Foundations of Electronics
Circuits and Devices

Foundations of Electronics
Circuits and Devices

3rd Edition

Russell L. Meade

Delmar Publishers

an International Thomson Publishing company I(T)P®

Albany • Bonn • Boston • Cincinnati • Detroit • London • Madrid
Melbourne • Mexico City • New York • Pacific Grove • Paris • San Francisco
Singapore • Tokyo • Toronto • Washington

NOTICE TO THE READER

DEDICATION

Once again, to my wonderful wife Betty for her love, support, and help during the project, and additionally to our four children: David, Stephen, Rebecca, and Timothy Meade.

Cover design by Gelia, Wells & Mohr

Delmar Staff:

Publisher:	Michael McDermott
Acquisitions Editor:	Gregory L. Clayton
Developmental Editor:	Michelle Ruelos Cannistraci
Senior Project Editor:	Christopher Chien
Production Manager:	Larry Main
Art Director:	Nicole Reamer
Editorial Assistant:	Amy E. Tucker
Marketing Manager:	Kitty Kelly
Marketing Coordinator:	Paula Collins

COPYRIGHT ©1999
By Delmar Publishers
a division of International Thomson Publishing Inc.

The ITP logo is a trademark under license.

Printed in the United States of America
For more information, contact:

Delmar Publishers
3 Columbia Circle, Box 15015
Albany, New York 12212-5015

International Thomson Publishing Europe
Berkshire House 168-173
High Holborn
London, WC1V 7AA
England

Thomas Nelson Australia
102 Dodds Street
South Melbourne, 3205
Victoria, Australia

Nelson Canada
1120 Birchmont Road
Scarborough, Ontario
Canada, M1K 5G4

International Thomson Editores
Campos Eliseos 385, Piso 7
Col Polanco
11560 Mexico D F Mexico

International Thomson Publishing GmbH
Konigswinterer Strasse 418
53227 Bonn
Germany

International Thomson Publishing Asia
60 Albert Street
#15-01 Albert Complex
Singapore 189969

International Thomson Publishing—Japan
Hirakawacho Kyowa Building, 3F
2-2-1 Hirakawacho
Chiyoda-ku, Tokyo 102
Japan

1 2 3 4 5 6 7 8 9 10 XXX 03 02 01 00 99 98

ISBN: 07668–0424–0 Vol I
       07668–0427–5 Vol II

Meade, Russell L.
    Foundations of electronics : circuits and devices / Russell L. Meade. — 3rd ed.
      p. cm.
    Includes index.
    ISBN 0–7668–0427–5
    1. Electronics. I. Title.
TK7816.M358 1998
821.381—dc21

98-41841
CIP

CONTENTS

v

PART 2 BASIC CIRCUIT ANALYSIS

4 SERIES CIRCUITS   100

5 PARALLEL CIRCUITS   146

PART 3 PRODUCING & MEASURING ELECTRICAL QUANTITIES

PART 4 REACTIVE COMPONENTS

PART 5 INTRODUCTORY DEVICES AND CIRCUITS

APPENDIXES

THE BOOK'S PURPOSE

The purpose of this book is to provide initial training in the fundamentals of electricity and electronics in a technically sound, "student-friendly," and easy-to-understand style. Because a solid foundation in fundamentals is critical to success in any facet of the electrical or electronics professions, this text provides the vital facts, concepts, and principles you will need for further progress in this field. Not only does *Foundations of Electronics* provide this critical training, it also promotes your ability to apply this knowledge to "real-life" situations. The instructional design of the book stresses the development of logical thinking patterns and practical applications of knowledge, and does not just present sterile theory. Ideally, those who study this book will have had some training in basic algebra and basic trigonometry; however, enough data are provided to enable even novices in these areas to successfully learn the material presented.

THE BOOK'S PATTERN

Key instructional design strategies used in the book's presentation include factors such as the following:

- Helping you understand, rather than just memorize, important formulas and concepts
- Giving meaningful previews and overviews to provide a context for content details
- Furnishing important learning objectives and clear targets for learning
- Offering clear and frequent examples
- Providing immediate application of knowledge gained
- Using frequent practice problems to reinforce concepts
- Supplying strategically located checkpoints
- Making liberal use of realistic pictorial and diagrammatic information
- Using color to enhance the instructional value of illustrations and diagrams
- Moving from the known to the unknown in a logical manner
- Progressing from the simple to the more complex
- Including effective summaries that provide the essence of what should be learned
- Presenting many troubleshooting, safety, and other practical hints
- Providing a useful troubleshooting technique we call the SIMPLER method to sharpen critical thinking skills and teach troubleshooting
- Furnishing Chapter Troubleshooting Challenge circuit problems for practicing the SIMPLER method of troubleshooting
- Formatting that allows for either group or individualized instruction delivery systems

THE BOOK'S PATH

The text begins with a brief overview of the wonders of today's electronic technology and the great opportunities afforded in this field. Next, crucial terms and elemental concepts are presented as foundations for the topics addressed

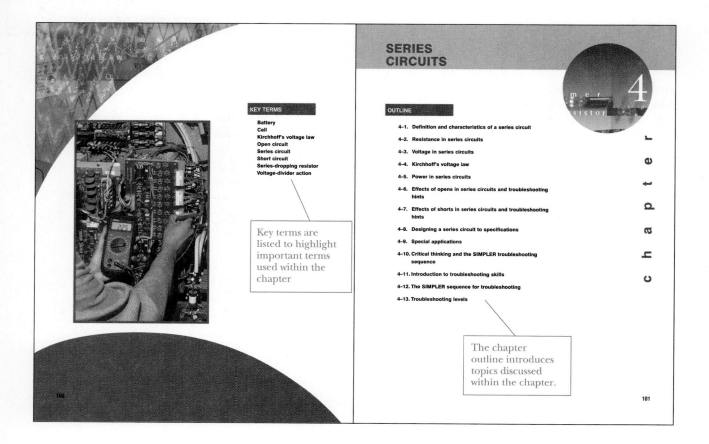

throughout the remainder of the book. Concepts such as the electron theory, basic electrical units and symbols, and how electronic circuitry can be illustrated with diagrams provide the stepping stones to what follows.

Essential circuit fundamentals are then presented so that students can develop the ability to analyze various circuit configurations. In the process of learning these fundamentals, several passive devices are studied, including resistors, inductors, and capacitors. Also, important test instruments and their uses are discussed. The book then applies the fundamental concepts, principles, and theories, first to dc circuits, and then to ac circuits.

Once these foundations have been laid, you will see several practical applications of these theories in circuits. For example, you will see how rectifiers and filters work to provide a universally significant system called the electronic power supply. Power supplies are found in virtually all electronic equipment.

Finally, you will get a glimpse of some of today's technological marvels. These include semiconductor diodes, transistors, and integrated circuits.

It is our hope that as you move from the introductory concepts of electronics through the principles of the circuit analysis to practical circuit applications and on to the introduction of modern electronic devices, you will find the book interesting, informative, and very useful in your training.

USING THE BOOK

The format of *Foundations of Electronics: Circuits and Devices* is "user-friendly." Each chapter features key terms, an introductory outline, a preview of the chapter's

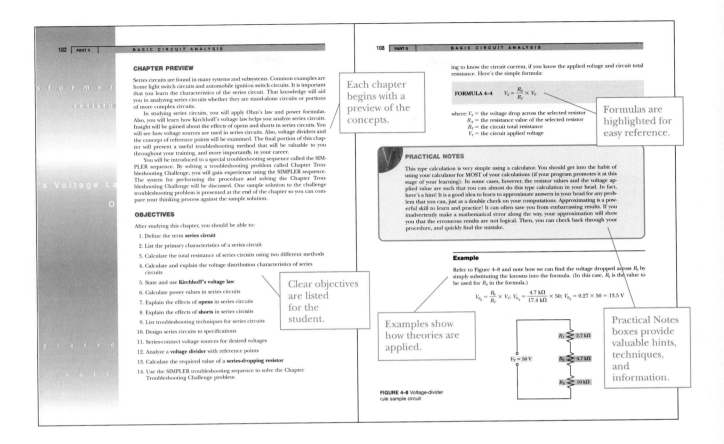

Each chapter begins with a preview of the concepts.

Formulas are highlighted for easy reference.

Clear objectives are listed for the student.

Examples show how theories are applied.

Practical Notes boxes provide valuable hints, techniques, and information.

contents, and learning objectives. The body of the chapter includes frequent step-by-step examples and practice problems, and In-Process Learning Checks to reinforce concepts. Formulas are highlighted throughout the text. Laboratory Manual projects are cross-referenced to the text. The chapter's formulas and some sample calculator sequences are summarized at the end of each chapter for easy reference. Each chapter closes with a detailed summary of important points as well as Review Questions, Problems, and Analysis Questions. (Answers to the practice problems, In-Process Learning Checks, and odd-numbered review, problems, and analysis questions may be found in Appendix B.)

KEY TERMS, CHAPTER OUTLINE, PREVIEW, OBJECTIVES—Each chapter begins by highlighting a few key terms that will be discussed in the chapter. Following that is a clear outline of topics that will be taught in the chapter. Next, a meaningful chapter preview is provided, and finally, the critical learning objectives are listed.

IN-PROCESS LEARNING CHECKS provide an opportunity for you to check your understanding of the subject each step of the way. Answers are provided in Appendix B.

PRACTICAL NOTES provide special hints, practical techniques, and information to the reader.

SAFETY HINTS highlight requirements for safety, both while learning and when applying electronic principles in "hands-on" situations.

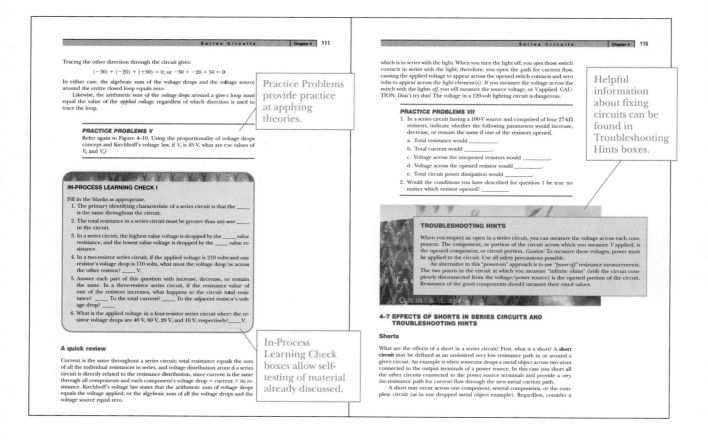

TROUBLESHOOTING HINTS offer experience-based information about circuit problems—what to look for and how to fix them.

NEW! GOOD IDEA boxes reinforce key practical ideas.

EXAMPLE and PRACTICE PROBLEMS—Numerous examples and practice problems show how theory is applied step by step and allow you to practice immediately after learning concepts.

FORMULAS and SAMPLE CALCULATOR SEQUENCES are summarized at the end of each chapter.

PERFORMANCE PROJECTS CORRELATION CHART—Projects from the Laboratory Manual are directly tied and cross-referenced to topics and sections in each chapter in the textbook.

SUMMARY, REVIEW QUESTIONS, PROBLEMS, ANALYSIS QUESTIONS—Each chapter concludes with a comprehensive Summary, Formula List, Sample Calculator Sequences, and a complete set of problems from quick review questions to critical thinking problems.

SIMPLER SEQUENCE and CHAPTER TROUBLESHOOTING CHALLENGES—The book uses a unique troubleshooting approach called the SIMPLER sequence. This successful troubleshooting method integrates critical thinking all along the way

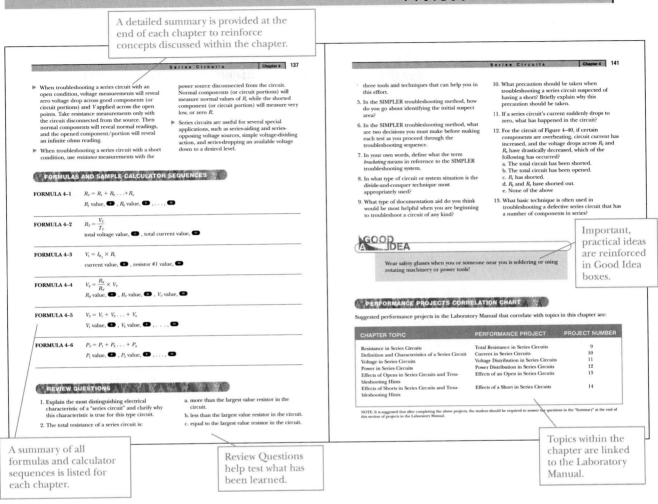

A detailed summary is provided at the end of each chapter to reinforce concepts discussed within the chapter.

Series Circuits | Chapter 4 | 137

▶ When troubleshooting a series circuit with an open condition, voltage measurements will reveal zero voltage drop across good components (or circuit portions) and V applied across the open points. Take resistance measurements only with the circuit disconnected from the source. Then normal components will reveal normal readings, and the opened component/portion will reveal an infinite ohms reading.

▶ When troubleshooting a series circuit with a short condition, use *resistance* measurements with the power source disconnected from the circuit. Normal components (or circuit portions) will measure normal values of R, while the shorted component (or circuit portion) will measure very low, or zero R.

▶ Series circuits are useful for several special applications, such as series-aiding and series-opposing voltage sources, simple voltage-dividing action, and series-dropping an available voltage down to a desired level.

FORMULAS AND SAMPLE CALCULATOR SEQUENCES

FORMULA 4–1 $R_T = R_1 + R_2 \ldots + R_n$

R_1 value, ⊞ , R_2 value, ⊞ , . . . , ⊟

FORMULA 4–2 $R_T = \dfrac{V_T}{I_T}$

total voltage value, ⊟ , total current value, ⊟

FORMULA 4–3 $V_1 = I_{R_1} \times R_1$

current value, ⊠ , resistor #1 value, ⊟

FORMULA 4–4 $V_x = \dfrac{R_x}{R_T} \times V_T$

R_x value, ⊟ , R_T value, ⊠ , V_T value, ⊟

FORMULA 4–5 $V_T = V_1 + V_2 \ldots + V_n$

V_1 value, ⊞ , V_2 value, ⊞ , . . . , ⊟

FORMULA 4–6 $P_T = P_1 + P_2 \ldots + P_n$

P_1 value, ⊞ , P_2 value, ⊞ , . . . , ⊟

A summary of all formulas and calculator sequences is listed for each chapter.

REVIEW QUESTIONS

1. Explain the most distinguishing electrical characteristic of a "series circuit" and clarify why this characteristic is true for this type circuit.

2. The total resistance of a series circuit is:

a. more than the largest value resistor in the circuit.

b. less than the largest value resistor in the circuit.

c. equal to the largest value resistor in the circuit.

Review Questions help test what has been learned.

Series Circuits | Chapter 4 | 141

three tools and techniques that can help you in this effort.

5. In the SIMPLER troubleshooting method, how do you go about identifying the initial suspect area?

6. In the SIMPLER troubleshooting method, what are two decisions you must make before making each test as you proceed through the troubleshooting sequence.

7. In your own words, define what the term *bracketing* means in reference to the SIMPLER troubleshooting system.

8. In what type of circuit or system situation is the divide-and-conquer technique most appropriately used?

9. What type of documentation aid do you think would be most helpful when you are beginning to troubleshoot a circuit of any kind?

10. What precaution should be taken when troubleshooting a series circuit suspected of having a short? Briefly explain why this precaution should be taken.

11. If a series circuit's current suddenly drops to zero, what has happened in the circuit?

12. For the circuit of Figure 4–40, if certain components are overheating, circuit current has increased, and the voltage drops across R_2 and R_4 have drastically decreased, which of the following has occurred?
a. The total circuit has been shorted.
b. The total circuit has been opened.
c. R_1 has shorted.
d. R_2 and R_4 have shorted out.
e. None of the above

13. What basic technique is often used in troubleshooting a defective series circuit that has a number of components in series?

GOOD IDEA

Wear safety glasses when you or someone near you is soldering or using rotating machinery or power tools!

Important, practical ideas are reinforced in Good Idea boxes.

PERFORMANCE PROJECTS CORRELATION CHART

Suggested performance projects in the Laboratory Manual that correlate with topics in this chapter are:

CHAPTER TOPIC	PERFORMANCE PROJECT	PROJECT NUMBER
Resistance in Series Circuits	Total Resistance in Series Circuits	9
Definition and Characteristics of a Series Circuit	Current in Series Circuits	10
Voltage in Series Circuits	Voltage Distribution in Series Circuits	11
Power in Series Circuits	Power Distribution in Series Circuits	12
Effects of Opens in Series Circuits and Troubleshooting Hints	Effects of an Open in Series Circuits	13
Effects of Shorts in Series Circuits and Troubleshooting Hints	Effects of a Short in Series Circuits	14

NOTE: It is suggested that after completing the above projects, the student be required to answer the questions in the "Summary" at the end of this section of projects in the Laboratory Manual.

Topics within the chapter are linked to the Laboratory Manual.

and encourages you to solve problems logically and efficiently. Instructors may use and teach other sequences, and the concepts outlined in the SIMPLER sequence are easily used with a wide variety of approaches.

APPENDIXES—The end of the book has extensive appendixes that include color codes, answers, schematic symbols, math review, how to use a scientific calculator, a comprehensive glossary, and more.

CHANGES TO THE THIRD EDITION

The development of this text began with an extensive review of *Foundations of Electronics*, second edition. Reviewers made detailed analyses of the second edition, and their suggestions for improving content, instructional quality, and art presentation formed the core of our plan for developing the third edition.

Two versions are available

There are two versions of *Foundations of Electronics*, third edition.

Foundations of Electronics, third edition, includes 22 dc/ac chapters. This book can be used for dc/ac courses.

Foundations of Electronics: Circuits and Devices, third edition, includes 22 dc/ac chapters plus 9 additional chapters on devices. This book can be used for both dc/ac and devices courses.

Using color to add learning value

As you use this book, you will notice that color has been used systematically to clarify the illustrations. For example, each resistor pictorial displays an actual color code, adding a practical, real-world dimension to these illustrations. In addition, voltage sources are always highlighted with yellow, resistance values are highlighted with brown, capacitive elements with blue, and so forth. We intend for this planned system to improve tangibly the value of the illustrations for students.

Calculator sequences

Sample calculator sequences have been provided for each formula to demonstrate the keystrokes. These are given at the end of each chapter to make study and homework more convenient.

Expanded content

In response to readers' requests, we have added important content that will be helpful to all users of the text.

Additional End-of-Chapter Questions: More than 400 additional new questions have been added in the text. These are found in the Review Questions, Problems, and Analysis Questions to provide even more in-depth practice and learning experiences for the student. The questions are comprehensive in sytle. Multiple-choice, short-answer, long-answer, problem-solving, and analysis questions provide a wide spectrum of practice and learning experiences.

New Topic Areas: Several new topic areas have been addressed, including superconductivity, surface mount technology (SMT), and components such as chip capacitors and chip resistors; additional component coding systems are also discussed.

TROUBLESHOOTING WITH THE SIMPLER SEQUENCE

To become logical troubleshooters, technicians must develop good critical thinking skills. This book introduces a useful troubleshooting approach called the SIMPLER sequence. This successful troubleshooting method integrates critical thinking all along the way and encourages you to solve problems logically and efficiently. Instructors may use and teach other sequences, and the concepts outlined in the SIMPLER sequence are easily used with a wide variety of approaches.

Chapter troubleshooting challenge

As you move through this book, you will encounter a series of troubleshooting problems called Chapter Troubleshooting Challenge. Each Chapter Troubleshooting Challenge includes a challenge circuit schematic and starting point symptoms information. Using the SIMPLER sequence, you will step through and solve the simulated troubleshooting problems.

Using the SIMPLER troubleshooting method, you will be able to guide yourself through each Chapter Troubleshooting Challenge circuit and think critically about the problems you need to solve. One possible approach to solving each Chapter Troubleshooting Challenge is provided in step-by-step illustrations following each challenge. The SIMPLER sequence and accompanying challenges are explained fully in Chapter 4.

The illustrations on pages xxiv–xxv show a Chapter Troubleshooting Challenge and give instructions on how to complete the challenge. We hope that you and your instructor will have great success with the SIMPLER sequence technique and that these Chapter Troubleshooting Challenges will make learning and teaching troubleshooting skills easier and more enjoyable

Selecting tests and finding the results

To simulate making a parameter test on a particular portion of the challenge circuit, you may select a test from the test listing provided with the Chapter Troubleshooting Challenge (for an example of this feature, see pages xxiv–xxv). You may then look up the results of the test you selected by looking in Appendix C. Look in the appendix for the identifier number assigned to the test on the Chapter Troubleshooting Challenge page. Next to that number, you will see the parameter value or condition that would be present if you made that same test on the actual circuit simulated by the Chapter Troubleshooting Challenge. *Please note:* The tolerances of components and of test instruments will create differences in the theoretical values and those actually shown in the answers. The theoretical values would be achieved only if all components were precisely rated and the test instruments had zero percent error.

THE LEARNING PACKAGE

The complete ancillary package was developed to achieve two goals:

1. To assist students in learning the essential information needed to prepare for the exciting field of electronics
2. To assist instructors in planning and implementing their instructional programs for the most efficient use of time and other resources

Laboratory Manual—*Foundations of Electronics,* third edition, has a well-correlated laboratory manual that can multiply the effectiveness of the learning process. The manual has 89 easy-to-use projects that provide hands-on experience with the concepts and principles you study in the theory text.

The format of the manual ensures that you will not miss the forest for the trees. You will not perform a whole series of "cookbook" steps, or the whole project, only to find out at the end you did not understand what you should have been learning in the project. Instead, you can analyze and learn important concepts each step of the way via the "programmed" conclusion column.

The projects are arranged as stand-alone, single-concept projects; however, they are grouped logically to allow maximum instructional flexibility. Each project can be performed individually or several related projects can be performed during one session.

Russ Meade wrote the laboratory manual, resulting in an optimally correlated package not commonly seen with other textbooks and laboratory manuals.

Order #: 0-7668-0430-5

Instructor's Guide—This comprehensive Instructor's Guide provides all the answers for the text and laboratory manual problems and projects, along with instructional strategies, sample course schedules, and transparency masters.

Order #: 0-7668-0429-1

Troubleshooting DC/AC Circuits with Electronics Workbench® (with enclosed Circuits Data Disk) by Richard Parker—This workbook teaches students how to

THE SIMPLER
SEQUENCE FOR
TROUBLESHOOTING

Follow the seven-step SIMPLER troubleshooting sequence, as outlined below, and see if you can find the problem with this circuit. As you follow the sequence, record your circuit test results for each testing step on a separate sheet of paper. This will aid you and your instructor to see where your thinking is on track or where it might deviate from the best procedure.

step 1 Symptoms
—(Gather, verify, and analyze symptom information) To begin this process, read the data under the "Starting Point Information" heading. Particularly look for circuit parameters that are not normal for the circuit configuration and component values given in the schematic diagram. For example, look for currents that are too high or too low, voltages that are too high or too low, resistances that are too high or too low, etc. This analysis should give you a first clue (or symptom information) that will aid you in determining possible areas or components in the circuit that might be causing this symptom.

step 2 Identify
—(Identify and bracket the initial suspect area) To perform this first "bracketing" step, analyze the clue or symptom information from the "Symptoms" step, then either on paper or in your mind, bracket, circle, or put parentheses around all of the circuit area that contains any components, wires, etc., that might cause the abnormality that your symptom information has pointed out. NOTE: Don't bracket parts of the circuit that contain items that could not cause the symptom!

step 3 Make
—(Make a decision about the first test you should make: what type and where) Look under the TEST column and determine which of the available tests shown in that column you think would give you the most meaningful information about the suspect area or components, in the light of the information you have so far.

step 4 Perform
—(Perform the first test you have selected from the TEST column list) To simulate performing the chosen test (as if you were getting test results in an actual circuit of this type), follow the dotted line to the right from your selected test to the number in parentheses. The number in parentheses, under the "Results in Appendix C" column tells you what number to look up in Appendix C to see what the test result is for the test you are simulating.

step 5 Locate
—(Locate and define a new "narrower" area of uncertainty) With the information you have gained from the test, you should be able to eliminate some of the circuit area or circuit components as still being suspect. In essence, you should be able to move one bracket of the bracketed area so that there is a new smaller area of uncertainty.

step 6 Examine
—(Examine available information and determine the next test: what type and where) Use the information you have thus far to determine what your next test should be in the new narrower area of uncertainty. Determine the type test and where, then proceed using the TEST listing to select that test, and the numbers in parentheses and Appendix C data to find the result of that test.

step 7 Repeat
—(Repeat the analysis and testing steps until you find the trouble) When you have determined what you would change or do to restore the circuit to normal operation, you have arrived at your solution to the problem. You can check your final result against ours by observing the pictorial step-by-step sample solution on the pages immediately following the "Chapter Troubleshooting Challenge"—circuit and test listing page.

step 8 Verify
NOTE: A useful 8th Step is to operate and test the circuit or system in which you have made the "correction changes" to see if it is operating properly. This is the final proof that you have done a good job of troubleshooting.

142

Troubleshooting with the SIMPLER sequence

1

Review the steps in the SIMPLER sequence.

2

Study the challenge circuit diagram and the starting point information.

3

Following the instructions, work through the SIMPLER sequence to find the problem with the circuit.

4

As you work through the SIMPLER sequence, choose tests (one at a time) from the list that you think might lead you to the problem solution.

5

To find the result of each test, look at the number across from the test you chose in the Results column.

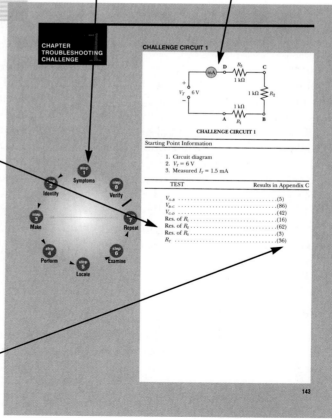

CHAPTER
TROUBLESHOOTING
CHALLENGE

CHALLENGE CIRCUIT 1

CHALLENGE CIRCUIT 1

Starting Point Information

1. Circuit diagram
2. $V_T = 6$ V
3. Measured $I_T = 1.5$ mA

TEST	Results in Appendix C
V_{A-B}	(5)
V_{B-C}	(86)
V_{C-D}	(42)
Res. of R_1	(16)
Res. of R_2	(62)
Res. of R_3	(3)
R_T	(36)

143

APPENDIX C

Chapter Challenge Test Results

Find the number listed next to the test you chose and record the result.

1. 1.4 V	31. 5 V	61. 14 V	89. 2.3 V
2. signal normal	32. 400 Ω	62. 1 kΩ	90. ∞ Ω
3. 2 kΩ	33. ∞ Ω	63. ∞ Ω	91. 0 V
4. 2 V	34. 6.4 V	64. 1.5 V	92. normal
5. 1.5 V	35. noticeably high	65. ≈6 VDC	93. 10 kΩ
6. high	36. 4 kΩ	66. 0 V	94. ≈5 V
7. 1.5 V	37. 10 kΩ	67. 7.5 V	95. 10 V
8. 0 V	38. 3.3 V	68. 10 kΩ	96. ≈11 V
9. 0 V	39. 14 V	69. low	97. 0 V
10. 3 V	40. 0 V	70. signal normal	98. low
11. 0 V	41. ≈1.5 mH	71. no change in opera-	99. 1.5 V
12. 7 V	42. 3.0 V	tion	100. ∞ Ω
13. ∞ Ω	43. 7 V	72. 0 Ω	101. 10 V
14. 0 Ω	44. 1.5 V	73. 1.5 V	102. 0 V
15. ≈0 V	45. high		103. 0 V
16. 1 kΩ	46. 0 Ω	74. (120 VAC)	104. ≈5 kΩ
17. 400 Ω	47. 0 V	75. 10 V	105. 0 V
18. 1.5 V	48. 2.3 V	76. slightly low	106. no signal
19. ∞ Ω	49. 14 V	77. 5.9 V	107. 50 pF
20. 1.5 V	50. 0.5 V	78. 3.1 V	108. ∞ Ω
21.	51. 1.5 V	79. 7.5 V	109. 12 kΩ
22. 10 V	52. 10 V	80. 7 V	110. 0 V
23. no signal		81. 2 V	111. no change in opera-
24. circuit operates nor-	53. (14 VAC)	82. 0 V	tion
mally	54. 10 kΩ	83. 10 V	112. 1.5 V
25. 10 kΩ	55. max value	84. 0 Ω	113. signal normal
26. slightly high	56. 10 V	85. slightly below max	114. 297 Ω
27. 0 V	57. signal normal	value	115. 0 V
28. 0 V	58. ∞ Ω	86. 1.5 V	116. ≈5 V
29. 0 Ω	59. 1.5 kΩ	87. 5 Ω	117. within normal range
30. greatly below max	60. 275 pF	88. 14 V	
value			

6

Turn to Appendix C and find that number from the results column. Next to the number in the Appendix you will find the result this test would yield. As you locate the number of the test you chose, record the result listed next to it on a sheet of paper.

7

When you have completed the challenge and pinpointed the circuits problem, check your work against the data in the answer pages. This answer offers one possible sequence of steps that might have been used to get to the solution, along with step-by-step color illustrations.

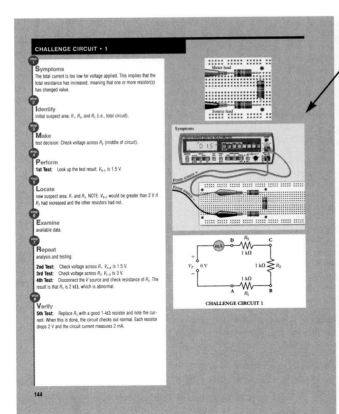

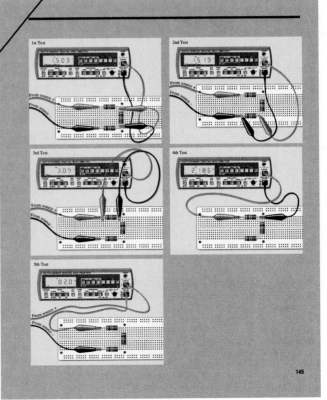

troubleshoot circuits with the help of Electronics Workbench.® Students will learn how to make measurements as they would in the lab. Students will replace components and then test the results to see how changes affect the operation of the circuit, all on a computer. The manual includes a disk with 100 circuits for troubleshooting with Electronics Workbench® taken directly from the Meade text.

Order #: 0-7668-1132-8

Instructor's Teaching System (ITS)—An innovative teaching system composed of tools that are designed to correlate, reinforce each other, and provide synergy, but that can also be used individually. All components of the Instructor's Teaching System (ITS) are neatly organized for your use in a three-ring binder. The binder will include the following components:

- **Instructor's Guide**—This softcover, comprehensive Instructor's Guide provides answers for the text and laboratory manual, along with instructional strategies, sample course schedules, and transparency masters.
- **Electronic Instructor's Resource Kit (EIRK)**—Available all on one **CD-ROM,** this custom-designed EIRK acts as an organizer and launch pad for all of the following applications:
  - **Computerized Testbank**—More than 1,500 questions provide mix-and-match capabilities for designing tests for any portion of the training.
  - **Electronic Gradebook**—Tracks student performance, prints student progress reports, organizes assignments, and more; simplifies administrative tasks.
  - **OnLine Testing**—This is a feature of the computerized testbank, which allows exams to be administered online via a school network or stand-alone PC.
  - **Power Point® Lecture Assistant**—Provides customizable lecture presentation outlines for every chapter in the textbook. Written by Russ Meade, they are closely tied to the textbook. Graphics from the Image library or your own images can be imported to create individualized classroom presentations.
  - **Image Library**—Selected color illustrations from the textbook provide the instructor with another means of promoting student understanding. The Image Library allows the instructor to display or print images for classroom presentations.
- **Circuits Data Disk**—Diskette includes troubleshooting circuits from the textbook created in Electronics Workbench®. Instructors may copy and distribute these circuit files to students free of charge.
- **Instructor's Teaching System Documentation**—Additional print material provides information, instruction, and teaching hints on how to use all of the tools together for maximum benefit.

Order #: 0-7668-1028-3

Technology

- **OnLine Companion™**—This text has a companion website, which will have high appeal to both educators and students. Features include the following:
  - Technology Updates

- Internet Activities
- Periodic Discussion Forums
- Ask the Author: Frequently Asked Questions
- Comprehensive listing of links to electronics industy and educational sites
- Instructor Resources
- Real Audio® broadcasts from author and contributors on new technology and educational strategies in electronics
- Free Internet-based online syllabus through Thomson's World Class Course (further details available from your sales representative)
- Come visit our site at **www.electronictech.com**

- **Electronics Workbench®**—New to this edition are optional Electronics Workbench® projects. They are available in the accompanying *Troubleshooting DC/AC Circuits with Electronics Workbench®* workbook by Richard Parker. This workbook includes troubleshooting projects using more than 100 circuits taken directly from the textbook. Students will learn how to make measurements as they would in the lab. Students will replace components and then test the results to see how changes affect the operation of the circuits, all on a computer.

  Using Electronics Workbench® to learn electronics still permits the learner to work a problem in student-sized steps, yet gives a deeper understanding of the subject. The teaching of electronics has to change because today's students have much more to learn but the learning time available has not increased. Understanding a powerful yet intuitive tool like Electronics Workbench® is essential if students are to really master all the concepts in the time available.

  This provides a new and exciting dimension to learning electronics by providing you with working circuits. The methods you will use are the same as you would use with real circuits in the lab. Circuits are stored on the disk that comes with this workbook. With this manual and Electronics Wrokbench®, learning how to troubleshoot should be no trouble!

  Order #: 0-7668-1132-8

- **Electronics into the Future: Circuit Fundamentals CD**—Electronics into the Future is a new interactive CD-ROM–based product from Delmar Publishers that is the perfect technology accompaniment to your dc/ac circuits text or as a stand-alone learning tool.

  Electronics into the Future offers students and instructors a range of multimedia presentations and interactive simulations designed to develop and expand major concepts in electronics.

  Electronics into the Future puts the power of nonlinear technology into the hands of students and instructors. As a classroom learning tool, *Electronics into the Future* gives instructors the power to use technology to assist in clearly illustrating difficult concepts. As a stand-alone learning tool, *Electronics into the Future* gives students the power to learn theory and troubleshooting from interactive practical applications.

  Features:

  - Available in either a student version or network version.
  - Contains six interactive modules featuring elements such as presentation video, interactive conceptualization, troubleshooting, and much more.

- The content modules cover the following topics: Ohm's law, amps, volts, Ohms, power, series circuits, parallel circuits, series-parallel circuits, network theorems (Kirchhoff's, superposition, Thevenin's), magnetism, and electromagnetism.

- Mathematics for Electronics content module. This module can be accessed as a stand-alone module and is linked logically into each content module.

- Recognizing that troubleshooting is one of the key learning objecitves for students in DC/AC circuits, special emphasis is placed on troubleshooting. *Electronics into the Future* has circuits for troubleshooting created in Electronics Workbench®.

- In addition, *Electronics into the Future* has the ability to launch the student edition of Electronics Workbench®, and the network version allows instructors to add links to Electronics Workbench® files.

- Runs in an easy-to-navigate browser environment.

- Includes free copy of Netscape Navigator®.

Order #: 0-7668-0657-X

Please visit our website at **www.electronictech.com** for more details and a demonstration. Please contact your sales representative for a full demonstration and for pricing details.

Scheduling instructions

Ideally, your program can use all the material provided in the text and the correlated laboratory manual to maximize learning and practice opportunities. The following suggestions are provided as possible scenarios for those whose time is more limited:

- Certain portions of the Network Theorems and Network Analysis Techniques in Chapters 7 and 8 may receive light treatment, as appropriate to your course.

- Also, some programs prefer to cover cells and batteries in briefer fashion as simply a lecture, which does not include all the data in Chapter 9.

- In addition, not all programs require the exhaustive treatment on analog meters provided in Chapter 11.

- If vector algebra and manipulation of complex numbers are not required in your program, the last half of Chapter 21 may be omitted.

- Further customizing of the program to your program's needs may be done by appropriate selection of the specific laboratory experiments provided in the *Foundations of Electronics* laboratory manual.

- For programs that must cover the dc and ac fundamentals in a two-quarter timespan, use the "trimming" ideas listed.

- In addition, limit assignments and discussions in Magnetism and Electromagnetism (Chapter 10) to the "fundamental principles" portions, deleting the discussion of the many SI magnetic units and presentation of "motor" and "generator" effects.

ACKNOWLEDGMENTS

REVIEWERS FOR THE THIRD EDITION:

Alex Breaux
SE College of Technology
Metairie, LA

Gary Cardinale
DeVRY Institute of Technology
Woodbridge, NJ

Sam Chafin
Dekalb Technical Institute
Clarkston, GA

Susan Clark
DeVRY Institute of Technology
Kansas City, MO

Tom Cress
Belleville Area College
Belleville, IL

Donald Enick
Electronics Institute
Kansas City, MO

Arnie Garcia
Texas State Technical College
Harlingen, TX

John Giancola
DeVRY Institute of Technology
Columbus, OH

Toby Givens
Orangeburg—Calhoun Technical
College
Orangeburg, SC

David Harris
Southeast College of Technology
Memphis, TN

Timothy Haynes
Haywood Community College
Wayneville, NC

David Heiserman
SweetHaven Publishing Services
Columbus, OH

Craig Hill
Erie Institute of Technology
Erie, PA

George Johnson
DeVRY Institute of Technology
Irving, TX

Peter Kerckhoff
DeVRY Institute of Technology
Kansas City, MO

Bill Kist
New England Institute
Pal Beach Gardens, FL

Bruce Koller
Diablo Valley College
Pleasant Hill, CA

Hector Lopez
City College
Ft. Lauderdale, FL

Patrick O'Connor
DeVRY Institute of Technology
Chicago, IL

JW Roberts
Western GA Tech
La Grange, GA

Rich Wagner
Dunwoody Industrial Institute
Minneapolis, MN

Saeed Shaikh
Miami Dade Community College
Miami, FL

I would like to express grateful appreciation to the publishing team at Delmar for patiently working with me in the attempt to produce the best books possible. For the publishing endeavors, this team includes Michael McDermott, Gregory L. Clayton, Michelle Ruelos Cannistraci, Larry Main, Christopher Chien, Nicole Reamer, and Amy E. Tucker. Special thanks are deservedly given to Michelle Ruelos Cannistraci, Christopher Chien, and Nicole Reamer, who spent a great amount of time, care, and effort in directly communicating with me concerning the innumerable details involved in this project.

Recognition is also gratefully given to the marketing team, Kitty Kelly and Paula Collins, whose efforts aid in getting the books into the hands of users. And, of course, this is where the books can realize their purpose in helping to facilitate significant technical training in the extraordinary field of electronics.

Russ Meade

Delmar Publishers wishes to thank the professional faculty at Hudson Valley Community College, with a special thanks to Gerard McEnearney, for their help, hospitality, and assistance in obtaining many of the new photos for this edition. Their expertise and professional advice contributed greatly to the technical accuracy and relevancy to many of our chapter opening photos. In addition, the Publisher would like to thank the following for providing chapter opening photos.

AMP (Chapter 2)
Cleveland Institute of Electronics (Chapter 7)
DeVRY Institute of Technology (Chapters 1, 8, 13, 15, 17)
Hewlett-Packard (Chapter 16, 19)
IBM (Chapter 23)
John Fluke Mfg. Co (Chapter 20)
Sencore (Chapter 21)

THE MAGNIFICENT GROWTH OF ELECTRONICS

Electronics has grown until it permeates almost everything we do! Whether we are at home, at work, or in our automobile, electronic systems, devices, and controls are all around us. The inventors of the telegraph, the telephone, and radio transmitters and receivers would not be able to believe the communications systems used today. Satellite communications, cable TV, the Internet, cell phones, computerized data communications, and Global Positioning systems would astound these inventors, probably even more than they astound us.

Today our *automobiles* are "loaded" with electronics:

- Electronic control units that automatically control all critical parameters that are involved in engine operations, transmissions, and emissions control
- Keyless entry systems
- Memory systems that automatically set seats, mirrors, and environmental settings based on individual preferences
- Computer-controlled suspension systems that automatically adjust to road conditions
- Electronic security systems
- Digital trip computers that provide an electronic message center with readouts to indicate remaining fuel and fuel consumption projections, current miles-per-gallon fuel use, compass direction, etc.
- Satellite-linked "navigation systems" that let you know "where on Earth" you are . . . and also may provide specific details on how to get where you want to go
- Awesome sound systems
- Other (such as "variable-assist" steering control)

Manufacturing uses a myriad of industrial controls, computer-aided drafting and design (CADD), and computer-aided manufacturing systems. Automation in industry includes such things as the following:

- Robots for welding, positioning, fastening, etc.
- Automated drilling, punching, and milling
- Automated shaping and bending
- Counting and sorting
- Process control

Business transactions involve everything from point-of-sale computers that take care of receipting, inventory, and change-making to on-line computerized transfers of money, documentation, and endless other transactions.

Our *homes* are filled with electronic wizardry that helps us take care of heating and cooling our homes, cooking, washing, cleaning, and providing security. The entertainment systems in our homes today could not have even been dreamed about just a few short years ago.

It is impossible to imagine what tomorrow will bring! It is obvious you have chosen a good field!

WHY ELECTRONICS FUNDAMENTALS ARE SO IMPORTANT!

To become a competent technician in the field of electronics demands that you get a thorough grasp of the fundamentals. In any field of endeavor, whether it be nursing, law, a craft, or other, there are fundamental terms, symbols, concepts, and principles that must be initially learned to acquire the more advanced knowledge and skills required in the profession. In electricity and electronics, it is vital that these basic principles be learned. They are the foundation upon which you build an understanding of all that is required for your career.

Develop the mindset to work diligently at learning these new terms, concepts, and the important foundational truths! You will reap large benefits later as you get further into this study.

Mastering the terms, symbols, formulas, and basic principles taught in this book will launch you into this stimulating field and will allow you to pursue it as far as your desire will carry you.

We wish you the best of success as you engage in the learning process . . . and beyond!

KEY TERMS

Ampere
Atom
Circuit
Compound
Conductor
Coulomb
Current
Electrical charge
Electron
Element
Energy
Force
Free electrons
Insulator
Ion
Load
Matter
Mixture
Molecule
Neutron
Ohm
Polarity
Potential
Proton
Resistance
Semiconductor
Source
Valence electrons
Volt

BASIC CONCEPTS OF ELECTRICITY

1

chapter

CHAPTER PREVIEW

This chapter will introduce you to a variety of concepts and theories proven useful as "thinking tools" for your study of electricity and electronics. These seemingly divergent foundational concepts and terms will come together in a meaningful way as you proceed in your studies. Consider each item as an important "knowledge capsule" upon which you can build in your continuing learning process.

OBJECTIVES

After studying this chapter, you should be able to:

1. Describe the types of tasks performed by electronics technicians, technologists, and engineers

2. Define the term **matter** and list its physical and chemical states

3. Describe the difference between **elements** and **compounds**

4. Discuss the characteristics and structure of an **atom, molecule,** and **ion**

5. Define the electrical characteristics of an **electron, proton,** and **neutron**

6. Explain the terms **valence electrons** and **free electrons**

7. List the methods used to create electrical imbalances

8. Describe the characteristics of **conductors, semiconductors,** and **insulators**

9. State the law of electrical charges

10. Discuss the terms **polarity** and **reference points**

11. Define charge and its unit of measure the **coulomb**

12. Define **potential** (emf) and give its unit of measure

13. Define **current** and explain its unit of measure

14. Calculate current when magnitude and rate of charge motion is known

15. Define **resistance** and give its unit of measure

16. List the typical elements of an electrical circuit

17. Describe the difference between closed and open circuits

1-1 GENERAL INFORMATION REGARDING THE FIELD OF ELECTRONICS

You are about to start your preparation in a field of work that is both interesting and almost limitless in opportunity. There are three broad categories of work in this field:

1. *Technicians* install, adjust, troubleshoot, repair and maintain a wide variety of electronic equipment and systems. See Figure 1–1.

2. *Technologists* assist in the design, development, and testing of electronic equipment and systems. See Figure 1–2.

3. *Engineers* design electronic equipment and systems. Engineers may also be involved as "customer engineers." These engineers serve in a technical service role for industry, providing expertise in making correct choices of equipment or systems for specific applications. See Figure 1–3.

Within the field of electronics, you can specialize in a wide variety of areas. A small sampling of areas is shown below:

- computers (e.g., CPUs, monitors, peripherals)
- communications (e.g., radio, television, telecommunications)
- consumer electronics (e.g., audio, video, alarm systems)
- medical electronics (e.g., x-ray, magnetic imaging, monitoring devices)

FIGURE 1–1 Electronics technician making tests on equipment *(Courtesy of Cleveland Institute of Electronics, Inc.)*

FIGURE 1–2 Electronics technologist performing computer calculations related to equipment design *(Courtesy of DeVry Institutes)*

FIGURE 1–3 Electronics engineer working with industrial customer to determine customer's system needs *(Courtesy of Cleveland Institute of Electronics, Inc.)*

- aerospace electronics (e.g., computing, navigation, satellite communications)
- marine electronics (e.g., radar, sonar, navigation, radio)
- automotive electronics (e.g., systems control, user conveniences)
- industrial electronics (e.g., process control, system monitoring)

As you can see, you have chosen a great field of study!

1-2 DEFINITION, PHYSICAL AND CHEMICAL STATES OF MATTER

Definition of matter

Everything that we see, touch, or smell represents some form of matter. In Figure 1–4 the workbench, the test equipment, and the air in the room represent **matter** in various forms. Although many definitions have been used, matter may be defined as anything that has weight and occupies space. We can also say that matter is what all things are made of and what our senses can perceive.

The basic building block of all matter is the atom. Later in this chapter, atoms will be thoroughly discussed because of their importance to the understanding of electronics.

Physical states of matter

Matter exists in three *physical* states. Matter is either a **solid,** such as the chair you are sitting on, a **liquid,** such as water, or a **gas,** such as oxygen.

When matter exists in the liquid or gaseous state, its dimensions are determined by the container.

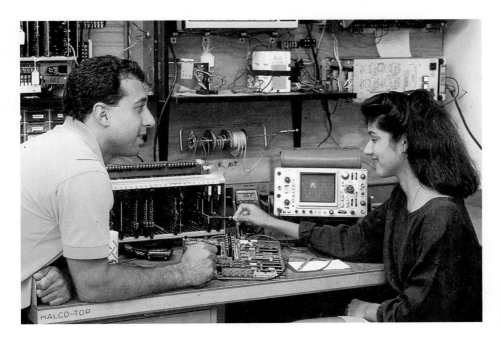

FIGURE 1–4 A scene full of matter *(Photo by Bruce Parker/Parker Productions)*

FIGURE 1–5 Some familiar metallic elements *(Photo by Michael A. Gallitelli)*

Chemical states of matter

The chemical states of matter are **elements, compounds,** and **mixtures.** An element, Figure 1–5, is a substance that cannot be chemically broken into simpler substances. In fact, an element has only one kind of atom. Examples of chemical elements are gold, iron, copper, silicon, oxygen, and hydrogen.

A compound is formed by a chemical combination of two or more elements. In other words, compounds are built from two or more atoms that combine into molecules. Also, a compound has a definite structure and characteristic (i.e., same weight and atomic structure). Figure 1–6 shows two examples of compounds. Examples of compounds are water, which is the chemical union of hydrogen and oxygen, and sugar, which contains carbon (a black, tasteless solid) plus hydrogen and oxygen (two gases).

Mixtures are a combination of substances where the individual elements possess the same properties as when they are alone. There is no chemical change resulting from the combination as there is with compounds. In Figure 1–7, mixing gold dust and sand would not yield a new, chemically different entity or compound. The substances would simply be mixed or combined.

FIGURE 1–6 Some commonly used compounds *(Photo by Michael A. Gallitelli)*

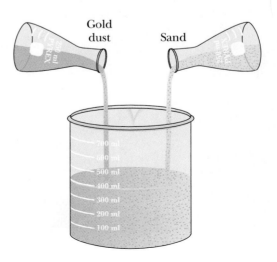

FIGURE 1–7 Mixtures are combinations of substances in which the individual elements possess the same properties as when they are alone. A mixture is different from a compound.

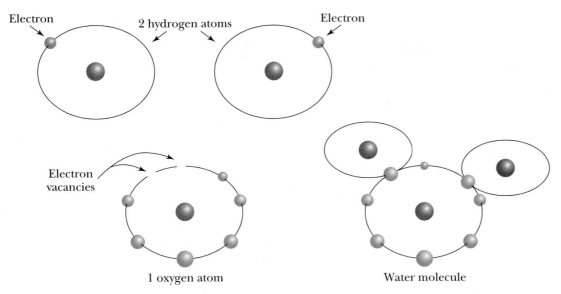

FIGURE 1–8 A molecule of water is formed from oxygen and hydrogen atoms.

1–3 COMPOSITION OF MATTER

The building blocks of all matter are atoms. Atoms combine chemically to form molecules, and the new compound (matter) created is different than each element that went into the compound's makeup, Figure 1–8. Also, a specific substance's molecules are different than the molecules of other types of matter.

In summary, *all matter is composed of atoms and molecules.* The smallest particle into which a compound can be divided but retain its physical properties is the molecule. The smallest particle into which an element can be divided but retain its physical properties is the atom, Figure 1–9.

1–4 STRUCTURE OF THE ATOM

The particles

The particles of the atom that interest electronics students are the **electron, proton,** and **neutron.** Although science has identified other particles (mesons, positrons, neutrinos), studying these particles is not necessary to understand the electron theory.

A *molecule* of water will exhibit the same physical characteristics as a drop of water.

An *atom* of gold will show the same chemical properties as a gold bar.

FIGURE 1–9 The smallest particle of a compound that has the same characteristics as the compound is a molecule. The smallest part of an element that will show the same characteristics as the element is an atom.

The model

A Danish scientist, Niels Bohr, developed a model of atomic structure that explains the electron theory. In his model the atom consists of a nucleus in the center of the atom with electrons orbiting the nucleus. The nucleus has two types of particles: protons, which are electrically positive in charge, and neutrons, which are neutral in electrical charge. The orbiting electrons are negatively charged particles, Figure 1–10. The common analogy for this model is our solar system: Planets orbit the sun like electrons orbit the nucleus of the atom.

The net charge of the atom is neutral because the orbiting electrons' total negative charge strength equals the total positive charge strength of the protons in the nucleus. Also, the number of protons and electrons in an electrically balanced atom are equal, Figure 1–11.

Some interesting characteristics of these three atomic particles are listed in Figure 1–12.

1. *The electron:*
    * has a negative electrical charge;
    * has small mass or weight (9×10^{-28} grams);
    * travels in orbits outside the nucleus;

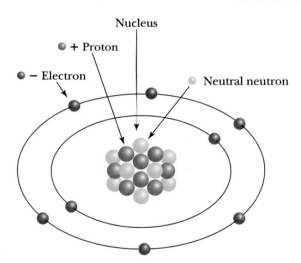

FIGURE 1–10 Bohr's model of the atom. The nucleus (center) contains protons and neutrons. Electrons orbit the nucleus.

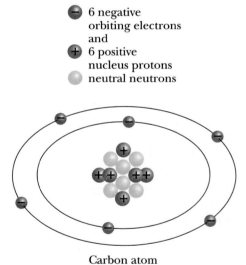

Carbon atom

FIGURE 1–11 Typically, atoms are electrically balanced.

ELECTRONS	PROTONS	NEUTRONS
Negative charge	Positive charge	No charge
Small mass/weight (about 9×10^{-28} grams)	1,836 times heavier than electron	Similar in mass/ weight to proton
Travel in orbits outside of nucleus	Located in the nucleus	Located in the nucleus
Rates of speed (trillions of orbits per sec.)	Equal in number to atom's electrons	

FIGURE 1–12 Some facts about atomic particles

- travels around the nucleus at unbelievable speed (trillions of times a second); and
- helps determine the atom's chemical characteristics.

2. *The proton:*
   - has a positive electrical charge;
   - is located in the nucleus of the atom;
   - is approximately 1,800 times heavier than an electron; and
   - is equal in number to the atom's electrons.

3. *The neutron:*
   - has no electrical charge;
   - is located in the nucleus of the atom;
   - is about the same mass or weight as a proton; and
   - may vary in number for a given element to form different "isotopes" of the same element; for example, hydrogen has three isotopes: protium, deuterium, and tritium, Figure 1–13.

Atomic number and weight

Although it is not necessary to thoroughly understand atomic numbers and atomic weights, it is appropriate to mention them.

The atomic number of an element is determined by the number of protons in each of its atoms. For example, the atomic number for copper is 29 and the atomic number for carbon is 6, Figure 1–14.

The atomic weight of an element is determined by comparing the weight of its atoms to an atom of carbon-12. For example, hydrogen, the simplest atom, has an atomic weight of about 1.007 "atomic mass units," and copper's atomic weight is 63.54.

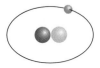

Deuterium
(heavy hydrogen)
nucleus =
1 proton
1 neutron

Tritium
(heavy, heavy hydrogen)
nucleus =
1 proton
2 neutrons

Protium
(light hydrogen)
nucleus =
1 proton

FIGURE 1–13 What atomic difference is there between these isotopes?

Carbon

Copper

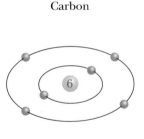
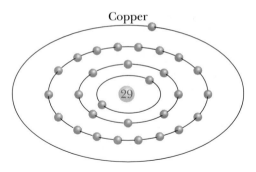

FIGURE 1–14 Carbon and copper atoms

IN-PROCESS LEARNING CHECK I

It will be a good idea at this point to initiate you into a technique that will be used from time to time throughout the book: the "In-Process Learning Check." Rather than wait until the end of the chapter to find out whether you're missing any of the key points, these special "checkups" will provide you with a chance to assure that you're learning the key points. Fill in the blanks for the following statements. If you have trouble with any of them, simply move back to that topic and refresh your memory. Answers are in Appendix B.

1. Matter is anything that has _____ and occupies _____.
2. Three physical states of matter are _____, _____, and _____.
3. Three chemical states of matter are _____, _____, and _____.
4. The smallest particle that a compound can be divided into but retain its physical properties is the _____.
5. The smallest particle that an element can be divided into but retain its physical properties is the _____.
6. The three parts of an atom that interest electronics students are the _____, _____, and _____.
7. The atomic particle having a negative charge is the _____.
8. The atomic particle having a positive charge is the _____.
9. The atomic particle having a neutral charge is the _____.
10. The _____ and _____ are found in the atom's nucleus.
11. The particle that orbits the nucleus of the atom is the _____.

Concept of atomic shells

As our illustrations indicate, all electrons traveling around the nucleus of the atom do not travel in the same path or at the same distance from the nucleus. Electrons align themselves in a structured manner, Figure 1–15.

Each ring, or shell, of orbiting electrons has a maximum number of electrons that can locate themselves within that shell if the atom is stable. The for-

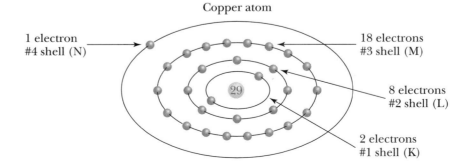

FIGURE 1–15 Electrons align at different distances from the nucleus.

Copper atom

1 electron
#4 shell (N)

18 electrons
#3 shell (M)

8 electrons
#2 shell (L)

2 electrons
#1 shell (K)

mula to use is $2n^2$, where n equals the number of the shell. For example, the innermost shell (closest to the nucleus) may have a maximum of 2 electrons; the second shell, 8 electrons; the third shell, 18 electrons; and the fourth shell, a maximum of 32 electrons. The outermost shell, whichever number shell that is for the given atom, can never contain more than 8 electrons.

1-5 CONCEPT OF THE ELECTRON THEORY

Purpose

The electron theory helps visualize atoms and electrons as they relate to electrical and electronic phenomena, and the previous descriptions of atomic structure are part of this electron theory. The next discussion of valence and free electrons illustrates some practical aspects of the electron theory.

Valence electrons

Valence electrons are those electrons in the outermost shell of the atom. The number of valence electrons in the atom determines its stability or instability, both electrically and chemically, Figure 1–16.

For all atoms, this outermost shell is *full* when it has eight *electrons*. If there are eight outermost ring electrons, then the material is stable and does not easily combine chemically with other atoms to form molecules. Also, these electrons are not easy to move from the atom. Examples of stable atoms are inert gases, such as neon and argon.

If the material's atoms have fewer than eight outermost ring electrons, the material is chemically and electrically active. Often the material will chemically combine with other atoms to gain stability and to form molecules and/or atomic bonds. Electrically, these valence electrons can be easily moved from their original home atom and are sometimes referred to as free electrons. Examples of such materials are copper, gold, and silver.

Also, there are materials that have four outermost ring electrons. They are halfway between being stable and very unstable. Germanium and silicon are two examples. These materials are used in most of today's solid-state, "semiconductor" devices, such as transistors and integrated circuits.

In summary, the concepts of the electron theory are based on atomic structure and help explain the various electrical phenomena discussed in this book.

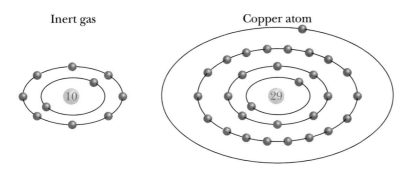

FIGURE 1–16 The number of valence electrons affects the chemical and/or electrical stability of an atom.

1-6 IONS

When an electron leaves its original home atom because of chemical, light, heat, or other types of energy, it leaves behind an atom that is no longer electrically neutral. An **ion** is any atom that is not electrically balanced (or neutral) and that has gained or lost electrons. A positive ion is an atom having fewer electrons than protons (a deficiency of electrons). A negative ion is an atom having more electrons than protons (an excess of electrons), Figure 1–17.

When an electron is torn from a neutral atom, leaving a positive ion, or when an electron is added to a neutral atom, producing a negative ion, the process is called ionization. Later in your study of electronics, you will see how this process can be useful in various electronic devices.

1-7 ENERGIES THAT CHANGE ELECTRICAL BALANCE

Why should anyone want to change the electrical balance of atoms, or control electron movement? As your study of electronics continues, you will see that the ability to control the movement of electrons, or electron flow, is the basis of electronics.

Figure 1–18 displays some common sources of energy causing electron movement and/or separation of charges. These sources are:

- friction (static electricity);
- chemical energy (batteries);
- mechanical energy (a generator or alternator);
- magnetic energy;
- light energy; and
- heat energy.

By appropriately using one or more of these energies, sources of electrical energy may be developed. These sources develop electrical potential, or the potential to move electrons through whatever circuits are electrically connected to these sources. Electrical **circuits** are closed paths designed to carry, manipulate, or control electron flow for some purpose. Later in this chapter, electrical potential and movement of electrons are discussed to show how the concepts presented thus far are foundations for the study of electronics.

Electron loss   Electron gain

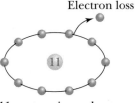   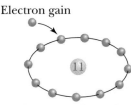

FIGURE 1–17 Ions are atoms that have lost or gained electrons.

11 protons in nucleus with only 10 orbiting electrons = "+ ion"

11 protons in nucleus with 12 orbiting electrons = "– ion"

(a)

(b)

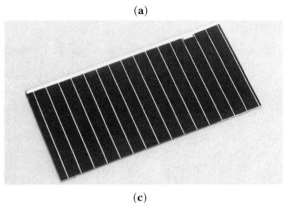

(c)

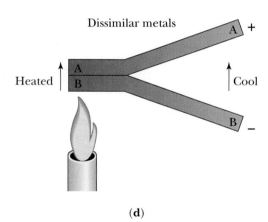

(d)

FIGURE 1–18 Some energy sources used to create useful energy: **(a)** A battery uses the chemical energy in its cell to create voltage. **(b)** A generator converts mechanical energy into electricity. *(Courtesy of AC-Delco)* **(c)** Solar cells convert light to electrical energy. **(d)** A thermocouple uses heat energy to produce an electric current.

1–8 CONDUCTORS, SEMICONDUCTORS, AND INSULATORS

Conductors, such as gold, silver, and copper, have many free electrons. These materials conduct electron movement easily because their outermost ring electrons are loosely bound to the nucleus. In other words, their outermost ring contains one, two, or three electrons rather than the eight electrons needed for atomic stability, Figure 1–19a.

Semiconductors, Figure 1–19b, are sample devices that use materials that are halfway between the conductors' characteristic of few outermost shell electrons and the stable, inert materials with eight valence electrons. Semiconductor materials have four outermost ring electrons. Germanium and silicon, as used in the pictured devices, are examples of semiconductor materials.

Insulator materials, Figure 1–19c, do not easily allow electron movement because their five to eight outermost shell electrons are tightly bound to the atom.

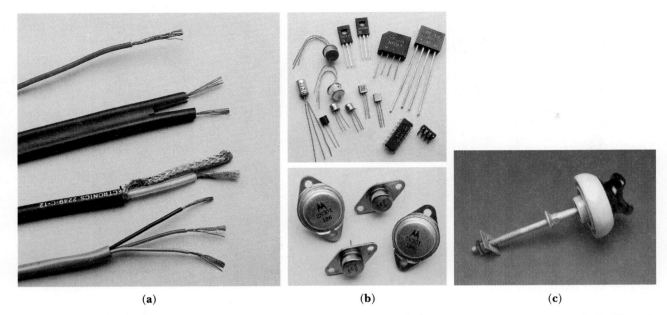

(a)   (b)   (c)

FIGURE 1–19 (a) Conductors *(Photo by Michael A. Gallitelli)*; **(b)** semiconductors *(Photos by Michael A. Gallitelli); and* **(c)** insulators

Insulators have few free electrons. Examples of insulator materials are glass and ceramic.

Electrical and electronic circuits are made up of a variety of components and interconnections consisting of conductors, insulators, and semiconductors.

1-9 SAMPLE OF AN ELECTRICAL SYSTEM

An electrical system typically has a source of electrical energy, a way to transport that electrical energy from one point to another, and an electrical load, Figure 1–20.

The source supplies energy through the transporting means to the load. The load may convert that electrical energy into another form of electrical energy or into another type of energy, such as heat, light, or motion.

Figure 1–21 illustrates an electrical lighting circuit. Observe that the method to transport the electrical energy to the light bulb (the load) is the conductor wires. The load then converts the electrical energy into another useful form, light.

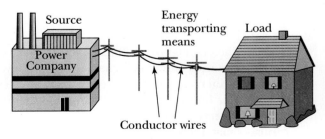

FIGURE 1–21 Components of a commonly used electrical system

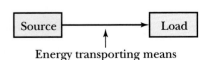

FIGURE 1–20 Parts of a basic electrical system

The rest of this chapter will discuss basic electrical laws and principles to help you in understanding what is happening in this basic light circuit. You should seek answers to the following questions: How does the source cause electrical energy or electricity to move through the conductors to the load? In what manner and/or form does the electricity move from one point to another? What determines the quantity of electricity moved from the source to the load?

1-10 BASIC PRINCIPLES OF STATIC ELECTRICITY

Before studying the movement of electrons in electrical circuits, it will be helpful to learn some basic principles about charges and static electricity as a frame of reference.

What is static electricity?

Everyone has been electrically shocked after walking on a thick pile rug and touching a door knob or after sliding across an automobile seat and touching the door handle. Where did that electricity come from? Sometimes this electricity is called static electricity. This term is somewhat misleading because static implies stationary and electrons are in constant motion around the nucleus of the atoms. Think of this type of electricity as being associated with nonconductors or insulator materials.

The basic law of electrical charges

In many science courses, the experiment of rubbing a rubber rod with fur or a glass rod with silk to develop charges is frequently performed, Figure 1–22. After doing this experiment, paper or other light materials are sometimes attracted to the charged body or object. What is meant by a charged body? This means that the object has more or less than its normal number of electrons. In the case of the rubber rod, it gained some electrons from the fur, and the fur lost those electrons to the rod. Thus, the net charge of the rubber rod is now negative, since it has an excess of electrons. Conversely, the fur is positively charged since it lost some of its electrons but retains the same number of protons in its nuclei.

   With the silk and glass rod experiment, the glass rod loses electrons and becomes positively charged. Furthermore, another experiment would show there is some attraction between the charged rubber rod (negative) and the charged glass rod (positive). This leads to a conclusion that has been accepted as the ba-

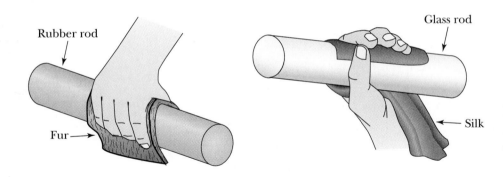

Rubber rod

Fur

Glass rod

Silk

FIGURE 1–22 Static electricity is usually associated with nonconductive materials.

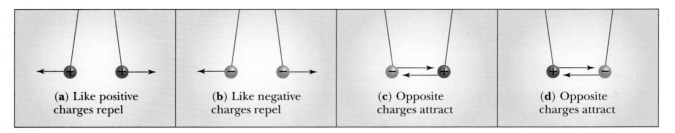

(a) Like positive charges repel

(b) Like negative charges repel

(c) Opposite charges attract

(d) Opposite charges attract

FIGURE 1–23 The basic law of charges is that like charges repel and unlike charges attract.

sic electrical law: Unlike charges attract each other, and like charges repel each other. Remember this important electrical law, Figure 1–23.

Polarity and reference points

You might have noticed that a common way of showing the difference between charges is by identifying them with a minus sign for negative or a plus sign for positive. Using positive (+) or negative (−) signs frequently indicates electrical **polarity.** Polarity denotes the relative electrical charge of one point in an electrical circuit "with reference to" another point, Figure 1–24. The notion of something "with reference to" something else is not really mysterious. Everyone has heard statements such as "John is taller than Bill," or "Bill is shorter than John." In the first statement, Bill is the reference point. In the second statement, the reference point is John.

Another familiar example of polarity and reference points is that the upper tip of the earth's axis is called the North Pole and the lower tip, the South Pole. These terms refer to their geographic locations.

Review some key facts:

1. Electrons are negatively charged particles and protons are positively charged particles.

2. Any substance or body that has excess electrons is negatively charged.

3. Any substance or body that has a deficiency of electrons is positively charged.

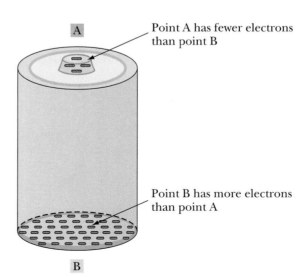

Point A has fewer electrons than point B

Point B has more electrons than point A

FIGURE 1–24 In a flashlight cell, point A is positive with respect to point B, and point B is negative with respect to point A.

4. Unlike charges attract and like charges repel.

Refer to Figure 1–25a, b, and c and answer the following questions:

1. In Figure 1–25a, which ball has a deficiency of electrons?
2. In Figure 1–25b, will the balls attract or repel each other?
3. In Figure 1–25c, if ball B has an equal number of electrons and protons, what is the polarity of ball A with respect to B?

The answers are 1) **B;** 2) **repel;** and 3) **negative.** In number one, the positive polarity indication on ball B indicates a deficiency of electrons. In number two, both balls have the same polarity charge, so they repel each other. In number three, if ball B has an equal number of electrons and protons, it is electrically neutral. However, ball A remains negative with respect to ball B, since ball A is not electrically neutral. In this case, ball A is negatively charged with respect to anything that does not have an equal or greater negative charge.

Coulomb's formula relating to electrical charges

Examining the forces of attraction or repulsion (repelling force) between charged bodies, Charles Coulomb, a French physicist, determined that the amount of attraction or repulsion between two charged bodies depended on the amount of charge of each body and the distance between the charged bodies. He expressed this relationship with the formula:

$$\text{Force} = \frac{\text{Charge on body \#1} \times \text{Charge on body \#2}}{\text{Distance between them squared}}$$

FORMULA 1–1 $F = k \dfrac{Q_1 \times Q_2}{d^2}$

where F = force (newtons)
   k = the constant 9×10^9 (for air or vacuum)
   Q = charge in units of charge (coulombs)
   d = distance between charged bodies (meters)

PRACTICAL NOTES

Wherever you see formulas throughout the book, be aware that "sample calculator sequences" are shown at the end of the chapter for each formula in the chapter. You will find these in the end-of-chapter feature called: "FORMULAS AND SAMPLE CALCULATOR SEQUENCES."

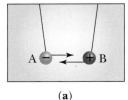

(a)

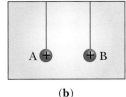

(b)

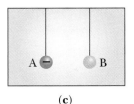

(c)

FIGURE 1–25 Some arrangements of electrical charges

Coulomb's law of charges states that the force of attraction or repulsion between two charged bodies is directly related to the product of their charges and inversely related to the square of the distance between them.

Unit of charge

In honor of Charles Coulomb, the unit of charge is designated the **"coulomb."** Many electrical units that define quantity or magnitude of electrical parameters are named after famous scientists who performed related experiments.

Electrical units of measure are powerful reference values used every day in electronics. Even as someone had to determine liquid quantities as a gallon, quart, pint, or cup, scientists define units of measure for electrical parameters so that each has its own "reference point" with respect to quantity or value.

In the case of electrical charge, the unit is the coulomb. This is the amount of electrical charge represented by 6.25×10^{18} electrons, or 6,250,000,000,000,000,000 electrons. Now you know the unit for charge used for the "Qs" in Formula 1–1.

Fields of force

You are familiar with several types of "fields," for example, gravitational fields and magnetic fields. Fields of force are represented by imaginary lines, which represent the field of influence for the force involved. Figure 1–26 shows electrostatic fields between unlike charges, Figure 1–26a, and between like charges, Figure 1–26b. This only shows a visualization of the nature of these fields, as represented by the lines. It does not attempt to indicate the strength of the charges, the strength of the field of force, or the distance between the charges. This representation helps us visualize the tangible forces between electrical charges.

1–11 ELECTRICAL POTENTIAL

What is electrical potential?

The implication drawn from these tangible forces is that they can be harnessed to do work. In this case, the difference of charge levels at two points has the potential to move electrons from a point of excess electrons to a point of electron deficiency, if a suitable path is provided. (Remember, the protons are in the nucleus and are not free to move.) This difference between two points having dif-

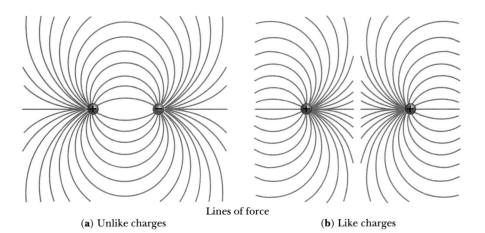

FIGURE 1–26 Electrostatic fields represented by lines of force

Lines of force

(**a**) Unlike charges   (**b**) Like charges

ferent charge levels is often termed a **potential** *difference*. This force, which moves electrons from one place to another, is called an **electromotive force** (emf). Remember these important terms because you will use them often.

The unit of electrical potential or electromotive force

Scientists have established a measure for the unit of charge (the coulomb), and the unit of measure for this difference of potential. They have named this unit of potential difference the **volt** in honor of Alessandro Volta who invented the electric battery and the electric capacitor. Often, this potential difference between two points, which is measured in volts, is called voltage. An in-depth discussion will come later. For now mentally picture that charges and/or different points are positive or negative relative to each other (that is, have polarity); that there are different quantities of charge depending on the number of excess electrons or electron deficiency; and that there is a potential difference between such points, which provides an electromotive force capable of causing electron movement from one point to another point.

What means can produce electrical potential?

Since electrical potential difference (electromotive force), or voltage, performs the electrical work of moving electrons from place to place, what establishes and maintains this voltage?

Earlier in this chapter, we discussed various types of energy that can cause "electrical imbalance," which you now know as electromotive force, or voltage. Static electricity, chemical energy, mechanical energy, magnetic energy, light energy, and heat energy were all mentioned. Look again at Figure 1–18 to review methods of using these various energies to establish and maintain differences of potential (voltage) between two points.

1-12 CHARGES IN MOTION

What happens if the two balls in Figure 1–27 touch each other? If you said that the excess electrons on ball A move to ball B and attempt to overcome the deficiency of electrons on ball B, you are right. In fact, electrons would continue to move until ball A and ball B had equal charges, or were neutral with respect to each other.

If the balls do not touch each other but are connected with a copper conductor wire, Figure 1–28, what happens? If you surmise that some electrons move through the conductor wire until the charges on ball A and ball B are equal, you are right.

Current

Are the electrons that move to ball B through the conductor the same electrons that left ball A? Probably not. As you have already studied, the conductor wire has many free electrons moving within it. Having a positive charge at one end of the conductor and a negative charge at the other end (a potential difference between its ends) causes the electrons to move from the negative end to the positive end. This movement of electrons is known as electrical **current flow,** or **current.**

An analogy of current flow

Figure 1–29 illustrates the concept of the movement of electrons (current flow) from atom to atom within the conductor (from negative to positive ends). Bin A

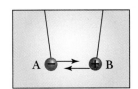

FIGURE 1–27 Two different charges

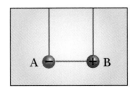

FIGURE 1–28 If a conductor wire is connected to ball A and ball B, what happens?

has a large quantity of rubber balls (representing electrons) and bin B has a few. The row of people between the two bins represent the conductor. They pass the balls from person to person, similar to the conductor material passing electrons from atom-to-atom when a current flows through the conductor. The progressive movement of balls from bin A to bin B simulates the movement of electrons from bin A through the conductor to bin B. In the conductor, as a free electron leaves its original atom and moves to an adjacent atom, it is replaced by another electron from another atom. This process is multiplied millions of times when there is current through a conductor. In Figure 1–28, as an electron leaves one end of the conductor (attracted by ball B's positive charge), it is replaced by an electron entering the conductor from ball A. Figure 1–30 also illustrates current flow.

Obviously, these analogies are oversimplifications of the physics involved in current flow, but they will aid your understanding as we discuss current flow and the specific electrical phenomena involved.

The unit of current

Recall that the unit of charge, the coulomb, was established in honor of Charles Coulomb, and the unit of potential difference, the volt, was named after Alessandro Volta. In like manner, the unit of measure for current was named in honor of the French mathematician and physicist, Andre Ampere.

An **ampere** of current is the quantity of electron movement represented by a flow rate of one coulomb of charge per second. Restated, *a flow of one coulomb per second = one ampere.*

FIGURE 1–29 An analogy relating to electron movement

Bin A   Bin B

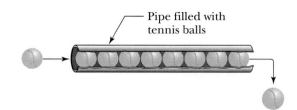

Pipe filled with tennis balls

FIGURE 1–30 Another analogy relating to electron movement

A formula that relates current to charge movement and time

A common formula that relates current (in amperes) to coulombs of charge and time (in seconds) is:

FORMULA 1–2 $\quad I = \dfrac{Q}{T}$

where I = current in amperes
$\quad Q$ = charge in coulombs
$\quad T$ = time in seconds

This formula allows you to calculate the current in amperes, if you know the charge in coulombs and the time in seconds. For example, if 10 coulombs of charge move from one point to another point in an electrical circuit over 2 seconds, current equals 5 amperes ($I = 10/2$). Also, this formula can be transposed to calculate the charge if the current and time are known. This transposition yields $Q = I \times T$. If at this point you do not understand how the formula was transposed, be encouraged by knowing that some techniques of transposing will be discussed again later in the book. Look at this formula and the examples, and use your knowledge about direct and inverse relationships to understand the formula. Is current *directly* or *inversely* related to the quantity of charge moved? Is current *directly* or *inversely* related to the duration of time taken to move the given charge? Your answers should have indicated a *direct* relationship between current and quantity of charge moved and an *inverse* relationship between current in amperes and the time it took to move the given charge. Recall that the direct relationship between quantities means that as one quantity increases, the other quantity increases. The inverse relationship indicates that as one quantity increases, the other quantity decreases, or vice versa.

IN-PROCESS LEARNING CHECK II

Fill in the blanks as appropriate.

1. An electrical system (circuit) consists of a source, a way to transport the electrical energy, and a _____.
2. Static electricity is usually associated with _____-type materials.
3. The basic electrical law is that _____ charges attract each other and _____ charges repel each other.
4. A positive sign or a negative sign often shows electrical _____.
5. If two quantities are *directly* related, as one increases the other will _____.
6. If two quantities are *inversely* related, as one increases the other will _____.
7. The unit of charge is the _____.

continued

8. The unit of current is the _____.

9. An ampere is an electron flow of one _____ per second.

10. If two points have different electrical charge levels, there is a difference of _____ between them.

11. The volt is the unit of _____ force, or _____ difference.

1-13 THREE IMPORTANT ELECTRICAL QUANTITIES

This chapter has discussed the magnitude of electrical charges (amount or value of charge); polarity (either negative or positive); and the difference of charge between two points that creates a potential difference (electromotive force or voltage) which can move electrons from one point to another (a current flow).

Also, you have learned that scientists have defined units of measure for the quantities of charge, potential difference, and current. Recall that the unit of charge is the coulomb (6.25×10^{18} electrons); the unit of electromotive force, or potential difference, is the volt; and the unit of current flow, or current, is the ampere. Of the electrical parameters described thus far, you will most frequently use the units for current and electromotive force (i.e., the ampere and the volt).

A third electrical quantity you will often use is **resistance.** If you have tried to sandpaper a piece of wood, you have experienced *physical resistance.*

Current flow through a conductor encounters molecular resistance to its flow. The amount of resistance to current flow depends on a number of factors. One factor is the type of material through which the current is flowing. Other factors are the dimensions of the conductor wire, including its cross-sectional area and its length, and temperature. An electrical pump (emf or voltage) causes current to flow through the resistance in its path.

As you have probably reasoned, electrical resistance is the opposition shown to current flow. Again, an electrical unit has been named after a scientist, Georg Simon Ohm; thus, the unit of resistance is the **ohm.**

The ohm of resistance has been defined in several ways. One definition states an ohm is the resistance of a column of pure mercury with a cross-sectional dimension of 1 millimeter squared and a length of 106.3 centimeters at a temperature of zero degrees centigrade. Another definition is an ohm is the amount of resistance that develops 0.24 calories of heat when one ampere of current flows through it. *The definition for ohm that is most useful is:*

An ohm of resistance is the amount of electrical resistance that limits the current to one ampere when one volt of electromotive force is applied.

1-14 BASIC ELECTRICAL CIRCUIT

Generalized description

Earlier in the chapter, a simple description was given of an electrical circuit. Recall that an electrical circuit has three basic ingredients: the source, the means to carry electricity from one point to another point, and a load. Before thoroughly examining this basic circuit, we will discuss the difference between a closed and an open circuit.

The closed circuit

You have learned that if a current path is provided when a potential difference exists between two points, electrons move from the point of excess electrons (negative polarity point) to the point of electron deficiency (positive polarity point). One method of providing this path is to connect a conductor wire between the points, thus providing a route through which electrons move and establish current flow. A closed circuit is a complete, unbroken path through which electrical current flows whenever voltage is applied to that circuit, Figure 1–31.

The open circuit

You probably already understand that if a closed circuit is an unbroken path for electron flow, then an open circuit has a break in the path for current flow, Figure 1–32. This break may be either a desired (designed) break or an undesired (unplanned) break. The most common method for purposely opening a circuit is an electrical switch, such as the one you use to turn the lights on or off. The switch is a fourth element of the basic electrical circuit, and many circuits contain such a control component, or circuit. In Figure 1–32 the switch is the control component.

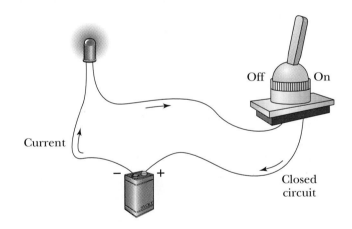

FIGURE 1–31 A closed circuit provides an unbroken path for current flow.

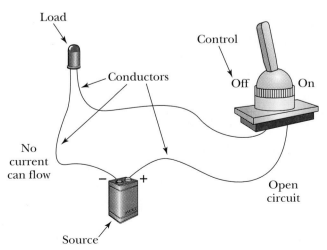

FIGURE 1–32 An open circuit has a break somewhere in the current path.

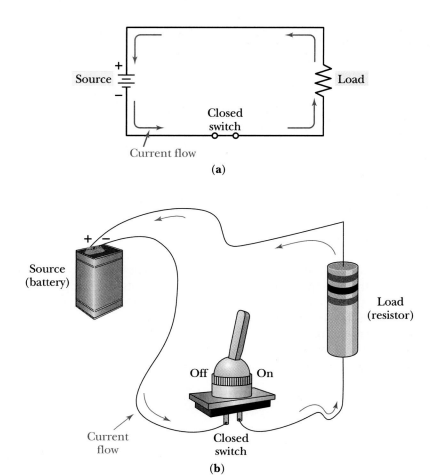

(a)

(b)

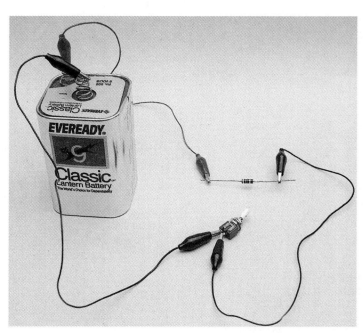

FIGURE 1–33 Direction of current flow through a circuit: **(a)** schematic; **(b)** pictorial; and **(c)** photo *(Photo by Michael A. Gallitelli)*

(c)

PARTS OF A CIRCUIT	EXAMPLE OF EACH	IMPORTANT ELECTRICAL QUANTITIES	
1) Source	Power company generators	EMF	(Volts)
		Resistance	(Ohms)
2) Transporting means	Conductor wires	Current	(Amperes)
3) Load	Light bulb(s)		
4) (Control)	(Switch)		

FIGURE 1–34 Important facts about a basic electrical circuit. As emf increases, current increases. As resistance increases, current decreases.

The basic electrical circuit summarized

The basic electrical circuit has 1) a source, 2) a means of conducting electron movement (current), 3) a load, and may have 4) a control element, such as a switch.

In order for current to flow (electron movement) through the circuit, there must be a voltage source and a closed circuit or a complete current-conducting path. If the conductor wire(s) were accidentally connected across the source with no other load component, the wire (with its small resistance) would become the load. Depending on the resistance of the wire and the amount of source voltage, this undesired low resistance condition (sometimes called a short circuit) would probably result in the wire overheating and melting, thus breaking the circuit.

Applying a voltage (electromotive force) to a closed circuit causes current to flow through the circuit conductors and load from the source's negative side through the circuit and back to the source's positive side, Figure 1–33. The amount of current (amperes) that flows depends on the value of voltage (volts) applied to the circuit and how much resistance (ohms) the circuit offers to current flow, Figure 1–34. Remember these points because they will become more important as your studies proceed.

SUMMARY

► Matter occupies space, has weight, and sometimes can be tangible to one or more of our five senses.

► Matter's physical states include solids, liquids, and gases.

► Matter's chemical states include elements, compounds, and mixtures.

► Elements are comprised of only one type of atom. But compounds are unique combinations of different elements' atoms that result in a new matter or substance.

► A mixture does not cause chemical or physical changes in the elements being mixed.

► The smallest particle into which a compound can be divided and retain its basic characteristics is a molecule. Elements can be divided to the atom level and retain their inherent characteristics.

► The atom is the fundamental building block of matter. Special combinations of atoms, called molecules, are the building blocks of compounds.

► Bohr's model of an atom pictures the central portion of the atom, the nucleus, being comprised of protons and neutrons. Around this nucleus is a particle called the electron, which orbits the nucleus like planets orbit the sun.

▶ The electron theory helps illustrate electricity and behavior of electrons in matter and relates to the atom model with its protons, neutrons, and electrons.

▶ The atom's proton has a positive electrical charge, while the orbiting electrons have a negative electrical charge. The neutrons in the center of the atom add mass to the atom but are electrically neutral.

▶ Protons and neutrons are approximately 1,800 times heavier than electrons.

▶ The net charge of an atom is typically neutral, since there are equal numbers of electrons and protons and their electrical charges offset each other.

▶ The atomic number of an element is the number of protons in the nucleus of each of its atoms.

▶ The atomic weight of an element is a comparison of its weight to that of carbon-12.

▶ Electrons travel around the nucleus of an atom in shells that are at different distances from the nucleus. The first ring (shell) carries a maximum of 2 electrons; the second shell outward a maximum of 8; the next shell, 18, and the fourth shell, 32. The outermost electron ring, valence electrons, determines if the atom is electrically and chemically active or stable and is full when it has 8 electrons.

▶ If the outer shell of an atom contains eight electrons, it is stable. If the outer ring has less than four electrons, it is usually a conductor. If the valence electrons number four, it is typically a semiconductor material.

▶ Materials, such as copper, that have only one outermost ring electron per atom are good conductors. These electrons, sometimes called free electrons, are easily moved within the material from atom to atom.

▶ An atom that has lost or gained electrons is called an ion. If it has lost electron(s), it is a positive ion. If it has gained electron(s), it is a negative ion.

▶ Ions and/or movement of electrons within a material are caused by various external and internal energies or forces. Several examples of energy that can electrically unbalance atoms or move electrons are friction, chemical, heat, and light.

▶ The law of electrical charges is that unlike charges attract and like charges repel each other.

▶ Static electricity is usually associated with nonconductor materials and is often caused by friction between them.

▶ Polarity designates differences between two points or objects relative to each other, or opposites. For example, electrical polarities of "−" or "+" show which point has excess electrons and electron deficiency relative to each other. Magnetic polarity indicates differences between magnetic poles (North and South). Also, electrical polarity shows direction of current flow through an electrical component or circuit. For example, electron movement (current) moves from the source's negative side to the source's positive side through the components and circuitry connected to that source.

▶ Coulomb's law of charges states that the force of attraction or repulsion between two charges is directly related to the product of the two charges and is inversely related to the square of the distance between them.

▶ The unit of charge is the coulomb, which represents 6.25×10^{18} electrons of charge.

▶ Electrical potential is the potential to perform the electrical work of moving electrons. It is sometimes termed electromotive force.

▶ The unit of measure for potential difference (electromotive force) is the volt.

▶ Several types of energy that establish and maintain potential differences between two points (i.e., create a voltage source) are chemical, heat, light, magnetic, and mechanical energies.

▶ Organized electron movement, or electron flow, is known as current flow or current.

▶ Current flow caused by electron movement is from the negative side of the voltage source to the positive side of the voltage source through the circuit connected to the source, if the circuit is a closed circuit.

▶ The unit of current is the ampere, which represents a rate of electron flow of one coulomb per second.

▶ A formula that relates the amount of current flow (in amperes) to the amount of charge (in coulombs) moved in a given amount of time (in seconds) is $I = Q/T$. The formula may also be transposed to read $Q = I \times T$.

▶ Three important electrical quantities of measure are:

  a. volt, the unit of potential difference, or amount of potential difference causing one ampere of current to flow through a resistance of one ohm;

  b. ampere, the unit of current flow, or amount of current flow represented by a rate of charge flow of one coulomb per second; and

  c. ohm, the unit of resistance, or amount of electrical resistance limiting the current to one ampere with one volt applied.

▶ A basic electrical circuit has an electrical source (emf), a way to transport electrons from one point to another point, and a load, which uses the electrical energy. Another part of many basic circuits is a control device, such as a switch.

▶ A closed circuit has an unbroken path for current flow: from one side of the source, through the circuitry and back to the other side of the source.

▶ An open circuit does not provide a continuous path for current flow. The path is broken, either by design (using a switch) or by accident. The open circuit presents an infinite resistance path for current.

▶ A short circuit is an undesired, low resistance path for current flow. If the short is across both terminals of a voltage source, it frequently causes undesired effects, such as melted wires, fire, and other damage to the voltage source.

▶ In the basic electrical circuit, the voltage source provides the electromotive force that moves electrons and causes current flow. The circuit conductors and components provide resistance to current, thereby limiting the current. The various techniques of controlling electrons and electron flow to perform desired effects form the basis of electronic science.

FORMULAS AND SAMPLE CALCULATOR SEQUENCES

NOTE: The calculator sequences shown throughout the book are based on a popular "Algebraic Notation" type of calculator. That is, the sequencing of calculations performed by the device are closely allied to the hierarchy system of operations used in Algebra. For example, operations such as reciprocal, square, and square root are higher priority than powers and roots. Powers and roots are higher priority than multiplication and division. Multiplication and division are higher priority than addition and subtraction, and the equal sign becomes the lowest priority. With algebraic operating systems such as this, lower-priority operations are delayed until higher-priority operations are complete.

Some training programs may have students use calculators that utilize what is known as "Reverse Polish Notation." These calculators use a system that evolved from an earlier system of simplifying arithmetic expressions, as conceived by a Polish logician and scientist named Lukasiewicz. The evolved system somewhat reverses the Polish logician's notation technique, hence the term *Reverse Polish Notation,* or RPN. In some cases the RPN-type calculators can accomplish a given calculation with slightly fewer keystrokes. For this reason, some instructors prefer having their students use this type of device.

No matter which type of calculator your program requires, let us encourage you to begin learning to use the calculator to your advantage at the earliest possible time. Calculators are wonderful tools for the technician and engineer alike!

FORMULA 1–1 $\quad F = k \dfrac{Q_1 \times Q_2}{d^2}$

k value, **⊗** , Q_1 value, **⊗** , Q_2 value, **═** , **÷** , distance value, **x²** , **═**

FORMULA 1–2 $\quad I = \dfrac{Q}{T}$

charge in coulombs, ➗ , time in seconds, 🟰

REVIEW QUESTIONS

1. Define the term *matter*.

2. List the three physical states of matter and give examples of each.

3. List the three chemical states of matter and give examples of each.

4. Define the term *element* and give two examples of elements.

5. Define the term *compound* and give two examples of compounds.

6. Sketch a hydrogen atom. Identify each particle.

7. On your hydrogen atom sketch, designate the electrical charge of each particle.

8. a. Which has more weight or mass, the electron or the proton?

   b. Approximately how many times heavier is the one compared to the other?

9. Define the term *free electron*.

10. a. Define the term *ion*.

    b. What causes an ion to be classified as a positive ion?

11. Name four energy types that can be used to purposely create electrical imbalances in atoms of materials.

12. How many valence electrons do each of the following materials have?

    a. A conductor.

    b. A semiconductor.

    c. An insulator.

13. Describe the basic law of electrical charges.

14. Describe Coulomb's law of electrical charges.

15. If the distance between two charges is tripled and the charge strength of each charge remains the same, what happens to the force of attraction or repulsion between them?

16. In an electrical circuit, what is potential difference?

17. What unit of measure expresses the amount of electromotive force present between two identified points in a circuit?

18. In an electrical circuit, what is current flow?

19. a. What basic unit of measure expresses the value of current in a circuit?

    b. What is the value of current flow if 20 coulombs of charge move past a given point in 5 seconds?

20. Describe the term *resistance* in terms of voltage and current, and give the unit of measure.

21. Describe the difference between a closed circuit and an open circuit.

22. List three or more elements of a basic electrical circuit.

23. What sign or symbol is used to denote a negative polarity point in an electrical diagram?

24. In a "closed" electrical circuit, electron movement (or electron current) is from the more (negative? positive?)_____ point in the circuit toward the more (negative? positive?)_____ point in the circuit.

25. In the circuit shown in Figure 1–35, is Point B representing the negative or the positive side of the source?

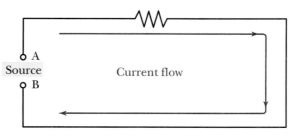

FIGURE 1–35

ANALYSIS QUESTIONS

1. Draw a diagram illustrating an electrically balanced atom that has 30 protons in its nucleus and whose first three shells are full, with the fourth (last) shell having two electrons.

2. Label all electrical charges in your diagram.

3. What kind of material is pictured in your diagram?

4. Is this material a conductor, semiconductor, or insulator?

5. What is its atomic number?

6. Define the term *polarity* as used in electrical circuits.

7. Draw an electrical circuit showing voltage source, conductor wires, and load. Show the polarity of that source, and, by using an arrow, show the direction of current flow through the circuit. (Assume a closed circuit.)

8. Referring to Figure 1–33b, list three ways this circuit might be made to become an open circuit.

9. In your own words, define the term *short circuit*.

10. What are some possible consequences of an undesired short circuit?

KEY TERMS

Ampere (A; mA; μA)
Block diagrams
Conductance *(G)*
Current *(I)*
Ohm (Ω; kΩ; MΩ)
Resistance *(R)*
Resistor (R)
Resistor color code
Schematic diagrams
Siemens (S; mS)
Switch (S)
Volt (V; mV; μV)
Voltage *(V)*

ELECTRICAL QUANTITIES AND COMPONENTS

2

chapter

CHAPTER PREVIEW

This chapter will further your study of the basic electrical quantities which are vital to your knowledge as a technician. For example, commonly used metric prefixes and powers of ten related to electrical quantities are presented. Information regarding basic characteristics and types of conductors and resistors is presented. You will also study and learn the resistor color code. (NOTE: Interpreting the color code is one of the most practical and most frequently used skills ever acquired by technicians.)

How to use meters to measure voltage, current, and resistance is also presented. Of course, application of this knowledge, and practice at doing this, is acquired by performing laboratory exercises that require making actual meter measurements.

Finally, the chapter introduces you to the technicians' and engineers' "shorthand" method of presenting electrical circuits. Both block diagram and schematic diagram techniques are presented. (NOTE: These "knowledge capsules" that may seem unrelated at this point will all begin coming together for you in the next chapter.)

OBJECTIVES

After studying this chapter, you should be able to:

1. List the units of measure for charge, potential (emf), current, resistance, and conductance and give the appropriate abbreviations and symbols for each

2. Use metric system terms and abbreviations to express subunits or multiple units of the primary electrical units

3. List the factors that affect the resistance of a conductor

4. Recognize common types of conductors

5. Use a wire table to find conductor resistance for given lengths

6. Recognize and/or draw the diagrammatic representations for conductors that cross and electrically connect, and that cross and do not connect

7. Define the term *superconductivity*

8. Give the characteristics of several common types of resistors

9. Explain the characteristics of surface-mount "chip" resistors

10. Use the resistor color code

11. Use other special resistor coding systems

12. Explain how to connect meters to measure voltage, current, and resistance

13. Recognize and/or draw the diagrammatic symbols for elemental electronic components or devices

14. Interpret basic facts from block and schematic diagrams

15. List key safety habits to be used in laboratory work

2-1 BASIC ELECTRICAL UNITS AND ABBREVIATIONS

Charge

Recall that the basic unit of charge is the coulomb and that the letter "Q" (or q) represents charge. However, you need to know that the abbreviation for the coulomb is **C.** (Remember that one C equals 6.25×10^{18} electrons.) Notice that this is a capital "C" and not a lower case "c." If the unit abbreviation or symbol for electronic measurement is named after an individual, such as coulomb, volt, ampere, or ohm, the abbreviation is a capital letter and not a lower case letter. Keep this in mind and get in the habit of correctly writing unit abbreviations or symbols.

Potential

The unit for potential difference, or electromotive force, is the **volt,** named after Alessandro Volta. The abbreviation, or symbol, for this unit is **V. Voltage** is expressed in volts. Recall that one volt equals the amount of electromotive force (emf) that moves a current of one ampere through a resistance of one ohm.

Current

The unit of measure for current flow is the **ampere.** The abbreviation, or symbol, for this basic unit of measure is **A.** Don't confuse the letter I (thought of as intensity of electron movement) that represents the general term **current** with the A used for the unit of current. Remember that one ampere equals an electron flow of one coulomb per second past a given point.

Resistance

Resistance is another electrical parameter that uses two letters: "R" represents the general term resistance and the Greek letter omega (Ω) represents the unit of resistance, the **ohm.** Remember that one ohm equals the resistance that limits the current to one ampere with one volt applied. (NOTE: Figure 2–1 shows some "multimeters" used to measure V, A, and Ω, the units of measure we have just discussed.)

Conductance

Another electrical parameter is **conductance.** It sometimes is defined as the comparative ease with which current flows through a component or circuit. Conductance is the opposite of resistance. The unit of conductance is the **siemens** (**S**) named after the scientist Ernst von Siemens. The abbreviation for the general term conductance is G, and the unit of measure is the siemens (S). Conductance, in siemens, is defined as the reciprocal of resistance in ohms.

$$G(\text{in siemens}) = \frac{1}{R} \text{ or } G = \frac{1}{R}(\text{S})$$

(a)

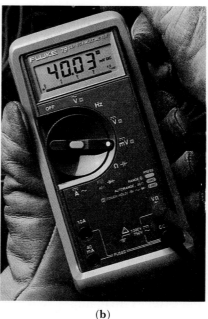

(b)

These units of measure and abbreviations, or symbols, are important, so a chart has been provided, Figure 2–2. Study this chart until you have memorized all the units and their abbreviations.

2–2 USING THE METRIC SYSTEM TO HELP

Some familiar metrics

Many people have been exposed to the metric system. The meter (approximately 39 inches), centimeter (one one-hundredth of a meter), millimeter (one one-thousandth of a meter), and kilometer (1,000 meters) are not uncommon terms. As you can see, the prefix to "meter" identifies a multiple or subdivision of the basic unit. In other words, the meter is the basic unit and the prefix centi equals 1/100, milli equals 1/1,000, and kilo equals 1,000 times the basic unit.

This technique of using metric prefixes is also useful in identifying multiples or subdivisions of basic electrical units. For example, in electronic circuits it is common to use thousandths or millionths of an ampere, or thousands of ohms of resistance. Although the subunits and multiple units in Figure 2–3 (page 38) do not represent all prefixes encountered in electronics, they do represent those you will be immediately applying. Study Figure 2–3 until you learn each of the metric terms and their abbreviations or symbols.

There is one other bit of related knowledge that can be of great help to you in practical work. That is, mathematical ways to represent each of these metric prefixes. These mathematical representations are helpful in studying electronic circuits, because simple calculations can be used to analyze, verify, or predict circuit behavior.

FIGURE 2–2 Important electrical units' abbreviations and symbols

BRIEF DEFINITION OF QUANTITY	ELECTRICAL QUANTITY OR PARAMETER	BASIC UNIT OF MEASURE	ABBREVIATION OR SYMBOL FOR UNIT
Excess or deficiency of electrons	Charge (Q)	Coulomb	C (6.25×10^{18} electrons)
Force able to move electrons	Potential difference (emf)	Volt (Force that moves one coulomb of charge per second through one ohm of resistance)	V
Progressive flow of electrons	Current (I)	Ampere (An electron flow rate of one coulomb per second)	A
Opposition to current flow	Resistance (R)	Ohm (A resistance that limits current to a value of one ampere with one volt applied)	Ω
Ease with which current can flow through a component or circuit	Conductance (G)	Siemens (The reciprocal of resistance, or, $\frac{1}{R}$)	S

FIGURE 2–2 Important electrical units' abbreviations and symbols

   Because the subunit and multiple-unit prefixes in Figure 2–3 are based on a decimal system (multiples or submultiples of 10), it is convenient to express these prefixes in powers of ten. Recall that:

10^0 = (ten to the zero power) = 1

10^1 = (ten to the 1st power) = 10

10^2 = (ten to the 2nd power) = 10×10 = 100

10^3 = (ten to the 3rd power) = $10 \times 10 \times 10$ = 1,000

10^{-1} = (ten to the negative 1st power) = 1/10 = 0.1

10^{-2} = (ten to the negative 2nd power) = 1/100 = 0.01

10^{-3} = (ten to the negative 3rd power) = 1/1,000 = 0.001

FIGURE 2–3 Some commonly used metric units

METRIC TERM	SYMBOL	MEANING	TYPICAL USE WITH ELECTRONIC UNITS
pico	p	One millionth of one millionth of the unit	Picoampere (pA)
nano	n	One thousandth of one millionth of the unit	Nanoampere (nA) Nanosecond (ns)
micro	μ	One millionth of the unit	Microampere (μA) Microvolt (μV)
milli	m	One thousandth of the unit	Milliampere (mA) Millivolt (mV)
kilo	k	One thousand times the unit	Kilohms (kΩ) Kilovolts (kV)
mega	M	One million times the unit	Megohm(s) (MΩ)

Figure 2–4 shows how powers of ten relate to the electrical units and prefixes discussed earlier. **Learn these because you will use them frequently.**

Study Figure 2–4. You can surmise that powers of ten can simplify writing very large or very small numbers. For example, instead of writing 3.5 microamperes (μA) as 0.0000035 amperes, you can state it as 3.5×10^{-6} amperes. The power of 10 tells us how many decimal places and which direction the decimal moves to state the number as a unit. Thus, 3.5×10^{-6} indicates the decimal moves six places to the left when stated as the basic unit (i.e., 0.0000035 amperes). This is three and one-half millionths of an ampere.

Another example is 5.5 megohms (MΩ) of resistance simplified as 5.5×10^{6} ohms. How would this be stated as a number representing the basic unit? If you said the power of 10 indicates moving the decimal six places to the right, you are correct. This yields 5,500,000 ohms.

Obviously, it is easier to manipulate numbers using powers of 10 than using the bulky multiple decimal places involved in the small and large numbers typically associated with electrical/electronic units. You will get more practice using powers of 10 in the next chapter.

FIGURE 2–4 Powers of ten related to metric and electronic terms

NUMBER	POWER OF 10	TERM	SAMPLE ELECTRONIC TERM
0.000000000001	10^{-12}	pico	pA (1×10^{-12} ampere)
0.000000001	10^{-9}	nano	nA (1×10^{-9} ampere)
0.000001	10^{-6}	micro	μA (1×10^{-6} ampere)
0.001	10^{-3}	milli	mA (1×10^{-3} ampere)
1,000	10^{3}	kilo	kΩ (1×10^{3} ohms)
1,000,000	10^{6}	mega	MΩ (1×10^{6} ohms)

IN-PROCESS LEARNING CHECK I

Fill in the blanks as appropriate.

1. Charge is represented by the letter _____. The unit of measure is the _____, and the abbreviation is _____.

2. The unit of potential difference is the _____. The symbol is _____.

3. The abbreviation for current is _____. The unit of measure for current is the _____, and the symbol for this unit is _____.

4. The abbreviation for resistance is _____. The unit of measure for resistance is the _____, and the symbol for this unit is _____.

5. Conductance is the _____ with which current can flow through a component or circuit. The abbreviation is _____. The unit of measure for conductance is the _____, and the symbol for this unit is _____.

6. How many microamperes does 0.0000022 amperes represent? _____ How is this expressed as a whole number times a power of 10? _____

7. What metric prefix represents one-thousandth of a unit? _____ How is this expressed as a power of 10? _____

8. Using a metric prefix how would you express 10,000 ohms? _____ How is this expressed as a whole number times a power of 10? _____

2-3 CONDUCTORS AND THEIR CHARACTERISTICS

Functions of conductors

You have already learned that the function of conductors is to carry electrical energy from one point to another. More explicitly, conductors provide a path for the flow of electrical current between components or circuits. As you can observe at home, at work, or at school, there are numerous types of conductors. One type is used to transport power into your home. Another type brings the TV signal from the antenna to your TV set. Still another style of conductors delivers telephone signals. The heavy battery cables in your automobile represent yet another type of conductor.

Types of conductors

The type of conductor required for a given job is determined by the nature of the application and the amount of electrical energy it must carry. Some ways of classifying conductors include the:

- type of metal used (silver, copper, gold, aluminum);
- physical dimensions of the wire itself (its "size");
- form of the metal conductor(s) (solid, stranded, or braided);
- number of conductors used or packaged together; and
- insulation characteristics (uninsulated, insulated, and type of insulation used).

Several examples of conductor types are shown in Figure 2–5.

FIGURE 2–5 Examples of several types of conductors: **(a)** auto battery cable, **(b)** coaxial cable, **(c)** TV twin lead, **(d)** stranded, **(e)** solid *(Photo by Michael A. Gallitelli)*

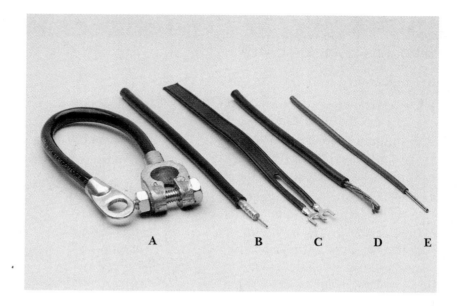

Resistance of solid conductors

The ideal conductor would have zero resistance, and thus would not dissipate any of the electrical energy being transported. In reality, however, all conductors do have some resistance to the flow of electrons at normal operating temperatures.

Superconductivity

Typically, the resistance of most metal conductors increases as their operating temperature is increased. In 1911, a unique discovery was made that if you cool (really cool) certain metal conductors' temperature to almost absolute zero (0° Kelvin, or −273° C), virtually all their resistance to current flow disappears. In other words, their resistance approaches zero ohms. This phenomenon only occurs below some critical, very cold temperature. The production and study of these super cold temperatures is often referred to as "cryogenics." These very low temperature requirements make it difficult to take advantage of this super-conductivity characteristic. Recently, scientists have begun experimenting with unique ceramic materials that can have very low resistance at temperatures considerably higher than absolute zero. If, in the future, scientists can find practical materials and methods to move that useable temperature range closer to normal room temperatures, the sky is the limit in terms of applications. Everything from electromagnetically operated devices (motors, generators, etc.) to computers would be greatly impacted.

Physical factors influencing conductor resistance

The resistance of a given conductor is related to the following factors:
1. The type material making up the conductor: The more conductive the material, the lower its resistance per given physical dimensions.
2. The length of the conductor: The longer it is, the higher its total resistance from one end to the other.

3. The cross-sectional area of the conductor: The greater the cross-sectional area, the lower its resistance per given length.

4. The temperature of the conductor: Typically, for metal conductors, as temperature increases, the resistance increases.

Resistivity

The "resistivity" (or "specific resistance") of a given material refers to its characteristic resistance (in ohms) per standard length and cross-sectional area, at a specified temperature. In SI units (International System of Units), resistivity is expressed in terms of meters for length, and square meters for cross-sectional area. The International System of Units (referred to as SI, after the initials of Systeme International) is basically the metric system. Seven basic categories of units are described in the International System: **length**—meter; **mass**—kilogram; **time**—second; **electric current**—ampere; **thermodynamic temperature**—Kelvin; **amount of substance**—mole; and **luminous intensity**—candela. Different materials each have their own resistivity characteristic. For example, in SI units, silver has a resistivity of 16 nano-ohms per meter; copper, a resistivity of 17 nano-ohms per meter; gold, a resistivity of 24 nano-ohms per meter; and aluminum, a resistivity of 27 nano-ohms per meter. These are expressed in nano-ohms per meter for simplicity of reading, rather than having to state that silver has a resistivity of 0.000000016 ohms/meter, and so on. The symbol used for resistivity is ρ (the Greek letter rho).

In everyday technician work, you are more likely to be dealing with conductors in terms of diameter in mils (thousandths of an inch), area in circular mils (CM), and length in feet. A circular mil describes the cross-sectional area of a round conductor having a diameter of one-thousandth of an inch (1 mil diameter). The cross-sectional area in circular mils is thus found as:

FORMULA 2–1 $A = d^2$

where A = cross-sectional area in circular mils
       d = diameter of wire in mils
See Figure 2–6.

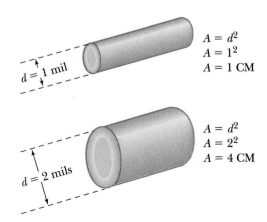

$A = d^2$
$A = 1^2$
$A = 1\ CM$

$A = d^2$
$A = 2^2$
$A = 4\ CM$

FIGURE 2–6 Cross-sectional area of conductors in circular mils (CM)

Example

If a wire has a diameter of 4 mils (0.004″), what is its cross-sectional area in circular mils?

Answer: $A = d^2$. $A = 4^2$. $A = 16$ CM

PRACTICE PROBLEMS I

What is the circular-mil cross-sectional area of a conductor whose diameter is 2.5 mils?

Because resistivity (ρ) of a material is a measure of a material's resistance per standard length and cross-sectional area (at a given temperature), ρ can be expressed in terms of circular-mil ohms per foot. That is, CM-Ω/ft.

Calculating the resistance of a conductor

A general formula for finding the resistance of a wire (assuming a given temperature) is:

FORMULA 2–2 $R = \dfrac{\rho l}{A}$

where R = resistance in ohms
  ρ = resistivity in CM-Ω/ft (for the type material in conductor)
  l = length in feet
  A = cross-sectional area of conductor in CM

Example

What is the resistance of a copper wire that is 250 feet in length and has a cross-sectional area of 39.75 CM? (NOTE: ρ for copper is 10.4 CM-Ω/ft.)

Answer: $R = \dfrac{\rho l}{A} = \dfrac{(10.4 \text{ CM-}\Omega/\text{ft})\,(250 \text{ ft})}{39.75\,CM} = 65.4\ \Omega$

PRACTICE PROBLEMS II

What is the resistance of a copper wire that is 300 feet in length and has a cross-sectional area of 2,048 CM? (Remember: ρ for copper is 10.4 CM-Ω/ft.)

Summary

The resistance a given conductor exhibits to current flow is directly related to its resistivity characteristic and its length, and inversely related to its diameter squared (its cross-sectional area). That is, the larger the wire (of a given type material), the lower its resistance per given length. The longer the wire, the higher its resistance per given cross-sectional area. See Figure 2–7. These are important facts to remember as you deal with conductors and conductive paths as a technician.

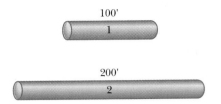

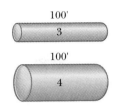

- Same type wire

- Same diameter for 1 and 2

- Conductor 2 is twice the length of conductor 1

- Conductor 2's resistance is twice that of Conductor 1, from one end to the other

- Same type wire

- Conductors 3 and 4 are the same length

- Conductor 4 has twice the diameter of 3 or ≈4 times the cm area

- Conductor 4's resistance is one-fourth that of conductor 3 from end to end

FIGURE 2–7 Relationship of wire resistance to conductor diameter and length

Wire sizes

There is a standardized sizing of wires by "wire gage numbers." This standard is referred to as the American Standard Wire Gage, or AWG. Figure 2–8 shows a tool some technicians use to measure the AWG size of wires. Wire sizes quickly communicate relative size information about conductors. For example, if a wire is categorized as an AWG #14 wire, its cross-sectional area in CM will be 4,107 CM. If the wire is sized as a #18 wire, its cross-sectional area in CM will be close to 1,624 CM. (NOTE: The higher the AWG number, the smaller the wire is in diameter [and hence, the less the cross-sectional area it has].) It is interesting to know that for each change of three wire-gage sizes, the CM area doubles or

FIGURE 2–8 A wire-gage tool *(Photo by Michael A. Gallitelli)*

halves, depending on which direction you move. For example, a #11 wire has approximately twice the cross-sectional area of a #14 wire. This means its resistance per unit length is approximately half that of #14 wire.

Wire table

A very useful source of information regarding conductor sizes and their resistances is a wire table, Figure 2–9. The wire table displays data regarding AWG wire numbers, their cross-sectional areas and their resistance in ohms per given length into one concise list or matrix. Most wire tables show the standard resistance per given length in terms of ohms per thousand feet.

PRACTICE PROBLEMS III

1. Refer to Figure 2–9 and determine the resistance of 500 feet of #12 wire.

2. What gage wire has a cross-sectional area of about 10,380 CM?

3. What is the resistance in ohms per thousand feet of the wire in question 2?

AWG WIRE SIZE	DIAMETER IN MILS	CIRCULAR-MIL AREA	OHMS PER 1,000 FT AT 25° C	CURRENT CAPACITY @ 700 CM/A
—	—	—	—	—
10	101.9	10,380	1.018	14.8
11	90.7	8,234	1.284	11.8
12	80.8	6,530	1.619	9.33
13	72.0	5,178	2.042	7.4
14	64.1	4,107	2.575	5.87
15	57.1	3,257	3.247	4.65
16	50.8	2,583	4.094	3.69
17	45.3	2,048	5.163	2.93
18	40.3	1,624	6.510	2.32
19	35.9	1,288	8.210	1.84
20	32.0	1,022	10.35	1.46
21	28.5	810	13.05	1.16
22	25.3	642	16.46	0.918
23	22.6	510	20.76	0.728
24	20.1	404	26.17	0.577
25	17.9	320	33.00	0.458
26	15.9	254	41.62	0.363
27	14.2	202	52.48	0.288
28	12.6	160	66.17	0.228
29	11.3	127	83.44	0.181
30	10.0	101	105.2	0.144
31	8.9	80	132.7	0.114
32	8.0	63	167.3	0.090
etc.	etc.	etc.	etc.	etc.

NOTE: Current-carrying capacity varies greatly, depending on environmental conditions. For example, a single #14 wire in open air can carry up to 32 amperes. In conduit, it can only carry a maximum of 17 amperes. At 700 CM/ampere, it can only carry about 5.87 amperes.

FIGURE 2–9 Sample partial wire table

As you might reason, the size of the conductor also dictates the maximum current it can safely carry without overheating and being destroyed. Some wire tables will also show this maximum current-carrying capacity for stated conditions, as well as providing the resistance-per-given length data. For a given type conductor, the larger the diameter of the wire, the greater current-carrying capacity it has. For example, as stated earlier, the cross-sectional area of the #11 wire is about double that of #14 wire. This means that the #11 wire can safely handle about double the current of the #14 wire. The general rule of thumb is that for every three wire sizes changed, the current-carrying capacity of the wire will double, or halve, depending on which direction you move in the wire sizes.

IN-PROCESS LEARNING CHECK II

1. The greater the resistivity of a given conductor, the (higher, lower) _____ its resistance will be per given length.

2. The circular-mil area of a conductor is equal to the square of its _____ in _____.

3. The smaller the diameter of a conductor, the (higher, lower) _____ its resistance will be per given length.

4. The higher the temperature in which a typical metal conductor must operate, the (higher, lower) _____ will be its resistance.

5. Wire tables show the AWG wire _____, the cross-sectional area of wires in _____, the resistance per given length of copper wire, and the operating _____ at which these parameters hold true.

2-4 RESISTORS

Uses

Resistors limit current and help determine current values for a given circuit. This current-controlling function is one of the important roles that resistors perform in electronics. Another function, which you'll study later, is dividing voltage. Both of these purposes are important.

Types

Resistors are designed to perform these functions under a variety of operating conditions. These include numerous voltage, current, and power level requirements. Also, various circuits have different demands related to accuracy of resistance, and stability of resistance value related to temperature variation. Add to these variables the fact that the physical size of a given resistor dictates its physical mounting method and electrical connection method.

Because of this diversity of requirements, several types of resistors have been created to meet these conditions. Two prominent types that you will commonly encounter are the carbon type resistor, Figure 2–10a and the higher power-handling wire-wound type, Figure 2–10b. As you might assume, the power handling capability of any resistor is related to its physical size and ability to dissipate

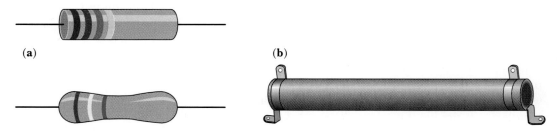

FIGURE 2–10 (a) Carbon resistor physical appearance; **(b)** wire-wound resistors

heat. It is likely that you will have many chances to see and use these two common types of resistors during your training.

Look at the chart in Figure 2–11 and study some basic features and characteristics of these, and several other useful type resistors. More in-depth knowledge of each of these type resistors will be gained as your training proceeds. As a technician or engineer, the considerations you will use in selecting or replacing resistors in circuits and systems will typically focus on power rating, voltage rating, accuracy of value, physical dimensions, and the component connection and mounting method requirements. In some cases, the temperature characteristics of the resistor also may be important.

Some other special-type resistor devices

In addition to the devices described in Figure 2–11, there are some special-application components or devices that might be considered in the resistor device family. One of these is called a fusible link, or a fusible resistor. These are resistive elements that are designed to open up the circuit should the current through them exceed a specified amount. (Much like your fuses or circuit breakers do at your home.) Of course, these must be replaced, should they open up. Their purpose is to limit current and also to provide circuit safety should the current get too high.

Because of today's robotics and automated production systems, a "zero-ohm" device has been created. This device allows industry to automate the insertion of a "glorified jumper wire" between two points on a printed circuit board, should the need arise. The very low resistance device (close to zero ohms) is packaged in a way that the automated machinery can insert it into the circuit board, rather than requiring a person to jumper the points with a wire, then solder, and so on.

Resistor types and symbols

You have been looking at some "fixed" resistance devices. Resistors may also come in "semifixed," or "variable" resistance forms. A fixed resistor is one in which the resistance between its terminals or component leads cannot be changed externally. See Figure 2–12a. A semifixed resistor has one contact, or terminal, that can be adjusted to any given spot along the resistance element, and affixed to that position, as shown in Figure 2–12b. Thus the resistance value can be set between this movable terminal and the end terminals at a desired value. Once adjusted, the movable element is not normally moved again. The variable resistor, however, has a movable contact that can be moved anywhere

Carbon-Composition Resistors

Key Characteristics: Values available from 1 ohm to several million ohms. Economical. Power ratings from $1/10$ to 2 watts. Use leads coming straight out each end (axial leads). Two designs are popular: the type for wiring to terminals and the type designed for automated insertion into printed circuit boards.

Construction: Resistive element is composed of ground up carbon and insulating material, held together by a resin binder. Resistive element and lead connections to that element are encapsulated in an insulating enclosure.

Precision Film-Type Resistors

Key Characteristics: Values can be accurately controlled. Quite immune to temperature changes. Have little undesired "noise" generation as well as other good electronic characteristics. Typically are more expensive than standard carbon composition resistors.

Construction: There are two types of film resistors. One uses a coating of carbon on a cylindrical ceramic substrate; the other uses a thin metal coating sprayed on the substrate. Both types generally have their resistive material cut in a spiral around the tubular form. The amount of resistance depends on the substances used and how much material is left in the spiral resistive element.

Surface Mount "Chip" Resistors

Key Characteristics: Tiny in size, good for compact applications. Have virtually zero lead length, which yields many electronic advantages. Ruggedly built and durable. Have good temperature characteristics. Because of their small size, these devices are typically available only in small power ratings ($1/8$ to $1/4$ watt).

Construction: Carbon layer deposited on a ceramic base or substrate. Contact to the resistive element is via some metal ends or terminals, which create virtually zero lead length. In application, these contacts then are soldered directly to a circuit board's conductive paths usually by automated soldering techniques.

Wire-Wound Resistors

Key Characteristics: In comparison to the other types of resistors, may be relatively large in physical size. Have more power-handling capability because of size and construction (e.g., 5 to 100 W). Can have good accuracy in resistance value. Have good temperature stability. Generally used in high-current applications.

Construction: A special resistance wire is wound around a tubular-shaped ceramic insulator form. The resistance characteristic of the wire and its length dictate the resistor's ohmic value.

FIGURE 2–11 Several common types of "fixed" resistors

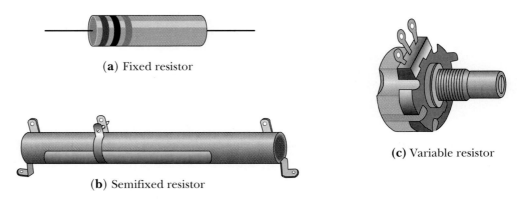

(**a**) Fixed resistor

(**b**) Semifixed resistor

(**c**) Variable resistor

FIGURE 2–12 Fixed, semifixed, and variable resistor constructions

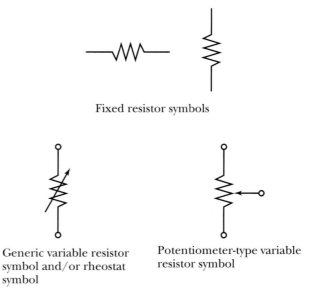

Fixed resistor symbols

Generic variable resistor symbol and/or rheostat symbol

Potentiometer-type variable resistor symbol

FIGURE 2–13 Schematic symbols for fixed and variable resistors

along the resistance element. In the variable resistor, the movable contact has a shaft attached to purposely enable easy changing of the contact's setting along the resistance element, Figure 2–12c.

Figure 2–13 shows the "schematic symbols" used to represent two of these types. As you will soon learn, schematic symbols are the technician's shorthand for representing electrical components and circuits on paper. You will become very familiar with these, and other electronic diagram symbols as you continue in your studies.

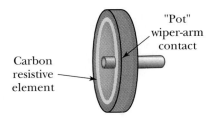

"Pot"
wiper-arm
contact

Carbon
resistive
element

FIGURE 2–14 Carbon resistor with color coding

FIGURE 2–15 Inside view of a potentiometer carbon element and wiper-arm contact

Resistor construction comparisons

Recall that cabon-composition resistors use a carbon-based resistive element and have a fixed resistance, depending on the mix of carbon and insulating material. See Figure 2–14. Please note the colored bands. You'll be studying about them later in this chapter.

The wire-wound resistors, on the other hand, use a special resistance wire for the resistive element. Refer again to Figures 2–10b and 2–11. One obvious difference between the "fixed" carbon resistor and the wire-wound types is that wire-wound resistors can be "fixed," "semifixed," or "variable."

Some variable resistors, called "potentiometers," have resistive elements that can be either carbon or wire-wound elements. Look at Figures 2–15 and 2–16 and observe that the position of the "wiper-arm" contact touching the circular carbon or wire-resistive element determines the amount of resistance from the center (wiper-arm) terminal to the terminals at each end of the potentiometer. You will learn in your further studies that the potentiometer is a valuable component for voltage-dividing functions. Volume controls for radios, audio amplifiers, and TVs are common applications for potentiometers.

Wire-wound resistors are often used where very low values of resistance are needed or where high current-carrying capacity is required. In Figure 2–16, you see a large wire-wound variable resistor, typically used as a rheostat.

2-5 RESISTOR COLOR CODE

A skill used continuously by electronics technicians is that of decoding the **resistor color code.** Because many resistors are so physically small, it is difficult to print their value in a readable size. A system of color coding has been adopted and standardized by the Electronics Industries Association (EIA); thus, technicians everywhere understand the code.

You should be aware that there are also resistors which can, and do, have their values labeled on them, and do not therefore need to have color coding. For those that do use the color code, the band system of color coding consists of three, four, or five bands (or stripes) of color around the tubular-shaped resistors. The first three stripes, when decoded, indicate the resistor's rated resistance value, in ohms. The fourth stripe, when present, indicates the tolerance percentage, or amount of deviation from the rated value that the resistor's actual value may vary in manufacture. If there is not a fourth band, the tolerance is automatically assumed to be ±20%.

FIGURE 2–16 High current-carrying capacity wire-wound variable resistor. This resistor can be wired as a two-terminal rheostat or as a three-terminal potentiometer.

FIGURE 2–17 Color band significance

A B C D E

A = First significant figure of resistance value

B = Second significant figure of resistance value

C = Decimal multiplier for resistance value

D = Tolerance rating (in percentage)

E = Reliability factor (in percentage, based on 1,000 hours of operation)

On certain types of resistors, when a fifth band is present, it is used to indicate a reliability factor. The rating is based on the percentage failure rate for 1,000 hours of service.

Also, many precision resistors (precision in terms of precise resistance values with very low tolerances) use a five-band color-coding system. In this case the first three stripes represent the first three digits of the resistance value. The fourth stripe is the decimal multiplier and the fifth stripe gives the tolerance rating.

During your training, you will probably be dealing mostly with the four-band, carbon-composition resistor type of color-coding.

Significance of color band placement

Look at Figure 2–17 and observe the placement of these color bands. The first color is considered to be the one closest to the end of the resistor. The other color bands progress toward the center of the resistor body. The key to decoding the color code is to learn the significance of each band position and of each color.

Refer again to Figure 2–17 and observe that the band positions are interpreted as follows:

1. The first band (closest to the end of the resistor) is the first significant digit in the resistor's ohmic value rating.

2. The next, or second, band represents the second significant digit in the resistor's ohmic value rating.

3. The third band indicates the multiplier, or number of zeros that should follow the first two numbers in order to know the resistor's rated ohmic value.

4. The fourth band provides the percentage tolerance information. That is how much, in percentage, the resistor can acceptably vary from its rated, color-coded value and still be within the manufacturer's specifications.

5. The fifth band, when present, indicates the failure rate (in percentage) per 1,000 hours of service. This is sometimes called the reliability factor.

Significance of colors

For resistors that are greater than 10 ohms in value, the 10 colors used to represent the values zero through nine (0–9) are used and are fairly easy to remember by means of a memory-aid sentence. As you can see in Figure 2–18, the color

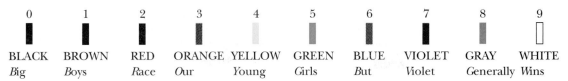

FIGURE 2–18 Sequence of colors memory aid

code starts with the dark colors at zero, and progresses toward lighter colors as the numbers increase. The figure shows how the sentence "*Big Boys Race Our Young Girls, But Violet Generally Wins*" can be used to remember the color code. These colors represent the sequential numbers zero through nine. The first letter of each word in the sentence is the same as the first letter of each color. Learn these!

Tolerance band colors

Gold and silver are frequently used tolerance band colors. Gold indicates ±5% tolerance from rated value. For example, if a resistor is color coded as a 100-ohm resistor with ±5% tolerance rating, its actual value could be anywhere between 95 and 105 ohms, and still be within the manufacturer's rating limits.

Silver represents a tolerance of ±10%. In this case, a resistor color coded 100 ohms could actually have a value between 90 ohms and 110 ohms and still be considered good.

As stated earlier, if there is no fourth band, the tolerance rating is ±20% of rated value.

Use of gold and silver colors as multipliers

Low value resistors (below 10 ohms) require a special multiplier value for the third band. This indicates that the numbers decoded from the first two color bands are to be multiplied by 0.1 or 0.01 to find the value for the resistor. A gold third band indicates the multiplier is one-tenth (0.1). For example, if the first two color bands are red and violet (27), then a gold third band would indicate a resistor value of 0.1 × 27, or 2.7 ohms.

A silver third band indicates a multiplier of one-hundredth (0.01). This means if the resistor is color coded brown-black-silver (in that order), the resistor value is 0.1 ohm (0.01 × 10 = 0.1).

A summary chart showing the significance of various colors used in the color code is shown in Figure 2–19.

Examples

Let's see how to apply the combination of color information and band position to decode some resistor values and tolerances. Also, we'll look at one case where a reliability factor is given. Notice in Figure 2–20a that the first band is yellow, representing the number 4. The second band is violet, representing 7. The third band is orange, indicating a multiplier of 1,000, which adds three zeros to the 47. Therefore, the resistor color-coded value is 47,000 ohms (or 47 kΩ). The fourth band is silver, indicating a 10% tolerance rating. This means that the resistor's actual value can be 47,000 ohms, plus or minus 4,700 ohms. That is, its

COLOR	NUMBER COLOR REPRESENTS (IN 1ST AND 2ND BANDS)	DECIMAL MULTIPLIER COLOR REPRESENTS (IN 3RD BAND)	TOLERANCE COLOR REPRESENTS (IN 4TH BAND)	RELIABILITY FACTOR COLOR REPRESENTS (IN 5TH BAND)	PRECISION RESISTOR TOLERANCE (IN 5TH BAND)
Black	0	1 (no zeros)	—	—	—
Brown	1	10 (1 zero)	—	1%	±1%
Red	2	100 (2 zeros)	—	0.1%	±2%
Orange	3	1,000 (3 zeros)	—	0.01%	—
Yellow	4	10,000 (4 zeros)	—	0.001%	—
Green	5	100,000 (5 zeros)	—	—	±0.5%
Blue	6	1,000,000 (6 zeros)	—	—	±0.25%
Violet	7	10,000,000 (7 zeros)	—	—	±0.1%
Gray	8	100,000,000 (8 zeros)	—	—	—
White	9	1,000,000,000 (9 zeros)	—	—	—
Gold	—	0.1	5%	—	—
Silver	—	0.01	10%	—	—

FIGURE 2–19 Color significance summary chart

value should be between 47,000 + 4,700 ohms (51,700 ohms) and 47,000 ohms − 4,700 ohms (42,300 ohms).

In Figure 2–20b, the color code is brown, black, orange, gold, in that order. Decoded, this value is 10,000 ohms, ±5%. In Figure 2–20c, the five-band color code is red, red, red, silver, brown. The value of the resistor is therefore 2, 2, and two zeros, or 2,200 ohms. The tolerance is ±10%. The reliability factor is 1%. (See Figure 2–19.)

Now you try a couple!

PRACTICE PROBLEMS IV

1. What is the resistor value and tolerance of the resistor shown?

Orange White Orange Silver

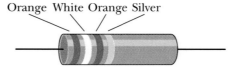

2. What is the resistor value, tolerance, and reliability rating of the resistor shown?

Brown Gray Orange Gold Red

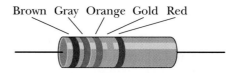

Special precision resistors

Although most resistors you will work with use the color coding as presented thus far, there are some precision resistors with tolerances of 2%, or less, that use a

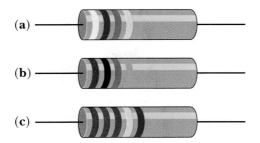

(a)

(b)

(c)

FIGURE 2–20 (a) 47 kΩ, 10% tolerance resistor; **(b)** 10 kΩ, 5% tolerance resistor; **(c)** 2.2 kΩ, 10% tolerance, 1% reliability factor

special five-band color-code technique. In these cases, the first three bands are value digits and the fourth band is the multiplier. The fifth band, then, provides the tolerance rating. For these special resistors, the fifth band tolerance rating uses the colors: brown = ±1%, red = ±2%, green = ±0.5%, blue = ±0.25%, and violet = ±0.1%; see Figure 2–19.

Other exceptions

We mentioned earlier that some resistors are labeled rather than color-coded. For example, the tiny resistors designed for surface mount technology (or chip resistors), often use a three-digit code to provide the device's resistance information. These three numbers are used in the same way that the first three color bands on carbon composition resistors are used. For example, the first digit on the chip resistor tells you the first number in the resistor's value. The second digit indicates the second number in its value. The third digit, then is the "multiplier" number, telling you how many zeros to add to the first two numbers to determine the resistor's value. If the chip resistor were coded with the digits 103, its value would be 1, then 0, followed by three zeros, or 10,000 ohms (10 kΩ).

There are also some other types of resistors that use a cryptic numbering and lettering system. The letter R indicates the decimal point position for resistors under 100 ohms. For example, if a resistor is labeled 2R7, its value is 2.7 ohms. Letters are also used to designate tolerance ratings, where the letter F means ±1%, G = ±2%, J = ±5%, K = ±10%, and M = ±20%. So if the label shows 2R7K, the resistor is a 2.7-ohm resistor with 10% tolerance.

For resistor values requiring more than three numbers, plus a multiplier, the first three labeled numbers are followed by a fourth number, which is the multiplier, indicating the number of zeros to follow the preceding numbers. An example of this might be a resistor labeled 4702K. In this case, the value is 47,000 ohms and the tolerance is ±10%.

2–6 MEASURING VOLTAGE, CURRENT, AND RESISTANCE WITH METERS

The importance of knowing how to measure V, I, and R

As a technician, you will frequently measure voltage, current, and resistance in electrical or electronic circuits. These measurements are sometimes used to set up or adjust circuit conditions for proper operation. Measurements are often used to check that a circuit is operating properly or to troubleshoot circuits that are not operating properly. The skill to make measurements correctly will become a valuable tool to you. In this section, we will introduce you to the most

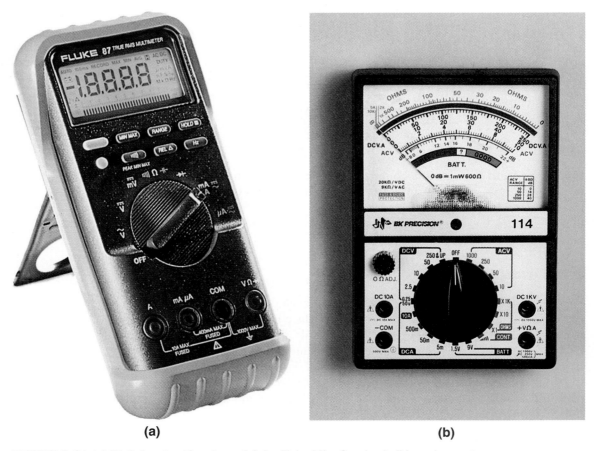

(a)   (b)

FIGURE 2–21 (a) Digital meter *(Courtesy of John Fluke Mfg. Co., Inc.)*; **(b)** analog meter *(Courtesy of B & K Precision)*

elemental concepts of measuring *V, I,* and *R.* Mastery of this measurement skill will only come with actual practice.

Types of meters

Two general categories of meters you will use in making measurements of *V, I,* and *R* are analog meters and digital meters, Figure 2–21. You will learn more about the significance of the terms analog and digital as you progress in your studies. For now, you can see that the scale on the digital meter is directly readable, just by reading the numbers in the readout portion of the meter. The analog meter, however, takes some practice in interpreting values in terms of where the pointer rests along the continuous scale. You should strive to become adept at using either type of meter, since both kinds are found in industry.

Symbols for meters

In Figure 2–22 you can see both the general schematic symbols and the pictorial symbols we will use in this book to represent analog and digital voltmeters,

ANALOG METER PICTORIALS

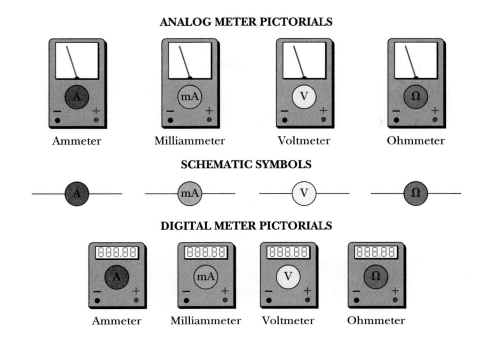

Ammeter   Milliammeter   Voltmeter   Ohmmeter

SCHEMATIC SYMBOLS

DIGITAL METER PICTORIALS

Ammeter   Milliammeter   Voltmeter   Ohmmeter

FIGURE 2–22 Pictorial and schematic meter symbols

ammeters (ammeters and milliammeters), and the ohmmeter (which measures resistance).

Making a voltage measurement with a voltmeter

To measure the voltage between two points, you simply connect the voltmeter across (or in parallel with) the two points at which you wish to measure the potential difference, Figure 2–23. Notice, when measuring dc voltage, it is necessary to connect the negative test lead from the meter to the more negative point

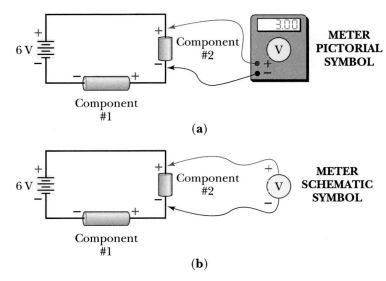

FIGURE 2–23 Measuring voltage "across" a component

in the circuit and the positive test lead to the more positive point in the circuit. You can see in the figure that the meter is reading the value of voltage across one component in the circuit (in this case, component number 2).

> When preparing to make a circuit test with an electrical meter, turn the power off (if this is both possible and reasonably practical) while connecting the meter's test leads to the circuit!

Making a current measurement with an ammeter

An illustration of how to connect an ammeter or a milliammeter to measure circuit current is shown in Figure 2–24. Observe that the circuit has to be broken for the meter to be put in the circuit's current path for measurement purposes. You can also see that in this dc circuit, the meter leads must be connected so the current enters the meter's negative terminal, travels through the meter, and then exits the meter from the meter's positive terminal. You will learn more details about the concept of polarity a little later.

> When measuring an unknown level of current or voltage with a meter that is not auto-ranging, always set the meter on its highest range and "work down" to the best range for reading the meter.

Making a resistance measurement with an ohmmeter

Very important *cautions* to be observed when using an ohmmeter are that *power should be off the circuit and the component to be measured should be disconnected from the remainder of the circuit!* For example, to measure the resistance of a component, as shown in Figure 2–25:

1. turn the power off the circuit by opening the switch;
2. disconnect one resistor lead from the remainder of the circuit; and
3. make the resistance measurement by connecting the ohmmeter across both leads of the resistor.

Key points regarding using meters

As indicated earlier, you can only gain practical experience in using meters to make measurements by actually making measurements. When learning and practicing this skill, be sure to observe all the safety precautions to protect yourself and the test equipment.

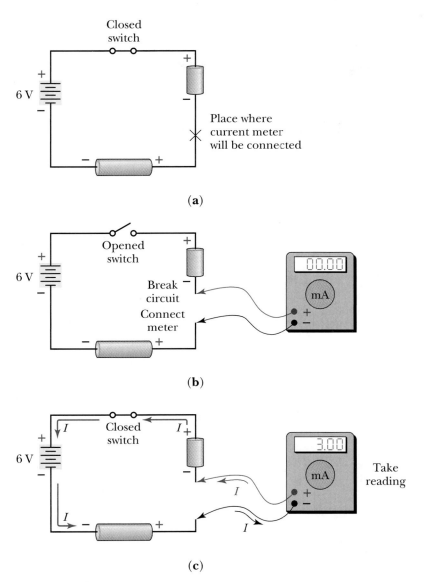

(a)

(b)

(c)

FIGURE 2–24 Measuring current "through" a circuit

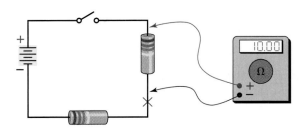

NOTICE: Power is off due to open switch. Component
being measured is isolated from the rest of the
circuit by disconnecting one of its leads (at "×").

(a)

FIGURE 2–25 Ohmmeter measurement

//

SAFETY HINTS

1. Turn power off in circuit to be tested when connecting meter leads.

2. Appropriately connect meter test leads to circuit. Observe proper polarity if measuring dc voltage or current.

3. Make sure meter is in correct mode (ac, dc, V, A, Ω).

4. Make sure range selected on meter is high enough to handle the value that is about to be tested.

5. a. If measuring V or I, turn on power and take reading.

   b. If measuring R, *do not turn on power.*

6. Turn off power and remove meter test leads from circuit.

7. If measuring I, reconnect circuit, as appropriate.

1. When measuring voltage, connect the voltmeter test leads to the two points at which you wish to know the potential difference, in volts. (That is, connect across, or in parallel, with the component or points to be checked.)

2. When measuring current, break the circuit and insert the current meter leads in line (in series) with the current path through which you are trying to measure the current.

3. If measuring dc voltage or current, be sure to observe the direction of current through the circuit, and connect the positive and negative test leads accordingly. (Observe polarity.)

4. When measuring resistance, be sure to *turn off the power* to the circuit, isolate the component to be measured, then make the resistance measurement, as appropriate.

IN-PROCESS LEARNING CHECK III

1. In the diagram shown, if you wanted to measure the dc voltage across component #1, at which point in the circuit would you attach the meter's negative test lead? At which point would you connect the meter's positive test lead?

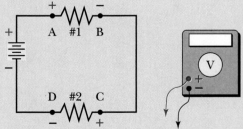

Meter negative lead connected to Point _____.

Meter positive lead connected to Point _____.

2. In the diagram shown, if you wanted to measure the dc circuit current and you were going to break the circuit at point "X," to which point would you connect the ammeter's negative lead? At which point would you connect the ammeter's positive lead?

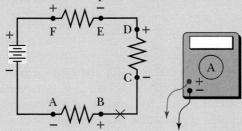

Meter negative lead connected to Point _____.

Meter positive lead connected to Point _____.

3. In the diagram shown, if you wanted to measure resistance, what is the first thing you would do, before connecting the meter across the component to be measured?

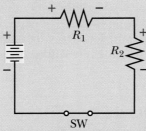

2-7 DIAGRAMS USED FOR ELECTRONIC SHORTHAND

Types of diagrams

During your experiences in electronics you will encounter various types of diagrams which can be helpful. Several of these types include **block diagrams, schematic diagrams,** chassis layout diagrams, and others. For now, we'll focus on the block diagram and the schematic diagram since these are the kind you will be using most.

Some basic symbols

Electronic circuits are commonly represented by either block diagrams, schematic diagrams, or both. In order to interpret and use schematic diagrams, it is necessary to learn some basic electronic symbols used in schematics. Because diagrams can be made to represent an infinite variety of circuits and configurations in a concise manner, schematic diagrams are the electronic technician's or engineer's shorthand.

At this point, we'll introduce you to some of the common schematic symbols you will be using almost immediately. Additional schematic symbols will be introduced throughout the text, as you need to know them. Become familiar with the symbols; the details of the represented components' operation and applications will unfold as your learning continues.

Examine Figure 2–26 and observe a few of the important electronic symbols. Don't be worried about committing these to rote memory at this time. You will automatically learn these by repeated usage as you continue in your studies.

Block diagrams

Block diagrams are useful because they are simple and conveniently portray the "functions" relating to each block. Recall that we used a simple block diagram in an earlier chapter when a simple electrical system was discussed.

Refer to Figure 2–27 (see page 63) and see if you can understand the concept of a block diagram. In essence, a block diagram is created by analyzing a system: breaking it into meaningful subsystems, which are drawn as individual blocks; then, interconnecting the subsystem blocks in a way that portrays flow and interaction between the blocks.

Note that the diagram quickly conveys what the system does. That is, this system distributes electrical power from a source to several loads. Also, it graphically shows the direction of flow of the electrical energy between blocks. Additionally, it very simply displays what each part of the system does. The power source supplies power, the distribution components control and allocate portions of the power to the various loads, and the loads use the power delivered to them.

Now, see if you can interpret the block diagram of a system most everyone is familiar with.

PRACTICE PROBLEMS V
Look at Figure 2–28 (see page 63) and answer the following questions:

1. What is the purpose of this system?
2. In which direction does the TV signal flow on the diagram, from right-to-left or from left-to-right?
3. What subsystem receives signals from the TV antenna?
4. What are the two types of outputs from the TV receiver circuitry?

Schematic diagrams

Now that you have briefly seen the usefulness of block diagrams, let's move on to the schematic diagrams. In schematic diagrams you apply the knowledge of electrical symbols. Because the symbols are standardized, it is easy for others working in electronics to "decode" the information that the original designer of the schematic is conveying. The information is concise and useful in communicating the detailed electrical connections in the circuitry, the types and values of components, and other important information.

Look at the sample pictorial and schematic diagrams in Figure 2–29 (see page 63) and notice the following facts:

1. Each component has an individualized callout. For example, one resistor is called R_1, another is called R_2, and so on. The switch is labeled S_1. If there were more switches, the series would carry on with S_2, and so on.

COMPONENT TYPES	PICTORIAL REPRESENTATIONS	SCHEMATIC SYMBOLS
Conductors (connected)		
Conductors (not connected)		
Cell		
Battery		
Switch (SPST) Single-pole Single-throw		
Switch (SPDT) Single-pole Double-throw		
Switch (DPST) Double-pole Single-throw		
Switch (DPDT) Double-pole Double-throw		
Switch Rotary type		
Switch (NOPB) Push-button type (normally open)		NOPB
Switch (NCPB) Push-button type (normally closed)		NCPB

FIGURE 2–26 Various components and their schematic symbols

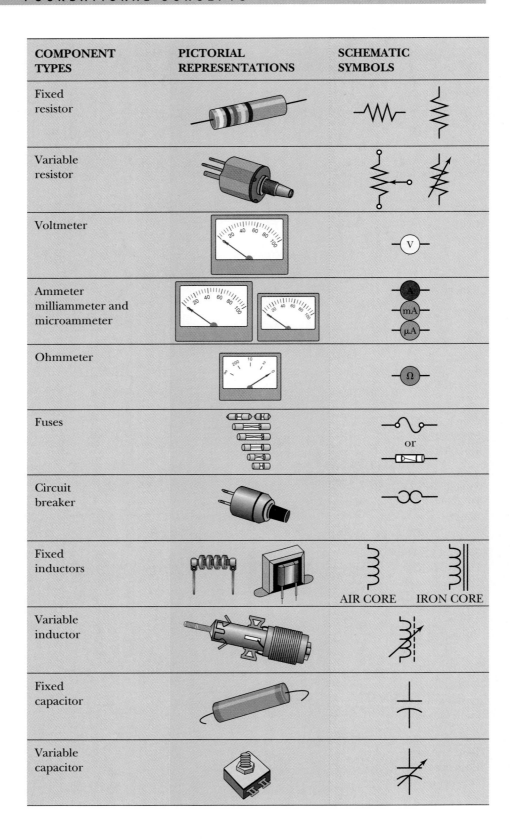

COMPONENT TYPES	PICTORIAL REPRESENTATIONS	SCHEMATIC SYMBOLS
Fixed resistor		
Variable resistor		
Voltmeter		
Ammeter milliammeter and microammeter		
Ohmmeter		
Fuses		or
Circuit breaker		
Fixed inductors		AIR CORE IRON CORE
Variable inductor		
Fixed capacitor		
Variable capacitor		

FIGURE 2–26 (cont.)
Various components and their schematic symbols

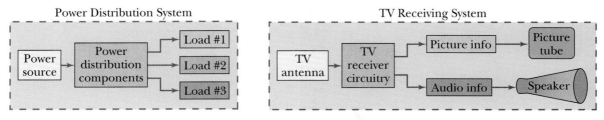

FIGURE 2–27 Sample block diagram   **FIGURE 2–28** Sample block diagram problem

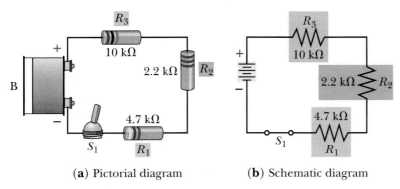

(**a**) Pictorial diagram   (**b**) Schematic diagram

FIGURE 2–29 Sample pictorial and schematic diagrams

2. In the pictorial, you can see that the resistor color-code bands are shown. In this schematic, the resistor values are given. If there were other components with important value data, their values would also be shown on the diagram.

3. In both the pictorial and the schematic diagrams, the battery terminals are marked with + and − , as appropriate. This marking of polarity can be given at each component; however, since you will study more about polarity in the next chapter, we will wait until then to use these polarity markings throughout the diagram.

4. Finally, an extremely important feature of all schematic, pictorial, or other wiring diagrams is that every circuit interconnection is shown, allowing the reader to understand every component's electrical relationship to every other component in the circuit.

Also, on many schematic diagrams, the values of electrical parameters such as voltage, current, resistance, and so on are called out. These data help immensely when technicians are trying to troubleshoot a circuit.

PRACTICE PROBLEMS VI

Look at Figure 2–30 and answer the following questions:

1. How many different kinds of components are there in the circuit? Name them.

2. What is the highest value resistor?

3. What kind of switch is used, a SPST or a SPDT?

4. Is the voltage source a cell or a battery?

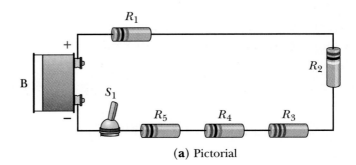

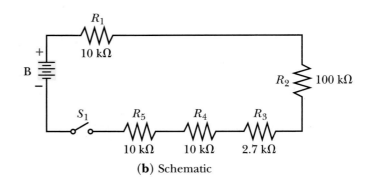

(a) Pictorial

FIGURE 2–30 Diagrams for Practice Problems VI

(b) Schematic

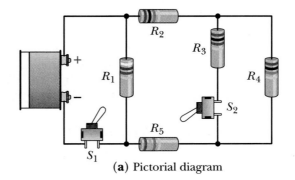

(a) Pictorial diagram

FIGURE 2–31 Sample pictorial and schematic diagrams for Practice Problems VII

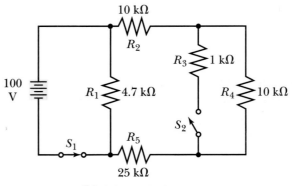

(b) Schematic diagram

PRACTICE PROBLEMS VII

Look at Figure 2–31 and answer the following questions:

1. How many resistors are in the circuit?
2. Which resistor has the lowest resistance value? What is its value?
3. How many switches are in the circuit?
4. What is the value of the circuit applied voltage?
5. Which switch is opened and which is closed?

Did you do well? As you can see, schematics can provide a large amount of information in a little space. Your appreciation for the value of schematics will undoubtedly continue to grow with your experiences in electronics.

SAFETY HINTS

You may have already been exposed to some safety hints about electricity and electronics. Nevertheless, here are a few facts and tips that can serve either as information or as reinforcement of earlier learning.

Background facts

Electrical shock occurs when your body provides a path for current between two points having a difference of potential. Generally, the amount of injury a person receives from a shock depends on the amount of current forced through the path of current. A range of effects from minor discomfort, severe muscular contractions, severe burns, ventricular fibrillation (erratic heart action), and death can occur. Many times the injury caused by involuntary jerking or muscular contractions is worse than the shock. Even very low values of current, if they flow through your body under the right conditions, can be fatal. Therefore, use extreme caution, no matter what value of voltage or current you are working with.

A higher voltage causes more current to flow through a circuit. The higher the resistance path, the lower the current. Therefore, anything that lowers the voltage on those circuits or any means of increasing resistance of possible current paths through the body is helpful.

Some general shop or laboratory safety tips

1. *Remove power* from circuits before working on them, as appropriate.
2. If you must work on a circuit with power applied, *work with only one hand at a time.* This will prevent touching two points in the circuit simultaneously and providing a current path from one hand, through your body, to the other hand. Also, avoid touching two points with any part of your body. For example, your hand touching a point in the circuit, and your arm leaning against the circuit chassis, ground, or another point in the circuit can provide "shocking results."
3. *Insulate your body* from ground so a path is not provided from one point (through your body) to ground.

continued

//

4. Use only *tools with insulated handles.*

5. *Know where the closest switch or main breaker switch is located* that will turn off power to the system. Make sure everyone working near you also knows this.

6. *Remove metal jewelry.* Dangling jewelry can not only make electrical contact, but, if you are working on "rotating machinery," can entangle you with the machinery.

7. When working on circuits with capacitors that store charge even after the power has been removed, *discharge the capacitors* before working on the circuit.

8. *Ground all equipment that should be grounded,* including equipment chassis, cabinets, and so forth.

Safety when working with hot soldering irons

1. *Position hot soldering irons properly* to prevent accidentally laying your arm, hands, or other objects on them.

2. *Wear safety glasses* when soldering.

3. *Be careful when wiping hot solder off a soldering iron.* Hot or molten solder splatters easily and causes serious burns.

NOTE: Other safety hints will be provided at strategic times during your training. Don't just read these safety hints; *follow them!*

SUMMARY

▶ Charge (Q or q) is measured in coulombs. The abbreviation for coulomb is C. A typical subunit of the coulomb is the μC, or microcoulomb, meaning one-millionth of a coulomb.

▶ Potential difference, or voltage, is represented by the letter V. The unit of measure is the volt, also abbreviated "V." The submultiples and multiple of voltage commonly found include mV (millivolt), μV (microvolt), and kV (kilovolt).

▶ Current (I) is measured in amperes. The abbreviation for the unit of measure is A. Typical submultiples of this unit are mA (milliampere), μA (microampere), and occasionally pA (picoampere).

▶ Conductance uses the letter G and the unit of measure is the siemens (S). The commonly found submultiples are mS (millisiemens) and μS (microsiemens).

▶ Resistance uses the letter R and the unit of measure is the ohm (Ω). Typical multiples of this unit are the kΩ (kilohm) and the MΩ (megohm).

▶ Several components found in many circuits are resistors, switches, batteries (or voltage sources), and the conductor wires. Know the standard symbols used for each.

▶ Commonly used metric subunit prefixes are pico, or micro micro, meaning one millionth of one millionth (one trillionth) of a unit; nano, meaning one thousandth of a millionth (a billionth) of a unit; micro, meaning one millionth of a unit; and milli, meaning one thousandth of a unit.

▶ Powers of ten for the above subunit prefixes are 10^{-12} for pico (or micro micro), 10^{-9} for nano, 10^{-6} for micro, and 10^{-3} for milli units.

► Common metric multiple-unit prefixes are kilo (1,000 times the unit) and mega (1,000,000 times the unit).

► Symbols for all of the above metric prefixes are p (pico), n (nano), μ (micro), m (milli), k (kilo), and M (mega).

► Conductors can vary in the materials they are made from, in physical size, in length, and in overall makeup. Solid and stranded wires are two basic types of conductors. Other differences between conductor types may relate to how they are packaged and the type of insulation used.

► There is a standard wire-gaging system called the American Standard Wire Gage, abbreviated AWG. The lower the gage number, the larger the wire.

► The cross-sectional area of conductors is measured in circular mils (CM). One mil equals one-thousandth of an inch. A one-CM wire has a diameter of 1 mil, or 0.001 inch.

► The larger a conductor's size (i.e., diameter and cross-sectional area) the lower its resistance, and the higher its current-carrying capacity. The greater the length of any given conductor, the greater its resistance from end to end.

► Resistivity (specific resistance) of any given type material is a measure of that material's resistance per standard length and cross-sectional area. The formula for resistivity is $R = \rho l / A$.

► A wire table is a convenient source of information regarding a conductor's resistance per given length and size, as well as its current-carrying capacity for specified conditions.

► Resistors perform several useful functions in electronic circuits: They limit current and divide voltage.

► Several types of resistors are carbon or composition resistors, film resistors, surface mount "chip" resistors, and wire-wound resistors.

Another way to categorize resistors is fixed, semifixed, and variable resistors.

► Carbon or composition resistors normally are manufactured with power dissipation ratings from ¼ to 2 watts. Wire-wound resistors generally are rated to handle power levels from 5 watts to over 100 watts.

► Two other types of resistors commonly used are precision resistors (often film-type resistors) and the surface-mount technology "chip" type. Because of their small size, chip resistors come only in very small power ratings from about ⅛ to ¼ watt; however, can have resistance values ranging from less than an ohm to in excess of a million ohms.

► Because carbon resistors are typically very small, a color-coding system has been devised to indicate resistor values.

► The sentence *"Big Boys Race Our Young Girls, But Violet Generally Wins"* can help you remember the ten colors in the code: Black, Brown, Red, Orange, Yellow, Green, Blue, Violet, Gray, and White.

► Block diagrams illustrate the major functions within a circuit or system and convey the flow of signals, electrical power, and so forth. General flow is typically shown from left to right.

► Schematic diagrams provide more information than block diagrams about types of components in the circuit(s), their values, and specific connections and related electrical parameter values.

► Block diagrams provide an overview of the system, whereas schematic diagrams provide the information needed to enable circuit analysis, circuit repairs or maintenance, and/or desired circuit modifications.

► Safety rules relating to electrical circuits, hot soldering irons, and rotating machinery should always be followed.

FORMULAS AND SAMPLE CALCULATOR SEQUENCES

FORMULA 2–1 $A = d^2$

diameter value (in mils), $\boxed{x^2}$

FORMULA 2–2 $R = \dfrac{\rho l}{A}$

Resistivity (CM-Ω/ft), **⊗**, length (ft), **÷**, area (CM), **⊜**

REVIEW QUESTIONS

1. Convert 3 mV to an equivalent value in V.

2. If a value of current is 12 μA, this can be stated as a value of 12 × 10 to what power amperes? What is this same value of current when stated as a decimal value of amperes?

3. What submultiple of a milli unit is a micro unit?

4. How many kΩ are there in a MΩ?

5. Draw a schematic diagram illustrating a voltage source (battery), two resistors, and a SPST switch all connected end-to-end. Illustrate the first resistor connected to the source's negative side, the other resistor connected to the source's positive side, and the switch between the resistors.

6. is the symbol for a fixed _____.

7. is the symbol for a variable _____.

8. is the symbol for a _____.

9. is the symbol for a _____.

10. Convert the following units:
    a. 15 milliamperes to amperes.
    b. 15 milliamperes to 15 times a power of 10.
    c. 5,000 volts to kilovolts.
    d. 5,000 volts to 5.0 times a power of 10.
    e. 0.5 megohms to ohms.
    f. 100,000 ohms to megohms.
    g. 0.5 ampere to milliamperes.

11. Name two or more types of resistors and discuss their physical construction.

12. Give the resistance value and tolerance rating for the following color-coded resistors: (NOTE: The first color shown is the first band; the second color represents the second band, and so forth.)
    a. Yellow, violet, yellow, and silver.
    b. Red, red, green, and silver.
    c. Orange, orange, black, and gold.
    d. White, brown, brown, and gold.
    e. Brown, red, gold, and gold.

13. Name at least three important parameters, or factors, to consider when selecting a resistor for a given circuit.

14. If a precision resistor has a fifth color band, what does this band designate?

15. What is the value and tolerance rating of a precision film-type resistor having the sequence of color bands: red, red, red, red, red?

16. What is the tolerance rating of a carbon-composition resistor having only three color-code bands?

17. If you had to replace a resistor having a power rating of 25 watts; what type of resistor would you likely be trying to procure?

18. What is the value of a surface-mount, chip resistor having coding numbers of 443?

19. If a resistor has a "reliability" coding, which band on the color-coded resistor would you use to determine that rating? If that band's color were orange, what percentage failure rate would you expect of that type resistor over 1,000 hours?

20. What is a very basic procedure you should use to prevent electrical shock when you are about to work on an electrical circuit?

21. List at least two physical precautions you can use to minimize the danger of shock if you *must* work on an energized electrical circuit?

22. What added precaution is used when working on circuits containing charged "capacitors"?

23. Name two possible dangers of wearing loose metal jewelry when working with electrical circuits or when using rotating machinery.

24. Name three special precautions you should use when working with hot soldering irons.

25. What is the basic situation that causes someone to get electrically shocked?

ANALYSIS QUESTIONS

1. In your own words, describe why a resistor color code is necessary.

2. If the environmental temperature conditions for a certain conductor's operation increase, describe what happens to the following parameters:
   a. Its resistance
   b. Its current-carrying capacity

3. Explain in your own words what the reliability rating on certain resistors means.

4. Describe in your own words the procedures and precautions for:
   a. Measuring dc voltage across a specified component in a multicomponent circuit.
   b. Measuring circuit current.
   c. Measuring the value of a specified resistor in a multiresistor circuit.

5. Draw a block diagram illustrating a simple power distribution system. Use at least three blocks, and label all blocks and diagram features.

6. Draw a schematic diagram showing two resistors, one meter, two switches, and a battery. Label all components.

7. List the color coding for the following resistors:
   a. 1,000 ohms, 10% tolerance.
   b. 27 kilohms, 20% tolerance.
   c. 1 megohm, 10% tolerance.
   d. 10 ohms, 5% tolerance.

8. List at least four safety precautions that should typically be used when connecting *any* type of measurement meter to an existing circuit.

9. Referring to Figure 2–26, if you were wiring a "fixed" resistor into a circuit, how many electrical connection points would you need to connect?

10. Referring to Figure 2–26, if you were wiring a "variable" resistor into a circuit, how many electrical connection points would you need to connect?

PERFORMANCE PROJECTS CORRELATION CHART

Suggested performance projects that correlate with topics in this chapter are:

CHAPTER TOPIC	PERFORMANCE PROJECT		PROJECT NUMBER
Making a Voltage Measurement with a Voltmeter	Voltmeters	(Use & Care of Meters series)	1
Making a Current Measurement with an Ammeter	Ammeters	(Use & Care of Meters series)	2
Making a Resistance Measurement with an Ohmmeter	Ohmmeters	(Use & Care of Meters series)	3
Resistor Color Code	Resistor Color Code	(Ohm's law series)	4

NOTE: It is suggested that after completing the above projects, the student should be required to answer the questions in the "Summary" at the end of this section of projects in the Laboratory Manual.

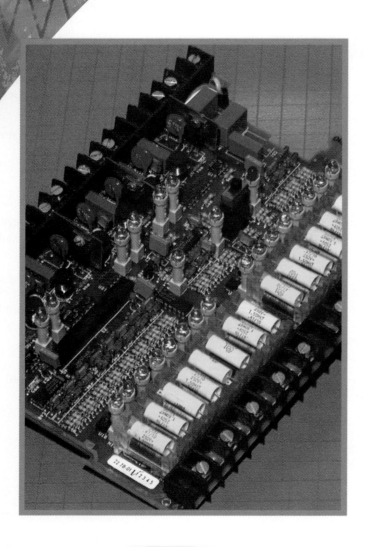

OHM'S LAW

3

chapter

CHAPTER PREVIEW

This chapter introduces you to Ohm's law. Ohm's law is fundamental to your career as a technician or engineer. It is the basis for calculations of electrical quantities in all types of electrical and electronic circuits.

You will have opportunity to use the knowledge you have already acquired and combine it with some new and very practical circuit analysis information. For example, you will practice your previously learned knowledge and skills in working with equations, using metrics, using powers of 10, and using a scientific calculator. For those who need a brief review in some of these skills, Appendices D and E provide a means for doing that.

OBJECTIVES

After studying this chapter, you should be able to:

1. Explain the relationships of current, voltage, and resistance
2. Use Ohm's law to solve for unknown circuit values
3. Illustrate the direction of current flow and polarity of voltage drops on a schematic diagram
4. Use metric prefixes and powers of 10 to solve Ohm's law problems
5. Use a calculator to solve circuit problems
6. Explain power dissipation
7. Use appropriate formulas to calculate values of power

3-1 OHM'S LAW AND RELATIONSHIPS BETWEEN ELECTRICAL QUANTITIES

Georg Simon Ohm, after whom the unit of resistance is named, observed and documented important relationships between three fundamental electrical quantities. He verified a relationship that indicated that electrical current in a circuit was *directly related* to the voltage applied and *inversely related* to the circuit's resistance. Recall that if two items are *directly related,* as one item increases, the other item also increases, and it increases proportionately to the first item's increase. If one item goes down, the other items goes down, proportionately. *Inversely related* means that as one item increases, the other item decreases, proportionately, and vice versa. Also recall that in defining some of the electrical quantities, the relationships in **Ohm's law** were implied in the definitions. For example, an ohm is that amount of resistance that limits current to one ampere when one volt is applied.

The relationship of current to voltage with resistance constant

These relationships are stated in a formula known as Ohm's law, where:

Current (amperes) = Potential difference (volts) / Resistance (ohms)

FORMULA 3–1 $I = \dfrac{V}{R}; \left(\text{Amperes} = \dfrac{\text{Volts}}{\text{Ohms}} \right)$

Example

If a circuit has 10 volts applied voltage and a resistance of 5 ohms, what is the value of current in amperes?

$$I = \frac{V}{R}, \text{ where } I = \frac{10 \text{ V}}{5 \text{ }\Omega} = 2 \text{ A (See Figure 3–1, page 74)}$$

As you can see from the formula, if V increases and R remains unchanged, then a larger value V divided by the same value R will produce a larger number for I. This means that with a given resistance value, if voltage is increased, then current must increase also, and it will *increase* in direct proportion to the voltage change. Compare the current in Figure 3–1a with the current in Figure 3–1b. As you can see, when voltage increased from 10 V to 15 V and the resistance remained at 5 Ω, the current increased from 2 A to 3 A. That is directly proportional to the voltage change, isn't it? As voltage increased to one and one-half times the original value, current also increased to one and one-half times its original value.

On the other hand, if voltage is decreased while the resistance remains the same, the current will *decrease* proportionately with the voltage change. Compare the current in Figure 3–1c with the current shown for the circuit in Figure 3–1a. Again, you can see a direct proportion. The circuit current in Figure 3–1c is half the value of 2 A, shown in Figure 3–1a. This is due to the fact the applied voltage in the circuit of Figure 3–1c (5 V) is half of the 10-V applied voltage, shown in Figure 3–1a. This type relationship may be referred to as a "linear proportionality."

We conclude, then, that *current is directly related to voltage* for a given resistance value. If the circuit applied *voltage increases* for a given resistance value, *current increases* in direct proportion to the increase in voltage. If *voltage decreases* for a given resistance value, *current* through the circuit will *decrease* in direction proportion to the voltage change.

The relationship of current to resistance with voltage constant

In Figure 3–2a (page 75), the voltage is 10 V and the resistance value is 10 Ω. Thus, the current value is 1 A.

Recall, in Figure 3–1a, when the resistance was 5 Ω, and the applied voltage was 10 V, current was 2 A. By *increasing* the resistance to 10 Ω, with the same value of voltage (that is, 10 V), the circuit current *decreases* to 1 A. In other words, with V remaining the same and R increasing, the I decreases proportionately, showing an inverse relationship between I and R. Current is *inversely related to resistance.*

In Figure 3–2b, you can see that if voltage is kept constant at 10 V and the resistance is decreased to 5 Ω, the current will increase proportionately. Like-

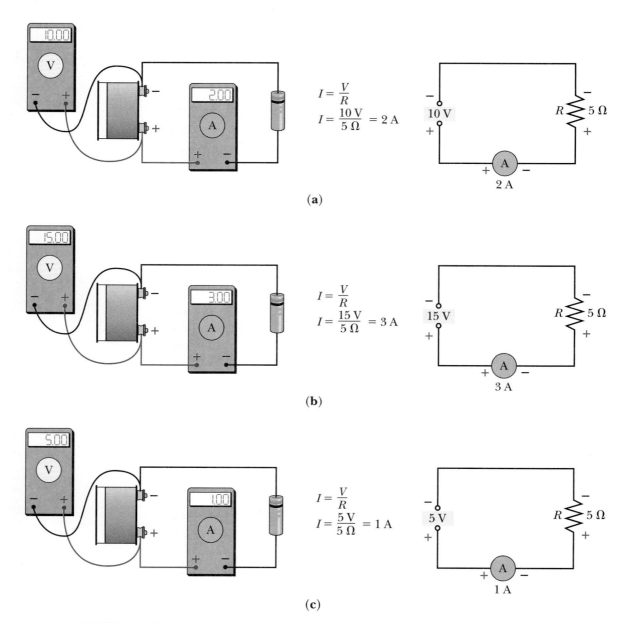

FIGURE 3–1 Ohm's law and the relationship of current to voltage with resistance constant

wise, in Figure 3–2c, it is obvious that if the resistance is increased in value, for a given applied voltage, the circuit current will decrease in proportion to the resistance change.

The direct relationship between voltage *(V)* and *(I)*, and the inverse relationship between resistance *(R)* and current *(I)* are illustrated with the Ohm's law formula in Figure 3–3 (page 76).

Now that we've shown you some examples, you can use the knowledge you have gained regarding Ohm's law, and direct and inverse relationships on the following problems.

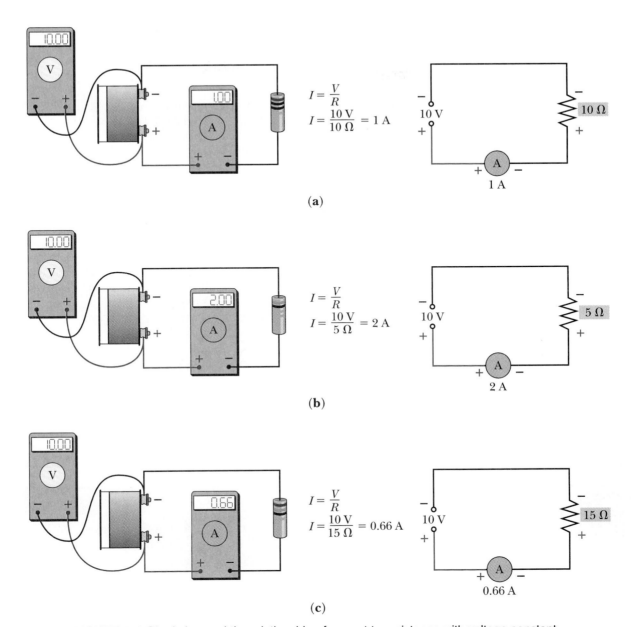

FIGURE 3–2 Ohm's law and the relationship of current to resistance with voltage constant

---

PRACTICE PROBLEMS I

1. Determine the circuit current for a circuit that has 150 V of applied voltage and a resistance of 10 Ω. (To solve, write down the Ohm's law formula, then substitute values in the formula as appropriate.)

2. If the voltage in problem 1 decreased to 75 V, but the resistance remained at 10 Ω, what would be the new circuit current?

FIGURE 3–3 Direct and inverse relationships of voltage, current, and resistance: **(a)** direct; **(b)** inverse

If $\quad \dfrac{V}{R}\,\substack{\uparrow \\ \rightarrow}\quad$ then I ↑ $\quad$ (I directly related to V. If V doubles, I doubles, if R is not changed)

If $\quad \dfrac{V}{R}\,\substack{\downarrow \\ \rightarrow}\quad$ then I ↓ $\quad$ (I directly related to V. If V halves, I halves, if R is not changed)

(a)

If $\quad \dfrac{V}{R}\,\substack{\rightarrow \\ \uparrow}\quad$ then I ↓ $\quad$ (I inversely related to R. If R doubles, I halves, if V is not changed)

If $\quad \dfrac{V}{R}\,\substack{\rightarrow \\ \downarrow}\quad$ then I ↑ $\quad$ (I inversely related to R. If R halves, I doubles, if V is not changed)

(b)

3. Suppose that both voltage and resistance doubled for the circuit in problem 1:

a. What would happen to the circuit current? Would it increase, decrease, or remain the same?

b. Use Ohm's law and calculate the actual value resulting from these changes.

PRACTICAL NOTES

To solve problems, get in the habit of writing down the appropriate formula, substituting the knowns into the formula, then solving for the unknown. Also, if the problem is stated in words (without a diagram), draw a simple diagram illustrating the circuit in question. It will help you think clearly!

3-2 THREE COMMON ARRANGEMENTS FOR THE OHM'S LAW EQUATION

How can the $I = V/R$ equation be transposed to solve for V or R?

Example

Use the formula for Ohm's law and solve for V and for R.

$$I = \frac{V}{R}$$

To isolate V, multiply both sides of the equation by R, and cancel factors, as appropriate.

Simply look at (or cover up) the desired quantity (I, V, or R) and observe the relationships of the other two quantities. To remember the correct version of the formula:

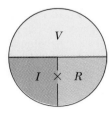

Look at I, the aid shows $\dfrac{V}{R}$

$$I = \dfrac{V}{R}$$

Look at V, the aid shows $I \times R$
$$V = I \times R$$

Look at R, the aid shows $\dfrac{V}{I}$

$$R = \dfrac{V}{I}$$

FIGURE 3–4 Ohm's law memory aid

Therefore, $I \times R = \dfrac{V}{R} \times R$. Then $I \times R = \dfrac{V\cancel{R}}{\cancel{R}}$.

FORMULA 3–2 $V = I \times R$; (Volts = Amperes $\times$ Ohms)

How can we isolate R? If both sides are divided by I, then the I on the right side of the equation will cancel because I divided by $I = 1$.

$$V = IR. \text{ Then, } \dfrac{V}{I} = \dfrac{\cancel{I} \times R}{\cancel{I}}$$

FORMULA 3–3 $R = \dfrac{V}{I}$; $\left(\text{Ohms} = \dfrac{\text{Volts}}{\text{Amperes}} \right)$

You have now been exposed to all three variations of Ohm's law formula.

$$I = \dfrac{V}{R}$$

$$V = I \times R$$

$$R = \dfrac{V}{I}$$

A useful memory aid often helps students remember these three variations of Ohm's law, Figure 3–4.

Restating these three Ohm's law variations:

$$I = \dfrac{V}{R} \text{ or Amperes} = \dfrac{\text{Volts}}{\text{Ohms}}$$

$$V = I \times R \text{ or Volts} = \text{Amperes} \times \text{Ohms}$$

$$R = \dfrac{V}{I} \text{ or Ohms} = \dfrac{\text{Volts}}{\text{Amperes}}$$

FIGURE 3–5 Powers of 10
related to metric and
electronic terms

NUMBER	POWER OF 10	TERM	SAMPLE ELECTRONIC TERM
0.000000000001	10^{-12}	pico	pA (1×10^{-12} ampere)
0.000000001	10^{-9}	nano	nA (1×10^{-9} ampere)
0.000001	10^{-6}	micro	μA (1×10^{-6} ampere)
0.001	10^{-3}	milli	mA (1×10^{-3} ampere)
1,000	10^{3}	kilo	kΩ (1×10^{3} ohms)
1,000,000	10^{6}	mega	MΩ (1×10^{6} ohms)

Knowing any two of the three electrical quantities, you can solve for the unknown value by using the appropriate version of the Ohm's law formula. This capability is important!

In the practical world of electronics, you usually will not be dealing with just the basic units of voltage, amperes, and ohms. You will often work with millivolts, milliamperes, microamperes, kilohms, and megohms. By way of review, observe Figure 3–5, which again shows the relationships between decimal numbers, powers of 10, metric terms, and electronic terms that are expressed using metric prefixes. Recall that in the preceding chapter, it was indicated that very small and very large electrical quantities can more easily be written and worked with by using powers of 10. In this chapter, we want you to begin to use the powers-of-10 mathematical tool because you will have some actual circuit problems to solve.

Example

Using powers of 10, the decimal number 250 can be expressed as:

250×10^{0}, or
25×10^{1}, or
2.5×10^{2}, or
0.25×10^{3}, and so forth.

PRACTICE PROBLEMS II

Convert the following decimal numbers to a number times a power of 10, as indicated:

1. $25,000 =$
   25×10 to what power?

2. $12.335 =$
   1.2335×10 to what power?

3. $0.100 =$
   1.0×10 to what power?

4. $0.0015 =$
   1.5×10 to what power?

Convert the following numbers to their decimal equivalents:

5. 235×10^{2}

6. 0.001×10^{6}

7. 10×10^{3}

8. 15×10^{-3}

Using the EE or EXP key

Now let's try to use the exponent EE or EXP key on a scientific calculator in conjunction with powers of 10. See Figure 3–6.

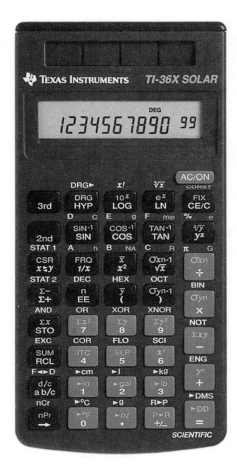

FIGURE 3–6 Photo close up of a typical scientific calculator *(Courtesy of Texas Instruments)*

1. Input a number with the calculator keypad.
2. Push the **EE** or **EXP** button.
3. Input the power of 10 you want to raise that number to.
4. Press the = sign.

You will see that number expressed as a decimal value.

Examples

Determine the value of (a) 235×10 to the 3rd power and (b) 235×10 to the minus 2nd power.

(a) Input 235, **EE** or **EXP**, input 3, **=**

The readout indicates that 235×10^3 equals 235,000.

If you want to use a negative power of 10, you simply input the negative sign prior to expressing the number for the power of 10.

(b) Input 235, **EE** or **EXP**, plus/minus **+/−** or **±**, input 2, **=**

The result indicates that 235×10^{-2} equals 2.35. Now you try the following problems.

PRACTICE PROBLEMS III

Use the calculator and the **EE** or **EXP** key to find the decimal values for the following:

1. Determine the value of the number 1.456 times 10 to the 6th power.

2. Determine the value of the number 333 times 10 to the minus 4th power.

3. Determine the value of the number 1,500,000 times 10 to the minus 6th power.

4. Determine the value of the number 10 times 10 to the 5th power.

3-3 SAMPLE APPLICATION OF METRIC PREFIXES AND POWERS OF 10

Suppose that a circuit has 10,000 ohms of resistance with 150 volts of applied voltage. What is the current?

$$I = \frac{V}{R}$$

$$I = \frac{150 \text{ V}}{10,000 \ \Omega} \left(\text{or } \frac{150 \text{ V}}{10 \text{ k} \ \Omega} \right)$$

$$I = 0.015 \text{ A (or 15 mA)}$$

It is cumbersome to divide units by thousands of units (kilounits), isn't it? It turns out that when you divide units (volts) by kilounits (kΩ), the answer is in milliunits (mA). That is a convenient thing to remember, and you will use it often.

However, how would this work if we used powers of 10?

$$I = \frac{V}{R}$$

$$I = \frac{150 \text{ V}}{10 \times 10^3 \ \Omega}$$

$$I = 15 \times 10^{-3} \text{ A (or 15 mA)}$$

Did you notice that when 10^3 is brought from the denominator to the numerator, it becomes 10^{-3}? Suppose we had divided one by one thousand. How many times would one thousand go into one? The answer is one one-thousandth time. Recall that one thousand equals 10^3 and one one-thousandth equals 10^{-3}. Therefore, when 150 volts are divided by 10,000 ohms, the answer is in thousandths of an ampere. Stated mathematically:

$$\frac{1}{1,000} = 0.001 \text{ or one thousandth (or } 1 \times 10^{-3}) \text{ and}$$

$$\frac{150}{10,000} = 0.015 \text{ or 15 thousandths (or } 15 \times 10^{-3})$$

See if you can apply Ohm's law, metric prefixes, and powers of 10.

IN-PROCESS LEARNING CHECK I

1. Refer to the diagram below and determine the circuit current. $I =$ _____ amperes, or _____ milliamperes.

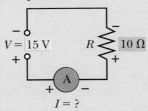

2. Refer to the circuit below and determine the circuit applied voltage. $V =$ _____ volts.

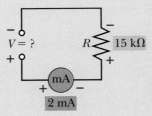

3. Refer to the circuit below and determine the value of circuit resistance. $R =$ _____ ohms, or _____ kilohms.

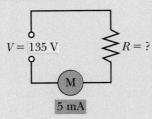

4. a. If R increases and V remains the same in Problem 1, does I increase, decrease, or remain the same?

   b. If I increases and R remains the same in Problem 2, would the voltage applied be higher, lower, or the same as stated in the problem?

   c. In Problem 3 if I doubles but the V remains the same, what must be true regarding R?

3–4 DIRECTION OF CURRENT FLOW

Electron flow approach

In Chapter 1, there was a brief discussion about electrons (negative in charge) moving to a positively charged point in a circuit. When circuit diagrams are shown, it is quite common to show the direction of current flow through the circuit by means of arrows. In the cases shown in Figure 3–7, the arrows are used to show the direction of the electron current flow. *For this text, this is the standard* that

FIGURE 3–7 Method of showing direction of electron current flow with arrows

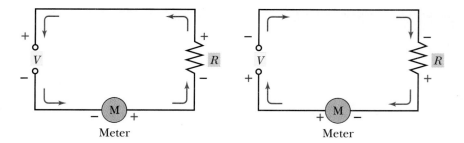

will be used for indicating direction of currents through circuits. This is coherent with our discussion of the fact that current flow is comprised of the movement of free electrons. Because electrons are negative in charge, they are repelled from negative points in circuits and are attracted toward more positive points; thus, electron current flow moves from negative-to-positive locations in components or circuits external to the voltage source.

Conventional current flow approach

Before the development of the electron theory, from which we derive our electron-flow approach to analyzing electron movement, there was a theory that assumed that current flow was in the direction that a positive charge would drift or move through a circuit. Obviously, this assumption would show current flowing from positive-to-negative points in a circuit. This difference in representing direction of current flow is referred to as **conventional current.** See Figure 3–8 for examples of how diagrams would represent this type current.

Because many engineering formulas and thought processes were created back when this theory was prevalent, numerous engineering texts represent current flow in this manner. Also, this is the manner that has been used in creating certain electronic component symbols. For these reasons, we want you to be aware of the term "conventional current." Be assured that regardless of whether electron current flow or conventional current flow is used in circuit analysis, *the answers come out the same!* As indicated earlier, this text will adhere to the electron current flow approach, as indicated in Figure 3–7.

3–5 POLARITY AND VOLTAGE

The concept of polarity, or one point in a component or circuit being considered more negative or positive than another point, was discussed earlier. Recall that the more negative point was labeled with a negative sign, and the less nega-

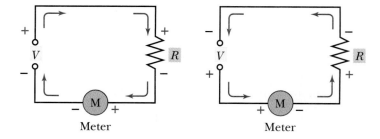

FIGURE 3–8 Arrows showing direction of "conventional current"

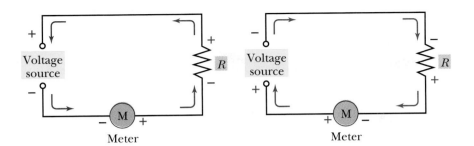

FIGURE 3–9 Illustrating polarity of voltages

tive (or positive point) was labeled with a positive sign. This technique indicates there is a potential difference, or voltage, between the two points.

To label the polarity of voltages across components in a circuit diagram, note the direction of current flow, then label the point where the current enters the component as the negative end. The other end of the component (where the current exits) is labeled as the positive point. This process is true for all components in the circuit, except the voltage source. In that case, the negative side of the source is the source of the electron flow, and the positive end of the source is the end where the electrons return after traveling through the external circuit, Figure 3–9.

PRACTICAL NOTES

You should have observed throughout our discussions that current flows through a component and/or circuit. Voltage does **not** "go" or "flow." It is a difference of potential between points, so think of voltage as being **across,** or **between** two points.

3-6 COMPARISON OF CIRCUIT CURRENT DIRECTIONS FROM DC AND AC SOURCES

So far, we have only mentioned dc (direct current) sources, and the resulting direction of current flow through circuits connected to them. At this point, we want to introduce you to the fact that there is also another type of voltage and current known as "ac." The ac stands for alternating current. An example of this type of voltage and current is that supplied to your home from the power company. An ac source provides voltage that periodically alternates in direction. The output waveform from such a source is known as a "sine wave." See Figure 3–10a. A complete "cycle" of a sine wave has occurred at the moment the waveform starts repeating itself. For example, in the waveform shown in Figure 3–10a, the voltage starts from zero level, moves to a maximum positive level, and returns back to zero. *Then,* the source's output voltage *reverses* direction, or polarity. The output voltage now moves to a maximum negative point, and then, back to zero. At this point, the voltage waveform starts the same sequence of events all over again. This begins the second cycle. See Figure 3–10b.

One-half of the complete cycle is called an "alternation"; therefore, there are two alternations in a cycle of ac, Figure 3–10a. As you can see, there is one positive alternation and one negative alternation in a cycle. You will be studying this type waveform in more depth in a later chapter. For now, we just want you to see the general differences between dc and ac.

FIGURE 3–10
Representation of ac sine-wave voltages

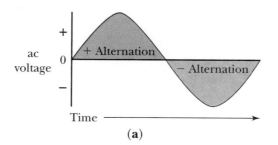

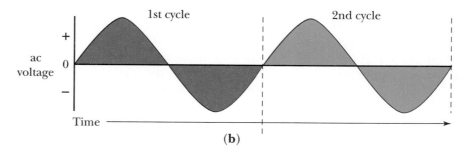

Notice the contrast between dc and ac voltage as shown in Figure 3–11. The dc voltage remains at one polarity. This means the current is always traveling through the circuit in one direction, Figure 3–11a. In the ac circuit, you see that during one alternation of the ac cycle, current travels through the circuit in one direction, Figure 3–11b. (Incidentally, you probably noticed that the electrons are traveling from the negative side of the source, through the external circuit, back to the positive side of the source, just like they did in the dc circuit.) Obviously, when the source reverses polarity (on the next alternation), the electrons will reverse their direction of travel through the circuit. However, once again, the electrons leave the negative side of the source, travel through the external circuit, and return to the positive side of the source, Figure 3–11c. This indicates that the polarity of the ac source voltage applied to the circuit determines the direction of the electron current flow through the circuit. As the polarity of source voltage reverses, so does the direction of current flow through the circuit, Figure 3–11b and c.

Later in your studies you will learn some more specific facts and terms related to ac voltages and signals. These facts will be presented at the particular time you will be best able to apply them.

3–7 WORK, ENERGY, AND POWER

Work is the expenditure of energy. **Energy** is the ability to do work, or that which is expended in doing work. **Power** is the rate of using energy. A formula for mechanical work is:

FORMULA 3–4   Work (Foot pounds) = Force (Pounds) × Distance (Feet)

The basic unit of mechanical work is the foot poundal. At sea level, 32.16 foot poundals equals 1 foot pound. Mechanical power is the rate of doing work. For example, the "horsepower" equals a rate of 550 foot pounds per second.

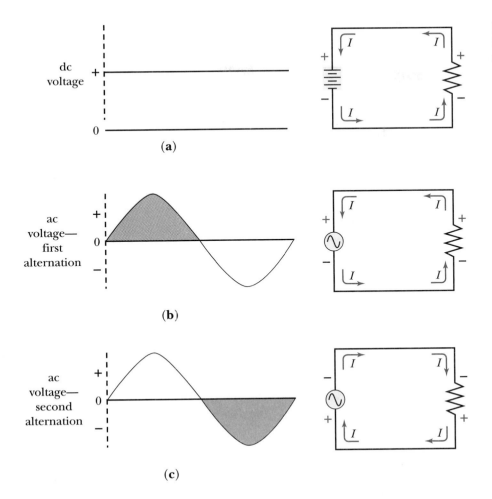

FIGURE 3–11

FIGURE 3–11 Contrasts between dc and ac voltages and circuit currents

dc voltage

(a)

ac voltage— first alternation

(b)

ac voltage— second alternation

(c)

Electrical energy is the ability to do electrical work. The unit of energy used for electrical energy is the joule (J). For electrical applications, the joule is defined as the energy required to move one coulomb of electrical charge between two points having a difference of potential of one volt.

Electrical power is the rate at which electrical energy is used. The unit of electrical power is the **watt,** named after the scientist James Watt. The abbreviation for watt or watts is **W.** *A watt of electrical power is defined as electrical energy expended at the rate of one joule per second.* Since joules can be related to energy, and energy can be related to force multiplied by distance, it can be determined that 746 watts is equivalent to one horsepower of mechanical power. As we already stated, electrical power in watts represents the time rate of expending electrical energy in performing electrical work; for example, the work of causing current flow (moving electrons) through an electrical resistance. Power can be related to electrical energy in the following manner:

FORMULA 3–5 $\quad \text{Power (Watts)} = \dfrac{\text{Energy (Joules)}}{\text{Time (Seconds)}}$

Example

If 100 joules of energy are used in 10 seconds, determine the power used.

$$P = \frac{100}{10} = 10 \text{ W}$$

PRACTICE PROBLEMS IV

1. What is the power, in watts, for a circuit in which 25 joules of energy are being used in 2 seconds?
2. What is the power, in milliwatts, in a circuit where 1 joule is used in 100 seconds?

3–8 MEASURING ELECTRICAL ENERGY CONSUMPTION

By finding how much power has been used over time, we can determine the total electrical energy that has been consumed. You should be aware that the symbol for both work and energy is W; however, the SI unit for both is the joule (J).

(NOTE: Don't get the symbol W for work or energy confused with the abbreviation for the unit of power, the watt [W].)

The electric power company charges you for the amount of electrical energy you use for a given period of time. For example, they typically charge you for the total number of "kilowatthours" of electrical energy you have used for that month. When you use one kilowatthour of energy, you are using 1,000 watts for one hour. For example, if you burn ten 100-watt bulbs for 1 hour, you are using one kilowatthour's worth of energy. If you use a 500-watt iron for 2 hours, you are using one kilowatthour of energy, and so on.

> **FORMULA 3–6**   Energy (Watthours) = Power (Watts) × Time (Hours)

The time can be measured in seconds or hours. When the time is expressed in seconds, the unit of energy consumption is called the **wattsecond (Ws).** When the time is expressed in hours, the unit of energy consumption is called the **watthour (Wh),** or, when convenient due to the amount of power involved (as in the case of the power company), the kilowatthour (kWh).

Examples

1. A power consumption of 200 watts for 3 hours = 600 Wh of electrical energy consumption.

$$W = P \times T; \ W = 200 \times 3 = 600 \text{ Wh (or, 0.6 kWh)}$$

2. If 450 watts are being used over 9 hours, find the energy consumption.

$$W = P \times T; \ W = 450 \times 9 = 4,050 \text{ Wh (or 4.05 kWh)}$$

PRACTICE PROBLEMS V

How many kilowatthours of energy are used by a 250-watt bulb if it is left on for 8 hours?

The basic power formula

Relating electrical power to practical circuits:

- A joule is the energy used to move one coulomb of charge between two points with a one volt potential difference.
- An ampere is the rate of charge movement of one coulomb per second.
- A watt is the rate of doing work at one joule per second.

  Thus, one watt of power equals one ampere times one volt. The formula is:

  Power (in watts) = Voltage (in volts) $\times$ Current (in amperes)

FORMULA 3–7 $P = V \times I$ (Watts = Volts $\times$ Amperes)

Example

If a circuit has 50 milliamperes of current flow and an applied voltage of 10 volts, how much power does the circuit dissipate? Using the basic formula for power:

$$P = V \times I$$

$$P = 10\ V \times 50\ mA$$

$$P = 10 \times 50 \times 10^{-3} = 500 \times 10^{-3}\ W \text{ or } 500\ mW$$

PRACTICE PROBLEMS VI

1. If a circuit has an applied voltage of 150 V, and the circuit resistance limits the current to 10 mA; how much power does the circuit dissipate?

2. If the circuit voltage for Problem 1 remained at 150 V, but the circuit resistance were doubled:

   a. What would happen to circuit current?

   b. What would the circuit power dissipation be?

IN-PROCESS LEARNING CHECK II

1. Electron current flow is considered to flow from the (negative, positive) _____ side of the source, through the external circuit, back to the (negative, positive) _____ side of the source.

continued

2. "Conventional" current is considered to flow from the (negative, positive) _____ side of the source, through the external circuit, back to the (negative, positive) _____ side of the source.

3. This text uses the (electron, conventional) _____ current flow approach.

4. One of the basic differences between dc and ac is that current in a dc circuit always flows in _____ direction; whereas, current flow in an ac circuit _____ direction for each alternation of the ac voltage.

5. Work is the _____ of energy; energy is the ability to do _____, or that which is expended in doing _____. Power is the rate of using _____.

6. Power in watts is equal to energy in _____ divided by time in _____.

7. Energy = power (in watts) × time. A common measure of energy consumption used by power companies for billing customers is the _____.

Variations of the basic power formula

Just as the Ohm's law formula can be rearranged, so can the power formula. The formula we started with was $P = V \times I$. Two other arrangements are $V = P/I$ and $I = P/V$. Incidentally, the same style of visual aid used for the Ohm's law formula will work here, Figure 3–12.

Other practical variations of the power formula are developed by combining knowledge of Ohm's law and the power formulas. For example, since $V = I \times R$ (Ohm's law) and $P = V \times I$, the formula can be combined to show that $P = I^2 \times R$ or I^2R. You have substituted for V in the Ohm's law formula $P = V \times I$; therefore $P = (I \times R) \times I$.

FORMULA 3–8 $P = I^2R$ (Watts = Amperes squared × Ohms)

Using this same idea, you can derive the formulas that state:

$$P = \frac{V^2}{R}, \qquad P = VI, \qquad P = V \times \left(\frac{V}{R}\right)$$

FORMULA 3–9 $P = \dfrac{V^2}{R} \left(\text{Watts} = \dfrac{\text{Volts squared}}{\text{Ohms}}\right)$

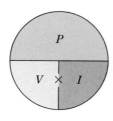

Again, by looking at the desired parameter (or covering it up) and seeing the arrangement of the other two parameters, the correct formula is evident. For example, if you know V and I, to find power take $V \times I$. If you know the power and the current, simply divide P by I to get V, and so forth.

FIGURE 3–12 Power formula memory aid

You have simply substituted the Ohm's law formula for I into the $P = V \times I$ power formula. In order to solve for any desired electrical quantity, each of the power formulas can be rearranged using the transposition and cancelling techniques learned earlier in this chapter.

$$P = V \times I, \qquad I = \frac{P}{V}, \qquad V = \frac{P}{I}$$

$$P = I^2 \times R, \qquad I = \sqrt{\frac{P}{R}}, \qquad R = \frac{P}{I^2}$$

$$P = \frac{V^2}{R}, \qquad V = \sqrt{PR}, \qquad R = \frac{V^2}{P}$$

3-9 COMMONLY USED VERSIONS OF OHM'S LAW AND POWER FORMULAS

The most important equations to remember are:

$$I = \frac{V}{R}, \qquad V = IR, \qquad \text{and } R = \frac{V}{I} \text{ (Ohm's law formulas)}$$

$$P = VI, \qquad P = I^2R, \qquad \text{and } P = \frac{V^2}{R} \text{ (power formulas)}$$

By understanding and knowing these Ohm's law and power formulas and by becoming adept at applying them to practical electronic problems, you will be on the way to becoming a good technician.

For a change of pace from simply reading about these formulas, we now want you to see how these formulas can be applied to practical problems. Follow the steps carefully as several sample problems and solutions are shown. You will be doing this sort of thing on your own throughout the remainder of your training.

Example

Refer to Figure 3–13 for the following.

1. In Diagram 1, find the voltage (V). Ohm's law formula for voltage is

$V = I \times R.$
Substitute the known values for I and R.
Therefore, $V = 5 \times 10^{-3}$ ampere $\times 10 \times 10^3$ ohms. $V = 50$ volts

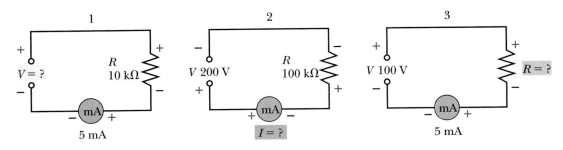

FIGURE 3–13 Ohm's law example problems

Remember: When multiplying powers of ten, add their exponents algebraically:

10^{-3} and 10^3 add to 10^0, which equals 1. Also, milliunits multiplied by kilounits result in units, and, in this case, $mA \times k\Omega$ result in volts.

2. In Diagram 2, find the circuit current (I). Ohm's law formula for current is $I = V/R$.

Substitute the known values for V and R.

$$I = \frac{200 \text{ V}}{100 \text{ k}\Omega} = \frac{200 \times 10^0}{100 \times 10^3} = 2 \times 10^{-3} = 2 \text{ mA}$$

Remember: When dividing powers of ten, subtract the exponents algebraically. Since the voltage did not show a power of 10, show it as 200×10^0 $[10^0 = 1]$. To subtract the denominator 10^3 from the numerator 10^0, change the sign of the denominator power of 10, then add 10^{-3} to 10^0, which equals 10^{-3}. Also, units divided by kilounits result in milliunits. In this case, $V/k\Omega$ results in milliamperes (mA).

3. In Diagram 3, find the resistance (R). Ohm's law formula for resistance is $R = V/I$.

Substitute the known values for V and I.

$$R = \frac{100 \text{ V}}{5 \text{ mA}} = \frac{100}{5 \times 10^{-3}} = 20 \text{ k}\Omega$$

Remember: When moving a power of 10 in the denominator by dividing it into the numerator, change its sign and add the exponents as appropriate. The 10^{-3} for the milliamperes becomes the 10^3 for kilohms in the answer. Also, in this problem units divided by milliunits result in kilounits; thus, volts divided by milliamperes result in the answer as kilohms.

PRACTICE PROBLEMS VII

1. Find the applied voltage in the circuit of Figure 3–14, Diagram 1.
2. Find the circuit current for the circuit of Figure 3–14, Diagram 2.
3. Find the circuit resistance for the circuit of Figure 3–14, Diagram 3.

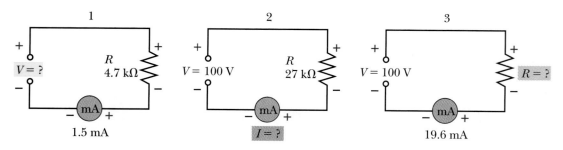

FIGURE 3–14 Ohm's law practice problems

In summary, when applying Ohm's law and power formulas:
- look at the circuit to determine the knowns and unknown(s);
- select and write the appropriate formula to use;
- substitute the knowns into the formula;
- use metrics and powers of 10 to manage large and small numbers;
- remember that units divided by milliunits equals kilounits and that units divided by kilounits equals milliunits;
- remember that milliunits multiplied by kilounits equals units; and
- recall various expressions for Ohm's law.

Solving problems using both the Ohm's law and power formulas

Technicians are often required to combine their knowledge of Ohm's law and the power formulas in order to find the electrical quantities of interest. The following examples provide an idea about how you need to logically address problems of this nature.

Example

Refer to Figure 3–15. Find the unknown values shown.

1. In Figure 3–15, Diagram 1, which formula would be easiest to use first? Observe that values for V and R are known.

$$I = \frac{V}{R} = \frac{100 \text{ V}}{10 \text{ }\Omega} = 10 \text{ A}$$

Knowing the current, you can solve the remaining unknown.

$$P = V \times I = 100 \text{ V} \times 10 \text{ A} = 1{,}000 \text{ watts (or 1 kilowatt)}$$

Also, since V, I, and R are known, you can use the other power formulas, as well.

2. In Figure 3–15, Diagram 2, which formula might be used first? In this case, voltage and current are known values, but resistance is unknown. If the current meter does not drop any appreciable voltage, you could assume that V_R equalled the applied voltage and first solve for power. In many problems,

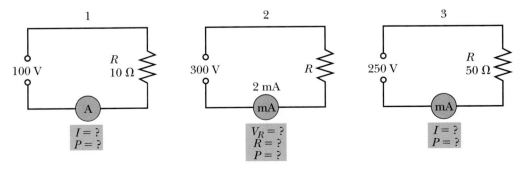

FIGURE 3–15 Combination Ohm's law power formula problems

you can start at more than one place, depending on the known value(s). Let's start with the power formula.

$$P = V \times I = 300 \text{ V} \times 2 \times 10^{-3} \text{ A} = 600 \times 10^{-3} \text{ W (or 600 mW)}$$

Now to solve for the other unknowns, assume 300 V across R, where $V_R = 300$ V.

$$R = \frac{V}{I} = \frac{300 \text{ V}}{2 \times 10^{-3} \text{ A}} = 150 \times 10^3 \text{ }\Omega \text{ (or 150 k}\Omega)$$

3. In Figure 3–15, Diagram 3, a good starting point is to first solve for I since you know the values of V and R.

$$I = \frac{V}{R} = \frac{250 \text{ V}}{50 \text{ k}\Omega} = \frac{250}{50 \times 10^3} = 5 \times 10^{-3} \text{ A (or 5 mA)}$$

Now that voltage and current are known, it is easy to solve for the other unknown.

$$P = V \times I = 250 \text{ V} \times 5 \times 10^{-3} \text{ A} = 1{,}250 \times 10^{-3} \text{ W or 1.25 W}$$

In summary, now you should know:

• how to find the known value(s) from the given diagram or facts;
• where there are several starting places it is best to look for the formula that might be the easiest to use and start there; and
• how to combine Ohm's law knowledge with power formulas.

PRACTICE PROBLEMS VIII

Try the problems in Figure 3–16 to see if you can apply logic and appropriate formulas. Don't cheat yourself by looking ahead.

Now check yourself to see how well you did! You will be getting more practice, both in lab and within this text, as you proceed in your training.

If you have difficulty in solving these, take a few minutes to review appropriate parts of this chapter to find the formulas and the needed guidance or hints.

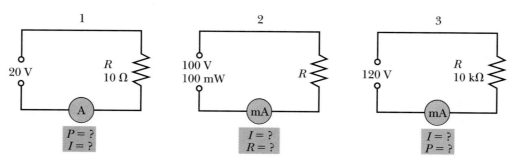

FIGURE 3–16

SUMMARY

▶ Current is directly related to voltage and inversely related to the circuit resistance. Ohm's law states this as $I = V/R$.

▶ If the other factors in the equation are known, the techniques of transposition allow rearrangement of the equation to solve for an unknown factor. For example, by rearranging the Ohm's law formula, two other useful equations can be derived, namely $V = I \times R$ and $R = V/I$.

▶ Powers of 10 are useful when working with either large or small numbers in formulas. When using powers of 10, the rules relating to exponents are applied.

▶ Scientific calculators are very useful to simplify working electrical problems. The **EE** or **EXP** keys are helpful aids when using calculators to solve problems.

▶ Two good habits to form when solving electrical circuit problems are to 1) draw a diagram of the circuit, if one is not given, and 2) write down any pertinent formulas that will be used; then, substitute knowns into them appropriately to solve for the unknowns.

▶ The direction of electron current flow through a circuit, external to the voltage source, is from the source's negative side, through the circuit, and back to the source's positive side. Arrows are frequently drawn on schematic diagrams to illustrate this. The direction electrons would move through a circuit is the standard being used in this textbook.

▶ Conventional current, as used in some engineering textbooks, is considered to flow in the direction that a positive charge would move through the circuit. That is, from positive to negative.

▶ When labeling the polarity of voltage across a given component in a circuit, the component end where the electron current enters is labeled the negative end, and the end where the electron current exits is labeled the positive end.

▶ Electrical power is the rate of using electrical energy to do electrical work. Power is generally dissipated in a circuit or in a component in the form of heat. One watt of power is the performance of electrical work at the rate of 1 joule per second.

▶ A joule is the energy used in moving one coulomb of charge between two points that have a difference of potential of one volt between them.

▶ The formulas commonly used to calculate electrical power, or power dissipation are: $P = V \times I$; $P = I^2R$, and $P = V^2/R$

▶ Electrical energy usage is computed by multiplying the power used times the amount of time it was used ($W = P \times T$). Units of measure for usage of electrical energy are: wattsecond (Ws); watthour (Wh); and for large amounts of energy usage, kilowatthour (kWh).

FORMULAS AND SAMPLE CALCULATOR SEQUENCES

FORMULA 3–1 $\quad I = \dfrac{V}{R}; \left(\text{Amperes} = \dfrac{\text{Volts}}{\text{Ohms}}\right)$

voltage value, **÷**, resistance value, **=**

FORMULA 3–2 $\quad V = I \times R$; (Volts = Amperes × Ohms)

current value, **×**, resistance value, **=**

FORMULA 3–3 $R = \dfrac{V}{I}; \left(\text{Ohms} = \dfrac{\text{Volts}}{\text{Amperes}} \right)$

voltage value, ➗, current value, ▭

FORMULA 3–4 Work (Foot pounds) = Force (Pounds) × Distance (Feet)

force value, ✖, distance value, ▭

FORMULA 3–5 Power (Watts) $= \dfrac{\text{Energy (Joules)}}{\text{Time (Seconds)}}$

energy value, ➗, time value, ▭

FORMULA 3–6 Energy (Watthours) = Power (Watts) × Time (Hours)

power value, ✖, time value, ▭

FORMULA 3–7 $P = V \times I$ (Watts = Volts × Amperes)

voltage value, ✖, current value, ▭

FORMULA 3–8 $P = I^{2}R$ (Watts = Amperes squared × Ohms)

current value, x^2, ✖, resistance value, ▭

FORMULA 3–9 $P = \dfrac{V^{2}}{R} \left(\text{Watts} = \dfrac{\text{Volts squared}}{\text{Ohms}} \right)$

voltage value, x^2, ➗, resistance value, ▭

REVIEW QUESTIONS

1. State the three forms of:
   a. the Ohm's law formula
   b. the formula for electrical power

2. State the following numbers as a number between 1 and 10 times a power of 10:
   a. 33,456
   b. 25
   c. 1,055,000
   d. 10

3. State the following numbers in the appropriate number times 10^3:
   a. 5,100
   b. 47,000
   c. 0.001
   d. 39

4. Define the term *work*.

5. Define the term *energy*.

6. Define the term *power.*

7. Give the unit used to measure work.

8. Give the unit used to measure energy.

9. Give the unit used to measure power.

10. In your own words, describe the relationships and/or proportionalities between the following electrical parameters:

a. power related to current (for a given resistance)
b. voltage related to current (for a given resistance)
c. energy (W) related to power (for a given time)

PROBLEMS

1. If voltage is doubled and resistance is halved in a circuit, what is the relationship of the new current to the original current?

2. If current triples in a circuit, but resistance remains unchanged, what happens to the power dissipated by the circuit?

3. a. Draw a schematic diagram showing a voltage source, an ammeter, and a resistor.
   b. Assume the voltage applied is 30 V and the ammeter indicates 2 mA of current. Calculate the circuit resistance and circuit power dissipation.
   c. Label all electrical parameters, the polarity of voltage across the resistor and across the voltage source, and show the direction of current flow.

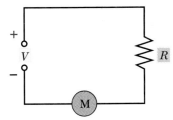

FIGURE 3–17

Refer to Figure 3–17 when answering questions 4 through 21:

4. In Figure 3–17, if the value of *V* is assumed to be 50 V and the value of *R* is assumed to be 33 kΩ, what value of current will the meter M read?

5. If the current meter (M) in Figure 3–17 is reading 3 mA, and *R* has a value of 12 kΩ, what is the value of circuit applied voltage?

6. Indicate whether the following items would increase, decrease, or remain the same if

the value of voltage in question 5 were doubled:
a. *V* has _____
b. *R* will _____
c. *I* will _____

7. If current through the circuit of Figure 3–17 is 50 mA and circuit voltage is 41 V, what is the value of *R*?

8. In Figure 3–17, what is the voltage applied if the circuit power dissipation is 100 mW and the circuit current is 12.5 mA?

9. For the circuit of Figure 3–17, what is the circuit current if the power dissipated by the 10-Ω resistor is 100 W?

10. If current in the circuit of Figure 3–17 decreases to one-third its original value, and the original power dissipation was 180 W, what is the new power dissipation?

11. Assume an applied voltage of 100 V for the circuit of Figure 3–17. What value of *R* is needed to limit circuit current to 8.5 mA?

12. Referring to Figure 3–17, assume an original applied voltage of 100 V and a resistance of 25 kΩ. What will happen to circuit current if the applied voltage is doubled and the circuit resistance is tripled?

13. Would the power dissipated by the circuit of Figure 3–17 increase, decrease, or remain the same if both the circuit resistance and the applied voltage were doubled?

14. For the circuit shown in Figure 3–17, if the value of *V* is 150 V and the value of *R* is 8 kΩ, what is the value of circuit current?

15. What is the power dissipation of the circuit described in question 14?

16. For the conditions expressed in question 14, express the amount of energy (in Watthours) that the circuit would consume over 3 hours.

17. Referring to Figure 3–17, if the value of V is 250 V and the value of R is 12kΩ, what is the value of circuit current?

18. What is the power dissipation of the circuit described in question 17?

19. What is the energy (in Watthours) consumed by the circuit of question 17 over 6.5 hours?

20. For the circuit of Figure 3–17, assume an applied voltage of 100 volts and a power dissipation of 50mW. What is the value of R?

21. For the circuit of Figure 3–17, assume a power dissipation is 100 mW when the value of R is 10 kΩ. What is the value of applied voltage?

22. Draw a schematic diagram showing a source, a resistor, and an ammeter. Assume the power dis-

sipation is 1,000 watts and the resistance is 10 ohms. Transpose the $P = I^2R$ formula and solve for the value of I.

23. Calculate the applied voltage for the circuit in question 22.

24. Label all the parameters on the circuit diagram you drew for question 22 and indicate the direction of current flow throughout the circuit, and the polarity of voltages across each component.

25. Using powers of 10, change the following as directed:

    a. $\dfrac{10^7 \times 10^{-4}}{10^5 \times 10^2 \times 10^{-2}}$ equals what whole number?

    b. $10^3 \times 10^5 \times 10^6 \times 10^{-2}$ equals 10 to what power?

    c. Add: $(7.53 \times 10^4) + (8.15 \times 10^3) + (225 \times 10^1)$

    d. Subtract: $(6.25 \times 10^3) - (0.836 \times 10^2)$

ANALYSIS QUESTIONS

1. Explain what the **EE** (or **EXP**) keys on a scientific calculator do when you input a number that you want to express times some power of 10.

2. If someone were to use a calculator and input: 222, **EE** (or **EXP**), 4, in that order, what would the calculator indicate as the result?

3. Describe the basic difference between current flow in a circuit having a dc source and that same circuit having an ac source.

4. Describe the difference between a dc voltage and an ac voltage in terms of polarity characteristics.

5. As contrasted with "electron-flow current," "conventional current" is considered to flow in a direction that a (negative, positive) _____ charge would move through the circuit.

6. Determine the electrical power equivalent to 3.5 horsepower.

7. In a given circuit, if power dissipation has increased to 8 times the original power, while resistance has decreased to one-half its original value, what must have happened to the circuit applied voltage?

A GOOD IDEA

Never set a soldering iron where you might accidentally lay your hand on it or easily touch the hot end of the iron!

A GOOD IDEA

Remove loose jewelry when working on electrical equipment or when using rotating machinery or power tools!

PERFORMANCE PROJECTS CORRELATION CHART

Suggested performance projects that correlate with topics in this chapter are:

CHAPTER TOPIC	PERFORMANCE PROJECT	PROJECT NUMBER
The Relationship of Current to Voltage with Resistance Constant	Relationship of I & V with R Constant (Ohm's law series)	5
The Relationship of Current to Resistance with Voltage Constant	Relationship of I & R with V Constant (Ohm's law series)	6
The Basic Power Formula	Relationship of P to V with R Constant (Ohm's law series)	7
Variations of the Basic Power Formula	Relationship of P to I with R Constant (Ohm's law series)	8

NOTE: It is suggested that after completing the above projects, the student should be required to answer the questions in the "Summary" at the end of this section of projects in the Laboratory Manual.

capacitor

V

n's Law

Kirchhoff's Current Law

KEY TERMS

Battery
Cell
Kirchhoff's voltage law
Open circuit
Series circuit
Short circuit
Series-dropping resistor
Voltage-divider action

SERIES CIRCUITS

4

chapter

CHAPTER PREVIEW

Series circuits are found in many systems and subsystems. Common examples are home light switch circuits and automobile ignition switch circuits. It is important that you learn the characteristics of the series circuit. That knowledge will aid you in analyzing series circuits whether they are stand-alone circuits or portions of more complex circuits.

In studying series circuits, you will apply Ohm's law and power formulas. Also, you will learn how Kirchhoff's voltage law helps you analyze series circuits. Insight will be gained about the effects of opens and shorts in series circuits. You will see how voltage sources are used in series circuits. Also, voltage dividers and the concept of reference points will be examined. The final portion of this chapter will present a useful troubleshooting method that will be valuable to you throughout your training, and more importantly, in your career.

You will be introduced to a special troubleshooting sequence called the SIMPLER sequence. By solving a troubleshooting problem called Chapter Troubleshooting Challenge, you will gain experience using the SIMPLER sequence. The system for performing the procedure and solving the Chapter Troubleshooting Challenge will be discussed. One sample solution to the challenge troubleshooting problem is presented at the end of the chapter so you can compare your thinking process against the sample solution.

OBJECTIVES

After studying this chapter, you should be able to:

1. Define the term **series circuit**

2. List the primary characteristics of a series circuit

3. Calculate the total resistance of series circuits using two different methods

4. Calculate and explain the voltage distribution characteristics of series circuits

5. State and use **Kirchhoff's voltage law**

6. Calculate power values in series circuits

7. Explain the effects of **opens** in series circuits

8. Explain the effects of **shorts** in series circuits

9. List troubleshooting techniques for series circuits

10. Design series circuits to specifications

11. Series-connect voltage sources for desired voltages

12. Analyze a **voltage divider** with reference points

13. Calculate the required value of a **series-dropping resistor**

14. Use the SIMPLER troubleshooting sequence to solve the Chapter Troubleshooting Challenge problem

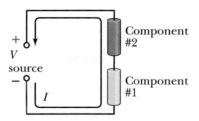

FIGURE 4–1 Components connected end-to-end in a two-component series circuit

4–1 DEFINITION AND CHARACTERISTICS OF A SERIES CIRCUIT

A **series circuit** is any circuit having only one path for current flow. In other words, two or more electrical components or elements are connected so the same current passes through all the connected components.

This situation exists when components are connected end-to-end in the circuit external to the source, Figure 4–1. This is termed a two-component series circuit. Also, notice there is only *one possible path* for current flow from the source's negative side, through the external circuit, and back to the source's positive side.

Key characteristics

Excluding the source, how many components are in series in Figure 4–2? If you said four, you are correct! The circuit in Figure 4–3 is a six-component series circuit. In all these cases, the two important characteristics to remember are that there is only one path for current flow and that current is the same through all parts of a series circuit. As a technician, you will have many opportunities to apply this knowledge.

4–2 RESISTANCE IN SERIES CIRCUITS

Knowing the characteristics that identify a series circuit helps you analyze the important electrical quantities or parameters of these circuits.

First, let's analyze the series circuit in terms of its resistance to current flow. In previous chapters we have limited circuits to a single resistor, and the circuit's

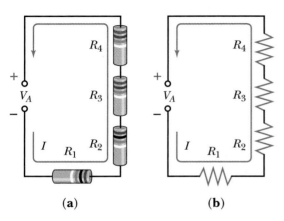

(a)   **(b)**

FIGURE 4–2 Four-resistor series circuit

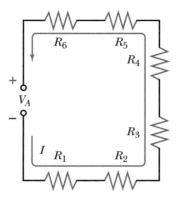

FIGURE 4–3 Six-resistor series circuit

resistance to current flow was obvious. Now, we are connecting two or more resistors in tandem or in series. How does this affect the circuit resistance?

Series circuit total resistance formula

If the circuit current must sequentially flow through all the resistors since there is only one path for current, then the total resistance to current flow will equal the *sum* of all the resistances in series. Once again, in series circuits, the total resistance (R_T or R total) equals the sum of all resistances in series.

FORMULA 4–1 $R_T = R_1 + R_2 \ldots + R_n$

R_n represents the last of the remaining resistor values.

EXAMPLE

Refer to Figure 4–4 for an application of this formula. This illustrates that to find a series circuit's total resistance, simply add all the individual resistance values.

PRACTICE PROBLEMS I

1. Referring back to Figure 4–2, what is the circuit's total resistance if R_1 and R_2 are 10 kΩ resistors and R_3 and R_4 are 27 kΩ resistors?
2. The values of the resistors in Figure 4–4 are changed so that $R_1 = 2.7$ kΩ, $R_2 = 5.1$ kΩ, and $R_3 = 8.2$ kΩ. What is the new total resistance value for the circuit?

The Ohm's law approach

Another important method to determine total resistance *in any circuit,* including series circuits, is Ohm's law. Recall that the Ohm's law formula to find resistance is $R = V/I$. To find total resistance, this formula becomes:

FORMULA 4–2 $R_T = \dfrac{V_T}{I_T}$

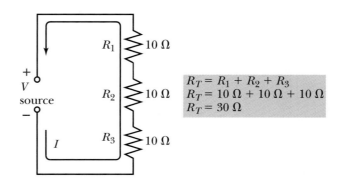

$$R_T = R_1 + R_2 + R_3$$
$$R_T = 10\ \Omega + 10\ \Omega + 10\ \Omega$$
$$R_T = 30\ \Omega$$

FIGURE 4–4 Application of series-circuit total resistance formula

If there is any means of determining the value of total voltage applied to the series circuit *and* the total current (which is the same as the current through any part of the series circuit), Ohm's law can be used to find total resistance by using $R_T = V_T/I_T$.

Example

Look at Figure 4–5 and use Ohm's law to find total resistance. Your answer should be 30 ohms because 60 volts divided by 2 amperes equals 30 ohms. Does this show the value of each of the resistors in the circuit? No, it only reveals the circuit's total resistance.

Example

Now look at Figure 4–6. Again, it is possible to use Ohm's law to determine the circuit's total resistance, where total resistance equals total voltage divided by total circuit current.

$$R_T = \frac{V_T}{I_T} = \frac{70 \text{ V}}{2 \text{ A}} = 35 \text{ } \Omega$$

Because we know the voltage drop across R_1 and know that the current through all parts of a series circuit is the same value, we can also determine the value of R_1. In this case, R_1 equals the voltage across R_1 divided by the 2 amperes of current through R_1. Thus, R_1 equals 30 volts divided by 2 amperes. $R_1 = 15$ ohms.

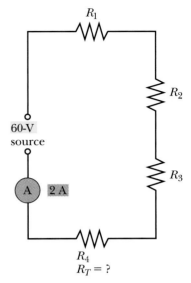

FIGURE 4–5 Finding R_T by Ohm's law

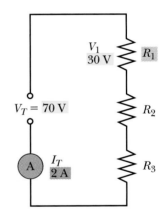

FIGURE 4–6 Finding one *R* value by Ohm's law

PRACTICE PROBLEMS II

1. Again refer to Figure 4–6, if $V_T = 100$ V, $I_T = 1$ A, and $V_I = 47$ V, what is the value of R_1? What is the value of R_T?

2. Assume that the electrical parameters in the circuit of Figure 4–6 are changed so that $V_T = 42$ V and $I_T = 4$ mA. What is the new value of R_T?

3. With the circuit V_T and I_T described in problem 2, if R_2 and R_3 are equal value resistors and $R_1 = 2.7$ kΩ, what is the value of R_2?

We have discussed and illustrated two important facts about series circuits. First, *current* is the same throughout all parts of a series circuit. Second, the total circuit *resistance* equals the sum of all the resistances in series, which indicates total resistance must be greater than any one of the resistances. We will now study voltage, a third important electrical parameter in series circuits.

4–3 VOLTAGE IN SERIES CIRCUITS

To help you understand how voltage is distributed throughout a series circuit, refer to Figure 4–7.

Individual component voltages

Because there is only one path for current, the current (*I*) through each resistor (R_1, R_2, and R_3) must have the same value since it is the same current, Figure 4–7.

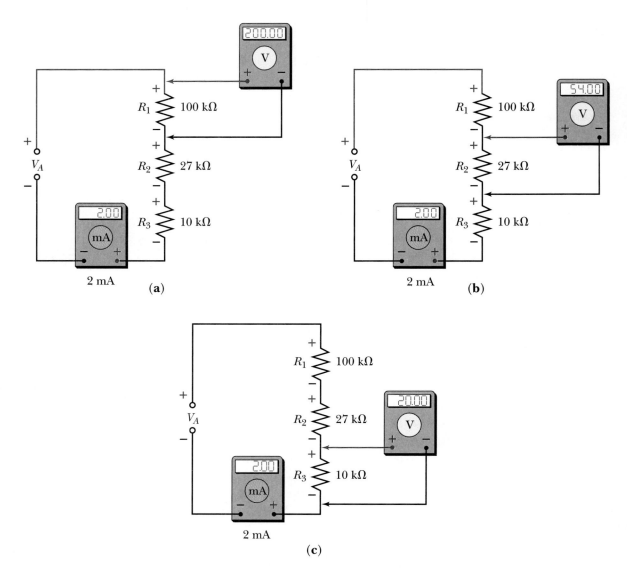

FIGURE 4–7 Voltage distribution in series circuits

From the study of Ohm's law, you know the value of voltage dropped by R_1 must equal its I times its R, or

FORMULA 4–3 $V_1 = I_{R_1} \times R_1$

Example

In this case, the current is 2 mA and the resistance of R_1 is 100 kΩ; therefore, $V_1 = 200$ volts, Figure 4–7a. (Remember that milliunits times kilounits equals units.) Voltage dropped by R_2 must equal 2 mA times 27 kΩ, or 54 volts, Figure

4–7b. The voltmeter, which is measuring the voltage across R_3, would indicate 20 volts because 2 mA $\times$ 10 kΩ = 20 volts, Figure 4–7c.

PRACTICE PROBLEMS III

1. In the circuit shown in Figure 4–7c, if the voltmeter reads 30 V, what are the values of V_A, I_T, V_{R_2}, and V_{R_1}?

2. In the circuit shown in Figure 4–7c, if the voltmeter reads 25 V, what are the values of V_A, I_T, V_{R_2}, and V_{R_1}?

PRACTICAL NOTES

In review, several observations are made.

1. The largest value resistor in a series circuit drops the largest voltage and the smallest value resistor drops the least voltage. These drops occur because the current is the same through each of the resistors and each resistor's voltage drop equals I times its R.

2. Since the same current passes through all components in series, a given component's voltage drop equals *the same percentage or fraction* of the total circuit applied voltage as its resistance value is of the total circuit resistance.

3. Relating the above concepts, when comparing two specific components' voltage drops, the ratio of voltage drops is the same as the ratio of their individual resistances.

Refer back to the voltages calculated for the circuit in Figure 4–7. Did the largest resistor value drop the highest voltage? Yes, R_1 dropped 200 volts compared with 54 volts for R_2 and 20 volts for R_3. Is the comparative ratio of voltage drops between resistors the same as the ratio of their resistances? Yes, R_1 resistance value is 10 times that of R_3 and V_1 is 10 times V_3. Also note that R_2 voltage drop is 2.7 times that of R_3, which is the same as their resistance ratios.

The concept of ratios of voltage drops equaling the ratios of resistances can be used to compare the voltage drops of any two components in the series circuit. For example, if the R values and applied voltage are known, it is possible to find all of the voltages for each component throughout the circuit using the proportionality technique. Naturally, you can also determine individual voltages around the circuit by solving for total resistance, then total current (V_T/R_T), and then determining each $I \times R$ drop with $I \times R_1$ for V_1, $I \times R_2$ for V_2, and so forth.

The voltage divider rule

You have just seen that the amount of voltage dropped by a given resistor in a series circuit is related to its resistance value compared with the others in the circuit. The voltage divider rule shows how in a series circuit, any given resistor's voltage drop relates to its resistance compared with the circuit's total resistance. Using this rule, you can determine a series resistor's voltage drop, without having to know the circuit current, if you know the applied voltage and circuit total resistance. Here's the simple formula:

FORMULA 4–4 $V_X = \dfrac{R_X}{R_T} \times V_T$

where: V_X = the voltage drop across the selected resistor
R_X = the resistance value of the selected resistor
R_T = the circuit total resistance
V_T = the circuit applied voltage

PRACTICAL NOTES

This type calculation is very simple using a calculator. You should get into the habit of using your calculator for MOST of your calculations (if your program promotes it at this stage of your learning). In some cases, however, the resistor values and the voltage applied value are such that you can almost do this type calculation in your head. In fact, here's a hint! It is a good idea to learn to approximate answers in your head for any problem that you can, just as a double check on your computations. Approximating is a powerful skill to learn and practice! It can often save you from embarrassing results. If you inadvertently make a mathematical error along the way, your approximation will show you that the erroneous results are not logical. Then, you can check back through your procedure, and quickly find the mistake.

Example

Refer to Figure 4–8 and note how we can find the voltage dropped across R_2 by simply substituting the knowns into the formula. (In this case, R_2 is the value to be used for R_X in the formula.)

$$V_{R_2} = \frac{R_2}{R_T} \times V_T;\ V_{R_2} = \frac{4.7\ \text{k}\Omega}{17.4\ \text{k}\Omega} \times 50;\ V_{R_2} = 0.27 \times 50 = 13.5\ \text{V}$$

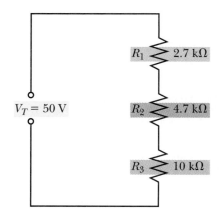

FIGURE 4–8 Voltage-divider rule sample circuit

PRACTICE PROBLEMS IV
Now you use the voltage divider formula to find the following answers.

1. For the circuit of Figure 4–8, what is the voltage dropped by R_1?
2. For the circuit of Figure 4–8, what is the voltage dropped by R_2 and by R_3?
3. Sum the rounded voltages of R_1, R_2, and R_3. Do they add up to close to the applied voltage of 50 volts?

As you can see, the voltage divider rule can be very handy! You will get more opportunities to use this rule as you proceed in your training.

Finding the value of applied voltage

Again referring to Figure 4–7 on page 106, let's examine some ways to find the circuit applied voltage. One obvious way is the Ohm's law expression where $V_T = I_T \times R_T$. In this circuit, V total (or V applied) = 2 mA times R_T. As you can easily determine, R_T in this circuit equals 100 kΩ + 27 kΩ + 10 kΩ = 137 kΩ total resistance. Therefore, circuit applied voltage equals 2 mA $\times$ 137 kΩ = 274 volts.

Another way to find applied voltage is to add all the individual voltage drops to find total voltage, just as we add all the individual resistances to find total resistance. This yields the same answer where 200 volts + 54 volts + 20 volts = 274 volts. This method suggests an important concept called Kirchhoff's voltage law.

4–4 KIRCHHOFF'S VOLTAGE LAW

Kirchhoff's voltage law states the *arithmetic sum* of the voltages around a single circuit loop (any complete closed path from one side of the source to the other) *equals V applied* (V_A). It also says the *algebraic sum* of all the loop voltages, including the source or applied voltage, must *equal zero*. That is, if you observe the polarity and value of voltage drops by the circuit elements and the polarity and value of the voltage source, and add the complete loop's values algebraically, the result is zero. For our purposes, the arithmetic sum approach will be most frequently used. It illustrates how Kirchhoff's voltage law helps determine unknown circuit parameters. Refer to Figure 4–9 as you study the following example.

Example

If the V_A is 50 volts and V_2 is 20 volts, how can Kirchhoff's voltage law help determine V_1? If the sum of voltages (*not* counting the source) must equal V applied, then:

$$V_A = V_1 + V_2$$

Since we know V_A and V_2, transpose to solve for V_1.

$$\text{Therefore, } V_1 = V_A - V_2 = 50\text{ V} - 20\text{ V} = 30\text{ V}$$

If individual voltage drops were known but not the applied voltage, find V applied by adding the individual voltage drops. In this case 30 V + 20 V = 50 V. This agrees with Kirchhoff's voltage law that the arithmetic sum equals the applied voltage. Kirchhoff's voltage law can help you find unknown voltages in

FIGURE 4–9 Using Kirchhoff's voltage law where $V_A = V_1 + V_2$

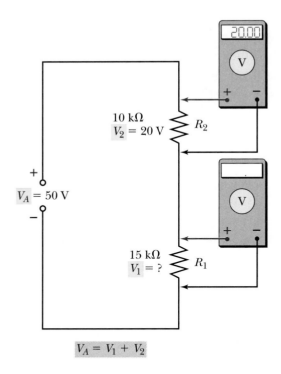

$$V_A = V_1 + V_2$$

series circuits through either addition or subtraction, as appropriate. And if there are more than two series components:

FORMULA 4–5 $V_T = V_1 + V_2 \ldots + V_n$

---

where V_n represents the last of the remaining voltage values.

In Figure 4–10, observe the polarity of voltages around the closed loop. To indicate polarities used in Kirchhoff's voltage law, trace the loop from the source's positive side (Point A), through the resistors and back to the source's negative side (Point B). Consider any voltage a positive voltage whose + point is reached first, and vice versa. In this case, the first voltage reached is +20 V; the next voltage is +30 V, and the source negative terminal is −50 V. Adding these voltages yields:

$$(+20) + (+30) + (-50) = 0; \text{ or } 20 + 30 - 50 = 0$$

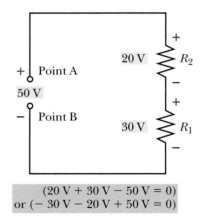

FIGURE 4–10 Kirchhoff's algebraic sum example

Tracing the other direction through the circuit gives:

$$(-30) + (-20) + (+50) = 0; \text{ or } -30 + -20 + 50 = 0$$

In either case, the algebraic sum of the voltage drops *and* the voltage source around the entire closed loop equals zero.

Likewise, the arithmetic sum of the *voltage drops* around a given loop must equal the value of the *applied voltage,* regardless of which direction is used to trace the loop.

PRACTICE PROBLEMS V

Refer again to Figure 4–10. Using the proportionality of voltage drops concept and Kirchhoff's voltage law, if V_1 is 45 V, what are the values of V_2 and V_A?

IN-PROCESS LEARNING CHECK I

Fill in the blanks as appropriate.

1. The primary identifying characteristic of a series circuit is that the _____ is the same throughout the circuit.

2. The total resistance in a series circuit must be greater than any one _____ in the circuit.

3. In a series circuit, the highest value voltage is dropped by the _____ value resistance, and the lowest value voltage is dropped by the _____ value resistance.

4. In a two-resistor series circuit, if the applied voltage is 210 volts and one resistor's voltage drop is 110 volts, what must the voltage drop be across the other resistor? _____ V.

5. Answer each part of this question with increase, decrease, or remain the same. In a three-resistor series circuit, if the resistance value of one of the resistors increases, what happens to the circuit total resistance? _____ To the total current? _____ To the adjacent resistor's voltage drop? _____

6. What is the applied voltage in a four-resistor series circuit where the resistor voltage drops are 40 V, 60 V, 20 V, and 10 V, respectively?_____V.

A quick review

Current is the same throughout a series circuit; total resistance equals the sum of all the individual resistances in series, and voltage distribution around a series circuit is directly related to the resistance distribution, since current is the same through all components and each component's voltage drop = current × its resistance. Kirchhoff's voltage law states that the arithmetic sum of voltage drops equals the voltage applied, or the algebraic sum of all the voltage drops and the voltage source equal zero.

PRACTICAL NOTES

As you continue to work more electronic circuit problems, it is good to remember that:

- Units divided by Kilo units = Milli units. This is often used when finding current: $I = V/R$; where V/kΩ = mA.

- Units divided by Milli units = Kilo units. This is often used when finding resistance: $R = V/I$; where V/mA = kΩ.

- Milli units times Kilo units = Units. This is often used when finding voltage: $V = I \times R$; where mA $\times$ kΩ = V.

4–5 POWER IN SERIES CIRCUITS

The last major electrical quantity to be discussed in relation to series circuits is power. Recall that power dissipated by a component or circuit (or supplied by a source) is calculated with the formulas $P = V \times I$, $P = I^2R$, or $P = V^2/R$. You have learned that in series circuits the current (I) is the same through all components. Therefore, the I^2 factor in the $P = I^2R$ formula is the same for each component in series and the largest R value dissipates the most power. Conversely, the component with the least R value dissipates the least power (I^2R), Figure 4–11.

Learn to use a calculator's various modes to ease the work involved in circuit calculations!

Individual component power calculations

Individual component power dissipations can be found if any two of the three electrical parameters for the given component are known. For example, if the

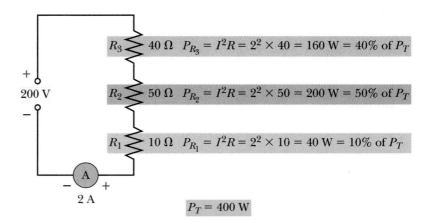

FIGURE 4–11 Largest R dissipates most power, and smallest R dissipates least power.

R_3 — 40 Ω $P_{R_3} = I^2R = 2^2 \times 40 = 160$ W = 40% of P_T

R_2 — 50 Ω $P_{R_2} = I^2R = 2^2 \times 50 = 200$ W = 50% of P_T

R_1 — 10 Ω $P_{R_1} = I^2R = 2^2 \times 10 = 40$ W = 10% of P_T

200 V

A

2 A

$P_T = 400$ W

component's resistance and the current through it are known, use the $P = I^2R$ formula. If the component's voltage drop and the current are known, use the $P = V \times I$ formula. Knowing the voltage and resistance allows you to use the $P = V^2/R$ version of the power formula.

As implied above, the individual power dissipations around a series circuit are *directly related* to the resistance of each element, just as the voltage distribution is directly related to each element's resistance. A specific resistor's power dissipation is the same percentage of the circuit's total power dissipation as its resistance value is of the total circuit resistance (R_T). For example, if its resistance is one-tenth (1/10) the total circuit resistance, it dissipates 10% of the total power. If it is half the R_T, it dissipates 50% of the total power, and so forth, Figure 4–11.

Total circuit power calculations

Total power dissipation (or power supplied to the circuit by the power source) is determined by adding all the individual power dissipations, if they are known. See Formula 4–6.

FORMULA 4–6 $P_T = P_1 + P_2 \ldots + P_n$

EXAMPLE

Refer to Figure 4–11. Using Formula 4–6:

$$P_T = P_{R_1} + P_{R_2} + P_{R_3} = 40 \text{ W} + 200 \text{ W} + 160 \text{ W} = 400 \text{ W}$$

PRACTICE PROBLEMS VI
Refer to Figure 4–12.

Find the values for:

$P_1 =$ _____ watts		$V_1 =$ _____ volts	
$P_2 =$ _____ watts		$V_2 =$ _____ volts	
$P_3 =$ _____ watts		$V_3 =$ _____ volts	
$P_T =$ _____ watts		$R_3 =$ _____ % of R_T	
$I_T =$ _____ amperes		$P_3 =$ _____ % of P_T	

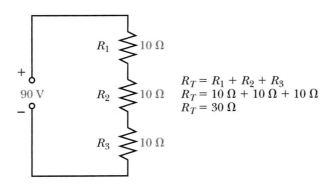

$$R_T = R_1 + R_2 + R_3$$
$$R_T = 10 \text{ } \Omega + 10 \text{ } \Omega + 10 \text{ } \Omega$$
$$R_T = 30 \text{ } \Omega$$

FIGURE 4–12 Diagram for Practice Problem VI

4–6 EFFECTS OF OPENS IN SERIES CIRCUITS AND TROUBLESHOOTING HINTS

Opens

Thus far, you have studied how to analyze current, resistance, voltage, and power in series circuits. At this point we will introduce some practical situations that might occur in *abnormal* series circuits. The two extreme conditions are opens and shorts.

First, what are the effects of an open in a series circuit? An **open circuit** occurs anytime a break in the current path occurs somewhere within the series circuit. Because there is only one path for current flow, if a break (or open) occurs *anywhere* throughout the current path, there is no current flow. In effect, the circuit resistance becomes "infinite" and I_T (circuit current) must decrease to zero.

If an open causes circuit resistance to increase to infinity and circuit current to decrease to zero, how does this affect the circuit voltages and power dissipations?

Example

In Figure 4–13, assume that R_3 is physically broken in half, causing an open or break in the current path. Since there is no complete path for current flow, what would the current meter read? The answer is zero mA.

If there is no current, what is the $I \times R$ drop of R_1? The answer is zero volts because zero times any value of R equals zero. And the answer for V_2 is also zero for the same reason. This means that the difference of potential across each of the good resistors (R_1 and R_2) is zero. That is, R_1 has zero voltage drop and R_2 has zero voltage drop. In effect, the positive side of the source potential is present anywhere from the source itself around the circuit to point A. The negative side of the source potential is present at every point along the circuit on the opposite side of the break all the way to point B.

What is the potential difference across the break, or open? If your answer is 100 volts (or V_A), you are correct! This indicates there will be no $I \times R$ drops across the *good* components in the circuit, and the circuit applied voltage appears across the open portion of the circuit, regardless of where the open appears in the circuit.

An example of a purposely opened circuit is the light switch circuit in your home. When you turn the light on, you *close* an open set of contacts on the switch,

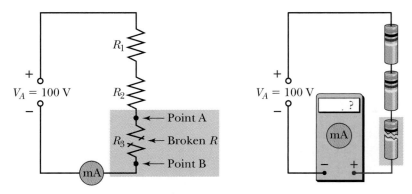

FIGURE 4–13 An open in a series circuit

which is in series with the light. When you turn the light off, you *open* those switch contacts in series with the light; therefore, you open the path for current flow, causing the applied voltage to appear across the opened switch contacts and zero volts to appear across the light element(s). If you measure the voltage across the switch with the lights *off*, you will measure the source voltage, or *V* applied. CAUTION: Don't try this! The voltage in a 120-volt lighting circuit is dangerous.

PRACTICE PROBLEMS VII

1. In a series circuit having a 100-V source and comprised of four 27-kΩ resistors, indicate whether the following parameters would increase, decrease, or remain the same if one of the resistors opened.

   a. Total resistance would _____.

   b. Total current would _____.

   c. Voltage across the unopened resistors would _____.

   d. Voltage across the opened resistor would _____.

   e. Total circuit power dissipation would _____.

2. Would the conditions you have described for question 1 be true no matter which resistor opened? _____

TROUBLESHOOTING HINTS

When you suspect an open in a series circuit, you can measure the voltage across each component. The component, or portion of the circuit across which you measure *V* applied, is the opened component, or circuit portion. *Caution:* To measure these voltages, power must be applied to the circuit. Use all safety precautions possible.

An alternative to this "power-on" approach is to use *"power-off"* resistance measurements. The two points in the circuit at which you measure "infinite ohms" (with the circuit completely disconnected from the voltage/power source) is the opened portion of the circuit. Resistance of the good components should measure their rated values.

4–7 EFFECTS OF SHORTS IN SERIES CIRCUITS AND TROUBLESHOOTING HINTS

Shorts

What are the effects of a short in a series circuit? First, what is a short? A **short circuit** may be defined as an undesired very low resistance path in or around a given circuit. An example is when someone drops a metal object across two wires connected to the output terminals of a power source. In this case you short all the other circuits connected to the power source terminals and provide a *very low* resistance path for current flow through the new metal current path.

A short may occur across one component, several components, or the complete circuit (as in our dropped metal object example). Regardless, consider a short to be a virtually zero resistance path that disrupts the normal operation of the circuit.

Example

Observe Figure 4–14. Notice that the undesired low resistance path is a piece of bare wire with a virtually zero resistance that has fallen across the leads of R_1. *Without* the short, the circuit total resistance would equal 20 kΩ ($R_1 + R_2$). Therefore, with 20 V applied voltage, the current is 1 mA (20 V/20 kΩ). *With* the short, the resistance from point A to point B is close to zero ohms because R_1 is shorted. As far as the voltage source is concerned, "it sees" only 10 kΩ resistance of R_2 as total circuit resistance. Thus, current is 20 V/10 kΩ, or 2 mA (double what it would be without the short). The short causes R_T to decrease to half its normal value and current to double since the *V* applied is unchanged.

This circuit analysis points to some important generalizations: 1) If any portion of a series (or any other type circuit) is shorted (or simply decreases in resistance), the total circuit resistance decreases; and 2) the circuit's total current increases, assuming the applied voltage remains unchanged.

What happens to the other circuit parameters if a short (or decreased resistance) occurs in a series circuit? Using our sample circuit in Figure 4–14: Let's think through it.

1. R_1 becomes zero ohms.

2. R_T decreases, causing circuit current (I_T) to increase.

3. Because I has increased and R_2 is still the same value, the voltage across R_2 increases. In this case, it will increase to *V* applied (20 V) because there is no other resistor in the circuit to drop voltage.

4. V_1 (the voltage across R_1) decreases to zero volts because any value of I times zero ohms equals zero volts.

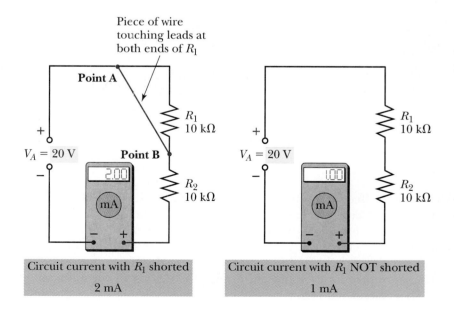

FIGURE 4–14 Effects of a short in a series circuit

5. The total circuit power dissipation increases because I has doubled and V applied has remained unchanged ($P = V \times I$).

6. Power dissipated by the good resistor (R_2) increases because its R is the same, but the current through it has doubled ($P = I^2R$).

7. Of course, the power dissipated by R_1 decreases to zero.

What happens if the *total* circuit is shorted instead of part of it? This means both R_1 and R_2 are shorted and the short is across the power supply terminals. The voltage source is looking at zero ohms, and it would try to supply "infinite" current. In reality this cannot happen. Probably the power supply fuse would blow, and/or the power supply and circuit wires/conductors might be damaged. However, the generalizations for the partially shorted circuit would also apply. That is, circuit resistance goes down, total current increases (until a fuse blows or the power supply "dies"), and the voltage across the shorted portion of the circuit decreases to zero.

PRACTICE PROBLEMS VIII

1. If a series circuit having a 200-V source is composed of a 10-kΩ, a 27-kΩ, a 47-kΩ and a 100-kΩ resistor, and the 47-kΩ resistor shorts, what will the voltages be across each of the resistors listed? (Don't forget to use the "draw-the-circuit" hint to help you in your analysis.)

2. With the short present across the 47-kΩ resistor, do the voltage drops across the other resistors increase, decrease, or remain the same?

TROUBLESHOOTING HINTS

If circuit fuses blow, components are too hot, or smoke appears, there is a good chance that a component, circuit, or portion of a circuit may have acquired a short. Again, measuring resistances helps locate the problem that requires removing power from the circuit to be tested. It is best to remove power from the circuit until the fault can be cleared; therefore, voltage measurement techniques must be used judiciously, if at all. Applying power to circuits of this type often causes other components or circuits to be damaged because excessive current may pass through the nonshorted components or circuits.

With power removed from the circuit, it is possible to use an ohmmeter to measure the resistance values of individual components, or of selected circuit portions. If the normal values are known, it is obvious which component(s), or portion(s) of the circuit have the very low resistance value. Often, there will be visual signs of components or wires that have overheated, leaving a trail that indicates where the excessive current flowed as a result of the short.

In special cases, if voltage measurements are used, voltage drops across the good components are higher than normal (due to the higher current), and voltage drops across the shorted component(s) or circuit portion(s) are close to zero.

If there are current meters or light bulbs in the circuit, the meter readings are higher than normal and/or the unshorted light bulbs glow brighter than normal.

A special troubleshooting hint

A simple but useful technique used in troubleshooting circuits that have sequential "in-line" components or subcircuits is called the divide-and-conquer approach, or split technique. This technique can save many steps in series situations where one of many components or subcircuits in the sequential line is the malfunctioning element.

The technique is to make the first test *in the middle* of the circuit. The results tell the technician which half of the circuit has the problem. For example, Figure 4–15 represents a twelve-light series Christmas tree circuit with an open bulb. Recall, if one light becomes open, all the lights go out since the only path for current has been interrupted. Note in Figure 4–15, the first test is made from one end of the "string" to the "mid-point," in this case between bulbs 6 and 7. If making a voltage test with power on and the meter indicates V applied, then the open bulb must be between bulb 1 through bulb 6. If V measures 0 volts, then the problem is in the other half of the circuit between bulb 7 through bulb 12. With *one* check, half of the circuit has been eliminated as the possible area of trouble.

Incidentally, this technique can be used with the power-off resistance measurement approach. With the circuit *disconnected* from the power source, if the first R measurement shows infinite resistance between bulbs 1 and 6, the problem is in that half of the circuit. If the reading is a low resistance, then the problem is in the other half of the circuit between bulb 7 through bulb 12.

After dividing the abnormally operating circuit's suspect area in half by making the first measurement made at the circuit mid-point, the next step is to divide the remaining suspect area of the circuit in half again by making the second check from one end of the suspect section to its mid-point; thus, narrowing down to a quarter of the circuit the portion that remains in question, Figure 4–16. This splitting technique can continue to be used until only two components or subcircuits remain. Then each is checked, as appropriate, to find *the* malfunctioning component.

The divide-and-conquer approach can be used in any in-line (linear) system of components or circuits where current flow, power, fluids, or signals must flow sequentially from one component or subcircuit to the next and is useful in trou-

FIGURE 4–15 Divide-and-conquer troubleshooting technique—1st step (NOTE: Meter "M" may be either a voltmeter, if voltage is applied to the circuit under test, or an ohmmeter, if the circuit is disconnected from the power source.)

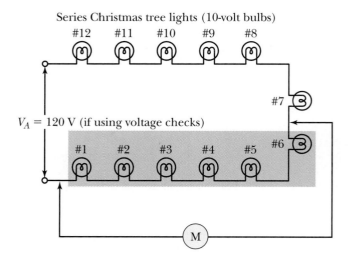

Series Christmas tree lights (10-volt bulbs)
#12   #11   #10   #9   #8

#7

$V_A = 120$ V (if using voltage checks)

#1   #2   #3   #4   #5   #6

M

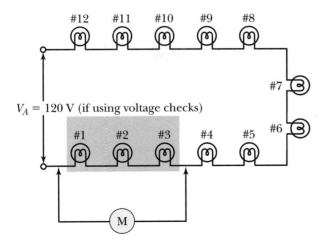

FIGURE 4–16 Divide-and-conquer troubleshooting technique—2nd step (NOTE: Meter "M" may be either a voltmeter, if voltage is applied to the circuit under test, or an ohmmeter, if the circuit is disconnected from the power source.)

$V_A = 120$ V (if using voltage checks)

bleshooting electrical, electronic, hydraulic, and many other systems having the in-line condition.

PRACTICAL NOTES

In review, solve circuit problems with the following steps:

1. Collect all known values of electrical parameters.
2. If no circuit diagram is given, draw a diagram and write known quantities on the diagram.
3. Solve the first part of the problem at a point where you have, or can easily find, sufficient knowns to solve for an unknown. Then proceed by solving the remaining portions of the problem, as appropriate.

4–8 DESIGNING A SERIES CIRCUIT TO SPECIFICATIONS

Let's apply this technique in a sample design problem, then you can try one to see whether you have learned the process.

Example

Design a three-resistor series circuit where two of the resistors are 10 kΩ, the total circuit current equals 2 mA, and the applied voltage equals 94 V.

The first step is to collect the knowns, which were given to you. The next step is to draw the circuit and label the knowns, Figure 4–17. The third step is to start at a point where sufficient knowns are available to solve for a desired unknown.

To apply the third step, observe that in the circuit we drew, we know both the I and the R values for both R_2 and R_3. This makes it easy to find their voltage drops.

$$V_2 = I \times R_2 = 2 \text{ mA} \times 10 \text{ k}\Omega = 20 \text{ V}$$

$$V_3 = I \times R_3 = 2 \text{ mA} \times 10 \text{ k}\Omega = 20 \text{ V}$$

The unknown is the value of R_1, which completes our design problem.

FIGURE 4–17

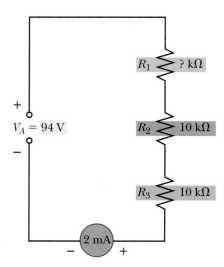

If the voltage drop of R_1 can be found, it is easy to find its resistance value. This is because we know the current must be 2 mA and its resistance must equal its voltage drop divided by 2 mA. It is convenient to use Kirchhoff's voltage law to find V_1. According to Kirchhoff, V_A must equal $V_1 + V_2 + V_3$. Thus, transposing to solve for the unknown V_1:

$$V_1 = V_A - (V_2 + V_3) = 94 \text{ V} - (20 \text{ V} + 20 \text{ V}) = 54 \text{ V}$$

$$\text{Therefore, } R_1 = \frac{54 \text{ V}}{2 \text{ mA}} = 27 \text{ k}\Omega.$$

Another way to solve R_1's value is to find R_T, where $R_T = V_T/I_T$. Thus, $R_T = 94 \text{ V}/2 \text{ mA} = 47 \text{ k}\Omega$. Knowing total resistance equals the sum of all individual resistances in a series circuit and R_2 plus R_3 equals 20 kΩ, R_1 must make up the difference of 27 kΩ to provide a total resistance of 47 kΩ. The more formal way to illustrate this is to write the R_T formula, then transpose it to solve for R_1. Thus,

$$R_T = R_1 + R_2 + R_3$$

$$R_1 = R_T - (R_2 + R_3) = 47 \text{ k}\Omega - (10 \text{ k}\Omega + 10 \text{ k}\Omega) = 27 \text{ k}\Omega$$

Does our design meet the specifications? Yes, it is a three-resistor series circuit where two of the resistors are 10 kΩ, the total circuit current is 2 mA, and the applied voltage equals 94 V.

PRACTICE PROBLEMS IX

Try this next design problem yourself. Complete it on a separate sheet of paper.

Problem: Design a three-resistor series circuit where R_1 is 20 kΩ and drops two-fifths of V applied, R_2 drops 1.5 fifths of V applied, and R_3 has a voltage drop equal to V_2. Assume V applied is 50 V. Draw the circuit and label all V, R, P, and I parameter values.

4–9 SPECIAL APPLICATIONS

Voltage sources in series

Voltage sources can be connected in series to provide a higher or lower total (resultant) voltage than one of the sources provides alone. The *resultant* voltage of more than one voltage in series depends on the values of each voltage *and* whether they "series-aid" or "series-oppose" each other. Figure 4–18a is a series-aiding type connection. Figure 4–18b is an example of a series-opposing arrangement.

Note that in Figure 4–18, you see the terms **cell** and **battery.** A cell is a single voltaic device which converts chemical energy to electrical energy. The symbols associated with a single cell are shown in each circuit in the figure as the 1.5 V sources. A familiar example of cells are the cells you put into your flashlight. Typically, these are 1.5-V cells.

A battery is simply two or more cells interconnected and put into one package. In the figure, the battery symbol is shown in each circuit related to the 6.0-volt sources. A familiar example of a battery is the 12-volt automobile battery. It is comprised of six 2-volt cells, interconnected and packaged in one battery case.

Note, in Figure 4–18a, that the "resultant" voltage applied to the circuit from the series-aiding sources equals the sum of the two sources. In this case 6.0 V + 1.5 V = 7.5 volts. The resulting current through the resistor is 7.5 amperes,

$$I = \frac{V}{R} = \frac{7.5\text{ V}}{1\ \Omega} = 7.5\text{ A}$$

Ways to know these sources are series-aiding are 1) the negative terminal of one source is connected to the positive terminal of the next, and 2) both sources try to produce current in the same direction through the circuit.

In Figure 4–19, what is the total voltage applied to the circuit? What is the current value? If you answered 27 V for the resultant voltage and 0.5 mA for the current, you are correct.

Series-aiding sources

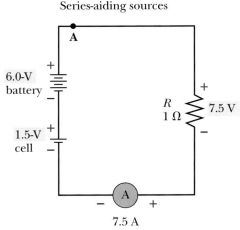

7.5 A

Kirchhoff's algebraic sum around closed loop = 0 V
Going clockwise from point A =
+ 7.5 V − 1.5 V − 6.0 V = 0 V

(a)

Series-opposing sources

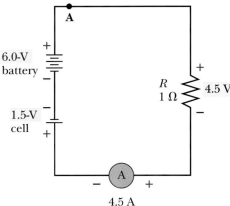

4.5 A

Kirchhoff's again
from point A (clockwise)
+ 4.5 V + 1.5 V − 6.0 V = 0 V

(b)

FIGURE 4–18 Series-connected voltage sources

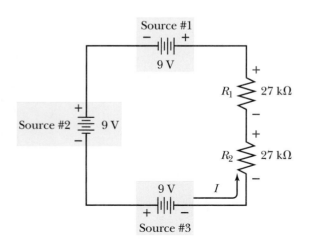

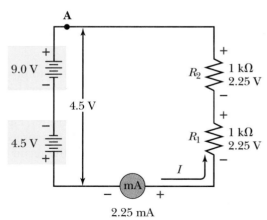

Pt. **A** (clockwise) = + 2.25 V + 2.25 V + 4.5 V − 9.0 V = 0 V

FIGURE 4–19 Series-aiding sources have the following features: (a) Sources are in series. (b) Negative terminal of one source connects to positive terminal of next, and so on. (c) All three sources try to produce current in the same direction through the circuit. (d) They are verifiable using Kirchhoff's law.

FIGURE 4–20 Series-opposing sources have the following features: (a) Sources are in series. (b) Negative terminal of one source connects to negative terminal of next (or positive connects to positive), and so on. (c) Both sources try to produce current in opposite directions through the circuit. (d) They are verifiable using Kirchhoff's law.

As you can see in Figure 4–20, series-opposing sources are connected so that: 1) the negative terminal of one source is connected to the negative terminal of the next, or positive-to-positive; and 2) the sources try to produce current through the circuit in opposite directions.

To determine the resultant or equivalent voltage of series-opposing sources, *subtract* the smaller voltage from the larger voltage.

Example

In Figure 4–20, 9.0 V minus 4.5 V equals a resultant of 4.5 V applied to the circuit connected to the series-opposing sources.

A final point about the concept of series-aiding and series-opposing voltages is that voltage drops can be series-aiding or series-opposing just like voltage sources are series-aiding or series-opposing. If the voltage drops are such that series-connected polarities are − to +, or + to −, *and* the current through the series components is in the same direction, they are series-aiding voltage drops. For example, V_{R_1} and V_{R_2} in Figure 4–20 are series-aiding. Conversely, if the polarities of series component voltage drops are − to −, or + to +, they are series-opposing voltage drops.

PRACTICE PROBLEMS X

1. Refer to Figure 4–19. If Source #1 were reversed in polarity (turned around in this circuit), what would be the value of V_{R_1}?

2. Refer to Figure 4–20. If the 4.5-V battery were replaced by a 6-V battery (using the same polarity connections into the circuit), what would be the value of V_{R_2}?

Simple voltage dividing action and reference points

Voltage dividing

As you have learned, voltage drops around a series circuit are proportional to the resistance distribution since the current is the same through all components. You can *select* the values of resistors in series to "distribute" or "divide" the applied voltage in any desired fashion. Therefore, combining this concept with the series-aiding voltage drop idea, we can create a simple **voltage-divider action** using a series circuit, Figure 4–21. Notice the polarity of voltage drops, and the difference of potential between the various points identified in the circuit.

Equal resistance values are used for R_1, R_2, and R_3, and the voltage source is equally divided by the three resistors. Also, since the voltage drops are series-aiding, the voltages at the various points are cumulative.

PRACTICE PROBLEMS XI

Referring again to Figure 4–21, assume the parameters are changed so source voltage = 188 V, R_1 = 20 kΩ, R_2 = 27 kΩ, and R_3 = 47 kΩ. Indicate the voltages between the following points: D to C; C to B; B to A; and D to B.

Sample applications for voltage dividers

Volume controls

You will recall that in an earlier chapter we mentioned that potentiometers were often used as voltage-dividing devices. In Figure 4–22, you can see that if the wiper-arm on the potentiometer is set at point A, the total voltage coming from the preceding stage in an amplifier or radio circuit is fed to the next stage. If the wiper arm is midway on the resistive element, half the voltage is applied to the next stage, and so on. This is the technique used in radio and TV receivers, and

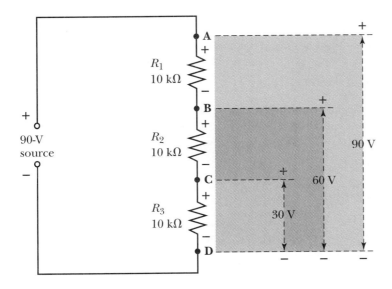

FIGURE 4–21 Example of voltage-divider action

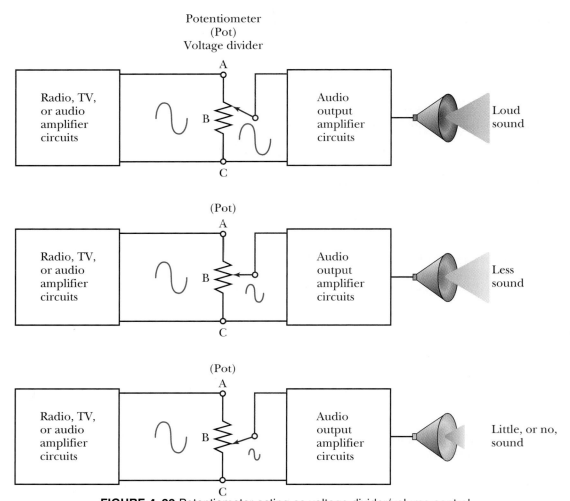

FIGURE 4–22 Potentiometer acting as voltage divider/volume control

in audio amplifiers to control the volume you hear from the system speaker(s). Obviously, the higher the "divided voltage" that is fed to the next (audio amplifier) stage after the potentiometer, the louder the sound you hear.

Transistor bias circuit

Later in your studies you will learn appropriate details about transistor amplifier stages. For now, you simply need to see that one very practical and commonly used application for a voltage divider is to supply appropriate voltage levels to various parts of a transistor, Figure 4–23.

Reference point(s)

Recall in an earlier chapter we discussed the idea of something "with respect to" or "in reference to" something else. The example used was "John is taller than Bill," or "Bill is shorter than John."

In electronic circuits it is common to relate electrical parameters, such as voltage, to a common reference point in the circuit. Most electronic circuits have some sort of common "return" path or connection point to which many of the circuits in the system are connected. This point becomes the circuit "common" reference point. When the circuits are mounted on a metal chassis, the chassis is often used as this common connection-point path. When the circuits are mounted on an epoxy or glass substrate, such as that used by printed circuit (PC) boards, a copper path around the PC board becomes the common connection point, or return path. This common point on chassis or PC boards is often called "chassis ground." The electrical schematic symbol for "chassis ground" is as shown here. /77

In many books and diagrams, the chassis ground symbol is not used. The "earth ground" symbol, $\equiv$, is commonly used to represent the common reference point for circuits. The chassis ground, or common circuit path may or may not be connected to an actual earth ground, as is the power-line system at your home. Throughout this book, we will use the more commonly used ground symbol ($\equiv$) to represent the common reference point in circuits where we do show a common ground reference.

In Figure 4–24, point C is where the ground reference is connected. How is this used as a reference point to describe the voltages along the "divider"? The voltages with respect to this ground reference point can be described as follows:

Point D with respect to ground reference = -30 V
Point C with respect to ground reference = $\ \ \ 0$ V
Point B with respect to ground reference = $+30$ V
Point A with respect to ground reference = $+60$ V

Suppose that the ground reference point is moved from point C to point B, Figure 4–25. How are the voltages described with respect to the reference point?

Point D with respect to ground reference = -60 V
Point C with respect to ground reference = -30 V

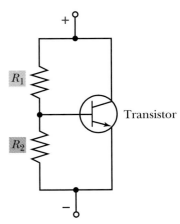

FIGURE 4–23 The voltages applied to the various elements of the transistor depend on the value of the applied voltage and the ratio of the voltage divider resistances R_1 and R_2.

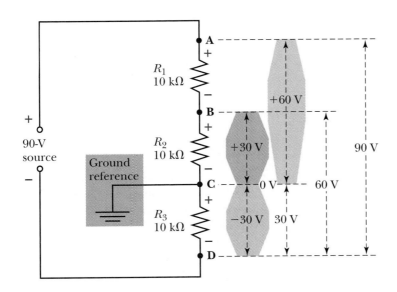

FIGURE 4–24 Ground as a reference point at point C. The 30 V, 60 V, and 90 V indicators show the cumulative voltage drops. The -30 V, $+30$ V, and $+60$ V indicators show voltages with respect to point C, where ground reference is connected.

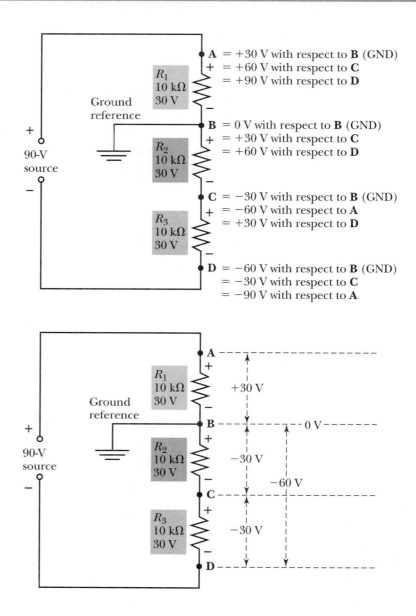

FIGURE 4–25 Ground as a reference point at point B

Point B with respect to ground reference = 0 V

Point A with respect to ground reference = +30 V

Polarity

Again, the polarity and value of voltage are described in terms of its polarity and value with respect to the designated reference point. For example, in Figure 4–24, the voltage at point C with respect to the voltage at point D is +30 V. Conversely, the voltage at point D with respect to the voltage at point C is −30 V. (Note the direction and amount of current is the same in both of the examples.)

A more in-depth study of various types of voltage dividers will be given in the chapter on series-parallel circuits. The idea of reference points, as just discussed, will be used throughout your studies and your career as a technician.

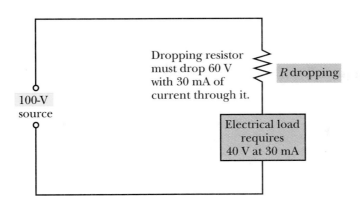

FIGURE 4–26 Example of a series-dropping resistor

PRACTICE PROBLEMS XII

Assume the circuit components and parameters are changed in Figure 4–25 so that V_A now equals 50 V; R_1 now equals 12 kΩ; R_2 now equals 33 kΩ; and R_3 now equals 15 kΩ. Assume that the ground reference point is still point B and that the polarity of the source is the same as in Figure 4–25. Using the voltage-divider rule, and your recently gained knowledge of reference points, determine the following: (*Hint:* Redraw the circuit so you will have something to look at. Use a calculator, if possible.)

1. What is the voltage across R_2?
2. What is the voltage at the top of R_2 with respect to point C?
3. What is the voltage across R_1?
4. What is the voltage at point C with respect to point A?

The voltage-dropping resistor

One final topic relating to some special applications of series circuits is the **series-dropping resistor.**

Often, there is a need to supply a specific voltage at a specific current to an electrical load. If the "fixed" power supply voltage available is higher than the desired voltage value, a series-dropping resistor is used that drops the voltage to the desired level, Figure 4–26.

Example

The current through the electrical load and the series-dropping resistor must be the same since they are in series. To calculate the value of the dropping resistor, the value of voltage it must drop, and the value of current the load requires, provide the knowns to solve the problem. In this case, the dropping resistor must drop 60 V, and the load requires 30 mA of current. Therefore, the dropping resistor must have 30 mA passing through it *when* it is dropping 60 V. Ohm's law indicates the resistor value must be $R = V/I$, or the dropping resistor value = 60 V/30 mA = 2 kΩ. Naturally, this design calculation technique can be used to select resistor values for an infinite variety of dropping-resistor applications and is another example of the many special applications for series circuits.

PRACTICAL NOTES

The power rating of the dropping resistor must be greater than the power it is required to dissipate. Usually, a rating *of at least two times* the actual dissipation is chosen.

It should be pointed out that a voltage drop is the difference in voltage between two points caused by a loss of pressure (or emf) as current flows through a component offering opposition to current flow. An example is the potential difference caused by a voltage drop across a resistor.

PRACTICE PROBLEMS XIII

1. What is the minimum power rating for the dropping resistor, Figure 4–26?

2. What should the value of the dropping resistor be if the applied voltage were 150 V and the load requirements remained 40 V at 30 mA? What is the minimum power rating for the dropping resistor?

4–10 CRITICAL THINKING AND THE SIMPLER TROUBLESHOOTING SEQUENCE

Your most valuable asset as a technician is the ability to think critically and logically. The ability to mentally move from a general principle or general case to a specific situation (deductive reasoning) is valuable. Likewise, moving from a specific case to a probable general case is worthwhile (inductive reasoning).

Throughout the remainder of this text, we will help you develop and enhance your logical reasoning skills. To get started, study the following section called "Introduction to Troubleshooting Skills."

4–11 INTRODUCTION TO TROUBLESHOOTING SKILLS

All electronic technicians and engineers perform troubleshooting to some degree throughout their careers! Whether designing, installing, testing, or repairing electronic circuits and systems, the skill of locating problem areas in the circuit or system by a *logical* narrowing down process becomes a valuable asset.

To enhance your skills of logical troubleshooting, we will introduce a simple sequence of troubleshooting steps you can easily learn. We'll call it the "SIMPLER" sequence (or method) for troubleshooting.

In many of the remaining chapters, you will have an opportunity to practice this sequence and solve "Chapter Troubleshooting Challenge" applications problems. By the time you finish the book, this SIMPLER sequence should be second nature; thus, enhancing your skills and making you an even more valuable technician.

4–12 THE SIMPLER SEQUENCE FOR TROUBLESHOOTING

Symptoms — Gather, verify, and analyze symptom information.

Identify — Identify initial suspect area for location of trouble.

Make — Make decisions about: "What type of test to make" and "Where to make it."

Perform — Perform the test.

Locate — Locate and define new "narrower" area in which to continue troubleshooting.

Examine — Examine available information and again determine "what type test" and "where."

Repeat — Repeat the preceding analysis and testing steps until the trouble is found.

Further information about each step in the sequence

1. **SYMPTOMS** may be provided or collected from a number of sources, such as:

   a. What *the user* of the circuit or system tells you.

   b. *Your five senses* frequently can give you strong clues to the trouble. Sight and hearing allow evaluation of the circuit or system output indicators; e.g., normal or abnormal picture, normal or abnormal sound, system

SIMPLER

Symptoms

Gather, verify, and analyze symptoms information.

Identify

the initial suspect area.

Make

a decision about "What type of test to make" and "Where to make it."

Perform

the test.

Locate

the new "narrower" suspect area.

Examine

the available information and again determine tests to be made.

Repeat

the analysis and testing steps until the trouble is found.

Verify

Test and verify that the circuit operates normally.

meter readings. Sight also allows you to see obviously overheated/burned components. Smell helps find overheated or burned components. (Once you have smelled certain types of overheated or burned components, you will never forget the smell!) Touch can also be used but should be used *very cautiously!* Bad burns result from touching a component that is running exceedingly hot! (Some components run that way normally, others do not.) Also, you can be dangerously shocked by touching "live" circuits!

c. By using your evaluation of the *easy-to-check indicators.* For example, if a TV has sound but no picture, or a radio receiver has hum, but you can't tune in radio stations. Most systems have easy-to-get-at switches and controls that aid in narrowing down possible trouble areas before you begin in-depth troubleshooting. For example, TVs have channel selectors, radios have tuning and volume controls, and electronic circuits frequently have switches and variable resistances.

d. By *comparing actual operation to normal operation and "norms."* "Norms" are normal characteristics of signals, voltages, currents, etc., that are known by experience or determined by referring to diagrams and documentation about the circuit or system being checked. This means you, as a technician, should know how to read and thoroughly understand block and schematic diagrams. (In specific cases, you may have to trace the actual circuit and draw the diagrams yourself.)

2. **IDENTIFY** the initial suspect area by:

a. analyzing all the symptom data;

b. from the analysis of the symptom data, determining all the possible sections of a system, or components within a circuit, that might cause or contribute to the problem; and

c. circling, or in some way marking, the initial suspect area.

3. **MAKE** a decision about "what kind of test" and "where" to make your first test by:

a. Looking at the initial suspect area, and determining where a test should be made to narrow down the suspect area most efficiently. And, determining what type of test would be most appropriate to make. (NOTE: Many times the "where" will dictate what type of test is to be made. At other times, the easiest type of test to make, which yields useful information, determines both the "where" and "what kind" of tests information.)

b. Typically, you begin with general tests, such as looking at obvious indicators, manipulating switches and controls, and similar measures. As you narrow down the suspect area, your tests become more precise, such as voltage, current, or resistance measurements. Finally, substituting a known good component for the suspected bad one confirms your analysis. (In some cases, eradicating the problem is done by soldering, or removing a short, or moving a wire rather than changing a component.)

4. **PERFORM** the first test based on what your analysis has indicated (i.e., what kind of test and where), and you are provided with new information and insight to use in your quest to narrow down and find the problem.

5. **LOCATE** and bracket the new narrower questionable area of the circuit or system using information you have. Again, you will identify this new smaller area by circling or bracketing it so you won't make needless checks outside the logical area to be analyzed.

6. **EXAMINE** the collected symptom information, test results, and other data you have. Now make a decision about the next test that will provide further meaningful information.

7. **REPEAT** the analysis, testing, and narrowing down process as many times as required, and you will eventually find the fault in the circuit or system. At this point, you have successfully used the SIMPLER sequence!

Using the SIMPLER sequence: an example

The information given to you is:

A basic circuit is comprised of a voltage source, conducting wires, a switch, and a light bulb (with its socket). It is known that the voltage source and conductor wires have been previously tested and they are good.

The complaint (which reveals symptom information) is the light bulb does not light, even when the switch is flipped to the "on" position.

1. The **symptom** information gathered is that the light bulb doesn't light! You verify this by connecting the circuit. Analysis of the symptom information can be done by mentally visualizing the circuit, or drawing the circuit diagram. It is then apparent that if the source and the conductor wires are good, the initial suspect areas are the switch, light bulb, and socket.

2. Now you mentally (or on paper) **identify** the portion of the circuit containing the switch, the light bulb, and the socket (by circling or bracketing), Figure 4–27. This becomes the "initial suspect area" for locating the trouble.

3. You now **make** a decision about "what type of test to make" and "where to make it." Some of the possibilities are to a) make a voltage test with a voltmeter across the light bulb socket; b) turn the switch to the "off" position and replace the bulb with a known good bulb, then turn the switch to the "on" position to see if the replacement bulb lights; c) disconnect the power source, unscrew the bulb, and make an ohmmeter test of the suspect bulb to see if it has continuity; d) measure the voltage across the switch terminals with the switch open and with it closed to see if the voltage changes; and e) disconnect the power source and use an ohmmeter to check the switch in both the "off" and "on" positions to see if it is operating properly. Because changing the bulb would be reasonably easy to do and bulbs do fail frequently, you decide to make the light bulb test first, Figure 4–28.

4. When you **perform** this test, you find that the "good" replacement bulb does not light, even with the power source connected to the circuit and the switch in the "on" position. Or if you tested the "suspect" bulb using the ohmmeter

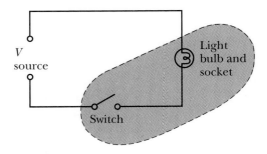

FIGURE 4–27 Initial suspect area

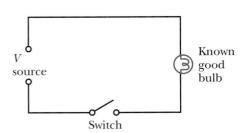

FIGURE 4–28 Checking the light bulb first

SIMPLER

FIGURE 4–29 Results of test:
(a) Good bulb does not light.
(b) Suspect bulb has
continuity.

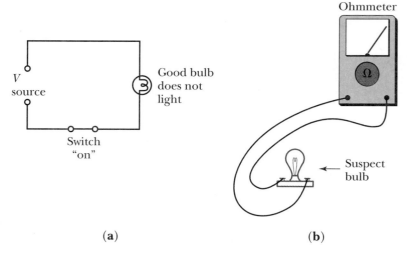

(a) (b)

check approach, you would find the bulb checks good, Figure 4–29.

5. You can now **locate** the new smaller suspect area that includes only the socket and switch, Figure 4–30.

6. When you **examine** all the accumulated information, you decide the most likely suspect between the socket and switch might be the switch since it is somewhat of a mechanical device which fails more frequently. Three ways you can check into this possibility are to a) disconnect the power source from the circuit and use an ohmmeter to check the switch continuity in both the "off" and "on" positions; b) measure the voltage at the light socket to see if voltage is present with the switch in the "on" position; and c) measure voltage across the switch in both the "off" and "on" positions to see if there is voltage across the switch when in the "off" position and zero volts present when the switch is in the "on" position. After examining the information, your decision about "what type of test" and "where" is to check voltage at the light socket since this is an easy check to make, Figure 4–31.

7. **Repeating** the analysis *and* testing procedures, you make the test at the bulb socket with the switch "on." Voltage is present, Figure 4–32. This indicates the switch is good. With this analysis, your remaining suspect is the light socket. In testing the socket (either by ohmmeter continuity testing with the power off or by replacing with a known good socket), you find the socket is indeed bad. You have found the problem! The solution is obvious: Replace the socket. After replacing the socket, you verify the accuracy of your troubleshooting by testing the circuit. Your reward is that the circuit works properly, Figure 4–33.

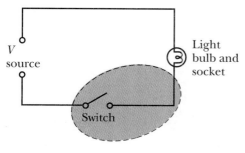

FIGURE 4–30 Narrowed suspect area

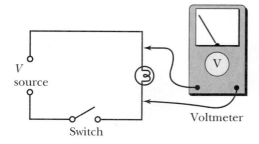

FIGURE 4–31 Voltage check at light socket

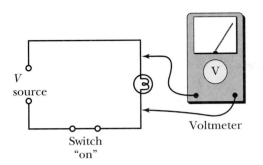

FIGURE 4–32 Voltage present at socket

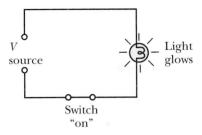

FIGURE 4–33 The circuit works

Although this hypothetical case is simple, it illustrates the concepts of the SIMPLER troubleshooting sequence. Practice using the sequence because it will help you become a logical troubleshooter!

4–13 TROUBLESHOOTING LEVELS

There are two basic levels of troubleshooting technicians may have to perform. One is troubleshooting to the "block" or "module" level. The other is troubleshooting to the single component level.

The block level requires knowledge of the normal "inputs" and "outputs" used for each block or module in the total system. These inputs and/or outputs may be audio or video signals, certain voltage levels, certain current values, and so forth.

To troubleshoot to the block level, we'll refer to the "inputs/outputs" theory, where the "inputs" refer to block inputs and the "outputs" refer to the block output(s). If the input quantities (signals, voltages, etc.) check normal for a given block but the output quantities are abnormal, the problem is probably in that block.

On the other hand, if the input quantities are not normal, you trace backward to where the input is supposed to come from until the place of abnormality is found. In some cases, the abnormality can be caused by the input circuitry of the block being tested, rather than from blocks feeding this block. Your job is to isolate the block causing the problem and to replace it. See "A Block-level Troubleshooting Example" on page 134.

Single-component level troubleshooting requires knowledge of normal parameters throughout the circuitry within each block of the system. By isolating and narrowing down (using the SIMPLER sequence approach), you will eventually narrow the problem to the bad component(s) within a module or block. Replacing the bad component(s) solves the problem. See "A Component-level Troubleshooting Example" on page 135. Study these two examples; then, proceed in the chapter, as appropriate.

The SIMPLER sequence and troubleshooting levels

The SIMPLER sequence can be applied to both the block and component levels of troubleshooting. You will have a chance to try both troubleshooting levels in the Chapter Challenge problems you will find throughout this book.

After studying the examples on pages 134 and 135, try the first Chapter Troubleshooting Challenge, found at the end of this chapter, to begin practicing these concepts.

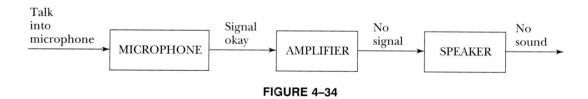

FIGURE 4–34

If you were told there was no sound from the speaker, Figure 4–34, even though someone was talking into the microphone, the block-level troubleshooting scenario might be as follows:

step 1 **Symptoms**   No sound from speaker.

step 2 **Identify**   initial suspect area anywhere in the system for this situation. The trouble could be in the speaker block, amplifier block, or microphone block.

step 3 **Make**   decisions about test (what kind of test and where). Probably make test about in the middle of the system (at the amplifier input). That way you have cut the area of possible trouble in half with your first test. (Remember the "divide-and-conquer" technique explained with Figure 4–16.)

step 4 **Perform**   the test: As you can see from the diagram, the signal into the amplifier is OK (the "input" is all right for this block).

step 5 **Locate**   the new smaller suspect area. Your first test has eliminated the microphone as a possible trouble spot. The remaining possible trouble area is the amplifier block and the speaker block.

step 6 **Examine**   available information and make new test decisions. Now see if the amplifier output is normal. The "type" of check is a signal check; the "where" is at the amplifier output.

step 7 **Repeat**   testing and analysis procedure. A check at the amplifier output shows no signal. (The "output" signal is bad.) If the input to the amplifier is good and the output of the amplifier is bad, the chances are the trouble is in the amplifier block. This is troubleshooting to the block level. If we were to go inside the block to find the component in the block causing the trouble, that would be troubleshooting to the component level.

When performing "block-level troubleshooting," if the output of a block is abnormal, check the condition of all the block's inputs before assuming that the trouble is in this block!

A COMPONENT-LEVEL TROUBLESHOOTING EXAMPLE

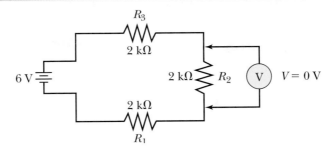

FIGURE 4-35

If you were given the facts in Figure 4-35, and told the voltage source had been checked as being OK, component-level troubleshooting might be done as follows:

step 1 **Symptoms**   No voltage across R_2, even though circuit input voltage is normal.

step 2 **Identify**   initial suspect area. The trouble could be with any one of the resistors or with the interconnecting wires in the circuit, since the symptom could be caused by R_2 being shorted or by an open elsewhere in the circuit.

step 3 **Make**   decisions about test (what type of test and where). Since voltage tests are easy and yield much information for small effort, the "what kind of test" is answered with a voltage test. Since you were told that the voltage source is OK, you decide to check the voltage across either one of the two remaining resistors to determine if current is flowing anywhere in the circuit.

step 4 **Perform**   the test. You decide to make the voltage test across R_1. The result of the test is that V_{R_1} is found to be 6 volts. This is abnormal for this circuit!

step 5 **Locate**   the new smaller suspect area. The voltage test across R_1 leads you to believe it may be the suspect! (Recall in a series circuit applied voltage appears across the open portion of the circuit, and zero volts appear across the good components.)

step 6 **Examine**   available information and make new test decisions. The logical new test might be to disconnect the power source and use an ohmmeter to check the value of resistance exhibited by R_1. The "what kind of test" is a resistance check. The "where to make the test" decision is across R_1.

step 7 **Repeat**   testing and analysis procedure. When R_1 is tested with the ohmmeter, it measures infinite resistance, or "open." R_1 is indeed the culprit.

step 8 **Verify**   This is verified by replacing R_1 and energizing the circuit. When this is done, the voltage drops measured across each of the resistors is 2 volts. This is troubleshooting down to the component level.

SUMMARY

▶ The definition and characteristics of a series circuit state that the circuit components are connected so that there is only one current path; therefore, the current is the same through all components in series.

▶ Current in a series circuit is calculated by Ohm's law, using either total voltage divided by total resistance:

$$I = \frac{V_T}{R_T}$$

or the voltage drop across one of the components divided by its resistance; for example:

$$I = \frac{V_1}{R_1}$$

▶ Total resistance in a series circuit equals the sum of the individual resistances ($R_T = R_1 + R_2 \ldots + R_n$). R_T can also be calculated if V_T and I are known using:

$$R_T = \frac{V_T}{I_T}$$

▶ Because I is the same throughout all components in series, each component's voltage drop ($I \times R$) is proportional to its resistance compared to the total resistance. The largest value resistance drops the highest percentage or fraction of V applied; the smallest value resistance drops the smallest portion of the total voltage.

▶ The voltage-divider rule is a convenient way of finding individual voltage drops in series circuits where the applied voltage and total resistance are known. The formula states:

$$V_X = \frac{R_X}{R_T} \times V_T$$

▶ Calculators are very useful for most calculations in electronics. Calculators may be used in several different modes of operation. One method is simply entering the literal numbers. Other methods use the advantages of a "scientific calculator" by application of powers of 10. The engineering mode deals with numbers times powers of 10 with exponents having multiples of three. The scientific notation mode translates every number into a number between 1 and 10 times a power of 10. The engineering mode is the mode you will probably use most.

▶ V applied must equal the sum of the voltage drops around the closed loop. That is, if there are three components around the series string, V applied must equal the sum of the three components' voltage drops. If there are four components, V applied equals the sum of the four individual voltage drops.

▶ Kirchhoff's voltage law states the arithmetic sum of the voltages around a single circuit loop equals V applied. Also, Kirchhoff's voltage law can be stated as the algebraic sum of all the loop voltages, including the source, must equal zero. It is important when using Kirchhoff's law(s) to notice the voltage polarities.

▶ The power distribution throughout a series circuit is directly related to the resistance distribution. For example, if a component's R is 25% of the total R, it dissipates 25% of the total power supplied by the source.

▶ Total power dissipated by a series circuit (or any other type circuit) equals the sum of all the individual power dissipations. Total power can be calculated by $P_T = V_T \times I_T$, or $P_T = P_1 + P_2 \ldots + P_n$.

▶ A break or open anywhere along the path of current in a series circuit causes R_T to increase to infinity; I_T to decrease to zero; voltage drops across unopened components or circuit portions to decrease to zero; and circuit applied voltage to be felt across the open portion of the circuit. P_T decreases to zero since I equals zero.

▶ A short (very low resistance path) across any component, or portion, of a series circuit causes R_T to decrease; I_T to increase; voltage drops across the normal portions of the circuit to increase; and voltage dropped across the shorted portion of the circuit to decrease to virtually zero. A short also causes P_T to increase and the power dissipated by the remaining unshorted components to increase.

▶ When designing a circuit or solving for unknown quantities in an electrical circuit, it is good to 1) collect all known values; 2) draw a diagram and label known value(s); and 3) solve at any point where enough knowns are available to solve for an unknown. (NOTE: Usually, Ohm's law and/or Kirchhoff's law[s] will be used in the solution.)

▶ When troubleshooting a series circuit with an open condition, voltage measurements will reveal zero voltage drop across good components (or circuit portions) and V applied across the open points. Take resistance measurements only with the circuit disconnected from the source. Then normal components will reveal normal readings, and the opened component/portion will reveal an infinite ohms reading.

▶ When troubleshooting a series circuit with a short condition, use *resistance* measurements with the

power source disconnected from the circuit. Normal components (or circuit portions) will measure normal values of R, while the shorted component (or circuit portion) will measure very low, or zero R.

▶ Series circuits are useful for several special applications, such as series-aiding and series-opposing voltage sources, simple voltage-dividing action, and series-dropping an available voltage down to a desired level.

FORMULAS AND SAMPLE CALCULATOR SEQUENCES

FORMULA 4–1 $R_T = R_1 + R_2 \ldots + R_n$

R_1 value, ⊕ , R_2 value, ⊕ , . . . , ⊜

FORMULA 4–2 $R_T = \dfrac{V_T}{I_T}$

total voltage value, ÷ , total current value, ⊜

FORMULA 4–3 $V_1 = I_{R_1} \times R_1$

current value, ✕ , resistor #1 value, ⊜

FORMULA 4–4 $V_X = \dfrac{R_X}{R_T} \times V_T$

R_X value, ÷ , R_T value, ✕ , V_T value, ⊜

FORMULA 4–5 $V_T = V_1 + V_2 \ldots + V_n$

V_1 value, ⊕ , V_2 value, ⊕ , . . . , ⊜

FORMULA 4–6 $P_T = P_1 + P_2 \ldots + P_n$

P_1 value, ⊕ , P_2 value, ⊕ , . . . , ⊜

REVIEW QUESTIONS

1. Explain the most distinguishing electrical characteristic of a "series circuit" and clarify why this characteristic is true for this type circuit.

2. The total resistance of a series circuit is:

a. more than the largest value resistor in the circuit.

b. less than the largest value resistor in the circuit.

c. equal to the largest value resistor in the circuit.

3. Explain why your answer for question 2 is true.

4. In a series resistive circuit, the largest voltage drop appears across:
   a. the smallest R value resistor.
   b. the largest R value resistor.
   c. neither the smallest or largest R value resistor.

5. In a series resistive circuit, the applied circuit voltage is equal to the:
   a. difference between the smallest and largest voltage drops around the circuit.
   b. sum of all the individual voltage drops around the circuit.
   c. product of the smallest and largest voltage drops around the circuit.

6. Kirchhoff's voltage law indicates that the *mathematical sum* of the voltage drops around a complete circuit loop must equal:
   a. zero.
   b. V_A.
   c. $V_1 + V_2$.
   d. none of the above.

7. The power distribution throughout a series circuit is:
   a. directly related to the resistance distribution around the circuit.
   b. inversely related to the resistance distribution around the circuit.
   c. has no relationship to the resistance distribution around the circuit.

8. If a break occurs in a series circuit:
   a. R_T decreases to zero ohms.
   b. R_T increases to infinite ohms.
   c. neither of the above are true.

9. If a short is placed across part of a series circuit:
   a. R_T decreases to zero ohms.
   b. R_T does not change.
   c. R_T decreases.
   d. R_T increases.

10. What type measurement would you make in troubleshooting a series circuit containing a short?
    a. Resistance measurements with the power on
    b. Resistance measurements with the power off
    c. Voltage measurements with the power off
    d. None of the above

PROBLEMS

NOTE: While working on these problems, don't forget to draw and label diagrams, as appropriate. It is a good habit to form.

1. If a series string of two equal resistors draws 5 amperes from a 10-volt source, what is the value of each resistor?

2. What is the current through a series circuit comprised of one 10-kΩ, one 20-kΩ, and one 30-kΩ resistor? Assume the circuit applied voltage is 40 V.

3. Draw a diagram of a series circuit containing resistors having values of 50 Ω, 40 Ω, 30 Ω, and 20 Ω, respectively. Label the 50-Ω resistor as R_1, the 40-Ω resistor as R_2, and so forth. Calculate the following and appropriately label your diagram:
   a. What is the value of R_T?
   b. What is V applied if I equals 2 A?
   c. What is the voltage drop across each resistor?
   d. What is the value of P_T?
   e. What value of power is dissipated by R_2? by R_4?

   f. What fractional portion of V_T is dropped by R_4?
   g. If R_3 increases in value while the other resistors remain the same, would the following parameters increase, decrease, or remain the same?
      (1) Total resistance
      (2) Total current
      (3) V_1, V_2, and V_4
      (4) P_T
   h. If R_2 shorted, would the following parameters increase, decrease, or remain the same?
      (1) Total resistance
      (2) Total current
      (3) V_1, V_3, and V_4

4. According to Kirchhoff's voltage law, if a series circuit contains components dropping 10 V, 20 V, 30 V, and 50 V, respectively, what is V applied? What is the algebraic sum of voltages around the complete closed loop, including the source?

5. Draw a diagram showing how you would connect three voltage sources to acquire a circuit applied

voltage of 60 V, if the three sources equaled 100 V, 40 V, and 120 V, respectively.

6. a. Calculate the P_T and P_1 values for the circuit shown in Figure 4–36.
   b. What is the V_T?
   c. What is the R_T?
   d. How many times greater is P_4 than P_1?
   e. If R_4 shorted what would the value of P_T become? of P_1?

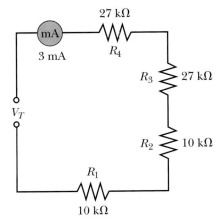

FIGURE 4–36

7. a. Find the value of V_A for the series circuit shown in Figure 4–37.
   b. Assume that R_3 and R_4 are equal. What is the value of R_3?
   c. What is the value of R_2?

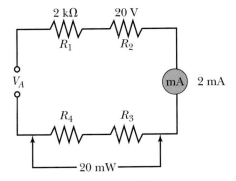

FIGURE 4–37

8. Draw a simple three-resistor series voltage divider that provides equal voltage division of a 180-V source and draws 2 mA of current.

Referring to Figure 4–38, calculate and answer questions 9–16:

9. What value of voltage will be read by the voltmeter?

10. What value of current will be read by the milliammeter?

11. What is the voltage drop across R_2?

12. What is the voltage drop across R_3?

13. What is the voltage drop across R_4?

14. If R_3 were to short out, what would be the voltage reading on the voltmeter?

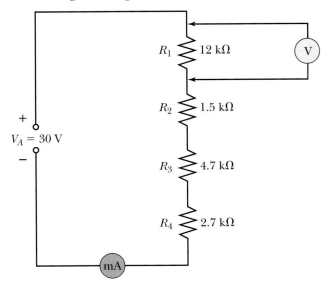

FIGURE 4–38 Review questions problem circuit

15. Which resistor in the circuit of Figure 4–38 dissipates the most power? The least power?

16. If V_A doubled for the original circuit of Figure 4–38, would the percentage of applied voltage dropped by R_3 increase, decrease, or remain the same?

17. In a three-resistor series circuit, if V applied is 200 V, V_1 is 40 V, and V_2 is 90 V, what is the value of V_3?

18. In a three-resistor series circuit, if V_1 is 25 V, V_2 is 50 V, and V_3 is twice V_1, what is the value of V applied?

19. For the conditions shown in the circuit in Figure 4–39.
   a. To what value is R_1 adjusted?
   b. What is the value of I_T?
   c. What percentage of total power is dissipated by R_1?
   d. If R_1 were set to the middle of its R range, what would be the value of I_T?

e. If R_1 were set at the least value possible, what value of voltage would be indicated by the voltmeter in the circuit?

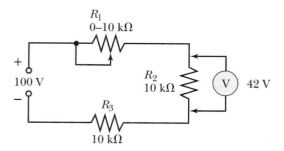

FIGURE 4-39

20. a. Find R_T and V_T for the circuit shown in Figure 4-40.
    b. What value will meter M_1 indicate?
    c. What value will meter M_3 indicate?
    d. What is the value of I_T for this circuit?
    e. If all R values remain the same, but P_T doubles, what is the new value of V_T? What is the new I_T?

21. Refer to Figure 4-40. If all R values are as shown, but P_{R_5} equals 45 mW:
    a. What is the value of P_{R_2}?
    b. What is the value of P_T?
    c. What is the value of I_T?

22. Draw a circuit diagram of a series circuit where V_3 is three times the value of V_1 and V_2 is twice V_1. Assume V applied is 60 V, R_T is 120 k Ω, and calculate:
    a. value of R_1.
    b. values of V_1, V_2, and V_3.
    c. values of I and P_T.

23. Determine the value of a series-dropping resistor needed to drop a source voltage of 100 V to the appropriate value for a load rated as 50 V at 5

mA. Draw the circuit and label all components and electrical parameters.

Refer to Figure 4-40 to answer questions 24-28.

24. If R_5 were to be shorted out by a jumper wire, would the voltage drop across R_1 increase, decrease, or remain the same?

25. If R_2 increased in value, because of age, would the meter reading of meter M_3 increase, decrease, or remain the same?

26. If R_4 were shorted, would the circuit applied voltage increase, decrease, or remain the same?

27. With the parameters shown in Figure 4-40, determine the value of circuit applied voltage.

28. All other factors (parameters) remaining the same, if the power dissipated by R_3 quadruples, what must have happened to the value of circuit applied voltage and the circuit current? (Be specific!)

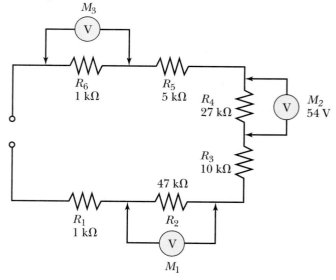

FIGURE 4-40

ANALYSIS QUESTIONS

1. In your own words, explain the difference between calculator scientific-mode notation and calculator engineering-mode notation.

2. List at least two other possible applications for simple voltage dividers, other than those mentioned in this chapter.

3. In your own words, express the key facts you have learned in this chapter regarding the "divide-and-conquer" rule.

4. In the SIMPLER troubleshooting sequence of steps, the first step relates to gathering and analyzing symptom information. Name at least

three tools and techniques that can help you in this effort.

5. In the SIMPLER troubleshooting method, how do you go about identifying the initial suspect area?

6. In the SIMPLER troubleshooting method, what are two decisions you must make before making each test as you proceed through the troubleshooting sequence.

7. In your own words, define what the term *bracketing* means in reference to the SIMPLER troubleshooting system.

8. In what type of circuit or system situation is the divide-and-conquer technique most appropriately used?

9. What type of documentation aid do you think would be most helpful when you are beginning to troubleshoot a circuit of any kind?

10. What precaution should be taken when troubleshooting a series circuit suspected of having a short? Briefly explain why this precaution should be taken.

11. If a series circuit's current suddenly drops to zero, what has happened in the circuit?

12. For the circuit of Figure 4–40, if certain components are overheating, circuit current has increased, and the voltage drops across R_2 and R_3 have drastically decreased, which of the following has occurred?
a. The total circuit has been shorted.
b. The total circuit has been opened.
c. R_1 has shorted.
d. R_2 and R_3 have shorted out.
e. None of the above

13. What basic technique is often used in troubleshooting a defective series circuit that has a number of components in series?

A GOOD IDEA

Wear safety glasses when you or someone near you is soldering or using rotating machinery or power tools!

PERFORMANCE PROJECTS CORRELATION CHART

Suggested performance projects in the Laboratory Manual that correlate with topics in this chapter are:

CHAPTER TOPIC	PERFORMANCE PROJECT	PROJECT NUMBER
Resistance in Series Circuits	Total Resistance in Series Circuits	9
Definition and Characteristics of a Series Circuit	Current in Series Circuits	10
Voltage in Series Circuits	Voltage Distribution in Series Circuits	11
Power in Series Circuits	Power Distribution in Series Circuits	12
Effects of Opens in Series Circuits and Troubleshooting Hints	Effects of an Open in Series Circuits	13
Effects of Shorts in Series Circuits and Troubleshooting Hints	Effects of a Short in Series Circuits	14

NOTE: It is suggested that after completing the above projects, the student should be required to answer the questions in the "Summary" at the end of this section of projects in the Laboratory Manual.

THE SIMPLER SEQUENCE FOR TROUBLESHOOTING

Follow the seven-step SIMPLER troubleshooting sequence, as outlined below, and see if you can find the problem with this circuit. As you follow the sequence, record your circuit test results for each testing step on a separate sheet of paper. This will aid you and your instructor to see where your thinking is on track or where it might deviate from the best procedure.

step 1 Symptoms

—(Gather, verify, and analyze symptom information) To begin this process, read the data under the "Starting Point Information" heading. Particularly look for circuit parameters that are not normal for the circuit configuration and component values given in the schematic diagram. For example, look for currents that are too high or too low; voltages that are too high or too low; resistances that are too high or too low, etc. This analysis should give you a first clue (or symptom information) that will aid you in determining possible areas or components in the circuit that might be causing this symptom.

step 2 Identify

—(Identify and bracket the initial suspect area) To perform this first "bracketing" step, analyze the clue or symptom information from the "Symptoms" step, then either on paper or in your mind, bracket, circle, or put parentheses around all of the circuit area that contains any components, wires, etc., that might cause the abnormality that your symptom information has pointed out. NOTE: Don't bracket parts of the circuit that contain items that could not cause the symptom!

step 3 Make

—(Make a decision about the first test you should make: what type and where) Look under the TEST column and determine which of the available tests shown in that column you think would give you the most meaningful information about the suspect area or components, in the light of the information you have so far.

step 4 Perform

—(Perform the first test you have selected from the TEST column list) To simulate performing the chosen test (as if you were getting test results in an actual circuit of this type), follow the dotted line to the right from your selected test to the number in parentheses. The number in parentheses, under the "Results in Appendix C" column tells you what number to look up in Appendix C to see what the test result is for the test you are simulating.

step 5 Locate

—(Locate and define a new "narrower" area of uncertainty) With the information you have gained from the test, you should be able to eliminate some of the circuit area or circuit components as still being suspect. In essence, you should be able to move one bracket of the bracketed area so that there is a new smaller area of uncertainty.

step 6 Examine

—(Examine available information and determine the next test: what type and where) Use the information you have thus far to determine what your next test should be in the new narrower area of uncertainty. Determine the type test and where, then proceed using the TEST listing to select that test, and the numbers in parentheses and Appendix C data to find the result of that test.

step 7 Repeat

—(Repeat the analysis and testing steps until you find the trouble) When you have determined what you would change or do to restore the circuit to normal operation, you have arrived at *your* solution to the problem. You can check your final result against ours by observing the pictorial step-by-step sample solution on the pages immediately following the "Chapter Troubleshooting Challenge"—circuit and test listing page.

step 8 Verify

NOTE: A useful *8th Step* is to operate and test the circuit or system in which you have made the "correction changes" to see if it is operating properly. This is the final proof that you have done a good job of troubleshooting.

CHALLENGE CIRCUIT 1

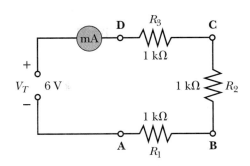

CHALLENGE CIRCUIT 1

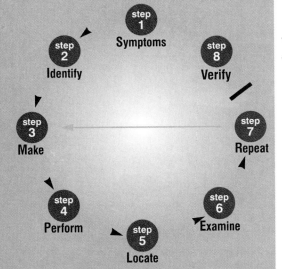

Starting Point Information

1. Circuit diagram
2. $V_T = 6$ V
3. Measured $I_T = 1.5$ mA

TEST	Results in Appendix C
$V_{A–B}$	(5)
$V_{B–C}$	(86)
$V_{C–D}$	(42)
Res. of R_1	(16)
Res. of R_2	(62)
Res. of R_3	(3)
R_T	(36)

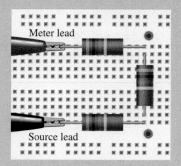

 step 1

Symptoms

The total current is too low for voltage applied. This implies that the total resistance has increased, meaning that one or more resistor(s) has changed value.

 step 2

Identify

initial suspect area: R_1, R_2, and R_3 (i.e., total circuit).

 step 3

Make

test decision: Check voltage across R_2 (middle of circuit).

 step 4

Perform

1st Test: Look up the test result. V_{B-C} is 1.5 V.

 step 5

Locate

new suspect area: R_1 and R_3. NOTE: V_{B-C} would be greater than 2 V if R_2 had increased and the other resistors had not.

 step 6

Examine

available data.

 step 7

Repeat

analysis and testing:

2nd Test: Check voltage across R_1. V_{A-B} is 1.5 V.
3rd Test: Check voltage across R_3. V_{C-D} is 3 V.
4th Test: Disconnect the V source and check resistance of R_3. The result is that R_3 is 2 kΩ, which is abnormal.

 step 8

Verify

5th Test: Replace R_3 with a good 1-kΩ resistor and note the current. When this is done, the circuit checks out normal. Each resistor drops 2 V and the circuit current measures 2 mA.

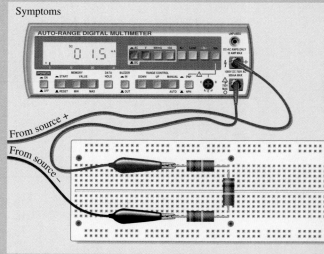

Symptoms

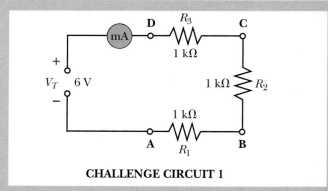

CHALLENGE CIRCUIT 1

1st Test

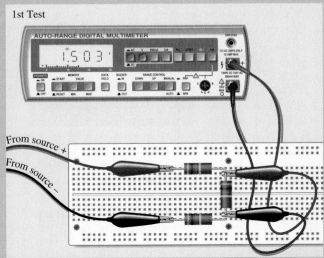

From source +

From source −

2nd Test

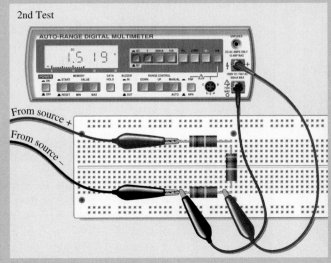

From source +

From source −

3rd Test

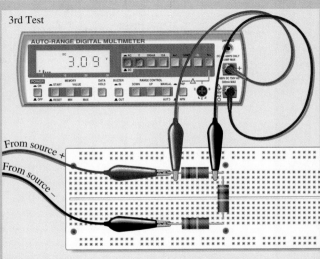

From source +

From source −

4th Test

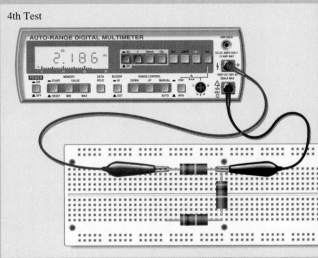

5th Test

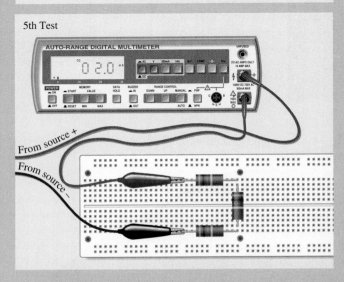

From source +

From source −

PARALLEL CIRCUITS

5

chapter

CHAPTER PREVIEW

Most electrical and electronic circuits, no matter how simple or complex, contain portions that can be examined using parallel circuit analysis. Of course, some circuits contain only parallel circuit arrangements. For example, the lights and wall outlets in your home generally use parallel circuitry to connect several lights or outlets on one circuit, Figure 5–1. Many automobile accessories (heaters and radios) are connected to the battery using parallel circuit methods, Figure 5–2.

In this chapter, you will again apply Ohm's law and the power formulas. Also, Kirchhoff's current law and some *vital contrasts* between parallel and series circuits will be examined. Several approaches to solve for total resistance in parallel circuits will be learned and applied. Finally, information on circuit troubleshooting techniques will be presented.

OBJECTIVES

After studying this chapter, you should be able to:

1. Define the term **parallel circuit**

2. List the characteristics of a parallel circuit

3. Determine voltage in parallel circuits

4. Calculate the total current and branch currents in parallel circuits

5. Compute total resistance and branch resistance values in parallel circuits using at least three different methods

6. Determine conductance values in parallel circuits

7. Calculate power values in parallel circuits

8. List the effects of opens in parallel circuits

9. List the effects of shorts in parallel circuits

10. Describe troubleshooting techniques for parallel circuits

11. Use current divider formulas

12. Use the SIMPLER troubleshooting sequence to solve the Chapter Troubleshooting Challenge problem

5–1 DEFINITION AND CHARACTERISTICS OF A PARALLEL CIRCUIT

Recall from the last chapter the important characteristic defining a series circuit is only one path for current exists. Therefore, the current through all components is the same. Because of this fact, the voltage distribution throughout the circuit (voltage across each of the components) depends on each component's resistance value (i.e., the highest R value dropping the highest voltage or the lowest R value dropping the lowest voltage).

In contrast to the series circuit, the significant features of **parallel circuits** are 1) *the voltage across all parallel components must be the same*, Figure 5–3; and 2) *there are two or more paths (branches) for current flow*, Figure 5–4. Because the voltage is

Lights in a home are connected in parallel.

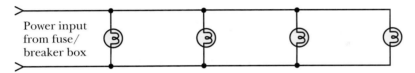

FIGURE 5–1 Parallel circuits are common in a home.

Wall outlets in a home are connected in parallel.

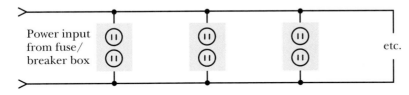

etc.

FIGURE 5–2 Automotive accessories are often connected in parallel across the battery. NOTE: Series switches in circuits are not shown.

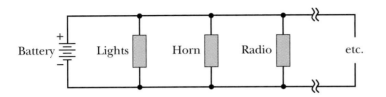

FIGURE 5–3 Voltage in parallel circuits is the same across all components, where $V_{R_1} = V_A$ and $V_{R_2} = V_{R_1} = V_A$.

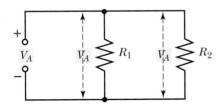
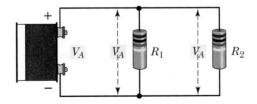

Series circuit

Parallel circuit

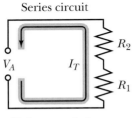
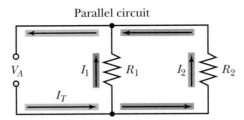

Only *one* path for current flow

Two (or more) paths for current flow

FIGURE 5–4 A basic difference between series and parallel circuits

the same across each **parallel branch,** the current distribution throughout the circuit depends on each branch's resistance value. Each branch current is *inverse* to its resistance value, which means the higher the resistance value, the lower the branch current, Figure 5–5.

Thus, a parallel circuit is defined as one where the voltage across all (parallel branch) components is the same and where there are two or more branches for current flow.

FIGURE 5–5 In parallel circuits, current through each branch is *inversely* proportional to each branch's resistance.

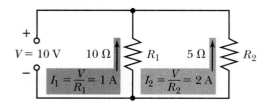

5–2 VOLTAGE IN PARALLEL CIRCUITS

If you know the voltage applied to a parallel circuit or the voltage across any one of the branches, you automatically know the voltage across any one or all of the parallel branches. This voltage is the same across all branches of a parallel circuit, Figure 5–3.

PRACTICE PROBLEMS I

Again refer to Figure 5–3. If R_2 is 27 kΩ and current through R_2 is 2 mA, what is the value of V_{R_1}? Of V_A?

5–3 CURRENT IN PARALLEL CIRCUITS

Branch currents and total current

Refer to Figure 5–6 and note the following points as current and Kirchhoff's current law are discussed:

- Total circuit current leaves from the source's negative side and travels along the conductor until it reaches the circuit "junction" at point A.
- At point A, a portion of the total current travels through branch resistor R_1 (current path I_1) on its way back to the source's positive side. The remaining portion travels on to the circuit junction at point B.
- At point B, current "splits" and a portion travels through R_2 (current path I_2) and the remainder through R_3 (current path I_3) on its way back to the source's positive side.
- At point C, currents I_2 and I_3 rejoin and travel to junction D. Thus, current value from point C to D must equal $I_2 + I_3$.

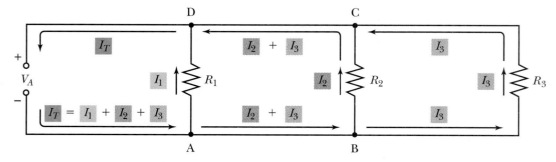

FIGURE 5–6 Kirchhoff's current law states that current coming to a point must equal the current leaving that same point.

• At point D, current $I_2 + I_3$ joins current I_1 and this total current travels back to the source's positive side. NOTE: *The total current equals the sum of the branch currents.* This is an important fact; remember it! Mathematically:

FORMULA 5–1 $I_T = I_1 + I_2 + I_3 \ldots + I_n$

PRACTICE PROBLEMS II

Assume a five-branch parallel circuit. What is the formula for I_T? Which resistor passes the most current? Which resistor passes the least current?

5–4 KIRCHHOFF'S CURRENT LAW

The previous detailed statements express **Kirchhoff's current law.** Simply stated, this law is *the value of current entering a point must equal the value of current leaving that same point.*

    In parallel circuit branch currents, since V across each branch is the same, the current divides through the branches in an inverse relationship to the individual branch resistances. That is, the highest R value branch allows the least current and the lowest R value branch allows the highest branch current.

Example

In using Ohm's law to solve for branch currents, this fact becomes apparent, Figure 5–7. It suggests an interesting proportionality approach in thinking about branch currents in parallel circuits, Figure 5–8. You can see that if branch 1 has twice the resistance of branch 2, it will have half the current value through it.

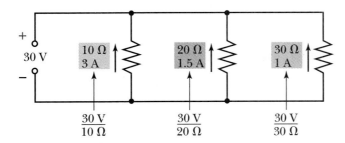

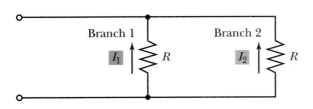

FIGURE 5–7 Current is inverse to branch R value.

FIGURE 5–8 Looking at branch 1 via the proportionality approach: If branch 1 has twice the R of branch 2, then I_1 = half of I_2 (or $I_2 = 2 \times I_1$). If branch 1 has one-fourth the R of branch 2, then $I_1 = 4 \times I_2$ (or I_2 is one-fourth of I_1).

Conversely, branch 2 will have twice the current value of branch 1. If branch 1 were to have one-fourth the R value of branch 2, then it would have four times the current of branch 2, and so forth.

At this point, you have learned that voltage is the same across each parallel branch, current to and from any given circuit point must be the same, branch currents are inverse to their branch resistances, and total current equals the sum of the branch currents. It's time to see if you can apply this knowledge and Ohm's law and Kirchhoff's law to solve some problems.

PRACTICE PROBLEMS III

Look at Figure 5–9 and, without looking ahead, determine the values for V_2, V_1, V_A, I_1, and I_T. Now, read the following explanation.

To find V_2 use Ohm's law, where $V_2 = I_2 \times R_2 = 1 \text{ A} \times 50 \text{ }\Omega = 50 \text{ V}$. Since the voltage must be the same across all parallel branches, then V_1 and V_A must also equal 50 V in this circuit. Since you know V_1 and R_1, then Ohm's law will again enable you to find the answer for I_1. That is, I_1 must equal V_1/R_1, or 5 A. And I_T equals the sum of the branch currents; thus, 1 A + 5 A = 6 A.

Using Ohm's law, what must the value of R_T be? The answer is

$$R_T = \frac{V_T}{I_T} = \frac{50 \text{ V}}{6 \text{ A}} = 8.33 \text{ }\Omega$$

It is interesting to note the total circuit resistance is *less* than either branch's resistance. With respect to the power source, the two branch resistances of 10 Ω and 50 Ω could be replaced by one resistor with a value of 8.33 Ω and it would still supply the same total current and circuit power. Later in the chapter you will study several methods to determine total circuit resistance for parallel circuits. However, at this point, remember that total resistance of parallel circuits is always less than the least branch resistance.

PRACTICE PROBLEMS IV

1. Refer to Figure 5–10 and with your knowledge of parallel circuits, solve for the unknowns. Don't cheat yourself by looking up the answers before working the problems.

2. Again referring to Figure 5–10, assume the parameters are changed to: $V_A = 125 \text{ V}$, $I_T = 10.64 \text{ mA}$, $I_1 = 4.63 \text{ mA}$, and $R_3 = 56 \text{ k}\Omega$. Find R_1, I_2, I_3, R_2, and the value of R_T.

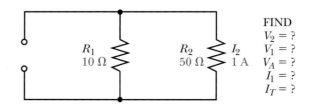

FIGURE 5–9

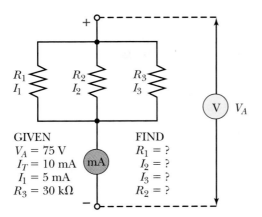

FIGURE 5–10

GIVEN
$V_A = 75$ V
$I_T = 10$ mA
$I_1 = 5$ mA
$R_3 = 30$ kΩ

FIND
$R_1 = ?$
$I_2 = ?$
$I_3 = ?$
$R_2 = ?$

5–5 RESISTANCE IN PARALLEL CIRCUITS

Again, total resistance of parallel circuits is always *less than the least branch resistance.* Let's examine the reason for this and learn several useful ways to solve for total resistance in parallel circuits.

Since each parallel branch provides another current path where the voltage/power source must supply current, then the source must provide more current for each additional branch in a parallel circuit. Since the *V* is the same from the source, but the *I* supplied by the source is higher for every branch added in parallel, then Ohm's law ($R = V/I$) tells us that the total or **equivalent circuit resistance** must decrease every time another branch (or current path) is added to the parallel circuitry. For example, if a circuit comprises only the lowest *R* value branch, the total resistance equals that value. If *any* value *R* branch is added in parallel with that branch, another current path has been provided. Therefore, total current supplied by the source increases, source voltage remains constant, and total or equivalent circuit resistance decreases. Moving forward from that thought, let's look at several methods available to determine parallel circuit total resistance.

5–6 METHODS TO CALCULATE TOTAL RESISTANCE (R_T)

The Ohm's law method

Recall the formula $R_T = V_T/I_T$ to find total resistance.

Example

Look at Figure 5–11. The total resistance according to Ohm's law is

$$R_T = \frac{100 \text{ V}}{20 \text{ mA}} = 5 \text{ k}\Omega$$

A logical extension of this law for parallel circuits is that each branch resistance can be solved by knowing branch voltage and branch current. This is stated as

$$R \text{ branch} = \frac{V \text{ branch}}{I \text{ branch}}$$

FIGURE 5–11 Ohm's law method to find R_T

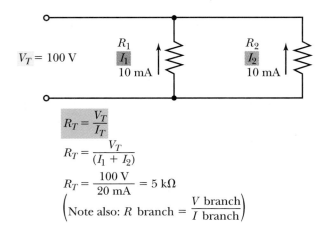

$$R_T = \frac{V_T}{I_T}$$

$$R_T = \frac{V_T}{(I_1 + I_2)}$$

$$R_T = \frac{100\ V}{20\ mA} = 5\ k\Omega$$

$$\left(\text{Note also: } R \text{ branch} = \frac{V \text{ branch}}{I \text{ branch}}\right)$$

Again, refer to Figure 5–11. Since we know the voltage is the same across all branches and must equal V applied, then

$$R_1 = \frac{100\ V}{10\ mA} = 10\ k\Omega$$

$$R_2 = \frac{100\ V}{10\ mA} = 10\ k\Omega$$

$$R_T = \frac{100\ V}{20\ mA} = 5\ k\Omega$$

Note this result of $R_T = 5\ k\Omega$ agrees with our earlier statement that total circuit R must be less than any one of the branch resistances in parallel.

Assume the following changes in parameters for the circuit of Figure 5–11. $V_T = 50\ V$, $I_1 = 1.06\ mA$, $I_2 = 1.85\ mA$. Again, total resistance can be found using Ohm's law. (NOTE: Rounding values to two decimal places, the answers will be close, but not absolutely precise in value.)

$$R_T = \frac{V_T}{I_T} = \frac{50\ V}{(1.06\ mA + 1.85\ mA)} = \frac{50\ V}{2.91\ mA} = 17.18\ k\Omega$$

$$R_1 = \frac{V_1}{I_1} = \frac{50\ V}{1.06\ mA} = 47\ k\Omega$$

$$R_2 = \frac{V_2}{I_2} = \frac{50\ V}{1.85\ mA} = 27\ k\Omega$$

PRACTICE PROBLEMS V

1. Again, refer to Figure 5–11 and apply Ohm's law. If $I_1 = 15\ mA$ and $R_1 = 2\ k\Omega$, what are the values of V_{R_1} and V_T?
2. If $R_2 = 10\ k\Omega$, what is the value of R_T?

The reciprocal method

Since total current in a parallel circuit equals the sum of the branch currents, we can derive a reciprocal formula for resistance using this fact plus Ohm's law.

Since $I_T = I_1 + I_2 \ldots + I_n$, and Ohm's law states $I_T = V/R_T$ and each branch current equals its V over its R, then:

$$\frac{V}{R_T} = \frac{V}{R_1} + \frac{V}{R_2} + \frac{V}{R_3} \ldots + \frac{V}{R_n}$$

Recall R_n represents the last resistor value in the sequence. Since V is common in parallel circuits and is the same across each of the resistances, we divide by V in each case and derive the formula for $\frac{1}{R_T}$.

FORMULA 5–2 $\quad \dfrac{1}{R_T} = \dfrac{1}{R_1} + \dfrac{1}{R_2} + \dfrac{1}{R_3} \ldots + \dfrac{1}{R_n}$

This formula, plus a couple simple math steps, are used to find total or equivalent resistance of the circuit (R_T or R_e) for any number of branch resistances.

Example

If a three-resistor circuit has a 10-Ω branch, a 15-Ω branch, and a 20-Ω branch, the total, or equivalent resistance, is found as follows:

$$\frac{1}{R_T} = \frac{1}{10} + \frac{1}{15} + \frac{1}{20} = \frac{6}{60} + \frac{4}{60} + \frac{3}{60} = \frac{13}{60}$$

To find R_T:

$$\frac{1}{R_T} = \frac{13}{60}$$

Cross-multiply.

$$60 = 13\,R_T$$

Then divide.

$$R_T = \frac{60}{13} = 4.62 \ \Omega$$

In the real world of electronics, rarely will the resistor values be as convenient to deal with as in our example. We chose those values to show you the concepts of using the reciprocal method. We now want to give you an example with more typical resistor values, and also promote the idea of your using a calculator to find R_T (or $R_{equivalent}$) for a parallel circuit. Since your calculator has a reciprocal function by using the "$1/x$" button (or x^{-1} button on some calculators), it makes calculations of this sort very easy.

Example

If a four-resistor parallel circuit is comprised of resistors having values of 22 Ω, 39 Ω, 56 Ω, and 82 Ω, the total resistance of the circuit may be found, using a calculator, as follows:

Step 1: Clear the calculator by pressing the `AC/ON` button. The readout should show 0.

Step 2: Enter 22, then press the ⟨1/x⟩ button, then the ⟨+⟩ button. The readout should show 0.04545.

Step 3: Enter 39, then the ⟨1/x⟩ button, then the ⟨+⟩ button. The readout should show 0.07109.

Step 4: Enter 56, then the ⟨1/x⟩ button, then the ⟨+⟩ button. The readout should show 0.0889.

Step 5: Enter 82, then the ⟨1/x⟩ button, then the ⟨=⟩ button. The readout should show 0.1011. This number is the $\dfrac{1}{R_T}$ value.

Step 6: Press the ⟨1/x⟩ button to find the circuit total resistance. $R_T = 9.886\ \Omega$ (rounded to two places, $R_T = 9.89\ \Omega$, or approximately 10 Ω).

(Is this less than the least branch R? Yes, and it does seem reasonable, using the mental approximating technique.) Again, R equivalent is less than the least branch. Another variation of the reciprocal method is:

FORMULA 5–3
$$R_T = \dfrac{1}{\dfrac{1}{R_1} + \dfrac{1}{R_2} + \dfrac{1}{R_3} \ldots + \dfrac{1}{R_n}}$$

(NOTE: This reciprocal of the reciprocals formula is the technique you just used in steps 1–6 of the "Example.")

PRACTICAL NOTES

As you have seen in the example, the reciprocal method is very easy with calculators. Practice this method until it is second nature to you. In most cases, it is the easiest method to use.

PRACTICE PROBLEMS VI

1. Assume a parallel circuit has branch resistances of 10 kΩ, 27 kΩ, and 20 kΩ, respectively. Use the reciprocal method and find R_T.

2. Assume a parallel circuit has branch resistances of 1.5 kΩ, 3.9 kΩ, 4.7 kΩ, and 5.6 kΩ, respectively. Again, use the reciprocal method and find the circuit R_T.

The conductance method

Recall resistance is the opposition shown to current flow, and conductance, conversely, is the ease with which current passes through a component or circuit. To calculate conductance use the reciprocal of resistance, $G = 1/R$, where G is conductance in siemens (S) and R is resistance in ohms (Ω).

Example

If a circuit has a resistance of 10 Ω, it exhibits one-tenth S of conductance. Conversely, if a circuit has one-tenth S of conductance, it must have 10 Ω of resis-

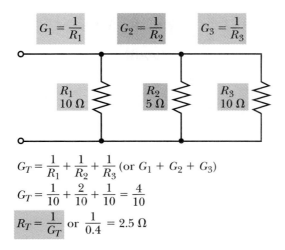

FIGURE 5–12 The conductance method to find R_T

$$G_T = \frac{1}{R_1} + \frac{1}{R_2} + \frac{1}{R_3} \text{ (or } G_1 + G_2 + G_3)$$

$$G_T = \frac{1}{10} + \frac{2}{10} + \frac{1}{10} = \frac{4}{10}$$

$$R_T = \frac{1}{G_T} \text{ or } \frac{1}{0.4} = 2.5 \ \Omega$$

tance. That is, $R = 1/G$, where $R = 1/0.1 = 10 \ \Omega$. Total conductance of parallel branches is found by adding all conductances of parallel branches, Figure 5–12.

FORMULA 5–4 $G_T = G_1 + G_2 + G_3$

In other words, $\dfrac{1}{R_T}$ (or G_T) $= \dfrac{1}{R_1} + \dfrac{1}{R_2} + \dfrac{1}{R_3}$

PRACTICE PROBLEMS VII

Refer to Figure 5–13, and use the conductance method to find the circuit total resistance.

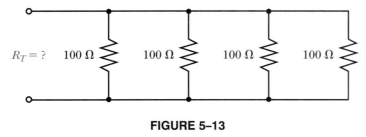

FIGURE 5–13

The product-over-the-sum method

One popular method to solve for total resistance in parallel circuits is the **product-over-the-sum method.** This method is limited to solving total resistance values for two-resistor circuits, or when circuits have more than two resistors, solving them, two resistors at a time. The mathematical formula is derived from the basic reciprocal resistance (conductance) formulas.
That is:

$$G_T = \frac{1}{R_T} = \frac{1}{R_1} + \frac{1}{R_2}$$

Since resistance is the reciprocal of conductance, the formula can be restated as:

$$R_T = \frac{1}{G_T}, \text{ or } R_T = \frac{1}{\dfrac{1}{R_1} + \dfrac{1}{R_2}}$$

FORMULA 5–5 $R_T = \dfrac{R_1 \times R_2}{R_1 + R_2}$

Example

An illustration of this formula is in the two-resistor parallel circuit of Figure 5–14.

What would be the total circuit resistance if $R_1 = 100$ ohms and $R_2 = 50$ ohms? In this case, the product is $100 \times 50 = 5{,}000$, and the sum is $100 + 50 = 150$. Therefore, $R_T = 5{,}000/150$, and total resistance $= 33.33$ ohms. This product-over-the-sum method works for any two branch values.

Example

Let's assume the resistor values in Figure 5–14 are changed so that one resistor's value is 3.9 kΩ and the other resistor's value is 6.8 kΩ.

$$R_T = \frac{R_1 \times R_2}{R_1 + R_2} = \frac{(3.9 \times 10^3) \times (6.8 \times 10^3)}{(3.9 \times 10^3) + (6.8 \times 10^3)} = \frac{26.52 \times 10^6}{10.7 \times 10^3}$$

$$R_T = 2.48 \times 10^3 \text{ or } 2.48 \text{ k}\Omega$$

Our mental approximation, quick-check technique here might be $4 \times 7 = 28$, and $28 \div 4 + 7$ (or 11) would be approximately 2.5 or so. Our answer makes sense! Now you try a couple of practice problems!

PRACTICE PROBLEMS VIII

1. If the circuit in Figure 5–14 has resistances of 30 Ω and 20 Ω, respectively, solve for R_T using the product-over-the-sum method.

2. Assume the values of resistors in Figure 5–14 are changed to 10 kΩ and 33 kΩ, respectively. Use the product-over-the-sum approach and determine the new value of R_T or (or R_e) for the new circuit.

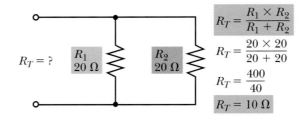

FIGURE 5–14 The product-over-the-sum method to find R_T

This method works well for two-branch circuits. How might a similar technique be used for a three-branch circuit? The answer is simple: Use the formula on two branches, *then use these results* from that calculation as being one equivalent resistance in parallel with the third resistor, and repeat the process. Of course, you can also use the reciprocal or conductance approach, which was recommended earlier, where

$$\frac{1}{R_T} = \frac{1}{R_1} + \frac{1}{R_2} + \frac{1}{R_3} \text{ or, better yet, } R_T = \frac{1}{\dfrac{1}{R_1} + \dfrac{1}{R_2} + \dfrac{1}{R_3}}$$

The following example shows how the product-over-the-sum approach might be used for a three-branch circuit.

Example

In Figure 5–15 three branch resistances are in parallel. Since each branch has a resistance of 3 Ω, what is the circuit total resistance?

Using the product-over-the-sum formula for the first two branches yields:

$$\frac{R_1 \times R_2}{R_1 + R_2} = \frac{3 \times 3}{3 + 3} = \frac{9}{6} = 1.5 \ \Omega$$

indicating that the equivalent resistance of branches 1 and 2 = 1.5 Ω. However, there is another branch involved! The equivalent resistance of 1.5 Ω is in parallel with branch R_3 (3 Ω). Repeating the product-over-the-sum formula for this situation gives:

$$\frac{R_e \times R_3}{R_e + R_3} = \frac{1.5 \times 3}{1.5 + 3} = \frac{4.5}{4.5} = 1 \ \Omega$$

Note that using the reciprocal approach would also yield:

$$\frac{1}{R_T} = \frac{1}{3} + \frac{1}{3} + \frac{1}{3} = \frac{3}{3} = \frac{1}{1} = 1 \ \Omega$$

PRACTICE PROBLEMS IX
If the resistances in Figure 5–15 are 47 Ω, 100 Ω, and 180 Ω, respectively, what is the value of R_T? Use the product-over-the-sum approach to calculate.

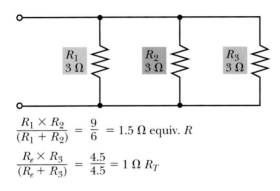

$$\frac{R_1 \times R_2}{(R_1 + R_2)} = \frac{9}{6} = 1.5 \ \Omega \text{ equiv. } R$$

$$\frac{R_e \times R_3}{(R_e + R_3)} = \frac{4.5}{4.5} = 1 \ \Omega \ R_T$$

FIGURE 5–15 Using product-over-the- sum method with more than two branches

What would happen if there were four branches? To find the equivalent resistance of any two branches, use the product-over-the-sum formula. Then find the equivalent resistance of the other two branches in the same way, and use the product-over-the-sum formula for the two equivalent resistances to find total circuit resistance.

A useful simplifying technique

You have undoubtedly observed in Figure 5–14 described earlier, which had two 20-Ω resistors in parallel, that the total $R = 10\ \Omega$. This number is half the value of one of the equal branch resistances. What do you suppose the total resistance of a three-branch parallel circuit is if the circuit has three equal branches of 30 Ω each? If you said 10 Ω again, you are right. That is, two equal parallel branches are equivalent to half the R value of one of the equal branches, three equal branches are equivalent to one-third the value of one of the equal branches, and so on. The idea is to mentally *combine* equal resistance branches to find their equivalent resistance. Doing this mental process reduces the complexity of the problem.

Look at Figure 5–16. By realizing that R_2 and R_4 can be combined as an equivalent resistance of 10 Ω, then combined with the two remaining 10-Ω branches, you can see the total circuit resistance will be one-third of 10 Ω, or 3.33 Ω.

This same concept can be used whenever branches of a multi-branched parallel circuit have equal values. Combine equal branches in a way to simplify to the fewest remaining branches. Then use the most appropriate method to solve for total circuit resistance, such as product-over-the-sum method, reciprocal method, Ohm's law, etc.

NOTE: The unequal values of resistances used in many circuits do not readily lend themselves to using this simplifying technique. However, whenever you can use it to help, do it! It will simplify the computation efforts.

The assumed voltage method

One technique to solve for total resistance of parallel circuits containing miscellaneous "mismatched" resistance values is the **assumed voltage method.** You'll often find problems of this sort on FCC radio licensing tests.

The technique uses a two-step approach:

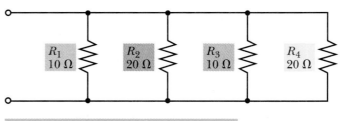

Step 1 = Combine R_2 and R_4 to yield 10 Ω.
Step 2 = Knowing that three 10 Ω resistors
         in parallel yields 1/3 of 10 Ω
         as their R_T. Then $R_T = 3.33\ \Omega$.

FIGURE 5–16 A useful simplifying technique

1. Assume an uncomplicated or easy circuit applied voltage value to determine each branch current, should the "assumed voltage" actually be applied to the circuit. For example, choose a value that is an even multiple of all the branch resistor values (the "least common denominator" value).

2. Determine the branch currents by dividing the assumed voltage by each branch R value. Then add branch currents to find total current. Now *divide* *"assumed voltage" by calculated total current to find total circuit resistance.*

Example

Refer to Figure 5–17. Following the preceding steps, the solution for this problem is as follows:

1. Assume 300 V applied voltage since it is easily divided by each of the branch R values.

2. Solve each branch current using Ohm's law. Branch currents are 12 A, 6 A, 3 A, and 2 A, respectively.

Total current = sum of branch currents = 23 A

Thus, $R_T = \dfrac{300 \text{ V}}{23 \text{ A}} = 13.04 \ \Omega$

In essence, this technique often allows you to mentally perform the first step without paper and pencil to find the total current. All that remains is one division problem, V_T/I_T, to find total circuit resistance.

In Figure 5–17, if the R values were changed to 10, 13, 20, and 26 Ω, respectively, what would be the total circuit resistance? Try using the assumed voltage method to find the answer.

Your answer should be 3.77 Ω. One value of voltage that could have been assumed is 260 V, since it is evenly divisible by all the R values in the circuit. NOTE: It does not matter if you assumed another value of voltage applied, your answer will be the same!

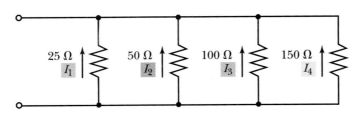

A good assumed voltage is 300 V due to R values.

$I_1 = 12 \text{ A}$ $I_2 = 6 \text{ A}$ $I_3 = 3 \text{ A}$ $I_4 = 2 \text{ A}$

I_T = sum of branch currents = 23 A

$R_T = \dfrac{300 \text{ V}}{23 \text{ A}} = 13.04 \ \Omega$

FIGURE 5–17 The assumed voltage method to find R_T

A useful parallel circuit resistance design formula

As a technician, there may be times when you will want to purposely decrease the resistance of an existing circuit by placing a resistor in parallel with it. Or you may need a resistor of some unavailable specific value. In either case, you may have a number of resistors, but none are the right value. This happens frequently! How can we determine the unknown resistor value to be paralleled with a given (known) value to achieve the desired resultant resistance? (NOTE: We interchangeably use the terms R_T [R total] and R_e [R equivalent] to indicate the resultant total resistance of paralleled resistances. Be familiar with both terms as you will see them used frequently.)

The answer is a formula derived from the product-over-the-sum method:

FORMULA 5–6 $R_u = \dfrac{R_k \times R_e}{R_k - R_e}$

Where R_u = R unknown
     R_e = equivalent R of paralleled resistors
     R_k = known R value that will be paralleled by R unknown achieving the desired R equivalent.

Example

What value of R is needed in parallel with a 10-ohm resistor to achieve a total resistance of 6 ohms? See the solution in Figure 5–18.

PRACTICE PROBLEMS X
Use this technique and solve the problem in Figure 5–19.

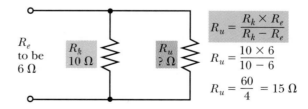

FIGURE 5–18 A practical circuit design formula

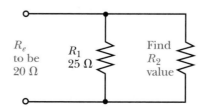

FIGURE 5–19

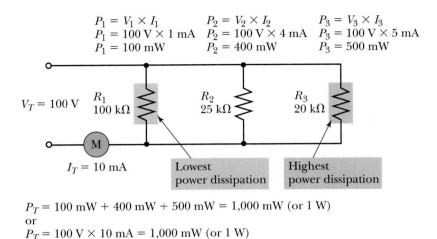

FIGURE 5–20 Power in parallel circuits

$$P_1 = V_1 \times I_1 \qquad P_2 = V_2 \times I_2 \qquad P_3 = V_3 \times I_3$$
$$P_1 = 100 \text{ V} \times 1 \text{ mA} \qquad P_2 = 100 \text{ V} \times 4 \text{ mA} \qquad P_3 = 100 \text{ V} \times 5 \text{ mA}$$
$$P_1 = 100 \text{ mW} \qquad P_2 = 400 \text{ mW} \qquad P_3 = 500 \text{ mW}$$

$P_T = 100 \text{ mW} + 400 \text{ mW} + 500 \text{ mW} = 1,000 \text{ mW (or 1 W)}$
or
$P_T = 100 \text{ V} \times 10 \text{ mA} = 1,000 \text{ mW (or 1 W)}$

5–7 POWER IN PARALLEL CIRCUITS

Total power

Total power for *any* resistive circuit (series, parallel, or combination) is computed by adding all individual power dissipations in the circuit, or by using one of the power formulas, $P_T = V_T \times I_T$; $P_T = I_T^2 \times R_T$; or $P_T = V_T^2 / R_T$. Likewise, power dissipated by any one resistive component is calculated using that particular component's parameters. That is, its voltage drop, current through it, and its resistance value.

Look at the circuit in Figure 5–20 as you read the next section about power in parallel circuits.

Power dissipated by each branch

To find a branch power dissipation, use the parameters applicable to that branch. In the case of branch 1 (R_1), $P = 100 \text{ V} \times I_1$. To find I_1, divide branch V (100 V) by branch R (100 kΩ). The answer is 1 mA. Thus, $P_1 = 100 \text{ V} \times 1 \text{ mA} = 100 \text{ mW}$.

$$\text{Branch 2: Branch } I = \frac{100 \text{ V}}{25 \text{ k}\Omega} = 4 \text{ mA}$$

$$\text{Branch } P = 100 \text{ V} \times 4 \text{ mA} = 400 \text{ mW}$$

$$\text{Branch 3: Branch } I = \frac{100 \text{ V}}{20 \text{ k}\Omega} = 5 \text{ mA}$$

$$\text{Branch } P = 100 \text{ V} \times 5 \text{ mA} = 500 \text{ mW}$$

Example

If we added all the individual power dissipations in the circuit in Figure 5–20, the total power = 100 mW + 400 mW + 500 mW = 1 W. Using the power formula, since V_T equals 100 volts and I_T equals 10 mA, the total power is 100 V × 10 mA =

1,000 mW (or 1 W). In either case, the total power supplied to and dissipated by the circuit is calculated as 1 W.

PRACTICE PROBLEMS XI

What is the total power in the circuit of Figure 5–20 if $V_T = 150$ V?

Relationship of power dissipation to branch resistance value

Refer again to Figure 5–20 and note that the *lowest* branch R dissipates the *most* power; the *highest* value R branch dissipates the *least* power. This is because V is the same for all branches, and current through each branch is *inverse* to its R value. **This is the opposite of what you learned about series circuits.** In *series circuits,* the largest value R dissipates the most power because its I is the same as all others in the circuit; however, its V is directly proportional to its R value.

   Soon we will look at other interesting similarities and differences between series and parallel circuits that you should learn. However, before we do that, let's look at some variations from the norm when parallel circuits develop opens or shorts.

5–8 EFFECTS OF OPENS IN PARALLEL CIRCUITS AND TROUBLESHOOTING HINTS

Opens

Analysis of one or more opened branches in a parallel circuit is simple. Refer to Figure 5–21.

1. Normal operation for this circuit is: $V_T = 100$ V; $I_T = 40$ mA; $P_T = 4,000$ mW (4 W); each branch $I = 10$ mA; and each branch $P = 1,000$ mW (1 W).

2. If branch 2 opens (R_2), then total current decreases to 30 mA, since there is no current through branch 2. Under these conditions, the following are true:

   • Total resistance increases.

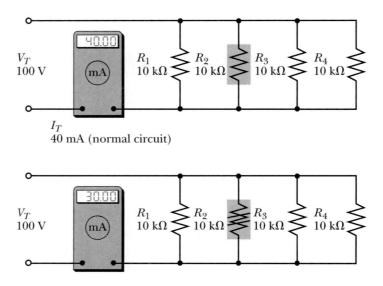

FIGURE 5–21 Effects of an open in a parallel circuit

- Total current decreases (30 mA rather than 40 mA).
- Total voltage and branch voltages remain unchanged.
- Total power decreases (100 V × 30 mA = 3,000 mW, or 3 W).
- Current through the good (unopened) branches remains unchanged because both their *V* and *R* values stay the same.
- Also, since voltage and current through unopened branches are unaffected, individual branch power dissipations stay the same, *except* for the opened branch or branches.
- Since current decreases to zero through the opened branch, its power dissipation also drops to zero. The voltage across the opened branch, however, stays the same.

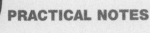

PRACTICAL NOTES

Some generalizations worth noting about an open in *any* circuit (series, parallel, or combination) are the following:

1. An open causes total circuit resistance to increase.
2. An open causes total circuit current to decrease.
3. An open causes total circuit power to decrease.

Incidentally, opens can be caused either by *components* opening, or by *conductor paths* opening. On printed circuit boards, a cracked conductor path or poor solder joint can cause an open. On "hard-wired" circuitry, a broken wire (or poor solder joint) can cause an open conductor path, Figure 5–22.

FIGURE 5–22 Open circuits can be caused by broken conductor paths or poor solder joints.

TROUBLESHOOTING HINTS

Since an open in a parallel branch causes total resistance to increase, total current to decrease, and current through the opened branch to decrease to zero, troubleshooting hints help isolate and identify these facts.

With power removed from the circuit, you can measure its total resistance with an ohmmeter. If the measured value indicates a higher than normal value for the circuit, then there is a good chance an open has occurred or at least some part of the circuit has opened or increased in R value. If it is convenient to electrically "lift" (electrically isolate) one end of each branch one at a time from the remainder of the circuit as you watch the meter, you can spot the troubled branch. For example, if lifting one end of a branch from the circuit causes a change in R (increase), the branch just lifted is probably good. If lifting of the branch causes no change in the R reading, you have probably found the branch with the problem. Because it was already opened, opening it again didn't change anything!

If you prefer using "power-on" current readings, the first indication of trouble you will see is a lower-than-normal total circuit current. If you can conveniently connect the meter to measure each branch's current individually, the branch with zero current is the open branch. Other branches should measure normal values.

5-9 EFFECTS OF SHORTS IN PARALLEL CIRCUITS AND TROUBLESHOOTING HINTS

Shorts

A shorted branch in a parallel circuit is a serious condition. Since each branch is connected directly across the voltage source, a shorted branch puts a resistance of zero ohms, or close to zero ohms, directly across the source. Refer to Figure 5–23.

1. Normal operation for this circuit is the same as described earlier for Figure 5–21, where $V_T = 100$ V; $I_T = 40$ mA; $P_T = 4,000$ mW (4 W); each branch $I = 10$ mA; and each branch $P = 1,000$ mW (1 W).

2. If branch 2 (R_2) shorted, the following things would probably happen:

   • I_T would try to drastically increase until a fuse blew, or the power supply "died"!

   • V_T would drastically decrease! Because of greatly increased $I \times R$ drop across the source's internal resistance, the output terminal voltage would decrease. It would probably drop to nearly zero volts with most power sources.

   • P_T would drastically increase until the power source quit, or a fuse blew!

   • Current through the other branches would decrease to zero, or close to it, because the V has dropped to almost zero, or zero.

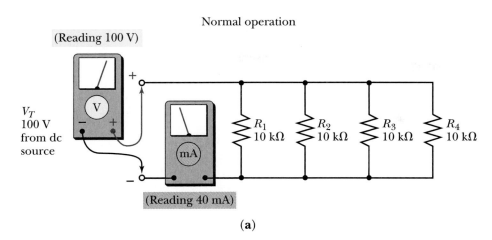

Normal operation

FIGURE 5–23 Effects of a short in a parallel circuit. R_T decreases and I_T increases until the fuse blows or the power source quits.

(a)

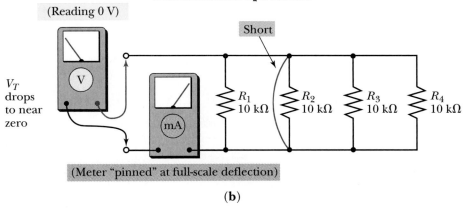

First instant after R_2 is shorted

(b)

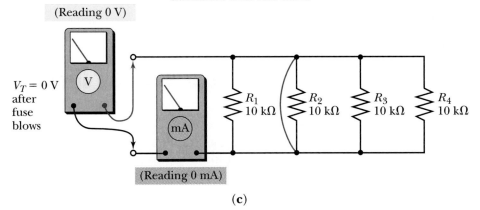

Conditions after fuse blows

(c)

- Total circuit resistance would drastically decrease! Remember the total resistance will be less than the least branch R.

- If there were any power supply fuses or house circuit fuses associated with the faulty circuit, they would probably blow. Also, there might be some smoke rising from some sections of the circuit or power supply.

PRACTICAL NOTES

Some generalizations about a short in *any* type of circuit (series, parallel, or combination) are the following:

1. A short causes total circuit resistance to decrease.
2. A short causes total circuit current to increase. (At least momentarily!)
3. A short causes total circuit power to increase. (At least momentarily!)
4. A short likely causes circuit voltage to decrease. (Probably to zero volts!)

TROUBLESHOOTING HINTS

First *disconnect* the circuit in question from the power source! Resistance checks can then be made on the circuitry without further damage either to the circuitry, power source, connecting wires, and so forth. Power-on troubleshooting should *not* be done for shorted conditions.

Again, the best method to find the troubled area is to isolate each branch one at a time. By electrically lifting one end of each branch from the remaining circuitry, an ohmmeter check of the branch resistance can be made. Normal *R* values will be found for the good branches. Close to zero ohms will be measured across the shorted branch or component.

Once the branch is located, a visual inspection of components, wires, printed circuit paths, and so on generally reveals the problem. If there is an internal short within the component, an *R* measurement of the component reveals this problem.

5–10 SIMILARITIES AND DIFFERENCES BETWEEN SERIES AND PARALLEL CIRCUITS

Some interesting and important similarities and differences between series and parallel circuits that you should know and remember are in the chart on the adjacent page. Use the chart on page 169 as an In-Process Learning Check for this chapter.

5–11 DESIGNING A PARALLEL CIRCUIT TO SPECIFICATIONS

Example

Assume you want to create a four-branch parallel circuit so that branches 1 and 3 have equal current values, and branches 2 and 4 also have equal current values but their current values are twice that of branches 1 and 3. Further assume that branch 1 has an *R* value of 10 kΩ and its current value is 10 mA.

COMPARING SERIES AND PARALLEL CIRCUITS

SERIES CIRCUIT CHARACTERISTICS	PARALLEL CIRCUIT CHARACTERISTICS
R_T = Sum of Rs	R_T = Less than lowest R in parallel
$I_T = \dfrac{V_T}{R_T}$	$I_T = \dfrac{V_T}{R_T}$
I_T = Same current throughout the circuit	I_T = Sum of branch currents, where branch current $$(I_b) = \frac{V_b}{R_b}$$ Branch I is inverse to branch R.
$V_T = I_T \times R_T$	$V_T = I_T \times R_T$
V_T = Sum of all IR drops	V_T = Same as across each branch
$V_{R_X} = I_{R_X} \times R_X$; Any given R's voltage drop is related to the ratio of its R to R total. $$V_{R_X} = \left(\frac{R_X}{R_T}\right) \times V_T$$	Voltage across all resistors is same value and equals the circuit applied voltage.
P dissipated by a given R is *directly related* to its R value. Largest R in circuit dissipates most power; smallest R dissipates least power.	P dissipated by a given R is *inverse* to its R value. Largest R dissipates least power, smallest R value dissipates most power.
$P_T = V_T \times I_T$; or sum of all Ps	$P_T = V_T \times I_T$; or sum of all Ps
Opens cause 0 current; 0 V across good components; and V_T (applied circuit voltage) to appear across open.	*Opens* cause no change in good branches; opened branch has zero current, but V across all branches stays the same.
Shorts cause I to increase; V to decrease across short; V to increase across remaining components; and total circuit power to increase.	*Shorts* cause I_T and P_T to increase (at least momentarily); V to decrease across all branches; and current through good branches to decrease.

The design problem involves determining all the circuit parameters and drawing the schematic for this circuit. Label all component values and all V, R, and I electrical parameters for each branch and for the total circuit.

Recall the way to begin is to draw a diagram and label all the knowns. Then proceed in the most appropriate manner to solve the unknowns based on the knowns.

See Figure 5–24a for this first step, which is drawing a diagram and labeling all knowns.

The second step is to use the knowns to solve for the unknowns, then label the diagram with the appropriate parameters.

- Since branch 1's R and I values are known, it's easy to find its voltage value. Thus, $V = I \times R = 10$ mA $\times$ 10 kΩ = 100 V.

- Because this is a parallel circuit, 100 V is the voltage across all the branches as well as the V applied value.

Step 1: Draw diagram. Label and/or list "knowns."   Step 2: Use knowns to solve unknowns. Label diagram.

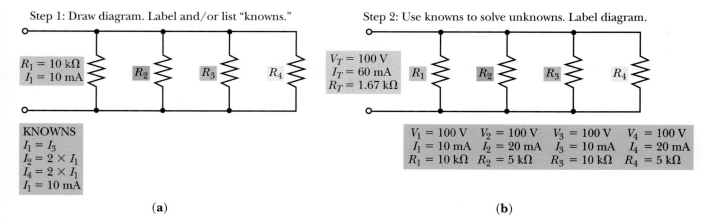

$R_1 = 10\ k\Omega$
$I_1 = 10\ mA$

KNOWNS
$I_1 = I_3$
$I_2 = 2 \times I_1$
$I_4 = 2 \times I_1$
$I_1 = 10\ mA$

$V_T = 100\ V$
$I_T = 60\ mA$
$R_T = 1.67\ k\Omega$

$V_1 = 100\ V$	$V_2 = 100\ V$	$V_3 = 100\ V$	$V_4 = 100\ V$
$I_1 = 10\ mA$	$I_2 = 20\ mA$	$I_3 = 10\ mA$	$I_4 = 20\ mA$
$R_1 = 10\ k\Omega$	$R_2 = 5\ k\Omega$	$R_3 = 10\ k\Omega$	$R_4 = 5\ k\Omega$

(a)   (b)

FIGURE 5–24

- Because branches 3 and 1 have the same current value, and since we know the V applied to both branches is the same, it is obvious branch 3 resistance value must be the same as branch 1. So we now know branch 3 $R_3 = 10\ k\Omega$, $I_3 = 10\ mA$, and $V_3 = 100\ V$.

- Because branches 2 and 4 have twice the current value as branches 1 and 3, it is obvious their current values must be 20 mA each. Since the V across each branch is 100 V, each branch's R value is

$$\frac{100\ V}{20\ mA} = 5\ k\Omega$$

- Solving the total circuit current and resistance are simple, and you already know the V applied value. Total current (I_T) equals the sum of the branch currents = 60 mA.

$$R_T = \frac{V_T}{I_T} = \frac{100\ V}{60\ mA} = 1.67\ k\Omega$$

Incidentally, you can use the R values and solve for the total resistance. The two 10-kΩ branches are the equivalent of 5 kΩ. This 5 kΩ in parallel with two other 5-kΩ branches is equivalent to 5 kΩ divided by 3, or 1.67 kΩ. Remember the technique with equal value branches?

- The last task is to label all appropriate parameters on the diagram, Figure 5–24b.

PRACTICE PROBLEMS XII

Try this problem yourself. Design and draw a three-resistor parallel circuit so that:

1. V applied to the circuit is 120 V.

2. The total circuit power dissipation is 1.44 W.

3. Branch 1 has half the current value of branch 2 and one-third the current value of branch 3.

Label your diagram with all V, I, R, and P values for the circuit and for each branch.

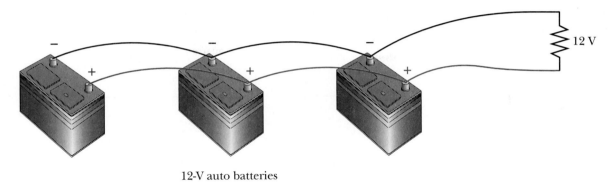

12-V auto batteries

FIGURE 5–25 Equal voltage sources in parallel can deliver more current and power than one source alone.

5–12 SOURCES IN PARALLEL

Voltage and power sources may be connected in parallel to provide higher current delivering capacity (and power) to circuits or loads (at a given voltage) than can be delivered by only one source. The sources should be equal in output terminal voltage value when using this technique. A common application of this method is the "paralleling" of batteries or cells, Figure 5–25.

5–13 CURRENT DIVIDERS

You are aware that parallel circuits cause the total current to divide through that circuit's branches. You also know the current divides in inverse proportion to each branch resistance value. These characteristics of parallel circuits offer the possibility of current dividing "by design," where one purposefully controls the ratios of branch currents by choosing desired values of resistances. Also, this current-dividing trait gives another method to find current through branches if only the total resistance, branch resistances, and total current are known.

General formula for parallel circuit with any number of branches

A general formula to find any given branch's current is:

FORMULA 5–7 $I_X = \dfrac{R_T}{R_X} \times I_T$

   where I_X is a specified branch's current value; R_T is the parallel circuit total resistance; and R_X is the value of resistance of the specified branch. Refer to Figure 5–26 for example applications of this formula.

PRACTICE PROBLEMS XIII

1. Refer again to Figure 5–26. Assume that $R_1 = 27$ kΩ, $R_2 = 47$ kΩ, $R_3 = 100$ kΩ, $R_T = 14.64$ kΩ, and $I_T = 3.42$ mA. Find each of the branch currents using the divider formula, as appropriate.

FIGURE 5–26 A general
current-divider example

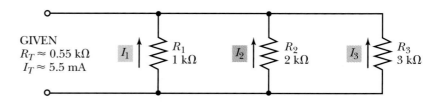

GIVEN
$R_T \approx 0.55 \text{ k}\Omega$
$I_T \approx 5.5 \text{ mA}$

Find approximate values for I_1, I_2, and I_3

$$I_1 \approx \left(\frac{R_T}{R_1}\right) \times I_T = \left(\frac{0.55 \text{ k}\Omega}{1 \text{ k}\Omega}\right) \times 5.5 \text{ mA} \approx 3.0 \text{ mA}$$

$$I_2 \approx \left(\frac{R_T}{R_2}\right) \times I_T = \left(\frac{0.55 \text{ k}\Omega}{2 \text{ k}\Omega}\right) \times 5.5 \text{ mA} \approx 1.5 \text{ mA}$$

$$I_3 \approx \left(\frac{R_T}{R_3}\right) \times I_T = \left(\frac{0.55 \text{ k}\Omega}{3 \text{ k}\Omega}\right) \times 5.5 \text{ mA} \approx 1.0 \text{ mA}$$

2. Assume the circuit of Figure 5–26 is changed so that $I_T = 4$ A, $R_1 = 10 \Omega$, $R_2 = 5.6 \Omega$, and $R_3 = 2.7 \Omega$.

   a. Find R_T.

   b. Use the general current divider formula to find the current through each branch.

Simple two-branch parallel circuit current divider formula

If there are only two branches, Figure 5–27, a formula is available to solve for each branch current, knowing I_T and branch R values:

FORMULA 5–8 $I_1 = \dfrac{R_2}{R_1 + R_2} \times I_T$

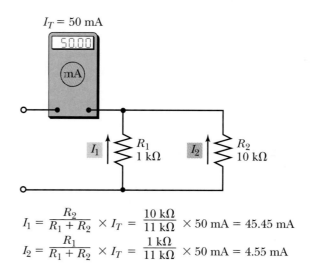

$$I_1 = \frac{R_2}{R_1 + R_2} \times I_T = \frac{10 \text{ k}\Omega}{11 \text{ k}\Omega} \times 50 \text{ mA} = 45.45 \text{ mA}$$

$$I_2 = \frac{R_1}{R_1 + R_2} \times I_T = \frac{1 \text{ k}\Omega}{11 \text{ k}\Omega} \times 50 \text{ mA} = 4.55 \text{ mA}$$

FIGURE 5–27 Two-resistor
current divider example

$$I_1 = \frac{10 \text{ k}\Omega}{11 \text{ k}\Omega} \times 50 \text{ mA} = 0.909 \times 50 \text{ mA} = 45.45 \text{ mA}$$

FORMULA 5–9 $I_2 = \dfrac{R_1}{R_1 + R_2} \times I_T$

$$I_2 = \frac{1 \text{ k}\Omega}{11 \text{ k}\Omega} \times 50 \text{ mA} = 0.0909 \times 50 \text{ mA} = 4.55 \text{ mA}$$

NOTE: There is an inverse relationship of current to branch Rs. Thus, when solving for I_1, R_2 is in the numerator; when solving for I_2, R_1 is in the numerator.

PRACTICE PROBLEMS XIV

1. Refer again to Figure 5–27 and assume the values of R_1 and R_2 change to 100 kΩ and 470 kΩ, respectively. Also, assume the total current is 2 mA. Find the values of current through each branch using the appropriate two-branch divider formula.

2. Assume that the circuit in Figure 5–27 is changed so that total current measures 20 mA, R_1 equals 1.8 kΩ, and R_2 equals 8.2 kΩ. Using the two-resistor current divider formula, determine the currents through each branch.

PRACTICAL NOTES

You have learned that whenever *any* resistance value is placed in parallel with an existing circuit, it alters the circuit resistance and current because another branch has been added; thus providing another current path.

In a later chapter, we will discuss how this feature impacts the types and uses of measurement devices, such as voltmeters. For now, we only want to alert you to an important observation about parallel circuits.

As already stated, no matter what value of R is paralleled with (or "bridged across") an existing circuit, the circuit R will decrease. But the question is *how much* change will there be?

Look at Figure 5–28. Note the impact (change) that adding a 1 megohm resistor branch has on the circuit resistance. As you can see, adding a *high R* in parallel causes a small change in total circuit R.

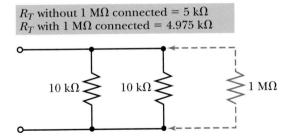

R_T without 1 MΩ connected = 5 kΩ
R_T with 1 MΩ connected = 4.975 kΩ

10 kΩ 10 kΩ 1 MΩ

FIGURE 5–28 Change in R_T of a given circuit when adding 1-MΩ branch

FIGURE 5–29 Change in R_T of a given circuit when adding 1-kΩ branch

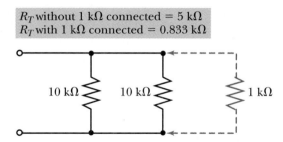

R_T without 1 kΩ connected = 5 kΩ
R_T with 1 kΩ connected = 0.833 kΩ

10 kΩ   10 kΩ   1 kΩ

Look at Figure 5–29. Note the change in R_T caused by introducing a 1-kΩ resistor in parallel with the same original circuit. Adding a *low R* in parallel with an existing circuit causes a much more significant change in the circuit total resistance (and total current) than paralleling an existing circuit with a high *R* value.

The amount of change caused by adding a new branch *R* is related to *both* the *R* value of the "new" *R* branch *and* to the total circuit resistance of the original circuit. If the original circuit's equivalent resistance is low initially, adding a low value *R* in parallel will not have nearly the effect that it does when the original circuit has a high R_T.

SUMMARY

▶ A parallel circuit is one in which there are two or more paths (or branches) for current flow, and all the branches have the same voltage applied across them.

▶ Total circuit current divides through the parallel branches in an inverse ratio to the branch resistances.

▶ Total circuit current in a parallel circuit equals the sum of the branch currents.

▶ Kirchhoff's current law states the value of current entering a point must equal the value of current leaving that same point in the circuit.

▶ The total resistance of a parallel circuit must be less than the least resistance branch *R* value.

▶ There are several methods to find the total resistance of parallel circuits:

1. *Ohm's law*

$$R_T = \frac{V_T}{I_T}$$

2. *Reciprocal resistance method*

$$\frac{1}{R_T} = \frac{1}{R_1} + \frac{1}{R_2} \cdots + \frac{1}{R_n} \text{ or } \cfrac{1}{\cfrac{1}{R_1} + \cfrac{1}{R_2} \cdots + \cfrac{1}{R_n}}$$

3. *Conductance method*

$$R_T = \frac{1}{G_T}$$

4. *Product-over-the-sum method*

$$R_T = \frac{R_1 \times R_2}{R_1 + R_2}$$

5. *Assumed voltage method*

▶ One technique to solve for parallel circuit total resistance is to combine equal branch *R* values into their equivalent resistance value, whenever possible. For example, two 10-kΩ resistors in parallel equal 5-kΩ equivalent resistance (or half of one equal branch *R* value). Three 10-kΩ resistors in parallel equal 3.33-kΩ equivalent resistance (or one-third of one equal branch *R* value).

▶ A useful formula to find the R value needed in parallel with a given R value to obtain a desired equivalent (total resistance) value is:

$$R_u = \frac{R_k \times R_e}{R_k - R_e}$$

where R_u is the unknown R value being calculated; R_e is the resultant total resistance desired from the "paralleled" resistors; and R_k is the known value resistor across which the unknown value resistor is paralleled to achieve the desired equivalent, or total parallel circuit resistance. Also, you can use:

$$\frac{1}{R_e} - \frac{1}{R_k} = \frac{1}{R_u}$$

▶ Total power supplied to, or dissipated by, a parallel circuit is found with the power formulas $P_T = V_T \times I_T$, $I_T^2 \times R_T$, or V_T^2 / R_T, or by summing the power dissipations of all the branches.

▶ Power dissipated by any branch of a parallel circuit is computed with the power formulas by substituting the specified branch V, I, and/or R parameters into the formulas, as appropriate.

▶ Branch power dissipations are such that the highest R value branch dissipates the least power, and the lowest R value branch dissipates the most power.

▶ An open branch in a parallel circuit causes total circuit current (and power) to decrease; total circuit R to increase; current through the unopened branches to remain the same, and current through the opened branch to decrease to zero. The circuit voltage remains the same.

▶ A shorted branch in a parallel circuit may cause damage to the circuit wiring and/or to the circuit power supply and/or may cause fuse(s) to blow. Until a fuse blows or the circuit power source fails, the short causes total circuit current to drastically increase; circuit R to decrease to zero ohms; and power demanded from the source to greatly increase. When a fuse blows or the source fails, circuit V, I, and P decrease to zero.

▶ When troubleshooting parallel circuits with an open branch problem, isolate the bad branch by power-off resistance checks or by power-on branch-by-branch current measurements. Another technique is to lift each branch electrically from the remainder of the circuit while monitoring circuit resistance or circuit current. The branch that causes no change is the bad branch.

▶ When troubleshooting parallel circuits with a shorted branch problem, *turn the power off and remove the circuit from the power source terminals before performing power-off checks to isolate the problem.* Branch-by-branch resistance measurements usually locate the problem. Also, you can electrically lift and reconnect each branch sequentially while monitoring the total circuit resistance.

▶ There are several significant *differences* between parallel and series circuits. Current is *common* in series circuits; voltage is *common* in parallel circuits. *Voltage* divides around a series circuit in *direct* relationship to the R values. *Current* divides throughout a parallel circuit in *inverse* relationship to the branch R values. In series circuits, the largest R dissipates the most power; in parallel circuits, the smallest R dissipates the most power. In series circuits, the total resistance equals the sum of all resistances. In parallel circuits, the total circuit resistance is *less* than the smallest branch resistance value.

▶ Voltage/power sources can be connected in parallel to provide higher current/power at a given voltage level.

FORMULAS AND SAMPLE CALCULATOR SEQUENCES

FORMULA 5–1 $\quad I_T = I_1 + I_2 + I_3 \ldots + I_n$

I_1 value, ⊕ , I_2 value, ⊕ , I_3 value, . . . , ⊜

FORMULA 5–2
$$\frac{1}{R_T} = \frac{1}{R_1} + \frac{1}{R_2} + \frac{1}{R_3} \ldots + \frac{1}{R_n}$$

R_1 value, ⟨**1/x**⟩, ⟨**+**⟩, R_2 value, ⟨**1/x**⟩, ⟨**+**⟩, R_3 value, ⟨**1/x**⟩, . . ., ⟨**=**⟩

FORMULA 5–3
$$R_T = \frac{1}{\dfrac{1}{R_1} + \dfrac{1}{R_2} + \dfrac{1}{R_3} \ldots + \dfrac{1}{R_n}}$$

R_1 value, ⟨**1/x**⟩, ⟨**+**⟩, R_2 value, ⟨**1/x**⟩, ⟨**+**⟩, R_3 value, ⟨**1/x**⟩, . . ., ⟨**=**⟩, ⟨**1/x**⟩

FORMULA 5–4
$$G_T = G_1 + G_2 + G_3$$

G_1 value, ⟨**+**⟩, G_2 value, ⟨**+**⟩, G_3 value, ⟨**=**⟩

FORMULA 5–5
$$R_T = \frac{R_1 \times R_2}{R_1 + R_2}$$

R_1 value, ⟨**×**⟩, R_2 value, ⟨**÷**⟩, ⟨**(**⟩, R_1 value, ⟨**+**⟩, R_2 value, ⟨**)**⟩, ⟨**=**⟩

FORMULA 5–6
$$R_u = \frac{R_k \times R_e}{R_k - R_e}$$

R known value, ⟨**×**⟩ , R equivalent value, ⟨**÷**⟩ , ⟨**(**⟩ , R known value, ⟨**−**⟩ , R equivalent value, ⟨**)**⟩ , ⟨**=**⟩

FORMULA 5–7
$$I_X = \frac{R_T}{R_X} \times I_T$$

R_T value, ⟨**÷**⟩, R_x value, ⟨**×**⟩, I_T value, ⟨**=**⟩

FORMULA 5–8
$$I_1 = \frac{R_2}{R_1 + R_2} \times I_T$$

R_2 value, ⟨**÷**⟩, ⟨**(**⟩, R_1 value, ⟨**+**⟩, R_2 value, ⟨**)**⟩, ⟨**×**⟩, I_T value, ⟨**=**⟩

FORMULA 5–9
$$I_2 = \frac{R_1}{R_1 + R_2} \times I_T$$

R_1 value, ⟨**÷**⟩, ⟨**(**⟩, R_1 value, ⟨**+**⟩, R_2 value, ⟨**)**⟩, ⟨**×**⟩, I_T value, ⟨**=**⟩

REVIEW QUESTIONS

1. What is the minimum number of current paths required for a circuit to be considered a parallel circuit?
   a. 1
   b. 2
   c. 3
   d. 4

2. Total current in a parallel circuit divides itself through various branches:
   a. in direct relationship to each branch resistance value.
   b. in inverse relationship to each branch resistance value.
   c. according to the power rating of each branch resistor.
   d. none of the above.

3. Total current in a parallel circuit is equal to:
   a. the sum of the two highest branch currents.
   b. the product of the lowest branch current times the highest branch current.
   c. the applied voltage divided by the circuit equivalent resistance.
   d. none of the above

4. Total resistance in a parallel circuit is always:
   a. the sum of the branch resistances.
   b. higher than the highest branch resistance.
   c. lower than the lowest resistance branch.
   d. equal to the lowest resistance branch.

5. Power distribution in a parallel circuit is such that:
   a. the lowest value resistor dissipates the most power.
   b. the highest value resistor dissipates the most power.
   c. since V is the same across all R's, all R's dissipate the same power.
   d. none of the above is true.

6. Give the formula for using the Ohm's law method to find total (or equivalent) resistance of a parallel circuit.

7. Give the formula(s) for using the reciprocal resistance method to find total resistance of a parallel circuit.

8. Give the formula for using the conductance method to find total resistance of a parallel circuit.

9. Give the formula for using the product-over-the-sum approach to find total resistance of a two-resistor parallel circuit.

10. List the steps required to find the total resistance of a parallel circuit using the "assumed voltage" method.

11. Give the formula for finding the value of resistor needed to be placed in parallel with a known value resistor in order to achieve a specified total resistance.

12. When troubleshooting a parallel circuit, the circuit power should always be removed before troubleshooting if the symptom information indicates that:
    a. the circuit resistance is too high and the circuit current too low.
    b. the circuit resistance is too low and the circuit current too high.

PROBLEMS

1. Refer to Figure 5–30 and find R_T.
   Circle the method you used to find the answer:
   Ohm's law
   reciprocal method
   product-over-the-sum method
   assumed voltage method

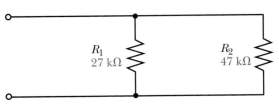

FIGURE 5–30

2. Refer to Figure 5–31 and find R_T.
   Describe the method(s) you used to solve this problem.

Recall that crossed wires are not connected since there is not a dot!

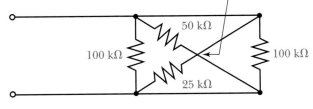

FIGURE 5–31

3. Refer to Figure 5–32 and find I_2.
Describe the method(s) you used to solve this problem.

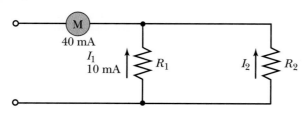

FIGURE 5–32

4. Assume in Figure 5–33 that each of the six resistors is 150 kΩ and determine the following circuit parameters:
R_T, I_T, and P_T
Current reading of meter 1
Current reading of meter 2

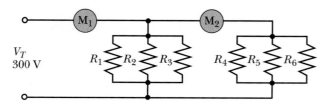

FIGURE 5–33

5. Refer to Figure 5–34 and using the information provided, answer the following with:
"I" for increase
"D" for decrease
"RTS" for remain the same
a. If R_2 opens,
I_1 will _____ P_T will _____
I_2 will _____ V_1 will _____
I_3 will _____ V_T will _____
I_T will _____

b. Assume that R_3 shorts internally. Before fuse or power supply failure,
I_1 will _____ total circuit current
I_2 will _____ will _____
I_3 will _____

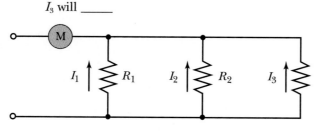

FIGURE 5–34

6. Refer to Figure 5–35 and find the value of R_1.

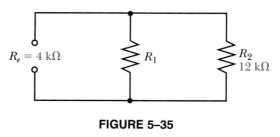

FIGURE 5–35

7. Refer to Figure 5–36. If I_T is 18 mA, what are the values of I_1 and I_2?

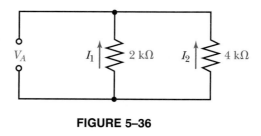

FIGURE 5–36

8. Refer to Figure 5–37. What is the value of R_1?

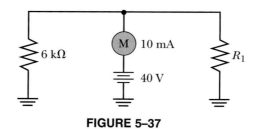

FIGURE 5–37

9. Refer to Figure 5–38. What is the value of P_{R_1}?

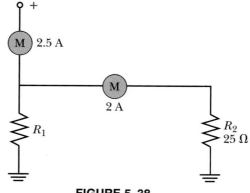

FIGURE 5–38

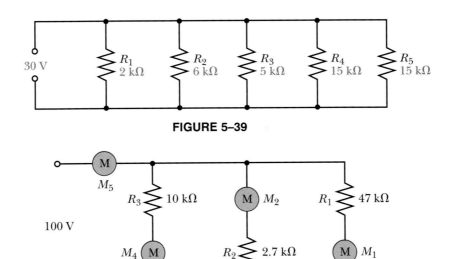

FIGURE 5–39

FIGURE 5–40

10. Refer to Figure 5–39 and find the following parameters:

$R_T =$ _____
$I_T =$ _____
$P_T =$ _____
$I_1 =$ _____
$I_2 =$ _____
$I_3 =$ _____
$I_4 =$ _____
$I_5 =$ _____
$P_1 =$ _____
$P_2 =$ _____
$P_3 =$ _____
$P_4 =$ _____
$P_5 =$ _____

11. Refer to Figure 5–40 and answer the following:
    a. What is current reading of M_1?
    b. What is current reading of M_2?
    c. What is current reading of M_3?
    d. What is current reading of M_4?
    e. What is current reading of M_5?
    f. What is the power dissipation of R_1?

g. What is the power dissipation of R_2?
h. What is the power dissipation of R_3?
i. What is total power dissipation of the circuit?
j. If R_2 opens, what is new value of P_T?

Refer to this circuit, Figure 5–41, when answering questions 12 through 16.

12. With nothing connected between points A and B, show your "step-by-step" work and solve for R_T using:
    a. the reciprocal method.
    b. the conductance method.
    c. the assumed voltage method.
    d. the Ohm's law method.

13. What resistance (R_X) needs to be connected between points A and B for the circuit total resistance to equal 10 kΩ?

14. Assume the R_X of the preceding problem is left connected to the circuit. What happens to the circuit R_T if R_4 is doubled in value and R_2 is halved in value? Would the answer be the same if R_X were removed?

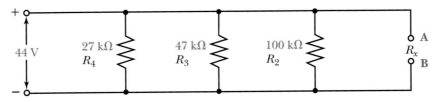

FIGURE 5–41

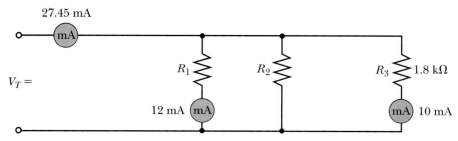

FIGURE 5–42

15. With the R_X of question 13 still connected:
    a. Which resistor in the circuit dissipates the most power?
    b. What is the value of power dissipated by this resistor?

16. With the R_X of question 13 *disconnected:*
    a. Which resistor in the circuit dissipates the least power?
    b. What is the value of power dissipated by this resistor?

    Refer to Figure 5–42 as you answer questions 17 through 25.

17. What is the value of V_T?

18. What is the value of R_1?

19. What is the value of current through R_2?

20. What is the value of R_2?

21. Which resistor dissipates the most power?

22. Which resistor dissipates the least power?

23. What would the total current be if R_2 were changed to a 1.5-kΩ resistor (assuming V_T remained the same)?

24. If R_1 were to increase in value (because of aging and use), what would happen to the total power supplied by the source?

25. For the conditions described in question 24, what would happen to the power dissipated by R_3?

    Refer to Figure 5–43 when answering questions 26–30. (NOTE: Round off answers to nearest whole number, as appropriate.)

26. What value resistors are R_2 and R_4?

27. What is the value of the circuit applied voltage?

28. If R_1 were changed to a resistance value of 27 kΩ, what would the total circuit power dissipation be?

29. What is the total equivalent resistance of the original circuit?

30. What value resistance would need to be added in parallel with the original circuit to have the total circuit resistance equal 4 kΩ?

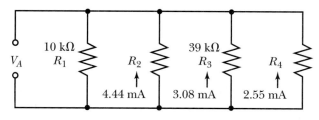

FIGURE 5–43

ANALYSIS QUESTIONS

1. What is the purpose of performing mental approximations, whenever possible, when performing circuit calculations?

2. What is one possible application of a current divider circuit?

3. Assume a circuit has three parallel resistors. If $R_1 = 12$ kΩ, $R_2 = 15$ kΩ, and $R_3 = 18$ kΩ, determine the total resistance using a calculator and the techniques defined below:

a. Use a calculator in the *engineering mode* and determine R_T by the reciprocal method. Show the answer as it is expressed in the calculator readout.

b. Use a calculator in the *scientific notation mode* and determine R_T by the reciprocal method. Show the answer as it is expressed in the calculator readout.

4. Because of the resistor values given, what approach might have been easily used for mental approximation of the R_T value?

5. Draw a diagram of a three-branch parallel circuit with the largest R having a value of 100 kΩ. The current divider circuit should have values that cause the total circuit current to divide as follows:

Branch 1's current is half of branch 2. Branch 2's current is one-fifth of branch 3. Label all R values on your diagram.

6. What value of R is needed in parallel with a 50-kΩ resistor to have a resultant equivalent resistance of 10 kΩ?

7. Describe a good method of isolating parallel components, or an individual branch, when troubleshooting a parallel circuit?

A GOOD IDEA

When troubleshooting, look up, find, or know the circuit's "norms" before and/or during detailed testing! This enables being alert to abnormalities that are clues to the trouble.

PERFORMANCE PROJECTS CORRELATION CHART

Suggested performance projects in the Laboratory Manual that correlate with topics in this chapter are:

CHAPTER TOPIC	PERFORMANCE PROJECT	PROJECT NUMBER
Voltage in Parallel Circuits	Voltage in Parallel Circuits	17
Current in Parallel Circuits & Kirchhoff's Current Law	Current in Parallel Circuits	16
Resistance in Parallel Circuits	Equivalent Resistance in Parallel Circuits	15
Power in Parallel Circuits	Power Distribution in Parallel Circuits	18
Effects of Opens in Parallel Circuits & Troubleshooting Hints	Effects of an Open in Parallel Circuits	19
Effects of Shorts in Parallel Circuits & Troubleshooting Hints	Effects of a Short in Parallel Circuits	20

NOTE: It is suggested that after completing the above projects, the student should be required to answer the questions in the "Summary" at the end of this section of projects in the Laboratory Manual.

THE SIMPLER SEQUENCE FOR TROUBLESHOOTING

Follow the seven-step SIMPLER troubleshooting sequence, as outlined below, and see if you can find the problem with this circuit. As you follow the sequence, record your circuit test results for each testing step on a separate sheet of paper. This will aid you and your instructor to see where your thinking is on track or where it might deviate from the best procedure.

step 1 Symptoms

—(Gather, verify, and analyze symptom information) To begin this process, read the data under the "Starting Point Information" heading. Particularly look for circuit parameters that are not normal for the circuit configuration and component values given in the schematic diagram. For example, look for currents that are too high or too low; voltages that are too high or too low; resistances that are too high or too low, etc. This analysis should give you a first clue (or symptom information) that will aid you in determining possible areas or components in the circuit that might be causing this symptom.

step 2 Identify

—(Identify and bracket the initial suspect area) To perform this first "bracketing" step, analyze the clue or symptom information from the "Symptoms" step, then either on paper or in your mind, bracket, circle, or put parentheses around all of the circuit area that contains any components, wires, etc., that might cause the abnormality that your symptom information has pointed out. NOTE: Don't bracket parts of the circuit that contain items that could not cause the symptom!

step 3 Make

—(Make a decision about the first test you should make: what type and where) Look under the TEST column and determine which of the available tests shown in that column you think would give you the most meaningful information about the suspect area or components, in the light of the information you have so far.

step 4 Perform

—(Perform the first test you have selected from the TEST column list) To simulate performing the chosen test (as if you were getting test results in an actual circuit of this type), follow the dotted line to the right from your selected test to the number in parentheses. The number in parentheses, under the "Results in Appendix C" column tells you what number to look up in Appendix C to see what the test result is for the test you are simulating.

step 5 Locate

—(Locate and define a new "narrower" area of uncertainty) With the information you have gained from the test, you should be able to eliminate some of the circuit area or circuit components as still being suspect. In essence, you should be able to move one bracket of the bracketed area so that there is a new smaller area of uncertainty.

step 6 Examine

—(Examine available information and determine the next test: what type and where) Use the information you have thus far to determine what your next test should be in the new narrower area of uncertainty. Determine the type test and where, then proceed using the TEST listing to select that test, and the numbers in parentheses and Appendix C data to find the result of that test.

step 7 Repeat

—(Repeat the analysis and testing steps until you find the trouble) When you have determined what you would change or do to restore the circuit to normal operation, you have arrived at *your* solution to the problem. You can check your final result against ours by observing the pictorial step-by-step sample solution on the pages immediately following the "Chapter Troubleshooting Challenge"—circuit and test listing page.

step 8 Verify

NOTE: A useful *8th Step* is to operate and test the circuit or system in which you have made the "correction changes" to see if it is operating properly. This is the final proof that you have done a good job of troubleshooting.

CHALLENGE CIRCUIT 2

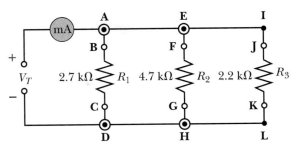

CHALLENGE CIRCUIT 2

Starting Point Information

1. Circuit diagram
2. V_T = 10 V
3. Measured I_2 = 2.1 mA
4. Measured I_T = 6.7 mA
5. Rs measured with remainder of circuit isolated from points being tested

TEST	Results in Appendix C
V_{A-D}	(83)
V_{B-D}	(11)
V_{E-H}	(22)
V_{F-G}	(95)
V_{I-L}	(52)
V_{J-K}	(75)
V_{A-B}	(8)
V_{C-D}	(56)
V_{I-J}	(103)
R_{A-B}	(46)
R_{C-D}	(90)
R_{I-J}	(14)
R_{K-L}	(29)

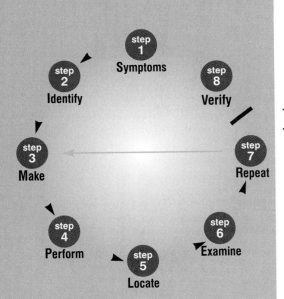

Symptoms

The current through R_2 is normal, but the total current is low for this circuit. This suggests that the total R is higher than normal.

Identify

initial suspect area: Branch 1 (points A to D) and branch 3 (points I to L) are suspect areas since branch 2 seems to be operating normally.

Make

test decision: Check the voltage between points A and D (branch 1).

Perform

1st Test: Look up the test result. V_{A-D} is 10 V, which is normal.

Locate

new suspect area: Points A to B and C to D in branch 1 and all of branch 3 are the new suspect areas.

Examine
available data.

Repeat

analysis and testing:

2nd Test: The V_{B-D}. V_{B-D} is 0 V, which is abnormal. This could result from a bad connection between points A and B or between points C and D.

3rd Test: Check V_{A-B}. V_{A-B} is 0 V, which is normal.

4th Test: Check V_{C-D}. V_{C-D} is 10 V, which is abnormal.

5th Test: Isolate branch 1 and check the resistance between points C and D. R_{C-D} is infinite ohms, indicating an open. (NOTE: This meter shows a blinking 30.00 to indicate infinite ohms. Other digital multimeters may use different readouts to indicate infinite ohms.)

Verify

6th Test: Make a solid connection between points C and D and note the total current reading. It should be 10.4 mA if all the branches are operating properly. That is, the result of a good connection is that total current is approximately 10.4 mA. When this is done, the circuit operates normally.

Symptoms

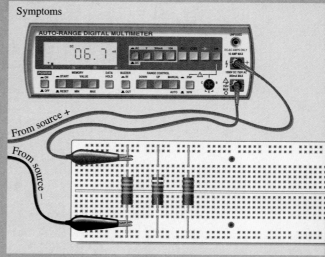

1st Test

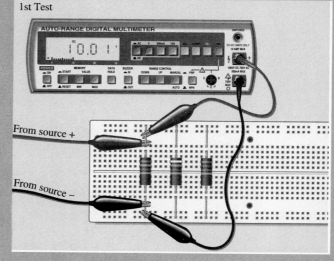

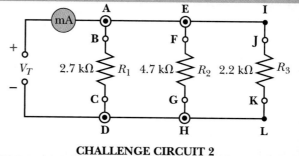

CHALLENGE CIRCUIT 2

2nd Test

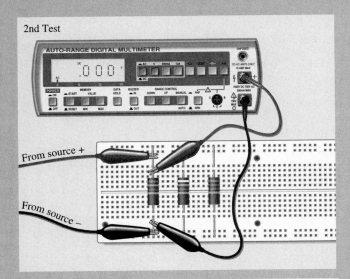

From source +

From source –

3rd Test

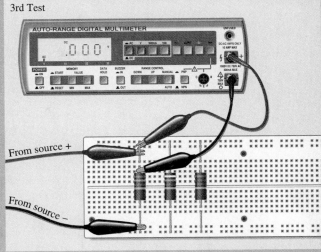

From source +

From source –

4th Test

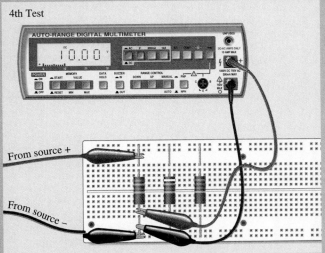

From source +

From source –

5th Test

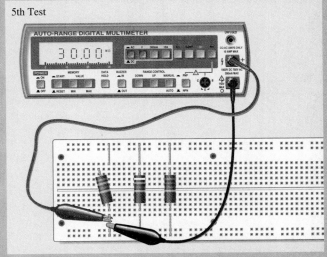

6th Test

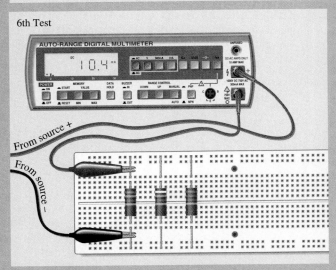

From source +

From source –

SPECIAL NOTE: You could have found the problem more quickly by calculating what the total current should be, then determining the amount of current missing was equal to the value branch 1 should have. Then you would have immediately known the problem was with branch 1. You still would have to perform isolation tests within that branch to locate the open.

185

SERIES-PARALLEL CIRCUITS

CHAPTER PREVIEW

Most electronic devices, equipment, and systems are comprised of series-parallel circuitry. Consumer products, such as high fidelity amplifiers, televisions, computers, and virtually all electronic products are filled with series-parallel circuits. A combination of both series and parallel connections form circuits that are coupled, or combined, to perform a desired task or function.

In this chapter, you will examine various configurations of these "combinational" circuits and will become familiar with analyzing them, using the principles you have already learned about series and parallel circuits. Also, you will see how a change in one part of a circuit has a greater or lesser effect on another portion of the circuit, depending on total circuit configuration and component values. Troubleshooting concepts will be examined. Your understanding of these circuits will be used by designing a simple series-parallel circuit to specifications. Finally, special applications of series-parallel circuitry will be studied.

OBJECTIVES

After studying this chapter, you should be able to:

1. Define the term **series-parallel circuit**

2. List the primary characteristic(s) of a series-parallel circuit

3. Determine the total resistance in a series-parallel circuit

4. Compute total circuit current and current through any given portion of a series-parallel circuit

5. Calculate voltages throughout a series-parallel circuit

6. Determine power values throughout a series-parallel circuit

7. Analyze the effects of an open in a series-parallel circuit

8. Analyze the effects of a short in a series-parallel circuit

9. Design a simple series-parallel circuit to specifications

10. Explain the **loading effects** on a series-parallel circuit

11. Calculate values relating to a **loaded voltage divider**

12. Make calculations relating to **bridge circuits**

13. Use the SIMPLER troubleshooting sequence to solve the Chapter Troubleshooting Challenge problem

6-1 WHAT IS A SERIES-PARALLEL CIRCUIT?

Definition

A **series-parallel circuit** contains a combination of both series-connected and parallel-connected components. There are both **in-line** series current paths and "branching-type" parallel current paths, Figure 6–1.

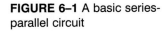

FIGURE 6–1 A basic series-parallel circuit

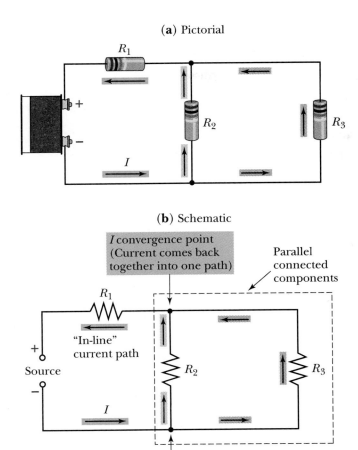

(**a**) Pictorial

(**b**) Schematic

I convergence point
(Current comes back
together into one path)

Parallel
connected
components

"In-line"
current path

Source

I branch point
(Current branches into
two different paths)

Characteristics

Series components in a series-parallel circuit may be in series with other *individual* components, or with other *combinations* of components, Figure 6–2.

Parallel components in a series-parallel circuit may be in parallel with other *individual* components, or with other *combinations* of components, Figure 6–3.

6–2 APPROACHES TO RECOGNIZE AND ANALYZE SERIES AND PARALLEL CIRCUIT PORTIONS

1. Start analysis at the portion of the circuit farthest from the source and work back toward the source to identify those components or circuit portions that are in series and in parallel.

2. Trace common current paths to identify components in series, Figure 6–4. Components, or combinations of components with common current are in series with each other.

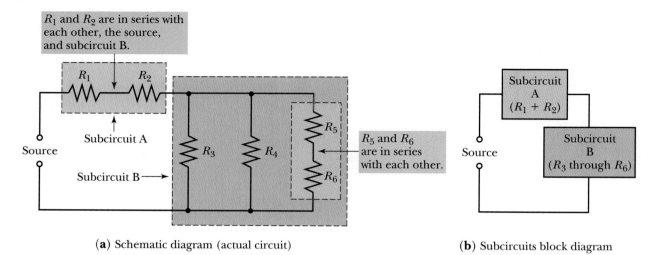

R_1 and R_2 are in series with each other, the source, and subcircuit B.

R_1 R_2

Subcircuit A

Source

Subcircuit B →

R_3 R_4 R_5

R_5 and R_6 are in series with each other.

R_6

(a) Schematic diagram (actual circuit)

Subcircuit A $(R_1 + R_2)$

Source

Subcircuit B $(R_3$ through $R_6)$

(b) Subcircuits block diagram

FIGURE 6–2 Series components in a series-parallel circuit

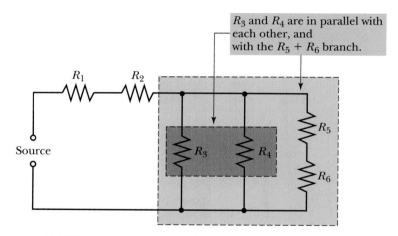

R_3 and R_4 are in parallel with each other, and with the $R_5 + R_6$ branch.

R_1 R_2

Source

R_3 R_4 R_5

R_6

FIGURE 6–3 Parallel components in a series-parallel circuit

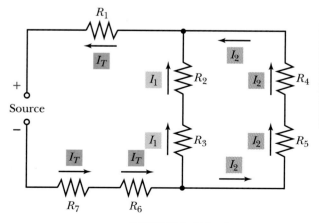

R_1

I_T

I_2

I_1 R_2 I_2 R_4

+

Source

−

I_1 R_3 I_2 R_5

I_T I_T I_2

R_7 R_6

Starting "away from the source" and working back:
1. R_4 and R_5 have common current (I_2) and are in series with each other.
2. R_2 and R_3 have common current (I_1) and are in series with each other.
3. R_1, R_6, and R_7 have common current (I_T) and they are in series with each other, and with the total parallel combination of the $R_2 + R_3$ branch in parallel with the $R_4 + R_5$ branch.

FIGURE 6–4 Tracing common current paths to identify series circuit portions

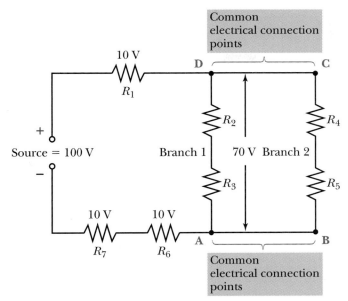

NOTE: Points **A** and **B** are electrically the same point and points **C** and **D** are electrically the same point. Therefore, the voltage from point **B** to point **C** is the same voltage as the voltage from point **A** to point **D**.

Branch 2 ($R_4 + R_5$) is across the same voltage points as branch 1 ($R_2 + R_3$). Therefore, branch 1 and branch 2 are in parallel with each other.

FIGURE 6–5 Observing common voltage points to identify parallel circuit portions

3. Observe voltages shared (in common) to identify components in parallel, Figure 6–5. Components, or combinations of components with the same voltage connection points (at both ends) are in parallel with each other.

4. Observe current branching and converging points to identify components, or combinations of components that are in parallel with each other, Figure 6–6.

   a. A point where current splits, or branches, is one end of a parallel combination of components, Figure 6–6, points A and B.

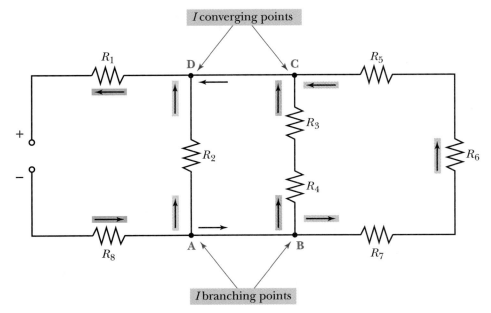

FIGURE 6–6 Observing current branching and converging points to identify parallel circuit portions

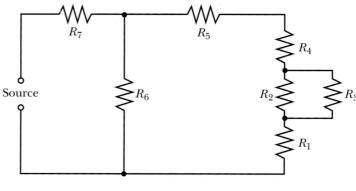

FIGURE 6–7

   b. The point where those same currents converge (or rejoin) is the other
       end of that same parallel combination, Figure 6–6, points C and D.

5. In summary, in the series portions of the circuit, current is common and the
   voltages (and resistances) are additive. In the parallel portions of the circuit,
   voltage is common and the branch currents are additive.

PRACTICE PROBLEMS I

Refer to Figure 6–7 and identify *individual* or single components that:

1. carry total current;

2. are in parallel with each other; that is, have the same voltage across
   them; and

3. are in series with each other; that is, have the same current through
   them.

Let's move to some analyses of series-parallel circuit electrical parameters.

6–3 TOTAL RESISTANCE IN SERIES-PARALLEL CIRCUITS

If total circuit current and circuit applied voltage are known, then solving for total circuit resistance is easy.

Using Ohm's law

Simply use Ohm's law, where $R_T = V_T / I_T$. If source voltage (V_S) and total current
are unknown, then apply the concepts discussed in the preceding section.

Using the "outside toward the source" approach

1. Start at the end of the circuit farthest from the source and work toward the
   source, identifying and solving the series, the parallel, and the series-parallel
   combinations, Figure 6–8. You will apply series circuit rules for series portions, and parallel circuit rules for parallel portions in the circuitry.

When given a circuit analysis problem without a diagram, sketch a simple circuit diagram to help you visualize the problem more clearly!

2. First, identify and solve parallel component segments. The common voltage and the current branching and converging concepts are true for R_2 and R_3; therefore, these two components are in parallel, Figure 6–8a. As you recall, the equivalent resistance of two equal value resistors in parallel is equal to half the value of one equal branch R. Thus, R_e of R_2 and $R_3 = 5\ \Omega$. Naturally, you can also solve it with the product-over-the-sum formula. That is:

$$R_e = \frac{R_2 \times R_3}{R_2 + R_3} = \frac{(10\ \Omega) \times (10\ \Omega)}{(10\ \Omega) + (10\ \Omega)} = \frac{100\ \Omega}{20\ \Omega} = 5\ \Omega$$

3. Next, work toward the source to solve the series and/or series-parallel combinations. Total current splits and rejoins as it flows through the parallel combination of R_2 and R_3. Total current also flows through R_1. Thus, R_1 and the parallel combination of R_2 and R_3 are in series with each other, Figure 6–8b. Therefore, total resistance of R_1 in series with the parallel combination of R_2 and R_3 equals $10\ \Omega + 5\ \Omega = 15\ \Omega$. And the circuit total resistance is $15\ \Omega$.

Example

Referring to Figure 6–9, let's look at another example of using the "outside-in" approach. This time we'll use more typical resistor values that aren't quite as easy to analyze as our previous example.

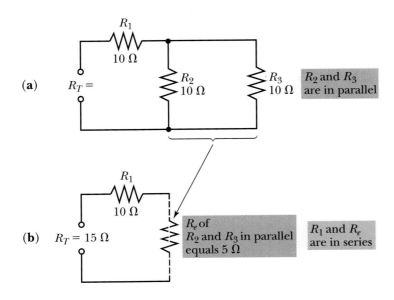

FIGURE 6–8 Analyzing a circuit to find R_T

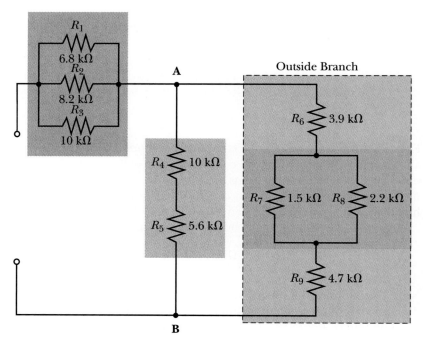

FIGURE 6–9

1. Starting with the outside branch, you can see that R_7 and R_8 are in parallel with each other, and their R_e is in series with both R_6 and R_9. Using the calculator's reciprocal function and several simple key strokes, (1.5, **EE** (or **EXP**), 3, **1/x**, **+**, 2.2, **EE** (or **EXP**), 3, **1/x**, **=**, **1/x**), we quickly find that the 1.5-kΩ and 2.2-kΩ parallel resistors have an R_e = 891.9 Ω or 0.892 kΩ. Let's round that to its approximate value of 0.9 kΩ. Now, adding the values of $R_6 + R_e$ (of R_7 and R_8 in parallel) + R_9, gives us 3.9 kΩ + 0.9 kΩ + 4.7 kΩ = 9.5 kΩ, which is the resistance of the outer branch.

2. Working into the next branch, we can see that the resistance between points A and B will be the equivalent resistance of the outer branch in parallel with the series resistors R_4 and R_5. $R_4 + R_5$ = 10 kΩ + 5.6 kΩ = 15.6 kΩ. The equivalent resistance of this 15.6 kΩ in parallel with the outer branch value of 9.5 kΩ can quickly be calculated with the calculator: 15.6, **EE** (or **EXP**), 3, **1/x**, **+**, 9.5, **EE** (or **EXP**), 3, **1/x**, **=**, **1/x** . The answer is 5.9 kΩ for the value of resistance between points A and B.

3. Now, all that remains is to find the equivalent resistance of R_1, R_2, and R_3, which are in parallel with each other, and add that R_e to the resistance between points A and B, and we will have solved the total resistance for this circuit. Once again, using our calculator and the reciprocal method: 6.8, **EE** (or **EXP**), 3, **1/x**, **+**, 8.2, **EE** (or **EXP**), 3, **1/x**, **+**, 10, **EE** (or **EXP**), 3, **1/x**, **=**, **1/x** . The R_e of the three resistors = 2.7 kΩ. This 2.7 kΩ is in series with $R_{A–B}$, which is 5.9 kΩ. Thus, R_T for this circuit equals 2.7 kΩ + 5.9 kΩ = 8.6 kΩ.

The reduce-and-redraw approach

A useful method to analyze series-parallel circuits is the reduce-and-redraw approach. This technique simplifies circuit analysis.

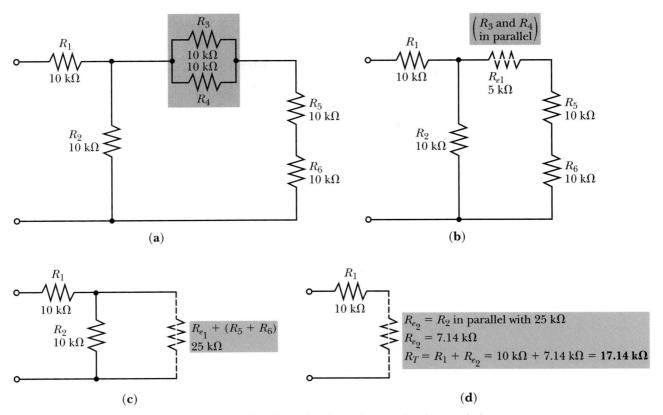

FIGURE 6–10 Simplifying by the reduce-and-redraw technique

Example

Look at Figure 6–10 as R_T is discussed for this series-parallel combination circuit.

1. R_3 and R_4 are in parallel with each other and have an equivalent resistance that we'll call R_{e_1}. Thus, $R_{e_1} = 5$ kΩ. R_{e_1}, the resistance of R_3 and R_4 in parallel equals:

$$\frac{(10 \text{ k}\Omega)(10 \text{ k}\Omega)}{(10 \text{ k}\Omega) + (10 \text{ k}\Omega)} = \frac{100 \times 10^6}{20 \times 10^3} = 5 \times 10^3 = 5 \text{ k}\Omega \text{ (Figure 6-10b)}$$

Of course, a simpler way is to remember that two equal branch resistances in parallel have an equivalent resistance equal to half of one of the branch resistances.

2. R_5 and R_6 are in series with each other and with R_{e_1} of parallel resistors R_3 and R_4. See Figure 6–10c and the calculations below.

The resistance of $R_{e_1} = 5$ kΩ

The resistance of $R_5 + R_6$ in series $= 10$ k$\Omega + 10$ k$\Omega = 20$ kΩ

The resistance of $R_{e_1} + R_5 + R_6 = 25$ kΩ

3. R_2 (10 kΩ) is in parallel with the total 25-kΩ combination of components R_3 through R_6 ($R_{e_1} + R_5 + R_6$). See Figure 6–10c and 6–10d and the following calculations.

$$R_{e_2} = \frac{(10 \text{ k}\Omega)(25 \text{ k}\Omega)}{(10 \text{ k}\Omega) + (25 \text{ k}\Omega)} = \frac{250 \times 10^6}{35 \times 10^3} = 7.14 \times 10^3 = 7.14 \text{ k}\Omega$$

4. The overall combination of components R_2 through R_6 (R_{e_2}) is in series with R_1. See Figure 6–10d and the following calculations.

Total circuit resistance $(R_T) = R_1 + R_{e_2} = 10 \text{ k}\Omega + 7.14 \text{ k}\Omega = 17.14 \text{ k}\Omega$

PRACTICE PROBLEMS II

Try practicing the reduce-and-redraw method as you perform the following tasks.

1. Find the total resistance of the circuit shown in Figure 6–11.
2. Find the total resistance of the circuit shown in Figure 6–12.

IN-PROCESS LEARNING CHECK I

Try solving the total resistance problem in Figures 6–13a and 6–13b. Use the outside toward the source approach to solve Figure 6–13a and use the reduce-and-redraw technique to solve the circuit in Figure 6–13b.

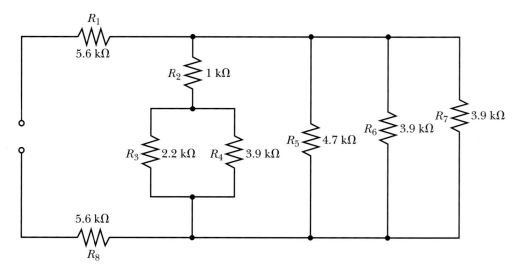

FIGURE 6–11

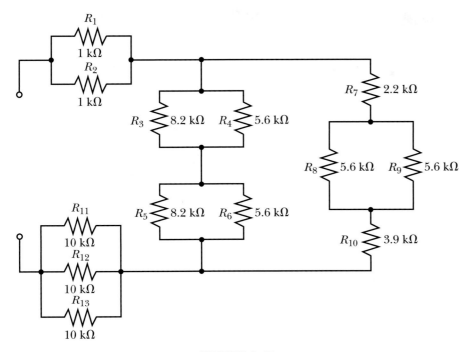

FIGURE 6–12

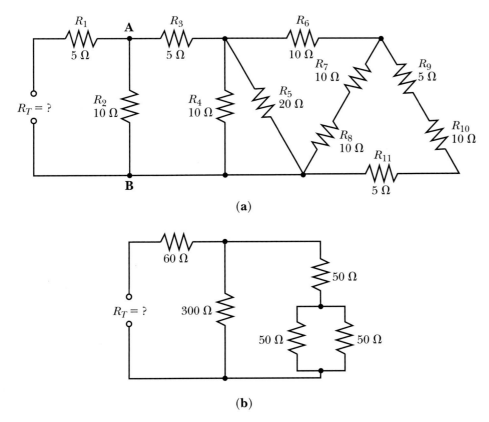

(a)

(b)

FIGURE 6–13 Solve for R_T values in each circuit.

6–4 CURRENT IN SERIES-PARALLEL CIRCUITS

Examine Figure 6–14 as we now discuss current in a series-parallel circuit. One can start by finding the total circuit current, then work from that point to determine current distribution throughout the circuit using Ohm's law, Kirchhoff's laws, current divider formulas, and/or other techniques.

It is easier if Ohm's law is used to find total current, where $I_T = V_S/R_T$. In this circuit, the source voltage (V_S) is 175 volts and the total resistance is 175 Ω. For practice, verify the R_T value using the techniques you learned in the preceding section.

Now that R_T and V_S are known, it's easy to solve for I_T.

$$I_T = \frac{V_S}{R_T} = \frac{175 \text{ V}}{175 \text{ }\Omega} = 1 \text{ A}$$

What about the distribution of current throughout the circuit? For this circuit, distribution of current is easy to find because of the values used. One example of a thought process that can be used is as follows:

1. R_1 and R_7 must have 1 ampere through them, since they are in series with the source and the remaining circuitry. In other words, I_T passes through both these resistors, Figure 6–14.

2. Since R_2 (250 Ω) and the outer branch (250 Ω) are equal resistance value branches, the total current equally divides between them. Thus, current through R_2 = 0.5 ampere, and current through the outer branch must equal 0.5 ampere, Figure 6–14.

3. It is apparent in tracing current through the outer branch that all of the 0.5 ampere must pass through both R_5 and R_6. However, this 0.5 ampere evenly splits between R_3 and R_4 because they are equal value parallel resistances. This means current through R_3 must be 0.25 ampere, and current through R_4 is also 0.25 ampere. In other words, the 0.5 ampere evenly splits at point C into two 0.25-ampere branch currents and rejoins at point D, Figure 6–14.

To find circuit resistance, you frequently start at a point farthest from the source and work back toward the source. *But when analyzing circuit current, you often start at the source with total current, then work outward through the circuit, analyzing the current distribution.* In this case, the circuit currents are determined to be:

I_T; I through R_1; and I through R_7 = 1 ampere.

I through R_2; I through R_5; and R_6 = 0.5 ampere.

I through R_3 and I through R_4 each = 0.25 ampere.

Of course, several other methods, or combinations of methods can be used to provide solutions, such as Ohm's law, Kirchhoff's voltage and current laws, and current divider formulas. In those cases where the values are not as convenient, you will probably have to use these other methods.

Example

Having solved for total current (1 ampere), you can refer to Figure 6–14 and apply known values to solve for the unknown parameters.

1. I_T passes through both R_1 and R_7. We can use Ohm's law to find the voltage drops. Thus,

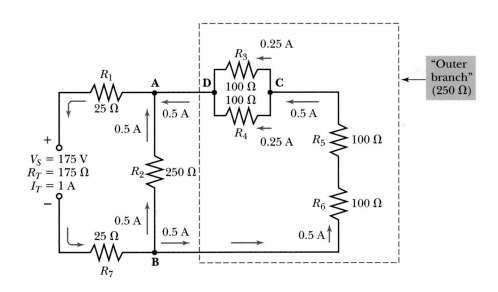

FIGURE 6–14 Solve for current values throughout a series-parallel circuit.

$$V_1 = I_T \times R_1 = 1 \text{ A} \times 25 \text{ } \Omega = 25 \text{ V}$$

$$V_7 = I_T \times R_7 = 1 \text{ A} \times 25 \text{ } \Omega = 25 \text{ V}$$

2. According to Kirchhoff's voltage law, the sum of the voltages around any closed loop must equal V applied. This means the sum of $V_1 + V_{A-B} + V_7$ must equal 175 V.

Thus, $25 \text{ V} + V_{A-B} + 25 \text{ V} = 175 \text{ V}$

$$50 \text{ V} + V_{A-B} = 175 \text{ V}$$

$$V_{A-B} = 175 \text{ V} - 50 \text{ V} = 125 \text{ V}$$

3. Kirchhoff's current law indicates the current leaving point A must equal the current entering point A. Stated another way, the total of the currents through R_2 and the outer branch must equal I_T in this case.

   a. Use Ohm's law to find current through R_2.

$$I \text{ through } R_2 = \frac{V_2}{R_2} = \frac{125 \text{ V}}{250 \text{ } \Omega} = 0.5 \text{ A}$$

   b. Using Kirchhoff's current law:

$$I \text{ through outer branch} = I_T - I \text{ (through } R_2);$$

$$= 1 \text{ A} - 0.5 \text{ A} = 0.5 \text{ A}$$

4. Ohm's law and Kirchhoff's voltage law are now used to solve the voltage drops across the resistors in the outer branch and the currents through R_3 and R_4.

   Knowing the outer branch current is 0.5 A:

   a. $V_6 = 0.5 \text{ A} \times R_6 = 0.5 \text{ A} \times 100 \text{ } \Omega = 50 \text{ V}$

   b. $V_5 = 0.5 \text{ A} \times R_5 = 0.5 \text{ A} \times 100 \text{ } \Omega = 50 \text{ V}$

   Kirchhoff's voltage law indicates that:

   c. $V_{C-D} = V_{A-B} - (V_5 + V_6) = 125 \text{ V} - (50 \text{ V} + 50 \text{ V}) = 25 \text{ V}$

Ohm's law can now be used to solve for the currents through R_3 and R_4.

d. I through $R_3 = \dfrac{V_3}{R_3} = \dfrac{25 \text{ V}}{100 \text{ } \Omega} = 0.25 \text{ A}$

e. I through $R_4 = \dfrac{V_4}{R_4} = \dfrac{25 \text{ V}}{100 \text{ } \Omega} = 0.25 \text{ A}$

PRACTICE PROBLEMS III

1. For practice in solving for current distribution throughout a series-parallel circuit, refer to Figure 6–15. Solve for I_T, I_1, I_2, and I_3.
2. Refer to Figure 6–16 and find the values of I_T, I_1, I_2, and I_3.

6–5 VOLTAGE IN SERIES-PARALLEL CIRCUITS

Voltage distribution throughout a series-parallel circuit is such that 1) voltage across components or circuit portions that are in series with each other are additive and 2) voltage across components or circuit portions that are in parallel must have equal values.

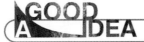

GOOD IDEA

Use the "approximation" technique to help prevent needless errors when performing calculations!

As you would expect, a given component's voltage drop in a series-parallel circuit is contingent on its circuit location and its resistance value. The location dictates what portion of the circuit's total current passes through it. Consequently, location of any given component within the series-parallel circuit has direct influence on its $I \times R$ drop.

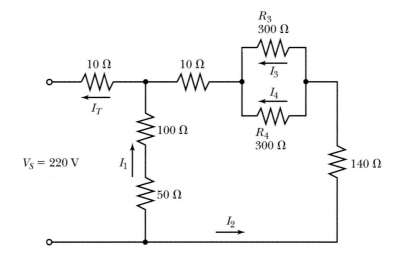

FIGURE 6–15

FIGURE 6–16

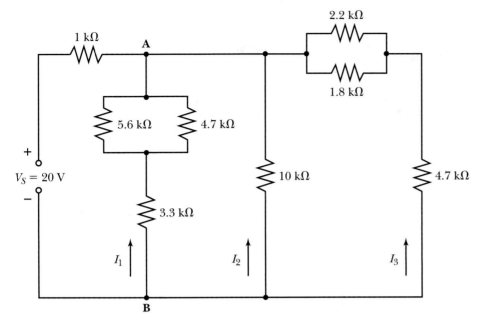

Refer to Figure 6–17 as the voltage distribution characteristics are analyzed with two methods: the Ohm's law method and resistance and voltage divider techniques.

Example

Ohm's law method

Solve circuit R and I values, then use Ohm's law to compute voltages.

1. $R_{C-D} = \dfrac{R_4 \times R_5}{R_4 + R_5} = \dfrac{(10 \times 10^3) \times (10 \times 10^3)}{(10 \times 10^3) + (10 \times 10^3)}$

$\qquad = \dfrac{100 \times 10^6}{20 \times 10^3} = 5 \times 10^3 = 5\ k\Omega$

2. $R_3 + R_{C-D} + R_6 = 5\ k\Omega + 5\ k\Omega + 5\ k\Omega = 15\ k\Omega$

3. $R_{A-B} = \dfrac{(R_2)(R_3 + R_{C-D} + R_6)}{(R_2) + (R_3 + R_{C-D} + R_6)}$

$\qquad = \dfrac{(15 \times 10^3) \times (15 \times 10^3)}{(15 \times 10^3) + (15 \times 10^3)} = \dfrac{225 \times 10^6}{30 \times 10^3} = 7.5 \times 10^3 = 7.5\ k\Omega$

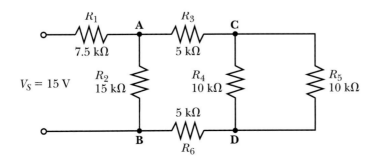

FIGURE 6–17 Analyzing voltages in a series-parallel circuit

4. $R_T = R_1 + R_{A-B} = 7.5 \text{ k}\Omega + 7.5 \text{ k}\Omega = 15 \text{ k}\Omega$

5. $I_T = \dfrac{V_T}{R_T} = \dfrac{15 \text{ V}}{15 \text{ k}\Omega} = 1 \text{ mA}$

6. $V_1 = I_T \times R_1 = 1 \text{ mA} \times 7.5 \text{ k}\Omega = 7.5 \text{ V}$

7. Current splits evenly between branches R_2 and the R_3 through R_6 branch since each branch R equals 15 kΩ. This means that:

$$V_2 = 0.5 \, I_T \times R_2 = 0.5 \text{ mA} \times 15 \text{ k}\Omega = 7.5 \text{ V}$$

$$V_3 + V_{C-D} + V_6 \text{ equals } 7.5 \text{ V}$$

8. $V_3 = 0.5 \, I_T \times R_3 = 0.5 \text{ mA} \times 5 \text{ k}\Omega = 2.5 \text{ V}$

9. Current splits evenly between R_4 and R_5; thus, each carries 0.25 mA of current.

$$V_4 = 0.25 \, I_T \times R_4 = 0.25 \text{ mA} \times 10 \text{ k}\Omega = 2.5 \text{ V}$$

$$V_5 = 0.25 \, I_T \times R_5 = 0.25 \text{ mA} \times 10 \text{ k}\Omega = 2.5 \text{ V}$$

10. $V_6 = 0.5 \, I_T \times R_6 = 0.5 \text{ mA} \times 5 \text{ k}\Omega = 2.5 \text{ V}$

Example

Resistance and voltage divider techniques

1. Observing the circuit, you can see that:

   a. R_4 and R_5 are in parallel; thus, $V_4 = V_5$.

   b. R_e of $R_{4-5} = 5 \text{ k}\Omega$.

   c. R_3, R_{4-5}, and R_6 are in series and together form a 15-kΩ branch of the circuit. Furthermore, each of these drops one-third the voltage present from point A to point B, since each of the three entities represents 5 kΩ.

   d. This 15-kΩ branch is in parallel with R_2, another 15-kΩ branch. Thus, R_e from point A to point B equals 7.5 kΩ.

   e. R_1 is also 7.5 kΩ and is in series with the equivalent 7.5 kΩ found from point A to point B. Thus V_1 and V_{A-B} are equal, and each equals half the voltage applied (Kirchhoff's voltage law).

2. Applying the principles of voltage dividers:

   a. $V_{A-B} = \dfrac{R_{A-B}}{R_T} \times V_T = \dfrac{7.5 \text{ k}\Omega}{15 \text{ k}\Omega} \times 15 \text{ V} = 7.5 \text{ V}$

   b. $V_{C-D} = \dfrac{R_{4-5}}{R_3 + R_{4-5} + R_6 \times V_{A-B}} = \dfrac{5 \text{ k}\Omega}{15 \text{ k}\Omega} \times 7.5 \text{ V} = 2.5 \text{ V}$

   c. $V_3 = V_{4-5} = V_6$

   d. $V_2 = V_{A-B}$

   e. $V_1 = \dfrac{R_1}{R_T} \times V_T = \dfrac{7.5 \text{ k}\Omega}{15 \text{ k}\Omega} \times 15 \text{ V} = 7.5 \text{ V}$

PRACTICAL NOTES

An approach often used to simplify notation for indicating that components are in parallel with each other is to use two vertical bars, or short vertical lines between component designators. For example, in the circuit of Figure 6–17 to show that R_4 and R_5 are in parallel, we could state $R_4 \| R_5$. If we were going to use this type of notation, we could indicate the resistors and their connections, which make up the resistance at points A–B, as follows:

$$R_{A-B} = R_2 \| [R_3 + (R_4 \| R_5) + R_6]$$

Example

Let's look at one more example of analyzing a series-parallel circuit, using the Ohm's law method. Solve for circuit R and I values, then compute voltages, as appropriate, for the circuit of Figure 6–18.

1. $R_{C-D} = R_4 \| (R_5 + R_6)$

$$R_{C-D} = \frac{R_4 \times (R_5 + R_6)}{R_4 + (R_5 + R_6)} = \frac{(3.9 \times 10^3) \times (4.9 \times 10^3)}{(3.9 \times 10^3) + (4.9 \times 10^3)}$$

$$R_{C-D} = \frac{19.11 \times 10^6}{8.8 \times 10^3} = 2.17 \text{ k}\Omega$$

2. $R_3 + R_{C-D} = 4.7 \text{ k}\Omega + 2.17 \text{ k}\Omega = 6.87 \text{ k}\Omega$

3. $R_{A-B} = R_2 \| (R_3 + R_{C-D}) = \frac{(5.6 \times 10^3) \times (6.87 \times 10^3)}{(5.6 \times 10^3) + (6.87 \times 10^3)} = \frac{38.47 \times 10^6}{12.47 \times 10^3}$

$= 3.09 \text{ k}\Omega$

4. $R_T = R_1 + R_{A-B} = 12 \text{ k}\Omega + 3.09 \text{ k}\Omega = 15.09 \text{ k}\Omega$

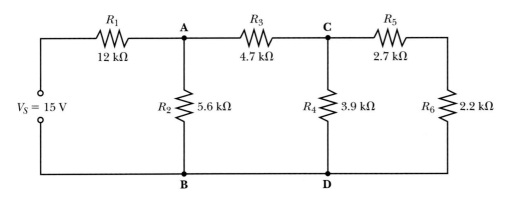

FIGURE 6–18

5. $I_T = \dfrac{V_S}{R_T} = \dfrac{15\text{ V}}{15\text{ k}\Omega} \approx 1\text{ mA}$

6. $I_{R_1} = 1\text{ mA}; \; V_{R_1} = I_{R_1} \times R_1 = 1\text{ mA} \times 12\text{ k}\Omega = 12\text{ V}$

7. $I_{R_2} = \dfrac{V_{R_2}}{R_2} = \dfrac{3\text{ V}}{5.6\text{ k}\Omega} \approx 0.54\text{ mA}$

8. $I_{R_3} = I_T - I_{R_2} = 1\text{ mA} - 0.54\text{mA} \approx 0.46\text{ mA}$
$V_{R_3} = I_{R_3} \times R_3 = 0.46\text{ mA} \times 4.7\text{ k}\Omega \approx 2.16\text{ V}$

9. $I_{R_4} = \dfrac{V_{R_4}}{R_4} = \dfrac{V_S - (V_{R_1} + V_{R_3})}{R_4} = \dfrac{15\text{ V} - 14.16\text{ V}}{3.9\text{ k}\Omega}$

$I_{R_4} = \dfrac{0.84\text{ V}}{3.9\text{ k}\Omega} \approx 0.22\text{ mA}$

10. $I_{R_5} = I_{R_6} = \dfrac{V_{\text{C–D}}}{R_5 + R_6} = \dfrac{0.84\text{ V}}{4.9\text{ k}\Omega} \approx 0.17\text{ mA}$

$V_{R_5} = I_{R_5} \times R_5 = 0.17\text{ mA} \times 2.7\text{ k}\Omega \approx 0.46\text{ V}$

$V_{R_6} = I_{R_6} \times R_6 = 0.17\text{ mA} \times 2.2\text{ k}\Omega \approx 0.37\text{ V}$

PRACTICE PROBLEMS IV

1. Analyze the voltage and current parameters throughout the circuit of Figure 6–19. List the voltage dropped by each resistor and the current through each resistor.

2. Analyze the voltage and current parameters throughout the circuit of Figure 6–20. Again, list the voltage dropped by each resistor and the value of current through each resistor.

6–6 POWER IN SERIES-PARALLEL CIRCUITS

Analysis of power dissipations throughout a series-parallel circuit employs the same techniques used in series or parallel circuits. Important points to remember are:

1. total power $= V_T \times I_T$ (or $I_T^2 \times R_T$; or, V_T^2/R_T);

2. total power also equals the sum of all individual power dissipations; and

3. individual component power dissipations are calculated using the individual component's parameters. That is, $V \times I$; I^2R; or V^2/R values.

Example

Notice the power dissipations throughout the circuit shown in Figure 6–21. If you take time to analyze this circuit completely, you find that the circuit's total resistance is 30 Ω. This means that the circuit's total current is 30 V/30 Ω, or 1 A. Current divides evenly in the two outer branches. Therefore, current through R_4 and $R_5 = 0.5$ A, and current through R_6 and R_7 is also 0.5 A. As you can see

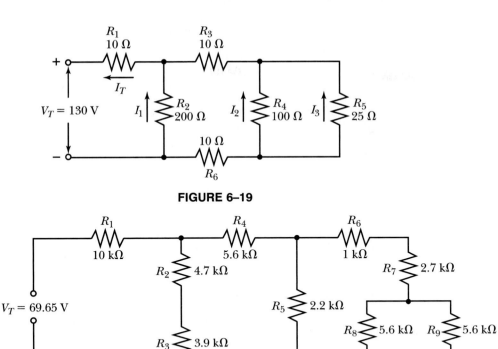

FIGURE 6–19

FIGURE 6–20

from the circuit diagram, the total 1 A current also evenly splits through R_2 and R_3, with each carrying 0.5 A.

With the current through each of the resistors known, you can then find the voltage drops across each component, and the power dissipated by each component, Figure 6–21. Since we know current, resistance, and voltage, any one of the power formulas ($P = V \times I$; $P = I^2 R$; or $P = V^2/R$) might be used to compute each of the component's power dissipations. For example: $P_{R_7} = V \times I = 10$ V $\times$ 0.5 A = 5 W; $P_{R_1} = V \times I = 10$ V $\times$ 1 A = 10 W; etc.

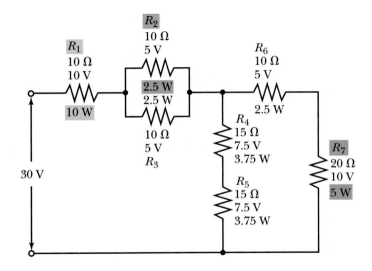

FIGURE 6–21 Effect of location on power, current, and voltage

PRACTICAL NOTES

Note the following:

1. In series circuits, the largest value resistor dissipates the most power, and the lowest value resistor dissipates the least power.
2. In parallel circuits, the smallest resistance branch dissipates the most power, and the largest resistance branch dissipates the least power.
3. In series-parallel circuits, the resistor dissipating the most power *may or may not* be the highest value resistor in the circuit. The resistor dissipating the lowest power *may or may not* be the lowest value resistor in the circuit.

    The power dissipated is determined by the position of the given resistor in the total circuit configuration. An illustration of this is in Figure 6–21. Note the following:

- R_7 is twice the value of R_1 but dissipates only half the power.
- R_1 and R_2 are equal value resistors but each dissipates a different value of power.

    To make sure you are learning to analyze the key parameters for a series-parallel circuit, perform the following In-Process Learning Check.

IN-PROCESS LEARNING CHECK II

Refer to the diagram below and find the following:

$I_1 = $ _____   Which resistor value in this circuit
$I_T = $ _____   is dissipating the most power? _____
$V_T = $ _____
$R_T = $ _____
$P_3 = $ _____

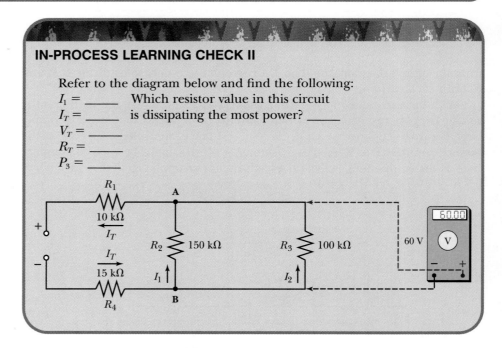

6–7 EFFECTS OF OPENS IN SERIES-PARALLEL CIRCUITS AND TROUBLESHOOTING HINTS

Effects

As you have seen, the electrical location of a given component in a series-parallel circuit has great influence on all its electrical parameters. The location of an open in the circuit configuration also has a strong influence on all the parameters throughout the circuit.

Looking at the circuit used in In-Process Learning Check II, you can see that if either R_1 or R_4 opens, then the path for total current is broken. Thus, all other components' voltage drops, currents, and power dissipations decrease to zero!

On the other hand, if R_3 opens, total current decreases, since the total resistance increases from 85 kΩ to 175 kΩ. But current does not decrease to zero. Thus, individual component voltages, currents, and power dissipations change; however V applied remains the same.

Here is a **special thinking exercise** for you. Again refer to the In-Process Learning Check circuit and assume R_2 opens. Will the following parameters increase, decrease, or remain the same?

V_T will _____ I_T will _____

V_1 will _____ I through R_1 will _____

V_2 will _____ I through R_2 will _____

V_3 will _____ I through R_3 will _____

V_4 will _____ I through R_4 will _____

                 P_T will _____ P_3 will _____ P_4 will _____

Total power and total current should decrease because total resistance increases while V applied remains the same.

Voltage drops across R_1 and R_4 also decrease, since they now have less current through them and their resistances remain unchanged.

Voltage across R_2 and R_3 increases because the series components (R_1 and R_4) are dropping less of the applied voltage, leaving more voltage to be dropped across the parallel combination of R_2 and R_3, points A–B.

Current decreases through R_1 and R_4 (remember I_T decreased); however, current through the unopened parallel branch, R_3, increases since current is no longer splitting between R_2 and R_3. Obviously, current through R_2 decreases to zero, since it is an open path.

Finally, power dissipations change. P_T, P_1, and P_4 decrease due to the drop in I_T, and P_3 increases because its current increases.

By now you should realize logical thinking is required when analyzing the sequence of events and the effects of an open in a series-parallel circuit. There is not just one simple approach. Rather, think of the interaction between the various portions of the circuit on each other and on the total circuit parameters.

TROUBLESHOOTING HINTS

Because the component's position in the circuit is important due to the interaction it has on the remaining circuitry, the hints given here are somewhat general.

If an open occurs in *any* component in the circuit, total resistance increases; consequently, total current decreases. If you measure (or monitor) total current and it is lower than normal with normal applied voltage, an open, or at least an increased resistance value, has occurred somewhere in the circuit.

continued

NOTE: Two typical ways to determine total current in a circuit are shown in Figure 6–22.
a. Insert current meter in series with source and the circuitry.
b. Measure voltage across a known resistor value that carries total current, and use Ohm's law (V/R) to determine current.

PRACTICAL NOTES

A convenient resistance value for this approach is a 1-kΩ resistor, if available, because its voltage drop equals the number of mA through it. In other words,

$$\frac{10 \text{ V}}{1 \text{ k}\Omega} = 10 \text{ mA}$$

(See Figure 6–22.)

1. If an open occurs in any component in series with the source, total R increases to ∞, and total I decreases to zero. As you learned in series circuits, V applied appears across the opened series component and zero volts are dropped across the good components.

2. If an open occurs in any component other than those in series with the source, current measurements and *common sense* will isolate the problem.

3. If current measurements cannot be easily performed, you can make a number of voltage checks throughout the circuit. Observe where voltages are higher and lower than normal, then logically determine where the open is located.

6–8 EFFECTS OF SHORTS IN SERIES-PARALLEL CIRCUITS AND TROUBLESHOOTING HINTS

Effects

Again, the location of the faulty component or circuit portion in the total circuit configuration controls what happens to the parameters of the remaining good components.

Refer to Figure 6–23. If the shorted component is in a line that carries total current, that is, in series with the source, then 1) total current increases, 2) voltage across the shorted component drops to zero, and 3) all other voltages throughout the circuit increase (except V_T).

Refer to Figure 6–24. If the shorted component is elsewhere in the circuit, then 1) total current still increases since total R decreases; however, 2) voltage across the shorted component *and* components *directly* in parallel with the shorted component decreases to zero, since the shorted component is shunting the current around components in parallel with it; and 3) other components' voltage drops may increase or decrease, depending on their location.

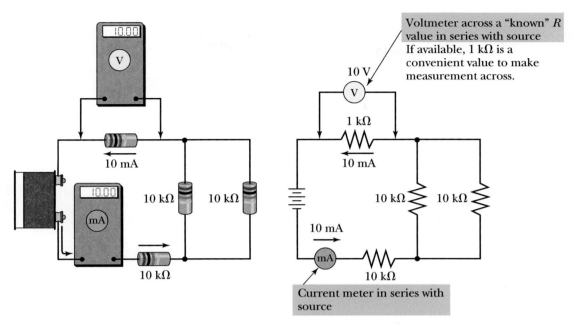

FIGURE 6–22 Two ways of checking total current

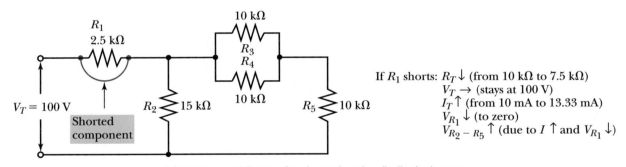

If R_1 shorts: $R_T \downarrow$ (from 10 kΩ to 7.5 kΩ)
$\qquad\qquad\quad V_T \rightarrow$ (stays at 100 V)
$\qquad\qquad\quad I_T \uparrow$ (from 10 mA to 13.33 mA)
$\qquad\qquad\quad V_{R_1} \downarrow$ (to zero)
$\qquad\qquad\quad V_{R_2 - R_5} \uparrow$ (due to $I \uparrow$ and $V_{R_1} \downarrow$)

FIGURE 6–23 Effects of a shorted series (in-line) element

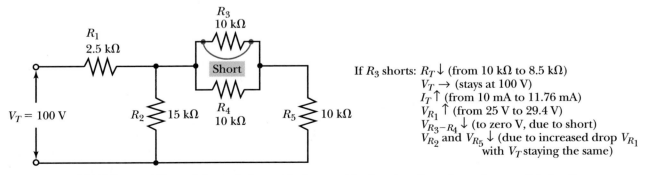

If R_3 shorts: $R_T \downarrow$ (from 10 kΩ to 8.5 kΩ)
$\qquad\qquad\quad V_T \rightarrow$ (stays at 100 V)
$\qquad\qquad\quad I_T \uparrow$ (from 10 mA to 11.76 mA)
$\qquad\qquad\quad V_{R_1} \uparrow$ (from 25 V to 29.4 V)
$\qquad\qquad\quad V_{R_3 - R_4} \downarrow$ (to zero V, due to short)
$\qquad\qquad\quad V_{R_2}$ and $V_{R_5} \downarrow$ (due to increased drop V_{R_1}
$\qquad\qquad\qquad\qquad$ with V_T staying the same)

FIGURE 6–24 Effect of shorted element in the combinational portion of a series-parallel circuit

TROUBLESHOOTING HINTS

Some general thoughts about troubleshooting shorts in series-parallel circuits are as follows:

1. A short of *any* component in the circuit causes total resistance to decrease and total current to increase.
2. A higher than normal total current (or lower than normal R_T) implies that either a short or a lowered R has occurred somewhere in the circuit.
3. Voltage measurements across the circuit components reveal close to zero volts across the shorted component and those directly in parallel with it. The good components may have either higher or lower than normal voltage levels, depending on their individual positions within the circuit. Voltage measurement is one simple way to locate the component or combination of components that are shorted.
4. Once the shorted portion of the circuit is located, resistance measurements will isolate the exact component or circuit portion containing the short. Resistance measurement involves turning the power off during measurement and lifting one end of each branch from the circuit.

Let's see if your analysis skills are developing. Try the troubleshooting problems in the following In-Process Learning Check.

IN-PROCESS LEARNING CHECK III

1. Refer to Figure 6–25. Determine the defective component and the nature of its defect.
2. Refer to Figure 6–26. Indicate which component or components might be suspect and what the trouble might be.
3. Refer to Figure 6–27. Indicate which voltages will change and in which direction, if R_2 increases in value.

6–9 DESIGNING A SERIES-PARALLEL CIRCUIT TO SPECIFICATIONS

For a design problem, let's assume we are designing and drawing the circuit diagram for a simple lighting circuit that contains four light bulbs, each of which normally operates at 10 volts. Using four lights and three SPST switches, configure the circuit so switch 1 acts as an on-off switch for all the lights; switch 2 controls two of the lights' on-off modes (assuming switch 1 is on), and switch 3 controls the remaining lights' on-off conditions (assuming switch 1 is on). The source voltage is 20 volts.

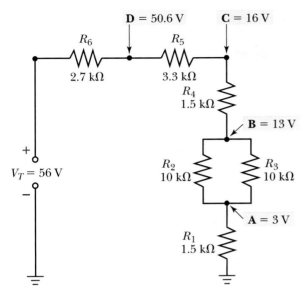

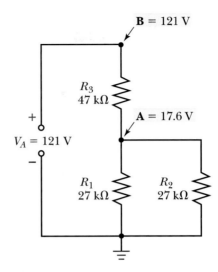

NOTE: Voltages indicated are from specified test points to ground reference.

FIGURE 6–25

NOTE: Voltages indicated are from specified test points to ground reference.

FIGURE 6–26

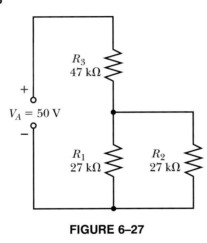

FIGURE 6–27

Example

The thought processes to solve this problem are as follows:

1. Since the source equals 20 V and the bulbs are 10-V bulbs, it is logical to assume two bulbs are in series across the 20-V source.

2. Since there are four lights, it is also logical to assume there are two, two-bulb branches connected to the 20-V source voltage.

3. Since switch 1 controls the on-off condition of the total circuit connected to the source, it is logical that this switch is in series with the source and connected between the source and the remaining circuitry.

4. Since the circuit specification is to control each of the two bulb branches independently, switches 2 and 3 must be placed in series with the bulbs in each branch, respectively.

Using the preceding knowns, it is possible to draw the circuit diagram, Figure 6–28.

FIGURE 6–28 Sample design problem circuit

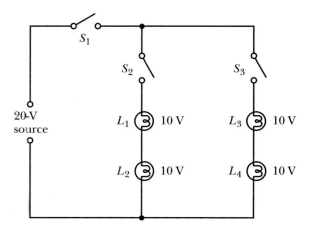

PRACTICE PROBLEMS V

See if you can design a series-parallel circuit that meets the following conditions:

Given six 10-kΩ resistors and a 45-V source, design and draw the circuit diagram where:

$V_1 = 20$ V; $V_2 = 5$ V; $V_3 = 5$ V; $V_4 = 5$ V; $V_5 = 5$ V; $V_6 = 20$ V

6–10 LOADED VOLTAGE DIVIDERS

Refer to Figure 6–29. Recall in the Series Circuits chapter we learned voltage values at different points along the series circuit are controlled by the ratios of the resistances throughout the series string. These voltage dividers are called unloaded voltage dividers because no external circuit loads (current-demanding components or devices) are connected to the voltage divider circuit.

Two applications of series-parallel circuits commonly encountered by electronic technicians are **loaded voltage dividers** and the **Wheatstone bridge circuit.**

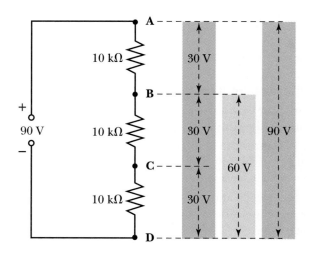

FIGURE 6–29 Unloaded voltage divider

Frequently in electronic circuits, a requirement is to provide different levels of voltage for various portions of the circuit or system. Also, these different portions may demand different load currents. Voltage dividers are often used with power supplies to provide and distribute these different levels of voltages at the different current values demanded.

To help in our discussion of loaded voltage dividers, it is appropriate to define some important terminology used in our discussion. Refer to Figure 6–30 as you study these definitions.

1. **Load** is a component/device or circuit drawing current from a power source.

2. **Load current** is the current required by the components or circuits that are connected to the power source and/or its output voltage divider.

3. R_L is the resistance value of the load resistor or component.

4. **Ground reference** is an electrical *reference point* used in defining or measuring voltage values in a circuit.

Frequently the electrical ground reference point in a circuit is chassis ground, which is the common metal conducting path through which many components or circuits are electrically connected to one side of the power supply. In printed circuits, it is a common metal conductor *path* where many components are electrically connected, or "commoned." In circuits built on a metal chassis, the metal chassis is often the "return" path, or common conductor path; hence, the term "chassis ground." The symbol for chassis ground is $\overline{/\!/\!/}$. In electrical wiring circuits, such as house wiring, "earth ground" is a common reference and connecting point. The symbol for earth ground is $\stackrel{\perp}{=}$. Since this symbol is often used to represent the "common" in many texts and many diagrams, we will also use it to represent the circuit common reference point, as well.

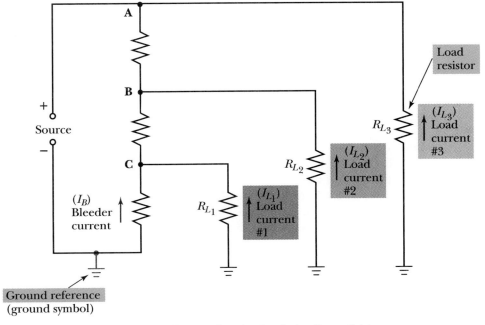

FIGURE 6–30 Terms related to loaded voltage dividers

PRACTICAL NOTES

Safety must be observed when dealing with equipment where the equipment chassis is used as the "common" circuit return path. In many (or most) cases, it is advisable to make sure that the chassis is also connected to a good earth ground via some means to prevent inadvertent shock or life-threatening danger caused by differences of potential between equipment chassis and earth ground. Earth ground connections are frequently made via electrical connection made to a water pipe, or a metal rod driven into the ground. The ac electrical power system normally has one of the conductor paths "neutralized" by its being connected to an electrical earth ground.

There are some pieces of equipment that are designed to be operated with a "floating ground," which is not connected to earth ground. That is the chassis is NOT to be connected to earth ground. In those cases, it is a desirable safety habit to use an "isolation transformer" to isolate the ac power source earth ground from the chassis common circuit return path. You will learn more about transformers in a later chapter. We simply wanted to CAUTION you to be careful when working with circuits and equipment that might shock (or kill) you if you touch both the chassis and an earth ground at the same time. Operate the equipment properly, and be aware of whether the chassis is properly grounded or isolated.

You must be aware of the ground conditions that exist for the equipment you are using, working on, or testing. It is always a good idea to check for differences of potential between equipment and earth grounds using a voltmeter. If the chassis is "hot" with respect to earth or power wiring grounds, you must determine whether you should even work on it at all, if it needs to be repaired before working on it, or whether the power plug needs to be reversed.

5. **Bleeder current** is considered the fixed current through a resistive network, or a bleeder resistor connected across the output of the power supply. Many power supplies use a bleeder resistor that draws a fixed minimum current from the supply to help regulate the power supply output voltage; that is, keeping it more constant under varying load conditions. Because loads connected to voltage dividers are often functional circuits rather than fixed-value resistors, load requirements (i.e., current demands by the loads) often change during operation. For this reason voltage dividers need to be designed so that reasonable variations in load current demands do not unreasonably upset the voltage level at the various divider output points. The higher the bleeder current is, with respect to the load currents drawn, the "stiffer" the voltage supply acts; that is, the less the output voltage will vary with changes in load currents. Bleeder currents of approximately 10% to 25% of load current demands are quite typical. If higher percentages are used, it is in cases where a stiffer supply is needed.

6. **Bleeder resistor** is the resistor, or resistor network, connected in parallel with the power supply circuitry, that draws the bleeder current from the source. Another common function of a bleeder resistor is to discharge power supply capacitors after the circuit is turned off, for safety purposes. You will learn more about this in later studies regarding power supply circuits.

7. **Potentiometer** (Figure 6–31) is a three-terminal resistive device used as a voltage-dividing component. One terminal is at one end of the resistive element,

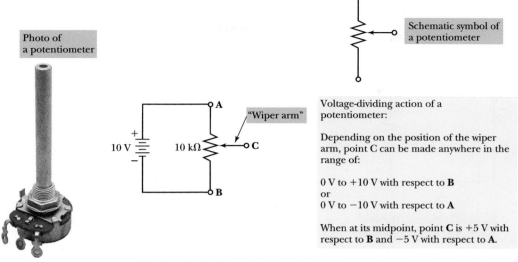

Photo of a potentiometer

Schematic symbol of a potentiometer

A

"Wiper arm"

10 V

10 kΩ

C

B

Voltage-dividing action of a potentiometer:

Depending on the position of the wiper arm, point C can be made anywhere in the range of:

0 V to +10 V with respect to **B**
or
0 V to −10 V with respect to **A**

When at its midpoint, point **C** is +5 V with respect to **B** and −5 V with respect to **A**.

Resistance characteristics:

$R_{AC} + R_{CB} = 10\ k\Omega$
$R_{AC} = 0 - 10\ k\Omega$
$R_{CB} = 0 - 10\ k\Omega$

FIGURE 6–31 The potentiometer *(Photo courtesy of Bourns, Inc.)*

a second contact is at the other end of the resistive element, and the third contact is a "wiper arm" that can move to any position along the resistive element. The position of the wiper arm along the resistive element determines the resistance, and thus the voltage from it to either end contact.

With this background information in mind, let's look at a brief analysis of a typical loaded voltage divider circuit.

Two-element divider with load

Observe Figure 6–32a and 6–32b as you study the following discussion. Let's think through the circuit.

What causes voltage between point A and ground reference to decrease from 25 volts to about 16.7 volts when load resistor R_L is connected? Before the load was connected, the two-series-connected equal value resistors divided the input voltage equally (25 volts each) since they were passing the same current.

When the load resistor was connected, a new branch path for current was created through R_L. The current through R_L must also pass through R_1 to return to the source's positive side. Now R_1 has a higher amount of current through it.

If the current through R_1 increases and its R value is unchanged, then its $I \times R$ drop must increase. In this case it increased from 25 volts to about 33.33 volts. Thus, only 16.66 volts remain to be dropped from point A to ground compared with 25 volts previously dropped before the load was connected.

Three-element divider with multiple loads

Observe in Figure 6–33a and 6–33b the changes caused by connecting a load to a three-element voltage divider are similar to those changes that occur in a two-element divider.

Let's look at the three-element loaded voltage divider in Figure 6–33b from the perspective of Ohm's law and Kirchhoff's voltage law.

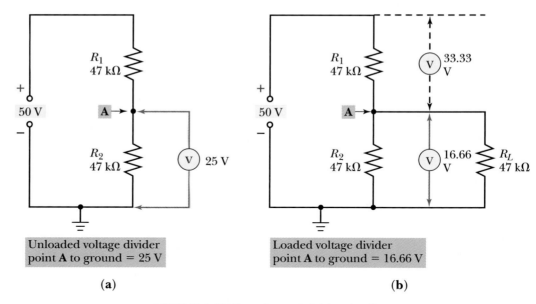

FIGURE 6–32 Two-element divider circuit

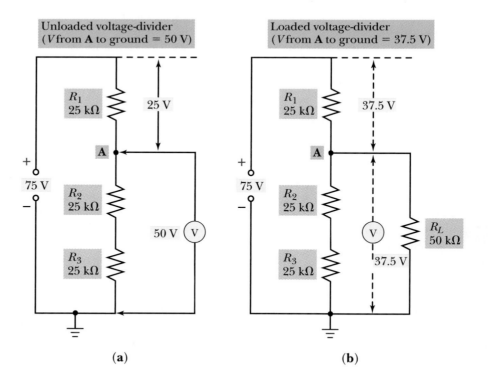

FIGURE 6–33 Comparison of loaded and unloaded voltage dividers (effect of connecting a load to a voltage divider circuit)

R_T is solved by using your knowledge of series and parallel circuits. R_L is in parallel with the series combination R_{2-3}. This indicates the R_L value of 50 kΩ is in parallel with 50 kΩ so the equivalent resistance from point A to ground equals 25 kΩ. This 25 kΩ is in series with the R_1 value of 25 kΩ yielding a total circuit resistance of 50 kΩ.

Solve I_T using Ohm's law, where

$$I_T = \frac{V_T}{R_T} = \frac{75 \text{ V}}{50 \text{ k}\Omega} = 1.5 \text{ mA}$$

Since I_T passes through R_1, $V_1 = I_T \times R_1 = 1.5 \text{ mA} \times 25 \text{ k}\Omega = 37.5$ volts.

Since point A to ground is composed of two equal 50-kΩ branches (R_L and R_{2-3}), total current evenly divides between the branches. This means 0.75 mA passes through R_L and 0.75 mA passes through R_{2-3}. The R_L voltage drop = 0.75 mA $\times$ 50 kΩ = 37.5 V; R_2 voltage drop = 0.75 mA $\times$ 25 kΩ = 18.75 V; R_3 voltage drop also = 0.75 mA $\times$ 25 kΩ = 18.75 V; therefore, V_{2-3} = 37.5 V.

Of course, by Kirchhoff's voltage law (the sum of the voltage drops around a closed loop must equal V applied), you know the voltage from point A to ground has to be 37.5 V. Since the calculated R_1 drop is 37.5 V, the remainder of V applied has to drop from A to ground reference.

Let's move one step further in studying loaded voltage dividers by performing a design and analysis problem to achieve specified voltage levels and load currents.

Example

Refer to Figure 6–34 for the desired parameters of the circuit to be designed. Note the voltage-dividing system provides voltage and current levels as follows:

$$\text{Load 1 } (R_{L_1}) = 25 \text{ V at } 12.5 \text{ mA}$$

$$\text{Load 2 } (R_{L_2}) = 75 \text{ V at } 25 \text{ mA}$$

$$\text{Load 3 } (R_{L_3}) = 225 \text{ V at } 50 \text{ mA}$$

Bleeder current (I_B) = 400 mA (to provide a fairly "stiff" divider)

For additional information we'll make one task to determine the resistance values of loads 1, 2, and 3. Using the parameters provided, this is not critical to the design tasks but will be of interest.

Since you know the desired voltage across each load and the currents passing through them, simply apply Ohm's law to find load resistance values.

$$R_{L_1} = \frac{25 \text{ V}}{12.5 \text{ mA}} = 2 \text{ k}\Omega$$

$$R_{L_2} = \frac{75 \text{ V}}{25 \text{ mA}} = 3 \text{ k}\Omega$$

$$R_{L_3} = \frac{225 \text{ V}}{50 \text{ mA}} = 4.5 \text{ k}\Omega$$

The second task, a design task, is to determine the values of voltage divider resistors R_1, R_2, and R_3 that will provide the appropriate voltages to the loads.

Again refer to Figure 6–34 as we examine the design task as follows:

1. Looking at the circuit, you can see that none of the load currents pass through R_3. This means only the given bleeder current (I_B) of 400 mA is passing through R_3. Because its voltage must equal the load 1 voltage of 25 V since they are in parallel, then

$$R_3 = \frac{25 \text{ V}}{400 \text{ mA}} = 62.5 \text{ }\Omega$$

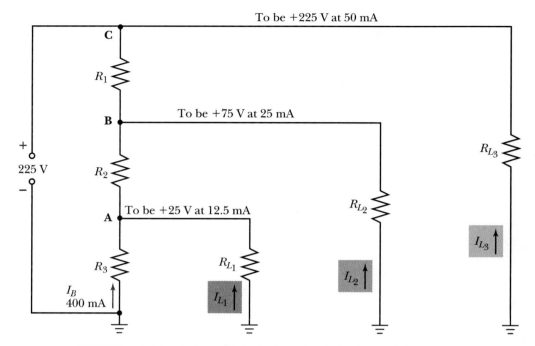

FIGURE 6–34 Analyzing and designing a loaded voltage divider circuit

2. The R_2 current must equal the 400-mA bleeder current, *plus* the 12.5-mA load 1, or 412.5 mA. The voltage dropped by R_2 must equal 75 V minus the voltage dropped by R_3. Thus, V_{R_2} equals 50 V. Therefore,

$$R_2 = \frac{50 \text{ V}}{412.5 \text{ mA}} = 121 \text{ } \Omega$$

3. The R_1 current must equal the bleeder current plus load 1 current plus load 2 current. Therefore, the current through R_1 equals 400 mA (bleeder current) plus 12.5 mA (I_{L_1}) plus 25 mA (I_{L_2}) equals 437.5 mA. Thus, R_1 voltage drop must equal 225 V minus the drop from point B to ground, or 225 V − 75 V = 150 V. This means

$$R_1 = \frac{150 \text{ V}}{437.5 \text{ mA}} = 343 \text{ } \Omega$$

Look at Figure 6–35 for the completed design.

PRACTICE PROBLEMS VI

1. Draw the circuit of a three-resistor voltage divider with three loads connected as they were in our example. Let the top divider resistor be R_1, the middle one R_2, and the bottom one R_3.

2. Determine the value of each of the load resistances and find the values for the divider resistors R_1, R_2, and R_3 assuming the following:

   a. source voltage = 100 V;

   b. bleeder current = 200 mA;

   c. load #1 needs 10 V and demands 10 mA of current;

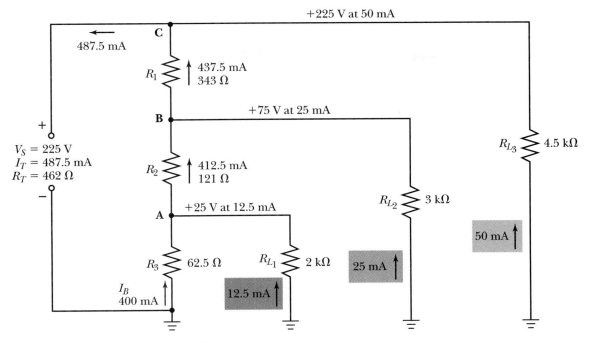

FIGURE 6-35 Voltage divider problem solution

    d. load #2 needs 30 V and demands 20 mA of current; and

    e. load #3 needs 100 V and demands 40 mA of current.

3. Completely label your drawing showing all resistances, voltages, and currents.

A practical voltage divider variation

Look at the circuit in Figure 6–36. With the potentiometer's wiper arm at the top (position 1), the 20-kΩ R_L is in parallel with total potentiometer resistance of 30 kΩ, yielding an equivalent resistance of 12 kΩ from point A to ground. The

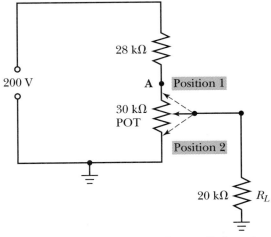

The potentiometer enables varying V across R_L from 0 to 60 V

FIGURE 6-36 Example of potentiometer varying voltage output

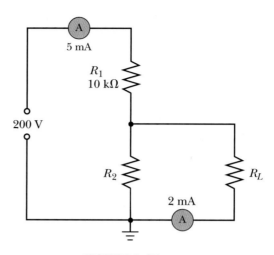

FIGURE 6–37

circuit voltage-dividing action is dividing the 200 V across the 40-kΩ total circuit resistance consisting of the 28-kΩ resistor and the 12-kΩ parallel circuit equivalent resistance in series. Therefore, R_L drops twelve-fortieths of 200 V, or 60 V. The 28-kΩ resistor drops the remaining 140 V.

On the other hand, if the wiper arm is at position 2 (or ground), both ends of R_L will connect to ground. Therefore, the voltage across R_L is zero V. This means that with the potentiometer, we can vary the voltage across R_L from 0 to 60 volts. As you will see throughout your electronics experience, this "variable-voltage-divider" capability of the potentiometer is convenient.

PRACTICE PROBLEMS VII

Refer to Figure 6–37 and answer the following:

1. What is the value of voltage across R_L? How much power is dissipated by R_L?

2. What is the value of R_L?

3. What is the value of R_2?

4. If V_{R_L} suddenly decreases due to a change in the R_2 value, does this indicate R_2 has increased or decreased in value?

5. If the decrease in V_{R_L} in question 4 was caused by a change in R_1 value, would it indicate R_1 has increased or decreased in value?

6. If R_L is disconnected from the voltage divider in the circuit, what is the new value of V_{R_2}?

7. What happens to V_{R_L} if R_1 opens? if R_2 shorts?

6–11 THE WHEATSTONE BRIDGE CIRCUIT

Various types of special series-parallel circuits, called **bridge circuits,** are used to make measurements in electronic circuits, Figure 6–38.

Note that zero volts appear between points A and B when the bridge is balanced. You will see this balance *only* when the ratio of the resistances in the left arm and in the right arm of the bridge are equal.

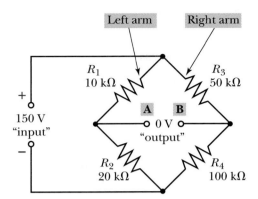

For example, observe Figure 6–38. In the left arm of the bridge, $R_1 = 10$ kΩ and $R_2 = 20$ kΩ. In the right arm of the bridge, $R_3 = 50$ kΩ and $R_4 = 100$ kΩ. If V applied is 150 V, then R_1 drops one-third the voltage across the left arm of the bridge, or 50 V (since it is one-third of the left arm's total resistance). R_2 drops the other two-thirds of the 150 volts, or 100 V.

What about the other arm of the bridge? The same 150 V is applied across that arm's 150 kΩ, with R_3 (50 kΩ) dropping one-third the applied voltage, or 50 V, and R_4 dropping two-thirds the applied voltage ($R =$ two-thirds the right arm's total R), or 100 V.

Note the voltage differential between points A and B is zero. This is because point A is at a potential of + 100 volts with respect to the power supply's negative terminal. Point B is also at a + 100 volts with respect to the power supply's negative terminal. Thus, the difference of potential between points A and B is zero volts, since they are at the same potential.

This balanced condition (zero volts between points A and B) exists whenever the resistance ratios of the top resistor to the bottom resistor in the left and right arm of the bridge are the same. That is,

FORMULA 6–1   Bridge balanced when: $\dfrac{R_1}{R_2} = \dfrac{R_3}{R_4}$

NOTE: The values of the resistors do not matter. It's the *ratio* of the resistances in each arm that determines the voltage distribution throughout that arm. Recall the series circuit concepts. If both arms have equal voltage distribution, then the midpoint of each arm is at the same potential and the bridge is balanced.

The **Wheatstone bridge circuit,** a special application of the bridge circuit, measures unknown resistance values. When the bridge circuit is balanced, the output terminals (points A and B) have zero potential difference between them. On the other hand, if the bridge circuit is unbalanced, there is a difference of potential between the output terminals. Refer to Figure 6–39 as you read the following discussion.

The following is one method of using the Wheatstone bridge circuit to determine an unknown resistance:

1. Carefully select R_1 and R_2 to be matched (equal) values.

2. Place a sensitive "zero-center-scale" current meter between points A and B, which are the output terminals of the bridge circuit.

FIGURE 6–39 Concepts of the Wheatstone bridge circuit

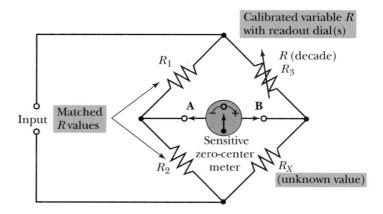

3. Use a calibrated variable resistance, such as a "resistance decade box," as R_3. NOTE: This type variable resistance has calibrated dials to enable you to read its resistance setting, Figure 6–40.

4. Place the unknown resistor (R_X) in the same arm of the circuit as the calibrated variable resistance.

5. Since R_1 and R_2 are equal value in the left arm, then the bridge is balanced when the calibrated variable resistance and the unknown resistance equal each other in the right arm. In this case, the sensitive current meter's pointer is pointing at "0" in the center of the scale. If the bridge is unbalanced, the pointer points to the left of center, indicating current through the meter in one direction, or to the right of center, indicating current in the other direction. The pointer's direction is dependent on the polarity of the voltage difference between points A and B, as determined by the comparative voltages at points A and B in the bridge network.

6. Once the calibrated variable resistance is adjusted for zero current reading (balanced), the value of the unknown resistor is read directly from the calibrated variable resistance's dial(s).

FIGURE 6–40 Decade resistor box used as a calibrated R value *(Courtesy of Clarostat Mfg. Company, Inc.)*

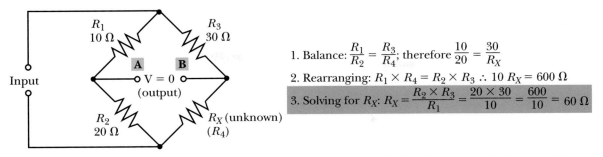

FIGURE 6–41 The calculation approach may be used to solve for the R_X value in a bridge circuit.

Another method to determine the unknown resistance value is to use the known resistance values and the concepts of the balanced bridge circuit.

1. Recall balance occurs when $\dfrac{R_1}{R_2} = \dfrac{R_3}{R_4}$

2. With these relationships, it is mathematically shown that:

$$R_1 \times R_4 = R_2 \times R_3 \text{ (cross-multiplying factors in formula)}$$

3. If R_4 is the unknown resistance, (R_X), the formula can be rearranged to solve R_X as follows:

FORMULA 6–2 $\qquad R_X = \dfrac{R_2 \times R_3}{R_1}$

4. Look at Figure 6–41 with the values shown to see if the steps described above will work.

5. Verifying results by substituting our answer into the original balanced bridge formula: $\dfrac{R_1}{R_2} = \dfrac{R_3}{R_X} = \dfrac{10}{20} = \dfrac{30}{60} = \dfrac{1}{2}$. (It checks out!)

Some applications for bridge circuits

In industry, there are numerous types of sensors and control circuits used to control the operation of machinery and manufacturing systems. For example, sensors for detecting changes in temperature, flow, pressure, or changes in electrical conditions are all quite common. The output of one of these sensors can be used as one leg of a bridge circuit, with the bridge circuit "zeroed" at the desired reference level for temperature, pressure, flow, and so on, Figure 6–42. Any change in the condition being sensed will then cause the bridge to become unbalanced. This differential at the output terminals of the bridge is then typically used as an input to circuits or devices that eventually translate the change into a readable output indication, such as a digital readout, for monitoring purposes, or a feedback control signal, Figure 6–42. This system condition change can then be managed, either manually, or in many cases automatically, to bring the system back to the desired operational characteristics.

FIGURE 6–42 Bridge circuit used for industrial control

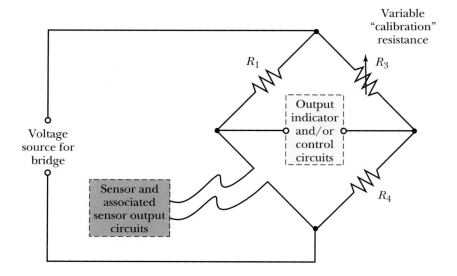

FIGURE 6–43 Bridge circuit used to measure unknown capacitor value

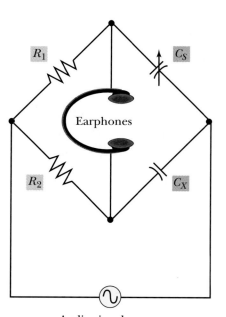

Audio signal source

We have already named one common application of the Wheatstone bridge; that is, measuring the value of an unknown resistance. Variations of the bridge circuit can also be used to measure the values of other types of components, such as capacitors, about which you will learn details in a later chapter. Again, the component to be measured is put into one leg of the bridge, Figure 6–43. When the bridge is balanced, by adjusting the "standard capacitor" (C_S) to the same value as the unknown capacitor's value (C_X), the audio tone in the earphones is "nulled" or minimized.

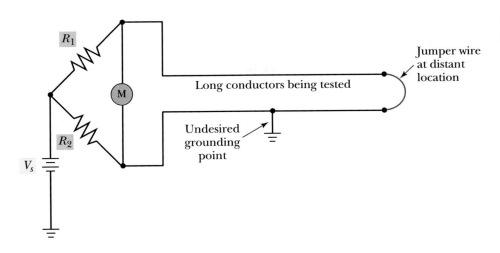

FIGURE 6–44 Example of the Murray loop

Still another application of a modified bridge circuit, which has been used for years by the power and telephone companies, is the Murray loop. The Murray loop is used to locate where an undesired conductor ground has occurred along the miles of lines used to transport power or telephone messages, Figure 6–44. By knowing the resistance per foot of the conductors, and using the outgoing and return wires in a loop that forms part of the bridge circuit, the precise location of the ground fault can be determined. A service person can then be sent to the location to perform the necessary repair.

PRACTICAL NOTES

Because series-parallel circuits are combinational circuits that are connected in an infinite variety of configurations, no single analysis technique is applicable. Rather, it is necessary to break the circuit down into its series and parallel portions for analysis.

In studying this chapter, you have learned the series portions of series-parallel circuits are analyzed using series-circuit concepts, where 1) current is the same through series components and/or circuit portions, and 2) voltage drops on series components are proportional to their resistance ratios.

Also you have learned that the parallel portions of the series-parallel circuits abide by the rules of parallel circuits, where 1) voltage is the same across parallel components or combinations, and 2) current divides in inverse proportion to the parallel branch resistances.

Furthermore, you have observed that the position (location) of a given component or circuit portion in the circuit influences all the parameters throughout the circuit. Therefore, to troubleshoot series-parallel circuits, be sure to isolate individual circuit portions using concepts appropriate to that circuit portion. Then move to the next portion, and so forth, until the faulty circuit portion is found. Isolation techniques are then used in that portion to isolate the specific faulty component or device within that portion.

SUMMARY

▶ A series-parallel circuit is a combination of series and parallel connected components and/or circuit portions.

▶ Series components or circuit portions are analyzed using the rules of series circuits.

▶ Parallel components or circuit portions are analyzed using the rules of parallel circuits.

▶ Points throughout the circuit where current divides and rejoins help identify parallel portions.

▶ Points throughout the circuit that share common voltage also help identify components or circuit portions that are in parallel.

▶ Paths in the circuitry that carry the same current help identify the series circuit elements.

▶ Analysis of total circuit resistance is approached by starting away from the source, then solving and combining results as one works back toward the source. (NOTE: If V_T and I_T are known, this is unnecessary. Use $R_T = V_T/I_T$ formula.)

▶ Analysis of current throughout the circuit is approached by finding the total current value. Then working outward from the source, analyze currents through various circuit portions.

▶ Power dissipations throughout the circuit are analyzed by finding the appropriate I, V, and R parameters. Then use the appropriate power formulas, such as $V \times I$, $I^2 R$, or V^2/R.

▶ A short, or decreased value of resistance, *anywhere* in the circuit causes R_T to decrease and I_T to increase.

▶ An open, or increased value of resistance, *anywhere* in the circuit causes R_T to increase and I_T to decrease.

▶ Two special applications of series-parallel circuits are voltage dividers and bridge circuits used for various purposes.

FORMULAS AND SAMPLE CALCULATOR SEQUENCES

FORMULA 6–1   Bridge balanced when: $\dfrac{R_1}{R_2} = \dfrac{R_3}{R_4}$

R_1 value, ⊝, R_2 value, ⊜, (check for equaling) R_3 value, ⊝, R_4 value, ⊜

FORMULA 6–2 $R_X = \dfrac{R_2 \times R_3}{R_1}$

R_2 value, ⊗, R_3 value, ⊝, R_1 value, ⊜

REVIEW QUESTIONS

1. In your own words, describe the key characteristics of series-parallel circuits.

2. Name two methods of identifying portions of combination (series-parallel) circuits in which components or branches are in parallel.

3. Name two methods of identifying portions of combination (series-parallel) circuits in which components are in series.

4. Describe a generally useful approach for analyzing a combination (series-parallel) circuit in terms of the circuit's total resistance.

5. Describe a generally useful approach for analyzing a combination (series-parallel) circuit in terms of the circuit's current distribution.

6. In series-parallel circuits, the largest resistor value drops the highest voltage.
   a. True all the time, regardless of the resistor's location in the circuit.
   b. False all the time, regardless of the resistor's location in the circuit.
   c. True sometimes, depending on the resistor's location in the circuit.
   d. Never true, no matter the location of the resistor in the circuit.

7. In series-parallel circuits, total power dissipation always equals the sum of all the individual component power dissipations.
   a. True all the time, no matter how many components are involved.
   b. False all the time, because total power equals $V_T \times I_T$.
   c. True only if all the components have the same value.
   d. Only true for the series portions of the circuit.

8. When troubleshooting a series-parallel circuit:
   a. always start with the series portions first when beginning the analysis.

b. always start with the parallel portions first when beginning the analysis.
   c. it doesn't matter where you start to begin the analysis.
   d. break the circuit down into series and parallel portions when beginning the analysis.

9. If a series-parallel-circuit's total circuit resistance has increased, then:
   a. a circuit portion or a component has definitely opened.
   b. a circuit portion or a component has definitely shorted.
   c. a component has decreased in value or a short has occurred in the circuit.
   d. a component has increased in value or an open has occurred in the circuit.

10. Adding a resistor in parallel with any series portion component, or any parallel portion components in a series-parallel circuit will cause the circuit total reistance to:
   a. increase.
   b. decrease.
   c. remain the same.

PROBLEMS

1. Refer to Figure 6–45 and determine R_T.

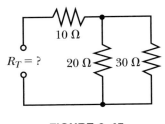

FIGURE 6–45

2. Refer to Figure 6–46 and find:
   I_T _____ V_3 _____ I_2 _____
   V_2 _____ V_5 _____ I_3 _____

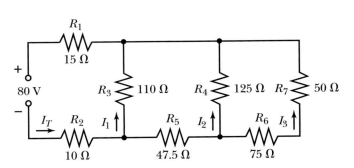

FIGURE 6–46

3. Refer to Figure 6–47 and answer the following with I, for increase; D, for decrease; or RTS, for remains the same. Assume that if R_5 is shorted:

R_T will _____ V_3 will _____
R_4 will _____ P_T will _____
V_1 will _____

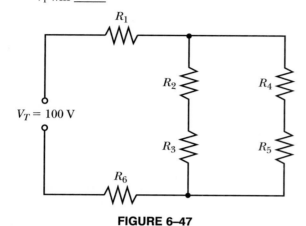

FIGURE 6–47

4. In Figure 6–48, what is the maximum total current possible?

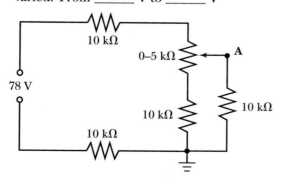

FIGURE 6–48

5. In Figure 6–49, what are the minimum and maximum voltages possible from point A to ground, if the setting of the potentiometer is varied? From _____ V to _____ V

FIGURE 6–49

6. In Figure 6–50, I_T is 10 amperes. Find R_1 and P_{R_2}.

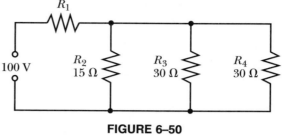

FIGURE 6–50

7. In Figure 6–51, find the value of R_4 and I_2.

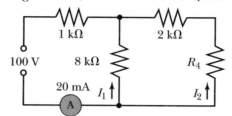

FIGURE 6–51

8. In Figure 6–52, find the resistance between points A and B.

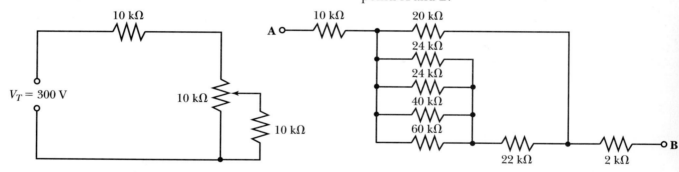

FIGURE 6–52

9. In Figure 6–53, will the light get brighter, dimmer, or remain the same after the switch is closed?

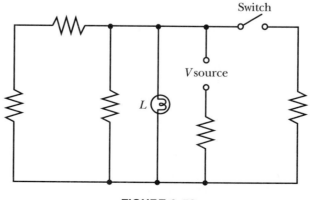

FIGURE 6–53

10. Find the applied voltage in Figure 6–54.

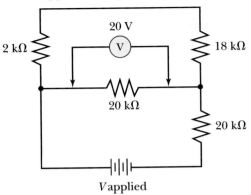

FIGURE 6–54

11. Accurately copy the diagram, Figure 6–55, on a separate sheet of paper. When you are done, draw arrows on the diagram showing the direction of current through each resistor. Using this circuit, assume all resistors have a value of 100 kΩ. Find the following values. (Round answers to whole numbers.)

 a. Find R_T
 b. Find I_T
 c. Find V_{R_1}
 d. Find V_{R_2}
 e. Find V_{R_3}
 f. Find V_{R_4}
 g. Find V_{R_5}
 h. Find V_{R_6}
 i. Find V_{R_7}
 j. Find V_{R_8}

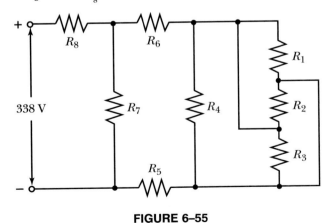

FIGURE 6–55

12. Refer to the circuit in Figure 6–56 and answer the following questions:

 a. If R_1 increases, what happens to the current through R_5?

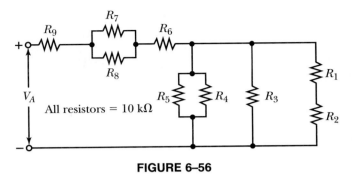

FIGURE 6–56

 b. If R_7 decreases, what happens to the voltage across R_9?
 c. If R_4 shorts, what happens to the voltage across R_2? Across R_8?
 d. If all resistors are 10 kΩ, what is the value of R_T? (All conditions are normal!)
 e. If the circuit applied voltage is 278.5 volts, what is the voltage drop across R_1?
 f. If the circuit applied voltage is reduced to half the value shown in question e, what is the power dissipation of R_9?
 g. If the circuit applied voltage is 557 volts, which resistor(s) dissipate the most power?
 h. Regardless of the applied voltage, which resistor(s) dissipate the least power?
 i. For the conditions defined in question g, what power is dissipated by the resistor dissipating the most power? What power is dissipated by the resistor dissipating the least power?

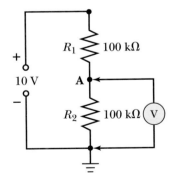

FIGURE 6–57 Circuit for Review Questions 13–21

Refer to Figure 6–57 and answer questions 13–21.

13. Before connecting the voltmeter between point A and ground, what is the voltage across R_1?

14. Before connecting the voltmeter between point A and ground, what is the voltage across R_2?

15. Assume the voltmeter is being used on its 10-volt range and the meter's resistance on that range is 200,000 ohms. When the meter is connected to the circuit to measure the voltage between point A and ground, what voltage will the meter indicate across R_2?

16. For the conditions described in question 15, and based on Kirchhoff's voltage law, what must be the voltage drop across R_1 at that time?

17. Does connecting the meter to the circuit change the circuit's electrical parameters? Explain.

18. Explain the circuit conditions if the voltmeter were connected across R_1 rather than R_2.

19. With the meter connected across just one of the resistors, what is the total circuit current?

20. What is the total circuit current without the meter connected?

21. At what points in the circuit could a meter be connected and not affect the voltage drops across R_1 and R_2?

Refer to Figure 6–58 and answer questions 22–31.

22. Determine the values of R_{L_1}, R_{L_2}, and R_{L_3}.

23. Determine the values for the divider resistors R_1, R_2, and R_3.

24. Would the voltage divider used in this circuit be considered a loaded or unloaded divider?

25. If the required conditions for R_{L_2} were changed so that the desired parameters were 150 V at 10 mA, would the value of R_1 in the divider need to be increased, decreased, or kept the same?

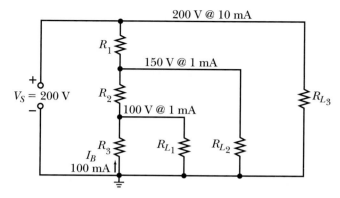

FIGURE 6–58 Circuit for Review Questions 22–31

26. For the conditions described in question 25, what would the new value of R_1 need to be for R_{L_2}'s parameters to be 150 V at 10 mA?

27. If all loads were removed from the circuit and R_1's value is changed to that calculated in question 26, what would the value of divider bleeder current (I_B) become?

28. To maintain a minimum resistor power rating of two times the power the resistor will be dissipating, what should the minimum power rating be for resistor R_3 in the circuit shown in Figure 6–58?

29. To maintain the two times power rating safety margin for resistor R_2 in the circuit of Figure 6–58, what minimum power rating should R_2 have?

30. To maintain the two times power rating safety margin for resistor R_3 in the circuit of Figure 6–58, what minimum power rating should R_3 have?

31. What is the total power being supplied by the power supply in the circuit of Figure 6–58?

ANALYSIS QUESTIONS

Perform appropriate analysis of the circuit shown in Figure 6–59 and answer the following statements with I for increase, D for decrease, or RTS for remain the same:

1. If R_2 decreased in value by a great amount:
   a. The voltage across R_3 would _____
   b. The voltage across R_1 would _____
   c. The current through the load connected in parallel with R_4 would _____

2. If R_2 "opened":
   a. The voltage across the load resistor connected to the bottom of R_1 (the "outer loop" load) would _____
   b. The voltage across R_1 would _____

3. Briefly explain how an industrial sensor's output may be used in conjunction with a bridge circuit to help control industrial processes.

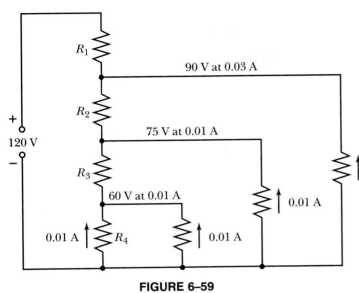

FIGURE 6–59

4. Determine the values of R_1, R_2, R_3, and R_4 for the circuit, Figure 6–59.

5. Determine the value of R_X in the circuit, Figure 6–60: Assume the bridge is balanced by the variable R set at 100 kΩ.

6. For the circuit of Figure 6–60, if R_1 is 47 kΩ, R_2 is 27 kΩ, and the variable R is set at 100 kΩ when the bridge is balanced, what is the value of R_X?

7. For the circuit of Figure 6–60, if the variable R is set at 1.5 kΩ and the resistor in the R_X position in the circuit is 27 kΩ, what must be the ratio of the resistances in the left arm of the bridge circuit?

8. For the circuit of Figure 6–60, assume that the input V is positive toward the top side of the circuit and negative toward the bottom. If the dc meter's negative lead is connected to the top of R_2 and its positive lead is connected to the top of R_X and the meter's pointer is pointing toward the left side of zero, does this indicate that R_2 or R_X has the greater voltage drop?

9. Briefly define the sequence of keystrokes or steps needed to perform calculation of $R_{equivalent}$ of three parallel resistors, using the reciprocal function of a calculator. Assume that you want to use the *engineering mode* of the calculator for this task.

10. Explain how using a voltmeter to measure voltage across part of a series portion of a series-parallel circuit causes circuit conditions to change.

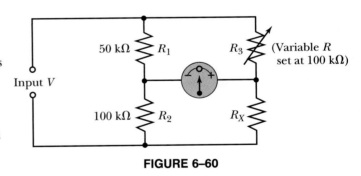

FIGURE 6–60

Suggested performance projects in the Laboratory Manual that correlate with topics in this chapter are:

CHAPTER TOPIC	PERFORMANCE PROJECT	PROJECT NUMBER
Total Resistance in S-P Circuits	Total Resistance in S-P Circuits	21
Current in S-P Circuits	Current in S-P Circuits	22
Voltage in S-P Circuits	Voltage Distribution in S-P Circuits	23
Power in S-P Circuits	Power Distribution in S-P Circuits	24
Effects of Opens in S-P Circuits and Troubleshooting Hints	Effects of an Open in S-P Circuits	25
Effects of Shorts in S-P Circuits and Troubleshooting Hints	Effects of a Short in S-P Circuits	26

NOTE: It is suggested that after completing the above projects, the student should be required to answer the questions in the "Summary" at the end of this section of projects in the Laboratory Manual.

THE SIMPLER SEQUENCE FOR TROUBLESHOOTING

Follow the seven-step SIMPLER troubleshooting sequence, as outlined below, and see if you can find the problem with this circuit. As you follow the sequence, record your circuit test results for each testing step on a separate sheet of paper. This will aid you and your instructor to see where your thinking is on track or where it might deviate from the best procedure.

 Symptoms

—(Gather, verify, and analyze symptom information) To begin this process, read the data under the "Starting Point Information" heading. Particularly look for circuit parameters that are not normal for the circuit configuration and component values given in the schematic diagram. For example, look for currents that are too high or too low; voltages that are too high or too low; resistances that are too high or too low, etc. This analysis should give you a first clue (or symptom information) that will aid you in determining possible areas or components in the circuit that might be causing this symptom.

 Identify

—(Identify and bracket the initial suspect area) To perform this first "bracketing" step, analyze the clue or symptom information from the "Symptoms" step, then either on paper or in your mind, bracket, circle, or put parentheses around all of the circuit area that contains any components, wires, etc., that might cause the abnormality that your symptom information has pointed out. NOTE: Don't bracket parts of the circuit that contain items that could not cause the symptom!

 Make

—(Make a decision about the first test you should make: what type and where) Look under the TEST column and determine which of the available tests shown in that column you think would give you the most meaningful information about the suspect area or components, in the light of the information you have so far.

 Perform

—(Perform the first test you have selected from the TEST column list) To simulate performing the chosen test (as if you were getting test results in an actual circuit of this type), follow the dotted line to the right from your selected test to the number in parentheses. The number in parentheses, under the "Results in Appendix C" column tells you what number to look up in Appendix C to see what the test result is for the test you are simulating.

 Locate

—(Locate and define a new "narrower" area of uncertainty) With the information you have gained from the test, you should be able to eliminate some of the circuit area or circuit components as still being suspect. In essence, you should be able to move one bracket of the bracketed area so that there is a new smaller area of uncertainty.

 Examine

—(Examine available information and determine the next test: what type and where) Use the information you have thus far to determine what your next test should be in the new narrower area of uncertainty. Determine the type test and where, then proceed using the TEST listing to select that test, and the numbers in parentheses and Appendix C data to find the result of that test.

 Repeat

—(Repeat the analysis and testing steps until you find the trouble) When you have determined what you would change or do to restore the circuit to normal operation, you have arrived at *your* solution to the problem. You can check your final result against ours by observing the pictorial step-by-step sample solution on the pages immediately following the "Chapter Troubleshooting Challenge"—circuit and test listing page.

step 8 **Verify**

NOTE: A useful *8ᵗʰ Step* is to operate and test the circuit or system in which you have made the "correction changes" to see if it is operating properly. This is the final proof that you have done a good job of troubleshooting.

CHALLENGE CIRCUIT 3

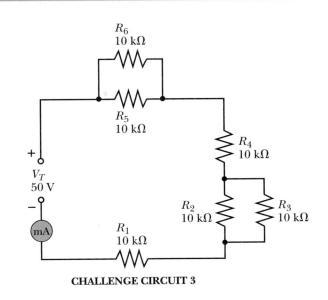

CHALLENGE CIRCUIT 3

Starting Point Information

1. Circuit diagram
2. I_T measures 1.4 mA
3. V_{R_1} measures 14 V
4. Resistances are measured with remainder of circuit being isolated from points being tested.

TEST	Results in Appendix C
V_{R_1}	(49)
V_{R_2}	(80)
V_{R_3}	(12)
V_{R_4}	(39)
V_{R_5}	(61)
V_{R_6}	(88)
Res. of R_1	(25)
Res. of R_2	(54)
Res. of R_3	(93)
Res. of R_4	(37)
Res. of R_5	(68)
Res. of R_6	(33)

step 1 Symptoms
step 2 Identify
step 3 Make
step 4 Perform
step 5 Locate
step 6 Examine
step 7 Repeat
step 8 Verify

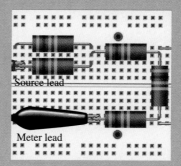

Symptoms

The total current is a little low. Since current should be close to 1.7 mA, R_T is higher than it should be.

Identify

initial suspect area: All resistors are initial suspect areas since if one or more resistors increased in value, the symptom would be present. However, the most likely suspect resistors and associated circuitry are related to R_2 and R_3, plus R_5 and R_6. This is because an open condition for any resistor branch would change the circuit resistance just enough to cause the symptom.

Make

test decision. **1st Test:**   Check the resistance of R_2 (with it isolated).

Perform

Look up the test result. R_2 is 10 kΩ, which is normal.

Locate

new suspect area: R_3, R_5, and R_6.

Examine

available data.

Symptoms

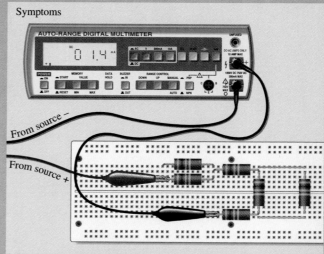

Repeat

analysis and testing:
2nd Test: Measure the resistance of R_3. R_3 is 10 kΩ, which is normal.
3rd Test: Measure the resistance of R_5. R_5 is 10 kΩ, which is normal.
4th Test: Measure the resistance of R_6. R_6 is infinite ohms. NOTE: This meter shows 30.00 MΩ when measuring an open (or an R value from 30 MΩ to ∞).

Verify

5th Test: Replace R_6 and note the circuit parameters. When this is done, the circuit operates properly.

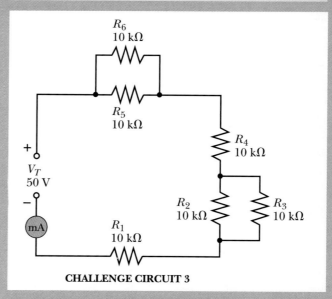

CHALLENGE CIRCUIT 3

1st Test

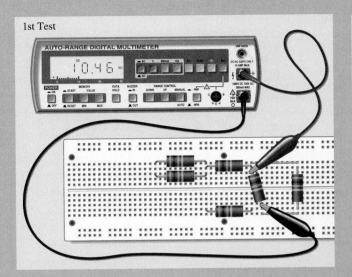

2nd Test

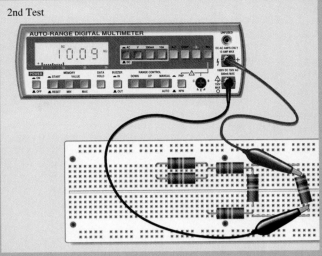

3rd Test

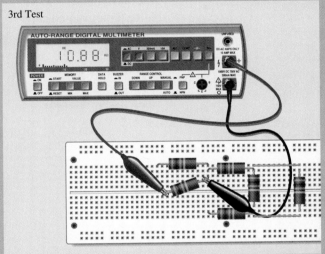

4th Test

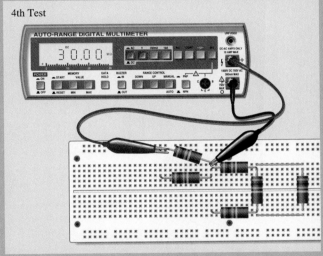

5th Test

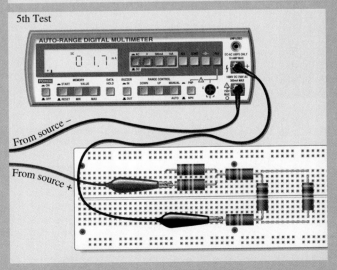

From source –

From source +

SPECIAL NOTE: Another way to check is to lift one end of each resistor and observe whether total current changed. The resistor you lift without having the total current change is the open one, since opening an open does not change the circuit conditions.

KEY TERMS

Bilateral resistance
Constant current source
Constant voltage source
Efficiency
Impedance
Linear network
Maximum power transfer theorem
Multiple-source circuit
Network
Norton's theorem
Steady-state condition
Superposition theorem
Thevenin's theorem

BASIC NETWORK THEOREMS

7

chapter

CHAPTER PREVIEW

In this chapter, several commonly used electrical network theorems will be discussed. These include **maximum power transfer, superposition,** and **Thevenin's** and **Norton's** theorems. For our purposes, a network is defined as a combination of components that are electrically connected. Theorems are ideas or statements that are not obvious at first but that we can prove using some more accepted premises. When the theorems are proved this way, they also become laws. The theorems you will learn about in this chapter can be proven in theory and in practice.

The theorems in this chapter deal with **bilateral resistances** that are in **linear networks** under **steady-state conditions,** not during transient switch on or off moments.

Using the maximum power transfer theorem, you can know what value of dc circuit load resistance or ac circuit load impedance allows maximum transfer of power from a source to its load (e.g., what speaker impedance in your stereo allows transfer of maximum power from amplifier to speaker). Using the superposition theorem, you can readily analyze circuits having more than one source.

One of the advantages in using Thevenin's theorem is once the Thevenin equivalent circuit has been determined, you can easily predict electrical parameters for a variety of load resistances (R_L) without having to recalculate the whole network each time.

Norton's theorem provides another approach to simplify complex networks and make circuit parameter analysis easier.

OBJECTIVES

After studying this chapter, you should be able to:

1. State the **maximum power transfer theorem**

2. Determine the R_L value needed for maximum power transfer in a given circuit

3. State the **superposition theorem**

4. Solve circuit parameters for a circuit having more than one source

5. State **Thevenin's theorem**

6. Determine V_L and I_L for various values of R_L connected across specified points in a given circuit or network

7. State **Norton's theorem**

8. Apply Norton's theorem in solving specified problems

9. Convert between Norton and Thevenin equivalent parameters

7–1 SOME IMPORTANT TERMS

In the Chapter Preview we mentioned several terms that should be defined at this point. We indicated that the theorems you will study in this chapter relate to circuits that are linear networks having bilateral resistances that are being operated under steady-state conditions. These are rather high-sounding words, but their meanings are quite simple. Let's briefly look at definitions for each term:

- **Bilateral resistance**—a resistance having equal resistance in either direction. That is, the resistance is the same for current passing either way through the component. This type of component produces the same voltage drop across it, no matter which direction the current flows through it.

- **Linear network**—a circuit whose electrical behavior does not change with different voltage or current levels or values.

- **Steady-state condition**—the condition where circuit values and conditions are stable or constant. The opposite of transient conditions, such as switch-on or switch-off conditions, when values are changing from one state or level to another.

7–2 MAXIMUM POWER TRANSFER THEOREM

In many electronic circuits and systems, it is important to have maximum transfer of power from the source to the load. For example, in radio or TV transmitting systems, it is desired to transfer the maximum power possible from the last stage of the transmitter to the antenna system. In your high fidelity amplifier, you want maximum power transferred from the last amplifier stage to the speaker system. This is usually accomplished by proper "**impedance** matching," which means the load resistance or impedance matches the source resistance or impedance. NOTE: In dc circuits, such as you have been studying, the term for opposition to current flow is *resistance.* In ac circuits, which you will soon study, the term for opposition to current flow is impedance.

If it were possible to have a voltage source with zero internal resistance but able to maintain constant output voltage (even if a load resistance of zero ohms were placed across its output terminals), the source theoretically could supply infinite power to the load, where $P = V \times I$. In this case, V would be some value and I would be infinite. Thus, $P = V$ value $\times$ infinite current = infinite power in watts.

Since these conditions are impossible, the practical condition is that there is an optimum value of R_L where maximum power is delivered to the load. Any higher or lower value of R_L causes less power to be delivered to the load.

The basic theorem

The **maximum power transfer** theorem states maximum power is transferred from the source to the load (R_L) when the resistance of the load *equals* the resistance of the source. In other words, it is a series circuit. In practical circuits, the internal resistance (r_{int}) of the source causes some of the source's power to be dissipated within the source, while the remaining power is delivered to the external circuitry, Figure 7–1.

FIGURE 7–1 Maximum power transfer theorem (NOTE: R_T seen by source is the series total of $r_{int} + R_L$.)

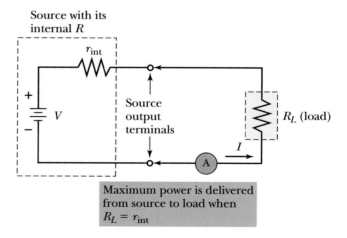

Source with its internal R

r_{int}

Source output terminals

R_L (load)

$+$ V $-$

I

A

Maximum power is delivered from source to load when $R_L = r_{int}$

PRACTICAL NOTES

It is important to realize that all power sources have some internal resistance, for which we are using the designation r_{int} in our discussions. We are showing you the r_{int} as an internal source series resistance, which alters the source output voltage delivered to the load, and limits the maximum amount of current that can be delivered by the source, Figure 7–1.

Refer to the circuits in Figure 7–2a, b, and c. Assuming the same source for all three circuits, notice the power delivered to the load in each case.

Observe in the first circuit, Figure 7–2a, the load resistance (R_L) equals the internal resistance of the source (r_{int}). The power delivered to the load is 10 watts.

In the circuit of Figure 7–2b, the load resistance is less than the source's internal resistance. The power delivered to the load is 8.84 watts.

In the third circuit, Figure 7–2c, R_L is greater than r_{int}, and the power delivered to the load is 7.5 watts. You can make calculations for a large number of R_L values to confirm that maximum power is delivered when $R_L = r_{int}$.

Efficiency factor

FORMULA 7–1   Efficiency (%) $= \dfrac{P_{out}}{P_{in}} \times 100$

If you calculate the **efficiency** factor, based on what percentage of total power generated by the source is delivered to the load, you will find efficiency equals 50% when R_L equals r_{int}.

In Figure 7–2a, $V = 20$ V and $I = 1$ A. Thus, P_T (or P_{in}) $= 20$ V $\times 1$ A $= 20$ W. The load is dissipating $(1 \text{ A})^2 (10 \ \Omega) = 10$ W, or half of the 20 watts.

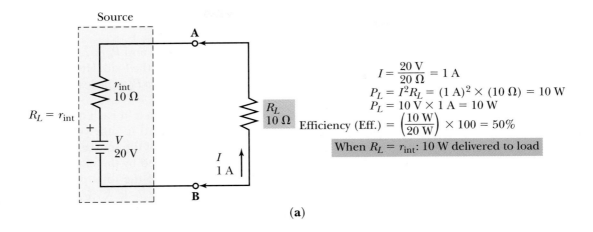

$$I = \frac{20 \text{ V}}{20 \text{ }\Omega} = 1 \text{ A}$$
$$P_L = I^2 R_L = (1 \text{ A})^2 \times (10 \text{ }\Omega) = 10 \text{ W}$$
$$P_L = 10 \text{ V} \times 1 \text{ A} = 10 \text{ W}$$
Efficiency (Eff.) $= \left(\frac{10 \text{ W}}{20 \text{ W}}\right) \times 100 = 50\%$

When $R_L = r_{int}$: 10 W delivered to load

(a)

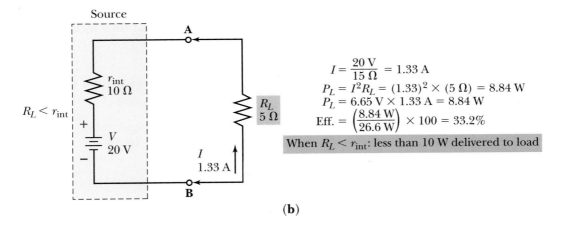

$$I = \frac{20 \text{ V}}{15 \text{ }\Omega} = 1.33 \text{ A}$$
$$P_L = I^2 R_L = (1.33)^2 \times (5 \text{ }\Omega) = 8.84 \text{ W}$$
$$P_L = 6.65 \text{ V} \times 1.33 \text{ A} = 8.84 \text{ W}$$
Eff. $= \left(\frac{8.84 \text{ W}}{26.6 \text{ W}}\right) \times 100 = 33.2\%$

When $R_L < r_{int}$: less than 10 W delivered to load

(b)

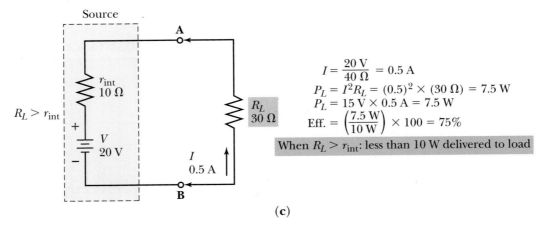

$$I = \frac{20 \text{ V}}{40 \text{ }\Omega} = 0.5 \text{ A}$$
$$P_L = I^2 R_L = (0.5)^2 \times (30 \text{ }\Omega) = 7.5 \text{ W}$$
$$P_L = 15 \text{ V} \times 0.5 \text{ A} = 7.5 \text{ W}$$
Eff. $= \left(\frac{7.5 \text{ W}}{10 \text{ W}}\right) \times 100 = 75\%$

When $R_L > r_{int}$: less than 10 W delivered to load

(c)

FIGURE 7–2 Verification of maximum power transfer theorem

In Figure 7–2b, $V = 20$ V and $I = 1.33$ A. Thus, $P_T = 20$ V $\times$ 1.33 A = 26.6 W, but only 8.84 W is delivered to the load. Therefore,

$$\text{Efficiency} = \left(\frac{8.84}{26.6}\right) \times 100 = 33.2\%$$

On the other hand, in Figure 7–2c, $P_T = 10$ W and $P_L = 7.5$ W. Thus, 75% of the power generated by the source is delivered to the load. However, the smaller value of total power produced by the source causes P_L to be less than when there was 50% efficiency (R_L matched r_{int}).

Where is the remainder of the power consumed in each case? If you said by the internal resistance of the source, you are right! In other words, when there is maximum power transfer, half the total power of the source is dissipated by its own internal resistance; the other half is dissipated by the external load. When the efficiency is about 33%, Figure 7–2b, 67% of the total power is dissipated within the source. When efficiency is 75%, Figure 7–2c, 25% of the total power is dissipated within the source, while 75% is delivered to the load.

Summary of the maximum power transfer theorem

- Maximum power transfer occurs when the load resistance equals the source resistance.
- Efficiency at maximum power transfer is 50%.
- When R_L is greater in value than r_{int}, efficiency is greater than 50%.
- When R_L is less than r_{int}, efficiency is less than 50%.

To visually see the effects on load power and efficiency when R_L is varied, using a given source, observe the graphs in Figures 7–3 and 7–4.

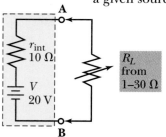

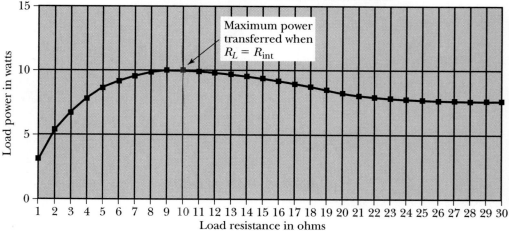

FIGURE 7–3

PRACTICAL NOTES

You may or may not have noticed the amount of voltage at the output terminals of the source varies with load current. This is due to the internal *IR* drop across r_{int}. That is, higher currents result in greater internal *IR* drop and less voltage available at the output terminals of the source. In fact, one way to obtain an idea of a source's internal resistance is to measure its open circuit output terminal voltage, then put a load that demands current from the source across its terminals. The change in output terminal voltage from no-load (zero current drain) to loaded conditions allows calculation of r_{int}.

FORMULA 7–2 $$r_{int} = \frac{V \text{ (no load)} - V \text{ (loaded)}}{I \text{(load)}}$$

If V (no-load) is 13 V and V (loaded) is 12 V with 100 A (load) current, then:

$$\frac{13 \text{ V} - 12 \text{ V}}{100 \text{ A}} = 0.01 \ \Omega$$

A practical example of this approach to determine if a source's internal resistance is higher than it should be is the automotive technician's technique of checking a battery's output voltage at no-load, then under heavy load. If the

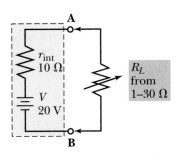

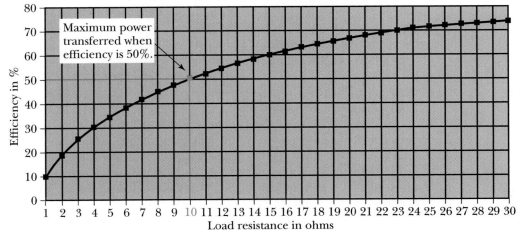

FIGURE 7–4

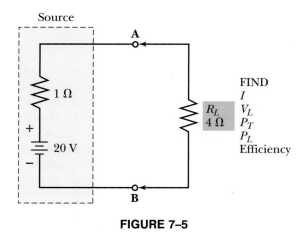

FIGURE 7–5

terminal voltage of the battery drops too much under load, its internal resistance is too high, and the battery is either defective or needs charging.

PRACTICE PROBLEMS I

Try solving the maximum power transfer problems in Figure 7–5.

7–3 SUPERPOSITION THEOREM

Often, you will analyze circuitry having more than one source. The superposition theorem is a useful tool in these cases. The **superposition theorem** states that in linear circuits having more than one source, the voltage across or current through any given element equals the algebraic sum of voltages or currents produced by each source acting alone with the other sources disabled, Figure 7–6. NOTE: To disable voltage sources for purposes of calculations, consider them shorted. To disable current sources, consider them opened.

EXAMPLE

A circuit containing two voltage sources—**a multiple-source circuit**—is illustrated in Figure 7–6a. In Figure 7–6b through 7–6d, the step-by-step application of the superposition theorem is shown. *Important:* **Noting the direction of current flow and polarity of voltage drops is important for each step.** This allows you to find the actual resultant current through a given component, or the resultant voltage drop of that component by adding or subtracting currents or voltages calculated from each source, as appropriate. When currents through a given component from the individual sources pass through that component in the same direction, the resultant current is calculated as the sum of the two individually computed currents. If currents pass through that component in opposite directions, subtract them. When computed voltage drops across a given component from each source are the same polarity, the resultant voltage equals the sum. If the sources cause opposite polarity voltage drops across the component, subtract them to find the resultant voltage drop across that component.

To get proper results from the final algebraic additions, use the same *reference point* in every case to determine polarity of voltage drop across a given component. Then you can combine the individual results properly to get the final resultant value.

In Figure 7–6b, the 60-V source is considered disabled (shorted out). Notice this effectively places R_3 in parallel with R_2. This means the R_1 value of 2 kΩ is in series with the parallel combination of $R_2 - R_3$ across the 30-V source. Thus, total R is 3 kΩ. Total I is 10 mA. R_1 passes the total 10 mA, whereas, the 10 mA is equally divided through R_2 and R_3 with 5 mA through each one. Using Ohm's law, the voltages are computed as: $V_1 = +20$ V with respect to point A; $V_2 = -10$ V with respect to point A; and $V_3 = +10$ V with respect to point B.

In Figure 7–6c, the 30-V source is considered disabled. In this case, R_1 and R_2 are in parallel with each other and in series with R_3 across the 60-V source. Notice the voltage drops are such that $V_1 = +20$ V with respect to point A; $V_2 = +20$ V with respect to point A; and $V_3 = +40$ V with respect to point B.

Circuit with two sources

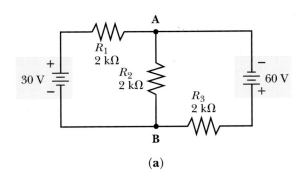

(a)

Assuming 60-V source shorted out

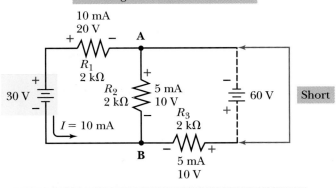

NOTE: If you were going to physically do this, you would *remove* the 60-V source and replace it with a conductor.

Step 1: Assume 60-V source is replaced by a short (0 Ω).
Step 2: Analyze R_T of circuit:

  a. $R_2 \| R_3 = \dfrac{2\text{ k}\Omega \times 2\text{ k}\Omega}{2\text{ k}\Omega + 2\text{ k}\Omega} = \dfrac{4 \times 10^6}{4 \times 10^3} = 1\text{ k}\Omega$

  b. R_e of $R_2 \| R_3$ is in series with R_1
  $R_T = R_1 + R_e = 2\text{ k}\Omega + 1\text{ k}\Omega = 3\text{ k}\Omega$

Step 3: Solve for total current and branch currents:

  a. $I_T = \dfrac{30\text{ V}}{3\text{ k}\Omega} = 10\text{ mA}$

  b. Since each branch = 2 kΩ, current divides equally:
  $I_{R_2} = 5\text{ mA}; I_{R_3} = 5\text{ mA}$

Step 4: Use Ohm's law and solve the voltage drops:

  a. $V_1 = I_T \times R_1 = 10\text{ mA} \times 2\text{ k}\Omega = +20$ V (with respect to point **A**)

  b. $V_2 = I_{R_2} \times R_2 = 5\text{ mA} \times 2\text{ k}\Omega = -10$ V (with respect to point **A**)

  c. $V_3 = I_{R_3} \times R_3 = 5\text{ mA} \times 2\text{ k}\Omega = +10$ V (with respect to point **B**)

(b)

FIGURE 7–6 Verification of superposition theorem

Assuming 30-V source shorted out

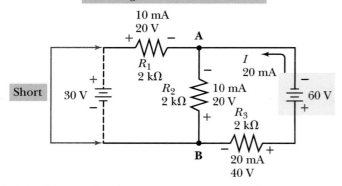

Resultant circuit Vs and Is with two sources

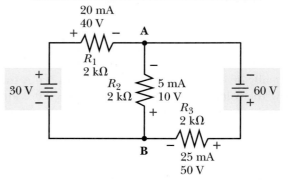

NOTE: To physically perform this step, you would *remove* the 30-V source and replace it with a "short."

Step 5: Assume 30-V source is replaced by a short (0 Ω).
Step 6: Analyze R_T of circuit:

a. $R_1 \| R_2 = \dfrac{2\ k\Omega \times 2\ k\Omega}{2\ k\Omega + 2\ k\Omega} = \dfrac{4 \times 10^6}{4 \times 10^3} = 1\ k\Omega$

b. R_e of $R_1 \| R_2$ is in series with R_3

c. $R_T = R_3 + R_e = 2\ k\Omega + 1\ k\Omega = 3\ k\Omega$

Step 7: Solve for total current and branch currents:

a. $I_T = \dfrac{60\ V}{3\ k\Omega} = 20\ mA$

b. Since each branch = 2 kΩ, current divides equally:
$I_{R_2} = 10\ mA;\ I_{R_3} = 10\ mA$

Step 8: Use Ohm's law and solve the voltage drops:

a. $V_1 = I_1 \times R_1 = 10\ mA \times 2\ k\Omega = +20\ V$ (with respect to point **A**)

b. $V_2 = I_{R_2} \times R_2 = 10\ mA \times 2\ k\Omega = +20\ V$ (with respect to point **A**)

c. $V_3 = I_T \times R_3 = 20\ mA \times 2\ k\Omega = +40\ V$ (with respect to point **B**)

(c)

Step 9: Algebraically add voltages and currents from the two sets of computations to find resultant component electrical parameters:

a. $V_1 = +20\ V + 20\ V = +40\ V$ (with respect to point **A**)

b. $V_2 = -10\ V + 20\ V = +10\ V$ (with respect to point **A**)

c. $V_3 = +10\ V + 40\ V = +50\ V$ (with respect to point **B**)

d. I_{R_1} resultant current = 20 mA (away from point **A**)

e. I_{R_2} resultant current = 5 mA (away from point **A**)

f. I_{R_3} resultant current = 25 mA (away from point **B**)

(d)

FIGURE 7–6 Continued

In Figure 7–6d, you see the algebraically computed results. Recall V_1 is +20 V with respect to A with the 60-V source disabled and +20 V with the 30-V source disabled. This means the voltages are additive and the resultant voltage across R_1 equals +40 V with respect to point A. This same type addition also determines the resultant value of V_2 equals +10 V with respect to point A, and the resultant value of V_3 equals +50 V with respect to point B.

In a similar manner, directions and values of the currents through each component are determined, using the algebraic summation technique. For example, note in Figure 7–6b, current flows from point B toward point A through resistor R_2 and has a value of 5 mA. In Figure 7–6c, current flows from point A toward point B through R_2 and has a value of 10 mA. This means the resultant current through R_2 with both sources active is 5 mA traveling from point A toward point B. You get the idea! Refer to Figure 7–6d to see the other resultant values, as appropriate.

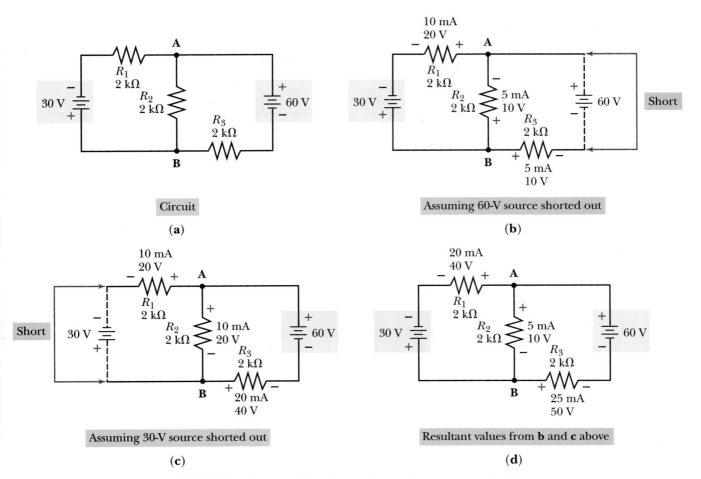

FIGURE 7–7 Superposition theorem example: reversed polarities

---

EXAMPLE

If the polarity of each of the sources in Figure 7–6 is reversed (while the circuit configuration remains the same), the results are like those in Figure 7–7.

The same analysis is used. That is, disable each voltage source, one at a time by viewing them as shorted. Then determine the currents and voltages throughout the circuit, using Ohm's law and so on.

With the 60-V source assumed as shorted, V_1 equals -20 V with respect to point A; V_2 equals $+10$ V with respect to point A; and V_3 equals -10 V with respect to point B, Figure 7–7b.

With the 30-V source assumed as shorted, V_1 equals -20 V with respect to point A; V_2 equals -20 V with respect to point A; and V_3 equals -40 V with respect to point B, Figure 7–7c.

Combining the results of the two assumed conditions, V_1 equals -40 V with respect to point A; V_2 equals -10 V with respect to point A; and V_3 equals -50 V with respect to point B, Figure 7–7d.

As you can see, when only the polarity of the sources changes, the resultant direction of currents and polarities of the voltage drops reverses but their values remain the same. Compare Figures 7–6d and 7–7d.

Summary of the superposition theorem

- Ohm's law is used to analyze the circuit, using one source at a time.
- Then the final results are determined by algebraically superimposing the results of all the sources involved.

PRACTICE PROBLEMS II

For practice in using the superposition theorem, try the following problems:

1. Solve for the resultant current(s) and voltage(s) for the circuit in Figure 7–8.
2. Assume only the 150-V source is reversed in polarity and solve for the resultant current(s) and voltage(s). **Caution: Only one source is being reversed!**

IN-PROCESS LEARNING CHECK I

Fill in the blanks as appropriate.

1. For maximum power transfer to occur between source and load, the load resistance should _____ the source resistance.
2. The higher the efficiency of power transfer from source to load, the _____ the percentage of total power is dissipated by the load.
3. Maximum power transfer occurs at _____ % efficiency.
4. If the load resistance is less than the source resistance, efficiency is _____ than the efficiency at maximum power transfer.
5. To analyze a circuit having two sources, the superposition theorem indicates that Ohm's law (can, cannot) _____ be used.
6. What are two key observations needed in using the superposition theorem to analyze a circuit with more than one source? _____

   _____

7. Using the superposition theorem, if the sources are considered "voltage" sources, are these sources considered shorted or opened during the analysis process? _____
8. When using the superposition theorem when determining the final result of your analysis, the calculated parameters are combined, or superimposed, (arithmetically, algebraically) _____ .

7–4 THEVENIN'S THEOREM

Thevenin's theorem is used to simplify complex **networks** so the calculation of voltage(s) across any two given points in the circuit (and consequently, current[s] flowing between the two selected points) is easily determined.

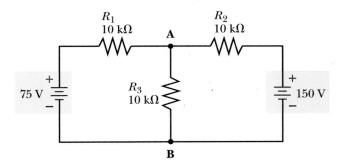

FIND
I through R_1
V across R_1, with respect to Point **A**
I through R_2
V across R_2, with respect to Point **A**
I through R_3
V across R_3, with respect to Point **A**

FIGURE 7–8

Once the Thevenin equivalent circuit parameters are known, solving circuit parameters for *any value* of R_L merely involves solving a two-resistor series circuit problem. A **constant voltage source** is assumed. This is a source whose output voltage remains constant under varying load current (or current demand) conditions.

As you recall, voltage distribution was considered when you studied series circuits. Thevenin's theorem deals with a simple equivalent series circuit to simplify voltage distribution analysis. Let's see how this is applied.

The theorem defined

One version of **Thevenin's theorem** states that any **linear** two-terminal **network** (of resistances and source[s]) can be replaced by a simplified equivalent circuit consisting of a single voltage source (called V_{TH}) and a single series resistance (called R_{TH}), Figure 7–9.

V_{TH} is the open-terminal voltage felt across the two load terminals (A and B) with the load resistance removed. R_{TH} is the resistance seen when looking back into the circuit from the two load terminals (without R_L connected) with voltage sources replaced by their internal resistance. For simplicity, we assume them shorted.

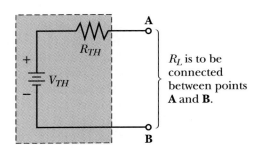

FIGURE 7–9 Thevenin's theorem equivalent circuit

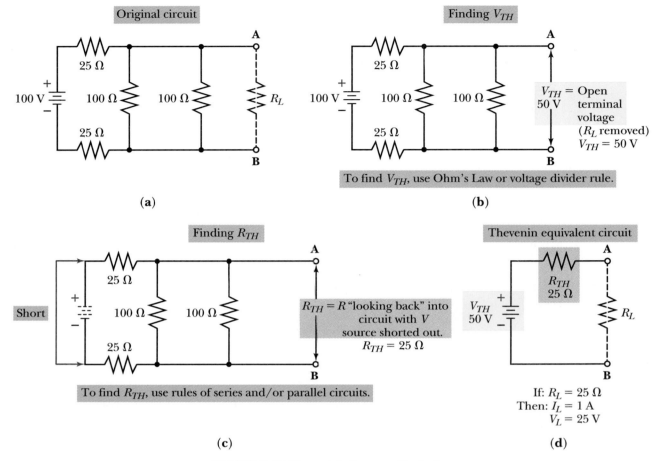

FIGURE 7–10 Thevenin's theorem illustration

Refer to Figure 7–10a, b, c, and d. Figure 7–10a shows the original circuit from which we will determine the Thevenin equivalent circuit, a process called "Thevenizing" the circuit.

Figure 7–10b illustrates how the value for V_{TH} is determined. Notice R_L is considered opened (between points A and B). V_{TH} is considered the value of voltage appearing at network terminals A and B (with R_L removed). For example, if Ohm's law is used, R_T is 100 Ω; V_T is 100 V; thus $I_T = 1$ A. Each of the 25-Ω series resistors drops 25 V (1 A × 25 Ω). The 1 A evenly splits between the two 100-Ω parallel resistors. Thus, voltage between points A and B = 0.5 A × 100 Ω, or 50 V. Therefore, $V_{TH} = 50$ V.

Figure 7–10c shows the process to find R_{TH}. Mentally short the source and determine the circuit resistance by looking toward the source from the two network points A and B. With the source shorted, the two 100-Ω parallel resistors are effectively in parallel with 50 Ω (25 Ω + 25 Ω). The equivalent resistance of two 100-Ω resistors is 50 Ω, and that 50 Ω in parallel with the remaining 50-Ω branch yields an R_T of 25 Ω. This is the value of R_{TH}.

Figure 7–10d shows the Thevenin equivalent circuit of a 50-V source (V_{TH}) and the series R_{TH} of 25 Ω. It is thus possible to determine the current through any value of R_L and the voltage across R_L by analyzing a simple two-resistor series circuit.

EXAMPLE

If R_L has a value of 25 Ω, what are the values of I_L and of V_L? If you answered 1 A and 25 V, you are correct!

Using formula 7–3, solve this way:

$$I_L = \frac{50 \text{ V}}{25 \text{ Ω} + 25 \text{ Ω}} = \frac{50 \text{ V}}{50 \text{ Ω}} = 1 \text{ A}$$

FORMULA 7–3 $I_L = \dfrac{V_{TH}}{R_{TH} + R_L}$

V_L is calculated by using Ohm's law ($I_L \times R_L$), or the voltage divider rule where:

FORMULA 7–4 $V_L = \dfrac{R_L}{R_L + R_{TH}} \times V_{TH}$

Using formula 7–4, solve this way:

$$V_L = \frac{25 \text{ Ω}}{25 \text{ Ω} + 25 \text{ Ω}} \times 50 \text{ V} = \frac{25 \text{ Ω}}{50 \text{ Ω}} \times 50 \text{ V} = 0.5 \times 50 \text{ V} = 25 \text{ V}$$

Summary of Thevenin's theorem

The Thevenin equivalent circuit consists of a source, V_{TH} (often called the Thevenin generator), and a single resistance, R_{TH} (often called the Thevenin internal resistance), in series with the load (R_L), which is across the specified network terminals. Therefore, this equivalent circuit is useful to simplify calculations of electrical parameters for various values of R_L since it reduces the problem to a simple two-resistor series circuit analysis.

Example

Let's assume that we are to put a load resistance (R_L) across the output terminals (points A and B) of a Wheatstone bridge circuit, Figure 7–11. Further, we would like to determine the V and I parameters associated with the load resistor. By using Thevenin's theory, we can define the simple Thevenin equivalent circuit for the bridge. Once that is found, simple two-resistor series circuit analysis techniques make it easy to find the V and I parameters for any given R_L value.

Refer to Figure 7–11 as you study the following steps.

1. Find V_{TH} (the difference of potential between points A and B, with the load resistor removed):

$$V_{TH} = V_A - V_B = \left(\frac{R_2}{R_1 + R_2}\right) \times V_S - \left(\frac{R_4}{R_3 + R_4}\right) \times V_S$$

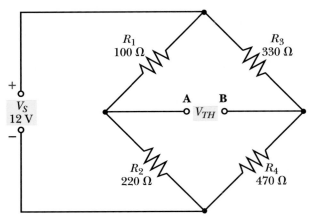

$$V_{TH} = V_A - V_B,$$

where (using the voltage divider method): $V_A = \left(\dfrac{R_2}{R_1 + R_2}\right) \times V_S$

$$V_B = \left(\dfrac{R_4}{R_3 + R_4}\right) \times V_S$$

$$V_{TH} = 1.2 \text{ V}$$

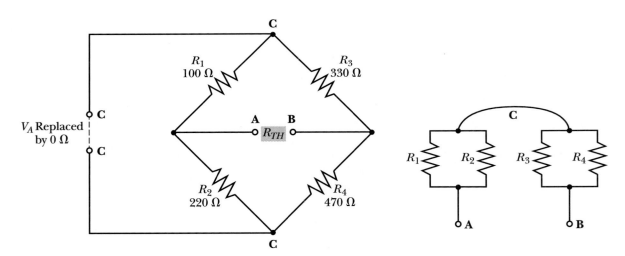

FIGURE 7–11 Analyzing a bridge circuit with Thevenin's theorem

$$V_{TH} = \left(\frac{220 \text{ }\Omega}{320 \text{ }\Omega}\right) \times 12 \text{ V} - \left(\frac{470 \text{ }\Omega}{800 \text{ }\Omega}\right) \times 12 \text{ V}$$

$$V_{TH} = 8.25 \text{ V} - 7.05 \text{ V} = 1.2 \text{ V}$$

2. Find R_{TH} by assuming source to be zero ohms:

$$R_{TH} = R_e \text{ of } (R_1 \parallel R_2) + R_e \text{ of } (R_3 \parallel R_4) \cong 263 \text{ }\Omega, \text{ as shown below.}$$

$$R_{TH} = \frac{R_1 \times R_2}{R_1 + R_2} + \frac{R_3 \times R_4}{R_3 + R_4}$$

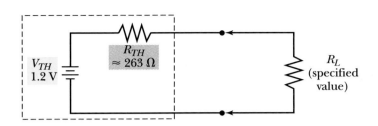

FIGURE 7–12 Thevenin equivalent circuit

$$R_{TH} = \frac{100\ \Omega \times 220\ \Omega}{100\ \Omega + 220\ \Omega} + \frac{330\ \Omega \times 470\ \Omega}{330\ \Omega + 470\ \Omega}$$

$$R_{TH} = 68.75\ \Omega + 193.875\ \Omega = 262.625\ \Omega$$

3. Draw the equivalent circuit. See Figure 7–12.

Let's assume that an R_L value of 270 Ω is to be connected between points A and B of the bridge circuit. What is the value of V_{R_L} and of I_L?

$$V_{R_L} = \left(\frac{R_L}{R_{TH} + R_L}\right) \times V_{TH} = \left(\frac{270\ \Omega}{263\ \Omega + 270\ \Omega}\right) \times 1.2\ \text{V} \approx 0.61\ \text{V}$$

$$I_L = \frac{V_{TH}}{R_T} = \frac{1.2\ \text{V}}{533\ \Omega} = 2.25\ \text{mA}$$

PRACTICAL NOTES

In a practical circuit, you can remove the load resistor (R_L) and measure the open-terminal voltage at those two points with a voltmeter to find V_{TH}. Then you can remove the power supply from the circuit and short the points where its output terminals are connected. Next, you can use an ohmmeter to measure the circuit resistance (at the open-terminal points) by looking back toward where the source points are shorted. Thus, the measured value of R is R_{TH}.

As you have seen, one example of a practical application of Thevenin's theorem is in analyzing an unbalanced bridge circuit. Try applying your Thevenin's theorem knowledge to the following practice problems.

PRACTICE PROBLEMS III

1. Refer again to Figure 7–10. What are the values of I_L and V_L, if R_L is 175 Ω?

2. In Figure 7–10, what value of R_L must be present if I_L is 125 mA?

3. Assume that the parameters in our sample bridge circuit, Figure 7–11, were changed so that $V_A = 18$ V, $R_1 = 220\ \Omega$, $R_2 = 330\ \Omega$, $R_3 = 470\ \Omega$, and $R_4 = 560\ \Omega$. If R_L is 470 Ω, what are the values of I_L and V_{R_L}?

4. In Figure 7–13, what are the values of V_{TH} and R_{TH}? What are the values of I_L and V_L, if R_L equals 16 kΩ?

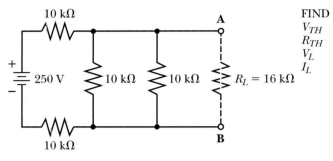

FIGURE 7–13

7–5 NORTON'S THEOREM

Norton's theorem is another method to simplify complex networks into manageable equivalent circuits for analysis. Again, the idea is to devise the equivalent circuit so parameter analysis at two specified network terminals (where R_L is connected) is easy.

Thevenin's theorem uses a voltage-related approach with a voltage source/series resistance equivalent circuit where voltage divider analysis is applied. On the other hand, Norton's theorem uses a current-related approach with a current source/shunt resistance equivalent circuit where a current-divider analysis approach is applied.

The theorem defined

Norton's theorem says any linear two-terminal network can be replaced by an equivalent circuit consisting of a single (**constant**) **current source** (called I_N) and a single shunt (parallel) resistance (called R_N), Figure 7–14.

Refer to Figure 7–15, points A and B. I_N is the current available from the Norton equivalent current source. I_N is determined as the current flowing between the load resistance terminals, if R_L were shorted (i.e., if R from points A to B is zero ohms). R_N is the resistance of the Norton equivalent shunt resistance. This value is computed as the R value seen when looking into the network with the load terminals open (R_L not present) and sources of *voltage* replaced by their r_{int}, Figure 7–16.

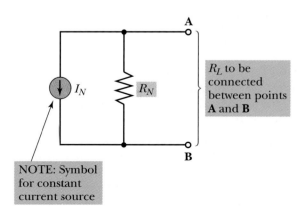

FIGURE 7–14 Norton's
theorem equivalent circuit

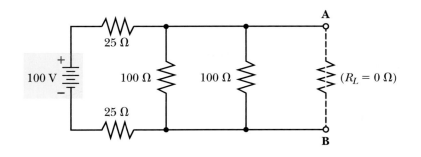

FIGURE 7–15 Norton's I_N = Current from A to B, if R_L shorted (0 Ω). (In this case, I_N would equal 100 V/50 Ω = 2 A.)

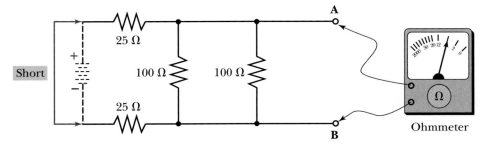

FIGURE 7–16 Norton's R_N = R_{A-B} with R_L removed and source considered as a short. (In this case, R_N = 25 Ω.)

This method is the same one used to calculate R_{TH} in Thevenin's theorem. In fact, the R_{TH} and R_N values are the same. The only difference is that in Norton's theorem, this resistance value is considered in parallel (or shunt) with the I_N equivalent current source. In Thevenin's theorem, this resistance value is considered in series with the V_{TH} equivalent voltage source.

Observe the similarities and differences between the two types of equivalent circuits in Figure 7–17b and 7–17c. Note these equivalent circuits are derived from the same original circuit used to explain Thevenin's theorem earlier in the chapter. (See Figure 7–10a [see page 250]. Also see Figure 7–17a.)

Correlating the prior thinking of the Thevenin analysis to the Norton analysis, let's assume the same 25-Ω R_L (load resistance) used in Figure 7–10d, and apply the Norton equivalent circuit approach.

First, let's see how the Norton equivalent circuit parameters, Figure 7–17c, are derived. Analysis starts with the original circuit, Figure 7–18a (see page 257).

I_N equals current between points A and B with R_L shorted. This short also shorts the two 100-Ω parallel resistors, so the source sees a total circuit R of 50 Ω (the two 25-Ω resistors in series, through the short). Thus, current between points A and B is 100 V/50 Ω, or 2 A (I_N = 2 A), Figure 7–18b (see page 257).

R_N equals R looking back into the network with points A and B open and with *voltage* source(s) shorted. R_N thus equals the equivalent resistance of the two 100-ohm parallel resistors (or 50 Ω) in parallel with the two 25-Ω series resistors' branch (or 50 Ω). This yields a resultant circuit equivalent resistance of 25 Ω (R_N = 25 Ω), again see Figure 7–18b.

Now, refer to Figure 7–18c to see how this equivalent circuit is used to determine electrical parameters associated with various values of R_L.

If an R_L value of 25 Ω is placed from terminal A to B, it is apparent the Norton equivalent circuit current of 2 A equally divides between R_N and R_L, since they are equal value parallel branches. This means I_L = 1 A. V_L is easily calculated using Ohm's law where, $I_L \times R_L$ = 1 A × 25 Ω = 25 V. It is interesting to note the I_L and V_L computed using Norton's theorem are exactly the same values as those

FIGURE 7–17 Comparisons of Thevenin and Norton equivalent circuits

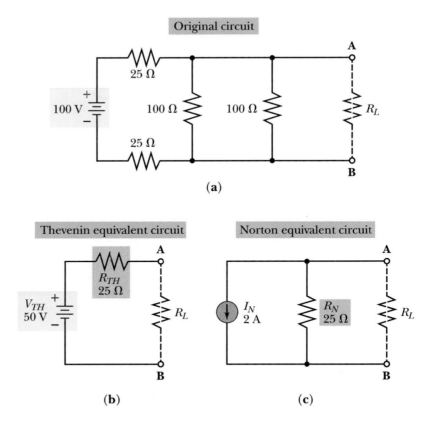

Original circuit

(a)

Thevenin equivalent circuit

(b)

Norton equivalent circuit

(c)

calculated for a load resistance of 25 Ω when using Thevenin's theorem for the same original circuit, Figure 7–10. It makes sense that the same R_L connected to the same circuit has the same operating parameters, regardless of the method used to analyze these parameters!

EXAMPLE

Suppose the R_L value changes to 75 Ω. The Norton equivalent circuit is the same. That is, $I_N = 2$ A and $R_N = 25$ Ω. With the higher value of R_L, I_L obviously is a smaller percentage of the I_N value.

Using the current divider rule, which is derived from the fact current divides inverse to the resistance of the branches, we can determine I_L from the following formula:

FORMULA 7–5
$$I_L = \frac{R_N}{R_N + R_L} \times I_N$$

With $R_L = 75$ Ω and $R_N = 25$ Ω,

$$I_L = \frac{25\ \Omega}{25\ \Omega + 75\ \Omega} \times 2\ \text{A} = 0.25 \times 2\ \text{A} = 0.5\ \text{A}$$

To find V_L, again use Ohm's law where $V_L = 0.5$ A $\times$ 75 Ω = 37.5 V

Original circuit

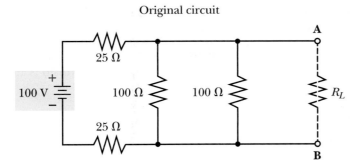

(a)

FIGURE 7–18 Application sample of Norton's theorem

Norton equivalent circuit $25 \ \Omega \ R_L$

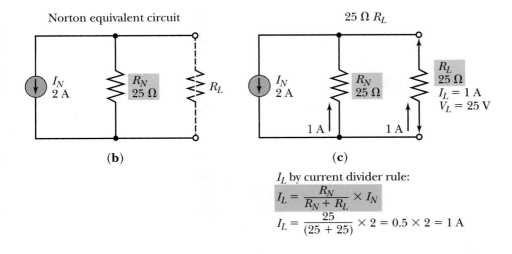

(b)   (c)

I_L by current divider rule:

$$I_L = \frac{R_N}{R_N + R_L} \times I_N$$

$$I_L = \frac{25}{(25 + 25)} \times 2 = 0.5 \times 2 = 1 \ \text{A}$$

PRACTICE PROBLEMS IV
In Figure 7–18a, assume R_L is 60 Ω. What are the values of I_L and V_L?

Summary of Norton's theorem

Norton's equivalent circuit is comprised of a constant current source (I_N) and a shunt resistance called R_N. When a load (R_L) is connected to this equivalent circuit, the current (I_N) divides between R_N and R_L according to the rules of parallel circuit current division (i.e., inverse to the resistances). The simple two-resistor current divider rule is applied to determine the current through the load resistance. Try the following Norton's theorem problems.

PRACTICE PROBLEMS V
1. Use Norton's theorem and determine the values of I_L and V_L for the circuit in Figure 7–18a, if R_L has a value of 100 Ω.

2. Refer again to Figure 7–18a. Will I_L increase or decrease if R_L changes from 25 to 50 Ω? Will V_L increase or decrease?

7-6 CONVERTING NORTON AND THEVENIN EQUIVALENT PARAMETERS

Since we have been discussing the similarities and differences between the two theorems, it is appropriate to show you some simple conversions that can be done between the two.

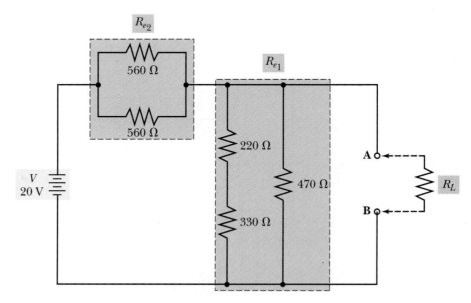

Thevenin's theorem:

$$V_{TH} = \left(\frac{R_{e_1}}{R_{e_1} + R_{e_2}}\right) \times 20\text{ V} = \left(\frac{253\ \Omega}{253\ \Omega + 280\ \Omega}\right) \times 20\text{ V} \cong 9.5\text{ V}$$

$$R_{TH} = \frac{R_{e_1} \times R_{e_2}}{R_{e_1} + R_{e_2}} = \frac{253\ \Omega \times 280\ \Omega}{253\ \Omega + 280\ \Omega} = 133\ \Omega$$

Norton's theorem:

$$I_N = \frac{V}{R_{e_2}} = \frac{20\text{ V}}{280\ \Omega} = 71\text{ mA}$$

$$R_N = \frac{R_{e_1} \times R_{e_2}}{R_{e_1} + R_{e_2}} = 133\ \Omega$$

CONVERSIONS
Thevenin from Norton:

$$R_{TH} = R_N$$
$$133\ \Omega = 133\ \Omega$$
$$V_{TH} = I_N \times R_N$$
$$V_{TH} = 71 \times 10^{-3} \times 133\ \Omega \cong 9.5\text{ V}$$

Norton from Thevenin:

$$R_N = R_{TH}$$
$$133\ \Omega = 133\ \Omega$$
$$I_N = \frac{V_{TH}}{R_{TH}}$$
$$I_N = \frac{9.5\text{ V}}{133\ \Omega} \cong 71\text{ mA}$$

FIGURE 7–19

You already know R_{TH} and R_N are the same value for a given network or circuit. You may have noticed if you multiply $I_N \times R_N$, the answer is the same as the value of V_{TH}.

Let's summarize these thoughts to show you the conversions possible.

Finding Thevenin parameters from Norton parameters:

$$R_{TH} = R_N$$

$$V_{TH} = I_N \times R_N$$

Finding Norton parameters from Thevenin parameters:

$$R_N = R_{TH}$$

$$I_N = \frac{V_{TH}}{R_{TH}}$$

Example

Let's look briefly at a sample circuit, analyze it with Thevenin's theorem and with Norton's theorem, then plug values into our simple conversion formulas. This will serve as a quick review of the two theorems, plus illustrate the simplicity of conversion from one set of parameters to the other. See Figure 7–19.

SUMMARY

▶ The theorems studied apply to two-terminal linear networks that have bilateral resistances or components.

▶ The maximum power transfer theorem indicates maximum power is transferred from source to load when the load resistance equals the source resistance. Under these conditions the efficiency is 50%.

▶ The superposition theorem says electrical parameters are computed in linear circuits with more than one source by analyzing the parameters produced by each source acting alone, then superimposing each result by algebraic addition to find the circuit parameters with multiple sources.

▶ Thevenin's theorem simplifies any two-terminal linear network into an equivalent circuit representation of a voltage source (V_{TH}) and series resistance (R_{TH}).

▶ Steps for Thevenizing a circuit are as follows:

1. Open (or remove R_L) from the network circuitry.

2. Determine the open-circuit V at points where R_L is to be connected. This is the value of V_{TH}.

3. Mentally short the circuit voltage source and determine resistance by looking into the network (from R_L connection points, with R_L disconnected). This is the value of R_{TH}.

4. Draw the Thevenin equivalent circuit with source = V_{TH} and series resistor of the R_{TH} value.

5. Make calculations of I_L and V_L for the desired values of R_L (where R_L is assumed to be across the equivalent circuit output terminals) by using two-resistor series circuit voltage analysis.

▶ Voltage distribution in a Thevenin equivalent circuit is computed using Ohm's law and/or the voltage divider rule.

▶ Norton's theorem simplifies any two-terminal linear network into an equivalent circuit representation of a current source (I_N) and a parallel or shunt resistance (R_N).

▶ Steps for Nortonizing a circuit are as follows:

1. Determine the value for I_N by finding the value of current flowing through R_L if it were zero ohms.

2. Find the value for R_N by assuming the R_L is removed from the circuit (R_L = infinite ohms) and any voltage sources are zero ohms. Calculate the circuit resistance by looking from the R_L connection points toward the source(s).

3. Draw the Norton equivalent circuit as a current source of a value equal to I_N and a resistance in parallel with the source equal to R_N.

4. Compute R_L parameters, as appropriate, by assuming the load resistance at the output terminals of the equivalent circuit, using concepts of two-resistor parallel circuit current division analysis.

▶ Current division in a Norton equivalent circuit is computed using Ohm's law and/or the current divider rule.

FORMULAS AND SAMPLE CALCULATOR SEQUENCES

FORMULA 7–1

$$\text{Efficiency (\%)} = \frac{P_{out}}{P_{in}} \times 100$$

P_{out} value, ➗ , P_{in} value, ✖ , 100, 🟰

FORMULA 7–2

$$r_{int} = \frac{V\,(\text{no load}) - V\,(\text{loaded})}{I\,(\text{load})}$$

V (no load) value, ➖ , V (loaded) value, 🟰 , ➗ , I (load) value, 🟰

FORMULA 7–3

$$I_L = \frac{V_{TH}}{R_{TH} + R_L}$$

V_{TH} value, ➗ , ⟮ , R_{TH} value, ➕ , R_L value, ⟯ , 🟰

FORMULA 7–4

$$V_L = \frac{R_L}{R_L + R_{TH}} \times V_{TH}$$

R_L value ➗ , ⟮ , R_L value, ➕ , R_{TH} value, ⟯ , 🟰 , ✖ , V_{TH} value, 🟰

FORMULA 7–5

$$I_L = \frac{R_N}{R_N + R_L} \times I_N$$

R_N value, ➗ , ⟮ , R_N value, ➕ , R_L value, ⟯ , 🟰 , ✖ , I_N value, 🟰

REVIEW QUESTIONS

1. Name at least two applications for the maximum power transfer theorem.

2. What is the efficiency of a circuit where there is maximum transfer of power?

3. How do you calculate circuit efficiency when evaluating power transfer in a circuit?

4. Define the term *bilateral linear network*.

5. State the sequence of steps used in "Thevenizing" a circuit.

6. State the sequence of steps used in "Nortonizing" a circuit.

7. Describe the most important application of the superposition theorem.

8. One of the benefits of using Thevenin's theorem is that once you have computed the equivalent circuit, load voltage and current parameters are easily calculated for ANY load resistance value:
   a. using a simple two-resistor voltage divider analysis.
   b. using a simple three-resistor voltage divider analysis.
   c. using a simple two-resistor current divider analysis.
   d. using a simple three-resistor current divider analysis.

9. One of the benefits of using Norton's theorem is that once you have computed the equivalent circuit, load voltage and current parameters are easily calculated for ANY load resistance value:
   a. using a simple two-resistor voltage divider analysis.
   b. using a simple three-resistor voltage divider analysis.
   c. using a simple two-resistor current divider analysis.
   d. using a simple three-resistor current divider analysis.

10. When using the superposition theorem to solve a network's parameters that has two voltage sources, the basic technique used is:

   a. analyzing parameters produced by each source acting alone, then superimposing each result by mathematical addition to find the resultant circuit parameters.
   b. analyzing parameters produced by each source acting alone, then superimposing each result by algebraic addition to find the resultant circuit parameters.
   c. removing both sources and considering them zero ohms each, then calculating circuit total resistance, then voltages and currents using Ohm's law, using the sum of or difference between the two supply voltages.
   d. none of the above.

11. Two methods for determining the voltage distribution in a network by using Thevenin's theorem are:
   a. Ohm's law and the current-divider rule.
   b. Ohm's law and Kirchhoff's voltage law.
   c. Ohm's law and Kirchhoff's current law.
   d. Ohm's law and the voltage-divider rule.

12. Two methods for determining the current distribution in a network by using Norton's theorem are:
   a. Ohm's law and the current-divider rule.
   b. Ohm's law and Kirchhoff's voltage law.
   c. Ohm's law and Kirchhoff's current law.
   d. Ohm's law and the voltage-divider rule.

PROBLEMS

1. In Figure 7–20, use the superposition theorem, and determine the voltage across R_1 and R_2.

2. Refer again to Figure 7–20. Assume the polarity of V_{S_2} is reversed. What is the value of V_{R_1} and V_{R_2}?

3. In Figure 7–21, use Thevenin's theorem and determine V_{TH}, R_{TH}, I_L, and V_L. Assume R_L is 10 kΩ.

4. Refer again to Figure 7–21. Assume R_L changes to 27 kΩ. What are the values of V_{TH}, R_{TH}, I_L, and V_L?

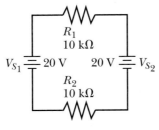

FIGURE 7–20

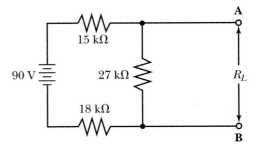

FIGURE 7–21

5. Draw and label the values for the Norton equivalent circuit for the circuit in question 3.

6. Draw and label the values for the Norton equivalent circuit for the circuit in question 4.

7. What is the maximum power that can be delivered to the load in Figure 7–22?

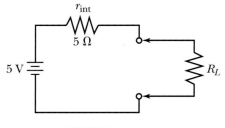

FIGURE 7–22

8. If the voltage at the source terminals of a circuit is 100 V without load and 90 V when 10 A are drawn from the source, what value is the load resistance to afford maximum power transfer?

9. For the circuit in Figure 7–23, what are the values of R_{TH} and V_{TH}?

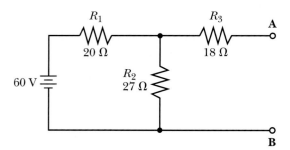

FIGURE 7–23 Find R_{TH} and V_{TH}.

10. For the circuit in Figure 7–24, what are the values of R_N and I_N?

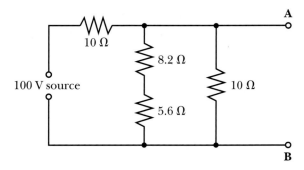

FIGURE 7–24 Find R_N and I_N.

11. Convert Norton parameters to Thevenin parameters, given $I_N = 3$ A and $R_N = 10\ \Omega$.

12. Convert Thevenin parameters to Norton parameters, given $R_{TH} = 15\ \Omega$ and $V_{TH} = 25$ V.

13. What is the percentage of efficiency for a circuit whose 100-V source has an internal resistance of $5\ \Omega$ and a load resistance of $25\ \Omega$?

14. Would the percentage efficiency be the same for the circuit described in question 13 if the source voltage were 200 V? Does the value of source voltage have any impact on the circuit's efficiency factor?

15. What load resistance value is needed for the circuit described in question 13 to have maximum power transfer?

ANALYSIS QUESTIONS

1. Explain why it is important for a voltage source to have as low an internal resistance (r_{int}) as possible.

2. For a given power source, why is it important for the load's opposition to current flow to be equal to that of the source?

3. If the efficiency of power transfer from source to load is 60%, should the load resistance be increased or decreased to enable maximum power transfer?

4. Which theorem relates most closely to voltage divider analysis techniques?

5. Which theorem relates most closely to current divider analysis techniques?

6. a. Give the main reason it is important to observe the direction of current flow through components when using the superposition theorem to analyze a two-source circuit.

   b. Describe which component(s) in the circuit

are especially impacted by the fact described in part (a) of this question.

7. In your own words, explain the significance of R_{TH} in using Thevenin's theorem, and explain how it is determined.

8. In your own words, explain the significance of V_{TH} in using Thevenin's theorem, and explain how it is found.

9. In your own words, explain the significance of R_N in using Norton's theorem, and explain how it is found.

10. In your own words, explain the significance of I_N in using Norton's theorem, and explain how it is found.

11. a. Perform appropriate research to enable listing at least one type circuit or application in which there may be two or more sources.

b. Which network theorem would you use to help you in analyzing such a circuit?

12. For the circuit of Figure 7–23, explain how you could use a voltmeter and an ohmmeter to determine the Thevenin equivalent circuit parameters of R_{TH} and V_{TH}.

13. For the circuit of Figure 7–23, is it possible to use a meter to measure the Norton equivalent circuit R_N value? If so, what is the procedure to do this?

14. Explain how to determine the internal resistance (r_{int}) of a voltage source.

15. What would be the internal resistance of a battery whose output terminal voltage was 12 V under no load conditions and 11 V when the load current was 50 A?

PERFORMANCE PROJECTS CORRELATION CHART

Suggested performance projects in the Laboratory Manual that correlate with topics in this chapter are:

CHAPTER TOPIC	PERFORMANCE PROJECT	PROJECT NUMBER
Maximum Power Transfer	Maximum Power Transfer Theorem	29
Thevenin's Theorem	Thevenin's Theorem	27
Norton's Theorem	Norton's Theorem	28

NOTE: It is suggested that after completing the above projects, the student should be required to answer the questions in the "Summary" at the end of this section of projects in the Laboratory Manual.

NETWORK ANALYSIS TECHNIQUES

chapter

8

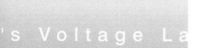

CHAPTER PREVIEW

As you have already learned, circuit and network analysis enable you to find the various currents and voltages throughout the circuit, or network, being analyzed. In this chapter, two other approaches will be introduced for solving network parameters. One method uses Kirchhoff's voltage law as a basis for writing equations for electrical quantities around a closed loop (also sometimes called a mesh). These equations then enable finding current and voltage parameters throughout the network. Another method uses Kirchhoff's current law as a basis for writing "nodal" equations, which also enable solving network parameters. Each method has advantages and disadvantages. The techniques presented in this chapter enable you to analyze complex networks in which connections between components are such that you cannot simply use basic series-parallel circuit analysis methods.

This chapter will also introduce some techniques for converting two specific types of networks from one form to the other when this would be advantageous. One type network is called the wye (Y) or (T) network configuration. The other is known as the delta (Δ) or pi (π) configuration. These networks, or circuit configurations, are used in a number of places in electrical power systems and in electronic circuitry. Knowledge regarding these circuit configurations is therefore very useful, and has value to you as a technician.

OBJECTIVES

After studying this chapter, you should be able to:

1. Define the terms mesh, loop, and node

2. Analyze a single-source circuit using a loop procedure

3. Use the assumed mesh currents approach to find voltage and current parameters for each component in a network having two sources

4. Use the nodal analysis approach to find voltage and current parameters for each component in a network having two sources

5. Convert from delta (Δ) circuit configuration parameters to wye (Y) circuit configuration parameters, and vice versa

8–1 ASSUMED LOOP OR MESH CURRENT ANALYSIS APPROACH

Background information

A **loop** is considered to be a closed, or complete, path within a circuit. A **mesh** is also considered to be an unbroken, or closed, loop. Observe in Figure 8–1a that two closed loops are shown in the circuit with arbitrary mesh current assignments and identifiers. Figure 8–1b illustrates the commonly accepted "single-window frame" definition for meshes and mesh currents.

For each mesh, or loop, we will use an **assumed mesh current** direction in which the mesh current does NOT branch, as an actual circuit current would. These arbitrarily assigned currents, for analysis purposes, are considered to flow only in their own loop, or mesh. (This nonbranching aspect simplifies matters,

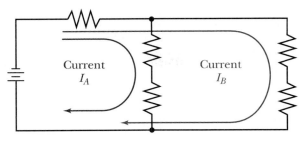

(**a**) Mesh currents—generalized loop definition

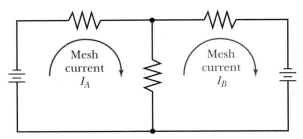

(**b**) Mesh currents—single "window-frame" definition

FIGURE 8–1 Examples of assumed mesh currents

and is one advantage of the mesh analysis method.) Kirchhoff's voltage law equations will then be written, using the assumed mesh current designators.

The version of Kirchhoff's voltage law that will be used stipulates that the sum of the voltage drops around a closed loop must equal the voltage source(s). That is: $V_s = V_1 + V_2 \ldots + V_n$. From the results of these equations, we will then be able to compute actual circuit voltages and currents throughout the network.

Polarity assignments

It is important to note directions of currents and polarities of voltages in the analysis steps. The general rules to follow during the analysis are to:

1. Use the arbitrary convention that assumed loop currents will be shown circulating in a clockwise direction through each loop;

2. Assume any voltage drop created across a resistor by its own assumed loop or mesh current to be positive in that mesh's equation. Any voltage drop across a resistor due to another loop or mesh current is considered negative for that mesh's equation; and

3. For source voltage polarity assignment (the algebraic sign to be used in equations), use the arbitrary convention that if the assumed mesh current is returning to the source into its positive terminal, the source voltage will be considered positive. If the assumed mesh current is returning (going into) the negative terminal, it will be considered negative.

Steps to use the loop/mesh analysis approach

1. Label the assumed mesh currents on the diagram. (Use the arbitrary clockwise direction convention.)

2. When assigning polarities to sources in the Kirchhoff equations, consider the source voltage positive if the assumed mesh current is entering its positive terminal, and negative if the assumed mesh current enters its negative terminal.

3. Write the Kirchhoff equation for each loop's voltage drops. Assume voltages across resistors in a given loop to be positive when caused by that mesh's assumed current, and negative when caused by an adjacent mesh's assumed current.

4. Solve the resulting equations using simultaneous equation methods.

5. Verify the results using Ohm's law, Kirchhoff's law, etc., as appropriate.

Example

Let's try this system out with a single-source bridge circuit, and see how it works. For the given circuit, let's find the value of voltage between points A and B.

Step 1. Draw assumed loop/mesh currents on the diagram. See Figure 8–2.

Step 2. Determine polarity of source polarity assignment(s). In this case, both mesh current I_A and mesh current I_B are shown entering the positive terminal of V_S. Thus, the source voltage will have a positive sign in our equations.

Step 3. Write the Kirchhoff equation for each loop's voltage drops:

a. Loop A:

$$(I_A + I_B)R_1 + I_A R_2 + I_A R_3 = V_S$$

Switching positions of terms for ease of notation and manipulation:

$$10(I_A + I_B) + 10I_A + 15I_A = 25$$

$$35I_A + 10I_B = 25 \text{ (equation 1)}$$

b. Loop B:

$$(I_A + I_B)R_1 + I_B R_4 + I_B R_5 = V_S$$

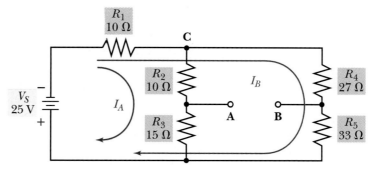

FIGURE 8–2 Illustration showing results of several key steps in the single-source "Example" problem

Kirchhoff's voltage law equation 1: $35\,I_A + 10\,I_B = 25$ V
Kirchhoff's voltage law equation 2: $10\,I_A + 70\,I_B = 25$ V
$$I_A = 0.638 \text{ A}$$
$$I_B = 0.266 \text{ A}$$

Switching positions of terms for ease of notation and manipulation:

$$10(I_A + I_B) + 27(I_B) + 33(I_B) = 25$$

$$10I_A + 70\,I_B = 25 \text{ (equation 2)}$$

Step 4. 　　　　　Equation 1:　$35I_A + 10I_B = 25$

　　　　　　　　Equation 2:　$10I_A + 70I_B = 25$

Multiplying equation 1 by -7 to make equation 1's "$10I_B$" become "$-70I_B$," and thus cancel the I_B term and solve for I_A:

$$-245I_A - 70I_B = -175$$

$$10I_A + 70I_B = 25$$

Adding to cancel the I_B term:

$$-235I_A = -150$$

$$I_A = 0.638 \text{ A}$$

Substituting 0.638 for I_A into equation 2 to solve for I_B:

$$10(0.638) + 70I_B = 25$$

$$6.38 + 70I_B = 25$$

$$70I_B = 18.62$$

$$I_B = 0.266 \text{ A}$$

Thus:

$$I_A = 0.638 \text{ A}$$

$$I_B = 0.266 \text{ A}$$

Using mesh currents to solve voltage drops:

$$V_1 = (I_A + I_B) \times 10 = 0.904 \times 10 = 9.04 \text{ V}$$

$$V_2 = I_A \times 10 = 0.638 \times 10 = 6.38 \text{ V}$$

$$V_3 = I_A \times 15 = 0.638 \times 15 = 9.57 \text{ V}$$

$$V_4 = I_B \times 27 = 0.266 \times 27 = 7.18 \text{ V}$$

$$V_5 = I_B \times 33 = 0.266 \times 33 = 8.78 \text{ V}$$

Point A is $+ 6.38$ V with respect to point C in the circuit. Point B is $+ 7.18$ V with respect to point C in the circuit. Thus, point A is $- 0.8$ V with respect to point B.

Step 5. Checking the voltage drops around the loops to see if it verifies our calculations:

Loop A: $V_1 + V_2 + V_3 = 9.04 + 6.38 + 9.57 = 24.99$ V, which checks out against our 25-V source.

Loop B: $V_1 + V_4 + V_5 = 9.04 + 7.18 + 8.78 = 25$ V, which also checks out against our 25-V source.

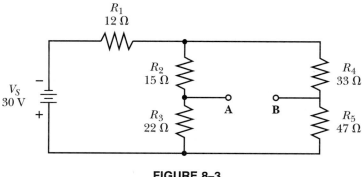

FIGURE 8–3

PRACTICE PROBLEMS I

Refer to Figure 8–3, and using the mesh analysis approach, solve the following:

1. I_A 5. V_3

2. I_B 6. V_4

3. V_1 7. V_5

4. V_2 8. V_{A-B}

Example

Now, let's try the mesh approach with a circuit containing two sources. Refer to Figure 8–4 as you follow the steps for this example.

Step 1. Label the assumed mesh currents on the diagram, using the arbitrary clockwise convention. See Figure 8–4.

Step 2. Determine polarity of sources. By observation you can see that the mesh currents are such that current goes into the positive terminal of the 15-V source, thus it will be designated as a positive number. The mesh current goes into the negative terminal of the 12-V source, thus it will be signed as a negative number in equations.

Step 3. Write the Kirchhoff voltage equation for each loop's voltage drops:

$$(3.9I_A + 5.6I_A) - 5.6I_B = 15 \text{ V}$$

$$9.5I_A - 5.6I_B = 15 \text{ V} \qquad \text{(equation 1)}$$

$$-5.6I_A + (5.6I_B + 6.8I_B) = -12 \text{ V}$$

$$-5.6I_A + 12.4I_B = -12 \text{ V} \qquad \text{(equation 2)}$$

Step 4: To cancel the I_B terms and solve for I_A, we multiply equation 1 by the factor 2.214 (which is derived from 12.4 being divided by 5.6). This yields:

$$21I_A - 12.4I_B = 33.21$$

$$-5.6I_A + 12.4I_B = -12$$

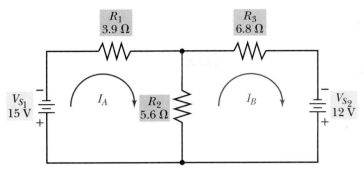

$$\text{Kirchhoff's voltage law equation 1: } 9.5\,I_A - 5.6\,I_B = 15 \text{ V}$$
$$\text{Kirchhoff's voltage law equation 2: } -5.6\,I_A + 12.4\,I_B = -12 \text{ V}$$
$$I_A = 1.377 \text{ A}$$
$$I_B = 0.346 \text{ A}$$

FIGURE 8–4 Illustration showing results of several key steps in the two-source "Example" problem

Adding to cancel the I_B term produces:

$$15.4 I_A = 21.21$$

$$I_A = 1.377 \text{ A}$$

To find I_B we substitute 1.377 A for I_A in the second equation, which yields:

$$-5.6(1.377) + 12.4 I_B = -12$$

$$-7.71 + 12.4 I_B = -12$$

$$12.4 I_B = -4.29$$

$$I_B = -0.346 \text{ A}$$

The negative sign of the outcome tells us that the actual current is in the opposite direction from that which we assumed in creating the mesh currents.

Now using the values of 1.377 A for I_A and 0.346 A for I_B, we can calculate circuit parameters quite easily.

$$V_1 = I_A \times 3.9 = 5.37 \text{ V}$$

$$V_2 = (I_A + I_B) \times 5.6 = (1.377 + 0.346) \times 5.6 = 9.65 \text{ V}$$

$$V_3 = I_B \times 6.8 = 0.346 \times 6.8 = 2.35 \text{ V}$$

Step 5. Checking the loop voltages as a verification of our calculations, we find:

Loop A: $V_1 + V_2$ should equal V_{S_1}: 5.37 + 9.65 = 15.02 V (which verifies loop *A*)

Loop B: $V_2 + V_3$ should equal V_{S_2}: 9.65 + 2.35 = 12 V (which verifies loop *B*)

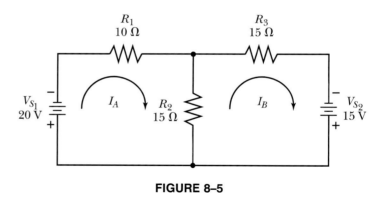

FIGURE 8–5

PRACTICE PROBLEMS II

For further practice in using the mesh approach, refer to Figure 8–5 and perform the required work to find the following circuit parameters.

1. I_A
2. I_B
3. V_1 and the current through R_1
4. V_2 and the current through R_2
5. V_3 and the current through R_3

IN-PROCESS LEARNING CHECK I

1. A loop is considered to be a closed, or complete _____ within a circuit.

2. In the loop or mesh approach, it is important to label the assumed _____ currents on the diagram you use for analysis.

3. When assigning polarities to sources in the Kirchhoff equations, a source voltage is considered (negative, positive) _____ if the assumed mesh current enters the positive terminal of the source.

4. When writing Kirchhoff equations for each loop's voltage drops, the component voltage is considered (negative, positive) _____ when caused by its own mesh current, and (negative, positive) _____ when caused by an adjacent mesh's current flow.

8–2 NODAL ANALYSIS APPROACH

Background information

A **node** is a current junction point of two or more components in a circuit; that is, a point where currents enter and leave. Kirchhoff's current law can be used to write equations related to these nodes in circuits. Nodes where three or more components are joined are sometimes identified as "major nodes."

For nodal analysis, one major node is specified as the "reference node." Typically, a node that is common to as many parts of the network as possible is chosen as the reference node. Voltages within the circuit are then designated for each of the other nodes in the circuit, relative to the reference node. A Kirchhoff current equation is written for each of the nodes, other than the reference node. For each current term, we use the appropriate V_R/R expression. This is in contrast to our having used the appropriate $I \times R$ expression for voltage terms in the mesh/loop analysis approach.

Steps to use the node analysis approach

1. Find the number of major nodes that are in the circuit, then select the one major node you will use as the reference node.

2. Designate currents into and out of all major nodes, with the exception of the reference point node.

3. Write Kirchhoff Current Law (KCL) equations for currents into and out of each major node, using the V_R/R approach to represent each current. The number of equations required is generally one less than the number of major nodes.

4. Incorporate known values of voltage and resistance in the equations, and solve for the unknown node voltages.

5. Based on the results of step 4, solve for current and voltage values for each component, as appropriate.

6. Verify results by using Kirchhoff's voltage law for each loop in the circuit.

Example

Refer to the circuit of Figure 8–6 as you observe the following steps in a sample nodal analysis.

Step 1. Find the major nodes and select the reference node. Observe in the circuit that only two major nodes exist. One is at the top of R_2, and the other is at other end of R_2, which happens to be ground reference in this case. Let's let the ground reference be the reference node. The voltage across R_2 will then be our voltage of interest, against which other parameters will be determined.

Step 2. Designate currents into and out of major nodes. See Figure 8–6 again.

Step 3. Write the Kirchhoff current equations for currents into and out of these major nodes. For the loop with the 15-V source:

$$I_1 + I_2 = I_3$$

(NOTE: the way our circuit components are labeled, current I_2 is the current through R_3, and current I_3 is the current through R_2 and the reference node.)

(Term for I_1) $\dfrac{V_{R_1}}{R_1}$ + (Term for I_2) $\dfrac{V_{R_3}}{R_3}$ = (Term for I_3) $\dfrac{V_{R_2}}{R_2}$ (which is the current through R_2)

(NOTE: V_{R_2} is the voltage of special interest, since it is associated with the reference node; V_{R_2} is the voltage drop of R_2 caused by I_3.)

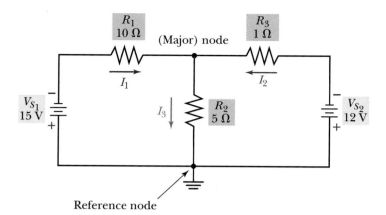

NOTE: I_1 is expressed as V_{R_1}/R_1
I_2 is expressed as V_{R_3}/R_3
I_3 is expressed as V_{R_2}/R_2

FIGURE 8–6 Illustration showing results of several key steps for first nodal analysis "Example" problem

Step 4. Since we know the V_{S_1} value, we can state that:

$$V_{R_1} + V_{R_2} = 15 \text{ V or } V_{R_1} = 15 - V_{R_2}$$

For the loop with the 12-V source, we can state that:

$$V_{R_3} + V_{R_2} = 12 \text{ V or } V_{R_3} = 12 - V_{R_2}$$

Now, incorporating these terms into our *V/R* current statements:

$$\frac{15 - V_{R_2}}{10} + \frac{12 - V_{R_2}}{1} = \frac{V_{R_2}}{5}$$

Multiplying each term by 10 to remove the denominators:

$$\left(\frac{\cancel{10}}{1}\right) \times \frac{15 - V_{R_2}}{\cancel{10}} + \left(\frac{10}{1}\right) \times \frac{12 - V_{R_2}}{1} = \left(\frac{\cancel{10}}{1}\right) \times \frac{V_{R_2}}{5}$$

$$15 - V_{R_2} + 10(12 - V_{R_2}) = 2\,V_{R_2}$$

$$15 - V_{R_2} + 120 - 10\,V_{R_2} = 2\,V_{R_2}$$

$$135 = 13\,V_{R_2}$$

$$V_{R_2} = \frac{135}{13}$$

$$V_{R_2} = 10.38 \text{ V}$$

Step 5. Therefore:

$$I_{R_2} = \frac{10.38}{5} = 2.076 \text{ A}$$

From this, we can continue with:

$$V_{R_3} = 12 - 10.38 = 1.62 \text{ V}$$

$$I_{R_3} = \frac{1.62}{1} = 1.62 \text{ A}$$

$$V_{R_1} = 15 - 10.38 = 4.62 \text{ V}$$

$$I_{R_1} = \frac{4.62}{10} = 0.462 \text{ A}$$

Step 6. Verifying results:

$$V_{R_1} + V_{R_2} = V_{S_1}$$

$$4.62 + 10.38 = 15 \text{ V (It checks!)}$$

$$V_{R_2} + V_{R_3} = V_{S_2}$$

$$10.38 + 1.62 = 12 \text{ V (It checks!)}$$

Example

The same procedures we have been showing you will work even if the component values are not as convenient to work with as the ones in our first example. Follow along in this example, which does not use a least common denominator to eliminate the denominators, but rather simply uses one of the denominators as the factor to move things from the denominator to the numerator. The steps used in the solution process are essentially the same as those used in the preceding example, so we will not detail the step numbers.

Refer to Figure 8–7 as you follow the procedure shown.

$$I_1 + I_2 = I_3$$

(Again, note that I_2 is R_3's current, and I_3 is R_2's current, where the R_2 parameters are the ones of special interest, since this is the major reference node part of the circuit.)

$$\text{Currents restated: } \frac{V_{R_1}}{R_1} + \frac{V_{R_3}}{R_3} = \frac{V_{R_2}}{R_2}$$

$$\text{15-V loop: } V_{R_1} + V_{R_2} = 15 \text{ V or } V_{R_1} = 15 - V_{R_2}$$

$$\text{12-V loop: } V_{R_3} + V_{R_2} = 12 \text{ V or } V_{R_3} = 12 - V_{R_2}$$

$$I_1 + I_2 = I_3$$

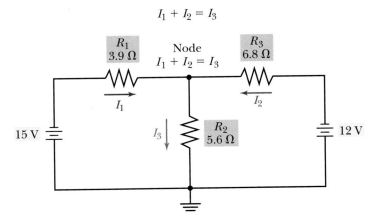

FIGURE 8–7 Reference diagram for nodal analysis second "Example" problem

Substituting these terms into the $I_1 + I_2 = I_3$ equation:

$$\frac{15 - V_{R_2}}{3.9} + \frac{12 - V_{R_2}}{6.8} = \frac{V_{R_2}}{5.6}$$

Multiplying each term by 6.8 to move the denominators:

$$1.74(15 - V_{R_2}) + (12 - V_{R_2}) = 1.214\ V_{R_2}$$

$$(26.1 - 1.74\ V_{R_2}) + (12 - V_{R_2}) = 1.214\ V_{R_2}$$

$$38.1 = 3.95\ V_{R_2}$$

$$V_{R_2} = 9.64\ \text{V}$$

Using the reference node voltage, we find the other parameters as follows:

$$V_{R_1} = 15 - 9.64 = 5.36\ \text{V}$$

$$I_{R_1} = \frac{5.36\ \text{V}}{3.9\ \Omega} = 1.37\ \text{A}$$

$$V_{R_3} = 12 - 9.64 = 2.36\ \text{V}$$

$$I_{R_3} = \frac{2.36\ \text{V}}{6.8\ \Omega} = 0.347\ \text{A}$$

$$\text{Current through } R_2 = \frac{9.64\ \text{V}}{5.6\ \Omega} = 1.72\ \text{A or}$$

$$I_1 + I_2 = I_3 : 1.37\ \text{A} + 0.347\ \text{A} = 1.717\ \text{A}$$

PRACTICE PROBLEMS III

Refer to Figure 8–8 and find the following circuit parameters, using the nodal analysis method.

1. V_{R_1}
2. V_{R_2}
3. V_{R_3}
4. Current through R_1
5. Current through R_2
6. Current through R_3

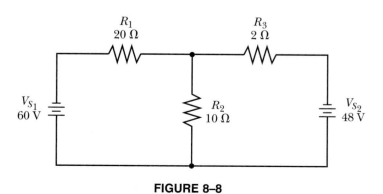

FIGURE 8–8

8–3 CONVERSIONS OF DELTA AND WYE NETWORKS

As a technician, you will occasionally deal with two circuit configurations found in numerous applications: the Wye or Tee configuration, shown in Figure 8–9, and the Delta, or Pi configuration, shown in Figure 8–10. For example, both types of connections are found in ac power applications. The pi configuration is found in many filter circuits used in electronic systems. As you can see, the names of these configurations are derived from the shapes they form. The delta configuration has the triangular shape of the Greek letter delta (Δ). The wye configuration is often represented in the shape of the letter Y.

There are times when it is convenient and helpful to be able to convert from one circuit configuration to the other. One example of simplifying calculations using this technique is the analysis of a bridge circuit, which we will show you a little later. To do this conversion, or transformation, we need to be able to derive equivalent circuits from one configuration to the other. So, let's look a little closer at each of these common network connection configurations, and how they might be converted, or transformed.

For ease of reference, we will label the components and their connection points as shown in Figure 8–11. Notice that we have superimposed one configuration on the other, so you can easily see the cross-referencing between components when transformation formulas are used. The wye components are labeled as R_1, R_2, and R_3; the delta components are labeled as R_A, R_B, and R_C, respectively.

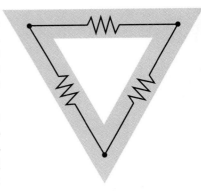

(**a**) Delta [Δ] configuration

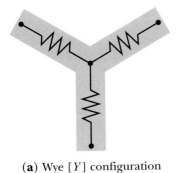

(**a**) Wye [Y] configuration

(**b**) Tee [T] configuration

FIGURE 8–9 Commonly used Y and T circuit configurations

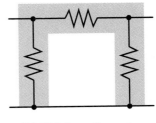

(**b**) Pi [π] configuration

FIGURE 8–10
Commonly used Δ and π configurations

FIGURE 8–11 Superimposed Δ and Y configurations with components labeled

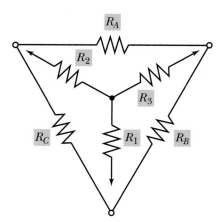

Converting a wye network to an equivalent delta network

The object of our conversion is to find out what values the delta components must be in order to provide the same electrical load to the source as the original wye network does. In other words, we are looking for the delta equivalent to the wye circuit. The equivalent circuit will present the same resistance, and draw the same current from the source as the original wye network does. It should be noted, however, that voltages between connection points will differ for each configuration.

The equations used to convert from wye to delta networks (**Wye-Delta Conversion**) are:

1. To find R_A's value:

FORMULA 8–1 $R_A = \dfrac{(R_1 \times R_2) + (R_2 \times R_3) + (R_3 + R_1)}{R_1}$

2. To find R_B's value:

FORMULA 8–2 $R_B = \dfrac{(R_1 \times R_2) + (R_2 \times R_3) + (R_3 \times R_1)}{R_2}$

3. To find R_C's value:

FORMULA 8–3 $R_C = \dfrac{(R_1 \times R_2) + (R_2 \times R_3) + (R_3 \times R_1)}{R_3}$

Observe that the numerator is the same in all three equations, and it is comprised of the sum of the products of all the wye resistors, two at a time. The denominator in each case is the resistor in the wye configuration that is across from, or opposite to, the resistor value being sought for the delta equivalent circuit.

Example

Find the delta equivalent circuit to the Y configuration shown in Figure 8–12. To make it a little easier to follow, we have again used our superimposed diagrams

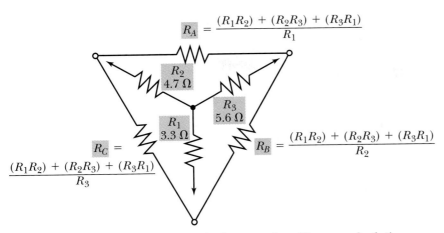

FIGURE 8–12
Reference diagram for wye-delta conversion "Example" problem

technique. NOTE: Get out your calculator and verify our calculations as you track our presentation.

Assume that $R_1 = 3.3\ \Omega$, $R_2 = 4.7\ \Omega$, and $R_3 = 5.6\ \Omega$.

$$R_A = \frac{(R_1 R_2) + (R_2 R_3) + (R_3 R_1)}{R_1}$$

$$R_A = \frac{(3.3 \times 4.7) + (4.7 \times 5.6) + (5.6 \times 3.3)}{3.3} = \frac{60.31}{3.3} = 18.28\ \Omega$$

The numerator is the same for all three equations. All we need to do is divide the 60.31 by the appropriate resistor value from the Y network to find the resistor of interest in our equivalent delta network. Thus:

$$R_B = \frac{60.31}{4.7} = 12.83\ \Omega$$

$$R_C = \frac{60.31}{5.6} = 10.77\ \Omega$$

PRACTICAL NOTES

On occasion, you will find that the same value resistors are used for R_1, R_2, and R_3 in the wye configuration. This simplifies conversion calculations. For example, if each of the wye resistors is a 10-Ω resistor, then the resulting numerator in our conversion formula (the sum of the products part) is 300. The denominator is 10. The resulting answer is that the delta resistor must equal 30 Ω, or three times the Y resistor. This is true for all three legs. It is also true that the equivalent delta resistor will always be three times its related Y resistor value, as long as all the wye network resistors are equal in value.

This fact can also be used in reverse order. When converting from a symmetrical value delta configuration to a wye configuration, the wye resistors will each be one-third the value of their related delta resistors. In other words, if, in our previous example, the delta resistors had been 30 Ω each, the wye equivalent circuit would have had resistors equal to 10 Ω each.

In summary, when converting from wye to delta (where all resistors are equal in value), use three times each wye resistor value for each delta resistor. When converting from delta to wye (where all resistors are equal in value), make each wye resistor one-third the value of each of the delta resistors.

Converting a delta network to an equivalent wye network

As an interesting way to find if the conversion techniques will work, let's look at the formulas for converting from delta to wye, insert the values we just calculated for the delta network, and see if, after conversion, we arrive back at the same values given in our original wye network. The equations used to convert from delta to wye networks (**Delta-Wye Conversion**) are:

FORMULA 8–4 $\qquad R_1 = \dfrac{R_B \times R_C}{R_A + R_B + R_C}$

FORMULA 8–5 $\qquad R_2 = \dfrac{R_C \times R_A}{R_A + R_B + R_C}$

FORMULA 8–6 $\qquad R_3 = \dfrac{R_A \times R_B}{R_A + R_B + R_C}$

Notice in these equations, the denominator is the same in every case. That is, the denominator equals the sum of the resistances in the delta network. Notice also that the numerator for each equation shows the product of the two delta resistors that connect to one end of the wye resistor of interest in our superimposed networks drawing, or the product of the two delta resistors that are adjacent to the wye network resistor of interest.

Example

Let's insert our computed delta resistor values into these formulas and see if we come close to the original wye circuit values.

$$R_1 = \frac{R_B \times R_C}{R_A + R_B + R_C} = \frac{12.83 \times 10.77}{18.28 + 12.83 + 10.77} = \frac{138.18}{41.88} = 3.3 \ \Omega$$

For these equations, the denominator will be constant, so we only need to determine the numerator in each case, then divide by 41.88. Therefore,

$$R_2 = \frac{R_C \times R_A}{41.88} = \frac{10.77 \times 18.28}{41.88} = \frac{196.88}{41.88} = 4.7 \ \Omega$$

$$R_3 = \frac{R_A \times R_B}{41.88} = \frac{18.28 \times 12.83}{41.88} = \frac{234.53}{41.88} = 5.6 \ \Omega$$

Observe that our values do check out and verify the techniques!

PRACTICE PROBLEMS IV

1. Perform the necessary conversion from wye to delta configurations for Figure 8–13.

FIGURE 8–13

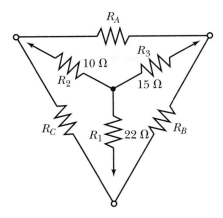

2. Use the values you computed for the delta resistor values in question 1 and determine the values of the wye network resistors, using the delta-wye conversion equations.

3. Did the answers you got in problem 2 appropriately correspond with the original circuit parameters shown in Figure 8–13?

IN-PROCESS LEARNING CHECK III

1. Because of their visual shapes and electrical connections:

   a. The wye circuit configuration is sometimes also called the _____ configuration.

   b. The delta circuit configuration is sometimes also called the _____ configuration.

2. Delta and wye circuit configurations are often found in _____ power applications.

3. The value in knowing how to determine equivalent circuits, or convert from delta-to-wye and wye-to-delta circuits is that the equivalent circuits will present the same resistance and draw the same _____ from the source feeding the circuit.

4. The voltage between connection points in wye and delta configurations (is, is not) _____ the same.

Bridge circuit analysis using conversion techniques

Observe the bridge circuit of Figure 8–14. When shown as a bridge circuit, it appears that the network is essentially two delta networks, back-to-back. Simple series-parallel circuit analysis techniques cannot be used here. If we convert the bottom delta portion of this circuit to a wye configuration, it will simplify the analysis of the circuit considerably.

FIGURE 8–14 Sample bridge circuit

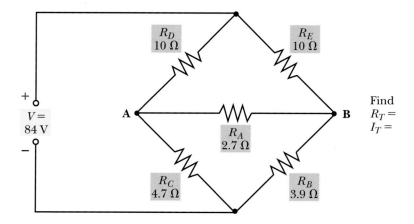

In Figure 8–15, you see that we have superimposed the proposed wye network in the bottom delta of the bridge diagram. By converting this delta circuit portion to an equivalent wye circuit, we can redraw the circuit as shown in Figure 8–16. As stated earlier, this lends itself to much easier analysis.

For simplification in transferring the concepts of the examples given earlier to this situation, we have assigned the resistors designations similar to those used in our preceding examples.

Example

If it were required that we find the R_T and I_T of the circuit, we might use the transformation techniques as follows.

Converting the bridge circuit delta portion to an equivalent wye:

$$R_1 = \frac{R_B \times R_C}{R_A + R_B + R_C} = \frac{3.9 \times 4.7}{2.7 + 3.9 + 4.7} = \frac{18.33}{11.3} = 1.62 \ \Omega$$

$$R_2 = \frac{R_C \times R_A}{11.3} = \frac{4.7 \times 2.7}{11.3} = \frac{12.69}{11.3} = 1.12 \ \Omega$$

$$R_3 = \frac{R_A \times R_B}{11.3} = \frac{2.7 \times 3.9}{11.3} = \frac{10.53}{11.3} = 0.93 \ \Omega$$

Now the bridge circuit can be redrawn as shown in Figure 8–17. The analysis is now that of a series-parallel circuit.

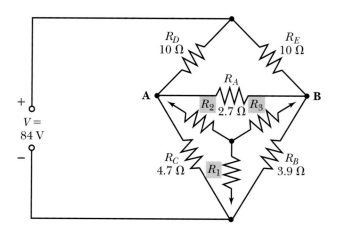

FIGURE 8–15 Sample bridge circuit analysis

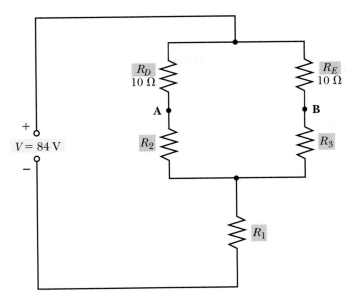

FIGURE 8–16 Redrawn bridge network equivalent with bottom delta transformed to wye

$$R_T = (R_D + R_2)\|(R_E + R_3) + R_1$$

$$R_T = (11.12\|10.93) + 1.62$$

$$R_T = 5.5 \ \Omega + 1.62 \ \Omega = 7.12 \ \Omega$$

$$I_T = \frac{V}{R_T} = \frac{84 \ V}{7.12 \ \Omega} = 11.8 \ A$$

PRACTICE PROBLEMS V

For the circuit of Figure 8–15, assume that the values are changed so that $R_A = 7 \ \Omega$, $R_B = 14 \ \Omega$, and $R_C = 21 \ \Omega$. Use the delta-to-wye transformation techniques to find the following parameters:

1. R_T
2. I_T

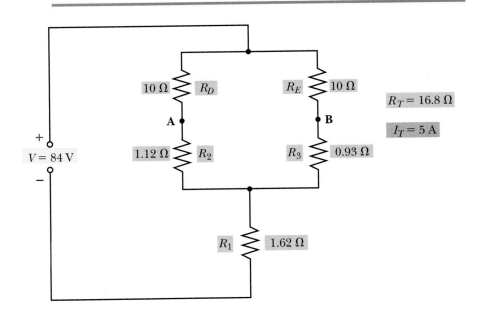

FIGURE 8–17 Reference diagram for bridge circuit conversion "Example" problem

SUMMARY

▶ A loop is a complete path in a circuit, from one side of the source, through a series of circuit components, back to the other side of the source.

▶ A mesh is an unbroken, closed loop within a network.

▶ Mesh currents are assumed currents (and current directions) used in conjunction with Kirchhoff's voltage law to analyze networks. Mesh currents are different than actual circuit currents, in that they are assumed to not branch.

▶ A common convention regarding labeling mesh currents on a diagram is that they are shown in a clockwise direction.

▶ Voltages across resistors in a network that are caused by their own mesh current are considered positive. Voltage drops due to another mesh current are considered negative.

▶ An arbitrary convention regarding assignment of source polarities in a network is that if a mesh current returns to the positive side of the source, the source is denoted as a positive term in the equations. If the mesh current returns to the negative terminal of the source, the source is signed as negative in the equations.

▶ For mesh analysis, voltages are expressed as $I \times R$ terms.

▶ Simultaneous equation techniques are used to eliminate one unknown from the resulting Kirchhoff equations. This enables solving for the other unknown factor.

▶ A node is a point where two or more circuit components join and where when circuit current is flowing, circuit currents may branch. A major node is where three or more components join.

▶ Nodal analysis uses Kirchhoff's current law to define parameters in the network. Currents to be used in the analysis, however, are expressed in terms of V/R.

▶ Delta and wye circuit configurations are quite common to many electrical and electronic circuits. The ability to analyze these circuits is enhanced by knowing how to convert, or transform, values of one type network to the other.

▶ In the conversion formulas for converting from wye to equivalent delta circuit values, the numerator is the common factor used in determining each of the three resistor values.

▶ In the conversion formulas for converting from delta to equivalent wye circuit values, the denominator is the common factor used in determining each of the three resistor values.

▶ In symmetrical wye circuits (those where all three legs of the wye have equal value resistors), the delta equivalent resistances will be three times the value of the wye resistors.

▶ In symmetrical delta circuits (those where all three sides of the delta have equal value resistors), the wye equivalent resistances will be one-third the value of the delta resistors.

FORMULAS AND SAMPLE CALCULATOR SEQUENCES

FORMULA 8–1 $R_A = \dfrac{(R_1 \times R_2) + (R_2 \times R_3) + (R_3 \times R_1)}{R_1}$

⬤, R_1 value, ⬤, R_2 value, ⬤, ⬤, ⬤, R_2 value, ⬤, R_3 value, ⬤, ⬤, ⬤, R_3 value, ⬤, R_1 value, ⬤, ⬤, ⬤, R_1 value, ⬤

FORMULA 8–2 $R_B = \dfrac{(R_1 \times R_2) + (R_2 \times R_3) + (R_3 \times R_1)}{R_2}$

⬤, R_1 value, ⬤, R_2 value, ⬤, ⬤, ⬤, R_2 value, ⬤, R_3 value, ⬤, ⬤, ⬤, R_3 value, ⬤, R_1 value, ⬤, ⬤, ⬤, R_2 value, ⬤

FORMULA 8–3

$$R_C = \frac{(R_1 \times R_2) + (R_2 \times R_3) + (R_3 \times R_1)}{R_3}$$

(, R_1 value, ✕, R_2 value,), +, (, R_2 value, ✕, R_3 value,), +, (, R_3 value, ✕, R_1 value,), =, ÷, R_3 value, =

FORMULA 8–4

$$R_1 = \frac{R_B \times R_C}{R_A + R_B + R_C}$$

R_B value, ✕, R_C value, ÷, (, R_A value, +, R_B value, +, R_C value,), =

FORMULA 8–5

$$R_2 = \frac{R_C \times R_A}{R_A + R_B + R_C}$$

R_C value, ✕, R_A value, ÷, (, R_A value, +, R_B value, +, R_C value,), =

FORMULA 8–6

$$R_3 = \frac{R_A \times R_B}{R_A + R_B + R_C}$$

R_A value, ✕, R_B value, ÷, (, R_A value, +, R_B value, +, R_C value,), =

REVIEW QUESTIONS

1. Define the term *loop.*

2. Define the term *mesh.*

3. Explain the basic difference between a mesh current and an actual circuit current.

4. A common convention regarding labeling mesh currents on a diagram is that they are shown in a:
   a. counter-clockwise direction.
   b. clockwise direction.
   c. different direction for each mesh window.

5. Voltages across resistors in a network that are caused by their own mesh current are considered:
   a. negative.
   b. neutral.
   c. positive.
   d. self-defining.

6. Voltages across resistors in a network that are caused by mesh currents other than their own mesh current are considered:
   a. negative.
   b. neutral.
   c. positive.
   d. self-defining.

7. When assigning source polarities in a network, an arbitrary convention is that if a mesh current returns to the positive side of the source:
   a. the source is indicated as positive in equations.
   b. the source is indicated as negative in equations.
   c. it makes no difference which assignment you use.
   d. sources are not considered in the equations.

8. The purpose of using "simultaneous equations" when using Kirchhoff equations is to:
   a. view two sets of parameters, simultaneously.
   b. eliminate one unknown in order to solve for another.
   c. neither of the above reasons.

9. A "node" is a point in a network where:
   a. two or more circuit components join, and circuit currents branch.
   b. no less than three circuit components join.
   c. where voltages are equal and cancel.
   d. none of the above.

10. A "delta" circuit configuration is the same as a:
    a. wye circuit configuration.
    b. pi circuit configuration.
    c. tee circuit configuration.
    d. none of these configurations.

11. A "wye" circuit configuration is the same as a:
    a. pi circuit configuration.
    b. tee circuit configuration.
    c. delta circuit configuration.
    d. none of these configurations.

12. In ac power circuitry it is common to find these two different circuit configurations:
    a. Wye and tee
    b. Delta and pi
    c. Wye and delta

PROBLEMS

Refer to Figure 8–4. Assume that all three resistors now equal 10 Ω. Assume V_{S_1} is changed to 20 V and V_{S_2} is changed to 16 V. Use the mesh analysis approach to determine the parameters called for in the following questions.

1. I_A

2. I_B

3. V_{R_1}

4. V_{R_2}

5. V_{R_3}

6. I_{R_1}

7. I_{R_2}

8. I_{R_3}

Refer to Figure 8–6. Assume that all three resistors equal 4.7 Ω. Assume V_{S_1} is changed to 21 V and V_{S_2} is changed to 14 V. Use the nodal analysis approach to determine the parameters called for in the following questions.

9. I_1

10. I_2

11. I_3

12. V_{R_1}

13. V_{R_2}

14. V_{R_3}

Refer to Figure 8–11. Assume that $R_1 = 10$ Ω, $R_2 = 12$ Ω, and $R_3 = 15$ Ω. Use the wye-to-delta conversion formulas to find the values of:

15. R_A

16. R_B

17. R_C

18. For the circuit of Figure 8–11, assume that $R_A = 150$ Ω, $R_B = 150$ Ω and $R_C = 150$ Ω. What would be the values of R_1, R_2, and R_3 in an equivalent wye circuit?

ANALYSIS QUESTIONS

1. When would mesh or nodal analysis be used rather than standard series-parallel analysis of a network?

2. How many equations must be written for a three-mesh circuit in order to find the circuit parameters?

3. Perform research and find where you might find delta and wye configurations of electrical circuits in an industrial plant setting.

4. Perform research and find what type(s) of electronic circuitry might commonly contain the pi circuit configuration.

PERFORMANCE PROJECTS CORRELATION CHART

CHAPTER TOPIC	PERFORMANCE PROJECT	PROJECT NUMBER
Assumed Loop or Mesh Current Analysis Approach	Loop/Mesh Analysis	30
Nodal Analysis Approach	Nodal Analysis	31
Conversions of Delta and Wye Networks	Wye-Delta and Delta-Wye Conversions	32

NOTE: It is suggested that after completing the above projects, the student should be required to answer the questions in the "Summary" at the end of this section of projects in the Laboratory Manual.

PRODUCING & MEASURING ELECTRICAL QUANTITIES

CELLS AND BATTERIES

9

chapter

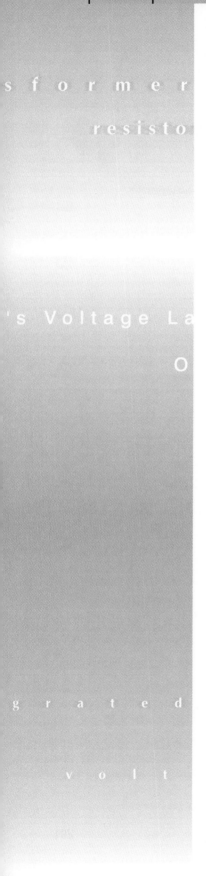

CHAPTER PREVIEW

In this chapter you will study common types of cells and batteries, learn about their operation; and examine some of their basic features. Also, some important facts about their care and usage are presented.

Cells and **batteries** are important sources of dc voltage and current that operate a variety of electrical and electronic circuits and devices. Everything, from automated cameras, burglar alarms, and emergency lighting systems, to portable radio and TV receivers is made possible by various types of cells and batteries, Figure 9–1.

A battery is two or more electrically connected cells usually contained in a single unit. Cells and batteries depend on the process of changing chemical energy into electrical energy by separating and/or producing charges.

Two general categories of these dc sources are nonrechargeable primary cells and rechargeable secondary cells. The common flashlight battery is an example of a nonrechargeable primary cell and the cells in a rechargeable automobile storage battery are examples of secondary cells.

OBJECTIVES

After studying this chapter, you should be able to:

1. Describe the difference between primary and secondary cells

2. Briefly explain chemical action in a cell

3. List physical and operating features of dry cells

4. List physical and operating features of secondary cells

5. Describe precautions to take with dry cells and wet cells

6. List a variety of primary and secondary cells

7. List typical electrical parameters of at least three primary and secondary cells

8. Use the SIMPLER troubleshooting sequence to solve the Chapter Troubleshooting Challenge problem

9-1 CHEMICAL ACTION IN A CELL

An early version of a cell that converts chemical energy to electrical energy is called a voltaic cell, named in honor of Alessandro Volta. A voltaic cell consists of two different metal plates (copper and zinc) immersed in a diluted acid solution (like a weak sulfuric acid solution), Figure 9–2. These metal plates are the **electrodes** and the acid solution is the **electrolyte.**

The chemical action occurring under these conditions is briefly described as follows:

1. When sulfuric acid dissolves in water, it splits into positive hydrogen ions and negative sulfate ions. (Recall from Chapter 1 that positive ions are atoms that have lost electrons and are positively charged entities. Negative ions are atoms that have gained electrons and are negatively charged entities.)

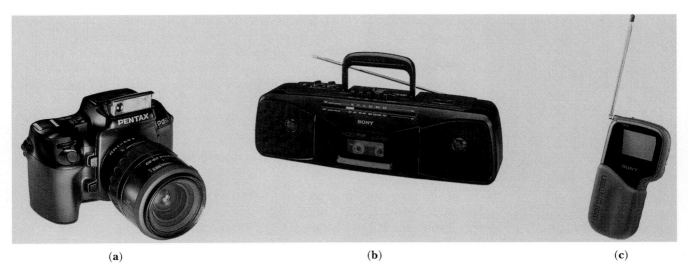

(a)   (b)   (c)

FIGURE 9–1 Some common items that use batteries as a power source: **(a)** 35-mm camera *(Courtesy of Pentax Corp);* **(b)** portable radio *(Courtesy of SONY Corp. of America);* **(c)** portable television *(Photo by Michael A. Gallitelli)*

2. When the zinc plate is immersed in the acid solution, the acid attacks the zinc. This action causes zinc ions to separate from the plate, leaving the zinc plate with an excess of electrons. The zinc plate has a net negative charge. NOTE: When the zinc plate wears away, the cell can no longer operate, and the cell is dead.

3. The positive hydrogen ions draw electrons from the copper plate, leaving the copper plate positively charged.

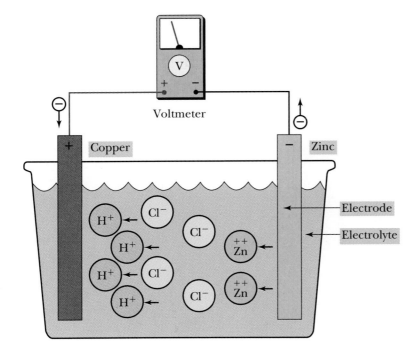

FIGURE 9–2 A simple voltaic cell consists of two different metal plates (electrodes) immersed in a diluted acid solution (electrolyte).

4. Because of the chemical actions described, the zinc plate is negative with respect to the copper plate (and/or the copper plate is positive with respect to the zinc plate). Therefore, a difference of potential exists between the electrodes. This electromotive force (emf) causes current to flow between the electrode terminals, if a conducting path is connected between them, Figure 9–3.

5. The open circuit difference of potential in primary cells depends on the metals used as the electrodes.

6. The amount of deliverable current depends upon the physical size of the electrodes.

In summary, the concepts discussed show that chemical reactions produce electromotive force. This principle of changing chemical energy to electrical energy is fundamental to the operation of all cells and batteries.

9-2 THE COMMON DRY CELL

Using the materials previously described, the useful life of the voltaic cell is very short. Therefore, it is not a practical type of cell.

A more useful primary cell is the carbon-zinc cell (also known as the LeClanche cell). This cell is commonly termed a dry cell.

Construction

The basic construction of a dry cell is shown in Figure 9–4. The key components are:

1. the positive electrode, which is carbon;
2. the negative electrode, which is zinc; and

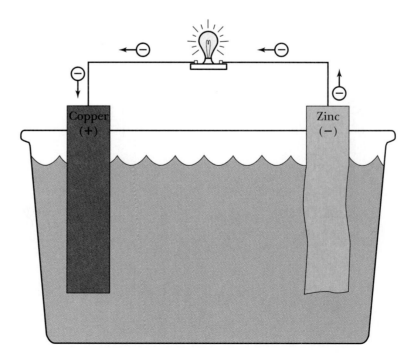

FIGURE 9–3 A cell produces current through the conducting path between electrodes.

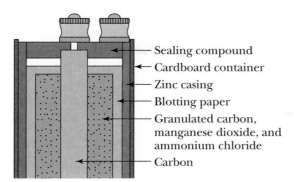

FIGURE 9–4 The construction of a carbon-zinc (LeClanche) cell, commonly termed a dry cell

Sealing compound
Cardboard container
Zinc casing
Blotting paper
Granulated carbon, manganese dioxide, and ammonium chloride
Carbon

3. the electrolyte, which is a solution of ammonium chloride called sal ammoniac that forms a paste-like substance.

Polarization

As the zinc electrode dissolves during the chemical process, hydrogen bubbles form on the positive carbon electrode that is detrimental to the cell's operation. This process is called **polarization.**

To overcome polarization, a **depolarizer** agent is introduced that combines its oxygen with the hydrogen bubbles to form water. This prevents the buildup of the undesired bubbles. A common depolarizing agent is manganese dioxide.

Local action

Another operational problem confronted in the manufacture of dry cells is **local action.** Because of the manufacturing process, most zinc has impurities (such as carbon). Such impurities are suspended in the electrolyte close to the zinc and create independent cell actions. This undesired "voltaic" action dissolves the zinc.

To reduce local action, a thin coating of mercury is put on the zinc by a process called **amalgamation.** Amalgamation seals the impurities, separating them from the electrolyte to prevent local action.

Shelf life

An unused dry cell has an indeterminant shelf life, since some local action occurs and the electrolyte paste will eventually dry up, depending on storage conditions.

The cell becomes worn-out when the zinc, electrolyte, or depolarizer is consumed. This is observed when output voltage, under normal load current conditions, is appreciably lower than it should be.

9–3 OPERATING CHARACTERISTICS OF DRY CELLS

During a dry cell's useful life, voltage output ranges from about 1.4 volts to 1.6 volts with 1.5 volts being the nominal voltage.

Current capability typically ranges from a few milliamperes for small cells with small electrodes to 1/8 ampere or more for large cells, such as a #6 dry cell.

As a cell or battery ages, its internal resistance increases. Consequently, at a given current level, its terminal voltage decreases accordingly. Recall in the previous chapter we described how you can determine the internal resistance of a cell or battery by dividing the difference between no-load and full-load terminal voltage by the load current at full load.

PRACTICE PROBLEMS I

1. What is the internal resistance of a dry cell having a no-load terminal voltage of 1.5 V and a full-load terminal voltage of 1.25 V with a load current of 100 mA?
2. According to our nominal normal range values discussed earlier, is this cell considered good, fair, or bad in performance?

PRACTICAL NOTES

As indicated earlier, drying of the sal ammoniac electrolyte paste and local action shorten the shelf life of dry cells. Also, trying to draw too much current and/or drawing high current for extended periods will shorten operating life.

To lengthen shelf life, dry cells and batteries should not be stored in hot environments and should not be operated at currents that are not within normal specified operating ranges.

If terminal voltage appreciably drops from the nominal rated value, the internal resistance of the cell has increased because of its deteriorated operating condition. To ensure proper operation of the circuit(s) using the cell as a source, replace the cell when this condition is apparent.

IN-PROCESS LEARNING CHECK I

Fill in the blanks as appropriate.

1. The carbon-zinc type dry cell is composed of a negative electrode made of _____ and a positive electrode made of _____. The electrolyte solution in this type cell is _____ _____, called sal _____.
2. The process which causes hydrogen bubbles to form on the positive electrode, and which is detrimental to the cell's operation, is called _____.
3. Manganese dioxide is often added to the cell to prevent the process mentioned in question 2. This agent is sometimes known as a _____ agent.
4. Another problem that can occur in dry cells is that of little batteries being formed because of impurities suspended in the cell's electrolyte near the zinc electrode. This type activity is known as _____ _____. To reduce this undesired activity, mercury is put on the zinc by a process called _____.
5. The shelf life of dry cells is (enhanced, hurt) _____ by storing the cells in hot conditions.
6. Nominal cell terminal voltage of dry cells is about _____ V.

9-4 RECHARGEABLE (SECONDARY) CELLS AND BATTERIES

Lead-acid batteries

The most commonly used rechargeable battery is, of course, the lead-acid type, used in automotive and other services, Figure 9–5. In its automotive application, it partially discharges at a high rate (during starting of the automobile), then is recharged by the automobile's alternator system while the car is running. While the car is not running, certain accessories may operate off the battery. For example, the clock runs continuously. Radios, tape decks, and so on may also be operated when the car's engine is off. The nominal cell voltage for the common 12-V lead-acid type battery is 2 V per cell.

Deep-cycle batteries

Another popular, high-power, lead-acid type battery is called the "deep-cycle" battery. These are typically used for powering electrical fishing motors and for other recreational devices. They are typically the size of a medium automotive battery and may furnish from 1,000 to 1,200 watt-hours per charge in normal operating temperatures. If cared for properly, this type battery may operate satisfactorily for more than 200 charge/discharge cycles.

Some other applications

Lead-acid batteries are often used in other applications. They provide power for mobile and portable transmitters and transceivers, and for electric vehicles and

FIGURE 9–5 Automotive lead-acid battery

FIGURE 9–6 The nickel-cadmium battery *(Courtesy of Eveready Battery Co., Inc.)*

wheel chairs. Special, glass-encased, stationary lead-acid batteries, or banks of batteries, provide backup power for telephone company installations. You will find this type battery being used almost anywhere considerable dc power is required, particularly where a portable or mobile source is needed.

Nickel-cadmium (ni-cad) cells

The nickel-cadmium (ni-cad) cell is probably the most commonly used, all-purpose, rechargeable dry cell available today, Figure 9–6. The nominal voltage, per cell, for this type is 1.2 V per cell. Properly used, these cells/batteries are capable of more than 500 charge/discharge cycles. This type cell or battery should never be discharged to less than 1.0 V per cell, as a cell can reverse polarity. This can cause a short circuit and/or ruptured seals.

Catalog listings for this battery indicate capacities in the range of 110 to more than 1,000 milliampere-hour ratings. This type of battery has a long shelf life, and is typically used in applications where there is cyclic rather than continuous use. A slow, "trickle" charging method is recommended for the recharging action.

Nickel-metal-hydride cell

This newer type rechargeable cell has some features that differ from the more commonly known ni-cad cells. Its main advantage is that it can deliver about 50% to 60% greater capacity per given size. However, it has several disadvantages at our present state of technology. Some of its disadvantages are that it is more costly than ni-cads, its cycle life is shorter, and it cannot be fast-charged. For example, you cannot charge it at a 1C rate, and there are some low-temperature charging concerns. For this reason, you cannot use the same charger to charge this type cell as you would use to charge a ni-cad.

9–5 BATTERY CAPACITY RATINGS

The capacity ratings of cells and batteries are related to the continuous current that can be supplied by the cell or battery over a specified period of time. This common rating is often referred to as the **ampere-hour (Ah) rating.** This rating is often expressed as the C rating. For example $C/12$ would indicate the current available continuously for 12 hours. The value for this C rating is dependent upon the discharge rate. Figure 9–7 shows a graph that illustrates the relationship of discharge rate to the ampere-hour (C) rating for two sizes of lead-acid storage batteries. As you can see, the higher the discharge rate, the lower the ampere-hour, or C rating for a given battery.

Typical capacity ratings may be in the range of a few milliampere-hours for very small batteries (such as used in hearing aids and other low-power devices) to more than 100 ampere-hours for deep-cycle storage batteries.

Example

A battery is rated as $C/15$. In this case, the C indicates the current continuously available for 15 hours. As shown in the following graph, if the discharge rate is changed, then the ampere-hour rating (C) changes.

From the graph shown in Figure 9–7, we can see that at a discharge rate of 10 amperes, the C rating for the type 27 lead-acid battery is about 85. The

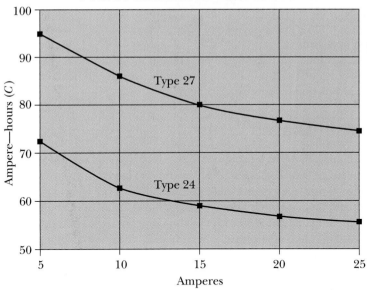

FIGURE 9–7 Data related to battery ratings

product of amperes and hours (or *C* rating), then, would indicate that this battery could deliver 10 amperes for about 85/10, or for 8.5 hours. That is 10 A × 8.5 h = 85 Ah.

PRACTICE PROBLEMS II

1. What continuous level of current can be delivered over a 20-hour period by a battery rated at 150 ampere-hours?

2. Using the graph, what would be the *C* rating for a type 24 battery being dischaged at a rate of 12.5 amperes?

9–6 DATA ON RECHARGING RECHARGEABLE BATTERIES

The full-charge rating *(C)* of a battery is expressed in ampere-hours. Since there is not 100% efficiency in recharging a battery, it will take more hours to fully charge the battery than that simply indicated by the *C* rating. For example, if you are charging a battery at one-tenth its *C* rating current (0.1 C, or the 10-hour rate), it may take more than 12 hours to fully charge the battery, rather than just the 10 hours indicated by the 0.1 C charging rate.

You should also be aware that charging methods used for lead-acid and ni-cad type cells and batteries should be different. To fully charge a ni-cad battery, it should be charged using a "constant-current" method at a rate of about 0.1 C (one-tenth full current rating) level. Typically, this would be about 50 mA for an AA-size cell. This charge rate is high enough to recharge the cells, yet not so high as to create overcharging damage. NOTE: Many ni-cads are manufactured so as to accept a certain level of overcharging, but overcharging still should be minimized.

For gel-filled (nonspill, sealed, any-mounting-position) lead-acid batteries, voltage-limited charging is best, since this type cannot accept overcharging.

Charging should typically be done at C/10 or C/20 until each cell's voltage reaches about 2.3 V. At this point, the charging current should taper to simply a maintenance of charge level. Deep-cycle, lead-acid batteries are also best charged at a slow rate.

Some automotive lead-acid batteries, and some ni-cad cells/batteries may be safely "quick-charged" at higher charging rates, but the manufacturer's recommendations should be followed in doing this.

9–7 SMART CHARGERS

Many quick-charge systems simply use a "timing" system, without regard to the condition of the cells or batteries being charged. There are occasions where this technique can actually harm the cell or battery.

New *Smart Chargers*, which *sample* and *test* the condition of the cell or battery being charged, are now on the market in the communications field. The charger then cycles on and off, appropriately, to optimally charge the cell or battery without overcharging or damaging it.

9–8 CONSTRUCTION AND OPERATIONAL CHARACTERISTICS OF THE LEAD-ACID BATTERY

The key components include:

1. the positive plate(s), which is lead peroxide (PbO_2) (brown);
2. the negative plate(s), which is spongy lead (Pb) (gray);
3. the electrolyte, which is dilute sulphuric acid (H_2SO_4);
4. nonconducting separators; and
5. hard rubber or plastic enclosure.

Charged condition(s)

When the automotive lead-acid battery is fully charged, the conditions include:

1. positive plate(s) that are lead peroxide and dark brown;
2. negative plate(s) that are spongy lead and gray;
3. electrolyte specific gravity that is between 1.275 and 1.30; and
4. closed-circuit (normal load) voltage per cell that is about 2 V. (Open-circuit voltage per cell is about 2.2 V without load when the battery is fully charged.)

Discharging condition(s)

As the lead-acid battery is discharging, the battery's chemical actions cause:

1. the positive plates to deteriorate, resulting in formulation of lead sulfate and water, which further dilutes the electrolyte;
2. the negative plates to deteriorate, resulting in formulation of lead sulfate; and
3. the electrolyte to weaken and decrease in specific gravity.

Discharged condition(s)

When the lead-acid battery has fully discharged, typical conditions are:

1. both positive and negative plates are changed to lead sulfate;
2. the creation of water weakens the sulfuric acid electrolyte so the specific gravity (measured with a hydrometer) drops below 1.150;
3. closed-circuit voltage per cell drops to about 1.75 V; and
4. if the battery remains in an uncharged condition for too long, the sulfate hardens and the chemical reversal (via charging action) becomes impossible. This is called sulfation.

Common testing methods

Two common methods for checking the condition of charge or discharge of a lead-acid battery are 1) checking the specific gravity of the battery's electrolyte and 2) checking the battery's output voltage under load. Most of us have seen both methods used or may have used these methods ourselves.

The "hydrometer" is a commonly used testing device used to check the specific gravity of the electrolyte in lead-acid batteries. By determining the specific gravity of the electrolyte, the "charge" condition of the battery can be observed.

Numerous devices and testing systems are used to put a load on a battery to check its output voltage under load. Everything from simply turning on the headlights and measuring the battery voltage to computerized loading and monitoring systems are in use today.

9-9 BATTERY CHARGING METHODS

Overview

Charging any partially or fully discharged secondary battery involves forcing current through it in the reverse direction (from the normal current flow) when it is acting as a source.

Look at Figure 9–8. Notice the difference in current flow directions when the battery is a source (providing current to a load) and when the battery is the

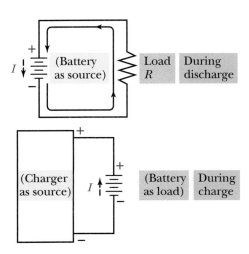

FIGURE 9–8 Directions of current through battery during charge and discharge

FIGURE 9–9 Constant voltage method of charging a battery

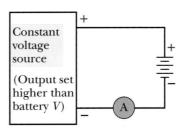

Higher charging current initially when battery's series-opposing voltage is lower.

Charging current tapers downward in value as battery charges, and its series-opposing voltage nears that of the charger voltage.

NOTE: A constant voltage source is one that has very low internal R; thus, varying values of load current won't cause source output voltage to change much.

load for the charging device or circuit. NOTE: To accomplish the charging action, the charger voltage must be greater than the battery voltage; the negative output terminal of the charger is connected to the negative battery terminal; and the positive charger output terminal is connected to the battery positive terminal. That is, the connections are *negative-to-negative* and *positive-to-positive*.

Two charging methods

The constant voltage method, Figure 9–9, applies a constant charging voltage to the battery. When the battery voltage, which is series-opposing to the charger voltage, is low, the charging current is high. As the battery increases its charge, thus its terminal voltage, the battery voltage almost equals the charging source. Thus, the charging current's value tapers downward as the battery charges and its series-opposing voltage nears that of the charger voltage.

The constant current method, Figure 9–10, essentially charges the battery at a constant current rate (usually the battery's normal discharge rate) for about 90% of the charging time. During the remaining time, the rate reduces to about 40% of normal discharge rate. For example, a 100-ampere-hour battery initially

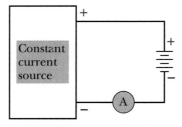

Charged at near normal discharge rate for about 90% of charging cycle, then reduced to about 40% of normal discharge rate for the remainder of the charging time.

FIGURE 9–10 Constant current method of charging a battery

NOTE: A constant current source has a high internal R compared to R of load; thus, load has little effect on circuit current.

charges at a rate of about 12 amperes, then reduces to about 5 amperes for the remaining period of charge.

9–10 EXAMPLES OF OTHER TYPES OF PRIMARY AND SECONDARY CELLS

Electronic parts catalogs list a variety of primary and secondary cells and batteries. We will briefly mention several common types whose application depend on inherent features, such as size, weight, current capacity, voltage, longevity, and durability.

Examples of rechargeable secondary cells and batteries

The Edison cell

The Edison (nickel-iron) cell is commonly used in heavy-duty industrial applications. This cell is lighter and more durable than the lead-acid cell. Although it endures many rechargings, it is more expensive than the lead-acid cell.

Other types

Other rechargeable secondary cells and batteries are lead-calcium, silver-zinc, and silver-cadmium.

Examples of nonrechargeable primary cells and batteries

Primary cells and batteries are used in a variety of applications, such as watches, cameras, computer memory protection circuits, and radios.

Carbon-zinc cell

The common dry cell used in flashlights, children's toys, and numerous other applications is the carbon-zinc cell. Note in Figure 9–11a a cell with its own built-in tester is shown.

Alkaline cell

The alkaline cell has a longer shelf life than the standard carbon-zinc dry cell, Figure 9–11b.

(a)

(b)

FIGURE 9–11 Two common types of nonrechargeable cells

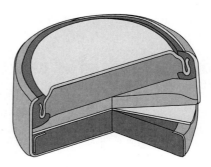

FIGURE 9–12 The mercury cell

Mercury cell

The mercury cell is often used in voltage reference applications because of the stable output voltage (typically 1.35 V per cell) during its lifetime, Figure 9–12.

Zinc-chloride cell

The main feature of the zinc-chloride cell is high current capacity.

Lithium cell

Figure 9–13 shows an example of a lithium cell. Because of their long life, lithium cells are commonly used in watches, cameras, calculators, computers, and similar devices. Although they are expensive, their long life and energy density make them practical.

FIGURE 9–13 The lithium cell *(Courtesy of Eveready Battery Co., Inc.)*

Solar cell

The solar cell, sometimes called a photovoltaic cell, is a device composed of a semiconductor diode that has the ability to convert light into electric current. See Figure 9–14. As light strikes the active surface, some electrons are moved, or released from their locations in the crystalline structure, and some of these can then become useful in establishing current through an external circuit. Typical voltage for these cells is about 0.26 V per cell. Because their current-producing capability is very small and present efficiency factors are as low as about 10% to 15% for the economically produced versions of these cells, they are usually used in large series-parallel arrays, rather than as single cells. Common applications include everything from hand-held calculators to microwave relay stations and space craft.

Refer to Figure 9–15 for a listing of other primary and secondary cells and batteries and their nominal output voltage per cell.

Final comments about cells and batteries

Selection of the right cell or battery for the job depends on the requirements of the application relative to *size, current drains, voltage, frequency of replacement,* and *cost.*

Recall from an earlier chapter the possibilities of connecting cells or batteries, Figure 9–16a, in series to increase voltage, Figure 9–16b, and of connecting like-value voltage sources in parallel to increase current delivery, Figure 9–16c.

FIGURE 9–14 A solar cell

PRIMARY CELLS/BATTERIES (NONRECHARGEABLE)		SECONDARY CELLS/BATTERIES (RECHARGEABLE)	
TYPE	**NOMINAL V**	**TYPE**	**NOMINAL V**
Carbon-Zinc	1.5	Lead-Acid	2.2
Alkaline	1.5	Nickel-Iron	1.4
Mercury	1.35	Nickel-Cadmium	1.2
Silver-Oxide	1.5		
Lithium	3.0		

FIGURE 9–15 Chart of primary and secondary cells/batteries with nominal voltage values

When performing either process, it is recommended that the cells and batteries have the same manufacturer's electrical ratings.

Also, recall that as a cell or battery deteriorates, its internal resistance increases. This means that for a given current drain, its output terminal voltage decreases because of the increased internal IR drop at the given current value. NOTE: Some *newer types* of batteries typically *maintain* their *voltage* output then *suddenly die*. Examples are mercury (primary) cells and nickel-cadmium (secondary) cells.

Shelf life of cells is greatly affected by temperature. Most dry cells operate near 70° F, so long exposure to higher temperatures drastically shortens their life. Conversely, batteries stored for long periods at low temperatures (e.g., 0° F) have a good shelf life.

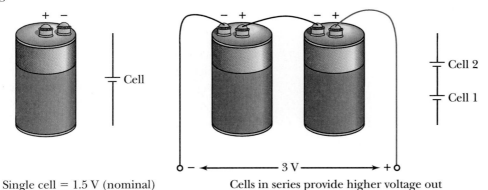

Single cell = 1.5 V (nominal)

(a)

Cells in series provide higher voltage out (series-aiding connections)

(b)

Cells in parallel can provide higher current to the load at a voltage = to that of one cell

(c)

FIGURE 9–16 Advantages of multiple-cell connections

PRACTICAL NOTES

Caution: *Do not leave dry cells or batteries in equipment for long periods without checking them!* With age, most cells and batteries leak electrolyte or acid that damages battery holder contacts, circuitry, or equipment. Therefore, remove a dead or degenerated cell or battery, even if a replacement is not available. Do not leave a dead cell or battery in the equipment.

Finally, do not recharge batteries at too high a charging rate; be cautious with sparks or flames near a charging battery; and, of course, be careful to prevent spillage of the electrolyte acid!

SUMMARY

▶ Cells and batteries (combinations of cells) are important sources of dc voltage and current, especially for portable equipment (e.g., emergency back-up equipment).

▶ Primary cells are *not rechargeable,* have a variety of voltage ratings, current capabilities, expected shelf life ratings, and costs, and are produced in many sizes and shapes.

▶ Secondary cells and batteries *are rechargeable.* Secondary cells or batteries (such as lead-acid batteries and ni-cad batteries) are available in a variety of ratings, sizes, shapes, and costs.

▶ The common dry cell is composed of a positive carbon electrode, a negative zinc electrode, and an electrolyte of ammonium chloride (and water), called sal ammoniac.

▶ During a dry cell's normal operating chemical process, hydrogen bubbles form around the positive electrode. This polarization degenerates the battery's output. To reduce polarization, a depolarizing agent is placed in the dry cell during manufacture. A common depolarizing agent is manganese dioxide.

▶ Local action also causes degeneration in dry cells. Impurities in the negative zinc electrode suspend in the electrolyte near the zinc, creating their own voltaic action and eating away the zinc. To reduce local action, the inside of the zinc container (negative electrode) is covered with a thin coating of mercury; thus preventing most potential for local action.

▶ Shelf life of cells and batteries depends on storage conditions and chemical properties.

▶ Voltage ratings of cells and batteries depend on the electrode material. Current ratings largely depend on the size (active area) of the electrodes.

▶ When using dry cells, care should be taken with their storage. Also, frequently check their condition, and make replacements, as appropriate.

▶ One common secondary (rechargeable) cell is the lead-acid cell. In this cell, the positive plates are lead peroxide, the negative plate(s) are spongy lead, and the electrolyte is diluted sulfuric acid.

▶ Batteries are rated according to the amount of current they deliver for a specified time. For example, a battery rated at 120 ampere-hours can nominally deliver 15 amperes for 8 hours, or 6 amperes for 20 hours. Small batteries and cells are rated in milliampere-hours.

▶ Rechargeable batteries are charged at high rates for short periods, or at lower rates for longer periods to restore them to full charge. The preferred method is to charge at lower rates for longer periods, decreasing the risk of cell or battery damage as well as preventing excessive gassing.

▶ In addition to the common carbon-zinc primary cell, many other types exist, such as alkaline, mercury, zinc-chloride, and lithium cells.

▶ Common types of rechargeable batteries are the lead-acid, nickel-iron (Edison), and nickel-cadmium.

REVIEW QUESTIONS

1. What is a positive ion?

2. In a carbon-zinc cell, which element is the positive electrode and which is the negative electrode?

3. Briefly describe the process of amalgamation.

4. What is the term used to describe the undesired collection of hydrogen bubbles around a dry cell's positive electrode?

5. Name a common depolarizing agent.

6. Name a common rechargeable cell used for many small portable electronic devices.

7. What type of cell is commonly used as a reference voltage source because of its stable output voltage?

8. Name two advantages and one disadvantage of the lead-acid cell.

9. What are some common applications of the modern ni-cad cell?

10. What are some common applications of solar cells?

PROBLEMS

1. What is the internal resistance of a battery whose no-load voltage is 12 V, but whose terminal voltage drops to 11 V when a load current of 20 A is demanded?

2. If the load current demand from the battery in question 1 were increased to 30 A, what would be the voltage delivered to the load?

3. What is the ampere-hour rating of a battery that delivers 12.5 A for 8 hours without its output voltage excessively dropping?

4. What circuit configuration of 1.5 V cells would you use to obtain an output of 6 V at a current demand equal to the current rating of one of the cells?

5. What circuit configuration of 1.5-V cells would you use to obtain an output of 3 V with the capability of demanding a load current twice the current-rating of one of the cells?

ANALYSIS QUESTIONS

1. Using the constant voltage method, describe the process and features of charging action when a battery is recharged.

2. Using the constant current method, explain the process when a battery is recharged.

3. Explain the polarity relationships between charger and battery when recharging a battery.

4. Refer to an electronic parts catalog and find the typical ampere-hour or milliampere-hour ratings for:
   a. ni-cad cells.
   b. alkaline dry cells.

5. Refer to an electronic parts catalog. Find and record the voltage and current ratings for the following cells:
   a. AAA   d. C
   b. AA   e. D
   c. A

6. What primarily determines the voltage-delivering capability of a single cell?

7. What primarily determines the current-delivering capability of a single cell?

8. Why does a cell's output voltage decrease under load, particularly when the cell is going bad?

9. a. If the value of load resistance connected to a battery or cell is increased in value, will the load current demand upon the battery increase, decrease, or remain the same?

   b. Will the battery or cell be able to deliver normal output voltage for a longer, shorter, or equal period?

10. What are some common applications of the modern lithium cell?

11. What makes the lithium cell so useful for the applications you mentioned in question 10?

THE SIMPLER SEQUENCE FOR TROUBLESHOOTING

Follow the seven-step SIMPLER troubleshooting sequence, as outlined below, and see if you can find the problem with this circuit. As you follow the sequence, record your circuit test results for each testing step on a separate sheet of paper. This will aid you and your instructor to see where your thinking is on track or where it might deviate from the best procedure.

step 1 **Symptoms**

—(Gather, verify, and analyze symptom information) To begin this process, read the data under the "Starting Point Information" heading. Particularly look for circuit parameters that are not normal for the circuit configuration and component values given in the schematic diagram. For example, look for currents that are too high or too low; voltages that are too high or too low; resistances that are too high or too low, etc. This analysis should give you a first clue (or symptom information) that will aid you in determining possible areas or components in the circuit that might be causing this symptom.

step 2 **Identify**

—(Identify and bracket the initial suspect area) To perform this first "bracketing" step, analyze the clue or symptom information from the "Symptoms" step, then either on paper or in your mind, bracket, circle, or put parentheses around all of the circuit area that contains any components, wires, etc., that might cause the abnormality that your symptom information has pointed out. NOTE: Don't bracket parts of the circuit that contain items that could not cause the symptom!

step 3 **Make**

—(Make a decision about the first test you should make: what type and where) Look under the TEST column and determine which of the available tests shown in that column you think would give you the most meaningful information about the suspect area or components, in the light of the information you have so far.

step 4 **Perform**

—(Perform the first test you have selected from the TEST column list) To simulate performing the chosen test (as if you were getting test results in an actual circuit of this type), follow the dotted line to the right from your selected test to the number in parentheses. The number in parentheses, under the "Results in Appendix C" column tells you what number to look up in Appendix C to see what the test result is for the test you are simulating.

step 5 **Locate**

—(Locate and define a new "narrower" area of uncertainty) With the information you have gained from the test, you should be able to eliminate some of the circuit area or circuit components as still being suspect. In essence, you should be able to move one bracket of the bracketed area so that there is a new smaller area of uncertainty.

step 6 **Examine**

—(Examine available information and determine the next test: what type and where) Use the information you have thus far to determine what your next test should be in the new narrower area of uncertainty. Determine the type test and where, then proceed using the TEST listing to select that test, and the numbers in parentheses and Appendix C data to find the result of that test.

step 7 **Repeat**

—(Repeat the analysis and testing steps until you find the trouble) When you have determined what you would change or do to restore the circuit to normal operation, you have arrived at *your* solution to the problem. You can check your final result against ours by observing the pictorial step-by-step sample solution on the pages immediately following the "Chapter Troubleshooting Challenge"—circuit and test listing page.

step 8 **Verify**

NOTE: A useful *8th Step* is to operate and test the circuit or system in which you have made the "correction changes" to see if it is operating properly. This is the final proof that you have done a good job of troubleshooting.

CHALLENGE CIRCUIT 4

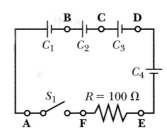

CHALLENGE CIRCUIT 4

Starting Point Information

1. Circuit diagram
2. Cells are 1.5 V cells
3. With S_1 closed, $V_R = 5$ V

TEST	Results in Appendix C
(With S_1 Open)	
V_{A-B}	(18)
V_{B-C}	(1)
V_{C-D}	(51)
V_{D-E}	(99)
V_{A-F}	(77)
(With S_1 Closed)	
V_{A-B}	(7)
V_{B-C}	(50)
V_{A-C}	(81)
V_{C-D}	(112)
V_{D-E}	(73)
V_{C-E}	(10)
V_{A-F}	(105)

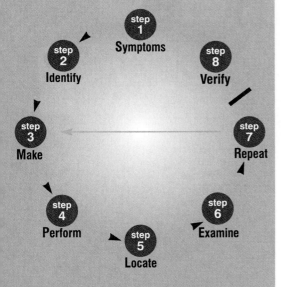

Symptoms

With the switch closed, V_R is 5 volts. However, if all the 1.5-volt cells were operating properly, it should be 6 volts. Since the resistor drops 5 volts with the switch closed, there is circuit current and circuit continuity. This implies a worn cell in the group.

Identify

initial suspect area: All the cells.

Make

test decision: If we use the divide-and-conquer (splitting) technique, we can check the voltage from A to C or from C to E. This narrows the suspect area to two cells, either C_1 and C_2, or C_3 and C_4. Let's start with the first pair by checking voltage from point A to C with S_1 closed to cause circuit current. That is, the batteries are being tested under load.

Perform

1st Test: The result of measuring V_{A-C} is 2 V, which is an abnormal reading for two 1.5-V cells in series.

Locate

new suspect area: Cells C_1 and C_2.

Examine

available data.

Repeat

analysis and testing: Check the voltage across either C_1 or C_2 with S_1 closed.

2nd Test: Check voltage across C_1. V is 1.5 V, which is normal.
3rd Test: Check voltage across C_2. V is 0.5 V, which is abnormal, suggesting that cell C_2 is probably the bad cell.

Verify

4th Test: Replace cell C_2 with a good cell, and observe voltage across R with S_1 closed. V_R is 6 V, which is normal.

Meter lead

Meter lead

Symptoms

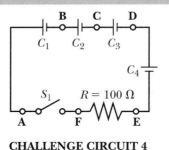

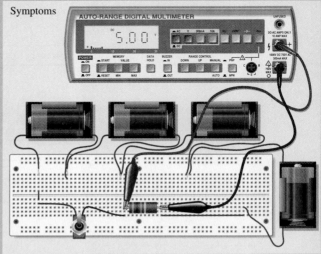

CHALLENGE CIRCUIT 4

1st Test

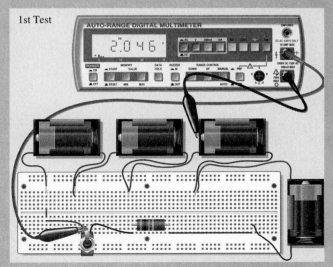

2nd Test

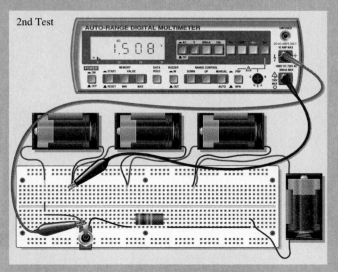

3rd Test

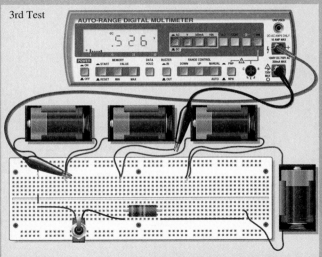

4th Test

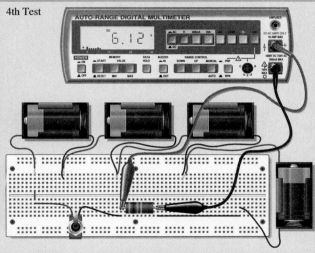

Ampere-turns (AT)
Electromagnetism
Faraday's law
Flux
Flux density
Induction
Left-hand rules
Lenz's law
Lines of force
Magnetic field
Magnetic field intensity
(H)
Magnetic polarity
Magnetism
Magnetomotive force
Permeability (μ)
Relative permeability (μ_r)
Reluctance
Residual magnetism
Saturation
Tesla
Weber

MAGNETISM AND ELECTROMAGNETISM

10

chapter

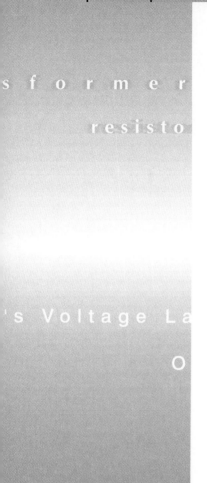

CHAPTER PREVIEW

Of all the phenomena in the universe, one of the most fascinating is magnetism and its unique relationship to electricity.

Magnetic storage media have become extremely important in today's world. Computers use magnetic "floppy disks" and "hard disks" to store tremendous amounts of data. Audio and video recorders use magnetic tape to store magnetic patterns that are transformed to sound and visual intelligence, which we can enjoy at our choosing.

Knowledge of magnetism is important for understanding a variety of industrial measurement and control devices and systems. Some examples include electrical meters, relays, solenoids, magnetic switches, inductors, transformers, and motors and generators, just to name a few.

In achieving your goal to become a technician, you will have the opportunity to study a number of these devices and components. In this chapter, you will focus on basic principles related to magnetism, some units of measure used with magnetic circuits, and important relationships between magnetic and electrical phenomena.

OBJECTIVES

After studying this chapter, you should be able to:

1. Define **magnetism, magnetic field, magnetic polarity,** and **flux**

2. Draw representations of magnetic fields related to permanent magnets

3. State the magnetic attraction and repulsion law

4. State at least five generalizations about magnetic lines of force

5. Draw representations of fields related to current-carrying conductors

6. Determine the polarity of electromagnets using the **left-hand rule**

7. List and define at least five magnetic units of measure, terms, and symbols

8. Draw and explain a *B-H* curve and its parameters

9. Draw and explain a hysteresis loop and its parameters

10. Explain motor action and generator action related to magnetic fields

11. List the key factors related to induced emf

12. Briefly explain the relationships of quantities in **Faraday's law**

13. Briefly explain **Lenz's law**

10-1 BACKGROUND INFORMATION

Early history

Early discovery of magnetism happened by people noticing the unique characteristics of certain stones called lodestones. These stones (magnetite material), which were found in the district of Magnesia in Asia Minor, attracted small bits of iron. It was also noted that certain materials stroked by lodestones became magnetized.

Materials or substances that have this unique feature are called magnets, and the phenomenon associated with magnets is called **magnetism.** Materials that are attracted by magnets are called magnetic materials.

As early as the eleventh century, the Chinese used magnets as navigational aids, since magnets that are suspended, or free to move, align approximately in a north-south geographic direction.

Today we have materials, called permanent magnets, that retain their magnetism for long periods. Examples of permanent magnets include hard iron or special iron-nickel alloys.

Magnetic materials that lose their magnetism after the magnetizing force is removed are termed temporary magnets.

Later discoveries

Centuries later the relationship between electricity and magnetism was discovered. In the 1800s several scientists made important observations. Hans Oersted noted that a free-swinging magnet would react to a current-carrying wire, if they were in close proximity. As long as current passed through the wire the reaction occurred. But when there was no current, there was no interaction between the magnet and wire. This indicated a steady current caused a steady magnetic effect near the current-carrying conductor. This "field of influence" would eventually be called a **magnetic field.**

Later in the same century, Joseph Henry and Michael Faraday discovered interesting phenomena that occurred when changing current levels or moving magnetic fields were present. Henry determined when current changed levels in a conductor, current was induced in a nearby conductor (having a closed current path). Faraday determined a permanent magnet moving in the vicinity of a conductor induced current in the wire (if there existed a complete path for current through the conductor). It should also be stated that even without a closed current path, voltage is induced.

All of these discoveries are very important! You will see these scientists' names again since several units of magnetic measure are named in their honor.

10-2 FUNDAMENTAL LAWS, RULES, AND TERMS TO DESCRIBE MAGNETISM

When a magnet is created, there is an area of magnetic influence near the magnet called a magnetic field. A magnetic field is established either by alignment of internal magnetic forces within a magnetic material, Figure 10–1, or by organized movement of charges (current flow) through conductor materials. This magnetic field is composed of magnetic **lines of force,** sometimes called **flux lines.** Thus, magnetic flux refers to all the magnetic lines of force associated with that magnet. The symbol for flux is the Greek letter Phi (ϕ). Each magnetic field line is designated by a unit called the maxwell (Mx). That is, one magnetic field line equals one maxwell, or 50 magnetic field lines equal 50 maxwells.

Another basic unit associated with flux is the **weber.** Whereas the maxwell equals one line of flux, the weber (Wb) equals 10^8 lines of flux, or 100 million flux lines. Thus, a microweber equals 100 lines, or 100 Mx.

Another fundamental term related to flux is **flux density,** which refers to the number of lines per given unit area. We'll discuss flux density later in the chapter.

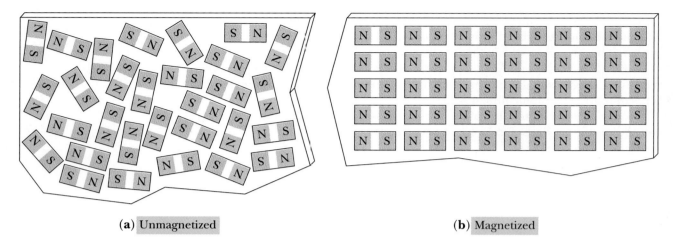

(a) Unmagnetized   **(b)** Magnetized

FIGURE 10–1 Magnetized and unmagnetized conditions

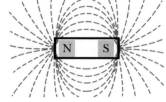

FIGURE 10–2 Iron filings displaying magnetic field pattern of a bar magnet

If a nonmagnetic material, like cardboard, is laid on top of a bar magnet (or other magnets), sprinkled with small iron filings, then tapped or vibrated, the iron filings position themselves in a pattern illustrating the effect of the magnetic field around the magnet, Figure 10–2.

A diagram frequently used to show this effect is in Figure 10–3. Notice that at each end of the bar magnet, the lines of force are concentrated, but as they get farther from the ends, they spread. As you can see, the ends of the bar magnets (where the lines of force are concentrated) are labeled "N" and "S" and are the "poles" of the magnet. That is, the N and S indicate North-seeking and South-seeking poles.

To represent directions of the lines of force, it is customary to say the flux lines exit the North pole and reenter the South pole of the magnet. Inside the magnet, of course, the flux lines travel from the South pole to the North pole. Each flux line is an unbroken loop, or ring.

In Figure 10–4 you can see two bar magnets used to illustrate that *like poles repel each other, and unlike poles attract each other.* This is an important law, so remember it!

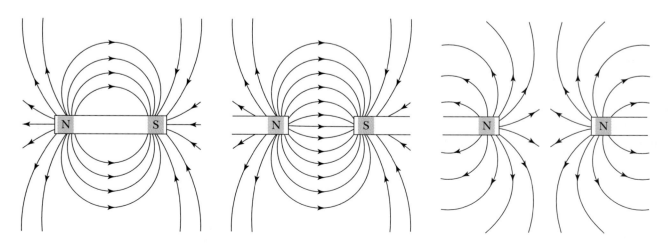

FIGURE 10–3 Lines of force

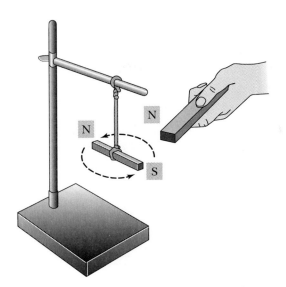

FIGURE 10–4 Like poles repel each other, and unlike poles attract each other.

Rules concerning lines of force

Along with the law defined above, you need to know the following important generalizations related to *lines of force* (flux lines), Figure 10–5.

1. Magnetic lines are continuous, Figure 10–5a.
2. Magnetic lines flow from North to South Pole outside the magnet and from South to North inside the magnet, Figure 10–5a.
3. Magnetic lines take the shortest or easiest path, Figure 10–5b.
4. Lines in the same direction repel each other but are additive, strengthening the overall field. Lines in opposite directions attract and cancel each other, weakening the overall field, Figure 10–5c.
5. Magnetic lines penetrate nonmagnetic materials, Figure 10–5d.
6. Lines of force do not cross each other, Figure 10–5e.

Some practical points about permanent magnets

Before discussing electromagnetism, here are some practical points about permanent magnets.

1. They are not really permanent because they lose their magnetic strength over time, or if heated to a high temperature, or if physically pounded upon, and so on.
2. They can be configured into different shapes in addition to bar magnets. One common shape is termed a horseshoe magnet because of its appearance, Figure 10–6.
3. The field lines' shape depends on the easiest path for flux, Figure 10–7.
4. Appropriate methods to store magnets help preserve their field strengths for longer periods. For example, store bar magnets so opposite poles are together, or use a "keeper bar" of magnetic material across the ends of a horseshoe magnet, Figure 10–8.

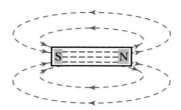

Magnetic lines are continuous and are assumed to flow from North to South poles external to the magnet and from South to North poles inside the magnet.

(a)

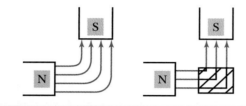

Magnetic lines tend to take the shortest or easiest path.

(b)

Lines in same direction repel each other but are additive, strengthening overall field.

Lines in opposite directions attract and cancel each other, weakening overall field.

(c)

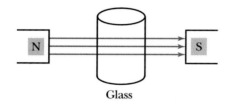

Glass

Magnetic lines penetrate nonmagnetic materials.

(d)

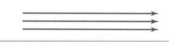

Lines of force don't cross each other.

(e)

FIGURE 10–5 Rules about lines of force

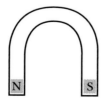

FIGURE 10–6
Horseshoe magnet

FIGURE 10–7
Shape of field lines depends on easiest path for flux.

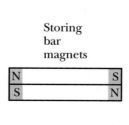

Storing bar magnets

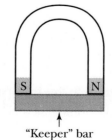

"Keeper" bar

FIGURE 10–8 Preserving the strength of magnets

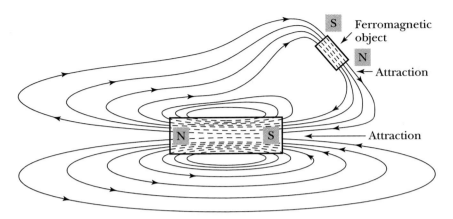

FIGURE 10–9 Induced magnetism

NOTE: polarity of "induced" magnetism places "unlike" poles next to each other. This explains how magnets attract metal objects.

5. A magnet induces magnetic properties in a nearby object of magnetic material, if the object is in the path of the original magnet's field lines, Figure 10–9.

IN-PROCESS LEARNING CHECK I

Fill in the blanks as appropriate.

1. Small magnets that are suspended and free to move align in a _____ and _____ direction.

2. Materials that lose their magnetism after the magnetizing force is removed are called _____ _____. Materials that retain their magnetism after the magnetizing force is removed are called _____ _____.

3. Can a wire carrying dc current establish a magnetic field? _____ If it does, is this field stationary or moving? _____

4. A law of magnetism is like poles _____ each other and unlike poles _____ each other.

5. The maxwell is a magnetic unit that represents how many lines of force? _____

6. The weber is a magnetic unit that represents how many lines of force? _____

7. Are magnetic lines of force continuous or noncontinuous? _____

8. Lines of force related to magnets exit the _____ pole of the magnet and enter the _____ pole.

9. Do nonmagnetic materials stop the flow of magnetic flux lines through themselves? _____

10. Can magnetism be induced from one object to another object? _____

FIGURE 10–10 (a) Field around current-carrying conductor; **(b)** left-hand rule for current-carrying conductor

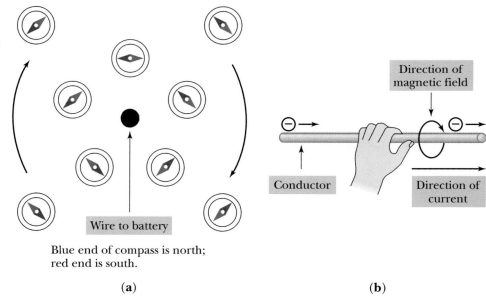

Blue end of compass is north; red end is south.

(a)   **(b)**

10–3 ELEMENTAL ELECTROMAGNETISM

Direction of field around a current-carrying conductor

Recall that Hans Oersted discovered a current-carrying wire has an associated magnetic field. A relationship between the current direction through the wire and the magnetic field direction around the current-carrying conductor is established while moving the compass around the conductor and noting the effect on the compass needle.

Note in Figure 10–10a you are looking at the end of wire (black dot), and the current is coming toward you. In Figure 10–10b you get an overview of the relationship of current direction with the magnetic field around the current-carrying conductor. Both of these figures illustrate one of the **left-hand rules.**

The left-hand rule for current-carrying conductors

Grasping the wire with your left hand so your extended thumb points in the direction of current flow, the magnetic field around the conductor is in the direction of the fingers that are wrapped around the conductor. (NOTE: For those who use "conventional current," where the direction of current flow is considered in the opposite direction, the "right-hand" rule is used instead of this left-hand rule.)

Example

Another way to illustrate this rule is shown in Figure 10–11. In this illustration the current (and your thumb) is coming toward you from the conductor with the dot, Figure 10–11a. The dot is like looking at the tip of an arrow coming toward you. In this case, the magnetic field is in a clockwise direction.

In the other picture, Figure 10–11b, the conductor has a plus sign, indicating current is going away from you into the page. This is similar to looking at the tail feathers of an arrow. In this case, the direction of magnetic field is counterclockwise.

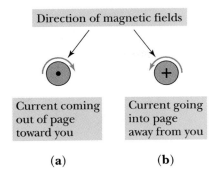

FIGURE 10–11 Magnetic fields around current-carrying conductors

(a)   *(b)*

PRACTICE PROBLEMS I

1. Draw a picture of two parallel conductors with the direction of current going away from you in both cases.

2. Using arrows, show the directions of the magnetic fields around each conductor.

Force between parallel current-carrying conductors

Look at Figure 10–12a. Notice that when two parallel current-carrying wires have current in the same direction, the flux lines between the wires are in opposite directions. This situation causes the flux lines to attract and cancel each other, and the magnetic field pattern is modified since lines of force cannot cross. This indicates the two adjacent wires move toward the weakened portion of the field or toward each other.

On the other hand, if the two wires have current in opposite directions, Figure 10–12b, the lines of force between the two conductors move in the same direction. Hence, the flux lines repel each other and the wires move apart.

PRACTICE PROBLEMS II

Indicate whether the conductors move together (toward each other) or apart in Figure 10–13a, b, and c.

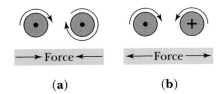

(a)   **(b)**

FIGURE 10–12 Force between adjacent current-carrying conductors

(a)   **(b)**   **(c)**

FIGURE 10–13

FIGURE 10–14 Stronger "combined" field

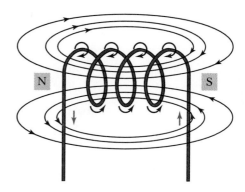

FIGURE 10–15 Strong field of coil and establishment of magnetic polarity

Field around a multiple-turn coil

Individual fields of conductors carrying current in the same direction modify to a larger and stronger field, as shown in Figure 10–14.

When a single current-carrying wire forms multiple loops (such as in a coil), there are a number of fields that combine from adjacent turns of the coil, producing a large field through and around the coil, Figure 10–15.

Also notice in Figure 10–15 the magnet established by this coil has North and South magnetic poles, like a permanent bar magnet. The **magnetic polarity** of this electromagnet is determined by another left-hand rule.

Left-hand rule to determine polarity of electromagnets

Grasping the coil so your fingers are pointing in the same direction as the current passing through the coil, the extended thumb points toward the end of the coil that is the North pole, Figure 10–16.

PRACTICE PROBLEMS III
See if you can determine the magnetic polarity of the coils in Figure 10–17a, b, and c.

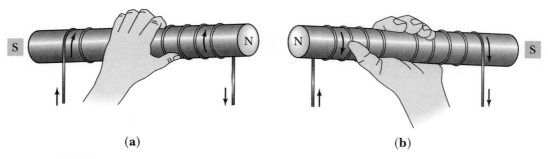

(a)   (b)

FIGURE 10–16 Left-hand rule to determine magnetic polarity of an electromagnet

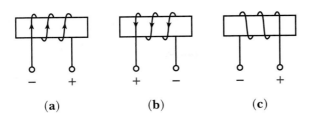

FIGURE 10–17

Factors influencing the field strength of an electromagnet

From the previous discussions, you should surmise the number of turns influences the strength of the electromagnet. That is, the higher the number of turns per unit length of coil, the greater the strength of the magnetic field, Figure 10–18a.

Also, you might have reasoned the amount of current through the conductor directly affects the magnetic strength, since a higher current indicates more moving charge per unit time, Figure 10-18b. These are correct assumptions. Furthermore, the larger the cross-sectional area of the coil, the greater the strength, Figure 10–18c.

Another factor influencing strength is the characteristics of the material in the flux-line path. If the material in the flux path is magnetic, the strength is greater compared with a path using nonmagnetic material, Figure 10–18d.

The length of the magnetic path also influences strength. For example, suppose we have a 100-turn coil with a given length of "X." Now if we spread that 100 turns over a length twice as great (2X), the strength reduces because the

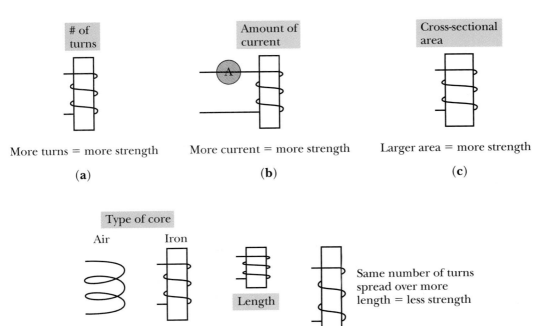

FIGURE 10–18 Some factors that determine strength of an electromagnet

turns are farther apart, thus the effective magnetizing force is less, Figure 10–18e.

To summarize, field strength is influenced by:

1. number of turns on the coil;
2. value of current through coil;
3. cross-sectional area of coil;
4. type of core (magnetic characteristics of material in flux path); and
5. length of the coil (number of turns per unit length).

IN-PROCESS LEARNING CHECK II

1. For a given coil dimension and core material, what two factors primarily affect the strength of an electromagnet? _____ and _____.

2. The left-hand rule for determining the polarity of electromagnetics states that when the fingers of your left hand point (in the same direction, in the opposite direction) _____ as the current passing through the coil, the thumb points toward the (North, South) _____ pole of the electromagnet.

3. Adjacent current-carrying conductors that are carrying current in the same direction tend to (attract each other, repel each other) _____.

4. If you grasp a current-carrying conductor so that the thumb of your left hand is in the direction of current through the conductor, the fingers "curled around the conductor" (will indicate, will not indicate) _____ the direction of the magnetic field around the conductor.

5. When representing an end view of a current conductor pictorially, it is common to show current coming out of the paper via a (dot, a cross, or plus sign) _____.

Highlighting magnetism and electromagnetism facts discussed thus far:

1. Like poles repel, and unlike poles attract, see Figures 10–4 (page 317) and 10–19a.

2. Amount of repulsion or attraction depends on the strength and proximity of the poles involved. Repulsion or attraction is directly related to the poles' strength and inversely related to the *square* of their separation distance, Figure 10–19a.

FORMULA 10–1 $\text{Force} = \dfrac{\text{Pole 1 Strength} \times \text{Pole 2 Strength}}{d^2}$

3. Strength of magnet is related to number of flux lines in its magnetic field per unit cross-sectional area.

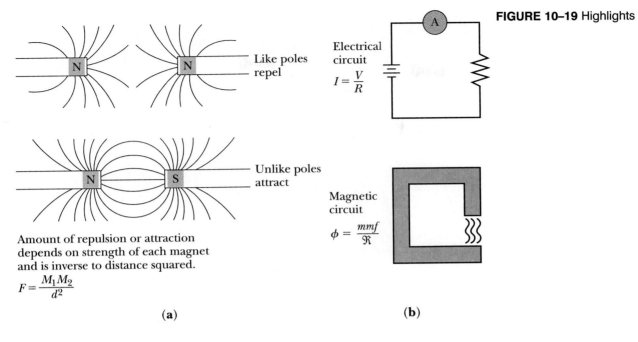

FIGURE 10–19 Highlights

Like poles repel

Unlike poles attract

Amount of repulsion or attraction depends on strength of each magnet and is inverse to distance squared.

$$F = \frac{M_1 M_2}{d^2}$$

Electrical circuit

$$I = \frac{V}{R}$$

Magnetic circuit

$$\phi = \frac{mmf}{\mathcal{R}}$$

(a)　　　　　　　　　　　　　　**(b)**

4. Magnetic lines outside of the magnet travel from North pole to South pole, see Figures 10–5a (page 318) and 10–19a.

5. Characteristics of the flux path determine how much flux is established by a given magnetizing force. Therefore, if the path is air, vacuum, or any non-magnetic material, less flux is established. If the path is a magnetic material, flux lines are more easily established. These concepts indicate that when air gaps are introduced into a magnetic path, the magnetic strength of the field is diminished.

6. The shape or configuration of the magnetic field flux lines is controlled by features of the materials in their path. That is, lines of force follow the path of least opposition, called **reluctance** in magnetic circuits, see Figure 10–5b.

7. An analogy can be drawn between Ohm's law in electrical circuits and the relationships of a magnetic circuit. In Ohm's law, the amount of current through a circuit is directly related to the amount of electromotive force *(V)* applied to the circuit and inversely related to the circuit's opposition to current flow *(R)*. In a magnetic circuit, the number of flux lines through the magnetic circuit is directly related to the value of the magnetizing force establishing flux and inversely related to the circuit's opposition to those flux lines. This opposition is the reluctance of the path, and the symbol is $\mathcal{R}$, Figure 10–19b.

10–4 SOME PRACTICAL APPLICATIONS OF MAGNETISM

Magnetic storage media

In Figure 10–20 you see several commonly used magnetic storage media elements.

Information on the computer diskettes is stored in structured track and sector formats, Figure 10–21. This allows the computer to selectively access and

FIGURE 10–20 Several magnetic media

Don't lay magnetic storage devices, such as computer disks or audio tapes, on or near equipment that may have strong magnetic fields! (Examples of these might be motors, generators, and TVs.)

recover data at any desired location in the "magnetically memorized" data, by virtue of that data's location on the diskette.

On audio cassette tapes, of course, sound waves have been translated into electrical signals at the recording tape head, which in turn causes magentic patterns to be induced onto the magnetic tape, Figure 10–22a. When the recorded tape is run over the playback head of the recorder, the reverse process happens. That is, the magnetic patterns on the tape induce electrical signals into the pickup head, Figure 10–22b. These signals are then processed, amplified, and eventually become sound waves, which activate your ear drums.

Video cassette tape recording systems use a technique similar to, but somewhat more complicated than, the audio recording system. Both audio and video signals are magnetically stored on a videotape's magnetic coating. Once recorded, you can enjoy both seeing and hearing the results by playing them back on your VCR.

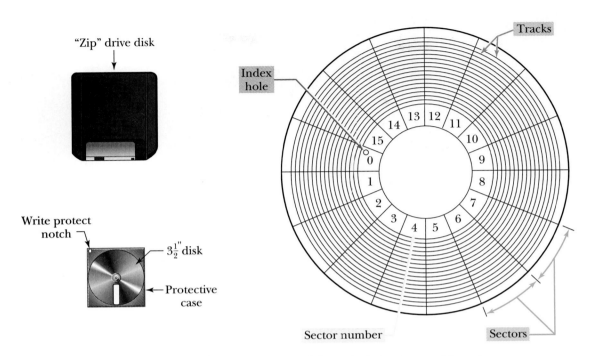

FIGURE 10–21 Computer "floppy" disks typical structure and format

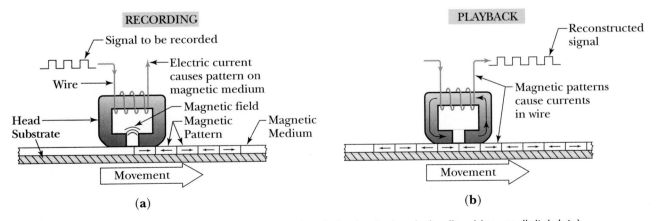

FIGURE 10–22 Using magnetism to record and play back signals (audio, video, or digital data)

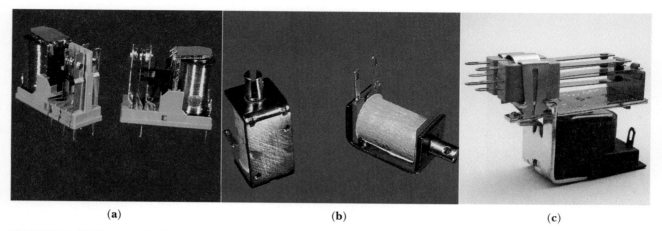

(a)   (b)   (c)

FIGURE 10–23 Common industrial electromagnetic devices *(Photos **a** and **b** courtesy of Guardian Electric Mfg. Co.; photo **c** by Michael A. Gallitelli)*

Industrial components and devices

In industry (and in many home appliances, as well), there are a myriad of applications for relays, solenoids, and magnetic switches, Figure 10–23. Although solid-state switching devices have begun replacing electromechanical relays and solenoids in low-power circuitry, relays and solenoids are still in wide use. The contacts on relays are used to make and break (open and close) circuits as the relay is energized by current flowing through its electromagnetic coil, or is deenergized by current ceasing to flow through its coil, Figure 10–24.

The movable element (plunger, or movable core) in a solenoid is typically used to physically move something. It changes electrical energy into linear, or straight-line, motion, Figure 10–25. The travel of this movable core, then, can be used to do everything from shifting cycles in your washing machine, to operating valves, or positioning items on a production line in industry.

As a technician in industry, you will frequently use metering devices or instruments, which also depend on magnetism and electromagnetism for their operation. At home or at work, you will hear speakers. Most speakers depend upon the interaction of a fixed (permanent magnet) and a changing current through a movable electromagnet (the voice coil) to cause the speaker cone to move the air in the room, Figure 10–26 (page 330). This movement of air, of course, sets up the sound waves you hear.

Motors and generators represent another large application area for the use of magnetism and electromagnetism. A motor changes electrical energy into rotary motion. In contrast, recall that the solenoid changes electrical energy into straight-line motion.

Later in this book, you will be studying the basic operation of transformers. Transformers are widely found in consumer products, delivery of electrical power, and industrial applications. They represent another important application of magnetism and electromagnetism.

In depth details of all these magnetic devices will be taught at the appropriate times in your studies. For now, we wanted you to see just a few of the applications of permanent and electromagnets and realize how important the knowledge of magnetics is to you as a technician or engineer.

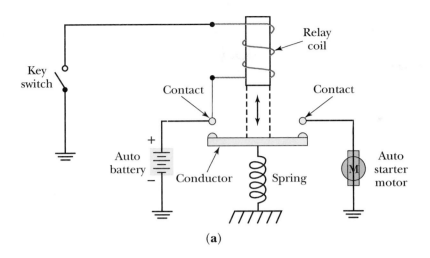

(a)

FIGURE 10–24 Simplified diagrams of relays in use: **(a)** automotive starter system contactor relay; **(b)** single-pole, single-throw relay operated by low power but controlling higher-power device

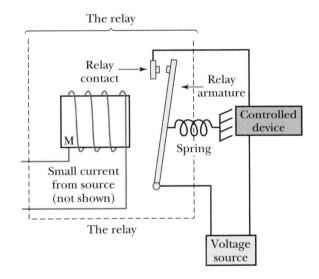

(b)

FIGURE 10–25 Several frame-style solenoids *(Courtesy of Guardian Electric Mfg. Co.)*

FIGURE 10–26 Permanent
magnet dynamic speaker

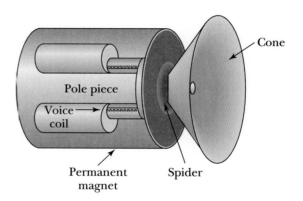

10–5 IMPORTANT MAGNETIC UNITS, TERMS, SYMBOLS, AND FORMULAS

Now, let's study some units of measure for the magnetic characteristics discussed earlier. Recall a single flux line (or magnetic line of force) is a maxwell, and 10^8 flux lines equal a Weber (Wb). Also, although magnetizing force was presented, we did not specifically define it.

Before thoroughly discussing magnetic units, some background about systems of units would be helpful. Most of us daily use units of measure pertaining to dimensions, volume, weight, speed, and so on. In the field of electronics, it is critical that some meaningful system of units exists.

Before 1960 there were several systems of units related to electricity, electronics, and magnetism. One system was the common unit of the volt, ampere, and watt; another system dealt with electrostatics; and a third system explained magnetics. Both electrostatics and magnetics used a system called the *cgs* system, where *c* was distance in centimeters, *g* was weight in grams, and *s* was time in seconds.

However, since the 1960s, a system has been accepted that uses the *Meter* for distance, the *Kilogram* for weight, the *Second* for time, and the *Ampere* (necessary for electrical systems) for current measure. This system is abbreviated as the *MKSA*. Since this system has been adopted internationally, it is called the International System, abbreviated "SI," *Système internationale d'unités*.

Examples of the units, symbols, and formulas are in Figure 10–27. You will need to refer to this chart several times as you study the next section.

The basic magnetic circuit

Since technicians do not typically have to deal with all the systems of units presented in Figure 10–27, we will not dwell on this subject. However, you should be aware of these units and have an understanding of the magnetic circuit. Repeating some of the ideas stated earlier, the magnetic equivalent of Ohm's law, sometimes called Rowland's law, is shown in Figure 10–28 (see page 332).

Relationships among B, H, and permeability factors

From our previous discussion, it is possible to note *H* (magnetic field intensity or mmf per unit length) is the factor producing flux per unit area (or flux density, *B*) within a given magnetic medium.

TERM OR QUANTITY	SYMBOL OR ABBREVIATION	SI UNIT AND FORMULA	CGS UNIT AND FORMULA
Flux (lines)	ϕ	weber (Wb) $= \dfrac{\text{number lines}}{10^8}$	Maxwell (Mx) number lines in field
Flux Density (Magnetic flux per unit cross-sectional area at right angles to the flux lines)	B	$\dfrac{\text{webers}}{\text{sq meter}} = $ **tesla** (T) $B = \dfrac{\phi \text{ (mks)}}{A \text{ (mks)}}$ or teslas $= \dfrac{\text{Wb}}{\text{sq mtr}}$	$\dfrac{\text{lines}}{\text{sq cm}}$ $B = \dfrac{\phi \text{ (cgs)}}{A \text{ (cgs)}}$ or gauss $= \dfrac{\text{Mx}}{\text{sq cm}}$
Magnetomotive Force (That which forces magnetic lines of force through a magnetic circuit)	MMF	Ampere-turns or $AT = NI$	F (cgs) $= 0.4\pi NI$ or F (gilberts) $= 1.26 \times N \times I$
Magnetic Field Intensity (Magnetomotive force per unit length)	H	Ampere-turns per meter $\dfrac{NI}{\text{length}}$ or $\dfrac{AT}{\text{meter}}$	H (oersteds) $= \dfrac{0.4\pi \, NI}{\text{length}}$ or $\dfrac{1.26 \, NI}{1 \text{ cm}}$
Permeability (Ability of a material to pass, conduct, or concentrate magnetic flux; analogous to conductance in electrical circuits)	μ	Vacuum or air; $\mu_o = B/H$. B is teslas; H is $AT/$meter where free space is considered to have an "absolute" permeability of: $4\pi \times 10^{-7}$ or 12.57×10^{-7}	Free space: $\mu_o = B/H$. B is in gauss; H is in oersteds; In cgs system, $\mu_o = 1$
Relative Permeability (Not constant because it varies with the degree of magnetization)	μ_r	Relative permeability of a material is a ratio. Thus, $\mu_r = \dfrac{\text{Flux density with core material}}{\text{Flux density with vacuum core}}$ where, flux density in the core material is: $B = \mu_o\mu_r H$ teslas, and absolute permeability of core materials is: $\mu = B/H = \mu_o\mu_r$ (SI units)	Same concept in both systems

FIGURE 10–27 Magnetic units, symbols, and formulas chart

It is further observed the amount of B a given H produces is directly related to the permeability of the flux path. For a vacuum only, it is stated that:

FORMULA 10–2 $\mu_0 = \dfrac{B}{H}$

FORMULA 10–3 $B = \mu_0 \times H$

OHM'S LAW	ROWLAND'S LAW	EXPLANATION
$R = \dfrac{V}{I}$, where: Resistance in ohms equals electromotive force in volts divided by current in amperes.	$\Re = \dfrac{F(mmf)}{\phi}$, where: Reluctance (the magnetic circuit's opposition to the establishment of flux) equals the magnetomotive force (F) divided by the flux (ϕ). Also, it can be shown that: $\Re = \dfrac{l}{\mu A}$ where l = length of the path in meters μ = permeability of the path A = cross-sectional area of the path perpendicular to flux, in meters	Reluctance equals magnetomotive force divided by flux. Flux equals magnetomotive force divided by reluctance. $F\,(mmf) = \phi \times \Re$ or flux times reluctance.

FIGURE 10–28 Rowland's Law

FORMULA 10–4 $H = \dfrac{B}{\mu_0}$

Referring again to Figure 10–27 and other related discussions, we will now apply this information to solve some problems dealing with magnetic parameters. This will give you a better understanding of the relationships.

Examples

Problem 1: What is the **magnetomotive force,** if a current of 5 amperes passes through a coil of 100 turns?

Answer:

FORMULA 10–5 $mmf = NI$

Thus, $mmf = 100 \times 5 = 500$ ampere-turns

Problem 2: If the length of the coil in problem 1 is 0.1 meters, what is the **magnetic field intensity?**

Answer:

FORMULA 10–6 $H = \dfrac{AT}{\text{meter}}$

Thus, $H = \dfrac{500}{0.1} = 5,000$ ampere-turns per meter

Problem 3: What is the **flux density** (SI units) if there are 10,000 lines of force coming from a North magnetic pole face (surface) having a cross-sectional area of 0.002 square meters?

Answer:

FORMULA 10–7 $B = \dfrac{\phi \ (\text{Wb})}{A \ (\text{m}^2)}$

$$\text{Thus, } B = \left[\dfrac{\dfrac{10^4}{10^8}}{2} \times (10^{-3}\text{m}^2)\right]$$

$$= \dfrac{10^{-4}}{0.002} = \dfrac{0.0001}{0.002} = 0.05 \ \text{T}$$

NOTE: Stated in micro units, $\dfrac{100 \ \mu \ \text{Wb}}{2,000 \ \mu\text{m}} = 0.05 \ \text{T}$

Problem 4: What is the **"absolute" permeability** (SI) of a core material if its relative permeability (μ_r) is equal to 2,000?

Answer:

FORMULA 10–8 $\mu = \mu_0 \times \mu_r$

$$\text{Thus, } \mu = (12.57 \times 10^{-7}) \times 2,000 = 2,514 \times 10^{-6}$$

Problem 5: What is the **flux density** *(B)* in tesla (T) if there are 100 webers per square meter?

Answer:

$$B = \dfrac{100 \ \text{Wb}}{1 \ \text{m}^2} = 100 \ \text{T}$$

Problem 6: If a magnetic circuit has an air gap with a length of 2 mm and a cross-sectional area of 5 cm², what is the gap's **reluctance?**

Answer:

Since $\mu_r = 1$ for air, then:

$$\Re = \dfrac{l}{\mu A} = \left[\dfrac{2 \times 10^{-3}}{12.57 \times 10^{-7} \times 5 \times 10^{-4}}\right]$$

$$= \dfrac{2 \times 10^{-3}}{62.85 \times 10^{-11}}$$

$$= 31.82 \times 10^5 \ \text{SI units}$$

Problem 7: If a given coil has a flux density of 3,000 webers per square meter using unknown core material and it has a flux density of 1,000 webers per square

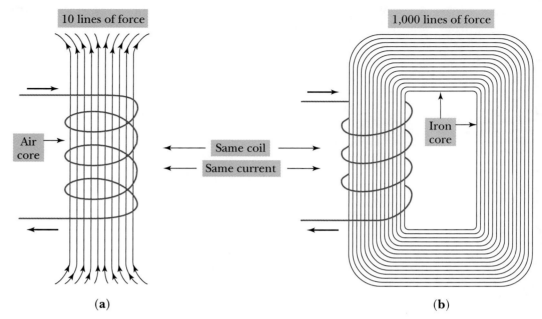

FIGURE 10–29 Effects of core material(s)

meter using air as its core, what is the **relative permeability** of the "brand x" material?

Answer:

$$\mu_r = \frac{\text{flux density with unknown core}}{\text{flux density with air core}}$$

$$= \frac{3,000}{1,000}$$

$$= 3$$

10–6 PRACTICAL CONSIDERATIONS ABOUT CORE MATERIALS

Since many of the components and devices you will be studying use principles related to electromagnetism, we will now consider the practical aspects of the core materials used in these components. You know that for a given magnetizing force, a coil's magnetic field strength is weaker if its core is air than if its core has a ferromagnetic substance, Figure 10–29. Also, we have indicated that certain magnetic properties of a material are not constant but vary with degrees of magnetization. Therefore, there is not a constant relationship between the magnetizing force field intensity *(H)* and the flux density *(B)* produced.

10–7 THE B-H CURVE

Manufacturers of magnetic materials often provide data about their materials to help select materials for a given purpose. One presentation frequently used is a

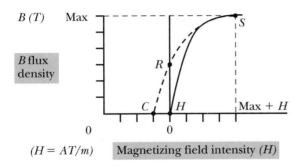

FIGURE 10–30 *B-H* curve

B-H curve, or a magnetization curve, Figure 10–30. Notice in Figure 10–30 that as the magnetizing intensity *(H)* increases, the flux density *(B)* increases but not in a linear fashion. This graph provides several important points.

1. During magnetization (the solid line), a point is reached when increasing *H* (by increasing current through the coil) does not cause a further significant flux density increase. This is due to the core reaching "saturation," represented at point S on the graph. At this point, it can't contain more flux lines.

2. During demagnetization (with current decreasing to zero through the coil), the dotted line indicates some other interesting characteristics:

   a. When the *H* reaches zero (no current through coil), some flux density *(B)* remains in the core due to **residual** or "remanent" magnetism. The amount is indicated by the vertical distance from the graph baseline to point *R*.

   b. To remove the residual magnetism and reduce the magnetism in the core to zero, it takes a magnetizing intensity in the opposite direction equal to that represented horizontally from point 0 to point C. This value indicates the amount of "coercive" force required to remove the core's residual magnetism.

An example of how the *B-H* curve is used to illustrate important magnetic features of different materials is shown in Figure 10–31. For example, notice different materials saturate at different levels of flux density *(B)*. **Saturation** occurs when a further increase in *H* does not result in an appreciable increase in *B*.

10-8 THE HYSTERESIS LOOP

Another useful graph is the hysteresis loop. It portrays what happens when a core is saturated with current in one direction through its coil, then the current direction is reversed and increased until the core is saturated in the opposite polarity. If this is continually repeated, a hysteresis loop graph, similar to that in Figure 10–32, can be created that enables meaningful analysis of the material in question. This is typically done by application of an alternating current source to the coil. NOTE: "Hysteresis" means "lagging behind." As this graph demonstrates, the flux buildup and decay lags behind the changes in the magnetizing force.

The following are important points provided by this graph:

1. Points A to B represent the initial magnetization curve from zero *H* until the core reaches saturation.

FIGURE 10–31
Magnetization curves for
different metals

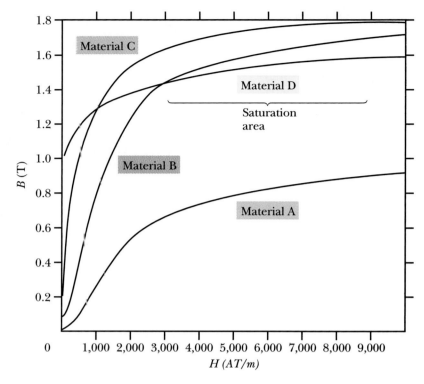

2. Points B to C represent the flux density decay as *H* changes from its maximum level back to zero. The vertical distance from A to C is the amount of residual magnetism in the core.

3. Points C to D represent the flux density decay back to zero as the opposite polarity *H* is applied. The horizontal distance from A to D represents the coercive force.

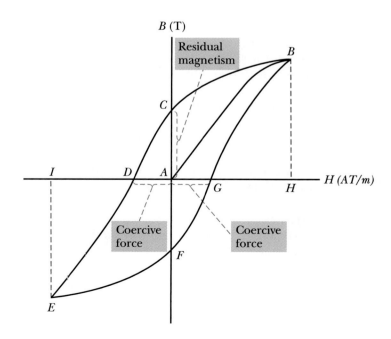

FIGURE 10–32 Hysteresis
loop

4. Points D to E represent the buildup of flux density to core saturation in the opposite polarity from the initial magnetization. Of course, the total horizontal distance from point A to I shows the H required for this saturation. The vertical axis of the graph shows the amount of flux density (B) involved.

5. Points E to F show decay of flux density as H changes from its maximum level of this polarity back to zero. The residual magnetism is represented from points A to F.

6. Points F to G illustrate the decay of flux density back to zero due to the coercive force of the reversed H overcoming the residual magnetism. The coercive force is from points A to G.

7. Points G back to B complete the loop, displaying the increase of B to saturation in the original direction. For this segment of the loop, the vertical distance from point B to H is the flux density value at saturation.

8. The area inside the hysteresis loop represents the core losses from this cycle of magnetizing, demagnetizing, magnetizing in the opposite polarity, and demagnetizing the core. These losses are evidenced by the heating of the core as a result of the "magnetic domains" continuously having to be re-aligned with alternating current (ac) applied to the coil. We'll discuss core losses in greater detail at a later point in this book.

Incidentally, you may wonder how material that has been magnetized can be demagnetized. One method is to put the material through this hysteresis loop cycle while gradually decreasing the H swing. The result would be shown as a shrinking area under the hysteresis loop. One practical way to accomplish a decrease in H swing is to use a large ac demagnetizing coil so the material is placed within the coil. Gradually pull the material away from the ac energized coil, Figure 10–33. You may have seen a TV technician use a "degaussing" coil on color

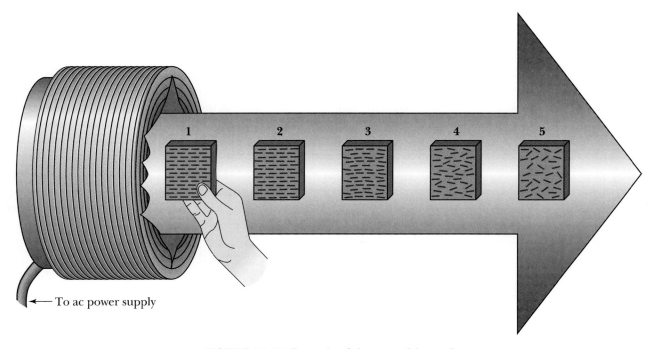

To ac power supply

FIGURE 10–33 Example of demagnetizing coil

picture tubes to demagnetize them. A color picture tube that has become magnetized will have a degraded picture. Demagnetizing it usually helps.

IN-PROCESS LEARNING CHECK III

1. A *B-H* curve is also known as a _____ curve.
2. *B* stands for flux _____.
3. *H* stands for magnetizing _____.
4. The point where increasing current through a coil causes no further significant increase in flux density is called _____.
5. The larger the area inside a "hysteresis loop" the (smaller, larger) _____ the magnetic losses represented.

10–9 INDUCTION AND RELATED EFFECTS

Motor action

When talking about motor action, we refer here to converting electrical energy to mechanical energy. You have studied the law that states lines of force in the same direction repel each other, and lines of force in opposite directions attract and cancel each other. One practical application of this phenomenon is motor action. The simplest example of this is a current-carrying conductor placed in a magnetic-field environment.

Look at Figure 10–34 and note the following:

1. The direction of the magnet's flux lines—North to South.
2. The direction of the current-carrying conductor's flux lines. (Use the left-hand rule for current-carrying conductors.)
3. The interaction between the conductor's flux lines and the magnet's lines of force is such that the lines at the top of the conductor are repelled (lines in the same direction repel); the lines at the bottom of the conductor are attracted (lines in opposite directions attract); and the field is weakened at the bottom. Therefore, the conductor will be pushed down. In other words, the lines at the top aid each other and the lines at the bottom cancel each other.
4. The amount of force that pushes the conductor down depends on the strength of both fields. The higher the current through the conductor, the stronger its field. And the higher the flux density of the magnet, the stronger its field.

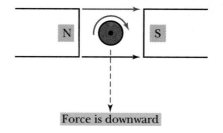

FIGURE 10–34 Motor action on a single conductor

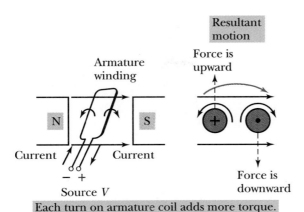

FIGURE 10–35** Concept of stronger motor action with an armature coil

Each turn on armature coil adds more torque.

5. Another method to cause stronger motor action is to make the single conductor into a coil (in the form of an armature) with many turns, thus increasing the ampere-turns and the strength of its field, Figure 10–35.

6. Refer to Figure 10–36b and c. When the current through the conductor is reversed or the magnetic poles of the magnet are switched, the force (motor

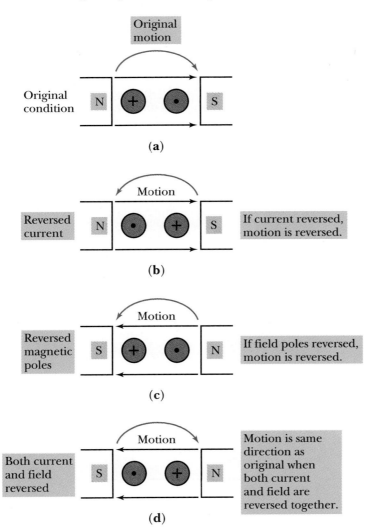

FIGURE 10–36 Effects of changing direction of current or field

action) is in the opposite direction of that in Figure 10–36a. When both current and poles are simultaneously reversed, Figure 10–36d, the force remains in the original direction, prior to being reversed, as that shown in Figure 10–36a.

Generator action

When talking about generator action, we refer to converting mechanical energy to electrical energy. Throughout the chapter, we have alluded to a relationship between electricity and magnetism. Electromagnetic **induction** clearly illustrates this linkage. As you will recall, Faraday observed there is an induced emf (causing current if there is a path) when a conductor cuts across magnetic flux, or conversely, when the conductor is cut by lines of force. In essence, *when there is relative motion* between the two (i.e., the conductor and the magnetic lines of flux), an emf is induced.

Observe Figure 10–37 and note the following:

1. The direction of the magnet's field: North to South.

2. The direction that the conductor is being physically moved with respect to the magnetic field of the magnet: Upward.

3. The fact that when the conductor cuts the field in this direction, the direction of induced current is toward you, Figure 10–37a. In fact, the left-hand rule for current-carrying conductors (studied earlier) would indicate the direction of current, as shown. For example, if you aim your fingers in the direction of the flux (N to S) and bend your fingers in the same direction the magnet's field is bent, then the induced current is in the same direction as your extended thumb.

4. If the conductor is moved down into the field, rather than up, the induced current is in the opposite direction, Figure 10–37b. (Try the left-hand rule again!)

5. The faster we move the conductor relative to the field, the higher the induced current (and/or emf), since more lines are being cut per unit time.

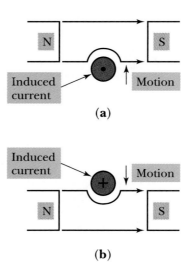

FIGURE 10–37 Converting mechanical energy into electrical energy

You may be wondering how cutting flux lines with a conductor causes induced emf, thus producing current flow (assuming a closed current path). Conceptually, it is briefly explained by indicating the magnetic fields involved with each of the electrons in the wire are caused to align, causing electron movement, which is current flow. When the conductor is open (with no complete path for current flow), one end of the conductor will be caused to have an excess of electrons (negatively charged); the other end has a deficiency of electrons (positively charged). Thus, we have an induced emf. If a closed path for current is provided, this induced emf causes current to flow.

What factors affect the amount of induced voltage? The key factors are:

1. amount of flux;

2. number of turns linked by the flux;

3. angle of cutting the flux; and

4. rate of relative motion.

All of these factors relate to the number of flux lines cut per unit time.

10–10 FARADAY'S LAW

Faraday's law states *the amount of induced emf depends on the rate of cutting the flux (ϕ)*. The formula is:

FORMULA 10–9 $V_{ind} = \dfrac{\Delta \phi}{\Delta T}$

where, V_{ind} is the induced emf, and $\Delta\phi/\Delta T$ indicates the *rate* of cutting the flux. Using the SI system, if a single conductor cuts one weber of flux (10^8 lines) in one second, the induced voltage is 1 volt, a two-turn coil would have 2 volts induced, and so forth. (NOTE: The symbol Δ is the fourth letter in the Greek alphabet called delta, which, for our purposes, represents the amount of change in, or a small change in.) Sometimes, the lower case letter "d" is used to indicate delta rather than the triangular symbol. Thus:

FORMULA 10–10 $V_{ind} = N \text{ (\# of turns)} \times \dfrac{d\phi \text{ (Wb)}}{dt \text{ (s)}}$

Example

If a 100-turn coil is cutting 3 webers per second ($d\phi/dt$ in formula), then the induced voltage is $100 \times 3 = 300$ V. In essence, 3×10^8 lines of flux are cutting the 100-turn coil each second, resulting in 300 V of induced voltage.

Although we will not thoroughly discuss generators, it is important you know that if a loop of wire is rotated within a magnetic field, Figure 10–38, the maximum voltage is induced when the conductor(s) cut at right angles to the flux lines. Voltage will incrementally decrease as the loop rotates at less than right angles, reaching zero when the loop conductor moves parallel to the lines of flux (in the vertical position), Figure 10–39. The key point is when the conductor

FIGURE 10–38 Regions of maximum and minimum induced voltage

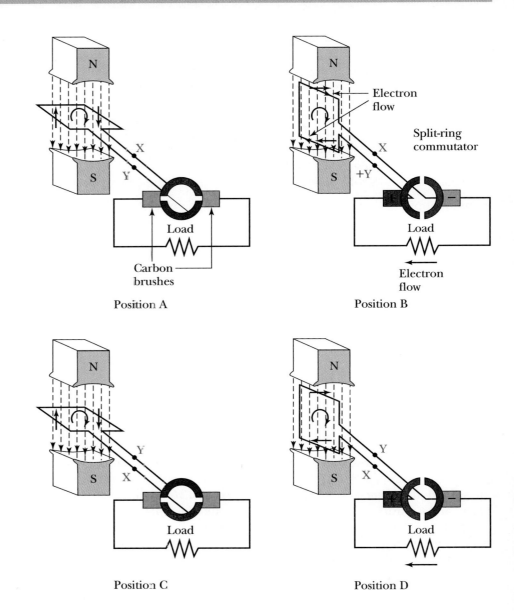

moves parallel to the lines of flux (an angle of 0°), no voltage is induced; when the conductor moves at right angles to the flux lines, maximum voltage is induced; and when the conductor moves at angles between 0° and 90°, some value between zero and the maximum value is induced.

10–11 LENZ'S LAW AND RECIPROCAL EFFECTS OF MOTORS AND GENERATORS

Lenz's law

As you have already noted, the law of energy conservation is you don't get something for nothing! In the case of motor action in the generator, to generate electricity with the generator we have to supply mechanical energy that turns the ar-

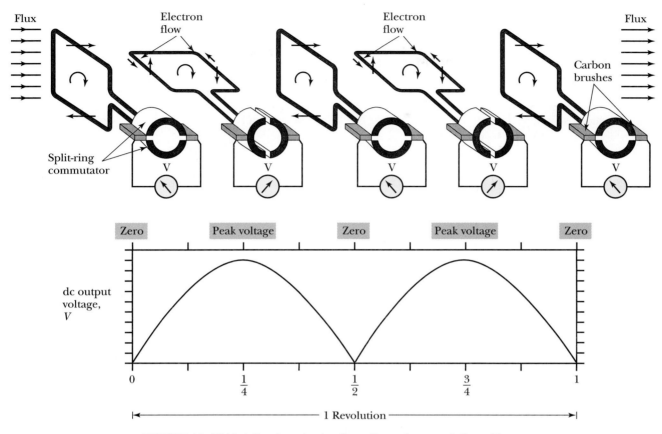

FIGURE 10–39 Variation in output voltage through one rotation of loop

mature shaft and overcomes the force of the opposing motor effect that results from the induced current and its resultant field(s).

Lenz's law states *the direction of an induced voltage, or current, is such that it tends to oppose the change that caused it.* Another way of stating this law is when there is a change in the flux linking a circuit, an emf/current is induced that establishes a field that opposes the change.

Look at Figure 10–40 and use the left-hand rules for current-carrying conductors *and* for electromagnets. Notice when the magnet is moved *down* in the coil, the induced current establishes a North pole at the top end of the coil, facing (and repelling) the North pole of the magnet, whose flux linking the coil caused the original induced current. In other words, as we force the magnet down into the coil, the induced current produces a field to oppose the motion, as stated by Lenz. Conversely, when we pull the magnet out of the coil, the induced current is in the reverse direction, establishing a South pole at the top of the coil and tending to oppose retraction of the magnet. Again, the magnetic polarity can be defined in accordance with Lenz's law.

Motor effect in a generator

The generator converts mechanical energy (armature shaft rotation) into electrical energy (generator's output voltage). As you know, when the generator's

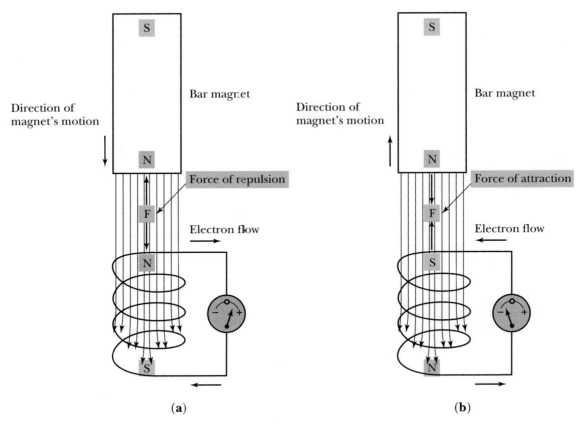

FIGURE 10–40 Lenz's law

conductors (armature windings) move through the magnetic field of the generator's field magnets, voltage is induced according to Faraday's law. This induced voltage causes current through the conductors, assuming a closed path for current. The *direction* of the resulting induced current is such that the magnetic field established around the conductor(s), resulting from the induced current, *opposes* the motion causing it, Figure 10–41.

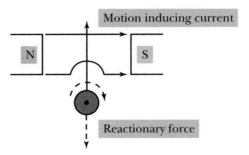

Induced current (coming toward you) results from upward mechanical movement.

Induced current sets up magnetic field around conductor which opposes upward motion that induced it in the first place.

FIGURE 10–41 Motor effect in a generator

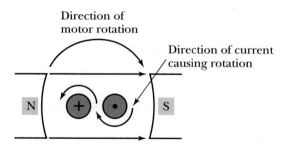

Direction of
motor rotation

Direction of current
causing rotation

N   +   ·   S

FIGURE 10–42 Generator effect in a motor is such that the induced current due to cutting flux will be in opposite direction of current causing motor to rotate.

Generator effect in a motor

In Figure 10–42, you can see the direction of current we are passing through the conductor causes the loop to rotate clockwise because of appropriate motor action. You can also observe when the conductor rotates clockwise, the wire loop's left side moves up into the flux and the wire loop's right side moves down into the flux. Since we have movement between a conductor and flux lines, there is an induced voltage/current.

This induced current is in the opposite direction from the current that is causing the motor's armature to move clockwise. This phenomenon is often termed generator effect in a motor and is related to Lenz's law.

10–12 OTHER TOPICS RELATED TO MAGNETISM

Several topics not previously discussed are now presented since you now have some general understanding of magnetism. These topics are:

1. the toroidal coil form (closed magnetic path);

2. magnetic shields;

3. shaping fields across a gap via pole pieces;

4. the Hall effect; and

5. special classifications of materials.

Toroidal coil form

Notice in Figure 10–43 the toroidal coil form resembles a doughnut and is made of soft iron, ferromagnetic, or other low-reluctance, high-permeability material. The key feature is this coil effectively confines the lines of flux within itself

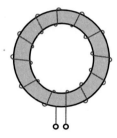

Flux is virtually contained within core
with very little "leakage" flux

FIGURE 10–43 Toroid coil

FIGURE 10–44 Concept of the magnetic shield

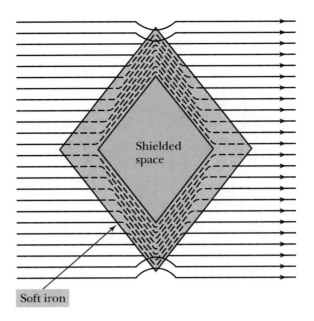

Shielded space

Soft iron

(because there are no air gaps); therefore, it is a very efficient carrier of flux lines. It allows little "leakage flux" (flux outside the desired path, generally in the surrounding air), thus does not create magnetic effects on nearby objects. At the same time, its internal field is affected very little by other magnetic fields near it.

It is interesting to note there are no poles, as such, in the toroid configuration. However, if one takes a slice out of the doughnut, creating an air gap, North and South poles are established at opposite sides of the air gap.

Magnetic shields

Remember magnetic lines of force penetrate nonmagnetic materials. To protect components, devices, and so forth from stray or nearby magnetic fields, a magnetic shield is used, Figure 10–44. A high-permeability, soft iron enclosure is useful for the shielding function. Actually, it is simply diverting the lines of flux through its low-reluctance path, thus protecting anything contained inside of it. One application for this type shielding is in electrical measuring instruments or meter movements.

Shaping fields via pole pieces

Notice in Figure 10–45 the use of pole pieces. These pieces control the shape of the field across the air gap to afford a more linear field with almost equal flux density across the area of the gap. This control is beneficial for many applications.

The Hall effect

Figure 10–46 illustrates the Hall effect. Named after its discoverer, E. H. Hall, the phenomenon states when a current-carrying material is in the presence of an external magnetic field, a small voltage develops on two opposite surfaces of the conductor.

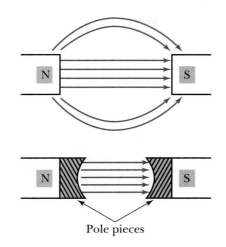

FIGURE 10–45 Concept of pole pieces linearizing a field

Pole pieces

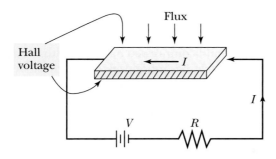

FIGURE 10–46 The Hall effect

As the illustration indicates, this effect is observed when the positional relationships of the current, the field's flux lines, and the surfaces where the difference of potential develops are such that the following are true:

1. Flux is perpendicular to direction of current flow through the conductor. (NOTE: The amount of voltage developed is directly related to the flux density, *B*.)

2. If current is traveling the length of the conductor, the Hall voltage (v_H) develops between the sides or across the width of the conductor.

Indium arsenide is a semiconductor material used in devices, such as gaussmeters, to measure flux density via the Hall effect. This material develops a relatively high value of v_H per given flux density. Because the amount of Hall-effect voltage developed directly relates to the flux density where the current-carrying conductor is immersed, a meter connected to this probe can be calibrated to indicate the flux density of the probed field, Figure 10–47.

Special classifications of materials

In this chapter, we have primarily been discussing ferromagnetic materials that typically have high permeabilities and can be strongly magnetized. Examples include iron, steel, nickel, and various alloys. Other classifications of materials worth noting are:

1. *paramagnetic* materials that have weak magnetic properties. (Examples are aluminum and chromium.)

FIGURE 10–47 Using the Hall effect to measure flux density

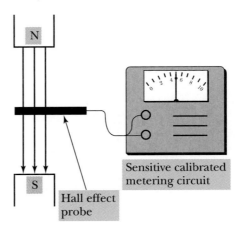

Sensitive calibrated metering circuit

Hall effect probe

2. *diamagnetic* materials that have permeabilities of less than one and align perpendicular to the direction of magnetic fields. (Examples are mercury, bismuth, antimony, copper, and zinc.); and

3. *ferrites* that are powdered and compressed materials with high permeabilities but high electrical resistance that minimizes I^2R/eddy current losses when used in high-frequency applications. (Examples are nickel ferrite, nickel-cobalt ferrite, and yttrium-iron garnet.) Note that eddy currents are those induced into the core material that cause undesired I^2R losses within the core.

SUMMARY

▶ A magnet is an object that attracts iron, steel, or certain other materials. It can be natural (magnetite or lodestone) or produced in iron, steel, and other materials in a form called a permanent magnet. Also, a magnet can exist in soft iron and other materials as a temporary magnet that requires a current-carrying coil around it.

▶ Magnetic materials, such as iron and nickel, are attracted by magnets and can be made to have magnetic properties.

▶ Magnets that are suspended and free to move align in an approximate North/South direction. The end of the magnet seeking the earth's North pole is the North-seeking pole. The other end of the magnet is the South-seeking pole.

▶ A magnetic field composed of magnetic lines of force called flux surrounds a magnet. This region is where the magnetic forces act.

▶ Lines of force are continuous; travel outside the magnet from the magnet's North pole to the magnet's South pole; and flow inside the magnet

from the South pole to the North pole. They travel the shortest path or the path with least opposition to the flow of flux, and when traveling in the same direction, repel each other. When traveling in opposite directions, they attract each other. Furthermore, lines of force do not cross each other and will penetrate nonmagnetic materials.

▶ A current-carrying conductor or wire is surrounded by a magnetic field. The strength of the magnetic field is directly related to the value of current passing through it.

▶ A coil of wire carrying current develops a magnetic field that is stronger than a straight conductor with the same current. The magnetic field becomes stronger if an iron core is inserted in the coil, making it an electromagnet.

▶ A basic law of magnetism is like poles repel each other and unlike poles attract each other.

▶ Permanent magnets come in a variety of shapes; for example, bar-shaped (rectangular), horseshoe-shaped, and disc-shaped. Permanent magnets are not really permanent since they lose their magnetism over time.

▶ Induction of magnetism occurs when a magnetized material is brought in proximity to or rubbed across some magnetic material.

▶ The left-hand rule to determine direction of the magnetic field surrounding a current-carrying conductor states when the wire is grasped in the left hand with the extended thumb in the direction of current flow through the wire, then the fingers are wrapped around the conductor in the same direction as the magnetic field.

▶ The left-hand rule to determine the polarity of electromagnets (i.e., which end is the North pole and which end is the South pole) states when the coil is grasped with the fingers pointing in the same direction as the current, the extended thumb points toward the end of the coil that is the North pole.

▶ Factors influencing the strength of an electromagnet are: the number of turns on the coil; the coil's length; the coil's value of current; the type of core; and the coil's cross-sectional area.

▶ Several important terms and units relating to magnetism include: flux; weber; tesla; magnetomotive force; magnetic field intensity; permeability; and relative permeability.

▶ Comparing magnetic circuit parameters with electrical circuits and Ohm's law can be performed by stating that the amount of flux flowing through a magnetic circuit is directly related to the magnetomotive force and inversely related to the path's reluctance (opposition) to flux.

▶ Core materials differ in their magnetic and electrical features. For example, some materials saturate at lower levels of magnetization.

▶ An ideal core material has very high permeability; loses all its magnetism when there is no current flow in the coil; does not easily saturate; and has low I^2R loss due to eddy currents.

▶ An important relationship exists between flux density produced by a given magnetic field intensity and the permeability of the material involved. This relationship stated for the permeability of a vacuum is:

$$\mu_0 = \frac{B}{H}, \text{ and } B = \mu_0 \times H, \text{ and } H = \frac{B}{\mu_0}$$

where: μ_0 is the absolute permeability of a vacuum
B is flux density and
H is the magnetic field intensity

▶ When an ac passes through an electromagnet's coil, there is a lag between the magnetizing force and the flux density produced. This is illustrated by the *B-H* curve and the hysteresis loop. These graphs also illustrate residual magnetism and the coercive force needed to overcome it.

▶ Faraday's law states the amount of voltage induced when there is relative motion between conductor(s) and magnetic flux lines is such that when 10^8 lines of flux are cut by one conductor (or vice versa) in one second, one volt is induced. In other words, induced voltage relates to the number of lines of force cut per unit time.

▶ Lenz's law states when there is an induced voltage (or current), the direction of the induced voltage or current opposes the change causing it.

▶ A motor converts electrical energy into mechanical energy.

▶ A generator converts mechanical energy into electrical energy.

FORMULAS AND SAMPLE CALCULATOR SEQUENCES

FORMULA 10–1 $\text{Force} = \dfrac{\text{Pole 1 Strength} \times \text{Pole 2 Strength}}{d^2}$

pole 1 value, ⊗, pole 2 value, ⊝, ⓘ, distance value, x^2, ⓘ, ⊜

FORMULA 10–2 $\mu_0 = \dfrac{B}{H}$

B value, ⊖, H value, ⊜

FORMULA 10–3 $B = \mu_0 \times H$

μ_0 value, ⊗, H value, ⊜

FORMULA 10–4 $H = \dfrac{B}{\mu_0}$

B value, ⊖, μ_0 value, ⊜

FORMULA 10–5 $mmf = NI$

number of turns value, ⊗, current value, ⊜

FORMULA 10–6 $H = \dfrac{AT}{meter}$

value of current (amperes), ⊗, number of turns, ⊖, number of meters, ⊜

FORMULA 10–7 $B\ (\mathrm{T}) = \dfrac{\phi(\mathrm{Wb})}{A(\mathrm{m}^2)}$

(, number of lines, ⊖, 10^8,), ⊖, area in meters squared, ⊜

FORMULA 10–8 $\mu = \mu_0 \times \mu_r$

μ_0 value, ⊗ , μ_r value, ⊜

FORMULA 10–9 $V_{ind} = \dfrac{\Delta\phi}{\Delta T}$

change in flux value, ⊖ , amount of time value, ⊜

FORMULA 10–10 $V_{ind} = N\ (\#\ \text{of turns}) \times \dfrac{d\phi\ (\mathrm{Wb})}{dt\ (\mathrm{s})}$

number of turns, ⊗ , (, number of webers cut, ⊖ , time to cut,) , ⊜

REVIEW QUESTIONS

1. Magnetism is:
   a. a material with unique characteristics.
   b. the phenomenon associated with magnetic materials.
   c. the property that attracts metals, woods, and papers.
   d. none of the above.

2. Flux is:
   a. magnetic lines of force related to magnetism.
   b. the space between magnetic poles.
   c. magnetic poles that are identified as N or S.
   d. none of the above.

3. A magnetic field is:
   a. the space at the ends of magnetic poles.
   b. the field of magnetic influence surrounding a magnet.
   c. the space found only within the magnet itself.
   d. none of the above.

4. Magnetic polarity refers to:
   a. a convention that defines the direction of flux relative to a magnet's end poles.
   b. the conventional means of defining a magnet's end faces as either North-seeking or South-seeking.
   c. neither a or b.
   d. both a and b.

5. The law of attraction and repulsion states that:
   a. like poles attract each other.
   b. like poles are neutral to each other.
   c. unlike poles repel each other.
   d. unlike poles attract each other.

6. List five statements about the behavior of magnetic lines of force.

7. What term indicates the ease with which a material passes, conducts, or concentrates magnetic flux?

8. What term indicates the opposition to magnetic lines of force?

9. What is the symbol for flux and what is its SI unit of measure?

10. What is the symbol for flux density and what is its SI unit of measure?

11. What is the symbol for magnetomotive force and what is its SI unit of measure?

12. What is the symbol for magnetic field intensity and what is its SI unit of measure?

13. How many lines of force are represented by a Weber?

14. What is meant by saturation in a magnetic path?

15. How does a magnetic shield give the effect of shielding something from magnetic lines of force?

16. What does a gaussmeter measure?

17. Draw a typical hysteresis loop and explain how saturation, residual magnetism, and coercive force are shown on the graph.

18. The magnetic field surrounding a current-carrying conductor is:
    a. inversely related to the material from which the conductor is made.
    b. directly related to the material from which the conductor is made.
    c. inversely related to the current passing through the conductor.
    d. directly related to the current passing through the conductor.

19. An ideal core material:
    a. has low permeability, low resistance, and saturates easily.
    b. has high permeability, high resistance, and does not easily saturate.
    c. has high eddy currents, and low I^2R losses.
    d. none of the above.

20. Hysteresis occurs:
    a. because of ac coercive force.
    b. because of dc magnetizing force.
    c. because of magnetic lagging effects.
    d. none of the above.

21. State Lenz's law in your own words.

22. State Faraday's law in your own words.

23. What is meant by the statement "motor effect in a generator"?

24. What is meant by the statement "generator effect in a motor"?

25. Why is a "toroidal" coil form such an efficient conductor of magnetic flux?

PROBLEMS

NOTE: Redraw illustrations on separate paper to answer questions involving diagrams.

1. Draw a typical field pattern for the magnetic situations in Figure 10–48. Indicate direction(s) of flux lines with arrows.

FIGURE 10–48

2. Draw magnetic field lines around the conductors in Figure 10–49 and indicate direction with arrows.

FIGURE 10–49

3. Indicate the North and South poles of the current-carrying coil in Figure 10–50.

FIGURE 10–50

4. If an air gap is 2 mm in length and has a cross-sectional area of 10 cm², what is the air gap's reluctance?

5. If a pole piece has a cross-sectional area of 0.005 m² and is emitting 1 μWb of flux:
   a. How many lines of flux are involved?
   b. What is the flux density, expressed in teslas?

6. How much induced emf results if a coil of 500 turns is cut at a reate of 10 μWb per second?

7. How many tesla units are represented by 1,750 gauss?

8. How many lines of flux per square centimeter are represented by 2,000 gauss?

9. What is the magnetic field intensity (H) of a coil having 1,500 turns per meter when the current through that coil is 3 amperes?

10. How many lines of flux are represented by 4 Wb?

ANALYSIS QUESTIONS

1. What is the difference between magnetomotive force and magnetic field intensity?

2. If the permeability of a given flux path is unchanged and H doubles, what happens to B?

3. What happens to the reluctance of a given flux path if its length triples while the magnetic circuit's μ and A remain unchanged?

4. If the current doubles through a given electromagnet's coil, does reluctance increase, decrease, or remain the same? (Assume the magnet is in its linear operating range.)

5. For the situation described in question 4, will flux density increase, decrease, or remain the same?

6. What is meant by "saturation" in a magnetic path?

7. Is the earth's geographic North pole actually the north or south magnetic pole of the earth relative to the way we mark bar magnet poles as being North-seeking and South-seeking?

DC MEASURING INSTRUMENTS

11

chapter

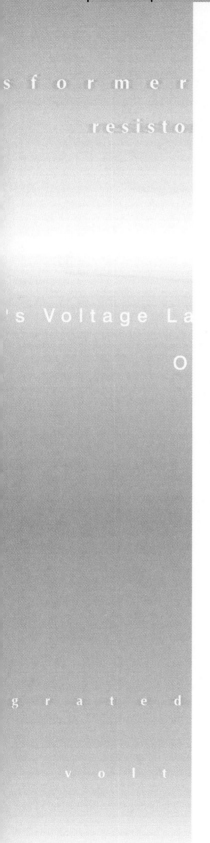

CHAPTER PREVIEW

Lord Kelvin can be paraphrased as saying, "If you can measure what you are speaking about, you know something about it. If you can't, your knowledge is very limited." Someone else has indicated that measuring instruments are like special eyes to technicians, engineers, and scientists. The dc measuring instruments you will study in this chapter are invaluable aids to people working in electronics. These instruments are indispensable for designing, troubleshooting, servicing, and monitoring equipment operation. Some samples of these instruments are in Figure 11–1.

There is a technical distinction between the terms instrument and meter. Instruments measure the value of present quantities. Meters, such as the watt-hour meter, measure and register quantities with respect to time. For convenience, we will refer to the measurement devices discussed using either term.

The digital multimeter has become the most used device by technicians and engineers for measuring voltage, current, resistance, and other electrical quantities. There are still a number of areas in which the analog-type meter has advantages over the digital meters. Advantages and disadvantages of each type of instrument will be discussed further in the chapter.

At a later time in your studies, you will study the theory and principles of operation of the digital meter. However, a major portion of this chapter is dedicated to helping you gain experience and practice in applying the electrical principles and laws you have studied thus far. You will use and practice your basic knowledge of magnetism, series circuits, parallel circuits, and series-parallel circuits. In doing this, you will see practical applications of this knowledge, and also gain a basic understanding of how dc analog meters operate.

In this chapter, you will examine the most frequently used moving-coil movement, called the **d'Arsonval movement**. You will see how this movement is incorporated in measuring instrument circuits that measure current, voltage, and resistance. Also, you will study the proper use of these instruments; the instruments' effects on the circuit(s) being tested; and some hints about their practical applications.

OBJECTIVES

After studying this chapter, you should be able to:

1. Describe the **d'Arsonval movement**

2. List key parts and functions of a movement

3. Define linear deflection, full-scale sensitivity, shunt, multiplier, **sensitivity rating, loading effect,** VOM, and DMM

4. Explain the proper methods to connect and use ammeters, voltmeters, and ohmmeters

5. Calculate the values required for **ammeter shunts**

6. Calculate the values required for **voltmeter multipliers**

7. Briefly explain calibration of series-type ohmmeters

8. Describe at least two special-purpose measuring devices

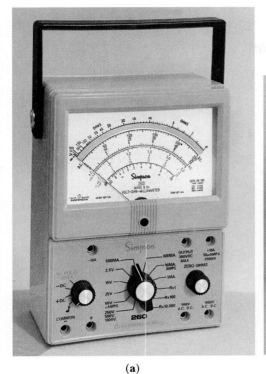

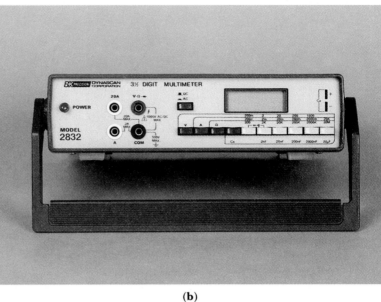

(a)   (b)

FIGURE 11–1 Examples of measuring instruments: **a.** Volt-ohm-milliammeter *(Courtesy of Simpson Electronic Company)*; **b.** Benchtop digital multimeter *(Courtesy of B & K Precision)*

9. List troubleshooting applications for ammeters, voltmeters, and ohmmeters

10. Use the SIMPLER troubleshooting sequence to solve the Chapter Troubleshooting Challenge problem

11-1 GENERAL REQUIREMENTS OF BASIC METER MOVEMENTS

Three essentials most analog meter movements contain are:

1. a moving element that reacts to the value of the electrical quantity being measured;

2. an indicating function that shows how much reaction has occurred; and

3. a damping function that aids the moving element to reach its appropriate resting place without undue overshoot or back and forth mechanical oscillation.

11-2 THE D'ARSONVAL (MOVING COIL) MOVEMENT

The **d'Arsonval movement** contains parts meeting the above requirements and is a device that reacts to the value of current through it.

FIGURE 11–2 Key parts of a moving-coil movement

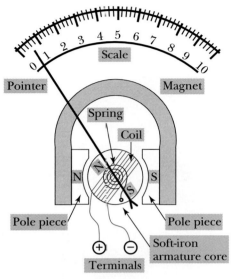

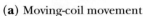

(**a**) Moving-coil movement

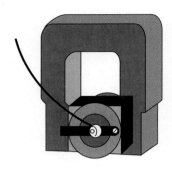

(**b**) "Phantom" view of commercial permanent-magnet moving-coil movement

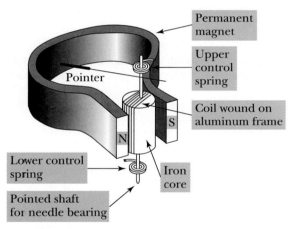

(**c**) Moving element with pointer and springs

Refer to Figure 11–2 and note the following key parts and their functions in this movement.

1. *Permanent magnet*—establishes magnetic field.

2. *Pole pieces*—linearize and strengthen magnet's field

3. *Core*—soft iron core concentrates, strengthens magnetic field and, along with pole pieces, produces a uniform air gap between the permanent magnet's pole pieces and itself.

4. *Movable coil and support frame*—light aluminum frame provides support and damping action for coil; movable coil creates electromagnetic field when current passes through it.

5. *Spindle and jewel bearings*—spindle supports movable coil and coil frame; jewel bearings provide a low-friction bearing surface to allow turning action of movable coil mechanism. (NOTE: Sapphire bearings are not shown.)

6. *Springs*—provide electrical contact or current path to coil and mechanical torque, restoring coil to starting position when current is not present; also provide a force against which the movable coil acts.

7. ***Pointer*** *and counterbalance*—provide means of indicating amount of deflection.

8. ***Scale***—calibrated in appropriate units gives meaning to pointer deflection.

9. ***Stops***—provide left and right limits to pointer deflection. (NOTE: Stops are not shown.)

Principle of operation

The operation of the **moving-coil movement** is based on the motor action occurring between the permanent magnet's field and the field established by the movable coil when current passes through it, Figures 11–2 and 11–3, as appropriate.

The moving element function, described earlier, involves many parts— the moving coil, the light aluminum coil frame, the pole pieces, the magnetic core piece, the spindle, the springs, and the sapphire bearings on which the spindle rotates. You will recall the strength of an electromagnet depends on the number of turns involved (per given length), the current through those turns, and the permeability features of the flux path. Since the number of turns and the flux path are known values, the only variable, in this case, is the amount of current through the movable coil. The higher current produces a stronger motor action; thus, the moving coil moves further.

Observe in Figures 11–2a and 11–3 the polarity of the electromagnetic field set up by current passing through the moving coil is such that the electromagnet's North pole is adjacent to the permanent magnet's North pole, and the electromagnet's South pole is adjacent to the permanent magnet's South pole. This means there is a force of repulsion. Thus, the movable coil moves in a clockwise direction when current passes in the correct direction.

It also is evident that current in the opposite direction through the coil reverses the polarity of the coil's field, causing attraction rather than repulsion. Thus, the movable coil moves in a counterclockwise direction, possibly damag-

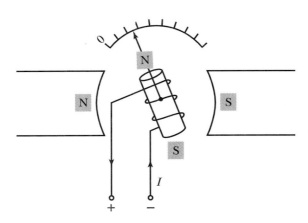

FIGURE 11–3 Polarity required for proper direction of motor action

ing the fragile meter movement. Note that proper operation of this movement depends on dc current (current in only one direction) and proper polarity, or direction of that current. The indicating function is performed by the combination of the pointer and calibrated scale.

The **damping** function is really an electrical damping provided by the light aluminum coil frame. Recall that when a conductor moves within or links a magnetic field and has relative motion, a current-producing emf is induced. The direction of the induced current establishes a magnetic field that opposes the motion causing the induced current in the first place (Lenz's law). This action causes a mechanical damping effect on the movement of the movable-coil-pointer assembly. The result is the pointer does not significantly overshoot its appropriate resting place, nor does it oscillate back and forth or above and below the appropriate deflection location for long periods.

Advantages and disadvantages of the moving-coil movement

Primary advantages of the moving-coil movement are as follows:

1. Linear scale, including scale divisions, is easy to read.
2. Construction provides a self-contained magnetic shielding effect.
3. It can be very accurate.
4. It can be quite sensitive because it has a large number of turns on coil and a strong permanent magnet.

Some disadvantages of the moving-coil movement are as follows:

1. It is quite fragile so will not stand rough treatment.
2. It measures only dc parameters.
3. It is very expensive.

IN-PROCESS LEARNING CHECK I

Fill in the blanks as appropriate.

1. Three essential elements in an analog meter movement include a _____ element, an _____ means, and some form of _____.
2. The purpose of the permanent magnet in a d'Arsonval meter movement is to establish a fixed _____ _____.
3. The movable element in a moving coil meter moves because of the interaction of the fixed _____ _____ and the magnetic field of the movable coil when current passes through the coil.
4. Electrical conduction of current in a movable coil meter movement is accomplished through the _____, which also provides a desired (mechanical, magnetic) _____ force against which the movable coil must act.
5. The element in a moving-coil movement producing electrical damping is the _____ _____.

11-3 AMMETERS

How ammeters are connected

The moving-coil movement is primarily a current indicator. Its basic action depends on the value of the current through its moving coil. Thus, to measure current, the circuit to be tested must be broken, and the current meter inserted in series with the circuit path where it is desired to measure current value (being sure to observe proper polarity), Figure 11–4. As you already know, current is the same throughout a series circuit path. Hence, the meter indicates the current through that portion of the circuit where it is inserted.

Sensitivity

Ammeters come in a variety of measurement ranges, Figure 11–5. Two common methods to define the **sensitivity** of current meters are the amount of current required to cause full-scale deflection and the number of millivolts dropped by the coil at a current equal to full-scale deflection. Sensitive basic movements (for example, the laboratory galvanometer) measure current values in the range of microamperes; other movements measure current in milliamperes.

///

SAFETY HINTS

Turn off power before connecting meter!

However, there is a limit to the range of measurement for which a basic movement can be designed. Physical factors, such as the size of wire, the weight of the coil frame, and the size of springs, limit the range when designing a basic movement. As you will see in a later discussion, using meter shunts helps solve this problem.

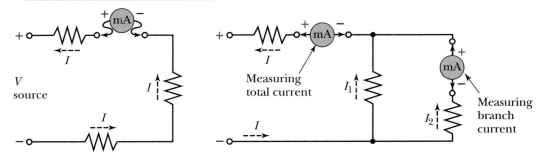

1. Turn power off.
2. Break part of circuit to be tested.
3. Insert current meter in series.
4. Observe polarity.

FIGURE 11–4 Proper connection of current meters

FIGURE 11–5 Meters come in different ranges. *(Photos courtesy of Simpson Electric Co.)*

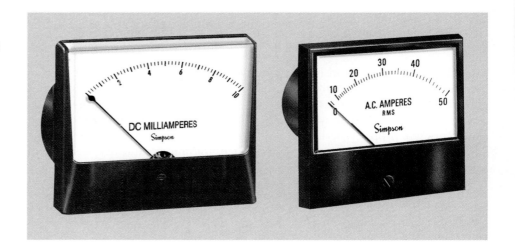

What factors cause a meter to be more or less sensitive? The primary factors are the number of turns on the coil and the permanent magnet's field strength. For example, when a meter deflects full-scale with 10 microamperes of current through the coil, the meter has a full-scale sensitivity (*fs*) of 10 microamperes. When it takes 100 microamperes to cause full-scale deflection, the full-scale (*fs*) sensitivity rating is 100 microamperes and so on.

Another physical factor influencing meter sensitivity is the force of the hairsprings. In order to have a very sensitive movement (one taking a small current value causing full-scale deflection of the pointer), it takes a strong permanent magnet and a coil with many turns of fine wire, and hairsprings without much force.

On the other hand, to have a less sensitive meter, a movable coil using less turns of larger diameter wire is used. Also, using a weaker permanent magnet and/or hairsprings with more force produces a less sensitive meter.

Typically, basic movements are produced in ranges from a few microamperes to a few milliamperes. When measuring currents higher than this, the use of a parallel (current bypass) path, called a **meter shunt,** is necessary.

The number of millivolts dropped by the meter movement at full-scale current adds yet another useful rating because it involves *both* the **full- scale current (I_M)** and the resistance of the wire in the movable coil **(R_M)** of the meter movement. For example, if the meter is a 1-mA meter (*fs* = 1 mA) and the meter resistance (R_M) is 10 Ω, then when the meter deflects a full-scale value of 1 mA, the meter movement is dropping a voltage equal to $I_M \times R_M = 1$ mA $\times$ 10 Ω $= 10$ mV. Therefore, this movement is rated as a 10 mV, 0–1-mA movement.

Trade-offs

What are the trade-offs of basic meter movements? Two important trade-offs are the desired sensitivity and *the desire not to disturb the tested circuit more than necessary.*

A more sensitive meter has a smaller wire and greater number of turns on the movable coil. Small diameter wire has a higher resistance per unit length than larger wire. This means that the more sensitive the meter, the higher is its own internal resistance called R_M. Consequently, a meter with higher resistance has more effect on the tested circuit, since its R_M is a larger percentage of the test circuit's total resistance. Ideally, a current meter should have zero internal resistance so as not to disturb the circuit tested. But zero resistance is impossible.

PRACTICAL NOTES

A common rule of thumb about ammeter resistance versus the resistance of the circuit tested is the ammeter should not be more than about 1% of the resistance of the tested circuit.

Example

For example, if the circuit through which you are trying to measure current has a resistance of 10,000 Ω, the ammeter should not have a resistance greater than 100 ohms to ensure that the tested circuit is not altered too much when the ammeter is inserted.

11-4 AMMETER SHUNTS

The basic shunt

Ammeter shunts extend the range of current that a given meter movement measures. The shunt is simply a low-resistance current path connected in parallel (or shunt) with the meter movement (the moving coil). The shunt effectively causes some percentage of the current to bypass the meter through the parallel path with the remaining percentage passing through the meter to cause deflection, Figure 11–6.

Important parameters

To calculate the required shunt resistance value that extends the meter's range, several electrical parameters are of importance. These parameters are:

1. the full-scale current for the meter movement (I_{fs}, or I_M);

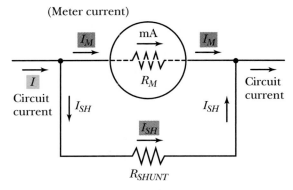

Circuit current "splits." A percentage of current passes through meter while remaining percentage passes through shunt. Thus, $I_M + I_{SH}$ = circuit current.

FIGURE 11–6 Concept of a "shunt" resistor path to extend meter range

2. the resistance of the meter movement (R_M);

3. the desired maximum current, or the extended meter range value to be indicated by full-scale deflection (I);

4. the current that passes through the shunt (I_S or $I - I_M$); and

5. the resistance of the shunt (R_S).

Methods used to calculate R_{SHUNT} value

More than one approach can be used to calculate the R value needed to shunt a given meter movement to increase its measurement range. However, all the techniques evolve from variations of Ohm's law, Figure 11–7.

One method is to use *Ohm's law* in its basic form, where:

FORMULA 11–1 $R_S = \dfrac{V_S}{I_S}$

NOTE: V_S is the same as the parallel voltage V_M (or $I_M R_M$).
I_S is the difference between the highest current to be measured (the extended range full-scale value) and the movement's full-scale current value ($I_S = I - I_M$). Substituting in Formula 11–1:

FORMULA 11–2 $R_S = \dfrac{I_M R_M}{(I - I_M)}$

FORMULA 11–3 $R_S = \dfrac{I_M R_M}{I_S}$

Incidentally, the $I_M R_M$ product is the same as the meter's millivolt rating spoken of earlier, assuming full-scale current value in milliamperes. The ($I - I_M$) portion in Formula 11–2 is really the current value that must pass through the shunt (I_S)

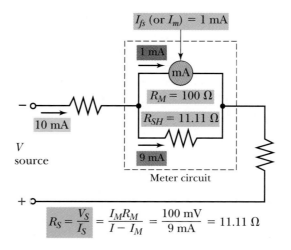

FIGURE 11–7 Basic shunt resistance calculation

in Formula 11–3. Therefore, if you know the millivolt rating of the meter and the current value passing through the shunt,

$$R_S = \frac{\text{millivolt rating}}{I_S}$$

Example

Given a 0–1 mA movement with a meter resistance of 100 Ω, what value of shunt R is needed to make the meter act as a 0–10 mA instrument?

$$\textit{Answer: } R_S = \frac{I_M R_M}{(I - I_M)}$$

$$R_S = \frac{(1 \times 10^{-3}) \times 100}{(10 \times 10^{-3}) - (1 \times 10^{-3})}$$

$$R_S = \frac{(100 \times 10^{-3})}{(9 \times 10^{-3})}$$

$$R_S = 11.11 \ \Omega$$

Another approach is to use the *inverse proportionality concept*. That is, branch currents are inverse to branch resistances.

For example, if it is desired to have nine times the current through branch 2 as there is current through branch 1, then make the R of branch 2 equal to one-ninth that of branch 1. The inverse relationship for our case is stated as:

FORMULA 11–4	$\dfrac{R_S}{R_M} = \dfrac{I_M}{I_S}$

NOTE: The ratio of the *shunt* resistance *to* the *meter* resistance is the same as the ratio of the *meter* current *to* the *shunt* current. Notice the inverse relationship on the two sides of the equation.

Another formula expressing the ratio concept is:

FORMULA 11–5	$R_S = \dfrac{R_M}{N - 1}$

where $R_S = R$ of shunt, $R_M = R$ of meter, and $N =$ number of times range is being multiplied. In other words, 1 mA meter whose increased range is 10 mA, so N equals 10.

Example

Using the same meter and desired current range as in the previous problem, substitute values in the inverse relationship equation as follows:

$$\frac{R_S}{R_M} = \frac{I_M}{I_S}$$

$$\frac{R_S}{100 \ \Omega} = \frac{1 \text{ mA}}{9 \text{ mA}} \text{ or } R_S = \text{one-ninth of } 100 \ \Omega$$

$$R_S = 11.11 \ \Omega$$

FIGURE 11–8 Switchable shunts provide multiple ranges with one meter.

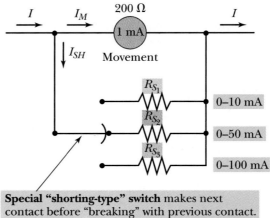

Special "shorting-type" switch makes next contact before "breaking" with previous contact.

Using this same concept, if you want to double the basic range of a given meter, what is the shunt resistance value? If you said the shunt resistance (R_S) equals the meter resistance (R_M), you are correct! In this case, we want the total current to equally split between the shunt path and the meter movement path; thus, the R values must be equal. Since you now understand the inverse proportionality method, try the following problem.

PRACTICE PROBLEMS I

What shunt resistance value is needed to extend the range of a 0–1-mA meter to 0–100 mA. Assume the meter movement's R is 100 Ω.

A basic multiple-range meter circuit

Notice in Figure 11–8 we have several shunts that can be selectively switched and connected with the meter. Obviously, since each shunt is a different value, the meter circuit can measure different values of current, depending on which shunt is connected to the circuit.

PRACTICAL NOTES

1. *A special shorting-type switch is used to switch ranges.* This type of switch, with its very long wiper contact, assures the meter is never without a shunt! In other words, when changing ranges, this switch makes contact with the next switch contact before it breaks contact with the contact it is leaving. Of course, the reason we don't want to leave the meter without a shunt when switching between ranges is that when measuring current in a circuit having current greater than the basic movement's full-scale current rating, the meter is damaged or ruined by excessive current through the movement during that moment when it is unshunted.

2. *The shunt value for each range is determined in the same manner as you determined the single shunt.* See Figure 11–9 for sample calculations.

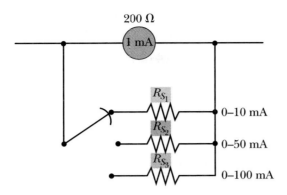

200 Ω
1 mA

R_{S_1}
0–10 mA
R_{S_2}
0–50 mA
R_{S_3}
0–100 mA

FIGURE 11–9 Sample calculation of shunts for three ranges

$$R_{S_1} = \frac{200 \text{ mV}}{9 \text{ mA}} = 22.22 \ \Omega$$

$$R_{S_2} = \frac{200 \text{ mV}}{49 \text{ mA}} = 4.08 \ \Omega$$

$$R_{S_3} = \frac{200 \text{ mV}}{99 \text{ mA}} = 2.02 \ \Omega$$

By way of general information, we want to simply mention (not study in detail) one other type of popular shunting system known as the Universal (Ayrton or Ring) Shunt. This type of shunt is also used to accomplish multiple ranges with a single meter. It is a variation from the single-shunt concept, using several series resistors across (in shunt with) the meter. Electrical contacts (taps) are connected between each of the series resistors. By switching between contacts, the various ranges are selected.

Ammeter key points capsule

1. Moving-coil movements are essentially dc current indicators.

2. The minimum current measured is based on the sensitivity of the basic meter movement.

3. The sensitivity of a basic meter movement is determined by the number of turns in the moving coil, the strength of the magnetic field, and other related mechanical factors, such as the springs, the weight of the coil frame, and so forth.

4. Currents higher than the basic movement's full-scale current rating can be measured if appropriate values of resistance are connected in shunt (or parallel) with the basic movement.

5. Ammeters are generally linear devices. That is, when 50% of full-scale current value flows through the meter movement, deflection is mid-scale on the meter. When 25% of full-scale current is measured, the pointer moves one-fourth full-scale distance on the scale.

6. To measure current, meters must be connected in series, observing proper polarity, and must have adequate range.

7. Because ammeters are connected in series, the lower the meter resistance (R_M) used in making the measurement, the less the actual circuit operating conditions will be changed and the more useful the reading will be.

FIGURE 11–10 The multiplier resistor

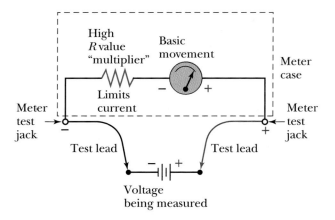

11-5 VOLTMETERS

The purpose of the multiplier resistor

Another extremely useful application for which movements can be adapted is that of measuring voltage, Figure 11–10. Remember that the basic meter movement is a current indicator. By adding select value resistors, called **multiplier resistors, R_{Mult}**, in series with the meter movement, it is possible to measure voltage. The multiplier resistor limits the current through the movement. The amount of meter deflection is directly related to the voltage applied to the meter circuit. For example, a voltage equaling one-third the voltage range causes one-third scale deflection. As you will soon see, different multiplier values provide different voltage ranges.

Connecting the voltmeter to the circuit being tested

Refer to Figure 11–11. Notice that although the voltmeter circuit is a series circuit ($R_M + R_{Mult}$), the voltmeter is always connected *in parallel* with the two points across which it is desired to measure the potential difference, or voltage. Because

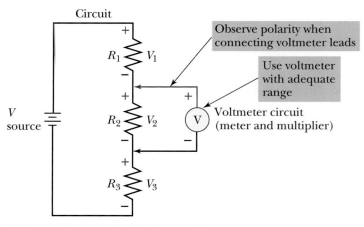

FIGURE 11–11 Connecting a voltmeter to make voltage measurements

Voltmeter is measuring V_2 value and is connected in parallel with voltage to be measured.

the voltmeter is connected in parallel with the portion of the circuit being tested, it is **not** *necessary to break (open) the circuit,* as is done when inserting a standard current-measuring instrument for measuring current.

It is necessary, however, to *observe polarity* when measuring dc voltages, or the meter deflection will be in the wrong direction, as it is for the basic current meter.

As indicated earlier in the chapter, *a sufficiently high range* is needed to assure no more than full-scale deflection of the meter; otherwise, damage will result. Put another way, you are asking for bent pointers and maximum smoke if an adequate range is not used! Again, look at Figure 11–11 to see these points illustrated.

11-6 CALCULATING THE MULTIPLIER VALUES

Using our familiar 0–1-mA meter with a resistance (R_M) of 100 Ω, let's see how we can design a voltmeter that will measure 10 V, 50 V, and 100 V.

Because the multiplier resistor's purpose is to limit the current through the meter, it makes sense that the multiplier resistor (R_{Mult}) value should limit the current through the meter to its full-scale current rating at the voltage range we want to design. That is, to have a 10-V range, the series value of $R_M + R_{Mult}$ should limit current to 1 mA when it is measuring 10 V. This is a calculation for a two-resistance series circuit using Ohm's law. As you read the following examples, refer to Figure 11–12.

Examples

For the 10-V range, then:
(Total R of meter circuit)

FORMULA 11–6 $\qquad R_{Mult} + R_M = \dfrac{V \,(\text{range})}{I \,(\text{full-scale})}$

(Multiplier R)

FORMULA 11–7 $\qquad R_{Mult} = \dfrac{V \,(\text{range})}{I(fs)} - R_M$

Therefore,

$$R_{Mult} = \frac{10\ \text{V}}{1\ \text{mA}} - 100\ \Omega = 10{,}000 - 100 = 9{,}900\ \Omega$$

For the 50-V range:

$$(\text{Total } R \text{ of meter circuit}) \quad R_{Mult} + R_M = \frac{V}{I}$$

$$(\text{Multiplier } R) \quad R_{Mult} = \frac{V}{I} - R_M$$

$$R_{Mult} = \frac{50\ \text{V}}{1\ \text{mA}} - 100\ \Omega = 50{,}000 - 100 = 49{,}900\ \Omega$$

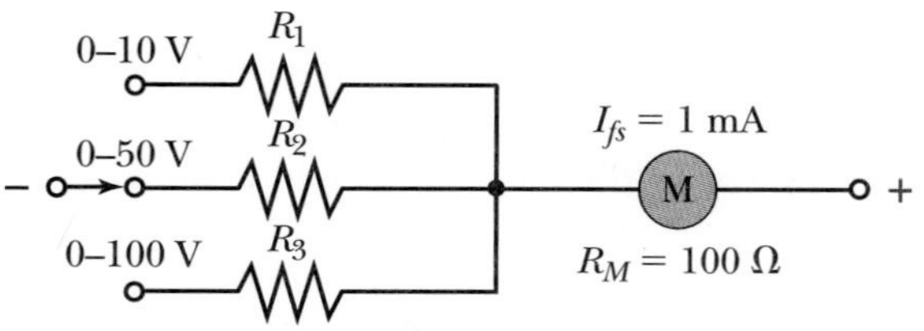

FIGURE 11–12 Examples of multiplier calculations

$$R_1 = \frac{10\text{ V}}{1\text{ mA}} - 100\ \Omega = 10{,}000\ \Omega - 100\ \Omega = 9{,}900\ \Omega$$

$$R_2 = \frac{50\text{ V}}{1\text{ mA}} - 100\ \Omega = 50{,}000\ \Omega - 100\ \Omega = 49{,}900\ \Omega$$

$$R_3 = \frac{100\text{ V}}{1\text{ mA}} - 100\ \Omega = 100{,}000\ \Omega - 100\ \Omega = 99{,}900\ \Omega$$

For the 100-V range:

$$\text{(Total } R \text{ of meter circuit) } R_{Multi} + R_M = \frac{V}{I}$$

$$\text{(Multiplier } R) \ R_{Multi} = \frac{V}{I} - R_M$$

$$R_{Multi} = \frac{100\text{ V}}{1\text{ mA}} - 100\ \Omega = 100{,}000 - 100 = 99{,}900\ \Omega$$

PRACTICE PROBLEMS II

Given a 50-μA movement with R_M value of 2,000 Ω, what value of multiplier resistor is needed to create a voltmeter range of 250 V?

Approximating multiplier values

An interesting point you have probably observed is that the meter's resistance is a very small percentage of the total meter circuit R ($R_M + R_{Mult}$). Therefore, not much error is introduced if we indicate that R_{Mult} is essentially equal to

$$\frac{V \text{ (desired range)}}{I(fs)}$$

Range selector switch

Observe in Figure 11–13a and 11–13b that a **range selector switch** selects the desired voltage measurement range by switching in the appropriate multiplier resistor(s). No matter what range is selected, the multiplier resistance limits current to the meter's full-scale current value when the voltage applied to the total metering circuit (meter plus multiplier) equals the voltage range value.

Also note that either the individual R multiplier or the cumulative R multipliers approach can be used. Most commercial multi-range meters use the cumulative R type circuitry. Refer again to Figure 11–13b.

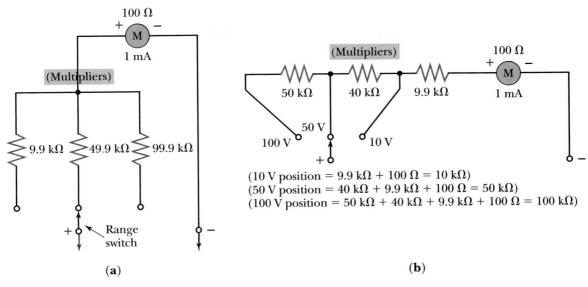

Individual range *R* multipliers approach

Cumulative *R* multipliers approach

(10 V position = 9.9 kΩ + 100 Ω = 10 kΩ)
(50 V position = 40 kΩ + 9.9 kΩ + 100 Ω = 50 kΩ)
(100 V position = 50 kΩ + 40 kΩ + 9.9 kΩ + 100 Ω = 100 kΩ)

FIGURE 11–13 The range selector switch

PRACTICAL NOTES

Caution about accuracy

There are many times when measurement accuracy is *critical!* In certain sensitive circuits, a percentage point of inaccuracy can make a meaningful difference—the difference between operating or not operating, in some cases. For that reason, we want to remind you: *Always be aware of the test instrument's accuracy as related to the acceptable level of accuracy needed to measure the circuit being tested!*

Meter scales have calibrations that either directly match or are a convenient multiple or submultiple of the various selectable ranges. Refer to Figure 11–14.

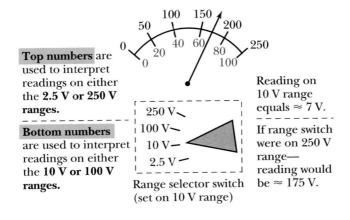

Top numbers are used to interpret readings on either the **2.5 V or 250 V** ranges.

Bottom numbers are used to interpret readings on either the **10 V or 100 V** ranges.

Range selector switch (set on 10 V range)

Reading on 10 V range equals ≈ 7 V.

If range switch were on 250 V range—reading would be ≈ 175 V.

FIGURE 11–14 Meter scales correlate to selectable ranges.

Voltmeter ohms-per-volt (Ω/V) ratings

Previously we discussed the full-scale current- and millivolt-type ratings used with the basic current meter movements.

Another useful rating, related particularly to voltmeters, is the **ohms-per-volt (Ω/V) rating**, Figure 11–15. This rating indicates how many ohms of meter circuit resistance ($R_M + R_{Mult}$) must be present to limit current to the meter's full-scale current rating value *for each volt* applied. In other words,

$$R = \frac{1 \text{ V}}{I(fs)}$$

or the reciprocal of the full-scale current rating provides the ohms-per-volt rating.

Example

A 1-mA movement has an Ω/V rating of $1/1 \times 10^{-3}$, or 1,000 ohms per volt. A sensitive 50-μA movement has an ohms-per-volt rating of $1/50 \times 10^{-6}$, or 20,000 ohms per volt.

A 10-mA movement has a rating equal to the reciprocal of 0.010, or $1/10 \times 10^{-3} = 100$ ohms per volt.

Greater sensitivity of the basic movement means a higher ohms-per-volt rating. Two observations are worth noting related to this rating.

One, if you know the ohms-per-volt rating, it's easy to calculate the multiplier R value for any desired voltmeter range, neglecting the small value of R_M. However, *be careful!* If a 20,000-Ω/V movement has a 2,000-Ω meter resistance, which is typical, a 10% error is introduced on the 1-V range by this technique. One example of using this ballpark or approximating technique is as follows. If the meter rating is 10,000 Ω/V, and you want to create a 100-V range, multiply the rating by 100. The answer is $100 \times 10,000$, or 1 MΩ. You can see the trend!

Another more important fact is this rating gives you a good idea of how much meter loading effect will occur when you make voltage measurements. The next section discusses this concept in more detail. It is important, so study it!

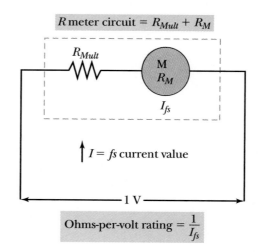

FIGURE 11–15 The ohms-per-volt rating

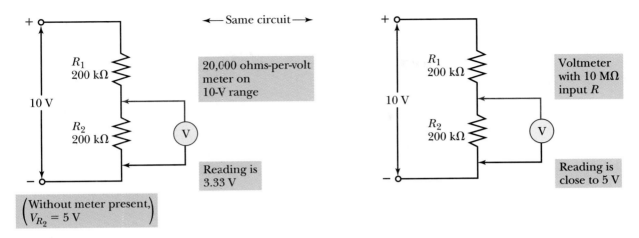

FIGURE 11–16 Voltmeter loading effect

PRACTICE PROBLEMS III

1. What is the ohms-per-volt rating of a meter that uses a 20-microampere movement?
2. What is the full-scale current of a meter rated at 2,000 ohms per volt?

Voltmeter loading effect

Because a voltmeter is connected in parallel with the component or the circuitry it is measuring, it would be ideal if the voltmeter circuitry had *infinite* resistance. This would mean none of the normal circuit operating conditions is altered with the voltmeter connected. But this is not the case! The voltmeter circuit has some finite value of resistance, and it alters circuit conditions! This situation is called **loading effect,** Figure 11–16.

It is apparent that the higher the voltmeter circuit's resistance, the less loading effect it has on the tested circuit. That is, the higher the ohms-per-volt rating of the voltmeter circuit *or* the higher the voltage range, the less the loading effect. Again, greater sensitivity of the basic movement means a higher ohms-per-volt rating. This situation correlates with the voltmeter causing less loading effect on the tested circuit.

Also, a higher voltage range results in a higher multiplier resistance, which causes less voltmeter loading effect, Figure 11– 17.

The greater the voltmeter circuit resistance ratio to the component or circuit resistance being tested, the more accurately the measurement reflects circuit conditions; that is, the less the circuit conditions change with the voltmeter attached. Stated another way, not only is the meter circuit resistance important, *but* the tested circuit's resistance characteristics are equally important! A higher circuit resistance before voltmeter connection results in greater disturbance of normal operating parameters when the meter is connected! In other words, low-resistance circuits are disturbed less by the meter connection than high-resistance circuits, Figure 11–18.

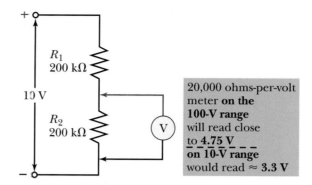

FIGURE 11–17 Sample conditions of loading effect

Using a 20,000 ohms-per-volt VOM—loading effect changes parameters as shown.

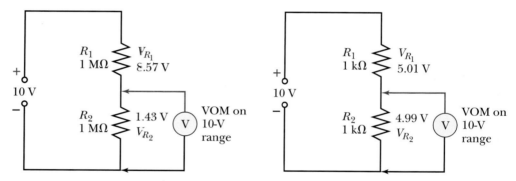

Without the voltmeter, V_{R_1} and V_{R_2} = 5 V each.
Note the loading effect difference between the two circuits!
Large loading effect is on high R circuit.
Small loading effect is on low R circuit.

FIGURE 11–18 Variation in loading effect caused by differences in circuits being tested

PRACTICAL NOTES

One practical rule of thumb is to be sure the resistance of the voltmeter circuit (on the range being used) is at least 10 times the resistance of the circuit portion being tested by the voltmeter. For practical purposes, commercial voltmeters range from 20,000-Ω/V sensitivity ratings for common multimeters, or VOMs (volt-ohm-milliammeters), to 10 megohms, or higher input circuit resistances for the voltmeter circuitry common in digital measuring devices, Figure 11–19. Obviously, meters with multi-megohm input resistances cause negligible loading effect on lower-resistance circuits to be tested.

Voltmeter key points capsule

1. A voltmeter is a combination of a current meter and a series current-limiting resistor, called a multiplier. The multiplier resistance limits current through the voltmeter when voltage is applied to the circuit.

2. A common sensitivity rating system for voltmeters is the ohms-per-volt rating that expresses how much resistance is needed to limit current through the meter to full-scale value when one volt is applied to the voltmeter circuitry.

3. The multiplier resistance value needed to achieve a given range voltmeter is determined by using Ohm's law and simple series-circuit techniques.

$$(R_M + R_{Multi}) = \frac{V\,\text{desired range}}{I_{fs}}$$

4. Linear scales are usually present on dc voltmeters.

5. To measure voltage, the voltmeter is connected in parallel with the component or circuit portion where the difference of potential is to be measured. When measuring dc voltage, it is necessary to observe polarity. Also, care must be taken to assure that the voltage range of the metering circuit is sufficient to handle the voltage to be measured.

6. Because the voltmeter is connected in parallel, the higher the metering circuit resistance, the less it changes the circuit conditions, and the more accurately the voltmeter reading shows the actual circuit operating conditions.

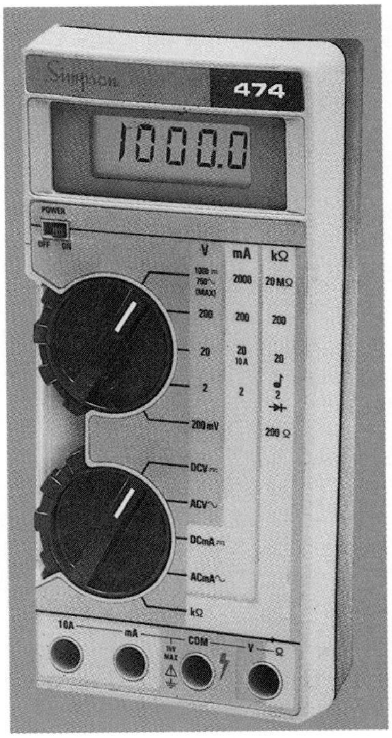

FIGURE 11–19 A DMM's higher meter circuitry resistance means less loading effect. *(Courtesy of Simpson Electric Co.)*

11-7 OHMMETERS

NOTE: Although **series-type ohmmeter circuits** are becoming less common, because of the availability of digital multimeter devices, we will study the series-type ohmmeter briefly so you can gain a basic understanding of ohmmeter concepts.

The ohmmeter is yet another application for dc meter movements. An ohmmeter circuit has three main elements—a battery (generally housed in the meter case), the meter movement, and some current-limiting resistance, Figure 11–20.

Because resistance measurements are made with the power off on the tested component or circuit, the internal battery becomes the current source during testing. Obviously, the meter reacts to the amount of current flowing, and the current-limiting resistance controls the amount of current flowing through the meter.

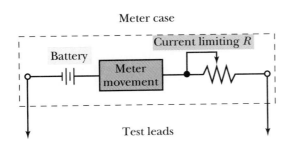

FIGURE 11–20 Typical ohmmeter elements

FIGURE 11–21 Series-type ohmmeter circuit characteristics

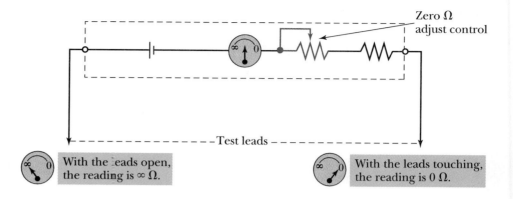

With the leads open, the reading is ∞ Ω.

With the leads touching, the reading is 0 Ω.

Adjusting, using, and reading the ohmmeter

The most common ohmmeter (used in analog instruments) uses the series-type ohmmeter circuit. Refer to Figure 11–21 and notice that when the test leads are not touching each other or are not connected to an external current path, there is an "open" circuit. Therefore, there is no current through the meter, and the **pointer** is at its left-hand resting place.

On the ohmmeter scale, zero current indicates infinite resistance between the test probes. The scale is marked with the infinity symbol (∞) on the left end of the scale. When the test leads are touched together, there are zero ohms between the test leads, and there is a complete current path. That is, current flows through the meter. The amount of current is controlled by adjusting the variable resistor (called the **zero-adjust control**) so there is exactly full-scale deflection on the current meter. The right end of the ohmmeter scale is marked with a "0" calibration mark, indicating that when full-scale current is flowing, there are zero ohms between the test probes.

It is apparent when a resistance value between 0 and ∞ ohms is placed between the test leads, some amount of current flows that is less than full-scale current value and more than zero current. Therefore, the scale is calibrated to show this resistance value. Series circuit analysis is used to show what percentage of full-scale deflection occurs for various resistance values, Figure 11–22.

Note the ohmmeter scale is a nonlinear scale! This nonlinear scale contrasts with most ammeter and voltmeter scales that have evenly spaced or linear scales. A nonlinear ohmmeter scale is sometimes termed a **backoff ohmmeter scale,**

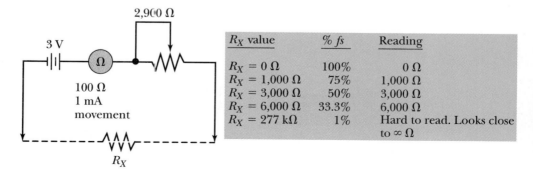

R_X value	% fs	Reading
$R_X = 0\ \Omega$	100%	$0\ \Omega$
$R_X = 1,000\ \Omega$	75%	$1,000\ \Omega$
$R_X = 3,000\ \Omega$	50%	$3,000\ \Omega$
$R_X = 6,000\ \Omega$	33.3%	$6,000\ \Omega$
$R_X = 277\ k\Omega$	1%	Hard to read. Looks close to ∞ Ω

FIGURE 11–22 Ohmmeter deflection for various R_x values

since it is a "right-to-left" scale. In other words, zero is on the right end of the scale, and higher resistance readings are to the left. Again, see Figure 11–21.

Typically, commercial multimeters (volt-ohm-milliammeters or VOMs) have more than one range for measuring resistance that enables reading both low and high R values. Notice in Figure 11–23 the ohmmeter range switch shows ranges expressed in terms of R times some factor. For example, in the $R \times 1$ range, the ohmmeter scale is read directly. In the $R \times 100$ range position, the technician should multiply the scale reading by 100 to interpret the value of resistance being measured. The method for multiple ranges involves switching to different metering circuit shunt resistances, which is necessary because when measuring high resistances, little current flows; thus high meter sensitivity is needed. When measuring low resistances, much current flows; thus, the meter can be shunted, as appropriate. Also, it is common to find a switching of internal voltage source(s), so when measuring high resistances the voltage supplied to the circuit is higher.

Ohmmeter key points capsule

1. The series-type ohmmeter is a dc movement, an internal dc source, and a series current-limiting resistance connected in series. The test probes, when connected to the external component(s) being tested, place the unknown value of R being measured in series with the metering circuit.

2. The commonly used series-type ohmmeter uses a backoff (right-to-left) scale that is nonlinear. The lower value resistance scale calibrations (on the right end of the scale) are spread apart more than the higher resistance value calibrations, which appear on the left end of the scale.

3. *Ohmmeters should NEVER be used to measure components or circuit portions that have power applied.* (Most of the time, if that is done, the meter will be damaged.) The ohmmeter's internal battery supplies the necessary voltage and current to make the measurement(s).

4. When using the series-type ohmmeter, the test probes are touched together and the zero-adjust control is adjusted so there is full-scale current through the meter and the ohmmeter scale reads "zero" resistance. After "zeroing" the

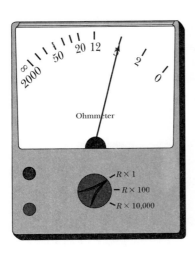

If the pointer is pointing at "5" on the scale, then
$R \times 1 = 5\ \Omega$
$R \times 100 = 500\ \Omega$
$R \times 10,000 = 50,000\ \Omega$

FIGURE 11–23 Switchable ohmmeter ranges

meter and with the power off, the test probes are then connected across the component or components being tested to read their resistance value(s).

5. Most ohmmeters have more than one resistance range to measure a wider range of resistance values. The range selector switch indicates R times some factor at each switch position. For example, if the meter indicates 100 on the scale and the range selector switch is in the $R \times 10$ position, the actual resistance being measured is 10×100, or 1,000 Ω. When changing ranges, be sure to re-zero the meter on the new range setting.

11-8 VOMS, DMMS, AND OTHER RELATED DEVICES

Background information

As indicated throughout the chapter, most test instruments are designed so they have multiple ranges for the electrical quantity they are measuring. Figure 11–24 shows an example of a panel meter designed to measure one type of electrical quantity. This type of meter is found in electronic equipment, such as radio transmitters and power supplies, and frequently has only one range.

As an electronics technician, you will probably use devices called **multimeters** in your troubleshooting and designing tasks. These meters not only have multiple ranges for each type of electrical quantity to be measured, but can measure more than one of the basic electrical quantities (I, V, or R). Let's take a brief look at several types of popular multimeters.

VOMs

The VOM or volt-ohm-milliammeter, Figure 11–25, is a popular multimeter used for many years by technicians. The advantages and disadvantages of the VOM are:

ADVANTAGES	DISADVANTAGES
• Compact and portable	• Can cause significant meter loading effects on circuit being tested
• Does not require an external power source	
• Measures dc current, dc voltage resistance, and ac voltage	• Less accurate than a DMM
• Typically has multiple ranges for each electrical quantity it measures	• Analog readout scale can introduce "parallax" reading errors if pointer is viewed from an angle
• Is analog in nature, so is *good* for adjusting, tuning, or peaking tuned circuits	

FIGURE 11–24 Single-range panel meter

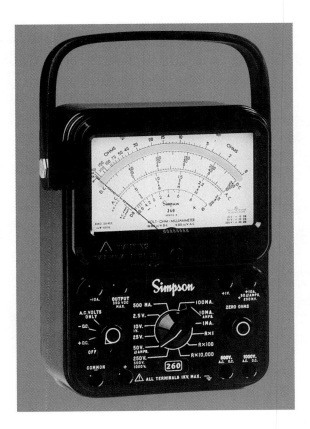

FIGURE 11–25 The popular VOM (volt-ohm-milliammeter) *(Courtesy of Simpson Electric Co.)*

11-9 THE DIGITAL MULTIMETER (DMM)

The digital multimeter (DMM), Figure 11–26, is probably the piece of test equipment most used by technicians today. There are a number of applications where the analog multimeter (VOM) is still the best device to use. However, the DMM's accuracy, expanded measurement capabilities, ease of reading, and lighter loading of the circuit under test make it a prime choice in most cases.

Some DMMs provide polarity-indicated readouts and autoranging features. If the polarity being measured is negative, the readout indicates this. In contrast, a VOM would have its pointer hitting and/or pushing against the scale's left-side limit peg, if the polarity were opposite the way you connected the test leads. Certainly, meter damage could result. When used in the autoranging mode, a DMM will automatically select the range with the best resolution, without the technician having to manually select the correct range.

Many DMMs also offer measurement capabilities beyond simply measuring voltage, current, or resistance. Some allow measurement of ac frequency, capacitor values, semiconductor diode condition, and so on. You will better understand the significance of these capabilities after you have studied ac signals and components used in ac.

Some popular DMMs also are capable of remembering important measurement data that you want to read at a later time. For example, there may be a "MIN MAX" mode on the meter. This means that the meter will remember the lowest and highest readings over a period of time, take the average, and provide that data in its readout. Also, some meters can store a reading in memory, then compare it to subsequent readings, and display the difference between the mem-

FIGURE 11–26 (a) The DMM (digital multimeter); **(b)** a portable DMM *(a. Courtesy of John Fluke Mfg. Co., Inc.* **b.** *Courtesy of Simpson Electric Co.)*

(a)   (b)

orized reading and the subsequent reading. For safety's sake, some meters have a "hold" the reading feature. This allows you to keep your eyes on the test probes when working in dangerous or difficult places for making measurements. You can then read the display when you have removed the test probes and are better able to fix your eyes on the meter readout.

Some of the advantages and disadvantages of the DMM are:

ADVANTAGES	DISADVANTAGES
• Digital readout usually prevents a reading error	• Can be physically larger than equivalent VOM (not in all cases)
• Can have "autoranging" that prevents destroying a meter by having it on the wrong range	• Can be more costly to buy and operate
• Can have "autopolarity" that prevents problems from connecting meter to test circuit with wrong polarity	• Because of "step-type" digital feature rather than "smooth" analog action, are *not as good* for adjusting "tuned" circuits or "peaking" tunable responses
• Can have excellent accuracy (plus or minus 1%)	• Frequency limitations
• Portable and uses internal batteries	

Related Devices

There are special probes used with multimeters and special self-contained metering devices that you should know about.

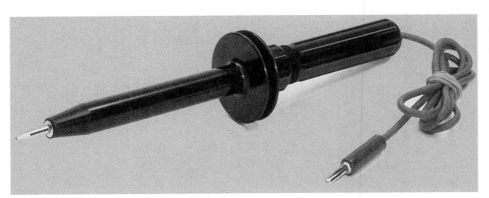

FIGURE 11–27 The high-voltage probe *(Courtesy of Amprobe Instruments)*

High-voltage probe

Basic multimeters cannot generally measure very high voltages in the multikilo-volt range; however, a high-voltage probe (Figure 11–27) will enable this to be done. In essence, the probe consists of a high-value resistance that acts as a multiplier resistance, external to the meter. The probe's resistor(s) are designed to withstand high voltages. One typical application of this probe is to measure high voltages on TV picture tubes.

Clamp-on current probe

Another accessory found on some multimeters is the clamp-on current probe that measures ac current, Figure 11–28a. When ac current levels are in the am-

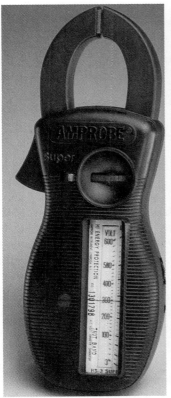

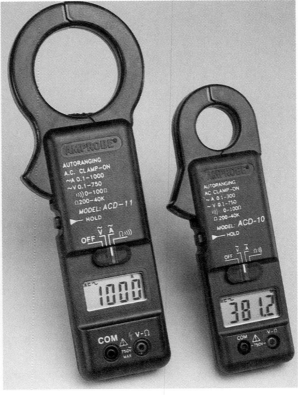

(a)   **(b)**

FIGURE 11–28 Clamp-on probes *(Photos courtesy of Amprobe Instruments)*

pere ranges, this device reacts to the magnetic fields surrounding the current-carrying conductors and translates this activity into current readings. The key advantage to this device is that it measures current *without having to break the circuit* when inserting the metering circuit. The clamp-on probe simply surrounds the current-carrying conductor without breaking the circuit. NOTE: This device *does not measure dc* current!

Other devices

There are a variety of special clamp-on measurement devices described in many electronic distributor catalogs. You will find the clamp-on meters we've been describing, as well as clamp-on volt-ohmmilliammeters (Figure 11–28b), clamp-on digital wattmeters, and "power factor" meters. (You will learn about "power factor" meters in later chapters.)

PRACTICAL NOTES

Measuring instruments are a vital tool to technicians when they are troubleshooting circuits that are operating abnormally. Electrical parameters indicate whether a component, circuit portion, or a total circuit is operating normally, abnormally, or is dead.

In a *qualitative* sense, the measurements indicate whether the values measured are normal, higher than normal, or lower than normal.

In a *quantitative* sense, the measurements indicate specific values for the parameters being measured. These values are then interpreted as normal or abnormal.

To troubleshoot any circuit or system, the technician must know "what should be." That is, he or she *must know the norms.* This is gained through experience, knowledge, or reference to appropriate technical documentation, such as schematics.

Although the following discussion is not exhaustive or comprehensive, the hints and techniques can be useful to you during your training and throughout your career as a technician. Study and apply these items, as appropriate.

11–10 TROUBLESHOOTING HINTS

General information

Use test instruments properly:

1. **Current meters**—*Turn power off circuit.* Insert meter in series, making sure to observe proper range and polarity. Turn power on circuit to be tested. Be sure the current metering circuit (meter or meter and shunt) does not have too high a resistance, causing undue or unaccounted-for changes in circuit conditions.

2. **Voltmeters**—Using appropriate safety precautions to prevent shock, connect the meter in parallel with desired test points, making sure to observe proper range and polarity. For safety's sake, only use one hand at a time in making voltage measurements. Be sure the meter and range do not cause unaccounted for meter loading because of too low a meter resistance.

3. **Ohmmeters**—*Turn power off circuit* where component or circuit portion to be measured resides. Make appropriate zero-adjust calibration with test probes shorted. Disconnect at least one component lead. Connect test probes

across component or circuit portion to be measured. Be sure ohmmeter batteries are good.

Techniques for voltmeter measurements

1. **Connecting one lead to common or ground**—Many electronic circuits have a common or ground reference point, such as the chassis or a common conductor "bus" where all voltages are referenced. By connecting one lead of the voltmeter (frequently the black or negative lead) to this ground reference, measurements are made throughout the circuit by moving the other lead to the various points where voltage measurements are desired, Figure 11–29. Frequently, the black lead on a meter has an alligator clip, making it easy to attach the lead to an electrically common point in the circuit.

2. **Using voltage checks to determine current**—Because it is often awkward and time consuming to break a circuit when installing a current meter, technicians frequently use voltage measurements to determine current. For example, if the measured voltage drop across a 1-kΩ resistor is 15 V, it is simple (Ohm's law) to deduce the current through that resistor must be 15 mA. That is,

$$I = \frac{V}{R} = \frac{15 \text{ V}}{1 \text{ k}\Omega} = 15 \text{ mA}$$

Obviously, Ohm's law enables you to determine current regardless of the R value where you are measuring the voltage. This technique prevents you from having to a) turn off the circuit, b) break the circuit, c) insert a current meter, d) make the measurement, then turn circuit off again, and e) reconnect the circuit when the current meter is removed.

3. **Using voltage measurements to find opens**—Recall from a previous chapter that when an open is in a series circuit, the applied voltage appears across

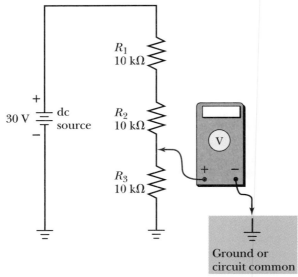

When + voltmeter lead is at:
Top of R_3, reading = 10 V
Top of R_2, reading = 20 V
Top of R_1, reading = 30 V

FIGURE 11–29 Making voltage measurements with respect to ground reference

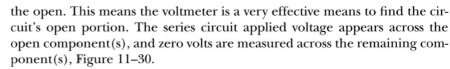

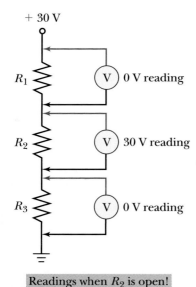

+ 30 V

R_1 (V) 0 V reading

R_2 (V) 30 V reading

R_3 (V) 0 V reading

Readings when R_2 is open!

FIGURE 11–30 Voltage measurements can find "opens."

the open. This means the voltmeter is a very effective means to find the circuit's open portion. The series circuit applied voltage appears across the open component(s), and zero volts are measured across the remaining component(s), Figure 11–30.

4. **Using voltage measurements to find shorts**—Again recall that a zero-volt *IR* drop appears across a shorted component or circuit portion. Voltage measurements quickly show this.

Techniques for ammeter measurements

1. **Total current measurement as a circuit condition indicator**—Total current value gives a general indication of circuit condition. If the total current is higher than normal, a lower-than-normal total circuit resistance is indicated. If the total current is lower than normal, a higher-than-normal total circuit resistance is indicated. Additional isolation techniques are then used to find the problem.

2. **Special jacks for current measurement**—Some circuits or systems have built-in closed-circuit jacks where the ammeters are connected to measure current in the circuit, Figure 11–31.

   When the ammeter is not present, the jack is shorted, providing a closed-current path. But when the ammeter is present, the current must pass through the meter. (NOTE: This current-jack situation is uncommon.) More typically, voltage measurements determine current(s). In transmitters and other similar circuits where monitoring current is important, permanently mounted and installed ammeters are used.

3. **Clamp-on metering for ease of measurement**— Previously we discussed the advantage of not breaking the circuit when using this type of current measurement. The limitations of this type instrument are the limited available ranges and the limited physical situations where it is possible to connect such a meter.

(a)

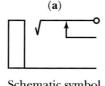

Schematic symbol for this type phone jack

(b)

FIGURE 11–31 Shorting-type jack

Techniques for ohmmeter measurements

1. **Continuity checks to find opens**—One frequently used check is the continuity check. The ohmmeter indicates ∞ ohms resistance if there is no "continuity," or an open between the points where the probes are connected. Of course, an open is one extreme condition that can be measured.

   A continuity check is very useful for checking fuses, light bulbs, switches, circuit wires, and circuit board paths. For example, all of these items will show very low or almost zero ohms resistance across their terminals, if they are normal. If the measurement shows infinite ohms, the item is obviously open or blown.

   In fact, sometimes visual checks are used to confirm where the open is occurring. You can see the broken filament in an automobile light bulb and a broken copper path on the circuit board!

2. **Ohmmeter checks to find shorts**—Ohmmeter checks are valuable for finding shorted components or circuit portions. It is apparent when measuring a component or circuit resistance that when the resistance is zero ohms, there is a short. In the case of a closed switch, this is the normal situation. However, in

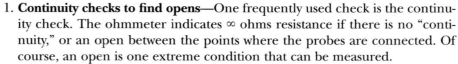

cases where there should be a measurable resistance value, you have located a problem. By physically lifting one end of each component or circuit portion from the remaining circuitry in order to electrically isolate it, and making appropriate measurements, you can determine which component is shorted.

3. **Checking continuity of long cables or power cords**—Frequently, it is handy to verify whether a cable has a break in one of its conductors, or a short between conductors. Examples are TV twin leads, coaxial cables for antenna installations, and cables in computer network systems. Note Figure 11–32 and see one popular technique used to verify continuity in such cables. Rather than having an ohmmeter with 200-foot-long test leads, the technician creates temporary shorts on the conductors at one end of the cable. Then, using the ohmmeter at the other end, determines if there is continuity throughout the cable length for both conductors. After making the measurement, the technician removes the temporary short to establish normal operation. Incidentally, when the temporary short is removed, there should *not* be continuity. If there is, this indicates an undesired low resistance, or shorted path between conductors.

4. **Measuring resistances with the ohmmeter**—Reiterating, *never* connect an ohmmeter to a circuit with power on it! Remove the power plug!

Watch out for "sneak paths"!

One frequent error made by technicians is not sufficiently isolating the component or circuit portion to be tested from the remaining circuitry. Any unaccounted current path that remains electrically connected to the specific path you want to measure can change the measured R values from the "expected" to

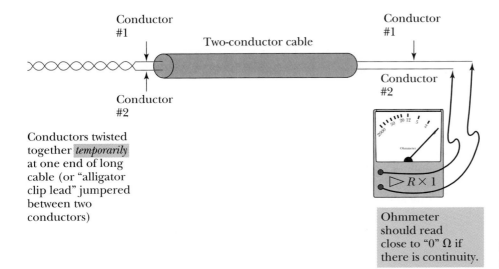

Conductor #1

Two-conductor cable

Conductor #1

Conductor #2

Conductor #2

Conductors twisted together *temporarily* at one end of long cable (or "alligator clip lead" jumpered between two conductors)

$\triangleright R \times 1$

Ohmmeter should read close to "0" Ω if there is continuity.

FIGURE 11–32 Checking continuity of long cables

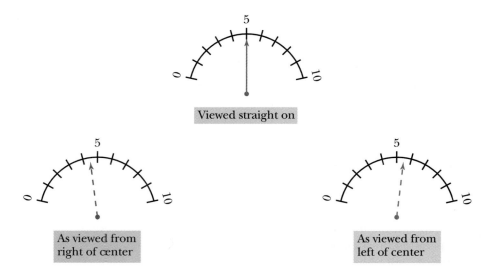

FIGURE 11–33 "Parallax" error

the "unexpected." Where possible, it is best to completely remove, or at least lift one end of the component being tested from the remaining circuit.

Another common "sneak path" occurs when the technician inadvertently allows his or her fingers to touch the test probes. This puts their body resistance in parallel with whatever is being measured. When it is impossible to isolate the component or circuit portion to be tested, but "in-circuit" tests are required, have a schematic diagram, or good knowledge about the circuitry so you can accurately interpret the measurements.

Naturally, when measuring resistances, use all of the normal operating procedures. That is, perform the zero-adjust calibration for the range to be used, remove power from the circuit or component being tested, and use the proper range. If resistance measurements indicate improper values, proceed with additional isolating procedures, as appropriate.

Closing comments about measuring instruments

In this chapter, we have focused on dc measurement instruments. We did mention, however, that most multimeters have additional built-in circuitry enabling the measurement of ac voltages.

Also you should know there are several other types of meter movement designs that react to ac quantities, which were not discussed in this chapter. These movements include the attraction-type moving-iron meter movement, the repulsion-type moving-iron movement, and the electrodynamometer movement (frequently used in wattmeters). It is beyond the scope of this chapter to discuss how each of these types operate, but we did want you to be aware of them.

As indicated throughout the chapter, meter movements are very fragile instruments. It is important that you give them the proper care to prevent physical abuse, and that you use proper operating procedures to prevent electrical damage.

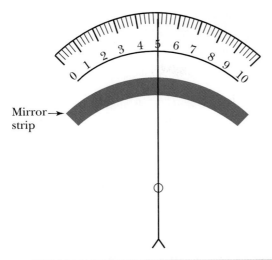

The technician makes sure the "real pointer" is superimposed over the reflection of the pointer when reading the meter to minimize the parallax reading error.

FIGURE 11–34 Mirrored scale for minimizing parallax error

Many instruments are manufactured with electrical protection circuitry built in. A variety of semiconductor circuits, from simple diode protection to more elaborate circuits, are examples of these circuits.

We have mentioned that technicians should be aware of factors affecting accuracy when making measurements. Meter loading effects were mentioned as one of these factors. Use simple series-parallel circuit analyses to determine the *voltmeter's* effect on the tested circuit. Use series circuit analyses to assess the *current meter's* effect on a given circuit.

Another error is that of parallax when reading scales on analog meters, Figure 11–33. Some instruments have a mirror on the scale so the technician will not read the pointer from an angle that might cause a misinterpreted reading, Figure 11–34. By making sure the real pointer is superimposed over the reflection, the technician will be reading the scale at the angle that yields the best accuracy.

Another point is the basic accuracy rating of the instrument itself. Typically, meters are rated from 1% to 3% accuracy at full-scale deflection for analog instruments and from 0.1% to 1% (or plus/minus one digit) on digital meter readouts. It is essential technicians understand the limitations of the instruments used and have knowledge of the circuits or systems they are testing to properly interpret what is being measured. As a technician, you will find that *properly used and properly interpreted* circuit parameter measurements (enabled by the test instruments we have been discussing) are essential to the performance of your job. This is true whether the task be in designing, troubleshooting, servicing, or monitoring the operation of circuits and systems.

SUMMARY

▶ A moving-coil meter movement (often called the d'Arsonval movement) contains several key elements—permanent magnet, pole pieces, soft iron core, movable coil and support frame, spindle and bearings, springs, pointer and counterbalance, calibrated scale, and stops.

▶ The moving-coil movement depends on a force of repulsion between the permanent magnet's fixed magnetic field and the movable coil's magnetic field when current passes through the movable coil.

▶ Electrical damping prevents the instrument's pointer from oscillating above and below the actual reading; develops from the effects of induced current in the movable coil frame; and is analyzed by Lenz's law.

▶ Moving-coil meters are current indicators and can be manufactured with sensitivities ranging from microamperes to milliamperes.

▶ Movement sensitivity relates to the strength of the permanent magnet, the number of turns on the moving coil, and other factors.

▶ Current meter sensitivities are usually expressed in full-scale current, but are also expressed in the amount of voltage dropped by the coil with full-scale current value passing through it, (e.g., 1-mA, 50-mV movement).

▶ When using the moving-coil movement as a current meter, the instrument is connected in series with the circuit portion where it is desired to measure current flow. Thus, breaking or opening the circuit, inserting the meter, and observing proper polarity and range are required. (*Caution:* Turn power off circuit before breaking circuit.)

▶ The minimum current value an instrument measures is determined by its basic movement full-scale current sensitivity. An instrument can measure higher currents by using current meter shunts. A shunt is a parallel-resistance path that divides the circuit current. Thus, a percentage of

the current passes through the meter and the remainder passes through the shunt.

▶ Parallel circuit analysis (current-divider) techniques determine the shunt resistance value for a desired current range. Inverse proportionality of current dividing in parallel circuits is a useful technique for this analysis, where

$$\frac{R_S}{R_M} = \frac{I_M}{I_S}$$

▶ When multiple ranges are available on the instrument for current measurements (selectable shunts), the switch is normally a shorting-type switch to assure the basic meter movement is never without a shunt between switch positions.

▶ A moving-coil meter movement can measure voltage by adding a series multiplier resistance. The multiplier R value and the meter's resistance must limit the current to the movement's full-scale current value for the voltage range desired. Ohm's law and series-circuit techniques determine the multiplier value. (For all practical purposes, the value of

$$R_{Multi} = \frac{V \text{ (desired range)}}{I \text{ (full-scale)}}$$

To be accurate, subtract the meter's resistance value from this answer.)

▶ Multiple ranges for voltage measurement are obtained by using a switch and contacting different multiplier resistance values at each switch position.

▶ Voltmeters are often rated by ohms-per-volt sensitivity. This rating indicates how many ohms must be in the metering circuit to limit current to the meter's full-scale current value when one volt (1 V) is applied. The formula for this is

$$\frac{\Omega}{V} \text{ (sensitivity)} = \frac{1 \text{ V}}{I_{fs}}$$

▶ Because voltmeters are connected in *parallel* with the portion of circuitry under test, the meter circuit's resistance alters the test circuit's operating conditions. This is called meter loading. The higher the meter circuit's input resistance, or Ω/V rating, the less loading effect it will cause.

▶ Most analog ohmmeters use series-type circuitry. The meter's resistance, the internal voltage source, and the resistance under test are all in series. Therefore, the ohmmeter scales read from right to left, with low R values appearing on the right side and higher R values appearing on the left side. This scale is sometimes called a backoff scale. Also, the scale calibrations for ohms are nonlinear, but the scale calibrations for current and voltage are linear.

▶ Factors to be observed when using dc instruments include using the proper range, connecting the meter to the circuit in a proper fashion, observing proper polarity, and, with the ohmmeter, making sure power is removed from the component or circuit being tested.

▶ Multimeters measure more than one type of electrical parameter, such as voltage, current, and resistance.

▶ A VOM typically has lower R input than a DMM. Therefore, a VOM causes more meter loading effects when measuring voltage.

▶ A number of special probes and devices are available for electrical measurements. Some of these include clamp-on current probes, thermocouples, clamp-on multimeters, and high-voltage probes.

▶ Using dc measuring instruments to troubleshoot actually means using common sense, properly applying Ohm's law, and having some knowledge of how the tested circuit or system operates.

FORMULAS AND SAMPLE CALCULATOR SEQUENCES

FORMULA 11–1 $R_S = \dfrac{V_S}{I_S}$

voltage across shunt, ⊟, current value to pass through shunt, ▬

FORMULA 11–2 $R_S = \dfrac{I_M R_M}{(I - I_M)}$

I_M value, ✖ , R_M value, ⊟ , ◖ , I (extended range) value, ▬ , I_M , ◗ , ▬

FORMULA 11–3 $R_S = \dfrac{I_M R_M}{I_S}$

I_M value, ✖ , R_M value, ⊟ , I_S value, ▬

FORMULA 11–4 $\dfrac{R_S}{R_M} = \dfrac{I_M}{I_S}$

R_S value, ⊟ , R_M value, ▬ , (ratio should compare to), I_M value, ⊟ , I_S value, ▬

FORMULA 11–5 $R_S = \dfrac{R_M}{N - 1}$

R_M value, ⊝ , ◖ , no. of times range is to be multiplied, ⊖ , 1 , ◗ , ⊜

FORMULA 11–6 $R_{Multi} + R_M = \dfrac{V\,(\text{range})}{I\,(\text{full-scale})}$

V range value, ⊝ , I full-scale value, ⊜

FORMULA 11–7 $R_{Multi} = \dfrac{V\,(\text{range})}{I(\text{fs})} - R_M$

V range value, ⊝ , I full-scale value, ⊖ , R_M value, ⊜

REVIEW QUESTIONS

1. Name at least five of the ten key elements or parts of a d'Arsonval moving-coil meter movement.

2. A typical moving-coil type meter movement depends on the action of magnetic:
   a. attraction.
   b. repulsion.
   c. neutralization.
   d. all of the above.

3. Electrical damping in a meter movement refers to:
   a. moisture buildup in the meter if the meter is not sealed well.
   b. current passing through the meter's coil.
   c. induced current in the coil frame retarding oscillation of the pointer.
   d. induced current in the coil frame assisting the meter point movement.

4. The basic moving-coil meter movement is:
   a. a voltage indicator.
   b. a current indicator.
   c. a resistance indicator.
   d. none of the above.

5. Two primary factors affecting the sensitivity of a basic meter movement are:
   a. the physical dimensions of the magnet and the coil.
   b. the strength of the permanent magnet and number of turns on the coil.
   c. the type bearings used and the kind of wire used in the springs.
   d. the calibrations on the meter scale and the linearity of the scale.

6. Basic current meter sensitivities may be expressed in terms of:
   a. full-scale current and/or meter voltage drop at full-scale current.
   b. full-scale current and/or meter voltage drop at half-scale current.
   c. resistance of the wire in the moving coil.
   d. millivolts drop per milliampere of current.

7. List the procedural steps you would use to safely measure current in a circuit.

8. To use a basic meter movement as a voltmeter, it is necessary to:
   a. add a shunt resistance to bypass some of the current around the meter.

b. add a series resistor to limit current through the meter.

c. add a range selector switch.

d. none of the above.

9. To have a multiple-range voltmeter, the range switch:

a. switches in a different series resistance value at each switch position.

b. switches in a different shunt resistance value at each switch position.

c. switches the meter's mode of operation.

d. none of the above.

10. A commonly used sensitivity rating for voltmeters is:

a. the full-scale current rating of the basic meter movement.

b. the ohms-per-volt rating.

c. the meter movement voltage drop rating.

d. none of the above.

11. Name at least two principal advantages of the DMM over analog-type meters.

12. Most analog-type ohmmeters use:

a. series-parallel-type circuitry.

b. parallel-type circuitry.

c. series-type circuitry.

13. The power source for ohmmeters is:

a. internal.

b. external.

c. both.

14. Typically, ohmmeters have:

a. linear scales.

b. nonlinear, left-to-right scales.

c. back-off nonlinear scales.

d. none of the above.

15. Why is it easier to use voltmeter checks, rather than current meter checks, when troubleshooting?

16. Why are pole pieces used in dc moving-coil movements?

17. What is meant by the term *damping* relative to dc moving-coil meter movements?

18. For what type situation is an analog-type instrument easier to use than a digital-type instrument?

19. The type of measuring instrument that causes less circuit "loading effect" for voltage measurements is:

a. the VOM.

b. the basic meter movement.

c. the DMM.

20. When measuring current, it is desirable for the meter circuit to have:

a. high resistance.

b. low resistance.

c. medium resistance.

d. it doesn't matter.

21. The advantage of a "clamp-on" type meter is that:

a. it is small.

b. you can measure with the power off.

c. it is not necessary to break the circuit to use it.

d. none of the above.

22. Current meters are connected in _____ with the component or circuit to be tested. Voltmeters are connected in _____ with the component or circuit to be tested. For ohmmeters, using the backoff scale, the resistance under test is in _____ with the meter, its current-limiting resistor, and the internal voltage source.

PROBLEMS

REMINDER: Don't forget to use your calculator for these problems. The more you use it, the more proficient and efficient you will become in performing calculations.

1. A current meter is rated as a 2-mA, 200-mV meter movement. What is its full-scale current rating? What is the resistance of the movement?

2. It is desired to increase the range of a 5-mA movement to 500 mA. What value of shunt resistance should be used? (Assume $R_M = 20\ \Omega$.)

3. A current meter is rated at 3 mA, 150 mV. What multiplier resistance value is needed to use this meter as a 0–30-V meter? Draw and label the circuit.

4. If the full-scale current rating of a meter is 40 μA, what is its ohms-per-volt sensitivity rating as a voltmeter? What is the value of $R_M + R_{Mult}$ if the voltage range selected is the 500-volt range?

5. Refer to Figure 11–35 and indicate what the voltmeter reading will be if the meter has Ω/V sensitivity rating of 20,000 Ω/V. (Assume that you are using the 0–10-V range.) What would the R_2 voltage drop be without the voltmeter present?

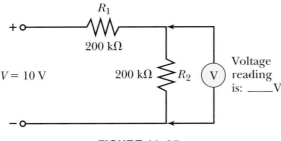

R_1

200 kΩ

$V = 10$ V

200 kΩ R_2

V

Voltage reading is: ___V

FIGURE 11–35

6. What is the value of the metering circuit resistance for a current meter having a 1-mA, 100-mV rating that is shunted to measure 0–2 mA?

7. If a voltmeter is reading 22 V on the 0–50-V range, what percentage of full-scale current is flowing through the meter movement? What would the reading be if 75% of full-scale current were flowing?

8. Draw the basic circuit of a series-type ohmmeter. Show the test probes and an unknown value resistor to be measured (R_x).

9. For the circuit in question 8, if the meter is a 1-mA, 100-Ω movement and the ohmmeter voltage source is a 1.5-V cell, what would be the calibration value at mid-scale on the ohmmeter scale? At quarter scale?

10. If a voltmeter with a sensitivity rating of 1,000 Ω/V that is set on the 0–50-V range is connected across one resistor of a series circuit comprised of two 100-kΩ resistors, what will be the voltmeter reading? (Assume a 100-V dc source.)

11. In question 10, what would be the reading if the meter had a sensitivity of 100,000 Ω/V and the 0–100-V range were used?

12. In either question 10 or 11, if two identical meters, each set on the same voltage range were simultaneously connected across R_1 and R_2 respectively (i.e., one meter measuring V_1, the other measuring V_2), what voltage reading would each show? (Still assume 100 V applied voltage.)

ANALYSIS QUESTIONS

1. Briefly explain the action of a shorting-type switch. Under what circumstances should it be used in dc measuring instruments?

2. Explain why a current meter circuit should have a low resistance.

3. Draw a three-resistor parallel circuit with a dc source. Show where and how current meters are connected to measure I_T and the current through the middle branch. Indicate polarities, as appropriate.

4. Should a voltmeter circuit have a high or low resistance? Why?

5. What indication would you have if an analog meter were connected to a dc circuit with the wrong polarity?

PERFORMANCE PROJECTS CORRELATION CHART

Suggested performance projects in the Laboratory Manual that correlate with topics in this chapter are:

CHAPTER TOPIC	PERFORMANCE PROJECT	PROJECT NUMBER
Ammeter Shunts	The Basic Ammeter Shunt	33
Voltmeters	The Basic Voltmeter Circuit	34
Ohmmeters	The Series Ohmmeter Circuit	35

NOTE: It is suggested that after completing the above projects, the student should be required to answer the questions in the "Summary" at the end of this section of projects in the Laboratory Manual.

THE SIMPLER SEQUENCE FOR TROUBLESHOOTING

Follow the seven-step SIMPLER troubleshooting sequence, as outlined below, and see if you can find the problem with this circuit. As you follow the sequence, record your circuit test results for each testing step on a separate sheet of paper. This will aid you and your instructor to see where your thinking is on track or where it might deviate from the best procedure.

step 1 **Symptoms**

—(Gather, verify, and analyze symptom information) To begin this process, read the data under the "Starting Point Information" heading. Particularly look for circuit parameters that are not normal for the circuit configuration and component values given in the schematic diagram. For example, look for currents that are too high or too low; voltages that are too high or too low; resistances that are too high or too low, etc. This analysis should give you a first clue (or symptom information) that will aid you in determining possible areas or components in the circuit that might be causing this symptom.

step 2 **Identify**

—(Identify and bracket the initial suspect area) To perform this first "bracketing" step, analyze the clue or symptom information from the "Symptoms" step, then either on paper or in your mind, bracket, circle, or put parentheses around all of the circuit area that contains any components, wires, etc., that might cause the abnormality that your symptom information has pointed out. NOTE: Don't bracket parts of the circuit that contain items that could not cause the symptom!

step 3 **Make**

—(Make a decision about the first test you should make: what type and where) Look under the TEST column and determine which of the available tests shown in that column you think would give you the most meaningful information about the suspect area or components, in the light of the information you have so far.

step 4 **Perform**

—(Perform the first test you have selected from the TEST column list) To simulate performing the chosen test (as if you were getting test results in an actual circuit of this type), follow the dotted line to the right from your selected test to the number in parentheses. The number in parentheses, under the "Results in Appendix C" column tells you what number to look up in Appendix C to see what the test result is for the test you are simulating.

step 5 **Locate**

—(Locate and define a new "narrower" area of uncertainty) With the information you have gained from the test, you should be able to eliminate some of the circuit area or circuit components as still being suspect. In essence, you should be able to move one bracket of the bracketed area so that there is a new smaller area of uncertainty.

step 6 **Examine**

—(Examine available information and determine the next test: what type and where) Use the information you have thus far to determine what your next test should be in the new narrower area of uncertainty. Determine the type test and where, then proceed using the TEST listing to select that test, and the numbers in parentheses and Appendix C data to find the result of that test.

step 7 **Repeat**

—(Repeat the analysis and testing steps until you find the trouble) When you have determined what you would change or do to restore the circuit to normal operation, you have arrived at *your* solution to the problem. You can check your final result against ours by observing the pictorial step-by-step sample solution on the pages immediately following the "Chapter Troubleshooting Challenge"—circuit and test listing page.

step 8 **Verify**

NOTE: A useful *8th Step* is to operate and test the circuit or system in which you have made the "correction changes" to see if it is operating properly. This is the final proof that you have done a good job of troubleshooting.

CHALLENGE CIRCUIT 5

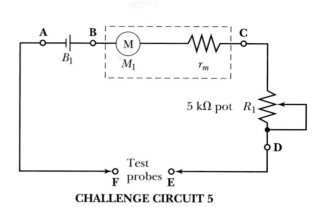

CHALLENGE CIRCUIT 5

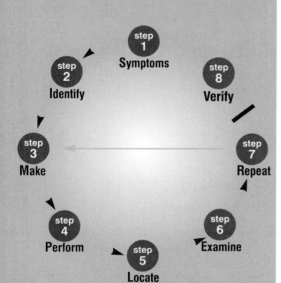

Starting Point Information

1. Circuit diagram
2. B_1 is a 1.5-V cell
3. Meter is a 1-mA movement
4. There is no meter response when the test probes are touched together to "zero" the ohmmeter

TEST	Results in Appendix C
(With test probes not touching)	
$V_{A\text{-}B}$	(44)
$V_{B\text{-}C}$	(115)
$V_{C\text{-}D}$	(9)
$V_{D\text{-}E}$	(91)
$V_{E\text{-}F}$	(40)
$V_{F\text{-}A}$	(101)
$R_{C\text{-}E}$	(13)
$R_{C\text{-}D}$	(59)
$R_{D\text{-}E}$	(108)
$R_{F\text{-}A}$	(84)
(With test probes touching)	
$V_{A\text{-}B}$	(64)
$V_{B\text{-}C}$	(27)
$V_{C\text{-}D}$	(97)
$V_{D\text{-}E}$	(20)
$V_{E\text{-}F}$	(110)
$V_{F\text{-}A}$	(47)

step 1 Symptoms
step 2 Identify
step 3 Make
step 4 Perform
step 5 Locate
step 6 Examine
step 7 Repeat
step 8 Verify

Symptoms

Information. When the ohmmeter test probes are touched to zero the ohmmeter, there is no meter response. This indicates a bad meter, a lack of circuit continuity somewhere, or a dead battery.

Identify

initial suspect area: The total circuit, since a discontinuity can be caused anywhere in the circuit.

Make

test decision: Since the battery is the only item in the circuit having a definite shelf life, let's first check the battery. Measure V between points A and B with the probes open.

Perform

1st Test: The result of measuring V_{A-B} is 1.5 V, which is normal. NOTE: There is no picture shown measuring V_{A-B}. With the test probes open, there is no load current drawn from the battery, thus it might measure normal without load and not completely good when under load. However, this tells us the battery is probably not totally dead, which it would have to be to match our symptom information.

Locate

new suspect area: The remainder of the circuit, since discontinuity can be anywhere along the current path.

Examine
available data.

Repeat

analysis and testing: By checking continuity between points C and E (with test leads not touching), we can check the continuity of the two circuit elements, R_1 and the test-lead/test-probe.

2nd Test: Check R_{C-E}. R is infinite ohms, indicating a problem. (NOTE: This meter shows "1" when measuring infinite ohms.)

3rd Test: Check R_{C-D} to verify that R_1's resistive element or wiper arm is not causing the discontinuity. R is about 1,500 ohms. However, moving the wiper arm causes R to change, which implies everything is normal.

4th Test: Check R_{D-E}. R is infinite ohms, indicating the discontinuity is between these points. A further check shows the test-lead wire has

Symptoms

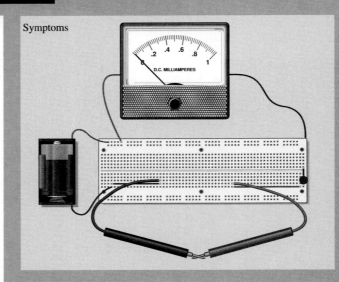

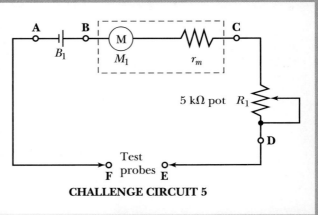

CHALLENGE CIRCUIT 5

become disconnected from the test-probe metal tip inside the test-probe plastic handle.

Verify

5th Test: Repair the wire connection to the test probe. Then, try to zero the ohmmeter. Now it works.

2nd Test

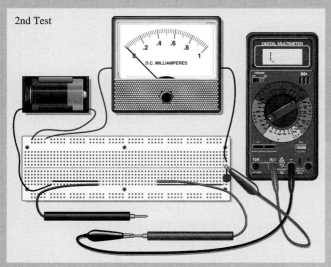

3rd Test

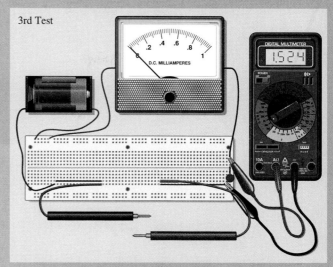

4th Test

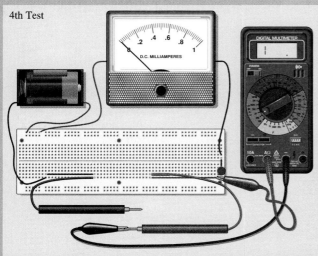

5th Test

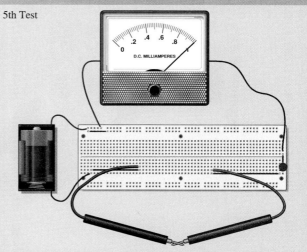

BASIC AC QUANTITIES

chapter

CHAPTER PREVIEW

Previous chapters have discussed dc voltages, currents, and related quantities. As you have learned, direct current (dc) flows in one direction through the circuit. Your knowledge of dc and your practice in analyzing dc circuits will help you as you now investigate alternating current (ac)—current that periodically reverses direction.

In this chapter, you will become familiar with some of the basic terms used to define important ac voltage, current, and waveform quantities and characteristics. Knowledge of these terms and values, plus their relationships to each other, form the critical foundation on which you will build your knowledge and skills in working with ac, both as a trainee and as a technician.

Studying this chapter, you will learn how ac is generated; how it is graphically represented; how its various features are described; and how specific important values, or quantities are associated with these features.

In the next chapter, you will study a powerful test instrument, called the oscilloscope, which will enable you to visually observe the ac characteristics you will study in this chapter.

OBJECTIVES

After studying this chapter, you should be able to:

1. Draw a graphic illustrating an **ac** waveform

2. Define **cycle, alternation, period, peak, peak-to-peak,** and **effective value (rms)**

3. Compute effective, peak, and peak-to-peak values of ac voltage and current

4. Explain **average** with reference to one-half cycle of sine-wave ac

5. Explain **average** with reference to nonsinusoidal waves

6. Define and calculate **frequency** and period

7. Draw a graphic illustrating phase relationships of two **sine waves**

8. Describe the phase relationships of V and I in a purely resistive ac circuit

9. Label key parameters of nonsinusoidal waveforms

12-1 BACKGROUND INFORMATION

The difference between dc and ac

Whereas direct current is unidirectional and has a constant value, or magnitude, alternating current, as the name suggests, alternates in direction and is bidirectional. Also, ac differs from dc in that it is continuously changing in amplitude or value, as well as periodically changing in polarity.

In Figure 12–1, you see how both a simple dc circuit and a simple ac circuit may be represented diagrammatically. You are already familiar with the symbols in the dc circuit shown in Figure 12–1a. Notice the symbol used to represent the ac source in the ac circuit shown in Figure 12–1b. You will be seeing this symbol

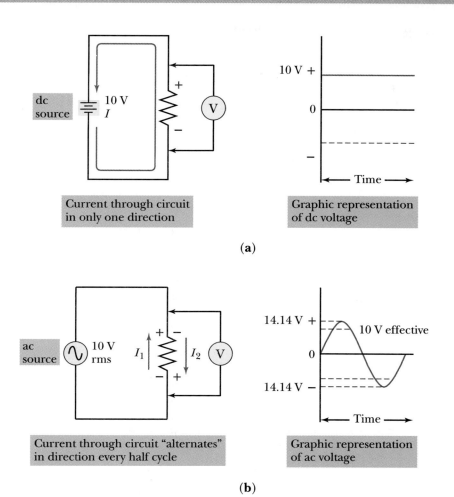

FIGURE 12–1 Graphic comparing dc and ac waveforms

Current through circuit in only one direction

Graphic representation of dc voltage

(a)

Current through circuit "alternates" in direction every half cycle

Graphic representation of ac voltage

(b)

used throughout the remainder of the book. The wiggly line in the center of the symbol represents a sine-wave waveform, which you will study shortly.

Figure 12–1 also highlights the comparisons and contrasts between dc and ac voltage characteristics, using their graphical representations. As you proceed in the chapter, you will be focusing your attention on learning important specifics about the sine wave and other ac waveforms.

Alternating current used for power transmission

Alternating current (**ac**) has certain advantages over dc when transporting electrical power from one point to another point. A key advantage is its ability to be "stepped up" or "stepped down" in value. Again, you will study more about this later in the text. These advantages are why power is brought to your home in the form of ac rather than dc.

Some interesting contrasts between dc and ac are shown in Figure 12–2. This ac electrical power is produced by alternating-current generators. A simple discussion about generating ac voltage will be covered shortly. However, some groundwork will now be presented.

FIGURE 12–2 Similarities and differences between dc and ac power

CHARACTERISTIC DESCRIPTOR	IS CHARACTERISTIC OF:	
	AC	DC
Current in one direction only	No	Yes
Current periodically alternates in direction one-half the time in each direction	Yes	No
Voltage can be "stepped up" or "stepped down" by transformer action	Yes	No
Is an efficient means to transport power over long distances	Yes	No
Useful for all purposes	Yes	No

Defining angular motion and the coordinate system

As we deal with ac quantities, it is helpful to understand how angular relationships are commonly depicted. Refer to Figures 12–3a (angular motion) and 12–3b (the coordinate system) as you study the following discussions.

Mathematical and graphic concepts about ac

1. Angular motion is defined in terms of a 360° circle or a four-quadrant coordinate system.
2. The zero degree (starting or reference plane) is a horizontal line or arrow extending to the right, Figure 12–3a. NOTE: Other terms you might see for this line or arrow are a rotating radius and a vector called a **phasor.** A **vector** identifies a quantity that has (or illustrates) both magnitude *and* direction.
3. Angles increase in a positive, counterclockwise (CCW) direction from the zero reference point. In other words, the horizontal phasor to the right is rotated in a CCW direction until it is pointing straight upward, indicating an angle of 90° (with respect to the 0° reference point).
   - When the phasor continues in a CCW rotation until it is pointing straight to the left, that point represents 180° of rotation.
   - When rotation continues until the phasor is pointing straight downward, the angle at that point is 270°.
   - When the phasor completes a CCW rotation, returning to its starting point of 0°, it has rotated through 360°.
   - When the phasor is halfway between 0° and 90°, the angle is 45°. When the phasor is halfway between 90° and 180°, the angle is 135°.
4. The **coordinate system,** Figure 12–3b, consists of two perpendicular lines, each representing one of the two axes. The horizontal line is the x-axis. The vertical line is the y-axis. It is worth your effort to learn the names and directions of these two axes—they will be referenced often during your technical training. Remember the x-axis is horizontal and the y-axis is vertical.

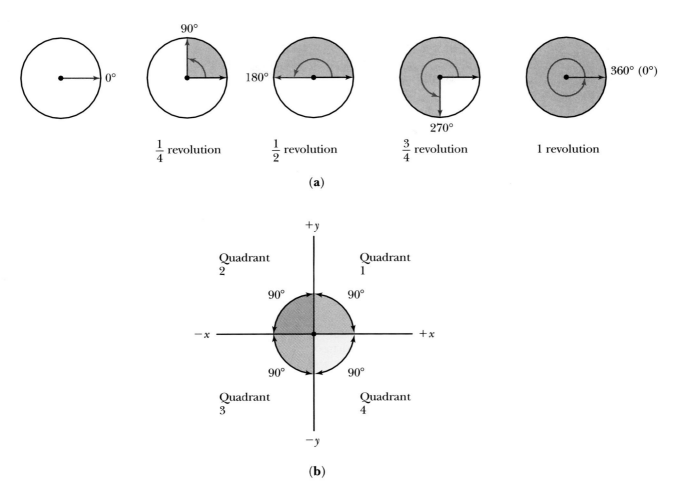

FIGURE 12–3 (a) Describing angular motion; (b) the coordinate system

5. There are four **quadrants** associated with this coordinate system. Again, look at Figure 12–3b and note that the angles between 0° and 90° are in quadrant 1. Angles between 90° and 180° are in quadrant 2. Angles between 180° and 270° are in quadrant 3; and angles between 270° and 360° (or 0°) are in quadrant 4.

6. When the phasor moves in a clockwise (CW) direction, the various angle positions are described as negative values. For example, the location halfway between 360° (0°) and 270° (−90°) in quadrant 4 (the same location as the +315° point) can be described as either −45° or +315°.

IN-PROCESS LEARNING CHECK I

1. Define the difference between dc and ac. _____

2. Which direction is the reference, or 0° position when describing angular motion? _____

3. Is the y-axis the horizontal or vertical axis? _____

continued

4. Is the second quadrant between 0° and 90°, 90° and 180°, 180° and 270°, or 270° and 360°? _____

5. Define the term *vector.* _____

Keeping in mind these thoughts about defining angles in degrees and a reference point on the coordinate system, let's move to a brief explanation of how an ac sine-wave voltage is generated.

12-2 GENERATING AN AC VOLTAGE

As a quick review, recall that generating a dc voltage was accomplished by means of rotating conductors (e.g., armature coil wires) cutting or linking lines of flux in a magnetic field. Remember that by means of a split-ring commutator and brushes, dc output voltage was delivered from the rotating armature to the load, and the dc was in the form of a pulsating dc, Figure 12–4a and b.

Refer to Figure 12–4c and note the elementary concepts of an ac generator. The simplified generator consists of a magnetic field through which a conductor is rotated. As you would expect when the maximum number of magnetic field flux lines are being cut (or linked) by the rotating conductor, maximum voltage is induced. Recall that if 10^8 flux lines are cut by a single conductor in one second, the induced voltage is one volt. Conversely, when minimum (or zero) flux lines are cut, minimum (or zero) voltage is induced.

Slip rings (sometimes called collector rings), are electrical contacting devices designed to make contact between stationary and rotating electrical junctures. As you can see in Figure 12–4c, the brushes are in a stationary position, and make contact between the external load resistance and the slip rings. The slip rings are connected to, and rotate along with, the rotating conductors. The outside surfaces of the slip rings, however, are in continuous sliding electrical contact with the brushes. The conductors, which are rotating through the magnetic field, are connected so that the same end of each conductor is always connected to the same slip ring.

Also, observe that the slip rings and brushes that connect the external circuit load to the rotating loop so each end of the load is always connected to the same end of the rotating conductor. Thus, if the polarity of the induced voltage in the conductor changes, the voltage felt by the load through the brushes and the slip rings also changes.

Figure 12–4d shows the graphical representation of the changing magnitude and polarity of generator output felt by the load throughout one complete 360° rotation of the conductor loop(s). As we stated earlier, this graphic representation of the changing amplitude and polarity of the output voltage over time is called a waveform. In this case, the waveform is that of one full **cycle** of an ac sine-wave signal. A cycle is often defined as one complete sequence of a series of recurring events. There are many examples of cycles. The 24-hour day, the 12-hour clock face, one revolution of a wheel, and so on. In Figure 12–4d, you can see how at any given instant or point throughout the cycle, the signal's amplitude

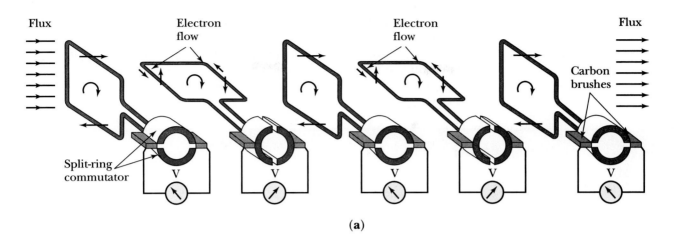

Flux   Electron flow   Electron flow   Flux

Carbon brushes

Split-ring commutator

(**a**)

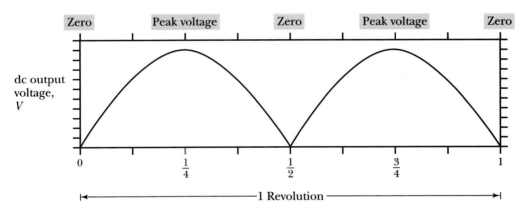

Zero   Peak voltage   Zero   Peak voltage   Zero

dc output voltage, V

0    $\frac{1}{4}$ $\frac{1}{2}$ $\frac{3}{4}$ 1

1 Revolution

(**b**)

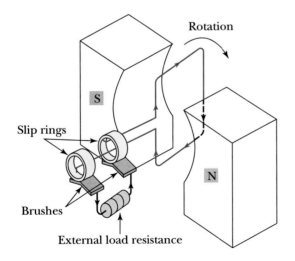

Rotation

S

Slip rings

N

Brushes

External load resistance

(**c**)

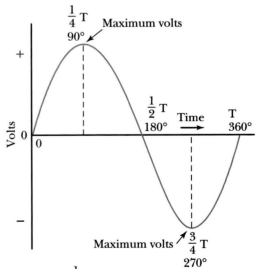

$\frac{1}{4}$ T   Maximum volts
90°

+

$\frac{1}{2}$ T   Time   T
180°   360°

Volts   0

0

−

Maximum volts $\frac{3}{4}$ T
270°

T = Period = $\frac{1}{f}$ = Time for 1 cycle

(**d**)

FIGURE 12–4 Basic concepts of an ac generator

can be related to the angular degrees of rotation of the generator. Figure 12–5 again relates sine-wave amplitude and polarity to degrees throughout 360°. For a sine wave, you can see that maximum positive amplitude occurs at 90°. Maximum negative amplitude occurs at 270°. And, the sine-wave signal is at zero level at 0°, 180°, and 360°. (Assuming we start the waveform analysis at the zero point, with the signal progressing in a positive direction.)

PRACTICAL NOTES

As you will see in upcoming studies, we do not always have to analyze an ac sine-wave cycle as starting at zero amplitude level, and a cycle being completed at zero amplitude. We can define one complete cycle as starting at any point along the sine wave, and being completed at the next point along the waveform that is at the same amplitude and changing in the same direction. (That is, the point at which the sequence of recurring events starts again.)

The sine wave

This unique, single-frequency waveform reflects a quantity that constantly changes in amplitude and periodically reverses in direction.

Radians of angular measure

You have been introduced to the fact that sine-wave signals are often analyzed in terms of angular measure. The radian is another angular measure of importance, as many of the mathematics associated with ac analysis often use this measure.

As you might surmise from the word *radian* itself, there is a relationship between the term *radius* (of a circle) and the term *radian* (used in circular angular measure). One definition for a radian (sometimes abbreviated "rad") is: "In a cir-

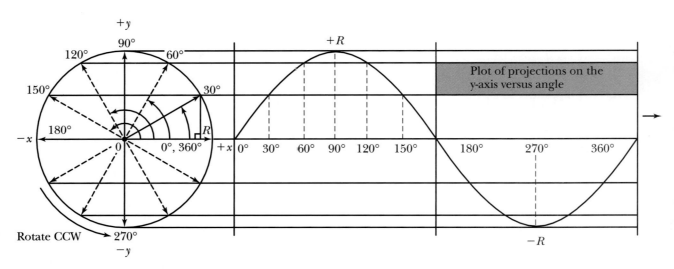

FIGURE 12–5 Projections of sine-wave amplitudes in relation to various angles

cle, the angle included within an arc equal to the radius of the circle." In Figure 12–6, you can see that the angle included by an arc of this length is about 57.3°.

The thing that makes the radian of particular interest, mathematically, is that there are exactly 2π radians in 360°. If you divide 360° by 57.3°, you will see that it would take 6.28 (or 2π) radians to make up a complete 360° circle. Observe Figure 12–7, and note the relationships between the degrees and radians around the circle. It is important to remember that one complete cycle of ac (360°) equals 2π radians, that half a cycle (180°) equals 1π radian, and so on. See Figure 12–7b. The waveform in Figure 12–7b is called a **sine wave** because the amplitude (amount) of voltage at any given moment is directly related to the sine function, used in trigonometry. That is, the trigonometric sine (sin) value for any given angle (θ or theta), represented along the sine wave, relates directly to the voltage amplitude at that same angle. NOTE: The sine of a given angle, using a right triangle as the basis of analysis, is equal to the ratio of the opposite side to the hypotenuse, Figure 12–8.

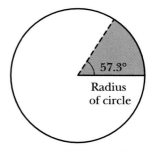

Arc = Radius =1 Radian

FIGURE 12–6 Illustration of the radian angular measurement

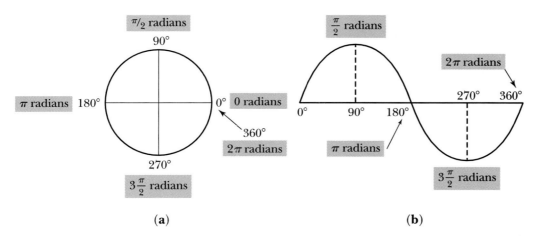

(a)   (b)

FIGURE 12–7 Degrees and radians of angular measure

For angle *A*:
1. The side opposite the right angle is called the *hypotenuse* (*c*).
2. The side opposite angle *A* is called the *opposite side* (*a*).
3. The side of angle *A* which is not the hypotenuse is called the *adjacent side* (*b*).

Sine (sin) of angle $= \dfrac{\text{opposite side}}{\text{hypotenuse}}$	; $\sin A = \dfrac{a}{c}$
Cosine (cos) of angle $= \dfrac{\text{adjacent side}}{\text{hypotenuse}}$	; $\cos A = \dfrac{b}{c}$
Tangent (tan) of angle $= \dfrac{\text{opposite side}}{\text{adjacent side}}$	; $\tan A = \dfrac{a}{b} = \dfrac{\sin A}{\cos A}$

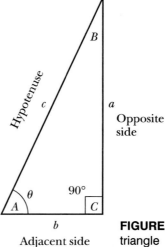

FIGURE 12–8 The right triangle

Also observe in Figure 12–8 that we have shown the side relationships that are important in computing the cosine (cos) and the tangent (tan) trig function values for the selected acute angle in the right triangle. As you can see, the cosine (cos) equals the ratio of the side that is adjacent (to the angle of interest) to the hypotenuse. The tangent (tan) equals the ratio of the side opposite to the side adjacent for the angle of interest. You will be studying and using these trig functions in later studies in this text, along with the sine (sin) function, which has been introduced in this chapter.

PRACTICAL NOTES

To find the sine of an angle using your calculator, be sure your calculator is in the degrees (DEG) mode of operation, rather than the RAD or GRAD mode. Recall that the units for angular measurement are as follows: a degree = 1/360 of a circle; a radian = $1/2\pi$ or approximately 0.159 of a circle (about 57.3°); and a grad = 1/400 of a circle, or about 0.9°. (This means that 90° = 100 grads.)

Once your calculator is in the degrees mode, simply input the value of the angle (e.g., 30, for 30°), press the **SIN** function button, and read the value of the sine. In our example of 30°, the value of the sine is 0.5. The key sequence again is as follows: (the angle value), then the trig function key. The answer shows the value of the trig function.

To go the other way, that is, if you know that the ratio of the opposite side to the hypotenuse equals 0.5, and you want to find the angle whose sine value is 0.5: input 0.5, then press the **2nd** and **SIN⁻¹** keys. On some calculators, you press **INV**, then **SIN**. By pressing the **2nd** and **SIN⁻¹** (or **INV**, **SIN**) keys, you are telling the calculator you want to know the angle whose sin is 0.5. When you do this, you should see 30 in the readout of the calculator. This means the angle whose sine is 0.5 is 30°. The key sequence again is: (the trig function value), the **2nd** or **INV** key, then the trig function key (i.e., the **SIN⁻¹** or **SIN** key). The answer shows the value of the angle.

You can use the same techniques to find the value for cosines and tangents and/or the angles involved. That is, to find the trig function value when the angle is known; input the value of the angle, then press the correct trig function key to get your answer. In the case of the cosine, if you input 30 (for 30°), then press the **COS** button, the readout will indicate that the value of the cosine for an angle of 30° is 0.866. If you input 30, then press the **TAN** button, you find the tangent of an angle of 30° is 0.577. Going the other way, input the trig function value, then the **2nd** or **INV** key, then the appropriate trig function key (e.g., **SIN⁻¹** or **SIN** key, as appropriate). The answer is the angle involved.

Refer to Figure 12–9 as you read the following discussion. For simplicity, we will equate the angle of rotation from the starting point of the ac generator's mechanical rotation (where no flux lines are being cut by the rotating conductor) to the angle of the sine-wave output voltage waveform. For example, at the 0° point of rotation, no flux lines are being cut and the output voltage is zero (position a). If you find the sine value of 0° on a "trig" table, or use the "sin" function on a calculator, you will find it to be zero, and the output voltage at this time is zero volts.

At 45° into its rotation, the conductor cuts sufficient flux lines to cause output voltage to rise from 0 to 70.7% of its maximum value. The sine value for 45° is 0.707.

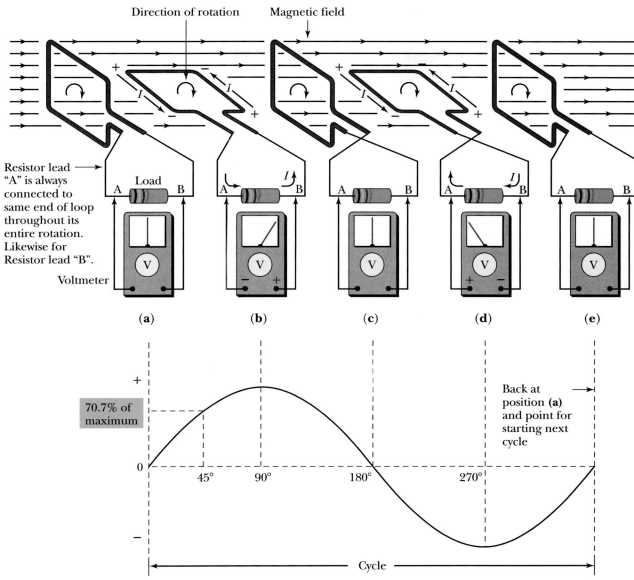

FIGURE 12–9 Relationship of flux lines cut in relation to output

As the conductor rotates to the 90° position (position b), it cuts maximum lines of flux per unit time; hence, has maximum induced voltage. The sine function value for 90° is 1.00. For a sine wave, 100% of maximum value occurs at 90°.

As the conductor rotation moves from 90° toward 180°, fewer and fewer magnetic lines of force per unit time are being cut; hence, induced voltage decreases from maximum (positive) value back to zero (position c).

Continuing the rotation from 180° to 270°, the conductor cuts flux in the opposite direction. Therefore, the induced voltage is of the opposite polarity. Once again, voltage starts from zero value and increases in its opposite polarity **magnitude** until it reaches maximum negative value at 270° of rotation (position d).

From 270° to 360°, the output voltage decreases from its maximum negative value back to zero (position e). This complete chain of events of the sine wave is a cycle. If rotation is continued, the cycle is repeated.

(NOTE: A cycle can start from *any* point on the waveform. That particular cycle continues until that same polarity and magnitude point is reached again. At that point, repetition of the recurring events begins again, and a new cycle is started.)

12-3 RATE OF CHANGE

Starting at one of the zero points, voltage increases from zero to maximum positive, then back to zero, on to maximum negative value, and then back to zero. As you can see, the sine wave of voltage continuously changes in amplitude and periodically reverses in polarity.

Let's look at another interesting aspect of the sine wave of voltage. The sine-wave waveform is nonlinear. That is, there is not an equal change in amplitude in one part of the waveform for a given amount of time, compared with the change in amplitude for another equal time segment at a different portion of the waveform. Therefore, the *rate of change* of voltage is different over different portions of the sine wave.

Example

Notice in Figure 12–10 the voltage is either increasing or decreasing in amplitude at its most rapid rate near the zero points of the sine wave (i.e., the largest amplitude *change* for a given amount of time).

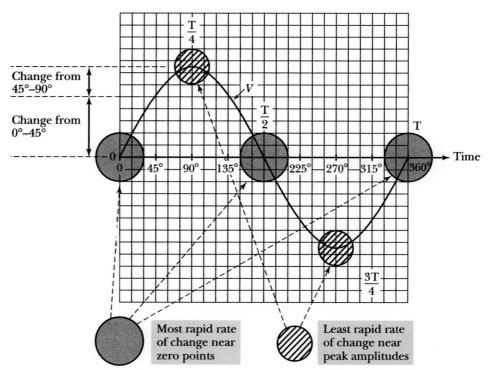

FIGURE 12–10 Different rates of change of voltage at various parts of sine wave

T = time for one cycle. This is known as the "period" of the waveform.

Also, notice that near the maximum positive and negative areas of the sine wave, specifically, near the 90° and 270° points, the rate of change of voltage is minimum. This rate of change has much less *change* in amplitude per unit time. Knowledge of this rate of change (with respect to time) will be helpful to you in some later discussions, so try to retain this knowledge.

12-4 INTRODUCTORY INFORMATION SUMMARY REGARDING AC SINE-WAVE SIGNALS

1. A sine-wave signal is an ac signal whose amplitude characteristics can be related to the trigonometric sine function.

2. An ac sine-wave signal continuously varies in amplitude and periodically changes polarity.

3. One complete cycle of an ac signal can be measured in terms of 360 degrees and/or 2π radians (a radian = 57.3°).

4. When a signal's 0° starting point is considered to be at zero amplitude, it is common to think of that signal's maximum positive point occurring at 90°, the next zero point being at 180°, the maximum negative point being at 270°, and the cycle being completed at 360°, at zero amplitude, ready to start the next cycle.

5. One cycle of a sine-wave signal can be measured from any starting point along the waveform to the next point along the waveform that matches it, and starts another similar waveform sequence in terms of amplitude and direction variations.

6. The "rate of change" of the sine-wave waveform is greatest at the zero amplitude points, and least at the maximum amplitude points of the waveform.

PRACTICE PROBLEMS I

1. Use a calculator or trig table and determine the sine of 35°.

2. Indicate on the following drawing at what point cycle 1 ends and cycle 2 begins, if the starting point for the first cycle is point A.

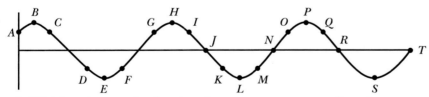

3. Which points on the drawing above represent areas of maximum rate of change of the sine wave(s)?

12-5 SOME BASIC SINE-WAVE WAVEFORM DESCRIPTORS

Refer to Figure 12–11. The horizontal line (x-axis) divides the sine wave into two parts—one above the zero reference line, representing positive values, and one below the zero reference line, representing negative values. Each half of the sine wave is an **alternation.** Naturally, the upper half is the "positive alternation," or positive half cycle. The lower half of the wave is the "negative alternation," or

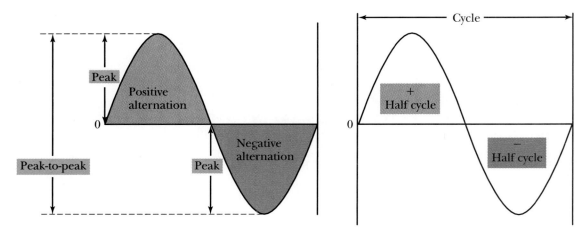

FIGURE 12–11 Several important descriptors of sine-wave ac

negative half cycle. With zero as the starting point, an alternation is the variation of an ac waveform from zero to a maximum value, then back to zero (in either polarity).

Examine Figure 12–11 again, and learn what is meant by positive peak, negative peak, and peak-to-peak values of the sine wave. Also, note in this illustration how a cycle is depicted.

Generalizing, a cycle, whether in electronics or any other physical phenomena, is one complete set of recurring events. Examples of this are one complete revolution of a wheel, and the 24-hour day (representing one complete revolution of the earth).

12-6 PERIOD AND FREQUENCY

Periodic waves

Periodic waves are those that repeat themselves regularly in time and form. Most ac signals and waves that you will study as a technician fall into this class. The period, or time of interest, for such waves is the time it takes for one complete cycle of that waveform. Let's see how the sine wave we have been discussing is defined in this regard.

Period

The ac sine-wave waveform has been described in terms of amplitude and polarity and has been referenced to various angles. It should be emphasized that the sine-wave occurs over time—it does not instantaneously vary through all its amplitude values. The household ac, with which you are familiar, takes 1/60 of a second to complete a cycle. Look at Figure 12–12 and observe how the sine-wave illustrations have been labeled. Notice that time is along the horizontal axis. In Figure 12–12a, a one-cycle-per-second signal is represented. If this waveform represented one cycle of the 60-cycle ac house current, the time, or **period** (*T*) of a cycle would be 1/60 of a second, rather than the one second shown for the one-cycle signal. Learn the term period! Again, period is the time required for one cycle.

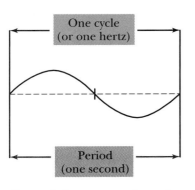

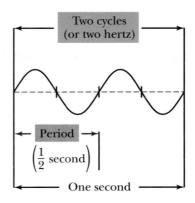

One cycle in one second equals
a "frequency" of one hertz

Two cycles in one second equals
a "frequency" of two hertz

(a)

(b)

FIGURE 12–12 Comparing
two frequencies

Frequency

The number of cycles occurring in one second is the **frequency** (*f*) of the ac. Refer again to Figure 12–12. The period of one cycle is inversely related to frequency. That is, the period equals the reciprocal of frequency, or $1/f$. The letter *T* represents period; that is, *T* is time of one cycle. Therefore:

FORMULA 12–1 $T(s) = \dfrac{1}{f(Hz)}$

PRACTICAL NOTES

Use the **1/x** (or x⁻¹) reciprocal function of your calculator for this type of calculation.

Restated, the period of one cycle (in seconds) equals 1 divided by the frequency in cycles per second. Rather than saying "cycles per second" when talking about frequency, we use a unit called the **Hertz (Hz),** named for the famous scientist Heinrich Hertz. For example, 1 Hz equals one cycle per second; 10 Hz equal 10 cycles per second; 1 kHz equals 1,000 cycles per second; and 1 MHz equals 1 million cycles per second. Notice the abbreviation for Hertz (Hz) *starts* with a capital H and is abbreviated Hz to prevent confusion with another electrical quantity you will study later—the henry, which is abbreviated with a capital H.

Example

1. The period of a 100-Hz sine wave is $\dfrac{1}{100}$ of a second, $\left(T = \dfrac{1}{f}\right)$.

2. The period of a 1,000-Hz signal is $\dfrac{1}{1,000}$ of a second, $\left(T = \dfrac{1}{F}\right)$.

Relationship of frequency and period

Converse to calculating the period of one cycle by taking the reciprocal of the frequency (i.e., $T = 1/f$), the formula is transposed to find the frequency, if you know the time it takes for one cycle. The formula is:

FORMULA 12–2 $\quad f(\text{Hz}) = \dfrac{1}{T\,(s)}$

This indicates the period is inverse to the frequency, and vice versa. That is, the higher (or greater) the frequency, the less time required for one cycle (the shorter the period), or the longer (or greater) the period, the lower the frequency must be. In other words, if $T\uparrow$, then $f\downarrow$; and if $f\uparrow$, then $T\downarrow$. NOTE: again the **1/x** function on your calculator is useful.

Example

1. A signal with a period of 0.0001 seconds indicates a signal with a frequency of 10,000 Hz, $\left(f = \dfrac{1}{T}\right)$.

2. A signal with a period of 0.005 seconds indicates a signal with a frequency of 200 Hz, $\left(f = \dfrac{1}{T}\right)$.

PRACTICAL NOTES

Another meaningful measurement related to frequency is wavelength. (In this case, it is an inverse relationship; i.e., the higher the frequency the shorter the wavelength, etc.) In your later studies, you will be likely to deal more with this measurement. You will be learning more details of the various frequency spectrums, such as audio frequencies, radio frequencies, optics, and x rays.

Wavelength is the distance a given signal's energy travels in the time it takes for one cycle of that energy. Incidentally, the symbol used to represent wavelength is the Greek letter lambda (λ). Basically, two factors determine the wavelength of any given frequency. The speed at which the signal's energy moves through its propagation medium, and the length of time for one cycle (or the period) of the specific frequency. For example, radio waves and light waves travel through space at a speed of about 186,000 miles per second, which translates to about 300 million meters per second. Signals in the audio spectrum, however, travel at around 1,130 feet per second. As you might surmise, the higher the frequency, the less time there is in one period (one cycle); hence, the less distance the wave travels in that time, and the shorter its wavelength.

The formula which describes the above data is:

$$\lambda = \text{velocity/frequency, or } \lambda = v/f$$

For frequencies (radio frequencies, etc.) that travel at the speed of light, the formula is:

$$\lambda \text{ (wavelength)} = \frac{300 \times 10^6}{\text{Freq. (Hz)}} \text{ or } \frac{300}{f\text{(MHz)}}$$

For example, the wavelength of a 2-MHz signal would be $300/2 = 150$ meters. You can also find a frequency, if you know its wavelength. That is, Frequency $= v/\lambda$. In our example, Frequency (in MHz) $= 300/150 = 2$ MHz.

IN-PROCESS LEARNING CHECK II

1. Determine the frequency of an ac signal whose period is 0.0001 second. _____

2. What is the period for a frequency of 400 Hz? _____

3. What time does one alternation for a frequency of 10 kHz take? _____

4. Does the y-axis represent time or amplitude in waveforms? _____

5. For waveforms, at what *angular points* does the amplitude equal 70.7% of the positive peak value? At what angular points is the sine wave at a 70.7% level during the negative alternation? _____

6. In the coordinate system, in which quadrants are all the angles represented by the positive alternation of a sine wave? _____

7. What is the period of a frequency of 15 MHz? _____

8. What frequency has a period of 25 ms? _____

9. As frequency increases, does T increase, decrease, or remain the same? _____

10. The longer a given signal's period, the _____ the time for each alternation. _____

12–7 PHASE RELATIONSHIPS

Alternating current (ac) voltages and currents can be *in phase* or *out of phase* with each other by a difference in angle. This is called **phase angle** and is represented by the Greek letter theta (θ). When two sine waves are the same frequency and their waveforms pass through zero at different times, and when they do not reach maximum positive amplitude at the same time, they are out of phase with each other.

On the other hand, when two sine waves are the same frequency, and when their waveforms pass through zero and reach maximum positive amplitude at the same time, they are in phase with each other.

Look at Figure 12–13 and note the following:

1. Sine waves a and b are in phase with each other because they cross the zero points and reach their maximum positive levels at the same time (shown on

FIGURE 12–13 Phase relationships

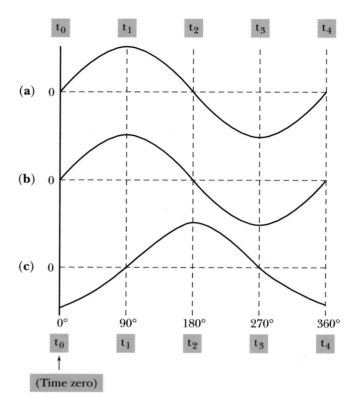

the x-axis). Zero points are at times t_0, t_2, and t_4, and the positive peak at time t_1.

2. Sine waves b and c are out of phase with each other because they do not pass through the zero points and reach peak values at the same time. In fact, sine wave b reaches the peak positive point (at t_1) of its waveform 90° before sine wave c does (at t_2). It can be said that sine wave c "lags" sine wave b (and also sine wave a) by 90°. Conversely, it can be said that sine wave b "leads" sine wave c by 90°. In other words, when viewing waveforms of two *equal-frequency* ac signals, the signal reaching maximum positive first is leading the other signal, in phase. "First" refers to the peak that is closest to the y-axis; thus, showing an occurrence at an earlier time.

Obviously, ac signals can be out of phase by different amounts. This is illustrated by sine-wave waveforms and **phasor** diagrams, Figure 12–14. Figure 12–14a shows two signals that are 45° out of phase. Which signal is leading the other one by 45°? The answer, of course, is *A* is leading *B* by 45°.

In Figure 12–14b, waveforms *A* and *B* are 180° out of phase. If these signals were of equal amplitude, they would cancel each other out.

In Figure 12–14c, you see two signals that are 90° out of phase.

Notice how the phasors in Figure 12–14 illustrate both magnitude and relative phase for several conditions. Observe that the *length* of the phasors can represent *amplitude* values of the two waveforms. Both phasors must represent the same relative point or quantity for each of the two waveforms being compared.

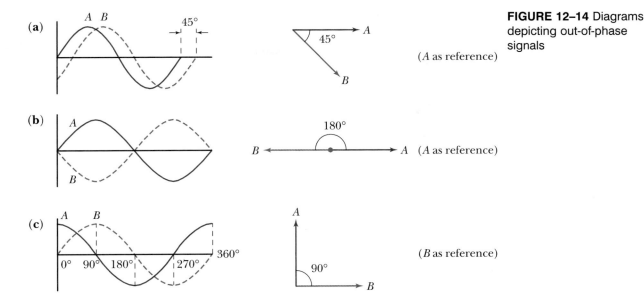

FIGURE 12–14 Diagrams depicting out-of-phase signals

PRACTICE PROBLEMS II

1. Illustrate two equal-frequency sine waves that are 60° out of phase with each other. Label the leading wave as *A*, and the lagging wave as *B*. Also, show the *A* waveform starting at the 0° point!

2. Assume the signals in problem 1 are equal in amplitude. Draw and label a diagram using phasors to show the conditions described in problem 1. Also, show the *B* phasor as the horizontal vector!

12-8 IMPORTANT AC SINE-WAVE CURRENT AND VOLTAGE VALUES

The importance of learning the relationships between the various values used to measure and analyze ac currents and voltages cannot be overemphasized! These values are defined in this section. Learn these relationships well, since you will use them extensively in your training and work from this point on.

Brief review of terms

Look at Figure 12–15 and recall the maximum positive or negative amplitude (height) of one alternation of a sine wave is called **peak value.** For voltage, this is called V_p, or V_M; for current, this is called I_p, or I_M.

Also, remember the total amplitude from the peak positive point to the peak negative point is called **peak-to-peak** value, sometimes abbreviated as V_{p-p} when speaking of voltage values.

In other words, for a symmetrical waveform:

$$V_p = \frac{1}{2} \times V_{p-p}, \text{ and } V_{p-p} = 2 \times V_p$$

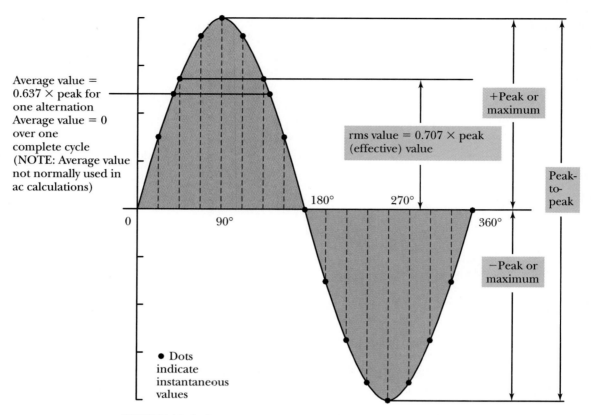

FIGURE 12–15 Important values related to sine-wave waveforms

Some new terms and values

Refer again to Figure 12–15 as you study the following discussions:

1. Since a given amount of current through a component in either direction causes the same amount of voltage drop, power dissipation, heating effect, and so on, the effects of an ac voltage or current are analyzed on a single alternation. For example, the positive alternation causes the same amount of current through a component (like a resistor) as the negative alternation, and vice versa.

2. The most frequently used value related to ac is called the **effective value (rms).** This is *that value of ac voltage or current producing the same heating effect as an equal value of dc voltage or current.* That is, it produces the same power:

$$I^2R \text{ or } \frac{V^2}{R}$$

For example, an effective value of ac current of two amperes through a given resistor produces the same heating effect as two amperes of dc current through that same resistor.

Effective value is also commonly termed the root-mean-square value. The root-mean-square term comes from the mathematical method used to find its value. It is derived from taking the square root of the mean (average) of all the squares of the sine values. Thus, effective value (rms) is 0.707 times maximum or peak value. Again, the reason effective value is greater than half of the maximum value relates to the nonlinearity of the rate of change of voltage or current throughout the alternation.

FORMULA 12–3   Effective value (rms) = 0.707 × Peak value

Example

Again, assuming 1-V peak value for the waveform in Figure 12–15, $V_{rms} = 0.707 \times 1 = 0.707$ V

3. Although **average value** only has meaningful significance as it relates to the subject of rectification of ac (changing of ac to dc), we will simply introduce you to the concepts here. This awareness will then become useful in your later studies. If you compute the average amplitude or level under the wave-form outline of *one-half cycle* or one alternation of an ac waveform, the result is the **average value** equals 0.637 (63.7%) of the maximum (or peak) value. Again, the reason the average value is more than 50% of the peak value is because the rate of change is slower near peak value than it is near the zero level. That means the voltage level is above 50% of maximum level for a longer time than it is below 50% of maximum level.

In other words, the sine wave is a nonlinear waveform. Therefore, the average is not half of the peak value. *Incidentally, when you "average" over the entire cycle, rather than for one alternation, the result is zero,* since the positive and negative alternations are symmetrical, equal in amplitude and time, and opposite in polarity. However, computing only one alternation or half-cycle, the formula is:

FORMULA 12–4   Average value = 0.637 × Peak value

Example

If the peak value in Figure 12–15 were 1 V, V_{avg} (one-half cycle) = 0.637 × 1.0 = 0.637 V.

PRACTICAL NOTES

Most voltmeters and current meters used by technicians are calibrated to measure effective, or rms, ac values. This includes VOMs, DMMs, and clip-on current and volt meters.

In ac, power (heating effect) is computed in purely resistive components or circuits by using the effective voltage or current values. That is, $P = V_{rms} \times I_{rms}$, or $P = I^2R$, or $P = V^2/R$.

Summary of relationships for common ac values (for sinusoidal waveforms)

1. Effective value (rms) = 0.707 times peak value, or 0.3535 times peak- to-peak value.

2. Average value = 0.637 times peak value. (NOTE: it also = 0.9 times effective value, since 0.637 is approximately 9/10 of 0.707.)

3. Peak value = maximum amplitude within one alternation. Since effective value (rms) is 0.707 of peak value, it follows that peak value equals 1/0.707, or is 1.414 times effective value. That is, $V_p = 1.414 \times$ rms value. (NOTE: it also equals 0.5 times peak-to-peak value.)

4. Peak-to-peak value = two times peak value and is the total sine wave amplitude from positive peak to negative peak. It follows then, that effective value (rms) is 0.3535 of the peak-to-peak value (or half of 0.707). That is, V_{p-p} equals 1/0.3535, or peak-to-peak value equals 2.828 times effective value.

With the knowledge of these various relationships, try the practice problems below.

PRACTICE PROBLEMS III

1. The peak value of a sine-wave voltage is 14.14 V. What is the rms voltage value?

2. For the voltage described in question 1, what is the V_{avg} value over one alternation?

3. Approximately what percentage of effective value computed in question 1 is the average value computed in question 2?

4. Does frequency affect the rms value of a sine-wave voltage?

5. Are the results from calculating ac values from the negative alternation the same as ac values calculated from the positive alternation?

6. The peak-to-peak voltage of a sine-wave signal is 342 V. Calculate the following:
   a. ac rms (effective) voltage value _____
   b. peak voltage value _____
   c. instantaneous voltage value at 45° _____
   d. instantaneous voltage value at 90° _____
   e. ac average voltage value (over one-half cycle) _____

12–9 THE PURELY RESISTIVE AC CIRCUIT

Earlier, we discussed in-phase and out-of-phase ac quantities. In purely resistive ac circuits, the ac voltage across a resistance is in phase with the current through the resistance, Figure 12–16. It is logical that when maximum voltage is applied to the resistor, maximum current flows through the resistor. Also, when the instantaneous ac voltage value is at the zero point, zero current flows and so forth. Ohm's law is just as useful in ac circuits as it has been in dc circuits. (NOTE: **instantaneous values** are expressed as e or v for instantaneous voltage and i for instantaneous current.)

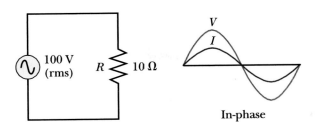

FIGURE 12–16 Basic ac resistive circuit

In-phase

Example

Applying Ohm's law to the circuit in Figure 12–16, if the rms ac voltage is 100 V, what is the effective value of ac current through the 10-Ω resistor? If you answered 10 A, you are correct! That is,

$$I = \frac{V}{R} = \frac{100 \text{ V}}{10 \text{ }\Omega} = 10 \text{ A}$$

In ac circuits where voltage and current are in phase, the calculations to find V, I, and P are the same as they are in a dc circuit. Generally, effective (rms) values are used in these computations, unless otherwise indicated.

If you want to determine the instantaneous voltage at the 30° point of the sine wave, the value would equal the sine of 30° times the maximum value. In our sample case, the sine of 30° = 0.5. Therefore, if the maximum voltage (V_{max} or V_p) is 141.4 volts, the instantaneous value at 30° = 0.5 × 141.4 = 70.7 V.

FORMULA 12–5 $v = V_p \sin \theta$

where v = instantaneous voltage value
  V_p = maximum voltage value
  $\sin \theta$ = sine of the angle value

The analyses you have been using for dc circuits also applies to purely resistive circuits with more than one resistor. The only difference is that you typically will deal with ac effective values (rms). When you know only peak or peak-to-peak values, you have to use the factors you have been studying to find the rms value(s). In later chapters, you will study what happens when out-of-phase quantities are introduced into circuits via "reactive" components.

Try to apply your new knowledge of ac quantities and relationships to the following practice problems.

PRACTICE PROBLEMS IV

1. Refer to the series circuit in Figure 12–17 and determine the following parameters:
   a. Effective value of V_1
   b. Effective value of V_2
   c. I_p
   d. Instantaneous peak power
   e. P_1

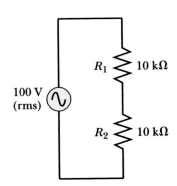

FIGURE 12–17 Purely resistive series ac circuit

FIGURE 12–18 Purely resistive parallel ac circuit

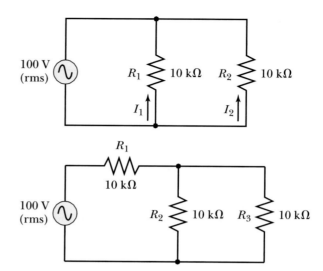

FIGURE 12–19 Purely resistive series-parallel ac circuit

2. Refer to the parallel circuit in Figure 12–18 and determine the following parameters:
   a. Effective value of I_1
   b. Effective value of I_2
   c. I_p
   d. V_{p-p}

3. Refer to the series-parallel circuit in Figure 12–19 and determine the following parameters:
   a. Effective value of V_1
   b. Effective value of V_2
   c. Effective value of V_3
   d. P_1 (NOTE: Use appropriate rms values.)

12–10 OTHER PERIODIC WAVEFORMS

Important waveshapes

Although the sine wave is a fundamental ac waveform, there are nonsinusoidal waveforms that are important for the technician to know and understand. Modern digital electronics systems, such as computers, data communications, radar, and pulse systems, and circuits requiring ramp waveforms, require that the technician become familiar with their features. It is not necessary in this chapter to deal in-depth with these various waveforms, but it is important to at least introduce you to them.

The most important waveforms include the **square wave,** the **rectangular wave,** and the **sawtooth** (ramp-shaped) **wave.** Refer to Figure 12–20 and note the comparative characteristic of each waveform with the sine wave in terms of period, wave shape, and peak-to-peak values.

How nonsinusoidal waveshapes are formed

It is interesting to note that a pure ac sine wave is comprised of one single frequency. Nonsinusoidal signal waveforms can be mathematically shown to be

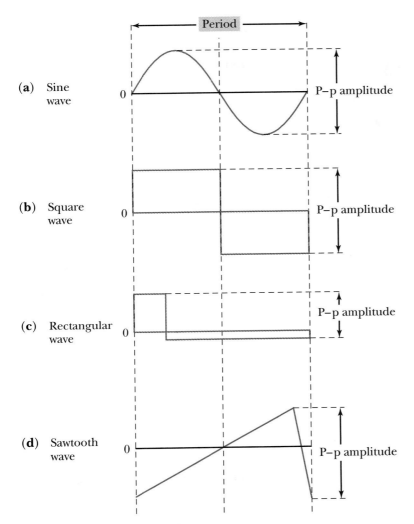

FIGURE 12–20 Comparison of sinusoidal and nonsinusoidal waveforms in terms of waveshape and peak-to-peak values

composed of a "fundamental" frequency sine wave *plus* a number of multiples of that frequency, called "harmonics." In other words, if the waveform is not sinusoidal, it indicates the presence of harmonics. For example, the square wave is composed of a fundamental sine-wave frequency and a large number of odd harmonic signals. The greater the number of odd harmonic signals that are added, or present, the closer to an ideal square wave the resultant complex waveform becomes (see Figure 12–21). Odd harmonics are those that represent frequencies of 3 times, 5 times, 7 times, and so on, the fundamental frequency. For example, if the fundamental frequency is 100 Hz, then the third harmonic is 300 Hz, the fifth harmonic is 500 Hz, and so on.

Other nonsinusoidal waveforms may be created by the addition of even harmonic signals, or a mixture of even and odd harmonics, Figure 12–22. Even harmonics are even multiples of the fundamental frequency involved. That is, 200 Hz is the second harmonic of a fundamental frequency of 100 Hz; 400 Hz is the fourth harmonic frequency, and so on.

FIGURE 12–21 Graphic illustration of forming of a square wave

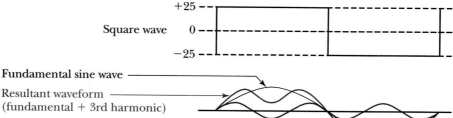

Square wave

Fundamental sine wave

Resultant waveform (fundamental + 3rd harmonic)

3rd harmonic

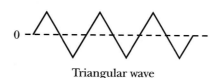

Resultant waveform (fundamental + 3rd + 5th harmonic)

5th harmonic

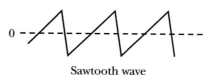

FIGURE 12–22 Sample of other nonsinusoidal waveforms

Triangular wave

Sawtooth wave

Important parameters for nonsinusoidal waveforms

As you have already seen, the period of a nonsinusoidal waveform is determined in the same way the period is determined for a sine-wave signal; that is, the time it takes for one complete cycle of the signal, Figure 12–20. You have also seen that the peak-to-peak amplitude of nonsinusoidal signals is determined in similar fashion as for the sine-wave signal. That is, from the top-most point of the waveform to the bottom-most point on the waveform, Figure 12–20. Let's look at some more parameters of interest for these nonsinusoidal waveforms.

Pulse-width, duty cycle, and average values

Notice how the **pulse width** is identified in the series of cycles of rectangular waves shown in Figure 12–23. By comparing the pulse width with the time for one period, we can determine the **duty cycle.** That is, the *duty cycle equals the pulse width divided by the time for one period.* Since duty cycle is generally expressed in percentage, we then take this decimal number times 100% to find the percentage duty cycle:

FORMULA 12–6 $\% \text{ duty cycle} = \dfrac{t_w \text{ (Pulse width)}}{T \text{ (Time for 1 period)}} \times 100\%$

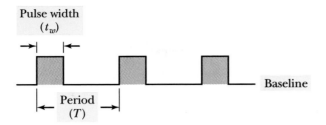

FIGURE 12–23 Pulse width and period of a pulse-type waveform

Example

If the time for one period of this repetitive pulse signal is 20 μs and the pulse width is 4 μs, then the percentage duty cycle is:

$$\% \text{ duty cycle} = \frac{4 \text{ μs}}{20 \text{ μs}} \times 100\% = 0.2 \times 100 = 20\% \text{ duty cycle}$$

PRACTICE PROBLEMS V

What is the duty cycle of a repetitive pulse waveform if $T = 15$ μs and the $t_w = 1$ μs?

Finding average value for a nonsinusoidal waveform

In a number of situations in electronic circuits, these nonsinusoidal waveforms are not symmetrically balanced above and below the zero reference line. Many times a dc component is purposely added to "offset" the waveform, for operational purposes. To find the average value of voltage (V_{avg}) of such waveforms, it is necessary to understand where the "baseline" of the waveform is considered to be, and be able to determine the duty cycle and amplitude of the signal. The basic formula for finding V_{avg} is:

FORMULA 12–7 V_{avg} = Baseline value + (duty cycle × p-p amplitude)

Examples

In Figure 12–24a, you can see that the baseline has been defined as being at the 0 V level. Also, you can observe that the peak-to-peak amplitude of the waveform is 5 volts. The duty cycle can be calculated as being 0.1 (or 10%). Thus,

$$V_{avg} = 0 + (0.1 \times 5) = 0 + 0.5 = 0.5 \text{ V}$$

In Figure 12–24b, the conditions are changed so that the baseline is now at +2 V. What happens to V_{avg} in this case?

$$V_{avg} = 2 + (0.1 \times 5) = 2 + (0.1 \times 5) = 2 + 0.5 = 2.5 \text{ V}$$

FIGURE 12–24 Finding V_{avg} for a rectangular waveform

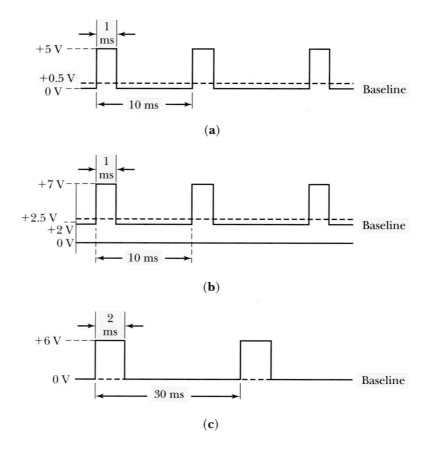

(a)

(b)

(c)

PRACTICE PROBLEMS VI

1. Refer to Figure 12–24c and determine the V_{avg} for these conditions.
2. What would V_{avg} be if the waveform were symmetrical above and below the reference point (i.e., if the baseline were −3 V, in this case)?

Later in your studies, you will be studying more details about pulse-type signals. Topics such as waveshaping and the waveform characteristics of ramp, slope, and so on will be studied at the time you can apply them to circuits you are studying. At this time, we simply wanted to introduce you to the fact that there are ac signals other than sine waves, and familiarize you with some of their basic characteristics.

SUMMARY

▶ Alternating current (ac) varies continually in magnitude and periodically reverses in direction (polarity).

▶ Whereas ac is bidirectional, dc is unidirectional. Also, ac alternates in polarity, but dc has only one polarity. Furthermore, ac can be transformed or

stepped up or down via transformer action, whereas dc cannot.

▶ The sine wave is related to angular motion in degrees, and any given point's value throughout the sine wave relates to the trigonometric sine function. Maximum values occur at 90° and 270°

($-90°$) points of the sine wave. Zero values occur at $0°$, $180°$, and $360°$ points of each cycle. (Assuming the wave is started at the zero level.)

▶ A sine-wave voltage is generated by relative motion between conductor(s) and a magnetic field, assuming the conductor(s) links the flux lines via a rotating motion. The ac generator output is typically delivered to the load via slip rings and brushes.

▶ Quantities having magnitude and direction are frequently represented by lines or arrows called vectors. A vector's length and relative direction illustrate the quantity's value and angle.

▶ When representing angles relative to ac quantities, the zero reference point is horizontal and to the right. Angles increase positively from that reference plane in a counterclockwise (CCW) direction.

▶ When representing angles relative to the coordinate system, angles between $0°$ and $90°$ are in quadrant 1; angles between $90°$ and $180°$ are in quadrant 2; angles between $180°$ and $270°$ are in quadrant 3; and angles between $270°$ and $360°$ are in quadrant 4. These positive angles apply with CCW rotation of the phasor.

▶ A cycle represents one complete set of recurring events or values. In ac, the completion of a positive alternation (half-cycle) and a negative alternation (or vice-versa) completes a cycle.

▶ A period is the time it takes for one cycle. It is calculated by the formula $T = 1/f$.

▶ Frequency is the number of cycles per second, and Hertz (Hz) indicates cycles per second. Frequency is calculated using the formula $f = 1/T$, if the time for one cycle (period) is known. Frequency and period are inversely related.

▶ Referring to a sine wave, rate of change indicates how much amplitude changes for a given amount of time. This rate of change is maximum near the zero level points and minimum near the maximum level points.

▶ In purely resistive ac circuits, voltage and current are in phase with each other.

▶ When ac quantities are out of phase, they do not reach the same relative levels at the same time. A generalization is the waveform that reaches maximum positive level first is leading the other quantity in phase.

▶ Phase angle is the difference in angle (time difference) between *two equal-frequency* sine waves. If sine waves A and B are in phase, there is zero phase angle between them. If sine wave A reaches maximum positive level 1/4 cycle before sine wave B, sine wave A is leading sine wave B by $90°$. Thus, the two signals are out of phase by that amount. If sine wave A reaches maximum positive level 1/8 cycle before sine wave B, sine wave A is leading sine wave B in phase by $45°$.

▶ Average value of ac voltage or current is the average level under one alternation's waveform. Average value is equal to 0.637 times peak value. (NOTE: Average value of a symmetrical waveform over a complete cycle equals zero.)

▶ Effective value (rms) of ac voltage or current is the value that causes the same heating effect as an equal value of dc voltage or current. Effective value is computed as 0.707 times peak value, or 0.3535 times peak-to-peak value, or 1.11 times average value.

▶ Peak value (maximum positive or negative value) of ac voltage or current is 1.414 times the effective value. It can also be calculated as being half the peak-to-peak value.

▶ Peak-to-peak value is the total difference in voltage or current between the positive and negative maximum values. Peak-to-peak value is calculated as two times peak value, or 2.828 times effective value.

▶ In ac resistive circuits, average power is calculated with the effective values (rms) of I and V.

▶ Several nonsinusoidal waveforms are the square wave, rectangular wave, and sawtooth wave.

▶ The period, frequency, and peak-to-peak amplitude values of nonsinusoidal waveforms (square, triangular, rectangular, and sawtooth) are determined in the same fashion as the period, frequency, and peak-to-peak amplitude of a sine wave are identified.

▶ The duty cycle of repetitive pulse-type waveforms is found by dividing the pulse width by the time for one period. This is typically multiplied by 100 to find the percentage duty cycle.

▶ The average voltage value (V_{avg}) of pulse-type waveforms is determined by adding the baseline voltage to the product of the duty cycle and the peak-to-peak amplitude of the waveform.

FORMULAS AND SAMPLE CALCULATOR SEQUENCES

FORMULA 12–1 $T(s) = \dfrac{1}{f(Hz)}$

frequency in hertz, [1/x]

FORMULA 12–2 $f(Hz) = \dfrac{1}{T(s)}$

time for one cycle in seconds, [1/x]

FORMULA 12–3 Effective value (rms) $= 0.707 \times$ Peak value

0.707, [×] , peak value of voltage, [=]

FORMULA 12–4 Average value $= 0.637 \times$ Peak value

0.637, [×] , peak value of voltage, [=]

FORMULA 12–5 $v = V_p \sin \theta$

V peak value, [×] , angle value, [SIN] , [=]

FORMULA 12–6 $\%\ \text{duty cycle} = \dfrac{t_w\ (\text{Pulse width})}{T\ (\text{Time for 1 period})} \times 100\%$

pulse width time value, [÷], time for one cycle, [×], 100, [=]

FORMULA 12–7 $V_{avg} = \text{Baseline value} + (\text{duty cycle} \times \text{p-p amplitude})$

baseline value, [+] , [(] , duty cycle decimal value, [×] , p-p value, [)] , [=]

REVIEW QUESTIONS

1. AC generator output is delivered via:
   a. brushes alone.
   b. commutator segments and brushes.
   c. magnetic linkage.
   d. slip rings and brushes.

2. The zero reference point often used to graphically represent angles for ac quantities on the coordinate system is:
   a. vertical $-90°$ upward from the horizontal plane.

b. horizontal, to the right.

c. horizontal, to the left.

d. vertical −90° downward from the horizontal plane.

3. Angles represented on the coordinate system increase positively from the zero reference point in a:

a. clockwise direction.

b. counter-clockwise direction.

c. either CW or CCW direction.

4. When using the coordinate system, angles between 90° and 180° are represented:

a. in quadrant 1.

b. in quadrant 2.

c. in quadrant 3.

d. in quadrant 4.

5. In an ac sine-wave, an alternation represents:

a. a full cycle.

b. a half cycle.

c. a quarter cycle.

d. two cycles.

6. A period is the time it takes for:

a. one-fourth cycle.

b. one-half cycle.

c. one cycle.

d. none of the above.

7. The formula for calculating a period is:

a. $T = \frac{1}{2}/f$.

b. $T = \frac{1}{4}/f$.

c. $T = 1/f$.

d. none of the above.

8. Frequency may be defined as:

a. the number of cycles per second, and is expressed in radians.

b. the number of cycles per second, and is expressed in Hertz.

c. the number of cycles per second, and is expressed in CPS.

d. none of the above.

9. As frequency is increased:

a. period increases.

b. period decreases.

c. period does not change.

d. none of the above.

10. The rate of change of an ac sinusoidal waveform is maximum at:

a. waveform maximum points.

b. waveform median points.

c. waveform zero points.

d. none of the above.

11. By convention, when two graphically represented sine-wave waveforms are not in phase:

a. the waveform that reaches maximum positive level first is considered leading.

b. the waveform that reaches maximum negative level first is considered leading.

c. the waveform that reaches the zero-point first is considered leading.

d. none of the above.

12. Phase angle between two equal-frequency sine waves represents the:

a. difference in amplitude.

b. difference in time.

c. both difference in amplitude and time.

d. none of the above.

13. The average value of ac voltage or current is its average level over:

a. one complete cycle.

b. two complete cycles.

c. one alternation.

d. two alternations.

14. The effective value of ac voltage or current may be calculated as:

a. 0.707 × peak value.

b. 0.3535 × peak-to-peak value.

c. both of the above.

d. none of the above.

15. Peak-to-peak value represents the total difference between:

a. zero and positive peak value.

b. zero and negative peak value.

c. positive peak to negative peak value.

d. both (a) and (b) above.

16. Explain the terms *fundamental* and *harmonic*.

17. Define the term *periodic wave*.

18. Wavelength of a signal is affected by:

a. the speed the signal travels and the frequency of the signal.

b. the speed the signal travels and the amplitude of the signal.

c. the time of day the signal is propagated and the signal strength.

d. none of the above.

19. A square wave is composed of:
    a. a fundamental sine wave and a great number of even harmonics.
    b. a fundamental sine wave and a great number of odd harmonics.
    c. a distorted sine wave.
    d. none of the above.

20. The period, frequency, and peak-to-peak amplitude values of square waves, triangular waves, and rectangular waveforms can be determined:
    a. in a slightly different manner than for regular sine waves.
    b. in a completely different manner than for regular sine waves.
    c. in the same manner as for regular sine waves
    d. cannot be determined.

PROBLEMS

1. What is the period of a 2-kHz voltage?

2. What is the frequency of an ac having a period of 0.2 μsec?

3. How much time does one alternation of 60-Hertz ac require?

4. The peak value of a sine wave is 169.73 volts. What is the rms value?

5. What is the frequency of a signal having one-fourth the period of a signal whose T equals 0.01 seconds?

6. What is the peak-to-peak voltage across a 10-kΩ resistor that dissipates 50 mW?

7. If the rms value of an ac voltage doubles, the peak-to-peak value must _____.

8. The ac effective voltage across a given resistor doubles. What happens to the power dissipated by the resistor?

9. The period of an ac signal triples. What must have happened to the frequency of the signal?

10. The rms value of a given ac voltage is 75 volts. What is the peak value? What is the peak-to-peak value?

11. What time is needed for one alternation of a 500-Hz ac signal?

12. One alternation of a signal takes 500 μs. What is its frequency? What is the time of two periods? (Use a calculator to find answers.)

13. Draw the diagram of a three-resistor series-parallel circuit having a 10-kΩ resistor (labeled R_1) in series with the source, and two 100-kΩ resistors (labeled R_2 and R_3) in parallel with each other and in series with R_1. Assume a 120-V ac source. Calculate the following parameters, and appropriately label them on the diagram.
    a. V_{R_1}
    b. I_{R_2}
    c. P_{R_1}
    d. I_{R_2}
    e. V_{R_3}
    f. θ between V applied and I total

14. What is the wavelength of a radio signal with a frequency of 1 MHz?

15. If the frequency of a signal is lowered to one-third its original frequency, what happens to its wavelength? Be specific.

ANALYSIS QUESTIONS

1. What are two differences between ac and dc?

2. Draw an illustration showing one cycle of a 10-V (rms), 1-MHz ac signal. Identify and appropriately label the positive alternation, the period, the peak value, the peak-to-peak value, the effective value, and points on the sine wave representing angles of 30°, 90°, and 215°.

3. The rms value of a given ac voltage is 100 V. What is the peak value? What is the peak-to-peak value?

Refer to Figure 12–25 and answer questions 4–8.

4. Which waveform illustrates the signal with the longest period?

5. What is the peak-to-peak voltage value of waveform B?

6. What is the effective value of voltage waveform A?

7. If the period of waveform A equals 0.002 ms, what is the frequency represented by waveform B?

8. If the peak-to-peak value of waveform A changes to 60 V, what is the new effective voltage value?

9. a. Draw a graphic illustration of one and one-half cycles of a rectangular wave. Assume the amplitude is 10 V, the baseline is +3 V, the period is 25 ms, and the duty cycle is 25%. Label all appropriate parameters on the drawing.

b. Determine the V_{avg} for the signal described in question 9a.

10. Draw one cycle of a sine wave and a cosine wave on the same "baseline." Assume each maximum value is 10 units. Appropriately label the diagram to show the following information:

a. Angles of 0°, 45°, 90°, 135°, 180°, 225°, 270°, 315°, and 360°, marked on the x-axis.

b. Amplitude values of the sine wave at 45° and 90°, respectively. (Use a calculator to determine values.)

c. Amplitude values of the cosine wave at 0°, 45°, and 90°, respectively. (Use a calculator to determine values.)

d. Indicate which wave is leading in phase, and by how much.

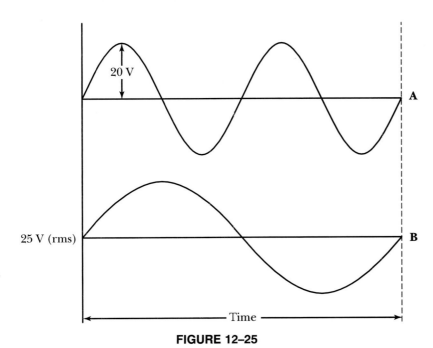

FIGURE 12–25

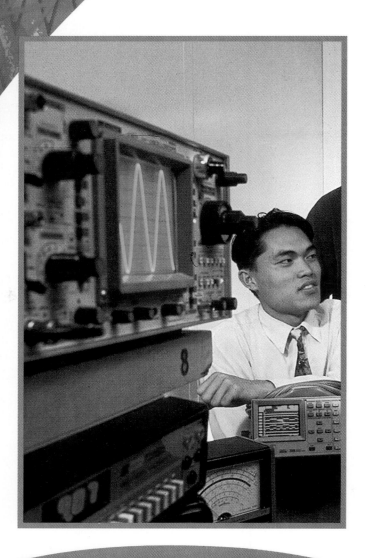

THE OSCILLOSCOPE

13

chapter

CHAPTER PREVIEW

The **oscilloscope,** often called a scope, is without doubt one of the most versatile tools available to technicians and engineers. It can display waveforms; help in determining voltages, currents, and phase relationships; and assist in computing and comparing frequencies.

In your studies, you have been introduced to other measuring instruments and to some important ac quantities. This chapter will increase your knowledge about measuring techniques. This chapter will also reinforce your knowledge about the important ac quantities you will measure and analyze throughout your remaining studies and in your career as a technician.

The purpose of this chapter is not to discuss the details of circuitry involved in making a scope function, nor to instruct you on subjects that are best learned with actual experience. However, this chapter will, perhaps, inspire you to become familiar with this valuable tool and will be a good background for you to become proficient in using this device.

OBJECTIVES

After studying this chapter, you should be able to:

1. List the key parts of the oscilloscope

2. List precautions when using scopes

3. List procedures when measuring voltage with a scope

4. List procedures to display and interpret waveforms

5. List procedures relating to phase determination

6. List procedures when checking frequency with a scope

7. Use the SIMPLER troubleshooting sequence to solve the Chapter Troubleshooting Challenge problem

13-1 BACKGROUND INFORMATION

Scopes come in a variety of brands and complexity. Two common categories of scopes include those used for general-purpose bench work applications, and the more complex (and expensive) laboratory-quality scopes, Figure 13–1a and b. Also, portable scopes, Figure 13–1c, and the "digital storage" oscilloscope (DSO), Figure 13–2, are now fairly commonplace. These scopes have the ability to store signal waveforms in memory for either immediate or later retrieval and display. This type of scope makes possible such applications as waveform comparisons at different times or locations, complex, long-duration and/or short-duration signal-change analysis, and other unique types of signal comparisons and analyses. These are analyses that would be difficult or impossible to perform using standard analog "real time" (what you see is happening right now) scopes. To store signals, this scope "digitizes" the analog signals by sampling signal levels many times over the duration of the signal waveform. The digital data is then placed in memory for retrieval at the desired time.

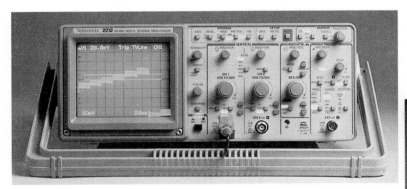

(a)

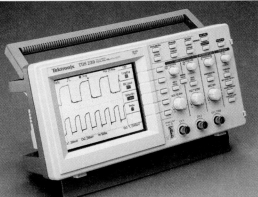

(c)

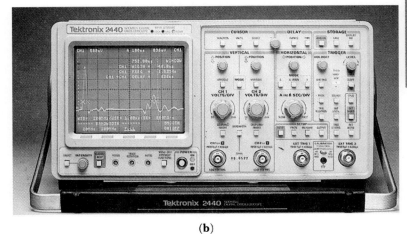

(b)

FIGURE 13–1 (a) General-purpose oscilloscope; **(b)** lab-quality oscilloscope; **(c)** portable oscilloscope *(Photos courtesy of Tektronix, Inc.)*

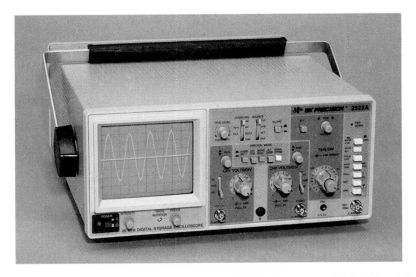

FIGURE 13–2 A digital storage oscilloscope (DSO) *(Courtesy of B & K Precision)*

Two general areas of commonality between scopes include the following:

1. All scopes have a cathode-ray tube (CRT) where the displays are viewed.
2. All scopes have controls and circuits that adjust the display to help you analyze voltage, time, and frequency parameters of the signal(s) under test.

Let's take a brief look at these two critical areas related to scopes.

13–2 KEY SECTIONS OF THE SCOPE

The key sections of a scope that will be discussed are:

1. cathode-ray tube;
2. intensity and focus controls;
3. position controls;
4. sweep frequency controls;
5. vertical section;
6. horizontal section; and
7. synchronization controls.

A typical scope front panel layout is shown in Figure 13–3. Use this figure while you read the following descriptions of the various controls and circuits.

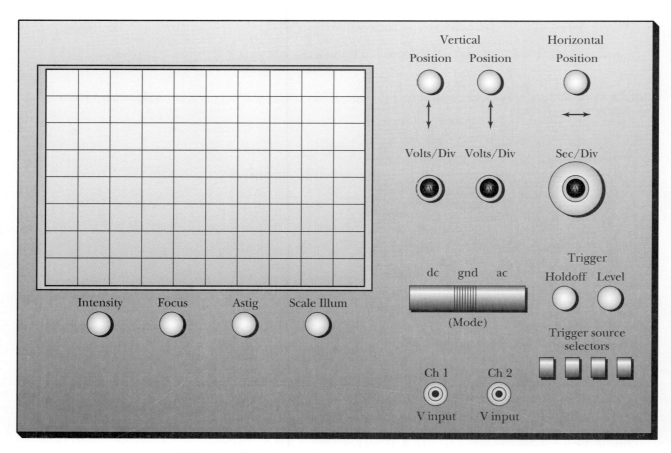

FIGURE 13–3 Typical controls on a dual-trace scope (simplified)

13-3 THE CRT

All oscilloscopes have a **cathode-ray tube (CRT),** which has a "screen." The screen end of the CRT is the part that displays visual information that is interpreted by the technician or engineer when using the scope. An "electron gun," which we will explain shortly, shoots an electron stream, or beam, at the inside coating of the CRT screen. The screen, in turn, lights up wherever the electron beam is striking the CRT screen coating.

The screen

As indicated, the purpose of the screen is to convert the electron beam energy into light. This conversion is accomplished by the impact of the electron beam on the fluorescent screen material. The screen material has two characteristics, which are defined as follows:

1. Luminescence—the property that converts the energy of the electron beam striking the screen's surface into light.
2. Phosphorescence—the property that causes the screen to continue to glow for a short time after the actual impact of the electron beam.

Without going into too much detail, it is interesting to know that the type and nature of the coating material used can provide long persistence time (such as used in radar screens), medium persistence time (such as used in most oscilloscopes), and short persistence time (such as used in televisions).

The electron gun

The electron beam, about which we have been talking, is created and controlled by various elements along the length of the cathode-ray tube. Figure 13–4 illus-

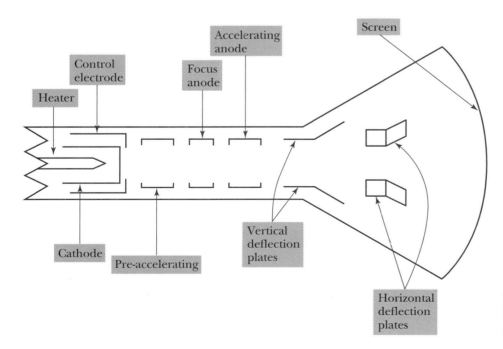

FIGURE 13–4 Elements in a typical cathode-ray tube (CRT)

trates these electrodes within the CRT. Refer to the figure as you study the following brief discussion of the function of each of these electrodes.

The heater

Functions to heat the cathode to a hot enough temperature so that the cathode will emit some electrons from its surface.

The cathode

A cylindrically shaped element that has a special coating on the side toward the CRT screen. When heated, it emits electrons, or supplies the electrons to be formed into and controlled as an electron stream, or beam.

The control electrode

A metal sleeve, open at the cathode end and closed, except for a small aperature, at the end facing the CRT screen. It concentrates the electrons into the shape of a beam, and controls the number of electrons moving toward the screen.

The accelerating anodes

The first (pre-accelerating) anode and second (accelerating) anode work together to perform two functions. One function is to give high velocity to the electron beam as it travels toward the screen. The second function is to combine with the focusing anode to determine the focus characteristics, or fineness of definition, of the beam. NOTE: There is a conductive coating on the inside of the glass CRT tube that extends almost to the screen. This aquadag (coating) is electrically connected to the second accelerating anode, and aids in the acceleration process.

The focusing anode

A cylindrically shaped electrode with a small opening at either end. It is used to concentrate, or focus, the electron stream so that when it strikes the screen, it is a fine point.

The deflection plates

Function to deflect the electron stream in accordance with the signal voltage being observed. Obviously, the vertical deflection plates control the upward and downward movement of the electron beam, and the horizontal deflection plates control the left-to-right, or horizontal, movement of the electron beam.

Now, let's see how these various control electrodes translate to some of the front-panel controls on an oscilloscope.

PRACTICAL NOTES

Caution! One thing you should learn when operating a scope is that it is not good to leave a bright spot at one position on the CRT. The CRT screen material can be burned or damaged if this occurs for any great length of time. Never have the trace bright enough to cause a "halo" effect on the screen.

Position controls

In Figure 13–4 (page 437), you can see this CRT has "deflection plates" that control the position and movement of the electron beam coming from the opposite end of the tube and striking the back of the screen. The position where this electron beam strikes the back of the screen (CRT) is controlled by the electrostatic field set up between the plates when there is a difference of potential. (It is worth mentioning that positional control of the electron beam [or stream] in televisions is normally accomplished through *electromagnetic* fields, rather than *electrostatic* fields used by most oscilloscopes.)

The reactions of the electron beam to these fields are illustrated in Figure 13–5. Notice the electron beam is attracted to the plate(s) having a positive charge and repelled from any deflection plates having a negative charge, or potential. Therefore, the position where the electron beam strikes the screen can be controlled by the voltages present at the deflection plates. Knowing that deflection of the electron beam is toward positive deflection plates and away from negative deflection plates, you can understand the **position controls** simply make the appropriate deflection plates either more positive or more negative, depending on which way we want the electron beam to move on the CRT face.

Look again at Figure 13–3 (page 437). You can see these front-panel controls that influence both the vertical and horizontal position of the trace. Notice what happens when we apply ac to the deflection plates, Figure 13–6a, b, and c.

Horizontal sweep frequency control(s)

A sawtooth (or ramp-type) waveform applied to the horizontal plates, creates a horizontal trace on the screen, Figure 13–7. The voltage applied to the horizontal plates is sometimes called the horizontal "sweep voltage," because the voltage causes the electron beam to horizontally sweep across the screen, creating a horizontal trace, or line.

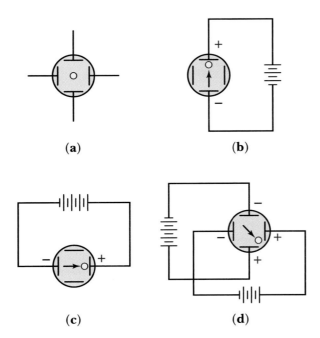

(a) (b)

(c) (d)

FIGURE 13–5 Electron-beam movement caused by dc voltages on deflection plates

FIGURE 13–6 Electron-beam movement caused by ac voltages on deflection plates

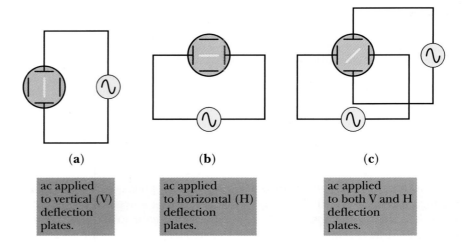

(**a**)

ac applied to vertical (V) deflection plates.

(**b**)

ac applied to horizontal (H) deflection plates.

(**c**)

ac applied to both V and H deflection plates.

FIGURE 13–7 Linear trace caused by sawtooth voltage on horizontal (H) plates

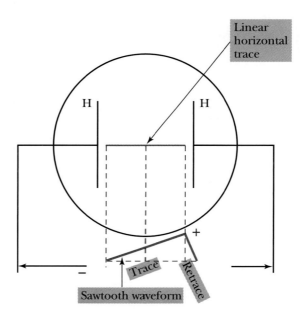

Linear horizontal trace

H   H

Trace   Retrace

Sawtooth waveform

The electron beam moves, or traces, at a constant rate (linear speed) across the screen from left to right, then quickly retraces, or "flies back," to its starting point on the left. During the very short retrace (or flyback) time, a blanking signal is applied to the system so that the scope screen will not display the electron beam's movement across the screen (from right to left) during retrace time.

Because the electron beam is moving at a constant speed across the screen during the left-to-right trace time, a horizontal "time base" has been established. In other words, we know how much time it takes for the beam to move from one horizontal point to another on the screen. Because of this, we can measure times and calculate frequencies of signal waveforms that are displayed, as you will see later.

The number of times the beam traces across the screen in a second is determined by the frequency of the sweep (sawtooth) voltage. You see a horizontal line on the screen when this happens at a rapid rate. Two reasons for this are that:

1. the CRT screen material has a persistence that causes continual emission of light from the screen for a short time after the electron beam has stopped striking that area; and

2. the retinas in our eyes also have a persistence feature. This is why you don't see flicker between frames as a movie film is projected, or flicker between frames in the TV "scanning" process.

The **sweep frequency controls** (horizontal time variable, "VAR", and horizontal sec/div) adjust the frequency output of the horizontal **sweep circuitry** in the oscilloscope. Naturally, this determines the number of times per second the electron beam is traced horizontally across the CRT screen. The horizontal frequency control allows us to view signals of different frequencies.

Having accumulated this information, let's now see how the electron beam is further controlled, amplified, or attenuated to provide meaningful displays on the CRT screen.

Vertical section

Key elements in the vertical section are the vertical input jack(s) (sometimes called Y input), and the **vertical attenuator** and **vertical amplifier,** with related controls. When the scope is a single-trace scope, there is only one vertical input jack. When the scope is a dual-trace scope, there are two vertical input jacks. (See the jacks in Figure 13–3 [page 436].)

The vertical attenuator and amplifier circuitry, and associated controls, decrease or increase the amplitude of the signals to be viewed; that is, the signals that are applied to the vertical section via the vertical input jack(s). View Figure 13–3 and note the calibrated **vertical volts/div** and the **variable control(s),** which are used for signal level control. Using these controls allows you to adjust for larger or smaller vertical deflection of the CRT trace with a given vertical input signal amplitude.

Horizontal section

We have briefly discussed some of the important horizontal deflection elements and controls. One of these controls adjusts the frequency of the internally generated **horizontal** sweep signal.

Other possible elements related to the horizontal trace may include a horizontal **gain control** that changes the length of the horizontal trace line; and a jack (often marked "Ext X" input) that enables inputting a signal from an external source to the horizontal deflection system in lieu of using the internally generated sweep signal.

Synchronization controls

The **synchronization controls** synchronize the observed signal with the horizontal trace; thus, the waveform appears stationary (assuming the signal has a periodic waveform). The effect is much like a strobe light that provides "stop-action" when timing an automobile, or the effect of making spinning fan blades appear to be standing still, using a light that is blinking at the appropriate rate. Without going into the technical details, suffice it to say that by starting or "triggering" the trace (left-to-right horizontal sweep) of the electron beam in proper time relationship with the signal to be observed on the scope (the signal fed to the vertical deflection plates), a stable waveform is displayed. If the horizontal trace and the vertical system signals are not **synchronized** (i.e., do not start from a given waveform point at a specific time), the waveform is a "moving target" and hard to view.

Some of the jacks and controls typically associated with this synchronizing function are also shown in Figure 13–3, and include:

1. trigger source selector switches (for selecting the synchronizing signal source); and

2. "trigger holdoff and level" controls, which help set the appropriate level for the triggering signal to work best.

IN-PROCESS LEARNING CHECK I

1. The part of the oscilloscope producing the visual display is the _____ _____ _____.

2. The scope control that influences the brightness of the display is the _____ control.

3. A waveform is moved up or down on the screen by using the _____ _____ control.

4. A waveform is moved left or right on the screen by using the _____ _____ control.

5. For a signal fed to the scope's vertical input, the controls that adjust the number of cycles seen on the screen are the horizontal _____ and the horizontal _____ control.

6. The control(s) that help keep the waveform from moving or jiggling on the display are associated with the _____ circuitry.

13–4 COMBINING HORIZONTAL AND VERTICAL SIGNALS TO VIEW A WAVEFORM

Having an electron beam cause a horizontal line across the CRT screen by applying voltage(s) to the horizontal deflection plates alone is not useful for displaying and analyzing waveforms. Causing an electron beam to vertically move by applying voltage(s) to the vertical deflection plates alone also is not useful for displaying and analyzing waveforms. However, useful waveforms appear when appropriate voltages are simultaneously applied to both the horizontal and vertical deflection plates.

Look at Figure 13–8. Notice how a sinusoidal waveform is displayed when a sine-wave voltage is applied to the vertical deflection plates, and when a sawtooth voltage of the same frequency (from internal sweep circuitry) is simultaneously applied to the horizontal deflection plates of the CRT. In this case, the frequency is 60 Hertz.

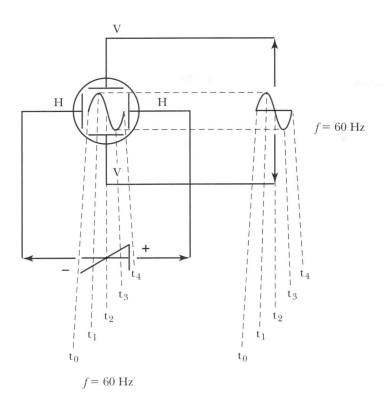

Again refer to Figure 13–8 and study the following discussion. Look at the conditions at times t_0, t_1, t_2, t_3, and t_4.

1. The sawtooth voltage is applied to the horizontal deflection plates so that at the beginning of the sawtooth, the electron beam is at the left side of the screen (viewed by the person looking at the front of the scope). Time is t_0.

2. As the sawtooth voltage increases in a positive direction, the right-hand deflection plate becomes increasingly positive. Thus, the beam moves from the screen's left side toward its right side. Time is t_1.

3. The 60-Hz sine-wave voltage is applied to the vertical deflection plates. At time zero (t_0) the beam is horizontally on the screen's left side. Since no vertical voltage is at t_0 of the sine wave, a spot appears on the screen's left side.

4. As the horizontal sawtooth signal causes the beam to begin its horizontal trace to the right, the voltage on the vertical deflection plates causes the beam to simultaneously move to its maximum upward position. Time is t_1.

5. By the time the vertical signal has decreased from maximum positive value back to zero (180° into the sine wave), the horizontal signal has caused the beam to arrive at the center of the screen. In other words, the positive alternation of the sine wave has been traced and displayed on the screen at t_2.

6. The same rationale is used when tracing the negative alternation's waveform. That is, as the beam continues its horizontal trek across the screen, the signal applied to the vertical deflection plates is causing the beam to be sequentially moved to the maximum downward position, then back to the zero voltage position, vertically (center of the screen vertically). During this sequence, the negative alternation of the sine wave has been displayed.

FIGURE 13–9 Result of two cycles of sine-wave signal per one horizontal sweep

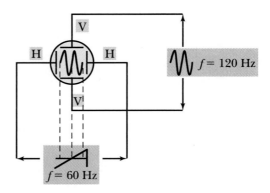

7. Throughout the cycle, one cycle of the sawtooth waveform applied to the horizontal plates has caused one horizontal trace. At the same time, the electron beam has been moved up and down, corresponding to the voltage throughout one sine-wave signal cycle applied to the vertical plates. The result is a sine-wave waveform viewed on the CRT.

8. Because these events have occurred 60 times in 1 second, our eyes perceive a single sine-wave waveform that traces over itself 60 times in 1 second.

Example

A horizontal sweep frequency of 60 Hz and a vertical signal of 120 Hz result in two cycles of vertical deflections occurring at the same period that one sweep occurred. The display shows two full cycles of a sine wave, Figure 13–9.

PRACTICE PROBLEMS I

Assume the vertical frequency is four times the horizontal frequency. What does the waveform look like?

By looking at the electron beam position at a number of instantaneous values of the applied horizontal and vertical signals, you can determine what type of waveform will appear on the scope screen, Figure 13–10. These concepts help

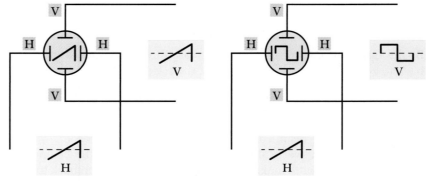

FIGURE 13–10 Illustration of how different waveforms are traced

NOTE: Horizontal and vertical signals are the same frequency.

interpret what we see on the CRT screen. Later you will see how these concepts help us determine frequency and phase of ac signals.

13–5 MEASURING VOLTAGE AND DETERMINING CURRENT WITH THE SCOPE

General information

Both dc and ac voltage measurements can be made with oscilloscopes. It is useful to know how to make these measurements and to interpret waveforms in terms of voltage amplitudes. We will discuss two common methods used to measure voltages with the scope. One method is to use a "built-in" calibration voltage to set up the scope for measuring voltages. This calibrating voltage method is generally found on older oscilloscopes; however, it is still used enough for us to take time to briefly discuss and illustrate the basic technique.

A second method is to interpret the scope display in reference to the "volts/division" control setting when the "V VAR" (variable) control is set at the "Cal" position. This technique is presently the more commonly used method.

As indicated earlier, learning to use the scope in a practical way *only comes through "hands-on" experience*. The descriptions in this chapter provide only conceptual knowledge. Therefore, you will have to expand on these concepts by actual laboratory experiences—be sure you do!

Calibration voltage approach

Many scopes have built-in calibrating voltage sources used to set up the scope for interpretation of voltage levels. Typically, the calibrating voltage output is 0.5-V or 1-V peak-to-peak value. Also, this calibrating voltage is often a square wave, Figure 13– 11.

Let's assume a 1-V peak-to-peak calibrating voltage and see how calibration enables us to measure voltages. Refer to Figure 13–12, as you read the following discussion.

1. With *no* signals connected to the vertical input jack, the horizontal sweep and gain controls, vertical and horizontal positioning controls, and "sync" controls are adjusted to produce a horizontal line trace across the screen, that is centered both vertically and horizontally on the screen, Figure 13–12a.

Scope screen

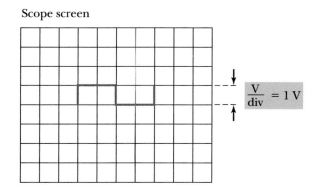

$$\frac{V}{div} = 1\ V$$

FIGURE 13–11 Typical square-wave calibration voltage

FIGURE 13–12 Calibrating for voltage measurement with calibration signal

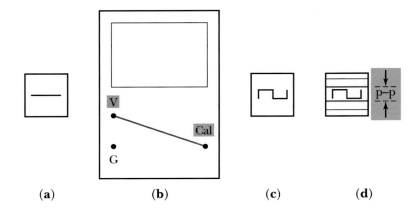

(a)　　　　　　　(b)　　　　　　(c)　　　　(d)

2. A "jumper" wire connects the calibrating voltage output terminal on the scope to the vertical input jack, Figure 13–12b.

3. Horizontal frequency controls are adjusted so one cycle of the calibration square wave is displayed on the screen, Figure 13–12c.

4. Vertical "variable" level control is set to the "Cal" position. The volts/div selector switch is then adjusted so the waveform is displayed with an "easy-to-read" amount of deflection on the screen calibrations (graticule). For example, one major division of vertical deflection appears, Figure 13–12d.

5. When this is done, each vertical square, or division of deflection caused by a voltage applied to the vertical input jack represents 1 V of input voltage.

6. The scope settings are now adjusted for one division to equal 1 V (peak-to-peak). Let's see how measurements are made.

AC voltage measurement

After removing the jumper wire from the vertical input jack (i.e., removing the calibrating voltage signal from the vertical input), there is a vertically and horizontally centered trace across the scope.

Let's now assume we want to measure a small ac (sine-wave) voltage that is connected between the vertical input jack and ground of the scope. Let's further assume a sine wave appears on the face of the scope. The sine wave is four divisions high, from top to bottom (peak-to-peak). What is the peak-to-peak value of the voltage being measured?

If you said four volts peak-to-peak, you are correct. Since each division in height represents one volt (peak-to-peak), then a four-divisions-high voltage must be four volts peak-to-peak.

Example

What is the rms value of the voltage just measured? (Here is where you apply your knowledge gained from the last chapter!)

1. If the voltage is 4 V peak-to-peak, then it must be 2 V peak.

2. If it is 2 V peak, then its rms value must be 0.707×2 V, or 1.414 volts. You can also use the peak-to-peak value $\times 0.3535$ and get the same results.

FIGURE 13-13

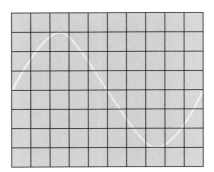

V_{p-p} = # of divisions deflection × volts/div. setting

PRACTICE PROBLEMS II

1. Suppose the voltage in our example had measured 10 V peak-to-peak on the scope. What would the rms value be? What would be the peak value of this voltage?

2. Assume the scope has been calibrated or set up so that a 1-V peak-to-peak signal causes *one-half division vertical deflection*. Refer to the scope pattern in Figure 13–13 and determine the following parameters:

   a. Peak-to-peak value of the ac signal displayed

   b. RMS value of the ac signal displayed

In summary, when measuring ac voltage with a scope, what you see on the screen is deflection caused by the peak-to-peak value of the voltage being measured. You must then use the appropriate calculation factors to interpret the waveform and determine the effective rms or other values of interest.

DC voltage measurement

Refer to Figure 13–14 as you read the following discussion on dc voltage measurement.

First, the "dc-gnd-ac" input mode switch is set to dc. For simplicity, assume the scope is set up so that a 1-V peak-to-peak signal causes a one-division-high waveform on the screen. With no voltage applied between the vertical input jack and ground, the scope shows a centered horizontal trace.

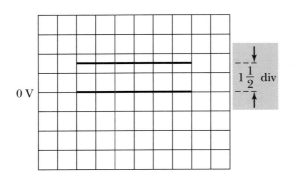

FIGURE 13–14 Direct current (dc) voltage measurement

FIGURE 13–15

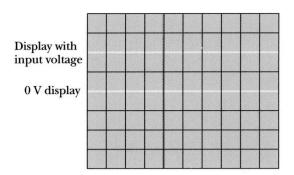

Display with
input voltage

0 V display

Assume when you connect the dc voltage to be measured to the vertical input, the trace jumps up one and one-half divisions from its starting point. What is the dc voltage being measured? If you said 1.5 V dc, you are right! Furthermore, you might have indicated the end of the source connected to the vertical input jack was positive at the jack side and negative at the ground connection side, because the trace moved up. Obviously, if you reversed the connections, the trace would move down from center by one and one-half divisions.

PRACTICE PROBLEMS III

1. Assume the scope is set up so that a 1-V peak-to-peak signal voltage causes two vertical divisions of deflection. Furthermore, assume that with 0 V applied to the vertical input jack, the horizontal trace line is centered on the screen. Refer to Figure 13–15 and determine the amount and polarity (with respect to ground) of the dc voltage being measured.

2. If 1-V peak-to-peak voltage causes 2.5 divisions of deflection, and the scope pattern looks like that in Figure 13–15, what is the value of the dc voltage being measured, and what is its polarity with respect to ground?

Volts-per-division control setting approach

By making sure that the "V-Variable" control is set in the "Cal" position, you can directly interpret voltage readings of scope patterns by reference to the vertical volts/div control setting. In Figure 13–16, you can see that our example control can be set at eight different positions. The beauty in this is that you can select the position that makes the signal display the most readable amplitude for interpreting the signal being measured. The signal voltage value equals the volts/div setting times the number of divisions of deflection.

Example

Observe in Figure 13–17 that the volts/div control is set at the 0.1 position. With the "V-Variable" control in the "Cal" position, each tenth-volt amplitude of the signal will cause one division displacement of the trace on the scope display. In our sample then, there are six divisions of deflection caused by the peak-to-peak amplitude of the signal. The interpretation is that the signal being measured

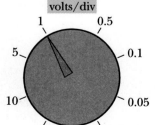

Vertical
volts/div

FIGURE 13–16 Using the calibrated volts/div switch setting to determine voltage

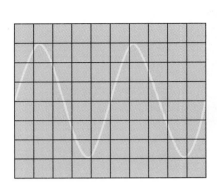

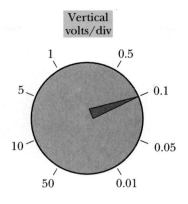

Vertical
volts/div

1   0.5

5   0.1

10   0.05

50   0.01

FIGURE 13–17 Example of volts/div voltage measurement approach

must be 0.6 volts peak-to-peak. The rms value is 0.3535×0.6, or 0.21 V. If the tested voltage were a dc voltage of 0.6 V, the trace would have jumped upward or downward by six divisions, depending on the polarity applied to the scope input jacks. This would cause the trace to be "off-screen." You would have to move the 0-V reference trace off center by two or more vertical divisions in order to see the 0.6-V dc deflection.

An alternative approach would be to set the **volts/div control** to the next higher setting (0.5 V/div). Then, the display would only jump one and one-fifth squares, which would be viewable. Of course, you would interpret the value based on the new volts/div setting.

PRACTICAL NOTES

Once you have calculated the rms value for a given measurement, you don't need to re-calculate to find the rms value for other measurements if you DO NOT change the vertical control settings. In the preceding example, the scope was set up so that 0.6 V peak-to-peak, or 0.21 V rms, caused six divisions of trace deflection. If the settings are not changed, you know that if you had three divisions of deflection, the rms must be one-half of 0.21 V rms, or 0.105 V rms. In other words, you would know that since 0.21 V rms caused six divisions of deflection, then each division must equal 0.21/6 V rms, 0.035 volts rms per division.

PRACTICE PROBLEMS IV

1. If the vertical volts/div control in Figure 13–17 were set on the "5" position and the scope display were the same as shown, what would be the peak-to-peak and rms values of the signal being checked?

2. What is the rms volts-per-division sensitivity setup used in question 1?

3. If you were going to measure a signal that you knew was about twice as great in amplitude as the one measured in question 1, which setting of the vertical volts/div control would you use?

FIGURE 13–18 An example of a "times ten" test probe *(Courtesy of B & K Precision)*

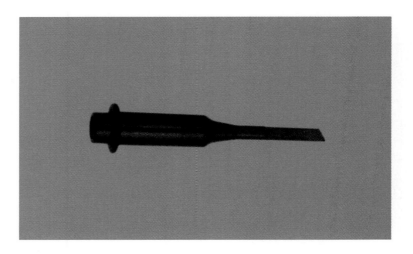

Attenuator test probes

Another technique often used with many scopes is a test probe that attenuates the signal by a given factor, for example 10 times attenuation, Figure 13–18. Thus, you multiply by 10 times what you are reading on the scope to find the actual value of voltage applied to the probe. This probe often has another purpose—it prevents the circuit being tested from being disturbed by the scope's input circuitry, or test cable. Also, the cable is a shielded cable, which minimizes stray signal pickup.

PRACTICAL NOTES

There are numerous types of test probes and associated cables available for use with oscilloscopes. Some are switchable and have one switch position for measuring low frequencies and another switch position for measuring high frequencies. Many have switchable attenuation; for example, X1 (times 1) and X10 (times 10). Sometimes you will see these written as 1:1 and 10:1 modes. As indicated earlier, the X10 mode indicates that the signal has been attenuated ten times; therefore, you would multiply your scope reading by 10 to find the actual signal level input to the test probe. There are also "current probes" available for use with scopes; however, the voltage probes are the more common types found in practice. As you proceed in your studies, you will be learning other parameters related to scope input circuit characteristics and matching of test probes/cables to the circuit conditions.

PRACTICE PROBLEMS V

1. Refer to the waveform shown in Figure 13–19 and determine the peak-to-peak and rms values of the signal. Assume that the volts/div setting of the vertical input signal controls is set at 0.5 V/div and that you are using a 1:1-type test probe.

2. Assume that the same waveform pattern is showing on the scope as shown in Figure 13–19; however, this time the volts/div setting is 100 mV/div and you are using a 10:1 test probe. What are the signal's peak-to-peak and rms voltage values?

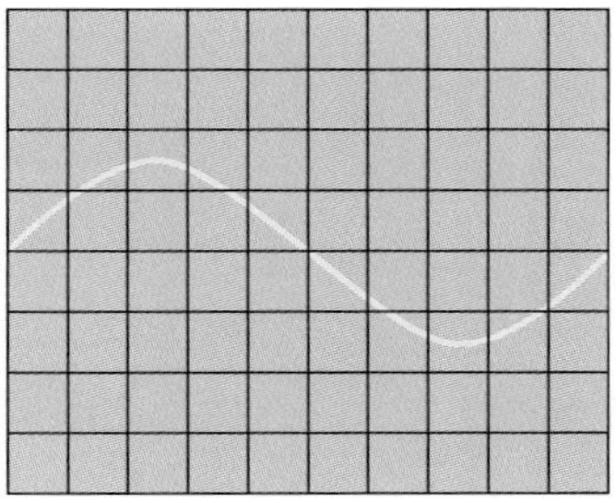

FIGURE 13–19

Summary of voltage measurement

Two basic techniques can be used. One method is to set up the scope controls using a calibrating voltage (either from an internal scope source or from a known-value external source). A second, more common technique is to use the scope's designed "volts/div" controls to obtain a known sensitivity to signal inputs.

Measuring voltage to find current

Current can be calculated by the Ohm's law technique you learned in previous chapters. If you use the scope to determine the rms voltage across a 1-kΩ resistor, for example, you automatically know the number of volts measured equals the number of milliamperes through the 1-kΩ resistor. That is, $I = V/R$, and V divided by kΩ equals mA. Obviously, Ohm's law applies to any value resistor. The 1-kΩ value is merely convenient to use, where possible.

13–6 USING THE SCOPE FOR PHASE COMPARISONS

In addition to exhibiting waveforms and measuring voltage (or determining current), the scope can compare the phase of two signals. Two methods for doing this are:

1. An overlaying technique where the two waveforms are visually superimposed on each other. This is accomplished by a dual-trace scope, or by an electronic switching device. Therefore, the displacement of the two signals' positive peaks, or the points where they cross the zero reference axis, determines the phase difference.

Example

Adjust the horizontal sweep frequency so exactly one cycle of the signal applied to the A (or #1) input covers the screen. (NOTE: Ten divisions are typical.)

The second signal is applied to the scope B (or #2) input. Since the screen is divided into 10 divisions, and 10 divisions represent 360°, then each division represents 36°. By noting the amount of offset between the two waveform traces and knowing that each division difference equals 36°, you can easily interpret the phase difference between the two signals.

PRACTICE PROBLEMS VI
Suppose the horizontal frequency in our example were adjusted so there were exactly 2 cycles across the 10 divisions of the screen. How many degrees would each major division on the screen represent?

2. Another method used to compare the phase of two signals is the **Lissajous** (pronounced liz-a-ju) pattern(s) technique. With this method, one signal is applied to the horizontal plates and the other signal is applied to the vertical plates. The resulting waveform (known as a Lissajous pattern) enables the technician to determine the phase differential between the two signals. (NOTE: This technique is not a frequently used method for phase measurement.)

The overlaying technique

If a dual-trace scope is used:

1. Two horizontal traces are obtained on the scope. They are then superimposed at the center of the screen by using the positioning controls, as appropriate.
2. Signal #1 is applied to one channel's vertical input jack. Scope sweep frequency is adjusted for one sine wave. Vertical level controls are adjusted to obtain a convenient size waveform for viewing purposes.
3. Signal #2 is connected to the other channel's vertical input jack. Again, appropriate adjustments are made to appropriately display signal #2. The signals should be adjusted to the same vertical size and should be the same frequency.
4. Look at Figure 13–20 to see one method to interpret the display in terms of phase difference between the signals.

13-7 DETERMINING FREQUENCY WITH THE SCOPE

Today's technology has given us devices called digital frequency counters that are excellent for making frequency measurements. However, there are still many

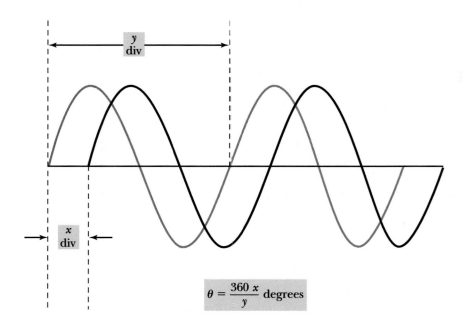

FIGURE 13–20 Overlay technique for phase measurement

$$\theta = \frac{360 \, x}{y} \text{ degrees}$$

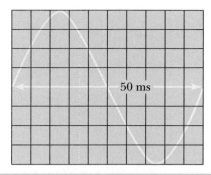

FIGURE 13–21 Example of frequency determination

50 ms

(Horizontal seconds/division set at 5 ms)

times when you have a scope connected to the circuit under test, and you do not want or need to pull out another piece of test equipment when you already have a pattern on the scope that you can interpret for frequency measurement.

Let's take a quick look at how this is done. You have already learned that the sweep frequency controls allow us to have a "linear horizontal time base sweep" whose time per division we can easily determine. You have learned in an earlier chapter that a signal's period (time for one cycle) and its frequency are inversely proportional. That is $f = 1/T$ and $T = 1/f$. Add your calculator's 1/x (reciprocal) function and you have all the tools you need to find frequency quickly from a scope pattern.

Example

Observe Figure 13–21. The sweep frequency control is set on 5 ms/div. This means the horizontal trace takes 5 milliseconds for each division of horizontal movement across the scope screen. With this knowledge, you can see that the signal being looked at is taking about 50 milliseconds for one cycle, or the period equals 50 ms. Using your calculator **1/x** function and inputting 50, **EE** (or **EXP**), **+/−**, 3, then using the **1/x** function key, the answer is shown as 20. This means the frequency is 20 Hz. Of course, you probably could have easily interpreted in your head that 50 milliseconds = 0.05 seconds and simply used the reciprocal function on your calculator for 0.05 to get to the answer of 20.

PRACTICE PROBLEMS VII

1. Assume the s/div control is set on the 0.1 s/div setting. Also assume that the scope pattern shows exactly 4 cycles of the signal to be measured across its 10 horizontal divisions.
   a. What is the period of the signal being measured?
   b. What is the frequency of the signal being measured?

2. Assume the s/div control is set on the 50 μs/div setting. This time, the pattern shows exactly 10 cycles of the signal under test.
   a. What is the signal's period?
   b. What is the signal's frequency?

PRACTICAL NOTES

Translating the $f = 1/T$ formula into practical terms relating to the scope settings and waveform being viewed:

$f = 1$ cycle display ÷ (number of horizontal squares for that 1 cycle × the "time per div" horizontal sweep setting)

Thus, for Figure 13–21:

$1/(5 \text{ ms} \times 10 \text{ divisions}) = 1/50 \text{ ms}$; therefore, $f = 1/0.05 \text{ s} = 20 \text{ Hz}$

Final comments about the oscilloscope

Certainly, reading this chapter has not made you a professional oscilloscope user! We only hope that by briefly discussing the remarkable versatility of the oscilloscope, you will desire to become knowledgeable and practiced in applying this instrument throughout your studies and career. The ability of the scope to display waveforms so that increments of time are measured and interpreted, and its ability to measure relative amplitudes makes this instrument one of the technician's most valuable assets. Learn to use it!

Figure 13–22 shows a very simplified block diagram of the elemental parts of a scope. As you can see, the vertical input signal is processed through the vertical amplifier/attenuator system before being applied to the vertical deflection plates of the CRT. The processing can either be in the form of amplification or attenuation of the signal to be displayed by the scope. Generally, the signal applied to the vertical input system is the signal you are trying to analyze or check.

The horizontal amplifier system has two possible inputs. The one most-often used is the input coming from the horizontal timebase generator (called the sweep system in our diagram). The output of this horizontal amplifier system is then fed to the horizontal deflection plates of the CRT. This produces the horizontal trace on the scope. Sometimes, it is desired to use a horizontal signal source other than the internal sweep generator circuitry. In that case, the horizontal signal is input to the horizontal amplifier system via the H input jack (sometimes labeled the "x" input jack).

Of course, the dc power supply shown in our diagram supplies the appropriate dc voltages to operate the CRT, scope amplifier circuits, timebase generator circuits, and so on throughout the oscilloscope circuitry.

When creating brackets for troubleshooting, place one of the brackets at the point where there is an abnormal "output indicator"! Place the other bracket at the first point in an earlier part of the flow path where things are known to be normal!

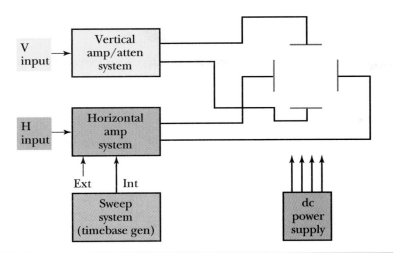

FIGURE 13–22 Simple oscilloscope functional block diagram

PRACTICAL NOTES

Some precautions: When using a scope, the technician should use common sense!

1. Do not leave a "bright spot" on the screen for any length of time.

2. Do not apply signals that exceed the scope's voltage rating.

3. Do not try to make accurate measurements on signals whose frequency is outside the scope's frequency specifications.

4. Be aware that the scope's input circuitry (including the test probe) can cause loading effects on the circuitry under test—use the correct probe for the job!

5. Check the attenuator probe setting (1:1 or 10:1) in order to interpret the scope display properly.

SUMMARY

▶ The oscilloscope (scope) is an instrument providing a visual presentation of electrical parameters with respect to time and amplitude.

▶ Intensity and focus controls adjust brightness and clarity of the visual presentation.

▶ Horizontal controls adjust the frequency of the internally generated horizontal sweep signal; select outside or inside sources for this sweep signal; horizontally position the visual presentation; and adjust the horizontal trace length.

▶ Vertical controls adjust the vertical amplitude and the vertical position of signals displayed on the screen.

▶ Synchronization control(s) adjust the start time of the horizontal sweep signal so the visual presentation of a signal is stable, or standing still.

▶ Scope presentations display the peak-to-peak values of the waveform. Other significant values are interpreted from the waveform by appropriate interpretation factors. For example, rms value = 0.707 × peak, or 0.3535 × peak-to-peak value.

▶ To measure voltage with an oscilloscope, the appropriate vertical controls are adjusted so the user knows the vertical sensitivity of the scope. That is, how many volts are represented by each division of vertical deflection on the calibrated screen.

▶ To determine the phase difference between two signals with an oscilloscope, a dual-trace scope is used so each signal is applied to a separate vertical input. Then, by appropriate adjustment, the two signals are superimposed on the visual presentation. Thus, the technician can interpret the phase differential in terms of the two signals.

REVIEW QUESTIONS

1. What is the name and abbreviation of the oscilloscope component that displays signals on its screen?

2. What are four practical uses of the oscilloscope?

3. Why should the intensity control be set to provide the minimum brightness, allowing good readability of the presentation?

4. What control(s) set the sweep frequency, therefore determining the number of cycles of a given signal viewed?

5. To what input terminal(s) is the signal to be viewed normally applied?

6. To what terminal(s) is an external sweep voltage applied?

7. To what terminal(s) is an external synchronizing signal applied?

8. Along what axis is deflection caused by a signal applied to the vertical deflection system?

9. Why should the setting of the scope vertical gain control and vertical input switch remain unchanged after initial calibration, if using the scope for voltage measurement?

10. What is an attenuator probe?

PROBLEMS

1. The time/div control and horizontal gain control settings are set so the horizontal sweep line represents 0.1 ms per division. What is the frequency of a signal if one displayed cycle uses four horizontal divisions on the horizontal sweep line?

2. If the controls are set so that the horizontal sweep line represents 2.5 ms per division, what is the frequency of a signal if a single displayed cycle uses 8 horizontal divisions on the horizontal sweep line?

3. For a frequency double that calculated in question 2, how many horizontal divisions would a single cycle take if the sweep controls were set so that each horizontal division represented 5 ms?

Refer to Figure 13–23 and answer questions 4–8:

4. What is the peak-to-peak value of voltage indicated by the waveform display?

5. What is the value of T for the signal displayed?

6. If you want the display to show only one cycle of the signal, which controls do you adjust on the scope?

7. What is the time-per-division setting to accomplish a one-cycle display?

8. The vertical sensitivity is set for 5 V per division (calibrated). What are the rms, peak, and peak-to-peak values of voltage of the signal?

9. If a scope is set with the vertical sensitivity at 2 V per division, and in measuring the voltage across a 4.7-kΩ resistor with the scope the peak-to-peak deflection is 2.5 divisions, what is the value of rms current through the resistor?

10. How many cycles of a 400-Hz signal would appear across a scope having 10 horizontal divisions if the horizontal sweep controls were set at 0.5 ms/div?

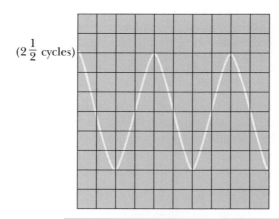

$(2\frac{1}{2}$ cycles)

Vertical sensitivity is set at 5 V/div
Horizontal time/div is set at 0.001 s/div

FIGURE 13–23

Since the best method to familiarize yourself with an oscilloscope is by seeing one, the following questions pertain to observing the oscilloscope you may have available in your training environment. Since the best method to learn how to use an oscilloscope is hands-on experience, we further recommend that you also perform "hands-on" laboratory projects(s) related to using the scope. (See the end-of-chapter "Performance Projects Correlation Chart" for suggestions along this line.)

1. List the precise names of the scope controls observed on your training program's oscilloscope that:
   a. control the trace focus.
   b. control the trace intensity.
   c. control the vertical position of the trace(s).
   d. control the horizontal position of the trace(s).
   e. allow for scope adjustment and measurement of vertical input signal amplitude(s).
   f. provide for measuring vertical input signal frequency(ies).

2. Observe the switches and controls on your training program's oscilloscope, and list the following data:
   a. Maximum possible vertical input signal sensitivity setting (the smallest volts/div setting)
   b. Minimum possible vertical input signal sensitivity

3. Refer to Figure 13–24 and for each pattern shown (except for the model waveform pattern) list the oscilloscope controls used to cause the waveform shown to match the model waveform. (NOTE: a = model waveform.)

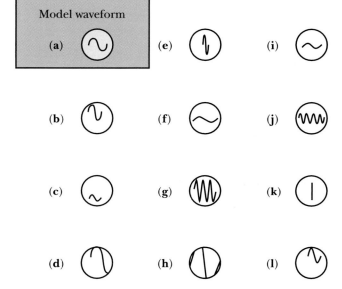

FIGURE 13–24

Suggested performance projects in the Laboratory Manual that correlate with topics in this chapter are:

CHAPTER TOPIC	PERFORMANCE PROJECT	PROJECT NUMBER
Key Sections of the Scope	Basic Operation	36
Measuring Voltage and Determining Current with the Scope	Voltage Measurements	37
Using the Scope for Phase Comparisons	Phase Comparisons (Part A)	38
Determining Frequency with the Scope	Determining Frequency	39

NOTE: It is suggested that after completing the above projects, the student should be required to answer the questions in the "Summary" at the end of this section of projects in the Laboratory Manual.

THE SIMPLER SEQUENCE FOR TROUBLESHOOTING

Follow the seven-step SIMPLER troubleshooting sequence, as outlined below, and see if you can find the problem with this circuit. As you follow the sequence, record your circuit test results for each testing step on a separate sheet of paper. This will aid you and your instructor to see where your thinking is on track or where it might deviate from the best procedure.

step 1 **Symptoms**

—(Gather, verify, and analyze symptom information) To begin this process, read the data under the "Starting Point Information" heading. Particularly look for circuit parameters that are not normal for the circuit configuration and component values given in the schematic diagram. For example, look for currents that are too high or too low; voltages that are too high or too low; resistances that are too high or too low, etc. This analysis should give you a first clue (or symptom information) that will aid you in determining possible areas or components in the circuit that might be causing this symptom.

step 2 **Identify**

—(Identify and bracket the initial suspect area) To perform this first "bracketing" step, analyze the clue or symptom information from the "Symptoms" step, then either on paper or in your mind, bracket, circle, or put parentheses around all of the circuit area that contains any components, wires, etc., that might cause the abnormality that your symptom information has pointed out. NOTE: Don't bracket parts of the circuit that contain items that could not cause the symptom!

step 3 **Make**

—(Make a decision about the first test you should make: what type and where) Look under the TEST column and determine which of the available tests shown in that column you think would give you the most meaningful information about the suspect area or components, in the light of the information you have so far.

step 4 **Perform**

—(Perform the first test you have selected from the TEST column list) To simulate performing the chosen test (as if you were getting test results in an actual circuit of this type), follow the dotted line to the right from your selected test to the number in parentheses. The number in parentheses, under the "Results in Appendix C" column tells you what number to look up in Appendix C to see what the test result is for the test you are simulating.

step 5 **Locate**

—(Locate and define a new "narrower" area of uncertainty) With the information you have gained from the test, you should be able to eliminate some of the circuit area or circuit components as still being suspect. In essence, you should be able to move one bracket of the bracketed area so that there is a new smaller area of uncertainty.

step 6 **Examine**

—(Examine available information and determine the next test: what type and where) Use the information you have thus far to determine what your next test should be in the new narrower area of uncertainty. Determine the type test and where, then proceed using the TEST listing to select that test, and the numbers in parentheses and Appendix C data to find the result of that test.

step 7 **Repeat**

—(Repeat the analysis and testing steps until you find the trouble) When you have determined what you would change or do to restore the circuit to normal operation, you have arrived at *your* solution to the problem. You can check your final result against ours by observing the pictorial step-by-step sample solution on the pages immediately following the "Chapter Troubleshooting Challenge"—circuit and test listing page.

step 8 **Verify**

NOTE: A useful *8th Step* is to operate and test the circuit or system in which you have made the "correction changes" to see if it is operating properly. This is the final proof that you have done a good job of troubleshooting.

CHALLENGE CIRCUIT 6

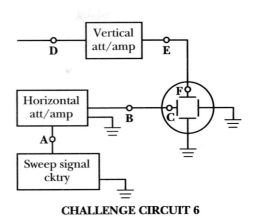

CHALLENGE CIRCUIT 6

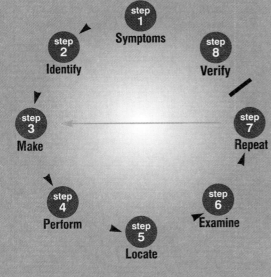

step 1
Symptoms

step 2
Identify

step 3
Make

step 4
Perform

step 5
Locate

step 6
Examine

step 7
Repeat

step 8
Verify

Starting Point Information

1. Simplified block diagram
2. Vertical input signal is 300-Hz ac sine-wave
3. Output of sweep signal circuitry is a 150-Hz sawtooth waveform
4. Horizontal sweep = 150 Hz
5. Trace seen on CRT is a horizontal line from left to right

TEST	Results in Appendix C

Signal/Voltage Checks
Point A to Ground ....(70)
Point B to Ground ....(2)
Point C to Ground ....(113)
Point D to Ground.......................................(57)
Point E to Ground ....(23)
Point F to Ground.......................................(106)

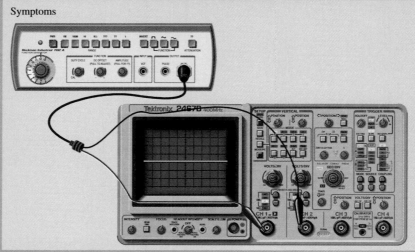

Symptoms

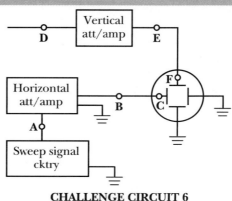

Vertical att/amp

D   E

Horizontal att/amp

B

A

Sweep signal cktry

F

C

CHALLENGE CIRCUIT 6

Symptoms

Although there should be a waveform on the CRT screen, there is only a straight horizontal line. This implies there is a horizontal signal being fed to the CRT horizontal plates, but no vertical signal is reaching the CRT vertical plates.

Identify

initial suspect area: The initial suspect area includes all elements involved in the vertical signal path to the CRT vertical plates. That is, the signal input to the vertical amplifier, the vertical amplifier, and the path from the vertical amplifier output to the CRT vertical plates.

Make

test decision: We'll use the "inputs-outputs" technique and first check the input of the vertical amplifier to see if an input signal is present.

Perform

1st Test: Check the presence of an input signal for the vertical amplifier at point D. The input signal is present.

Locate

Locate new suspect area: The new suspect area is the remainder of the vertical signal path to the CRT vertical plates.

Examine

available data.

Repeat

analysis and testing:
2nd Test: Check the signal "out of" the vertical amplifier at point E to ground. No signal is present, indicating a problem in the vertical amplifier circuitry. We have performed block-level troubleshooting. At this point, the vertical amplifier requires component-level troubleshooting.

Verify

3rd Test: After component-level troubleshooting has been completed and the needed repairs performed, test the system with the same signals. The result is that two sine-wave cycles are seen on the CRT screen.

1st Test

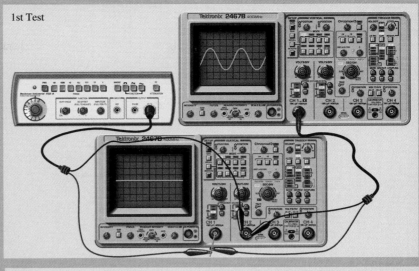

2nd Test

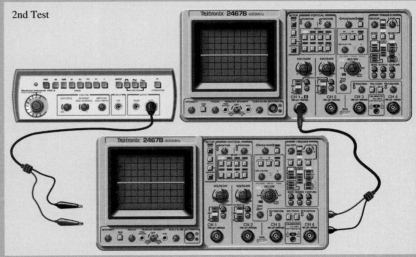

3rd Test

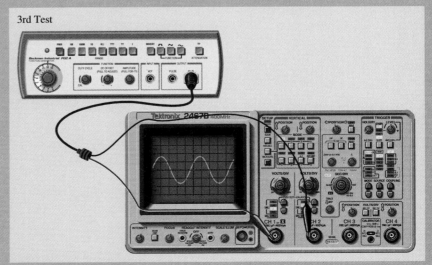

IV

capacitor

V

n's Law

I

Kirchhoff's Current Law

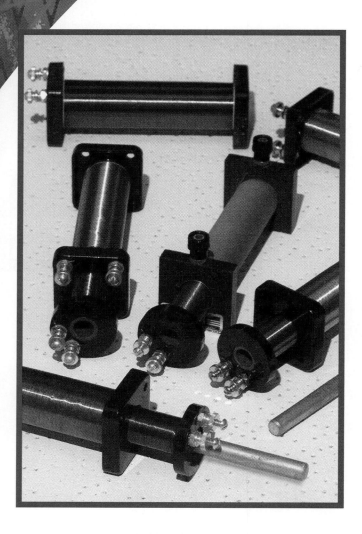

INDUCTANCE

14

chapter

CHAPTER PREVIEW

The electrical property of resistance and its component, a resistor, are important elements in your understanding of electronics. Likewise, the property of inductance and its related component, an inductor, are equally important. A significant contrast between resistors and inductors is that resistors dissipate electrical energy and inductors have the ability to temporarily store electrical energy. Even as you applied your newly gained insights about magnetism and electromagnetism as you studied dc meters and generator action, you will again apply your learning about electromagnetism, as you consider the characteristics of inductors.

Inductors have many uses in electronics. For example, they are used in "tuned" circuits, which enable selection of a desired radio or television station. They are used as filter elements to smooth waveforms or to suppress unwanted frequencies. Inductance is also used in transformers, which are devices that increase or decrease ac voltages, currents, and impedances. The inductor (or coil) in automobiles has been used for years to help develop high voltages, which create the spark across the spark plug gaps. These applications and many others use this unique electrical property of inductance.

In this chapter, we will discuss the property of inductance as it relates to dc circuit action. In the following chapters, you will study the action of this electrical property under ac conditions. An understanding of both will be very useful to you!

OBJECTIVES

After studying this chapter, you should be able to:

1. Define **inductance** and **self-inductance**

2. Explain **Faraday's** and **Lenz's laws**

3. Calculate induced **cemf** values for specified circuit conditions

4. Calculate inductance values from specified parameters

5. Calculate inductance in series and parallel

6. Determine energy stored in a magnetic field

7. Draw and explain time-constant graphs

8. Calculate time constants for specified circuit conditions

9. Use ϵ to aid in finding circuit quantities in dc RL circuits

10. List common problems of inductors

14–1 BACKGROUND INFORMATION

From previous chapters, you know that electrical resistance (R) is that property in a circuit that *opposes current flow*. It is also important for you to know that **inductance (L)** is the property in an electrical circuit that *opposes a change in current*.

Inductors come in a variety of physical and electrical sizes, Figure 14–1. The forms you see here are components that have been designed to oppose changes in current. It should be pointed out that even a straight piece of wire exhibits a

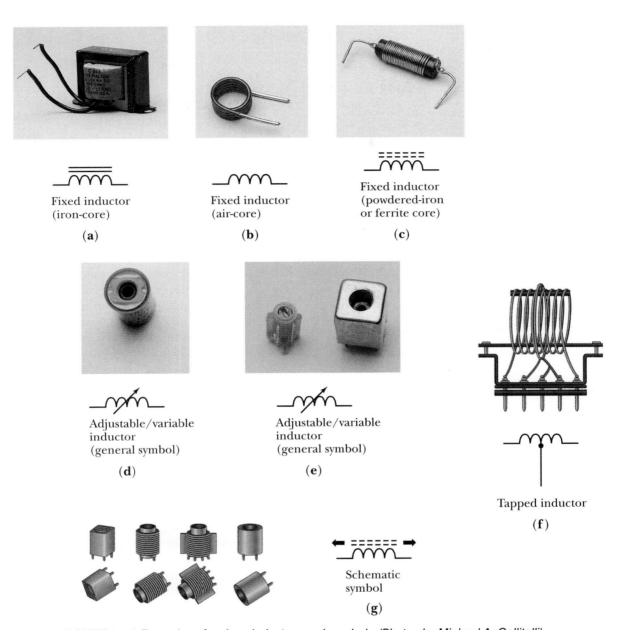

FIGURE 14–1 Examples of various inductors and symbols *(Photos by Michael A. Gallitelli)*

small amount of inductance. Sometimes this "stray" (many times undesired) inductance presents conditions that must be contended with in circuit designs and layouts. This is particularly true in high-frequency circuits, which you will study later in your training. Again, remember that inductance does not oppose current flow, as such, but opposes a *change* in current flow.

Another important characteristic of inductance and inductors is they store electrical energy in the form of the magnetic field surrounding them. As current increases through the inductor, a resulting magnetic field expands, surrounding the coil, Figure 14–2a. When current tries to decrease, the expanded field collapses and tries to prevent the current from decreasing, Figure 14–2b. (Recall

FIGURE 14–2 Expanding and collapsing fields

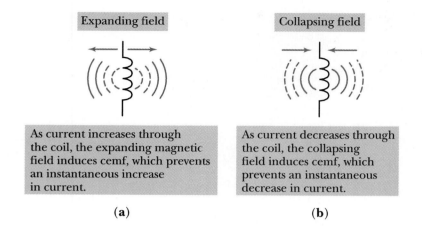

Expanding field

As current increases through the coil, the expanding magnetic field induces cemf, which prevents an instantaneous increase in current.

(a)

Collapsing field

As current decreases through the coil, the collapsing field induces cemf, which prevents an instantaneous decrease in current.

(b)

Lenz's law!) As the field expands, it absorbs and stores energy. When the field collapses, it effectively returns that stored energy to the circuit.

Another feature of inductance and inductors is that an emf (electromotive force or voltage) is induced when a current *change* occurs through them. This induced emf is due to the expanding or collapsing magnetic field as it cuts the conductors of the coil. This self-induced emf is sometimes called **counter-emf,** or **cemf,** because its polarity opposes the change that induced it, Figure 14–3. The term **self-inductance** is frequently used to refer to inductors that create this situation.

In a previous chapter, we discussed magnetism and electromagnetism. Additionally, you learned the factors determining the strength of this electromagnetic field around the coil—number of turns; coil length (number of turns per unit length); cross-sectional area of coil; amount of current through the coil; and the type of core material (characteristics involved in the flux path). Keep these factors in mind, as there is a relationship between these items and the property of inductance, which is addressed in this chapter.

In the earlier discussion of electromagnetism, we examined a constant level dc current through the coil. In this chapter, we will begin to examine what happens during the switch-on/switch-off "transient" (temporary condition) periods when dc is applied or removed from a circuit containing inductance.

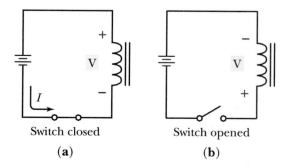

Switch closed

(a)

Switch opened

(b)

(**a**) The polarity of the induced cemf is opposite to that of the source as the magnetic field expands.
(**b**) The polarity of the induced voltage reverses as the field collapses.

FIGURE 14–3 The concept of cemf

14–2 REVIEW OF FARADAY'S AND LENZ'S LAWS

Faraday's law

Recall **Faraday's law:** The emf induced in a circuit is proportional to the time-rate of change in the magnetic flux linking the circuit. In other words, the amount of induced emf depends on the rate of cutting the flux (ϕ).

Also, recall that in the SI system of units, a weber is 10^8 lines of flux. The formula to determine the induced voltage due to cutting flux lines (or flux linkages) is defined as:

$$V_{ind} = \frac{d\phi}{dt}$$

where $\dfrac{d\phi}{dt}$ indicates rate of cutting or linking flux lines. For a coil then:

FORMULA 14–1 $V_{ind} = N \text{ (number of turns)} \times \dfrac{d\phi \text{ (Wb)}}{dt \text{ (s)}}$

Example

When a coil has 500 turns and the rate of flux change is 2 webers per second,

$$V_{ind} = 500 \times \frac{2}{1} = 1{,}000 \text{ V}$$

PRACTICE PROBLEMS I

What is the induced voltage (per turn) when 2×10^8 lines of flux are cut in 0.5 second?

Lenz's law

Recall **Lenz's law:** The direction of an induced voltage (or current) opposes the change that caused it. That is, when the amount of magnetic flux linking an electric circuit changes, an emf is induced that produces a current opposing this flux change.

To summarize, Faraday's law describes how much voltage is induced by the parameters affecting it, and Lenz's law describes the polarity and nature of this induced voltage and the current produced. Let's now begin to correlate these laws with some new information as we examine inductance and its dc characteristics.

14–3 SELF-INDUCTANCE

In addition to inductance being defined as the property of a circuit that opposes a change in current, another definition for self-inductance (or inductance) is *the property of an electric circuit where a voltage is induced when the current flowing in the circuit changes.* You have learned the symbol for inductance, *L*. The **unit of inductance** is the **henry.** Named in honor of Joseph Henry, this unit of inductance is abbreviated **H.**

As you might suspect, the amount or magnitude of induced emf produced is related to the amount of inductance *(L)* in henrys (H) and the rate of current change. A formula expressing this is:

FORMULA 14–2 $V_L = L\dfrac{di}{dt}$ (change in current, in amperes)
(change in time, in seconds)

where L = inductance in henrys

$$\dfrac{di}{dt} = \text{rate of change of current } \left(\dfrac{\text{A}}{\text{s}}\right)$$

It should be noted that the polarity of the induced voltage is opposite to the source voltage that caused the current through the coil.

From the above, a definition for a henry of inductance states that *a circuit has an inductance of one henry when a rate of change of 1 ampere per second causes an induced voltage of 1 volt.* The induced voltage is in the opposite direction to the source voltage; therefore, it is called **counter-emf** or **back-emf.** Again see Figure 14–3.

Example

When the rate of current change is 5 amperes per second through an inductance of 10 henrys, what is the magnitude of induced emf?

$$V_L = L\dfrac{di}{dt} \text{ (change in current, in amperes)}$$
(change in time, in seconds)

$$V_L = 10 \times 5 = 50 \text{ V} \text{ (Or, induced cemf} = 50 \text{ V.)}$$

PRACTICE PROBLEMS II

What is the counter-emf produced by an inductance of 5 henrys if the rate of current change is 3 amperes per second?

From these definitions and formulas, a useful formula to calculate inductance value is:

FORMULA 14–3 $L = \dfrac{V_L}{\dfrac{di}{dt}}$

where L = inductance in henrys
$$\dfrac{di}{dt} = \text{rate of current change } \left(\dfrac{\text{A}}{\text{s}}\right)$$

Example

What is the circuit inductance if a current change of 12 amperes in a period of 6 seconds causes an induced voltage of 5 volts?

$$L = \dfrac{V_L}{\dfrac{di}{dt}} = \dfrac{V_L}{1} \times \dfrac{dt}{di} = \dfrac{5}{1} \times \dfrac{6}{12} = 5 \times 0.5 = 2.5 \text{ H}$$

Inductance:

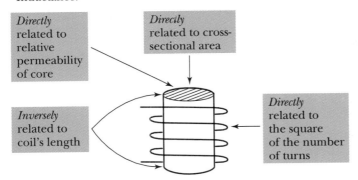

Directly related to relative permeability of core

Directly related to cross-sectional area

Inversely related to coil's length

Directly related to the square of the number of turns

FIGURE 14-4 Factors that affect the amount of inductance

14-4 FACTORS THAT DETERMINE INDUCTANCE

As implied earlier, some of the same factors that affect the strength of a coil's electromagnetic field are also related to the amount of coil inductance.

Look at Figure 14–4. The physical properties of an inductor that affect its inductance include the following:

1. *Number of turns.* The greater the number of turns, the greater the inductance value. In fact, inductance is proportional to the square of the turns. That is, when you double the number of turns in a coil of a given length and diameter, the inductance quadruples, assuming the core material is unchanged.

2. *Cross-sectional area (A) of the coil.* This feature relates to the square of the coil diameter. The greater the cross-sectional area, the greater the inductance value. When the cross-sectional area doubles, the inductance (L) doubles.

3. *Length of the coil.* The longer the coil for a given diameter and number of turns, the *less* the inductance value. This is due to a less concentrated flux, thus, less flux cut per unit time and the less the induced voltage produced.

4. *Relative permeability of the core.* For air-core inductors, the relative permeability (μ_r) is virtually one. The higher the permeability of the inductor's core material, the greater the inductance value. Recall that permeability relates to the magnetic path's ability to concentrate magnetic flux lines.

As a technician you probably will not design inductors. However, Formula 14–4 illustrates the relationship of inductance to the various factors we have been discussing. In SI units, the formula is:

FORMULA 14-4 $$L = 12.57 \times 10^{-7} \times \frac{\mu_r N^2 A}{l}$$

where L = inductance in henrys
μ_r = relative permeability (for air or vacuum = 1)
N = number of turns
A = cross-sectional area in square meters
l = length of coil in meters
12.57×10^{-7} = a constant representing the absolute permeability of air (in SI units)

IN-PROCESS LEARNING CHECK I

Fill in the blanks as appropriate.

1. All other factors remaining the same, when an inductor's number of turns increases four times, the inductance value _____ by a factor of _____.

2. All other factors remaining the same, when an inductor's diameter triples, the inductance value _____ by a factor of _____.

3. All other factors remaining the same, when an inductor's length increases, the inductance value _____.

4. When the core of an air-core inductor is replaced by material having a permeability of ten, the inductance value _____ by a factor of _____.

Example

An air-core coil that is 0.01 meters in length with a cross-sectional area of 0.001 square meters has 2,000 turns. What is its inductance in henrys?

$$L = 12.57 \times 10^{-7} \times \left(\frac{1 \times (2,000)^2 \times 0.001}{0.01} \right)$$

$$L = 12.57 \times 10^{-7} \times \left(\frac{40 \times 10^5) \times 1 \times 10^{-3}}{10 \times 10^{-3}} \right)$$

$$L = 12.57 \times 10^{-7} \times 4 \times 10^5 = \text{about 0.5 H, or 500 mH}$$

Since the μ_r of air is one, the formula is modified for *air-core* inductors and stated as:

FORMULA 14–5 $L = 12.57 \times 10^{-7} \times \dfrac{N^2 A}{l}$

14–5 INDUCTORS IN SERIES AND PARALLEL

For our purposes, we will assume the inductors in this section do not have any *coupling* between them. That is, there are no flux lines from one inductor linking the turns of another inductor. In technical terms, no *mutual inductance* exists between coils.

Inductance in series

Finding the total inductance of inductors in *series,* Figure 14–5, is simple. Since they are in series, the current change through each inductor is the same. Thus, when they are equal-value inductors, the induced voltage in each is equal, and the total inductance is two times the value of either inductor. This implies that

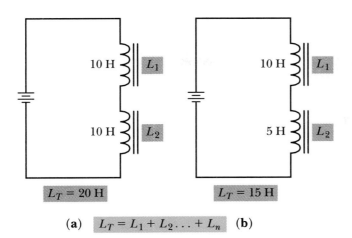

FIGURE 14–5 Total inductance of series inductors

(a) $L_T = L_1 + L_2 \ldots + L_n$ (b)

when the inductors are not equal in value, the total inductance (L_T) is the sum of the inductance values in series. That is, $L_T = L_1 + L_2$.

Stated another way: *Inductors in series add inductances in the same way that resistors in series add resistances!*

FORMULA 14–6 $L_T = L_1 + L_2 \ldots + L_n$

REMINDER: This formula holds true as long as there is no coupling of flux lines from one inductor to another; that is, there is no coupling between inductors.

Example

When the inductors in Figure 14–5a are tripled in value, the total inductance of the circuit on the left is 60 H, rather than 20 H, and the total inductance of Figure 14–5b is 45 H, rather than 15 H.

PRACTICE PROBLEMS III
Determine the total inductance of the circuit shown in Figure 14–6.

Inductance in parallel

As you might expect, the total inductance of inductors in parallel is less than the least value inductor in parallel. This is analogous to resistors in parallel. The general formulas used to compute total inductance for inductors in parallel, Figure 14–7, are the reciprocal formulas: (NOTE: Don't forget to use the reciprocal function of your calculator to advantage when dealing with these formulas.)

FORMULA 14–7 $\dfrac{1}{L_T} = \dfrac{1}{L_1} + \dfrac{1}{L_2} + \dfrac{1}{L_3} \ldots + \dfrac{1}{L_n}$

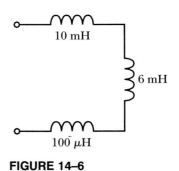

FIGURE 14–6

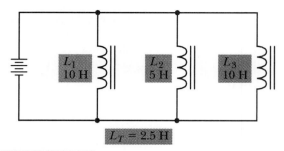

$$\frac{1}{L_T} = \frac{1}{L_1} + \frac{1}{L_2} + \frac{1}{L_3} = \frac{1}{10} + \frac{1}{5} + \frac{1}{10} = 0.1 + 0.2 + 0.1 = 0.4$$

$$\frac{1}{L_T} = 0.4; \ L_T = \text{Reciprocal of } \frac{1}{L_T} = \frac{1}{0.4} = 2.5 \text{ H}$$

FIGURE 14–7 Finding L_T using the reciprocal formula

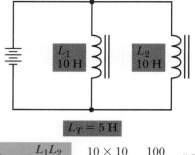

$$L_T = \frac{L_1 L_2}{L_1 + L_2} = \frac{10 \times 10}{10 + 10} = \frac{100}{20} = 5 \text{ H}$$

FIGURE 14–8 Finding L_T using the product-over-the-sum formula

FORMULA 14–8 $\quad L_T = \dfrac{1}{\dfrac{1}{L_1} + \dfrac{1}{L_2} + \dfrac{1}{L_3} \dots + \dfrac{1}{L_n}}$

PRACTICAL NOTES

Be aware that Formula 14–8 ("the reciprocal of the summed reciprocals") is simply displaying what you would need to do to find L_T, after you use Formula 14–7 and find $1/L_T$. That is, to find L_T, you take the reciprocal of $1/L_T$, or 1 over $1/L_T$. This then gives you the total inductance of the parallel inductors used in the calculations.

(NOTE: If all units are in henrys, the answer is in henrys.) Also, the product-over-the-sum approach, used in Figure 14–8, works for two inductors in parallel:

FORMULA 14–9 $\quad L_T = \dfrac{L_1 \times L_2}{L_1 + L_2}$

Example

You have probably already surmised that the same time-saving techniques can be used for inductance calculations as you used in resistance calculations. For example, if the two inductors in parallel are of equal value, the total inductance equals half of one inductor's value. This is similar to two 10-Ω resistors in parallel having an equivalent total resistance of 5 Ω.

PRACTICE PROBLEMS IV

1. What is the total inductance of three 15-H inductors that are connected in parallel and have no mutual inductance?

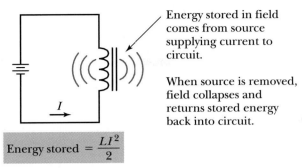

Energy stored in field comes from source supplying current to circuit.

When source is removed, field collapses and returns stored energy back into circuit.

Energy stored $= \dfrac{LI^2}{2}$

FIGURE 14–9 Concept of energy being stored in inductor's magnetic field

2. Use your calculator and the reciprocal method and find the total inductance of four inductors in parallel. Their individual values are: 10 mH, 12 mH, 15 mH, and 20 mH. Assume no coupling between them.

14–6 ENERGY STORED IN THE INDUCTOR'S MAGNETIC FIELD

The magnetic field surrounding an inductor when current passes through the inductor is a form of stored energy, Figure 14–9. The current producing the electromagnetic field is provided by the circuit's source. Hence, this stored energy is supplied by the source to the magnetic field.

When the source is removed from the circuit, the field collapses and returns the stored energy to the circuit in the form of induced emf. This induced emf tries to keep the current through the coil from decaying.

If the inductor were a perfect inductor with no resistance, no I^2R loss would occur; therefore, no power would be dissipated in this energy storing-and-returning process. NOTE: In reality, inductors have some resistance (the resistance of the wire), and they dissipate power; however, for the present discussion, we won't analyze the details of this fact. Note also that in dc circuits, the inductor's resistance (plus any other resistance in series with the inductor) and the amount of voltage applied determine the amount of current through the coil (Ohm's law).

Notice in Figure 14–10, if a coil has 10 ohms of resistance, and a dc voltage of 20 V is applied, the dc current through the coil will be 20 V/10 Ω, or 2 amperes, after the field has finished building up.

Computing the energy stored in an inductor's field is not a daily calculation for most technicians. Many technicians, however, take FCC examinations to obtain commercial radio licenses. Since this computation is required for these tests, we will show you how this calculation is performed.

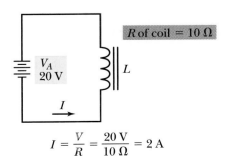

R of coil $= 10\ \Omega$

V_A 20 V

L

I

$$I = \frac{V}{R} = \frac{20\ \text{V}}{10\ \Omega} = 2\ \text{A}$$

FIGURE 14–10 Direct current (dc) through inductor found by using Ohm's law

In an earlier chapter on Ohm's law, we discussed the unit of electrical energy, the **joule.** A joule is sometimes defined as the amount of electrical energy required to move one coulomb of electrical charge between two points having a potential difference of one volt. Also remember the watt of power relates to the joule since the performance of electrical work at a rate of one joule per second equals one watt.

Relating all of these factors, it can be shown that the stored energy correlates to the average power supplied to the inductor circuit and the time involved. The amount of energy stored, in turn, relates to the amount of inductance and current involved.

Accumulating these facts, then:

FORMULA 14–10   Energy $= 0.5 \ LI^2$

where $L =$ inductance in henrys
$I =$ current in amperes
Energy $=$ joules (or watt-seconds)

Example

What is the energy stored by an inductor of 20 henrys that has a current of 5 amperes passing through it?
Substituting into the formula:

$$\text{Energy} = 0.5 \ LI^2 \text{ or } \frac{LI^2}{2}$$

$$\text{Energy} = \frac{20 \times (5)^2}{2} = \frac{20 \times 25}{2} = \frac{500}{2} = 250 \text{ J}$$

PRACTICE PROBLEMS V

A dc voltage of 100 V is applied to a 1-H inductor that has 20 Ω of dc resistance. (a) What is the dc current through the coil? (b) How many joules of energy are stored in the inductor's magnetic field?

14–7 THE L/R TIME CONSTANT

We earlier established that an inductor opposes a change in current flow due to the expanding or contracting magnetic field as it cuts the turns of the coil. The induced voltage opposes the change that originally induced it (Lenz's law). This implies that even when dc voltage is first applied to a coil, the current cannot instantly rise to its maximum value (the V/R value), which is true!

Let's examine the transient conditions, or the temporary state caused by a sudden change in circuit conditions in a simple inductive circuit when dc is first applied. Refer to Figures 14–11 and 14–12 as you study the following discussion.

1. In Figure 14–11, you see the conditions that exist in a purely resistive circuit. When the switch is closed, the current immediately reaches its dc (Ohm's law) value determined by the voltage applied and the resistance involved.

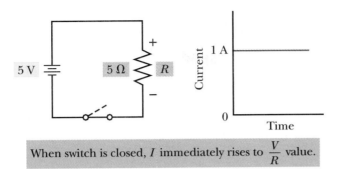

FIGURE 14–11 Direct current (dc) change in purely resistive circuit

When switch is closed, I immediately rises to $\dfrac{V}{R}$ value.

2. In Figure 14–12a, the inductor in the resistor-inductor *(RL)* circuit prevents current from rising to its maximum value during the first moments after the switch is closed. This is because the current increases from zero to some value, causing an expansion of the magnetic field in the inductor. The expanding field cuts the turns of the coil, inducing a back-emf (cemf) that opposes the current change. Therefore, the current does not abruptly rise from zero to the final (stable) dc value.

3. In Figure 14–12b, at time t_1, the current has arrived at only 63.2% of its final value. The time from t_0 to t_1 is equal to the value of *L/R*, where *L* is in henrys and *R* is the circuit resistance value in ohms. This value *(L/R)* is known as one **time constant.** The symbol for the time constant is the Greek letter tau (τ). In a circuit containing *L* and *R*, the formula for time constant is:

FORMULA 14–11 $\tau = \dfrac{L}{R}$

Let's think about Formula 14–11 for a moment. One time constant (τ) equals the value of the inductance *(L)* in henrys divided by the resistance *(R)*

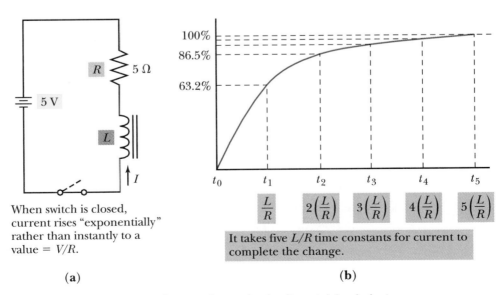

When switch is closed, current rises "exponentially" rather than instantly to a value = V/R.

(a)

It takes five *L/R* time constants for current to complete the change.

(b)

FIGURE 14–12 Current change in circuit containing inductance

in ohms. The amount of time for one time constant is *directly* related to the value of inductance. The time for one time constant is inversely related to the resistance value. Why might this be true?

You have already learned that the greater the inductance value, the greater will be the amount of counter-emf generated for a given amount of current change ($V_L = L \times d\phi/dt$). This indicates that the larger the L value, the greater the opposition to current change. This means it will take more time for the current (and field) to build up to their maximum, or "steady-state" level, which occurs after five time constants. In other words: As L increases, the time for one time constant (τ) increases. Conversely, as L decreases, τ decreases.

In accordance with Ohm's law, the smaller the resistance value, the higher the final current will be ($I = V/R$); thus, the greater the required change in current to arrive at the higher current value, and the longer the time involved for one time constant. In other words: As R decreases, the time for τ increases. Conversely, as R increases, τ decreases.

With these thoughts in mind, continue examining the graphing of the rise of current through the RL circuit, with dc applied. Incidentally, this type graph is often referred to as the "Universal Time-Constant Graph" or chart.

4. In Figure 14–12b, at time t_2 (where the time from t_1 to t_2 represents a second time constant), the current has risen 63.2% of the remaining distance to the final dc current level. That is, current rises 63.2% of the remaining 36.8% to be achieved, or 63.2% of 36.8% equals an additional 23.3%. Thus, in two time constants the current achieves 63.2% + 23.3%, or 86.5% of its final maximum value.

5. As time continues, the current rises another 63.2% of the remaining value for each time constant's worth of time. After *five time constants*, the current has virtually arrived at its final dc value. In this case,

$$\frac{5 \text{ V}}{5 \text{ }\Omega} = 1 \text{ A}$$

Therefore, the amount of time required for current to rise to maximum V/R value is directly related to the value of inductance (L) and inversely related to the circuit R, as indicated by the L/R relationship.

The inductor's expanding magnetic field and the resulting induced voltage cause current to exponentially increase (in a nonlinear fashion), rather than instantly as it would in a purely resistive circuit.

What happens when the inductor's field collapses, that is, when the source is removed after the field has fully expanded? In Figure 14–13, when the switch is in position A, the source is connected to the RL circuit.

When the switch is in position B, the source is disconnected from the RL circuit. However, a current path is provided for the RL circuit, even when the source is disconnected.

Assume the switch was in position A for at least five time constants and the current has reached its maximum level. Therefore, the magnetic field is fully expanded. When the switch is moved to position B, the expanded field collapses, inducing an emf into the coil, which attempts to keep the current from changing. The inductor's collapsing field and induced emf have now, in effect, become the source. Obviously, as the field fully collapses, current decays and eventually

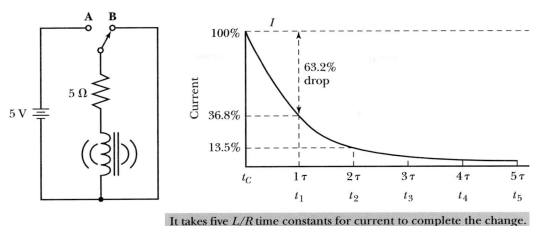

It takes five L/R time constants for current to complete the change.

FIGURE 14–13 Circuit with same L/R time constant for current decay as for current buildup

becomes zero. The following comments describe what happens during this declining current time. Figure 14–14 shows a graphic display of both the current build-up and decay. Refer to it while reading the following discussion.

1. The current decays 63.2% from maximum current value toward zero in the first time constant. In other words, it falls to 36.8% of maximum value.

2. During the second τ, it falls 63.2% of the remaining distance, or another 23.3% for a total of 86.5% decay. In other words, the current is now at 13.5% of its maximum value.

3. This continues, until at the end of the five L/R time constants, the current is virtually back at the zero level.

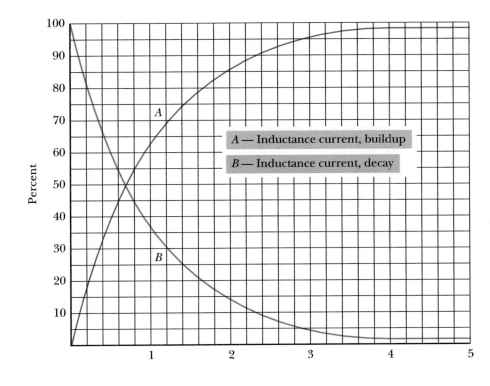

FIGURE 14–14 Universal time-constant chart

4. Given the same L/R conditions, the current decay is the mirror image (or the inverse) of the current-rise condition, described earlier.

Let's summarize. In a dc circuit comprised of an inductor and a series resistance (RL circuit), it takes five L/R time constants for current to change *from one static level to a "changed" new level*, whether it be an increase or a decrease in current levels. (NOTE: This is true whether the change is from 0 to 1 ampere, from 1 to 5 amperes, or from 4 amperes to 2 amperes, and so forth.) This is an important concept, so remember it!

Observe Figure 14–15 and notice that in a *dc RL* circuit, the sum of V_L and V_R must equal V applied. This means as current increases through the RL circuit, V_R is increasing and V_L is decreasing proportionately until V_R equals approximately V applied and V_L equals approximately 0 V when the current has finally reached a steady-state value. (Assuming the R of the inductor is close to zero ohms.) Conversely, when current decays, V_R also decays at the same rate to the steady-state minimum value the circuit is moving toward. The τ curves are one convenient method to determine what values V_R and V_L have after specified times.

Example

Refer to Figure 14–15. What is the value of V_R 1 second after the switch closed? Answer: First, compute how many time constants are represented by 1 second:

$$\tau = \frac{L}{R} = \frac{10 \text{ H}}{12 \text{ }\Omega} = 0.833 \text{ seconds}$$

Therefore, 1 second = 1/0.833 time constants, or 1 second = 1.2 τ.

Next, refer to the τ curves and note that at 1.2 τ, the current rises to 70% of maximum value. Therefore, $V_R = 0.7 \times V$ applied. Thus, V_R at that instant equals $0.7 \times 10 \text{ V} = 7 \text{ V}$ and V_L is 3 V.

PRACTICE PROBLEMS VI
Using the same circuit as in Figure 14–15, determine the V_R and V_L values two seconds after the switch is closed.

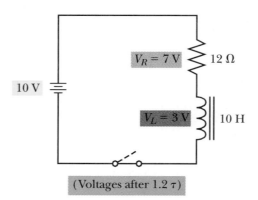

FIGURE 14–15 Simple dc *RL* circuit

14–8 USING EPSILON (ε) TO HELP CALCULATE CIRCUIT PARAMETERS

As you have seen, you can come fairly close in approximating circuit conditions and values from the exponential Universal Time Constant Chart. Another more precise way to calculate circuit parameters at any given instant is to use natural logarithms and epsilon, rather than graphical means.

The curves represented on the time-constant graph are sometimes called exponential curves. Many things in nature follow this exponential curve. The natural logarithms are mathematical values related to exponential changes. The base number of the natural logarithms is 2.71828. This number is often referred to by the Greek letter epsilon (ε). That is, ε equals 2.71828.

Graph based on ε raised to negative powers

Observe in Figure 14–16 that the descending curve on the graph is actually a plot that can be produced by raising ε (2.71828) to negative powers. That is, if we find ϵ^{-1}, it has a value of 0.3679, ϵ^{-2} equals 0.1353, and so on. The ascending curve on the graph represents 1 minus the other curve's value at each point along the curve. For example, when the descending curve is at 0.3679, the ascending curve is at $1 - 0.3679$, or 0.6321, and so on.

Relationship of curves to circuit parameters

For our purposes here, the ascending curve represents the increase in current over five time constants. It can also represent the increasing voltage across the resistor in the RL circuit as current increases to maximum, or "full," value (I_f) over five time constants. As I increases, $I \times R$ increases, in step. The descending curve represents the current decay when the external voltage source is removed from the circuit and the inductor's field is collapsing (assuming there is a current path when the external source is removed, Figure 14–13). Also, the descending curve can represent the decay in voltage drop across the resistor, since its $I \times R$ drop is going to be in step with the current conditions.

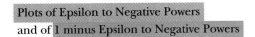

Plots of Epsilon to Negative Powers
and of 1 minus Epsilon to Negative Powers

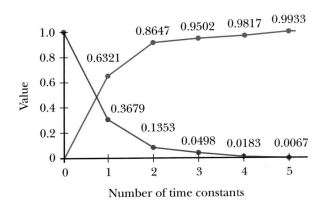

FIGURE 14–16 Graph based on ε raised to negative powers

As you know, in the dc circuit, the $V_R + V_L$ must add up to equal $V_{applied}$. At any given point along these curves, the decimal values of the descending and ascending curves must add up to 1, representing 100% of the maximum, or final, state value.

Relationship of negative powers for epsilon and L/R time constant

As stated, the negative power to which epsilon (2.71828) is raised, determines where we are on the descending exponential curve. How does this curve relate to time constants?

You have already learned that after five L/R time constants, the circuit parameters will have changed from wherever they are to the new "stable state" conditions. The relationship of importance to us is: *how much time is being allowed for the change versus the time for one time constant?*

If the time being allowed for the change is five time constants or greater, 1) current will be at its final value, if it is building up, or at zero value, if decaying; 2) V_R will be equal to $V_{applied}$ if current is building up, or at zero value, if current is decaying; 3) V_L will be at zero value in either case.

If the time being allowed for the change is only one time constant, then current will be at 63.2% of its new maximum, if current is building up, or at 36.8% of maximum, if current is decaying.

The bottom line is, if we know the amount of time being allowed and know, or can calculate, the L/R time constant, we know how many time constants are being allowed for the change. *This ratio of time allowed to time for one time constant is the number we want to use as the negative power of epsilon (ϵ) in calculating circuit parameters!*

The method of finding the number of time constants allowed is

$$\text{Number of time constants} = \frac{\text{Time allowed for change } (t)}{\text{Time for one } L/R \text{ time constant}}$$

$$\text{Number of } \tau = \frac{t}{\dfrac{L}{R}} = \frac{t}{1} \times \frac{R}{L} = \frac{Rt}{L}$$

So the negative power we will want to raise epsilon to in our calculations related to L/R time constants is Rt/L.

Of course, your scientific calculator has either the natural log function key e^x, or a key labeled $\boxed{\ln_x}$, which will greatly simplify calculations involving raising epsilon to the various powers of interest. In our examples, we will use the e^x key.

Armed with these facts, let's introduce the formulas used to solve for the instantaneous value of current. Of course, if you know the current, you can calculate the $I \times R$ resistor voltage drop. If you know the applied voltage, you can then subtract V_R from V_S to find V_L at that point in time.

The circuit conditions assumed here are as follows: On current buildup, we are increasing from zero to maximum or final value. On current decay, we are decreasing from the present level to zero current.

FIGURE 14–17

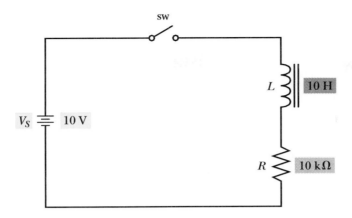

FORMULA 14–12 (Increasing I) $i = \dfrac{V_S}{R}(1 - \epsilon^{-Rt/L})$

where $i =$ instantaneous current value

Examples

Refer to Figure 14–17 and determine the circuit current 1 millisecond after the switch is closed.

$$i = \frac{V_S}{R}(1 - \epsilon^{-Rt/L})$$

$$i = \frac{10 \text{ V}}{10 \text{ k}\Omega}(1 - \epsilon^{-10 \text{ k}\Omega \times 1 \text{ ms}/10 \text{ H}})$$

$$i = 1 \text{ mA } (1 - \epsilon^{-1}) = 1 \text{ mA } (1 - 0.3678) = 1 \text{ mA} \times 0.6322$$

$$i = 0.0006322 \text{ A or } 0.6322 \text{ mA}$$

This checks out against our exponential graph. The time allowed for change was 1 time constant and the current increased from zero to about 63% of its maximum possible value of 1 mA.

$$V_R = 0.63 \text{ mA} \times 10 \text{ k}\Omega, \text{ or } 6.3 \text{ V}; \ V_L = 10 \text{ V} - 6.3 \text{ V} = 3.7 \text{ V}$$

For many calculators the calculator keystrokes for finding $\epsilon^{-Rt/L}$ for our example are: 10, **EE**, 3, **×**, 1, **EE**, **+/−**, 3, **÷**, 10, **=**, **+/−**, **2nd**, **e^x**. This yields the 0.3678, which easily plugs into the formula in place of the ϵ^{-1} from our example.

The complete calculator sequence for many calculators for this problem would be: 10, **EE**, 3, **×**, 1, **EE**, **+/−**, 3, **÷**, 10, **=**, **+/−**, **2nd**, **e^x**, **+/−**, **+**, 1, **=**, **×**, 10, **÷**, 10, **EE**, 3, **=**

Observe that this sequence indicates that first you find the product of $R \times t$, then divide it by the L value, yielding the value of Rt/L (the number of time constants being allowed for the change). Then use that result as the negative power to raise epsilon to in order to find out where we are on the exponential curve. Then, make that result negative with the ⊕/⊖ button in order to subtract it from a + 1. The resultant decimal value is then multiplied times the value of maximum current found by *V/R*. The result, for our example, is the circuit current present after the one time constant allowed for the current change from zero.

PRACTICE PROBLEMS VII

Suppose for the example circuit in Figure 14–17, we want to compute the current 2.5 milliseconds after the switch was closed. What would be the value of current? (Use the calculator sequence, then see if the result makes sense by checking against the τ chart.)

To find the value of decaying (decreasing) current, it is not necessary to subtract from 1, since we are not trying to find the value of the ascending curve, but rather we want the descending curve value. The formula for instantaneous value of current during decreasing current is:

FORMULA 14–13 (Decreasing *I*) $i = \dfrac{V_S}{R} \epsilon^{-Rt/L}$

Example

Refer to the circuit shown in Figure 14–18. Assume that the current is at maximum, final-state value, then the switch is moved from point A to point B. What is the value of circuit current after 68 microseconds?

$$i = \frac{V_S}{R} \times \epsilon^{-Rt/L}$$

$$i = \frac{6\ V}{2.2\ k\Omega} \times \epsilon^{-(2.2\ k\Omega\ \times\ 68\ \mu s)/50\ mH} = 2.72\ mA \times \epsilon^{-2.992} = 2.72\ mA \times 0.05$$

$$= 0.136\ mA$$

This checks out against our τ chart. After about 3 time constants, the current should have decayed to approximately 0.0498 times maximum current. Our cal-

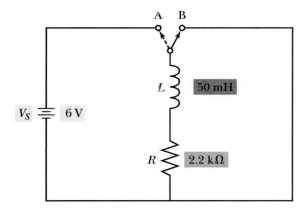

FIGURE 14–18

culations indicated that after 2.992 time constants, the current is at 0.05 times its original value. $V_R = 0.136$ mA $\times$ 2.2 k$\Omega = 0.299$ volts at this point in time.

PRACTICAL NOTES

Adaptations of the current formulas can be made to find resistor and inductor voltage at any instant during current rise or at any instant during current decay. The formulas that may be used are as follows:

Voltage across resistor during current buildup:

$$V_R = V_S \times (1 - \epsilon^{-Rt/L})$$

Voltage across resistor during current decay:

$$V_R = V_S \times (\epsilon^{-Rt/L})$$

Voltage across inductor during current buildup or decay:

$$V_L = V_S \times (\epsilon^{-Rt/L})$$

A rapidly collapsing field induces very high voltage(s)

In a circuit, such as in Figure 14–15 (see page 480), when the source voltage is disconnected, there is no low-resistance current path, as was shown earlier when explaining the current decay events. For purposes of explanation, let's assume the resistance across the open-switch contacts is extremely high. What happens to the total R in the current path when the switch opens? The R drastically increases (approaching infinity). What impact does this have on the L/R time constant? Since R drastically increases and L remains the same, the time constant must greatly decrease!

The result of all this activity is the expanded field collapses almost instantly. This means many flux lines cut the turns of the inductor in a very short time; thus, a very high voltage is induced in the coil. (NOTE: This phenomenon produces high voltage and high-voltage "arcs" jumping the gaps of spark plugs in your automobile.) In Figure 14–15, the induced voltage could be high enough to cause an arc across the open-switch contacts. The stored magnetic energy represented by the collapsing field is thus dissipated by the energy used in producing the arc.

14–9 SUMMARY COMMENTS ABOUT INDUCTORS

You have studied that inductors have properties of inductance (which oppose a *change* in current) and resistance in the wire from which the inductor is wound (which opposes current flow). The amount of inductance, L (measured in henrys), relates to the physical properties of the coil, including the length, cross-sectional area, number of turns, and type of core. Also, there is a relationship between the amounts of inductance and induced cemf that develops for any given rate of current change through the coil.

Also, you know when a dc voltage is first applied to a circuit with an inductor present, current does not instantly rise to the value dictated by the dc resistance of the circuit divided by the applied voltage. This delay is due to the inductor's opposing any current change. It takes five L/R time constants for the complete change from one current level (even if it is zero amperes) to a new current level. Under opposite conditions (i.e., if the dc source is removed from the inductor circuit) current will not abruptly decrease to zero, since the inductor opposes the current change and the collapsing field induces an emf in the coil to attempt to prevent the current from changing. When the time constant is very small during the decay sequence, the field quickly collapses, inducing a very high voltage in the coil. Wise circuit designers, therefore, make sure that current is not abruptly interrupted through the inductors in general dc circuits. When it must be abruptly interrupted, preventive measures are generally used to avoid unwanted voltage spikes. (NOTE: There are some special applications where high voltage is desired, and the component ratings and circuit design are appropriately chosen for this purpose.)

Laminated iron-core inductors, such as those used in power supplies and audio circuitry, are often in henrys of inductance. Powdered iron-core and variable ferrite-core inductors can have inductance ranges in millihenrys (mH). Air-core inductors are typically in the microhenrys (μH) range.

Also, inductors are wound with various sizes of wire and in varying numbers of turns which affect the dc resistance of the coil. Later, you will study how the resistance to inductance ratios affects inductors' operation in certain ac circuit applications.

Excellent information is available about the details of winding coils to desired specifications in *The Radio Amateur's Handbook,* published by the American Radio Relay League (ARRL). Also, there are numerous inexpensive, special slide rules that allow you to establish required parameters and to read the number of turns, coil length, coil diameter, and so on, needed for the desired results.

Watch out for "inductive kick" when breaking a dc circuit containing a large inductance! Use appropriate safety precautions to avoid being shocked!

14–10 TROUBLESHOOTING HINTS

Typical troubles

Troubles in inductors are usually associated with either opens or shorts. For this reason, the ohmmeter is a useful instrument to check the condition of an inductor—*provided* you know the normal resistance for the coil being checked.

Usual values range from 1 Ω to several hundred ohms for iron-core "chokes" and inductors with many turns and inductance values in the henrys range. Inductors in the millihenrys and microhenrys ranges typically have dc resistances from less than 1 Ω to something less than 100 Ω.

CAUTION! Care should be taken when removing ohmmeter test leads from large inductances. The "inductive kick" caused by the quick collapse of the inductor's field when the ohmmeter is removed can give you a nasty shock! This field is created by current produced through the inductor by the ohmmeter's dc source during testing. Be sure your hands or fingers are not touching or near the metallic tips of the test probes when removing the test leads from the inductor connections.

Open coil

If an inductor is open, it is obvious with an ohmmeter check that it measures infinite ohms. *CAUTION!* Be sure there are no sneak paths when you measure. If it is easy to do, it is good to isolate the component being measured from the rest of the circuitry when checking it. Probably the most common trouble is this open condition. Either there is a broken turn in the coil, or a bad connection to the coil, and so forth.

Shorted turns

Because each coil turn must be electrically insulated from all other coil turns for the inductor to operate properly, when the varnish or insulation on adjacent turns breaks down, these turns can short together. This is more difficult to detect with an ohmmeter test. The key is to know what R value the given coil should read. If there are shorted turns, the R value will read lower than normal. (NOTE: If only one or two turns are shorted, it may be impossible to detect by simple ohmmeter measurements!)

Obviously, the more turns that are shorted, the greater is the difference between the normal value and that which is measured. It is also possible in iron-core inductors to have turns shorting to the core, or being shorted by the core.

Another possibility is the whole inductor is being shorted from one end to the other. The reading will then be 0 Ω. This condition, however, is less apt to happen than simply having shorted turns. Many times, an inductor with shorted turns will give the technician a clue by the burned insulation smell or visual effects shown by an extremely overheated component.

PRACTICAL NOTES

When selecting inductors for replacement or new designs, be sure to notice the important parameter ratings. Four important ratings you should note are:

1. the inductance value;

2. the current rating;

3. the dc resistance of the inductor; and

4. any voltage and insulation ratings or limitations related to the inductor's use.

SUMMARY

▶ Inductance or self-inductance is the property of a circuit that opposes any *change* in current flow. Any change in current creates moving magnetic flux lines that, when linked (or cutting conductor[s]), induces an emf opposing the current change that initially produced it (Lenz's law).

▶ An inductor is a component that has the property of inductance. Any current-carrying conductor possesses some amount of inductance. However, most inductors are created by forming a conductor into loops of wire wound around some sort of structure. The amount of inductance created depends on the number of turns, the cross-sectional area, the length of the coil, and the permeability (magnetic characteristics) of the core material.

▶ Inductors come in a variety of physical and electrical sizes. Iron-core inductors typically have inductances in the henrys ranges. Powdered iron-core and variable ferrite-core inductors have values in the millihenrys range. Air-core inductors are typically in the microhenrys range.

▶ The amount of back-emf induced in an inductor relates to the amount of inductance involved and the rate of current change. That is,

$$V_L = L \frac{di}{dt}$$

where L = inductance in henrys

$$\frac{di}{dt} = \frac{\text{the rate of current change}}{\text{in amperes per second}}$$

▶ Inductance value is calculated when the inductor's physical characteristics and the core's magnetic properties (permeability) are known. That is,

$$L = (12.57 \times 10^{-7}) \times \frac{(\mu_r N^2 A)}{\text{length}}$$

where 12.57×10^{-7} is the absolute permeability of air, μ_r is the relative permeability of the core, N is the number of turns, and A is the core's cross-sectional area in square meters.

▶ Total inductance of inductors in series (with no mutual coupling) is calculated by adding the individual inductance values. That is, $L_T = L_1 + L_2 \ldots + L_n$ This method is similar to finding total resistance of series resistors.

▶ Total inductance of inductors in parallel (with no mutual coupling) is calculated in a similar fashion to that used when calculating equivalent resistance of parallel resistors. That is, either the reciprocal formula or the product-over-the-sum formula may be used. These methods are

$$\frac{1}{L_T} = \frac{1}{L_1} + \frac{1}{L_2} \ldots + \frac{1}{L_n}$$

$$\text{and } L_T = \frac{L_1 \times L_2}{L_1 + L_2}$$

▶ Energy is stored in the inductor magnetic field when there is current flowing through the inductor. The amount of energy stored is calculated in joules. That is,

$$E = \frac{LI^2}{2} \left(\text{or } \frac{1}{2} LI^2\right)$$

where E is energy in joules, L is inductance in henrys, and I is current in amperes.

▶ Current through an inductor does not instantly rise to its maximum value. Neither does it instantly fall from the value it is at to the minimum value where it decays. Current takes five time constants to change from the level it is at to a new stable level. A time constant (τ) is equal to the inductance in henrys, divided by the R in ohms, or $\tau = L/R$. (NOTE: The L/R time for the increasing current condition may be different than the L/R time for the decaying current condition depending on the circuitry involved.)

▶ The universal time-constant chart is a graph of the exponential growth and decay of many things in nature. Looking at the curve of increasing values, after one time constant (τ), the quantity reaches 63.2% of maximum value. After two time constants, the quantity has reached 86.5% of maximum value and so on. After five time constants, the quantity has virtually reached 100% of the maximum value to be achieved. Looking at the decay curve, after one time constant, the value has fallen from 100% of maximum to 36.8% of maximum (or a decay of 63.2%). After two time constants, the quantity has fallen to 13.5% of the original maximum value, and so forth. After five time constants, the quantity has fallen to the minimum value involved.

▶ Natural logarithms and epsilon can be used to mathematically determine instantaneous values of dc transient circuit quantities, such as instantaneous current (*i*) in dc *RL* circuits.

For increasing current: $i = \dfrac{V_S}{R} \times (1 - \epsilon^{-Rt/L})$

For decaying current: $i = \dfrac{V_S}{R} \times \epsilon^{-Rt/L}$

▶ When an inductor's magnetic field rapidly collapses, a high voltage is induced. Therefore, in dc circuits, precautions should be taken to prevent abrupt interruptions of current through

large inductances. Also, circuit design should minimize the effects of such changes.

▶ Typical problems with inductors are opens or shorts. The ohmmeter is useful to check for these problems, provided the technician has knowledge of normal resistance values for the component being tested. Shorted turns are more difficult to find with ohmmeter checks.

FORMULAS AND SAMPLE CALCULATOR SEQUENCES

FORMULA 14–1 $V_{\text{ind}} = N \text{ (number of turns)} \times \dfrac{d\phi \text{ (Wb)}}{dt \text{ (s)}}$

number of turns, ⊗, (❘❘, number of webers cut, ➗, amount of time, ❘❙), ▱

FORMULA 14–2 $V_L = L \dfrac{di}{dt}$

inductance value, ⊗, (❘❘, change in current, ➗, amount of time, ❘❙), ▱

FORMULA 14–3 $L = \dfrac{V_L}{\dfrac{di}{dt}}$

V_L value, ➗, (❘❘, change in current, ➗, amount of time, ❘❙), ▱

FORMULA 14–4 $L = 12.57 \times 10^{-7} \times \dfrac{\mu_r N^2 A}{l}$

12.57, **EE** or **EXP**, **+/−**, 7, ⊗, (❘❘, μ_r value, ⊗, number of turns, **x^2**, ⊗, area of coil in square meters, ❘❙), ➗, length of coil in meters, ▱

FORMULA 14–5 $L = 12.57 \times 10^{-7} \times \dfrac{N^2 A}{l}$

12.57, **EE** or **EXP**, **+/−**, 7, ⊗, (❘❘, number of turns, **x^2**, ⊗, area of coil in square meters, ❘❙), ➗, length of coil in meters, ▱

FORMULA 14–6 $L_T = L_1 + L_2 \ldots + L_n$

L_1 value, ⊕, L_2 value, . . . , ▱

FORMULA 14–7 $\dfrac{1}{L_T} = \dfrac{1}{L_1} + \dfrac{1}{L_2} + \dfrac{1}{L_3} \ldots + \dfrac{1}{L_n}$

L_1 value, **1/x**, ⊕, L_2 value, **1/x**, ⊕, L_3 value, **1/x**, . . . , ▱

FORMULA 14–8 $L_T = \dfrac{1}{\dfrac{1}{L_1} + \dfrac{1}{L_2} + \dfrac{1}{L_3} \ldots + \dfrac{1}{L_n}}$

L_1 value, **1/x** , **+** , L_2 value, **1/x** , **+** , L_3 value, **1/x** , **=** , **1/x**

FORMULA 14–9 $L_T = \dfrac{L_1 \times L_2}{L_1 + L_2}$

L_1 value, **×** , L_2 value, **÷** , **(** , L_1 value, **+** , L_2 value, **)** , **=**

FORMULA 14–10 Energy $= 0.5\, LI^2$

0.5, **×** , **(** , L value, **×** , I value, **x²** , **)** , **=**

FORMULA 14–11 $\tau = \dfrac{L}{R}$

inductance value (henries), **÷** , resistance value (ohms), **=**

FORMULA 14–12 $i = \dfrac{V_S}{R} \times (1 - \epsilon^{-Rt/L})$

(NOTE: This is the formula for increasing current.)

R value, **×** , time value, **÷** , inductance value, **=** , **+/−** , **2nd** , **eˣ** , **+/−** , **+** , 1, **=** , **×** , V_S value, **÷** , R value, **=**

FORMULA 14–13 $i = \dfrac{V_S}{R} \times \epsilon^{-Rt/L}$

(NOTE: This is the formula for decreasing current.)

R value, **×** , time value, **÷** , inductance value, **=** , **+/−** , **2nd** , **eˣ** , **×** , V_s value, **÷** , R value, **=**

REVIEW QUESTIONS

1. Define inductance.

2. In what form is electrical energy stored by an inductor?

3. Define the four important physical elements that affect the amount of inductance.

4. Explain Lenz's law.

5. Explain Faraday's law.

6. State the formula for one time constant. Define terms.

7. The cause of "cemf" is:
   a. the basic property of resistance.
   b. the basic property of inductance.
   c. changing current and the property of inductance.
   d. changing current and the property of resistance.

8. Iron-core inductors have typical inductance ranges:
   a. in the microhenrys range.
   b. in the henrys range.
   c. in the millihenrys range.
   d. none of the above.

9. Powdered iron-core inductors have typical inductance ranges:
   a. in the microhenrys range.
   b. in the henrys range.
   c. in the millihenrys range.
   d. none of the above.

10. Air-core inductors have typical inductance ranges:
    a. in the microhenrys range.
    b. in the henrys range.
    c. in the millihenrys range.
    d. none of the above.

11. The amount of back-emf induced in an inductor is related to:
    a. the length of the coil and the amount of current through it.
    b. the inductance of the coil and the rate of current change.
    c. the inductance of the coil and the coil's wire resistance.
    d. none of the above.

12. In terms of inductors, permeability relates to:
    a. the size of wire used in the inductor.
    b. the size of the core used in the inductor.
    c. the material used in the inductor core.
    d. all of the above.

13. Total inductance of magnetically noncoupled series inductors is calculated the same as:
    a. resistances in series.
    b. resistances in parallel.
    c. resistances in series-parallel.
    d. none of the above.

14. Total inductance of magnetically noncoupled parallel inductors is calculated the same as:
    a. resistances in series.
    b. resistances in parallel.
    c. resistances in series-parallel.
    d. none of the above.

15. If the voltage applied to a dc RL circuit changes level:
    a. the circuit conditions will instantly be at the new level parameters.
    b. the circuit will take one L/R time constant to arrive at the new parameters.
    c. it will take more than one L/R time constant to arrive at the new parameters.
    d. the circuit will take 10 time constants to arrive at the new level parameters.

16. The most common problems found with inductors are:
    a. too few turns, and loose turns.
    b. open connections and shorted turns.
    c. core decay and flux leakage.
    d. none of the above.

17. What physical precaution should be taken when checking a large inductance with an ohmmeter?

18. What electrical precaution should be taken when checking any inductor with an ohmmeter?

19. When viewing the universal time-constant chart (see Figure 14–14, page 479), for a dc RL circuit, it can be said that there is a 63.2% change in current rise in the time interval between the second and third time constant. Specifically explain what is referred to by this 63.2% term.

20. Is the 63.2% concept true between each of the 5 L/R time constants it takes for complete change?

PROBLEMS

1. When the number of turns on an inductor that has an inductance of 100 mH doubles, while keeping its length, cross-sectional area, and core material constant, what is its new inductance value? (Express your answer in three ways: µH, mH, and H.)

2. Current through an inductor is changing at a rate of 100 mA/s and the induced voltage is 30 mV. What is the value of inductance?

3. What is the new inductance for a 100-µH air-core coil whose core is replaced with one having a relative permeability of 500?

4. How much energy is stored in the magnetic field of a 15-H inductor that is carrying current of 2 A?

5. A 100 mH, a 0.2 H, and a 1,000-μH inductor are connected in series. Assuming no mutual coupling exists between them, what is the circuit's total inductance? (Express your answer in mH.)

6. A 10-H and a 15-H inductor are connected in parallel. Assuming no mutual coupling exists between them, what is the resultant total inductance?

7. A series *RL* circuit consists of a 250-mH inductor and a resistor having a resistance of 100 Ω. What is the value of one time constant? How long will it take for current to change completely from one level to another?

8. Refer to the Universal Time Constant (UTC) chart, Figure 14–14, and determine what percentage of applied voltage appears across the resistor in a series *RL* circuit after four time constants. What percentage of the applied voltage is across the inductor at this same time?

9. A 500-turn inductor has a total inductance of 0.4 H. What is the inductance when connections are made at one end of the coil and at a tap point connection at the 250th turn?

10. How long will it take the voltage across the resistance in a series *RL* circuit consisting of 5 H of inductance and 10 Ω of resistance to achieve 86.5% of *V* applied? (Assume the inductor has negligible resistance.)

Perform the calculations for questions 11 and 12 using your calculator and the ⒠ˣ function, as appropriate.

11. Refer to Figure 14–19. What is the voltage across the inductor after the switch has been closed for 0.6 μs?

12. Refer to Figure 14–20. How much time will it take for the voltage across the resistor to reach 38 V? (Round off, as appropriate.)

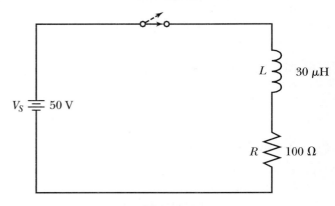

V_S ⎓ 50 V L 30 μH R 100 Ω

FIGURE 14–19

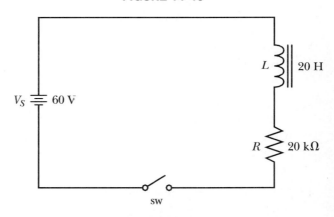

V_S ⎓ 60 V L 20 H R 20 kΩ sw

FIGURE 14–20

ANALYSIS QUESTIONS

1. If the iron core is removed from an inductor so that it becomes an air-core inductor, will the coil's inductance increase, decrease, or remain the same?

2. When the current through an inductor has reached its steady-state condition and the current through the coil remains the same, is there an induced back-emf?

3. When additional inductors are connected in series with a circuit's existing inductors, will the total inductance of the circuit increase, decrease, or remain the same? (Assume no mutual coupling exists between inductors.)

4. When additional inductors are connected in parallel with a circuit's existing inductors, will the total inductance of the circuit increase,

decrease, or remain the same? (Assume no mutual coupling exists between inductors.)

5. If a 1,000-turn inductor has an inductance of 4 H, how many additional turns does it take to increase the inductance to 6 H?

6. Draw a graph of the V_R and V_L voltages with respect to V applied for an inductor circuit consisting of a 10-H inductor, a series 10-Ω resistance, and a source voltage of 20 V. Identify the specific voltages across L and R at the 1τ, 2τ, 3τ, and 4τ points on your graph.

7. Explain why a 1,000 turn coil, consisting of wire that has a varnish insulation with a voltage breakdown rating of only 100 V, can safely handle an applied voltage across the coil at 2,000 V.

8. What is the difference between the current-limiting effects of an inductor's resistance and its property of inductance?

9. Explain why the joules of energy stored by an inductor is directly related to the amount of inductance the inductor has.

10. Explain why there is a direct relationship between the current through a coil and the amount of energy stored by that coil.

PERFORMANCE PROJECTS CORRELATION CHART

Suggested performance projects in the Laboratory Manual that correlate with topics in this chapter are:

CHAPTER TOPIC	PERFORMANCE PROJECT	PROJECT NUMBER
Self-Inductance The L/R Time Constant	L Opposing Change of Current	40
Inductors in Series and Parallel	Total Inductance in Series and Parallel	41

NOTE: It is suggested that after completing the above projects, the student should be required to answer the questions in the "Summary" at the end of this section of projects in the Laboratory Manual.

KEY TERMS

Inductive reactance
Leading in phase
Phase angle
Q **(figure of merit)**

INDUCTIVE REACTANCE IN AC

15

chapter

CHAPTER PREVIEW

In the previous chapter, we learned that an inductor opposes a current change but provides very little opposition to a steady (dc) current. The dc winding resistance is the only opposition to dc current; whereas the opposition to the current change is because an inductor reacts to changing current by producing a back-emf. You saw that energy is stored in the inductor's magnetic field during current buildup and returned to the circuit during current decay. If the inductor were "perfect" (without dc wire resistance or other losses), it would return all the stored energy to the circuit, dissipate no power, and have no effect on dc circuit parameters. Practical inductors have some resistance and dissipate some power, even in dc circuits. However, these losses are not caused by the property of inductance.

Inductors react to ac current, which is constantly varying. Also, they produce an opposition to ac current known as inductive reactance, the reactance that will be studied in this chapter.

Important concepts discussed earlier were the inductance value and rate of current change relationships to the induced voltage appearing across the coil. In this chapter, we will relate these concepts to the ac circuit and explore what happens when inductance is present in a sine-wave ac environment. We will show how these characteristics translate into how inductance influences parameters in ac circuits.

OBJECTIVES

After studying this chapter, you should be able to:

1. Illustrate *V-I* relationships for a purely resistive ac circuit

2. Illustrate *V-I* relationships for a purely inductive ac circuit

3. Explain the concept of **inductive reactance**

4. Write and explain the formula for inductive reactance

5. Use Ohm's law to solve for X_L

6. Use the X_L formula to solve for inductive reactance at different frequencies and with various inductance values

7. Use the X_L formula to solve for unknown *L* or *f* values

8. Determine X_L, I_L, and V_L values for series and parallel-connected inductances

9. Use the SIMPLER troubleshooting sequence to solve the Chapter Troubleshooting Challenge problem

15–1 V AND I RELATIONSHIPS IN A PURELY RESISTIVE AC CIRCUIT

For a quick review, observe Figure 15–1. Note in a purely resistive circuit, the voltage across a resistor is in-phase with the current through it. Both the voltage and current pass through the zero points and reach the maximum points of the same polarity at the same time.

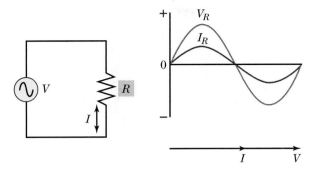

FIGURE 15–1 *V* and *I* relationships in a purely resistive ac circuit

15-2 V AND I RELATIONSHIPS IN A PURELY INDUCTIVE AC CIRCUIT

For purposes of explanation, let's assume the inductor is an ideal inductor, having only inductance and no resistance. What phase relationship will the current through the coil and the voltage across the coil have with respect to each other? Examine Figure 15–2 as you study the following explanations and comments.

1. In the circuit shown, the inductor is connected across the source output terminals. Therefore, recalling the rule about voltage in parallel circuits, you know the impressed voltage across the coil and the source applied voltage are the same.

2. In the circuit shown, there is only one current path. Therefore, recalling the rule about current in series circuits, you know the current through all circuit parts is the same. That is, the circuit current (I) and the current through the inductor (I_L) are the same.

3. Recall the induced voltage (cemf) across the inductor is maximum when the rate of current change is maximum and minimum when the rate of current change is minimum, Figure 15–3.

4. Also, remember for a sine wave the maximum rate of change occurs as the waveform crosses the zero points. The minimum rate of change occurs at the maximum positive and negative peaks of the sine wave, Figure 15–3.

5. The implication is that maximum induced V occurs across the inductor when the current sine wave is either increasing from (or through) zero in the positive direction or decreasing from (or through) zero in the negative direction. See Figure 15–3 again.

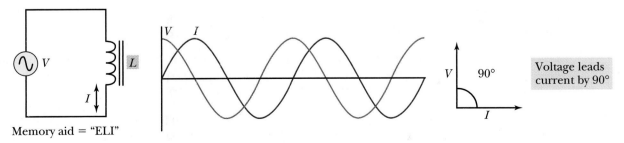

Memory aid = "ELI"

FIGURE 15–2 *V* and *I* relationships in a purely inductive ac circuit

FIGURE 15–3 Inductor voltage related to rate of change of current

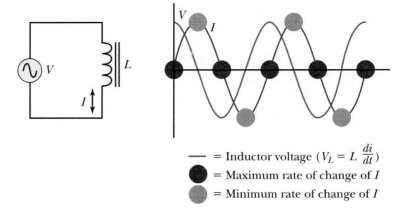

— = Inductor voltage ($V_L = L \frac{di}{dt}$)

● = Maximum rate of change of I

● = Minimum rate of change of I

6. The *amplitude* of the induced voltage is determined by the rate of current change and the value of L.

7. Two factors determine the *polarity* of the induced voltage:
   a. the direction of the current flow through the coil, and
   b. whether the current is trying to increase or decrease.

8. Because the induced voltage opposes any current change, when the current tries to increase, the cemf will oppose it. When the current tries to decrease, the cemf will again oppose it (Lenz's law).

9. Figures 15–2 and 15–3 illustrate the results of these actions and reactions. In essence, because the induced back-emf is maximum when the rate of current change is maximum and minimum when the rate of current change is minimum, the current lags behind the impressed voltage. The time lag, or delay of current behind voltage, has obviously been caused by the inductance's opposition to changes in current flow. The current and voltage are out of phase. The 90° **phase angle** is due to the fact that at the moment when the current increases or decreases, the inductor instantly develops the cemf, which opposes the change in current.

10. Since the V and I are one-fourth cycle apart, they are 90° out of phase with each other. A complete cycle represents 360° or 2π radians, where a radian equals 57.3°, or the angle described by an arc on the circle's circumference that is equal in length to the radius.

11. Using our rule of thumb from a previous chapter, you can see the voltage is at maximum positive value before the current reaches maximum positive value. Hence, the voltage is **leading** the current by 90°. Incidentally, it is also accurate to say current is lagging voltage by 90°. Recall the reference point concept—John is taller than Bill, or Bill is shorter than John.

At this point we will introduce a helpful phrase that will serve as a memory aid and that will help throughout your studies in ac and in your career. The understanding of the complete expression will come only after you have studied capacitors and inductors. However, the phrase can also be helpful to you now.

Here's the phrase: **Eli the ice man.** It's a short saying, so you should not have trouble memorizing it, particularly after you understand the implications!

You have just learned in a sine-wave ac circuit environment, voltage across an inductor leads the current through the inductor by 90° (in a pure inductance).

The word "Eli" helps you remember that voltage leads current since the E (emf or voltage) is before the I (current). Obviously the "L" indicates we are talking about an inductor. (You have probably already surmised that "ice" in this special phrase refers to current *[I]* leading voltage *[E]* in a capacitor *[C]*, but we'll discuss this in a later chapter.)

PRACTICAL NOTES

Although the memory aid can be helpful to some, it is more important that you learn and understand the reasons for the electrical quantity relationships. As you are learning in this chapter, current lags voltage in an inductor due to the opposition an inductor shows to a change in current. You will be learning in an upcoming chapter that the reason current leads voltage across a capacitor is that a capacitor opposes a change in voltage. The details of why this is true will become apparent as you continue studying these subjects. We simply want to indicate that the memory aid should not replace your understanding of these matters.

15-3 CONCEPT OF INDUCTIVE REACTANCE

Because an inductor reacts to the constantly changing current in an ac circuit by producing an emf that opposes current changes, inductance displays opposition to ac current, Figure 15–4. Recall the inductor did not show opposition to dc current, except for the small resistance of the wire. This lack of opposition is because current must be changing to cause a changing flux that induces the back-emf we've been discussing. In dc the current is not changing, except for the temporary switch-on and switch-off conditions discussed earlier.

This special opposition an inductor displays to ac current is **inductive reactance.** The symbol "X" denotes reactance. Because it is inductive reactance, the correct abbreviation or symbol is X_L. Because X_L is an opposition to current, the *unit of measure for inductive reactance (X_L) is the ohm!*

15-4 RELATIONSHIP OF X_L TO INDUCTANCE VALUE

We have already stated that the amount of opposition inductance shows to ac current is directly related to the amount of back-emf induced. For a given rate

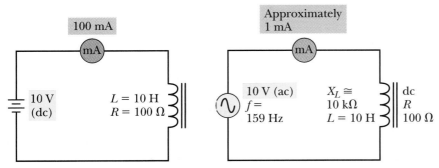

With ac applied, current is much less than with dc applied to same circuit. The additional opposition to current in the ac circuit is due to X_L of the inductor.

FIGURE 15–4 Concept of inductive reactance to ac

FIGURE 15–5 Relationship of X_L to L

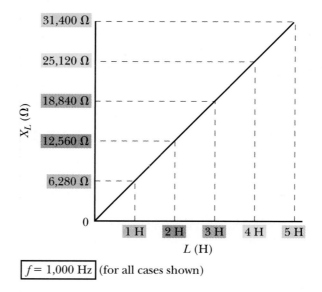

$f = 1{,}000$ Hz (for all cases shown)

of current change, the amount of back-emf induced is directly related to the value of coil inductance (i.e., the greater the L, the greater the back-emf produced). Therefore, the higher the value of L, the higher will be the inductive reactance (X_L) value. In other words, if the L value doubles, the number of ohms of X_L doubles; if the L in the circuit decreases to half its original value, X_L decreases to half, and so on, Figure 15–5.

Example

If a purely inductive circuit has a 5-H inductor with a second 5-H inductor connected in series, what happens to the circuit current? (Assume all other circuit parameters remain the same.)

    Answer: Current decreases to half the original value, since the total opposition to current (X_{L_T}) has doubled when inductance was doubled.

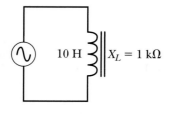

FIGURE 15–6

PRACTICE PROBLEMS I

1. For the circuit in Figure 15–6, what is the value of total X_L when two more equal value inductors are added in series with the existing inductor, thus making $L_T = 30$ H? (Assume all other circuit conditions remain the same.)

2. If the three 10-H inductors in question 1 are parallel-connected rather than series-connected, what is the total X_L?

15–5 RELATIONSHIP OF X_L TO FREQUENCY OF AC

Rate of change related to frequency

It has been previously established that the amount of back-emf produced in an inductor relates to both the value of inductance and to the rate of current change. As you know, the faster the rate of change, the greater the induced emf produced. And the greater the back-emf, the higher the value of opposition or X_L.

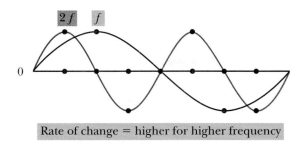

FIGURE 15–7 Comparison of rate of change for two frequencies

Let's take a look at how changing the frequency of an ac signal affects this rate of change. Refer to Figure 15–7 as you study the following discussion.

1. Only one cycle is shown for the lower of the two frequency signals displayed. Two cycles of the higher frequency signal are shown.

2. The two signals have the same maximum (peak) amplitude. You can see the higher frequency signal must have a higher rate of change of current to reach its maximum amplitude level in half the time. Carrying this thought throughout the whole cycle, it is apparent that the higher frequency signal has twice the rate of change compared to the lower frequency signal.

3. Because the back-emf is directly related to the rate of change, when frequency doubles (which doubles the rate of change), the inductive reactance also doubles. When frequency decreases, rate of current change decreases and, consequently, the inductive reactance decreases proportionately, Figure 15–8.

Rate of change related to angular velocity

In an earlier chapter, a rotating vector to describe a sine wave was introduced. At this point, we want to indicate a relationship exists between frequency and the angular velocity of the quantity represented by a rotating vector.

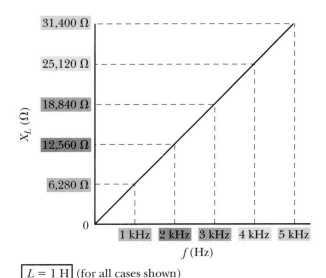

$L = 1\ \text{H}$ (for all cases shown)

FIGURE 15–8 Relationship of X_L to f

Angular velocity is the speed at which a vector rotates about its axis. The symbol representing angular velocity is the lowercase Greek letter omega, ω. Thus, the relationship of angular velocity to frequency is ω equals $2\pi f$ (radians per second). (Recall that 2π radians equal $360°$, or one complete cycle.) We introduce angular velocity and its mathematical representation at this point to acquaint you with the expression "$2\pi f$," which is used in formulas dealing with ac quantities. You will see this expression as you continue in your studies and in your work as a technician.

IN-PROCESS LEARNING CHECK I

Fill in the blanks as appropriate.

1. When current increases through an inductor, the cemf (helps, hinders) _____ the current increase.

2. When current decreases through an inductor, the cemf (helps, hinders) _____ the current decrease.

3. In a pure inductor, the _____ leads the _____ by $90°$.

4. What memory aid helps in recalling the relationship described in question 3? _____

5. The opposition that an inductor shows to ac is termed _____ _____.

6. The opposition that an inductor shows to ac (increases, decreases, remains the same) _____ as inductance increases.

7. The opposition that an inductor shows to ac (increases, decreases, remains the same) _____ as frequency decreases.

8. X_L is (directly, inversely) _____ related to inductance value.

9. X_L is (directly, inversely) _____ related to frequency.

15–6 METHODS TO CALCULATE X_L

The X_L formula

Two common methods are used to calculate the value of X_L. One method is to use Ohm's law. The other method is the inductive reactance formula,

FORMULA 15–1 $\qquad X_L = 2\pi fL$

where X_L = inductive reactance in ohms
$\qquad 2\pi$ = 6.28 (approximately)
$\qquad f$ = frequency in hertz
$\qquad L$ = inductance in henrys

You can see from the formula and previous discussions that inductive reactance is directly related to both frequency and inductance. That is, when *either* frequency *or* inductance increases, inductive reactance increases. When *either* frequency *or* inductance decreases, X_L decreases. That is, when one factor doubles while the other one halves, the reactance stays the same, and so on. You get the idea!

The X_L formula can also be transposed to solve for L or f, if the other two factors are known.

FORMULA 15–2 $L = \dfrac{X_L}{2\pi f}$

That is, $L = \dfrac{X_L}{6.28\,f}$ henrys

and

FORMULA 15–3 $f = \dfrac{X_L}{2\pi L}$

That is, $f = \dfrac{X_L}{6.28\,L}$

Example

What is the inductive reactance of an iron-core choke having 5 H of inductance that is in a 60-Hz ac circuit?

$$X_L = 2\pi f L = 6.28 \times 60 \times 5 = 1{,}884\ \Omega$$

Example

What is the inductance of an inductor that offers 2 kΩ of inductive reactance at a frequency of 1.5 MHz?

$$L = \frac{X_L}{2\pi f} = \frac{2 \times 10^3}{6.28 \times (1.5 \times 10^6)} = 212\ \mu\text{H}$$

Example

What frequency of operation would cause a 30-μH inductor to produce 4 kΩ of inductive reactance?

$$f = \frac{X_L}{2\pi L} = \frac{4 \times 10^3}{6.28 \times (30 \times 10^{-6})} = 21.23\ \text{MHz}$$

PRACTICE PROBLEMS II

1. What is the X_L of a 10-mH inductor that is operating at a frequency of 100 kHz?

2. What is the inductance in a purely inductive circuit, if the frequency of the applied voltage is 400 Hz and the inductive reactance is 200 Ω?

3. What frequency causes a 100-mH inductor to exhibit 150 Ω of inductive reactance?

Using the Ohm's law approach

The ac current through an inductor is also calculated by using the Ohm's law expression: $I_L = V_L/X_L$. We can transpose this formula to find any one of the three factors, provided we know the other two factors. (We have previously done this using Ohm's law expressions.) This means that:

$$X_L = \frac{V_L}{I_L}$$

$$V_L = I_L X_L$$

$$I_L = \frac{V_L}{X_L}$$

PRACTICAL NOTES

When using Ohm's law, as indicated, it is common to use the (rms) or effective values of voltage and current in your calculations, unless otherwise indicated!

PRACTICE PROBLEMS III

Use Ohm's law and answer the following.

1. What is the inductive reactance of an inductor whose voltage measures 25 V and through which an ac current of 5 mA is passing?

2. What is the voltage across a 2-H inductor passing 3 mA of current? Assume the frequency of operation is 400 Hz. (Hint: Calculate X_L using the reactance formula first.)

3. What is the current through an inductor having 250 Ω of inductive reactance if the voltage across the inductor measures 10 V?

15-7 INDUCTIVE REACTANCES IN SERIES AND PARALLEL

You are aware that inductances in series are additive just like series resistances are additive. Likewise, you have already learned that inductive reactance, or X_L, is directly related to the amount of inductance. Therefore, the ohmic values of inductive reactances of series inductors add like the ohmic values of series resistances add. In other words, $X_{L_T} = X_{L_1} + X_{L_2} \ldots + X_{L_n}$, Figure 15–9.

Inductance values in parallel are also treated in the same manner that parallel resistance values are treated. That is, the reciprocal or product-over-the-sum approaches are used to find the total or equivalent value(s). Because inductive reactance is directly related to inductance value, total X_L is again correlated to the total inductance of parallel-connected inductors like it was to the total inductance of series-connected inductors, Figure 15–10.

Useful formulas to find the total inductive reactance of parallel-connected inductors are:

FORMULA 15–4 $\quad \dfrac{1}{X_{L_T}} = \dfrac{1}{X_{L_1}} + \dfrac{1}{X_{L_2}} \ldots + \dfrac{1}{X_{L_n}}$

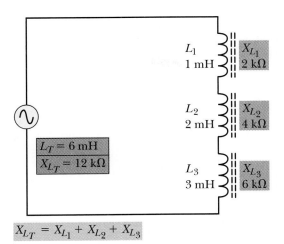

FIGURE 15–9 Inductive reactances in series

L_1
1 mH
X_{L_1}
2 kΩ

L_2
2 mH
X_{L_2}
4 kΩ

L_3
3 mH
X_{L_3}
6 kΩ

$L_T = 6\text{ mH}$
$X_{L_T} = 12\text{ k}\Omega$

$$X_{L_T} = X_{L_1} + X_{L_2} + X_{L_3}$$

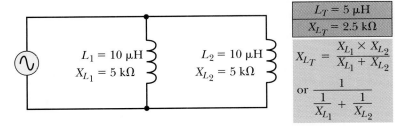

FIGURE 15–10 Inductive reactances in parallel

$L_1 = 10\ \mu\text{H}$
$X_{L_1} = 5\text{ k}\Omega$

$L_2 = 10\ \mu\text{H}$
$X_{L_2} = 5\text{ k}\Omega$

$L_T = 5\ \mu\text{H}$
$X_{L_T} = 2.5\text{ k}\Omega$

$$X_{L_T} = \frac{X_{L_1} \times X_{L_2}}{X_{L_1} + X_{L_2}}$$

or $\dfrac{1}{\dfrac{1}{X_{L_1}} + \dfrac{1}{X_{L_2}}}$

or

FORMULA 15–5 $\quad X_{L_T} = \dfrac{1}{\dfrac{1}{X_{L_1}} + \dfrac{1}{X_{L_2}}}$, etc.

PRACTICAL NOTES

Once again, the reciprocal ( ) function on your calculator is found to be very useful. You can solve for total inductive reactance of parallel-connected inductive reactances using the same techniques you have previously used to solve for total resistance and total inductance of parallel-connected inductances. (NOTE: Assume no coupling between inductances.)

For two inductors in parallel, you can use:

FORMULA 15–6 $\quad X_{L_T} = \dfrac{X_{L_1} \times X_{L_2}}{X_{L_1} + X_{L_2}}$

Where appropriate, you can use all the "shortcuts" you have previously learned about parallel circuits—two 25-ohm reactances in parallel yield a total equivalent reactance of 12.5 ohms, or half of either reactor's value, and so forth.

15–8 FINAL COMMENTS ABOUT INDUCTORS AND INDUCTIVE REACTANCE

Collecting what you now know about inductance and reactance, the following comments can be made:

1. The inductive reactance of an inductor limits or opposes the ac current flow through the inductor. That is, given an inductor in a circuit with a specified value of dc applied voltage, if that same inductor is put into a circuit with the same ac voltage (rms value), the current will be less in the ac circuit than in the dc circuit. This situation is due to the opposition of the inductive reactance to ac current, Figure 15–4.

2. Connecting inductances in series causes total inductance to increase, total inductive reactance to increase, and current to decrease (assuming a constant V applied), Figure 15–11.

3. Series-connected inductors may act as a series, inductive ac voltage divider with the greatest voltage across the largest inductance and the smallest voltage across the smallest inductance, and so on. NOTE: In a series circuit containing *nothing but inductors,* the total voltage equals the sum of the individual voltages across the inductors. The current through the inductors is 90° out of phase with the voltage across each inductor. Each of the inductor's voltages is in phase with the other inductors' voltages, thus are added to obtain the total voltage, Figure 15–12.

4. Connecting inductances in parallel causes total inductance to decrease, total inductive reactance to decrease, and current to increase (assuming a constant V applied), Figure 15–13.

5. Parallel-connected inductors can act as a parallel inductive ac current divider with the highest current through the lowest inductance and the lowest current through the highest inductance and so on. NOTE: In a parallel cir-

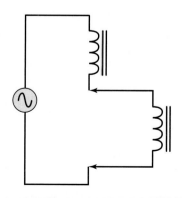

FIGURE 15–11 Effects of connecting inductances in series

Connecting inductances in series causes:
L_T ↑
X_{L_T} ↑
I_T ↓

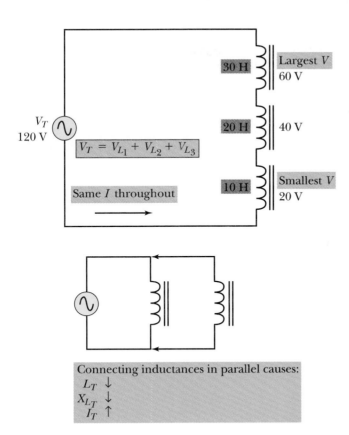

FIGURE 15–12 Voltage division across series-connected inductors

V_T 120 V

30 H — Largest V 60 V

20 H — 40 V

10 H — Smallest V 20 V

$$V_T = V_{L_1} + V_{L_2} + V_{L_3}$$

Same I throughout

FIGURE 15–13 Effects of connecting inductors in parallel

Connecting inductances in parallel causes:
L_T ↓
X_{L_T} ↓
I_T ↑

cuit containing *nothing but inductors,* the total current equals the summation of the individual inductor branch currents. Again, there is a 90° phase difference between the current through each inductor and the voltage across that inductor. But because all branch currents are lagging the circuit voltage by 90°, the branch currents are in phase with each other, thus are added to find total current, Figure 15–14.

15–9 INDUCTOR Q FACTOR

We have already implied that a perfect inductor would have zero resistance, and that the only opposition this inductor would show to ac current flow would be the inductive reactance. It has also been stated that every coil has some resistance in the wire making up the coil. In many cases, this resistance is so small that it has negligible effect on the circuit operation. In other cases, this is not true.

$$I_T = I_1 + I_2 + I_3$$

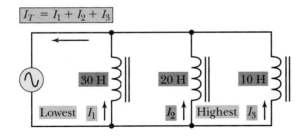

30 H — Lowest I_1

20 H — I_2

10 H — Highest I_3

FIGURE 15–14 Current division through parallel-connected inductors

In fact, under certain conditions it may have considerable effect. Therefore, inductors sometimes have a **Q** rating, which is a ratio describing the energy the inductor stores in its field compared to the energy it dissipates due to I^2R losses.

In essence, the ratio of the X_L the inductor shows (at a given frequency) to the R it has, depicts this ratio. That is:

FORMULA 15–7 $\qquad Q = \dfrac{X_L}{R}$

Example

The Q of a coil having 100-mH inductance and 50-Ω resistance that is operating at a frequency of 5 kHz would be:

$$Q = \frac{X_L}{R}$$

$$X_L = 6.28 \times 5 \times 10^3 \times 100 \times 10^{-3} = 3{,}140 \ \Omega$$

$$R = 50 \ \Omega$$

$$Q = \frac{X_L}{R} = \frac{3{,}140}{50} = 62.8$$

(Q has no units because it is a ratio of quantities with like units.)

The preceding discussion indicates that the lower the R that a given inductor (L) has, the higher is its **figure of merit** or **Q**. (Also, the higher the frequency of operation, the greater the X_L and the higher the Q for a given inductor.)

It is generally desirable to use high Q coils, particularly at radio frequencies. A coil is considered to be high Q if its Q value is greater than approximately 20 to 25. Of course, this is an arbitrary value. It does, however, give some indication of the inductance-to-resistance ratio of the inductor. Therefore, the value is the ratio of energy stored to energy dissipated or the X_L/R ratio we have been examining.

SUMMARY

▶ Inductive reactance is the opposition that inductance displays to ac (or pulsating) current.

▶ Inductive reactance is associated with the amount of back-emf generated by the inductor, which is associated with the value of inductance (L) and the rate of change of current.

▶ Inductive reactance is measured in ohms. The symbol for inductive reactance is X_L. X_L is calculated by the formula $X_L = 2\pi fL$, where X_L is in ohms, f is in hertz, and L is in henrys. The 2π is the constant equal to two times pi or approximately 6.28.

▶ Voltage leads current by 90° in a purely inductive circuit. (Voltage and current are in phase for a purely resistive circuit.)

▶ The first part of the memory-aid phrase helps you remember the phase relationship of voltage and current relative to inductors. The phrase is *Eli the ice man*. The word *Eli* indicates that E comes before I in an inductor (L).

▶ Inductive reactance is *directly proportional* to both inductance *(L) and* to frequency *(f)*.

▶ Quality (or the figure of merit) of an inductor is sometimes described as its Q. This factor equates the ratio of the inductor's X_L to its R. Thus, $Q = X_L/R$.

▶ Inductive reactances in series add just as resistances in series add. That is, $X_{L_T} = X_{L_1} + X_{L_2} \ldots + X_{L_n}$.

▶ Total (equivalent) inductive reactance of inductances in parallel is computed using similar methods to those used to find total or equivalent resistance of resistors in parallel. That is, the reciprocal formula or the product-over-the-sum formula is used.

The reciprocal formula(s) include:

$$\frac{1}{X_{L_T}} = \frac{1}{X_{L_1}} + \frac{1}{X_{L_2}} + \ldots \frac{1}{X_{L_n}} \text{ or } X_{L_T} = \frac{1}{\frac{1}{X_{L_1}} + \frac{1}{X_{L_2}}}$$

The product-over-the-sum formula is:

$$X_{L_T} = \frac{X_{L_1} \times X_{L_2}}{X_{L_1} + X_{L_2}}$$

FORMULAS AND SAMPLE CALCULATOR SEQUENCES

FORMULA 15–1 $X_L = 2\pi f L$

6.28, ⊗, frequency value, ⊗, inductance value, ⊜

FORMULA 15–2 $L = \dfrac{X_L}{2\pi f}$

X_L value, ÷, (, 6.28, ⊗, frequency value,), ⊜

FORMULA 15–3 $f = \dfrac{X_L}{2\pi L}$

X_L value, ÷, (, 6.28, ⊗, inductance value,), ⊜

FORMULA 15–4 $\dfrac{1}{X_{L_T}} = \dfrac{1}{X_{L_1}} + \dfrac{1}{X_{L_2}} \ldots + \dfrac{1}{X_{L_n}}$

X_{L_1} value, 1/x, +, X_{L_2} value, 1/x, . . . , ⊜

FORMULA 15–5 $X_{L_T} = \dfrac{1}{\dfrac{1}{X_{L_1}} + \dfrac{1}{X_{L_2}}}$, etc.

X_{L_1} value, 1/x, +, X_{L_2} value, 1/x, +, . . . , ⊜, 1/x

FORMULA 15–6 $X_{L_T} = \dfrac{X_{L_1} \times X_{L_2}}{X_{L_1} + X_{L_2}}$

X_{L_1} value, ⊗, X_{L_2} value, ÷, (, X_{L_1} value, +, X_{L_2} value,), ⊜

FORMULA 15–7 $Q = \dfrac{X_L}{R}$

X_L value, ⊕ , R value, ⊜

REVIEW QUESTIONS

1. Inductive reactance is the opposition that an inductor shows:
   a. to direct current.
   b. to dc and ac.
   c. to alternating current.
   d. all of the above.

2. The higher the back-emf produced by an inductor:
   a. the higher its X_L.
   b. the lower its X_L.
   c. the higher its resistance.
   d. back-emf has no relationship to either resistance or reactance.

3. X_L is:
   a. directly related to only frequency.
   b. directly related to only inductance.
   c. directly related to both frequency and inductance.
   d. none of the above.

4. Inductive reactance is indicated in:
   a. siemens.
   b. ohms.
   c. ampere turns.
   d. impedance.

5. In a purely inductive circuit:
   a. voltage lags current by 90°.
   b. current lags voltage by 90°.
   c. voltage and current are in phase.
   d. none of the above.

6. Total inductive reactance for series inductors:
   a. is calculated similarly to determining R_{equiv} of parallel resistors.
   b. is calculated similarly to determining total inductance of series inductors.
   c. is calculated using the reciprocal formula.
   d. none of the above.

7. The Q of an inductor is equal to:
   a. the ratio of its turns to its length.
   b. the ratio of its length to its turns.
   c. the ratio of its resistance to its inductive reactance.
   d. the ratio of its inductive reactance to its resistance.

8. The characteristic of an inductor that causes voltage to lead current is:
   a. its resistance and its related impedance.
   b. its inductance and its related impedance.
   c. its inductance and its related reactance.
   d. all of the above.

9. In a series circuit composed of only inductors:
   a. V_T equals the difference between the maximum and minimum V drops.
   b. V_T equals the sum of all the individual voltage drops.
   c. V_T equals the sum of the maximum and minimum V drops.
   d. none of the above.

10. In a parallel circuit composed of only inductors:
   a. total current is the same throughout each inductor.
   b. total current is the sum of all the branch currents.
   c. both (a) and (b) above.
   d. none of the above.

PROBLEMS

1. What is the inductance of an inductor that exhibits 376.8 Ω reactance at a frequency of 100 Hz? What is inductive reactance at a frequency of 450 Hz?

2. What is the X_L of a 5-H inductor at 120 Hz? What is the value of X_L at a frequency of 600 Hz?

3. What is the total inductance of the circuit in Figure 15–15? When the voltage applied is 100 V

and the current through L_1 is 100 mA, what is the frequency of the applied voltage? (Use one decimal place in your calculations.)

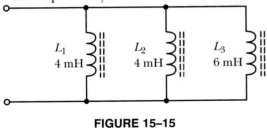

FIGURE 15–15

4. Find L_T for the circuit in Figure 15–16. At approximately what frequency does X_{L_T} equal 5,000 ohms?

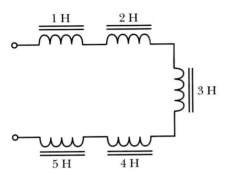

FIGURE 15–16

5. Find the Q of the coil in the circuit shown in Figure 15–17. When the frequency triples and the R of the inductor doubles, does Q increase, decrease, or remain the same? If the applied voltage were dc instead of ac, would Q change? Explain.

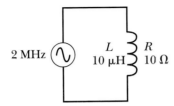

FIGURE 15–17

6. Draw a waveform showing two cycles that illustrates the circuit V and I for the circuit in Figure 15–18. Appropriately label the peak values of V and I, assuming the source voltage is shown in rms value.

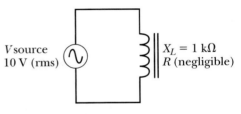

FIGURE 15–18

7. Find V_T, V_{L_1}, and I_T for the circuit in Figure 15–19.

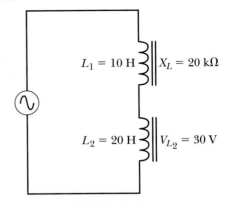

FIGURE 15–19

8. What is the branch current through L_2 for the circuit in Figure 15–20?

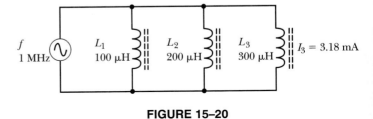

FIGURE 15–20

9. What is the value of the source voltage in the circuit of Figure 15–20?

10. Determine the total inductance of the circuit in Figure 15–21.

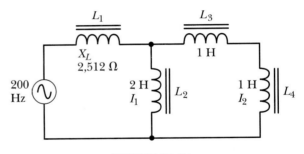

FIGURE 15–21

Refer again to Figure 15–21 and answer questions 11 and 12 with "I" for increase, "D" for decrease, and "RTS" for remain the same.

11. If frequency doubles and V applied stays the same,
    L_T will _____.
    X_{L_T} will _____.
    I_T will _____.
    V_{L_4} will _____.
    I_1 will _____.

12. If L_2 changes to a 1-H inductor,
    L_T will _____.
    X_{L_1} will _____.
    I_T will _____.
    V_{L_1} will _____.
    f will _____.

13. What value of inductor will produce 2,000 ohms of X_L at a frequency of 1 kHz?

14. Refer to the circuit in Figure 15–22. Assuming that $L_1 = 10$ H, $L_2 = 15$ H, and $L_3 = 20$ H, what is th total inductance of the circuit?

15. For the circuit of Figure 15–22, what is the total inductive reactance value?

16. For the circuit of Figure 15–22, what is:
    a. the circuit total current?
    b. the current through each inductor?

Referring to the circuit of Figure 15–22, answer questions 17–20 with increase, decrease, or remain the same, as appropriate for each question.

17. If L_2 were removed from the circuit, would the circuit total current increase, decrease, or remain the same?

18. If the source frequency were increased, would the current through L_3 increase, decrease, or remain the same?

19. What would have to happen to the circuit applied voltage to keep the circuit current constant if the frequency of the source were decreased to one-third its original value? Be specific.

20. If the three inductors were rewired to be in series, rather than in parallel, what would the new circuit total current be? (Assume f and V source remained the same.)

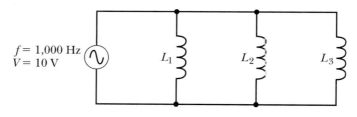

$f = 1,000$ Hz
$V = 10$ V

L_1 L_2 L_3

FIGURE 15–22

ANALYSIS QUESTIONS

1. If a circuit's parameters are changed so that frequency triples and the amount of inductance doubles, how many times greater will the new X_L be compared with the original value?

2. Assume circuit conditions change for a given circuit so that frequency is halved and inductance is tripled. What is the relationship of the new X_L to the original value?

3. In a given circuit containing inductance, when frequency decreases, will circuit current increase, decrease, or remain the same?

4. When the frequency of applied voltage for a given inductive circuit doubles and the circuit voltage increases by three times, will the circuit current increase, decrease, or remain the same?

5. When the frequency of applied to a given inductor decreases, will the Q of the coil change? If so, in what way?

6. Does the R of a given inductor change when frequency is changed?

7. Explain why you think, in most cases, it is best to have an inductor with a high Q factor.

8. For a theoretically "perfect" inductor:
   a. current and voltage are in phase.
   b. current and voltage are 45° out of phase.
   c. current and voltage are 90° out of phase.
   d. current and voltage are 180° out of phase.

9. When current through an inductor is increasing, the induced voltage in the inductor:
   a. aids the increase in current.
   b. opposes the increase in current.
   c. has no effect on the increase in current.
   d. none of the above.

PERFORMANCE PROJECTS CORRELATION CHART

Suggested performance projects in the Laboratory Manual that correlate with topics in this chapter are:

CHAPTER TOPIC	PERFORMANCE PROJECT	PROJECT NUMBER
Concept of Inductive Reactance	Induced Voltage	42
Relationship of X_L to Inductance Value	Relationship of X_L to L and Frequency	43
Relationship of X_L to Frequency of AC		
Methods to Calculate X_L	The X_L Formula	44

NOTE: It is suggested that after completing the above projects, the student should be required to answer the questions in the "Summary" at the end of this section of projects in the Laboratory Manual.

THE SIMPLER SEQUENCE FOR TROUBLESHOOTING

Follow the seven-step SIMPLER troubleshooting sequence, as outlined below, and see if you can find the problem with this circuit. As you follow the sequence, record your circuit test results for each testing step on a separate sheet of paper. This will aid you and your instructor to see where your thinking is on track or where it might deviate from the best procedure.

step 1 Symptoms

—(Gather, verify, and analyze symptom information) To begin this process, read the data under the "Starting Point Information" heading. Particularly look for circuit parameters that are not normal for the circuit configuration and component values given in the schematic diagram. For example, look for currents that are too high or too low; voltages that are too high or too low; resistances that are too high or too low, etc. This analysis should give you a first clue (or symptom information) that will aid you in determining possible areas or components in the circuit that might be causing this symptom.

step 2 Identify

—(Identify and bracket the initial suspect area) To perform this first "bracketing" step, analyze the clue or symptom information from the "Symptoms" step, then either on paper or in your mind, bracket, circle, or put parentheses around all of the circuit area that contains any components, wires, etc., that might cause the abnormality that your symptom information has pointed out. NOTE: Don't bracket parts of the circuit that contain items that could not cause the symptom!

step 3 Make

—(Make a decision about the first test you should make: what type and where) Look under the TEST column and determine which of the available tests shown in that column you think would give you the most meaningful information about the suspect area or components, in the light of the information you have so far.

step 4 Perform

—(Perform the first test you have selected from the TEST column list) To simulate performing the chosen test (as if you were getting test results in an actual circuit of this type), follow the dotted line to the right from your selected test to the number in parentheses. The number in parentheses, under the "Results in Appendix C" column tells you what number to look up in Appendix C to see what the test result is for the test you are simulating.

step 5 Locate

—(Locate and define a new "narrower" area of uncertainty) With the information you have gained from the test, you should be able to eliminate some of the circuit area or circuit components as still being suspect. In essence, you should be able to move one bracket of the bracketed area so that there is a new smaller area of uncertainty.

step 6 Examine

—(Examine available information and determine the next test: what type and where) Use the information you have thus far to determine what your next test should be in the new narrower area of uncertainty. Determine the type test and where, then proceed using the TEST listing to select that test, and the numbers in parentheses and Appendix C data to find the result of that test.

step 7 Repeat

—(Repeat the analysis and testing steps until you find the trouble) When you have determined what you would change or do to restore the circuit to normal operation, you have arrived at *your* solution to the problem. You can check your final result against ours by observing the pictorial step-by-step sample solution on the pages immediately following the "Chapter Troubleshooting Challenge"—circuit and test listing page.

step 8 Verify

NOTE: A useful *8th Step* is to operate and test the circuit or system in which you have made the "correction changes" to see if it is operating properly. This is the final proof that you have done a good job of troubleshooting.

CHALLENGE CIRCUIT 7

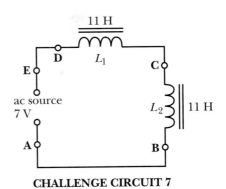

CHALLENGE CIRCUIT 7

Starting Point Information

1. Circuit diagram
2. Each inductor is rated as 11 H, 400 Ω dc resistance
3. The ac voltage measured across L_1 is significantly lower than the voltage measured across L_2

TEST	Results in Appendix C
$V_{A\text{-}B}$ . . . .(66)
$V_{B\text{-}C}$ . . . .(31)
$V_{C\text{-}D}$ . . . .(4)
$V_{D\text{-}E}$ . . . .(82)
$V_{E\text{-}A}$ . . . .(43)
$R_{B\text{-}C}$ . . . .(17)
$R_{C\text{-}D}$ . . . .(114)

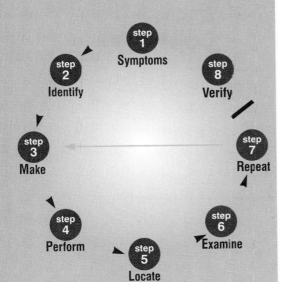

Symptoms

The series circuit inductors are rated at the same value, and in a series circuit the current through both inductors should be the same. However, there is a significant difference in the $I \times X_L$ drops across the two inductors. This implies one inductor has a problem.

Identify

initial suspect area: Both inductors until further checks are made.

Make

test decision: It is less likely that an inductor would increase rather than decrease in inductance (due to shorted turns or change in magnetic path). So let's start by examining the inductor with the lowest voltage drop (L_1). Our assumption is L_2 has remained at the rated value of L, while L_1 may have decreased. This would fit the symptom information regarding L_1's voltage drop being low.

Perform

1st Test: Check the dc resistance of L_1 versus its rated value to see if there might be a hint of shorted turns. (NOTE: Source is removed from the circuit.) R is 297 Ω and the rated value is 400 Ω, indicating the possibility of some shorted turns.

Locate

new suspect area: L_1 is now a strong suspect area. But we still need to check L_2's parameters to compare and verify our suspicions about L_1.

Examine

Examine the available data.

Repeat

analysis and testing: A resistance check of L_2 might be necessary.
2nd Test: Check R_{B-C}. R_{B-C} is 401 Ω, which agrees with the rating. Replace L_1 with the appropriately rated inductor. Then check the voltage across L_1.
3rd Test: V_{C-D} is now close to 3.5 V, which is half of the source voltage. This is normal for this circuit with equal inductor values.
4th Test: Check the voltage across L_2 to verify that it is approximately equal to the voltage across L_1.

Symptoms

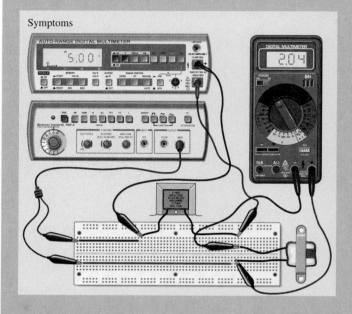

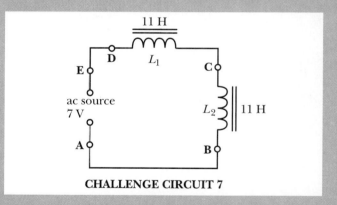

CHALLENGE CIRCUIT 7

Verify

When checked, the result is also about 3.5 V. The trouble is L_1 has shorted turns. Replacing L_1 restores the circuit to normal.

1st Test

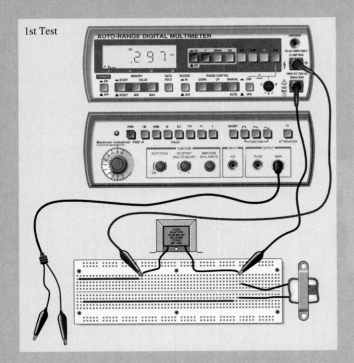

2nd Test

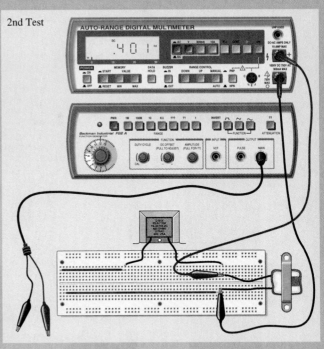

3rd Test

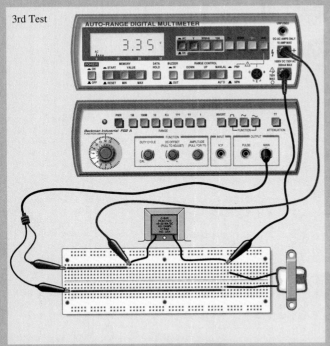

4th Test

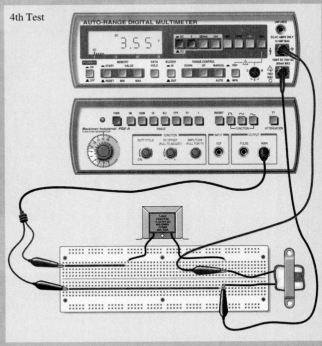

KEY TERMS

Cosine function
Impedance
Phase angle (θ)
Pythagorean theorem
Scalar
Sine function
Tangent function
Vector

RL CIRCUITS IN AC

16

chapter

CHAPTER PREVIEW

In the preceding chapters, you have studied the circuit characteristics of both dc and ac circuits containing only inductance. You have learned basic concepts related to inductors, such as inherent opposition to current changes, the self-induced back-emf, phase relationships of V and I, and other characteristics of inductance.

The purpose of this chapter is to introduce you to basic facts and circuit analysis techniques used to examine ac circuits containing both resistance and inductance (RL circuits). Impedance, a combination of resistance and reactance, will be discussed and analyzed. Further considerations about **phase angle** will be given. Primary methods of using **vectors** and other simple math techniques to analyze RL circuit parameters will also be introduced. Later in the text, you will study vector analysis techniques that involve polar and rectangular coordinate systems, and methods of expressing electrical quantities using these systems. For this chapter, however, we simply want to introduce you to basic concepts related to analyzing RL circuits in ac.

The concepts discussed in this chapter are important because they provide valuable knowledge for all your future ac circuit analyses endeavors. This chapter begins a series of chapters in which you will be performing calculator operations such as square roots, squares, trigonometric functions, and so on. We want to remind you that different models of scientific calculators may use slightly different keystrokes for certain operations. As you get involved in this chapter, study and/or refer back to the "Special Notations Regarding Scientific Calculators" chart in Appendix D, as required.

OBJECTIVES

After studying this chapter, you should be able to:

1. Use **vectors** to determine magnitude and direction
2. Determine circuit impedance using the **Pythagorean theorem**
3. Determine V_T and I_T using the Pythagorean theorem
4. Determine ac circuit parameters using trigonometry
5. Calculate ac electrical parameters for series RL circuits
6. Calculate ac electrical parameters for parallel RL circuits
7. List at least three practical applications of inductive circuits
8. Use the SIMPLER troubleshooting sequence to solve the Chapter Troubleshooting Challenge problem

16–1 REVIEW OF SIMPLE R AND L CIRCUITS

Simple R circuit

The ac current through a resistor and the ac voltage across a resistor are in phase, as you have already learned. You know this may be shown by a waveform display or phasor diagram, Figure 16–1.

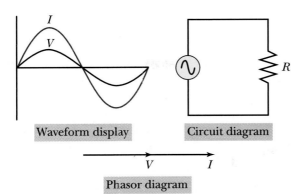

FIGURE 16–1 *V* and *I* in a simple *R* circuit

In the purely resistive *series* ac circuit, Figure 16–2a, you simply add the individual series resistances to find total circuit opposition to current. Or you add voltage drops across resistors to find total circuit voltage, since all voltage drops are in phase with each other.

In purely resistive *parallel* circuits, Figure 16–2b, total resistance is found by using the product-over-the-sum or reciprocal methods. The total circuit current is found by adding the branch currents, because all branch currents are in phase with each other.

Simple L circuit

In a purely inductive ac circuit, you know current through the inductor and voltage across the inductor are 90° out of phase with each other. Again, this is illustrated by a waveform display, Figure 16–3a or a phasor diagram, Figure 16–3b.

You know that in purely inductive series circuits, total inductance is the sum of the individual inductances; total inductive reactance (X_{L_T}) is the sum of all X_Ls; and total voltage equals the sum of all V_Ls, since each V_L is in phase with the other V_Ls although they are not in phase with the current, Figure 16–4a. In purely inductive parallel circuits, total circuit opposition to current (X_{L_T}) is found by using reciprocal or product-over-the-sum approaches. Also, total circuit current

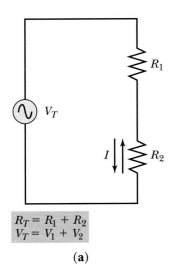

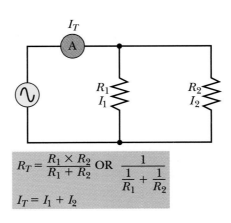

FIGURE 16–2 Purely resistive circuits

FIGURE 16–3 *V* and *I* in a simple *L* circuit

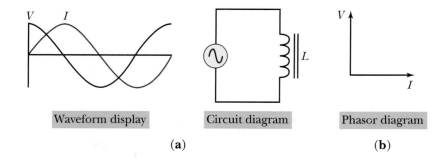

Waveform display   Circuit diagram   Phasor diagram

(a)   (b)

FIGURE 16–4 Purely inductive circuits

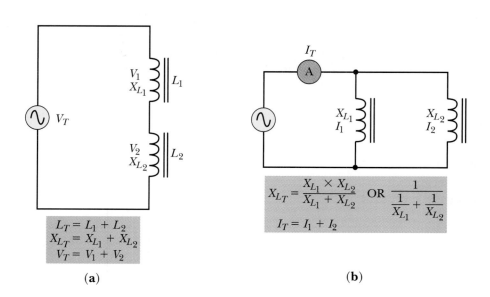

(a)   (b)

equals the sum of the branch currents, since branch currents are in phase with each other although not in phase with the circuit voltage, Figure 16–4b.

16–2 USING VECTORS TO DESCRIBE AND DETERMINE MAGNITUDE AND DIRECTION

Background

When ac circuits contain reactances and resistances, their voltage(s) and current(s) are not in-phase with each other. In this case, we cannot add these quantities together to find resultant totals as we did with dc circuits or ac circuits that are purely resistive or reactive. Thus, other methods are used to analyze ac circuit parameters.

One of these methods is to represent circuit parameters using **vectors.** A *vector quantity expresses both magnitude and direction.* Examples of quantities having both magnitude and direction that are illustrated using vectors include mechanical forces, the force of gravity, wind velocity and direction, and magnitudes and relative angles of electrical voltages and currents in ac.

Scalars, on the other hand, define quantities that exhibit only magnitude. Units of measure, such as the henry or the ohm, are scalar quantities. Even dc circuit parameters, or instantaneous values of ac electrical quantities are treated

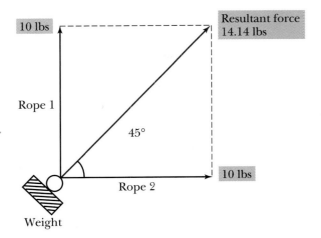

FIGURE 16–5 Vectors show both magnitude and direction.

as scalar quantities, since there are no time differences or angles involved. Simple addition is used to add scalar quantities, but this is not true for vector quantities.

Plotting and measuring vectors

Vectors can be plotted to scale to determine *both* magnitude and direction of given quantities, Figure 16–5. If a force of 10 pounds is exerted on Rope 1 and a force of 10 pounds is exerted on Rope 2, what is the total force exerted on the weight, and in which direction is this total force? The resultant force is 14.14 pounds in a direction that is 45° removed from either rope's direction.

Look at Figure 16–6. Notice these vectors are plotted to scale so you can then *measure* both the magnitude and direction of the resultant force. *Measure the length* of the resultant vector *to find magnitude,* and *measure the angle* between the original forces and the resultant force by a protractor *to find direction.*

Observe Figure 16–7a, which illustrates how vectors are laid end-to-end or head-to-tail to determine resultant values. Also, notice in Figure 16–7b that a parallelogram is drawn that projects the same magnitudes as the end-to-end approach. Although either method can be used to find resultant values, we will use the parallelogram approach most frequently.

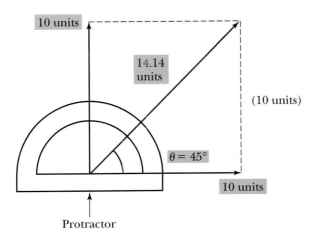

FIGURE 16–6 Plotting vectors to scale

FIGURE 16–7 Methods of
formatting vectors

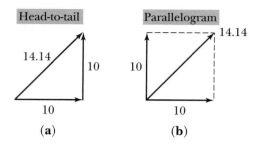

Later in the chapter you will see how this plotting and measuring technique
is used to analyze *RL* circuit parameters. Although it is not always convenient to
have graph paper and protractor handy, or to draw circuit parameters to scale
when analyzing a circuit, the plotting approach can be used. Also, it is a useful
tool to help introduce you to vectors.

Even when not plotted to scale for the direct measurement method, vectors
are useful to illustrate ac circuit parameters and show relative magnitudes and
angles. Sketching this diagram helps you understand the situation and verify
your computations.

PRACTICAL NOTES

Even when not precisely measuring vectors to determine resultant values, get in the habit
of approximating the *relative* lengths and directions of vectors when you draw them to
solve ac problems. This allows you to visually determine if your computations are making sense! A good practice is to draw simple vector diagrams whenever you are analyzing
ac circuits—they help visualize the situation.

PRACTICE PROBLEMS I

Assume the forces on the ropes in Figure 16–5 are 3 pounds for Rope 1
and 4 pounds for Rope 2. What are the magnitude and direction of the
resultant force on the weight? (NOTE: Use the plotting-to-scale approach and remember to keep things relative to the scale you use.)

IN-PROCESS LEARNING CHECK I

1. In a purely resistive ac circuit, the circuit voltage and current are _____
   phase.

2. In a purely inductive ac circuit, the current through the inductance and
   the voltage across the inductance are _____ degrees out of phase. In this
   case, the _____ leads the _____.

3. A quantity expressing both magnitude and direction is a _____ quantity.

4. The length of the vector expresses the _____ of a quantity.

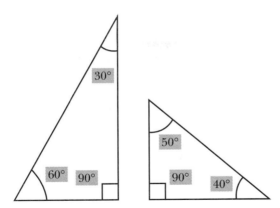

FIGURE 16–8 Examples of right triangles

16-3 INTRODUCTION TO BASIC AC CIRCUIT ANALYSIS TECHNIQUES

Most ac circuit analyses solve for quantities that are illustrated with vectors in right-triangle form. Recall from our previous discussions that a right triangle has one angle forming 90° and two angles that total another 90°, Figure 16–8. Many electrical parameters have this 90° phase difference between quantities. For example, in perfect inductors and capacitors, the voltages and currents are 90° out-of-phase. Also, ac circuit resistive and reactive voltage drops or currents are often represented as 90° out of phase with each other.

We will discuss two basic techniques used to analyze these right-triangle problems. The first method is the **Pythagorean theorem,** which is useful to determine the *magnitude* of quantities. The other method briefly addressed in this chapter is trigonometric functions that determine the circuit phase angles. Also, trigonometry can be used to find the magnitude of various electrical quantities, if the phase angle is known.

PRACTICAL NOTES

NOTE: The calculations are easy if you have a commonly used scientific calculator with square, square root, and "trig" function keys. You probably own or presently have access to such a calculator. If not, you might want to gain access to one and learn to use these functions as you proceed through the remainder of the chapter. If this device is not available, paper and pencil will work just fine! NOTE: *Be aware* that some calculators only accept radian measure! Therefore, to convert degrees to number of radians, divide the number of degrees by 57.3; and to convert radians to number of degrees, multiply the number of radians by 57.3. Recall that 2π radians equal 360°; that is, 6.28 times 57.3° equals 360°.

Pythagorean theorem

The Pythagorean theorem states the square of the hypotenuse of a right triangle equals the sum of the squares of the other two sides.

FORMULA 16–1 $c^2 = a^2 + b^2$

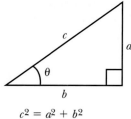

$$c^2 = a^2 + b^2$$
$$\sqrt{c^2} = \sqrt{a^2 + b^2}$$
$$c = \sqrt{a^2 + b^2}$$

FIGURE 16–9 Pythagorean theorem

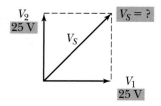

FIGURE 16–10

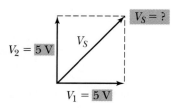

FIGURE 16–11

Refer to Figure 16–9 and review the equations related to the right triangle and Pythagorean theorem.

Now, let's transfer this information into finding the resultant value, or magnitude, of two voltages that are displaced from each other by 90° and vectorially added.

Example

Refer to Figure 16–10 and observe the following steps: In the diagram shown,

1. V_S represents the hypotenuse of the right triangle.
2. V_1 represents one of the other two sides of the right triangle.
3. V_2 represents the remaining side of the right triangle.
4. Transferring this information into the Pythagorean formula:

$$V_S = \sqrt{V_1^2 + V_2^2}$$
$$V_S = \sqrt{25^2 + 25^2}$$
$$V_S = \sqrt{1,250} = 35.35 \text{ V}$$

PRACTICE PROBLEMS II
Refer to Figure 16–11. Use the Pythagorean theorem and solve for the magnitude of side V_S.

Trigonometric functions

Trigonometric functions result from special relationships existing between the sides and angles of right triangles. "Trig" is a common shortened term for trigonometry or trigonometric. We will only discuss three basic trig functions used for your study of ac circuits—sine, cosine, and tangent functions.

Notice the side opposite (opp), the side adjacent (adj), and the hypotenuse (hyp) in Figure 16–12. As you can see, the relationship of each side to the "angle of interest," θ, determines whether it is called the opposite or the adjacent side. The hypotenuse is *always* the longest side of the triangle.

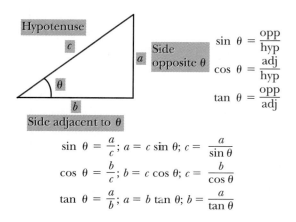

$$\sin \theta = \frac{\text{opp}}{\text{hyp}}$$
$$\cos \theta = \frac{\text{adj}}{\text{hyp}}$$
$$\tan \theta = \frac{\text{opp}}{\text{adj}}$$

$$\sin \theta = \frac{a}{c}; \, a = c \sin \theta; \, c = \frac{a}{\sin \theta}$$
$$\cos \theta = \frac{b}{c}; \, b = c \cos \theta; \, c = \frac{b}{\cos \theta}$$
$$\tan \theta = \frac{a}{b}; \, a = b \tan \theta; \, b = \frac{a}{\tan \theta}$$

FIGURE 16–12 Labeling sides of a right triangle

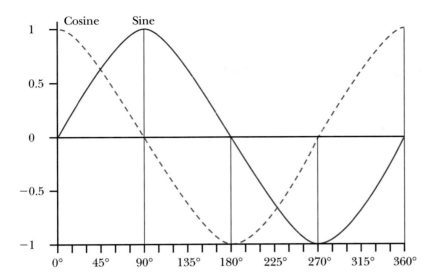

FIGURE 16–13 Sine and cosine waves

Refer to Figure 16–12 to review the basic trig functions.

If we assume a rotating vector with a unit length of one starts at 0° and rotates counterclockwise to the 90° position, the values of sine, cosine, and tangent vary within limits. That is:

sin θ increases from 0 to 1.0 (with angles from 0° to 90°)

cos θ decreases from 1.0 to 0 (with angles from 0° to 90°)

tan θ increases from 0 to ∞ (with angles from 0° to 90°)

See if you can relate this to the sine and cosine waves in Figure 16–13. Assume the maximum (peak) value is one volt for each waveform. Relate the zero points and the maximum points to the angles. Do they compare?

Figure 16–14 carries this idea one step further. Notice how these functions vary in the four quadrants of the coordinate system we discussed in an earlier chapter. Again, try relating the sine and cosine waves in Figure 16–13 to the statements in Figure 16–14.

Example

What is the described angle if the side opposite is 2 units and the hypotenuse is 2.828 units?

QUADRANT	SIN θ		COS θ		TAN θ	
	FROM	TO	FROM	TO	FROM	TO
I (0° to 90°)	0	1.0	1.0	0	0	∞
II (90° to 180°)	1.0	0	0	−1.0	−∞	0
III (180° to 270°)	0	−1.0	−1.0	0	0	∞
IV (270° to 360°)	−1.0	0	0	1.0	−∞	0

FIGURE 16–14 Variations in values of trig functions

FIGURE 16–15 Series *RL* circuit voltages

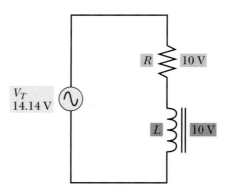

Referring to the trig formulas for sine, cosine, and tangent, the trig formula containing the side opposite and the hypotenuse is the formula for the sine of the angle. Therefore,

$$\sin \theta = \frac{2}{2.828} = 0.707$$

The $\sin^{-1}$ of $0.0707 = 45°$.
Now try the following practice problems.

PRACTICE PROBLEMS III

1. What is the described angle if the side adjacent has a magnitude of 5 units and the hypotenuse has a magnitude of 10 units?

2. What is the tangent of 45°?

3. What is the sine of 30°?

16–4 FUNDAMENTAL ANALYSIS OF SERIES RL CIRCUITS

Now we will take the knowledge you have acquired and apply it to some practical situations and problems, so you can see the benefit of this knowledge!

Refer to Figure 16–15. Notice if you added the voltages across the series resistor and the inductor to find total applied voltage, you would get 20 V. This answer is wrong! The actual applied voltage is 14.14 V.

Let's see why 20 V is the wrong value! Recall in series circuits, the current is the same through all components. Again referring to Figure 16–15 and from our previous discussions, we know the voltage across the resistor is in-phase with the current. We also know the voltage across the inductor leads the current by about 90°, which means the voltages across the resistor and the inductor are *not* in-phase with each other. Since the current through both is the same current, one voltage is in-phase with the current (V_R), and one voltage is not in-phase with the current. Therefore, we cannot add the voltages across R and L to find total voltage, as we would do in a simple dc circuit. Rather, we must use a special method of addition called vector addition to find the correct answer. The voltages must be added "vectorially," not "algebraically."

Let's try the two methods previously discussed to find total voltage in series *RL* circuits.

Using the Pythagorean theorem to analyze voltage in series RL circuits

Figure 16–16 illustrates how V_R, V_L, and V_T are typically represented in a right triangle by phasors or vectors.

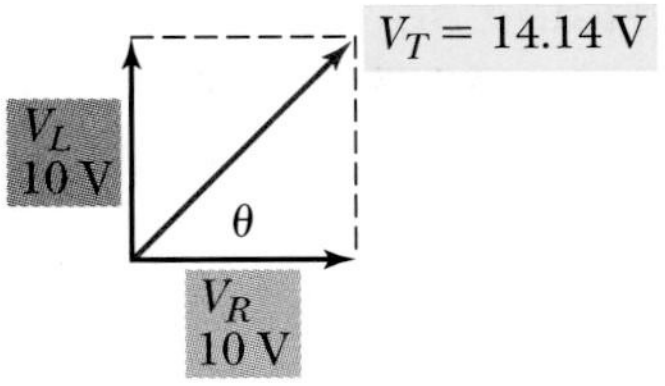

$V_T = 14.14$ V

V_L 10 V

θ

V_R 10 V

FIGURE 16–16 Voltage vector diagram

Example

Let's transfer the Pythagorean theorem formula in Figure 16–9, $c = \sqrt{a^2 + b^2}$, to this situation. As you can see, the longest side is V_T. This is equivalent to the c in the original Pythagorean formula. The other two quantities (V_R and V_L) are the other quantities in the formula. Thus, the formula to solve for V_T becomes:

FORMULA 16–2 $V_T = \sqrt{V_R^2 + V_L^2}$

Therefore, $V_T = \sqrt{10^2 + 10^2} = \sqrt{200} = 14.14$ volts.

Example

Assume the voltage values in Figure 16–15 change so that V_R is 30 volts and V_L is 40 volts. What is the new V applied value?
Answer: The V_T is 50 volts, Figure 16–17.

PRACTICE PROBLEMS IV
In a simple series *RL* circuit, V_R is 120 V and V_L is 90 V. What is the V applied value? SUGGESTION: Use your calculator.

You can see that using the Pythagorean theorem, it is easy to find the magnitude of total voltage in a series *RL* circuit—simply substitute the appropriate values in the formula. We are not able to find the resultant angular direction by this method. However, the trig functions help us.

Using trig to solve for angle information in series RL circuits

Look at Figure 16–18 as you study the following statements.

1. To find the desired angle information using trig functions, identify the sides of the right triangle in terms of the circuit parameters of interest. Note in Figure

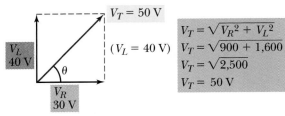

$V_T = 50$ V

($V_L = 40$ V) $V_T = \sqrt{V_R^2 + V_L^2}$
$V_T = \sqrt{900 + 1{,}600}$
$V_T = \sqrt{2{,}500}$
$V_T = 50$ V

V_L 40 V

θ

V_R 30 V

(NOTE: V_T is *not* equal to the sum of V_R and V_L. It is greater than either one alone, but not equal to their sum.)

FIGURE 16–17

FIGURE 16–18 Using trig to find angle information

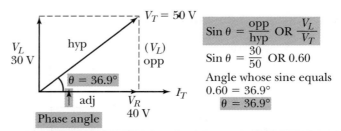

NOTE: Phase angle is considered the angular difference in phase between circuit total voltage and circuit total current. Get in the habit of looking at the sketch to see if the answer is logical. It is obvious that the angle in this case must be less than 45° because the vertical leg is shorter than the horizontal leg. The answer looks and seems logical!

16–18 that V_L is in the upward ($+90°$) direction, and V_R is at the 0° reference direction. In *series* circuits, the current is the same throughout the circuit and is in phase with the resistive voltage. Therefore, I is also on the 0° axis.

PRACTICAL NOTES

For series circuits since current is common to all parts, *current is commonly the reference phasor* and is plotted on the reference phasor 0° axis.

Because these directions are important, let's restate them! They are general rules of thumb that you will use from now on! V_L is vertical (up), V_R is horizontal to the right, and I is in-phase with V_R.

2. Referring to the values in Figure 16–18, let's find the *circuit phase angle* that is defined as *the phase angle between circuit applied voltage and circuit total current.*
   a. Consider the angle between V_T and I_T the angle of interest. Thus, the side opposite to this angle is the V_L value and the hypotenuse is V_T.
   b. Use the information relating these two sides in the sine function formula to find the angle. (Recall that sin θ = opp/hyp.) Therefore, in our example, sin $\theta = 30/50 = 0.60$.
   c. Using a calculator, or computer, (or trig table) to find the angle having a sine function value of 0.60, we find that it is 36.9°.

PRACTICAL NOTES

In ac circuits containing both resistance and reactance, the phase difference in degrees between applied voltage and total current must fall between 0° (indicating a purely resistive circuit) and 90° (indicating a purely reactive circuit). Since both resistance and reactance are present, the circuit cannot act purely resistive or purely reactive.

   d. In this example, since we know all sides, we can also use the other two trig functions. That is,

$$\cos \theta = \frac{\text{adj}}{\text{hyp}} = \frac{V_R}{V_T} = \frac{40}{50} = 0.8$$

The angle with a cosine value of 0.8 is 36.9°.

$$\tan \theta = \frac{\text{opp}}{\text{adj}} = \frac{V_L}{V_R} = \frac{30}{40} = 0.75$$

The angle with a tangent value of 0.75 is 36.9°.

PRACTICAL NOTES

Rather than having to say in words "the angle having a tangent value equal to . . . ," it is common to use the mathematical term "arc tangent," or "arctan." In other words, if I find the arctan of 0.75, it is an angle of 36.9°. You will also see arctan written as "$\tan^{-1}$." To find the arctan on a calculator, input the ratio of the opposite side to the adjacent side. That is: (opposite side value, ÷ , adjacent side value, =), then press the **2nd** and **TAN⁻¹** function keys (in that order). You will then see the $\tan^{-1}$ (arctan, or angle value) displayed in the readout.

Naturally, you can find the arc sine or arc cosine using the same technique. For the sine, the values to input are the opposite side divided by the hypotenuse. For the cosine, the values to input are the adjacent side divided by the hypotenuse. Once you have the sine or cosine value, use the **2nd** and **SIN⁻¹** function keys or the **2nd** and **COS⁻¹** function keys as appropriate to find the arc sine or arc cosine, respectively. (Note: This gives you the angle value.)

---

e. Use the sine, cosine, or tangent functions to solve for the unknown angle depending on the information available. Obviously, when you know the values for the opposite side and the hypotenuse, use the sine function. When you know the values for side adjacent and the hypotenuse, use the cosine function. When you know only the values for side opposite the angle of interest and the side adjacent to the angle of interest, use the tangent function.

Stated another way, select and use the formula containing the two known values and the value you are solving for.

PRACTICE PROBLEMS V

1. Draw the vector diagram for the voltages in Figure 16–19.
2. Use trig and solve for the phase angle of the circuit in Figure 16–19.
3. Which trig function did you use in this case?
4. If the voltages had been reversed (i.e., $V_R = 20$ V and $V_L = 30$ V), would the phase angle have been greater, smaller, or the same?

Using trig to solve for voltages in series RL circuits

You have seen how trig helps find angle information. You have also learned the Pythagorean theorem is useful to find magnitude or voltage value in series *RL* circuits. Can trig also be used to find magnitude? Yes, it can! Let's see how!

FIGURE 16–19

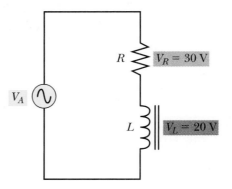

Example

If the angle information is known, determine the sine, cosine, or tangent values for that angle. Next, use the appropriate, available sides information, and find the value of the unknown side. Look at Figure 16–20 and let's see how to do this.

1. Start by making sure the formula for the trig function used includes the unknown side of interest, plus a known side. This allows us to solve for the unknown by using equation techniques.

2. To solve for V_L, use the sine function since the sin θ uses the opposite side divided by the hypotenuse. V_L is the unknown side of interest and is the side opposite the known angle. Since we know the hypotenuse value, we have our starting point.

3. Sin θ = opp/hyp, thus sin 60° = $V_L/100$. Now we determine the sine of 60° and substitute in the equation. Use the trig table, calculator, or computer and find the sine of 60° is 0.866. (NOTE: Round-off to 0.87.) Put this in the formula, thus 0.87 = $V_L/100$. Transpose and solve: V_L = 100 × 0.87 = 87 V.

4. To solve for V_R, use either the cosine or tangent functions with the information now known. Cosine uses the adjacent divided by the hypotenuse and tangent uses the opposite divided by the adjacent. In either case, a known and the desired unknown (the adjacent) values appear in the formulas.

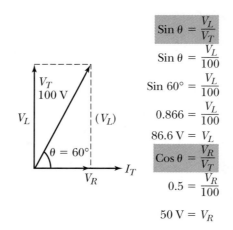

FIGURE 16–20

5. Assuming we do not know the V_L value, we use the cosine function. Cos θ = adj/hyp, thus cos 60° = $V_R/100$. Next, determine the cosine of 60°, which equals 0.5. Thus, substitute in the formula, 0.5 = $V_R/100$. Transpose and solve: $V_R = 100 \times 0.5 = 50$ V.

Now, let's see if these results make sense with respect to our vector diagram in Figure 16–20. Since the angle is greater than 45°, we know V_L must be greater than V_R and it is!

Now we can use the tangent function to check our results, since it uses the opposite divided by the adjacent sides. Insert the values found for V_L and V_R into the tan θ = opp/adj formula to find that 86.6/50 = 1.732. The angle whose tangent is 1.732 is 60°. It checks out! Now, you try some problems.

PRACTICE PROBLEMS VI

1. A series *RL* circuit has one resistor and one inductor. What is the V_R value if the applied voltage is 50 V and the circuit phase angle is 45°?

2. A series *RL* circuit comprised of a single resistor and inductor has an applied voltage value of 65 V and the circuit phase angle is 60°.
   a. What is the value of V_R?
   b. What is the value of V_L?

3. A series *RL* circuit comprised of a single resistor and inductor has values where $V_L = 36$ V and $V_R = 24$ V.
   a. What is the value of the angle between V_R and V_T?
   b. What is the value of V_T?

Analyzing impedance in series RL circuits

You have learned how to analyze various voltages in series *RL* circuits by measuring vectors, using the Pythagorean theorem, and applying basic trig functions. Now, let's look at another important ac circuit consideration, impedance.

Impedance is the total opposition a circuit offers to ac current flow at a given frequency. This means impedance is the combined opposition of all **resistances and reactances** in the circuit. Since impedance is sensitive to frequency, there must be some reactance—*pure* resistance is not sensitive to frequency. The symbol for impedance is "Z," and the unit of impedance is the ohm.

Before discussing impedance, let's review some important facts you already know, Figure 16–21.

1. The voltage drop across the resistor is in-phase with the circuit current through the resistor, and V_R equals *IR*. Note that this is a series circuit. Since current is the common factor in series circuits, the current is the reference vector and is at the 0° position of the coordinate system. Also V_R is at 0° since it is in phase with *I*.

2. The voltage drop across the inductor leads the circuit current by 90° (assuming a perfect inductor with no resistance), and V_L equals IX_L. Since V_L leads *I* by 90°, the IX_L vector is leading the *I* vector by 90°.

3. As you know, total circuit voltage is the *vector resultant* of the individual series voltages. Since the circuit is not purely resistive, V_T cannot be at 0°. Since the

FIGURE 16–21 Series *RL* circuit quantities

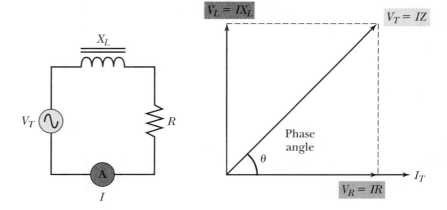

circuit is not purely inductive, V_T cannot be at 90°. Remember in circuits containing both resistance and reactance, the resultant circuit phase angle between circuit *V* and circuit *I* is between 0° and 90°. As you can see from the diagram, total circuit voltage equals *IZ*, where *Z* is the total opposition to ac current from the combination of resistance and reactance. Since *I* is at 0° and *IZ* is at an angle between 0° and 90°, when we plot *Z*, it also will be between 0° and 90°.

Let's pursue these thoughts. Refer to Figure 16–22 as you study the following discussion about how to plot a diagram of *R*, X_L, and *Z* (impedance diagram) for series *RL* circuits.

1. The V_R equals *IR*, and *I* is plotted at the 0° reference vector position. Also, since the resistor voltage is in-phase with the current through the resistor, *R* is plotted at the 0° reference position when plotting an impedance diagram of *R*, X_L, and *Z*.

2. Since V_L equals IX_L and V_L is plotted at 90° from the reference vector position, it is logical that X_L is plotted at the 90° position when plotting *R*, X_L, and *Z* on an impedance diagram. Thus, the result of *I* times X_L is plotted at 90°.

3. The V_T equals *IZ*, and V_T is plotted at an angle (depending on the vector resultant of V_R and V_L) between 0° and 90°. Then it is logical to plot *Z* at the same angle in an impedance diagram where V_T is plotted in the voltage diagram of the same series *RL* circuit. In fact, the angle between *Z* and *R* is the same as the circuit phase angle between V_T and I_T.

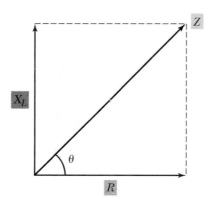

FIGURE 16–22 An impedance diagram

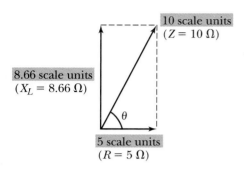

FIGURE 16–23

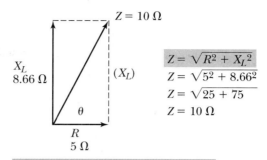

$$Z = \sqrt{R^2 + X_L{}^2}$$
$$Z = \sqrt{5^2 + 8.66^2}$$
$$Z = \sqrt{25 + 75}$$
$$Z = 10 \ \Omega$$

(NOTE: Z is not equal to the sum of R and X_L. It is greater than either one alone but not equal to their sum.)

FIGURE 16–24

Knowing all this, how can the techniques to analyze voltages in series RL circuits be applied to impedance? As you can see by comparing the diagrams of Figures 16–21 and 16–22, the layout of the impedance diagram looks just like the diagram used in the voltage vector diagram analysis. This implies the techniques used to find impedance (Z) are those techniques used to find V_T. Recall these methods include:

1. Draw to scale and measure to find magnitude and direction.
2. Apply the Pythagorean theorem to find magnitudes.
3. Apply trig functions to find angles or magnitudes.

Example

Figure 16–23 illustrates the drawing-to-scale approach. Figure 16–24 shows the Pythagorean theorem version used for impedance solutions. That is:

FORMULA 16–3 $Z = \sqrt{R^2 + X_L{}^2}$

Finally, Figure 16–25 solves for the angle using the simple trig technique. Now you try applying these techniques in the following situation.

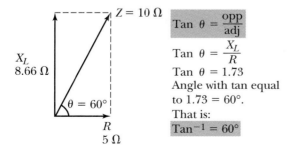

$$\text{Tan } \theta = \frac{\text{opp}}{\text{adj}}$$
$$\text{Tan } \theta = \frac{X_L}{R}$$
Tan $\theta = 1.73$
Angle with tan equal to $1.73 = 60°$.
That is:
$\text{Tan}^{-1} = 60°$

FIGURE 16–25

PRACTICE PROBLEMS VII

1. Draw the schematic diagram of a series RL circuit containing 15 Ω of resistance and 20 Ω of inductive reactance.

2. Draw an impedance diagram for the circuit described in question 1 (to scale). Measure and record the impedance value. Use a protractor and measure the angle between R and Z. Label this angle on the diagram.

3. Use the Pythagorean theorem formula where $Z = \sqrt{R^2 + X_L^2}$ and calculate and record the Z value. Does the answer agree with your measurement results?

4. Use appropriate trig formula(s) and solve for the angle where Z should be plotted. Show your work. Does this answer agree with the angle you measured using the protractor in question 2?

Summary of analyses of series RL circuits

1. The larger the R value compared with the X_L value, the more resistive the circuit will act. Conversely, the larger the X_L compared with the R, the more inductive the circuit will act.

2. The more resistive the circuit, the closer the circuit phase angle is to 0°. The more inductive the circuit, the closer the circuit phase angle is to 90° because circuit voltage is leading circuit current, which is the 0° reference vector in series circuits. Recall in a purely resistive series circuit, phase angle equals 0°. In a purely inductive series circuit, phase angle equals 90°. Refer to Figure 16–26 for some examples of R, X_L, Z, V_R, V_L, V_T, and phase angle (θ) relationships.

3. Since current (I) is the same through all parts of a series circuit, the ratio of R and X_L is the same as the relationship (or ratio) of V_R and V_L. That is, $V_R = IR$ and $V_L = IX_L$. For example, when the R value is 2 times the X_L value, the V_R value is twice the V_L value.

4. Total circuit impedance is the result of both factors, resistance and reactance. Impedance is greater than each factor but *not* equal to their sum. Right-triangle methods are used to find its value. Refer again to Figure 16–24.

R Ω	V_R (IR)	X_L Ω	V_L (IX_L)	$Z\Omega$ $(\sqrt{R^2 + X_L^2})$	V_T (IZ) VOLTS	θ (PHASE ANGLE) Arc Tan of $\dfrac{X_L}{R}$ $\left(\text{or } \dfrac{V_L}{V_R}\right)$
10	10 V	5	5 V	11.18	11.18	26.56°
10	10 V	10	10 V	14.14	14.14	45°
5	5 V	10	10 V	11.18	11.18	63.43°
100	100 V	10	10 V	100.5	100.5	5.71°
10	10 V	100	100 V	100.5	100.5	84.29°

FIGURE 16–26 Examples of parameter relationships in series *RL* circuits

Assume an I_T value of one ampere in each case. (This means V applied is changing.)

5. Total circuit voltage is the vector resultant of the resistive and reactive voltages in series. This voltage is greater than each factor but *not* equal to their sum. Right-triangle methods are used. Refer again to Figure 16–17.

6. Phase angle for series *RL* circuits is found using sine, cosine, or tangent functions. For example:

$$\sin \theta = \frac{V_L}{V_T}$$

$$\cos \theta = \frac{V_R}{V_T}$$

$$\tan \theta = \frac{V_L}{V_R}$$

NOTE: The tangent function is useful for many ac circuit problems because it does not require knowing the hypotenuse value.

For the impedance diagram:

$$\sin \theta = \frac{X_L}{Z}$$

$$\cos \theta = \frac{R}{Z}$$

$$\tan \theta = \frac{X_L}{R}$$

7. A greater *R* or X_L value results in greater circuit impedance and lower current value for a given applied voltage.

8. The circuit current value is found by using Ohm's law, where

$$I_T = \frac{V_T}{Z_T}$$

9. *Impedance diagrams can only be drawn and used to analyze series ac circuits!* Also, they can be used to analyze series portions of more complex circuits but *should not* be drawn to analyze parallel circuits.

10. An impedance diagram is not a phasor diagram, since the quantities shown are not sinusoidally varying quantities. However, because *Z* is the vector resulting from *R* and *X*, it can be graphed.

16–5 FUNDAMENTAL ANALYSIS OF PARALLEL RL CIRCUITS

Basically, the same strategies we have been using in the series circuit solutions can be applied to parallel circuits. That is, we can use vectors, the Pythagorean theorem, and trig functions. The differences are in *which* parameters are used on the diagrams and in the formulas.

Because *voltage* is the common factor in parallel circuits, it becomes the reference vector at the 0° position on our diagrams. Since I_T equals the vector sum of the branch currents in parallel circuits, our diagrams illustrate the branch currents in appropriate right-triangle form. (Recall that in series circuits, current was the reference and voltages are plotted in right-triangle format.)

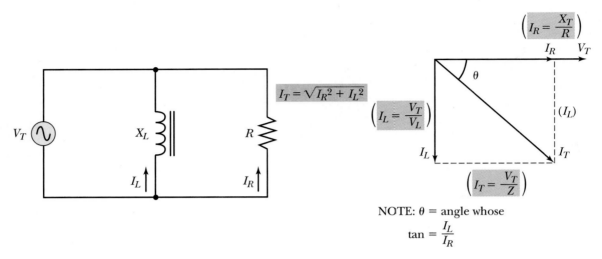

FIGURE 16–27 Parallel *RL* circuit analysis

Another contrast between parallel *RL* circuit analysis and series *RL* circuit analysis is that we *do not* use impedance diagrams to analyze parallel *RL* circuits.

With these facts in mind, let's apply the principles you have learned to some parallel *RL* circuit analysis. Refer to Figure 16–27 as you study the following comments.

1. V_T is at the reference vector position because it is common to both the resistive and the reactive branches.

2. Since I_R (resistive branch current) is in-phase with V_R (or V_T), it is also at the 0° position.

3. We know the current through a coil lags the voltage across the coil by 90°. Thus, I_L is lagging V_T (or V_L) by 90°. (Remember in the four-quadrant system that −90° is shown as straight down.)

4. The total current is the vector resultant of the resistive and the reactive branch currents. Thus, it is between 0° and −90°.

5. The higher the resistive branch current relative to the reactive branch current, the more resistive the circuit will act. This means the smaller phase angle is between V_T and I_T. Also, the greater the inductive branch current relative to the resistive branch current, the more inductive the circuit will act. This means the phase angle is greater between V_T and I_T.

6. In a parallel *RL* circuit, total impedance is the resultant opposition offered to current by the combined circuit resistance and reactance. That is,

$$Z = \frac{V_T}{I_T}$$

Because of the phase relationships in a circuit with both resistance and reactance, the impedance is less than the impedance of any one branch. However, the shortcut methods used in purely resistive or reactive circuits will not work. For example, if there are 10 Ω of resistance in parallel with 10 Ω of reactance, the impedance will *not* be 5 Ω but 7.07 Ω. This is because the total current is not the simple arithmetic sum of the branch currents, as it is in purely resistive or reactive circuits. Rather, total current is the vector sum.

7. Branch currents are found using Ohm's law, where

$$I_R = \frac{V_R}{R} \text{ and } I_L = \frac{V_L}{X_L}$$

8. Total current is found using the Pythagorean formula, where:

FORMULA 16–4 $I_T = \sqrt{I_R^2 + I_L^2}$

Example

Parameters for the circuit in Figure 16–28 are $V_T = 150$ V, $R = 50\ \Omega$, and $X_L = 30\ \Omega$. Find I_T, Z, and the circuit phase angle (θ).

1. Label the known values on the diagram, Figure 16–29.

2. Determine the branch currents:

$$I_R = \frac{V_T}{R} = \frac{150\text{ V}}{50\ \Omega} = 3\text{ A}$$

$$I_L = \frac{V_T}{X_L} = \frac{150\text{ V}}{30\ \Omega} = 5\text{ A}$$

3. Draw a current vector diagram and label knowns, Figure 16–30.

4. Use Pythagorean theorem and solve for I_T:

$$I_T = \sqrt{I_R^2 + I_L^2}$$
$$I_T = \sqrt{9 + 25}$$
$$I_T = \sqrt{34}$$
$$I_T = 5.83\text{ A}$$

5. Use Ohm's law and solve for Z:

$$Z = \frac{V_T}{I_T}$$

$$Z = \frac{150\text{ V}}{5.83\text{ A}}$$

$$Z = 25.73\ \Omega$$

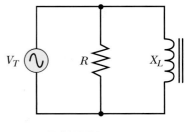

FIGURE 16–28

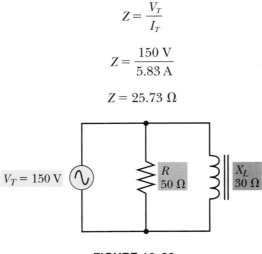

FIGURE 16–29

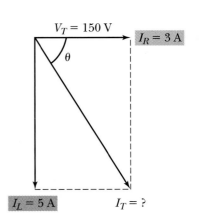

FIGURE 16–30

6. Use the trig formula and solve for phase angle:

$$\tan \theta = \frac{I_L}{I_R}$$

$$\tan \theta \approx 1.666$$

Angle whose tangent = 1.666 is 59° (i.e., $\text{Tan}^{-1} 1.666 = 59°$)

Answers: $I_T = 5.83$ A, $Z = 25.73$ Ω, and $\theta = -59°$

(NOTE: The angle is considered a negative angle because the circuit current is lagging the reference circuit voltage by 59°.)

Now you try the following practice problem.

PRACTICE PROBLEMS VIII

A circuit similar to the one in Figure 16–28 has the following parameters: $V_T = 300$ V, $R = 60$ Ω, and $X_L = 100$ Ω. Perform the same steps previously shown and find 1) I_T, 2) Z, and 3) the phase angle (θ).

A formula to find Z without knowing I

To find the impedance *(Z)* of a simple parallel *RL* circuit, you can use the following formula:

FORMULA 16–5 $\quad Z = \dfrac{RX_L}{\sqrt{R^2 + X_L^2}}$

Example

Using the same values of R and X_L just used in our preceding example:

$$Z = \frac{RX_L}{\sqrt{R^2 + X_L^2}} = \frac{50 \times 30}{\sqrt{50^2 + 30^2}} = \frac{1,500}{\sqrt{3,400}} = 25.72 \text{ Ω}$$

As you can see, it checks out with our previous solution in which we used Ohm's law to solve for *Z*. Now you try one using this *Z* formula technique.

PRACTICE PROBLEMS IX

Assume a simple parellel *RL* circuit in which the resistance equals 47 Ω and the inductive reactance equals 100 Ω. What is the impedance value of this circuit?

Summary of analysis of parallel RL circuits

1. The greater the resistive branch *current* compared to the inductive branch *current,* the more resistive the circuit acts. Conversely, the greater the inductive branch *current* compared with the resistive branch current, the more inductive the circuit acts.

R Ω	I_R $\left(\dfrac{V_T}{R}\right)$	X_L Ω	I_L $\left(\dfrac{V_T}{X_L}\right)$	I_T $(\sqrt{I_R{}^2 + I_L{}^2})$	Z $\left(\dfrac{V_T}{I_T}\right)$	θ (PHASE ANGLE) Arc Tan of $\dfrac{I_L}{I_R}$
10	5 A	5	10 A	11.18 A	4.47 Ω	−63.43°
10	5 A	10	5 A	7.07 A	7.07 Ω	−45°
5	10 A	10	5 A	11.18 A	4.47 Ω	−26.56°
100	0.5 A	10	5 A	5.02 A	9.96 Ω	−84.29°
10	5 A	100	0.5 A	5.02 A	9.96 Ω	−5.7°

NOTE: The larger the (−) phase angle, the more inductive the circuit is acting.

FIGURE 16–31 Examples of parameter relationships in parallel *RL* circuits

Assume a voltage applied (V_T) of 50 V for each case. (This means I_T will be changing).

2. The more resistive the circuit is, the closer the circuit phase angle is to 0°. The more inductive the circuit is, the closer the circuit phase angle is to −90°. These situations result from circuit current lagging behind the circuit voltage, which is the 0° reference vector in parallel circuits. Recall in a purely *resistive* parallel circuit $\theta = 0°$, but in a purely inductive *parallel* circuit $\theta = -90°$. Refer to Figure 16–31 to see some examples of the relationships of R, X_L, Z, I_R, I_L, I_T, and phase angle (θ) with selected circuit parameters purposely varied.

3. Branch current is *inverse* to the resistance or reactance of a given branch. For example, when R is two times X_L, current through the *resistive* branch is half the current through the *reactive* branch.

4. Total circuit impedance of a given parallel *RL* circuit is less than the opposition of any one branch. However, impedance cannot be found with product-over-the-sum or reciprocal formulas used in dc circuits. Neither can impedance be found using an impedance diagram approach. Ohm's law or other methods are used.

5. Total circuit current is the vector resultant of the resistive and reactive branch currents. Current is greater than any one branch current but *not* equal to their sum. Right-triangle methods are used. Refer again to Figure 16–27.

6. Phase angle for parallel *RL* circuits is found using sine, cosine, or tangent trig functions. For example:

$$\sin \theta = \frac{I_L}{I_T}$$

$$\cos \theta = \frac{I_R}{I_T}$$

$$\tan \theta = \frac{I_L}{I_R}$$

NOTE: The tangent function is useful for many ac circuit problems because it does not require knowing the hypotenuse value.

7. A greater R or X_L value results in greater circuit impedance and lower circuit current value for a given applied voltage.

8. The circuit impedance value is found by using Ohm's law, where

$$Z_T = \frac{V_T}{I_T}$$

9. *Impedance diagrams can only be drawn and used to analyze series ac circuits!* Also, they can be used to analyze series portions of more complex circuits but *should not* be drawn to analyze parallel circuits. Obviously, since the impedance of a parallel *RL* circuit is smaller than any one branch, the hypotenuse cannot be shorter than one of the sides when using the right-triangle techniques. Therefore, an impedance diagram cannot be drawn for parallel *RL* circuits.

16–6 EXAMPLES OF PRACTICAL APPLICATIONS FOR INDUCTORS AND RL CIRCUITS

Inductors and inductances have a variety of applications. Some of these applications use the property of inductance or mutual inductance. Other applications involve using inductance with resistance or capacitance, or both. Inductance can be used in circuits operating at power-line (low) frequencies; circuits operating in the audio-frequency range; and circuits operating at higher frequencies, known as radio frequencies.

The following is a brief list of applications.

1. Power-frequency applications, Figure 16–32.
   a. Power transformers
   b. Filter circuits (both power line and power supply filters)
   c. Solenoids, electromagnets, relays

2. Audio-frequency applications, Figure 16–33.
   a. Audio transformers (used for coupling and Z-matching)
   b. Audio chokes (as loads for amplifiers)
   c. Filter circuits

3. Radio-frequency applications, Figure 16–34.
   a. Tuning circuit applications
   b. Radio frequency chokes/filters
   c. Waveshaping applications (using L/R time constant characteristics)

Typically, inductance values used at power-line frequencies are in the henrys range. For audio frequencies, values are often in the tenths of a henry, or millihenry ranges. For radio frequency *(RF)* applications, values are typically in the microhenrys range.

One key characteristic of inductors allowing them to be used in these applications is their sensitivity to frequency. (Recall, X_L varies with frequency.) Their relationship to creating and reacting to magnetic fields is another very useful characteristic.

You have already studied the basic principles of inductors, which make the varied applications shown in Figure 16–33 possible. For example, you know inductive reactance varies with frequency. The application of inductors in filter and tuning circuits depends on this principle. Also, you studied the principle for waveshaping when you studied L/R time constants.

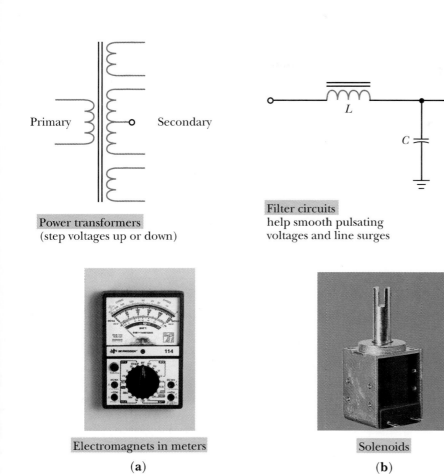

Power transformers
(step voltages up or down)

Filter circuits
help smooth pulsating
voltages and line surges

Electromagnets in meters

(a)

Solenoids

(b)

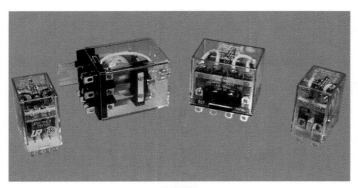

Relays
(often used to cause switching actions)

(c)

FIGURE 16–32 Examples of applications at low ac (or pulsating dc) frequencies
*(Photo **a** courtesy of B & K Precision; photos **b** and **c** courtesy of Guardian Electric Mfg. Co.)*

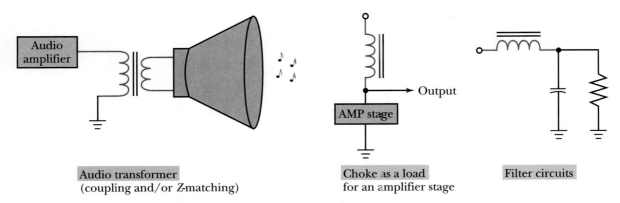

Audio transformer
(coupling and/or Z-matching)

Choke as a load
for an amplifier stage

Filter circuits

FIGURE 16–33 Examples of applications at audio frequencies

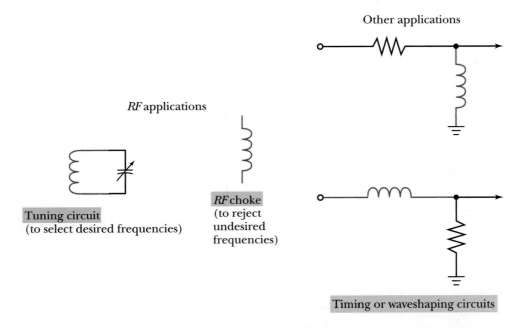

Other applications

RF applications

Tuning circuit
(to select desired frequencies)

RF choke
(to reject
undesired
frequencies)

Timing or waveshaping circuits

FIGURE 16–34 Examples of radio-frequency *(RF)* and other applications

As you can see, inductors, the property of inductance, and inductive reactance are valuable tools in the field of electronics. Your understanding of inductors will be of great help in your career.

SUMMARY

▶ Phase angle is the difference in phase between the circuit applied voltage (V_T) and the circuit current (I_T).

▶ Circuit current and voltage are in phase in purely resistive ac circuits; are 90° out of phase in purely inductive circuits; and are between 0° and 90° out of phase in circuits containing both resistance and inductance.

▶ For *series* ac circuits containing both resistance and inductive reactance, total circuit voltage and impedance cannot be found by adding voltage values or oppositions to current, as is done in dc circuits. Rather, values must be determined by vectors or right-triangle techniques.

▶ For voltage-current vector diagrams relating to *series RL* circuits, *current* is the *reference vector* positioned at the 0° location since it is common throughout the circuit. Resistive voltage drop(s) are also at the 0° position, since they are in-phase with current. Inductive voltages are shown at 90° (since V_L leads I by 90°). Total circuit voltage is plotted as the vector resultant of the resistive and inductive voltages.

▶ For impedance diagrams (used *only* for *series* circuits), the R values are plotted at the 0° reference position. The X_L value is plotted vertical (up) at 90° (assuming a perfect inductor). Impedance is plotted at a position that is the vector resultant of the R and X_L vectors.

▶ The Pythagorean theorem applied to *series RL* circuits is used to calculate the total voltage value where $V_T = \sqrt{V_R^2 + X_L^2}$. It is also useful to determine the impedance *(Z)* value where $Z = \sqrt{R^2 + X_L^2}$.

▶ The trig functions of sine, cosine, and tangent are used to determine phase angle, if two sides are known. Also, these functions can be used to determine voltage and impedance values of *series RL* circuits, if the phase angle and one side are known.

▶ For *parallel* ac circuits containing both resistance and inductive reactance, total circuit current cannot be found by adding branch currents, as is done in dc circuits. Right-triangle methods are again used.

▶ For voltage-current vector diagrams relating to *parallel RL* circuits, *voltage* is the *reference vector* at the 0° location since it is common throughout the circuit. Resistive branch current(s) are also at the 0° position, since they are in phase with the

voltage. Inductive branch currents are at −90° since I_L lags V by 90°. Total circuit current is plotted as the vector resultant of the resistive and inductive branch currents.

▶ The Pythagorean theorem applied to *parallel RL* circuits is used to calculate the total current value where $I_T = \sqrt{I_R^2 + I_L^2}$.

▶ A formula that may be used to find Z in a simple parallel *RL* circuit is:

$$Z = \frac{RX_L}{\sqrt{R^2 + X_L^2}}$$

▶ The trig functions (sine, cosine, and tangent) are also used to determine phase angle in parallel circuits, as well as series circuits, if two sides are known. They can be used to determine current values in *parallel RL* circuits, if the phase angle and one side are known.

▶ Plotting vectors to scale and then measuring lengths and angles is one method to determine parameters of interest in *RL* circuits. For series *RL* circuits, the resistive and inductive voltage values are plotted to find V_T and θ. For parallel *RL* circuits, the resistive and inductive branch current values are plotted to find I_T and θ.

▶ Impedance is the total opposition offered to ac current by a combination of resistance and reactance. The symbol for impedance is *"Z,"* and the unit of impedance is the ohm.

▶ Inductors, either alone or in conjunction with other components (R or C), are used in many applications. General uses include transformers, tuning circuits, filter circuits, and waveshaping cir-

FORMULAS AND SAMPLE CALCULATOR SEQUENCES

FORMULA 16–1 $c^2 = a^2 + b^2$

a value, ⬛x^2, ➕, b value, ⬛x^2, 🟰

FORMULA 16–2 $V_T = \sqrt{V_R^2 + V_L^2}$

V_R value, ⬛x^2, ➕, V_L value, ⬛x^2, 🟰, ⬛$\sqrt{x}$

FORMULA 16–3 $Z = \sqrt{R^2 + X_L^2}$

R value, ⬛x^2, ➕, X_L value, ⬛x^2, 🟰, ⬛$\sqrt{x}$

FORMULA 16–4 $I_T = \sqrt{I_R^2 + I_L^2}$

I_R value, $\boxed{x^2}$, $\boxed{+}$, I_L value, $\boxed{x^2}$, $\boxed{=}$, $\boxed{x}$

FORMULA 16–5 $Z = \dfrac{RX_L}{\sqrt{R^2 + X_L^2}}$

R value, $\boxed{\times}$, X_L value, $\boxed{\div}$, $\boxed{(}$, R value, $\boxed{x^2}$, $\boxed{+}$, X_L value, $\boxed{x^2}$, $\boxed{)}$, $\boxed{x}$, $\boxed{=}$

REVIEW QUESTIONS

1. Phase angle can be described as:
   a. the phase difference between circuit current and circuit impedance.
   b. the phase difference between circuit current and circuit applied voltage.
   c. the phase difference between circuit impedance and circuit applied voltage.
   d. none of the above.

2. If the phase angle of a series RL circuit is 36°, the circuit is:
   a. more resistive than reactive.
   b. more reactive than resistive.
   c. equally resistive and reactive.
   d. none of the above.

3. If the phase angle of a parallel RL circuit is −36°, the circuit is:
   a. more resistive than reactive.
   b. more reactive than resistive.
   c. equally resistive and reactive.
   d. none of the above.

4. Impedance vector diagrams should not be used for analyzing:
   a. series RL circuits.
   b. parallel RL circuits.
   c. series-parallel RL circuits.
   d. can be used for all the above circuits.

5. For analyzing series RL circuits, the reference vector is:
   a. current.
   b. voltage.
   c. impedance.
   d. none of the above.

6. For analyzing parallel RL circuits, the reference vector is:
   a. current.
   b. voltage.

   c. impedance.
   d. none of the above.

7. In series RL circuit analysis, resistive voltage drops are considered to be at:
   a. +90°.
   b. −90°
   c. 0°.
   d. none of the above.

8. In series RL circuit analysis, reactive voltage drops are considered to be at:
   a. +90°.
   b. −90°.
   c. 0°.
   d. none of the above.

9. In parallel RL circuit analysis, resistive branch currents are considered to be at:
   a. +90°.
   b. −90°.
   c. 0°.
   d. none of the above.

10. In parallel RL circuit analysis, reactive branch currents are considered to be at:
    a. +90°.
    b. −90°.
    c. 0°.
    d. none of the above.

11. For impedance diagrams:
    a. R is plotted at 0° and X_L is plotted at −90°.
    b. R is plotted at −90° and X_L is plotted at 0°.
    c. R is plotted at 0° and X_L is plotted at +90°.
    d. R is plotted at +90° and X_L is plotted at 0°.

12. What is the formula that can be used to find V_T in a series RL circuit?

13. What is the formula that can be used to find Z in a series RL circuit?

14. What is the formula that can be used to find I_T in a parallel RL circuit?

15. What is the formula that can be used to find Z for a simple parallel RL circuit?

16. Impedance is the:
    a. algebraic sum of resistance and reactance in an ac circuit.
    b. arithmetic sum of resistance and reactance in an ac circuit.
    c. vector sum of resistance and reactance in a series RL ac circuit.
    d. vector sum of resistance and reactance in a parallel RL ac circuit.

17. For a series RL circuit:
    a. the greater the resistance compared with the inductive reactance, the greater the circuit phase angle will be.
    b. the greater the resistance compared with the inductive reactance, the smaller the circuit phase angle will be.
    c. the greater the inductive reactance compared with the resistance, the smaller the circuit phase angle will be.
    d. none of the above.

18. For a parallel RL circuit:
    a. the greater the resistance compared with the inductive reactance, the greater the circuit phase angle will be.
    b. the greater the resistance compared with the inductive reactance, the smaller the circuit phase angle will be.
    c. the greater the inductive reactance compared with the resistance, the greater the circuit phase angle will be.
    d. none of the above.

19. For an impedance diagram:
    a. $\sin \theta = X_L / Z$.
    b. $\sin \theta = R/Z$.
    c. $\sin \theta = X_L/R$.
    d. none of the above.

20. For a parallel RL circuit:
    a. $\cos \theta = I_L/I_T$.
    b. $\cos \theta = I_L/I_R$.
    c. $\cos \theta = I_R/I_T$.
    d. none of the above.

PROBLEMS

1. A series RL circuit contains a resistance of 20 Ω and an inductive reactance of 30 Ω. Find the following parameters using the specified techniques.
    a. Use the plot-and-measure technique to find Z. (Round-off to the nearest whole number.) Measure the θ with a protractor. (Round-off to the nearest degree.)
    b. Use Pythagorean theorem to find Z.
    c. Use trig to find the angle between R and Z.

2. A parallel RL circuit has one branch consisting of a 200-Ω resistance, and another branch consisting of an inductor with X_L equal to 150 Ω. The applied voltage is 300 V. Find the values for the following parameters. Draw and label the schematic and vector diagrams.
    a. I_R
    b. I_L
    c. I_T
    d. θ
    e. Z

3. Draw the circuit diagram, voltage-current vector diagram and impedance diagram for a series RL circuit having one R and one L. Voltage drop across the 10-kΩ resistor is 5 V, and circuit phase angle is 45°. Label all components and vectors, and show all calculations.

4. Draw the circuit diagram and voltage-current vector diagram of a parallel RL circuit having one R and one L. Assume a total current of 12 A and a phase angle of $-75°$. Label all components and vectors in the diagrams, and show all calculations. (Round answers to the nearest whole number and the nearest degree values.)

5. Refer to Figure 16–35 and solve for the following:
    a. Z
    b. V_R
    c. θ

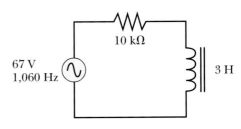

67 V
1,060 Hz

10 kΩ

3 H

FIGURE 16–35

6. Refer to Figure 16–36 and solve for the following:
   a. X_L
   b. V_L
   c. V_R
   d. θ

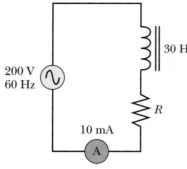

FIGURE 16–36

7. Refer to Figure 16–37 and solve for the following:
   a. I
   b. Z
   c. V_L
   d. X_L
   e. L

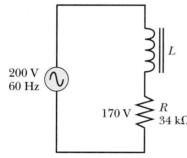

FIGURE 16–37

8. Draw the circuit diagram and all vector diagrams of the series RL circuit described below to find and illustrate all voltage, current, impedance, and phase-angle parameters. Show all calculations.
   Given: Inductance = 10 H
        Voltage leads current by 60°
        Frequency = 1 kHz
        Total voltage (V_T) = 72.5 V

9. Draw the circuit diagram and all vector diagrams for the parallel RL circuit described below to find and illustrate all voltage, current, and phase-angle parameters. Show all calculations. (NOTE: Rounding off is acceptable for this problem.)
   Given: Inductance = 1.99 H
        Frequency = 400 Hz
        Resistance = 7 kΩ
        V_T = 21 V

10. Draw the circuit diagram for a series RL circuit containing two 4-mH inductances and two 5-kΩ resistors. The circuit has a frequency of 200 kHz. Find I_T if the applied voltage is 140 V. (Round off to the nearest kilohm, volt, and milliampere, as appropriate.)

ANALYSIS QUESTIONS

Answer the following with "I" for increase, "D" for decrease, and "RTS" for remain the same.

1. If the frequency applied to a parallel RL circuit is doubled, and the circuit has one branch consisting of a 200-Ω resistance, and another branch consisting of an inductor with an X_L of 150-Ω, what will happen to the following circuit parameters?
   a. R will _____
   b. L will _____
   c. X_L will _____
   d. Z will _____
   e. θ will _____
   f. I_T will _____
   g. V_T will _____

2. If the inductance in a series RL circuit consisting of a 10-kΩ R and a 10-kΩ X_L doubles while the circuit R remains the same, what will happen to the following circuit parameters?
   a. R will _____
   b. L will _____
   c. X_L will _____
   d. Z will _____
   e. θ will _____
   f. I_T will _____
   g. V_T will _____

3. For the same original circuit as described in question 2, if the R value doubles while X_L remains the same, what will happen to the circuit parameters?
   a. R will _____
   b. L will _____
   c. X_L will _____
   d. Z will _____
   e. θ will _____

f. I_T will _____
g. V_T will _____

4. For the parallel RL circuit you drew in Problems, question 4, if inductance doubles while R remains the same, what will happen to the circuit parameters?
   a. R will _____
   b. L will _____
   c. X_L will _____
   d. Z will _____
   e. θ will _____
   f. I_T will _____
   g. V_T will _____

5. For the parallel RL circuit you drew in Problems, question 4, if the R value doubles while X_L remains the same, what will happen to the circuit parameters?
   a. R will _____
   b. L will _____
   c. X_L will _____
   d. Z will _____
   e. θ will _____
   f. I_T will _____
   g. V_T will _____

6. For the parallel RL circuit you drew in Problems, question 4, if both R and X_L values are halved, what will happen to the circuit parameters?
   a. R will _____
   b. L will _____
   c. X_L will _____
   d. Z will _____
   e. θ will _____
   f. I_T will _____
   g. V_T will _____

7. What calculator keystrokes would you use to find the sine of 57°?

8. What calculator keystrokes would you use to find the $\tan^{-1}$ related to a ratio of 0.80?

9. What calculator keystrokes would you use to find the cosine of 34°?

10. What calculator keystrokes would you use to find the arcsine for a ratio of 0.75?

11. What, if any, are the advantages of using the tangent function for ac circuit analysis?

12. How many radians are represented by 75°?

PERFORMANCE PROJECTS CORRELATION CHART

Suggested performance projects in the Laboratory Manual that correlate with topics in this chapter are:

CHAPTER TOPIC	PERFORMANCE PROJECT	PROJECT NUMBER
Fundamental Analysis of Series RL Circuits	Review of V, I, R, Z, and θ Relationships in a Series R Circuit	45
	V, I, R, Z, and θ Relationships in a Series RL Circuit	46
Fundamental Analysis of Parallel RL Circuits	Review of V, I, R, Z, and θ Relationships in a Parallel R Circuit	47
	V, I, R, Z, and θ Relationships in a Parallel RL Circuit	48

NOTE: It is suggested that after completing the above projects, the student should be required to answer the questions in the "Summary" at the end of this section of projects in the Laboratory Manual.

THE SIMPLER SEQUENCE FOR TROUBLESHOOTING

Follow the seven-step SIMPLER troubleshooting sequence, as outlined below, and see if you can find the problem with this circuit. As you follow the sequence, record your circuit test results for each testing step on a separate sheet of paper. This will aid you and your instructor to see where your thinking is on track or where it might deviate from the best procedure.

step 1 **Symptoms**

—(Gather, verify, and analyze symptom information) To begin this process, read the data under the "Starting Point Information" heading. Particularly look for circuit parameters that are not normal for the circuit configuration and component values given in the schematic diagram. For example, look for currents that are too high or too low; voltages that are too high or too low; resistances that are too high or too low, etc. This analysis should give you a first clue (or symptom information) that will aid you in determining possible areas or components in the circuit that might be causing this symptom.

step 2 **Identify**

—(Identify and bracket the initial suspect area) To perform this first "bracketing" step, analyze the clue or symptom information from the "Symptoms" step, then either on paper or in your mind, bracket, circle, or put parentheses around all of the circuit area that contains any components, wires, etc., that might cause the abnormality that your symptom information has pointed out. NOTE: Don't bracket parts of the circuit that contain items that could not cause the symptom!

step 3 **Make**

—(Make a decision about the first test you should make: what type and where) Look under the TEST column and determine which of the available tests shown in that column you think would give you the most meaningful information about the suspect area or components, in the light of the information you have so far.

step 4 **Perform**

—(Perform the first test you have selected from the TEST column list) To simulate performing the chosen test (as if you were getting test results in an actual circuit of this type), follow the dotted line to the right from your selected test to the number in parentheses. The number in parentheses, under the "Results in Appendix C" column tells you what number to look up in Appendix C to see what the test result is for the test you are simulating.

step 5 **Locate**

—(Locate and define a new "narrower" area of uncertainty) With the information you have gained from the test, you should be able to eliminate some of the circuit area or circuit components as still being suspect. In essence, you should be able to move one bracket of the bracketed area so that there is a new smaller area of uncertainty.

step 6 **Examine**

—(Examine available information and determine the next test: what type and where) Use the information you have thus far to determine what your next test should be in the new narrower area of uncertainty. Determine the type test and where, then proceed using the TEST listing to select that test, and the numbers in parentheses and Appendix C data to find the result of that test.

step 7 **Repeat**

—(Repeat the analysis and testing steps until you find the trouble) When you have determined what you would change or do to restore the circuit to normal operation, you have arrived at *your* solution to the problem. You can check your final result against ours by observing the pictorial step-by-step sample solution on the pages immediately following the "Chapter Troubleshooting Challenge"—circuit and test listing page.

step 8 **Verify**

NOTE: A useful *8th Step* is to operate and test the circuit or system in which you have made the "correction changes" to see if it is operating properly. This is the final proof that you have done a good job of troubleshooting.

CHALLENGE CIRCUIT 8

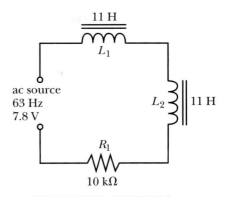

CHALLENGE CIRCUIT 8

Starting Point Information

1. Circuit diagram
2. V_{R_1} is higher than it should be
3. V source checked as being 7.8 V

TEST	Results in Appendix C
V_{R_1}	(34)
V_{L_1}	(89)
V_{L_2}	(48)
R_1	(109)

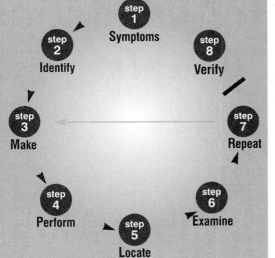

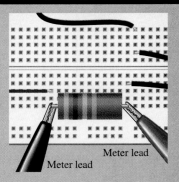

Meter lead

Meter lead

Symptoms

The voltage drop across the resistor is slightly higher than it should be. This could be due to an inductor that has decreased in value due to shorted turns or a resistor that has increased or opened since, at this point, we don't know how high the voltage is across the resistor.

Identify

initial suspect area: All components (resistor and two inductors) are in the initial suspect area.

Make

test decision: Since inductors typically open more frequently than they short and since we don't know how high the voltage is across the resistor, it makes sense to check the voltage across R_1.

Perform

1st Test: V_{R_1} is 6.4 V. If everything were normal, the resistor would drop 5.9 V, and each inductor would drop about 2.6 V, since their inductive reactances at this frequency would be close to 4.35 kΩ each. The total X_L is about 8.7 kΩ and the rated R is 10 kΩ. Z would then equal about 13.3 kΩ, and circuit current would be approximately 0.59 mA. Thus, the vector sums of the resistive and inductive voltage drops would equal approximately 7.8 V.

Locate

new suspect area: The new suspect area stays the same as the initial suspect area since we do not have enough information to eliminate the inductors or the resistor.

Examine

available data.

Repeat

analysis and testing: It would be good to look at the inductor voltage drops so we can draw some conclusions.

2nd Test: Measure V_{L_1}. V_{L_1} is close to 2.3 volts, which is slightly lower than normal.

3rd Test: Measure V_{L_2}. V_{L_2} is close to 2.36 volts, which is slightly lower than normal.

4th Test: Measure R_1. R_1 is measured with power removed and equals 12 kΩ, which is higher than the 10 kΩ it should be.

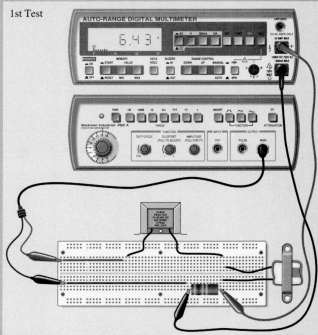

1st Test

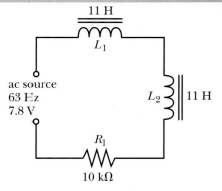

ac source
63 Hz
7.8 V

L_1 — 11 H

L_2 — 11 H

R_1 — 10 kΩ

CHALLENGE CIRCUIT 8

Verify

5th Test: Since R_1 resistance is high, replace R_1 with a new 10-kΩ resistor. After replacing the resistor, all the circuit parameters become normal.

2nd Test

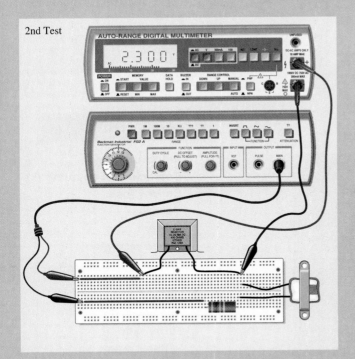

3rd Test

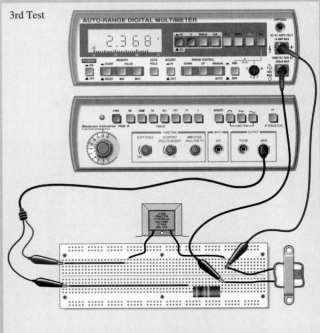

4th Test

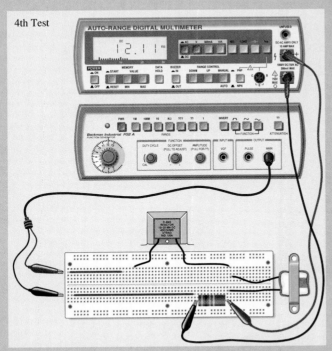

5th Test

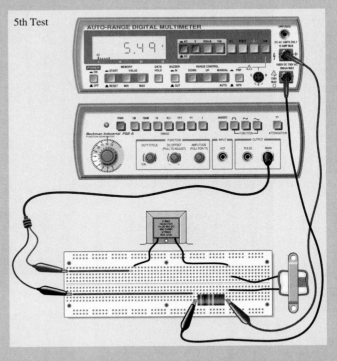

s Voltage Law

V

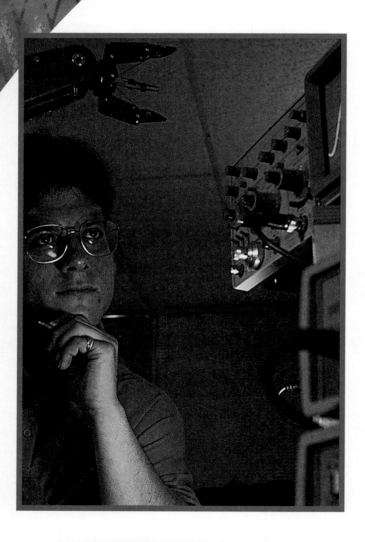

BASIC TRANSFORMER CHARACTERISTICS

chapter

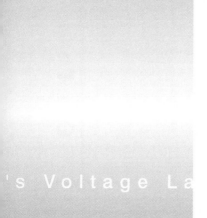

CHAPTER PREVIEW

We stated earlier that one application of electromagnetic induction is the transformer. Furthermore, you know these devices have application in different frequency ranges and for various purposes. That is, there are power transformers, audio transformers, and radio-frequency (rf) transformers, which all find various applications in many types of electronic circuits and systems. Power supplies, audio amplifiers, radio and TV receivers, transmitters, and other systems and subsystems use the unique characteristics of these devices.

In this chapter, you will study the basic features and types of transformers that a technician needs to understand. Practical information, such as schematic symbols and color codes, will be provided. Turns ratios, voltage ratios, current ratios, impedance ratios, and other major aspects about transformers will be studied. Finally, typical malfunctions and troubleshooting techniques will be examined.

OBJECTIVES

After studying this chapter, you should be able to:

1. Define **mutual inductance**

2. Calculate mutual inductance values

3. Calculate **coefficient of coupling** values

4. Calculate **turns, voltage, current,** and **impedance ratios**

5. List, draw, or explain physical, magnetic, electrical, and schematic characteristics of various **transformers**

6. List common transformer color codes

7. Define at least two types of **core losses**

8. List common problems found in transformers

9. List troubleshooting procedures

17–1 BACKGROUND INFORMATION

You know when a changing current passes through a conductor wire or a coil, an expanding or collapsing magnetic field is produced. This expanding or collapsing field, in turn, induces a counter-emf (cemf). This process of producing voltage through a changing magnetic field is known as electromagnetic induction. When the induced voltage is created across the current-carrying conductor or coil itself, it is called self-induction. As you know, the symbol for this self-inductance is L.

Another important kind of induction is **mutual induction.** The unit of mutual inductance is the henry and the symbol is M. (NOTE: Sometimes you will also see the abbreviation L_M.)

One definition for mutual inductance is the property of inducing voltage in one circuit by varying the current in another circuit. Refer to Figure 17–1 and note that Coil A is connected to an ac source. Coil B is not connected to any source. However, it is located close to coil A so that most of the flux produced in

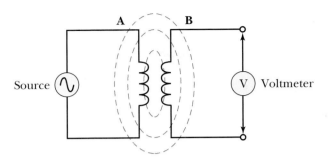

FIGURE 17–1 Mutal induction by flux linkage

Coil A by current from the source cuts or "links" Coil B. Since a *conductor, magnetic field,* and *relative motion* will induce voltage, a voltage is induced in Coil B because of the changing current in Coil A. By definition, there must be mutual inductance. These two circuits (Coil A and Coil B) are positioned so that energy is transferred by *magnetic linkage.* This coupling between circuits is also sometimes called *inductive coupling.* Inductive coupling is due to flux linkages.

17–2 COEFFICIENT OF COUPLING

In Figure 17–2 you can see smaller and greater amounts of flux linkage between coils. A term expressing this relationship by the fractional amount of total flux linking the two coils is **coefficient of coupling,** represented by the letter **k.** Coefficient of coupling and mutual inductance are *not* the same. However, they are related. Figure 17–2b shows the coils with the higher coefficient of coupling. The coefficient of coupling is therefore computed as the fractional amount of the total flux linking the two circuits. When *all* the flux from one circuit links the other, coefficient of coupling *(k)* equals one. When two-thirds of the flux links the other circuit, *k* equals 0.66, and so forth.

See Figure 17–3 for examples of factors that influence the coefficient of coupling. These factors include the proximity of coils (the closer they are the higher the *k*); the relative positions of coils with respect to each other (parallel coils at a given distance have greater *k* than perpendicular coils); and other factors, such as when the coils are wound on the same core. Coils wound on the same iron core have a coefficient of coupling almost equal to unity (1). Virtually all the flux produced by one coil links the other coil, with little leakage flux occurring. Air-

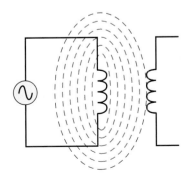

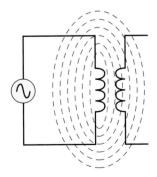

(**a**) Small amount of flux linkage: low coefficient of coupling (low *k*)

(**b**) Increased amount: higher coefficient of coupling (high *k*)

FIGURE 17–2 Coefficient of coupling related to degree of flux linkage

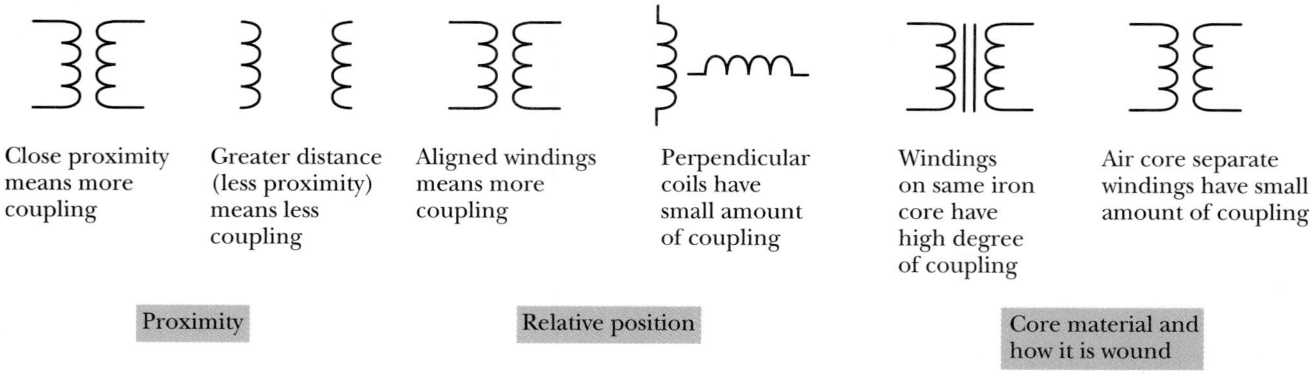

FIGURE 17–3 Factors influencing coefficient of coupling *(k)*

core coils (often used in rf circuits) have a *k* that indicates a percentage of coupling from less than 5% (*k* = 0.05) to about 35% (*k* = 0.35). Obviously, air-core coils have more leakage flux.

IN-PROCESS LEARNING CHECK I

Fill in the blanks as appropriate.

1. Producing voltage via a changing magnetic field is called electromagnetic _____.

2. Inducing voltage in one circuit by varying current in another circuit is called _____ _____.

3. The fractional amount of the total flux that links two circuits is called the _____ of coupling, which is represented by the letter _____. When 100% of the flux links the two circuits, the _____ of coupling has a value of _____.

4. The closer coils are, the _____ the coupling factor produced. Compared with parallel coils, perpendicular coils have a _____ degree of coupling.

17–3 MUTUAL INDUCTANCE AND TRANSFORMER ACTION

As stated, when two circuits are coupled magnetically, they have mutual inductance. Such is the case for transformers. **Transformers** are devices that transfer energy from one circuit to another by electromagnetic induction (mutual induction). Typically, transformers have their *primary* and *secondary* windings wound on a single core.

Refer to Figure 17–4. Notice the winding connected to the source is the primary winding. The winding connected to the load is the secondary winding. Some transformers can have more than one secondary winding. Also, observe the schematic symbol(s) used for various transformers.

Another feature shown on some diagrams, Figure 17–4, are sense dots. Sense dots indicate the ends of the windings that have the same polarity at any given moment. Usually, when sense dots are *not* shown, there is 180° difference between the primary and secondary voltage polarity.

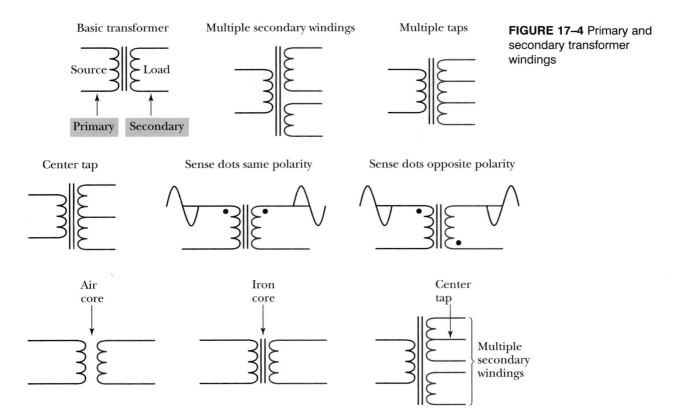

FIGURE 17–4 Primary and secondary transformer windings

PRACTICAL NOTES

Most transformers shift the phase of voltage by 180° from input to output due to the electromagnetic induction process. That is, when the top of the primary winding is positive, the top of the secondary winding is negative, and vice versa. When it is desired to produce a transformer that does not shift the phase, the manufacturer can reverse the direction the wire is wound on the iron core for one of the windings. In this case, the top of the secondary can be positive at the same time the top of the primary is positive. As stated earlier, the "sense dots" are often used to communicate this information.

Two circuits have a mutual inductance of 1 henry when a rate of current change of one ampere per second in the first circuit induces a voltage of one volt in the second circuit.

Mutual inductance between windings is directly related to the inductance of each winding and the coefficient of coupling. The formula is:

FORMULA 17–1 $M = k\sqrt{L_1 \times L_2}$

where M = mutual inductance in henrys
L_1 = self-inductance of the primary and L_2 = self-inductance of the secondary (for transformers) where Ls are in henrys

Example

A transformer has a k of 1 (100% coupling of flux lines), and the primary and secondary each have inductances of 3 H. The mutual inductance (M or L_M) is computed as $M = 1\sqrt{3 \times 3} = 1\sqrt{9} = 1\sqrt{9} = 3$ H. From our definition, this indicates a current change of one ampere per second in the primary induces 3 volts in the secondary.

PRACTICE PROBLEMS I

What is the mutual inductance of a transformer where the coefficient of coupling is 0.95, the primary inductance is 10 H, and the secondary inductance is 15 H?

17–4 MUTUAL INDUCTANCE BETWEEN COILS OTHER THAN TRANSFORMERS

We have been discussing mutual inductance in transformers. How does mutual inductance (coupled inductance) affect the total inductance of coils other than transformers that are located where their fields can interact?

Series coils

When two coils are connected in *series* so that their fields *aid* each other, total inductance is calculated from the formula:

FORMULA 17–2 $\qquad L_T = L_1 + L_2 + 2M$

where L_T = total inductance (H)
$\quad L_1$ = inductance of one coil (H)
$\quad L_2$ = inductance of other coil (H)
$\quad M$ = mutual inductance (H)

When two coils are connected in *series* so their fields oppose each other, the formula becomes:

FORMULA 17–3 $\qquad L_T = L_1 + L_2 - 2M$

Therefore, the general formula for series inductors is:

FORMULA 17–4 $\qquad L_T = L_1 + L_2 \pm 2M$

Example

Two inductors are series connected so their fields aid one another. Inductor 1 has an inductance of 8 H. Inductor 2 also has an inductance of 8 H. Their mutual inductance is 4 H. What is the total inductance of the circuit?

$$L_T = L_1 + L_2 + 2M = 8 + 8 + 8 = 24 \text{ H}$$

PRACTICE PROBLEMS II

What is the total inductance of two series-connected 10-H inductors whose mutual inductance is 5 H and whose magnetic fields are opposing?

Parallel coils

When two inductors are in *parallel* with their fields *aiding* each other, the formula is:

FORMULA 17–5
$$L_T = \frac{1}{\dfrac{1}{L_1 + M} + \dfrac{1}{L_2 + M}}$$

When two inductors are in *parallel* with their fields *opposing* each other, the formula becomes:

FORMULA 17–6
$$L_T = \frac{1}{\dfrac{1}{L_1 - M} + \dfrac{1}{L_2 - M}}$$

PRACTICE PROBLEMS III

1. What is the total inductance of two parallel coils whose fields are aiding, if $L_1 = 8$ H, $L_2 = 8$ H, and $M = 4$ H?
2. What is the total inductance of two parallel coils whose fields are opposing, if $L_1 = 8$ H, $L_2 = 8$ H, and $M = 4$ H?

Relationship of k and M

The coefficient of coupling and mutual inductance are related by the following formulas for k and M.

FORMULA 17–7
$$k = \frac{M}{\sqrt{L_1 \times L_2}}$$

Recall Formula 17–1 is $M = k\sqrt{L_1 \times L_2}$. As you can see from these two formulas, mutual inductance and coefficient of coupling are directly related.

17-5 IMPORTANT TRANSFORMER RATIOS

We will begin this section by making the following assumptions.

1. With *no load* connected to the secondary winding, the primary winding acts like a simple inductor.

2. Current flowing in the primary depends on applied voltage and inductive reactance of the primary.

$$I_P = \frac{V}{X_P}$$

3. The primary current produces a magnetic field that cuts both the primary and secondary turns.

4. A cemf is produced in the primary. This cemf almost equals the applied voltage because of self-inductance.

5. A cemf is produced in *each turn* of the secondary, which equals the voltage induced in *each turn* of the primary because of mutual inductance. This means when there are equal turns on the secondary and the primary, their voltages are equal.

Using these assumptions, let's look at the ratio between electrical output and input parameters under specified conditions. When 100% of the primary flux links the secondary winding (when $k = 1$), the ratio of the induced voltages (i.e., the ratio of primary voltage to the secondary voltage) is the same as the ratio of turns on the primary compared with the turns on the secondary. Thus,

$$\frac{V_P}{V_S} = \frac{N_P}{N_S}$$

where V_P is primary induced voltage; V_s is secondary induced voltage; N_P is the number of turns on the primary winding; and N_S is the number of turns on the secondary winding.

Observe Figure 17–5 and notice how the terms step-up and step-down relate to transformers. When the secondary (output) voltage is higher than the primary (input) voltage, the transformer is a step-up transformer. Conversely, when the secondary voltage is lower than the primary voltage, the transformer is a step-down transformer.

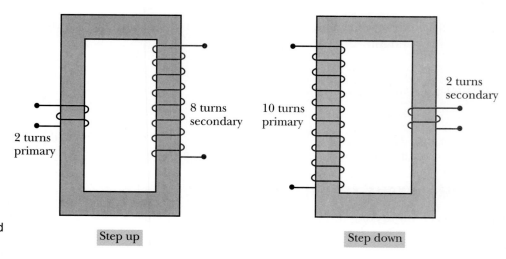

FIGURE 17–5 Step-up and step-down transformers

Turns ratio

You will often find the **turns ratio (TR)** of a transformer expressed in terms of the ratio of primary turns to secondary turns (N_P/N_S). For example, when the primary has 250 turns and the secondary has 1,000 turns, the turns ratio is:

$$\frac{N_P}{N_S} = \frac{250}{1,000} = 1{:}4$$

NOTE: A number of texts also show the opposite approach (i.e., secondary-to-primary ratio). *Either approach works as long as formulas are consistently set up using the same approach.* In any case, the technician should be careful in making transformer calcualtions! Be *consistent* in the way you set up the various ratios. We will use *both* ratios to familiarize you with both approaches.

Notice that in Figure 17–6a the turns ratio is shown for both approaches. Also, in Figure 17–6b, the **voltage ratio** is shown for both approaches.

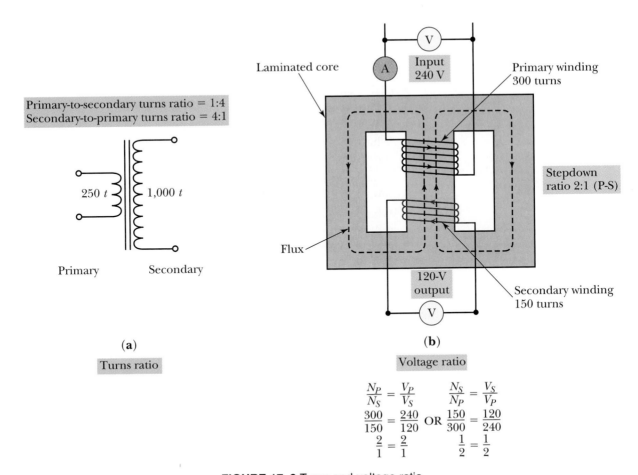

Primary-to-secondary turns ratio = 1:4
Secondary-to-primary turns ratio = 4:1

250 t   1,000 t

Primary   Secondary

(a)

Turns ratio

Laminated core

Input 240 V

Primary winding 300 turns

Stepdown ratio 2:1 (P-S)

Flux

120-V output

Secondary winding 150 turns

(b)

Voltage ratio

$$\frac{N_P}{N_S} = \frac{V_P}{V_S} \qquad \frac{N_S}{N_P} = \frac{V_S}{V_P}$$

$$\frac{300}{150} = \frac{240}{120} \quad \text{OR} \quad \frac{150}{300} = \frac{120}{240}$$

$$\frac{2}{1} = \frac{2}{1} \qquad\qquad \frac{1}{2} = \frac{1}{2}$$

FIGURE 17–6 Turns and voltage ratio

Voltage ratio

As you learned with the ideal transformer, the **voltage ratio** has the same ratio as the turns ratio; again see Figure 17–6b. That is:

FORMULA 17–8	$\dfrac{V_P}{V_S} = \dfrac{N_P}{N_S}$

Example

1. What is the primary-to-secondary turns ratio of a transformer having eight times as many turns on the primary as on the secondary?

$$TR = \frac{N_P}{N_S} = \frac{8}{1} \text{ or } 8{:}1$$

2. What is the secondary voltage on a transformer having a (p-s) turns ratio of 1:14 if the applied voltage is 20 V?

$$\frac{V_P}{V_S} = \frac{N_P}{N_S}; \frac{20}{V_S} = \frac{1}{14}; \ V_S = 20 \times 14 = 280 \text{ V}$$

PRACTICE PROBLEMS IV

1. What is the p-s turns ratio of an ideal transformer whose secondary voltage is 300 V when 100 V is connected to the primary?
2. What is the p-s turns ratio of an ideal transformer that steps-up the input voltage six times?
3. What is the secondary voltage of an ideal transformer whose p-s turns ratio is 1:5 and whose primary voltage is 50 V?
4. What is the voltage ratio of an ideal transformer whose p-s turns ratio is 4:1? Is this a step-up or a step-down transformer?

Current ratio

Up to now, our discussions have been about transformers with no load connected to the secondary(ies). Notice in Figure 17–7 that when a load is connected to the secondary, current flows through the load and the secondary. This current produces a magnetic field related to the secondary current. The polarity produced by the secondary current cancels some of the primary field. Hence, the cemf is reduced in the primary, which reduces the impedance of the primary to current flow. This reduced impedance causes primary current to increase. The interaction between the secondary and the primary is termed "reflected impedance." In other words, the impedance reflected across the input or primary results from current in the output or secondary.

In Figure 17–7, the power available to the load from the transformer secondary must come from the source that supplies the primary, since a trans-

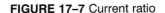

FIGURE 17–7 Current ratio

$V_P = 100$ V
$I_P = 0.4$ A
$P_P = 40$ W

$V_S = 20$ V
$I_S = 2$ A
$P_S = 40$ W

$10 \ \Omega$
40 W

A

0.4 A

A

2 A

$$\frac{I_S}{I_P} = \frac{N_P}{N_S} = \frac{V_P}{V_S}$$

Voltage = stepped down
Current = stepped up

former does not *create* power, but *transfers* power from the primary circuit to the secondary circuit. Transformers are not 100% efficient. Therefore, the power delivered to the primary by its source is slightly higher than the power delivered by the secondary to its load. Since transformers are highly efficient (generally ranging from 90% to 98% efficient), Figure 17–7 has assumed 100% efficiency to illustrate this principle. Later in this chapter, we will look at some losses that prevent actual 100% efficiency.

Also notice in Figure 17–7 the power delivered to the load is 40 W, where $V_S = 20$ V, $R_S = 10 \ \Omega$, and $I_S = 2$ A. The power delivered to the primary is 40 W, where $V_P = 100$ V and $I_P = 0.4$ A. This is a step-down transformer where the voltage has been stepped down five times. That is,

$$\frac{N_P}{N_S} = 5{:}1 \text{ and } \frac{V_P}{V_S} = 5{:}1$$

What about the **current ratio?** Observe the current in the secondary is five times higher than the current in the primary. Using our primary-to-secondary principle, the current ratio is 1:5, or the inverse of the voltage ratio.

This makes sense, if we assume the power in the secondary is the same as the power in the primary. In other words, $V_S I_S = V_P I_P$. If voltage were decreased by five times, the current would have to be increased by five times so the *V* times *I* (or *P*) would be the same in both primary and secondary.

(NOTE: Power supplied to the primary must equal power delivered to the secondary plus power lost.)

Now let's relate current ratio to the turns and voltage ratios. Thus,

$$\frac{I_S}{I_P} = \frac{N_P}{N_S} = \frac{V_P}{V_S}$$

Note the inversion in the current ratio. Since quantities equaling the same thing are equal to each other, we can list three useful equations from the preceding information.

FORMULA 17–9 $\dfrac{N_P}{N_S} = \dfrac{V_P}{V_S}$

FORMULA 17–10 $\dfrac{N_P}{N_S} = \dfrac{I_S}{I_P}$

FORMULA 17–11 $\dfrac{V_P}{V_S} = \dfrac{I_S}{I_P}$

If you know any three of the quantities in any of these equations, you can solve for the unknown quantity, assuming 100% efficiency.

Examples

1. A transformer has 250 turns on the primary, 1,000 turns on the secondary, and the primary voltage is 120 V. What is the secondary voltage value?

$$\frac{N_P}{N_S} = \frac{V_P}{V_S}$$

$$\frac{250}{1,000} = \frac{120}{V_S}$$

$$250 V_S = 120,000$$

$$V_S = \frac{120,000}{250} = 480 \text{ V}$$

2. A transformer has a primary to secondary turns ratio of 5:1. What is the primary current when the secondary load current is 0.5 A?

$$\frac{N_P}{N_S} = \frac{I_S}{I_P}$$

$$\frac{5}{1} = \frac{0.5}{I_P}$$

$$5 I_P = 1 \times 0.5$$

$$I_P = \frac{0.5}{5} = 0.1 \text{ A}$$

3. The primary-to-secondary current ratio of a transformer is 4:1, and the secondary voltage is 60 V. What is the voltage on the primary?

$$\frac{V_P}{V_S} = \frac{I_S}{I_P}$$

$$\frac{V_P}{60} = \frac{1}{4}$$

$$4 V_P = 60 \times 1$$

$$V_P = \frac{60}{4} = 15 \text{ V}$$

Impedance ratio

The **impedance ratio** is related to the turns ratio, Figure 17–8. Also, since inductance and thus the inductive reactance of a coil, relates to the square of the turns, the impedance also has a similar relationship. A load of a given impedance connected to the transformer secondary is transformed to a different value by what the source sees "looking into" the primary (unless the turns ratio is 1:1). In fact, for an ideal (zero loss) transformer, the impedance transformation (secondary-to-primary) is proportional to the square of the secondary-to-primary turns ratio. That is,

FORMULA 17–12 $\left(\dfrac{N_S}{N_P}\right)^2 = \dfrac{Z_S}{Z_P}$

where Z_S = the impedance of the load connected to the secondary

Z_P = the impedance looking into the primary from source

$\dfrac{N_S}{N_P}$ = the secondary-to-primary turns ratio

Another way of thinking about impedance ratios and their relationship to turns ratios is to consider that impedance can be regarded as the ratio of V to I. That is $Z = V/I$. For example, if we consider a step-up transformer that steps up voltage by two times, we understand the current is "stepped-down" by two times (or is one-half). Then our ratio of V/I has become $2V/0.5I$, which translates to a ratio of 4. Thus, the impedance of the winding with the higher turns is four times that of the winding with the lesser turns. Of course, the converse is true if the transformer is a step-down transformer. The winding with the lesser turns would have half the voltage and twice the current; thus, the $0.5V/2I$ ratio would indicate that the winding with the lower number of turns has an impedance that is one-fourth that of the winding with the greater number of turns. Therefore, we see again, that the impedance ratio is related to the square of the turns ratio.

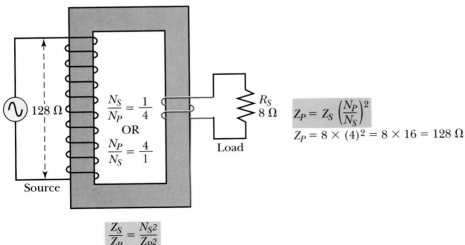

$$\frac{Z_S}{Z_P} = \frac{N_{S2}}{Z_{P2}}$$

FIGURE 17–8 Impedance ratio

Example

When the N_S/N_P (secondary-to-primary) turns ratio of a transformer is 1:4 and the impedance connected to the secondary is 8 Ω, the primary impedance is related to the square of the turns ratio. Hence, Z_P is 16 times Z_S. That is $Z_P = 8 \times 16 = 128$ Ω, Figure 17–8. A practical formula expressing this in terms of *primary-to-secondary* turns ratio is:

FORMULA 17–13
$$Z_P = Z_S \left(\frac{N_P}{N_S}\right)^2$$

Example

A transformer has a secondary-to-primary turns ratio of 0.5, and a load resistance of 2,000 ohms is connected to the secondary. What is the apparent primary impedance?

Since the secondary-to-primary turns ratio is 0.5, the secondary has half as many turns as the primary, or the primary has twice as many turns as the secondary. Appropriately substitute in the formula:

$$Z_P = Z_S \left(\frac{N_P}{N_S}\right)^2$$

$$Z_P = 2,000 \ \Omega \times (2)^2$$

$$Z_P = 2,000 \ \Omega \times 4 = 8,000 \ \Omega$$

PRACTICE PROBLEMS V

In the prior example, suppose the turns ratio reverses. In other words, the secondary has twice as many turns as the primary. What is the apparent primary impedance?

These discussions indicate the primary impedance of a very efficient transformer (as it looks to the power source feeding the primary) is determined by the load connected to the secondary and the turns ratio. Because this is true, transformers are often used to match impedances. That is, a transformer is used so that a secondary load of given value is transformed to the optimum impedance value into which the primary source should be looking. (The source may be an amplifier stage, another ac signal, or power source.) Therefore, a useful formula is:

FORMULA 17–14
$$\frac{N_P}{N_S} = \sqrt{\frac{Z_P}{Z_S}}$$

where $\dfrac{N_P}{N_S}$ = the required *primary-to-secondary* turns ratio

Z_P = the primary impedance desired (required)
Z_S = the impedance of the load connected to the secondary

Example

An amplifier requires a load of 1,000 ohms for best performance. The amplifier output is to be connected to a loudspeaker having an impedance of 10 ohms. What must the turns ratio *(primary-to-secondary)* be for a transformer used for impedance matching?

$$\frac{N_P}{N_S} = \sqrt{\frac{Z_P}{Z_S}}$$

$$\frac{N_P}{N_S} = \sqrt{\frac{1,000}{10}} = \sqrt{100} = 10$$

This indicates the primary must have 10 times as many turns as the secondary.

PRACTICE PROBLEMS VI

1. What is the primary-to-secondary impedance ratio of a transformer that has 1,000 turns on its primary and 200 turns on its secondary?

2. What impedance would the primary source see looking into a transformer where the primary-to-secondary turns ratio is 16 and the load resistance connected to the secondary is 4 Ω?

17–6 TRANSFORMER LOSSES

Because of energy losses, practical transformers do not meet all the criteria for the ideal, 100% efficient devices we have been discussing. These transformer losses consist primarily of two categories. One category is **copper loss,** caused by the I^2R copper wire losses that make up the primary and secondary windings. Current passing through the resistance of the wire produces energy loss in the form of heat.

The other category of transformer losses is termed **core loss.** Core losses result from two factors. One factor is called *eddy current* losses. Because of the changing magnetic fields and the iron-core conductivity, currents are produced in the iron core that do not aid transformer output. These eddy currents produce an I^2R core loss related to the core material resistance and the current amount. To reduce the current involved, the iron core is usually constructed of thin sheets of metal called *laminations.* These laminations are electrically insulated from each other by varnish or shellac. Reducing the cross-sectional area of the conductor by laminating significantly increases the core resistance, which decreases the current, and greatly reduces the I^2R loss.

The other core loss is called *hysteresis loss.* Recall magnetic domains in the magnetic core material require energy to rearrange. (Recall in an earlier chapter, the lagging effect of the core magnetization behind the magnetizing force was illustrated by the hysteresis loop.) Since ac current constantly changes in magnitude and direction, the tiny molecular magnets within the core are constantly being rearranged. This process requires energy that causes some loss.

Hysteresis loss increases as the frequency of operation is increased. For this reason, iron-core transformers, such as those used at power frequencies, are not used in radio frequency applications.

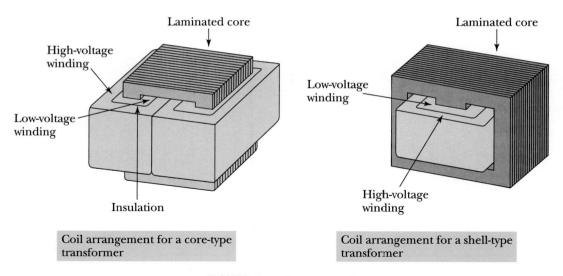

FIGURE 17–9 Power transformers

Copper losses, core losses, leakage flux, and other factors prevent a transformer from being 100% efficient. Efficiency percentage is calculated as output power divided by input power times 100.

FORMULA 17–15   Efficiency % $= \dfrac{P_{\text{out}}}{P_{\text{in}}} \times 100$

To minimize these various losses, the physical features of coils and transformers used for applications in power, audio, and rf frequency ranges slightly differ. We will investigate some of these characteristics in the next section.

17–7 CHARACTERISTICS OF SELECTED TRANSFORMERS

Power transformers

Power transformers, Figure 17–9, typically have the following features:

1. They are heavy (because of their iron core) and can be very large, like the transformers on telephone poles. However, power transformers in electronic devices are much smaller and vary considerably in size.

FIGURE 17–10
Autotransformer

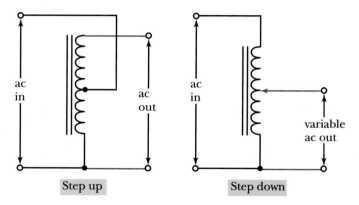

2. They have a laminated iron core to reduce eddy-current losses.

3. The flux path is as short as possible to reduce leakage flux and minimize magnetizing energy needed.

4. They use one of two core shapes—core type (primary and secondary windings on separate legs of the core) and shell type (both primary and secondary windings on the center leg), again see Figure 17–9.

5. They have one or more primary and secondary windings.

PRACTICAL NOTES

Some power transformers are used when it is necessary to be able to select either a 120- or 240-V primary input source for the transformer. Therefore, there are transformers with two primary windings. These primary windings may be connected in series, when 240 V is th source (i.e., the end of one winding is connected to the beginning of the other winding, being careful to pay attention to phasing of the two windings). Alternatively, by connecting the two primaries in parallel (again being careful of phasing), the transformer primary can be fed from a 120-V source. By using these techniques, the secondary voltage output will be the same with 120- or 240-V inputs, as appropriate.

6. They may have *one* tapped winding, such as the **autotransformer,** Figure 17–10. The autotransformer is a special single-winding transformer.

When the input source is connected from one end to the tap while the output is across the whole coil, the voltage is stepped up. If the input voltage is applied across the whole coil and the output taken from the tap to one end, the voltage is stepped down.

Variacs or *Powerstats* are trade names for autotransformers having a movable wiper-arm tap that allows varying voltage from 0 V to the maximum level. They are often used as a variable ac source in experimental setups.

SAFETY HINTS

Caution: Bad shock can occur if no **isolation transformer** is used on the autotransformer input side, because there is no isolation from the raw line voltage and ground connections in the autotransformer circuit. For this reason, do *not* use autotransformers with a movable wiper-arm tap unless you plug the autotransformer into an isolation transformer output, which isolates it from the raw ac power source.

7. They isolate the circuitry connected to its secondary from the primary ac source (e.g., the ac powerline). A 1:1 turns ratio transformer is specifically designed for isolation applications, Figure 17–11. (NOTE: all standard transformers isolate secondary from primary source regardless of turns ratios, except for autotransformers.)

When using this special 1:1 turns-ratio isolation transformer, the load on the secondary has no *direct* connection to the primary source. This provides a safety measure for people working with the circuitry connected to the isolation transformer secondary.

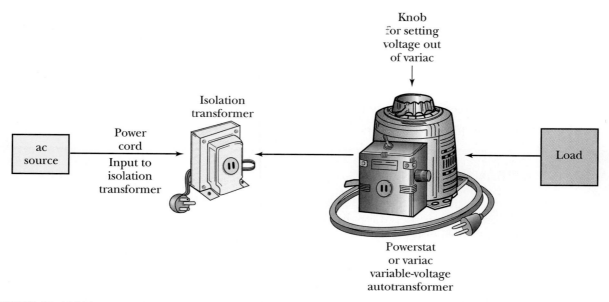

FIGURE 17–11 Using an isolation transformer for safety when using a powerstat or variac-type variable voltage autotransformer

8. They cannot operate at high frequency because of excessive losses that would be created.

9. Just two examples of the many color codes in use are shown in Figure 17–12a and b.

Audio transformers

Audio transformers are similar to power transformers in their construction. They typically are:

1. smaller and lighter than most power transformers; and

2. unable to operate at frequencies above the audio range without excessive losses.

RF transformers and coils

Because the flux is changing *very rapidly*, *RF* transformers or coils, Figures 17–13 and 17–14 (see page 574) have different features compared to power or audio transformers. Some features include the following:

1. Cores are air core, or powdered iron core, depending on frequency and application. Air-core coils have virtually zero hysteresis or eddy-current losses. Powdered iron cores are made by grinding metal materials and nonconductive materials into granules, which are pressed into a cylindrical-shaped "slug." Each metal particle is therefore insulated, greatly reducing losses such as eddy currents. See Figure 17–13a for examples of this coil.

Also, nonconductive ferrite materials can be used in *RF* coils or transformers. As you recall, ferrite materials are special magnetic materials that provide easy paths for magnetic flux but are nonconductive. Often, toroid-type cores are used in radio-frequency circuits and lower-frequency applications, Figure 17–13b.

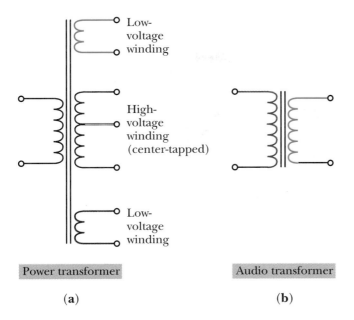

FIGURE 17–12 Typical power and audio transformer color-coding systems

Low-voltage winding

High-voltage winding (center-tapped)

Low-voltage winding

Power transformer

Audio transformer

(a)

(b)

2. *RF* coils and transformers typically have very low inductance values. Generally, these values are in the microhenry or millihenry ranges.

3. Figure 17–14 shows examples of air-core coils found in transmitter applications.

Color coding

Today, many power, audio, and *RF* transformers are designed and manufactured to be used and mounted on printed circuit (PC) boards. Consequently, their connector terminals are designed for that technique. However, some transformers that have wire leads by which connection is made to the various windings are still available and in use. In that case, the leads may be color coded to identify the various windings and their connection points. Although, there may be variations from these, frequently the colors used include the following:

- *Power transformers:* Typically primary leads are black, high-voltage secondary is red, and if center-tapped, the center-tap is red with yellow stripe; low-voltage windings may be yellow for one winding, green for another, brown for another and slate for another, depending on the number of separate low-voltage windings the transformer has, and the type circuitry involved.

- *Audio transformers:* Typically, one primary winding lead is blue and connects to the output electrode of the device feeding the transformer primary, the other end of the primary winding is red, and connects to the circuit source; the secondary has one lead that is green and feeds the input to the stage connected to the transformer secondary, the other secondary lead (return) is typically black.

- *Radio frequency* (*RF*) inter-stage transformers, such as intermediate-frequency (i.f.) transformers used in radio receivers, use colors similar to the audio transformers. That is, primary leads are blue and red, and secondary leads are green and black.

FIGURE 17–13 Examples of various types coils and cores: **(a)** Powdered iron-core type coils; **(b)** toroidal cores

(**a**)

(**b**)

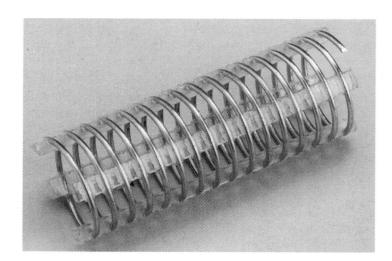

FIGURE 17–14 Example of an air-core *RF* coil

17-8 TROUBLESHOOTING HINTS

Transformer problems are virtually the same problems as those described in the chapter on inductance. Two basic problems are opens or shorts. The primary test to detect these defects is the ohmmeter.

//

SAFETY HINTS

Once again, we want to remind you that when checking large inductances (including transformer windings) with an ohmmeter, take precautions not to get shocked by the "inductive kick" of the winding's collapsing field when removing the ohmmeter leads from the winding under test.

Open winding(s)

With an open winding, the obvious result is zero secondary output voltage, even when source voltage is applied to the primary. When the primary is *not* open and only one secondary of a multiple-secondary transformer is open, only the open winding shows zero output. To check these conditions, remove the power and use the ohmmeter.

Refer to Figure 17–15a to see an open winding. In this case, the ohmmeter registers an infinite resistance reading, indicating the winding under test has an open. Both the primary and secondary windings should be checked! Normal readings are the specified dc resistance values for each winding.

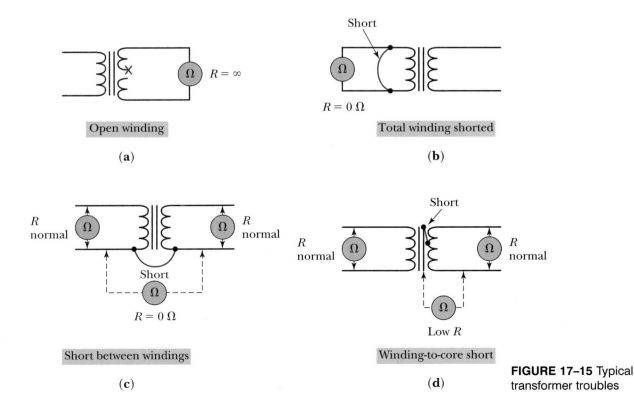

FIGURE 17-15 Typical transformer troubles

Shorted winding(s)

Shorted turns or a shorted winding can cause a fuse in the primary circuit to blow (open) from excessive current. Another possible clue to this condition is if the transformer operates at a hotter than normal temperature. Also, another hint of shorted turns is lower than normal output voltage. Actually, partially shorted windings or windings with very low *normal* resistances are hard to check only using an ohmmeter. Therefore, watch for these clues when troubleshooting transformer problems.

However, the ohmmeter is useful to detect certain types of transformer short conditions. Figure 17–15b shows a total-winding short condition. In this case, the ohmmeter reads virtually zero ohms when it should be reading the normal winding resistance value, if there were no short.

Figure 17–15c illustrates the condition when the primary and secondary windings are shorted together. In this case, each winding resistance might be normal; however, there would be a low resistance reading between windings. When conditions are normal, "infinite ohms" is measured between primary and secondary windings.

Another possible short condition is when one or both windings are shorted to the metal core, Figure 17–15d. Again, if things are normal, an infinite ohms reading will be present between either winding and the core material.

PRACTICAL NOTES

The following discussion describes one practical approach used to determine whether a short in a *power transformer* exists. This approach is a simple test involving the transformer in question and a light bulb. In Figure 17–16, you can see a light bulb is placed in series with the transformer primary winding. When a short is present in *either* the primary or secondary, the lamp glows brightly because of the higher than normal primary current. When no short exists, the lamp does *not* glow or may glow dimly.

The requirements for bulb size and its electrical ratings are as follows:

1. The voltage rating should equal the same voltage rating as the transformer primary.

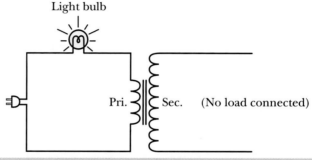

Light bulb

Pri.   Sec.   (No load connected)

If lamp glows dimly or not at all = no short(s)
If lamp glows brightly = short(s); (throw transformer away!)

FIGURE 17–16 One way to find shorted turns

NOTE: Use small bulb rating for small transformers; use larger bulb rating for large transformers.

2. For small transformers, a small-wattage lamp should be used.

3. For large transformers, a large-wattage lamp should be used.

Be aware that choosing the correct bulb size and interpreting the results come only with experience. This test is simply mentioned to show you some innovative ways technicians find to learn *what they need to know,* even in difficult situations.

Practical Transformer Ratings

POWER TRANSFORMERS

Input volts
Output V, I, and kVA ratings
Insulation ratings
Size
Mounting dimensions
Weight

AUDIO TRANSFORMERS

Primary impedance
Secondary impedance
Maximum V, I, and P
Size
Mounting dimensions

SUMMARY

▶ When the magnetic flux of two coils link and interact, mutual inductance and inductive coupling occur between the two circuits.

▶ When mutual inductance is present, current changing in one circuit causes an induced voltage in the other circuit. The symbol for mutual inductance is M or L_M, and the unit of measure is the henry.

▶ One henry of mutual inductance is present when a change of one ampere per second in one circuit induces one volt in the mutually coupled circuit.

▶ Expressing the fraction of total flux linking one circuit to another circuit is called coefficient of coupling. The abbreviation for coefficient of coupling is k. NOTE: Multiplying k by 100 yields the percentage of total flux lines linking the circuits.

▶ The relationship between mutual inductance and coefficient of coupling is shown by the formulas:

$$M = k\sqrt{L_1 \times L_2} \text{ and } k = \frac{M}{\sqrt{L_1 \times L_2}}$$

where M = mutual inductance in henrys
k = coefficient of coupling
L_1 = self-inductance of coil or transformer winding 1
L_2 = self-inductance of coil or transformer winding 2

▶ Mutual inductance can exist between coils that are not transformer windings but that are close enough for flux linkage to occur. This can affect total inductance for the circuit.

▶ Transformers are comprised of separate coils or windings, having inductive coupling. The primary winding is connected to an ac source. The load(s) are connected to the secondary winding(s).

▶ One feature of a two-winding transformer is that the primary and secondary are magnetically coupled, yet electrically isolated from each other. This feature can keep circuits and loads from being directly connected to the power lines, which increases safety.

▶ One key transformer parameter is the turns ratio. Turns ratio is expressed either by the primary-to-secondary ratio N_P/N_S, or secondary-to-primary ratio N_S/N_P.

▶ The ratio of voltages between primary and secondary is the same as the ratio of turns. That is,

$$\frac{N_P}{N_S} = \frac{V_P}{V_S} \text{ or } \frac{N_S}{N_P} = \frac{V_S}{V_P}$$

▶ The current ratio of a transformer is the inverse of the voltage ratio. When voltage is stepped up from primary to secondary, current is stepped down by a proportionate factor:

$$\frac{V_P}{V_S} = \frac{I_S}{I_P} \text{ or } \frac{V_S}{V_P} = \frac{I_P}{I_S}$$

▶ The impedance ratio of a transformer is related to the square of the turns ratio. That is,

$$\frac{Z_P}{Z_S} = \frac{N_P^2}{N_S^2}$$

▶ When a transformer has 100% efficiency, the power delivered to the load by the secondary equals the power supplied to the primary by the ac source. The efficiency of iron-core transformers is usually in the 90% to 98% range.

▶ Two categories of transformer losses are copper losses and core losses. Copper losses relate to the I^2R losses of the copper wire in the primary and secondary windings. Core losses result from eddy currents being induced in the core by the changing magnetic flux, and the energy wasted due to hysteresis losses. Eddy currents in the core material cause I^2R losses.

▶ Transformers have various applications at different frequencies. Three transformer categories include power transformers, audio transformers, and rf transformers.

▶ Power and audio transformers generally have laminated iron-cores, for efficiency's sake. Rf transformers usually use air cores, special powdered iron, or ferrite in their core material.

▶ Transformers can develop abnormal operations because of opens or shorts. Opens are evident by lack of output. Shorts between turns, from winding-to-winding or from winding-to-core, can be seen by overheating, blown fuses, and so forth.

FORMULAS AND SAMPLE CALCULATOR SEQUENCES

FORMULA 17–1 $M = k\sqrt{L_1 \times L_2}$

k value, ⊗, ⊙, L_1 value, ⊗, L_2 value, ⊙, ⊗, ⊜

FORMULA 17–2 $L_T = L_1 + L_2 + 2M$

L_1 value, ⊕, L_2 value, ⊕, 2, ⊗, M value, ⊜

FORMULA 17–3 $L_T = L_1 + L_2 - 2M$

L_1 value, ⊕, L_2 value, ⊖, 2, ⊗, M value, ⊜

FORMULA 17–4 $L_T = L_1 + L_2 \pm 2M$

L_1 value, ⊕, L_2 value, ⊕ or ⊖ as appropriate, 2, ⊗, M value, ⊜

FORMULA 17–5 $L_T = \dfrac{1}{\dfrac{1}{L_1 + M} + \dfrac{1}{L_2 + M}}$

L_1 value, ⊕, M value, ⊜, 1/x, ⊕, ⊙, L_2 value, ⊕, M value, ⊙, 1/x, ⊜, 1/x

FORMULA 17–6

$$L_T = \cfrac{1}{\cfrac{1}{L_1 - M} + \cfrac{1}{L_2 - M}}$$

L_1 value, ⊖, M value, ▬, 1/x, ⊕, (▐, L_2 value, ⊖, M value, ▐), 1/x, ▬, 1/x

FORMULA 17–7

$$k = \frac{M}{\sqrt{L_1 \times L_2}}$$

M value, ÷, (▐, L_1 value, ✕, L_2 value, ▐), ✔x, ▬

FORMULA 17–8

$$\frac{V_P}{V_S} = \frac{N_P}{N_S}$$

V_P value, ÷, V_S value, ▬, (should equal) N_P value, ÷, N_S value, ▬

FORMULA 17–9

$$\frac{N_P}{N_S} = \frac{V_P}{V_S}$$

N_P value, ÷, N_S value, ▬, (should equal) V_P value, ÷, V_S value, ▬

FORMULA 17–10

$$\frac{N_P}{N_S} = \frac{I_S}{I_P}$$

N_P value, ÷, N_S value, ▬ (should equal) I_S value, ÷, I_P value, ▬

FORMULA 17–11

$$\frac{V_P}{V_S} = \frac{I_S}{I_P}$$

V_P value, ÷, V_S value, ▬, (should equal) I_S value, ÷, I_P value, ▬

FORMULA 17–12

$$\left(\frac{N_S}{N_P}\right)^2 = \frac{Z_S}{Z_P}$$

N_S value, ÷, N_P value, ▬, x^2, (should equal) Z_S value, ÷, Z_P value, ▬

FORMULA 17–13

$$Z_P = Z_S\left(\frac{N_P}{N_S}\right)^2$$

Z_S value, ✕, (▐, N_P value, ÷, N_S value, ▐), x^2, ▬

FORMULA 17–14

$$\frac{N_P}{N_S} = \sqrt{\frac{Z_P}{Z_S}}$$

N_P value, ÷, N_S value, ▬, (should equal) (▐, Z_P value, ÷, Z_S value, ▐), √x

FORMULA 17–15

$$\text{Efficiency \%} = \frac{P_{\text{out}}}{P_{\text{in}}} \times 100$$

P_{out} value, ÷, P_{in} value, ✕, 100, ▬

REVIEW QUESTIONS

1. When current changing in one circuit causes induced voltage in another, it is called:
   a. transformation.
   b. mutual inductance.
   c. decoupling.
   d. flux leakage.

2. When a change of 1 A per second in one circuit induces 1 V in another circuit:
   a. there is no mutual inductance present.
   b. there is 1 henry of mutual inductance present.
   c. the coefficient of coupling is unity.
   d. there must be a transformer involved.

3. Coefficient of coupling expresses:
   a. a relationship to flux density and magnetizing force.
   b. a fractional value of the total flux linking one circuit to another.
   c. a percentage.
   d. none of the above.

4. In a transformer:
   a. the secondary is connected to the source.
   b. the secondary is always connected to a load.
   c. the primary is connected to the source.
   d. none of the above.

5. In a transformer:
   a. the primary and secondary are electrically connected to each other.
   b. the primary and secondary are only magnetically coupled.
   c. the primary and secondary are both electrically and magnetically coupled.
   d. none of the above.

6. Turn ratios for transformers may be expressed:
   a. only as a ratio of primary turns-to-secondary turns.
   b. only as a ratio of secondary turns-to-primary turns.
   c. either as primary-to-secondary or secondary-to-primary ratios.
   d. none of the above.

7. The voltage ratio of a transformer:
   a. is equal to the turns ratio.
   b. is not equal to the turns ratio.
   c. has no relationship to the turns ratio.
   d. none of the above.

8. The current ratio of a transformer:
   a. is directly related to the voltage ratio.
   b. is inversely related to the voltage ratio.
   c. is not related to the voltage ratio.
   d. none of the above.

9. The impedance ratio of a transformer is equal to:
   a. the voltage ratio.
   b. the current ratio.
   c. the square of the turns ratio.
   d. none of the above.

10. The typical efficiency of iron-core transformers is approximately:
   a. 10% to 20%.
   b. 80% to 88%.
   c. 90% to 98%.
   d. none of the above.

11. Copper losses in a transformer come from:
   a. resistance of the iron in the core material.
   b. eddy currents.
   c. resistance of the wire in the windings.
   d. none of the above.

12. Core losses in a transformer may come from:
   a. hysteresis and eddy currents.
   b. winding losses and metal core losses.
   c. leakage flux alone.
   d. all of the above.

13. Iron cores are typically laminated, rather than solid:
   a. to decrease their resistance to current.
   b. to increase their resistance to current.
   c. to make the cores lighter in weight.
   d. none of the above.

14. Iron cores are usually used for:
   a. audio and *RF* transformers.
   b. power and *RF* transformers.
   c. power and audio transformers.
   d. power and large *RF* transformers.

15. The most common problems in transformers are:
   a. shorted turns and lamination buzz.
   b. shorted turns and overheating.
   c. opens and shorts.
   d. opens between windings and core.

16. A blown fuse in a transformer circuit may indicate:
    a. an open transformer winding.
    b. a short between windings.
    c. an open in a single winding.
    d. none of the above.

17. A lack of output from a transformer may indicate:
    a. a low resistance.
    b. an open.

c. excessive leakage flux.
d. none of the above.

18. The ohmmeter can be used to find transformer troubles, such as:
    a. open windings.
    b. shorts between windings.
    c. shorts between core and windings.
    d. all of the above.

PROBLEMS

1. What is the primary current of a transformer whose source is 120 V and secondary is supplying 240 V at 100 mA?

2. What is the secondary voltage of a transformer having a secondary-to-primary turns ratio of 5.5:1, if the source voltage is 100 V?

3. What is the primary-to-secondary turns ratio of a transformer whose primary voltage is 120 V and secondary voltage is 6 V?

4. If a transformer has a primary-to-secondary turns ratio of 5:1 and the load resistance connected to the secondary is 8 Ω, what impedance is seen looking into the primary?

5. What is the mutual inductance of two coils having a coefficient of coupling of 0.6, if each coil has a self-inductance of 2 H?

6. What is the total inductance of two coils in series having 0.3 k, if each of the coils has a self-inductance of 10 mH? (Assume their fields are aiding.)

7. What is the total inductance of two coils in parallel having 0.5 k, if each coil has a self-inductance of 10 H. (Assume their fields are opposing.)

8. What is the secondary voltage of an ideal transformer whose secondary-to-primary turns ratio is 15 and primary voltage is 12 V?

9. How many volts-per-turn are in a transformer having a 500-turn primary and a 100-turn secondary, if the primary voltage is 50 V?

10. A 6:1 step-down transformer has a secondary current of 0.1 A. Assuming an ideal transformer, what is the primary current?

11. What is the number of turns in the primary winding of the transformer shown in Figure 17–17?

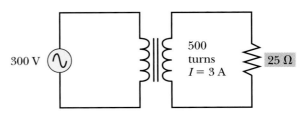

FIGURE 17–17

12. Assuming an ideal transformer, what is the power delivered to R_L in Figure 17–18?

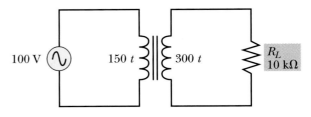

FIGURE 17–18

13. What is the primary-to-secondary turns ratio in Figure 17–19?

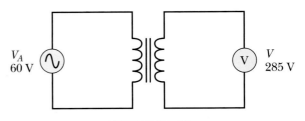

FIGURE 17–19

14. What is the primary current in the transformer circuit in Figure 17–20? (Assume an ideal transformer.)

15. What is the voltage applied to the primary of the circuit in Figure 17–21?

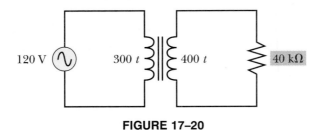

FIGURE 17–20

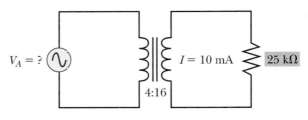

FIGURE 17–21

ANALYSIS QUESTIONS

1. Research the term "flyback transformer." Indicate where it is used, and whether its output is a high or low voltage. Also, list the frequency at which a TV flyback transformer is operating.

2. Why are the cores of transformers laminated iron rather than solid iron?

3. An amplifier having an output impedance of 5,000 Ω is feeding its output signal to a matching transformer. If the speaker connected to the secondary is a 16-Ω speaker and the number of turns on the secondary is 200 turns, what number of turns should the primary winding have to provide the best impedance match? (Assume a perfect transformer.)

4. The primary-to-secondary turns ratio of a transformer is 8:1. What is the primary current,

if the power delivered to a 200-Ω load resistance is 405 mW?

5. If the impedance ratio (primary-to-secondary) of an ideal transformer is 36:1 and the number of turns on the primary is 1,200, how many turns are on the secondary winding?

6. A TV receiver contains a transformer in its power supply. If you unplug the receiver from its 120-V, 60-Hz wall plug (ac source), turn the on/off switch to the on position, and connect an ohmmeter to each lead of the ac cord's plug, the ohmmeter reads 10 Ω. Using Ohm's law, you think the current supplied to the transformer is 12 A, since $I = V/R$, or 120 V/10 Ω. Explain why the circuit is fused with only a 5-A fuse, yet, when the TV receiver is turned on, the fuse does not blow!

PERFORMANCE PROJECTS CORRELATION CHART

Suggested performance projects in the Laboratory Manual that correlate with topics in this chapter are:

CHAPTER TOPIC	PERFORMANCE PROJECT	PROJECT NUMBER
Mutual Inductance and Transformer Action and	Turns, Voltage, and Current Ratios and	49
Important Transformer Ratios	Turns Ratio(s) versus Impedance Ratio	50

NOTE: It is suggested that after completing the above projects, the student should be required to answer the questions in the "Summary" at the end of this section of projects in the Laboratory Manual.

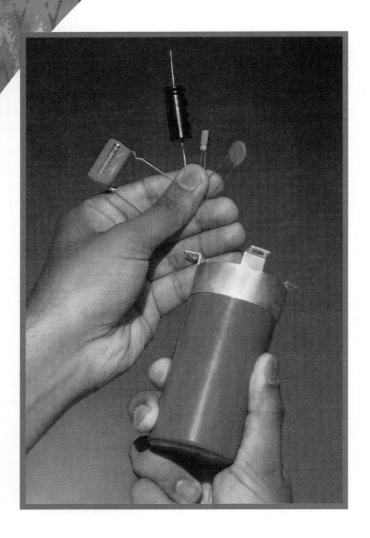

KEY TERMS

Capacitance
Capacitor
Charging a capacitor
Dielectric
Dielectric constant
Dielectric strength
Discharging a capacitor
Electrostatic (electric)
field
RC **time constant**

CAPACITANCE

18

chapter

CHAPTER PREVIEW

Three basic properties of electrical circuits are resistance, inductance, and capacitance. You have studied resistance (opposition to current flow) and inductance (property opposing a *change in current*). In this chapter, you will study capacitance (property opposing a *change in voltage*).

You will learn that capacitors have the unique ability to store electrical energy because of an electric field. Even as current produces a magnetic field, voltage establishes an electric field. This phenomenon and other important dc characteristics of capacitors will be analyzed.

Also, you will study how capacitors charge and discharge; the physical factors affecting capacitance values; some examples of capacitors; how to calculate total capacitance for circuits containing more than one capacitor; examples of capacitor coding systems; and some common capacitor problems and troubleshooting techniques.

OBJECTIVES

After studying this chapter, you should be able to:

1. Define **capacitor, capacitance, dielectric, dielectric constant, electric field,** farad, *RC* **time constant,** and leakage resistance

2. Describe **capacitor charging** action and **discharging** action

3. Calculate charge, voltage, **capacitance,** and stored energy, using the appropriate formulas

4. Determine total **capacitance** in circuits with more than one **capacitor** (series and parallel)

5. Calculate circuit voltages using appropriate *RC* **time constant** formulas

6. List and describe the physical and electrical features of at least four types of **capacitors**

7. List typical **capacitor** problems and describe troubleshooting techniques

8. Use the SIMPLER troubleshooting sequence to solve the Chapter Troubleshooting Challenge problem

18-1 DEFINITION AND DESCRIPTION OF A CAPACITOR

In *physical* terms, a **capacitor** is an electrical component, generally consisting of two conducting surfaces (often called capacitor plates) that are separated by a nonconductor (called the dielectric), Figure 18–1.

The nonconducting material, or **dielectric,** can be any of a number of materials. Some examples of dielectric materials are vacuum, air, waxed paper, plastic, glass, ceramic material, aluminum oxide, and tantalum oxide.

The schematic symbol for the capacitor plates is one straight line and one curved line, each perpendicular to the lines representing the wires or circuit conductors connected to the plates; again see Figure 18–1.

In *electrical* terms, a capacitor is an electrical component that stores electrical charge when voltage is applied.

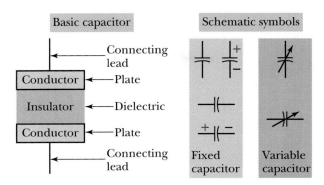

FIGURE 18–1 Basic capacitor and symbols

18–2 THE ELECTROSTATIC FIELD

You know that unlike charges attract and like charges repel. Refer to Figure 18–2 as you study the following discussion.

Lines are used to illustrate **electrostatic fields** between charged bodies, much like lines illustrate magnetic flux between magnetic poles. Direction of the electrostatic field lines is usually shown from the positive charged body to the negative charged body. (NOTE: An electron placed in this field travels in the opposite direction from the electrostatic field. In other words, the electron is repelled from the negative body and attracted to the positive body.)

The force between charged bodies is *directly* related to the product of the charges on the bodies and *inversely* related to the square of the distance between them.

$$F = \frac{kQ_1 \times Q_2}{d^2} \text{ (coulomb's law)}$$

This means the closer together a capacitor's plates are for a given charge, the stronger the electric field produced. Also, the greater the charge (in coulombs) stored on the plates, the stronger the field produced.

The electrostatic field represents the storage of electric energy. This energy originates with a source and can be returned to a circuit when the source is re-

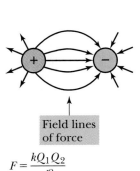

$$F = \frac{kQ_1 Q_2}{d^2}$$

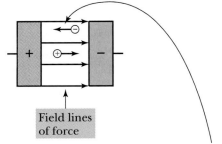

The direction an electron, or negative charge would move *if* placed in this field is in the opposite direction from that used to illustrate the electrostatic field. (NOTE: There is not an actual electron flow through a good dielectric.) A positive charge, placed in the field, would move in the direction used to represent the field lines.

FIGURE 18–2 The electrostatic field

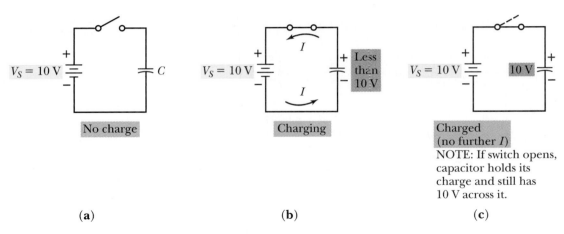

(a) No charge

(b) Charging

(c) Charged (no further *I*)
NOTE: If switch opens, capacitor holds its charge and still has 10 V across it.

FIGURE 18–3 Capacitor charging action

moved. (This is analogous to energy stored in the buildup of an inductor's magnetic field, which may be returned to the circuit when the magnetic field collapses.)

18–3 CHARGING AND DISCHARGING ACTION

Figure 18–3a illustrates a capacitor with no *charge,* that is, no electrostatic field exists between the plates. The plates and dielectric between them are electrically *neutral.*

Charging action

In Figure 18–3b, the circuit switch is closed and electrons move for a short time, **charging the capacitor.** Let's look at this action.

As soon as the switch is closed, neither capacitor plate has an electron excess or deficiency. Also, there is no voltage across the plates or electric field across the dielectric to oppose electron movement to or from the plates.

Electrons leave the negative terminal of the battery and travel to the capacitor plate, which is electrically connected to it. The electrons do not travel through the nonconductive dielectric material between the capacitor plates. Rather, the electrons accumulate on that side of the capacitor, establishing an electron excess and a negative charge at that location. Incidentally, the electron orbital paths in the atoms of the dielectric material (unless it is vacuum) are distorted, since they are repelled by the negative charged plate of the capacitor, Figure 18–4. (NOTE: The distorted orbits represent storage of some electrical energy.)

As electrons accumulate on one side of the capacitor, making that side negative, electrons are being simultaneously attracted to the source's positive terminal from the capacitor's opposite side. Recall in a series circuit, the current is the same through all circuit parts.

Obviously, as an electron surplus builds on one plate and an equal electron deficiency occurs on the other plate, a potential (voltage) difference is established between the capacitor plates.

Notice in Figure 18–3c the voltage polarity established across the capacitor plates is one that *series-opposes* the source voltage (V_S). When sufficient *charging current* has flowed to cause the capacitor voltage to equal the source voltage, no more current can flow. In other words, the capacitor has *charged* to the source voltage.

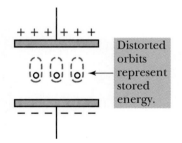

Distorted orbits represent stored energy.

FIGURE 18–4 Distorted orbits represent stored electrical energy.

Current flowing from the negative source terminal to one capacitor plate and to the positive source terminal from the other capacitor plate is maximum the moment the switch is closed. For a moment, the capacitor acts almost like a short. NOTE: Current does not flow *through* the capacitor via the nonconductive dielectric. Rather, current acts like it has flowed because of the electron movement on either side of the capacitor. Charging current is maximum the first instant the switch is closed and decreases to zero as the capacitor becomes *charged* to the source voltage value.

When the source voltage increases to a higher voltage, current again flows until the capacitor becomes charged to this new voltage level. Once the capacitor has charged to the source voltage, it acts like an open and blocks dc, again see Figure 18–3c.

In summary, maximum charging current flows when voltage is first applied to the capacitor. As the capacitor charges, voltage across the capacitor increases to the source voltage value and charging current decreases. When the capacitor is charged to V_S, no further current flows and the capacitor is holding a charge, which causes a voltage equal to V_S across its plates. The electrical energy has been stored in the electrostatic (dielectric) field of the capacitor.

When we open the switch in the charged capacitor circuit, Figure 18–5, and we assume the dielectric has infinite resistance to the current (i.e., no leakage current flows between the plates through the dielectric), the capacitor remains charged *as long* as no external *discharge* path for current is provided.

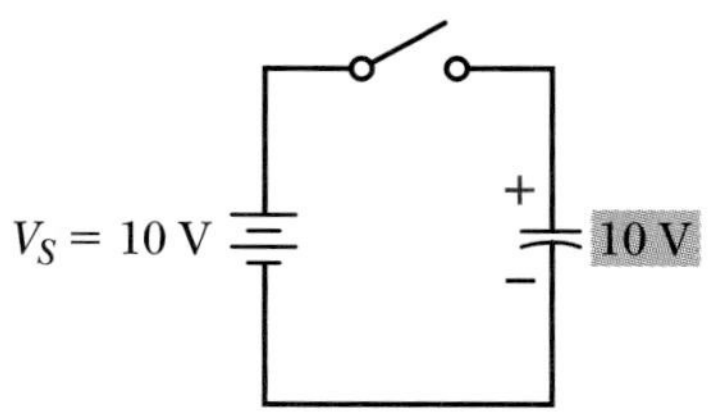

FIGURE 18–5 Capacitors retain charge until it bleeds off either through leakage *R* of dielectric or through some external current path.

//

SAFETY HINTS

Caution: **Always discharge circuit capacitors after power has been removed and before working on circuits containing them!** Technicians have been killed by accidentally becoming the discharge path! By providing an appropriate discharge path to bleed the charge from circuit capacitors before working on the circuit, this danger is avoided.

Many power supplies have a "bleeder resistor," which is intended to bleed off the charge from power supply capacitors when the power is turned off. IT IS NOT SAFE to depend on the bleeder working, as they occasionally become open. A number of technicians have been killed, assuming the bleeder resistor would protect them!

Some dangerous equipment has a "shorting stick" mounted inside the cabinet of the equipment. This is a long, nonconductive, wooden or plastic stick with a large braided conductor attached at one end. The other end of the braided conductor is permanently connected to the system ground. The technician holds the end of the stick at the opposite end from which the conductor is attached to the stick, then positions the conductor end of the stick to touch the capacitor terminals to be discharged. This of course discharges the capacitor through the braid to ground. Caution must be used in doing this due to the large arc that can result. All operator safety precautions recommended by the equipment manufacturer and common sense MUST be followed when using this procedure.

The point to be made here is: NEVER ASSUME LARGE CAPACITORS IN SYSTEMS ARE DISCHARGED JUST BECAUSE THE POWER HAS BEEN TURNED OFF!

FIGURE 18–6 Discharging a capacitor

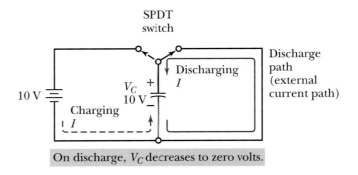

On discharge, V_C decreases to zero volts.

Discharging action

Figure 18–6 illustrates **discharging a capacitor.** An external path for electron movement has been supplied across the capacitor plates. The excess electrons on the bottom plate quickly move through this path to the top plate, which has an electron deficiency. The charge difference between the plates is neutralized, and the potential difference between the plates falls to zero. When full neutralization occurs and the potential difference between plates is zero, we have completely discharged the capacitor.

IN-PROCESS LEARNING CHECK I

Fill in the blanks as appropriate.

1. A capacitor is an electrical component consisting of two conducting surfaces called _____ that are separated by a nonconductor called the _____.

2. A capacitor is a device that stores electrical _____ when voltage is applied.

3. Electrons _____ travel through the capacitor dielectric.

4. During charging action, one capacitor plate collects _____, making that plate _____. At the same time, the other plate is losing _____ to become _____.

5. Once the capacitor has charged to the voltage applied, it acts like an _____ circuit to dc. When the level of dc applied increases, the capacitor _____ to reach the new level. When the level of dc applied decreases, the capacitor _____ to reach the new level.

6. When voltage is first applied to a capacitor, _____ charge current occurs.

7. As the capacitor becomes charged, current _____ through the circuit in series with the capacitor.

18–4 THE UNIT OF CAPACITANCE

You have seen that **capacitance** relates to the *capacity* of a *capacitor* to store electrical charge. As you might expect, the more charge a capacitor stores for a given voltage, the larger its capacitance value must be. The unit of capacitance is the farad, named in honor of Michael Faraday. The farad is that amount of capacitance where a charge of one coulomb develops a potential difference of one volt across the capacitor plates, or terminals. The farad is abbreviated as capital F.

From this definition, relationships between capacitance *(C)*, voltage *(V)*, and charge *(Q)* are:

FORMULA 18–1 $\quad C = \dfrac{Q}{V}$

FORMULA 18–2 $\quad Q = CV$

FORMULA 18–3 $\quad V = \dfrac{Q}{C}$

where V = volts
$\quad\quad Q$ = coulombs
$\quad\quad C$ = capacitance in farads

As you can see from Formula 18–2, the amount of charge a capacitor stores is *directly related to* the amount of *capacitance and* the *voltage* across the capacitor.

Example

When 10 coulombs of charge are stored on a capacitor charged to 2 V, the capacitance (in farads) is calculated as:

$$C \text{ (farads)} = \frac{Q \text{ (coulombs)}}{V \text{ (volts)}}$$

$$C(F) = \frac{10 \text{ C}}{2 \text{ V}}$$

$$C = 5 \text{ F}$$

NOTE: The unit of capacitance *(C)* is the farad (F). The unit of charge *(Q)* is the coulomb, also abbreviated as "C."

It should be noted that the farad is a large amount of capacitance. Practical capacitors are usually in the range of millionths of a farad (μF) or millionths of a millionth of a farad, formerly called micro-micro farad ($\mu\mu$F). This value is now called picofarad, pF. Using powers of 10, these values are expressed as:

$$\text{One microfarad} = 1 \ \mu\text{F} = 1 \times 10^{-6} \ \text{F}$$

$$\text{One picofarad} = 1 \ \text{pF} = 1 \times 10^{-12} \ \text{F}$$

PRACTICE PROBLEMS I

1. What is the capacitance of a capacitor that develops 25 V across its plates when storing a 100-μC charge? (Express answer in microfarads.)

2. What is the charge on a 10-μF capacitor when the capacitor is charged to 250 V? (Express answer in μC.)

3. What voltage develops across the plates of a 2-μF capacitor with a charge of 50 μC?

18–5 ENERGY STORED IN CAPACITOR'S ELECTROSTATIC FIELD

It was stated earlier that electrical energy supplied by the source during the capacitor charge is stored in the electrostatic field of the charged capacitor. This energy is returned to the circuit when the capacitor discharges.

For inductors, the energy (joules) stored in the magnetic field is related to the inductance and the current creating that field

$$W = \frac{1}{2}LI^2$$

However, for capacitors, the joules of stored energy are related to the capacitance and the voltage, or potential difference, between capacitor plates created by the electrostatic field. The formula for energy (in joules) is:

FORMULA 18–4   Energy (J) $= \frac{1}{2}CV^2$ or 0.5 CV^2

Example

How many joules are stored in a 5-μF capacitor that is charged to 250 V?

$$\text{Energy } (J) = \frac{1}{2}CV^2$$

$$\text{Energy} = \frac{5 \times 10^{-6} \times (250)^2}{2} = \frac{0.3125}{2} = 0.15625 \, \text{J}$$

PRACTICE PROBLEMS II

How much electrical energy is stored in a 10-μF capacitor that is charged to 100 V?

18–6 FACTORS AFFECTING CAPACITANCE VALUE

Capacitance value is related to how much charge is stored per unit voltage. The physical factors influencing the capacitance of parallel-plate capacitors are size of plate area, thickness of dielectric material (spacing between plates), and type of dielectric material, Figure 18–7. Let's examine these factors.

• **Plate area.** Obviously, the larger the area of the plates facing each other, the greater the number of electrons (and charge) that can be stored. Therefore

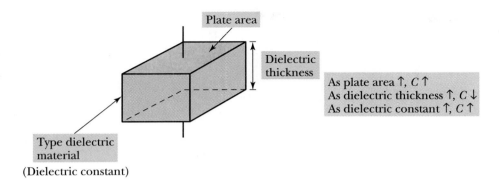

FIGURE 18–7 Physical factors affecting capacitance value

Plate area

Dielectric thickness

As plate area ↑, C ↑
As dielectric thickness ↑, C ↓
As dielectric constant ↑, C ↑

Type dielectric material
(Dielectric constant)

in Figure 18–7, C is directly related to plate area. For example, when the plate area doubles, there is room for twice as many electric lines of force with the same potential difference between plates. Thus, the amount of charge the capacitor holds per given voltage has doubled and the capacitance has doubled. The area of the plates directly facing each other equals the area of one plate. Using SI units, this area is expressed in square meters.

- **Distance between plates (dielectric thickness).** From previous discussions, it is logical to assume the further apart the plates, the less the electric field strength established for a given charge and potential difference between plates. Thus, the less the electrical energy or charge stored in this electric field. In other words, the *capacitance value is inverse to the spacing between plates.* Conversely, the closer the plates (the thinner the dielectric material), the stronger the electric field intensity per given voltage and the higher the capacitance value. In SI units, this distance (thickness of dielectric) is expressed in meters.

- **Type of dielectric material.** Different dielectric materials exhibit different abilities to concentrate electric flux lines. This property is sometimes called *permittivity,* which is analogous to conductivity of conductors. Vacuum (or air) has a permittivity of one. The relative permittivity of a given dielectric material is the ratio of the absolute permittivity (ϵ_o) of the given dielectric to the absolute permittivity of free space (vacuum) (ϵ_v). This relative permittivity is called the **dielectric constant** *(k)* of the material.

Dielectric constants

Refer to the chart in Figure 18–8 for examples of the average dielectric constant of various materials. The ratio reflecting the dielectric constant is shown as:

FORMULA 18–5 $\quad k = \dfrac{\epsilon_o \ \textit{(absolute permittivity of dielectric materials)}}{\epsilon_v \ \textit{(absolute permittivity of vaccum)}}$

(NOTE: Sometimes the symbol K_ϵ is used in lieu of k to represent dielectric constant.)

The higher the dielectric constant is, the greater the density of electric flux lines established for any given plate area and spacing between plates. Therefore, capacitance is directly related to the dielectric constant. For example, when the

FIGURE 18–8 Sample dielectric constants

KIND OF MATERIAL	APPROXIMATE K
Air	1.0
Glass	8.0
Waxed paper	3.5
Mica	6.0
Ceramic	100.0 +

NOTE: Values vary depending on quality, grade, environmental conditions, and frequency.

average dielectric constant (k or K_ϵ) of a given material is five, the capacitance is five times as great with this dielectric material compared with a dielectric of air or vacuum.

Dielectric strength

The electric field intensity is increased by increasing the voltage across a dielectric. As we continue increasing the potential difference across the capacitor plates, eventually a point is reached where the electron orbits in the dielectric material are stressed (distorted in their orbital paths), causing electrons to be torn from their orbits. At this time, the dielectric material breaks down and typically punctures or arcs, becoming a conductor. The material is no longer functioning as a nonconducting dielectric material (unless it is air or vacuum, which are self-healing). The breakdown value relates to the **dielectric strength** of the material. The practical measure of dielectric strength is called the *breakdown voltage*. The breakdown voltage depends on the kind of material and its thickness. Figure 18–9 lists some materials and their approximate average dielectric strength ratings in terms of volts per mil (thousandths of an inch) breakdown voltage ratings. Be aware these numbers will vary, depending on manufacture, frequency of operation, and so on.

18–7 CAPACITANCE FORMULAS

As you have learned, capacitance depends on the surface dimensions of plates facing each other, and the spacing and the dielectric constant of the material between plates. Using the appropriate factors to compute capacitance in farads us-

KIND OF MATERIAL	APPROXIMATE DIELECTRIC STRENGTH (VOLTS PER 0.001 INCH)
Air	80
Glass	200 + (up to 2kV)
Waxed paper	1,200
Mica	2,000
Ceramic	500 + (up to ≈ 1kV)

FIGURE 18–9 Examples of dielectric strengths

ing the SI units, the formula for capacitance of a parallel-plate capacitor with air (or vacuum) as the dielectric is:

FORMULA 18–6 $C = \dfrac{8.85A}{10^{12}s}$

where C = capacitance in farads
     A = area of the plates facing each other, in square meters
     s = spacing between plates, in meters
         the 8.85 constant relates to the absolute permittivity of air or vacuum

NOTE: If the plates have different dimensions, *use the smaller plate area for "A"* in the formula.

To calculate capacitance when dielectric materials other than air or vacuum are used, use this formula and insert the dielectric constant *(k)* in the formula. Thus:

FORMULA 18–7 $C = \dfrac{8.85kA}{10^{12}s}$

Again, C = capacitance in farads
     A = area of the plates facing each other in square meters
     k = the dielectric constant
     s = the spacing between the parallel plates in meters

For your information, a practical formula used to determine the capacitance in *microfarads* (μF) of multiplate capacitors is in Figure 18–10. NOTE: This formula uses inches and square inches, rather than meters and square meters.

FORMULA 18–8 $C = 2.25\dfrac{kA}{10^{7}s}(n - 1)$

where C = capacitance in μF
     k = dielectric constant in material between plates
     A = area of one plate in square inches
     s = separation between plate surfaces in inches
     n = number of plates

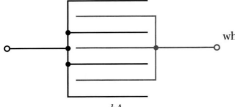

where C = capacitance in μF
     k = dielectric constant of dielectric
     A = area of one plate in square inches
     s = separation between plates in inches
     n = number of plates

$$C = 2.25\frac{kA}{10^{7}s}(n - 1)$$

FIGURE 18–10 Multiplate capacitor

Examples

1. A two-plate capacitor has a dielectric with a dielectric constant of 3. The plates are each 0.01 square meters in area and spaced 0.001 meter apart. What is the capacitance?

$$C = \frac{8.85kA}{10^{12}s} = \frac{8.85 \times 3 \times 0.01}{10^{12} \times 0.001}$$

$$C = \frac{0.2655}{1 \times 10^{9}} = 0.0000000002655 \text{ F}$$

$$C = 265.5 \text{ pF}$$

2. What is the capacitance (in microfarads) of a capacitor if the plate area of one plate is 2 square inches, and the dielectric has a k of 100 and is 0.005 inches thick? (Assume a two-plate capacitor.)

$$C = 2.25\frac{kA}{10^{7}s}(n - 1)$$

$$C = \frac{2.25 \times 100 \times 2}{10^{7} \times 0.005} \times (2 - 1) = \frac{450}{50,000} \times 1 = 0.009 \text{ μF}$$

PRACTICE PROBLEMS III

1. A capacitor has four plates of equal size. Each plate has an area of 1 square inch. This capacitor uses a dielectric having a k of 10. The dielectric thickness between each pair of plates is 0.1 inch. What is the capacitance value in microfarads?

2. Each plate of a two-plate capacitor has an area of 0.2 square meters. The plates are uniformly spaced 0.005 meters apart and the dielectric is air. What is the capacitance in picofarads?

PRACTICAL NOTES

As a technician, you will rarely, if ever, have to actually measure capacitor plate dimensions, plate-spacing, and so on, and use the formulas we have shown you to calculate capacitance value. These formulas were introduced as a reinforcement tool to help you understand the important physical factors that influence the value of capacitance.

Technicians do, however, have to check and measure capacitances quite routinely. There are several pieces of test equipment that can be used for doing this (see Figure 18–11):

1. DMMs that have capacity-checking range(s)

2. "Dedicated" digital capacitance meters or checkers

3. LCR meters that can check inductance, resistance, and/or capacitance

4. Other (such as bridges and dip-meters)

(a)

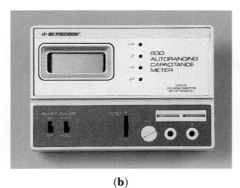

(b)

(c)

FIGURE 18–11 Test instruments used to measure capacitance *(Photo **a** courtesy of John Fluke Mfg. Co., Inc.; photos **b** and **c** courtesy of B & K Precision)*

18–8 TOTAL CAPACITANCE IN SERIES AND PARALLEL

As a technician, you may have to determine circuit parameters with capacitors connected in series or parallel. The following information will help you learn to perform such required circuit analysis.

Capacitance of series capacitors

Refer to Figure 18–12 as you study the following comments.

1. Connecting two capacitors in series increases the dielectric thickness, while the plate areas connected to the source remain the same, Figure 18–12. From our previous discussions, you can surmise that connecting capacitors in series *decreases* the resultant total capacitance.

2. Finding total capacitance of *capacitors in series* is analogous to finding total resistance of *resistors in parallel.*

3. The formulas to determine total capacitance (C_T) of series-connected capacitors are as follows:

   a. For two capacitors in series:

FORMULA 18–9 $C_T = \dfrac{C_1 \times C_2}{C_1 + C_2}$

FIGURE 18–12 Total capacitance of series capacitors

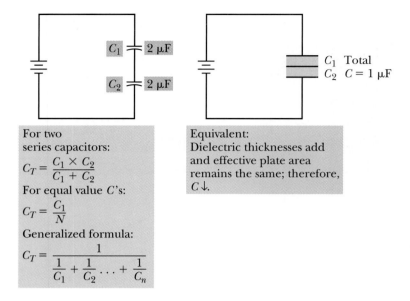

For two series capacitors:
$$C_T = \frac{C_1 \times C_2}{C_1 + C_2}$$
For equal value C's:
$$C_T = \frac{C_1}{N}$$
Generalized formula:
$$C_T = \frac{1}{\dfrac{1}{C_1} + \dfrac{1}{C_2} \cdots + \dfrac{1}{C_n}}$$

Equivalent:
Dielectric thicknesses add and effective plate area remains the same; therefore, $C\downarrow$.

b. For equal-value capacitors in series:

FORMULA 18–10 $\qquad C_T = \dfrac{C}{N}$

where C is the value of one of the equal-value capacitors and N is the number of capacitors in series.

c. The general formula for any number and value of capacitors in series:

FORMULA 18–11 $\qquad C_T = \dfrac{1}{\dfrac{1}{C_1} + \dfrac{1}{C_2} + \dfrac{1}{C_3} \cdots + \dfrac{1}{C_n}}$

Voltage distribution related to series capacitors

Because the current throughout a series circuit is the same at any point, capacitors connected in series have the same number of coulombs (C) of charge (Q). The potential difference between plates of equal-value capacitors for a given charge is equal. But for different-value capacitors, potential difference between plates is not equal. Recall the $V = Q/C$ formula, where V is directly related to Q and inversely related to C. This indicates in a series circuit (since the Q is the same), the voltage across any given capacitor is *inversely* proportional to its C value. For example, if we have a 5-μF capacitor and a 10-μF capacitor in series, the voltage across the 5-μF capacitor is two times the voltage across the 10-μF capacitor, Figure 18–13. To obtain capacitances in the ranges described (i.e., multi-microfarad), the electrolytic capacitor is typically used. You will study about this capacitor later in the chapter. However, you should take note of the preceding comments and following examples about voltage distribution.

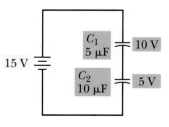

FIGURE 18–13 Voltage distribution across series capacitors

PRACTICAL NOTES

Because of the large tolerances typical of electrolytics, the voltage distribution may not "track" the theoretical values computed by using the "marked values" on the capacitors. If the capacitors' values are precisely as marked, the resulting voltage distribution will be as indicated in the preceding discussion.

Example

1. What is the total capacitance of the circuit in Figure 18–14?

$$C_T = \frac{C_1 \times C_2}{C_1 + C_2} = \frac{3 \ \mu F \times 6 \ \mu F}{3 \ \mu F + 6 \ \mu F} = \frac{18 \ \mu F}{9} = 2 \ \mu F$$

2. What is the total capacitance of three 10-μF capacitors connected in series?

$$C_T = \frac{C}{N} = \frac{10 \ \mu F}{3} = 3.33 \ \mu F$$

3. When the voltage applied to the circuit in Figure 18–14 is 90 V, what are the V_{C_1} and V_{C_2} values?

Because the voltage distribution is *inverse* to the capacitance values, V_{C_1} equals twice the value of V_{C_2}, or two-thirds of the applied voltage. Thus,

$$V_{C_1} = \frac{2}{3} \times 90 = 60 \text{ V, and } V_{C_2} = \frac{1}{3} \times 90 = 30 \text{ V}$$

An alternate solution is to find Q by using the total capacitance value in the formula $Q = C_T \times V_T$. Then solve for each capacitor voltage value using $V = Q/C$. Therefore, since the total capacitance is 2 μF:

$$Q = 2 \times 10^{-6} \times 90 = 180 \ \mu C$$

$$V_{C_1} = \frac{180 \ \mu C}{3 \ \mu F} = 60 \text{ V}$$

$$V_{C_2} = \frac{180 \ \mu C}{6 \ \mu F} = 30 \text{ V}$$

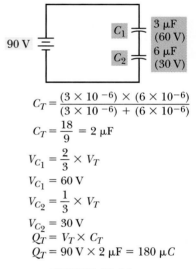

$$C_T = \frac{(3 \times 10^{-6}) \times (6 \times 10^{-6})}{(3 \times 10^{-6}) + (6 \times 10^{-6})}$$

$$C_T = \frac{18}{9} = 2 \ \mu F$$

$$V_{C_1} = \frac{2}{3} \times V_T$$

$$V_{C_1} = 60 \text{ V}$$

$$V_{C_2} = \frac{1}{3} \times V_T$$

$$V_{C_2} = 30 \text{ V}$$

$$Q_T = V_T \times C_T$$

$$Q_T = 90 \text{ V} \times 2 \ \mu F = 180 \ \mu C$$

FIGURE 18–14

PRACTICE PROBLEMS IV

1. What is the total capacitance of the circuit in Figure 18–15?

2. What is the total capacitance of four 20-μF capacitors connected in series?

3. What are the voltages across each capacitor in Figure 18– 15?

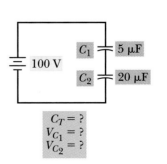

$$C_T = ?$$
$$V_{C_1} = ?$$
$$V_{C_2} = ?$$

FIGURE 18–15

15 μF C_X

V_S
60 V

30 μF

30 μF

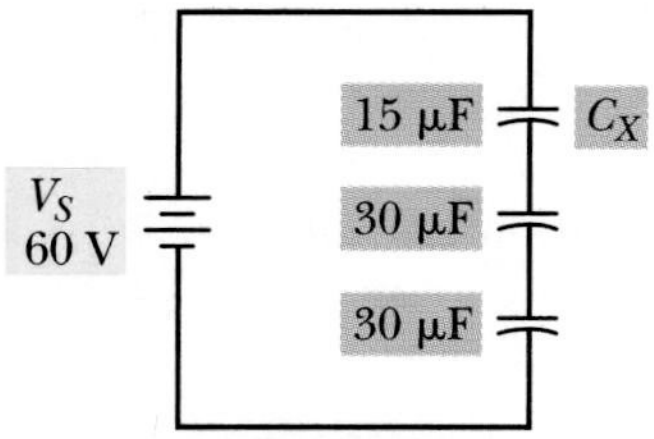

$$V_X = V_S \left(\frac{C_T}{C_X} \right)$$

$$V_X = 60 \left(\frac{7.5}{15} \right)$$

$$V_X = 60 \times 0.5$$

$$V_X = 30 \text{ V}$$

FIGURE 18–16 Series
capacitors voltage formula

18–9 FINDING VOLTAGE WHEN THREE OR MORE CAPACITORS ARE IN SERIES

A handy formula used to find voltage across a given capacitor when three or more capacitors are in series is (see Figure 18–16):

FORMULA 18–12 $V_X = V_S \dfrac{C_T}{C_X}$

where V_X = voltage across capacitor "x"
       V_S = the dc source voltage
       C_T = total series capacitance
       C_X = value of capacitor "x"

Capacitance of parallel capacitors

Refer to Figure 18–17 as you study the following information. Since the top and bottom plates of all the parallel capacitors are electrically connected to each other, the combined plate area equals the sum of the individual capacitor plate areas. However, the dielectric thicknesses do not change and are not additive.

Increasing the effective plate area while keeping the dielectric thickness(es) the same results in the total capacitance of the parallel-connected capacitors equaling the sum of the individual capacitances. In other words, *capacitances in parallel add like resistances add in series*. The formula for total capacitance (C_T) of parallel-connected capacitors is:

FORMULA 18–13 $C_T = C_1 + C_2 \ldots + C_n$

Charge distribution related to parallel capacitors

Because the voltage across components connected in parallel is the same, the charge (in coulombs) contained by each capacitor in a parallel-capacitor circuit is *directly* related to each capacitance value. That is, the larger the capacitance value, the greater the charge for a given voltage, $Q = CV$. For example, a 10-μF, 20-μF, and 30-μF capacitor are in parallel, and the voltage applied to this circuit is 300 V. Thus, the charge (Q) on the 10-μF capacitor is half that on the 20-μF ca-

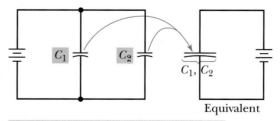

C_1 C_2 C_1, C_2
Equivalent

When plate areas are combined,
dielectric thickness remains unchanged.

FIGURE 18–17 Capacitance
of parallel capacitors

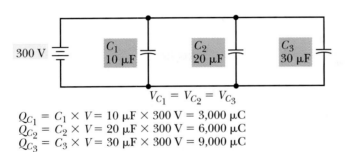

FIGURE 18–18 Charge distribution on parallel capacitors

$$V_{C_1} = V_{C_2} = V_{C_3}$$

$Q_{C_1} = C_1 \times V = 10 \ \mu F \times 300 \ V = 3{,}000 \ \mu C$
$Q_{C_2} = C_2 \times V = 20 \ \mu F \times 300 \ V = 6{,}000 \ \mu C$
$Q_{C_3} = C_3 \times V = 30 \ \mu F \times 300 \ V = 9{,}000 \ \mu C$

Charge on parallel capacitors is directly related to each one's capacitance value.

pacitor and one-third that on the 30-μF capacitor ($Q = CV$). The Q on the 10-μF capacitor equals 10 μF × 300 V = 3,000 μC. The charge on the 20-μF capacitor equals 20 μF × 300 V = 6,000 μC. The charge on the 30-μF capacitor equals 30 μF × 300 V = 9,000 μC.

NOTE: Because total charge equals the sum of the charges, it can also be computed from the $Q_T = C_T \times V_T$ formula. Therefore, total charge = 60 μF × 300 V, so Q_T = 18,000 μC, Figure 18–18.

Examples

1. What is the total capacitance of a parallel circuit containing one 15-μF capacitor, two 10-μF capacitors, and one 3-μF capacitor?

$$C_T = C_1 + C_2 + C_3 + C_4 = 15 + 10 + 10 + 3 \ \mu F = 38 \ \mu F$$

2. What is the charge on the 3-μF capacitor, if the voltage applied to the circuit in question 1 is 50 V?

$$Q = CV = 3 \times 10^{-6} \times 50 = 150 \ \mu C$$

PRACTICE PROBLEMS V

1. What is the total capacitance of five 12-μF capacitors connected in parallel?

2. If V_T equals 60 V, what is the charge on each capacitor in problem 1? What is the total charge on all capacitors?

18–10 THE RC TIME CONSTANT

At the beginning of the chapter, we said a capacitor opposes a *change* in voltage. You have seen how charge current is required to establish a potential difference between the capacitor plates. Obviously, the rate of charge for a given capacitor is directly related to the charge current. That is, the higher the current is, the quicker the capacitor is charged to V_S.

You also know current allowed to flow is *inversely* related to the resistance in the current path. Therefore, the larger the resistance value in the charge (or discharge) path of the capacitor, the longer the time required for voltage to reach V_S across the capacitor during charge, or to reach 0 V during discharge.

The larger the capacitance *(C)* value is, the more charge required to develop a given potential difference between its plates, $V = Q/C$. Because a capacitor prevents an instantaneous voltage change across its plates (if there is any resistance in the charge path or discharge path), a capacitor is said to oppose a *change* in voltage across itself.

Recall when inductors were discussed in a previous chapter, there was an L/R time constant relating the time required to change current through the coil. You learned that 5 L/R time constants are needed for a complete change from one circuit current value to the new level.

In like manner, an **RC time constant** relating the time required to charge (or discharge) a capacitor exists. *To completely change from one voltage level to a "new level" of source voltage,* the capacitor needs 5 *RC* time constants. *One RC time constant (τ)* = R (in ohms) $\times$ C (in farads). That is:

FORMULA 18–14 $\tau = R \times C$

Refer to Figure 18–19 as you study the following information about a typical series *RC* circuit.

The sum of the dc voltages across the capacitor and the resistor equals the source voltage (V_S) at all times.

As the capacitor is charged, the voltage across the capacitor is series-opposing the source voltage. Therefore, the voltage across the resistor at any given moment equals the source voltage minus the capacitor voltage. That is, $V_R = V_S - V_C$. Thus, as the capacitor charges to the V_S value, the resistor voltage decreases toward 0 V. When the capacitor is fully charged, V_C equals V_S and zero current flows, and no *IR* drop occurs across the resistor.

When the switch is initially closed, the capacitor has zero charge and zero voltage. Therefore, the source voltage is dropped by the resistor *(R)*. Because this is a series circuit, and the current is the same throughout a series circuit, Ohm's law reveals what the charging current is at that moment.

$$I = \frac{V_R}{R}$$

In other words, at that moment, the capacitor behaves like a short; therefore, only the resistor limits current.

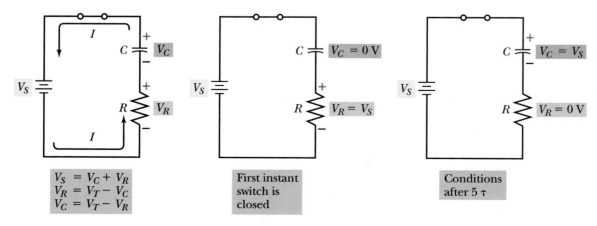

$$V_S = V_C + V_R$$
$$V_R = V_T - V_C$$
$$V_C = V_T - V_R$$

First instant switch is closed

Conditions after 5 τ

FIGURE 18–19 Voltages in an *RC* circuit

As the charging current flows, a series-opposing potential difference across the capacitor develops, and the charging current diminishes at a nonlinear rate (toward zero). The *IR* drop across the resistor drops accordingly. For example, after a certain interval of time, the capacitor is charged to half of V_S. At that time, half the source voltage is across the resistor, and one-half the source voltage is across the capacitor.

The percentage of voltage across the capacitor at any given moment relates to the *RC* time constant, where one time constant *(in seconds)* = R (Ω) × C (farads). Often the Greek lowercase letter tau (τ) is used to represent one time constant. Note also that τ (sec) = R (megohms) × C (microfarads). Also, observe that τ *(μsec)* = R (Ω) × C (μF).

Refer to Figure 18–20 (curve A) and notice that after one time constant *(τ)*, *the capacitor voltage (V_C)* equals 63.2% of V_S. After 2 τ, V_C equals 86.5% of V_S; after 3 τ, V_C equals 95%; after 4 τ, V_C equals 98.5%; and after 5 τ, V_C equals more than 99% of V_S; therefore, it is considered fully charged. Are these percentages and corresponding numbers familiar? Of course! They are the same percentages and numbers used when the *L/R* time constant was discussed. In other words, the same exponential curves (or time constant chart curves) are applicable for analyzing voltages in a capacitor circuit as were used to analyze current in an inductor circuit.

Note in Figure 18–20 that curve B illustrates what happens to the resistor voltage as the capacitor is charged. That is, at the first instant V_R equals V_S. As the capacitor charges, V_R decreases toward zero. In effect, the resistor voltage and

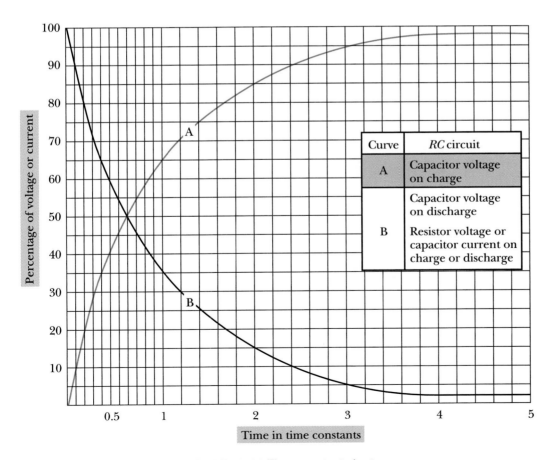

FIGURE 18–20 Time-constant chart

the circuit current are inverse to the capacitor voltage. As the capacitor voltage increases, the circuit current, and thus the voltage across the resistor, decrease.

Using exponential formulas to find voltage(s)

Values along the exponential curves on the time constant chart in Figure 18–20 are more precisely found by using some simple formulas. The values along the descending exponential curve relate to the base of *natural logarithms,* called epsilon (ϵ), which has a value of 2.718. (Recall this number of natural growth was discussed in an earlier chapter.)

When a capacitor is charging in a series RC circuit, the voltage across the resistor during the first instant is maximum, then decreases exponentially as the capacitor charges to the new voltage level. Thus, the resistor voltage follows the descending exponential curve shown in the time constant chart. The level of v_R for any given point of time relates to the length of time the capacitor has charged relative to the circuit's RC time constant. That is, the time allowed (t) divided by the time of 1 RC time constant (τ), or t/τ. Obviously, when the t (time allowed) equals 5 τ or more, then the capacitor has fully charged, no current is flowing, and v_R equals zero. When t is less than 5 τ, then v_R equals a value that is found by the formula:

FORMULA 18–15 $v_R = V_S \times \epsilon^{-t/\tau}$

where ϵ = epsilon (or 2.71828)
  t = time allowed in seconds
  $\tau = R \times C$ (or 1 time constant)

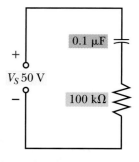

FIGURE 18–21

Example

For the circuit in Figure 18–21 find v_R after 10,000 µs.

$$v_R = V_S \times \epsilon^{-t/\tau}$$

$$v_R = 50 \times 2.718^{-10,000/(100,000 \times 0.1)}$$

$$v_R = 50 \times 2.718^{-1}$$

$$v_R = 50 \times 0.3679$$

$$v_R = 18.395 \text{ V}$$

PRACTICAL NOTES

Using a calculator to find the ϵ value raised to the $-t/\tau$ power is easy. You simply enter the result of dividing t by τ and then press the **+/−** key to make it negative. Following that, you press the **2nd** key, then the **e^x** key to get the natural antilogarithm of the displayed number.

For our preceding example then, the keystrokes to find v_R would be:
1, **+/−**, **2nd**, **e^x**, **×**, 50, **=**

Example

For the circuit in Figure 18–21, what is the V_R after 1.8 time constants?

$$v_R = V_S = \epsilon^{-t/\tau}$$

$$v_R = 50 \times 2.718^{-1.8}$$

$$v_R = 50 \times 0.1653$$

$$v_R = 8.265 \text{ V}$$

The calculator keystrokes that can be used to perform the calculations in this example are:

1.8, **+/−**, **2nd**, **eˣ**, **×**, 50, **=**

The 1.8, **+/−**, **2nd**, **eˣ**, sequence gives the result of raising 2.718 to the negative 1.8th power, which is 0.1653. This result times the 50 V (V_A) provides the value of voltage across the resistor (8.265 V).

The capacitor voltage is the difference between the resistor voltage and the new voltage applied or source voltage for any given instant. A formula for v_C is:

FORMULA 18–16 $v_C = V_S(1 - \epsilon^{-t/\tau})$

NOTE: The small letter v denotes an instantaneous voltage value for a given moment in time.

For the first example then:

$$v_C = 50 \times (1 - 0.3679)$$

$$v_C = 50 \times 0.6321$$

$$v_C = 31.605 \text{ V}$$

To check, add the V_R and the v_C (18.395 + 31.605); the result is 50 V (our applied voltage). It checks out!

For the second example (see Figure 18–21):

$$v_C = 50 \times (1 - 0.1653)$$

$$v_C = 50 \times 0.8347$$

$$v_C = 41.735 \text{ V}$$

When we add v_R (8.265 V) + v_C (41.735 V), we get 50 V. It checks!

PRACTICE PROBLEMS VI

Use either the time-constant chart or the formulas and a scientific calculator to solve the following.

1. What is the voltage across the capacitor of the circuit in Figure 18–22, 20 μs after the switch is closed?

2. What is the voltage across the resistor in Figure 18–22, 15 μs after the switch is closed? What is the charging current value at that moment?

3. What is the value of v_C 30 μs after the switch is closed?

4. What is the value of v_R 25 μs after the switch is closed?

FIGURE 18–22

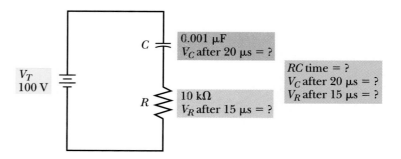

V_T 100 V

C — 0.001 µF
V_C after 20 µs = ?

R — 10 kΩ
V_R after 15 µs = ?

RC time = ?
V_C after 20 µs = ?
V_R after 15 µs = ?

18–11 TYPES OF CAPACITORS

General classification

There are two basic classifications of capacitors—*fixed* and *variable*. These capacitors come in various configurations, sizes, and characteristics. As the two names imply, fixed capacitors have one value of capacitance; variable capacitors have a range of capacitance values that may be adjusted, Figure 18–23. By changing the position of the *rotor* plates with respect to the fixed *stator* plates, Figure 18–23a, the plate areas facing each other are varied. This changes the capacitance value.

When the plates are fully meshed, capacitance is maximum. When the plates are fully unmeshed, capacitance value is minimum. The dielectric for this type of capacitor is air.

Another type of variable capacitor is shown in Figure 18–23b. This compression-type capacitor has only one movable plate. By turning the screw, the spacing between the movable plate and the fixed plate is varied thus changing the capacitance value. When the movable plate is screwed tight, capacitance is maximum. As the screw is loosened, capacitance decreases.

Capacitors and their characteristics

Capacitors are generally referred to in terms of their dielectric material. The following discussion describes several common capacitors and their features.

1. *Paper and plastic capacitors* use a variety of dielectric materials. Some of these include mylar, polystyrene and polyethylene for the plastic types, and waxed or oiled paper for the older, less expensive paper type. Typically, the plates

FIGURE 18–23 Variable capacitors

(a)   (b)

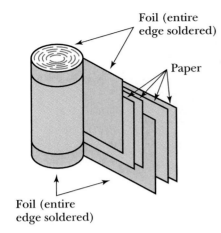

Foil (entire edge soldered)

Paper

Foil (entire edge soldered)

FIGURE 18–24 Paper-type capacitor

Typical available capacitance values: Approximately 0.001 to 1.0 μF

Typical working voltage dc (WVDC): 100 to 1,500 V

are long strips of tinfoil separated by the dielectric material. The foil and dielectric material are commonly rolled into a cylindrical component, Figure 18–24. It is interesting to note that because of their construction, each plate has two active surfaces. This means to calculate the plate areas, use twice the area of one plate, rather than only the area of one plate. Typical capacitance values for this type capacitor range from about 0.001 μF to 1 μF or more. Paper and plastic capacitors are typically used in fairly low- frequency applications such as in audio amplifiers.

2. *Mica capacitors* use mica for the dielectric material. Because the mica dielectric has a high breakdown voltage, these low-capacitance, high-voltage capacitors are frequently found in high-voltage circuits. Often, their construction is alternate layers of foil with mica that is molded into a plastic case, Figure 18–25. They are compact, moisture-proof and durable. Voltage ratings are in thousands of volts. Typical capacitance values range from about 5 to 50,000 picofarads, depending on voltage ratings.

3. *Ceramic capacitors* are typified by their small size and high-dielectric strength. Ceramic capacitors generally come in the shape of a flat disk (disk-ceramics), Figure 18–26a, or in cylindrical shapes. These capacitors are also compact, moisture-proof, and durable. Typical available ranges having 1,000 V ratings are from approximately 5 pF to about 5,000 pF. At lower voltage ratings, higher capacitance values are available. Because of good dielectric characteristics, both mica and ceramic capacitors can be used in applications from the audio-frequency range up to several hundred megahertz. Figure 18–26b shows a sample of typical coding used with disk ceramic capacitors.

4. *Electrolytic capacitors* have several prominent characteristics, including:
   a. high capacitance-to-size ratio;
   b. polarity sensitivity and terminals marked + and −;
   c. allow more leakage current than other types; and
   d. have their *C* value and voltage rating printed on them.

Refer to Figure 18–27 for the typical construction of electrolytic can-type capacitors. The external aluminum can or housing is typically the negative plate (or electrode). The positive electrode external contact is generally aluminum foil immersed or in contact with an electrolyte of ammonium borate (or equivalent). If the electrolyte is a borax solution, the capacitor is called

FIGURE 18–25 Mica capacitors *(Photo by Michael A. Gallitelli)*

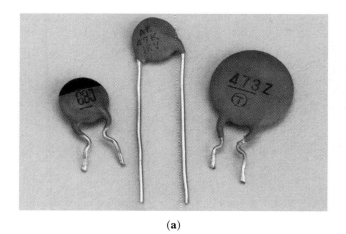

(a)

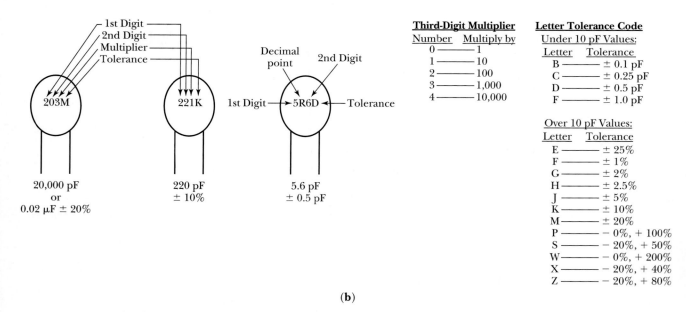

(b)

FIGURE 18–26 (a) Ceramic capacitors; **(b)** sample ceramic disk capacitors coded markings

a *wet electrolytic*. If the electrolyte is a gauze material saturated with borax solution, the capacitor is called a *dry electrolytic*. To create the dielectric, a dc current is passed through the capacitor. This causes a very thin aluminum oxide film (about 10 microcentimeters thick) to form on the foil surface. This thin oxide film is the dielectric. (NOTE: This type of capacitor is *not* named after its dielectric.) Because of the extremely thin dielectric, capacitance value per given plate area is very high. Also, voltage breakdown levels are relatively low. Furthermore, since the dielectric is very thin, some leakage current occurs. Typically, this leakage is a fraction of a milliampere for each microfarad of capacitance.

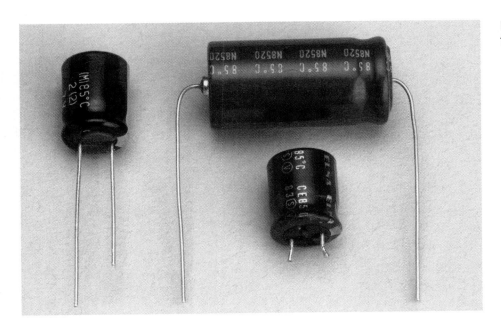

FIGURE 18–27 Aluminum-type electrolytic capacitors

SAFETY HINTS

A **very important** fact about electrolytics is you must *observe polarity* when connecting them into a circuit. If current is passed in the wrong direction through the capacitor, the chemical action that created the dielectric layer will be reversed and will destroy the capacitor. A shorted capacitor is actually created. Gas may build up in the component, causing an explosion! Depending on the voltage ratings, common values of electrolytic capacitors range from 2 µF to several hundred µF. Usual voltage ratings for this capacitor range from 10 V to about 500 V.

The main advantage of the electrolytic capacitor is the large capacitance-per-size factor. Two obvious disadvantages are the polarity, which must be observed, and the higher leakage current feature. Also, in many capacitors of this type, the electrolyte can dry out with age and depreciate the capacitor quality or render it useless.

PRACTICAL NOTES

Some specially made "nonpolarized" electrolytic capacitors are used in ac applications. One application for this type capacitor is as a motor-starting capacitor. This application requires a large capacitance value that must operate in an ac environment. These capacitors are actually constructed by putting two electrolytic capacitors "back-to-back" in one package. To function properly, they are connected with series-opposing polarities.

FIGURE 18–28 Tantalum
electrolytic capacitors

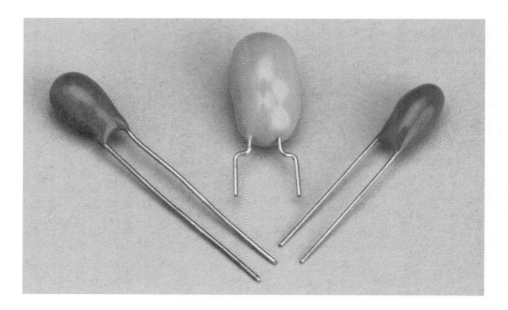

Because of the losses of the dielectric at higher frequencies, electrolytic capacitor applications are generally limited to power-supply circuits and audio-frequency applications.

5. *Tantalum capacitors,* Figure 18–28, are in the electrolytic capacitor family. They use tantalum instead of aluminum. One important quality of tantalum capacitors is they provide very high capacities in small-sized capacitors. They have lower leakage current than the older electrolytics. Also, they do not dry out as fast, thus have a longer shelf-life. Because they are generally only manufactured with low-voltage ratings, they are used in low-voltage semiconductor circuitry. Tantalum capacitors are expensive.

6. *Chip (SMT) capacitors,* Figure 18–29a and b. Very small in size, these devices are primarily used as compact components on PC board–type circuitry where space is at a premium. Also, the construction style, with virtually no lead length, enables these components to have minimal stray inductance or capacitance. This makes them useful in high-frequency applications. Materi-

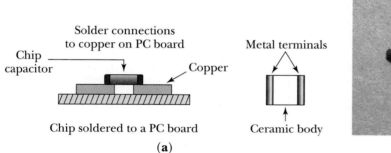

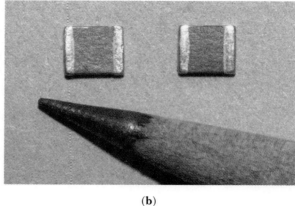

(a)

(b)

FIGURE 18–29 "Chip" capacitors

als commonly used in the construction of these devices include layers of conductive material, using ceramic as the dielectric between layers. As seen in Figure 18–29a, the general exterior portions of the chip capacitor are the ceramic-type body and the metallic end contacts. These metallic end pieces are the electrical contact points to the capacitor. Their miniature size makes marking their value difficult. Therefore, they are typically marked with a short two-character code (which you may need a magnifying glass to read). This two-letter code consists of a letter, which indicates the significant digits in its value, and a number, which indicates the multiplier (or how many zeroes should follow the significant digits). This coding indicates the capacitor value in pF.

Observe Figure 18–30 to see the meaning of the letters and numbers in this two-character coding system.

Surface-Mount Technology (SMT) "Chip" Capacitor Coding

SMT Capacitor Significant Figures Letter Code				SMT Capacitor Multiplier Code	
Character	Significant Figures	Character	Significant Figures	Number	Decimal Multiplier
A	1.0	R	4.3	0	1
B	1.1	S	4.7	1	10
C	1.2	T	5.1	2	100
D	1.3	U	5.6	3	1,000
E	1.5	V	6.2	4	10,000
F	1.6	W	6.8	5	100,000
G	1.8	X	7.5	6	1,000,000
H	2.0	Y	8.2	7	10,000,000
J	2.2	Z	9.1	8	100,000,000
K	2.4	a	2.5	9	0.1
L	2.7	b	3.5		
M	3.0	d	4.0		
N	3.3	e	4.5		
P	3.6	f	5.0		
Q	3.9	m	6.0		
		n	7.0		
		t	8.0		
		y	9.0		

(a)

Decoding Examples:

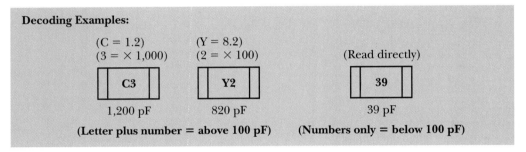

(C = 1.2)
(3 = × 1,000)

C3

1,200 pF

(Y = 8.2)
(2 = × 100)

Y2

820 pF

(Letter plus number = above 100 pF)

(Read directly)

39

39 pF

(Numbers only = below 100 pF)

(b)

FIGURE 18–30

Ratings

Several important factors should be observed when selecting or replacing capacitors in various applications. Some of these include:

- physical size and mounting characteristics;
- capacitance value (in microfarads or picofarads);
- capacitance tolerance (in percentage of rated value);
- voltage ratings (e.g., working volts dc or WVDC);
- safe temperature operating range;
- temperature coefficient (for temperature-compensating ceramic capacitors used in tuned rf circuits);
- power factor of the capacitor (expressing the loss characteristics of dielectric used in capacitor); and
- inductance characteristics (when used at high frequencies).

Taking into consideration these factors, Figure 18–31 shows examples of typical ratings for various types of capacitors.

18-12 TYPICAL COLOR CODES

Capacitors are color-coded with various systems. It is not critical to learn these systems at this time. However, you should be aware that they exist.

Appendix A displays some common methods of color-coding for mica, ceramic disc, ceramic tubular, and paper capacitors. As in other color-coding systems, the position and color of the dots or bands conveys the desired information. (NOTE: Electrolytic capacitors are generally large enough so that the information is printed on their body, rather than using color codes. Also, ceramic capacitors often have the critical information printed on them.)

TYPICAL RANGE OF RATINGS FOR VARIOUS CAPACITORS
(Shown in standard electronic parts catalogs)

CAPACITOR TYPE	CAPACITANCE RATINGS	VOLTAGE RATINGS	TOLERANCE RATINGS	OPER. TEMPS	TEMP. COEFF.
Paper	0.001–1.0 μF	100–1,500 V	±10% (common)	−40 to +85°C	N/A
Mica	5–50,000 pF	600 to several kV	±1–5%	−55 to +125°C	N/A
Ceramic	1–10,000 pF	1 kV–6 kV	±10–20%	−55 to +85°C	N or P 0–750
Electrolytic	10–1,000 μF	5–500 V	−10–+50%	−40 to +85°C	N/A

NOTE: Temperature coefficient ratings are generally given for ceramic-type capacitors that are designed to increase or decrease in capacitance with temperature change. The rating is normally given in "parts-per-million per degree Celsius." For example, if a capacitor decreases 500 ppm per degree Celsius, it would be labeled "N500."

FIGURE 18–31 Summary of typical ratings

18–13 TYPICAL PROBLEMS AND TROUBLESHOOTING TECHNIQUES

Capacitors develop problems more frequently than resistors or inductors. As with the other components, troubles may be related to a shorted capacitor, an open capacitor, or degenerated performance caused by excessive leakage, age, and other causes.

Causes of open capacitors

One reason a capacitor acts similar to an open is that a broken connection occurs between an external lead and a capacitor plate. In fact, this situation can occur with any capacitor! In electrolytic capacitors, the electrolyte resistance can drastically increase because of drying out, age, or operation at higher than normal temperatures.

Causes of shorted capacitors

Shorts can occur because of external leads touching or plates shorted internally. Internal shorts are most common with paper capacitors and electrolytic capacitors. Common causes for the dielectric material breaking down are age and usage. Again, operating under high temperature conditions accelerates this aging process. In paper electrolytic capacitors, the dielectric material sometimes deteriorates with age, which decreases leakage resistance. The resulting increase in leakage current is equivalent to a *partial short* condition.

Troubleshooting techniques

Two general methods for testing capacitors are *in-circuit* testing and *out-of-circuit* testing. There are capacitance testers that measure capacitance values, leakage characteristics, and so on; however, the ohmmeter is often useful for making out-of-circuit general-condition tests. Because of the capacitor charging action, there typically is a current surge when a voltage is first applied to the capacitor, and the current decreases to zero (or to leakage-current level) when the capacitor is fully charged. Often, this action is monitored with the ohmmeter mode on a VOM.

Before working on a circuit, make sure all appropriate power supply capacitors are discharged, even though the power has been removed! Sometimes the "bleeder resistor" fails, and someone can get injured or killed!

Using the ohmmeter for general tests

General precautions

When using the ohmmeter to check electrolytic capacitors, you should *observe polarity*. First, determine the voltage polarity at the test leads of the ohmmeter when in the ohms function. Frequently, the red lead is positive and the black lead is negative *but not* for all brands and models of instruments—check with the instrument's literature or another voltmeter. Next, observe the polarity markings on the capacitor, as appropriate. Also, to protect the meter, *make sure the capacitor to be checked is discharged* before testing.

Action with good capacitors

An out-of-circuit test is where one capacitor lead is disconnected from the circuit, or the capacitor is physically removed from the circuit. When checking a capacitor, use a high range (e.g., $R \times 1$ meg selector position) and connect the ohmmeter leads across the capacitor, Figure 18–32. As shown, the ohmmeter pointer initially deflects close to the zero ohms position. As the capacitor charges to the voltage provided by the ohmmeter battery, the pointer moves across the scale toward the infinite resistance scale position, and eventually rests at a very high resistance. For small-value capacitors, this value can be close to infinite. For electrolytic capacitors, the leakage resistance by this method varies from about 0.25 to 1 megohm, depending on the capacitor value and condition. When the R reading is less than this, the capacitor probably needs to be replaced.

For small-value capacitors, the charging deflection action is very quick and little pointer movement is seen. For larger values of capacitors, the charging action and resultant deflection are slower because of the larger RC time constant involved.

With leads first connected, Ω reading low

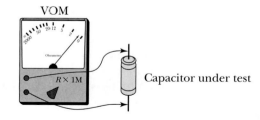

Capacitor under test

As capacitor charges, Ω reading increases

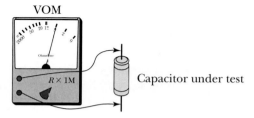

Capacitor under test

After good capacitor is charged, Ω reading stable at high R value

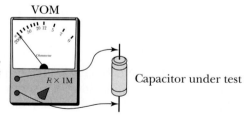

Capacitor under test

FIGURE 18–32 Checking capacitors with an ohmmeter

Open capacitors

When an ohmmeter is connected across an open capacitor, the pointer does not move (or an infinite Ω reading is shown), and no charging action is observed. Of course, the technician should be aware of the capacitor type and value being checked. For example, small-value capacitors (such as mica capacitors) may not provide an observable deflection, while large-value capacitors should provide some discernable charging action.

Shorted capacitors

When an ohmmeter is connected across a shorted capacitor, the pointer moves to 0 Ω on the scale and remains there. Sometimes, a variable capacitor short is remedied by locating the plate(s) that touch. In this case, simply bend the plates apart. Fixed capacitors indicating a shorted condition must be replaced.

Excessive leakage indication

When an ohmmeter is connected across a capacitor having excessively high leakage current (excessively low leakage resistance), the obvious result is a lower-than-normal resistance reading when the pointer finally comes to rest. The capacitor shows a charging action; however, the leakage resistance is lower than it should be.

Replacing electrolytic capacitors

Electrolytic capacitors definitely deteriorate with age. Their electrolyte may dry out, causing an increase in resistance and degeneration of operation. Also, the dielectric resistance greatly decreases from constant use causing excessive leakage current and reduced effectiveness in operation. When selecting an electrolytic capacitor for replacement, pay attention to the age of the capacitor—shelf-life is an important consideration.

Closing comments

Capacitors are tremendously useful components found in most electronic circuits. Their characteristics of opposing a change in voltage, of blocking dc current once charged, and of storing electrical energy make them useful in numerous applications.

You should be aware that resistance, inductance, and capacitance are present in each of the three basic components you have studied thus far. Also, there is stray, or "distributed," capacitance and inductance present in most circuits. For example, wires and conductors that carry current exhibit some inductance, even though it may be a small amount. The leads and foil of capacitors have some inductance, as well. There is capacitance between the turns of an inductor. Distributed capacitance is present between wires and chassis, or ground. Both inductance and capacitance are present along the length of antenna elements. Coaxial cables have capacitance between conductors and inductance along their length. A resistor has a small degree of both inductance and capacitance present because of its structure.

These stray, or distributed, capacitances and/or inductances normally do not affect circuits operating at low frequencies. However, in your later studies of high-frequency components and circuits, you will see they can play a part in proper or improper operation of circuits and systems. You will learn that the effects of some of these stray electrical quantities can be minimized by using shielding and/or using proper location of components and wise routing of wires. In summary, then, the frequency where components are expected to operate determines how critical these stray or distributed *R, L,* and *C* values are to circuit operation.

SUMMARY

▶ Capacitance is the circuit property that opposes a voltage *change* caused by the storing of electrical energy (charge) in an electrostatic field. The unit of capacitance is the farad (F). One farad of capacitance stores one coulomb of charge when 1 V is applied or is present across its plates. Practical subunits of the farad are the microfarad (μF) and the picofarad (μμF or pF). The symbol for capacitance is *C*.

▶ A capacitor is a device that possesses the property of capacitance. Capacitance is present when there are two conductors separated by an insulator (nonconductor or dielectric). Capacitance can be lumped as in the capacitor component, or distributed as the capacitance between wires and a chassis.

▶ Capacitors come in various sizes and shapes and in various capacitance values. Capacitors are often named after the type of dielectric used in their construction (e.g., paper, mica, and ceramic). Electrolytic capacitors are an exception and are *not* named after their aluminum oxide dielectric. This capacitor is also the only basic category that is polarity sensitive.

▶ Factors affecting the capacitance amount include conductive plate areas, spacing between the plates (dielectric thickness), and the type of dielectric material. The formula is

$$C = \frac{8.85kA}{10^{12}s}$$

The greater the plate areas, the greater the capacitance value. The greater the spacing between plates (the thicker the dielectric) is, the less the capacitance value. Also, the higher the dielectric constant *(k)* is, the higher the capacitance value.

▶ The capacitance, charge, and voltage relationships for a capacitor are expressed as:

$$Q = CV, \ C = \frac{Q}{V}, \text{ and } V = \frac{Q}{C}$$

where *Q* is charge in coulombs; *C* is capacitance in farads; and *V* is voltage across the capacitor in volts.

▶ Dielectric constant (*k* or K_e) is the comparative charge that different materials store relative to that charge stored in vacuum or air. That is, vacuum and air have a reference *k* of one.

▶ When capacitors charge, they are accumulating electrons on their negative plate(s) and losing electrons from their positive plate(s). As this happens, the electron orbits in the dielectric material are stressed, or distorted. This stored energy is returned to the circuit when the capacitor discharges. When the capacitor discharges, the excess electrons on the negative plate(s) travel to the positive plate(s), eventually neutralizing the charge and the potential difference between plates.

▶ Leakage resistance is related to the resistance of the dielectric material to current flow. If the dielectric were a perfect nonconductor, the leakage resistance would be infinite. Most capacitors have very high leakage resistance, unless their voltage ratings are exceeded. Electrolytic capacitors have a lower leakage resistance than the other types because of the thinness and type of dielectric.

▶ When capacitors are connected in series, the total capacitance is less than the least capacitance in series. That is, capacitances in series add like resistances in parallel.

▶ When capacitors are in parallel, total capacitance equals the sum of all the individual capacitances. That is, capacitances in parallel add like resistances in series, where $C_T = C_1 + C_2 + \ldots C_n$.

▶ Capacitors take time to charge and discharge. The time depends on the amount of capacitance storing the charge and the amount of resistance limiting the charging or discharging current. One RC time constant equals R (in ohms) times C (in farads). Five RC time constants are required for the capacitor to completely charge or discharge to a new voltage level.

▶ Basic types of capacitors are paper, mica, ceramic, and electrolytic types that come in a variety of combinations. Some of these variations are epoxy-dipped tantalum capacitors, epoxy-coated polyester film capacitors, dipped-mica capacitors, polystyrene film capacitors, dipped-tubular capacitors, "chip" capacitors, and so on.

▶ Capacitors have various color codes or coding systems. Electronic Industries Association (EIA) and MIL (military) color codes are common. Tubular capacitors often use a banded color-code system (typically six bands). Older types of mica capacitors are frequently color coded with either a five-dot, or six-dot system. Ceramic capacitors may sometimes be color coded with MIL color codes, using three-, four-, or five-color dots or bands. Chip capacitors use a two-character coding system.

▶ Typical problems of capacitors include opens, shorts, or degeneration in operation due to low leakage resistance, or high electrolyte resistance.

▶ An ohmmeter check of a good capacitor will indicate a charging action occurring when the ohmmeter is first connected to the capacitor. The capacitor is charging to the ohmmeter's internal voltage source value. This is displayed by the ohmmeter pointer initially moving to a lower resistance point on the scale then moving to a final resting point at very high resistance once the capacitor has charged.

▶ Typical R values after charging are 100 or more megohms. Electrolytic capacitors have a lower leakage resistance in the range of 1/2 to 1 megohm.

FORMULAS AND SAMPLE CALCULATOR SEQUENCES

FORMULA 18–1 $C = \dfrac{Q}{V}$

charge value in coulombs, ⊖, voltage value, ⊜

FORMULA 18–2 $Q = CV$

capacitance value in farads, ⊗, voltage value, ⊜

FORMULA 18–3 $V = \dfrac{Q}{C}$

charge value in coulombs, ⊕, capacitance value in farads, ⊜

FORMULA 18–4   Energy (J) $= \dfrac{1}{2}CV^2$ or $0.5\ CV^2$

capacitance value in farads, ⊗, voltage value, x^2, ⊕, 2, ⊜

FORMULA 18–5 $k = \dfrac{\epsilon_o \ (\textit{absolute permittivity of dielectric material})}{\epsilon_V \ (\textit{absolute permittivity of vaccum})}$

ϵ_o value, ⊘, ϵ_V value, ⊜

FORMULA 18–6 $C = \dfrac{8.85A}{10^{12}s}$

8.85, ⊗ area of plates, ÷, (, 1, EE, 12 , ⊗, spacing,), ⊜

FORMULA 18–7 $C = \dfrac{8.85kA}{10^{12}s}$

8.85, ⊗, k value, ⊗, area of plates, ÷, (, 1, EE, 12, ⊗, spacing,), ⊜

FORMULA 18–8 $C = 2.25\dfrac{kA}{10^7 s}(n-1)$ (where A is in square inches and s is in inches)

2.25, ⊗, k value, ⊗, A value, ÷, (, 1, EE, 7, ⊗, s value,), ⊗, (, n value, −, 1,), ⊜

FORMULA 18–9 $C_T = \dfrac{C_1 \times C_2}{C_1 + C_2}$ (series capacitors)

C_1 value, ⊗, C_2 value, ÷, (, C_1 value, +, C_2 value,), ⊜

FORMULA 18–10 $C_T = \dfrac{C}{N}$ (series capacitors)

capacitor value of one of the equal-value capacitors, ÷, number of capacitors, ⊜

FORMULA 18–11 $C_T = \dfrac{1}{\dfrac{1}{C_1} + \dfrac{1}{C_2} + \dfrac{1}{C_3} \ldots + \dfrac{1}{C_n}}$ (series capacitors)

C_1 value, 1/x, +, C_2 value, 1/x, . . . , ⊜, 1/x

FORMULA 18–12 $V_X = V_S \dfrac{C_T}{C_X}$

V_S value, ⊗, (, C_T value, ÷, C_X value,), ⊜

FORMULA 18–13 $C_T = C_1 + C_2 \ldots + C_n$ (parallel capacitors)

C_1 value, +, C_2 value, . . . , ⊜

FORMULA 18–14 $\tau = R \ (\text{ohms}) \times C \ (\text{farads})$

R value, ⊗, C value, ⊜

FORMULA 18–15 $v_R = V_S \times \epsilon^{-t/\tau}$

time allowed value, **÷**, **((**, *R* value, **×**, *C* value, **))**, **=**, **+/−**, **2nd**, **eˣ**, **×**, V_S value, **=**

FORMULA 18–16 $v_C = V_S(1 - \epsilon^{-t/\tau})$

time allowed value, **÷**, **((**, *R* value, **×**, *C* value, **))**, **=**, **+/−**, **2nd**, **eˣ**, **+/−**, **+**, 1, **=**, **×**, V_S value, **=**

REVIEW QUESTIONS

1. Capacitors store energy in the form of:
   a. an electromagnetic field.
   b. an electrostatic field.
   c. an electromagnetic charge.
   d. none of the above.

2. The property of capacitance requires:
   a. two or more insulators connected by conductors.
   b. two or more conductors with electrical connections.
   c. two or more conductors separated by insulator(s).
   d. all of the above.

3. Various types of capacitors are often named:
   a. after the type dielectric used in them.
   b. after the type conductors used in them.
   c. only after their size and shape.
   d. none of the above.

4. Capacitance value is:
   a. directly related to dielectric thickness.
   b. inversely related to dielectric thickness and plate area.
   c. inversely related to the dielectric constant.
   d. none of the above.

5. Dielectric constant relates to:
   a. comparative charge stored for a given material versus air.
   b. the fact that the dielectric doesn't change characteristics.
   c. the insulating properties of the nonconducting material.
   d. none of the above.

6. When a capacitor charges:
   a. it is returning energy to the circuit from which it got the energy.
   b. it is losing electrons to the source.
   c. it is building up an electrostatic field.
   d. none of the above.

7. Because dielectric materials are nonconductive:
   a. a capacitor's leakage resistance is infinite.
   b. a capacitor's leakage resistance is usually low.
   c. a capacitor cannot pass current through the circuit.
   d. none of the above.

8. Capacitances in series add like:
   a. resistances in series.
   b. inductances in series.
   c. resistances in parallel.
   d. none of the above.

9. The total capacitance of series capacitors is:
   a. greater than any one of the capacitors in series.
   b. less than any one of the capacitors in series.
   c. equal to the capacitance of one, divided by the number of capacitors.
   d. none of the above.

10. Capacitances in parallel:
    a. add like resistances in series.
    b. add like inductances in parallel.
    c. add like resistances in parallel.
    d. none of the above.

11. Charge on a capacitor is directly related to:
    a. voltage and inversely related to capacitance.
    b. capacitance and inversely related to voltage.
    c. both capacitance and voltage.
    d. none of the above.

12. The type of capacitor most likely to deteriorate with age is the:
    a. mica capacitor.
    b. ceramic capacitor.

c. paper capacitor.
d. electrolytic capacitor.

13. List one very impotant precaution that must be observed when connecting electrolytic capacitors into a circuit.

14. List the main advantage of an electrolytic capacitor.

15. What type of capacitor is noted for its high-voltage breakdown characteristic?

16. What are two basic classifications for capacitors?

17. List two advantages of "chip" capacitors.

18. What rule of thumb is used to indicate the amount of time it will take a capacitor to charge to the source voltage, if there is a resistor in series with the capacitor?

19. In a series dc circuit containing both a resistor and a capacitor, as the capacitor charges:
a. the voltage across the resistor increases.
b. the voltage across the resistor decreases.
c. the voltage across the resistor remains the same.
d. the voltage across the capacitor decreases.

20. Name at least two test devices that may be used to check capacitance value.

PROBLEMS

1. The capacitor plate area doubles while the dielectric thickness is halved. What is the relationship of the new capacitance value to the original capacitance value? (Assume the same dielectric material.)

2. What charge is a 100-pF capacitor storing, if it is charged to 100 V?

3. What is the capacitor value that stores 200 pC when charged to 50 V?

4. What voltage is present across a 10-μF capacitor having a charge of 2,000 μC?

5. What is the total capacitance of a circuit containing a 100-pF, 400-pF, and 1,000-pF capacitor connected in series? If the capacitors were connected in parallel, what would be the total capacitance?

6. A 10-μF, 20-μF, and 50-μF capacitor are connected in series across a 160-V source. What voltages appear across each capacitor after it is fully charged? What value of charge is on each capacitor?

7. A 10-μF, 20-μF, and 50-μF capacitor are connected in parallel across a 100-V source. What is the charge on each capacitor? What is the total capacitance and total charge for the circuit?

8. Using

$$V_X = V_S \frac{C_T}{C_X}$$

determine the voltage across the 20-μF capacitor in a series circuit containing a 20-μF, 40-μF, and 60-μF capacitor across a dc source of 100 V. (Show all work.)

9. Use the appropriate exponential formula(s) and find the v_R value in a series RC circuit after 1.2 τ if V_S is 200 V. What is the v_C value? (Use a calculator with the e^x or l_{nx} key, if possible.)

10. What is the total capacitance of the circuit in Figure 18–33?

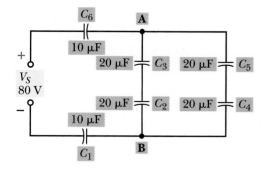

FIGURE 18–33

11. What is the voltage across C_5 in question 10?

12. What is the charge on C_6 in question 10?

13. Assume each capacitor in question 10 has a breakdown voltage rating of 300 V. What voltage should not be exceeded between points A and B?

14. For the voltage described between points A and B in question 13, what is the V_S value that causes that condition to exist?

15. What is the total energy stored in the circuit of question 10 with 80 V V_s?

16. Assume a 15-μF capacitor is charged from a 150-V source and another 15-μF capacitor is charged from a separate 300-V source. If the capacitors are disconnected from their sources after being charged and connected in parallel with positive plate to positive plate and negative plate to negative plate, what is the resultant *voltage* and *charge* on each capacitor?

17. For the conditions described in question 16, what is the total stored energy after the capacitors are connected?

ANALYSIS QUESTIONS

1. Resarch and describe:
   a. The procedure for using a DMM to measure capacitance values.
   b. For the particular meter you research, define the ranges of capacitance the meter can measure.

2. Research and describe:
   a. The procedure for using a dedicated capacitance meter to measure capacitance values.
   b. For the particular meter you research, define the ranges of capacitance the meter can measure.

3. Research and describe:
   a. The procedure for using an LCR meter to measure capacitance values.
   b. For the particular meter you research, define the ranges of capacitance the meter can measure.

4. Research and describe:
   a. Any other types of measurements, if any, that can be made relative to capacitors' characteristics or conditions.
   b. By what test equipment and/or means these other measurements might be made.

A GOOD IDEA

After making your first test in troubleshooting, move one of the brackets you used to define the "area of uncertainty" so that the bracketed area becomes smaller! (If at all possible!) Then, make new tests only in the new, smaller bracketed area!

PERFORMANCE PROJECTS CORRELATION CHART

Suggested performance projects that correlate with topics in this chapter are:

CHAPTER TOPIC	PERFORMANCE PROJECT	PROJECT NUMBER
Charging and discharging action The *RC* Time Constant	Charge and Discharge Action and *RC* Time	51
Total Capacitance in Series and Parallel	Total Capacitance in Series and Parallel	52

NOTE: It is suggested that after completing the above projects, the student should be required to answer the questions in the "Summary" at the end of this section of projects in the Laboratory Manual.

THE SIMPLER SEQUENCE FOR TROUBLESHOOTING

Follow the seven-step SIMPLER troubleshooting sequence, as outlined below, and see if you can find the problem with this circuit. As you follow the sequence, record your circuit test results for each testing step on a separate sheet of paper. This will aid you and your instructor to see where your thinking is on track or where it might deviate from the best procedure.

step 1 — Symptoms

—(Gather, verify, and analyze symptom information) To begin this process, read the data under the "Starting Point Information" heading. Particularly look for circuit parameters that are not normal for the circuit configuration and component values given in the schematic diagram. For example, look for currents that are too high or too low; voltages that are too high or too low; resistances that are too high or too low, etc. This analysis should give you a first clue (or symptom information) that will aid you in determining possible areas or components in the circuit that might be causing this symptom.

step 2 — Identify

—(Identify and bracket the initial suspect area) To perform this first "bracketing" step, analyze the clue or symptom information from the "Symptoms" step, then either on paper or in your mind, bracket, circle, or put parentheses around all of the circuit area that contains any components, wires, etc., that might cause the abnormality that your symptom information has pointed out. NOTE: Don't bracket parts of the circuit that contain items that could not cause the symptom!

step 3 — Make

—(Make a decision about the first test you should make: what type and where) Look under the TEST column and determine which of the available tests shown in that column you think would give you the most meaningful information about the suspect area or components, in the light of the information you have so far.

step 4 — Perform

—(Perform the first test you have selected from the TEST column list) To simulate performing the chosen test (as if you were getting test results in an actual circuit of this type), follow the dotted line to the right from your selected test to the number in parentheses. The number in parentheses, under the "Results in Appendix C" column tells you what number to look up in Appendix C to see what the test result is for the test you are simulating.

step 5 — Locate

—(Locate and define a new "narrower" area of uncertainty) With the information you have gained from the test, you should be able to eliminate some of the circuit area or circuit components as still being suspect. In essence, you should be able to move one bracket of the bracketed area so that there is a new smaller area of uncertainty.

step 6 — Examine

—(Examine available information and determine the next test: what type and where) Use the information you have thus far to determine what your next test should be in the new narrower area of uncertainty. Determine the type test and where, then proceed using the TEST listing to select that test, and the numbers in parentheses and Appendix C data to find the result of that test.

step 7 — Repeat

—(Repeat the analysis and testing steps until you find the trouble) When you have determined what you would change or do to restore the circuit to normal operation, you have arrived at *your* solution to the problem. You can check your final result against ours by observing the pictorial step-by-step sample solution on the pages immediately following the "Chapter Troubleshooting Challenge"—circuit and test listing page.

step 8 — Verify

NOTE: A useful *8th Step* is to operate and test the circuit or system in which you have made the "correction changes" to see if it is operating properly. This is the final proof that you have done a good job of troubleshooting.

CHALLENGE CIRCUIT 9

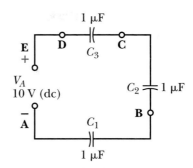

CHALLENGE CIRCUIT 9

Starting Point Information

1. Circuit diagram
2. V_{C_2} is lower than it should be

TEST	Results in Appendix C
V_{C_1}	(116)
V_{C_2}	(15)
V_{C_3}	(94)

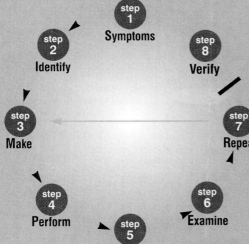

Symptoms

All we know is that the voltage across C_2 is lower than it should be. If things were normal, voltage across C_1 would be 3.3 V; the voltage across C_2 would be 3.3 V; and the voltage across C_3 would also be 3.3 V. Since Q is the same for all capacitors, the voltage distributions should be as described above. Also, we could predict that if C_2 were opened, its voltage would be 10 V (which does not agree with our symptom information); if C_2 were shorted or one of the other capacitors were opened, V_{C_2} would be 0 V; and if V_{C_2} were leaky, voltage would also be close to 0 V.

Identify

initial suspect area: C_2 is the initial suspect area. However, we can't disregard the other capacitors since we have limited information.

Make

test decision: Let's first check C_2's voltage to see where we are.

Perform

1st Test: Check V_2. V_{C_2} is close to 0 V, which is lower than normal.

Locate

new suspect area: If either one of the other capacitors were open, the voltage for C_2 would still be 0 V. If either one of the other two capacitors were leaky or shorted, C_2's voltage would be higher than normal.

Examine

available data.

Repeat

analysis and testing: A check of V_{C_1} or V_{C_3} might be necessary.

2nd Test: Check V_{C_3}. V_{C_3} is about 5 V, which is higher than normal, but C_3 is certainly not open or shorted. It looks like C_2 might be leaky. Let's take a look at C_2.

3rd Test: Lift one of C_2's leads and check the capacitor's dc resistance. The test reveals a lower than normal R.

Verify

4th Test: Replace C_2 and check the circuit operation. The circuit checks out normal. C_2 was leaky.

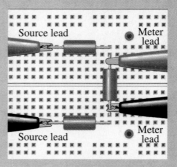

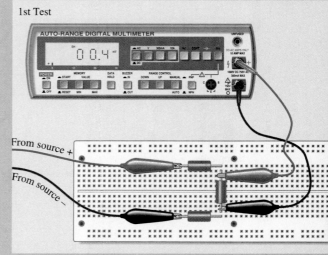

1st Test

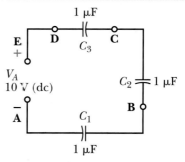

CHALLENGE CIRCUIT 9

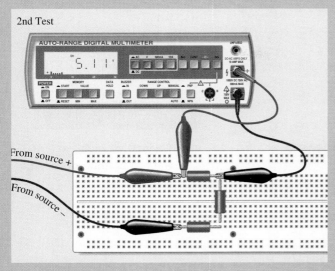

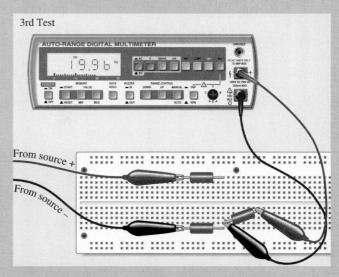

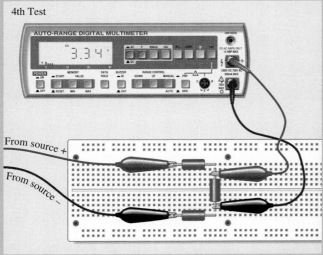

CAPACITIVE REACTANCE IN AC

19

chapter

CHAPTER PREVIEW

You have learned that a capacitor allows current flow, provided it is charging or discharging. However, a capacitor does not allow dc current to continue once charged to the source voltage. Also, you learned a capacitor does not instantly charge or discharge to a new voltage level; it requires current flow and time.

The unique characteristics of capacitors allow them to be used for a number of purposes in electronic circuits. Their ability to block dc current (once charged) makes them useful in coupling and decoupling applications for amplifiers, and so on. Their ability to store electrical energy provides great value when they are used for filtering circuits in power supplies and other circuits. Their charging and discharging characteristics allow them to be used in wave-shaping applications and for timing circuits of all kinds.

The characteristic of allowing ac current due to the charging and discharging current will be studied in this chapter. You will learn there is an inverse relationship between the amount of *opposition* a capacitor gives to ac current and the circuit frequency and capacitance values involved. Capacitor opposition to ac current is called **capacitive reactance.** (Recall inductive reactance describes inductor opposition to ac current.)

In this chapter, you will examine capacitance as it relates to the capacitor action in a sine-wave ac circuit. You will study the phase relationship of voltage and current for capacitive components; learn to compute capacitive reactance; discover some applications of this characteristic; and learn other practical information about this important electrical parameter.

OBJECTIVES

After studying this chapter, you should be able to:

1. Illustrate *V-I* relationships for purely resistive and purely capacitive circuits

2. Explain **capacitive reactance**

3. Use Ohm's law to solve for X_C value(s)

4. Use capacitive reactance formula to solve for X_C value(s)

5. Use X_C formula to solve for unknown C and f values

6. Use Ohm's law and **reactance** formulas to determine circuit reactances, voltages, and currents for series-connected and parallel-connected capacitors

7. List two practical applications for X_C

8. Use the SIMPLER troubleshooting sequence to solve the Chapter Troubleshooting Challenge problem

19-1 V AND I RELATIONSHIPS IN A PURELY RESISTIVE AC CIRCUIT

As a quick review, look at Figure 19–1. Note that V and I are in phase in a purely resistive ac circuit.

19-2 V AND I RELATIONSHIPS IN A PURELY CAPACITIVE AC CIRCUIT

Observe Figure 19–2 as you study the following statements about the relationship of V and I in a purely capacitive ac circuit.

1. To develop a voltage (potential difference or p.d.) across capacitor plates, charge must move. That is, charging current must flow first.

2. When a sine-wave voltage is applied across a capacitor that is effectively in parallel with the source terminals, voltage across the capacitor must be the same as the applied voltage.

3. For the voltage across the capacitor to change, there must be alternate charging and discharging action. This implies charging and discharging current flow. The value of instantaneous capacitor current (i_C) directly relates to the amount of capacitance and the rate of change of voltage. (This principle fits well with the concept learned earlier where $Q = CV$.)

4. The formula that expresses the direct relationship of i_C to C and rate of change of voltage is:

FORMULA 19–1 $i_C = C\dfrac{dv}{dt}$

where dv/dt indicates change in voltage divided by amount of time over which the change occurs, or the rate of change of voltage.

For example, if the voltage across a 100-pF capacitor changes by 100 V in 100 µs, then:

$$i_C = C\frac{dv}{dt}$$

$$i_C = 100 \times 10^{-12} \times \frac{100}{100 \times 10^{-6}}$$

$$i_C = \frac{10{,}000 \times 10^{-12}}{100 \times 10^{-6}}$$

$$i_C = 100 \times 10^{-6}, \text{ or } 100 \ \mu A$$

5. Because the amount of charging (or discharging) current directly relates to the rate of change of voltage across the capacitor, capacitor current, i_C, is maximum when voltage rate of change is maximum and minimum when voltage rate of change is minimum.

6. Since **rate of change of voltage** is maximum near the zero points of the sine wave and minimum at the maximum points on the sine wave, current is maximum when voltage is minimum and vice versa. In other words, I is 90° out of phase with V.

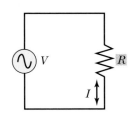

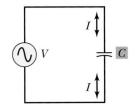

V and I are in phase

FIGURE 19–1 V and I phase relationships in a purely resistive circuit

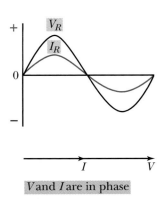

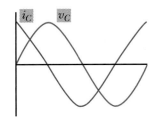

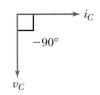

FIGURE 19–2 V and I phase relationships in a purely capacitive circuit

7. To illustrate these facts, Figure 19–2 shows that capacitor current *leads* the voltage across the capacitor by 90°. Remember our phrase mentioned in an earlier chapter—*"Eli, the ice man."* *Eli* represents voltage or emf *(E)* that comes before (leads) current *(I)* in an inductor *(L)* circuit. Obviously, the word "ice" represents that *current (I) leads voltage (E) in a capacitive (C) circuit.* If this phrase helps you remember phase relationships of voltage and current for inductors and capacitors, good! But if you understand the concepts related to inductors opposing a current change and capacitors opposing a voltage change, you will not have to rely on this phrase!

IN-PROCESS LEARNING CHECK I

Fill in the blanks as appropriate.

1. For a capacitor to develop a potential difference between its plates, there must be a _____ current.
2. For the potential difference between capacitor plates to decrease (once it is charged), there must be a _____ current.
3. When ac is applied to capacitor plates, the capacitor will alternately _____ and _____.
4. The value of instantaneous capacitor current directly relates to the value of _____ and rate of change of _____.
5. The amount of charging or discharging current is maximum when the rate of change of voltage is _____. For sine-wave voltage, this occurs when the sine wave is near _____ points.
6. Capacitor current _____ the voltage across the capacitor by _____ degrees. The expression that helps you remember this relationship is "Eli, the _____ man."

19-3 CONCEPT OF CAPACITIVE REACTANCE

As you have just seen, a sine-wave current develops in a capacitor circuit by the charge and discharge action of the capacitor when a sine-wave voltage is applied. This sine-wave capacitor current leads the sine-wave capacitor voltage by 90°.

In a purely capacitive circuit, the factor that limits the ac current is called capacitive reactance. The symbol used to represent this quantity is "X_C". It is measured in ohms. As you would anticipate, the higher the capacitive reactance, the lower the amount of current produced for a given voltage. That is,

$$I_C = \frac{V_C}{X_C}$$

which is simply an application of Ohm's law.

What factors affect this current-limiting trait called capacitive reactance? It is apparent that for a given voltage the larger the amount of charge moved per unit time, the higher the current and the lower the value of X_C produced and vice versa. Let's examine the relationship of X_C to the two factors that affect this movement of charge, or amount of current for a given voltage. These two factors are *capacitance* and *frequency*.

19–4 RELATIONSHIP OF X_C TO CAPACITANCE VALUE

The larger the capacitor, the greater the charge accumulated for a given V across its plates

$$V = \frac{Q}{C}$$

To match a given V applied, a higher charging current is required in a high C value circuit than is required with a smaller capacitance. This implies a lower opposition to ac current or a lower capacitive reactance (X_C). This means X_C *is inversely proportional to C.* In other words, the larger the C, the lower the X_C present; and the smaller the C, the higher the X_C present.

You will soon be introduced to a formula for calculating X_C; however, for now, notice in Figure 19–3a that for a given frequency of operation (100 Hz, in this case), the amount of capacitive reactance in ohms is *inversely related* to the capacitance value, which illustrates our previous statement. We will ask you to confirm the values shown in the figure after you study the X_C formula. For now, assume the values are valid. The point here is that the greater the C value, the less the X_C value, and vice-versa.

19–5 RELATIONSHIP OF X_C TO FREQUENCY OF AC

If the capacitor voltage is tracking the source V, as in the circuit of Figure 19–3b, then the higher the frequency and the less available time there is for the required amount of charge (Q) to move to and away from the respective capacitor plates during charge and discharge action. As you know, it takes a given amount of Q to achieve a specific voltage across a given capacitor $(V = Q/C)$.

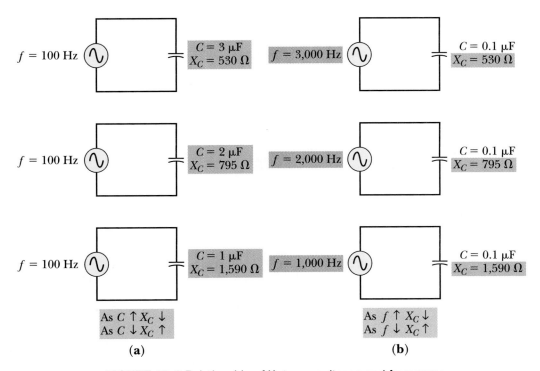

FIGURE 19–3 Relationship of X_C to capacitance and frequency

The higher the frequency of applied voltage, the faster is the rate of change of voltage for any given voltage level. As you learned earlier, the i_c directly relates to the voltage rate of change *(dV/dt)*. Since higher frequencies produce a greater charge and discharge current for any given voltage level, as frequency increases opposition to current decreases. More current for a given *V* applied indicates there is *less* opposition, meaning capacitive reactance *(X_C)* decreases as frequency increases. That is, X_C *is inversely proportional to f.* In other words, the higher the frequency, the lower the X_C present; and the lower the frequency, the greater the X_C present. See again Figure 19–3b.

19–6 METHODS TO CALCULATE X_C

Ohm's law method

Ohm's law is one practical method to calculate X_C, where

$$X_C = \frac{V_C}{I_C}$$

For example, if the effective ac voltage applied to a capacitive circuit is 100 volts, and the I_C is 1 ampere (rms), capacitive reactance *(X_C)* must equal 100 ohms. NOTE: Unless told otherwise, assume effective (rms) values of ac voltage and current in your measurements and computations relating to V_C, I_C, and X_C. As an electronics student and technician, you will frequently make such measurements and calculations. Remember, the Ohm's law formula can also be transposed to yield

$$V_C = I_C \times X_C, \text{ and } I_C = \frac{V_C}{X_C}$$

The basic X_C formula

From the previous discussions, it is easy to understand the concepts illustrated by the formula to calculate X_C when only C and f are known. The formula states that:

FORMULA 19–2 $\qquad X_C \text{ (ohms)} = \dfrac{1}{2\pi fC}$

where f is frequency in hertz
C is capacitance in farads

PRACTICAL NOTES

Some people find it convenient to alter the formula

$$X_C = \frac{1}{2\pi fC} \text{ to read } \frac{0.159}{fC}$$

This can be done because the value of 2π does not change. That is, it is a "constant." One divided by $2\pi = 0.159$; therefore, the formula can be simplified to read $0.159/(fC)$. When X_C calculations must be performed without a calculator, this modified formula may be easier to use. When a calculator is available, either formula format is easily used.

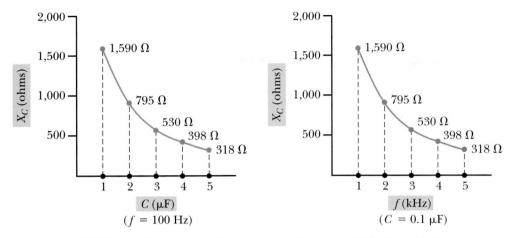

FIGURE 19–4 Examples of the inverse relationship of X_C to C and to f

Notice that X_C is *inversely* proportional to both f and C. That is, if either f or C (or both) are increased, the resulting larger product in the denominator divided into one results in a smaller value of X_C. In fact, if either f or C is doubled, X_C is half. If either f or C is halved, X_C doubles, and so forth, Figure 19–4.

Stated another way, larger C means more charging and discharging current flows to charge to and discharge from a given voltage. Thus, a larger C results in a lower X_C. Since a higher f means the voltage rate of change is faster, more charging and discharging current must flow in the time allowed by the higher frequency; hence, a higher f results in a lower X_C.

Recall the "$2\pi f$" in the formula relates to angular velocity, which relates to sine-wave frequency. The X_C formula is the second formula where you have found this angular velocity expression—the other formula was for inductive reactance where $X_L = 2\pi f L$.

Example

1. What is the capacitive reactance of a 2-μF capacitor operating at a frequency of 120 Hz?

$$X_C = \frac{1}{2\pi f C} = \frac{1}{6.28 \times 120 \times (2 \times 10^{-6})} = \frac{1}{1,507.2 \times 10^{-6}} = 663.5\Omega$$

2. The frequency of applied voltage doubles while the C remains the same in a capacitive circuit. What happens to the circuit X_C?

$$X_C = \frac{1}{2\pi f C}, \text{ where } X_C \text{ for stated conditions} = \frac{1}{2\pi(2f)\,C}$$

Therefore, the denominator's product is twice as large as that divided into the unchanged numerator. That is, the X_C is half of the original frequency X_C.

3. In a capacitive circuit, if X_C quadruples while the frequency of the applied voltage remains the same, what must have happened to the C value in the circuit?

Since X_C is inversely proportional to f and C and f remains unchanged, C must have decreased to one-fourth its original value. That is,

$$4 \times \text{original } X_C = \frac{1}{2\pi f(0.25C)}.$$

PRACTICE PROBLEMS I

1. Use the X_C formula and verify values of X_C shown in Figure 19–3. Do the formula results validate the values of X_C shown?

2. What is the capacitive reactance of a 5-μF capacitor operating at a frequency of 200 Hz?

3. What happens to the X_C in a circuit when f doubles and the capacitance value is halved at the same time?

Rearranging the X_C formula to find f and C

By rearranging the X_C equation, we can solve for frequency if both C and X_C are known, or solve for C if both f and X_C are known. When the equation is transposed, the new equations are as follows:

FORMULA 19–3 $f = \dfrac{1}{2\pi C X_C}$

and

FORMULA 19–4 $C = \dfrac{1}{2\pi f X_C}$

Try these formulas on the following practice problems.

PRACTICE PROBLEMS II

1. What is the frequency of operation of a capacitive circuit where the total X_C is 1,000 Ω and the total capacitance is 500 pF?

2. What is the total circuit capacitance of a circuit exhibiting an X_C of 3,180 Ω at a frequency of 2 kHz?

19–7 CAPACITIVE REACTANCES IN SERIES AND PARALLEL

In the last chapter, you learned that capacitances in series add like resistances in parallel, and capacitances in parallel add like resistances in series. It is interesting that when we consider capacitive *reactances* in series and parallel, the converse is true.

Capacitive reactances in series

Since capacitive reactance (X_C) is an opposition to current flow, measured in ohms, capacitive reactances (current oppositions) in series add like resistances in series. That is, $X_{C_T} = X_{C_1} + X_{C_2} + X_{C_3} \ldots + X_{C_n}$, Figure 19–5.

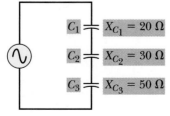

$X_{C_T} = X_{C_1} + X_{C_2} + X_{C_3}$
$X_{C_T} = 100 \ \Omega$

(X_Cs in series add like resistances in series)

FIGURE 19–5 X_C in series

Example

What is the total X_C of a circuit containing three series-connected 3-μF capacitors operating at a frequency of 100 Hz?

Approach 1: Find total C, then use the X_C formula:

C_T for three equal-value capacitors in series equals the C value of one capacitor divided by the number of equal-value capacitors ($C_T = C_1/N$).

$$C_T = \frac{3\ \mu\text{F}}{3} = 1\ \mu\text{F}$$

$$X_{C_T} = \frac{1}{2\pi f C_T} = \frac{1}{6.28 \times 100 \times 1 \times 10^{-6}} = 1{,}592\ \Omega$$

Approach 2: Find X_C of a 3-μF capacitor, then multiply by 3:

$$X_{C_1} = \frac{1}{2\pi f C_1} = \frac{1}{6.28 \times 100 \times 3 \times 10^{-6}} = 530.7\ \Omega$$

$$X_{C_T} = 3 \times 530.7\ \Omega = 1{,}592\ \Omega$$

PRACTICE PROBLEMS III

What is the total X_C of the circuit in Figure 19–6?

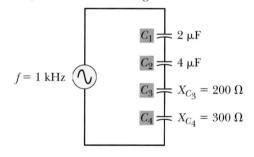

C_1	2 μF
C_2	4 μF
C_3	$X_{C_3} = 200\ \Omega$
C_4	$X_{C_4} = 300\ \Omega$

$f = 1$ kHz

FIGURE 19–6

Capacitive reactances in parallel

Because capacitive reactance is an opposition to ac current flow, when we supply more current paths by connecting X_Cs in parallel, the resulting total opposition decreases just as it does when we parallel resistances.

Useful formulas to solve for total capacitive reactance of capacitors in parallel are as follows:

• General formula for X_Cs in parallel:

FORMULA 19–5 $$X_{C_T} = \frac{1}{\dfrac{1}{X_{C_1}} + \dfrac{1}{X_{C_2}} + \dfrac{1}{X_{C_3}} \cdots + \dfrac{1}{X_{C_n}}}$$

• Formula for two X_Cs in parallel:

FORMULA 19–6　　$X_{C_T} = \dfrac{X_{C_1} \times X_{C_2}}{X_{C_1} + X_{C_2}}$

• Formula for equal-value X_Cs in parallel:

FORMULA 19–7　　$X_{C_T} = \dfrac{X_{C_1}}{N}$

where N = number of equal-value X_Cs in parallel

Example

What is the total X_C of a circuit containing three parallel-connected capacitor branches with values of 2 μF, 3 μF, and 5 μF, respectively? Assume the circuit is operating at a frequency of 400 Hz.

Two possible approaches are:

1. Find total C, then use X_C formula to find X_{C_T}.

2. Find X_C of each capacitor using X_C formula, then use the general formula to solve for X_{C_T}.

Of these two approaches, approach 1 is easiest for these circumstances. Therefore,

$$C_T = C_1 + C_2 + C_3$$

$$C_T = 2\ \mu F + 3\ \mu F = 5\ \mu F = 10\ \mu F$$

$$X_{C_T} = \frac{1}{2\pi f C_T} = \frac{1}{6.28 \times 400 \times 10 \times 10^{-6}} = \frac{1}{0.02512} = 39.8\ \Omega$$

PRACTICE PROBLEMS IV

1. What is the total capacitive reactance of a circuit containing three parallel-connected capacitors of 1 μF, 2 μF, and 3 μF, respectively? (Assume an operating frequency of 1 kHz.)

2. What is the total capacitive reactance of two parallel-connected capacitors where one capacitance is 5 μF and the other capacitance is 20 μF? (Circuit is operating at a frequency of 500 Hz.)

3. What is the total capacitive reactance of four 1,600-Ω capacitive reactances connected in parallel?

19–8 VOLTAGES, CURRENTS, AND CAPACITIVE REACTANCES

Our computations have concentrated on finding the capacitive reactances in capacitive circuits. Let's discuss voltage distribution throughout series capacitive circuits and current distribution throughout parallel capacitive circuits.

Capacitive reactances represent opposition to ac current (in ohms). When dealing with V, I, and X_C, we treat ohms of opposition like we treat resistances in resistive circuits. That is, voltage drops are calculated by using $I \times X_C$; and currents are calculated by using V/X_C, and so on.

Refer to Figure 19–7 to see series circuit parameters and computations illustrated. Then refer to Figure 19–8 and observe how parallel circuit parameters and computations are made. After studying these figures, try the following practice problems.

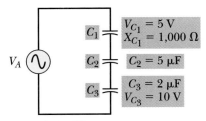

Find f; C_1; I; X_{C_2}; X_{C_3}; X_{C_T}; V_{C_2}; and V_A.

(Since we know two things about C_1, let's start there.)

$$I = \frac{V_{C_1}}{X_{C_1}} = \frac{5}{1,000} = 0.005 \text{ A, or } I = 5 \text{ mA.}$$

Since V_{C_3} is double V_{C_1}, its X_C must be double that of C_1 because they have the same I and $V = I \times X_C$. Thus, $X_{C_3} = 2,000 \ \Omega$.

Since V_{C_1} is half V_{C_3}, C_1 must have twice the capacitance and half the X_C of C_3. Thus, $C_1 = 4 \ \mu\text{F}$.

Since C_2 is 2.5 times the C value of C_3 and X_C is inversely related to C, X_{C_2} must equal $\frac{X_{C_3}}{2.5}$ or $\frac{2,000}{2.5}$. Thus, $X_{C_2} = 800 \ \Omega$.

$$V_{C_2} = I \times X_{C_2} = 5 \text{ mA} \times 800 \ \Omega$$
$$V_{C_2} = 4 \text{ V}$$
$$X_{C_T} = X_{C_1} + X_{C_2} + X_{C_3} = 1,000 + 800 + 2,000$$
$$X_{C_T} = 3,800 \ \Omega$$

$$C_T = \frac{1}{\left(\frac{1}{4} + \frac{1}{5} + \frac{1}{2}\right)}$$
$$C_T = 1.05 \ \mu\text{F}$$

$$f = \frac{1}{2\pi C_T X_{C_T}} = \frac{1}{6.28 \times 1.05 \times 10^{-6} \times 3,800}$$
$$f = \frac{1}{0.02505} = 39.9 \text{ Hz}$$
$$f = 39.9 \text{ Hz or about 40 Hz}$$
$$V_A = I \times X_{C_T} = 5 \text{ mA} \times 3.8 \text{ k}\Omega$$
$$V_A = 19 \text{ V}$$

(Also, $V_1 + V_2 + V_3 = 19$ V)

FIGURE 19–7 Series capacitive circuit analysis

FIGURE 19–8 Parallel capacitive circuit analysis

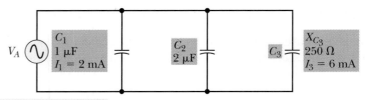

Find V_A; f; X_{C_2}; X_{C_1}; I_2; I_T.

Starting with information regarding C_3,

$$V_A = V_3 = I_3 \times X_{C_3}$$
$$V = 6 \text{ mA} \times 0.25 \text{ k}\Omega$$
$$V_A = 1.5 \text{V}$$

Since I_3 is triple I_1, X_{C_3} must be one-third X_{C_1}. Therefore, C_3 must be triple C_1.

$$C_3 = 3 \ \mu\text{F}$$
$$X_{C_1} = 3 \times X_{C_3} = 3 \times 250 \ \Omega$$
$$X_{C_1} = 750 \ \Omega$$

$$X_{C_2} = \frac{1}{2} X_{C_1} = \frac{1}{2} \text{ of } 750 \ \Omega$$
$$X_{C_2} = 375 \ \Omega$$
$$I_2 = \frac{V_2}{X_{C_2}} = \frac{1.5 \text{ V}}{375 \ \Omega}$$
$$I_2 = 4 \text{ mA}$$
$$I_T = I_1 + I_2 + I_3 = 2 \text{ mA} + 4 \text{ mA} + 6 \text{ mA}$$
$$I_T = 12 \text{ mA}$$

f is found by using any one of the branches information or by using the total C and X_C parameters. That is,

$$f = \frac{1}{2\pi C X_C} = \frac{1}{6.28 \times 1 \times 10^{-6} \times 750}$$
$$f = \frac{1}{0.00471}$$
$$f = 212 \text{ Hz}$$

PRACTICE PROBLEMS V

1. For the circuit in Figure 19–9, find X_{C_1}, X_{C_2}, X_{C_3}, X_{C_T}, C_2, C_3, V_2, and I_T.

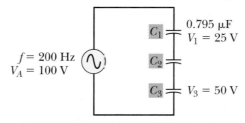

FIGURE 19–9

Find X_{C_1}; X_{C_2}; X_{C_3}; X_{C_T}; C_2; C_3; V_2; and I_T.

2. For the circuit in Figure 19–10, find C_1, X_{C_1}, X_{C_2}, X_{C_3}, I_2, and I_T.

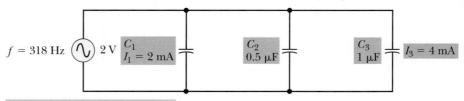

FIGURE 19–10

Find C_1; X_{C_1}; X_{C_2}; X_{C_3}; I_2; and I_T.

FIGURE 19–11 Comparisons of *C* and *L* circuit characteristics

PARAMETER INVOLVED	CAPACITIVE CIRCUIT	INDUCTIVE CIRCUIT
Opposition to *I*	X_C in ohms	X_L in ohms
As *f* increases	X_C decreases	X_L increases
As *f* decreases	X_C increases	X_L decreases
As *C* increases	X_C decreases	Not applicable
As *C* decreases	X_C increases	Not applicable
As *L* increases	Not applicable	X_L increases
As *L* decreases	Not applicable	X_L decreases
Voltage across component	$I \times X_C$	$I \times X_L$
Instantaneous values	$i_C = C \times dv/dt$	$v_L = L \times di/dt$
Phase of *V* and *I*	*I* leads *V* by 90 degrees	*V* leads *I* by 90 degrees
To calculate *X*	$X_C = 1/2\pi fC$	$X_L = 2\pi fL$
Series-connected	$C_T <$ least *C*	$L_T =$ sum of *L*'s
Parallel-connected	$C_T =$ sum of *C*'s	$L_T <$ least *L*
Series-connected	$X_{C_T} =$ sum of X_C's	$X_{L_T} =$ sum of X_L's
Parallel-connected	$X_{C_T} <$ least X_C	$X_{L_T} <$ least X_L

19–9 FINAL COMMENTS ABOUT CAPACITORS AND CAPACITIVE REACTANCE

Having studied capacitive reactance, you can more clearly see how capacitors are used. As you proceed in your studies and career, you should *keep in mind* the inverse relationship of capacitance and capacitive reactance. Also, you remember capacitances in series add like resistances in parallel. Keeping these ideas in mind will help prevent errors when analyzing circuits containing both inductances and capacitances.

A list of similarities and differences between capacitive and inductive circuits is shown in Figure 19–11. Review what you have learned about the operating features of these two electronic components. It is important that you thoroughly understand each component to analyze circuits that combine these components in various configurations.

SUMMARY

▶ Capacitive reactance is the opposition a capacitor displays to ac or pulsating dc current. The symbol is X_C and it is measured in ohms. Capacitive reactance is calculated by the formula

$$X_C = \frac{1}{2\pi fC}$$

where *f* is frequency in hertz and *C* is capacitance in farads.

▶ Factors influencing the amount of capacitive reactance (X_C) in a circuit are the amount of capacitance *(C)* and the frequency *(f)*. Voltage does *not* affect capacitive reactance. As *C or f* increases, X_C decreases. As *C or f* decreases, X_C increases proportionately.

▶ In a purely capacitive circuit, current leads voltage by 90°. (Charging current must flow before a potential difference develops across the capacitor plates.)

▶ Capacitive reactance is calculated by using Ohm's law or the capacitive reactance formula, where

$$X_C = \frac{V_C}{I_C}, \text{ or } X_C = \frac{1}{2\pi fC}$$

▶ The capacitive reactance formula can be rearranged to find either *f* or *C*, if all other factors are known. For example, $f = \dfrac{1}{2\pi CX_C}$, and $C = \dfrac{1}{2\pi fX_C}$

▶ Capacitive reactances in series add like resistances in series. That is, $X_{C_T} = X_{C_1} + X_{C_2} + X_{C_3} \ldots + X_{C_n}$.

▶ Capacitive reactances in parallel add like resistances in parallel. The reciprocal formula, product-over-the-sum formula, or equal-value parallel formulas are used, as appropriate, to find total capacitive reactance of parallel-connected capacitors.

▶ All three variations of Ohm's law may be used with relation to capacitor parameters. That is,

$$I_C = \frac{V_C}{X_C}; \; V_C = IX_C; \; \text{and} \; X_C = \frac{V_C}{I_C}$$

▶ Capacitance and capacitive reactance exhibit converse characteristics. In a *series* circuit, the *smallest C* has the largest voltage drop and the largest *C* the least voltage drop. In this same series circuit, the largest X_C drops the most voltage (because the smallest *C* has the highest X_C), and so forth. In *parallel* circuits, the largest *C* holds the largest amount of charge *(Q)* and the smallest *C* has the least *Q*.

FORMULAS AND SAMPLE CALCULATOR SEQUENCES

FORMULA 19–1 $i_C = C \dfrac{dv}{dt}$

Cap value, ⊗, voltage change value, ⊘, amount of time, ⊜

FORMULA 19–2 $X_C \text{ (ohms)} = \dfrac{1}{2\pi fC}$

6.28, ⊗, frequency value, ⊗, capacitance value, ⊜, 1/x

FORMULA 19–3 $f = \dfrac{1}{2\pi C X_C}$

6.28, ⊗, capacitance value, ⊗, X_C value, ⊜, 1/x

FORMULA 19–4 $C = \dfrac{1}{2\pi f X_C}$

6.28, ⊗, frequency value, ⊗, X_C value, ⊜, 1/x

FORMULA 19–5 $X_{C_T} = \dfrac{1}{\dfrac{1}{X_{C_1}} + \dfrac{1}{X_{C_2}} + \dfrac{1}{X_{C_3}} \ldots \dfrac{1}{X_{C_n}}}$

X_{C_1} value, 1/x, ⊕, X_{C_2} value, 1/x, ⊕, X_{C_3} value, 1/x, ⊜, 1/x

FORMULA 19–6 $X_{C_T} = \dfrac{X_{C_1} \times X_{C_2}}{X_{C_1} + X_{C_2}}$

X_{C_1} value, ⊗, X_{C_2} value, ⊘, (, X_{C_1} value, ⊕, X_{C_2} value,), ⊜

FORMULA 19–7 $X_{C_T} = \dfrac{X_{C_1}}{N}$

X_{C_1} value, ⊘, number of equal value X_Cs in parallel, ⊜

1. Capacitive reactance is:
   a. the opposition a capacitor shows to dc only.
   b. the opposition a capacitor shows to ac only.
   c. the opposition a capacitor shows to ac or pulsating dc.
   d. none of the above.

2. The amount of capacitive reactance shown by a capacitor can be influenced by:
   a. circuit voltage only.
   b. circuit voltage and current.
   c. frequency and resistance.
   d. frequency and capacitance.

3. Two commonly used methods used to compute capacitive reactance values are:
   a. Ohm's law and Kirchhoff's law.
   b. Ohm's law and the capacitive reactance formula.
   c. Ohm's law and epsilon natural log approach.
   d. none of the above.

4. The symbol used to represent capacitive reactance is:
   a. Z.
   b. R.
   c. X_C.
   d. none of the above.

5. Total capacitive reactance of parallel capacitors can be calculated like resistances in:
   a. series.
   b. parallel.
   c. series-parallel.
   d. none of the above.

6. Total capacitive reactance of series capacitors can be calculated like resistances in:
   a. series.
   b. parallel.
   c. series-parallel.
   d. none of the above.

7. In a series capacitive circuit, the smallest capacitance has:
   a. the lowest voltage drop.
   b. the highest voltage drop.
   c. the same voltage drop as all other capacitors.
   d. none of the above.

8. In a parallel capacitive circuit, the largest capacitance has:
   a. the lowest charge value.

b. the highest charge value.
   c. the same Q as the other capacitors in the circuit.
   d. none of the above.

9. If you know the frequency and capacitive reactance:
   a. the value of C can be found by the formula $1/2\pi fC$.
   b. the value of C can be found by the formula $0.159/fC$.
   c. the value of C can be found by the formula $1/2\pi fX_C$.
   d. the value of C can be found with just this amount of information.

10. If four capacitors of equal value are in series:
    a. C_T = four times one C's value, and X_{C_T} = one fourth each capacitor's reactance.
    b. X_{C_T} = four times one C's reactance and C_T = one-fourth of one of the Cs.
    c. total capacitance and total reactance are four times each individual C's values.
    d. none of the above.

11. If four capacitors of equal value are in parallel:
    a. C_T is four times one C's value, and X_{C_T} is one-fourth each capacitor's reactance.
    b. X_{C_T} = four times one C's reactance, and C_T = one-fourth each capacitor's C.
    c. total capacitance and total reactance are four times each individual C's values.
    d. none of the above.

12. The voltage rating of the smallest capacitor in a series C circuit needs:
    a. to be higher than the other capacitors' voltage ratings, if the others are operating at their voltage rating.
    b. to be lower than the other capacitor's voltage ratings, if the others are operating at their voltage rating.
    c. to be the same as the other capacitors' voltage ratings, if the others are operating at their voltage rating.
    d. no special consideration.

13. In a purely capacitive circuit, the current:
    a. leads the voltage by 45°
    b. lags the voltage by 45°.
    c. leads the voltage by 90°.
    d. lags the voltage by 90°.

14. Capacitive reactance and inductive reactance:
    a. vary in the same way as frequency of operation is changed.
    b. vary in opposite ways as frequency of operation is changed.
    c. both remain the same as frequency of operation is changed.
    d. none of the above.

15. Capacitive reactance is related to the rate of change of:
    a. current.
    b. voltage.
    c. frequency.
    d. none of the above.

16. When computing X_C, the constant 0.159, when used, is equivalent to:
    a. 2π.
    b. $1/2\pi$.
    c. $2\pi f C$.
    d. none of the above.

17. The process of developing a voltage across capacitor plates may be called:
    a. rate of change of voltage.
    b. charging.
    c. discharging.
    d. none of the above.

18. When the rate of change of voltage is maximum, capacitor charging or discharging current is:
    a. maximum.
    b. minimum.
    c. not changing.
    d. none of the above.

19. When the sine-wave voltage applied to capacitor circuit is near maximum level, i_C is
    a. maximum.
    b. minimum.
    c. not changing.

20. What formula would you use to find f if you know the C and X_C?

PROBLEMS

1. What is the capacitive reactance of a 0.00025-μF capacitor operating at a frequency of 5 kHz?

2. What capacitance value exhibits 250 Ω of reactance at a frequency of 1,500 Hz?

3. The frequency of applied voltage to a given circuit triples while the capacitance remains the same. What happens to the value of capacitive reactance?

4. Series-circuit capacitors are all exchanged for Cs, each having twice the capacitance of the capacitor it replaces. What happens to total circuit X_C and total C?

5. For the circuit in Figure 19–12 find C_T, X_{C_T}, V_A, C_1, C_2, and X_{C_2}.

6. For the circuit in Figure 19–12, if C_1 and C_2 were equal in value, what would be the new voltages on C_1 and C_2?

7. Assume the frequency doubles in the circuit in Figure 19–12. Answer the following statements with "I" for increase, "D" for decrease, and "RTS" for remain the same.
    a. C_1 will _____
    b. C_2 will _____
    c. C_T will _____
    d. X_{C_1} will _____
    e. X_{C_2} will _____
    f. X_{C_T} will _____
    g. Current will _____
    h. V_{C_1} will _____
    i. V_{C_2} will _____
    j. V_A will _____

8. For the circuit in Figure 19–13 find C_T, I_1, I_2, X_{C_T}, X_{C_1}, X_{C_2}, and I_T.

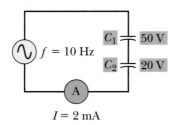

C_1 50 V
C_2 20 V
$f = 10$ Hz
A
$I = 2$ mA

FIGURE 19–12

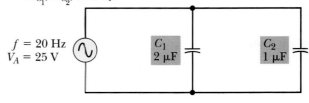

$f = 20$ Hz
$V_A = 25$ V
C_1 2 μF
C_2 1 μF

FIGURE 19–13

9. For the circuit in Figure 19–13, if C_2 is made to equal C_1, what is the I_T?

10. Assume the frequency doubles in the circuit in Figure 19–13. Answer the following statements with "I" for increase, "D" for decrease, and "RTS" for remain the same.
    a. C_1 will _____
    b. C_2 will _____
    c. C_T will _____
    d. X_{C_1} will _____
    e. X_{C_2} will _____
    f. X_{C_T} will _____
    g. Current will _____
    h. V_{C_1} will _____
    i. V_{C_2} will _____
    j. V_A will _____

Refer to Figure 19–14 and answer questions 11–15:

11. What is the X_C of C_1?

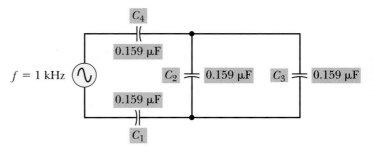

FIGURE 19–14

12. What is the X_{C_T}?

13. What is the value of C_T?

14. If f were changed to 5 kHz, what would be the new value of circuit X_{C_T}?

15. If C_2 were replaced by a larger capacitance value capacitor, would the voltage across C_4 increase, decrease, or remain the same?

ANALYSIS QUESTIONS

1. In your own words, explain when and why a capacitor might be said to be an "open circuit" with respect to dc, or that a capacitor "blocks" dc.

2. In your own words, explain how a capacitor can "pass" ac.

3. Draw a capacitive voltage divider circuit where the following conditions are found. (Label the diagram with all component and parameter values.)

    a. Three capacitors are used.
    b. The voltage across the middle capacitor will be double that found across the capacitors on either side of it.
    c. The frequency of operation is 2 kHz.
    d. The capacitor with maximum X_C has a reactance value of 79.5 Ω.
    e. $V_A = 159$ V.

4. Draw a circuit configuration where the total capacitance is 0.66 μF, using six 1-μF capacitors.

PERFORMANCE PROJECTS CORRELATION CHART

Suggested performance projects that correlate with topics in this chapter are:

CHAPTER TOPIC	PERFORMANCE PROJECT	PROJECT NUMBER
V & *I* Relationships in a Purely Capacitive AC Circuit	Capacitance Opposing a Change in Voltage	53
Relationship of X_C to Capacitance Value and Relationship of X_C to Frequency of AC	X_C Related to Capacitance and Frequency	54
Methods to Calculate X_C	The X_C Formula	55

NOTE: It is suggested that after completing the above projects, the student should be required to answer the questions in the "Summary" at the end of this section of projects in the Laboratory Manual.

THE SIMPLER SEQUENCE FOR TROUBLESHOOTING

Follow the seven-step SIMPLER troubleshooting sequence, as outlined below, and see if you can find the problem with this circuit. As you follow the sequence, record your circuit test results for each testing step on a separate sheet of paper. This will aid you and your instructor to see where your thinking is on track or where it might deviate from the best procedure.

step 1 Symptoms

—(Gather, verify, and analyze symptom information) To begin this process, read the data under the "Starting Point Information" heading. Particularly look for circuit parameters that are not normal for the circuit configuration and component values given in the schematic diagram. For example, look for currents that are too high or too low; voltages that are too high or too low; resistances that are too high or too low, etc. This analysis should give you a first clue (or symptom information) that will aid you in determining possible areas or components in the circuit that might be causing this symptom.

step 2 Identify

—(Identify and bracket the initial suspect area) To perform this first "bracketing" step, analyze the clue or symptom information from the "Symptoms" step, then either on paper or in your mind, bracket, circle, or put parentheses around all of the circuit area that contains any components, wires, etc., that might cause the abnormality that your symptom information has pointed out. NOTE: Don't bracket parts of the circuit that contain items that could not cause the symptom!

step 3 Make

—(Make a decision about the first test you should make: what type and where) Lock under the TEST column and determine which of the available tests shown in that column you think would give you the most meaningful information about the suspect area or components, in the light of the information you have so far.

step 4 Perform

—(Perform the first test you have selected from the TEST column list) To simulate performing the chosen test (as if you were getting test results in an actual circuit of this type), follow the dotted line to the right from your selected test to the number in parentheses. The number in parentheses, under the "Results in Appendix C" column tells you what number to look up in Appendix C to see what the test result is for the test you are simulating.

step 5 Locate

—(Locate and define a new "narrower" area of uncertainty) With the information you have gained from the test, you should be able to eliminate some of the circuit area or circuit components as still being suspect. In essence, you should be able to move one bracket of the bracketed area so that there is a new smaller area of uncertainty.

step 6 Examine

—(Examine available information and determine the next test: what type and where) Use the information you have thus far to determine what your next test should be in the new narrower area of uncertainty. Determine the type test and where, then proceed using the TEST listing to select that test, and the numbers in parentheses and Appendix C data to find the result of that test.

step 7 Repeat

—(Repeat the analysis and testing steps until you find the trouble) When you have determined what you would change or do to restore the circuit to normal operation, you have arrived at *your* solution to the problem. You can check your final result against ours by observing the pictorial step-by-step sample solution on the pages immediately following the "Chapter Troubleshooting Challenge"—circuit and test listing page.

step 8 Verify

NOTE: A useful *8th Step* is to operate and test the circuit or system in which you have made the "correction changes" to see if it is operating properly. This is the final proof that you have done a good job of troubleshooting.

CHALLENGE CIRCUIT 10

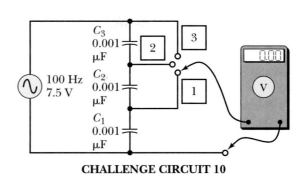

CHALLENGE CIRCUIT 10

Starting Point Information

1. Circuit diagram
2. When testing at test point "1," voltmeter reading is zero volts.
3. When testing at test point "2," voltmeter reading is 7.5 volts.
4. When testing at test point "3," voltmeter reading is 7.5 volts.

TEST	Results in Appendix C
V_{C_1}	.(28)
V_{C_2}	.(67)
V_{C_3}	.(102)
V_T	.(79)
"Bridging" with known good 0.001-μF capacitor:	
Bridge C_1	.(71)
Bridge C_2	.(24)
Bridge C_3	.(111)

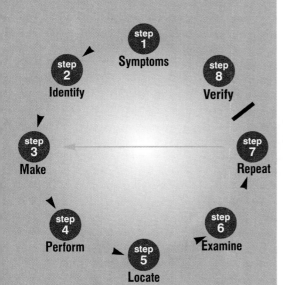

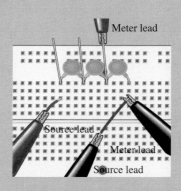

Meter lead

Source lead

Meter lead

Source lead

Symptoms

The capacitive voltage divider is not working correctly. The X_Cs of the equal value capacitors should be equal. The $I \times X_C$ drops should all be approximately equal. At test point 1, the voltmeter should read about 2.5 V; at test point 2 about 5.0 V; and at test point 3 about 7.5 V.

Identify

initial suspect area: If C_1 were shorted, the V at test point 1 would be 0 V. The V at test point 2 would be 3.75 V rather than 2.5 V. If C_3 were shorted, the V at switch position 1 would be 3.75 V, not 0 V. If C_1 were open, the V at switch position 1 would be 7.5 V. If C_3 were open, the V at switch position 2 would be 0 V. The normal voltage at switch position 3 is 7.5 V (the source voltage), which is what is measured. Without making our first test, we have narrowed the initial suspect area to C_2. If it were open, the circuit would operate similar to the symptom descriptions.

Make

test decision: Bridge the suspected open capacitor with a known good one of the same ratings and note any changes in circuit operation. We'll try the quick bridging technique. CAUTION: Make bridging connections with the power off. Then turn power on and make the necessary observations.

Perform

1st Test: Bridge C_2 while monitoring voltage at switch position 2. When this is done, V measures about 4.8 V, close to the nominal 5-V normal operation.

Locate

new suspect area: This is not necessary since we've found our problem.

Examine

available data.

Repeat

analysis and testing.

Verify

that the circuit operates normally by checking the voltage at each switch position **(2nd, 3rd, and 4th Tests)**. C_2 was open.

Symptoms

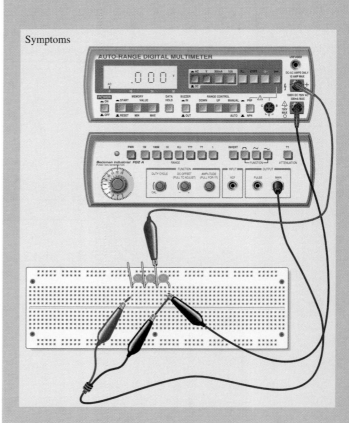

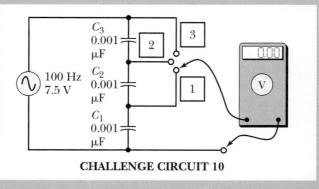

CHALLENGE CIRCUIT 10

1st Test

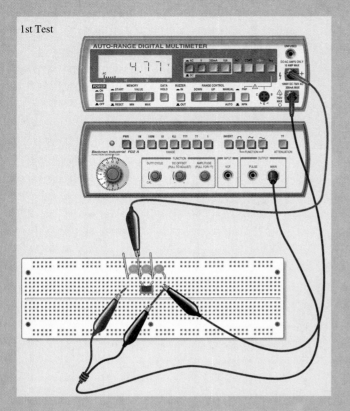

2nd Test

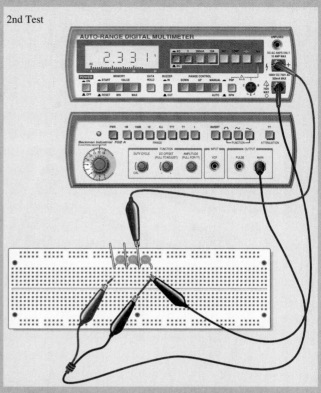

3rd Test

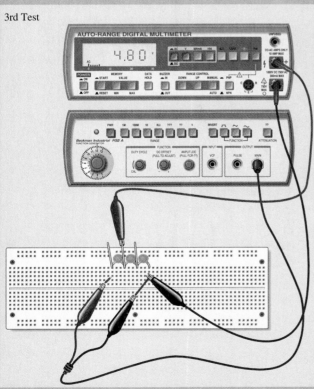

4th Test

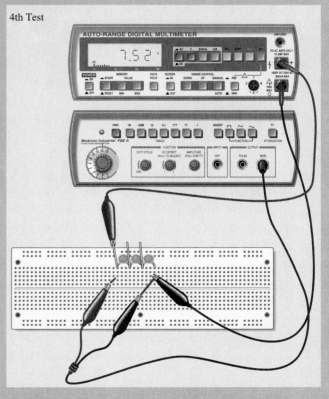

Current leading voltage
Long time-constant circuit
Short time-constant circuit

RC CIRCUITS IN AC

20

chapter

CHAPTER PREVIEW

Seldom does the technician find capacitors alone in a circuit. They are generally used with other components, such as resistors, inductors, transistors, and so on, to perform some useful function. In this chapter, we will investigate simple circuits containing both resistors and capacitors. Understanding *RC* circuits is foundational to dealing with practical circuits and systems, such as amplifiers, receivers, transmitters, computers, power supplies, and other systems and subsystems.

You will learn to analyze both series and parallel combinations of resistance and capacitance that operate under sine-wave ac conditions. An overview of how these components are used for waveshaping in nonsinusoidal applications will also be given. Also in this chapter, as in the earlier chapter "*RL* Circuits in AC," we will have you use the basic, or elemental, approaches of the Pythagorean theorem, and simple trig functions for circuit analysis. After completing this chapter, you will have had more practice in these techniques. Since after this chapter, you will have familiarity with basic vector notations for both *RL* and *RC* circuits, the next chapter will introduce additional powerful vector analysis and notation techniques. So be diligent in practicing the techniques and calculator usage afforded in this chapter. This will help you master the next chapter, which combines all your knowledge of *RL* and *RC* circuits, as well as teaching you some additional powerful analysis techniques. The concluding portion of this chapter will provide a summary and review of the similarities and differences between *RC* and *RL* circuits and supply some troubleshooting hints for *RC* circuits.

OBJECTIVES

After studying this chapter, you should be able to:

1. Draw or describe operation of simple *R* and *C* circuits

2. Analyze appropriate series and parallel *RC* circuit parameters using Pythagorean theorem

3. Use vector analysis to analyze series and parallel *RC* circuit parameters

4. List differences between *RC* and *RL* circuits

5. Predict output waveform(s) from a waveshaping network

6. List two practical applications for *RC* circuits

7. Apply troubleshooting hints to help identify problems

20–1 REVIEW OF SIMPLE R AND C CIRCUITS

Although you have studied purely resistive and capacitive circuits, review Figure 20–1 as a quick refresher.

20–2 SERIES RC CIRCUIT ANALYSIS

As Figure 20–1 has just revealed, current and voltage are in phase for purely resistive ac circuits. Circuit **current leads voltage** by 90° in purely capacitive circuits. Therefore, when resistance and capacitance are combined, the overall differ-

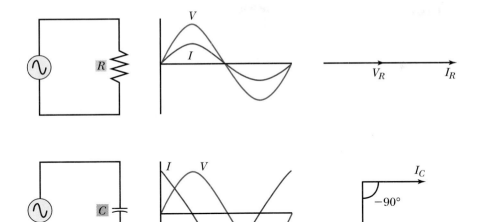

FIGURE 20–1 Voltage-current relationships in purely *R* and purely *C* circuits

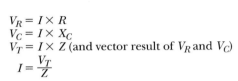

$$V_R = I \times R$$
$$V_C = I \times X_C$$
$$V_T = I \times Z \text{ (and vector result of } V_R \text{ and } V_C)$$
$$I = \frac{V_T}{Z}$$

FIGURE 20–2 Positional information for phasors in series *RC* circuits

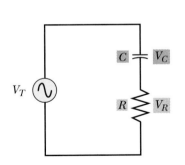

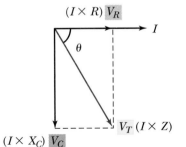

Angular difference between V_T and I (θ) is greater than 0° but less than 90°.

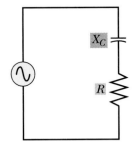

ence in angle between circuit voltage and current is an angular difference between 0° and 90°, Figure 20–2. The actual angle depends on the relative values of the components' electrical parameters and the frequency of operation.

Positional information for phasors

In series circuits, the current is used as the zero-degree reference vector for *V-I* vector diagrams because *current is common throughout a series circuit*. The *V-I* vector diagram, therefore, shows the resultant circuit voltage lagging behind the current by some angle between 0° and −90°, again see Figure 20–2. Remember, this is in the fourth quadrant of the coordinate system.

In *series RC* circuits (and *only* for *series* circuits), we can also develop an impedance diagram as well as a *V-I* vector diagram, Figure 20–3. Since the resistive voltage drop *(I × R)* vector must indicate the resistor voltage is in phase with the

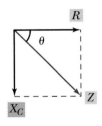

FIGURE 20–3 Impedance diagram for a series *RC* circuit

circuit current, the impedance diagram line representing R is shown at the $0°$ position. Because the capacitor voltage $(I \times X_C)$ lags the capacitor (and circuit) current by $90°$, the X_C on the impedance diagram is shown downward at the $-90°$ position.

Summarizing for *series RC* circuits, I and V_R are shown at $0°$ and V_C is shown at $-90°$ in *V-I* vector diagrams. The R is shown at $0°$ and X_C at $-90°$ for impedance diagrams. The angle between V_T and I (in *V-I* vector diagrams) and between Z and R (in impedance diagrams) is the value of the circuit phase angle. With this information, let's proceed with some analysis techniques.

Solving magnitude values by the Pythagorean theorem

As in the chapter on RL circuits, we can use the Pythagorean theorem to solve for magnitudes of voltages, currents, and circuit impedance in series RC circuits.

Recall the Pythagorean theorem deals with right triangles and the relationship of triangle sides. The hypotenuse (c) relates to the other sides by the formula $c^2 = a^2 + b^2$, or $c = \sqrt{a^2 + b^2}$. Observe Figure 20–4 as you study the following statements to see how this theorem is applied in phasor addition for *V-I* vector and impedance diagrams for a series RC circuit.

Solving for V$_T$ in the V-I vector diagram

1. The hypotenuse of the right triangle in the *V-I* vector diagram is V_T, or the vector resultant of combining V_R and V_C.

2. The side adjacent to the angle of interest (the phase angle) is V_R. The side opposite the angle is V_C.

3. Substituting into the Pythagorean formula, the formula to solve for the magnitude of the hypotenuse, or V_T, is:

FORMULA 20–1 $V_T = \sqrt{V_R^2 + V_C^2}$

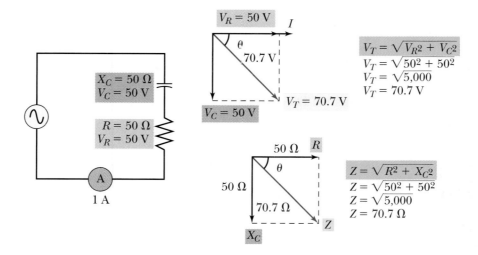

FIGURE 20–4 Using the Pythagorean theorem approach

Example

Thus, in Figure 20–4, $V_T = \sqrt{50^2 + 50^2} = \sqrt{2{,}500 + 2{,}500} = \sqrt{5{,}000} = 70.7$ V.

Solving for Z in the impedance diagram

1. The hypotenuse of the right triangle in the impedance diagram is Z (impedance).
2. The side adjacent to the angle representing the phase angle is R. The side opposite the angle represents X_C.
3. Substituting into the Pythagorean formula, the formula to solve for the magnitude of the hypotenuse, or Z, is:

FORMULA 20–2	$Z = \sqrt{R^2 + X_C^2}$

Example

Thus, in Figure 20–4, $Z = \sqrt{50^2 + 50^2} = \sqrt{2{,}500 + 2{,}500} = \sqrt{5{,}000} = 70.7$ Ω.
    Try applying the Pythagorean formula to the following problems.

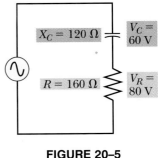

FIGURE 20–5

PRACTICE PROBLEMS I

1. For the circuit in Figure 20–5, use the Pythagorean approach and solve for V_T. Draw the V-I vector diagram and label all elements. (NOTE: You may want to try the drawing-to-scale approach learned in an earlier chapter to check your results.)
2. For the circuit in Figure 20–5, use the Pythagorean approach and solve for Z. Draw and label an appropriate impedance diagram.

PRACTICAL NOTES

Reminder: It is a good habit to sketch diagrams for any problem you are solving even if you don't draw them precisely. By sketching, you can approximate to see if the answers you get are logical.

Solving for phase angle using trigonometry

Recall from the RL circuits chapter that we can use simple trig functions to solve for the angles of a right triangle. Remember that

$$\sin \theta = \frac{\text{opp}}{\text{hyp}}, \cos \theta = \frac{\text{adj}}{\text{hyp}}, \text{ and } \tan \theta = \frac{\text{opp}}{\text{adj}}$$

Because we have already identified which sides are the side adjacent, side opposite, and the hypotenuse in the series RC circuit V-I vector and impedance diagrams, substitute into the trig formulas to solve for angle information.

Note the trig formula that does not require knowing the magnitude of the hypotenuse to solve for the angle information is the tangent function. Therefore, this function is often used.

PRACTICAL NOTES

With the accessibility of scientific calculators, finding phase angles (and other circuit parameters) in ac circuits has become increasingly simple. In this chapter, we are having you use, practice, and continue to learn the rudimentary concepts related to trig functions for phase-angle solutions.

In the next chapter, you will have the opportunity to learn rectangular and polar vector notation techniques and simple calculator keystrokes that allow conversion between these notation methods. As you will see, these notation methods and the calculator can take much of the "grief" out of ac circuit calculations.

The foundations you have built in previous chapters, and are building in this chapter, will help you to better understand and use these techniques as you progress in your studies. We encourage you to try to understand concepts, rather than simply memorizing formulas!

Solving the angle from the V-I vector diagram

Example

Refer again to Figure 20–4. Since we know all three sides, we can use any of the three basic trig formulas. However, let's use the tangent function.

1. To find the phase angle, or the angle between V_T and I, the V_C and V_R values can be used.
2. Substitute these values into the tangent formula.

$$\tan \theta = \frac{\text{opp}}{\text{adj}} = \frac{V_C}{V_R} = \frac{50}{50} = 1$$

3. Using a calculator, we find the arctan θ, or the angle whose tangent equals 1 is 45°.
4. Therefore, the phase angle is 45°. This makes sense since both legs of the triangle are equal in length. If you draw this diagram to scale and use a protractor to measure the angle, the result should verify a 45° angle.

Solving the angle from the impedance diagram

Example

Refer to Figure 20–6 and use the tangent function to find the phase angle.

1. The angle between Z and R is the same value as the phase angle (the angle between the circuit voltage and circuit current).

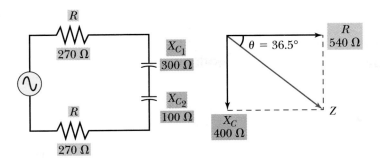

FIGURE 20–6

2. Using the tangent function to find this angle requires the X_C and R values.

   a. Add the resistances to get the R value.

$$R = 270 \ \Omega + 270 \ \Omega = 540 \ \Omega$$

   b. Add the X_Cs to get the X_C value.

$$X_C = 300 \ \Omega + 100 \ \Omega = 400 \ \Omega$$

3. Substitute appropriate values into the tangent formula.

$$\tan \theta = \frac{\text{opp}}{\text{adj}} = \frac{X_C}{R} = \frac{400}{540} = 0.74$$

4. Find the $\tan^{-1}$ of 0.74 to get the angle.

$$\tan^{-1} 0.74 = 36.5°; \text{ thus, } \theta = 36.5°$$

NOTE: When impedance for series RC circuits is plotted on an impedance vector diagram, the location of the Z vector is in the fourth quadrant, somewhere between $0°$ and $-90°$ removed from the reference vector (R). This means the phase angle is considered a negative angle. Thus, $\theta = -36.5°$, in this case.

Apply these principles to solve the following problems.

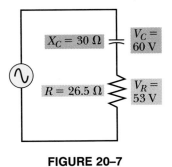

FIGURE 20–7

PRACTICE PROBLEMS II

1. Refer to Figure 20–7. Draw a V-I vector diagram for the circuit shown, and find the phase angle using a basic trig function. (Round-off to the nearest volt.)

2. Draw an impedance diagram for the circuit in Figure 20–7. Then solve for the value of θ, using an appropriate trig function.

20–3 PHASE-SHIFT APPLICATION OF SERIES RC NETWORK

Background information

Phase-shift networks are frequently needed and used in electronic circuits and electronically controlled systems. The series RC network is one of the simpler networks for achieving a desired phase shift (or phase difference). The network is used to create and/or control the phase difference between the voltage into the network, and the voltage being taken as output voltage from the network. These phase-shift networks are sometimes called *RC* "lead" or "lag" circuits.

The principle upon which these *RC* phase-shift networks operate is based upon the fact that a capacitor opposes a change in voltage. That is, it takes time for the capacitor to change from one voltage level to another (because of charging and discharging time required). You have already learned that in a series *RC* circuit, the circuit current may lead the circuit applied voltage (V_{in}) anywhere from 0° to 90°, depending on the relative values of R and X_C. Also, you have learned that the voltage across the resistor is in-phase with the circuit current.

We can control the relative R and X_C values by choosing or varying the component values. For example, we can use a variable resistor in the network to control the amount of phase shift. We can also select which component we will take the output voltage across. By selecting the appropriate component, we can choose whether the output voltage leads or lags the input voltage. A *"lead"* network is defined as one in which the *output voltage leads input voltage*. A *"lag"* network is one in which the *output voltage lags the input voltage*.

Example (an RC "Lead" Network)

Refer to Figure 20–8 as you study the following comments.

1. For the lead network, the output is taken across *R*.
2. For the lead network, then, V_R and V_{out} are the same voltage.
3. For our vector diagram, we will use V_{out} as the reference vector.
4. Since the capacitor causes the circuit current to "lead" the circuit voltage and the output voltage is taken across the resistor, the output voltage is leading the input voltage.
5. The amount of lead is controllable by the values of *C* and *R*, which control the relative values of X_C and *R*.

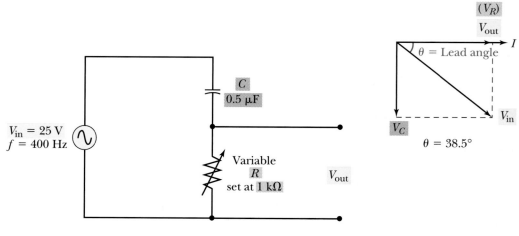

FIGURE 20–8 An *RC* "lead" network

6. The circuit phase angle $(\theta) = \arctan \dfrac{X_C}{R}$

$$X_C = \frac{1}{2\pi fC} = \frac{1}{6.28 \times 400 \times 0.5 \times 10^{-6}} = 796 \ \Omega$$

$$\frac{X_C}{R} = \frac{796}{1,000} = 0.796$$

$$\arctan 0.796 = 38.5°$$

7. For this network, the output voltage leads the input voltage by $38.5°$.

8. The value of V_{out} is determined by the voltage divider action of the *RC* network. In essence, the output voltage (in this case V_R) relates to the ratio of the *R* to the circuit impedance. That is, $V_{out} = V_{in} \times R/Z$. (Of course *Z* in this circuit equals $\sqrt{R^2 + X_C^2}$.)

$$V_{out} = V_{in} \ (R/Z)$$
$$\text{where } Z = \sqrt{R^2 + X_C^2} = \sqrt{1,000^2 + 796^2} = 1,278 \ \Omega$$
$$V_{out} = 25 \text{ V} \ (1,000/1,278) = 25 \text{ V} \times 0.78 = 19.5 \text{ V}$$

PRACTICE PROBLEMS III

What would be the angle of "lead" should the variable resistor in Figure 20–8 be set at 500 Ω?

Example (an RC "Lag" Network)

Refer to Figure 20–9 as you study the following information.

1. For the "lag" network, the output is taken across *C*.

2. For the lag network, then, V_C and V_{out} are the same voltage.

3. For our vector diagram, we will use V_R as the reference vector.

4. Since the capacitor causes the circuit current to lead the circuit voltage and the output voltage is taken across the capacitor, the output voltage is lagging the input voltage.

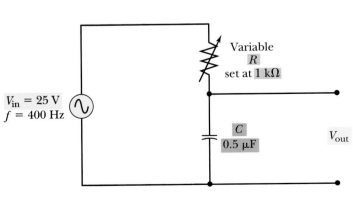
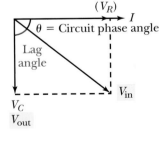

FIGURE 20–9 An *RC* "lag" network

5. The circuit phase angle and the amount of lag are NOT the same thing, in this case. The circuit phase angle is the phase difference between V_R and V_{in}. Since the output is taken across the capacitor, the lag angle is the difference in degrees between V_C and V_{in}.

6. The circuit phase angle and the amount of lag are controllable by the values of C and R, which, in turn, control the relative values of X_C and R.

7. The circuit phase angle (θ) = arctan $\dfrac{X_C}{R}$

$$X_C = \frac{1}{2\pi f C} = \frac{1}{\sqrt{6.28 \times 400 \times 0.5 \times 10^{-6}}} = 796 \ \Omega$$

$$\frac{X_C}{R} = \frac{796}{1,000} = 0.796$$

$$\text{arctan } 0.796 = 38.5°$$

8. For this network, V_{in} lags V_R by this 38.5° (the circuit phase angle); however, the amount of lag between V_{out} compared with V_{in} equals 90° minus 38.5° or 51.5° of lag.

$$\text{Lag angle} = 90° - \text{circuit phase angle}$$

9. To vary the amount of lag in this circuit, we can vary the R value. If the R value is decreased, the circuit becomes more capacitive and the phase angle will increase; however, the amount of lag will decrease. Remember that the lag angle is the angular difference between V_{in} and V_C (the output voltage). If the R value is increased, the circuit phase angle will decrease; however, the lag angle will increase.

10. The value of V_{out} is determined by the voltage-divider action of the RC network. In essence, the output voltage (in this case V_C) relates to the ratio of the X_C to the circuit impedance. That is, $V_{out} = V_{in} \times X_C / Z$. (Of course Z, in this circuit, equals $\sqrt{R^2 + X_C^2}$.

$$V_{out} = V_{in} \left(\frac{X_C}{Z} \right)$$

$$\text{where } Z = \sqrt{R^2 + X_C^2} = \sqrt{1,000^2 + 796^2} = 1,278 \ \Omega$$

$$V_{out} = 25 \text{ V} \left(\frac{796}{1,278} \right) = 25 \text{ V} \times 0.62 = 15.5 \text{ V}$$

PRACTICE PROBLEMS IV

What would be the angle of lag should the R be set at 750 Ω in the circuit of Figure 20–9?

Summary of series RC circuits

Figure 20–10 shows examples of series RC circuit parameters for various circuit conditions. These examples help reinforce the following generalizations about series RC circuits.

1. Ohm's law formulas can be used at any time to solve for individual component or total circuit voltages, currents, resistances, reactances, and impedances, as long as the appropriate parameter values are substituted into the formulas. That is,

$$V_C = I_C \times X_C, \; Z = \frac{V_T}{I_T}, \text{ and so forth}$$

2. The circuit voltage lags the circuit current by some angle between $0°$ and $-90°$. Conversely, the circuit current leads the circuit voltage by that same angular difference.

3. Since current is common throughout series circuits, current is the reference vector in phasor diagrams of V and I.

4. The larger the R value is with respect to the X_C value, the more resistive the circuit is. That is, the phase angle is smaller or closer to $0°$. Conversely, the larger the X_C value is with respect to the circuit R value, the greater is the phase angle and the closer it is to $-90°$.

5. Impedance of a series RC circuit is greater than either the R value alone or the X_C value alone but *not* as great as their arithmetic sum. Impedance equals the vector sum of R and X_C.

6. When either R or X_C increases in value, circuit impedance increases and circuit current decreases.

CAPACITANCE CHANGED BY A FACTOR OF 2 (FROM 2 μF TO 1 μF)									
f (Hz)	C (μF)	X_C (Ω)	R (Ω)	Z (Ω)	I (amps)	V_C (IX_C)	V_R (IR)	V_T (IZ)	θ deg
100	2	795	7,950	7,989.65	0.025	19.90	199.01	200	−5.71
100	1	1,590	7,950	8,107.44	0.0246	39.22	196.12	200	−11.31

(C to half; X_C doubled; Z increased; I decreased; V_C increased; V_R decreased; and θ increased)

(a)

RESISTANCE CHANGED BY A FACTOR OF 2 (FROM 7,950 Ω TO 3,975 Ω)									
f (Hz)	C (μF)	X_C (Ω)	R (Ω)	Z (Ω)	I (amps)	V_C (IX_C)	V_R (IR)	V_T (IZ)	θ deg
100	1	1,590	7,950	8,107.44	0.0246	39.22	196.12	200	−11.31
100	1	1,590	3,975	4,281.21	0.0467	74.28	185.70	200	−21.80

(R to half; Z decreased; I increased; V_C increased; V_R decreased; and θ increased)

(b)

FREQUENCY CHANGED BY A FACTOR OF 10 (FROM 100 Hz TO 10 Hz)									
f (Hz)	C (μF)	X_C (Ω)	R (Ω)	Z (Ω)	I (amps)	V_C (IX_C)	V_R (IR)	V_T (IZ)	θ deg
100	1	1,590	3,975	4,281.21	0.0467	74.28	185.7	200	−21.80
10	1	15,900	3,975	16,389.34	0.0122	194.03	48.51	200	−75.96

(f to $1/10$; X_C increased ten times; Z increased; I decreased; V_R decreased; and θ increased greatly)

(c)

FIGURE 20–10 Sample series *RC* circuit parameters charts

7. Total voltage of a series *RC* circuit is greater than either the resistive voltage drops or the reactive voltages alone but *not* as great as their arithmetic sum. The voltages must be added vectorially to find the resulting V_T value.

8. For series *RC* circuits, both *V-I* vector and impedance diagrams are used. The quantities used in the *V-I* vector diagram are the circuit current as a reference vector at 0°; V_R (also at the 0° reference position); and V_C at −90°. V_T (or V_A) is plotted as the vector resulting from the resistive and reactive voltage vectors. For the impedance diagram, *R* is the reference vector and X_C is plotted downward at the −90° position. Circuit impedance (*Z*) is the vector sum resulting from *R* and X_C.

IN-PROCESS LEARNING CHECK I

Fill in the blanks as appropriate.

1. In series *RC* circuits, the _____ is the reference vector, shown at 0°.

2. In the *V-I* vector diagram of a series *RC* circuit, the circuit voltage is shown _____ the circuit current.

3. In series *RC* circuits, an impedance diagram (can, cannot) _____ be drawn.

4. Using the Pythagorean approach, the formula to find V_T in a series *RC* circuit is _____.

5. Using the Pythagorean approach, the formula to find *Z* in a series *RC* circuit is _____.

6. The trig function used to solve for phase angle when you do not know the hypotenuse value is the _____ function.

20–4 PARALLEL RC CIRCUIT ANALYSIS

In parallel *RC* circuits, the voltage is common to all parallel branches. Therefore, *voltage* is the reference vector for the *V-I* vector diagrams, Figure 20–11a. Again, you *will not* draw impedance diagrams for *parallel* circuits.

In drawing a *V-I* vector diagram, branch currents are plotted relative to the voltage reference vector. I_R is plotted at the zero-degree position and is the side adjacent to the angle of interest. I_C is plotted at the +90° position because it leads the voltage by 90° and is the side opposite the angle for computational purposes. Again, refer to Figure 20–11a.

Solving for I$_T$ using the Pythagorean approach

For parallel *RC* circuits, the Pythagorean formula to solve for I_T becomes:

FORMULA 20–3 $I_T = \sqrt{I_R^2 + I_C^2}$

Example

Use the quantities in Figure 20–11b, and substitute into the preceding formula. Thus,

$$I_T = \sqrt{I_R^2 + I_C^2} = \sqrt{15^2 + 20^2} = \sqrt{625} = 25 \text{ A}$$

Solving for phase angle using trigonometry

The sine, cosine, and tangent functions, as related to a parallel RC circuit, are stated as:

$$\sin \theta = \frac{\text{opp}}{\text{hyp}} \text{ or } \sin \theta = \frac{I_C}{I_T}$$

$$\cos \theta = \frac{\text{adj}}{\text{hyp}} \text{ or } \cos \theta = \frac{I_R}{I_T}$$

$$\tan \theta = \frac{\text{opp}}{\text{adj}} \text{ or } \tan \theta = \frac{I_C}{I_R}$$

It should be noted that due to the fact that the tangent function does not require solving for the hypotenuse (I_T, in this case), it is often used to solve for phase angles.

Examples

Referring to the information in Figure 20–11b, trig functions can be used to solve for the phase angle (θ) as follows:

1. Using the tangent function:

$$\tan \theta = \frac{I_C}{I_R} = \frac{20}{15} = 1.33. \text{ The arctan } 1.33 = 53.1°$$

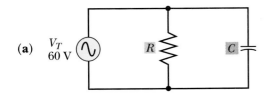

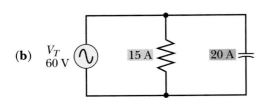

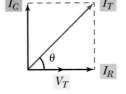

$$I_T = \sqrt{I_R{}^2 + I_C{}^2}$$
$$I_T = \sqrt{15^2 + 20^2}$$
$$I_T = \sqrt{625}$$
$$I_T = 25 \text{ A}$$

$$\tan \theta = \frac{I_C}{I_R}$$
$$\tan \theta = \frac{20}{15}$$

$$\tan \theta = 1.33$$
Angle whose tan = 1.33 is 53.1°

$$Z = \frac{V_T}{I_T} = \frac{60}{25} = 2.4 \ \Omega$$

$$R = \frac{V_R}{I_R} = \frac{60}{15} = 4 \ \Omega$$

$$X_C = \frac{V_C}{I_C} = \frac{60}{20} = 3 \ \Omega$$

(NOTE: Z is less than either of the branch ohms.)

FIGURE 20–11 Parallel RC circuit analysis

2. Using the sine function:

$$\sin \theta = \frac{I_C}{I_T} = \frac{20}{25} = 0.8. \text{ The arcsin } 0.8 = 53.1°$$

3. Using the cosine function:

$$\cos \theta = \frac{I_R}{I_T} = \frac{15}{25} = 0.6. \text{ The arccos } 0.6 = 53.1°$$

Looking at the diagram, the angle makes sense because the current through the reactive branch is greater than the current through the resistive branch. The circuit is acting more reactive than resistive. In other words, the angle is greater than 45°.

Solving for impedance in a parallel RC circuit

An impedance diagram CANNOT be used for illustrating impedance in parallel circuits as it is used for series circuits. The circuit impedance for a parallel circuit is less than the impedance of any one branch; therefore, it cannot equal the hypotenuse, or vector sum of the oppositions, as it does in series circuits. (For example, if the resistive branch and the reactive branch had equal ohmic values, the circuit impedance would equal 70.7% of one of the branches.) There are a couple of simple ways, however, to solve for the value of impedance in a basic parallel RC circuit.

1. Use Ohm's law, where $Z = V_T/I_T$.

2. Use the following formula:

FORMULA 20–4 $$Z = \frac{RX_C}{\sqrt{R^2 + X_C^2}}$$

Let's try these out!

Examples

Refer again to Figure 20–11b.

1. Using Ohm's law:

$$Z = \frac{V_T}{I_T}$$

$$Z = \frac{60 \text{ V}}{25 \text{ A}}$$

$$Z = 2.4 \text{ }\Omega$$

2. Using the special formula:

$$Z = \frac{RX_C}{\sqrt{R^2 + X_C^2}}$$

To use this formula, we need to know the R and X_C values. Since the voltage is known and each branch current is known, we can use Ohm's law to find these.

$$R = \frac{60 \text{ V}}{15 \text{ A}} = 4 \text{ }\Omega; \; X_C = \frac{60 \text{ V}}{20 \text{ A}} = 3 \text{ }\Omega$$

$$Z = \frac{4 \times 3}{\sqrt{4^2 + 3^2}} = \frac{12}{5} = 2.4 \text{ }\Omega$$

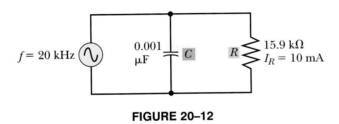

FIGURE 20–12

PRACTICE PROBLEMS V

Refer to Figure 20–12 and solve the following problems.

1. Use the capacitive reactance formula, Ohm's law, and the Pythagorean approach, as appropriate, to solve for I_T.

2. Draw the V-I vector diagram for the circuit and determine the phase angle, using appropriate trig formula(s). Label all the elements in the diagram.

3. Determine the circuit impedance.

IN-PROCESS LEARNING CHECK II

1. In parallel RC circuits, _____ is the reference vector in V-I vector diagram.

2. In parallel RC circuits, current through the resistor(s) is plotted (in phase, out of phase) _____ with the circuit current vector.

3. In parallel RC circuits, current through capacitor branch(es) is plotted at _____ with respect to circuit applied voltage.

4. The Pythagorean formula to find circuit current in parallel RC circuits is _____.

5. Is the Pythagorean formula used to find circuit voltage in a parallel RC circuit? _____

6. Can the tangent function be used to find phase angle in a parallel RC circuit? _____

7. Can an impedance diagram be used to find circuit impedance in a parallel RC circuit? _____

Summary of parallel RC circuits

Figure 20–13 provides several examples of parallel RC circuit parameters that illustrate the following facts. Study and understand these general principles before proceeding in the chapter.

1. Ohm's law formulas are used at any time to solve for individual component or overall circuit parameters, providing you substitute proper quantities into the formulas.

2. The applied voltage (V_T) is the reference vector in phasor diagrams of parallel RC circuits, since V is common to all branches.

CAPACITANCE CHANGED BY A FACTOR OF 2 (FROM 2 μF TO 1 μF)

f (Hz)	V_T (V)	C (μF)	X_C (Ω)	R (Ω)	I_C (amps)	I_R (amps)	I_T (amps)	Z (ohms)	θ deg
100	100	2	795	5,000	0.1258	0.0200	0.1274	758.14	80.97
100	100	1	1,590	5,000	0.0629	0.0200	0.0660	1,515.2	72.36

(*C* to half; X_C doubled; I_C halved; I_R = same; I_T decreased; and θ decreased)

(**a**)

RESISTANCE CHANGED BY A FACTOR OF 2 (FROM 5,000 Ω TO 2,500 Ω)

f (Hz)	V_T (V)	C (μF)	X_C (Ω)	R (Ω)	I_C (amps)	I_R (amps)	I_T (amps)	Z (ohms)	θ deg
100	100	1	1,590	5,000	0.0629	0.0200	0.0660	1,515.23	73.36
100	100	1	1,590	2,500	0.0629	0.0400	0.0745	1,341.64	57.54

(*R* to half; X_C = same; I_C = same; I_R doubled; I_T increased; and θ decreased)

(**b**)

FREQUENCY CHANGED BY A FACTOR OF 10 (FROM 100 Hz TO 10 Hz)

f (Hz)	V_T (V)	C (μF)	X_C (Ω)	R (Ω)	I_C (amps)	I_R (amps)	I_T (amps)	Z (ohms)	θ deg
100	100	1	1,590	2,500	0.0629	0.0400	0.0745	1,341.64	57.54
10	100	1	15,900	2,500	0.0063	0.0400	0.0405	2,469.66	8.94

(*f* to $^1/_{10}$; X_C increased 10x; I_C decreased to $^1/_{10}$; I_R = same; I_T decreased; and θ decreased greatly)

(**c**)

RESISTANCE CHANGED TO MAKE EQUAL WITH X_C

f (Hz)	V_T (V)	C (μF)	X_C (Ω)	R (Ω)	I_C (amps)	I_R (amps)	I_T (amps)	Z (ohms)	θ deg
10	100	1	15,900	2,500	0.0063	0.0400	0.0405	2,469.66	8.94
10	100	1	15,900	15,900	0.0063	0.0063	0.0089	11,243.00	45

(*R* made = X_C; X_C = same; I_C = same; I_R decreased greatly; I_T decreased greatly; and θ = 45° [since I_C and I_R are equal])

(**d**)

V_T IS DOUBLED

f (Hz)	V_T (V)	C (μF)	X_C (Ω)	R (Ω)	I_C (amps)	I_R (amps)	I_T (amps)	Z (ohms)	θ deg
10	100	1	15,900	15,900	0.0063	0.0063	0.0089	11,243.00	45
10	200	1	15,900	15,900	0.0126	0.0126	0.0178	11,243.00	45

(V_T doubled; all currents doubled; X_C, R, Z, and θ did not change)

(**e**)

FIGURE 20–13 Sample parallel *RC* circuit parameters charts

3. The total circuit current leads the circuit voltage by some angle between 0° and +90°.

4. The *smaller* the R value with respect to the X_C value, the more resistive the circuit is. That is, the closer to 0° the phase angle is. Conversely, the larger the R value is with respect to the X_C value, the more capacitive the circuit is.

5. Impedance in a parallel RC circuit is less than the ohmic opposition of any one branch and must be found through Ohm's law techniques or a special formula. Impedance for a parallel RC circuit **cannot** be found by drawing an impedance diagram.

6. When either R or X_C increases, impedance increases and circuit current decreases.

7. Total current in a parallel RC circuit is greater than any one branch current. However, total current is *not* as great as the sum of the branch currents. The branch currents must be added vectorially to find the resulting I_T value.

8. For a *V-I* vector diagram in a parallel RC circuit, voltage applied is the reference vector, I_R is plotted at 0°, and I_C is plotted at +90°. I_T is plotted as the vector resulting from the resistive and reactive branch currents.

20–5 SIMILARITIES AND DIFFERENCES BETWEEN RC AND RL CIRCUITS

So that you can see and remember the key points about RL and RC circuits, Figure 20–14 provides an important list of similarities and differences. If you understand the facts listed, you will find the following chapters about circuits containing all three quantities of resistance, inductance, and capacitance easy to understand.

ITEM DESCRIPTION	SERIES RC CIRCUIT	PARALLEL RC CIRCUIT	SERIES RL CIRCUIT	PARALLEL RL CIRCUIT
Formulas	$V_T = \sqrt{V_R^2 + V_C^2}$	$V_T = I_T \times Z$	$V_T = \sqrt{V_R^2 + V_L^2}$	$V_T = I_T \times Z$
	$I_T = I_R = I_C = \dfrac{V_T}{Z}$	$I_T = \sqrt{I_R^2 + I_C^2}$	$I_T = I_R = I_L = \dfrac{V_T}{Z}$	$I_T = \sqrt{I_R^2 + I_L^2}$
	$Z = \sqrt{R^2 + X_C^2}$	$Z = \dfrac{V_T}{I_T}$	$Z = \sqrt{R^2 + X_L^2}$	$Z = \dfrac{V_T}{I_T}$
	$\tan \theta = \dfrac{X_C}{R}$ or $\dfrac{V_C}{V_R}$	$\tan \theta = \dfrac{I_C}{I_R}$	$\tan \theta = \dfrac{X_L}{R}$ or $\dfrac{V_L}{V_R}$	$\tan \theta = \dfrac{I_L}{I_R}$
	$X_C = \dfrac{0.159}{fC}$ or $\dfrac{V_C}{I_C}$	$X_C = \dfrac{0.159}{fC}$ or $\dfrac{V_C}{I_C}$	$X_L = 2\pi fL$ or $\dfrac{V_L}{I_L}$	$X_L = 2\pi fL$ or $\dfrac{V_L}{I_L}$
As $f\uparrow$	$X_C\downarrow Z\downarrow I\uparrow \theta\downarrow$	$X_C\downarrow Z\downarrow I\uparrow \theta\uparrow$	$X_L\uparrow Z\uparrow I\downarrow \theta\uparrow$	$X_L\uparrow Z\uparrow I\downarrow \theta\downarrow$
As $C\uparrow$	$X_C\downarrow Z\downarrow I\uparrow \theta\downarrow$	$X_C\downarrow Z\downarrow I\uparrow \theta\uparrow$	N/A	N/A
As $L\uparrow$	N/A	N/A	$X_L\uparrow Z\uparrow I\downarrow \theta\uparrow$	$X_L\uparrow Z\uparrow I\downarrow \theta\downarrow$
As $R\uparrow$	$Z\uparrow I\downarrow \theta\downarrow$	$Z\uparrow I\downarrow \theta\uparrow$	$Z\uparrow I\downarrow \theta\downarrow$	$Z\uparrow I\downarrow \theta\uparrow$

FIGURE 20–14 Similarities and differences between RL and RC circuits

20–6 LINEAR WAVESHAPING AND NONSINUSOIDAL WAVEFORMS

Many electronic circuits depend on the ability to change nonsinusoidal circuit waveforms (such as square or rectangular waveforms) into different shapes for different purposes. For example, in radar systems it is not uncommon to need sharp or pointed waveforms to act as *trigger pulses* for circuits where timing action is performed.

In Figure 20–15, a square-wave input fed to a series *RC* network is shaped to be somewhat sharp or peaked across the resistor. This output is called the *differentiated output*. By definition, a differentiating circuit is one whose output relates to the rate of change of voltage or current of the input.

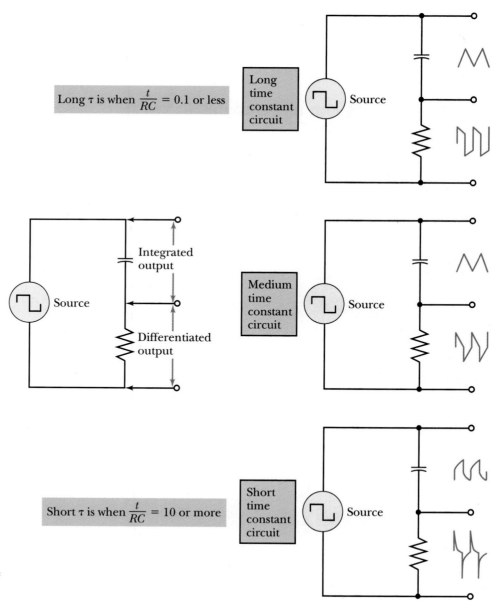

Long τ is when $\dfrac{t}{RC} = 0.1$ or less

Long time constant circuit — Source

Integrated output

Differentiated output

Source

Medium time constant circuit — Source

Short τ is when $\dfrac{t}{RC} = 10$ or more

Short time constant circuit — Source

FIGURE 20–15 Examples of *RC* circuit waveshaping

On the other hand, across the capacitor an output called an *integrated output* is taken. In this case, the waveform is rounded-off or flattened-out. This seems logical since a capacitor opposes a voltage change. Therefore, the voltage can't make sudden or "sharp" changes across the capacitor. By definition, an integrating circuit output relates to the integral of the input signal.

The amount of change of the input signal waveshape at the output relates to the amount of time allowed for the capacitor to charge or discharge during one alternation of the input signal. If the capacitor has more than ample time (over 5 time constants) to make its voltage change, the circuit is a **short time-constant circuit.** By definition, a short time-constant circuit is one where the time allowed (*t*) is 10 times or more one *RC* time constant. That is, $t/RC = 10$ or more. *RC* time is short compared with *t*.

Conversely, if the *RC* time constant is such that the time allowed by the input signal alternation is short compared with the time needed for complete change across the *RC* network, it is a **long time-constant circuit.** That is, $t/RC = 0.1$ or less. *RC* time constant is long compared with the time allowed to charge or discharge.

Refer again to Figure 20–15. Notice both the differentiated and integrated waveforms present across the resistor and capacitor for long, medium, and short time-constant conditions. You will see practical applications of such circuits as you study a variety of waveshaping circuit applications.

In Figures 20–16 and 20–17, you can again see how the *RC* time of resistive-capacitive networks is useful in integrating or differentiating signals, causing desired change in the waveshape of square or rectangular wave signals.

20-7 OTHER APPLICATIONS OF RC CIRCUITS

In addition to waveshaping and timing applications, *RC* circuits have other valuable uses. One common function is that of coupling desired signals from one circuit point or one circuit stage to another. Often they are used to block dc while passing along the ac component of a signal, Figure 20–18 (page 670).

Capacitors are also used to bypass or filter undesired signals. You will see examples of power-supply filtering when you study rectifier and filter circuits.

Also, capacitor networks are used for ac and dc voltage dividers, Figure 20–19 (page 670). Recall, voltage divides inversely to the capacitance values when capacitors are connected in series. That is, the smallest *C* has the highest *V* and the largest *C* has the smallest *V*.

By now, it is obvious that capacitors are found in almost every electronic circuit and system you will study or work on as a technician. The importance of learning about them and their operation cannot be overemphasized.

20-8 TROUBLESHOOTING HINTS AND CONSIDERATIONS FOR RC CIRCUITS

Based on information supplied earlier regarding typical troubles with capacitors, one can surmise some of the problems that can occur for the various applications where capacitors are used. Some examples follow.

When used as a filtering device—for example, in power supplies or in decoupling circuits—either an excessive leakage condition or excessive internal resistance condition will prevent proper filtering action. As you will learn, in such

These circuits will pass part of any dc component present in input.

Square-wave input

+5 V
0 V
−5 V

The longer the time constant, the more the output voltage approaches the dc average of the input. In other words, the low-frequency components are passed and the high-frequency components are filtered out.

Small R
Small C
+5 V
0 V
−5 V
Short time constant

Medium R
Medium C
+5 V
0 V
−5 V
Medium time constant

Large R
Large C
+5 V
0 V
−5 V
Long time constant

(20% duty cycle)

Short-pulse input

5 V
0 V

The dc output of the integrator is proportional to the duty cycle of the incoming waveform.

Small R
Small C
5 V
0 V
Short time constant

Medium R
Medium C
+5 V
0 V
Medium time constant

This average voltage is 20% of the maximum value of 5 V.

Large R
Large C
5 V
1 V
0 V
Long time constant

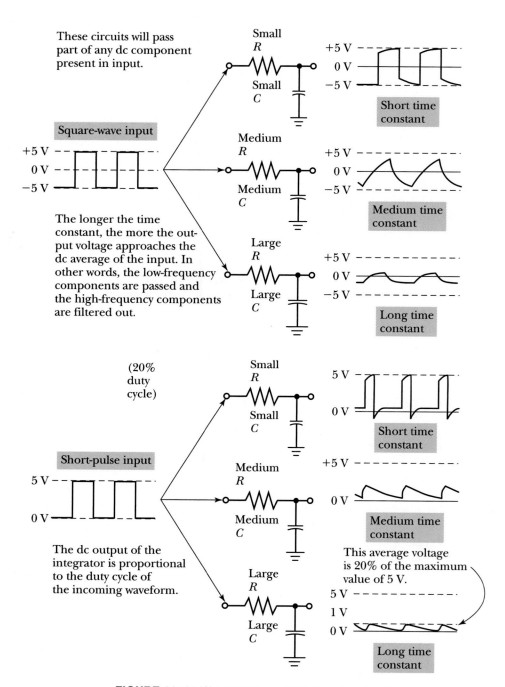

FIGURE 20–16 Waveshaping—*RC* integrator circuits

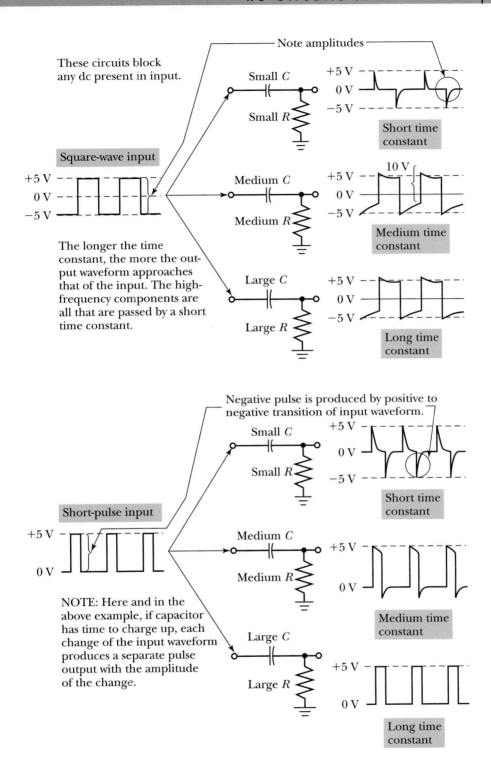

These circuits block any dc present in input.

Square-wave input

The longer the time constant, the more the output waveform approaches that of the input. The high-frequency components are all that are passed by a short time constant.

Negative pulse is produced by positive to negative transition of input waveform.

Short-pulse input

NOTE: Here and in the above example, if capacitor has time to charge up, each change of the input waveform produces a separate pulse output with the amplitude of the change.

FIGURE 20–17 Waveshaping—*RC* differentiator circuits

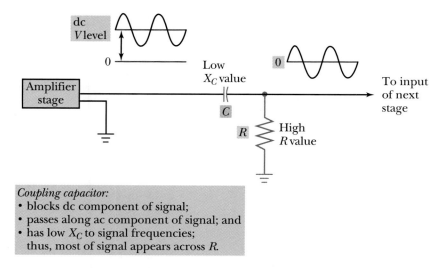

Coupling capacitor:
• blocks dc component of signal;
• passes along ac component of signal; and
• has low X_C to signal frequencies;
  thus, most of signal appears across R.

FIGURE 20–18 Example of an *RC* coupling network

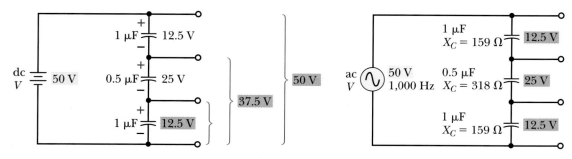

FIGURE 20–19 Capacitor voltage dividers

systems as amplifiers, radio or TV receivers a range from excessive hum to high-pitched squeals can be heard when such problems exist.

The only cure for this situation is to isolate the bad capacitor and *replace it!* Sometimes, bridging a degenerated capacitor with a good capacitor displays sufficient difference to help isolate the bad capacitor.

PRACTICAL NOTES

1. If a capacitor is used in a *timing circuit,* one symptom of a possible bad C (or R) is an improper time period.
2. If the capacitor is used in a waveshaping function, obvious clues are noted if an oscilloscope is used to analyze the normality or abnormality of the waveform(s).
3. If a capacitor is used in a *coupling circuit* where it is supposed to block dc and pass ac, the presence of measurable dc across the network resistor in series with the coupling capacitor shows the capacitor is probably leaking dc.

The primary tools used to help find defective capacitors in these various functions are the oscilloscope, digital multimeters, and capacitor testers. Another useful device is

a component *substitution box,* which enables selection of an appropriate-value component. This component is temporarily substituted in the suspected area of trouble to see
if the problem disappears with the substitute part.

As you study power supplies, amplifiers, oscillators, pulse and digital circuits, you will
gain more familiarity with common symptoms and troubleshooting techniques associated with *RC* circuits. Perhaps these few comments will whet your appetite to learn more.

SAFETY HINTS

1. To avoid shock, don't make connections with the circuit on.

2. In a defective circuit, capacitors may not always be discharged. BE CARE-
   FUL when inserting test equipment. Check that capacitors are discharged
   by making appropriate voltage measurements, and/or using safe discharge techniques.

SUMMARY

▶ Circuit current and circuit voltage are in phase in
purely resistive ac circuits. They are 90° out of
phase in purely capacitive circuits. They are
somewhere between 0° and 90° out of phase with
each other in circuits containing both resistance
and capacitance.

▶ To calculate total circuit voltage in a series *RC*
circuit, vectorially add the resistive and capacitive
voltages, as opposed to adding them with simple
arithmetic.

▶ Since current is common throughout any series
circuit, the reference vector for phasor diagrams
of series *RC* circuits is the circuit current. The
various circuit voltages (V_C and V_R) are then
plotted with respect to the reference vector. V_C is
shown lagging the current by 90°, and V_R is in
phase with the current.

▶ The reference vector for phasor diagrams of
parallel *RC* circuits is the circuit voltage. The
various parallel branch currents are plotted with
respect to the reference vector. I_C is shown
leading the voltage by 90°, and I_R is in phase with
the circuit voltage.

▶ Impedance is the opposition shown to ac current
by the combination of resistance and reactance.

▶ Impedance diagrams can only be used for *series*
circuits. They should not be used for analyzing
parallel *RC* (or *RL*) circuits.

▶ The Pythagorean theorem approach is used to
solve for the magnitudes of various phasors. In
series *RC* circuits, use the formula to calculate Z
and V_T. For parallel *RC* circuits, use the formula
to calculate I_T.

▶ Trig functions are used to solve for both phase
angles and magnitudes of various sides. One
commonly used trig function is the tangent
function, used for both series and parallel
resistive and reactive circuits.

▶ In series *RC* circuits, the larger the *C* is, the
smaller the X_C present. And the less the
capacitance controls the circuit parameters. In
series circuits, the smaller the X_C is with respect to
the *R*, the less the $I \times X_C$ drop is compared to the
$I \times R$ drop. Hence, the less capacitive the circuit
acts, and the smaller the phase angle will be.

▶ In parallel *RC* circuits, the larger the *C* is, the
smaller the X_C present. And the more the
capacitance controls the circuit parameters. In
parallel circuits, the smaller the X_C is with respect
to the *R*, the higher is the current through the

capacitive branch. Hence, the more capacitive the circuit acts, and the larger the phase angle will be.

▶ Any frequency change changes the X_C. Therefore, frequency change affects circuit current, component voltages, impedance, and phase angle for series RC circuits. Frequency changes also affect branch currents, impedance, and phase angle in parallel RC circuits.

▶ Capacitors and RC circuits have various

applications, including filtering, coupling, voltage-dividing, waveshaping, and timing. These uses take advantage of the capacitor's sensitivity to frequency, ability to block dc while passing ac and to store charge.

▶ Because electrolytic capacitors age faster than other components, such as resistors or inductors, periodically replace them to assure optimum operation of the equipment where they are used.

FORMULAS AND SAMPLE CALCULATOR SEQUENCES

FORMULA 20–1 $V_T = \sqrt{V_R^2 + V_C^2}$ (Series RC circuits)

V_R value, ⬛x^2, ⬛$+$, V_C value, ⬛x^2, ⬛$=$, ⬛x

FORMULA 20–2 $Z = \sqrt{R^2 + X_C^2}$ (Series RC circuits)

R value, ⬛x^2, ⬛$+$, X_C value, ⬛x^2, ⬛$=$, ⬛x

FORMULA 20–3 $I_T = \sqrt{I_R^2 + I_C^2}$ (Parallel RC circuits)

I_R value, ⬛x^2, ⬛$+$, I_C value, ⬛x^2, ⬛$=$, ⬛x

FORMULA 20–4 $Z = \dfrac{RX_C}{\sqrt{R^2 + X_C^2}}$ (Parallel RC circuits)

R value, ⬛$\times$, X_C value, ⬛$=$, ⬛$\div$, (R value, ⬛x^2, ⬛$+$, X_C value ⬛x^2,), ⬛x, ⬛$=$

NOTE: If the circuit capacitance value is in μF, use: Cap. value number, ⬛EE, ⬛$+/-$, 6 to express the "Cap. value" in microfarads for the calculator sequence.

TRIG FUNCTIONS:

sin θ opp/hyp $= V_C/V_T$, or X_C/Z (series RC circuits)

sin $\theta = $ opp/hyp $= I_C/I_T$ (parallel RC circuits)

opp side value, ⬛$\div$, hyp value, ⬛$=$

cos $\theta = $ adj/hyp $= V_R/V_T$, or R/Z (series RC circuits)

cos $\theta = $ adj/hyp $= I_R/I_T$ (parallel RC circuits)

adj side value, ⬛$\div$, hyp value, ⬛$=$

tan $\theta = $ opp/adj $= V_C/V_R$, or X_C/R (series RC circuits)

$\tan \theta = \text{opp}/\text{adj} = I_C/I_R$ (parallel *RC* circuits)

opp side value, ➗, adj side value, 🟰

$\arctan \theta$ = the angle represented by the tangent value

tan value, **2nd**, **TAN⁻¹**

IMPEDANCE OF PARALLEL RC CIRCUIT:

$$Z = \frac{RX_C}{\sqrt{R^2 + X_C^2}}$$

R value, ✖, X_C value, ➗, **((**, *R* value, **x²**, ➕, X_C value, **x²**, **))**, **x**, 🟰

REVIEW QUESTIONS

1. In RC circuits:
   a. circuit phase angle is 90°.
   b. circuit phase angle is 0°.
   c. circuit phase angle is 45°.
   d. circuit phase angle is between 0° and 90°.

2. Total voltage in a series *RC* circuit is found by:
   a. algebraic addition.
   b. vector addition.
   c. mathematical addition.
   d. none of the above.

3. The reference vector, when analyzing parallel *RC* circuits is:
   a. circuit current.
   b. circuit voltage.
   c. circuit impedance.
   d. none of the above.

4. The quantity(ies) that is(are) plotted with respect to the reference vector in parallel RC circuits:
   a. are the circuit currents.
   b. is the circuit voltage.
   c. are the circuit reactances.
   d. none of the above.

5. In a series *RC* circuit, the larger the *C* value for a given *R* value:
   a. the larger the circuit phase angle.
   b. the smaller the circuit phase angle.
   c. the value of *C* does not affect phase angle.
   d. the more "capacitive" the circuit acts.

6. In parallel *RC* circuits, the larger the *C* value for a given *R* value:
   a. the larger the circuit phase angle.

b. the smaller the circuit phase angle.
   c. the value of *C* does not affect phase angle.
   d. the less "capacitive" the circuit acts.

7. If frequency of operation is increased for a series *RC* circuit:
   a. the circuit will act more capacitively.
   b. the circuit will act less capacitively.
   c. the circuit conditions will remain the same.
   d. none of the above.

8. If the frequency of operation is decreased for a parallel *RC* circuit:
   a. the circuit will act more capacitively.
   b. the circuit will act less capacitively.
   c. the circuit conditions will remain the same.
   d. none of the above.

9. *RC* circuits can act as filters because:
   a. capacitors oppose a change in current.
   b. capacitors oppose a change in voltage.
   c. resistors react to changes in frequency.
   d. none of the above.

10. *RC* circuits can act as signal coupling networks because:
    a. capacitors block ac and pass dc.
    b. capacitors block both ac and dc.
    c. capacitors block dc and pass ac.
    d. none of the above.

11. A series *RC* circuit's impedance increases as:
    a. frequency increases.
    b. frequency decreases.
    c. phase angle increases.
    d. phase angle decreases.

12. A parallel *RC* circuit's phase angle increases as:
    a. frequency increases.
    b. frequency decreases.
    c. impedance increases.
    d. none of the above.

13. Capacitive by-pass filters:
    a. by-pass or filter desired signal frequencies.
    b. by-pass or filter undesired signal frequencies.
    c. by-pass or filter undesired electronic components.
    d. none of the above.

14. In a series *RC* circuit, as frequency is increased:
    a. X_C decreases, Z decreases, I decreases, and θ decreases.
    b. X_C increases, Z increases, I decreases, and θ increases.
    c. X_C decreases, Z decreases, I increases, and θ decreases.
    d. none of the above.

15. For a *V-I* vector diagram for a parallel *RC* circuit:
    a. V_A is the reference, I_R is at 0°, I_C is at −90°, and I_T is between 0° and +90°.
    b. V_A is the reference, I_R is at 0°, I_C is at +90°, and I_T is between 0° and +90°.
    c. I_T is the reference, V_R is at 0°, V_C is at −90°, and V_A is between 0° and +90°.
    d. I_T is the reference, V_R is at 0°, V_C is at +90°, and V_A is between 0° and +90°.

16. For a *V-I* vector diagram for a series *RC* circuit:
    a. V_A is the reference, I_R is at 0°, I_C is at −90°, and I_T is between 0° ad −90°.
    b. V_A is the reference, I_R is at 0°, I_C is at +90°, and I_T is between 0° and −90°.
    c. I_T is the reference, V_R is at 0°, V_C is at −90°, and V_A is between 0° and −90°.
    d. I_T is the reference, V_R is at 0°, V_C is at +90°, and V_A is between 0° and −90°.

17. For a series *RC* *"lead"* (phase shift) network, the output voltage is taken across:
    a. the capacitor.
    b. the resistor.
    c. both the resistor and the capacitor.
    d. neither the resistor nor the capacitor.

18. For a series *RC* *"lag"* (phase shift) network, the output voltage is taken across:
    a. the capacitor.
    b. the resistor.
    c. both the resistor and the capacitor.
    d. neither the resistor nor the capacitor.

19. What two formulas might be used to find total current in a parallel *RC* circuit?

20. Given a parallel *RC* circuit's R and X_C values, what formula would you use to find the circuit's impedance?

PROBLEMS

Answer questions 1–5 in reference to Figure 20–20.

1. What is the circuit impedance?

2. What are the V_C, V_R, and V_T values?

3. What is the C value?

4. Is circuit voltage leading or lagging circuit current? By how much?

Answer the following with "I" for increase, "D" for decrease, and "RTS" for remain the same.

5. If the frequency of the circuit in Figure 20–20 doubles and component values remain the same:
    a. R will _____   e. θ will _____
    b. X_C will _____   f. V_C will _____
    c. Z will _____   g. V_R will _____
    d. I_T will _____   h. V_T will _____

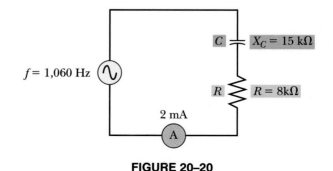

FIGURE 20–20

Answer questions 6–10 in reference to Figure 20–21.

6. What are the I_C and I_T values?

7. What is the R value?

8. What is the Z value?

9. What is the frequency of operation?

10. Answer the following with "I" for increase, "D" for decrease, and "RTS" for remain the same. If the frequency of the circuit in Figure 20–21 doubles and component values remain the same:
   a. R will
   b. X_C will
   c. Z will
   d. I_T will
   e. θ will
   f. I_C will
   g. I_R will
   h. V_T will

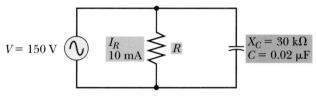

FIGURE 20–21

Answer questions 11–15 in reference to Figure 20–22.

11. Find R_T

12. Find X_{C_T}

13. Find Z

14. Find θ

15. If C_1 equals 0.1 µF, what is the frequency of the source voltage?

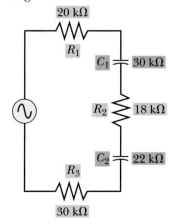

FIGURE 20–22

Answer questions 16–20 in reference to Figure 20–23.

16. Find I_R

17. Find I_C

18. Find I_T

19. Find θ

20. Find Z_T

21. A two-component series RC circuit is operated at a frequency of 3,975 Hz. The circuit is comprised of a 0.04-µF capacitor and an undisclosed-value resistance. The circuit impedance is 1,250 Ω. The applied voltage is 25 V. Determine the following circuit parameters: R, X_C, I, V_R, V_C, and θ.

22. A two-branch parallel RC circuit has a total current of 5 mA. The circuit's capacitive reactance is 2 kΩ. The applied voltage is 5 V. Determine the following circuit parameters: R, I_C, I_R, Z, and θ.

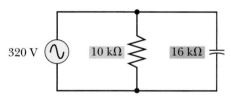

FIGURE 20–23

ANALYSIS QUESTIONS

1. In your own words, describe two basic circuit operational differences between a series *RC* circuit and a series *RL* circuit.

2. In your own words, describe two basic circuit operational differences between a parallel *RC* circuit and a parallel *RL* circuit.

3. Are there logical comparisons between a series *RC* circuit and a parallel *RL* circuit? Explain.

4. Draw the circuit diagram of a phase "lead" *RC* circuit that will produce a 45° lead angle. The components you can use include a variable resistor whose range of *R* is from 0 to 10,000 Ω, and a 0.25-μF capacitor. Assume the frequency of operation is 120 Hz.

PERFORMANCE PROJECTS CORRELATION CHART

Suggested performance projects that correlate with topics in this chapter are:

CHAPTER TOPIC	PERFORMANCE PROJECT	PROJECT NUMBER
Series *RC* Circuit Analysis	*V, I, R, Z,* and θ Relationships in a Series *RC* Circuit	56
Parallel *RC* Circuit Analysis	*V, I, R, Z,* and θ Relationships in a Parallel *RC* Circuit	57

NOTE: It is suggested that after completing the above projects, the student should be required to answer the questions in the "Summary" at the end of this section of projects in the Laboratory Manual.

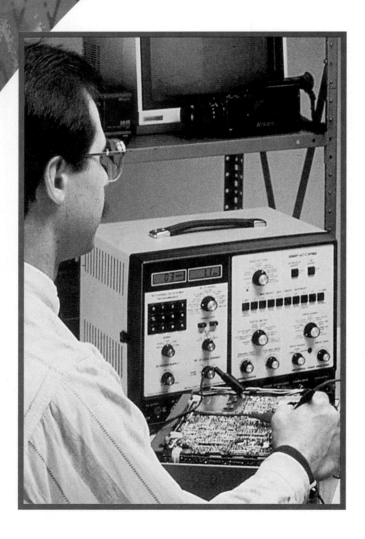

RLC CIRCUIT ANALYSIS

21

chapter

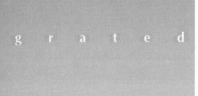

CHAPTER PREVIEW

As a technician, you will work with many circuits that contain combinations of R, L, and C. You have already investigated the property of inductance in ac circuits, in pure inductive circuits, and in circuits with inductance and resistance. Likewise, you have seen how capacitance behaves in purely capacitive circuits and in circuits with both capacitance and resistance. Furthermore, you have learned to use the Pythagorean theorem and simple trigonometric functions to analyze these circuits. In this chapter, you will analyze circuits with various combinations of R, L, and C.

The outline of this chapter accommodates either programs that teach vector algebra and its manipulation of complex numbers or only the approaches learned to this point. The first section of the chapter will present RLC circuits and the basic analysis techniques you have been using and provide a discussion of power in ac circuits. The second section introduces rectangular and polar form vector analysis, sometimes called vector algebra. The last portion provides an opportunity to apply vector algebra to various configurations of RLC circuits.

OBJECTIVES

After studying this chapter, you should be able to:

1. Solve RLC circuit problems using Pythagorean approach and trig functions

2. Define and illustrate ac circuit parameters using both **rectangular** and **polar form notation**

3. Define **real numbers** and **imaginary numbers**

4. Define real power, **apparent power, power factor,** and **voltampere-reactive**

5. Calculate values of real power, apparent power, and power factor, and draw the power triangle

6. Analyze RLC circuits and state results in rectangular and polar forms

7. Use the SIMPLER troubleshooting sequence to solve the Chapter Troubleshooting Challenge problem

Series Circuit

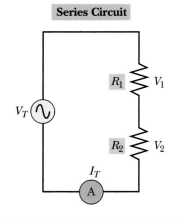

Parallel Circuit

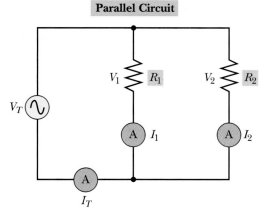

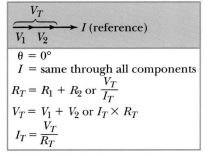

$$\begin{array}{l} V_T \\ \overrightarrow{} \ I \text{ (reference)} \\ V_1 \ \ V_2 \end{array}$$

$\theta = 0°$

I = same through all components

$R_T = R_1 + R_2$ or $\dfrac{V_T}{I_T}$

$V_T = V_1 + V_2$ or $I_T \times R_T$

$I_T = \dfrac{V_T}{R_T}$

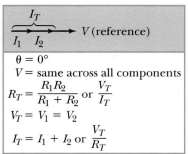

$$\begin{array}{l} I_T \\ \overrightarrow{} \ V \text{ (reference)} \\ I_1 \ \ I_2 \end{array}$$

$\theta = 0°$

V = same across all components

$R_T = \dfrac{R_1 R_2}{R_1 + R_2}$ or $\dfrac{V_T}{I_T}$

$V_T = V_1 = V_2$

$I_T = I_1 + I_2$ or $\dfrac{V_T}{R_T}$

FIGURE 21–1 Purely resistive circuits

21-1 BASIC RLC CIRCUIT ANALYSIS

Review of resistive circuits

In studying circuits containing only resistances, you learned resistance in ac is treated the same as resistance in dc. Key facts about series and parallel resistive circuits are in Figure 21–1. Take a few minutes to review these facts.

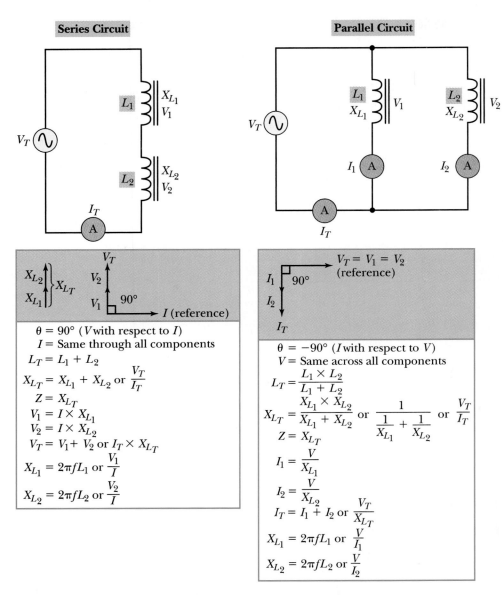

FIGURE 21–2 Purely inductive circuits

Review of inductive circuits

As you know, inductance opposes a change in current. This feature causes the voltage to lead current by $90°$ in purely inductive circuits and by some angle between $0°$ and $90°$ in circuits with both L and R. Examine Figures 21–2 and 21–3 as you review inductive circuits. Remember in vector or phasor diagrams of cir-

cuit parameters V_L is plotted upward at 90° for *series* circuits with I as the reference vector. V_R is plotted at the 0° position. For the impedance diagram, X_L is
plotted upward at 90°, and R is at the 0° position. For *parallel* circuits, I_L is plotted downward at −90°. When resistance exists, I_R is plotted at the 0° position with
circuit V, which is the reference vector.

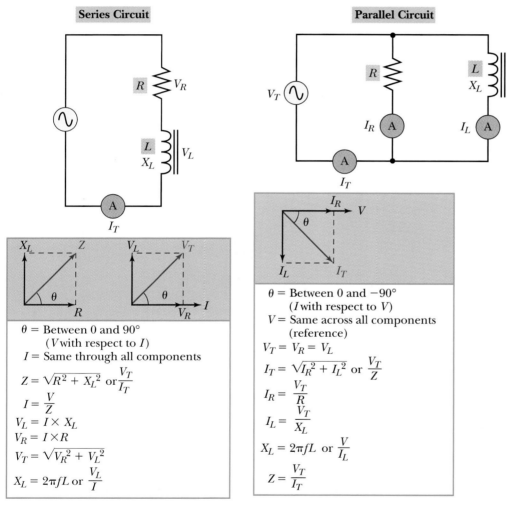

FIGURE 21–3 *RL* circuits

Review of capacitive circuits

Because it takes time for a potential difference to build up or change in value on a capacitor due to required charging or discharging current action, capacitance opposes a voltage change. This feature causes current to lead voltage by 90° in purely capacitive circuits and by some angle less than 90° in circuits with both C and R.

Refer to Figures 21–4 and 21–5. Notice for series circuit V-I vector diagrams, V_C is plotted downward at $-90°$. For Z diagrams, X_C is plotted downward at $-90°$.

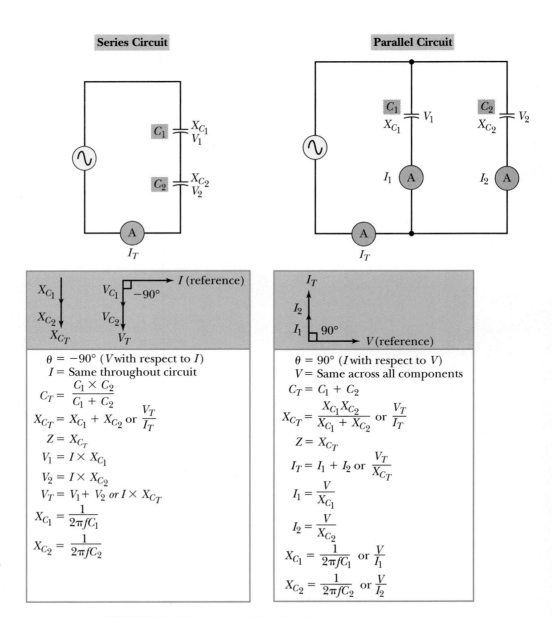

FIGURE 21–4 Purely capacitive circuits

When resistance exists, V_R is plotted at the 0° position with the circuit I, which is the reference vector. X_C is plotted downward and R is plotted at 0° for series circuit impedance diagrams.

For parallel circuit V-I diagrams, I_C is plotted upward at 90°, and I_R is plotted at the 0° position along with circuit V, which is the reference vector. Remember all of these facts as we move to the analysis of circuits containing all three components—R, L, and C.

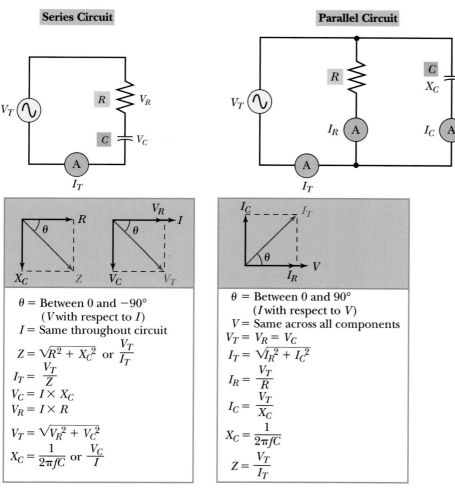

Series Circuit

θ = Between 0 and −90°
  (V with respect to I)
I = Same throughout circuit
$Z = \sqrt{R^2 + X_C^2}$ or $\dfrac{V_T}{I_T}$
$I_T = \dfrac{V_T}{Z}$
$V_C = I \times X_C$
$V_R = I \times R$
$V_T = \sqrt{V_R^2 + V_C^2}$
$X_C = \dfrac{1}{2\pi fC}$ or $\dfrac{V_C}{I}$

Parallel Circuit

θ = Between 0 and 90°
  (I with respect to V)
V = Same across all components
$V_T = V_R = V_C$
$I_T = \sqrt{I_R^2 + I_C^2}$
$I_R = \dfrac{V_T}{R}$
$I_C = \dfrac{V_T}{X_C}$
$X_C = \dfrac{1}{2\pi fC}$
$Z = \dfrac{V_T}{I_T}$

FIGURE 21–5 *RC* circuits

21-2 THE SERIES RLC CIRCUIT

A simple series LC circuit without resistance

Let's look at what happens when we combine inductive reactance (X_L) and capacitive reactance (X_C) in a simple series circuit. Refer to Figure 21–6 as you study the following statements.

1. Since X_L is plotted upward and X_C is plotted downward, they *cancel* each other.

2. The result of combining them is that any remaining, or *net* reactance is the difference between the two reactances. That is, if X_L is greater than X_C, the net reactance is some value of X_L. If X_C is greater than X_L, the net circuit reactance is some value of X_C.

3. Apply the same rationale to voltages. Since the voltages are plotted 180° out-of-phase with each other and calculated by multiplying the common current (I) times each reactance, the resulting voltage of the two opposite-reactive voltages is the difference between them. In other words, V_{X_T} = largest V_X − smaller V_X.

4. Since the resulting reactance ohms are less than either reactance (because of the cancelling effect), the current is higher than it would be with either reactance alone. This means the $I \times X_L$ or $I \times X_C$ voltages are higher than the V applied! Obviously, since the opposite reactive voltages also cancel each other, the net voltage equals the applied voltage V_T.

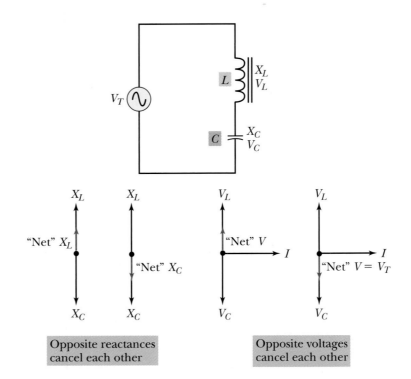

FIGURE 21–6 Series *LC* circuit without resistance

Opposite reactances cancel each other

Opposite voltages cancel each other

Series RLC circuit analysis

Refer to Figure 21–7 and note the *RLC* circuit analysis as you study the following procedures.

Computing impedance by Pythagorean approach

Using the concepts previously presented, the impedance must equal the vector sum of R and the remaining reactance. The formula is $Z = \sqrt{R^2 + (X_L - X_C)^2}$ when X_L is greater than X_C. If X_C is greater than X_L, then $Z = \sqrt{R^2 + (X_C - X_L)^2}$. In essence,

FORMULA 21–1 $Z = \sqrt{R^2 + X^2}$

where net reactance (X) depends on the relative reactance values, that is

$$X = X_L - X_C$$

$$X = X_C - X_L$$

In Figure 21–7, $R = 60\ \Omega$, $X_L = 160\ \Omega$, and $X_C = 80\ \Omega$. The net reactance is 80 Ω of inductive reactance in series with 60 Ω of resistance. Using the Pythagorean formula to find Z:

$$Z = \sqrt{R^2 + X^2}$$

$$Z = \sqrt{60^2 + 80^2}$$

$$Z = \sqrt{3{,}600 + 6{,}400}$$

$$Z = \sqrt{10{,}000}$$

$$Z = 100\ \Omega$$

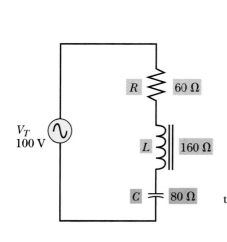

$$Z = \sqrt{R^2 + (X_L - X_C)^2}$$
$$Z = \sqrt{60^2 + 80^2}$$
$$Z = \sqrt{10{,}000}$$
$$Z = 100\ \Omega$$
$$I = \frac{V_A}{Z} = \frac{100\ \text{V}}{100\ \Omega} = 1\ \text{A}$$
$$V_R = 1 \times 60 = 60\ \text{V}$$
$$V_L = 1 \times 160 = 160\ \text{V}$$
$$V_C = 1 \times 80 = 80\ \text{V}$$
$$V_T = \sqrt{V_R^2 + (V_L - V_C)^2}$$
$$V_T = \sqrt{60^2 + 80^2}$$
$$V_T = \sqrt{10{,}000}$$
$$V_T = 100\ \text{V}$$
$$\tan\theta = \frac{X_T}{R} = \frac{80}{60} = 1.33$$
$$\theta = 53.1° \ (V \text{ with respect to } I)$$

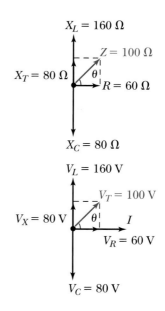

FIGURE 21–7 Series *RLC* circuit

Computing current by Ohm's law

In series circuits, current is common throughout the circuit. Circuit current is found through Ohm's law where $I = V/Z$. Since V_A is given as 100 V, then

$$I = \frac{V}{Z} = \frac{100\ V}{100\ \Omega} = 1\ A$$

Computing individual voltage by Ohm's law

The voltage across the resistor $= IR$, across the inductor $= IX_L$, and across the capacitor $= IX_C$.

$$V_R = 1\ A \times 60\ \Omega = 60\ V$$

$$V_L = 1\ A \times 160\ \Omega = 160\ V$$

$$V_C = 1\ A \times 80\ \Omega = 80\ V$$

Computing total voltage by Pythagorean approach

Since the resistive and reactive voltages are out-of-phase, we must again use vector summation. As you know, total voltage is found by the formula $V_T = \sqrt{V_R^2 + (V_L - V_C)^2}$, if inductor voltage is greater than capacitor voltage. If V_C is greater than V_L, the formula is $V_T = \sqrt{V_R^2 + (V_C - V_L)^2}$. Simplifying, the formula becomes

FORMULA 21–2　　$V_T = \sqrt{V_R^2 + V_X^2}$

where V_X is *net reactive voltage*. Thus, in Figure 21–7:

$$V_T = \sqrt{60^2 + 80^2}$$

$$V_T = \sqrt{3{,}600 + 6{,}400}$$

$$V_T = \sqrt{10{,}000}$$

$$V_T = 100\ V$$

Computing phase angle from impedance and voltage data

Using impedance data, the formula is:

FORMULA 21–3　　tangent of $\theta = \dfrac{X\ \text{(net reactance)}}{R}$

$$\tan \theta = \frac{80}{60}$$

$$\tan \theta = 1.33$$

The angle whose tangent $= 1.33$ is 53.1°. Since the net reactance is inductive reactance (plotted upward), the angle is +53.1°. NOTE: If the net reactance were capacitive reactance (plotted downward) of the same value, the phase angle would be −53.1° rather than +53.1°.

Using voltage data, the formula is:

FORMULA 21–4   tangent of $\theta = \dfrac{V_X}{V_R}$

$$\tan \theta = \frac{80}{60}$$
$$\tan \theta = 1.33$$

The angle whose tangent $= 1.33$ is $53.1°$.

Since the net reactive voltage is inductive voltage, which leads current (the reference vector), the angle is $+53.1°$. NOTE: If the net reactive voltage were V_C (plotted downward) of the same value, the phase angle would be $-53.1°$ rather than $+53.1°$.

PRACTICAL NOTES

With more than one resistor or one reactance of each kind in the circuit, combine resistances to one total R value; inductive reactances to one total X_L value; and capacitive reactances to one total X_C value. Then solve the problem with the standard procedures to solve series RLC circuits.

Example

Refer to the series RLC circuit in Figure 21–8 and calculate Z, V_T, V_{L_1}, V_{C_1}, V_{R_1}, and θ.

In our example circuit, $R_1 = 20\ \Omega$ and $R_2 = 10\ \Omega$, so circuit $R = 30\ \Omega$.

$X_{L_1} = 50\ \Omega$ and $X_{L_2} = 40\ \Omega$, so circuit $X_L = 90\ \Omega$.

$X_{C_1} = 80\ \Omega$ and $X_{C_2} = 40\ \Omega$, so circuit $X_C = 120\ \Omega$.

Net reactance $= 120\ \Omega (X_C) - 90\ \Omega\ (X_L) = 30\ \Omega$ of capacitive reactance.

$Z = \sqrt{R^2 + X^2} = \sqrt{30^2 + 30^2} = \sqrt{900 + 900} = \sqrt{1{,}800} = 42.42\ \Omega$

I is given as 1 ampere.

$V_R = IR = 1 \times 30 = 30\ \text{V}$

$V_L = IX_L = 1 \times 90 = 90\ \text{V}$

$V_C = IX_C = 1 \times 120 = 120\ \text{V}$

$V_T = \sqrt{V_R^2 + V_X^2} = \sqrt{30^2 + 30^2} = \sqrt{900 + 900} = \sqrt{1{,}800} = 42.42\ \text{V}$

$V_{L_1} = IX_{L_1} = 1\ \text{A} \times 50\ \Omega = 50\ \text{V}$

$V_{C_1} = IX_{C_1} = 1\ \text{A} \times 80\ \Omega = 80\ \text{V}$

$V_{R_1} = IR_1 = 1\ \text{A} \times 20\ \Omega = 20\ \text{V}$

$\tan \theta = \dfrac{X}{R}$, or $\dfrac{V_X}{V_R}$

$\tan \theta = \dfrac{30}{30} = 1$

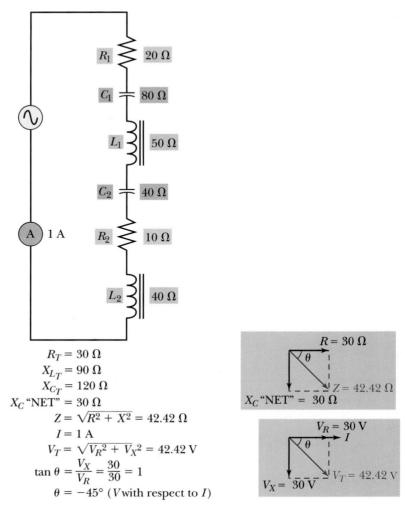

$$R_T = 30\ \Omega$$
$$X_{L_T} = 90\ \Omega$$
$$X_{C_T} = 120\ \Omega$$
$$X_C \text{ "NET"} = 30\ \Omega$$
$$Z = \sqrt{R^2 + X^2} = 42.42\ \Omega$$
$$I = 1\ \text{A}$$
$$V_T = \sqrt{V_R^2 + V_X^2} = 42.42\ \text{V}$$
$$\tan \theta = \frac{V_X}{V_R} = \frac{30}{30} = 1$$
$$\theta = -45° \ (V \text{ with respect to } I)$$

FIGURE 21–8

The angle whose tangent = 1 is 45°.

Now that you have seen how this is done, try the following practice problem.

PRACTICE PROBLEMS I

Refer to Figure 21–9 and calculate R, X_{C_T}, X_L, Z, V_T, V_{C_2}, V_L, V_R, and θ.

IN-PROCESS LEARNING CHECK I

Fill in the blanks as appropriate.

1. When both capacitive and inductive reactance exist in the same circuit, the net reactance is _____.

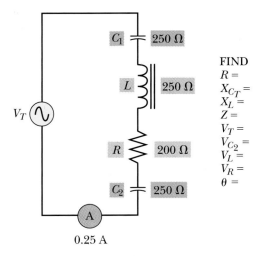

FIND
$R =$
$X_{C_T} =$
$X_L =$
$Z =$
$V_T =$
$V_{C_2} =$
$V_L =$
$V_R =$
$\theta =$

FIGURE 21–9

2. When plotting reactances for a series circuit containing L and C, the X_L is plotted _____ from the reference, and the X_C is plotted _____ from the reference.

3. In a series circuit with both capacitive and inductive reactance, the capacitive voltage is plotted _____ degrees out of phase with the inductive voltage.

4. If inductive reactance is greater than capacitive reactance in a series circuit, the formula to find the circuit impedance is _____.

5. In a series RLC circuit, current is found by _____.

6. True or False. Total applied voltage in a series RLC circuit is found by the Pythagorean approach or by adding the voltage drops around the circuit.

7. If the frequency applied to a series RLC circuit decreases while the voltage applied remains the same, the component(s) whose voltage will increase is/are the _____.

21-3 THE PARALLEL RLC CIRCUIT

A simple LC parallel circuit

Refer to Figure 21–10 as you study the following discussion on combining inductive and capacitive reactances in parallel.

1. Since I_L is plotted downward and I_C is plotted upward, obviously the reactive branch currents *cancel* each other.

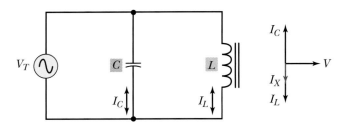

FIGURE 21–10 Parallel *LC* circuit without resistance

2. Combining the reactive branch currents produces the resulting reactive current that is the difference between the opposite-reactance branch currents. That is, the resulting reactive current (I_X) is the larger reactive branch current minus the smaller reactive branch current.

3. In a parallel circuit, the voltage is the same across all branches, that is, V_T.

Parallel RLC circuit analysis

As was just stated, the *opposite-direction phasors* cancelling effect is also applicable to parallel circuits. But this cancellation effect is different between series and parallel circuits because of the circuit parameters involved. In series circuits, the current is common and the reactive voltage opposite-direction phasors cancel. However, in parallel circuits, the voltage is common and the reactive branch currents are opposite-direction phasors. Look at Figure 21–11 as you study the following discussion.

1. I_T is the vector sum of the circuit resistive and reactive branch currents.

2. The resulting reactive branch current value used to calculate total current is found by subtracting the smaller reactive branch current value from the larger reactive branch current value. This value (I_X) is then used to calculate I_T.

3. The general formula for I_T is $I_T = \sqrt{I_R^2 + (I_L - I_C)^2}$, where I_L is greater than I_C. If I_C is greater than I_L, the formula becomes $I_T = \sqrt{I_R^2 + (I_C - I_L)^2}$. Using the *net reactive current,* the formula becomes

FORMULA 21–5 $I_T = \sqrt{I_R^2 + I_X^2}$

Let's apply this knowledge and Ohm's law to solve a problem.

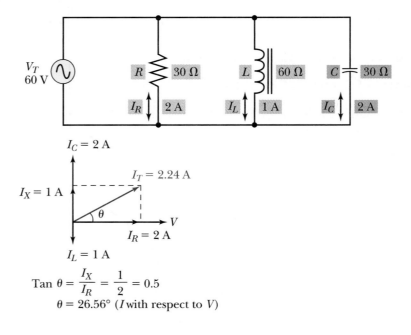

$$\text{Tan } \theta = \frac{I_X}{I_R} = \frac{1}{2} = 0.5$$

$$\theta = 26.56° \ (I \text{ with respect to } V)$$

FIGURE 21–11 Parallel *RLC* circuit

Example

Refer again to the circuit in Figure 21–11 and find I_R, I_L, I_C, I_X, I_T, Z, and θ.
Use Ohm's law to solve for each branch current:

$$I_R = \frac{V_T}{R} = \frac{60}{30} = 2 \text{ A}$$

$$I_L = \frac{V_T}{X_L} = \frac{60}{60} = 1 \text{ A}$$

$$I_C = \frac{V_T}{X_C} = \frac{60}{30} = 2 \text{ A}$$

Subtract the smaller reactive current from the larger to solve for I_X:

$$I_X = I_C - I_L = 2 - 1 = 1 \text{ A}$$

Use the Pythagorean approach to solve for I_T:

$$I_T = \sqrt{I_R^2 + I_X^2} = \sqrt{2^2 + 1^2} = \sqrt{4 + 1} = \sqrt{5} = 2.24 \text{ A}$$

Use Ohm's law to solve for Z:

$$Z = \frac{V_T}{I_T} = \text{approximately } 26.8 \ \Omega$$

Use the tangent function to solve for the phase angle:

$$\text{Tan } \theta = \frac{I_X}{I_R} = \frac{1}{2} = 0.5$$

The angle whose tangent = 0.5 is 26.56°. Since the net reactive current is capacitive, current leads the reference circuit voltage and the phase angle (I_T with respect to V_T) is +26.56°.
    Try solving the following practice problem.

PRACTICE PROBLEMS II
Refer to Figure 21–12 and solve for I_R, I_L, I_C, I_X, I_T, Z, and θ.

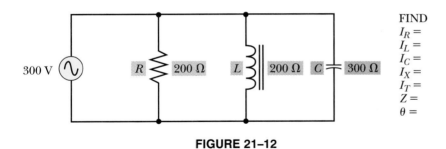

FIND
$I_R =$
$I_L =$
$I_C =$
$I_X =$
$I_T =$
$Z =$
$\theta =$

FIGURE 21–12

21–4 POWER IN AC CIRCUITS

Background

In dc circuits, you know how to find the power expended by a circuit through the resistive power dissipation. That is,

$$I^2R, \frac{V^2}{R}, \text{ and } V \times I$$

In ac circuits with both resistance and reactance, the apparent power supplied by the source is not the real or true power dissipated in the form of heat. Obviously, this is because the reactances take power from the circuit during one portion of the cycle and return it to the circuit during another portion of the cycle. As you know, circuits with reactances cause the voltage and current to be out of phase with each other. Therefore, the product of the **out-of-phase** circuit *v* and *i* does not indicate the actual circuit power dissipation.

Power in a purely resistive circuit

Look at Figure 21–13. The power is dissipated during both half cycles. During the positive alternation, the power graph is represented by the products of a series of instantaneous values of *i* and *v* along the waveform. During the negative alternation, the power graph represents the products of the series of instantaneous values of negative *i* and negative *v* along the waveform. The product of a negative times a negative, of course, comes out positive. In other words, the resistor dissipates power during both alternations. Over the whole cycle, an average power dissipation equal to the product of current and voltage results. It can be mathematically shown that the average power dissipated equals the effective (rms) voltage value times the effective (rms) current value.

Power in purely reactive circuits

Phase differences between voltage and current for purely inductive or capacitive circuits cause the power graphs, Figure 21–14. Because reactances take power from the circuit for one-quarter cycle and return it the next quarter cycle, you

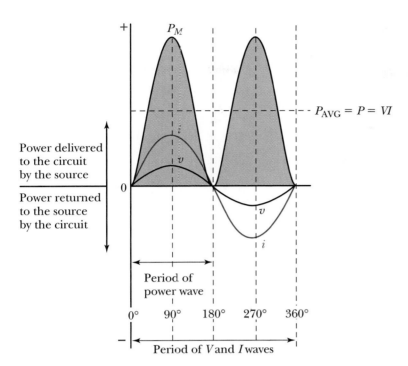

FIGURE 21–13
Instantaneous power graphs
in a purely resistive circuit

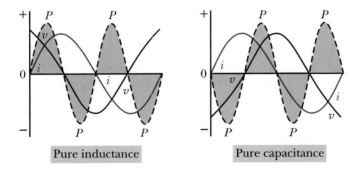

FIGURE 21–14
Instantaneous power graphs
in purely reactive circuits

see a power graph of twice the source frequency. Notice, too, the average power over the complete cycle is zero because the reactive instantaneous v times i products are positive half the time and negative half the time. This fact implies that purely inductive or capacitive reactances do not dissipate power. In reality, inductors (because of their wire resistance) have some resistive component, which dissipates some power in the coil. Good capacitors have very low effective resistance and typically dissipate almost zero power.

Combining resistive and reactive power features

View Figure 21–15 and note the curves for $v \times i$ products in an ac circuit with both resistance and reactance indicate power dissipation over one cycle. That is, the graphs resulting from positive $v \times i$ products and those resulting from negative $v \times i$ products are not equal. There is power dissipated. The amount of power represents the *true power* dissipated by the resistive share of the circuit.

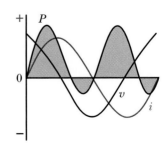

FIGURE 21–15
Instantaneous power
graphs—*R* and *X* combined

FIGURE 21–16 Power and power factor in a series ac circuit

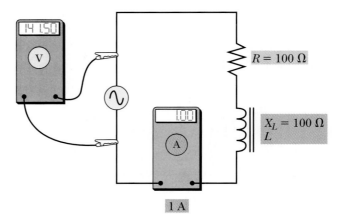

$$p.f. = \cos \theta = \cos 45° = 0.707$$
$$p.f. = \frac{R}{Z} \text{ or } \frac{V_R}{V_T}$$

$S = VI = 141.5 \text{ VA (apparent power)}$
$P = VI \cos \theta = I^2 R = 100 \text{ W (true power)}$
$Q = VI \sin \theta = 100 \text{ VAR (reactive power)}$

Apparent power

To someone not familiar with our preceding discussion, it would appear that by taking a voltmeter reading and a current meter reading for a given ac circuit, the power is found by multiplying their readings ($V \times I$). But the meters are *not* indicating the phase difference between the two quantities. Therefore, what is computed from this technique is **apparent power.** Apparent power is found by multiplying $V \times I$.

Refer to Figure 21–16 and notice that since R and X_L are equal, the circuit phase angle is 45°. The apparent power is found by the product of V applied times I. In this case, apparent power = 141.5 volts × 1 ampere = 141.5 VA. NOTE: Apparent power is expressed as **volt-amperes,** not watts! Also, note the abbreviation for apparent power is **S.**

With these thoughts, let's find how phase angle is used to find the real or true power supplied to and dissipated by the circuit.

True power

True power dissipation is determined by finding the amount of power dissipated by the resistance in the circuit. If we assume a perfect inductor in Figure 21–16, the power dissipated by the resistor equals the true power dissipation. Because the I_R and V_R are in phase with each other, the formula $V_R \times I_R$ is used. Therefore, true power = 100 V × 1 A = 100 W.

The I²R method

The preference is to say that true power equals $I^2 R$, where R represents all circuit resistive quantities. This also includes resistor components and resistance values in reactive components. These R values are easily measured. It is not as easy (for example) to separate reactive and resistive voltage components in an inductor by measuring V_L.

The power factor method

The power factor method considers the phase angle between current and voltage, created by the reactance. The value of the circuit **power factor** *(p.f.)* reflects how resistive or reactive the circuit is acting. A power factor of one indicates a purely resistive circuit. A power factor of zero represents a purely reactive circuit. Obviously, a power factor between zero and one indicates the circuit has both resistance and reactance, and true power is less than apparent power. Power factor is equal to the cosine of the phase angle.

PRACTICAL NOTES

A power factor of one indicates a circuit with a purely resistive load that uses all the energy delivered to it by the source. Power factors of less than one indicate some reactance is present in the circuit; and hence, the source must deliver more energy (or current) than is actually being put to work as "true power." In practice then, the power company likes industries and people (who are their "loads") to keep the power factor as close to one as possible. This means the power company's electrical generators have to produce less electrical energy, and their wires have to carry less current than when the power factors indicate reactive loads. Having smaller generator requirements and being able to transport the energy via smaller conductors can save the power company a lot of money.

Since many loads are inductive (such as motors), many industries use reactive compensating components to offset the reactance of their loads. For example, capacitors can be used to produce opposite reactances to help offset the inductance of inductive loads, such as large motors used in industry.

FORMULA 21–6 Power factor *(p.f.)* = $\cos \theta$

Then for our sample circuit in Figure 21–16:

FORMULA 21–7 Apparent power $(S) = V \times I$

$$S = 141.5 \times 1 = 141.5 \text{ volt-amperes}$$

FORMULA 21–8 True power $(P) = I^2 R$

$$P = 1^2 \times 100 = 100 \text{ watts OR}$$

FORMULA 21–9 $P =$ Apparent power $\times$ power factor

$$P = 141.5 \times \cos \theta \text{ (where } \cos 45° = 0.707)$$
$$P = 141.5 \times 0.707 = 100 \text{ watts}$$

Power factor formulas

Power factor really expresses the ratio of true power to apparent power (which is the same ratio expressed by the cos θ, Formula 21–6).

> **FORMULA 21–10** $p.f. = \dfrac{\text{True power } (P)}{\text{Apparent power } (S)}$

$$\frac{VI \times \cos \theta}{VI} \text{ OR } \frac{R}{Z} \text{ (Series)} = \frac{I_R}{I_T} \text{ (Parallel)}$$

Example

Recall in *series circuits*

$$\cos \theta = \frac{R}{Z}\left(or \; \frac{V_R}{V_T}\right)$$

For Figure 21–16, our series circuit example:

$$Z = \frac{V_T}{I_T} = \frac{141.5 \text{ V}}{1 \text{ A}} = 141.5 \; \Omega$$

$$\cos \theta = \frac{R}{Z} = \frac{100}{141.5} = 0.707. \text{ Therefore, } p.f. = 0.707$$

(Incidentally, the angle whose cos = 0.707 is 45°, as you know.)

In *parallel circuits* the *power factor* = cos θ = I_R/I_T. Refer to Figure 21–17 and notice the calculations for power factor using the resistive and total current values. In this case, the voltage applied is 100 V. Thus:

$$I_R = \frac{100}{100} = 1 \text{ A}$$

$$I_L = \frac{100}{100} = 1 \text{ A}$$

$$I_T = \sqrt{I_R^2 + I_L^2} = \sqrt{1^2 + 1^2} = \sqrt{2} = 1.414$$

$$p.f. = \frac{I_R}{I_T} = \frac{1}{1.414} = 0.707$$

Again, in this case $\theta = 45°$.

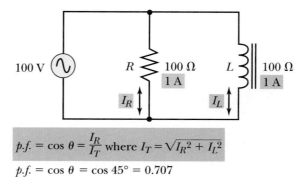

FIGURE 21–17 Power factor in a parallel ac circuit

$p.f. = \cos \theta = \dfrac{I_R}{I_T}$ where $I_T = \sqrt{I_R^2 + I_L^2}$

$p.f. = \cos \theta = \cos 45° = 0.707$

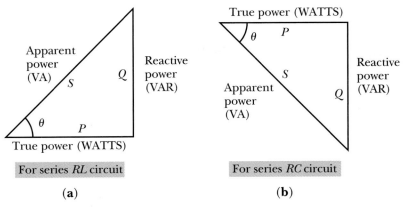

FIGURE 21–18 The power triangle

For series *RL* circuit

(a)

For series *RC* circuit

(b)

The power triangle

The relationship between true and apparent power is sometimes represented visually by a *power triangle* diagram.

Figure 21–18a is a power triangle for a series *RL* circuit where $R = X_L$. Apparent power (S) in volt-amperes is plotted on a line at an angle equal to θ from the *x*-axis. Its length, representing magnitude or value, is $V \times I$. Notice the apparent power is the result of two components:

1. The **true power *(P)***, sometimes called *resistive power,* is plotted on the *x*-axis (horizontal) and has a magnitude of $VI \times \cos \theta$ watts.

2. The **reactive power *(Q)***, often called **voltampere-reactive (VAR),** is plotted on the *y*-axis (vertical) and has a magnitude of $VI \times \sin \theta$.

Figure 21–18b is a power triangle for a circuit with resistance and capacitive reactance. From your knowledge of the Pythagorean theorem and right-triangle analysis, you can see apparent power is computed from the formula:

FORMULA 21–11 $S = \sqrt{P^2 + Q^2}$

where S = apparent power (measured in volt-amperes, VA)
P = true power (measured in watts, W)
Q = net reactive power (measured in volt-amperes reactive, VAR)

(NOTE: Power triangles are often used to analyze ac power distribution circuits.)

PRACTICAL NOTES

This formula is applicable to both series and parallel circuits. To summarize:

1. Knowing only the value of voltage or current does not indicate the amount of power actually being dissipated by the circuit.

2. A power triangle (different from *V-I* vector diagram) is used to diagram the various types of power in the ac circuit. Because the power waveform is a different frequency sinusoidal representation compared with the *V* and *I* quantities, we cannot plot the power values on the same vector diagram with *V* and *I*. Thus, the right-triangle (Pythagorean) approach displays the various types of power in the ac circuit.

continued

3. The vertical element in the power triangle is the reactive power. The horizontal element in the triangle is the true or resistive power. Therefore, the hypotenuse of the triangle is the apparent power, which is the result of reactive and resistive power.

IN-PROCESS LEARNING CHECK III

1. Power dissipated by a perfect capacitor over one whole cycle of input is _____.

2. The formula for apparent power is _____, and the unit expressing apparent power is _____.

3. True power in *RLC* circuits is that power dissipated by the _____.

4. The formula for power factor is _____.

5. If the power factor is one, the circuit is purely _____.

6. The lower the power factor, the more _____ the circuit is acting.

7. In the power triangle, true power is shown on the _____ axis, reactive power (VAR) is shown on the _____ axis, and apparent power is the _____ of the right triangle.

21–5 RECTANGULAR AND POLAR VECTOR ANALYSIS

Overview of rectangular and polar forms

In studying dc circuits, you learned that adding, subtracting, multiplying, and dividing electrical quantities was all that was needed to analyze circuits. When you studied ac circuits, you learned that out-of-phase quantities need a different treatment. You learned several approaches—graphic analysis (plotting and measuring vectors or phasors and angles); the Pythagorean theorem (to find resulting magnitudes or values); and simple trigonometry (to find related angles). Also, you learned trigonometry can be used to find magnitude information.

In this section, you will learn about algebraic vector analysis that involves both rectangular and polar forms to analyze vector quantities. You will learn how the *j* operator is used in complex numbers to distinguish the vertical vector from other components. Actually, these techniques are only a small extension of what you already know and use!

Phasors

You know that phasors represent sinusoidal ac quantities when they are the *same* frequency. Of course, they must be the same frequency so that when using phasor drawings the quantities compared maintain the same phase differential throughout the ac cycle. The normal direction of rotation of the constant-magnitude rotating radius vector (phasor) is counterclockwise from the 0° position (reference point). The relative position of the phasor shows its relationship to the reference (0°) position and to other phasors in the diagram. The phasor

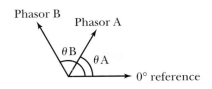

FIGURE 21–19 Phasor illustration

Length of phasor denotes magnitude or value.
Angle of phasor denotes relative direction.

length indicates the magnitude, or relative value of the quantity represented, Figure 21–19. In ac, this is the effective (rms) value of V, I, and so forth.

Rectangular notation

In **rectangular form notation,** the magnitude and direction of a vector relates to both the x-axis (horizontal) and the y-axis (vertical) of the rectangular coordinate system, Figure 21–20. One way to find the location and magnitude of a given vector on this coordinate system is by **complex numbers.** A complex number locates a positional point with respect to two axes.

A complex number has two parts—**real** and **imaginary.** The first part, real term or number, indicates the horizontal component on the x-axis. The second element, imaginary term or number, indicates the vertical component on the y-axis. For example, if it is stated that a specific vector $= 4 + j3$, this means the point at the end of that vector is located by projecting upward from 4 on the horizontal axis until a point even with 3 on the $+j$ (vertical) axis is reached. (Incidentally, in this case, the length of this vector from the origin is 5, and its angle from the reference plane is $36.9°$.)

Notice in Figure 21–20 the horizontal axis is also called the x-axis, the real axis, or the resistance axis. The vertical axis is also called the imaginary axis, or

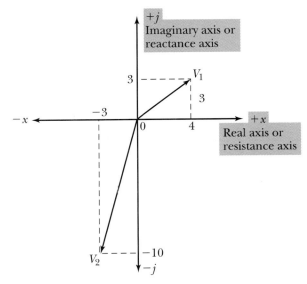

$V_1 = 4 + j3$
$V_2 = -3 - j10$

FIGURE 21–20 Illustration of rectangular notation form

FIGURE 21–21 The *j* operator

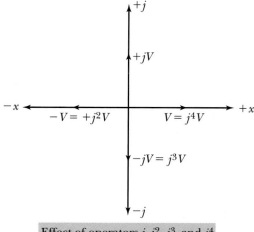

Effect of operators j, j^2, j^3, and j^4

the reactance axis. The *j* operator designates values on that axis. The $+j$ indicates an upward direction, while $-j$ signifies a downward direction. When a number in a formula is preceded by the letter *j*, the number is a vertical (or *y*-axis) component.

Since the *j* operator is so important in the rectangular form of vector algebra, let's define the *j* operator more specifically:

$$j = \sqrt{-1}$$

$$j^2 = (\sqrt{-1})(\sqrt{-1}) = -1$$

$$j^3 = (j^2)(j) = (-1)(j) = -j$$

$$j^4 = (j^2)(j^2) = (-1)(-1) = +1$$

Look at Figure 21–21 to relate these equations to the rectangular coordinate format. Notice the various powers of the *j* operator rotate a phasor as follows:

- $+j$ is a 90° operator, rotating a phasor 90° in the counterclockwise (CCW) direction, or +90°.
- $-j$ rotates a phasor 90° in the clockwise (CW) direction, or −90°.
- j^2 rotates the phasor 180°.
- j^3 rotates the phasor to the +270° position, or −90°.
- j^4 is +1, rotating the vector back to the starting point (0°).

Example

If the value of V_T for a series *RL* circuit (with a perfect inductor) is $30 + j40$, it might be illustrated as in Figure 21–22. Observe V_T is 50 V and θ is 53.1°. Suppose that V_T, expressed in rectangular form, is $4 + j3$. What is the value of V_T and the angle? This should be familiar to you.

Answer: 5 V at an angle of 36.9°.

In our previous discussions, you saw how this was done graphically by projecting from 4 (horizontal) and $+j3$ (vertical) points on the coordinate system. The resultant vector and angle could then be measured. You have also previously

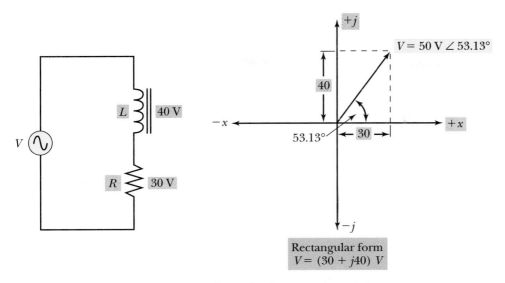

FIGURE 21–22 Example of rectangular notation

learned that the resultant of these two projections can be found by vectorial addition; that is, by using the Pythagorean theorem to find the length of the resulting vector (the hypotenuse of the right triangle). The angle can be found by using the tangent function. The scientific calculator offers still another method, which is the preferred method. We will introduce you to this method after you are familiarized with basic concepts of polar notation.

Polar notation

Refer to Figure 21–23 to see how vectors are represented on polar coordinates.

Whereas the rectangular form expresses vectors by the horizontal and vertical axes, the **polar form notation** expresses the length and angle of the resulting vector to define a vector. Previously, we stated one value of V_T is $30 + j40$ in rectangular notation. This same V_T is expressed in polar form as $50 \angle 53.1° \text{ V}$. (The resulting vector of the other two vectors described in the rectangular form.)

The advantage for using polar form notation is that the values are the same as those measured by instruments such as voltmeters, ammeters, and so on. That is, in our example above, a voltmeter would actually read 50 V for V_T.

Converting from rectangular to polar form

To convert a complex number in rectangular form to polar form simply find the vector resulting from the real term, j term, and angle.

Example

In a series RL circuit, what is the polar notation for a value expressed in rectangular notation as $60 \text{ V} + j70 \text{ V}$?
Use phasor addition to find the magnitude:

$$V_T = \sqrt{V_R^2 + V_L^2} = \sqrt{60^2 + 70^2} = \sqrt{8,500} = 92.2 \text{ V}$$

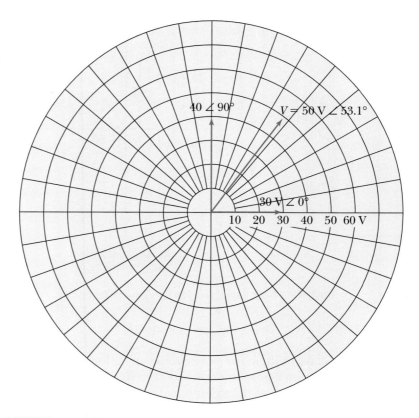

FIGURE 21–23 Illustration of vectors represented on polar coordinate system

Use the tangent function to solve for the angle:

$$\tan \theta = \frac{V_L}{V_R} = \frac{70}{60} = 1.166$$

The angle whose tangent = 1.166 is 49.3°

Polar form = 92.2 ∠49.3° V

You get the idea! To convert, solve for the magnitude by vectorially adding the real and the j terms; then solve for the angle, using the arctangent, as appropriate. The resulting magnitude and angle are then combined to express the polar notation for the vector.

Using a sample calculator to convert from rectangular to polar form

In Figure 21–24, you see one example of a scientific calculator. Let's see the steps you can use to perform the conversion of values expressed in rectangular form to find the equivalent in polar form.

Keys of particular interest will be the **x↔y** key, and the **R►P** key. Let's use the values given in our previous example and perform the conversion using the calculator.

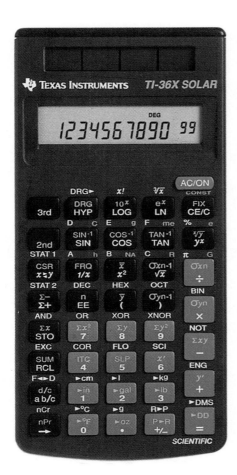

FIGURE 21–24 Sample scientific calculator with *P-R* and *R-P* conversion modes *(Courtesy of Texas Instruments)*

Example

1. Enter the real term value, in this case 60.

2. Press the **x↔y** key. (This enters the real term portion.)

3. Enter the imaginary term value, in this case 70.

4. Press the **3rd** key, then the **R►P** key (rectangular to polar key). By pressing the **3rd** and **R►P** key in sequence, you have invoked the rectangular-to-polar conversion mode of the calculator.

5. Read the number in the calculator display area (92.19). This number represents the vector magnitude for the polar notation. Observe that this number agrees with what we found using the Pythagorean formula.

6. Press the **x↔y** key again to get the angle information. The readout says 49.3, indicating the angle is 49.3°.

7. Combining the results of steps 5 and 6, you find that the polar notation indicates that $92.19 \angle 49.3°$ is the equivalent of the original $60 + j70$ rectangular notation form.

NOTE: If the imaginary number is a $-j$, then enter the number, followed by the **+/−** key, before pressing the **3rd** key (in step 4).

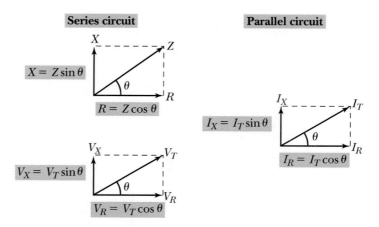

FIGURE 21–25 Converting from polar to rectangular form

Verify the following samples of rectangular and equivalent polar form notations.

Rectangular		Polar
$3 + j4$	$=$	$5\angle 53.1°$
$4 - j3$	$=$	$5\angle -36.9°$
$60 + j70$	$=$	$92.2\angle 49.3°$

Now you try a couple of conversions on your own.

PRACTICE PROBLEMS III

1. Convert $20 + j30$ to polar form.
2. Convert $30 - j20$ to polar form.

Converting from polar to rectangular form

Later in the chapter, you will see the rectangular form is most useful to *add and subtract* vector values and the polar form is useful to *multiply and divide* vector values. For this reason, it is convenient to know how to convert from rectangular to polar form and from polar to rectangular form, as appropriate.

Recall that trig is used to solve for the sides of a right triangle, if the angle and one side are known. Let's put this fact to use.

Since the polar form expresses the hypotenuse of the right triangle plus the angle, we can solve for the other two sides of the triangle by using the appropriate trig functions. Obviously, knowing the sides allows us to express the real and *j* terms appropriately.

Refer to Figure 21–25 and note the horizontal component; that is, the resistive or real term (R in series circuit impedance diagrams), V_R in series circuits, and I_R in parallel circuits is found by:

$$R = Z \cos \theta$$

$$V_R = V_T \cos \theta$$

$$I_R = I_T \cos \theta$$

Recall

$$\cos\theta = \frac{\text{adj}}{\text{hyp}}; \text{ therefore, adj} = \text{hyp} \times \cos\theta$$

In like manner, use trig to solve for the j (imaginary) term (i.e., X in series impedance diagrams, V_X in series circuits, and I_X in parallel circuits). Referring to Figure 21–25 again, the reactive term(s) are found by:

$$X = Z\sin\theta$$

$$V_X = V_T\sin\theta$$

$$I_X = I_T\sin\theta$$

Recall

$$\sin\theta = \frac{\text{opp}}{\text{hyp}}; \text{ therefore, opp} = \text{hyp} \times \sin\theta$$

Once the real and j terms are known, we can express the answer in rectangular form.

Using a sample calculator to convert from polar to rectangular form

You can use the trigonometry we have been discussing to solve for these values or, as you might guess, you can use the scientific caluclator to do the job. At this point, let's give you the calculator steps used in solving a sample problem. Let's assume that the polar form notation is $50\angle37°$. To find the rectangular notation equivalent, use the following steps:

Example

1. Enter 50, then press the **x↔y** key.
2. Enter 37.
3. Press the **2nd** key, then the **P►R** key. The display shows the value of the real term in the rectangular expression, 39.93.
4. Press the **x↔y** key. The display shows the value of the imaginary (j) component of the rectangular expression, $+j30.09$.
5. Combine the results of steps 3 and 4 to get the answer; that is, $50\angle37$ degrees in polar form $= 39.93 + j30.09$ in rectangular form.

Both the trig-related formulas and the related calculator sequences follow for your convenience.

FORMULA 21–12 $Z\angle\theta° = (Z\cos\theta) + (j\,Z\sin\theta)$

for a series circuit impedance diagram
Z value, **x↔y**, angle value (θ), **2nd**, **P►R**, **x↔y**
Answer is combination of last two readouts; that is, **P►R** gives the real term and **x↔y** gives the imaginary (j) term.

> **FORMULA 21–13** $V_T \angle \theta° = V_T \cos \theta + j\, V_T \sin \theta$

for a series circuit V-I vector diagram

V_T value, **x↘y**, angle value (θ), **2nd**, **P►R**, **x↘y**

Answer is combination of last two readouts; that is, **P►R** gives the real term and **x↘y** gives the imaginary (j) term.

> **FORMULA 21–14** $I_T \angle \theta° = I_T \cos \theta + j\, I_T \sin \theta$

for a parallel circuit V-I vector diagram

I_T value, **x↘y**, angle value (θ), **2nd**, **P►R**, **x↘y**

Answer is combination of last two readouts; that is, **P►R** gives the real term and **x↘y** gives the imaginary (j) term.

Examples

Convert $45 \angle 30°$ to rectangular form (trig method).

$$45 \angle 30° = 45\,(\cos 30°) + j45\,(\sin 30°)$$

$$45 \angle 30° = 45 \times 0.866 + j45 \times 0.5$$

$$45 \angle 30° = 38.97 + j22.5$$

Convert $45 \angle 30°$ to rectangular form (calculator method).

45, **x↘y**, 30, **2nd**, **P►R**, (gives R value). Then press **x↘y** again (gives reactive value). Verify the results by drawing a sketch to see if the results look logical.

> #### PRACTICE PROBLEMS IV
> 1. Convert $66 \angle 60°$ to rectangular form.
> 2. Convert $75 \angle 48°$ to rectangular form.
>
> Now that you are familiar with rectangular and polar notation and their relationships, we will show how these expressions are used mathematically.

21–6 ALGEBRAIC OPERATIONS

Addition (rectangular form)

Basic rule: Add in-phase (resistive) terms and add out-of-phase (reactive) j terms.

Example

Find the sum of V_1 and V_2.

Example 1:
$$\begin{aligned} V_1 &= 15 + j20 \\ V_2 &= \underline{20 - j10} \\ V_1 + V_2 &= 35 + j10 \end{aligned}$$

Example 2:
$$\begin{aligned} V_1 &= 20 + j20 \\ V_2 &= \underline{30 + j10} \\ V_1 + V_2 &= 50 + j30 \end{aligned}$$

Subtraction (rectangular form)

Basic rule: Change the sign of the subtrahend and add.

Example

Subtract branch current I_2 from branch current I_1, where

$$I_1 = 5 - j6 \text{ A}$$
$$I_2 = 4 + j7 \text{ A}$$

Change the sign of branch current I_2 and add:

$$
\begin{array}{r}
5 - j6 \\
-4 - j7 \\
\hline
I_1 - I_2 = 1 - j13
\end{array}
$$

Example

When given total current and one branch current (I_1), find the other branch current (I_2).

$$I_T = I_1 + I_2; \text{ therefore, } I_2 = I_T - I_1.$$

where

$$I_T = 13 - j5.5 \text{ mA}$$
$$I_1 = 6 - j4.0 \text{ mA}$$

Change the sign of branch current 1, and add:

$$
\begin{array}{r}
13 - j5.5 \\
-6 + j4.0 \\
\hline
I_2 = 7 - j1.5
\end{array}
$$

PRACTICAL NOTES

If addition or subtraction is needed to analyze a problem and the terms are given in polar form, convert them to rectangular form because you cannot use polar form to add or subtract.

On the other hand, even though multiplication and division are possible in rectangular form, polar form is easier for multiplying and dividing. If you know the rectangular form, you will first convert it to the polar form to multiply or divide. For example, $V_T = I_T \times Z_T$.

Multiplication (polar form)

Basic rule: Multiply the two magnitudes and add the angles algebraically.

Example

If a current of 5 $\angle 30°$ A flows through a Z of 20 $\angle -15°$ Ω, what is the voltage?

$$V = IZ$$

$$V = (5 \angle 30°) \times (20 \angle -15°)$$

$$V = 100 \angle 15° \text{V}$$

Division (polar form)

Basic rule: Divide one magnitude by the other and subtract the angles algebraically.

Example

Find the current in a circuit if the voltage = 35 $\angle 60°$ and the impedance = 140 $\angle -20°$.

$$I = \frac{V}{Z}$$

$$I = \frac{35 \angle 60°}{140 \angle -20°}$$

$$I = 0.25 \angle 80° \text{ A}$$

Raising to powers (polar form)

Basic rule: Raise the magnitude to the power indicated and multiply the angle by that power.

Example

If $I = 3 \angle 20°$, then $I^2 = 9 \angle 40°$.

Taking a root (polar form)

Basic rule: Take the root of the magnitude and divide the angle by the root.

Example

If $I^2 = 100 \angle 60°$, then $I = 10 \angle 30°$.

PRACTICE PROBLEMS V

1. Add $20 + j5$ and $35 - j6$
2. Add $34.9 - j10$ and $15.4 + j12$
3. Subtract $20 + j5$ from $35 - j6$
4. Subtract $22.9 - j7$ from $37.8 - j13$
5. Multiply $5 \angle 20° \times 25 \angle -15°$
6. Multiply $24.5 \angle 31° \times 41 \angle 23°$

7. Divide $5 \angle 20°$ by $25 \angle -15°$

8. Divide $33 \angle 34°$ by $13 \angle 45°$

9. Find I^2 if $I = 12 \angle 25°$

10. Find I if $I^2 = 49 \angle 66°$

11. Find the result of: $\dfrac{5 \angle 50° + (30 + j40)}{2.6 \angle -15°}$

12. Find the result of: $\dfrac{10 \angle 53.1° + (25 - j25)}{4.8 \angle 36°}$

21-7 APPLICATION OF RECTANGULAR AND POLAR ANALYSIS

Sample series RLC circuit analysis and applying vector analysis

Refer to Figure 21–26 as you study the following discussion on finding various circuit parameters. Note if addition and subtraction are needed, the calculations use rectangular notation; if multiplication or division is involved, the calculations use polar notation.

1. Find Z_T:

    Polar form:

    Rectangular form: $20 + j10$

    Z_T = vector sum of oppositions

    $Z_T = \sqrt{R_T^2 + X_T^2}$

    $Z_T = \sqrt{20^2 + 10^2}$

    $Z_T = \sqrt{500}$

    $Z = 22.36 \angle 26.56° \, \Omega$ $Z_T = 22.36 \, \Omega$

2. Find I_T (Assume I_T is at 0° because this is a series circuit.):

    Polar form: Rectangular form:

    $I_T = \dfrac{V_T}{Z_T}$

    $I_T = \dfrac{60.00 \angle 26.56}{22.36 \angle 26.56}$

    $I_T = 2.68 \angle 0° \, \text{A}$ $I_T = 2.68 + j0 \, \text{A}$

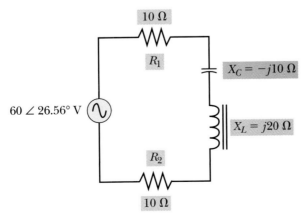

FIGURE 21–26 Sample *RLC* circuit analysis

3. Find individual voltages:
   a. $V_{R_1} = I \times R_1 = 2.68 \angle 0° \times 10 \angle 0° = 26.8 \angle 0°$ V
   b. $V_{R_2} = I \times R_2 = 2.68 \angle 0° \times 10 \angle 0° = 26.8 \angle 0°$ V
   c. $V_L = I \times X_L = 2.68 \angle 0° \times 20 \angle 90° = 53.6 \angle 90°$ V
   d. $V_C = I \times X_C = 2.68 \angle 0° \times 10 \angle -90° = 26.8 \angle -90°$ V

4. Verify total voltage via sum of individual drops. That is, convert each V from polar to rectangular form, then add:

Polar form:	Rectangular form:
$V_{R_1} = 26.8 \angle 0°$	a. $26.8 + j0$
$V_{R_2} = 26.8 \angle 0°$	b. $26.8 + j0$
$V_L = 53.6 \angle 90°$	c. $0.0 + j53.6$
$V_C = 26.8 \angle -90°$	d. $0.0 - j26.8$

   Total voltage = $53.6 + j26.8$ or $60 \angle 26.56°$

   V_T = approximately $60 \angle 26.56°$ V (polar) or $53.6 + j26.8$ (rectangular)

Sample parallel RLC circuit analysis

Remember for parallel circuits in dc, it is easier to use the conductance of branches or add branch currents, than to combine reciprocal resistances. Recall conductance is the reciprocal of resistance. That is, $G = 1/R$ siemens.

In ac circuits, we have to deal with pure resistance *and* reactance *and* the combination of resistance and reactance (impedance). The comparable term for conductance of a pure reactance is called **susceptance,** where susceptance = $1/\pm X$. The symbol for susceptance is **B.** The reciprocal of impedance is called **admittance,** which is equal to $1/Z$. Thus:

- For pure resistance, conductance $(G) = 1/R$.
- For pure reactance, susceptance $(B) = 1/\pm X$.
- For combined R and X admittance $(Y) = 1/Z$.

The phase angle designation for B or Y is opposite that expressed for X or Z, since it is a reciprocal relationship. This means reactance for an inductive branch is $+jX_L$, and inductive susceptance is $-jB_L$. Also, capacitive reactance is $-jX_C$ and capacitive susceptance is jB_C.

FORMULA 21–15   Inductive susceptance: $B_L = \dfrac{1}{X_L} = \dfrac{1}{2\pi f L}$

FORMULA 21–16   Capacitive susceptance: $B_C = \dfrac{1}{X_C} = 2\pi f C$

Notice the inversion of the X_L and X_C formulas. Important admittance relationships are:

$$Y = G \pm jB = G + j(B_C - B_L)$$

FORMULA 21–17 $Y = \sqrt{G^2 + B^2}$

$$\theta = \text{angle whose tangent} = \frac{B}{G}$$

$$\text{NOTE: } Y = \frac{I}{V}$$

$$I = V \times Y$$

$$V = \frac{I}{Y}$$

Example

Refer to Figure 21–27. Use the admittance approach and find the total current of a parallel circuit operating at 60 Hz with the following parameters: $R = 500\ \Omega$; $L = 1$ H; $C = 4\ \mu$F; and $V = 100$ V.

$$G = \frac{1}{R} = \frac{1}{500} = 0.002 \text{ S}$$

$$B_L = \frac{1}{2\pi fL} = \frac{1}{6.28 \times 60 \times 1} = 0.00265 \text{ S}$$

$$B_C = 2\pi fC = 6.28 \times 60 \times (4 \times 10^{-6}) = 0.00151 \text{ S}$$

$$Y = G + j(B_C - B_L) = 0.002 + j(0.00151 - 0.00265)$$

$$Y = 0.002 - j0.00114 = 0.0023\angle - 29.7° \text{ S}$$

$$I_T = VY = 100 \angle 0° \times 0.0023 \angle -29.7° = 0.23 \angle -29.7° \text{ A}$$

You can calculate the parameters using rectangular forms to add and subtract and polar forms to multiply and divide. When using these notations, it is a good idea to make appropriate conversions so each significant parameter is expressed in both forms. In this way, you use the form most applicable to the calculation you are performing.

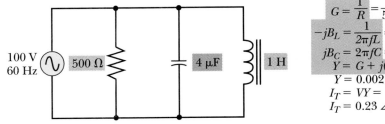

$$G = \frac{1}{R} = \frac{1}{500} = 0.002 \text{ S}$$
$$-jB_L = \frac{1}{2\pi fL} = \frac{1}{6.28 \times 60 \times 1} = -j0.00265 \text{ S}$$
$$jB_C = 2\pi fC = 6.28 \times 60 \times (4 \times 10^{-6}) = j0.00151 \text{ S}$$
$$Y = G + j(B_C - B_L) = 0.002 + j(0.00151 - 0.0026 \text{ S})$$
$$Y = 0.002 - j0.00114 = 0.0023 \angle -29.7° \text{ S}$$
$$I_T = VY = 100 \angle 0° \times 0.0023 \angle -29.7°$$
$$I_T = 0.23 \angle -29.7° \text{ A}$$

FIGURE 21–27

FIGURE 21–28 Sample combination circuit calculations

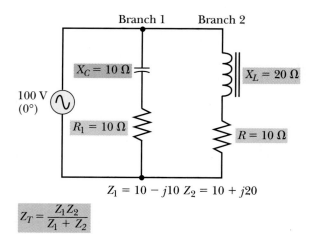

Branch 1   Branch 2

$X_C = 10 \ \Omega$ $X_L = 20 \ \Omega$

100 V (0°)

$R_1 = 10 \ \Omega$ $R = 10 \ \Omega$

$Z_1 = 10 - j10 \quad Z_2 = 10 + j20$

$$Z_T = \frac{Z_1 Z_2}{Z_1 + Z_2}$$

Sample combination circuit calculations

Refer to Figure 21–28 and note the following calculations.

1. Find Z_T. Since in a parallel circuit

$$Z_T = \frac{Z_1 Z_2}{Z_1 + Z_2}$$

we need to know both the rectangular and polar forms of each branch impedance. These forms are a good starting point.

$Z_1 = 10 - j10$ (rectangular) $14.14 \angle -45°$ (polar)

$Z_2 = 10 + j20$ (rectangular) $22.36 \angle 63.4°$ (polar)

$$Z_T = \frac{Z_1 \times Z_2}{Z_1 + Z_2} = \frac{14.14 \angle -45° \times 22.36 \angle 63.4°}{10 - j10 + 10 + j20} = \frac{316.17 \angle 18.4°}{20 + j10}$$

$$Z_T = \frac{316.17 \angle 18.4°}{22.36 \angle 26.56°} = 14.13 \angle -8.16° \ \Omega$$

2. Assume 100 V applied voltage at 0° and find the current.

$$\text{Branch 1 current} = \frac{V}{Z_1}$$

$$\text{Branch I current} = \frac{100 \angle 0°}{14.14 \angle -45°} = 7.07 \angle 45° \ A$$

$$\text{Branch 2 current} = \frac{V}{Z_2}$$

$$\text{Branch 2 current} = \frac{100 \angle 0°}{22.36 \angle 63.4°} = 4.47 \angle -63.4° \ A$$

3. To find I_T, convert each branch current to rectangular form and add the branch currents.

$$I_1 = 7.07 \angle 45° \text{ (polar)} \qquad 5 + j5 \text{ (rectangular)}$$

$$I_2 = \underline{4.47 \angle -63.4° \text{ (polar)}} \qquad \underline{2 - j4 \text{ (rectangular)}}$$

$$I_T = 7.07 \angle 8.13° \text{ A (polar)} \qquad 7 + j1 \text{ A (rectangular)}$$

4. Individual component voltages are calculated using $I \times R$, $I \times X_L$, and $I \times X_C$, as appropriate. You can practice rectangular and polar notation and appropriate conversion techniques by doing this. If you do this, remember to sketch vector diagram(s), as appropriate, to check the logic of your results.

PRACTICAL NOTES

The calculations and techniques introduced in this chapter will be used by technicians with varying frequency, depending upon their specific job functions. Regardless of your job function, having this knowlege provides you with a stronger foundation in ac analysis. Some specific examples of where these calculations and techniques are used include radio and TV transmission systems and power transmission systems. Certainly, many other areas of endeavor for technicians can and will take advantage of the knowledge gained in this chapter.

SUMMARY

▶ Vectors in opposite directions cancel each other. If one has greater magnitude than the other, there is a net resulting vector equal to the larger value minus the smaller value.

▶ Power in ac circuits with both resistive and reactive components is described in three ways:

1. Apparent power *(S)*—applied voltage times the circuit current, expressed in volt-amperes (VA) and plotted as the hypotenuse of the power triangle

2. Reactive power *(Q)* or voltampere-reactive (VAR)—equal to $VI \times \sin \theta$, and plotted on the vertical axis

3. True power *(P)* in watts—equal to $VI \cos \theta$, and plotted on the horizontal axis

Because of the right-triangle relationship:
$S = \sqrt{P^2 + Q^2}$.

▶ Vector quantities are analyzed in several ways—graphic means, algebraic expressions based on rectangular coordinates, and expressions based on polar coordinates.

▶ Separating a vector quantity into its components is easily accomplished by simple trigonometric functions. These functions are sine, cosine, and tangent.

▶ Complex numbers express vector quantities relative to the rectangular notation system. The complex number has a term that defines the horizontal (real or resistive) component and a j term that expresses the reactive component or out-of-phase element. The general term for impedance is $R \pm jX$, for applied voltage is $V_R \pm V_X$, and for total current in a parallel circuit is $I_R \pm jI_X$.

▶ Vectors expressed algebraically in rectangular coordinates are added, subtracted, multiplied, and divided. However, this form of notation is most easily used to add and subtract. To add, like terms are collected (i.e., resistive terms collected and j terms collected), and added algebraically. To subtract, change the sign of the subtrahend and add algebraically.

▶ Vectors expressed in polar form *cannot* be added or subtracted algebraically. However, this form is easily used to multiply, divide, and find square roots. To multiply polar expressions, multiply magnitudes and add the angles algebraically. To divide polar expressions, divide the magnitudes and subtract the angles algebraically. To raise a polar expression to a power, raise the magnitude by the given power and multiply the angle by the power. To take a root, take the root of the magnitude value and divide the angle by the root value.

▶ Because polar form facilitates multiplication and division and rectangular form expedites addition and subtraction, it is a good idea to convert from one form to the other, as appropriate.

▶ When two branches are in parallel, with each having resistive and reactive components, the resultant Z is found by the formula $Z = \dfrac{Z_1 Z_2}{(Z_1 + Z_2)}$

▶ It is convenient to use the polar form for the numerator (for multiplying) and the rectangular form for the denominator (for adding). Then, convert the denominator to the polar form to allow division into the numerator to get the final answer.

▶ The ease that current flows through a resistance is called conductance, where $G = 1/R$. A comparable term for reactance is called susceptance, where $B = 1/\pm X$. When resistance and reactance are combined in a circuit, the term for ease of current flow is admittance (Y). Admittance is the reciprocal of impedance, where $Y = 1/Z$. The unit of measure for all these terms is siemens (S).

▶ Some important relationships between admittance, susceptance, and conductance are $Y = G \pm jB$, $Y = \sqrt{G^2 + B^2}$, and $\theta =$ angle whose tangent $= B/G$.

FORMULAS AND SAMPLE CALCULATOR SEQUENCES

FORMULA 21–1 $Z = \sqrt{R^2 + X^2}$ (Series RLC circuits)

R value, $\boxed{x^2}$, $\boxed{+}$, X_{net}, $\boxed{x^2}$, $\boxed{=}$, $\boxed{\sqrt{x}}$

FORMULA 21–2 $V_T = \sqrt{V_R^2 + V_X^2}$ (Series RLC circuits)

V_R value, $\boxed{x^2}$, $\boxed{+}$, V_X value, $\boxed{x^2}$, $\boxed{=}$, $\boxed{\sqrt{x}}$

FORMULA 21–3   tangent of $\theta = \dfrac{X \text{ (net reactance)}}{R \text{ (Series } RLC \text{ circuits)}}$

X(net reactance) value, $\boxed{\div}$, R value, $\boxed{=}$

FORMULA 21–4   tangent of $\theta = \dfrac{V_X}{V_R}$ (Series RLC circuits)

$V_{X(\text{net})}$ value, $\boxed{\div}$, V_R value, $\boxed{=}$

FORMULA 21–5 $I_T = \sqrt{I_R^2 + I_X^2}$ (Parallel RLC circuits)

I_R value, $\boxed{x^2}$, $\boxed{+}$, $I_{X(\text{net})}$, $\boxed{x^2}$, $\boxed{=}$, $\boxed{\sqrt{x}}$

FORMULA 21–6 Power factor $(p.f.) = \cos \theta$

Angle value, **COS**

FORMULA 21–7 Apparent power $(S) = V \times I$

V value, **×**, I value, **=**

FORMULA 21–8 True power $(P) = I^2 R$

I value, **x²**, **×**, R value, **=**

FORMULA 21–9 $P =$ Apparent power $\times$ power factor

Apparent power value, **×**, power factor value, **=**

FORMULA 21–10 $p.f. = \dfrac{\text{True power } (P)}{\text{Apparent power } (S)}$

True power value, **÷**, apparent power value, **=**

NOTE: In series circuits $p.f. = \cos \theta = R/Z$. In parallel circuits $p.f. = \cos \theta = I_R/I_T$.

FORMULA 21–11 $S = \sqrt{P^2 + Q^2}$

True power value, **x²**, **+**, net reactive power value, **x²**, **=**, **√x**

FORMULA 21–12 $Z \angle \theta° = (Z \cos \theta) + (j Z \sin \theta)$ (For series circuits)

USING THE SCIENTIFIC CALCULATOR CONVERSION TECHNIQUE:
Z value, **x↔y**, angle value (θ), **2nd**, **P►R**, **x↔y**
Answer is combination of last two readouts; that is, **P►R** gives the real term and **x↔y** gives the imaginary (j) term.

FORMULA 21–13 $V_T \angle \theta° = V_T \cos \theta + j V_T \sin \theta$ (For series circuits)

V_T value, **x↔y**, angle value (θ), **2nd**, **P►R**, **x↔y**
Answer is combination of last two readouts; that is, **P►R** gives the real term and **x↔y** gives the imaginary (j) term.

FORMULA 21–14 $I_T \angle \theta° = I_T \cos \theta + j\, I_T \sin \theta$ (For parallel circuits)

I_T value, **x\y** , angle value (θ), **2nd** , **P►R** , **x\y**
Answer is combination of last two readouts; that is, **P►R** gives the real term and **x\y** gives the imaginary (j) term.

FORMULA 21–15   Inductive susceptance: $B_L = \dfrac{1}{X_L} = \dfrac{1}{2\pi fL}$

6.28, **x** , frequency value, **x** , inductance value, **=** , **1/x**

FORMULA 21–16   Capacitive susceptance: $B_C = \dfrac{1}{X_C} = 2\pi fC$

6.28, **x** , frequency value, **x** , capacitance value, **=**

FORMULA 21–17 $Y = \sqrt{G^2 + B^2}$

Value of conductance (G), **x²** , **+** , value of susceptance (B) **x²** , **=** , **x**

REVIEW QUESTIONS

1. In a series *RLC* circuit *V-I* vector diagram, voltage across capacitors is plotted:
   a. upward at 90°.
   b. downward at −90°.
   c. to the right at 0°.
   d. to the left at 180°.

2. In a series *RLC* circuit *V-I* vector diagram, circuit current is plotted:
   a. upward at 90°.
   b. downward at −90°.
   c. to the right at 0°.
   d. to the left at 180°.

3. In a series *RLC* circuit with unequal reactances, applied voltage for a *V-I* vector diagram is plotted:
   a. upward at 90°.
   b. downward at −90°.
   c. to the right at 0°.
   d. none of the above.

4. In a series *RLC* circuit impedance diagram, *R* is plotted:
   a. upward at 90°.
   b. downward at −90°.
   c. to the right at 0°.
   d. none of the above.

5. In a series *RLC* circuit impedance diagram, capacitive reactance is plotted:
   a. upward at 90°.
   b. downward at −90°.
   c. to the right at 0°.
   d. none of the above.

6. In a series *RLC* circuit with unequal reactances, impedance is plotted:
   a. upward at 90°.
   b. downward at −90°.
   c. to the right at 0°.
   d. none of the above.

7. In a parallel *RLC* circuit *V-I* vector diagram, inductor branch current(s) is(are) plotted:
   a. upward at 90°.
   b. downward at −90°.
   c. to the right at 0°.
   d. to the left at 180°.

8. In a parallel *RLC* circuit *V-I* vector diagram, capacitor branch current(s) is(are) plotted:
   a. upward at 90°.
   b. downward at −90°.
   c. to the right at 0°.
   d. to the left at 180°.

9. In a parallel *RLC* circuit with a net reactance involved, the total circuit current plot in a *V-I* vector diagram:
   a. is upward at 90°.
   b. is downward at −90°.
   c. may be upward or downward at some angle.
   d. none of the above.

10. In a parallel *RLC* circuit with a net reactance involved, a *V-I* vector diagram would illustrate total circuit voltage:
    a. upward at 90°.
    b. downward at −90°.
    c. to the right at 0°.
    d. at some angle other than 0°.

11. In a parallel *RLC* circuit *V-I* vector diagram, resistive branch current(s) is(are) plotted:
    a. upward at 90°.
    b. downward at −90°.
    c. to the right at 0°.
    d. to the left at 180°.

12. In parallel *RLC* circuits, inductive and capacitive branch currents are:
    a. in phase with each other.
    b. 90° out of phase with each other.
    c. 180° out of phase with each other.
    d. 270° out of phase with each other.

13. For a purely resistive circuit in ac, an instantaneous power graph would show that power is dissipated:
    a. only on the positive alternation from the source.
    b. only on the negative alternation from the source.
    c. during both alternations.
    d. none of the above.

14. For ideal reactive devices and a purely reactive circuit, an ac instantaneous power graph would show that average power dissipated over a cycle equals:
    a. 2 × Power in.
    b. *V* × *I.*
    c. zero.
    d. none of the above.

15. When a circuit contains both resistances and reactances, apparent power is:
    a. lower than the actual power dissipated by the circuit.

b. higher than the actual power dissipated by the circuit.
    c. equal to the actual power dissipated by the circuit.
    d. none of the above.

16. A "power triangle" is made up of:
    a. true power, apparent power, and *V* × *I.*
    b. true power, apparent power, and their mathematical sum.
    c. true power, reactive power, and apparent power.
    d. none of the above.

17. True power in *RLC* circuits is dissipated by:
    a. resistive elements.
    b. reactive elements.
    c. both resistive and reactive elements.
    d. none of the above.

18. The *j* operator is used in:
    a. polar notation.
    b. vector notation.
    c. rectangular notation.
    d. none of the above.

19. The advantage of using the polar form of notation is that:
    a. it uses the *j* operator to great advantage.
    b. it gives the same values for electrical quantities that measuring instruments show.
    c. it is easier to write down than the rectangular notation.
    d. it is easy to add and subtract with this type notation.

20. One advantage of rectangular notation over polar notation is that:
    a. it is easier to multiply and divide vector quantities in rectangular form.
    b. it is easier to add and subtract vector quantities in rectangular form.
    c. it provides the angle within its notation.
    d. none of the above.

PROBLEMS

(Be sure to make appropriate vector diagram sketches to help clarify your thinking.)

1. Refer to Figure 21–29a. Find Z, I, V_C, V_L, V_R, and θ using the Pythagorean theorem, Ohm's law, and appropriate trig functions. Draw impedance and V-I vector diagrams, labeling all parts, as appropriate.

2. Refer to Figure 21–29b. Find Z, I_C, I_L, I_R, I_T, and θ using the Pythagorean theorem, Ohm's law, and appropriate trig functions. Draw an appropriate V-I vector diagram and label all parts, as appropriate.

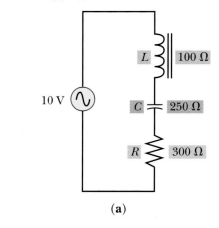

(a)

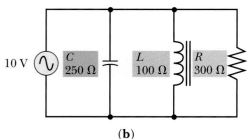

(b)

FIGURE 21–29

3. Calculate true power, apparent power, and reactive power for the circuit of question 1. What is the power factor?

4. Draw a power triangle for the circuit in question 1.

5. A series circuit has a 20-Ω resistance and a 40-Ω inductive reactance. How do you state the circuit Z in rectangular form? If the inductive reactance is changed to an equal value of capacitive reactance, how do you state Z?

6. A series RLC circuit is composed of 10-Ω resistance, 20-Ω inductive reactance, and 30-Ω capacitive reactance. How is the circuit Z stated in rectangular form?

7. A parallel circuit with a 20-V source has a resistive branch of 20 Ω and a capacitive reactance branch of 20 Ω. What is the total current stated in polar form? See Figure 21–30.

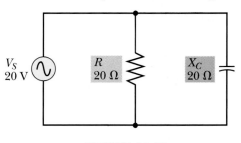

FIGURE 21–30

8. Convert $30 + j40$ to polar form.

9. Convert $14.14 \angle 45°$ to rectangular form.

10. A series RLC circuit has a 10-V, 1-kHz source, an R of 8 kΩ, an X_L of 6 kΩ, and an X_C of 4 kΩ. What is the total circuit current? (Express in polar form.) What is the true power?

11. What is the total impedance of a series RLC circuit that has a 20-kHz ac source, $C = 100$ pF, $L = 0.159$ H, and $R = 15$ kΩ? (Express in polar form and in rectangular form.)

12. A series RLC circuit has 100 kΩ of resistance, a capacitor with 0.0005 μF of capacity, and an inductor with 1.5 H of inductance. What is the value of voltage drop across the inductor, if the applied ac voltage is 100 V at a frequency of 10 kHz? (Express in polar form.)

13. If the frequency of the applied voltage increases, what happens to V_C?

14. What is the true power dissipated by a series RLC circuit with 100 V applied, if $R = 30$ Ω, $X_L = 40$ Ω, and $X_C = 10$ Ω?

15. A parallel RC circuit has a capacitor with 2 Ω of X_C and a resistor with 4 Ω of resistance. What is the impedance? (Express in polar form.)

16. A parallel RC circuit has a power factor of 0.5

and $I_T = 5$ A, $f = 1$ kHz, and $R = 60 \, \Omega$. What is the value of C?

17. What is the total impedance of a parallel *RLC* circuit comprised of $R = 6 \, \Omega$, $X_L = 12 \, \Omega$, and $X_C = 24 \, \Omega$?

18. A parallel *RLC* circuit has 400-Ω resistance, 500-Ω inductive reactance branch, and 800-Ω capacitive reactance branch. What is the circuit admittance value?

19. What are the Z and θ of the circuit in question 18? (Assume a voltage applied of 800 V.)

20. What value of X_C produces a $\theta = 0°$?

21. For the circuit in question 18, what happens to the following parameters if frequency doubles? (Answer with "I" for increase, "D" for decrease, and "RTS" for remain the same.)
    a. I_R will
    b. I_C will
    c. I_L will
    d. I_T will
    e. Z_T will
    f. θ will

Referring to the answers you obtain for selected Problems, use the appropriate calculator key sequences and solve for the following.

22. Refer to Problems question 6. State the Z in polar form.

23. Refer to Problems question 7. State the total current in rectangular form.

24. Refer to Problems question 10. State the total current in rectangular form.

25. Refer to Problems question 15. State the impedance in rectangular form.

26. Referring to Figure 21–31 and using rectangular and polar notation techniques, find Z_T, and express your answer in polar form.

27. Express the Z_T of the circuit in Figure 21–31 in rectangular form.

28. Find I_T for the circuit of Figure 21–31 and express your answer in polar form.

29. Express the I_T value for the circuit of Figure 21–31 in rectangular form.

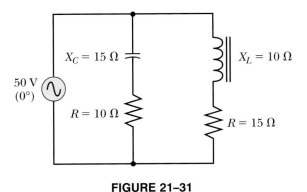

FIGURE 21–31

ANALYSIS QUESTIONS

1. In your own words, define what is meant by the terms *real number* and *imaginary number*.

2. In your own words, define what is meant by the term *complex number*.

3. Explain why using measurements of a circuit's voltage with a voltmeter and its current with a current meter does not provide the necessary information to find the actual power dissipated by the circuit.

4. Explain how you can make measurements to determine the actual power dissipated by an ac circuit containing inductors and/or capacitors.

THE SIMPLER SEQUENCE FOR TROUBLESHOOTING

Follow the seven-step SIMPLER troubleshooting sequence, as outlined below, and see if you can find the problem with this circuit. As you follow the sequence, record your circuit test results for each testing step on a separate sheet of paper. This will aid you and your instructor to see where your thinking is on track or where it might deviate from the best procedure.

step 1 Symptoms

—(Gather, verify, and analyze symptom information) To begin this process, read the data under the "Starting Point Information" heading. Particularly look for circuit parameters that are not normal for the circuit configuration and component values given in the schematic diagram. For example, look for currents that are too high or too low; voltages that are too high or too low; resistances that are too high or too low, etc. This analysis should give you a first clue (or symptom information) that will aid you in determining possible areas or components in the circuit that might be causing this symptom.

step 2 Identify

—(Identify and bracket the initial suspect area) To perform this first "bracketing" step, analyze the clue or symptom information from the "Symptoms" step, then either on paper or in your mind, bracket, circle, or put parentheses around all of the circuit area that contains any components, wires, etc., that might cause the abnormality that your symptom information has pointed out. NOTE: Don't bracket parts of the circuit that contain items that could not cause the symptom!

step 3 Make

—(Make a decision about the first test you should make: what type and where) Look under the TEST column and determine which of the available tests shown in that column you think would give you the most meaningful information about the suspect area or components, in the light of the information you have so far.

step 4 Perform

—(Perform the first test you have selected from the TEST column list) To simulate performing the chosen test (as if you were getting test results in an actual circuit of this type), follow the dotted line to the right from your selected test to the number in parentheses. The number in parentheses, under the "Results in Appendix C" column tells you what number to look up in Appendix C to see what the test result is for the test you are simulating.

step 5 Locate

—(Locate and define a new "narrower" area of uncertainty) With the information you have gained from the test, you should be able to eliminate some of the circuit area or circuit components as still being suspect. In essence, you should be able to move one bracket of the bracketed area so that there is a new smaller area of uncertainty.

step 6 Examine

—(Examine available information and determine the next test: what type and where) Use the information you have thus far to determine what your next test should be in the new narrower area of uncertainty. Determine the type test and where, then proceed using the TEST listing to select that test, and the numbers in parentheses and Appendix C data to find the result of that test.

step 7 Repeat

—(Repeat the analysis and testing steps until you find the trouble) When you have determined what you would change or do to restore the circuit to normal operation, you have arrived at *your* solution to the problem. You can check your final result against ours by observing the pictorial step-by-step sample solution on the pages immediately following the "Chapter Troubleshooting Challenge"—circuit and test listing page.

step 8 Verify

NOTE: A useful *8th Step* is to operate and test the circuit or system in which you have made the "correction changes" to see if it is operating properly. This is the final proof that you have done a good job of troubleshooting.

CHALLENGE CIRCUIT 11

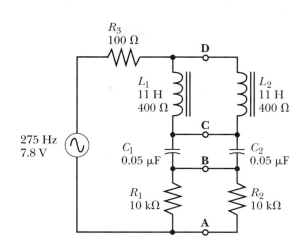

Starting Point Information

1. Circuit diagram
2. $V_{R_3} = 0.057$ V

TEST	Results in Appendix C
$V_{A\text{-}B}$	(78)
$V_{B\text{-}C}$	(38)
$V_{C\text{-}D}$	(96)
$R_{A\text{-}B}$	(104)
$R_{B\text{-}C}$	(63)
$R_{C\text{-}D}$	(32)

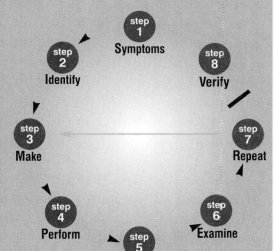

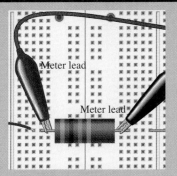

Symptoms

the voltmeter measures approximately 0.057 V. If the circuit were operating as prescribed by the diagram, the voltmeter should measure about 0.112 V. This implies circuit current is lower than it normally would be. Thus, circuit impedance is higher than normal.

Identify

initial suspect area: Between points C and D, the total inductive reactance of the parallel inductors should be about 9.5 kΩ. Between points B and C, the net capacitive reactance of the parallel capacitors should be about 5.8 kΩ. The net resistance between points A and B should be about 5 kΩ. If the circuit were operating correctly, the total impedance would be about 6.2 kΩ. However, it's acting like the circuit has about 14 kΩ total impedance. Suspect area includes the whole circuit.

Make

test decision: The upward shift in Z could be caused by any R, C, or L opening or any C or L shorting. We'll first look at the resistors. Let's look at V_{A-B}.

Perform

1st Test: Test V_{A-B}. V_{A-B} is about 3.11 V, again indicating lower than normal circuit current. We know the Rs are not shorted or the voltage would read zero volts. Also, since we know the circuit current is about 0.57 mA, then 0.57 mA times the normal equivalent R of approximately 5 kΩ comes close to that value. The resistors appear to be acting normally.

Locate

new suspect area: The Rs have been eliminated. Thus, the reactive components are the new suspect area.

Examine
available data.

Repeat

analysis and testing: Let's look at the inductive branches, isolating as appropriate.

2nd Test: Check V_{C-D}. V_{C-D} is about 11 V. An open inductor could cause this situation so let's check that out.

3rd Test: Check R_{C-D}. R_{C-D} is about 400 Ω. This should be about 200 Ω with the parallel circuit situation.

4th Test: Check R_{L_1}. R_{L_1} is about 400 Ω, which is normal.

Symptoms

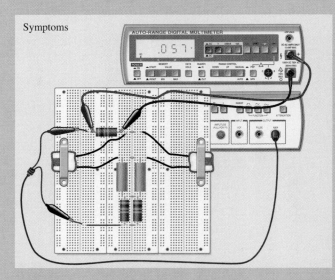

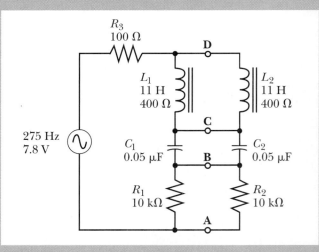

5th Test: Check R_{L_2}. R_{L_2} is infinite ohms, which is an open coil.

Verify

6th Test: Change L_2 and check the circuit operation. It's normal.

1st Test

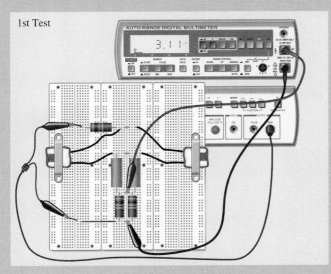

2nd Test

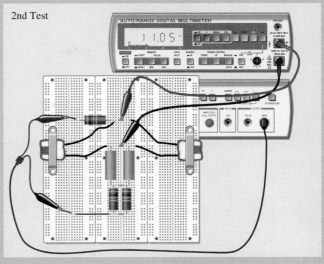

3rd Test

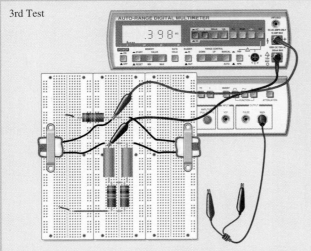

4th Test

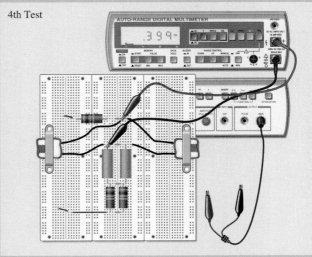

5th Test

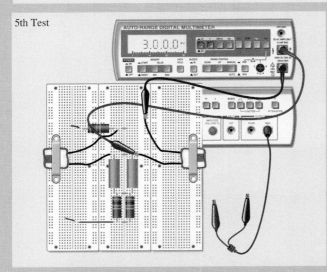

6th Test

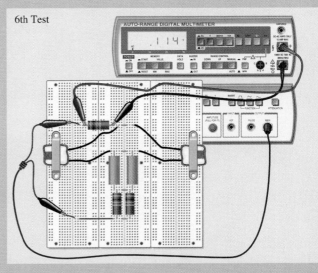

Bandpass
Bandpass filter
Bandstop filter
Bandwidth
High-pass filter
L/C **ratio**
Low-pass filter
Q **factor**
Resonance
Selectivity
Skirts

SERIES AND PARALLEL RESONANCE

22

chapter

CHAPTER PREVIEW

Series and parallel resonant circuits provide valuable capabilities in numerous electronic circuits and systems. When you select a radio or TV station, you are using the unique capability of a resonant circuit to select a narrow group of frequencies from a wide spectrum of frequencies that impact the receiver's antenna system. The transmitter that is transmitting the selected signals also has a number of "tuned" resonant circuits that assure only the desired frequencies are being transmitted and that spurious signals are not being transmitted. Both the transmitting and receiving antenna systems are further examples of special tuned or resonant circuits.

In addition to the valuable tuning application, resonance allows a related application called *filtering*. Filtering can involve either passing desired signals or voltages, or blocking undesired ones. Tuning and filtering are the most frequent applications for series and parallel resonance.

In this chapter, you will study the key characteristics of both series and parallel resonant circuits. Your knowledge of these circuits will be frequently used throughout your career.

OBJECTIVES

After studying this chapter, you should be able to:

1. List the key characteristics of series and parallel resonant circuits

2. Calculate the resonant frequency of circuits

3. Calculate **L** or **C** values needed for **resonance** at a given f_r

4. Calculate **Q** factor for series and parallel resonant circuits

5. Determine **bandwidth** and **bandpass** of resonant circuits

6. Draw circuit diagrams for three types of **filters**

7. Use the SIMPLER troubleshooting sequence to solve the Chapter Troubleshooting Challenge problems

22-1 X_L, X_C, AND FREQUENCY

Previously you learned that as frequency increases, X_L increases proportionately. The formula for inductive reactance expresses this by $X_L = 2\pi f L$. You also learned that as frequency increases, X_C decreases in inverse proportion, expressed in the formula

$$X_C = \frac{1}{2\pi f C}$$

Conversely, as frequency decreases, X_L decreases and X_C increases. In other words, X_L and X_C change in opposite directions as frequency changes.

To review, refer to Figure 22–1 and note the relationships between X_L, X_C, and frequency. Notice X_C is not illustrated at frequencies approaching 0 Hz. This is because X_C is approaching infinite ohms at these frequencies, thus is difficult to illustrate on such a graph.

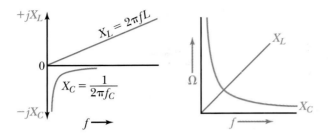

FIGURE 22–1 Relationships between X_L, X_C, and frequency

22-2 SERIES RESONANCE CHARACTERISTICS

The implication of these statements about the relationships of X_L, X_C, and frequency is that for a given set of L and C values, there is a specific frequency where the *absolute* magnitudes of X_L and X_C are equal. Look at Figure 22–2 and notice that for this series *RLC* circuit, there is one frequency where this occurs. In a series circuit containing L and C, the condition when the magnitude of X_L equals the magnitude of X_C is called **resonance** and the frequency where this occurs is the *resonant frequency*. At resonance, some interesting observations can be made about circuit electrical parameters.

Refer to Figure 22–3 as you study the following list of features.

The characteristics of a series *RLC* circuit at resonance are as follows:

1. $X_L = X_C$ (In this case, 100 Ω = 100 Ω.)
2. Z = minimum and equals R in circuit. (In this case, 10 Ω.)
3. I = maximum and equals V/R. (In this case, 50 mV/10 Ω = 5 mA.)
4. V_C and V_L are maximum. ($I \times X_C$ = 500 mV; $I \times X_L$ = 500 mV.)
5. $\theta = 0°$ (Since circuit is acting purely resistive.)

Let's examine these statements.

1. When X_L equals X_C, the reactances cancel, leaving 0 Ω of net reactance in the circuit. The remaining opposition to current, under these conditions, is any resistance in the circuit. Generally, most of this resistance is the resistance in the inductor windings (unless we have purposely added an additional discrete resistor component). The coil resistance is shown in Figure 22–3 as r_s, indicating series resistance.

2. Because the reactances cancel each other, the Z is minimum and equals the R in the circuit.

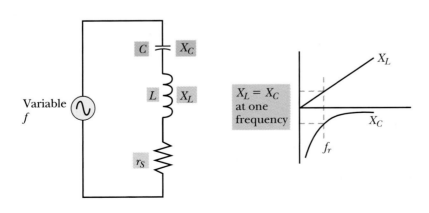

FIGURE 22–2 $X_L = X_C$ at resonant frequency (f_r)

FIGURE 22–3 Sample series circuit parameters at resonance

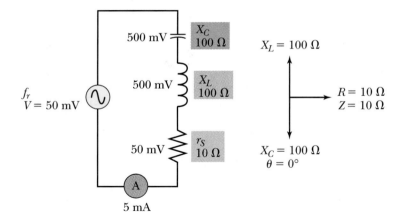

3. When Z is minimum ($Z = R$), I is maximum. That is, $I = V/R$.

4. V_C and V_L are maximum since I is limited only by the R value and is therefore maximum. The voltage across the capacitor equals $I \times X_C$, and the voltage across the inductor equals $I \times X_L$. As noted in our sample circuit, the voltage across the coil and the capacitor is 10 times V applied. There is a *resonant rise* in voltage across the reactive components. We will examine this fact again later in the chapter.

5. Again, because the reactances cancel each other, the circuit acts purely resistive; therefore, phase angle (θ) is zero.

IN-PROCESS LEARNING CHECK I

Fill in the blanks as appropriate.

1. In circuits containing both inductance and capacitance, when the source frequency increases, inductive reactance _____ and capacitive reactance _____.

2. In a series RLC (or LC) circuit, when X_L equals X_C, the circuit is said to be at _____.

3. In a series RLC (or LC) circuit, when X_L equals X_C, the circuit impedance is _____; the circuit current is _____; and the phase angle between applied voltage and current is _____ degrees.

4. In a series RLC (or LC) circuit, when X_L equals X_C, the voltage across the inductor or the capacitor is _____.

Since you now know some of the series resonance effects, you may wonder how the frequency of resonance is determined.

Let's deviate from examining circuit parameters to look at the basic formulas that allow you to determine the resonant frequency for a circuit with given component values, or to find the appropriate component values for a desired resonant frequency.

22–3 THE RESONANT FREQUENCY FORMULA

The basic formula

In our previous discussions, a fixed value of L and C were assumed, while f was changed until X_L equaled X_C. With fixed L and C values, there is only one possible resonant frequency. In many practical applications, either the C or L value or both can be varied to achieve resonance at various frequencies. An example is tuning in different radio stations on a receiver. Also, different **L/C ratios** for any given frequency can be obtained. You can see that a technician should know how to determine the resonant frequency of any given set of LC parameters. Let's look briefly at the derivation and formula for frequency of resonance. Starting with resonance $X_L = X_C$, where L is in henrys, C is in farads, and f is in Hz, the derivation is $X_L = X_C$.

$$2\pi fL = \frac{1}{2\pi fC}$$

$$2\pi f_r L = \frac{1}{2\pi f_r C}$$

$$2\pi L (f_r)^2 = \frac{1}{2\pi C}$$

$$f_r^2 = \frac{1}{(2\pi)^2 LC}$$

FORMULA 22–1 $\quad f_r = \dfrac{1}{2\pi\sqrt{LC}}$ OR $\dfrac{0.159}{\sqrt{LC}}$

Example

What is the resonant frequency of a series LC circuit having a 200-μH inductor and a 400-pF capacitor?

$$f_r = \frac{1}{2\pi\sqrt{LC}}$$

$$f_r = \frac{1}{6.28 \times \sqrt{200 \times 10^{-6} \times 400 \times 10^{-12}}}$$

$$f_r = \frac{1}{6.28 \times \sqrt{80{,}000 \times 10^{-18}}}$$

$$f_r = \frac{1}{1{,}776 \times 10^{-9}} = 562.98 \text{ kHz}$$

Now you try one!

PRACTICE PROBLEMS I

What is the resonant frequency of a series LC circuit containing an inductor with an inductance of 250 μH and a capacitor with a capacitance of 500 pF?

Variations of the f_R formula

As you can see from the basic resonant frequency formula, the L and C values determine the frequency of resonance. That is, the larger the LC product is, the lower is the f_r, and vice versa, Figure 22–4a. Note also in Figure 22–4b how various combinations of L and C can be resonant at one particular frequency (in this case, 1 MHz).

It is frequently useful to find the L value needed with an existing C value or the C value needed with an existing L value to make the circuit resonant at some specified frequency. By rearranging the resonant frequency formula (i.e., squaring both sides to delete the square root portion), we can find L or C, if the other two variables are known. When this is done, the following useful formulas are derived:

$$L = \frac{1}{(2\pi)^2 f_r^2 C} = \frac{1}{4\pi^2 f_r^2 C} =$$

FORMULA 22–2 $\quad L = \dfrac{0.02533}{f_r^2 C}$

$$C = \frac{1}{(2\pi)^2 f_r^2 L} = \frac{1}{4\pi^2 f_r^2 L} =$$

FORMULA 22–3 $\quad C = \dfrac{0.02533}{f_r^2 L}$

NOTE: These formulas are for your information—You need not memorize them!

L (μH)	C (pF)	f_r (MHz)
100	100	1.59
100	200	1.13
400	400	0.398
400	10	2.52

(a)

$$\left(\text{Using } f_r = \frac{1}{2\pi\sqrt{LC}} \text{ OR } \frac{0.159}{\sqrt{LC}}\right)$$

Different LC products cause different f_rs

L (μH)	C (pF)	f_r (MHz)
500	51	1
400	63.75	1
300	85	1
200	127.5	1

(b)

$$\left(\text{Using } f_r = \frac{1}{2\pi\sqrt{LC}} \text{ OR } \frac{0.159}{\sqrt{LC}}\right)$$

Different values having same LC product yield same f_r

FIGURE 22–4 (a) Values of L and C determine f_r and **(b)** different combinations of L and C can produce same f_r.

Other useful variations

When dealing in radio frequency circuits, it is sometimes inconvenient to manage henrys, farads, and hertz, as in the previous formulas. It is often more convenient to deal with microhenrys (μH), picofarads (pF), and megahertz (MHz). The following formulas are based on these convenient units.

$$f^2 = \frac{25{,}330}{LC} \text{ or}$$

FORMULA 22–4 $L = \dfrac{25{,}330}{f^2 C}$

where L is in microhenrys, C is in picofarads, f is in megahertz or

FORMULA 22–5 $C = \dfrac{25{,}330}{f^2 L}$

NOTE: Again, these formulas are for your information, but you may find occasion to use them.

Example

A circuit has an L value of 100 μH. What C value will make the circuit resonant frequency equal 1 MHz?

$$\text{Using } C = \frac{0.02533}{f_r^2 L}$$

where L is in henrys, C in farads, f in hertz.

$$C = \frac{0.02533}{(1 \times 10^6)^2 \times 100 \times 10^{-6}}$$

$$C = \frac{0.02533}{1 \times 10^{12} \times 100 \times 10^{-6}}$$

$$C = \frac{25{,}330 \times 10^{-12}}{100} = 253.3 \text{ pF}$$

$$\text{Using } C = \frac{25{,}330}{f^2 L}$$

where L is in microhenrys, C is in picofarads, and f is in megahertz.

$$C = \frac{25{,}330}{f^2 L} = \frac{25{,}330}{100} = 253.3 \text{ pF}$$

Incidentally, you can also use the X_C equals X_L approach. Since you know the frequency and the L, you can calculate X_L using the $X_L = 2\pi f L$ formula. Then,

Using basic units: $L =$ H, $C =$ F, and $f_r =$ Hz $$L = \frac{0.02533}{f_r^2 C} \text{ and } C = \frac{0.02533}{f_r^2 L}$$
Using convenient units: $L = \mu$H, $C =$ pF, and $f_r =$ MHz $$L = \frac{25{,}330}{f_r^2 C} \text{ and } C = \frac{25{,}330}{f_r^2 L}$$
Using the $X_L = X_C$ approach: If you know X_C, then $L = \dfrac{X_C}{2\pi f_r}$ If you know X_L, then $C = \dfrac{1}{2\pi f_r X_L}$

FIGURE 22–5 Some useful variations derived from the f_r and reactance formulas

knowing that X_C is that same value at resonance, you use the X_C formula to solve for C. That is,

$$C = \frac{1}{2\pi f X_C}$$

See Figure 22–5.

Now you apply the appropriate formula(s) to solve for an unknown L when C and f_r are known.

PRACTICE PROBLEMS II

What L value is needed with a 100-pF capacitor to make the circuit resonant frequency equal 5 MHz?

22–4 SOME RESONANCE CURVES

Figure 22–6 shows the resonance curves (graphic plots) of current, impedance, and phase angle versus frequency for a series RLC circuit.

Plot of current (I) versus frequency (f)

When frequency is considerably below the circuit resonant frequency, the current is low compared with current at resonance. This is because $Z = \sqrt{r^2 + (X_C - X_L)^2}$; therefore, circuit impedance (Z) is much higher than it is at resonance. For series RLC circuits below resonance, the net reactance is capacitive reactance. The same pattern is true above resonance, except the net reactance for a series RLC circuit is X_L. (NOTE: For series resonant circuits, plotting current rising to maximum value at resonance is the most significant aspect of the response curves.)

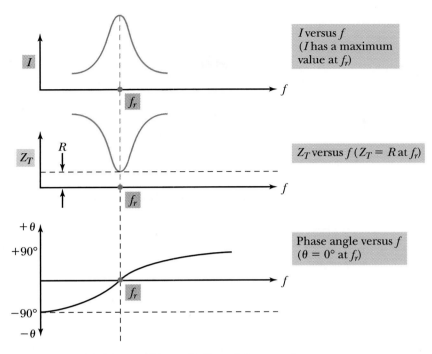

FIGURE 22–6 Plots of *I, Z,* and *θ* versus frequency

Plot of impedance (Z) versus frequency

As you would expect, impedance is lowest at the resonant frequency, since the reactances cancel each other at resonance. The impedance graph in Figure 22–6 shows that Z equals R (resistance) at resonance. Thus, for any circuit, when Z is low, I is high and when Z is high, I is low.

Plot of phase angle versus frequency

At resonance, the circuit acts resistively and the phase angle is $0°$. (This is because the reactances have cancelled.) Below resonance the circuit acts capacitively, since the net reactance is capacitive reactance. Above resonance, the circuit acts inductively since the net reactance is X_L.

22–5 Q AND RESONANT RISE OF VOLTAGE

Q and the response characteristic

The shape of a given circuit's resonance response relates to an important factor called the **Q factor,** or figure of merit, Figure 22–7. The higher the Q factor is, the sharper is the response characteristic. This sharp response curve is sometimes described as having steep **skirts,** or as indicating a circuit with high **selectivity** (i.e., one that is very frequency selective). The lower the Q factor is, the broader, or flatter the response curve and the less selective the circuit is.

FIGURE 22–7 Response curves versus circuit Q

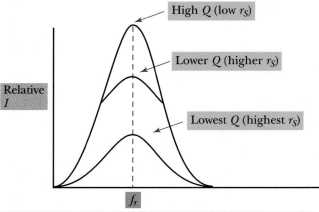

NOTE: r_S value does not change f_r but does affect response.

$$Q = \frac{X_L}{r_S}$$

Q equals X_L/R or X_C/R

The Q factor of a resonant circuit relates to two parameters. The first parameter is associated with the ratio of reactance to resistance at the resonant frequency. In formula form, $Q = X_L/r_s$ or X_C/r_s at resonance, where r_s is the series resistance in the circuit, which often is the coil resistance. In other words, the Q of the coil (X_L over R) is frequently the same as the Q of the resonant circuit. Obviously, the lower the series circuit resistance is, the higher the Q is and the sharper the resonant circuit response. The higher the r_s is, the lower the Q is and the flatter the response. Note that since I is the same throughout the circuit, the ratio

$$\frac{I \times X_L}{I \times R}, \text{ or } \frac{V_L}{V_R}$$

is also the same as the Q factor. Likewise, the ratio of reactive power to true power in the circuit also equals the Q factor.

Q related to L/C ratio

A second related factor that affects the Q is the ratio of L to C; that is L/C. Since both L and C values affect the frequency of resonance, a *given* resonant frequency can be achieved by various LC values, refer again to Figure 22–4b. However, different combinations of LC values have different reactances at resonance. Therefore, it is best to have a *high* L/C ratio to achieve a high X_L at resonance; hence, a high X_L/R or Q factor, Figure 22–8.

Resonant rise of voltage

Refer to Figure 22–9 and note the voltages around the series circuit. The term *magnification factor* describes the fact that voltage across each of the reactive components at resonance is Q times V applied. This is because of the resonant rise of I at resonance, which is due to the reactances cancelling each other at the resonant frequency. Since I is limited only by R at that frequency, and since each re-

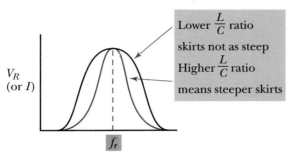

FIGURE 22–8 *Q* related to *L/C* ratios

Lower $\frac{L}{C}$ ratio skirts not as steep

Higher $\frac{L}{C}$ ratio means steeper skirts

V_R (or I)

f_r

(Same I since r_S assumed to be constant)

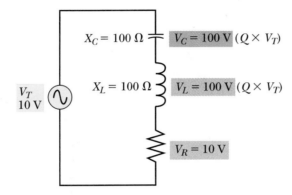

FIGURE 22–9 Magnification factor

$X_C = 100\ \Omega$ — $V_C = 100\ \text{V}$ $(Q \times V_T)$

$X_L = 100\ \Omega$ — $V_L = 100\ \text{V}$ $(Q \times V_T)$

V_T 10 V

$V_R = 10\ \text{V}$

active voltage drop is I times X with X being very high, the reactive voltage can be very high. In fact, at resonance, the reactive voltages are Q times V applied; thus, voltage has been magnified (Q times) across the reactive components. In other words, the higher the Q is, the greater the magnification is. As stated, the ratio of reactive voltage to applied voltage at resonance equals the Q. If we measure the voltage across one reactive component and call it V_{out} and call V applied V_{in}, then

$$Q = \frac{V_{\text{out}}}{V_{\text{in}}}$$

PRACTICE PROBLEMS III

1. What is the Q of the circuit in Figure 22–10?

2. What is the V_C value at resonance, if the applied voltage in Figure 22–10 is 12 V?

3. If r_s doubles in this circuit, what is the Q value?

4. If r_s is a smaller value than shown in Figure 22–10, are the response curve skirts more or less steep?

22–6 PARALLEL RESONANCE CHARACTERISTICS

A parallel resonant circuit is a frequently used circuit in receivers, transmitters, and frequency measuring devices. As you might expect, many parallel resonant

FIGURE 22-10

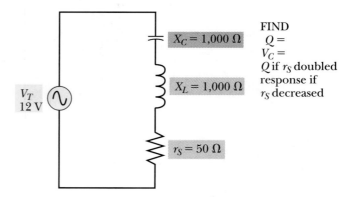

FIND
$Q =$
$V_C =$
Q if r_S doubled
response if
r_S decreased

circuit features are opposite to those of a series resonant circuit. Look at Figure 22–11 as you study the following discussion.

Highlighting the basic features of a high-Q parallel resonant circuit:

1. $Z =$ maximum (and is purely resistive)
2. Line current = minimum
3. Current inside *tank* = maximum
4. $X_L = X_C$ (ideal components)
5. $\theta = 0°$ (and *p.f.* = 1)

Let's examine these statements further.

1. Z is maximum at resonance because the inductive and capacitive branch currents are 180° out of phase, thus cancelling each other (assuming ideal L and C components). This means the vector sum of the branch currents (or I line) is very small. If the inductor and capacitor are perfect, the Z is infinite. Since

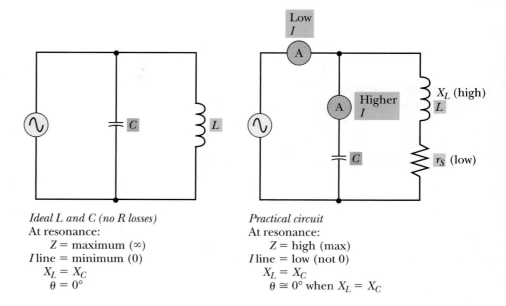

FIGURE 22-11 Parallel resonant circuit characteristics

Ideal L and C (no R losses)
At resonance:
  $Z =$ maximum (∞)
I line = minimum (0)
  $X_L = X_C$
  $\theta = 0°$

Practical circuit
At resonance:
  $Z =$ high (max)
I line = low (not 0)
  $X_L = X_C$
  $\theta \cong 0°$ when $X_L = X_C$

practical inductors and capacitors have resistances and losses, there is an effective R involved, Figure 22–11. (For your information, $Z_T = L/C_{r_s}$.)

2. Since Z is maximum at resonance, obviously, the line current (or total current supplied by the source) is minimum.

3. Within the LC (tank) circuit, the circulating current can be very high. In fact, the current inside the tank circuit equals $Q \times I$ line and the total circuit impedance is $Q \times X_L$.

4. When $X_L = X_C$, the preceding characteristics are exhibited. In practical parallel resonant circuits having less-than-perfect components, some r_s value associated with the coil exists. This means when $X_L = X_C$, the Z_L of the inductive branch is slightly greater than the X_C of the capacitive branch. In low-Q circuits, this means the frequency needs to be slightly lower to allow more I_L, so the inductive branch current equals the capacitive branch current. For circuits having a Q above 10, however, this effect is negligible.

5. At resonance (for high-Q circuits), circuit current and applied voltage are virtually in phase.

PRACTICAL NOTES

For low-Q circuits, the condition of unity power factor occurs at a slightly different frequency than when $X_L = X_C$. The frequency of $p.f. = 1$ is sometimes termed antiresonance, or the antiresonant frequency.

IN-PROCESS LEARNING CHECK II

1. In a high-Q parallel resonant circuit, Z is _____.
2. In a high-Q parallel circuit using ideal components, X_L _____ equal X_C at resonance.
3. The power factor of an ideal parallel resonant circuit is _____.
4. Phase angle in an ideal parallel resonant circuit is _____.
5. In a parallel resonant circuit, the current circulating in the tank circuit is _____ than the line current.

22–7 PARALLEL RESONANCE FORMULAS

For Q greater than 10

When the circuit Q is high (when the r_s is low, or the equivalent R_p is high), the resonant frequency is determined by the LC component values as it is in series resonant circuits. With Qs greater than 10, the resonant frequency is calculated with the same formula used for series resonance. That is,

$$f_r = \frac{1}{2\pi\sqrt{LC}}$$

For Q less than 10

When the circuit Q is low (i.e., when r_s is high or effective R_p is low, or when a reflected impedance from a load coupled to the tuned circuit is low causing an effective low value of equivalent R_p), the R, L, and C values enter in the formula for parallel resonance. That is, the resistance has a significant effect on the frequency of resonance so it should be considered. The following formula (only for your information) illustrates that resistance affects resonance in parallel LC circuits.

FORMULA 22–6
$$f_r = \frac{1}{2\pi\sqrt{LC}} \times \sqrt{1 - \frac{C_{r_S}^2}{L}}$$

NOTE: Formula 22–6 is given as information only to show that resistance has effect on resonant frequency in low-Q, parallel resonant circuits. No computations using this formula will be required in this text. In most cases, the Q of parallel tuned circuits is high enough so that the effect of the resistance on the resonant frequency can be generally neglected.

To check your grasp of parallel resonant circuits thus far, try to solve the following problems.

PRACTICE PROBLEMS IV

1. Calculate the resonant frequency of a parallel LC circuit with the following parameters: $L = 200$ µH having a resistance of 10 Ω and $C = 20$ pF.

2. What is the Q of the circuit in question 1?

3. What is the impedance at resonance of the circuit in question 1?

4. Is the resonant frequency of the circuit in question 1 lower or higher if $r_s = 1$ kΩ?

22–8 EFFECT OF A COUPLED LOAD ON THE TUNED CIRCUIT

Whenever a parallel resonant circuit delivers energy to a load (i.e., a load is coupled to the tuned circuit), the Q of the tank circuit is affected, Figure 22–12. If the power dissipated by the coupled load is greater than 10 times the power lost in the tank circuit reactive components, the tank impedance is high compared to the load impedance. Thus, for all practical purposes, the impedance of the parallel combination of the tank and load virtually equals the load resistance value. When a tank is heavily loaded by such a resistive load impedance, the Q of the system equals R over X, or $Q = R/X$, where R is the parallel load resistance and X is the reactance. (NOTE: This is the inverse of the X_L/R formula used for Q in series resonance and high-Q parallel resonant circuits.) For example, if a resistive load of 4,000 Ω is connected across a parallel resonant circuit where X_C (or X_L) = 400 Ω,

$$Q = \frac{R}{X_L} = \frac{4,000}{400} = 10$$

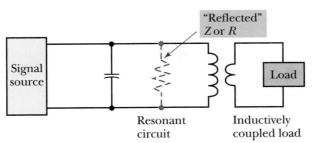

FIGURE 22–12 Effect of coupling a load to a parallel resonant circuit

The lower the reflected R value, the lower the Q. If load is "heavy" (i.e., much energy absorbed by load) $Z_T \approx Z$ of load and $Q = \dfrac{R_p}{X_L}$.

22–9 Q AND THE RESONANT RISE OF IMPEDANCE

The preceding discussions show us that the current characteristic is the most meaningful response factor in series resonant circuits, where the impedance characteristic is the most significant response factor in parallel resonant circuits. The response characteristics of a parallel tuned circuit are also related to the Q factor. In Figure 22–13 you can see there is a resonant rise of impedance as the resonant frequency is approached. As we previously stated, the impedance at resonance is $Q \times X_L$. You can also see from this illustration that the higher the Q is, the sharper the response characteristics, and the lower the Q is, the flatter and broader the response.

22–10 SELECTIVITY, BANDWIDTH, AND BANDPASS

Selectivity

The sharper the response curve of a resonant circuit, the more selective it is. This means that only frequencies close to resonance appear near the maximum response portion of the resonant circuit response. This indicates that in series resonant circuits, the current value *falls off* rapidly from its maximum value at frequencies slightly below or above the resonant frequency. In parallel resonant circuits, the impedance *falls off* rapidly from the maximum value obtained at resonance. Conversely, resonant circuits that have less selectivity allow a wider band of frequencies close to their maximum response levels. Selectivity, used to de-

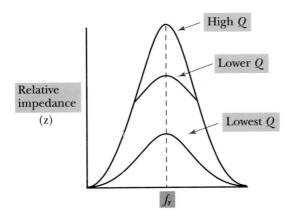

FIGURE 22–13 Resonant rise of Z for a parallel resonant circuit

FIGURE 22–14 Examples of various selectivities

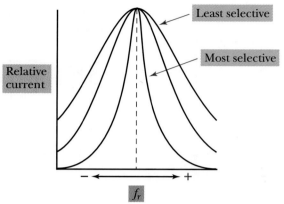

scribe a resonant circuit, refers to the capability of the circuit to differentiate between frequencies. Refer to Figure 22–14 for examples of various selectivities versus response characteristics.

Bandwidth and bandpass

You have looked at various response curves and have noticed that none have perfect vertical skirts. Therefore, to define **bandwidth,** some value or level on the response curve is chosen that indicates all frequencies below that level are rejected and frequencies above the chosen value are selected. The accepted level is 70.7% of maximum response level.

For series resonant circuits, we consider those two points on the response curve where current is 70.7% of maximum. For parallel resonant circuits, the points where Z is 70.7% of maximum are used.

Total bandwidth *(BW)* is calculated as:

FORMULA 22–7 $\Delta f = f_2 - f_1$

where: $\Delta f = BW$
$f_2 = $ higher frequency 0.707 point
$f_1 = $ lower frequency 0.707 point on response curve
Also, $f_1 = f_r - \dfrac{\Delta f}{2}$, and $f_2 = f_r + \dfrac{\Delta f}{2}$.

Notice that if $P = I^2 R$ and at f_1 and f_2 the current level is 0.707 of maximum, then at those points on the response curve, $P = (0.707\ I_{max})^2 R = 0.5\ P_{max}$. For this reason, these points (f_1 and f_2) are often called *half-power points,* Figure 22–15.

PRACTICAL NOTES

The response curves are not perfectly symmetrical above and below resonance. Therefore, the f_1 and f_2 expressions are approximations. They are close to true for very high Q circuits.

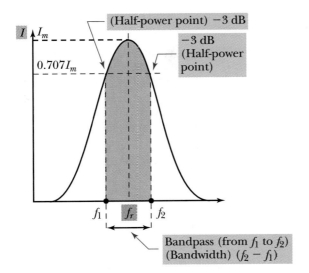

FIGURE 22–15 Bandwidth and bandpass

Bandwidth related to Q

You have learned the Q of a resonant circuit is the primary factor determining the response curve shape. Thus, BW (Δf) and Q are related. The formula showing this relationship is:

> **FORMULA 22–8** $\Delta f = \dfrac{f_r}{Q}$

Rearranging this formula, we can also say that:

> **FORMULA 22–9** $Q = \dfrac{f_r}{\Delta f}$

Parallel resonant circuit damping

Parallel resistance (R_p) can be used to "damp," lower Q, and broaden the bandwidth of a parallel resonant circuit.

Recall that series resistance (R_s), typically in the inductor branch, lowers Q. The higher the series resistance present, the lower the Q. In contrast, however, when parallel resistance is present across a parallel resonant circuit (R_p), the lower the parallel resistance, the lower the Q for the circuit. This parallel resistance may be from reflected impedance loading effects (mentioned earlier), or from purposely placing a resistance in parallel with the circuit, for damping purposes.

Bandpass

As has been discussed, the bandwidth is the width of the band of frequencies where a given resonant circuit responds within specified criteria. For example, a given resonant circuit responds over a 20-kHz bandwidth with levels at or above 70.7% of maximum response level.

On the other hand, **bandpass** is the specific frequencies at the upper and lower limits of the bandwidth. For example, if the resonant frequency is 2,000 kHz, and the circuit has a bandwidth of 20 kHz, the bandpass is from 1,990 to 2,010 kHz (assuming a symmetrical response). This indicates the specific band of frequencies passed is from 1,990 to 2,010 kHz, and frequencies outside this limit cause a response that falls below the 70.7% level, or below the half-power points (sometimes expressed as the -3 dB points, as you will learn later). Again, Figure 22–15 illustrates bandwidth and bandpass.

Example

1. What is the total bandwidth of a resonant circuit if f_r is 1,000 kHz and f_1 is 990 kHz?

   Since one half-power point is 10 kHz below the resonant frequency, it is assumed that if the response curve is symmetrical, the other half-power point is 10 kHz above the resonant frequency. This means the total $BW = 20$ kHz. That is, $f_1 = 990$ kHz and $f_2 = 1,010$ kHz.

2. What is the bandwidth of a resonant circuit if resonant frequency is 5 kHz and Q is 25?

$$\Delta f = \frac{f_r}{Q} = \frac{5,000}{20} = 250 \text{ Hz}$$

3. What is the bandpass of the circuit described in question 2?

Bandpass = from $5,000 - 125$ Hz to $5,000 + 125$ Hz = 4,875 to 5,125 Hz.

PRACTICE PROBLEMS V

1. In a resonant circuit, f_r is 2,500 kHz and Δf is 12 kHz. What are the f_1 and f_2 values?

2. In a resonant circuit, Q is 50 and bandwidth is 10 kHz. What is the resonant frequency and the bandpass?

3. What is the f_1 value in a circuit whose resonant frequency is 3,200 kHz and whose bandwidth is 2 kHz?

22–11 MEASUREMENTS RELATED TO RESONANT CIRCUITS

Because resonant circuits are so prevalent in electronics, it is worthwhile to know some of their practical measurements. These measurements include measuring resonant frequency; determining Q and magnification factor; determining maximum current (for series resonant circuits) and maximum impedance (for parallel resonant circuits); determining bandwidth and bandpass; measuring the *tuning range* of a tunable resonant circuit (with variable C or L); and determining the tuning ratio. Let's briefly look at each of these measurements.

1. Measuring resonant frequency
   There are several ways to determine resonant frequency. Of course, if you are trying to measure at what frequency a tunable signal source is operating

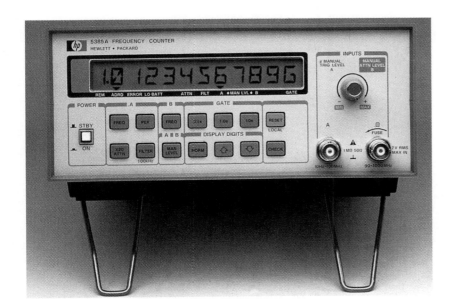

FIGURE 22–16 A device used to determine f_r. *(Courtesy of Hewlett-Packard Co.)*

(what frequency it is tuned to), a digital frequency counter coupled to the output circuit might be used for direct frequency measurement, Figure 22–16. One way to get an approximate resonant frequency measurement of a "stand-alone" *LC* circuit is to use a metering device that reacts to circuit conditions. One type of metering device is called a tunable "dip" meter. This type device is used with "power-off" conditions to determine the resonant frequency for a given *LC* circuit. When these tunable devices are near (usually inductively coupled) to *LC* circuits, as the tuned circuit in the dip meter is tuned near to the resonant frequency of the *LC* circuit, the *LC* circuit absorbs energy from the dip meter circuit and causes the meter reading to "dip" downward. The frequency at which the dip occurred is then read from the calibrated dial on the dip meter. Other means may include field strength meters or oscilloscopes to monitor frequency outputs of signal sources. It is imperative that the technician be aware of the possibility of "detuning" the circuit under test, and use the proper test probes and/or coupling methods with the scope, or other test equipment to minimize detuning effects.

Keep in mind the fact that there are various means for checking resonant frequencies of circuits. As you continue in your studies, you will probably have opportunity to get some practical experience in using some of these devices, and you will learn the proper techniques for making these measurements.

2. Determining Q and magnification factor

In a series resonant circuit, using a high input impedance DVM, or an oscilloscope allows the measurement of voltage across one reactive component (generally, the capacitor). When the circuit is tuned to resonance, the voltage is maximum. This voltage is then compared with the voltage source to determine the Q and the voltage magnification factor. That is,

$$Q = \frac{V_{\text{out}}}{V_{\text{in}}}$$

See Figure 22–17.

FIGURE 22–17 Determining
Q of series resonant circuit

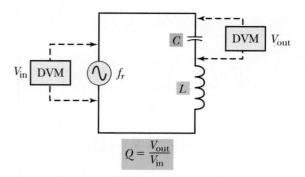

$$Q = \frac{V_{\text{out}}}{V_{\text{in}}}$$

3. Determining maximum current (I) for series and maximum impedance (Z) for parallel resonant circuits

In a series resonant circuit, a current meter or a voltmeter reading across a series resistor is used to determine maximum current. Read the current meter directly, or use Ohm's law where

$$\frac{V_R}{R} = I$$

to determine current when the circuit is tuned for maximum current or resonance, Figure 22–18.

In a parallel resonant circuit, use a variable R in series with the source and adjust it so that V_{TANK} across the tuned circuit and V_R across the variable R are equal (at resonance). Remove the R from the circuit and measure its value. Its value is the same as the Z of the tuned circuit at resonance since the V drops were equal. Since we can calculate X_L from the resonant frequency formula and we know that

$$Q = \frac{Z_T}{X_L}$$

we can determine the Q, Figure 22–19.

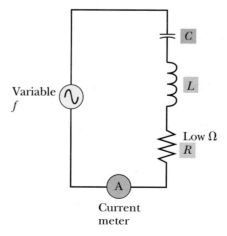

Current
meter

FIGURE 22–18 Determining
maximum I for a series
resonant circuit

Use a current meter, *or*, insert low R value, measure V_R, and then calculate I. $\left(I = \dfrac{V_R}{R}\right)$

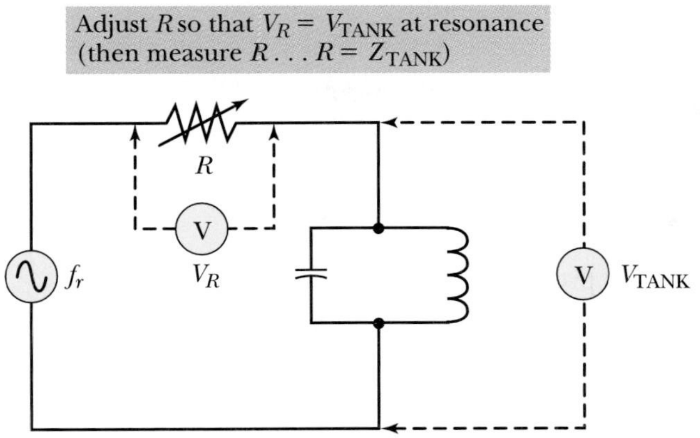

Adjust R so that $V_R = V_{TANK}$ at resonance
(then measure $R \ldots R = Z_{TANK}$)

FIGURE 22–19 Determining maximum Z for a parallel resonant circuit

4. Determining bandwidth and bandpass

For a series resonant circuit, tune the circuit through resonance and note the frequencies where current is 70.7% of maximum value. The difference between the lower frequency and the higher frequency is the bandwidth. The actual frequencies at those two points represent the bandpass.

For a parallel resonant circuit, you can use the Z determining system and find the frequencies where Z is 70.7% of maximum value. Again, the difference between those frequencies is the bandwidth, and the precise lower and upper frequencies are the bandpass, Figure 22–20.

5. Measuring the tuning range

If the tuned circuit has a variable C or L, or if both are variable, there are limits to the range of resonant frequencies to which the circuit can be tuned.

The various resonant frequency measurement devices previously mentioned can check the actual resonant frequencies that are capable of being tuned by the circuit. Simply measure the frequency at both ends of the tuning extremities, and you will know the circuit tuning range.

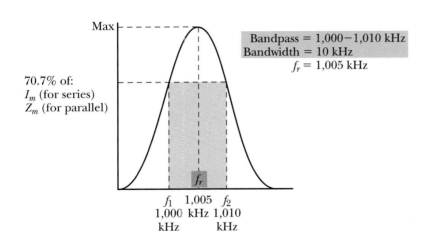

70.7% of:
I_m (for series)
Z_m (for parallel)

Bandpass = 1,000−1,010 kHz
Bandwidth = 10 kHz
f_r = 1,005 kHz

f_1 1,005 f_2
1,000 kHz 1,010
kHz   kHz

FIGURE 22–20 Determining bandwidth and bandpass

6. Determining the tuning ratio

From the resonant frequency formula, you can reason that the frequency change caused by a change in C or L is inversely proportional to the square root of the *change* in C or L. That is,

$$f_r = \frac{1}{2\pi\sqrt{LC}}$$

therefore, f_r is proportional to $1/\sqrt{LC}$.

If it is desired to tune through a range of frequencies from 1,000 to 6,000 kHz, this means this tuning ratio of 6:1 (highest to lowest frequency) requires a change of C or L in the ratio of 36:1. If C is the variable component and its minimum C value is 10 pF, it has to have a maximum C value of 36×10, or 360 pF to tune the desired range.

///

SAFETY HINTS

When working around equipment with resonant circuits, particularly in transmitter circuits, be aware that it is possible to get bad *rf burns* by touching or getting too near a live circuit or component. In addition, these circuits often have high dc voltages, as well as high radio frequency ac energy. If touched while the circuit is live, injury or death can result!

22–12 SPECIAL NOTATIONS ABOUT SINGLE COMPONENTS

Distributed capacitance

Distributed capacitance is present in most circuits. It exists anywhere there are two conductors separated by an insulator or dielectric material. For example, there is distributed capacitance between conductor paths on printed circuit boards. Distributed capacitance is also present between wires and chassis in "hard-wired" circuits, and so on.

Self-resonance of an inductor

It is important to know that inductors also have distributed capacitance, Figure 22–21a. Because each turn in an operating coil is at a slightly different potential, each pair of turns effectively creates a "parasitic" capacitor. The result of this effective distributed capacitance is that at some frequency, the effective capacitance of the coil will have a reactance equal to the inductive reactance of the inductor. At this frequency, the inductor will show "self-resonance." At frequencies above this self-resonant frequency, the coil will begin to act like a capacitor, rather than an inductor. Typically, this self-resonance effect does not hinder inductor operation except at higher frequencies.

Distributed or stray inductance

Anytime there are current-carrying conductors, inductance will be present. The changing flux caused by changing current levels will induce voltage into the conductor itself, or any conductors that are close to, and linked by, the magnetic field. This means that capacitors have some inductance, because their leads and plates are current-carrying conductors. Capacitors whose plates are wrapped in a coil, such as paper and electrolytic types, exhibit more inductance than other types, such as mica or ceramic capacitors. Resistors also have a measure of inductance, since their leads, plus the resistive body of the resistor, represent current-carrying conductors.

Component equivalent circuits

Observe the equivalent circuits for inductors, capacitors, and resistors, shown in Figure 22–21b, c, and d. The physical construction, size, and so on of these com-

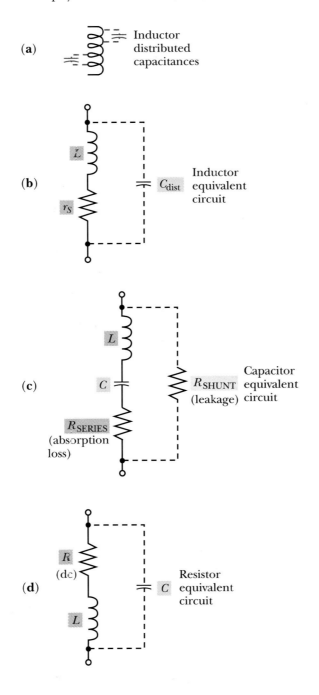

FIGURE 22–21

ponents determines how much of the various characteristics a given component will exhibit. Be aware that circuit designers must consider both circuit layouts and single-component characteristics when designing high-frequency circuits.

22–13 FILTER APPLICATIONS OF NONRESONANT AND RESONANT RLC CIRCUITS

Introduction

Since you have learned about the sensitivity to frequency of reactive components and various resonance effects, it is natural to provide an introduction to filters. This brief section about filters will introduce some basic filter networks and their common applications. Two general filter classifications that will be discussed are nonresonant filters and resonant filters.

Another way filters are categorized is by the frequency range where they are used. For example, power supply filters typically operate at frequencies of 60 Hz, 120 Hz, 400 Hz, and so forth. Audio filtering networks obviously operate in audio frequency environments. *Rf* filters are used at frequencies above the audio range in receivers, transmitters, and other *rf* systems.

Names of filter networks or individual filter components often reflect what they do; for example, low-pass, high-pass, bandpass, bandstop, smoothing, decoupling, and so on. Another common way to refer to filter networks is the way the components are *laid out;* for example, L-shaped, pi-shaped, T-shaped, and so on. With these few classifications in mind, let's look at some filter networks.

A general principle

Remember this general principle while analyzing any type filter: A device or circuit that has low opposition to a given frequency or band of frequencies and is in series with the output or load passes that frequency or band of frequencies along to the output or load. A device or circuit that has a low opposition to a given frequency or band of frequencies and is in parallel (or shunt) with the output or load bypasses or prevents that frequency or band of frequencies at the output or load. The converse of these two statements is also true. High oppositions in series with the signal path prevent signals from arriving at the load. High oppositions in parallel allow signals at the output or load.

Nonresonant filters

Low-pass filters

A **low-pass filter** passes along low frequency components of a given signal or waveform, while impeding the passage of higher frequency components. For example, look at the simple *RC* filter network in Figure 22–22. Notice that the output is being taken across the capacitor. The voltage distribution of the signal from the source will depend on the relative values of R and X_C of the capacitor. If the X_C is high compared to the R, then a greater portion of the signal will be across the capacitor, and the network "output" signal will be greater. If the X_C is low with respect to the R, the resistor drops most of the signal and the output is very low. Since the capacitor is frequency-sensitive, its X_C will be high at low frequencies and low at high frequencies.

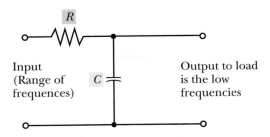

FIGURE 22–22 Example of a low-pass filter

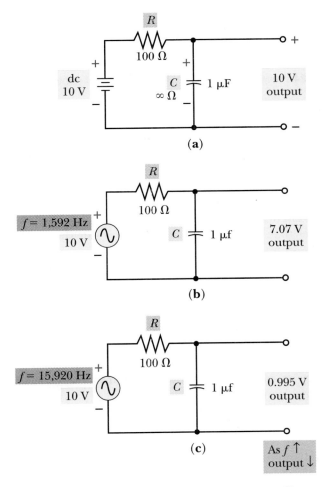

FIGURE 22–23 Example of simple *RC* low-pass filter operation

In Figure 22–23a, the source frequency is dc (or 0 Hz). The X_C is virtually infinite and almost all the input voltage appears across the capacitor or is present at the output. In Figure 22–23b, the source frequency is such that R and X_C are equal; thus, the output signal and the voltage drop across the R are equal. (Don't forget, these voltages are out-of-phase; they would have to be added vectorially to equal the source voltage value.) In Figure 22–23c, the frequency is high, thus the X_C is low, and the output voltage is low since the R drops most of the signal

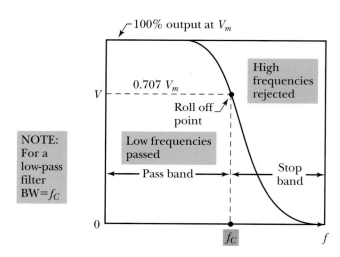

FIGURE 22–24 Frequency response characteristics of a low-pass filter

voltage. As you can see, this "low-pass" filter has caused the output voltage to be maximum at low frequencies and minimum at high frequencies. Thus, it is called a low-pass filter! The frequency response of a low-pass filter is shown in Figure 22–24. NOTE: f_c stands for "cutoff frequency." That is the 0.707 of V_{in} point where the output starts declining greatly. The steepness of signal fall off depends on the type of filter circuit and its components. Various circuit and component configurations, other than the simple RC circuit previously shown, that provide low-pass filtering are shown in Figure 22–25.

Several applications for low-pass filters

Power supply filters are low-pass filters used to pass along the dc voltage and current to power supply loads, which generally require as "clean" a dc as possible. A large part of this type filter's job is to filter out any 60-Hz ac components present before their getting to the filter's output and to the power supply loads.

Audio frequency low-pass filters are used in systems in which there are both rf and audio frequency signals present, and it is desired to pass along only the audio spectrum signals to succeeding stages in the system. A classic example of this appears in radio receivers, just before the audio amplifier stages in the receiver.

Low-pass filters may also be used to filter out undesired higher-frequency RF signals from desired lower-frequency RF signals. For example, a transmitter operating at several megahertz may use a low-pass filter to ensure that the desired frequencies from the transmitter arrive at the antenna, while not permitting higher frequency harmonics or spurious signals that might interfere with TV receivers, etc., to get past the filter circuit.

Other devices that may require filtering to prevent unwanted "high-frequency interference" include telephones, high-fi amplifiers, and p.a. systems. For these types of systems the interference may be coming from high-frequency electromagnetic interference (called EMI) from nearby radio stations, etc., or the unwanted signals may be getting into the system via the power lines, telephone lines, and so on. In many cases, good filters can alleviate the problem.

Some systems we use can be either the victims of interference, or the cause of interference, or both. Computers and all the microprocessor-controlled sys-

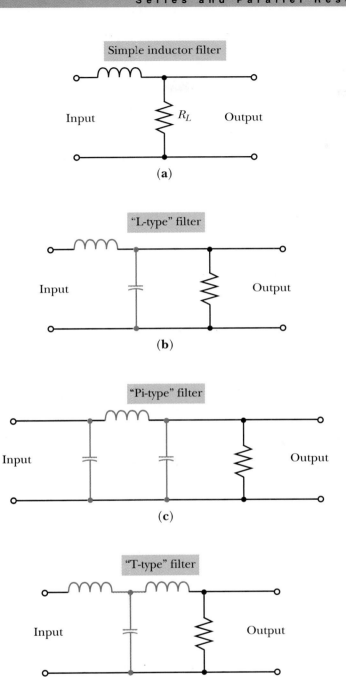

FIGURE 22–25 Examples of circuit configurations for low-pass filters

tems used around our homes are examples of this scenario. As you can see, filters have an important role in today's electronic society.

High-pass filters

A **high-pass filter** passes the high frequency components of a signal (or group of signals) to the output or load, while greatly attenuating or preventing, the low frequency components from being passed to the output or load.

Figure 22–26a illustrates a simple *RC,* high-pass filter. We can use the same concepts to explain this filter as we did in explaining the simple *RC* low-pass filter. That is, the relative values of *R* and X_C will determine the amount of the input signal that appears at the output. The difference between the two types of filters, is that in this case we take the output across the *R,* rather than the *C.* This means that the higher the X_C compared with the *R* value, the lower the output will be, and vice versa.

Let's assume the same values for *R* and *C* in Figure 22–26a as was used in our low-pass filter examples in Figure 22–23. If *R* = 100 Ω and the value of *C* = 1 μF, the following conditions will exist.

1. With a 10-V dc source as input, the output will be zero because the capacitor (once charged) blocks dc. This "ultra-low" frequency (dc) is not passed to the output.

2. At a frequency of 1,592 Hz, *R* and X_C are equal, and the output will be 7.07 V with a 10-V ac source as the input.

3. At a frequency of 15,920 Hz, *R* = 100 Ω and X_C = about 10 Ω; thus, the output will be about 9 V.

Since this filter passes along the higher frequencies to the output, while passing very little signal at low frequencies, it is called a high-pass filter!

Figure 22–26 shows several examples of configurations of *R, L,* and *C* components used as high-pass filters. Typically, the capacitors, which have a high opposition to low frequencies, are in series with the output or load. The inductors, which have a low opposition to low frequencies, are in parallel, or shunt with the output, thus effectively bypassing the low frequencies around the load. The result is the high frequency components of the input signal are passed to the load; and the low frequency components are greatly attenuated or prevented from reaching the load. Thus, the filter is called a high-pass filter.

Figure 22–27 shows the typical response characteristic of a high-pass filter.

Several applications for high-pass filters

Coupling circuits are one common use for networks that will attenuate frequencies below the desired frequency range (including dc) but will pass along the desired frequencies. For example, the circuit that couples a signal into the input of an amplifier effectively blocks dc and passes along the signal frequencies to be amplified.

Television receivers may use high-pass filters to prevent unwanted signals below the TV channel frequencies from causing interference. Filtering may also be used in other receiver antenna input circuits to prevent unwanted signals from getting into the system.

Some tone control circuits in radios and audio amplifiers use the concept of bypassing higher audio frequencies to alter the tone of the sound prescribed to the listener.

These are just a few of the many applications for filters in general, and high-pass filters, in particular.

NOTE: Low-pass and high-pass filter networks can be combined to pass a desired band of frequencies. Those frequencies passed along to the output or load are the frequencies that are not stopped by either filter network. This is one application of *bandpass filtering.* Typically however, this application is often fulfilled by resonant circuits.

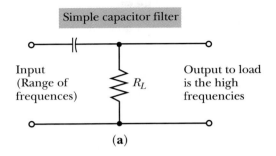

Simple capacitor filter

Input
(Range of
frequencies)

R_L

Output to load
is the high
frequencies

(a)

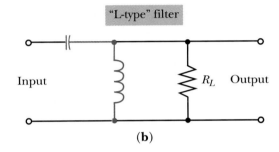

"L-type" filter

Input R_L Output

(b)

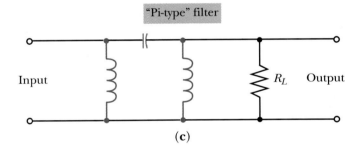

"Pi-type" filter

Input R_L Output

(c)

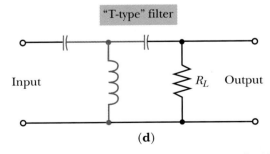

"T-type" filter

Input R_L Output

(d)

FIGURE 22–26 Examples of different configurations for high-pass filters

PRACTICAL NOTES

As you know, points falling below 70.7% or the half-power points of a filter response are out of the band of frequencies that the filter will pass. In low-pass filters, Figures 22–24 and 22–25, the high frequencies are rejected. In high-pass filters, Figures 22–26 and 22–27, the lower frequencies are rejected. The frequency where rejection begins is sometimes called the *cutoff frequency* (f_c). For a simple RC network, Figure 22–26a, $X_C = R$ at this cutoff frequency. The formula to find the cutoff frequency for this RC network is:

FORMULA 22–10 $\qquad f_c = \dfrac{1}{2\pi RC}$

To find the cutoff frequency where $X_L = R$ for an RL network similar to the one shown in Figure 22–25a, the formula is:

FORMULA 22–11 $\qquad f_c = \dfrac{1}{2\pi\left(\dfrac{L}{R}\right)}$

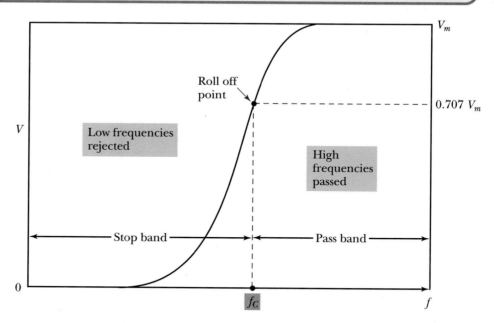

FIGURE 22–27 Typical response characteristic of a high-pass filter

Resonant filters

The bandpass filter

Refer to Figure 22–28 and notice a combination of series and parallel resonant circuits are used to pass a desired band of frequencies. The desired output band of frequencies finds little opposition from the series resonant circuits in series

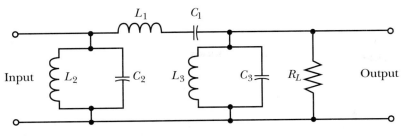

FIGURE 22–28 Resonant-type bandpass filter

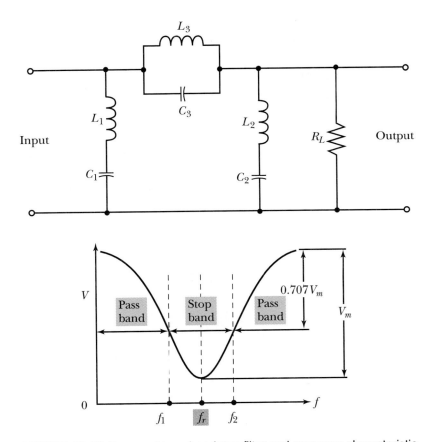

FIGURE 22–29 Resonant-type bandstop filter and response characteristic

with the output path and very high opposition from the parallel resonant circuits. Therefore, the desired band of frequencies appears at the output or load. This tunable filter is tuned with variable Cs or Ls to pass one desired band of frequencies while rejecting all other frequencies. An example of a tunable **bandpass filter** is the single tunable parallel resonant circuit that tunes different radio stations.

The bandstop filter

By placing the series resonant circuit(s) in shunt with the signal path and one or more parallel resonant circuit(s) in series with the signal path, a **bandstop filter** is developed, Figure 22–29. Also, note the frequency response characteristic.

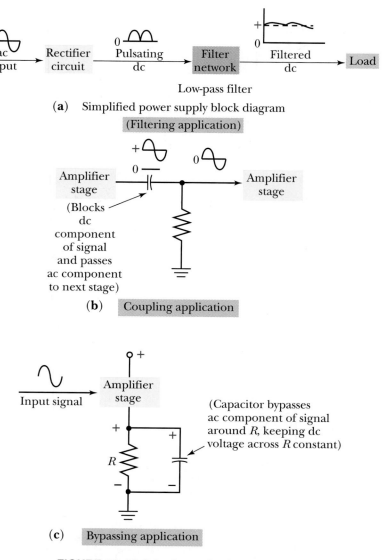

(a) Simplified power supply block diagram
(Filtering application)

(b) Coupling application

(c) Bypassing application

FIGURE 22–30 Sample applications of filters

The previous discussions have only introduced you to filters. We have not attempted to investigate filter design formulas (e.g., for *constant-k* or *m-derived filters*), or the wide spectrum of modern filters, such as YIG (yittrium-iron-garnet) filters, multichannel analog filters, and computer-controlled filters.

Perhaps your interest in studying filters was increased by this simple overview. Filter networks have numerous practical uses, from simple power supply filtering to signal processing in such high-technology areas as speech processing and medical instrumentation. As you continue in your studies, you will have the opportunity to study some of the basic applications of filters, coupling networks, and bypassing networks shown in Figure 22–30. The principles discussed in this chapter should help you in the more detailed studies.

SUMMARY

▶ X_L and X_C change in opposite directions for a given frequency change. In other words, as f increases, X_L increases and X_C decreases. Also, as f decreases, X_L decreases and X_C increases.

▶ For any LC combination, there is one frequency where X_L and X_C are equal: the resonant frequency or the frequency of resonance.

▶ At resonance, series RLC circuits exhibit the characteristics of: $X_L = X_C$, maximum current, minimum Z (and $Z = R$), θ of $0°$, and voltage across each reactance of Q times V applied.

▶ At resonance, parallel RLC circuits exhibit the characteristics of: $X_L = X_C$, minimum current, maximum Z ($Z = Q \times X_L$), $\theta = 0°$, and current within the tank circuit = Q times I line.

▶ See Figure 22–31 for comparison of series and parallel resonant circuits.

▶ The general formula to find the resonant frequency for series LC circuits and for high-Q parallel LC circuits is: $f_r = \dfrac{1}{2\pi\sqrt{LC}}$

▶ The Q of a resonant circuit is determined by the circuit resistance and the L/C ratio. The more

	SERIES RESONANCE	PARALLEL RESONANCE
SIMILARITIES	$X_L = X_C$	$X_L = X_C$
	$f_r = \dfrac{1}{2\pi\sqrt{LC}}$	$f_r = \dfrac{1}{2\pi\sqrt{LC}}$
	$\theta = 0°$	$\theta = 0°$ (High Q circuits)
	$p.f. = 1$	$p.f. = 1$ (High Q circuits)
	$\Delta f = \dfrac{f_r}{Q}$	$\Delta f = \dfrac{f_r}{Q}$
DIFFERENCES	$f_r = \dfrac{1}{2\pi\sqrt{LC}}$ no matter circuit R value	f_r can be affected by R
	I = maximum	I line = minimum
	Z = minimum ($=R$)	Z = maximum
	$Q = \dfrac{X_L}{r_s}$	$Q = \dfrac{R}{X_L}$ (low Q circuits)
	Resonant rise of I	Resonant rise of Z
	Resonant rise of reactive V ($Q \times V_A$)	($Z = Q \times X_L$)
	Above f_r = Inductive	Above f_r = Capacitive
	Below f_r = Capacitive	Below f_r = Inductive $p.f. = 1$ called "antiresonance" ... can be a different f than maximum Z point.

FIGURE 22–31 Similarities and differences between series and parallel resonance

the circuit resistive losses are, the lower the Q is. The higher the L/C ratio is, the higher the Q is for a given f_r.

▶ The higher the Q of a resonant circuit is, the greater the magnification factor is for voltage (series resonant circuit) and impedance (parallel resonant circuit).

▶ The higher the Q of a resonant circuit is, the sharper is its response curve and the more selective it is to a band of frequencies. High Q means narrow bandwidth and narrow bandpass.

▶ Bandwidth is often defined by the frequencies where the response curve for the circuit is above the 70.7% of the maximum level. This is sometimes called the half-power or −3 dB points on the response curve.

▶ The relationship of resonant frequency, Q, and bandwidth (BW) is shown by:

$$BW\,(\Delta f) = \frac{f_r}{Q}, \text{ or } Q = \frac{f_r}{\Delta f}$$

▶ R, L, and C components can be configured to produce filtering effects. Low-pass filters allow low frequencies to pass to the output, while attenuating high frequencies. High-pass filters pass high frequencies to the output while attenuating the low frequencies.

▶ Generally, to allow a certain frequency spectrum to pass to the output, a low impedance circuit or component is placed in series with the signal path, and/or a high impedance circuit or component is placed in shunt with the signal path.

▶ To prevent a certain frequency spectrum at the output of the filter, a high impedance circuit or component is placed in series with the signal path, and/or a low impedance circuit or component is placed in shunt with the signal path.

▶ Filters are categorized in various ways. Some classifications are physical layout shape (L, pi, T); usage (bypass, decouple, bandpass, bandstop, smoothing); component types (LC, RC, active, passive); number of sections they have (single-section, two-section); and the frequency spectrum in which they are used (power supply, audio, RF).

▶ Multiple-section filters generally provide more filtering than single-section filters.

FORMULAS AND SAMPLE CALCULATOR SEQUENCES

FORMULA 22–1 Resonant frequency: $f_r = \dfrac{1}{2\pi\sqrt{LC}}$

Inductance value, ⊗, capacitance value, ⊜, ⊗, ⊗, 6.28, ⊜, ⊙

FORMULA 22–2 Inductance value for given f_r and C: $L = \dfrac{0.02533}{f_r^2 C}$

0.02533, ⊕, ⊙, resonant frequency value, ⊗, ⊗, capacitance value, ⊙, ⊜

FORMULA 22–3 Capacitance value for given f_r and L: $C = \dfrac{0.02533}{f_r^2 L}$

0.02533, ⊕, ⊙, resonant frequency value, ⊗, ⊗, inductance value, ⊙, ⊜

FORMULA 22–4 Inductance value (in μH) for given f_r and C: $L = \dfrac{25{,}330}{f^2 C}$

25,330, ⊕, ⊙, resonant frequency value in MHz, ⊗, ⊗, capacitance value in pF, ⊙, ⊜

FORMULA 22–5   Capacitance value (in pF) for given f_r and L: $C = \dfrac{25{,}330}{f^2 L}$

25,330, ⊟, ⟨⟨, resonant frequency value in MHz, x^2, ⊠, inductance value in μH, ⟩⟩, ⊜

FORMULA 22–6 $f_r = \dfrac{1}{2\pi\sqrt{LC}} \times \sqrt{1 - \dfrac{C_{r_S}^{\,2}}{L}}$

FORMULA 22–7   Total bandwidth *(BW)*: $\Delta f = f_2 - f_1$

half-power point (higher frequency), ⊟, half-power point (lower frequency), ⊜

FORMULA 22–8   Bandwidth *(BW)* related to Q: $\Delta f = \dfrac{f_r}{Q}$

resonant frequency value, ⊟, Q value, ⊜

FORMULA 22–9 Q value: $Q = \dfrac{f_r}{\Delta f}$

resonant frequency value, ⊟, bandwidth value, ⊜

FORMULA 22–10   Cutoff frequency *(RC* filter): $f_c = \dfrac{1}{2\pi RC}$

6.28, ⊠, resistor value, ⊠ capacitance value, ⊜, 1/x

FORMULA 22–11   Cutoff frequency *(RL* filter): $f_c = \dfrac{1}{2\pi\left(\dfrac{L}{R}\right)}$

6.28, ⊠, ⟨⟨, inductance value, ⊟, resistance value, ⟩⟩, ⊜, 1/x

---

REVIEW QUESTIONS

1. As f increases:
   a. X_L decreases and X_C increases.
   b. X_L increases and X_C decreases.
   c. X_L and X_C increase.
   d. none of the above.

2. When L and C are both present in a circuit:
   a. there are two frequencies at which L and C resonante.
   b. there is only one frequency of resonance for given values of L and C.

c. series and parallel resonance occur at the same frequency in all cases.
   d. none of the above.

3. For series resonant circuits:
   a. Z is minimum, I is maximum, $X_L = X_C$, θ is 0°, and $V_C = Q$ times voltage applied.
   b. Z is maximum, I is minimum, and $V_L = Q$ times voltage applied.
   c. Z is minimum, I is maximum, and $\theta = 45°$.
   d. none of the above.

4. If an *LC* circuit has a high *Q*:
   a. the circuit is broad-band in nature.
   b. the circuit is not very selective in terms of frequency.
   c. the circuit has a narrow bandpass.
   d. none of the above.

5. If an *LC* circuit has a low *Q*:
   a. its bandwidth is narrow.
   b. its bandwidth is broad.
   c. its bandwidth equals its *Q*.
   d. none of the above.

6. By definition, bandwidth is defined as frequencies wherein the response curve is close to:
   a. 50% of maximum level.
   b. 100% of maximum level.
   c. 70% of maximum level.
   d. none of the above.

7. Bandwidth and *Q* are:
   a. directly related.
   b. not related.
   c. inversely related.
   d. none of the above.

8. "Half-power points" may be considered:
   a. 50%, of maximum response points.
   b. −3 dB points.
   c. 0.5 points.
   d. none of the above.

9. Resonant frequency of a series *RLC* circuit:
   a. can be affected by the amount of *R* in the circuit.
   b. cannot be affected by the amount of *R* in the circuit.
   c. cannot be affected by the amount *L* in the circuit.
   d. none of the above.

10. Resonant frequency of a parallel *RLC* circuit:
    a. can be affected by the amount of *R* in the circuit.
    b. cannot be affected by the amount of *R* in the circuit.
    c. cannot be affected by the amount of *C* in the circuit.
    d. none of the above.

11. Low-pass filters:
    a. attenuate low frequencies.
    b. amplify low frequencies.
    c. allow passage of low frequencies.
    d. allow passage of high frequencies.

12. Classification of filter circuits may include:
    a. physical layout and components used.
    b. use, physical layout, components used, and frequency spectrum.
    c. components used, physical layout, use, and number of sections.
    d. number of sections, components used, use, physical layout, and frequency.

13. A bandpass type filter provides:
    a. high opposition in series with the output.
    b. low opposition in parallel with the output.
    c. low opposition in series with the output.
    d. none of the above.

14. A band-stop type filter provides:
    a. high opposition in series with the output.
    b. high opposition in parallel with the output.
    c. low opposition in series with the output.
    d. none of the above.

15. Inductors:
    a. have inductance, resistance, and capacitance.
    b. have inductance and resistance only.
    c. must have an external *C* connected to produce a resonant condition.
    d. none of the above.

16. Resistors have the properties of:
    a. resistance and capacitance.
    b. resistance and inductance.
    c. resistance, inductance, and capacitance.
    d. resistance only.

17. Several ways of measuring the resonant frequency of a signal source include:
    a. a digital frequency counter and a tunable dip meter (with power on).
    b. a digital frequency counter, with power on, a dip meter (with power off).
    c. an oscilloscope, or a frequency counter, with power on.
    d. all of the above.

18. For a simple *RC* filter, the cutoff frequency is:
    a. at a point where *R* is double X_C.
    b. directly related to the value of *R* and *C*.
    c. inversely related to the value of *R* and *C*.
    d. only related to the value of *R*.

19. For a simple *RL* filter, the cutoff frequency will increase if:
    a. the value of *L* increases and *R* stays the same.
    b. the value of *R* increases and *L* stays the same.
    c. the value of *R* and *L* change by the same factor.
    d. none of the above.

20. To create a resonant bandstop filter, you would use:
    a. a series resonant circuit in series with the signal path.
    b. a series resonant circuit in parallel with the signal path.
    c. a parallel resonant circuit in parallel with the signal path.
    d. none of the above.

PROBLEMS

1. Disregarding the generator or source impedance, a series *LC* circuit is comprised of a 100-pF capacitor and a 1-μH inductor having 10 Ω of resistance.
   a. What is the impedance at resonance?
   b. What is the resonant frequency?
   c. What is the *Q*?
   d. What is the bandwidth?
   e. What is the bandpass?
   f. Below the resonant frequency, does the circuit act inductively or capacitively?

2. The voltage applied to the circuit in question 1 is 20 V.
   a. What is the voltage across the capacitor at resonance?
   b. What is the current at resonance?
   c. What is the true power at resonance?
   d. What is the net reactive power?

3. A parallel *LC* circuit has a capacitive branch with a 30-pF capacitor and an inductive branch with an inductor having 28 μH and a r_s of 30 Ω.
   a. What is the f_r?
   b. What is the *Q*? (Assume a *Q* greater than 10.)
   c. What is the bandwidth?
   d. What is the bandpass?
   e. At resonance, what is the *Z*?
   f. Below resonance, does the circuit act inductively or capacitively?

4. If the voltage applied to the circuit in question 3 is 20 V, what is the line current? What is the current through the capacitor branch?

5. A series *RLC* circuit consists of a 100-Ω resistor, a 75-H inductor, and a 300-pF capacitor. What is the resonant frequency?

6. A series *RLC* circuit has a resonant frequency of 40 kHz and the *Q* is 20. At what frequency will the phase angle be 45°?

7. A series *RLC* circuit is resonant at 150 kHz, and has *R* = 100 Ω and *C* = 25 pF. What is the inductance value?

8. A series *RLC* circuit is resonant at 21 kHz and has *R* = 40 Ω and inductance = 12 H. What is the *C* value?

9. Given a series *RLC* circuit that is resonant at 1.5 MHz, and knowing that *C* has a reactance of 500 Ω, what is the *R* value for the circuit to have a bandwidth of 10 kHz?

10. A series *RLC* circuit is comprised of a 200-Ω resistor, a 36-mH inductor, and a capacitor. What is the voltage across the *R* at resonance if *V* applied is 100 V?

11. What is the resonant frequency of a parallel *LC* combination having a 200-μH inductor and a 0.0002-μF capacitor?

12. If the frequency of the source in question 11 is adjusted to a new frequency above the resonance frequency, will I_T lead or lag V_A?

13. A 220-μH inductor has a distributed capacitance of 25 pF and an effective resistance of 80 Ω. What impedance will it exhibit at resonant frequency?

ANALYSIS QUESTIONS

Perform the required research to determine the answers to the following questions.

1. Is the tuned circuit at the input of a receiver typically a series or parallel *LC* circuit?

2. What is meant by a "double-tuned" transformer, as used between certain stages of radio or TV receivers?

3. Would the circuits described in questions 1 and 2 be considered resonant or nonresonant filters?

4. Match the frequency response curves, shown in the right column, to the types of filters, listed in the left column. Place the letter identifying each response curve in the blank identifying the filter or circuit type that would have that type response.

_____ Parallel resonant
_____ Low-pass   **(a)**
_____ Bandpass
_____ High-pass
_____ Bandstop
_____ Series resonant

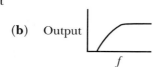

(b)   Output

(c)   Output

(d)   Output

(e)   Z

(f)   Output

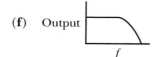

5. In your own words, define the meaning of the terms *selectivity, bandwidth,* and *bandpass.*

Answer the following questions with "I" for increase, "D" for decrease, and "RTS" for remain the same.

6. For a series resonant *RLC* circuit, if frequency is slightly increased:
   a. X_L will _____   i. V_R will _____
   b. X_C will _____   j. V_T will _____
   c. R will _____   k. BW will _____
   d. Z will _____   l. Q will _____
   e. I will _____   m. I_C will _____
   f. θ will _____ n. I_L will _____
   g. V_C will _____   o. I_T will _____
   h. V_L will _____

7. For a series resonant *RLC* circuit, if frequency is slightly decreased:
   a. X_L will _____   i. V_R will _____
   b. X_C will _____   j. V_T will _____
   c. R will _____   k. BW will _____
   d. Z will _____   l. Q will _____
   e. I will _____   m. I_C will _____
   f. θ will _____ n. I_L will _____
   g. V_C will _____   o. I_T will _____
   h. V_L will _____

8. For a parallel resonant *RLC* circuit, if frequency is slightly increased:
   a. X_L will _____   i. V_R will _____
   b. X_C will _____   j. V_T will _____
   c. R will _____   k. BW will _____
   d. Z will _____   l. Q will _____
   e. I will _____   m. I_C will _____
   f. θ will _____ n. I_L will _____
   g. V_C will _____   o. I_T will _____
   h. V_L will _____

9. For a parallel resonant *RLC* circuit, if frequency is decreased:
   a. X_L will _____   i. V_R will _____
   b. X_C will _____   j. V_T will _____
   c. R will _____   k. BW will _____
   d. Z will _____   l. Q will _____
   e. I will _____   m. I_C will _____
   f. θ will _____ n. I_L will _____
   g. V_C will _____   o. I_T will _____
   h. V_L will _____

10. Research the terms *m*-derived and constant-*k* filters and write the definition of each.

11. Assuming "perfect" reactive components, state the condition that must exist if a series *RLC* circuit is at resonance.

12. Draw a pi-type *LC* filter that can filter a power supply output. Is this a low-pass or a high-pass filter?

13. Draw a circuit showing a series resonant and a parallel resonant circuit connected so that the combination passes a band of frequencies close to the resonant frequencies of the circuits.

14. If a parallel *LC* circuit is acting inductively at a frequency of 150 kHz, tuning the circuit to resonance requires which of the following?
    a. Increasing X_C
    b. Decreasing X_L
    c. Decreasing L
    d. Increasing C

Be sure to use your five senses, plus, easy-to-manipulate "front-panel" controls in gathering and analyzing symptoms, before digging out the test equipment and beginning detailed testing!

PERFORMANCE PROJECTS CORRELATION CHART

Suggested performance projects that correlate with topics in this chapter are:

CHAPTER TOPIC	PERFORMANCE PROJECT	PROJECT NUMBER
X_L, X_C, and Frequency	X_L and X_C Relationships to Frequency	58
Series Resonance Characteristics	V, I, R, Z, and θ Relationships when $X_L = X_C$	59
The Resonant Frequency Formula		
Q and the Resonant Rise of Voltage	Q and Voltage in a Series Resonant Circuit	60
Selectivity, Bandwidth, and Bandpass	Bandwidth Related to Q	61
Parallel Resonance Characteristics	V, I, R, Z, and θ Relationships when $X_L = X_C$	62
Parallel Resonance Formulas		
Q and the Resonant Rise of Impedance	Q and Impedance in a Parallel Resonant Circuit	63
Selectivity, Bandwidth, and Bandpass	Bandwidth Related to Q	64

NOTE: It is suggested that after completing the above projects, the student should be required to answer the questions in the "Summary" at the end of this section of projects in the Laboratory Manual.

Follow the seven-step SIMPLER troubleshooting sequence, as outlined below, and see if you can find the problem with this circuit. As you follow the sequence, record your circuit test results for each testing step on a separate sheet of paper. This will aid you and your instructor to see where your thinking is on track or where it might deviate from the best procedure.

step 1 Symptoms

—(Gather, verify, and analyze symptom information) To begin this process, read the data under the "Starting Point Information" heading. Particularly look for circuit parameters that are not normal for the circuit configuration and component values given in the schematic diagram. For example, look for currents that are too high or too low; voltages that are too high or too low; resistances that are too high or too low, etc. This analysis should give you a first clue (or symptom information) that will aid you in determining possible areas or components in the circuit that might be causing this symptom.

step 2 Identify

—(Identify and bracket the initial suspect area) To perform this first "bracketing" step, analyze the clue or symptom information from the "Symptoms" step, then either on paper or in your mind, bracket, circle, or put parentheses around all of the circuit area that contains any components, wires, etc., that might cause the abnormality that your symptom information has pointed out. NOTE: Don't bracket parts of the circuit that contain items that could not cause the symptom!

step 3 Make

—(Make a decision about the first test you should make: what type and where) Look under the TEST column and determine which of the available tests shown in that column you think would give you the most meaningful information about the suspect area or components, in the light of the information you have so far.

step 4 Perform

—(Perform the first test you have selected from the TEST column list) To simulate performing the chosen test (as if you were getting test results in an actual circuit of this type), follow the dotted line to the right from your selected test to the number in parentheses. The number in parentheses, under the "Results in Appendix C" column tells you what number to look up in Appendix C to see what the test result is for the test you are simulating.

step 5 Locate

—(Locate and define a new "narrower" area of uncertainty) With the information you have gained from the test, you should be able to eliminate some of the circuit area or circuit components as still being suspect. In essence, you should be able to move one bracket of the bracketed area so that there is a new smaller area of uncertainty.

step 6 Examine

—(Examine available information and determine the next test: what type and where) Use the information you have thus far to determine what your next test should be in the new narrower area of uncertainty. Determine the type test and where, then proceed using the TEST listing to select that test, and the numbers in parentheses and Appendix C data to find the result of that test.

step 7 Repeat

—(Repeat the analysis and testing steps until you find the trouble) When you have determined what you would change or do to restore the circuit to normal operation, you have arrived at *your* solution to the problem. You can check your final result against ours by observing the pictorial step-by-step sample solution on the pages immediately following the "Chapter Troubleshooting Challenge"—circuit and test listing page.

step 8 Verify

NOTE: A useful *8th Step* is to operate and test the circuit or system in which you have made the "correction changes" to see if it is operating properly. This is the final proof that you have done a good job of troubleshooting.

CHALLENGE CIRCUIT 12

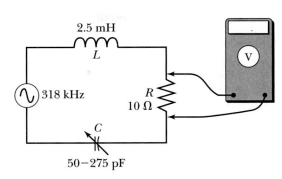

CHALLENGE CIRCUIT 12

Starting Point Information

---

1. Circuit diagram
2. V_R is lower than it should be when C is adjusted so that the LC resonance is below the 318 kHz source signal frequency.

TEST	Results in Appendix C

V_R (when C set at the position
for 318 kHz LC resonance) . . . .(55)
V_R (when C set at a position for
LC resonance above 318 kHz) . . . .(85)
V_R (when C set at a position for
LC resonance below 318 kHz) . . . .(30)
R_C (C set at a 318 kHz position) . . . .(58)
R_C (C set above 318 kHz position) . . . .(100)
R_C (C set below 318 kHz position) . . . .(72)
NOTE: R_C is resistance measured from stator-to-rotor plates across the capacitor.

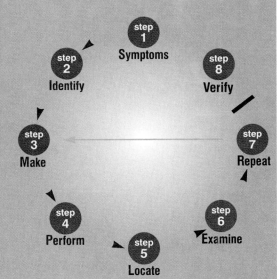

step 1 Symptoms
step 2 Identify
step 3 Make
step 4 Perform
step 5 Locate
step 6 Examine
step 7 Repeat
step 8 Verify

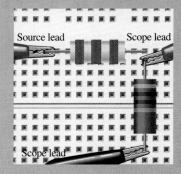
Source lead   Scope lead

Scope lead

step 1

Symptoms

When the variable C is adjusted so the LC circuit is resonant at about 318 kHz source signal (i.e., C is about 100 pF), the scope shows maximum deflection due to maximum I_R drop at series resonance (Figure 1). When the variable C is adjusted so the LC circuit resonance is slightly higher in frequency (plates more unmeshed), the V_R (and scope deflection) slightly decrease as would be expected (Figure 2). When the capacitor is adjusted so the LC resonance is slightly lower (plates more meshed), the V_R and scope deflection drop drastically (Figure 3). This indicates some drastic change in the circuit. The off-resonance impedance must have increased at an abnormal rate for this circuit.

step 2

Identify

initial suspect area: Since the only item being varied is the capacitance (because of our tuning efforts), the variable tuning capacitor is the initial suspect area.

step 3

Make

test decision: We can check the tuning capacitor's operation with an ohmmeter to make sure that it isn't shorting at some point in its tuning range. The ohmmeter check is an easy check to find any possible shorts.

step 4

Perform

1st Test: Perform a resistance check throughout the tuning range. As the plates approach being more fully meshed a short between the stator and rotor plates becomes evident. This appears to be our problem. (NOTE: No Figure shown for this test.)

step 5

Locate

new suspect area: There is no new area because the variable capacitor has proven to be the likely culprit.

step 6

Examine

available data.

step 7

Repeat

analysis and testing: Let's visually inspect the capacitor to see if we can see some shorted plates.
2nd Test: Inspect the capacitor with strong light. The result is you see two plates that are rubbing as the plates begin to more fully mesh. The problem is shorting plates on the tuning capacitor. (NOTE: No Figure shown for this test.)
3rd Test: Bend the plates (Figure 4) to eliminate the rubbing. Then,

Symptoms (1)   Figure 1

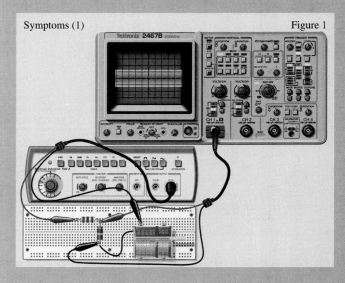

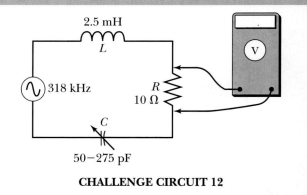

CHALLENGE CIRCUIT 12

2.5 mH
L

$\sim$ 318 kHz

R
10 Ω

C

50−275 pF

check with the ohmmeter again, while turning the rotor plates through their range.

step 8

Verify

4th Test: The result is everything is normal (Figure 5).

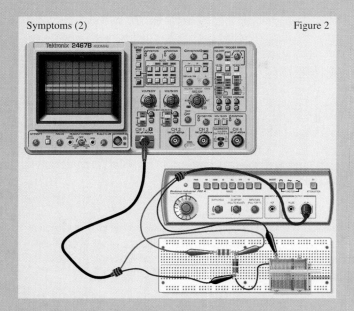

Figure 2

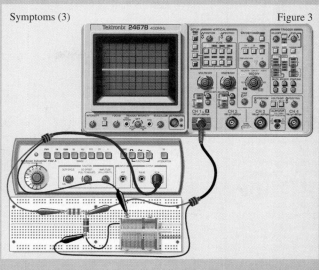

Figure 3

3rd Test

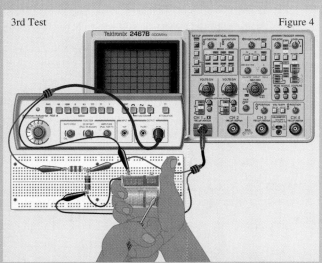

Figure 4

4th Test

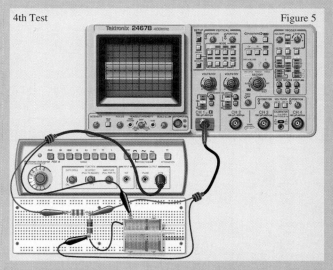

Figure 5

THE SIMPLER SEQUENCE FOR TROUBLESHOOTING

Follow the seven-step SIMPLER troubleshooting sequence, as outlined below, and see if you can find the problem with this circuit. As you follow the sequence, record your circuit test results for each testing step on a separate sheet of paper. This will aid you and your instructor to see where your thinking is on track or where it might deviate from the best procedure.

Symptoms

—(Gather, verify, and analyze symptom information) To begin this process, read the data under the "Starting Point Information" heading. Particularly look for circuit parameters that are not normal for the circuit configuration and component values given in the schematic diagram. For example, look for currents that are too high or too low; voltages that are too high or too low; resistances that are too high or too low, etc. This analysis should give you a first clue (or symptom information) that will aid you in determining possible areas or components in the circuit that might be causing this symptom.

Identify

—(Identify and bracket the initial suspect area) To perform this first "bracketing" step, analyze the clue or symptom information from the "Symptoms" step, then either on paper or in your mind, bracket, circle, or put parentheses around all of the circuit area that contains any components, wires, etc., that might cause the abnormality that your symptom information has pointed out. NOTE: Don't bracket parts of the circuit that contain items that could not cause the symptom!

Make

—(Make a decision about the first test you should make: what type and where) Look under the TEST column and determine which of the available tests shown in that column you think would give you the most meaningful information about the suspect area or components, in the light of the information you have so far.

Perform

—(Perform the first test you have selected from the TEST column list) To simulate performing the chosen test (as if you were getting test results in an actual circuit of this type), follow the dotted line to the right from your selected test to the number in parentheses. The number in parentheses, under the "Results in Appendix C" column tells you what number to look up in Appendix C to see what the test result is for the test you are simulating.

Locate

—(Locate and define a new "narrower" area of uncertainty) With the information you have gained from the test, you should be able to eliminate some of the circuit area or circuit components as still being suspect. In essence, you should be able to move one bracket of the bracketed area so that there is a new smaller area of uncertainty.

Examine

—(Examine available information and determine the next test: what type and where) Use the information you have thus far to determine what your next test should be in the new narrower area of uncertainty. Determine the type test and where, then proceed using the TEST listing to select that test, and the numbers in parentheses and Appendix C data to find the result of that test.

Repeat

—(Repeat the analysis and testing steps until you find the trouble) When you have determined what you would change or do to restore the circuit to normal operation, you have arrived at *your* solution to the problem. You can check your final result against ours by observing the pictorial step-by-step sample solution on the pages immediately following the "Chapter Troubleshooting Challenge"—circuit and test listing page.

Verify

NOTE: A useful *8th Step* is to operate and test the circuit or system in which you have made the "correction changes" to see if it is operating properly. This is the final proof that you have done a good job of troubleshooting.

CHALLENGE CIRCUIT 13

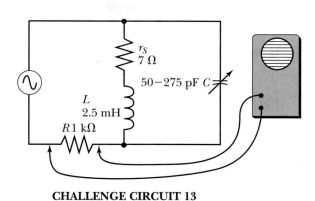

CHALLENGE CIRCUIT 13

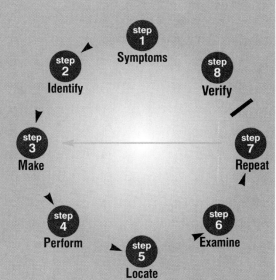

Starting Point Information

1. Circuit diagram
2. Using the scope to monitor when V across the 1-kΩ R is minimum (indicating that the parallel LC impedance is maximum), we find that the range of frequencies tuned from maximum C value to minimum C value is slightly higher in frequency than should be expected from the rated values shown on the diagram.

TEST	Results in Appendix C
$R_{inductor}$ ·	(87)
$R_{capacitor}$ ·	(19)
C of capacitor when "fully meshed" ·	(60)
C of capacitor when "fully unmeshed" · · · · · · · · · · · · · · · · · · ·	(107)
L of inductor ·	(41)

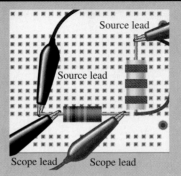

Source lead

Source lead

Scope lead   Scope lead

Symptoms

One would normally expect the tuning range of the *LC* circuit to be from about 190 kHz to 450 kHz as the *C* is varied between 50 and 275 pF. It is actually tuning from about 220 kHz (Figure 1) to 580 kHz (Figure 2).

Symptoms (1)   Figure 1

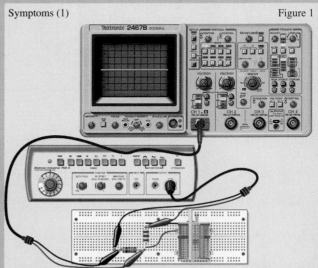

Identify

initial suspect area: Since the frequencies are higher than expected, it looks like the *LC* product must be less than the rated values. Either the capacitor has decreased by losing plates (which is highly unlikely) or the inductor has decreased in value for some reason. We can't be sure at this point. The inductor is the initial suspect. However, we'll keep both the *C* and *L* in the initial suspect area.

Make

test decision: What could make the inductor have less inductance than anticipated? Possibly, shorted turns which would make *L* decrease. Since the rated dc resistance of this coil is about 7 Ω, an ohmmeter check (with the inductor isolated from the rest of the circuit) might reveal if enough turns are shorted to make a difference in coil resistance.

Perform

1st Test: With *L* isolated and source removed, measure its resistance (Figure 3). *R* measures about 5 Ω. Even this little resistance change could be significant, since we're dealing with such a small rated resistance in the first place. A number of turns could be shorted.

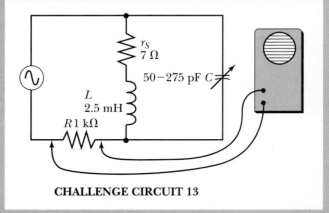

r_S 7 Ω

50–275 pF C

L 2.5 mH

R 1 kΩ

CHALLENGE CIRCUIT 13

Locate

new suspect area: The hot zone is still the inductor. However, we need to make another inductor check so we can exclude the capacitor.

Examine

available data.

Repeat

analysis and testing: Since the coil looks suspicious, let's substitute a known good coil of like ratings and see how the circuit operates.

Verify

2nd Test and **3rd Test:** Substitute an inductor of the same ratings. Then check the circuit tuning characteristics. The result is the component substitution worked. The circuit tunes with the expected range of frequencies, 190 kHz (Figure 4) to 450 kHz (Figure 5). The trouble was shorted turns on the inductor.

Symptoms (2) Figure 2

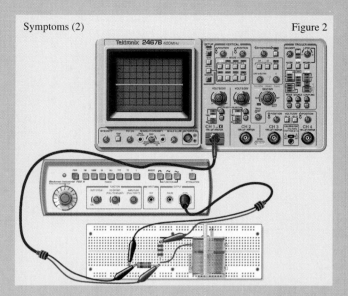

1st Test Figure 3

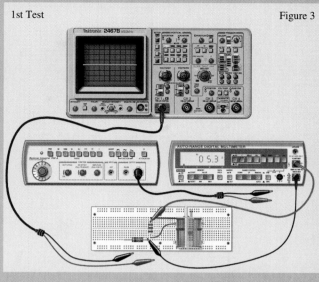

2nd Test Figure 4

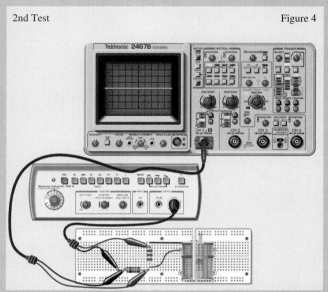

3rd Test Figure 5
Checking operation through tuning range

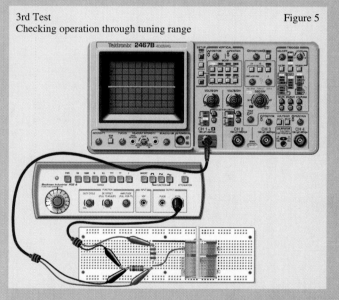

KEY TERMS

Barrier potential
Covalent bond
Depletion region
Doping material
Forward bias
Majority carriers
Minority carriers
N-type material
Pentavalent atoms
P-N junction
P-type material
Reverse bias
Reverse breakdown
voltage
Tetravalent atoms
Trivalent atoms
Valence electrons

SEMICONDUCTOR MATERIALS AND P-N JUNCTIONS

23

chapter

CHAPTER PREVIEW

Semiconductor devices have changed our world. Progress in electronics was once limited by having to use fragile, bulky, and power-gobbling vacuum tubes instead of tiny semiconductor devices. Transistors, diodes, and integrated circuits are the most common examples of modern-day semiconductors. Without them, we wouldn't have pocket calculators, desktop computers, digital watches, video games, VCRs, or portable electronic keyboards. Trying to imagine a world without the gadgets and appliances made possible with semiconductors is like trying to imagine modern society functioning without the automobile.

In this chapter, you get a look at the very basic inner workings that are common to all kinds of semiconductor devices. You will see why atoms of silicon are so important for making semiconductor devices, and you will learn about a few of the other important elements such as germanium, arsenic, gallium, and phosphorus that are applied in semiconductor technology. You will be introduced to the unique way that current flows through semiconductor materials, and you will get a first look at some graphs that are commonly used for describing the operation of simple semiconductor devices.

OBJECTIVES

After studying this chapter, you should be able to:

1. Describe the difference between **valence electrons** and **conduction-band electrons**

2. Describe the main difference between **N-type** semiconductor materials and **P-type** semiconductor materials

3. Draw a diagram of a **P-N junction,** including the **depletion region**

4. Draw a P-N junction that shows the polarity of applied voltage for **forward biasing** the junction

5. Draw a P-N junction that shows the polarity of applied voltage for **reverse biasing** the junction

6. Explain the difference between the **barrier potential** and **reverse breakdown voltage** for a P-N junction

7. Sketch the **I-V curve** for a typical P-N junction, showing both the forward and reverse bias parts of the curve

23-1 SEMICONDUCTOR MATERIALS

Recall from Chapter 1 that conductors, nonconductors, and semiconductors offer different levels of resistance to current flow. Also remember that conductors such as copper, gold, and silver offer little opposition to current flow, whereas good insulators such as glass, mica, and most plastics offer a great deal of opposition to current flow. Semiconductor materials, such as silicon combined with very small amounts of aluminum or arsenic, oppose current flow at levels somewhere between that of the best conductors and best insulators.

In this chapter, you will discover how it is possible to adjust a semiconductor's ability to pass electrical current. You will see how the same semiconductor device can be a fairly good conductor under certain conditions and yet function as a rather good insulator under a different set of conditions. This ability to control the conductance of semiconductor material makes it suitable for use in devices that can control the flow of current through a circuit and amplify electrical signals.

Review of atomic structure

Referring to Figure 23–1, remember that the three primary particles of atoms are protons, neutrons, and electrons. The protons (positive particles) and neutrons (neutral particles) are located in the center, or nucleus, of the atom. The electrons (negative charges) are arranged in orbits that surround the nucleus. In a normal, stable atom, the number of electrons in the orbits exactly equals the number protons in the nucleus. This means that the total electrical charge of the atom is zero.

There is a definite limit to the number of electrons that can be included in each electron orbit, or shell. If you number the shells such that 1 is the innermost, 2 is the second from the nucleus, etc., you can use the following formula to determine the maximum number of electrons that can occupy each shell, where n is the shell number.

FORMULA 23–1 $2n^2$

Formula 23–1 shows that the innermost shell (n = 1) can contain no more than 2 electrons, the second shell (n = 2) can hold up to 8, the third (n = 3) up to 18, and so on.

Each shell represents a different energy level for the electrons that occupy it. As shown in Figure 23–2, the energy levels increase with increasing distance from the nucleus. So you can see that electrons located in the innermost shell possess the least amount of energy, while those in the outermost shell have the greatest amount of energy. The electronic behavior of an atom mainly depends on the relative number of electrons in its outermost shell, or valence shell. Electrons included in the valence shell are called **valence electrons**.

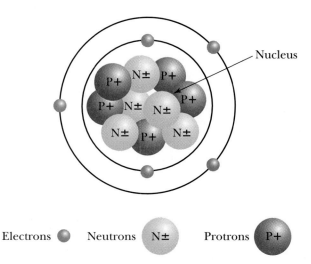

Electrons ●   Neutrons N±   Protrons P+

FIGURE 23–1 Atoms are composed of protons, neutrons, and electrons.

FIGURE 23–2 Energy levels for atoms

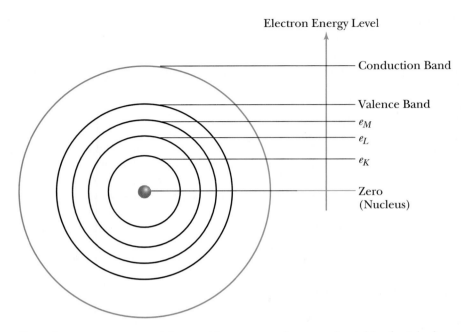

Exposing an atom to certain outside sources of energy (notably electrical energy) gives the valence electrons additional energy, which allows them to break away from their valence band and enter a **conduction band**. This is an important effect in electronics because electrons at the conduction-band energy levels are the only ones that are free to take part in the process of electron current flow. When a valence-band electron absorbs energy, it jumps up into a conduction band. Sooner or later that same electron has to fall back down to the valence-band level, and when that happens, the electron gives up its extra load of energy—usually in the form of heat, but sometimes as a particle of light energy.

Atomic structures for semiconductors

You will soon see that the most useful semiconductor materials are those made from atoms that have three, four, or five valence electrons. Atoms that normally have three valence electrons are called **trivalent atoms**, those normally having four valence electrons are called **tetravalent atoms**, and those normally having five outer electrons are called **pentavalent atoms.** Like any other atom, these are electrically neutral when they have their normal number of valence electrons. Figure 23–3 shows examples of each of these three kinds of atoms.

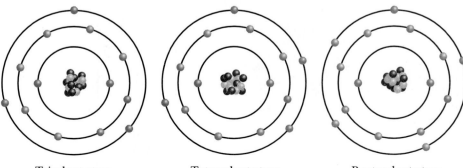

FIGURE 23–3 Trivalent, tetravalent, and pentavalent atoms

| Trivalent atom | Tetravalent atom | Pentavalent atom |
| Aluminum | Silicon | Phosphorus |

The valence electrons in these semiconductor materials are apt to absorb energy from heat and light sources as well as electrical sources. You will find through your studies of semiconductor devices that they respond to changes in temperature. In many instances, a relatively small increase in temperature can drive millions of valence electrons into conduction bands, thereby causing a dramatic decrease in electrical resistance.

These descriptions of atoms—shells, energy levels, valence electrons, and conduction-band electrons—are vital to your understanding of semiconductor materials and the behavior of semiconductor devices. Here is a brief summary of some key facts about the atomic structures of semiconductor materials.

1. "Tri-" means *three*, so a trivalent atom is one that has three electrons in the valence shell. Typical trivalent semiconductor materials are boron, aluminum, gallium, and indium.

2. "Tetra-" means *four*, so a tetravalent atom is one that has four electrons in its valence shell. Typical tetravalent semiconductor materials are germanium and silicon. Silicon is by far the more common (which explains why the center of semiconductor technology in northern California is nicknamed "Silicon Valley").

3. "Penta-" means *five*, so a pentavalent atom is one that has five electrons in its valence shell. Typical pentavalent semiconductor materials are phosphorus, arsenic, and antimony.

4. Valence electrons in semiconductor materials can absorb energy and jump to conduction bands by energy sources that include electricity, heat, and light.

5. Conduction-band electrons in semiconductor materials usually emit heat energy when they fall back to their valence bands. In certain semiconductors, particularly gallium arsenide, electrons dropping down to valence-band levels emit light energy.

Covalent bonds

You have already learned from an earlier lesson that matter is composed of molecules, which are, in turn, composed of atoms that represent the basic chemical elements. Although every kind of matter in the universe is composed of atoms, there are a number of different ways atoms can be assembled to create molecules. The materials used for making semiconductors form what is known as a **covalent bond** between the individual atoms. In a covalent bond, two or more atoms share valence electrons.

Figure 23–4 shows how a single tetravalent atom forms covalent bonds with four other atoms of the same type. The bonds are actually three-dimensional, thus forming a crystalline molecule, or lattice, that is shaped like a cube. A single grain of common beach sand is made up of millions upon millions of tetravalent silicon atoms that are bonded in this fashion.

Each atom of a trivalent material forms covalent bonds with three other atoms; and by the same token, pentavalent atoms form covalent bonds with five other identical atoms. Semiconductors are manufactured from these trivalent, tetravalent, and pentavalent elements that are first highly purified, then carefully combined to produce the necessary electronic effects. Because of their solid, crystalline makeup, you often hear semiconductor devices (diodes, transistors, and integrated circuits, for instance) called *solid-state devices*.

FIGURE 23–4 Covalent bonds for tetravalent atoms (silicon: Si)

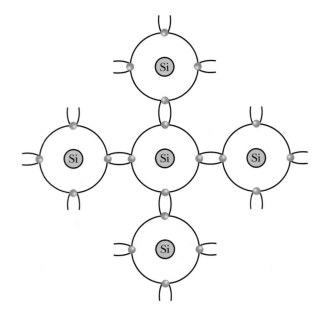

Practical semiconductor materials are not made from just one kind of atom, however. Most are made from a highly purified tetravalent atom (silicon or germanium), which is combined with extremely small amounts of trivalent and pentavalent atoms that are called the **doping material**. The atoms that are chosen for this task have the ability to combine with one another as though they were atoms of the same element. In fact, a tetravalent atom can be "fooled" into making a covalent bond with a trivalent or pentavalent atom.

Figure 23–5 shows covalent bonds between four tetravalent atoms and one trivalent atom. The bond between the leftmost tetravalent atom and the trivalent atom is missing one electron, but the covalent bonding takes place in spite of the shortage of one electron. Likewise, a pentavalent atom can bond with a tetrava-

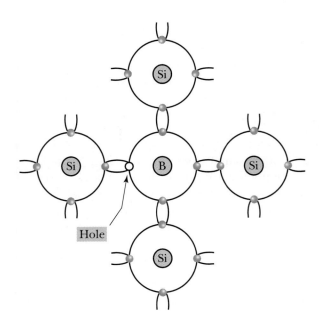

FIGURE 23–5 Covalent bonds between a trivalent atom (boron: B) and four tetravalent atoms (silicon: Si)

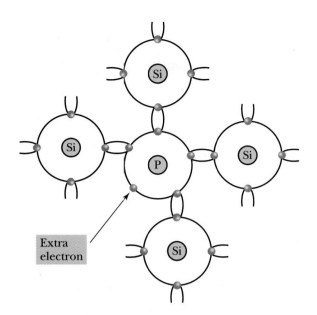

FIGURE 23–6 Covalent bonds between a pentavalent atom (phosphorus: P) and four tetravalent atoms (silicon: Si)

Extra electron

lent atom. As shown in Figure 23–6, this situation leaves an extra electron in the bonding arrangement.

The theory of operation of semiconductor devices rests heavily upon the notions of gaps, or *holes,* that are left when creating covalent bonds between trivalent and tetravalent atoms, and the excess electrons that result from covalent bonds between tetravalent and pentavalent atoms.

N-type semiconductors

An **N-type semiconductor material** is one that has an excess number of electrons. This means that one out of every million or so atoms has five electrons in covalent bonds instead of four. A block of highly purified silicon, for example, has four electrons available for covalent bonding. Arsenic is a similar material, but has five electrons available for covalent bonding. So when a minute amount of arsenic is mixed with a sample of silicon, the arsenic atoms move into places normally occupied by silicon atoms. The "fit" is a good one except for the fact that there is no place in the covalent bonds for the fifth electron. So the "fifth" electrons contributed by arsenic impurity are free to wander through the semiconductor material under external influences such as heat, light, and electrical energy.

Applying a source of electrical energy, or voltage, to a semiconductor material that has an excess number of electrons causes those electrons to drift through the material. These free electrons are repelled by the negative terminal of the applied potential and are attracted to the positive terminal. See Figure 23–7. Electron flow (current) is thus established through the semiconductor and any external circuitry connected to it. The current is carried through the semiconductor by electrons. Because electrons have a negative electrical charge, this type of semiconductor is called an N-type (negative-type) semiconductor.

The physical principle of conduction of electrons through an N-type semiconductor is slightly different from electron flow through a good conductor. For practical purposes, though, you can think of electron flow through an N-type semiconductor as ordinary electron flow through a common conductor.

FIGURE 23-7 Current flow through an N-type semiconductor material

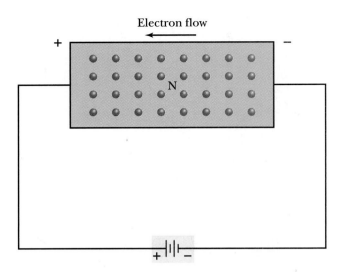

P-type semiconductors

A **P-type semiconductor material** is one that has a shortage, or deficiency, of electrons. As with N-type semiconductors, the basic material is a highly purified tetravalent semiconductor such as silicon or germanium. The impurity, or doping material, for a P-type semiconductor is a trivalent element such as gallium or antimony. A trivalent doping material contributes covalent bonds that are made of three electrons instead of four. Such bonds are missing one electron, leaving a hole where a fourth electron would normally reside.

Every covalent bond that contains a hole is an unstable bond. This means that a little bit of external energy can cause the hole to be filled by an electron from a nearby tetravalent bond. The original hole is thus filled and its bond made more stable, but a hole is then left in the bond that gave up its electron. See Figure 23-8. We can say that an electron moved to fill a hole, but left a hole behind. It is more proper to say, however, that the hole moved. Applying an external voltage to a P-type semiconductor causes holes to drift from the point of positive charge to the point of negative charge. Holes, in other words, behave as positive charges. This is why such semiconductors are called P-type (positive-type) semiconductors.

More about N- and P-type semiconductor materials

A highly purified semiconductor material that has not yet been doped is known as an *intrinsic semiconductor*. Once the doping material is added, the material becomes an *extrinsic semiconductor*.

N-type semiconductor materials are formed by adding minute amounts of a pentavalent element to the intrinsic semiconductor. This forces a few of the material's pentavalent atoms to provide a spare electron. The doping atoms in this instance are called *donor atoms* because they "donate" extra electrons to the material. The charge carriers in an N-type semiconductor are electrons, which are said to be the **majority carriers**, while holes are the **minority carriers**.

P-type semiconductors are formed by adding a tiny amount of one of the trivalent doping materials. This leaves a portion of the valence shells with a short-

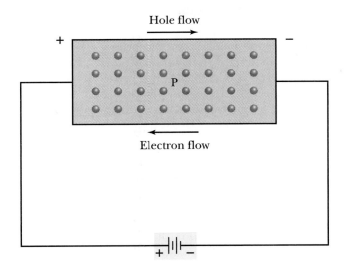

Hole flow

Electron flow

FIGURE 23–8 Current flow through a P-type semiconductor material

age of electrons (or excess holes). Because a trivalent doping material leaves holes that can subsequently accept electrons from other bonds, it is called an *acceptor atom*. The charge carriers in a P-type material are positively charged holes. Holes are the majority charge carriers in this instance, and electrons are said to be the minority carriers.

It is important to realize that N- and P-type materials are not electrically charged. One might suppose that an N-type material would possess a negative charge because it contains an excess number of electrons; and by the same token, a P-type material would have an inherent positive charge because of its shortage of electrons (or oversupply of positively charged holes). This is not so. Semiconductor materials cannot supply negative and positive charges to an external circuit as batteries do, for instance.

IN-PROCESS LEARNING CHECK I

Fill in the blanks as appropriate.

1. Before doping (in their pure form) semiconductor materials are sometimes called _____ semiconductors. Once they are doped with tiny amounts of an impurity atom, they are called _____ semiconductors.

2. N-type materials are formed by doping a _____ valent semiconductor material with a _____ valent material. P-type materials are formed by doping a _____ valent semiconductor material with a _____ valent material.

3. The doping atoms for N-type materials are called _____ atoms because they donate extra electrons to the covalent bonds. The doping atoms for P-type materials are called _____ atoms because they accept electrons that will fill the holes in the covalent bonds.

4. The majority carriers in an N-type material are _____, and the minority carriers are _____. The majority carriers in a P-type material are _____, and the minority carriers are _____.

23-2 THE P-N JUNCTION

N- and P-type semiconductor materials are rarely used alone. Practical semiconductor devices use both materials. Semiconductor *diodes,* for example, use a section of N-type material that is chemically and electrically fused to a section of P-type material (see Figure 23–9a). As you will discover in Chapter 26, common transistors are composed of one type of semiconductor sandwiched between sections of the other type: an N-type material between two sections of P material (called a *PNP transistor*), or a P-type material fit between two sections of an N material (called an *NPN transistor*). See the examples in Figures 23–9b and 23–9c.

The region where a P-type material is chemically and electrically fused with an N-type material is called a **P-N junction**. At the P-N junction within a semiconductor device, an excess number of electrons from the N-type material come into contact with an excess number of holes from the P-type material (see Figure 23–10a). Since there are free electrons in the N-type material near the junction of the P- and N-type materials, they are attracted to the positive holes near the junction in the P-type material. Some electrons leave the N-type material to fill

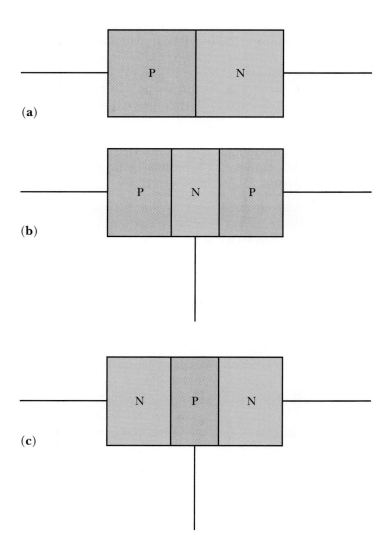

(a)

(b)

(c)

FIGURE 23–9 Practical devices made from combinations of N- and P-type semiconductor materials: **(a)** diode; **(b)** PNP junction transistor; **(c)** NPN junction transistor

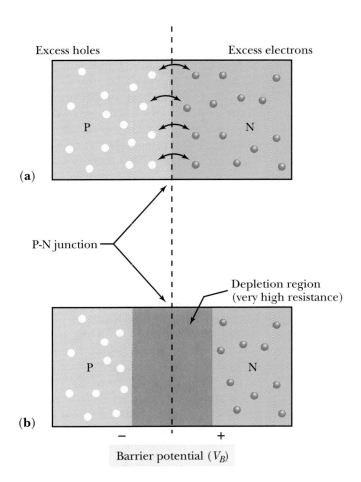

Excess holes   Excess electrons

P   N

(a)

P-N junction

Depletion region
(very high resistance)

P   N

(b)

−   +

Barrier potential (V_B)

FIGURE 23–10 The depletion region of a P-N junction: **(a)** Holes and electrons recombine at the junction; **(b)** the recombination of holes and electrons quickly forms the depletion region.

the holes in the P-type material. And while electrons are leaving the N-type material in this fashion, they leave behind positive hole charges. In other words, equal numbers of electrons and holes flow across the P-N junction. Electrons move from the N-type material into the P-type material, and holes from the P-type material to the N-type material.

The result of this activity is that the N-type material now has an electron deficiency and is positively charged because of positive ions near the P-N junction, and the P-type material has a surplus of electrons and is negatively charged because of negative ions near the junction. This causes a potential difference near the junction that is known as the **barrier potential** or *contact potential* (V_B). See Figure 23–10b. The barrier potential for silicon materials is 0.7 V, while that of germanium-based semiconductors is about 0.3 V.

Current flowing through the junction without an externally applied source of energy is called the **diffusion current**. Diffusion current flows only long enough to establish the barrier potential and maintain it when an external source of energy—such as heat, light, or an electrical potential—is applied.

The junction region where the current carriers of both materials have balanced out is called the **depletion region**. This indicates that the region has been depleted of majority carriers because of the diffusion current. And because there are virtually no majority carriers (excess holes or excess electrons) in this region of the semiconductor, the region acts as a good insulator. Study these details in Figure 23–10b.

The barrier potential between the N and P layers in a semiconductor device is usually between 0.3 V and 0.7 V, depending on the exact nature of the materials as well as outside influences such as temperature. Although it is not possible to measure the barrier potential directly, you will learn how it influences the operation of the semiconductor devices in later chapters. Also, it is important to realize that the depletion region is quite thin, measuring only micrometers (millionths of a meter) across. The actual thickness of this insulating region depends on a number of factors, but the most significant factor for our purposes is the polarity and amount of voltage that is being applied to the semiconductor.

Biasing P-N junctions

Figure 23–11 shows a voltage source applied to the ends of a two-layer (P-N) semiconductor device. In this instance, the positive terminal of the source is applied to the P-type region and the negative terminal to the device's N-type region. In the first figure, Figure 23–11a, the switch is open and no potential is applied to the material. There is no current flow in the depletion region. Any motion of majority carriers (electrons in the N-type material and holes in the P-type material) is low level and random, largely because of heating effects of the surrounding air.

The instant the switch is closed, Figure 23–11b, a negative potential is applied to the N-type material and a positive potential to the P-type material. Both the electron majority carriers in the N-type material and the hole majority carriers in the P-type material are forced toward the P-N junction.

If the applied voltage is greater than the contact (barrier) potential of that particular P-N junction, the oncoming flood of majority carriers eliminates the charges in the depletion region. See Figure 23–11c. Positive hole charges in the depletion region are fully occupied with electrons from the N-type material, and electrons in the depletion region are all fit into holes coming from the P-type material. The result is the total elimination of the non-conductive depletion region and the establishment of a steady-state electron current flow through the device.

A P-N junction that is conducting in this fashion—with electron current flow from the N-type material to the P-type material—is said to be **forward biased**. This is possible only when an external power supply is connected so that its negative supply is connected to the semiconductor's N-type material and the positive side of the power supply is connected to the P-type material.

If an external power source is connected with the polarities reversed from the direction just described, current cannot flow through the junction. Figure 23–12 shows a P-N junction that is **reverse biased**. When a positive potential is applied to the N-type material, electron majority carriers are drawn away from the junction's depletion region; and by the same token, a negative potential applied to the P-type material draws holes away from the junction. The overall result is that the depletion region is actually widened, Figure 23–12b, thus becoming a very good barrier to current flow through the device.

When you take a moment to compare the effects of forward and reverse bias on a P-N junction, you may begin to see one of the most important properties of P-N junctions: *A P-N junction conducts current in only one direction.*

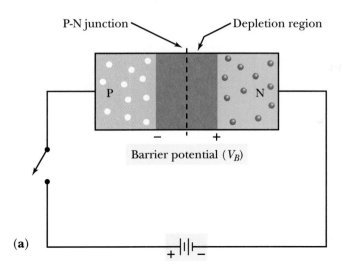

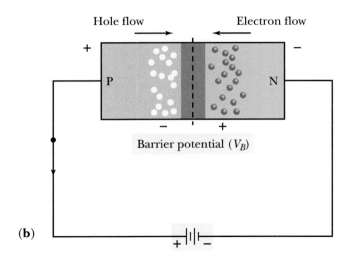

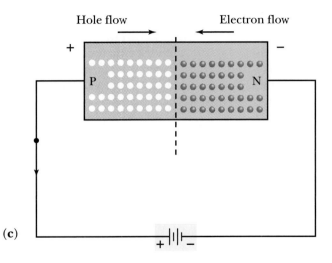

FIGURE 23–11 A forward-biased P-N junction: **(a)** no voltage applied to the semiconductor; **(b)** immediate reaction to closing the switch; **(c)** current flow established through the circuit

FIGURE 23–12 A reverse-biased P-N junction: **(a)** no voltage applied to the semiconductor; **(b)** current cannot flow through the device

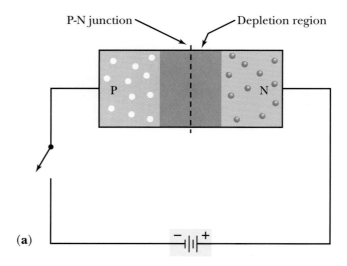

(a)

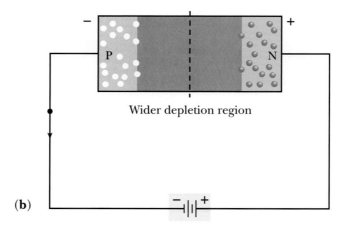

(b)

Characteristic curves for P-N junctions

From the previous discussion, you can conclude that forward bias occurs when the external source voltage causes the P-type material to be positive with respect to the N-type material. Under these circumstances, as bias voltage increases from zero, conduction similar to that shown in Figure 23–13 results. Little current flows until the barrier potential of the diode is nearly overcome. Then, there is a rapid rise in current flow. The typical barrier potential for a silicon P-N junction is about 0.7 V. Once you reach this point on the conduction curve, you can see that the voltage increases very little compared with large increases in forward current.

You have also seen that reverse biasing a P-N junction means we connect the negative terminal of the external source to the P-type material and the external positive terminal to the N-type material. As you increase the amount of applied voltage, only a minute amount of reverse current (leakage current) flows through the semiconductor by means of minority carriers. This reverse current

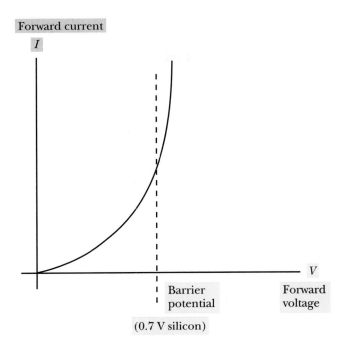

FIGURE 23–13 Forward-bias portion of the I-V curve for a typical P-N junction

remains very small until the reverse voltage is increased to a point known as the **reverse breakdown voltage** level, Figure 23–14. Before reaching this breakdown point, the leakage current in silicon P-N junctions is on the order of a few microamperes. For germanium junctions, the leakage current is less than a milliampere. The symbol for leakage current is I_{CO}.

An interesting effect occurs when you exceed the reverse breakdown voltage level of a P-N junction. The reverse current increases dramatically, but the voltage across the junction remains fairly constant once you reach the breakdown level.

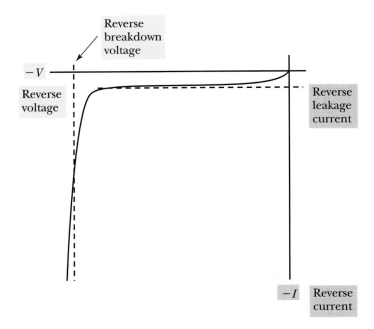

FIGURE 23–14 Reverse-bias portion of the I-V curve for a typical P-N junction

FIGURE 23–15 A single I-V curve showing both the forward- and reverse-bias characteristics of a typical P-N junction

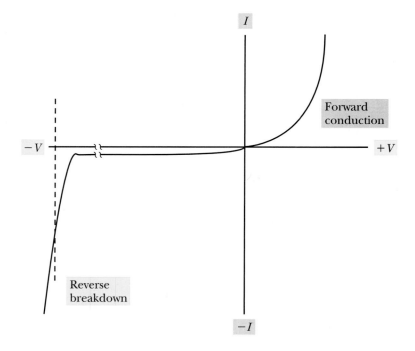

The reverse breakdown voltage level of a P-N junction is mainly determined at the time of its manufacture. These voltage specifications can be as low as 5 V for certain special applications, are frequently between 100 V and 400 V, and may be as high as 5,000 V for other special applications.

The curves for forward conduction current, Figure 23–13, and reverse leakage and breakdown current, Figure 23–14, are called *I-V curves*. This name is appropriate because the curves show the amount of current flow (I) as you vary the voltage (V) applied to the P-N junction. Forward and reverse I-V curves are usually combined into a single graph, as shown in Figure 23–15.

SUMMARY

▶ Semiconductor materials have conductivity levels falling in a range between good conductors and good insulators.

▶ Adding energy (electricity, heat, or light) to an atom causes its electrons to gain energy. In conductors and semiconductors, valence electrons that gain energy will jump up to a conduction band. When conduction-band electrons fall back to their valence bands, they give up the energy they had gained from an outside source. They usually give up their energy in the form of heat, but certain semiconductor materials emit light as well.

▶ Semiconductor materials such as silicon and

germanium have four electrons in their valence shell.

▶ Introducing selected impurities into the crystal structure of intrinsic semiconductor materials is called *doping*. Doping a tetravalent material with a pentavalent material introduces donor atoms in the structure that donate relatively free electrons. This produces an N-type semiconductor material. Doping a tetravalent material with a trivalent material introduces acceptor atoms into the structure that easily accept electrons to fill the holes where the covalent bonds lack an electron. This produces a P-type semiconductor material.

▶ The majority carriers in N-type semiconductor

materials are electrons. In P-type materials, the majority carriers are holes.

▶ When P- and N-type materials are formed together, a P-N junction is created at the area of contact. Near the junction, a depletion region is created by electrons from the N-type material moving in to fill holes in the P-type material, and holes moving in the opposite direction (from the P-type material) to combine with available electrons. The depletion region is electrically neutral, but separates the N- and P-type materials, which have a difference in potential called the barrier potential (or junction voltage). This potential is positive on the N-type side and negative on the P-type side of the depletion region.

▶ Applying an external voltage to a P-N junction causes the semiconductor to conduct current freely (forward bias) or act as a good insulator (reverse bias). This depends on the polarity of the applied voltage. The junction is forward biased when the applied voltage is positive to the P-type material and negative to the N-type material. On the other hand, the junction is reverse biased when the applied voltage is negative to the P-type material and positive to the N-type material.

▶ The I-V curve for a P-N junction shows that forward conduction begins when the applied voltage reaches the junction's barrier potential. The curve also shows that very little current flows through a P-N junction that is reverse biased until the applied voltage reaches the reverse breakdown voltage level.

FORMULAS AND SAMPLE CALCULATOR SEQUENCES

FORMULA 23–1 $2n^2$

shell number, ⬛x^2 , ⬛$\times$, 2, ⬛$=$

REVIEW QUESTIONS

1. Explain the meaning of the following terms:
   a. Trivalent
   b. Tetravalent
   c. Pentavalent

2. List the number of valence electrons that are found in:
   a. silicon.
   b. arsenic.
   c. gallium.

3. For an N-type semiconductor material, cite:
   a. the name of the majority carrier.
   b. the name of the minority carrier.

4. For a P-type semiconductor material, cite:
   a. the name of the majority carrier.
   b. the name of the minority carrier.

5. Indicate the typical barrier potential for:
   a. a silicon P-N junction.
   b. a germanium P-N junction.

6. Which type of semiconductor material occurs when you dope a tetravalent material with a pentavalent material?
   a. N-type material
   b. P-type material

7. Which type of semiconductor material occurs when you dope a tetravalent material with a trivalent material?
   a. N-type material
   b. P-type material

8. If the electrons in a P-type material are flowing from left to right, the holes in the same material are:
   a. flowing from left to right.
   b. flowing from right to left.
   c. not flowing at all because they are the minority carriers.

9. Draw a P-N junction semiconductor showing an external dc power supply connected for forward biasing.

10. Draw a P-N junction semiconductor showing an external dc power supply connected for reverse biasing.

11. Sketch the forward-bias portion of the I-V curve for a P-N junction. Indicate the barrier potential on the $+V$ axis.

12. Sketch the reverse-bias portion of the I-V curve for a P-N junction. Indicate the breakdown voltage level on the $-V$ axis.

PROBLEMS

1. What is the maximum number of electrons that can occupy the fourth shell from the nucleus in a normal atom?

2. What is the maximum number of electrons that can occupy the second shell from the nucleus in a normal atom?

3. Which atomic shell has a maximum of 18 electrons?

1. Explain the main difference between valence-band and conduction-band electrons.

2. Explain the main difference between intrinsic and extrinsic semiconductor materials.

3. Describe how increasing and decreasing the amount of forward-bias voltage affects the thickness of the depletion region of a P-N junction.

4. Explain why the conductance of a P-N junction is much greater when it is forward biased than when it is reverse biased.

5. Explain why it can be correctly said that holes follow conventional current flow and electrons follow electron flow.

6. Applying energy, such as heat and light, to a semiconductor increases carrier activity. Explain how this accounts for the fact that semiconductors tend to be more conductive as their temperature rises. Compare this effect with the reaction of a normal conductive material such as copper.

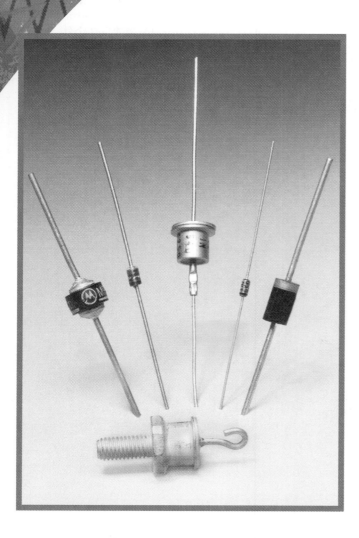

DIODES AND DIODE CIRCUITS

24

chapter

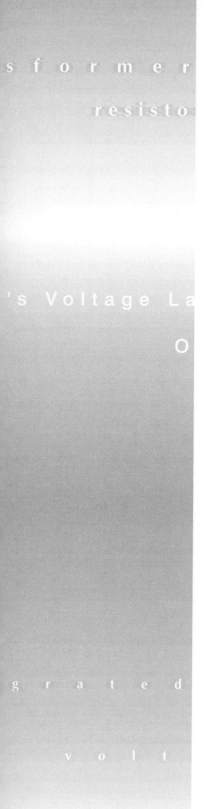

CHAPTER PREVIEW

You learned in Chapter 23 that a P-N junction normally allows current to pass in only one direction. This normal direction of current flow occurs when the junction is forward biased. This chapter introduces you to a practical application of P-N junctions in the form of semiconductor diodes.

You will discover that diodes are used for allowing or stopping electron flow and for controlling the direction of current flow through useful electronic circuits. Diode specifications include limitations on the amount of current they can carry and the amount of reverse-bias voltage they can withstand before breaking down. As long as you use diodes properly and within their specified limits, they are very reliable devices.

Rectifier diodes are typically used for power supply applications. You will see them introduced as elements that convert ac power to dc power. Switching diodes, on the other hand, have lower current ratings than rectifier diodes, but you will see that switching diodes can function better in clipping and clamping operations that deal with short-duration pulse waveforms.

Zener diodes are introduced in this chapter as a special type of P-N junction device. These diodes are very commonly used as voltage-level regulators and protectors against high voltage surges.

Optoelectronic devices combine the effects of light and electricity. Certain kinds of P-N junction diodes are used in optoelectronics. Light-emitting diodes (LEDs), for example, emit light energy when current flows through them. Photodiodes, on the other hand, vary in their P-N junction conduction as the intensity of light falling onto them varies. Another important optoelectronic diode is the laser diode.

Some high frequency applications of junction diodes include tunnel diodes (used for producing high frequency oscillators) and varactor diodes (used mainly as a voltage-controlled capacitor in high frequency tuning circuitry).

OBJECTIVES

After studying this chapter, you should be able to:

1. Describe how to connect a dc source to a **junction diode** for forward bias and for reverse bias

2. Determine the distribution of voltages in a series circuit consisting of a junction diode and resistor

3. Sketch the waveforms found in an ac circuit consisting of a junction diode and resistor

4. Explain the function of diode **clamping** and **clipper circuits**

5. Understand the meaning and importance of diode specifications

6. Describe the operation and specifications for **zener diodes**

7. Explain the function of a simple zener diode circuit

8. Describe the operation of **LEDs** and **photodiodes**

9. Determine the value of resistor to be placed in series with an LED for proper operation

10. Describe the purpose of **laser diodes, tunnel diodes,** and **varactor diodes**

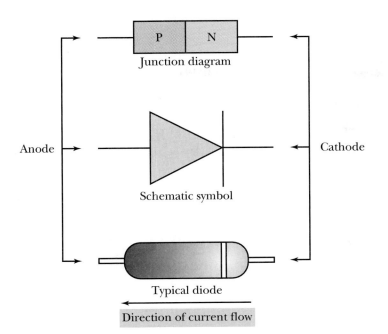

FIGURE 24–1 A P-N junction diode

24–1 P-N JUNCTION DIODES

A P-N junction is formed wherever a P-type semiconductor material comes into contact with an N-type semiconductor. You have seen that current flows easily only in one direction through a P-N junction. Figure 24–1 shows a very practical application of a P-N junction in the form of a semiconductor *diode.* Technically speaking, the diode is known as a **P-N junction diode**, or simply *junction diode.* The figure also shows the schematic symbol for a diode. The arrow represents the **anode**—the P-type material. The straight bar or line represents the **cathode**—the N-type material. Many diodes tend to resemble carbon resistors. Usually there is only one color band near one end of a diode that marks the cathode.

How a diode works in a circuit

The circuits in Figure 24–2 show basically how a diode behaves in a simple kind of circuit. In the first drawing, the diode is forward biased. This means the negative terminal of the power source is connected to the cathode (N-type material) and the positive terminal to the anode (P-type material). Because this diode is forward biased, current is allowed to flow freely through the circuit.

The second diagram in Figure 24–2 shows the same diode, but with the terminals of the power source reversed so that the diode is reverse biased. This means the positive terminal of the power source is connected to the cathode (N-type material) and the negative terminal to the anode (P-type material). Perhaps a very small amount of reverse leakage current will flow through this circuit, but for all practical purposes, we can say that there is no current flow.

Figure 24–3 (see page 801) demonstrates the operation of a semiconductor diode in a circuit that includes a resistor as well as a diode and dc power source. When the power source is such that the diode is forward biased as in Figure 24–3a, current flows through the circuit. The voltage you measure across the diode (from anode to cathode) is equal to the diode's P-N barrier potential—about 0.7 V for silicon diodes.

FIGURE 24–2 Current action in a junction diode that is forward biased and one that is reverse biased

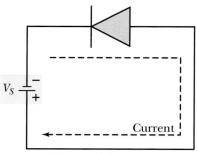

Forward-biased diode

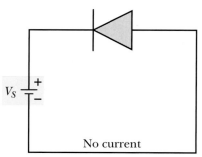

Reverse-biased diode

According to Kirchhoff's voltage law, the voltage across the resistor has to be equal to the source voltage minus the voltage across the diode: $10\,V - 0.7\,V = 9.3\,V$.

The diode in Figure 24–3b is reverse biased. You can see that the negative side of the power source is connected to the anode of the diode. In this instance, the diode blocks current from flowing through the circuit. In other words, the diode acts as an open switch. No current flows through the circuit, all of the source voltage is dropped across the diode, and there is no voltage across the resistor.

You can use some simple formulas to help understand the operation of diode circuit and determine the possible causes of troubles in simple diode circuits. Formula 24–1 determines the voltage across the resistor (V_R) as the power supply voltage (V_S) minus the voltage across the diode (V_D). The amount of voltage across the diode depends on the polarity of the power source (whether the diode is forward or reversed biased) and the type of semiconductor material (e.g., silicon or germanium). Here is a summary of these effects:

1. When the diode is reverse biased, the voltage across the diode is equal to the source voltage: $V_D = V_S$.

2. When the diode is forward biased, the voltage across the diode is equal to the diode's P-N junction barrier potential: about 0.7 V for a silicon diode, and about 0.3 V for a germanium diode.

FORMULA 24–1 $V_R = V_S - V_D$

Example

Suppose you have a forward-biased silicon diode connected through a series resistor to a 12.6-Vdc power supply. What is the voltage across the diode? across the

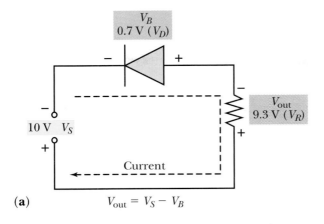

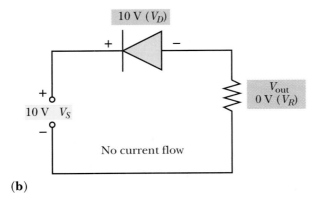

FIGURE 24–3 Distribution of voltages in a simple diode circuit

resistor? Because this is a forward-biased silicon diode, the voltage dropped across it is close to 0.7 V.

Answer:

According to Formula 24–1, the voltage across the resistor is:

$$V_R = V_S - V_D = 12.6 \text{ V} - 0.7 \text{ V} = 11.9 \text{ V}$$

PRACTICE PROBLEMS I

1. What is the voltage across silicon diode D_1 in Figure 24–4a?
2. What is the voltage across resistor R_1 in Figure 24–4a?
3. What is the voltage across silicon diode D_1 in Figure 24–4b?
4. What is the voltage across resistor R_1 in Figure 24–4b?

You have seen how you can determine the amount voltage dropped across a series circuit that is made up of a diode and resistor in series with a dc power source. If you know the value of the resistor, you can also calculate the amount of current flowing through the circuit. And because we are still dealing with a series circuit, you know that the amount of current flowing through the diode has to be equal to the amount of current flowing through the resistor.

Formula 24–2 is a form of Ohm's law that lets you calculate the resistor current (I_R) as the voltage across the resistor (V_R) divided by the value of the resistor (R).

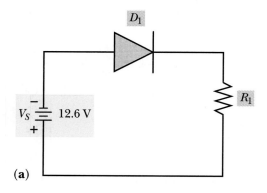

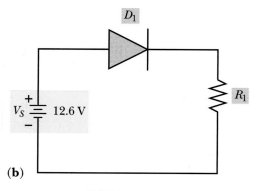

FIGURE 24–4

FORMULA 24–2 $I_R = V_R/R$

So if you have previously found that the voltage across the resistor is 5.5 V and the value of the resistor is 10 kΩ:

$$I_R = V_R/R$$

$$I_R = 5.5 \text{ V}/10 \text{ k}\Omega = 0.55 \text{ mA}$$

A more practical version of Formula 24–2 would be one that lets you calculate the current through the resistor in terms of the supply voltage (V_S), voltage across the diode (V_D), and the value of the resistor (R). From Formula 24–2:

$$I_R = V_R/R$$

From Formula 24–1:

$$V_R = V_S - V_D$$

Substituting Formula 24–1 into Formula 24–2:

$$I_R = (V_S - V_D)/R$$

So the value of the current flowing through a circuit consisting of a dc power source and a diode connected in series with a resistor is given by Formula 24–3.

FORMULA 24–3 $I_R = (V_S - V_D)/R$

FIGURE 24–5

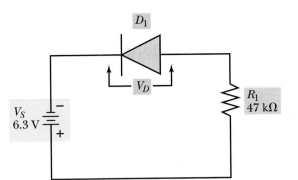

Example

Suppose you are working with the circuit shown in Figure 24–5. If D_1 is a silicon diode, determine the current through the circuit.

Answer:

From Formula 24–3:

$$I_R = (V_S - V_D)/R$$

$$I_R = (6.3\text{ V} - 0.7\text{ V})/47\text{ k}\Omega = 119\ \mu\text{A}$$

The current through the resistor, as well as the diode, is 119 μA.

If the diode in Figure 24–5 is turned around so that it is reverse biased, the voltage across the diode would be equal to the source voltage. So $V_S - V_D$ in Formula 24–3 would be 0 V and, as a result, the current through the resistor would also be 0 A:

$$I_R = (V_S - V_D)/R$$

$$I_R = (6.3\text{ V} - 6.3\text{ V})/47\text{ k}\Omega$$

$$I_R = 0/47\text{ k}\Omega = 0\text{ A}$$

PRACTICE PROBLEMS II

Referring to the circuit in Figure 24–6:

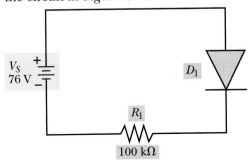

FIGURE 24–6

1. What is the voltage across silicon diode D_1?
2. What is the voltage across resistor R_1?
3. How much current is flowing through resistor R_1?

General diode ratings

Diodes have a number of ratings or specifications. Some ratings are more important than others, most often depending on the application for which the diode is designed. Typically, in most practical cases, if you give careful attention to the most important diode ratings, the ratings of lesser importance automatically fall into line.

There are three particular diode ratings that apply in one way or another to all types of diodes and applications.

1. *Forward voltage drop, V_F.* As you have seen many times through earlier discussions, the forward voltage drop (or barrier potential) is the forward-conducting junction voltage level: about 0.3 V for germanium diodes and 0.7 V for silicon.

2. *Average forward current, I_O.* This is the maximum amount of forward conduction that the diode can carry for an indefinitely long period. If the average current exceeds this value, the diode will overheat and eventually destroy itself.

3. *Peak reverse voltage, V_R.* The peak reverse voltage (PRV) is sometimes called the *reverse breakdown voltage.* This is the largest amount of reverse-bias voltage the diode's junction can withstand for an indefinitely long period. If a reverse voltage exceeds this level for more than a few seconds, the voltage will "punch through" the depletion layer and allow current to flow backwards through the diode.

Excessive forward current and reverse breakdown are the most common causes of diode failure. In both cases, the diode gets very hot first, then the P-N junction is destroyed by the heat. If the heating effect is not held in check, the diode eventually cracks in two and fizzles in a puff of smoke. If the excessive forward current or reverse voltage is removed within a few seconds, a diode can sometimes survive the ordeal and return to normal operation after a cooling-off period. But when the diode doesn't survive, it must be replaced.

IN-PROCESS LEARNING CHECK I

Fill in the blanks as appropriate.

1. To cause conduction in a semiconductor diode, the diode must be _____ biased.

2. To reverse bias a P-N junction, connect the negative source voltage to the _____ -type material and the positive source voltage to the _____ -type material.

3. The anode of a junction diode corresponds to the _____ -type material, and the cathode corresponds to the _____ -type material.

4. The forward conduction voltage drop across a silicon diode is approximately _____ V.

5. When a diode is reverse biased in a series circuit, it acts like a(n) _____ (open, closed) switch.

6. There will be no current flow through a series diode circuit when the diode is _____ biased.

7. When a diode is connected in series with a resistor, the voltage across the resistor is very nearly equal to the dc source voltage when the diode is _____ biased.

8. The three most general diode ratings are _____, _____, and _____.

24–2 RECTIFIER DIODES

Rectifier diodes are used where it is necessary to change an alternating-current power source into a direct-current power source. (You will study this important application of rectifier diodes in more detail in Chapter 25.) Rectifier diodes are the most rugged and durable of the semiconductors in the junction diode family. They are especially noted for their relatively large average forward current and reverse breakdown voltage ratings. Figure 24–7 shows some common rectifier diodes and their specifications.

PRACTICAL NOTES

Part identification numbers for diodes usually begin with the designator 1N. So if a part is labeled 1N5243, you can safely assume that it is some sort of diode. You might not be able to tell exactly what sort of diode it is—you need a catalog or data sheet to do that—but at least you can identify it as a diode.

Part Identification Code	I_O (A)	V_R (V)
1N4001	1.0	50
1N4002	1.0	100
1N4003	1.0	200
1N4004	1.0	400
1N4007	1.0	1000
1N5400	3.0	50
1N5401	3.0	100
1N5402	3.0	200
1N5404	3.0	400
1N5408	3.0	1000

FIGURE 24–7 Typical ratings for rectifier diodes

FIGURE 24–8 High-current diode *(Photo courtesy of International Rectifier)*

The physical size of a rectifier diode is an indication of its current-carrying capacity. Actually, heating is the only factor that limits the forward current-carrying capacity of a diode. So when a diode is made physically larger, it has more surface area for dissipating heat. Also, you can dissipate unwanted diode heat by mounting the diode onto a **heat sink**. A heat sink is a plate of aluminum or an arrangement of grooves and fins made of aluminum that carries away the heat generated by a semiconductor and delivers it efficiently into the surrounding air. Figure 24–8 shows a *stud mounted,* high current diode.

Figure 24–9 shows the ratings for some stud mounted, high current, rectifier diodes.

PRACTICAL NOTES

1. *The maximum voltage applied to a rectifier diode should not exceed the diode's reverse breakdown voltage level.* So if a diode is being used in a circuit that is operated from a 120-V (rms) source, the diode's reverse breakdown rating must exceed the peak value of the applied ac voltage:

$$V_{pk} = 1.414 \times V_{rms}$$

$$V_{pk} = 1.414 \times 120 \text{ V} = 170 \text{ V}$$

So the reverse breakdown rating of the diode must be greater than 170 V. A design rule of thumb suggests doubling the calculated reverse voltage value. In this example, the doubled value is 340 V. Because no diodes are rated at exactly 340 V, it is good practice to go to the next higher standard rating: 400 V.

2. *The maximum current through the diode portion of a circuit should not exceed the diode's average forward current rating.* If a silicon diode is connected (and forward biased) to a

Part Identification Code	I_O (A)	V_R (V)
1N1199A	12	50
1N1183A	40	50
1N1200A	12	100
1N1184A	40	100
1N1185A	35	150
1N1202A	12	200
1N1186A	40	200
1N1187A	40	300
1N1204A	12	400
1N1188A	40	400
1N1189A	40	500
1N1206A	12	600
1N1190A	40	600

FIGURE 24–9 Typical ratings for high current, stud mounted, rectifier diodes

12-Vdc source through a 100-Ω resistor, Formula 24–3 tells us the average forward current is:

$$I_R = (V_S - V_D)/R$$

$$I_R = (12 \text{ V} - 0.7 \text{ V})/100 \ \Omega$$

$$I_R = 11.3/100 = 0.113 \text{ A}$$

This shows that the diode should have an average forward current rating of at least 0.113 A. Of the common silicon diodes used as power supply rectifiers on the market today, the smallest have I_O ratings of 1.0 A. So using a 1-A rectifier diode in this circuit would be a practical choice.

You can use a VOM ohmmeter to test the condition of a diode that is removed from a circuit. Refer to Figure 24–10 as you take note of the following procedures. (NOTE: If a DDM is used, you would use the device's "diode checker" circuit rather than the ohmmeter function.)

1. Set the ohmmeter for a midrange scale: $R \times 100$ for an analog ohmmeter.

2. Connect the ohmmeter leads to the diode, positive lead to the anode and negative lead to the cathode. Note the resistance (you don't have to be exact).

3. Reverse the ohmmeter connections to the diode so that the positive lead is now to the cathode and the negative lead to the anode. Note the resistance (again, you don't have to be exact).

FIGURE 24–10 Ohmmeter
tests for diodes

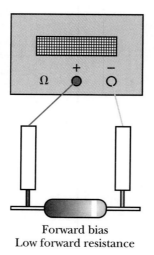
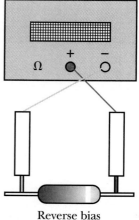

Forward bias
Low forward resistance

Reverse bias
High reverse resistance

If the diode is in good working order, you should find a much higher resistance
when the diode is reverse biased than when it is forward biased. The diode is defective (usually shorted) when the forward and reverse ohmmeter readings are
in the same general range.

24–3 SWITCHING DIODES

A second type of junction diode is designed for high frequency applications such
as the circuits commonly found in communications equipment and in computers. They are also used in circuits where diode action has to take place reliably
in a very short period. These diodes, called *switching diodes,* usually have low current and voltage ratings, and they are no larger than the smaller rectifier diodes.
(Most switching diodes are less than a quarter-inch long, and the surface mount
versions are only about 0.1 inch square.) Figure 24–11 shows the specifications
for several typical switching diodes.

Part Identification	V_R (V)	V_F	I_R	t_{rr} (ns)
1N4150	50	0.74 V @ 10 mA	100 nA @ 50 V	4
1N4153	75	0.88 V @ 20 mA	50 nA @ 50 V	2
1N4151	75	1.00 V @ 50 mA	50 nA @ 50 V	2
1N4148	75	1.00 V @ 10 mA	25 nA @ 20 V	4
1N914B	75	0.72 V @ 5 mA	25 nA @ 20 V	4
1N914A	75	1.00 V @ 20 mA	25 nA @ 20 V	4
1N914	75	1.00 V @ 10 mA	25 nA @ 20 V	4
1N4448	75	0.72 V @ 5 mA	25 nA @ 20 V	4

FIGURE 24–11 Typical
ratings for switching diodes

Switching diodes are made in such a way that they can respond very quickly to changes in the polarity of the signal applied to them. This *reverse recovery time* (t_{rr}) rating is less than 50 ns (nanoseconds), or 50×10^{-9} s. This means the diode will recover from a reverse-voltage state and begin forward conducting within 50 ns. By contrast, the bulkier construction of rectifier diodes gives them a longer recovery time. A typical reverse recovery time for a rectifier diode is on the order of 30 μsec, which is 600 times longer than a typical switching diode.

Diode clipper circuits

A diode **clipper circuit** removes the peaks from an input waveform. There are a number of reasons for using such a circuit. A diode clipper can, for example, be used in waveshaping circuits to get rid of unwanted spikes. Diode clippers are also called *limiter* circuits.

There are two basic kinds of diode clipper circuits: series and shunt (parallel) clippers. In a series diode clipper, the diode is connected in series between the circuit's input and output terminals. In a shunt diode clipper, the diode is connected in parallel between the input and output terminals of the circuit.

Series diode clipper

Figure 24–12 shows a pair of series diode clippers. The only difference between the two is that the circuit in Figure 24–12a clips off the negative portion of the input waveform, while the other circuit clips off the positive portion.

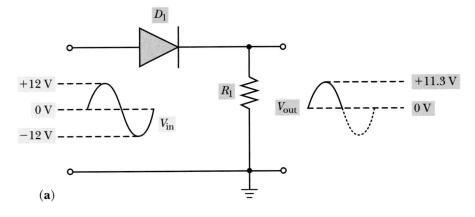

(a)

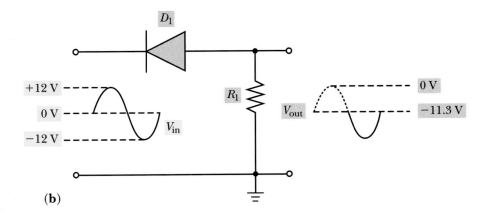

(b)

FIGURE 24–12 Series diode clippers: **(a)** negative clipper; **(b)** positive clipper

The clipping action of these series clippers works according to the principle that the diode will conduct current only when it is forward biased. In Figure 24–12a, the diode is forward biased on the positive portion of the input waveform, thereby allowing the positive portion of the waveform to be duplicated across the output resistor. During the negative half-cycle, on the other hand, the diode is reverse biased and cannot pass that part of the input waveform to the output resistor. In Figure 24–12b, the diode is turned around so that it is forward biased on the negative portion of the input waveform.

Shunt diode clipper

Figure 24–13 shows examples of shunt clippers. Note that the diodes are in parallel with this output. That is why they are called shunt clippers. In Figure 24–13a, the diode conducts while the input waveform is positive. When a diode is forward biased and conducting in this fashion, the voltage across it is fixed fairly close to the diode's barrier potential. When a silicon switching diode is used, this voltage is about 0.7 V. So during the positive half-cycle of the input waveform in Figure 24–13a, the output is fixed rather close to a steady 0.7-V level. When the input waveform goes negative, however, the diode is reverse biased and the entire negative half-cycle appears across the diode and output of the circuit. So in this case, the positive half-cycle is effectively clipped from the input waveform.

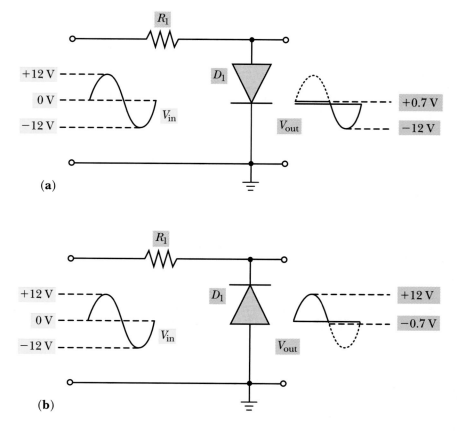

FIGURE 24–13 Shunt (parallel) diode clippers: **(a)** positive clipper; **(b)** negative clipper

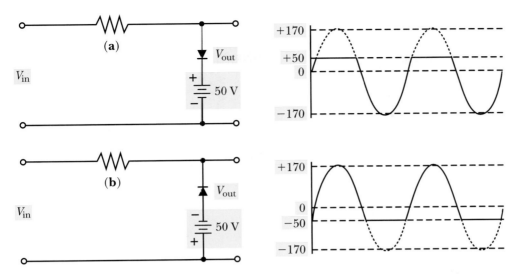

FIGURE 24–14 Biased shunt-diode clippers: **(a)** biased positive clipper; **(b)** biased negative clipper

The simple shunt clipper in Figure 24–13b is identical with the first circuit except that the diode is reversed in its place across the output. The input waveform is clipped, or limited, during the negative portion of the input cycle. The diode is reverse biased during the positive half-cycle, so that is the portion of the input waveform that appears at the output terminals of the circuit.

Biased diode clippers

An interesting variation of the basic diode clipping action is to bias the diodes purposely so they limit the waveform to a value other than zero. Examples of *biased diode clippers* are shown in Figure 24–14.

In both examples, the circuit is biased at 50 Vdc and the peak input voltage is 170 V. You can see that the positive limiter in Figure 24–14a is a +50-V clipper. Everything above that bias level is clipped off. The negative half of the input waveform is passed to the output without change. The clipping action is of the opposite polarity in Figure 24–14b. Here, the negative swing is clipped at −50 V, and the positive half-cycle is unchanged.

The dc source that is used for setting the bias level for these clipper circuits can be a battery, but you will find that a zener diode (which you will study very soon) is simpler and less expensive. Most electronic circuits that require clipping action of any kind now accomplish the job with operational amplifiers (which you will study in Chapter 29).

Diode clamping circuits

A **clamping circuit** or **clamper** is one that changes the baseline voltage level of a waveform. Unless stated otherwise, you can safely assume that the baseline level for a waveform is zero. A 160-V (peak) sine waveform, for instance, normally goes to +160 V and swings down through 0 V on its way to −160 V. The 0-V level is the exact middle of that voltage swing. It is the baseline voltage. Sometimes, however, it is desirable to change the baseline level.

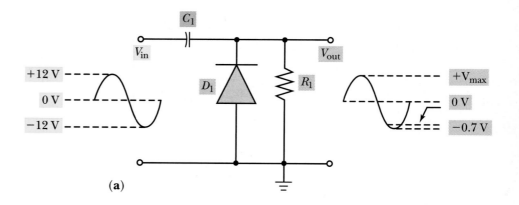

(a)

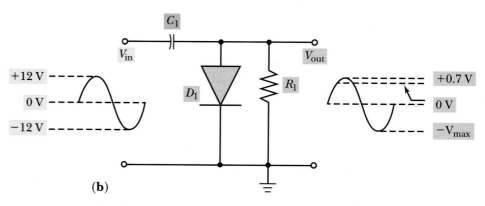

(b)

FIGURE 24–15 Diode clamping circuits: **(a)** positive clamping; **(b)** negative clamping

Figure 24–15 shows a pair of diode clamping circuits. In both cases, the peak input waveform is shown to be 12 V. Because of the combined action of the capacitor, resistor, and diode, the baseline of the output is radically shifted—positive in Figure 24–15a and negative in Figure 24–15b.

The output voltage levels of these diode clamping circuits depend on the values of C_1 and R_1, and the frequency of the input waveform. We will not be calculating those values in this lesson, but you should be able to recognize this kind of circuit when you see it on a schematic diagram.

IN-PROCESS LEARNING CHECK II

Fill in the blanks as appropriate.

1. Rectifier diodes are used where it is necessary to change _____ current power to _____ current power.

2. Where additional cooling is necessary, a rectifier diode can be connected to a(n) _____ to dissipate heat more efficiently.

3. When an ac waveform is applied to a rectifier diode, the diode's _____ rating must be greater than the peak voltage level.

4. The main current specification for rectifier diodes is _____.

5. To test the forward conduction of a diode with an ohmmeter, connect the _____ lead of the meter to the cathode and the _____ lead to the anode.

6. The forward resistance of a good diode should be much _____ than its reverse resistance.

7. Switching diodes have a(n) _____ rating that is hundreds of times less than most rectifier diodes.

8. The type of diode circuit that removes the peaks from an input waveform is called a(n) _____ circuit.

9. The type of diode circuit that changes the baseline level of an input waveform is called a(n) _____ circuit.

24–4 ZENER DIODES

You have already learned that every junction diode has a certain reverse breakdown voltage specification. For rectifier and switching diodes, it is essential to avoid circuit conditions that approach their reverse breakdown ratings. The zener diode is different, however. It is a diode that is designed to operate normally in its reverse-breakdown mode.

Figure 24–16 shows the I-V curve for a typical P-N junction diode. It is important you notice the part of the curve that represents the reverse-breakdown

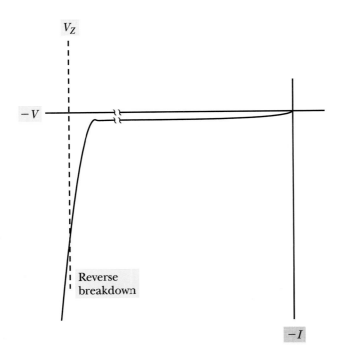

FIGURE 24–16 Reverse-bias I-V curve for a P-N junction diode

FIGURE 24–17 A zener diode test circuit

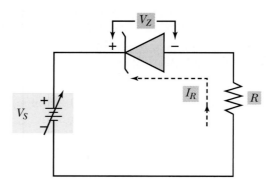

current. Once this voltage is reached, the voltage can hardly be increased any farther. If the reverse-breakdown voltage is specified at -100 V, for example, it is difficult to make the applied reverse voltage more negative beyond that level. You can see that a reverse-biased P-N junction can act as a *voltage regulator* device. Let's study this effect in more detail.

Zener diodes are designed for reverse breakdown at specified voltage levels. They can handle the reverse current without damage as long as there is a series resistance in the circuit to limit the reverse current below a specified maximum level. The diagram in Figure 24–17 is a simple demonstration circuit. Notice that the symbol for a zener diode looks like an ordinary P-N junction diode, but with the cathode-marker bar being bent on the ends.

Of course there will be no current flow through the circuit when the variable dc power supply, V_S, is set to 0 V. And even as you slowly increase the applied voltage level, there will be virtually no current flow (only reverse leakage current in the microampere range) because the diode is reverse biased. You can increase the supply voltage and see no appreciable reverse current flow (I_R) through the circuit until the value of V_S equals the zener's rated voltage level (V_Z). Once the source voltage exceeds V_Z, reverse current begins flowing through the circuit. The amount of reverse current flow depends on the value of the source voltage, the zener voltage rating, and the value of the series resistor (R).

FORMULA 24–4 $I_R = (V_S - V_Z)/R$

(NOTE: Formula 24–4 is valid as long as V_S is greater than V_Z. The source voltage must be greater than the zener voltage before any current can flow through the circuit.)

Example

In Figure 24–17, $V_S = 100$ V, $V_Z = 60$ V, and $R = 470$ Ω. How much current flows through the circuit?

Answer:

From Formula 24–4:

$$I_R = (V_S - V_Z)/R$$

$$I_R = 40 \text{ V}/470 \ \Omega = 85.1 \text{ mA}$$

A second rating for zener diodes is their maximum power dissipation (rather than current rating as for other types of diodes you have studied so far). It is very important that the power dissipation of an operating zener does not exceed its power rating. The actual power dissipation of a zener diode is found by multiplying the rated zener voltage by the measured or calculated amount of current flowing through it.

FORMULA 24–5 $P_Z = V_Z \times I_R$

Example

Refer to the circuit in Figure 24–18 to determine the current flow through the circuit and the power dissipated by the zener diode.

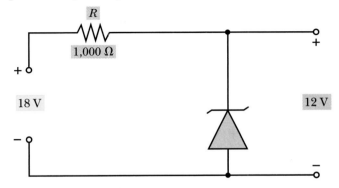

FIGURE 24–18 A simple zener diode regulator circuit

Answer:

From Formula 24–4:

$$I_R = (V_S - V_Z)/R$$

$$I_R = 6 \text{ V}/1,000 \text{ } \Omega = 0.006 \text{ A (or 6 mA)}$$

From Formula 24–5:

$$P_Z = V_Z \times I_R$$

$$P_Z = 12 \text{ V} \times 0.006 \text{ A} = 0.072 \text{ W (or 72 mW)}$$

Figure 24–19 lists some zener diode part designations along with their zener voltage and power ratings.

PRACTICAL NOTES

Zener diodes look exactly like rectifier diodes, a stripe near one end marks the cathode terminal, and they usually have part designations that begin with 1N. This means you cannot tell the difference between a rectifier diode and a zener diode without checking the specifications in a data book or catalog.

FIGURE 24–19 Typical
ratings for zener diodes

Part Identification	V_Z (V)	Max. power (W)
1N4728A	3.3	1.0
1N4731A	4.7	1.0
1N4733A	5.1	1.0
1N5233B	6.0	0.5
1N4742A	12.0	1.0
1N5252B	24.0	0.5

FIGURE 24–20 A working
zener diode voltage regulator
circuit

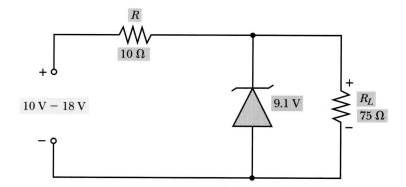

Figure 24–20 illustrates a practical voltage regulator circuit. The function of this circuit is to allow the input voltage to swing between 10 V and 18 V, yet maintain a steady 9.1 V across the load resistor, R_L. Whenever the supply voltage increases, the voltage has to remain fixed at 9.1 V across the zener diode and R_L, so series resistor R takes up the difference. In fact, the voltage drop across R varies along with the supply voltage (and the current through R and the zener diode vary accordingly). But the voltage and current for R_L remain regulated.

PRACTICE PROBLEMS III

Refer to the zener circuit in Figure 24–21.

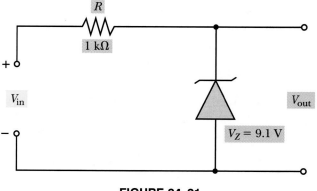

FIGURE 24–21

1. How much voltage is dropped across series resistor R when $V_{in} =$ 10 V? When $V_{in} = 20$ V?

2. What is the voltage at V_{out} when $V_{in} = 10$ V? When $V_{in} = 20$ V?

3. What is the current through R and the zener diode when $V_{in} = 10$ V? When $V_{in} = 20$ V?

4. What is the power dissipation of the zener diode when $V_{in} = 10$ V? When $V_{in} = 20$ V?

24–5 OPTOELECTRONIC DIODES

An **optoelectronic device** is an electronic device that somehow interacts with light energy. The optoelectronic devices discussed in this lesson are light-emitting diodes, photodiodes, and laser diodes.

Light-emitting diodes

Whenever electrical energy is applied to a semiconductor material, electrons in their valence bands absorb some of that energy and jump up to the higher energy conduction-band levels. As a result, the number of free electrons and holes increases. Conduction-band electrons cannot hold onto their extra energy levels for very long, however. Within a few nanoseconds after jumping up into a conduction band, an electron gives up its extra energy and falls back into a more stable valence band. This extra energy is usually given up in the form of heat, but certain kinds of semiconductor materials favor giving up this extra energy in the form of light. So it is possible to create semiconductor devices that actually emit light that can be used for practical purposes. These devices are **light-emitting diodes** (abbreviated **LEDs**).

The semiconductor materials used for making LEDs include gallium, arsenic, phosphorus, and carbon. The choice of semiconductor material determines the color of the light (red, amber, green, blue, orange, and infrared) that is emitted. For a given material the range of frequencies of light is fairly wide, but centers about one of these colors. The most common type of LED is made of a compound, gallium arsenide. These LEDs emit a bright red light that can be used for all kinds of indicators and displays for electronic equipment. Silicon carbide LEDs emit blue light.

Figure 24–22 shows the electronics symbol for an LED. This looks very much like the symbol for a common junction diode, but also includes arrows that point away from the diode. The idea of the arrows is to suggest that the diode emits light energy. LEDs are easy to identify because a prominent feature is the transparent lens assembly.

Another common configuration for LEDs is represented by seven-segment displays. A seven-segment display is made up of seven individual LEDs that have a long, narrow bar shape. These narrow bars are known as *segments*. Lighting certain combinations of these segments allows you to indicate decimal numbers 0 through 9.

An LED emits light whenever it is forward biased and conducting current. Refer to the simple circuit in Figure 24–23. If you are using a gallium arsenide LED, the barrier potential (forward voltage drop) is 1.5 V, and you need about 15 mA of forward current to generate a useful amount of light. The value of the current limiting resistor connected in series with the LED depends on the

FIGURE 24–22 The light-emitting diode (LED)

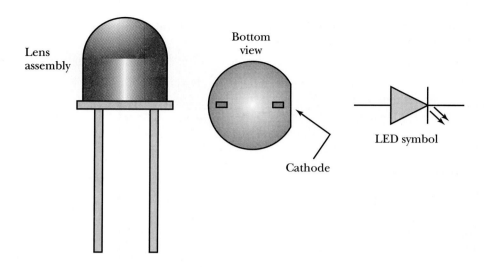

FIGURE 24–23 A practical LED circuit

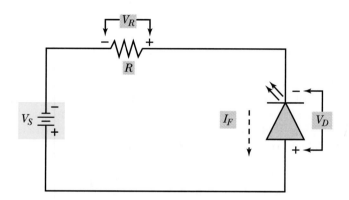

amount of supply voltage. Formula 24–1 lets you determine the amount of voltage that has to be dropped across the resistor (V_R) in terms of the source voltage (V_S) and the forward voltage drop across the diode (V_D).

Example

Referring to the circuit in Figure 24–23, let $V_S = 12$ V, and assume that the forward voltage drop for the LED (V_D) is 1.5 V. Calculate the voltage across the current limiting resistor (V_R).

Answer:

From Formula 24–1:

$$V_R = V_S - V_D$$

$$V_R = 12 \text{ V} - 1.5 \text{ V} = 10.5 \text{ V}$$

So the voltage across the current limiting resistor in this example is 10.5 V.

When you also know the amount of forward current (I_F) that is necessary for operating the LED, you can calculate the value of the series resistor (R). Formula

24–6 allows you to calculate the value of R on the basis of the voltage across resistor and the desired amount of LED current. Formula 24–7 lets you calculate the value of the resistor in terms of source voltage, diode voltage, and forward current.

FORMULA 24–6 $R = V_R/I_F$

FORMULA 24–7 $R = (V_S - V_D)/I_F$

Example

Referring again to Figure 24–23, let the voltage source be 9 Vdc, the forward voltage drop for the LED be 1.5 V, and the normal forward current for the LED be 15 mA. Calculate the value of the resistor and the amount of voltage you can expect to find dropped across it.

Answer:

Using Formula 24–7 to find the resistor value:

$$R = (V_S - V_D)/I_F$$

$$R = (9\text{ V} - 1.5\text{ V})/15\text{ mA} = 500\ \Omega$$

Using Formula 24–1 to find the voltage across the resistor:

$$V_R = V_S - V_D$$

$$V_R = 9\text{ V} - 1.5\text{ V} = 7.5\text{ V}$$

PRACTICAL NOTES

You can build a simple LED tester according to the example you have just studied. The circuit in Figure 24–24 uses an ordinary 9-V battery and a 470-Ω, 1/4-W resistor. (A 470-Ω resistor is used instead of the calculated value of 500 Ω because 470-Ω resistors are far more common and therefore easier to find and less expensive.)

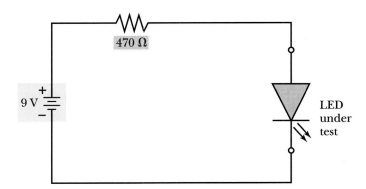

FIGURE 24–24 An LED tester

FIGURE 24–25 Typical ratings for LEDs

Color	V_D (V)	Max. I_F (mA)	Max. Reverse Voltage (V)	Max. Power (mW)
Red	1.5	25	4	70
Green	1.5	30	4	90
Amber	1.5	30	3	90
Orange	1.3	30	4	90
Blue	3.4	30	5	200

PRACTICE PROBLEMS IV

Use Formula 24–7 to calculate the value of the resistor in Figure 24–23 if $V_S = 9$ V, the LED's rated forward voltage drop is 1.2 V, and the required forward current is 24 mA.

Aside from emitting light energy, LEDs operate like common junction diodes. You have already seen, however, that the forward-bias voltage drop from LEDs tends to be larger than that of common silicon diodes (about 1.5 V for LEDs and about 0.7 V for silicon junction diodes). Also, the reverse-breakdown voltage for LEDs is much lower than for practical switching and rectifier diodes. Whereas the reverse-breakdown voltages for ordinary diodes are commonly on the order of 100, 200, and 300 V, the reverse breakdown voltage for LEDs may be typically in the range of 4 V or so. Because LEDs have such low reverse-breakdown voltage ratings, they cannot be used in practical rectifier circuits.

Figure 24–25 summarizes the specifications for several common LEDs.

Photodiodes

Earlier discussions of semiconductor materials and P-N junctions described these junctions as sensitive to light energy. Semiconductor devices are normally encased in light-tight enclosures, so they are unable to respond to changes in surrounding light level. In a **photodiode**, there is a lens that focuses light onto the P-N junction. This light creates available charge carriers—increasing the amount of light falling onto the junction increases the current flow. Figure 24–26 shows the typical enclosure and schematic symbol for the photodiode.

A photodiode is operated in its reverse-biased configuration, Figure 24–27. This means the amount of light falling onto the depletion region changes the amount of reverse leakage current. As the light intensity increases, the photodiode allows more current to flow through the circuit; and this means the voltage across the output resistor increases with increasing light level. Likewise, the voltage across the output resistor decreases with a decreasing light level.

Opto-isolators

Sometimes an LED (used as a source of light) and a photodiode (used as detector of light) are used to provide isolation between circuits or processes. The information is transmitted from the source to the detector by light waves. Such a device is called an *opto-isolator* and is illustrated in Figure 24–28.

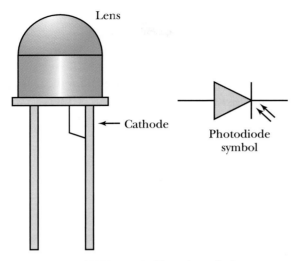

FIGURE 24–26 The photodiode

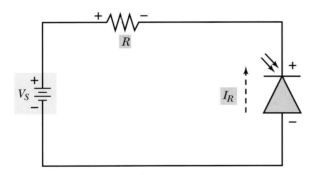

FIGURE 24–27 Elements of a photodiode circuit

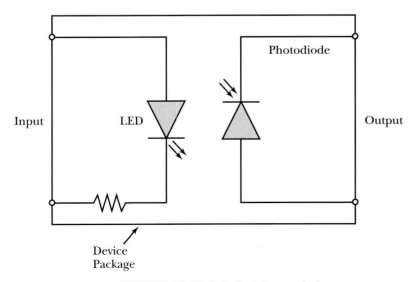

FIGURE 24–28 Opto-isolator symbol

Two common applications for opto-isolators include preventing a low impedance load from loading down a high impedance source, and enabling a low-current source to control a high-current detector. The distance between the source and detector diodes is usually small (a tenth of an inch or less), but the electrical isolation effect and efficiency are better than those of a transformer.

With this system, it is important that the source emit light frequencies to which the detector is sensitive. Therefore, sources and detectors are matched according to the frequencies emitted by the source and the most sensitive frequency range of the detector. The emitter and source diodes are usually enclosed in a lightproof package. However, some applications require that the package is open and exposed to outside light sources. In this case, the light from the source diode is sharply focused onto the photodiode, thereby drowning out the effects of outside light sources.

Other optoelectronic diode devices and systems

A fairly new type of semiconductor diode that is growing in popularity these days is the **laser diode**. Like an LED, a laser diode emits light energy. But the light energy from a laser diode is far more intense and is at a single frequency *(monochromatic)*. The effects of laser diodes are clearly seen as the criss-crossing pattern of ruby-colored light in the price scanners of modern supermarket checkout counters. Laser diodes are also used in compact disk (CD) players and optical disk drives for computers.

Laser diodes must be pulsed with short bursts of high current. The pulses occur so rapidly the light seems to be steady, however.

Another type of optoelectronic system uses an LED or laser diode as a source of light and a photodiode (or phototransistor) as the detector. The system works much like an opto-isolator except that light from the source is transmitted through a length of thin fiber-optic cable. A fiber-optic cable can transmit light energy over long distances with virtually no loss in light intensity—and certainly without interference from outside light sources or poor atmospheric conditions.

Optoelectronic/fiber-optic systems are already replacing ordinary copper wire for telephone communications. They are quickly becoming the preferred systems for communications between computers in modern office buildings.

IN-PROCESS LEARNING CHECK III

Fill in the blanks as appropriate.

1. An LED emits light energy when an electron in the _____ band falls to the _____ band.

2. The choice of _____ determines the color of the light from an LED.

3. The arrow on an LED symbol points _____ the diode.

4. An LED emits light when it is _____ biased.

5. The formula for calculating the value of the resistor to be used in series with an LED and its dc voltage source is $R = (V_S - V_D)/I_F$, where:

   V_S is _____

   V_D is _____

   I_F is _____

6. The reverse-breakdown voltage of an LED is generally _____ than that of a typical rectifier diode, and the forward voltage drop of an LED is _____ than that of a rectifier diode.

7. Light falling onto the depletion region of a reverse-biased diode _____ the amount of reverse leakage current.

8. The arrow on a photodiode symbol points _____ the diode.

9. A photodiode is connected into the circuit so that it is _____ biased.

10. In an opto-isolator, a(n) _____ is the source and a(n) _____ is the detector of light energy.

24-6 OTHER TYPES OF DIODES

The tunnel diode

The **tunnel diode** is a two-terminal semiconductor that is manufactured with a high proportion of the doping material and an especially thin depletion region. Because of the heavy doping, the barrier potential is relatively high, yet because of the thin depletion region, electrons can be forced to "tunnel" through the depletion region. At the biasing voltage levels, which cause this tunneling effect, there is a unique feature exhibited by the diode, called the *negative-resistance* region of the I-V curve (see Figure 24–29). This means there is a portion of the characteristic curve where an increase in forward voltage actually causes a decrease in forward current.

Negative resistance effects are advantageous in circuits known as *oscillators,* which you will learn about later. These oscillator circuits are commonly used to generate high frequency ac signals. The tunnel diode is also sometimes used as an electronic switch. Its key feature in this application is its high switching speed and low power consumption.

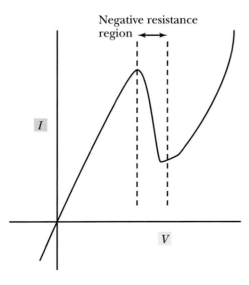

FIGURE 24–29 I-V curve for a tunnel diode

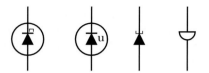

FIGURE 24–30 Tunnel diode schematic symbols

Figure 24–30 shows the different schematic symbols for tunnel diodes.

The varactor diode

Varactor diodes act like small variable capacitors. By changing the amount of reverse bias across their P-N junction, their capacitance can be varied. The greater the amount of reverse bias, the wider the depletion region becomes. Also, the greater the effective distance (depletion region) between the P- and N-type sections, the smaller the capacitance. On the other hand, the smaller the amount of reverse bias, the narrower the depletion region and the greater the amount of capacitance produced.

In effect, the varactor is a voltage-variable capacitance compared with a variable capacitor whose physical plates must be meshed or unmeshed to change capacitance. These devices are designed with capacitance values ranging from 1 pF to 2,000 pF or more. Frequency of operation affects the *Q* factor. The varactor symbol is shown in Figure 24–31.

Schottky diode

The Schottky diode, or Schottky barrier diode (SBD), is specially constructed and doped for ultra-fast switching speeds. Ordinary junction diodes have difficulty keeping up with changes in polarity greater than 1 MHz. It isn't unusual, however, to find Schottky diodes operating comfortably at 20 MHz and higher. There is an entire family of digital IC devices that use diodes and transistors built according to Schottky's hot-carrier junction effect.

Figure 24–32 shows the schematic symbol for a Schottky diode.

FIGURE 24–31 Varactor diode schematic symbol

FIGURE 24–32 Schematic symbol for a Schottky diode

SUMMARY

▶ The P-N junction creates a diode. (The prefix "di" indicates two parts.) By applying an external bias voltage to the diode, it can be made to conduct or not to conduct. Conduction is caused by forward biasing the P-N junction (i.e., negative voltage connected to the N-type material or cathode, and positive voltage connected to the P-type material or anode). When reverse biased, the diode will not conduct and acts as an open switch.

▶ There is a definite distribution of voltages in a circuit consisting of a dc power source, diode, and a resistor connected in series. When the dc power source is polarized such that it forward biases the diode, about 0.7 V (forward voltage drop) appears across the diode and the remainder of the source voltage appears across the series resistor. The current through the forward-biased circuit can be calculated by dividing the resistor value into the source voltage minus the forward drop across the diode. When the power source is connected so as to reverse bias the diode, there is virtually no current flow through the circuit; the full source voltage can be found across the diode, and zero voltage across the resistor.

▶ The three important diode specifications (or ratings) are forward voltage drop (V_F), average forward current (I_O), and peak reverse voltage (V_R). The forward voltage drop, also known as the barrier potential, is the voltage found across the diode when it is forward biased and fully conducting. This value is about 0.7 V for silicon diodes. Average forward current is the maximum amount of forward-bias current a diode can carry indefinitely. Operating a diode outside these specifications for extended periods will eventually destroy it.

▶ Rectifier diodes are designed for converting ac power to dc power. They can have relatively high current ratings (e.g., 10 A) and heat sinks are often used to cool the diodes where this is required.

▶ Rectifier and switching diodes can be tested with an ohmmeter. A diode in good working order will show a much higher resistance when it is reverse biased by the ohmmeter connection than when it is forward biased. A diode is defective whenever you find the forward and reverse ohmmeter readings in the same general range.

▶ Switching diodes usually have lower current and voltage ratings than rectifier diodes. However, they have much faster reverse recovery times. This means switching diodes recover much faster from reverse-bias conditions than rectifier diodes do.

▶ Diode clipper circuits (also called limiters) remove all or a portion of one polarity from an input waveform. In a series clipper, the diode is connected in series with the output of the circuit. In a shunt clipper, the diode is connected in parallel with the output of the circuit. Biased clippers allow clipping at levels other than zero or the forward voltage rating of the diode. The shape of the waveform is drastically changed by clipper action.

▶ Diode clamping circuits change the baseline of an input ac waveform from zero to some other value. The shape of the waveform is not changed by clamping action.

▶ Zener diodes are intentionally used in the reverse-breakdown mode. This action fixes the voltage across the zener diode at a specified level. Zener diodes are used where it is necessary to obtain a steady (regulated) voltage level despite swings in the input voltage levels. The two main specifications for zener diodes are their reverse voltage drop (V_Z) and their maximum continuous power.

▶ An optoelectronic device is one that interacts with light energy. Common optoelectronic diodes are light-emitting diodes (LEDs), photodiodes, and laser diodes. LEDs and laser diodes both emit light when they are forward biased and carrying current. The color of light from an LED depends on the semiconductor materials selected for its construction. The light from LEDs covers a fairly broad range of frequencies, but light from a laser is at a single frequency (monochromatic). A photodiode, when reverse biased, conducts in proportion to the amount of light focused onto its P-N junction.

▶ An LED and photodiode can be combined into a single unit that is called an opto-isolator.

▶ A tunnel diode has a negative-resistance I-V curve that makes it useful for generating high frequency waveforms.

▶ The varactor diode acts like a small capacitor that is variable by means of changing the amount of

reverse dc bias applied to it. The larger the reverse bias voltage, the smaller the amount of capacitance.

▶ The Schottky diode is specially constructed for ultra-fast switching times. It is found in high-frequency circuits and digital IC devices.

FORMULAS AND SAMPLE CALCULATOR SEQUENCES

FORMULA 24–1 $V_R = V_S - V_D$

V_S value, ⊖, V_D value, ⊜

FORMULA 24–2 $I_R = V_R/R$

V_R value, ÷, R value, ⊜

FORMULA 24–3 $I_R = (V_S - V_D)/R$

V_S value, ⊖, V_D value, ⊜, ÷, R value, ⊜

FORMULA 24–4 $I_R = (V_S - V_Z)/R$

V_S value, ⊖, V_Z value, ⊜, ÷, R value, ⊜

FORMULA 24–5 $P_Z = V_Z \times I_R$

V_Z value, ✕, I_R value, ⊜

FORMULA 24–6 $R = V_R/I_F$

V_R value, ÷, I_F value, ⊜

FORMULA 24–7 $R = (V_S - V_D)/I_F$

V_S value, ⊖, V_D value, ⊜, ÷, I_F value, ⊜

REVIEW QUESTIONS

1 Draw two diode circuits, one showing the diode as it is reverse biased and the second as it is forward biased. Label the drawings as appropriate.

2. Indicate the polarity of the voltages dropped across the resistor and diode in Figure 24–33.

3. Modify the circuit in Figure 24–33 by reversing the polarity of the dc source. Indicate on your drawing the polarity and amount of voltage across the diode and resistor. Show the amount of current flowing through the circuit.

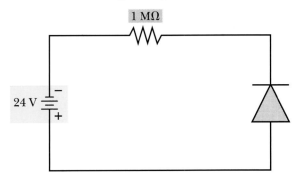

FIGURE 24–33

4. Sketch the waveform found across the resistor in Figure 24–34. What is the maximum amount of reverse voltage found across the diode?

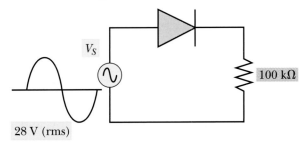

FIGURE 24–34

5. Sketch a circuit for a series diode clipper that clips the positive levels from the input ac waveform.

6. Sketch a circuit for a shunt diode clipper that clips the negative levels from the input ac waveform.

7. Describe how a diode clamp circuit affects an ac input waveform.

8. Explain the difference between the action of a biased diode clipper and a diode clamping circuit.

9. Explain the following diode specifications:
   a. Reverse breakdown voltage
   b. Forward conduction voltage
   c. Average forward current
   d. Reverse recovery time

10. Describe the condition necessary for a zener diode to hold its rated V_z level.

11. Suppose the source voltage in Figure 24–35 increases. What will happen to the voltage across the zener diode? The resistor? How will this increase in applied voltage affect the current through the circuit?

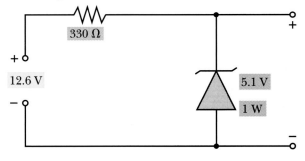

FIGURE 24–35

12. Describe the conditions required for getting an LED to emit light energy.

13. Sketch a simple LED circuit that consists of a dc power source, resistor, and LED. Arrange the polarities of the power source and LED so that the LED will light.

14. Explain how light falling onto the junction of a photodiode affects its conductance.

15. Draw a circuit showing a dc power source, resistor, and photodiode connected in series. Show the polarities and direction of the photodiode for normal operation.

16. Describe in words the difference between the schematic symbol for an LED and a photodiode.

17. Name the two types of diodes that can be found in an opto-isolator.

18. Describe how the light energy from an LED is different from that of a laser diode.

19. List at least two kinds of equipment that use laser diodes.

20. Define the term *optoelectronic device*.

21. What is the peculiar feature of the I-V curve for tunnel diodes?

22. Draw at least one schematic symbol for a tunnel diode.

23. Name two important applications of tunnel diodes.

24. Describe how a varactor changes capacitance with changing levels of reverse bias.

25. Describe how Schottky diodes are different from most other kinds of semiconductor diodes.

PROBLEMS

1. Referring to the circuit in Figure 24–36, let V_S = 6 V, R_S = 1 kΩ, and V_F = 0.7 V. Determine.
   a. the total current through the circuit.
   b. the voltage drop across resistor R_S.
   c. the forward current through the diode.

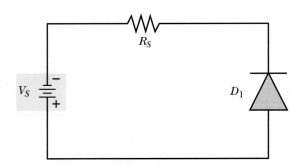

FIGURE 24–36

2. Referring to the circuit in Figure 24–36, let V_S = 12 V and V_F = 0.7 V. Determine:
   a. the value for R_S that will limit the diode current to 500 mA.
   b. the power dissipation of R_S.

3. If the maximum current rating for the diode in Figure 24–36 is 1 A and R_S = 100 Ω, calculate the maximum value of V_S that can be applied without exceeding the diode's current rating.

4. Determine the following value for the zener diode circuit in Figure 24–37:
   a. voltage drop across the zener diode
   b. voltage drop across the resistor
   c. current through the circuit

5. For the circuit in Figure 24–37, let V_S = 12 V, R = 120 Ω, and V_Z = 9 V. Determine:
   a. the voltage across R.

   b. the current through the circuit.
   c. the power dissipation of R.
   d. the power dissipation of Z.

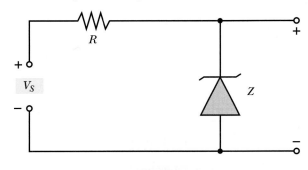

FIGURE 24–37

6. Referring to the circuit in Figure 24–37, let V_S = 12 V and V_Z = 9 V. If the current through the circuit is to be set at 1 A, determine:
   a. the value of R.
   b. the voltage across R.
   c. the power dissipation of R.
   d. the power dissipation of Z.

7. Suppose the zener diode in Figure 24–37 is rated at 6.3 V, the dc power source is 9 V, and the value of R = 680 Ω. Determine the amount of power dissipated by the zener diode.

8. Suppose the dc source voltage in a zener diode circuit (such as in Figure 24–37) increases by 2 V. How much will the voltage across the resistor increase? How much will the voltage across the zener diode increase?

9. For the LED circuit in Figure 24–38, let V_S = 6 V, R_S = 180 Ω, and V_F = 1.5 V. Determine:
   a. the amount of current through the circuit.
   b. the current through the LED.
   c. the power dissipation of R_S.

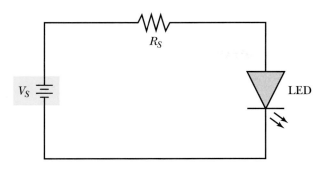

FIGURE 24–38

10. You want to operate an LED from a 9-V battery. The specifications for the LED list I_F at 150 mA and V_F at 1.5 V. Calculate the value of resistance you should connect between the battery and LED.

ANALYSIS QUESTIONS

1. Only one of the diodes in the circuit in Figure 24–39 will conduct. The conducting diode will carry the full current load, and the other will carry no current. Explain why this happens. (*Hint:* There are slight variations in the forward conduction voltage.) Explain why connecting a 5-Ω resistor in series with each diode will allow both to conduct, each carrying about half the current load.

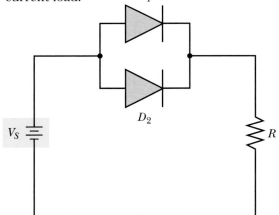

FIGURE 24–39

2. Use an electronics supply catalog to locate the part number for the rectifier diode that has the highest peak reverse voltage. Also locate the diode having the highest average current rating.

3. Consult a electronics supply catalog for the highest and lowest values for zener diode voltage and power.

4. Sketch the waveform you might see across the zener diode in Figure 24–37 if the zener is rated at 6 V, and the dc voltage source is replaced with an 8-V peak-to-peak ac source.

5. Explain how laser diodes are used in bar-code readers.

6. Research the basic circuits and range of operating frequencies for tunnel diode oscillators.

7. Use an electronics supply catalog to determine the specifications that are important for selecting Schottky diodes.

8. Use a dictionary or glossary of technical terms to describe the difference between discrete and integrated electronic devices.

PERFORMANCE PROJECTS CORRELATION CHART

CHAPTER TOPIC	PERFORMANCE PROJECT	PROJECT NUMBER
P-N Junction Diodes	Forward and Reverse Bias, and I vs. V	65
Rectifier Diodes	Ohmmeter Tests for Diodes	66
Diode Clipper Circuits	Diode Clipper Circuits	67
Zener Diodes	Zener Diodes	68
Light-Emitting Diodes	Light-Emitting Diodes	69

NOTE: It is suggested that after completing the above projects, the student should be required to answer the questions in the "Summary" at the end of this section of projects in the Laboratory Manual.

KEY TERMS

Bleeder resistor
Bridge rectifier
Full-wave rectifier
Half-wave rectifier
Power supply
Pulsating dc
Rectifier circuit
Regulator

POWER SUPPLY CIRCUITS

25

chapter

CHAPTER PREVIEW

The voltage supplied by power companies to our homes and businesses is ac voltage. As you learned earlier, this is because of the higher efficiency of transporting power from the power company to its customers by ac rather than dc. This efficiency is due to the step-up or step-down ability for ac voltages by using transformers, which minimize line losses.

Power supplies, which change ac source voltages to appropriate dc voltages, are found in most electronic equipment. Because of this, circuits associated with power supplies are a critical area that a technician should know. In this chapter, you will study each of the basic parts of a power supply. You will have a chance to apply directly what you have already learned about rectifier diodes as well as R, C, and L networks.

More complex power supplies have provisions for regulating their output voltage levels against swings in the input voltage. Also, there are instances where it is important to rectify and step up voltage levels with a voltage multiplier power supply.

OBJECTIVES

After studying this chapter, you should be able to:

1. List the basic elements of a power supply system

2. Draw the three basic types of rectifier circuits: **half-wave, center-tapped full-wave,** and **bridge circuits**

3. Explain the paths for current flow through the three basic types of rectifier circuits

4. Describe the waveforms found across the diode(s) and at the output of the three basic types of rectifier circuits

5. Determine the unfiltered dc output voltage of specified rectifier circuits

6. Briefly describe power supply **filter action**

7. Identify power supply filter configurations

8. Explain the purpose of a power supply **voltage regulator**

9. Recognize two **voltage multiplier** circuits

10. Briefly describe the most common types of power supply troubles

25-1 ELEMENTS OF POWER SUPPLY CIRCUITS

The purpose of a **power supply** circuit is to convert available ac power into usable dc power. Most of the circuits inside an ordinary TV receiver, for example, require dc voltage levels such as +12 V and +48 V. The source of power for the TV is 120 Vac from a standard wall outlet. A power supply inside the TV receiver is responsible for converting electrical power from the 120-Vac source into the lower dc level required for operating most of the circuits. Power supplies are required in any electronic device that requires dc and is operated from ac power sources.

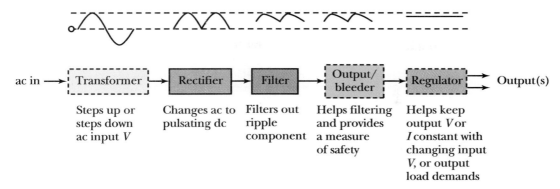

FIGURE 25–1 Power supply system block diagram

Figure 25–1 shows the main elements of a complete power supply system:

1. A transformer is used to step up or step down the ac voltage level and isolate the remainder of the electronic system from the ac power source.

2. A **rectifier circuit** is used to change the ac power into pulsating dc.

3. A **filter** removes the ac components from the pulsating dc waveform and provides a smooth dc power level.

4. An optional **bleeder resistor** or resistor divider circuit is used to aid filtering and discharge any capacitors in the power supply that might otherwise remain charged when ac power is removed.

5. An optional **regulator** provides an especially smooth and constant output voltage or current level.

You will learn more about these elements as the various power supply configurations are discussed in this chapter. You will also learn about power supply specifications, including:

- output voltage levels;
- output current levels;
- amount of voltage regulation; and
- ripple voltage levels.

And you will see how these specifications are met by the selection of components for the various sections in the power supply system (e.g., transformer, rectifier, filter, and regulator).

25-2 RECTIFIER CIRCUITS

Rectifier circuits transform ac waveforms into pulsating dc waveforms. This is an important first step in the process of obtaining useful, smooth, dc sources from commercial ac power sources. The most important part of a rectifier circuit is its diodes (the rectifier diodes you studied in Chapter 24).

Half-wave rectifier circuits

The **half-wave rectifier** introduced in this section is an ac version of the simple junction diode circuit described in the previous chapter. In Figure 25–2, note

FIGURE 25–2 Half-wave rectifier circuit and waveforms

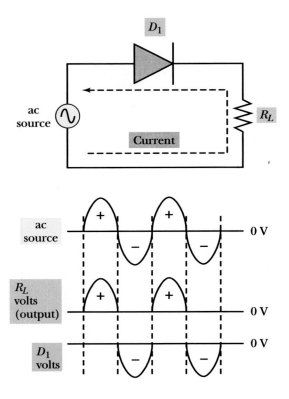

that the ac source is supplying power to a diode and resistor that are connected in series. The diode conducts current whenever the applied voltage forward biases the diode, and the diode blocks current flow through the circuit whenever the applied voltage reverse biases the diode. To put it another way, you will find the applied voltage dropped across resistor R_L whenever the diode conducts and zero voltage across the resistor whenever the diode is not conducting.

Refer to the circuit in Figure 25–2 while you consider the following points:

- The direction of electron flow through the diode is from cathode to anode (against the direction of the arrow in the diode symbol).

- The diode (D_1) conducts only on each ac alternation when its anode is positive with respect to the cathode. When this is the case, virtually all of the ac voltage is found across the load resistor (R_L).

- On ac input alternations when the diode is reverse biased, it cannot conduct because its anode is negative with respect to its cathode. When this happens, nearly all of the ac voltage is found across the diode and virtually none across the load resistor.

The waveforms in Figure 25–2 show the operation of this circuit through two full cycles of applied ac voltage. During the first half-cycle, the diode is forward biased. As a result, only the forward conduction voltage appears across the diode, and most of the waveform is applied to the resistor at the output. On the second half-cycle (the negative half-cycle), the diode is reverse biased. Under this condition, current cannot flow through the circuit and virtually all of the input voltage appears across the diode. (NOTE: This is indicated by the flat, 0 V, part of the R_L volts or output waveform.)

Because there is no filtering connected to this rectifier circuit, the output waveform is called **pulsating dc**. It is considered a pulsating waveform because the current flows in half-cycle bursts. It is a dc waveform because the output voltage is of just one polarity.

Recall for ac voltage waveforms, V_{rms} equals 0.707 times V_{pk} and that V_{avg} for each alternation = 0.637 times V_{pk}. Since 0.637 is about nine-tenths of 0.707, V_{avg} equaled 0.9 times V_{rms}.

Since a half-wave rectifier circuit is producing an output pulse only during one-half cycle, it is logical that the average dc output is only half the average ac voltage. That is, half of 0.9 = 0.45. Hence, the average dc output voltage of the half-wave rectifier circuit equals 0.45 times V_{rms}. Of course you can also compute the dc output as half the average ac voltage. That is, half of 0.637 times V_{pk}. This means the unfiltered dc output (half-wave rectifier) is calculated from either of the following formulas:

FORMULA 25–1 $V_{dc} = 0.45 \times V_{rms}$

FORMULA 25–2 $V_{dc} = 0.318 \times V_{pk}$

PRACTICAL NOTES

The reason for showing you Formula 25–1 is that you can measure the rms ac input voltage with any DMM or VOM—then easily compute average dc output. Not all meters can measure peak values, directly. (Of course, a scope will!)

The V_{dc} formulas shown don't take into account the very small dc voltage dropped across the diode, since it does not appreciably change the output voltage.

Example

What is the unfiltered dc output voltage of a half-wave rectifier circuit if the ac input voltage is 100 V peak?

Answer:

$$V_{dc} = 0.318 \times V_{pk} = 0.318 \times 100 \text{ V} = 31.8 \text{ V}$$

PRACTICE PROBLEMS I
What is the unfiltered dc output voltage of a half-wave rectifier circuit if the ac input is 200 V rms?

You already know that the voltage drop across a conducting silicon diode is about 0.7 V. And that's the voltage you will find across the diode in a half-wave rectifier circuit on the half-cycle when the diode is conducting. But what about the voltage across the diode during the half-cycle when it is reverse biased? More specifically, what is the maximum voltage across the diode when it is not conducting? This value is called the *peak inverse voltage, PIV.*

The PIV appears across the diode when the applied ac voltage has a polarity that reverse biases the diode. Because there is no current flowing through the diode in Figure 25–2, there can be no current flowing through the load resistor, R_L. The diode is acting like an open switch, so you will find the full amount of applied voltage dropped across it. The peak value of this reverse voltage waveform is PIV = $1.414 \times V_{rms}$.

FORMULA 25–3   PIV (H-W) = $1.414 \times V_{rms}$

As you have seen in Figure 25–2, there is one ripple in the output for each ac cycle of input.

FORMULA 25–4   Ripple Frequency (H-W) = ac input frequency

If the ac input frequency to a half-wave rectifier circuit is 60 Hz, the ripple frequency of the output across resistor R_L will also be 60 Hz.

The circuits in Figure 25–3 are both half-wave rectifiers. They are identical except for the direction the rectifier diode is connected. In Figure 25–3a, the

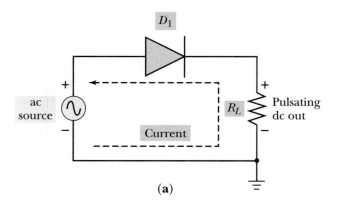

(a)

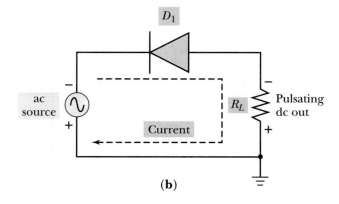

FIGURE 25–3 Changing output polarity from a half-wave rectifier: **(a)** circuit with positive output; **(b)** circuit with negative output

(b)

diode is connected so that the output waveform across R_L is positive with respect to ground. In Figure 25–3b, the diode is connected in such a way that the output waveform is negative with respect to ground.

PRACTICE PROBLEMS II
Refer to Figure 25–4 and find the dc output voltage, PIV, output ripple frequency, and output polarity with respect to ground.

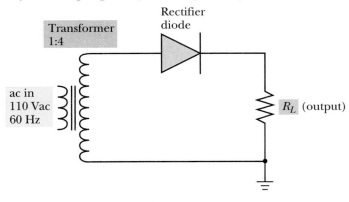

FIGURE 25–4

Full-wave rectifier circuits

Refer to the rectifier circuit in Figure 25–5 while you study the following discussion of the basic operation of a **full-wave rectifier** circuit. Notice that the secondary of the transformer is *center tapped*. The center-tapped transformer is absolutely necessary for the operation of this full-wave rectifier circuit.

For a transformer with a center-tapped secondary winding, when the top of the center-tapped secondary is positive, the bottom of the secondary winding is negative. It is important to note also that when the top of the tapped secondary is positive, the center tap is negative with respect to the top of the winding, and positive with respect to the bottom of the secondary winding. Of course, on the next alternation of ac input, when the top of the secondary is negative, the bottom of the full secondary is positive. In this case, the center tap is positive with respect to the top of the secondary, and negative with respect to the bottom of the secondary. It is obvious that from center tap to either end of the secondary, the voltage equals only half of the full secondary voltage. Keep these thoughts in mind as you follow the discussion of the operation of the full-wave rectifier circuit.

The rectifier diodes are connected in such a way that D_1 conducts on the first half-cycle of the input ac, when the top half of the center-tapped secondary is forward-biasing it. The current path is from center tap, up through R_L, through D_1, back to the top of the secondary winding. D_2 conducts on the opposite half-cycle of the input ac, when the bottom half of the secondary causes D_2 to be forward-biased. The two diodes thus conduct alternately, D_1 on one half-cycle of input ac, and D_2 on the other half-cycle.

The following four formulas allow you to calculate the average dc voltage from a full-wave rectifier circuit that uses a center-tapped transformer. The first two yield the output voltage when you know the rms voltage that is applied.

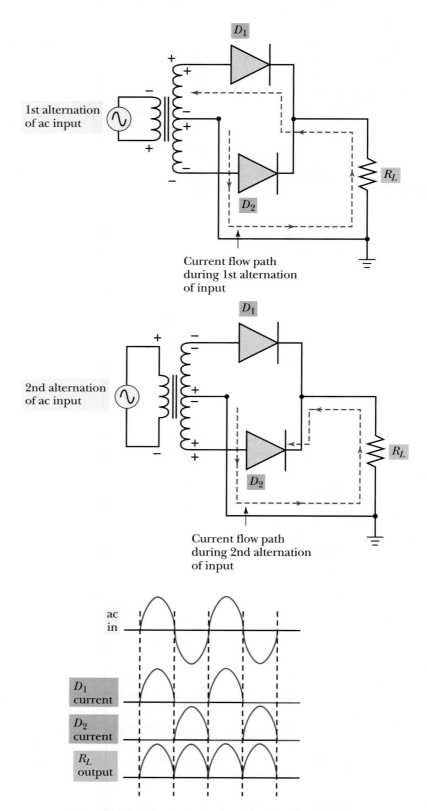

1st alternation of ac input

Current flow path during 1st alternation of input

2nd alternation of ac input

Current flow path during 2nd alternation of input

ac in

D_1 current

D_2 current

R_L output

FIGURE 25–5 Full-wave rectifier circuit and waveforms

Formula 25–5 lets you calculate V_{dc} on the basis of the rms voltage (V_{rms}) measured between the center tap and one end of the transformer secondary (as shown in Figure 25–6a):

FORMULA 25–5 $V_{dc} = 0.9 \times V_{rms}$ of half the secondary

Formula 25–6 also lets you calculate V_{dc}, but after measuring the rms voltage across the full transformer secondary (see Figure 25–6b):

FORMULA 25–6 $V_{dc} = 0.45 \times V_{rms}$ of the full secondary

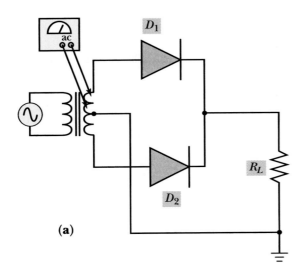

(a)

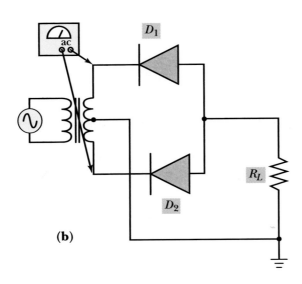

(b)

FIGURE 25–6 Center-tapped transformer voltages: **(a)** center tap to one end; **(b)** end-to-end

Formulas 25–7 and 25–8 allow you to calculate the average output dc voltage level on the basis of peak voltage measurements (V_{pk}) as measured from the center tap or across the full secondary:

FORMULA 25–7 $V_{dc} = 0.637 \times V_{pk}$ of half the secondary

FORMULA 25–8 $V_{dc} = 0.318 \times V_{pk}$ of the full secondary

The peak inverse voltage (PIV) across EACH diode in the full-wave rectifier circuit equals the peak voltage measured across the FULL secondary of the transformer (end-to-end).

FORMULA 25–9   PIV (F-W) $= 1.414 \times V_{rms}$ of full secondary

Example

The end-to-end (full secondary) voltage on the secondary of a center-tapped transformer is 260 V. What is the minimum reverse voltage rating for the diodes if they are used in a full-wave rectifier?

Answer:

From Formula 25–9:

$$\text{PIV} = 1.414 \times V_{rms} \text{ of full secondary}$$
$$\text{PIV} = 1.414 \times 260 = 368 \text{ V}$$

So the reverse voltage rating of the diodes has to be at least 368 V. You could select diodes having a 400-V peak reverse rating, but a 600-V version would be even better.

As far as the ripple frequency is concerned, a full-wave rectifier produces an output ripple for each half-cycle of ac input. This means the output ripple frequency equals two times the frequency of the ac input.

FORMULA 25–10   Ripple Frequency $= 2 \times$ ac input frequency

PRACTICE PROBLEMS III

Assume a full-wave rectifier circuit is connected to a transformer having a full-secondary rms voltage of 300 V. Further assume that the frequency of the ac source feeding the transformer is 400 Hz. Determine the dc output, ripple frequency, and PIV for each diode.

The output polarity of a full-wave rectifier with center-tapped transformer, can be reversed by reversing the cathode and anode connections for *both* diodes.

Caution: Both diodes must be reversed! If only one is reversed, the transformer winding will be shorted when both diodes are forward biased during one of the half-cycles of applied power. Compare the circuits in Figure 25–7. It is also possible to reverse the output polarity by taking the output from the center tap and grounding the common-connection point of the two diodes.

The bridge rectifier

A **bridge rectifier** is a type of full-wave rectifier circuit, but one that does not require a transformer having a center-tapped secondary. As shown in Figure 25–8, a bridge rectifier uses four rectifier diodes.

You can see from the diagrams that two of the diodes are forward biased on one half-cycle of the input ac waveform and the other two are forward biased on

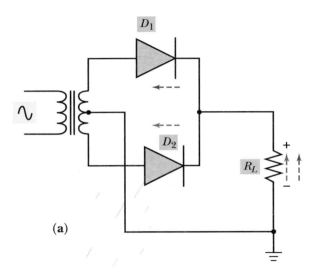

(a)

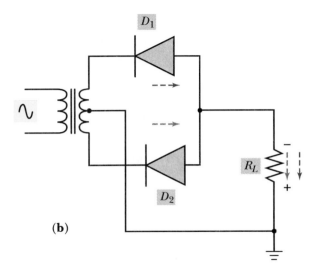

(b)

FIGURE 25–7 Diode direction affects output polarity: **(a)** positive output; **(b)** negative output

FIGURE 25–8 Bridge rectifier circuit: **(a)** first half-cycle; **(b)** second half-cycle

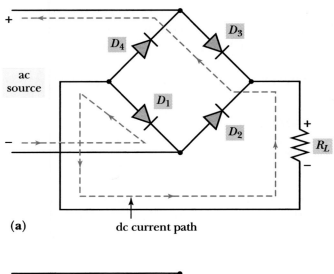

(a)   dc current path

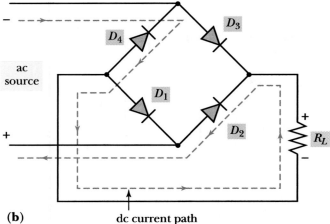

(b)   dc current path

the opposite half-cycle. When the input waveform has the polarity shown in Figure 25–8a, you can see that diodes D_1 and D_3 conduct. To trace the series path for electron flow during this half-cycle of applied voltage:

- start from the negative end of the ac power source;
- pass through diode D_1 (against the symbol arrow);
- pass through the output load resistor in the direction indicated on the diagram;
- pass through diode D_3; and
- return to the positive side of the ac power source.

Diodes D_2 and D_4 do not conduct on this half-cycle because they are reverse biased.

On the second half-cycle, Figure 25–8b, D_2 and D_4 are allowed to conduct, while D_1 and D_3 are reverse biased. To trace the series path for electron flow during this second half-cycle:

- start from the negative end of the ac power source;
- pass through diode D_4;

- pass through the output load resistor in the direction indicated on the diagram;
- pass through diode D_2; and
- return to the positive side of the ac power source.

Since there is a pulse of output current during each half-cycle of the ac input waveform, the frequency of the pulsating dc waveform at the output is twice that of the input frequency.

Formulas 25–11 and 25–12 allow you to calculate the average dc voltage level at the output on the basis of the input voltage. If you know the rms voltage (V_{rms}) at the input, you can use Formula 25–11 to calculate the average dc output. But if you know the peak voltage level at the input (V_{pk}), use Formula 25–12.

FORMULA 25–11 $V_{dc} = 0.9 \times V_{rms}$ input

FORMULA 25–12 $V_{dc} = 0.637 \times V_{pk}$ input

Example

What is the average dc voltage level at the output of the bridge rectifier circuit in Figure 25–9 if you measure 12.6 V (rms) across the secondary of the transformer?

Answer:

Using Formula 25–11:

$$V_{dc} = 0.9 \times V_{rms} \text{ input}$$
$$V_{dc} = 0.9 \times 12.6 \text{ V} = 11.34 \text{ V}$$

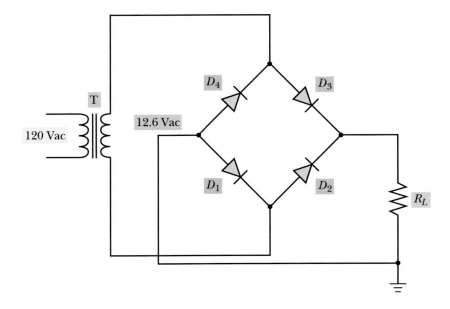

FIGURE 25–9

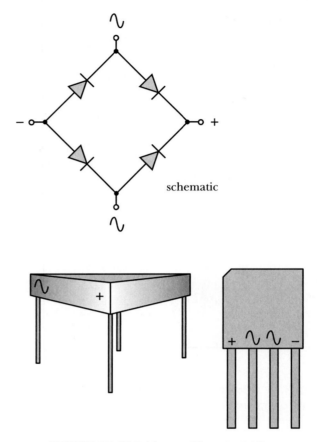

FIGURE 25–10 Bridge rectifier assemblies

The peak inverse voltage across each diode of an unfiltered bridge rectifier circuit equals the peak value of the ac input waveform. You can use Formula 25–13 to calculate the peak value, given the rms voltage present at the input:

FORMULA 25–13   PIV (Bridge) $= 1.414 \times V_{rms}$ input

As seen earlier in this chapter, the ripple frequency at the output of an un-filtered bridge rectifier circuit is twice the frequency of the input waveform:

FORMULA 25–14   Ripple Frequency (Bridge) $= 2 \times$ ac input frequency

You have seen that a bridge rectifier requires four separate diodes. This arrangement is so commonly used, however, that semiconductor manufacturers simplify the actual circuitry by providing the four-diode bridge assembly in a single package. Figure 25–10 shows two popular packages for bridge rectifier assemblies. The ac inputs are marked with a small sine-wave symbol, and the two dc outputs are labeled plus (+) and minus (−).

Bridge rectifier assemblies simplify the procedures for building, testing, and repairing power supplies. These assemblies are rated much like single rectifier diodes—by average forward current and maximum reverse breakdown voltage.

PRACTICE PROBLEMS IV

Refer to the circuit in Figure 25–11.

1. If the input rms voltage level is 120 V as shown in the figure, what is the average dc output voltage level?
2. If the load (R_L) is 50 Ω, what is the average dc current through the load? (*Hint:* Use Ohm's law, $I_{dc} = V_{dc}/R_L$.)
3. What is the minimum reverse breakdown voltage specification for the diodes?
4. What is the ripple frequency across the load?

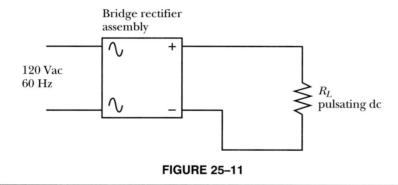

FIGURE 25–11

Summary of basic rectifier circuit features

Consider the three basic kinds of rectifier circuits you have just studied.

- A half-wave rectifier requires only a single diode in addition to the power source and output load. It is the simplest of the three basic kinds of rectifier circuits. Its main drawback is that there is no power delivered to the load during one half of each input ac cycle.

- A center-tapped full-wave rectifier circuit requires two diodes; but more important, it also requires the use of a center-tapped power transformer. The full-wave rectifier circuit provides output power to the load during both alternations of ac input, and that is its advantage over a half-wave rectifier.

- The full-wave bridge circuit requires four diodes, but has the power delivering capability of full-wave rectifiers and does not require the use of a center-tapped power transformer. Furthermore, for a given transformer, the bridge rectifier circuit delivers twice the voltage output, when compared with either the half-wave, or center-tapped transformer full-wave circuit. Although the need for four diodes might seem to be a disadvantage, the low cost and reliability of suitable rectifier diodes (especially bridge rectifier assemblies) make this circuit the first choice for most practical applications.

IN-PROCESS LEARNING CHECK I

Fill in the blanks as appropriate.

1. The simplest rectifier circuit is the _____ circuit.

2. The ripple frequency of a bridge rectifier circuit is _____ the ripple frequency of a half-wave rectifier.

3. The circuit having the highest output voltage for a given transformer secondary voltage is the _____ rectifier.

4. For a given full transformer secondary voltage, the full-wave rectifier circuit unfiltered dc output voltage is _____ the dc output of a half-wave rectifier that is using the same transformer.

5. To find the average dc output (unfiltered) of a center-tapped, full-wave rectifier, multiply the full secondary rms voltage by _____.

25–3 BASIC POWER SUPPLY FILTERS

The purpose of a power supply filter is to keep the ripple component from appearing in the output. It is designed to convert pulsating dc from rectifier circuits into a suitably smooth dc level.

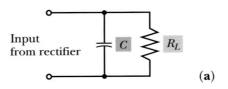

(a)

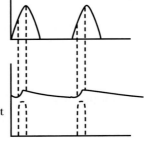

(b)

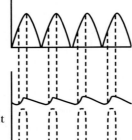

(c)

FIGURE 25–12 Simple C-type filter circuit: **(a)** circuit diagram; **(b)** half-wave rectification before and after C-type filter action; **(c)** full-wave rectification before and after C-type filter action

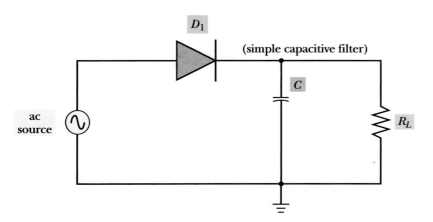

FIGURE 25–13 Half-wave power supply with C-type filter

Capacitance filters

Figure 25–12 shows a simple capacitance (C-type) filter. You can see that the output of the rectifier circuit is connected in parallel with a filter capacitor and the load resistance. Figure 25–12b shows how the C-type filter helps smooth out the valleys in the waveform from a half-wave rectifier, and Figure 25–12c shows the filtering when using a full-wave rectifier.

Figure 25–13 illustrates an example of a C-type filter that is used with a simple half-wave rectifier network. This is the simplest and most economical sort of power supply. It is used when simplicity and economy are important and when a relatively large amount of ripple does not seriously affect the operation of electronic devices that operate from it.

Inductance filters

Figure 25–14 shows an inductance (L-type) power supply filter. In this case, the output of a half- or full-wave rectifier (including a bridge rectifier) is connected to the load through an inductor, or choke. Figure 25–14b and 25–14c show how an L-type filter circuit affects the waveforms from half- and full-wave rectifier circuits, respectively.

During the time a rectifier is supplying power through the inductor in an L-type filter, energy is stored in the form of a magnetic field. As the circuit current tries to decrease, the inductor's energy is used to tend to keep current flowing; thus, providing a "smoothing" action on the output waveform. The quality of the filtering depends on the value of the inductance and the amount of current required to operate the load.

Figure 25–15 is an example of an L-type filter that is driven by a center-tapped full-wave rectifier. This is a relatively expensive circuit because of the need for electromagnetic components (namely the center-tapped power transformer and filter choke).

Filter networks

You have seen how capacitors and inductors, used separately, can filter pulsating dc waveforms. The best kinds of power supply filter circuits use combinations of

FIGURE 25–14 Simple L-type filter circuit: **(a)** circuit diagram; **(b)** half-wave rectification before and after L-type filter action; **(c)** full-wave rectification before and after L-type filter action

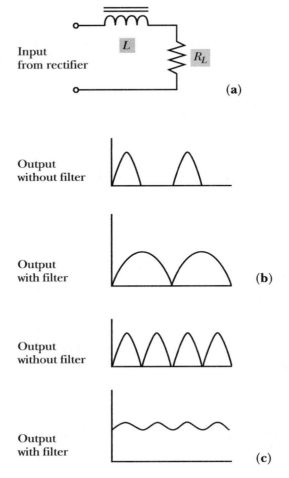

(a)

Output without filter

Output with filter

(b)

Output without filter

Output with filter

(c)

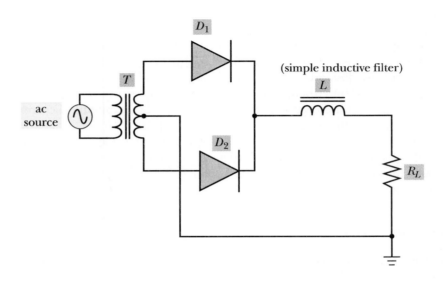

FIGURE 25–15 Full-wave power supply with L filter

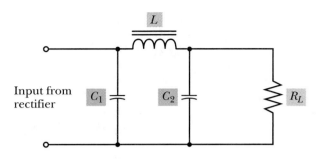

FIGURE 25–16 Capacitor-input pi-type LC filter

L, C, and R. The main considerations in the selection and quality of a power supply filter network are:

1. allowable output ripple;
2. allowable minimum regulation;
3. rectifier peak current limits;
4. load current; and
5. output voltage.

The filter circuit in Figure 25–16 is one of the most popular in equipment that requires a very high amount of filtering (low ripple) and good regulation. It is called a *capacitor-input pi filter*. It is called "capacitor input" because the output from the rectifier sees a shunt capacitor first. It is called a pi filter because the formation of the two capacitors and inductor on a schematic diagram resembles the Greek letter pi (π).

A slightly different capacitor-input pi filter is shown in Figure 25–17. The inductor is replaced with a resistor. The ripple and regulation qualities for this circuit are not as good as for the LC version, but the RC version is less expensive and good for many less-demanding applications.

Figure 25–18 shows the best kind of LCR power supply filter circuit. This is a *choke-input, two-section L filter*. It is a "choke-input" filter because inductor L_1 is

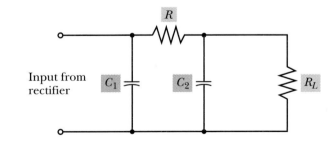

FIGURE 25–17 Capacitor-input pi-type RC filter

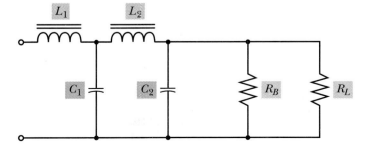

FIGURE 25–18 Choke-input, two-section L filter

the first component seen by the pulsating dc output from the rectifiers. It is called an "L" filter because the schematic drawing of L_1 and C_1 looks like an inverted letter L (as does L_2 and C_2). It is a two-section filter because there are two L sections connected one after the other.

This type of two-section L filter is by far the best kind of power supply filter network, both in terms of extremely low ripple and excellent regulation. It is very expensive, however, so it is used only in the most demanding kinds of equipment such as radar, medical instruments, and certain kinds of communications gear.

Power supply ripple

You have seen how filter circuits go a long way toward converting pulsating dc from a rectifier circuit to a reasonably steady dc level. In actual practice, no filter circuit is perfect—there is always some slight amount of variation left over. The part of the original ac waveform that remains at the output of a filtered rectifier circuit is called the *ripple*. You can see the ripple on the output waveforms for the circuits in Figures 25–12 and 25–14.

The higher the quality of a power supply, the smaller the amount of ripple in its output voltage. The amount of ripple, however, is not always the same value. Drawing more current from a power supply, for example, tends to increase the amount of ripple. The specifications for a power supply ought to cite the amount of ripple (usually measured in ac volts or millivolts) at a certain level of current.

A useful specification for power supply ripple is the percent of ripple. As shown in Formula 25–15, you can determine the percent of ripple of a power supply by using an oscilloscope to measure the ac ripple (peak-to-peak) and the amount of dc voltage below the ripple. You then divide the ac amount by the dc amount, and multiply by 100.

> **FORMULA 25–15**   % Ripple = ac/dc × 100

For example, if the ripple is found to be 1 $V_{\text{p-p}}$ and the dc level is 48 V, the percent of ripple is $1/48 \times 100 = 2.08\%$. That is a pretty good power supply, considering that many allow up to 10% ripple.

25–4 SIMPLE POWER SUPPLY REGULATORS

The purpose of a **power supply voltage regulator** is to provide a very steady, or well-regulated, dc output. Two different factors commonly cause the voltage from a rectifier circuit to swing away from the desired output level. One factor is the changes in the ac input voltage. Power supplies that are operated from 120-Vac, 60-Hz utility sources can be affected by fluctuations in that so-called 120-Vac level. This ac voltage can drop as low as 110 Vac at times and rise to 128 Vac or so at other times. Without the help of a voltage regulator, the output of a power supply will vary according to such changes in the ac input level.

Another factor that causes the dc output from a power supply to vary is the amount of current that is required to operate the load at any given moment.

This is not ordinarily a problem with very simple electronic devices, but the more complex an electronic system becomes, the more the current requirement varies as the equipment performs different tasks. The voltage from a power supply tends to decrease when the load "pulls" more current from it. Then when the load lightens, the power supply voltage tends to rise.

Electronic power supply regulators eliminate dc voltage swings caused by variations in input voltage and output loading. Whenever the dc output from a regulated power supply tries to change, the regulator instantly makes an internal adjustment that causes the dc output to remain steady. Modern regulators respond so rapidly that they can do much of the ripple filtering as well. So the output from a good power supply regulator is not only very steady, but virtually ripple free.

Some regulation is offered by the type of filter network that is used in a power supply. Traditionally, these filters were made up of passive components that have limited capabilities. In recent years, however, nearly ideal voltage regulation has become possible at a low cost with the help of small, integrated-circuit voltage regulator components.

Integrated-circuit (IC) components are made up of many semiconductors, resistances, and interconnections that are all included on a single chip of silicon. IC power supply regulators may have just three terminals on them. Figure 25–19 shows a typical IC voltage regulator used with a bridge rectifier circuit. The three terminals are for:

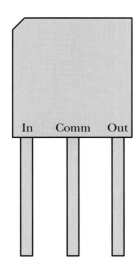

FIGURE 25–19 A typical power supply regulator IC device

- unregulated dc input;
- regulated dc output; and
- common ground connection.

Voltage regulator ICs are rated according to their output regulation voltage and maximum current capability. Typical regulators are rated at +5 V at 1 A, +12 V at 1 A, and +24 V at 1/2 A.

The quality of a regulator is expressed in terms of percent of regulation. This is a figure that expresses how well a power supply retains its dc output level under changes in the amount of current being drawn from its output. A power supply is said to have *no load* when no current is being drawn from it, and *full load* when the maximum rated current is being drawn from it. Formula 25–16 shows that the percent of regulation depends on:

- V_F, the output of dc voltage when maximum current is being drawn from it; and
- ΔV, the difference in dc output voltage when the supply is operated at full load and no load.

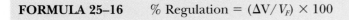

FORMULA 25–16   % Regulation = $(\Delta V / V_F) \times 100$

Suppose you are working with a power supply that has a dc output of 12.2 V at no load and 11.8 V at full load. What is the percent of regulation? In this instance, $V_F = 11.8$ V, and $\Delta V = 12.2 - 11.8$ V. So the regulation is $[(12.2 - 11.8)/12.2] \times 100 = 3.28\%$.

IN-PROCESS LEARNING CHECK II

Fill in the blanks as appropriate.

1. In a C-type filter, the output of the rectifier circuit is connected in _____ with a filter _____ and the load resistance.
2. In an L-type filter, the output of the rectifier circuit is connected in _____ with a filter _____ and the load resistance.
3. In an L-type filter, energy is stored in the form of a _____ field.
4. The five main considerations in the selection and quality of a power supply filter network are: _____, _____, _____, _____, and _____.
5. The two main causes of swings in the output voltage of a power supply are changes in the _____ voltage level and changes in the _____ current demand.
6. The three terminals on an integrated-circuit voltage regulator are the unregulated _____, regulated _____, and _____.
7. The (higher, lower) _____ the percent of ripple, the higher the quality of the power supply.
8. Full-load output voltage is taken when the current from the power supply is (maximum, minimum) _____.
9. The (higher, lower) _____ the percent of regulation, the higher the quality of the power supply.

25–5 BASIC VOLTAGE MULTIPLIER CIRCUITS

A **voltage doubler** circuit is a type of rectifier circuit that not only changes an ac input to a dc output, but doubles the voltage in the process. An ideal voltage doubler circuit provides an output voltage level that is equal to two times the peak value of the ac input voltage. There are no ideal voltage doubler circuits, so you can expect the output voltage to be slightly lower than twice the peak input voltage, especially under heavy current demands from the load circuit.

Example

A voltage doubler power supply is used with the normal household 120-Vac input. What is the ideal dc voltage level at the output of the doubler?

Answer:

The 120-Vac rating is an rms value. The peak value is $1.414 \times 120 \text{ V} = 169.7 \text{ V}$. The output of the multiplier is $2 \times 169.7 \text{ V} = 339.4 \text{ Vdc}$.

The circuit in Figure 25–20 is a full-wave voltage doubler circuit. The rectified input voltage charges the capacitors series-aiding, with each one charging to the input voltage. Therefore, the output voltage is about twice the input voltage (under light load). This is called a "full-wave" doubler because there are two output pulses for every ac input cycle.

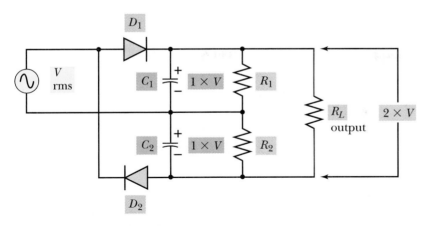

FIGURE 25–20 Full-wave voltage doubler

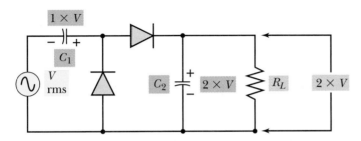

FIGURE 25–21 Half-wave cascaded voltage doubler

Figure 25–21 shows a different type of voltage doubler that is known as a half-wave cascaded voltage doubler. Capacitor C_2 is charged close to twice the ac input voltage. In this circuit, C_2 gets a pulse of charging current once during each ac input cycle, thus the "half-wave" designation. The reason C_2 charges to twice the input voltage is because the charge on C_1 is series-aiding with the source in the charge path for C_2. Thus, C_2 charges to the voltage across C_1 plus the peak value of the ac source.

Voltage doubler circuits are used where it is desirable to obtain a higher dc voltage without the use of a step-up transformer.

Figures 25–22 and 25–23 are the basic circuits for voltage tripler and quadrupler power supplies, respectively. The dc output from the tripler is about three times the peak input voltage, and the output from the quadrupler is approximately four times the peak input voltage (under light load).

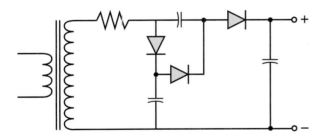

FIGURE 25–22 Voltage tripler circuit

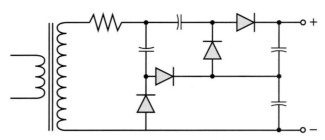

FIGURE 25–23 Voltage quadrupler circuit

Although it is possible to connect combinations of doublers and triplers to create voltage multiplication greater than four, it isn't practical to do so. The main problem is that the full-load current capability and percent of regulation drop off dramatically as you increase the number of times the input voltage is multiplied. Once you get past the quadrupler, voltage multiplier circuits have virtually no practical applications.

25–6 TROUBLESHOOTING POWER SUPPLIES

TROUBLESHOOTING HINTS

First, isolate a power supply problem to the general section of trouble by carefully considering the symptoms. Recall the main sections of a typical power supply are the transformer, the rectifier(s), the filter network, and possibly a regulator circuit. Excellent tools to help you with symptoms are the oscilloscope to look at waveforms, and/or a voltmeter to make voltage measurements. An ohmmeter can also be helpful for checking rectifier diodes as described in the previous chapter. Your eyes become a tool as you can see burned parts, blown fuses, and so forth, and you can use them to analyze the readings and scope patterns. Your nose is a great symptom locator to find overheated parts. If the system you are troubleshooting involves an audio output (such as a radio receiver or cassette tape player) your ears are a very helpful tool, since you can hear hum and can probably tell if it is 60 Hz or 120 Hz, and you can hear "motorboating."

By noting whether there is excessive ripple in the output and zero or too low an output voltage, you can isolate the section you might want to troubleshoot.

Troubles in the transformer section of a power supply can include power sources feeding the primary of the transformer (loose or broken connections), fuses in the primary and/or secondary, and the transformer itself. Transformers usually don't go bad by themselves. Usually some external circuit or component connected to the transformer causes too much current through the transformer circuit that either blows a fuse or damages the transformer. The best way to check a transformer is to measure the voltage on the primary and secondary windings and compare the results with known norms.

PRACTICAL NOTES

It is not unusual for power transformers to operate at high temperatures that seem a bit uncomfortable to the touch. Whenever a transformer contains shorted windings (a definite trouble that must be fixed), it runs much hotter than normal. Often, a strong odor of melted wax or burned insulation is given off, and/or some smoke, and areas showing discoloration may be seen.

A common trouble with rectifiers is short circuits. A shorted diode places a short across the ac input to the power supply, often blowing a fuse and overheating a transformer that might be connected at the input to the rectifier portion of the circuit. The voltage is lower than normal at the dc output of the power supply.

The best way to check a rectifier while it remains in the circuit is to measure the ac voltage across each diode. If the diode is shorted, the ac voltage reading across it is virtually zero. If the diode is open, the ac voltage across it is much higher than normal. You can also check a rectifier by removing it from the circuit and measuring its forward and reverse resistance with an ohmmeter. The diode is defective if the forward and reverse resistance readings are not vastly different.

Often, electrolytic capacitors are used in the filter section of power supplies. When a capacitor fails, an internal short circuit is usually the problem. Short circuits in capacitors are usually caused by voltage spikes that exceed the capacitor's rated value. Shorted capacitors quickly overheat, sometimes exploding or popping open with a sharp cracking sound. Capacitors in this condition are quite easy to locate by visual inspection. (They may give off a strong acrid odor as well.)

To check for open filter capacitors in an operating circuit, measure the *ac voltage* (the ripple level) across each capacitor. If the capacitor is good, the ripple level will be within tolerance for the power supply. If it is open, the ripple voltage is a large percentage of the rated dc output level.

If the power supply uses filter inductors, or chokes, they are usually connected in series with the load current. Excessive load current causes too much current to flow through the filter choke. Overheating is the result. Shorted windings within a filter choke also cause it to overheat. The symptoms are virtually identical with those you have seen earlier for shorted power transformers.

An open filter choke is easy to diagnose because 1) it will have full rectifier voltage at the input end and no voltage at the output end, and 2) an ohmmeter check will show infinite resistance across the choke when it is disconnected from the circuit.

Troubles in voltage regulators are easily detected by comparing the input and output voltages for the regulator circuit. If the voltage at the input of the regulator is within its normal range, but the output of the regulator is not, the trouble is most likely with the regulator.

PRACTICAL NOTES

Some integrated-circuit regulators have a built-in circuit protection feature that greatly reduces the output voltage in the event of excessive load current. The fact that the output of the regulator is lower than normal might indicate a short-circuit condition in the load rather than a faulty regulator circuit.

SUMMARY

▶ A typical power supply system takes existing ac voltage and converts it to dc voltage of an appropriate level. Many power supplies use a transformer to step up or to step down ac voltages, a rectifier to change ac to pulsating dc, a filter to smooth out the pulsating dc and eliminate the ac component of the waveform, an output network (in some cases) that can be a voltage divider, a bleeder, and so on, and, in some cases, a voltage regulator that keeps the voltage constant with varying loads.

▶ Common rectifier circuit configurations are the half-wave rectifier (using one diode), the conventional full-wave rectifier (using two diodes and a center-tapped transformer), and the bridge rectifier (using four diodes). Other special circuit configurations are used that act as voltage multipliers, namely the full-wave doubler and half-wave cascade doubler.

▶ Unfiltered dc output voltages for a given transformer can be computed for the various standard circuit configurations:

$$V_{dc} \text{ (H-W)} = 0.45 \times V_{rms} \text{ or } 0.318 \times V_{pk}$$

$$V_{dc} \text{ (F-W)} = 0.9 \times V_{rms} \text{ of half the secondary}$$

$$= 0.318 \times V_{pk} \text{ of full secondary}$$

$$V_{dc} \text{ (Bridge)} = 0.9 \times V_{rms} \text{ of full secondary}$$

$$= 0.637 \times V_{pk} \text{ of full secondary}$$

▶ Peak inverse voltage (PIV) is the maximum voltage across the diode during its non-conduction period of the applied ac waveform. PIV values for each rectifier diode (without filter) for the common circuit configurations are as follows:

$$\text{PIV (H-W)} = 1.414 \times V_{rms} \text{ of full secondary}$$

$$\text{PIV (F-W)} = 1.414 \times V_{rms} \text{ of full secondary}$$

$$\text{PIV (Bridge)} = 1.414 \times V_{rms} \text{ of full secondary}$$

(NOTE: With capacitor filters, peak inverse voltage can be as high as $2.828 \times V_{rms}$.)

▶ Capacitors, inductors, and resistors are often used in power supply filter networks. Common filters include simple C types, simple L types, and more common, combinations of L and C in L-type and pi-type configurations. Filters can be single section or multiple section. Generally, when there are more sections, filtering is greater.

▶ Ripple frequency relates to the number of pulses per cycle of ac input. Half-wave rectifiers have ripple frequency outputs equal to the ac input frequency. Full-wave and bridge rectifiers have ripple frequency outputs equal to twice the ac input frequency.

▶ The percentage of ripple of a power supply is an indication of how well the filter is performing under a given set of load condition. Generally, the greater the amount of loading on a power supply, the higher the percentage of ripple. A power supply operating under no-load conditions might have nearly zero percentage of ripple.

▶ The percentage of regulation of a power supply is an indication of how well it maintains its rated output voltage under changing load conditions. Generally, the greater the amount of loading, the lower the dc output voltage. Higher-quality power supplies tend to have low figures for percentage of regulation.

▶ Voltage multipliers use diodes and capacitors to produce dc output voltages that are roughly two times the peak ac input voltage (voltage doubler), three times the peak ac input voltage

(voltage tripler), and four times the peak ac input voltage (voltage quadrupler). The higher the multiple, however, the less useful the power supply becomes because the multiplying process reduces the power supply's current capacity and percentage of regulation.

▶ Oscilloscope waveforms and voltage measurements are commonly used for troubleshooting power supplies. Also, the ohmmeter is useful for checking fuses as well as the forward and reverse resistances of rectifier diodes.

FORMULAS AND SAMPLE CALCULATOR SEQUENCES

FORMULA 25–1 $V_{dc} = 0.45 \times V_{rms}$

.45, ✖, V_{rms} value, ▬

FORMULA 25–2 $V_{dc} = 0.318 \times V_{pk}$

.318, ✖, V_{pk} value, ▬

FORMULA 25–3   PIV (H-W) $= 1.414 \times V_{rms}$

1.414, ✖, V_{rms} value, ▬

FORMULA 25–4   Ripple Frequency (H-W) = ac input frequency

FORMULA 25–5 $V_{dc} = 0.9 \times V_{rms}$ of half the secondary

.9, ✖, V_{rms} value, ▬

FORMULA 25–6 $V_{dc} = 0.45 \times V_{rms}$ of the full secondary

.45, ✖, V_{rms} value, ▬

FORMULA 25–7 $V_{dc} = 0.637 \times V_{pk}$ of half the secondary

.637, ✖, V_{pk} value, ▬

FORMULA 25–8 $V_{dc} = 0.318 \times V_{pk}$ of the full secondary

.318, ✖, V_{pk} value, ▬

FORMULA 25–9   PIV (F-W) = $1.414 \times V_{rms}$ of the full secondary

1.414, ⊗, V_{rms} value, ⊜

FORMULA 25–10   Ripple Frequency = $2 \times$ ac input frequency

2, ⊗, input frequency value, ⊜

FORMULA 25–11 $V_{dc} = 0.9 \times V_{rms}$ input

.9, ⊗, V_{rms} value, ⊜

FORMULA 25–12 $V_{dc} = 0.637 \times V_{pk}$ input

.637, ⊗, V_{pk} value, ⊜

FORMULA 25–13   PIV (Bridge) = $1.414 \times V_{rms}$ input

1.414, ⊗, V_{rms} value, ⊜

FORMULA 25–14   Ripple Frequency (Bridge) = $2 \times$ ac input frequency

2, ⊗ , input frequency value, ⊜

FORMULA 25–15   % Ripple = ac/dc $\times$ 100

ac ripple, ÷, dc voltage, ⊗ , 100 ⊜

FORMULA 25–16   % Regulation = $(\Delta V / V_F) \times 100$

⦅, no-load voltage, ⊖, full-load voltage, ⦆, ÷, full-load voltage, ⊗, 100, ⊜

REVIEW QUESTIONS

1. Draw the basic diagram for a half-wave rectifier.

2. Sketch a diagram that compares the ac input for a half-wave rectifier with its output when: (a) there is no output filtering, (b) when there is a C-type output filter.

3. What happens to the output polarity of a half-wave rectifier if you turn around the rectifier

diode? Switch around the connections to the ac source?

4. If the input ac frequency is 60 Hz, what is the frequency (in pulses per second, pps) of the ripple from a half-wave rectifier circuit?

5. Draw the basic diagram for a typical full-wave rectifier circuit that uses a center-tapped transformer.

6. Sketch a diagram that compares the ac input (transformer primary) with the output of a full-wave, center-tapped transformer rectifier circuit when: (a) there is no output filtering, (b) when there is a C-type output filter.

7. If the input ac frequency is 60 Hz, what is the frequency (in pps) of the ripple from a full-wave, center-tapped, transformer rectifier circuit?

8. Draw the basic diagram for a bridge rectifier circuit.

9. Referring to your drawing in question 8, draw a circle around two diodes that conduct at the same time on the first half-cycle, then draw a square around the two diodes that conduct on the second half-cycle.

10. Sketch a diagram that compares the ac input for a bridge rectifier circuit with its output when: (a) there is no output filtering, (b) when there is a C-type output filter.

11. If the input ac frequency is 400 Hz, what is the frequency (in pps) of the ripple from a bridge rectifier circuit?

12. Draw the circuit for a full-wave power supply that fits the following description:
   a. Center-tapped power transformer
   b. *RC*, capacitor-input, pi-type filter
   c. Resistor load of 100-kΩ

   Is the dc output of this circuit equal to, greater than, or less than the same rectifier without filtering?

13. If the ripple output of a rectifier-filter system is excessive, is the problem most likely in the transformer, rectifier(s), or filter network? Indicate which component(s) could cause the problem.

14. Briefly describe the purpose of a voltage regulator in a power supply.

15. Sketch a block diagram of a complete dc power supply that includes a power transformer, bridge rectifier, integrated-circuit voltage regulator, C-type filter, and a bleeder resistance across the dc output.

16. Describe the difference between percent of ripple and percent of regulation.

17. Explain the difference between no-load and full-load conditions for a power supply.

18. What is the dc output voltage of a voltage tripler that has an ac input of 12.6 V_{rms}?

19. What rms voltage level should be applied to a voltage quadrupler if the dc output voltage is to be 1,200 V?

20. The percent regulation for a quadrupler voltage multiplier is (better than, worse than, the same as) _____ the regulation for a tripler multipler.

PROBLEMS

1. The ac input to a half-wave, unfiltered rectifier is 120 Vac (rms). Determine:
   a. the peak dc output voltage.
   b. peak inverse voltage applied to the rectifier diode.
   c. frequency of the output ripple.

2. What is the PIV of the diode in a half-wave rectifier circuit that uses a capacitor filter, if the transformer feeding the rectifier circuit is a 1:3 step-up transformer supplied by a 150-Vac source?

3. If the peak output voltage from a half-wave rectifier is 17 V, what is the peak output current when the load resistance is 10 Ω?

4. What is the unfiltered dc output of a full-wave rectifier circuit fed by a center-tapped, 1:2 step-up transformer that is connected to a primary voltage source of 200 V rms?

5. The output of an unfiltered full-wave rectifier is supposed to be 5.5 V peak dc. If the rectifier is

fed by a center-tapped transformer, what is the correct transformer ratio if the primary voltage is 120 Vac (rms)?

6. What is the unfiltered dc output of a bridge rectifier circuit fed by 120 Vac (rms)?

7. Calculate the percent of ripple when an oscilloscope shows a 500-mV peak-to-peak ac waveform riding on top of a 5-V dc level.

8. The percent of ripple for a high-quality power supply is supposed to be 5%. If it is a 12-Vdc

power supply, what is the maximum amount of ripple you should expect to see at the output?

9. Determine the percent of regulation of a power supply that has a no-load output of 26 V and a full-load output of 23.5 V.

10. The percent of regulation for a certain power supply is rated at 6%. If the no-load voltage is 5.5 V, what is the minimum full-load voltage?

ANALYSIS QUESTIONS

1. Draw the circuit for a half-wave rectifier that provides a negative output voltage.

2. Research how a center-tapped power transformer can be used with a bridge rectifier to produce a power supply that provides both a positive and a negative dc voltage.

3. Explain why using a larger filter capacitor improves the percent ripple and regulation for a power supply.

4. Research and write a paragraph that defines the term *critical inductance.*

5. Research and write a paragraph defining the term *optimum inductance.*

6. Explain why the voltage rating of a filter capacitor must be larger than the peak voltage level from the rectifier.

7. Explain why a shorted diode in a rectifier circuit will eventually destroy the power supply's electrolytic filter capacitor.

8. Describe how a leaky filter capacitor will affect the ripple and regulation of a power supply circuit.

PERFORMANCE PROJECTS CORRELATION CHART

CHAPTER TOPIC	PERFORMANCE PROJECT	PROJECT NUMBER
Half-Wave Rectifier Circuits Capacitance Filters	Half-Wave Rectifier	70
The Bridge Rectifier Capacitance Filters	Bridge Rectifier	71
Basic Voltage Multiplier Circuits	Voltage Multiplier	72

NOTE: It is suggested that after completing the above projects, the student should be required to answer the questions in the "Summary" at the end of this section of projects in the Laboratory Manual.

THE SIMPLER SEQUENCE FOR TROUBLESHOOTING

Follow the seven-step SIMPLER troubleshooting sequence, as outlined below, and see if you can find the problem with this circuit. As you follow the sequence, record your circuit test results for each testing step on a separate sheet of paper. This will aid you and your instructor to see where your thinking is on track or where it might deviate from the best procedure.

step 1 — Symptoms

—(Gather, verify, and analyze symptom information) To begin this process, read the data under the "Starting Point Information" heading. Particularly look for circuit parameters that are not normal for the circuit configuration and component values given in the schematic diagram. For example, look for currents that are too high or too low; voltages that are too high or too low; resistances that are too high or too low, etc. This analysis should give you a first clue (or symptom information) that will aid you in determining possible areas or components in the circuit that might be causing this symptom.

step 2 — Identify

—(Identify and bracket the initial suspect area) To perform this first "bracketing" step, analyze the clue or symptom information from the "Symptoms" step, then either on paper or in your mind, bracket, circle, or put parentheses around all of the circuit area that contains any components, wires, etc., that might cause the abnormality that your symptom information has pointed out. NOTE: Don't bracket parts of the circuit that contain items that could not cause the symptom!

step 3 — Make

—(Make a decision about the first test you should make: what type and where) Look under the TEST column and determine which of the available tests shown in that column you think would give you the most meaningful information about the suspect area or components, in the light of the information you have so far.

step 4 — Perform

—(Perform the first test you have selected from the TEST column list) To simulate performing the chosen test (as if you were getting test results in an actual circuit of this type), follow the dotted line to the right from your selected test to the number in parentheses. The number in parentheses, under the "Results in Appendix C" column tells you what number to look up in Appendix C to see what the test result is for the test you are simulating.

step 5 — Locate

—(Locate and define a new "narrower" area of uncertainty) With the information you have gained from the test, you should be able to eliminate some of the circuit area or circuit components as still being suspect. In essence, you should be able to move one bracket of the bracketed area so that there is a new smaller area of uncertainty.

step 6 — Examine

—(Examine available information and determine the next test: what type and where) Use the information you have thus far to determine what your next test should be in the new narrower area of uncertainty. Determine the type test and where, then proceed using the TEST listing to select that test, and the numbers in parentheses and Appendix C data to find the result of that test.

step 7 — Repeat

—(Repeat the analysis and testing steps until you find the trouble) When you have determined what you would change or do to restore the circuit to normal operation, you have arrived at *your* solution to the problem. You can check your final result against ours by observing the pictorial step-by-step sample solution on the pages immediately following the "Chapter Troubleshooting Challenge"—circuit and test listing page.

step 8 — Verify

NOTE: A useful *8th Step* is to operate and test the circuit or system in which you have made the "correction changes" to see if it is operating properly. This is the final proof that you have done a good job of troubleshooting.

CHALLENGE CIRCUIT 14

CHALLENGE CIRCUIT 14

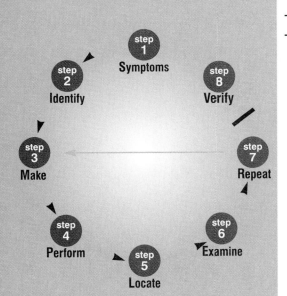

Starting Point Information

1. Circuit diagram
2. The output voltage at point D is lower than it should be.
3. You are to troubleshoot down to the "block" level.

TEST	Results in Appendix C
Waveform at point A shows	(74)
Waveform at point B shows	(53)
Waveform at point C shows	(21)
DC voltage at point D measures	(65)

Symptoms

The dc output voltage is considerably lower than it should be.

Identify

initial suspect area: The low voltage can be caused by a transformer problem, a rectifier problem, a filter problem, or an excessively high current drain by the load. For now, let's keep all of these possibilities in the initial suspect area.

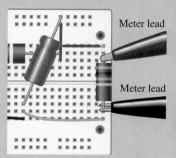

Meter lead

Meter lead

Make

test decision: The magnitude of change in the output voltage is large, and we would probably see smoke if the transformer were that faulty or if the load were that heavy. Therefore we'll make our first check at the midpoint between the rectifier circuit and the filter circuit. That is, at point C or the rectifier output.

Perform

1st Test: Using a scope, check the waveform at point C. The result is only one pulse per ac input cycle. Half of the full-wave rectifier output is missing.

Locate

new suspect area: it looks like we may have a rectifier problem. But, using our "input-output" technique, we'll have to verify that the ac input into the rectifier is normal before we exclude the transformer. Thus, the new suspect area includes the transformer and rectifier circuits.

Examine
available data.

Repeat

analysis and testing: The next test is to look at the ac input to the rectifier circuit at point B.

2nd Test: Check the waveform at point B. The scope check indicates a sine-wave voltage with peak voltage of about 39 V. This is what it should be. So we have isolated the trouble to the rectifier block.

Verify

3rd Test: When we restore the circuit by replacing the faulty component(s), the circuit works fine.

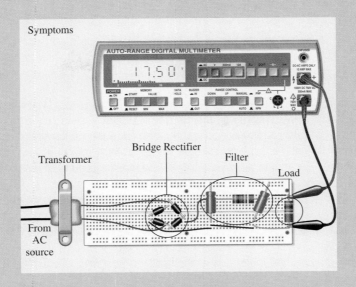

Symptoms

Transformer

Bridge Rectifier

Filter

Load

From AC source

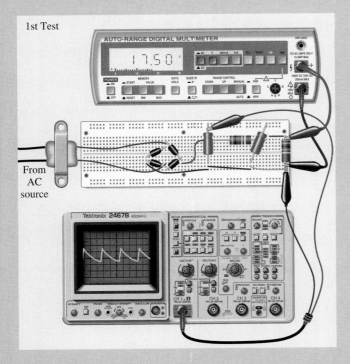

1st Test

From AC source

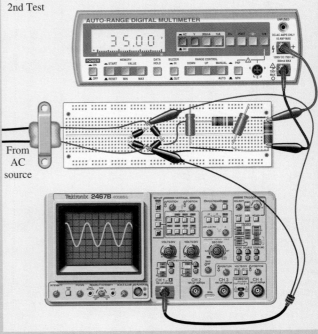

2nd Test

From AC source

THE SIMPLER SEQUENCE FOR TROUBLESHOOTING

Follow the seven-step SIMPLER troubleshooting sequence, as outlined below, and see if you can find the problem with this circuit. As you follow the sequence, record your circuit test results for each testing step on a separate sheet of paper. This will aid you and your instructor to see where your thinking is on track or where it might deviate from the best procedure.

 Symptoms

—(Gather, verify, and analyze symptom information) To begin this process, read the data under the "Starting Point Information" heading. Particularly look for circuit parameters that are not normal for the circuit configuration and component values given in the schematic diagram. For example, look for currents that are too high or too low; voltages that are too high or too low; resistances that are too high or too low, etc. This analysis should give you a first clue (or symptom information) that will aid you in determining possible areas or components in the circuit that might be causing this symptom.

 Identify

—(Identify and bracket the initial suspect area) To perform this first "bracketing" step, analyze the clue or symptom information from the "Symptoms" step, then either on paper or in your mind, bracket, circle, or put parentheses around all of the circuit area that contains any components, wires, etc., that might cause the abnormality that your symptom information has pointed out. NOTE: Don't bracket parts of the circuit that contain items that could not cause the symptom!

 Make

—(Make a decision about the first test you should make: what type and where) Look under the TEST column and determine which of the available tests shown in that column you think would give you the most meaningful information about the suspect area or components, in the light of the information you have so far.

 Perform

—(Perform the first test you have selected from the TEST column list) To simulate performing the chosen test (as if you were getting test results in an actual circuit of this type), follow the dotted line to the right from your selected test to the number in parentheses. The number in parentheses, under the "Results in Appendix C" column tells you what number to look up in Appendix C to see what the test result is for the test you are simulating.

 Locate

—(Locate and define a new "narrower" area of uncertainty) With the information you have gained from the test, you should be able to eliminate some of the circuit area or circuit components as still being suspect. In essence, you should be able to move one bracket of the bracketed area so that there is a new smaller area of uncertainty.

 Examine

—(Examine available information and determine the next test: what type and where) Use the information you have thus far to determine what your next test should be in the new narrower area of uncertainty. Determine the type test and where, then proceed using the TEST listing to select that test, and the numbers in parentheses and Appendix C data to find the result of that test.

 Repeat

—(Repeat the analysis and testing steps until you find the trouble) When you have determined what you would change or do to restore the circuit to normal operation, you have arrived at *your* solution to the problem. You can check your final result against ours by observing the pictorial step-by-step sample solution on the pages immediately following the "Chapter Troubleshooting Challenge"—circuit and test listing page.

step 8 **Verify**

NOTE: A useful *8th Step* is to operate and test the circuit or system in which you have made the "correction changes" to see if it is operating properly. This is the final proof that you have done a good job of troubleshooting.

CHALLENGE CIRCUIT 15

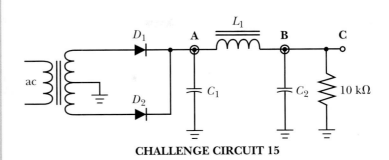

CHALLENGE CIRCUIT 15

Starting Point Information

1. Circuit diagram
2. Output voltage at C is lower than normal. A scope check also indicates higher than normal ac ripple present.

TEST	Results in Appendix C
DC voltage at pt C	(69)
AC voltage at pt C	(26)
DC voltage at pt B	(98)
AC voltage at pt B	(6)
DC voltage at pt A	(76)
AC voltage at pt A	(35)
R of L_1	(92)
C_1 R measurements	(45)
C_2 R measurements	(117)

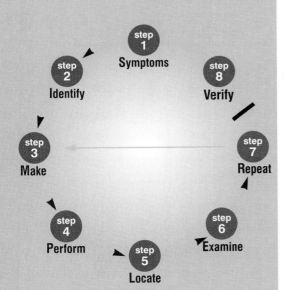

Symptoms

The dc voltage out seems low, and excessive ac ripple is present.

Identify

initial suspect area: The ripple symptom indicates the filter components are involved. If the inductor were shorted, the dc output might be slightly higher than normal. If the inductor were open, there would be no output. So this narrows our initial suspect area to the two filter capacitors.

Make

test decision: If either capacitor were shorted, the output voltage would be drastically lower than normal (virtually zero). So we need to see if a capacitor might be open, which would agree with our symptom information. Typically, an aging electrolytic capacitor that has dried out and become virtually open will show much more ac than it would if it were normal. First, let's check the ac voltage across the output filter capacitor (C_2).

Perform

1st Test: Check the ac voltage across C_2. It shows higher than the normal ac component. However, this could be caused by the input filter capacitor (C_1) not feeding a normal signal to the inductor and this capacitor.

Locate

new suspect area: The first test is not conclusive, therefore the suspect area remains the same at this point.

Examine

available data.

Repeat

analysis and testing.

2nd Test: Measure ac voltage across C_1. The result shows a much higher than normal ac component.

3rd Test: Make a front-and-back resistance check of capacitor C_1, with power removed. Since the capacitor checks open, we have probably found our problem—an open filter capacitor (C_1).

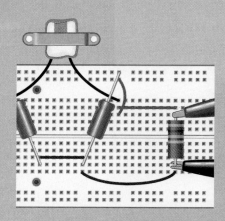

Symptoms

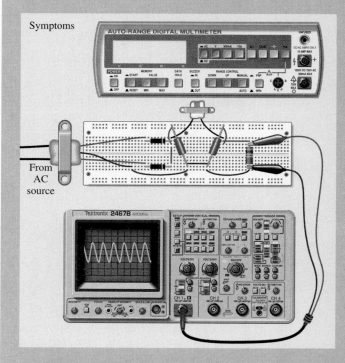

Verify

4th Test: Replace C_1 with a capacitor of appropriate ratings and restore normal opeation.

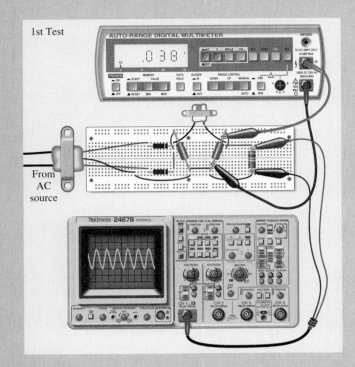

1st Test

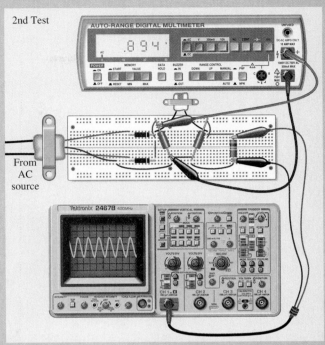

2nd Test

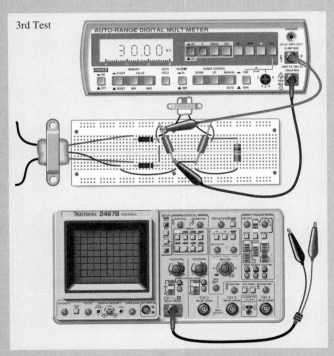

3rd Test

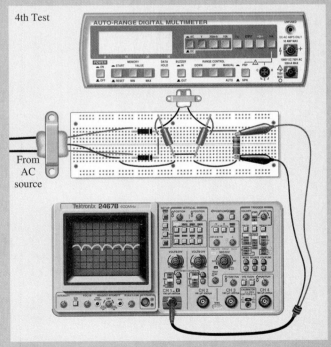

4th Test

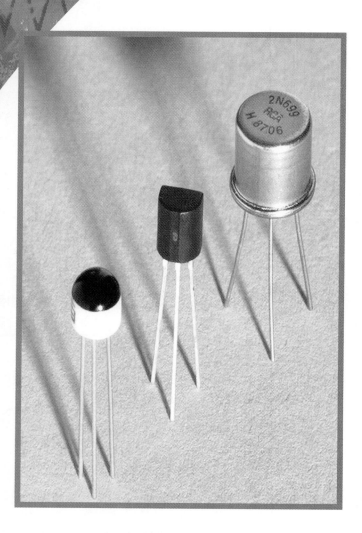

BIPOLAR JUNCTION TRANSISTORS

26

chapter

CHAPTER PREVIEW

The transistor was invented in the late 1940s. The term *transistor* is really a contraction of "transfer" and "resistor." Credit for its invention is given to three Bell Laboratories scientists, John Bardeen, Walter Brattain, and William Shockley.

In this chapter you will learn more about the basics of transistors. You will find out exactly what transistors are, you will see some of the things they can do, and you will find out how they work. You will also learn about the transistor specifications that are found in transistor data sheets.

OBJECTIVES

After studying this chapter, you should be able to:

1. Draw the symbols and identify the **emitter, base,** and **collector** leads for NPN and PNP transistors

2. Draw the symbols for **NPN** and **PNP transistors** and show the proper **voltage polarities** for the base-emitter terminals and for the collector-base terminals

3. Explain the meaning and cite the mathematical symbols for **emitter current, base current, collector current, base-emitter voltage,** and **collector-emitter voltage**

4. Describe how increasing the **forward-bias base current** in a **BJT amplifier** decreases the voltage between the emitter and collector

5. Describe the operation of a BJT when applied as a switch

6. Explain the meaning of the curves shown on a family of **collector characteristic curves**

7. Describe the meaning of the **maximum voltage, current,** and **power ratings** listed in BJT data sheets

26–1 BASIC TYPES OF TRANSISTORS

As you learned in the previous chapter, a forward-biased P-N junction is basically a low resistance path for current flow. Conversely, a reverse-biased P-N junction represents a high resistance path. This is the "resistance" part of transistor operation. If the current is almost the same through a given high resistance as it is through a given low resistance, the I^2R (power) in the high resistance is greater than in the low resistance. This is one basic principle on which power gain is produced by a transistor.

For example, if a signal is introduced in a forward-biased P-N junction and output is taken from a reverse-biased P-N junction with similar current in both the input and output circuits, there is *power gain*. In effect, the transistor has transferred the signal from the low resistance (input) to the high resistance (output). Thus, the word **transistor** relates the concepts of *transfer* and *resistance* as described earlier.

Resistance transfer—basic transistor operation—cannot be accomplished with a single P-N junction. Diode action can be accomplished with a single P-N junction, but transistor action cannot. Transistor action requires two separate

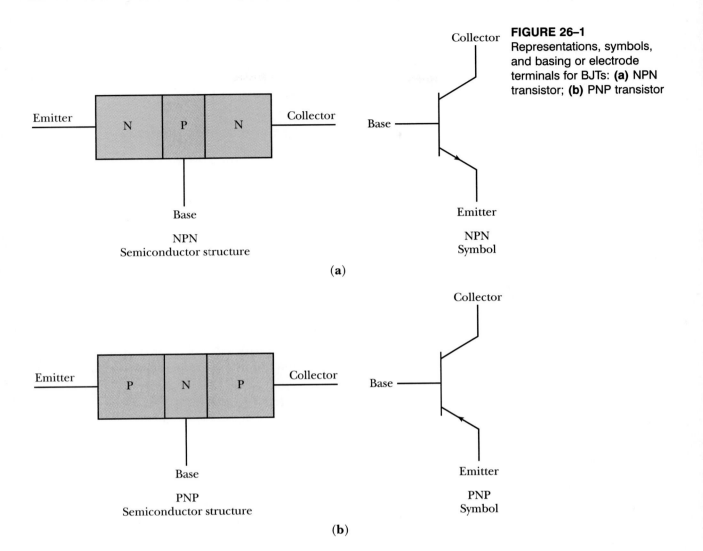

P-N junctions. Figure 26–1 shows two possible ways to create a two-junction semiconductor device. The example in Figure 26–1a uses a P-type semiconductor material located between two layers of an N-type material. This is known as an **NPN transistor**. In Figure 26–1b, an N-type material is sandwiched between two layers of a P-type material. This is a **PNP transistor**. Both NPN and PNP structures are valid and useful transistor types.

Although it is common practice to call these devices transistors, they are technically described as **bipolar junction transistors** (abbreviated as *BJT*s). They are "bipolar" because they use two different types of semiconductor materials, namely N-type and P-type materials, and because they use opposite polarity biasing voltages. They are "junction" transistors because they use current-carrying P-N junctions.

Note the similarities and difference for the symbols for NPN and PNP transistors, as shown in Figure 26–1. The external electrical connections (also called terminals or electrodes) are the same for both types. One of the outer connections is called the **emitter**, the opposite outer connection is called the **collector**, and the terminal for the middle material is called the **base**. The symbols differ only by the direction the arrow points in the emitter part of the symbol. For an NPN transistor, the arrow points outward and away from the base (remember that NPN is "*N*ot-*P*ointing-i*N*"). For a PNP transistor, the emitter arrow points in toward the base (remember that PNP is "*P*ointing-i*N*").

Some common BJT packages and basing schemes

In Figure 26–2 you see some common transistor lead basing layouts (i.e., bottom view of transistor package with lead connection layout). *E* stands for emitter, *B* stands for base, and *C* stands for the collector. The transistor outline (TO) numbers identify each specific packaging configuration used by manufacturers. Figure 26–3 shows BJTs mounted on heat sinks. The reasons for this are discussed later in this chapter.

BJTs are listed according to device identification codes. Recall that diodes are listed according to a code that begins with 1N (1N4001, for instance). BJTs are listed according to a code that usually begins with 2N. Some actual NPN transistors are 2N697, 2N222A, and 2N4401. Some actual PNPs are 2N2904, 2N4030, and 2N5400. There is no way to tell the difference between an NPN and PNP transistor just by looking at the part identification code. You should check the manufacturers' data sheets, a transistor data book, or a parts catalog in order to discover the specifications for a given transistor part identification number.

26–2 BJT BIAS VOLTAGES AND CURRENTS

Figure 26–4 is a demonstration circuit that is intended to show how currents flow through a BJT. It also introduces some labels you will find in nearly all discus-

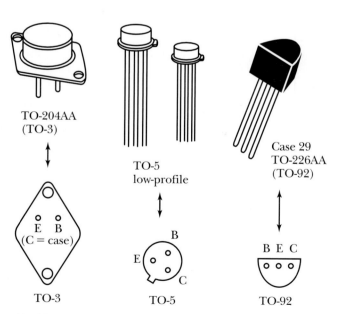

FIGURE 26–2 A sample of some common packaging and basing schemes

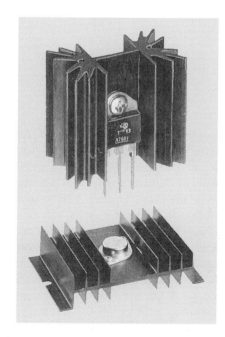

FIGURE 26–3 High-power BJTs mounted on heat sinks *(Photo courtesy of Thermalloy Inc.)*

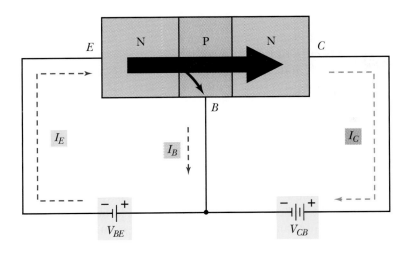

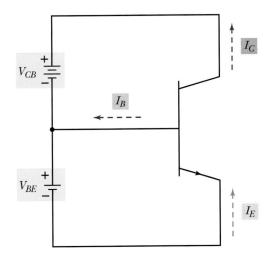

FIGURE 26–4 Bias voltages and currents in an NPN transistor

sions and descriptions of BJTs. First note this is an NPN transistor, then consider the following:

1. The dc source connected between the emitter and base of the transistor is called the **base-emitter voltage**, and is labeled V_{BE}.

2. The polarity of V_{BE} is such that it forward biases the base-emitter junction.

3. The dc source connected between the collector and base of the transistor is called the **collector-base voltage**, and is labeled V_{CB}.

4. The polarity of V_{CB} is such that it reverse biases the collector-base junction.

5. Current at the emitter connection (called **emitter current**) is labeled I_E.

6. Current at the base connection (called **base current**) is labeled I_B.

7. Current at the collector connection (called **collector current**) is labeled I_C.

The base-emitter junction of a BJT behaves very much like the P-N junction of an ordinary diode. For instance, you have to forward bias the base-emitter junction in order to allow any current flow through it. You can see from the diagram that forward-bias current will indeed flow from the negative terminal of V_{BE}, through the base-emitter junction, and back to the positive terminal of V_{BE}. Increasing the value of V_{BE} increases the value of I_B much the same way that increasing forward bias for a diode causes the diode current to increase. Base currents for BJTs, however, are usually limited to the microampere range. The forward-bias junction potential for the base-emitter junction of a BJT is identical with that of a diode, namely about 0.3 V for germanium transistors and 0.7 V for silicon transistors.

You will confuse yourself, however, if you try to compare the collector-base junction of a BJT with the P-N junction of an ordinary diode. Under normal operating conditions, the collector-base junction is reverse biased. Yet, it actually carries far more current than the forward-biased base-emitter junction. How can this be? The main clue is that the *base region in a BJT is extremely thin and very lightly doped.* This accounts for why base currents are usually quite small. What is more important, this also accounts for the fact that you can force a relatively large amount of current to flow between the emitter and collector of the BJT, provided the base-emitter junction is forward biased.

Let's take a closer look at the emitter-collector circuit. Note from the diagram that dc source V_{CB} and V_{BE} are in a series-aiding configuration. In other words, the total voltage applied between the emitter and collector terminals (**collector-emitter voltage, V_{CE}**) of the BJT is equal to the sum of the two dc sources: $V_{CE} = V_{CB} + V_{BE}$. If the base-emitter junction is forward biased and V_{CE} is sufficiently high, there will be a complete path for current flow that begins from the negative terminal of V_{BE}, goes to the emitter, passes through the extremely thin base region of the BJT, goes to the collector terminal, and returns to the positive terminal of V_{CB}. One of the vital characteristics of a BJT is that the collector current (I_C) is always much larger than the base current (I_B) that controls it.

The normal operating conditions for a BJT can be summarized this way:

- The base-emitter junction must be forward biased with a little bit of base current, I_B.
- The collector-base junction must be reverse biased.

The following occur as a result of setting up these conditions:

- The emitter current, I_E, is equal to the sum of the base and collector currents.
- The collector current, I_C, is always much larger than the base current.

Formula 26–1 expresses the fact that emitter current is exactly equal to the sum of the collector and base currents.

FORMULA 26–1 $I_E = I_C + I_B$

The fact that I_C is always much greater than I_B is expressed in terms of a ratio. Typically, I_C is between 40 and 300 times greater than I_B. This ratio is expressed as the β_{dc} (**dc beta**) or sometimes h_{FE}. (We will use the β_{dc} notation in our discussions, but you should expect to see the "h convention" often used in other places.) Formula 26–2 shows how you can calculate the value of dc beta when you know the values of collector and base current.

FORMULA 26–2 $\beta_{dc} = I_C/I_B$

Used in this way, β_{dc} represents the current gain of the transistor.

Example

In a test setup for a BJT, you find that the collector current is 1.2 mA when the base current is 24 μA. What is the current gain, β_{dc}, of the transistor?

Answer:

From Formula 26–2:

$$\beta_{dc} = I_C/I_B = 1.2 \text{ mA}/24 \text{ μA} = 50$$

The current gain of this BJT is 50.

Formulas 26–3 and 26–4 show how you can determine collector current and base current when you know the dc beta and one of the other terms.

FORMULA 26–3 $I_C = \beta_{dc} \times I_B$

FORMULA 26–4 $I_B = I_C/\beta_{dc}$

Example

The β_{dc} of a certain BJT is known to be approximately 180. How much base current is required in order to have 18 mA of collector current?

Answer:

From Formula 26–4:

$$I_B = I_C/\beta_{dc} = 18 \text{ mA}/180 = 100 \text{ μA}$$

There is another BJT specification that is important when you are working with certain transistor configurations. This is the α (alpha) ratio—the ratio of collector current to emitter current.

FORMULA 26–5 $\alpha = I_C/I_E$

Values of α for typical BJTs are always less than 1, but usually greater than 0.9.

Example

The collector current for a certain BJT is 9.8 mA when the emitter current is 10 mA. What is the BJT's α rating?

Answer:

From Formula 26–5:

$$\alpha = I_C/I_E = 9.8 \text{ mA}/10 \text{ mA} = 0.98$$

The α for this BJT is 0.98.

All of the BJT specifications described so far apply to the experimental setup for an NPN transistor in Figure 26–4. You might be pleased to know that the same kind of thinking, expressions, and formulas apply equally well to PNP transistors. See Figure 26–5. The only difference is the polarity of the applied bias voltages.

You can see that the base-emitter junction is forward biased by V_{BE}, and that the collector-base junction is reverse biased by V_{CB}. This is the same set of conditions required for operating an NPN transistor.

PRACTICE PROBLEMS I

1. How much emitter current flows in a BJT that has a base current of 100 μA and a collector current of 10 mA?

2. Suppose you measure the base current for a certain BJT and find it is 15 μA. The corresponding collector current is 2 mA. What is the dc current gain, β_{dc}, of the BJT if it is an NPN transistor? If it is a PNP transistor?

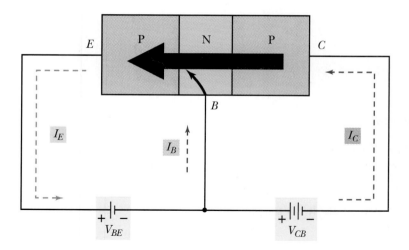

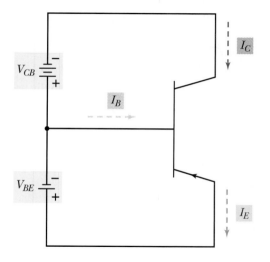

FIGURE 26–5 Bias voltages and currents in a PNP transistor

3. A certain BJT has a β_{dc} of 100. Calculate the amount of collector current, amount of emitter current, and the value of α when the base current is 10 μA.

IN-PROCESS LEARNING CHECK I

Fill in the blanks as appropriate.

1. Under normal operating conditions, the base-emitter junction of a BJT must be _____ biased, while the collector-base junction is _____ biased.

2. Current must be flowing through the _____ junction of a BJT before current can flow between the emitter and collector.

3. The symbol for an NPN transistor shows the emitter element pointing _____ the base element.

4. The symbol for a PNP transistor shows the emitter element pointing _____ the base element.

5. For BJTs:
   V_{BE} is the _____ voltage
   V_{CB} is the _____ voltage
   V_{CE} is the _____ voltage

6. For a typical BJT circuit:
   I_C stands for the _____ current
   I_B is the _____ current
   I_E is the _____ current

7. Stated in words, the dc beta of a BJT is the ratio of _____ divided by _____.

8. Stated in words, the alpha of a BJT is the ratio of _____ divided by _____.

9. The α of a BJT is _____ than 1, while the β_{dc} is _____ than 1.

26-3 BASIC USES OF THE BJT

Before we look at a few more details about the operation of BJTs, it is important to be able to describe basically how transistors are used. Transistors are controllers of current. The BJT uses a small amount of base current to control a larger amount of collector current. You will learn in a later lesson that field-effect transistors use a voltage level to control the flow of current. BJTs are basically used in two ways: as switches and as amplifiers. Often, these two applications are combined in one circuit.

The BJT as a switch and current amplifier

Figure 26–6 shows a typical circuit where an NPN transistor is used as a switch and as an **amplifier** of current. The rectangular waveform at the input switches between 0 V and +5 V. Let's examine how the BJT responds to the 0-V input and to the +5-V input.

FIGURE 26–6 BJT used as a switch and current amplifier

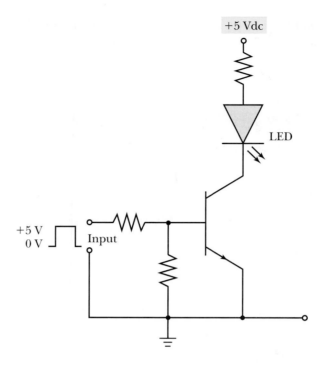

1. When the input is at 0 V, there is no base current flowing in the BJT, so:
   - there is no collector current in the BJT;
   - the collector voltage is equal to the V_{CC} level (+5 V); and
   - the LED does not light.

   The circuit is switched off.

2. When the input is at +5 V, base current is flowing in the BJT, so:
   - the collector current is flowing in the BJT (about 15 mA);
   - the collector voltage is equal to the forward voltage drop of the BJT (about +0.7 V); and
   - the LED is turned on.

In summary, the LED lights whenever the input waveform switches to +5 V, and it goes out whenever the input waveform switches to its 0-V level. The LED switches on and off at the frequency of the input waveform.

But why do we go to the trouble of using a BJT? Why not connect the input waveform direct to the LED and simplify the circuit? Well, the reason is that the input waveform probably comes from a source that cannot provide the 15-mA current level that is required for operating most LEDs. Maybe the waveform source provides a maximum of 5 mA. In this case, the BJT serves as a current amplifier as well as a switch.

The BJT as a voltage amplifier

The circuit in Figure 26–7 is a voltage amplifier. The input waveform in this example is a sinusoidal waveform that measures 2 V peak-to-peak. The values of the resistors at the base connection are selected so that there is some base-bias current flowing, even when the input signal is at 0 V. Let's study how the BJT reacts to one complete cycle of the waveform at 1/4-cycle intervals.

1. At 0°, the input is at 0 V.

   The only base current flowing is the base-bias current. The BJT is conducting, so the collector voltage is somewhere between 0 V and the $+V_{CC}$ level of +12 V. Let's suppose the base bias is adjusted so that the collector voltage is +6 V.

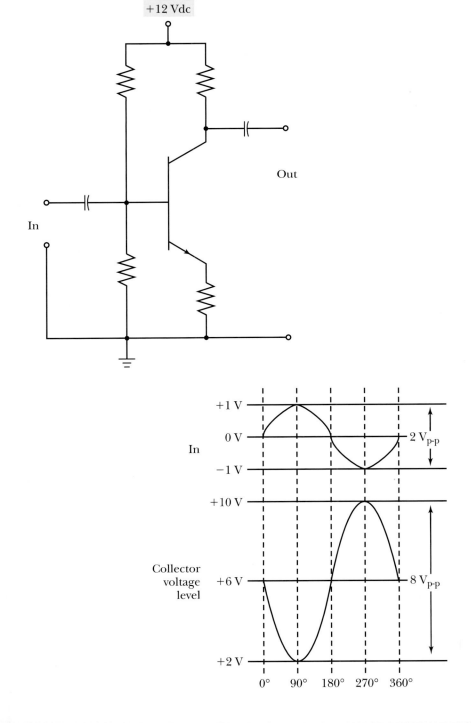

FIGURE 26–7 BJT used as a voltage amplifier

2. At 90°, the input is at +1 V.

   The positive polarity of the input signal now aids the base-bias current, so the BJT conducts more. This means the collector voltage (from collector to ground) drops to a lower level. This is because with higher I_C, the I × R drop across the resistor between the collector and V_{CC} source increases. Thus, less voltage is available to be dropped by the transistor. (Remember Kirchhoff's voltage law?) Let's say we find it drops down to +2 V.

3. At 180°, the input is at 0 V.

   The only base current flowing is the base-bias current. There is less base current than while the input waveform was positive; there is less collector current, so there is a larger collector voltage. The collector voltage returns to +6 V.

4. At 270°, the input is at −1 V.

   The negative polarity of the input signal bucks against base-bias current, so the BJT conducts even less. This means the collector voltage rises to a higher level, say, +10 V.

You can see from the resulting plot of the output waveform that it is a sinusoidal waveform that is 8 V peak-to-peak. This circuit is a voltage amplifier that increases the input signal from 2 V_{p-p} to 8 V_{p-p}. It is said that this amplifier has a voltage gain of 4 (8 V_{p-p} divided by 2 V_{p-p}). Also notice that this type of BJT amplifier inverts the input waveform or, in other words, shifts the waveform by 180°.

Later lessons will enable you to understand and analyze transistor switches and amplifiers in greater detail.

26–4 BJT CHARACTERISTIC CURVES

Figure 26–8 shows how the collector current (I_C) of an NPN transistor responds when:

- The forward-biasing base current (I_B) is fixed at some given level.
- The collector-emitter voltage (V_{CE}) is increased from 0 V to some upper positive level.

You can see from the graph that, as you begin increasing V_{CE} from 0 V, I_C increases sharply at first. But when V_{CE} reaches about +0.5 V, the increase in I_C begins to taper off, even though you continue increasing the value of V_{CE}. In this example, I_C very nearly levels off at 8 mA, even while V_{CE} increases from about 0.5 V to 20 V. Operating in this region of the curve, the transistor is said to be in a state of current saturation. This is a desirable and normal operating condition (as you will soon see). The graph of this effect is called a **collector characteristic curve.**

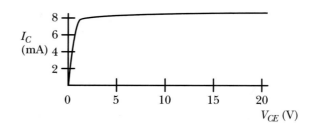

FIGURE 26–8 NPN collector characteristic curve

Saturation in a transistor occurs when the amount of current reaches a level where all available charge carriers (holes and electrons) are in use. So increasing the applied voltage causes only an insignificant increase in current flow through the device. However, there is a way to increase the population of available holes and electrons: increase the amount of base current. Figure 26–9b shows the *family of collector characteristic curves* for a typical NPN transistor. It is actually a series of collector characteristic curves, each generated by a different base current (I_B) level.

Let's take a closer look at how a family of collector characteristic curves is generated (refer to Figure 26–9a):

Step 1. Adjust V_{BE} to 0 V so that $I_B = 0$ μA.

Step 2. Slowly adjust V_{CE} from 0 V to +16 V, noting and plotting the values of I_C as you go along.
At this point, you have generated the collector characteristic curve for $I_B = 0$ μA.

Step 3. Adjust V_{BE} so that $I_B =$ is fixed at 40 μA.

Step 4. Slowly adjust V_{CE} from 0 V to +16 V, plotting the values of I_C as you go along.
This completes the second collector characteristic curve, the one where $I_B = 40$ μA.

The only difference between these two sets of steps is the value of base current. You can complete the entire family of collector characteristic curves, one at a time, by repeating these steps where $I_B = 40$ μA, 80 μA, 120 μA, 160 μA, 200 μA, and 240 μA.

The family of collector characteristic curves for a PNP transistor is the same as an NPN transistor except the PNP values for voltage and current are negative values.

Collector characteristic curves are very nearly the same for a given type of BJT. The 2N697 is a popular general-purpose NPN transistor. Every 2N697 that is

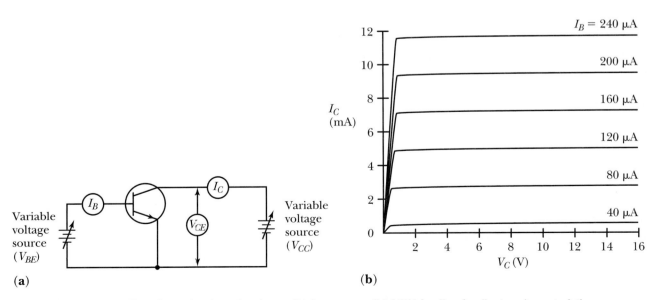

(a)   (b)

FIGURE 26–9 (a) Transistor circuit to develop collector curves; **(b)** NPN family of collector characteristic curves

tested should have nearly identical families of collector characteristic curves. The 2N222A is another popular NPN transistor, and each of them should have the same set of collector characteristic curves. But the family of collector characteristic curves for a 2N697 is different from that of a 2N222A. Transistor manufacturers publish the collector characteristic curves for each of their BJT products.

IN-PROCESS LEARNING CHECK II

Fill in the blanks as appropriate.

1. BJTs use a small amount of base _____ to control a larger amount of collector _____. It can be said that BJTs are _____ controllers.
2. When base current in a BJT is zero, the collector current is _____.
3. BJTs are basically used as _____ and _____.
4. When a BJT is being used as a switch (as in Figure 26–6), V_{CE} is maximum when I_B is _____, and V_{CE} is minimum when I_B is _____. In the same circuit, I_C is maximum when I_B is _____, and I_C is minimum when I_B is _____.
5. When a BJT is being used as voltage amplifier (as in Figure 26–7), an increase in base current causes a(n) _____ in V_{CE}.
6. On a family of collector characteristic curves, the horizontal axis represents the _____ and the vertical axis represents the _____. Each curve in the family represents a different level of _____.

26–5 OTHER BJT RATINGS

Maximum collector-emitter voltage

Recall there must be a voltage applied between the emitter and collector in order to operate a BJT properly. What's more, the polarity is important and depends on whether the BJT is an NPN or PNP type. This is the collector-emitter voltage, V_{CE}. There is a definite limit to the amount of collector-emitter voltage that a BJT can withstand without breaking down and destroying itself with an excess of heat-producing current flow between the emitter and collector. This upper voltage limit is called the maximum collector-emitter voltage rating, and it applies when the transistor is turned off—when $I_B = 0$ A (Figure 26–10). The symbol is BV_{CEO}. The BV stands for "*B*reakdown *V*oltage" and CEO stands for "*C*ollector-*E*mitter" voltage when the remaining terminal, the base, is "*O*pen" (no connection to it). Typical values of BV_{CEO} for BJTs are between 25 and 100 V.

Maximum collector-base voltage

You have already learned that the collector-base junction of a BJT must be reverse biased. This voltage is specified as V_{CB}. One of the important ratings, or specifications, for a BJT is the maximum allowable amount of collector-base voltage. If this reverse bias value is exceeded, the voltage will break down the collector-base junction and destroy the transistor. The symbol for the maximum collector-base voltage is BV_{CBO}. This signifies a breakdown voltage between the collector and base when the emitter is open. See Figure 26–11.

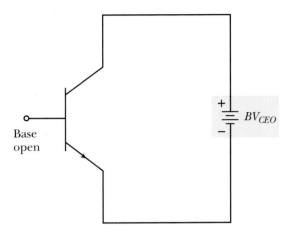

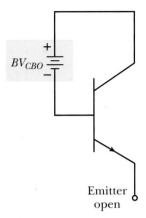

FIGURE 26–10 Diagram of maximum collector-emitter voltage

FIGURE 26–11 Diagram of maximum collector-base voltage

Values for BV_{CBO} run about 50 to 100 percent higher than the BV_{CEO} rating of the same transistor. So if a certain BJT has a collector-emitter breakdown voltage rating of 80 V, you can expect the collector-base rating to be on the order of 120 V to 160 V.

Maximum emitter-base voltage

The emitter-base voltage should either be 0 V (transistor turned off) or some forward-biasing value (transistor conducting). This P-N junction is rarely reverse biased. One reason is that the base portion of the BJT is extremely thin, which means it cannot withstand very much reverse-bias voltage without breaking down and destroying the junction.

The voltage between emitter and base, Figure 26–12, is usually shown as V_{EB}. The reverse breakdown voltage level for this parameter is shown as BV_{EBO}. This is the breakdown voltage (reverse) between the emitter and base terminals when the collector terminal is open. Typical values for BV_{EBO} are between 5 V and 10 V.

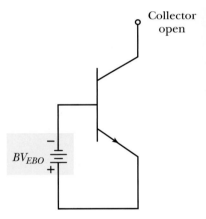

FIGURE 26–12 Diagram of maximum emitter-base voltage

PRACTICAL NOTES

Be careful when applying reverse dc levels and ac waveforms directly between the emitter and base of a BJT. The P-N junction between the emitter and base of a BJT is easily destroyed by seemingly harmless voltage levels.

Maximum collector current and power dissipation

The collector current for a BJT is usually denoted as I_C. The maximum allowable amount of continuous collector current is shown as $I_{C_{(max)}}$. For BJTs that are designed for small-signal and light-duty applications, $I_{C_{(max)}}$ is less than 1 A. For power BJTs $I_{C_{(max)}}$ is 1 A or more. Larger BJTs also have a maximum surge current rating. BJTs can withstand surges of current that are two or three times greater than their continuous current rating. The data sheet will list the allowable duration of the surges and their maximum duty cycle.

The power dissipation (P_D) of a BJT is figured by multiplying the collector-emitter voltage by the collector current. Power dissipation is related to the amount of heating the BJT will undergo. Whenever the power dissipation of a BJT exceeds the maximum power rating ($P_{D_{(max)}}$), the transistor's own heat will cause it to destroy itself. However, you can effectively increase the maximum power dissipation of a transistor by mounting it to a heat sink assembly (see Figure 26–3).

AC current gain

You have learned that a BJT has a certain amount of dc current gain that is called dc beta, β_{dc}. This is found in the transistor data sheets (sometimes as h_{FE}), and it is defined as the ratio of collector current to base current. There is a second version of current gain that appears on transistor data sheets. This is the **ac beta**, β_{ac} (sometimes shown as h_{fe}). Figure 26–13 will help you understand how the value of β_{ac} can be determined.

Notice that a change in base current between 120 μA and 160 μA (40 μA change) causes the collector current to change from 5.5 mA to 7.5 mA (2 mA change). Ac beta is calculated as the ratio of the change in collector current (2 mA in this example) to the change in base current (40 μA) in this example. This comes out to provide β_{ac} = 2 mA/40 μA = 50. In this formula, Δ means "a change in."

FORMULA 26–6 $\beta_{ac} = \Delta I_C / \Delta I_B$

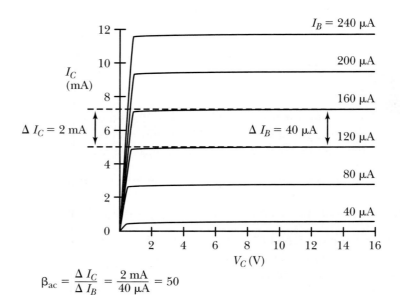

FIGURE 26–13 Calculating β_{ac} from a family of collector characteristic curves

$$\beta_{ac} = \frac{\Delta I_C}{\Delta I_B} = \frac{2 \text{ mA}}{40 \text{ μA}} = 50$$

26-6 INFORMATION FOR THE TECHNICIAN

PRACTICAL NOTES

Some transistor troubles are caused by physical factors. An example is when a solder connection fails to make contact between a transistor electrode and the conductive foil on a circuit board.

Many problems occur in transistors when their electrical ratings are exceeded. Damage results from excessive current through or excessive heating of the materials in the device. (You can damage a semiconductor device when you are soldering it to the circuit or circuit board if you do not appropriately heatsink the leads you are soldering.) Exceeding the voltage parameters for such devices is easy to do, and they "die silently" with no warning other than abnormal circuit or system operation. That is, there is no smoke or sound involved. Also, wrong polarity voltages applied to the electrodes cause problems.

There are some sophisticated *in-circuit testers* and *out-of-circuit testers,* which test transistors. An ohmmeter can also be used. When using an ohmmeter for transistor testing, troubles are found by checking the forward- and reverse-biased resistances from junction to junction. Many times, the problem is an internal short or open. Of course, the ohmmeter supplies the battery voltage for these tests. (NOTE: It helps to know the voltage polarity of the test probes when in the ohmmeter mode. Another *caution* is that currents caused by using ohmmeter on certain ranges can exceed ratings of the transistor being tested. Thus, you can make a transistor go from good to bad instantly.)

Refer to Figure 26–14 for some examples of how ohmmeter tests are used. Note that the values shown are only relative and can differ from device to device.

PNP-type transistor

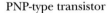

50 kΩ or more

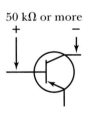

50 kΩ or more

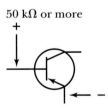

500 Ω or less

(To test NPN—simply reverse polarity in each case)

500 Ω or less

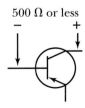

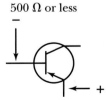

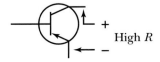

High *R*

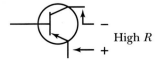

High *R*

FIGURE 26–14 VOM Ohmmeter testing of BJT junctions with R × 10 range

Another technique to check transistors during operation is to check the bias voltage from element to element. A bad transistor causes either too much or too little current through the various biasing networks, which can be detected by bias voltage measurements.

In summary, semiconductor devices can be damaged during manufacture, by mishandling during installation or replacement, or by improper currents or voltages. Normal precautions require using heat sinks to drain heat away from the transistor elements during soldering.

If a transistor or semiconductor fails, the device is replaced. But associated circuit components should also be checked prior to replacement to ensure that another malfunctioning part did not initially cause the failure (or that the failure of the transistor has not damaged an external part).

Common sense is the common ingredient in preventing problems and in solving them when they occur. In your further studies, you should become familiar with how semiconductor devices and circuits operate, what their normal parameters should be, and the various tools and techniques used to install, test, and replace these remarkable devices.

SUMMARY

▶ *Transistor* is derived from the words *transfer* and *resistor*.

▶ A transistor consists of two P-N junctions. Typically, one P-N junction is forward-biased and is a low resistance path for current; the other junction is reverse-biased and is a high resistance path for current.

▶ Because a transistor transfers signals from its low resistance input circuit to its high resistance output circuit, it exhibits power gain.

▶ Two basic BJTs are the NPN and PNP. The letters N and P designate the type of semiconductor material in the three sections. In the NPN device, P-type material is sandwiched between two areas of N-type material. In the PNP device, N-type material is placed between two outside areas of P-type material.

▶ Three elements of a BJT are the emitter, base, and collector.

▶ Forward bias is a voltage applied to a P-N junction so that positive polarity is connected to the P-type material and the negative polarity voltage is connected to the N-type material. In a BJT, the base-emitter junction is forward biased by V_{BE}.

▶ Reverse bias exists when a source's negative polarity voltage is connected to the P-type material, and the positive polarity voltage is connected to the N-type material. In a BJT, the collector-base junction is reverse biased by V_{CB}.

▶ Currents at the connections of a BJT are: emitter current, I_E; base current, I_B; and collector current, I_C.

▶ The base region in a BJT is extremely thin and very lightly doped. This is why it is possible to force current to flow between the emitter and collector whenever base-emitter junction is forward biased.

▶ Dc beta β_{dc} is an indication of the current gain of the BJT; it is the ratio of collector current to base current. Dc beta is typically between 40 and 300, depending on the design of the transistor. This same specification is often shown as h_{FE} as well as β_{dc}.

▶ The alpha (α) of a BJT is the ratio of collector current to emitter current. Alpha is always less than 1.

▶ A BJT is a current controlling device. It uses a relatively small amount of base current to control a larger amount of collector current. When there is no base current flowing in a BJT, there can be no collector current. Collector current increases as forward-bias base current increases.

▶ Even though a BJT is essentially a current-controlled device, it can be used as a voltage amplifier. Used in this way, changes in base current are said to cause changes in collector voltage.

▶ BJTs are basically used as switches and as amplifiers, and often as both at the same time.

▶ An I-V curve for a BJT shows how collector current increases with increasing collector voltage at first, but then levels off so that further increases in collector voltage cause only insignificant increases in collector current. A set of such curves for several different levels of base current is called a family of collector characteristic curves.

▶ Maximum ratings for BJTs are maximum collector-emitter voltage (BV_{CEO}), maximum collector-base voltage (BV_{CBO}), maximum emitter-base voltage (BV_{EBO}), maximum collector current ($I_{C_{(max)}}$), and maximum power dissipation ($P_{D_{(max)}}$).

▶ The power dissipation (P_D) of a BJT can be determined by the power formula as it applies to the collector current and voltage drop, i.e., $P_D = I_C \times V_{CE}$.

▶ Ac beta, β_{ac}, is a measure of dynamic current gain. This means the ratio of a change in collector current to the related change in base current that caused it. Ac beta can be determined from the BJT's family of collector characteristic curves. This same specification is also indicated by h_{fe}.

▶ Semiconductors are sensitive to overheating. Therefore, proper caution must be used when soldering and desoldering them.

▶ BJTs are tested in various ways. Test equipment is available to test them in circuit and out of circuit. Also, the ohmmeter, if properly used, is a good tool for general testing. Bias voltages are also good indicators of normal or abnormal circuit operation.

FORMULAS AND SAMPLE CALCULATOR SEQUENCES

FORMULA 26–1　　$I_E = I_C + I_B$

I_C value, ⊕, I_B value, ⊜

FORMULA 26–2　　$\beta_{dc} = I_C/I_B$

I_C value, ⊝, I_B value, ⊜

FORMULA 26–3　　$I_C = \beta_{dc} \times I_B$

β_{dc} value, ⊗, I_B value, ⊜

FORMULA 26–4　　$I_B = I_C/\beta_{dc}$

I_C value, ⊝, β_{dc} value, ⊜

FORMULA 26–5　　$\alpha = I_C/I_E$

I_C value, ⊝, I_E value, ⊜

FORMULA 26–6　　$\beta_{ac} = \Delta I_C/\Delta I_B$

ΔI_C value, ⊝, ΔI_B value, ⊜

REVIEW QUESTIONS

1. How many P-N junctions does a typical NPN transistor have? a typical PNP transistor?

2. Which type of semiconductor material is connected to the base of an NPN transistor? a PNP transistor?

3. Which type of semiconductor material is connected to the collector of an NPN transistor? a PNP transistor?

4. For proper operation of an NPN transistor, what polarity of voltage is applied to the base-emitter junction? to the collector-base junction?

5. For proper operation of a PNP transistor, what polarity of voltage is applied to the base-emitter junction? to the collector-base junction?

6. Draw the schematic symbols for the NPN and PNP transistors. Label all three terminals.

7. Approximately what percentage of emitter current appears in the collector circuit of a typical silicon BJT?

8. Explain forward biasing with respect to a P-N junction.

9. Explain reverse biasing with respect to a P-N junction.

10. For normal operation, the base-emitter circuit of a BJT is (forward, reverse) _____ biased.

11. For normal operation, the base-collector circuit of a BJT is (forward, reverse) _____ biased.

12. Is the base current in a BJT a high or low percentage of the emitter current?

13. Is it true that the emitter current and collector current in a BJT have vastly different values?

14. Explain the term *basing diagram*.

15. Describe how increasing the amount of base current affects the collector current in a BJT amplifier.

16. Explain why the voltage drop between emitter and collector of a BJT being used as a switch will be at its minimum level when the base current is at its maximum level.

17. Describe the relationship between forward-biasing current and collector-emitter voltage drop in a typical BJT voltage amplifier.

18. Define the following maximum ratings for a BJT:
   a. Maximum collector-emitter voltage (BV_{CEO})
   b. Maximum collector-base voltage (BV_{CBO})
   c. Maximum emitter-base voltage (BV_{EBO})

19. Define the following ratings for a BJT:
   a. $I_{C_{(max)}}$
   b. $P_{D_{(max)}}$

20. Explain the difference between dc beta and ac beta.

PROBLEMS

1. For a typical BJT amplifier circuit, the collector current is 40 mA and the base current is 500 μA. Determine the following:
   a. The value of the emitter current
   b. The value of β_{dc}

2. When I_E = 50 mA and I_B = 120 μA, determine the following:
   a. The value of I_C
   b. The value of β_{dc}

3. What is the dc beta of a BJT if the collector current is 10 mA when applying 20 μA of base current?

4. When β_{dc} = 100 and I_C = 800 mA, determine the following:

   a. The value of I_B
   b. The value of I_E

5. What is the value of I_C for a BJT that has a β_{dc} of 80 and an I_B of 20 μA?

6. What is the alpha of a BJT if you find 20 mA of emitter current and 19.5 mA of collector current?

7. Determine α_{dc} when I_C = 200 mA and I_E = 198 mA.

8. For a certain BJT circuit, α_{dc} = 0.95 and I_C = 400 mA. Determine the following:
   a. I_E
   b. I_B
   c. β_{dc}

9. For a certain BJT circuit, $\alpha_{dc} = 0.96$ and $I_E = 40$ mA. Determine the following:
   a. I_C
   b. I_B
   c. β_{dc}

10. For a certain BJT circuit, $\beta_{dc} = 200$ and $I_B = 40$ μA. Determine the following:
    a. I_C
    b. I_E
    c. α_{dc}

11. A change of 200 μA at the base causes a change of 20 mA at the collector. What is the β_{ac} for this transistor?

12. If the β_{ac} for a BJT is 400, how much change in collector current can you expect to see when the base current changes by 10 μA?

13. What is the power dissipation of a BJT were $V_{CE} = 4$ V and $I_C = 120$ mA?

14. If $P_{d_{(max)}}$ for a BJT = 600 mW, what is the maximum allowable collector current when $V_{CE} = 1.2$ V?

15. Determine the β_{ac} for $\Delta I_B = 80$ μA as shown on the family of collector characteristic curves in Figure 26–15.

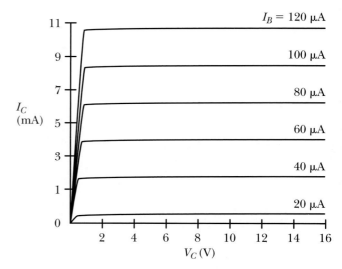

FIGURE 26–15

ANALYSIS QUESTIONS

1. Look up and draw the transistor outlines for the following case/basing configurations: TO-3, TO-5, TO-92, TO-220, and SOT-89. Label each electrode on your drawings.

2. Look up the part identification codes for five NPN and five PNP transistors.

3. Research and explain the features and use of heat sinks associated with power transistors.

4. Locate data sheets for several BJTs and find the ratings for β_{ac} (or h_{fe}), BV_{CEO}, BV_{CBO}, BV_{EBO}, $I_{C_{(max)}}$, and $P_{D_{(max)}}$.

5. Use Formulas 26–1 and 26–5 to prove that α is always a value that is less than 1.

PERFORMANCE PROJECTS CORRELATION CHART

CHAPTER TOPIC	PERFORMANCE PROJECT	PROJECT NUMBER
BJT Bias Voltages and Currents	BJT Biasing	73
BJT Bias Voltages and Currents	Transistor Alpha and Beta	74
Information for the Technician	Ohmmeter Tests for BJTs	75

NOTE: It is suggested that after completing the above projects, the student should be required to answer the questions in the "Summary" at the end of this section of projects in the Laboratory Manual.

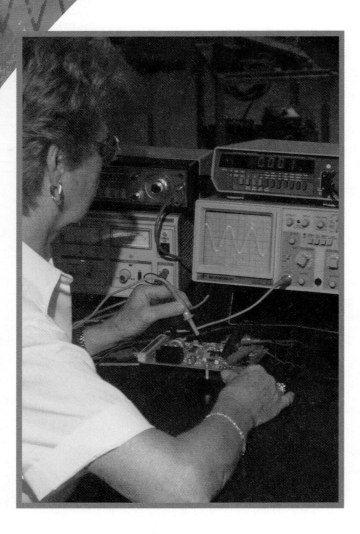

KEY TERMS

Class A amplifiers
Class B amplifiers
Class C amplifiers
Common-base amplifier (CB)
Common-collector amplifier (CC)
Common-emitter amplifier (CE)
Large-signal amplifiers
Load line
Push-pull amplifier
Q point
Small-signal amplifiers
Voltage follower

892

BJT AMPLIFIER CIRCUITS

OUTLINE

27

chapter

CHAPTER PREVIEW

Amplifier circuits are found in numerous systems and subsystems in electronics. For example, the ability to amplify enables applications such as building a very small signal from a microphone or recording into a large signal that drives the speakers in a stereo amplifier system.

This chapter provides an overview of some key operating features of BJT amplifier circuits. You will learn classifications of amplifiers relative to their circuit configuration, signal levels, and classes of operation. These circuits will provide information to help you in further studies of these important topics, including amplifiers built around field-effect transistors as summarized for you in a later study.

OBJECTIVES

After studying this chapter, you should be able to:

1. Explain the basic **transistor amplification process**

2. Describe the input and output characteristics of **common transistor amplifier stages**

3. List the advantages of each common type of transistor amplifier stage

4. Describe the difference between **small-signal** and **power amplifier circuits**

5. Classify amplifiers by **class of operation**

6. Describe the classification of amplifiers and their operation from their **load lines**

7. List typical applications for each classification of amplifier

8. Perform a basic **analysis** of a common-emitter, Class A BJT that uses voltage-divider biasing

27-1 HOW A TRANSISTOR AMPLIFIES

You learned in an earlier lesson that a transistor transfers current from a low resistance input circuit to a high resistance output circuit, thus exhibiting power gain. In actuality, the amplifier stage is not manufacturing power from nothing. Rather, the dc power source provides the power, but the amplifier's special transfer feature enables **amplification.** Likewise, by proper selection of transistor elements in the input and output circuits, voltage gain and/or current gain and changes in impedance values from input to output circuits can occur.

An analogy sometimes used to relate transistor operation is that a transistor acts like a variable resistor in a three-resistor voltage divider. See Figure 27–1. The transistor is in series with its collector resistor (R_C) and its emitter resistor (R_E). The collector voltage source (V_{CC}) is the voltage source for the voltage divider.

If the transistor operates so that its collector-to-emitter resistance is a very low resistance, the voltage drop across the transistor is fairly small compared with the voltages across the fixed resistors. If the transistor operates so that it acts like an open circuit (very high resistance) between the collector and emitter, then virtually all the source voltage falls across the transistor.

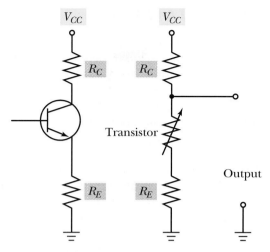

FIGURE 27–1 Analogy of a transistor as a variable resistor

The bias setting of the base-emitter junction and the input signal to the stage determines whether the transistor operates at a low resistance, high resistance, or somewhere in between.

There are several common ways to classify amplifiers:

1. By circuit configuration
2. By signal levels involved
3. By class of operation
4. By the type of transistor used

You will learn about the first three methods of classification in detail in this lesson. The type of semiconductor discussed in this chapter is the BJT. You will learn more about another type of transistor (the field-effect transistor, or FET) and its applications later.

27-2 AMPLIFIER CIRCUIT CONFIGURATIONS

There are only three basic configurations for BJT amplifier circuits. Any amplifier circuit found in electronics is one of these three basic configurations or, in some special instances, a combination of two of them. It is important you understand and recognize these basic circuit configurations. Without this understanding, you will become confused when attempting to analyze and troubleshoot more complex transistor amplifier circuits.

PRACTICAL NOTES

The amplifier examples in this lesson show only the use of NPN transistors. This is customarily done because the NPN is more commonly used than the PNP in actual practice. Even so, you can create PNP versions for study by drawing the circuits with a PNP transistor and reversing the polarity of the voltage sources.

The common-emitter (CE) amplifier

Figure 27–2 shows one of the simplest and most commonly used BJT amplifier circuits. Notice that the input signal is applied between the base and emitter terminals on the BJT, and that the output signal is taken between the collector and emitter terminals. (NOTE: In operating circuits, the transistor dc sources are bypassed as far as the signals are concerned.) Thus, the input signal is effectively between base and ground [emitter], and the output signal is effectively between collector and ground [emitter]. What BJT terminal is used for both the input and output connections? The emitter. For this reason, this circuit configuration is called the **common-emitter**, or CE, amplifier configuration.

The model in Figure 27–2 represents the CE amplifier in terms of signal flow, but it is not a practical circuit. You should never connect a BJT directly to voltage sources as shown in that figure. The model in Figure 27–3 is a more realistic model of a CE amplifier. You can see the series voltage-divider effect that will take place between the emitter-collector terminals on the BJT and the collector resistor, R_C. Let's study several things that happen whenever the input signal level is changed.

Referring to Figure 27–3, suppose the input signal changes and causes the base current to increase. Increasing the base current effectively lowers the resistance between the BJT's emitter and collector. As a result, 1) the voltage drop between the emitter and collector decreases, and 2) the current through the BJT and resistor R_C increases. Notice that an increase in base current at the input of this circuit causes a decrease in voltage at the output. On the other hand, causing the base current to decrease has the opposite effect on the output voltage. This is explained next.

Still referring to the working model of a CE amplifier in Figure 27–3, suppose there is a change in the input signal that causes a decrease in base current.

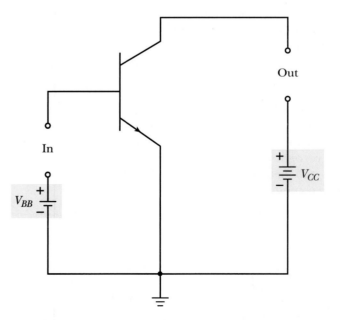

FIGURE 27–2 The theory model of a common-emitter amplifier

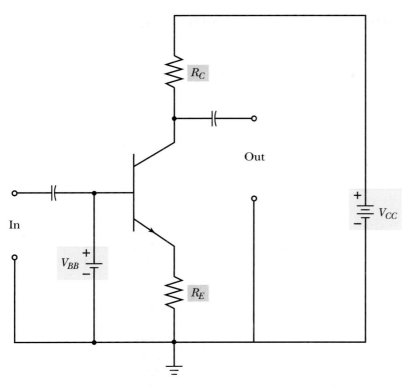

FIGURE 27–3 A voltage-divider model of a common-emitter amplifier

A decrease in base current causes a decrease in conductivity of the BJT's emitter-collector circuit. So 1) the voltage measured between emitter and collector increases, and 2) the current through the transistor and R_C decreases. Now you can see that a decrease in input signal causes an increase in output voltage level.

The dc source labeled V_{BB} in Figure 27–3 is used for forward biasing the NPN transistor. It acts as a clamping circuit that shifts the baseline of the input voltage more positive. In this way, it is ensured that a negative swing in the input ac signal will not reverse bias the base junction and thereby turn off the BJT altogether. It is usually impractical to use a separate battery for biasing the base circuits of CE amplifiers, so you will more commonly find the type of voltage-divider bias shown in the practical CE circuit in Figure 27–4.

There is a fixed amount of forward-bias base current in Figure 27–4 by means of a conduction path from ground, through resistor R_E, through the emitter-base junction of the BJT, and through resistor R_{B_1} to the positive collector supply voltage, V_{CC}. The amount of forward-bias current depends on the values of those two resistors and the amount of supply voltage.

PRACTICAL NOTES

In actual practice, Class A amplifier distortion occurs whenever the input signal is larger than the circuit design intends. Distortion (usually in the form of signal clipping) is also evident under a number of different types of circuit faults.

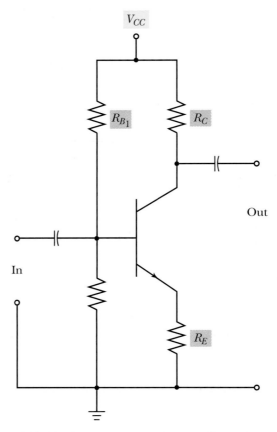

FIGURE 27–4 A practical, working model of a common-emitter amplifier

Features of CE amplifiers that use BJTs

Here is a summary of the basic features of practical CE amplifiers:

1. The input signal is introduced into the base circuit, and the output is taken from the collector circuit (the emitter is common to the input and output).

2. The input circuit is low impedance (because of the forward-biased emitter-base junction). Typically, the input impedance is in the range of 25 Ω to 5 kΩ.

3. The output circuit is medium to high impedance (approximately 50 Ω to 50 kΩ).

4. The circuit provides voltage, current, and power gain. Power gains range as high as 10,000. Current gain is always greater than 1.

5. There is a 180° phase reversal between the input and output signals.

Multiple-stage CE amplifier circuits

Figure 27–5 shows two separate CE amplifier circuits connected one after the other. This is a *two-stage amplifier* circuit. The original input signal is applied at the base circuit of the first stage, transistor Q_1. An amplified version is taken from the collector circuit of Q_1 and coupled by capacitor C_2 to the input of the second

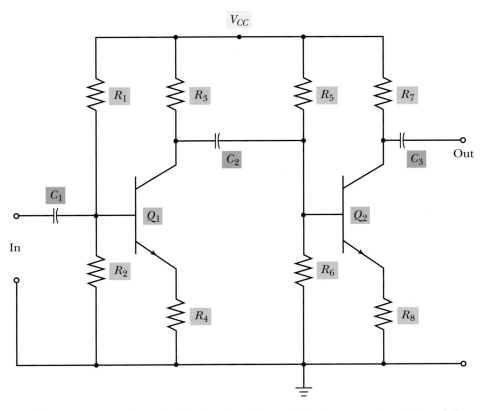

FIGURE 27–5 Two-stage, ac-coupled, common-emitter, amplifier circuit

amplifier stage, transistor Q_2. So the signal is amplified once again at Q_2 and the output taken from the collector of that stage. The signal is thus amplified twice.

In an amplifier that uses more than one stage, the overall gain multiplies from one stage to the next. So if the voltage gain of the input signal at the first stage is 10 and the voltage gain of the second stage is 6, then the overall gain is 10×6, or 60. Also, in each stage of a CE amplifier the signal is inverted. This means that the output signal from a two-stage CE amplifier is in phase with the input signal.

The two-stage circuit in Figure 27–5 is also considered *ac coupled* because the signal is introduced into the circuit through a capacitor (C_1), it is passed from the first to the second stage through a capacitor (C_2), and the signal leaves from the output through yet another capacitor (C_3). Recall that capacitors pass only ac signals. So in a multistage transistor circuit, gradual fluctuations in dc bias levels for the transistors are not amplified, and they do not affect the quality of the ac portion of the signal that is supposed to be amplified.

PRACTICAL NOTES

A disadvantage of using capacitor coupling between the stages of transistor amplifiers used as audio amplifiers is that the capacitors act as high-pass filters that block the low frequency response that is desirable in modern audio equipment. For this reason, many audio amplifiers are direct coupled (no C or L components between stages) using operational-amplifier circuits discussed in a later chapter.

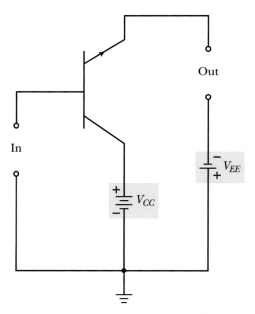

FIGURE 27–6 The theory model of a common-collector amplifier

The common-collector (CC) amplifier

Figure 27–6 is a simplified theory diagram of an amplifier configuration that has the input signal applied to the base terminal, and output taken from emitter terminal. If the input is applied to the base and the output taken from the emitter, it is apparent that the collector terminal is common to both the input and output. This circuit configuration is called the **common-collector**, or CC, amplifier configuration.

It is important to notice that the BJT is still biased to meet the basic operating conditions outlined in earlier discussions. First, you can see that dc source V_{CC} reverse biases the BJT's base-collector junction (imagine the input signal source is a short circuit to ac signals). Second, the V_{EE} source forward biases the base-emitter junction (if you imagine the input and output signal sources being short circuited to ac signals). No matter what BJT amplifier configuration you are working with, these two important conditions will be met.

The model of the CC amplifier in Figure 27–7 shows the components rearranged to give a clearer impression of how the circuit operates. This model shows that the voltage-divider effect will take place between the emitter resistor, R_E, and the emitter-collector connection of the BJT.

Suppose the input signal goes more positive. This will increase the amount of base current and decrease the resistance between the BJT's emitter and collector. As a result, 1) the voltage drop between the emitter and collector decreases, 2) the current through the BJT and resistor R_E increases, and 3) the voltage across R_E increases. Note that an increase in base current at the input of this circuit causes an increase in voltage at the output.

Whenever the input signal goes less positive, the amount of base current decreases, and the amount of emitter-collector current decreases as well. So 1) the voltage drop between the emitter and collector increases, 2) the current through the transistor and R_E decreases, and (3) the voltage across R_E decreases.

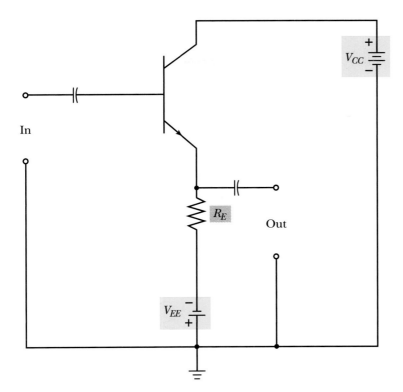

FIGURE 27–7 A voltage-divider model of a common-collector amplifier

Carefully notice that the output voltage "follows" the input voltage for a CC amplifier. As the input voltage goes more positive, so does the output; and as the input voltage goes less positive, so does the output. In other words, the input and output signals for a CC amplifier are in phase. For this reason a CC amplifier is often called a **voltage follower** amplifier.

If you study the little circuit comprised of input signal source, the base-biasing source, output resistor R_E, and the BJT's base-emitter junction, you will discover that the voltage dropped across the output resistor can never exceed the sum of the input signal and biasing sources. In fact, the voltage across R_E will always be less than the total input voltage because of the portion of the input that will be dropped across the emitter-base junction.

The CC circuit in Figure 27–8 has the form used in most electronic circuits. Forward-bias base current in Figure 27–8 is created by means of series-parallel paths. First consider the conduction path from ground through resistors R_{B_2} and R_{B_1} to V_{CC}. The other route is from ground through resistor R_E and the emitter-base junction of the BJT, then through resistor R_{B_1} to V_{CC}.

Features of CC amplifiers that use BJTs

Here is a summary of the basic features of practical CC amplifiers:

1. The input signal is introduced into the base circuit, and the output is taken from the emitter circuit (the collector is common to both the input and output).

2. The input circuit can have a very high impedance (determined by the β_{ac} value of the BJT and the values of resistors R_{B_2} and R_E).

3. The output impedance is relatively low (no greater than the value of R_E).

FIGURE 27–8 A practical model of a common-collector (voltage follower) amplifier

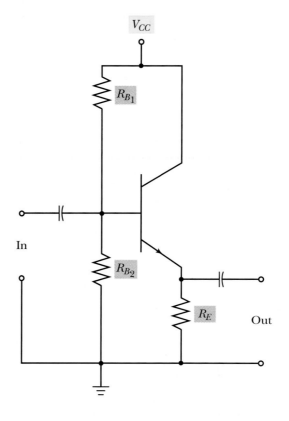

FIGURE 27–9 A CC amplifier followed by a CE amplifier

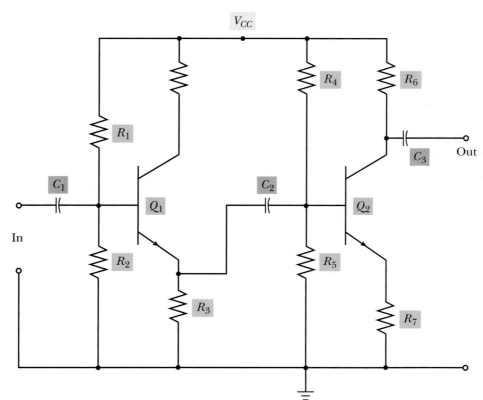

4. The circuit provides voltage gain less than 1.

5. The circuit provides good current gain, but less power gain than any of the three basic amplifier configurations.

6. There is no phase reversal between the input and output signals.

Application of the CC amplifier

You have seen that the input impedance of a CE voltage amplifier is relatively low. We often want to take advantage of the superior voltage-amplifying characteristic of a CE amplifier, but there is sometimes a requirement for a high impedance input. This problem is readily solved by a two-stage amplifying process as shown in Figure 27–9. The first stage uses a CC amplifier, which is noted for high input impedance but no voltage gain. By the time the signal passes through the CC stage, there is no longer a need for high input impedance, so the second stage can be a CE amplifier.

By using these two amplifiers in sequence, we get a result that cannot be achieved by either one alone. This principle is used frequently in modern electronic technology.

If two CC amplifiers are connected together as shown in Figure 27–10, the result is a circuit that has an extremely high input impedance. The input im-

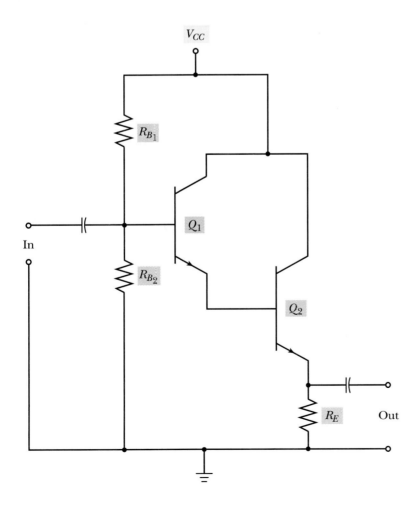

FIGURE 27–10 A Darlington amplifier circuit

FIGURE 27–11 The theory model of a common-base amplifier

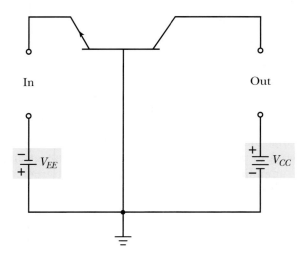

In

Out

V_{EE}

V_{CC}

pedance is created by the multiplication of the β_{ac} for Q_1, the β_{ac} for Q_2, and the value of R_E. It is not unusual for such a circuit to have an input impedance rated in megohms. This circuit is very commonly used in the input stages of sensitive measuring equipment. The circuit is known in the industry as a **Darlington amplifier**.

Besides very high input impedances, Darlington amplifiers are noted for high current gain (the product of the individual betas), but have a voltage gain less than 1. The circuit is so commonly used today that it is made available in integrated-circuit packages that include both BJTs and their interconnections.

The common-base (CB) amplifier

The theory circuit in Figure 27–11 shows an input signal applied to the emitter terminal and the output signal coming from the collector terminal. This is the definition of a **common-base** (CB) amplifier circuit. Both biasing requirements for BJT are met:

1. Dc source V_{EE} will forward bias the base-emitter junction.
2. Dc source V_{CC} will reverse bias the base-collector junction.

The voltage-divider model in Figure 27–12 is useful for stressing how changes in conductance of the BJT affects the output voltage level. The analysis is very nearly the same as for a CE amplifier. Whenever the signal goes more positive at the base circuit, the increasing amount of base current lowers the resistance between the BJT's emitter and collector. You can then expect 1) the voltage drop between the emitter and collector to decrease, 2) the current through resistor R_C to increase, and 3) the voltage across R_C to increase. Notice that a positive-going input signal causes a positive-going output signal. If you then consider the case where the input signal goes less positive, you will find that the output voltage follows along with it. For a CB amplifier, the output signal is in phase with the input signal.

A practical CB circuit is shown in Figure 27–13. If you trace through the circuitry, you will find that base-emitter forward bias is obtained by a path starting from ground, going through resistor R_E, through the base-emitter junction of the BJT, and through resistor R_{B_1} to the V_{CC} source.

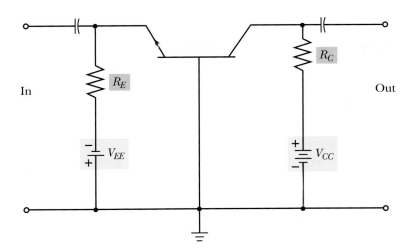

FIGURE 27–12 A voltage-divider model of a common-base amplifier

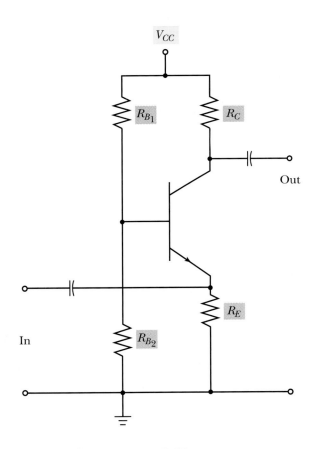

FIGURE 27–13 A practical model of a CB amplifier

Features of CB amplifiers that use BJTs

Here is a summary of the basic features of practical CB amplifiers:

1. The input signal is introduced into the emitter, and the output is taken from the collector circuit (the base is common to the input and output).
2. The input circuit is very low impedance (usually between 1 Ω and 50 Ω).
3. The output circuit is high impedance (about 1 kΩ to 1 MΩ).

4. The circuit provides good voltage and power gain. Current gain is always less than 1 (because collector current is always less than emitter current).

5. There is no phase reversal between the input and output signals.

Applications of CB amplifier circuits.

The very low input impedance of CB amplifier circuits seriously limit their use. In fact, this is the least common of the three basic amplifier configurations. Even so, CB amplifiers are consistently found in one application that requires low input impedances, namely the sensitive input stages of high frequency communications receivers.

IN-PROCESS LEARNING CHECK I

1. The BJT amplifier circuit having the base terminal common to both the input and output circuits is the _____ amplifier.

2. The BJT amplifier circuit providing both voltage and current gain is the _____ amplifier.

3. The BJT amplifier circuit having 180° phase difference between input and output signals is the _____ amplifier.

4. The BJT amplifier circuit having a voltage gain of less than 1 is the _____ amplifier.

5. The BJT amplifier circuit having an input to the base and output from the emitter is called a common-_____ amplifier.

6. In an amplifier that uses more than one stage, the overall gain is found by _____ the gains of the individual stages.

7. The voltage-follower amplifier is another name for a common-_____ amplifier.

8. The Darlington amplifier is made up of two common-_____ amplifiers.

27–3 CLASSIFICATION BY SIGNAL LEVELS

As its name implies, a **small-signal amplifier** is designed to operate effectively with small ac input signals and at power levels less than 1 W. What is a small signal? For our purposes, a small signal is one whose peak-to-peak ac current value is less than 0.1 times the amplifier's input bias current.

Examples of small-signal amplifiers are the first stages in radio and TV receivers. The small signals coming from the antennas are usually in the range of microvolts or millivolts. See the block diagram in Figure 27–14.

Large-signal amplifiers properly handle signals that have significantly larger input current levels and output power levels. The final **power amplifier** stages that drive the loudspeakers in an audio system are examples of large-signal amplifiers. See the example in the block diagram in Figure 27–15.

27–4 CLASSIFICATION BY CLASS OF OPERATION

The family of collector characteristic curves that you studied in an earlier lesson plays an important part in your understanding of classes of amplifier operation. By using these curves, we can explain the operating points and determine operating values for any class of amplifier operation.

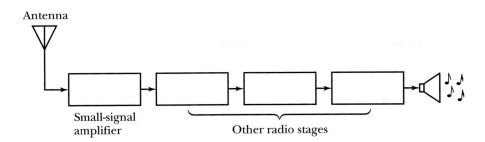

FIGURE 27–14 Small-signal amplifier fed by an antenna signal

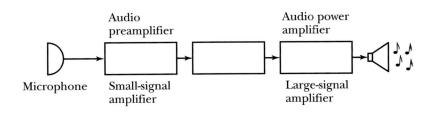

FIGURE 27–15 Large-signal amplifier feeding a loudspeaker

One of the important elements in this procedure is a **load line**. When drawn onto a set of collector characteristic curves, a load line describes in detail how the BJT will respond in a particular circuit. To construct a load line on a set of curves, you need to know two extreme values: 1) the collector-emitter voltage when the BJT is biased to its cutoff point, and 2) the amount of collector current when the BJT is at its saturation point. These values can always be determined from the amount of supply voltage and the values of resistors that are in series with the emitter-collector circuit.

When the base current for a BJT is zero, the collector current is always zero with a very small amount of collector leakage current. This condition occurs when the base-emitter junction is at 0 V or slightly reverse biased, and when the base-collector junction is reverse biased (as it usually is). Under these conditions, the collector-emitter voltage drop (V_{CE}) is maximum because of the very high impedance of the BJT. At that point, V_{CE} virtually equals the source voltage (V_{CC}). This is the cutoff point on the collector characteristic curves. See Figure 27–16.

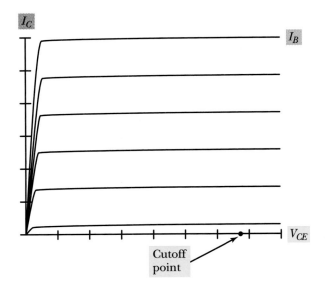

FIGURE 27–16 Plotting the cutoff point on the collector characteristic curves

FIGURE 27–17 Plotting the saturation point on the collector characteristic curves

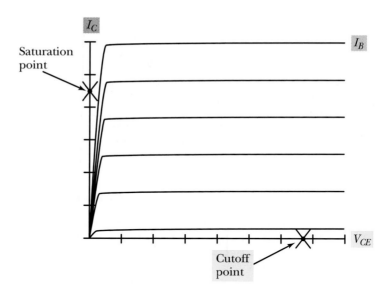

The opposite extreme operational condition occurs when I_C is maximum or the transistor operates at a point called *saturation*. This state exists when the forward base current increases to a point where a further increase in base current causes no additional increase in collector current. At this point, the collector-emitter voltage is minimum and is taken as virtually zero. See this saturation point plotted on the family of curves in Figure 27–17. The symbol for the amount of collector saturation current is $I_{C_{(sat)}}$.

Referring to the circuit in Figure 27–13 (page 905), you can calculate $I_{C_{(sat)}}$ with the following formula:

FORMULA 27–1 $I_{C_{(sat)}} = V_{CC}/(R_C + R_E)$

Example

Referring to the circuit in Figure 27–13, let $V_{CC} = 12$ V, $R_C = 10$ kΩ, and $R_E = 2.2$ kΩ. Determine the cutoff point and saturation current.

Answer:

The cutoff voltage is the same as V_{CC}, 12 V. From Formula 27–1:

$$I_{C_{(sat)}} = V_{CC}/(R_C + R_E)$$

$$I_{C_{(sat)}} = 12 \text{ V}/(10 \text{ k}\Omega + 2.2 \text{ k}\Omega) = 984 \text{ }\mu\text{A}$$

A load line is constructed by connecting the two extreme operating points with a straight line, Figure 27–18. From the load line, you can determine the voltage across the transistor and the amount of collector current for a given value of base current. As you will soon see, by setting the base-bias value, you set the operating points for the transistor.

To summarize this load line procedure:

1. Mark the cutoff point on the collector where $V_{CE} = V_{CC}$. Refer to Figure 27–16.

2. Determine the saturation, or maximum I_C, by calculating the collector current required to cause the total V_{CC} voltage value to be dropped across R_E and

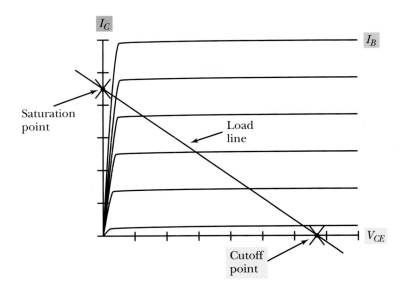

FIGURE 27–18 Plotting the load line on the collector characteristic curves

R_C (disregarding the forward voltage drop across the transistor). Refer to Figure 27–17.

3. Draw the load line by connecting the saturation and cutoff points on the characteristic curves with a straight line. Refer to Figure 27–18.

PRACTICE PROBLEMS I

Referring to the circuit in Figure 27–13, let $V_{CC} = 5$ V, $R_C = 470$ Ω, and $R_E = 47$ Ω. Determine the cutoff point and saturation current.

The **Q point**, or *quiescent operating point,* of a transistor is determined by the value of dc base bias. The intersection of the base-bias (I_B) line and the load line is called the Q point. By drawing a line down to the V_{CE} axis, we can see the V_{CE} value for that circuit operating at that particular bias point. By projecting a line horizontally from the Q point to the I_C axis, we can determine the collector current (I_C) value for that circuit operating with that base bias and the circuit's component values. See, for example, the load line and Q point in Figure 27–19.

Class A amplifier operation

When the Q point is set at the midpoint of the load line, Figure 27–19, the amplifier stage is operating at or near the definition for a **Class A amplifier**. By definition, a Class A amplifier is one where collector current flows throughout the entire input cycle. Under these conditions, the maximum applied base current does not drive the BJT into saturation, and the minimum base current does not drive the BJT into cutoff. So the transistor is always conducting above and below the Q point, but never hitting either extreme.

Because the output signal is not distorted by driving the amplifier into cutoff or saturation, Class A amplifiers are used where it is desirable to keep signal distortion as low as possible. This is especially important in the small-signal amplifiers for radio, TV, and audio equipment.

Class A amplifiers provide minimum signal distortion, but they are also known for their low operating efficiency. The low efficiency occurs because a sig-

FIGURE 27–19 Q point set at midpoint on the load line for Class A amplifier operation

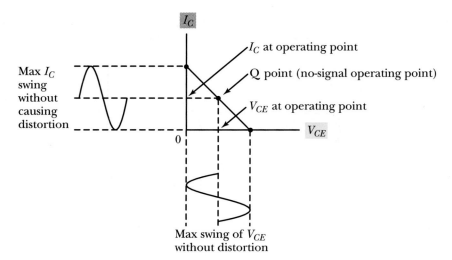

nificant amount of dc bias current flows when there is no signal applied to the amplifier.

Class B amplifier operation

When the BJT amplifier is biased so that its Q point is at cutoff, Figure 27–20, the stage is biased as a **Class B amplifier**. Collector current flows only as long as the input signal is positive. This is usually just 180° of a sinusoidal input waveform. Because of being biased at cutoff, half of the input cycle biases the transistor stage below cutoff. The output waveform resembles the half-wave rectified signal you studied in an earlier chapter.

Class B amplifiers have higher efficiency than Class A stages because the Class B transistors are in cutoff through half the input waveform. When no signal is applied, the transistor is turned off (recall that a Class A amplifier is conducting even when no signal is applied). However, Class B amplifiers distort the signal by clipping off any part of the signal that drives the transistor into cutoff (Class A amplifiers do not distort the signal).

This class of amplifier is used in radio frequency circuits and can be used in audio amplifiers by using a special two-transistor circuit known as a **push-pull amplifier**, Figure 27–21. The transistors conduct on opposite half-cycles, thus providing a full 360° signal at the output.

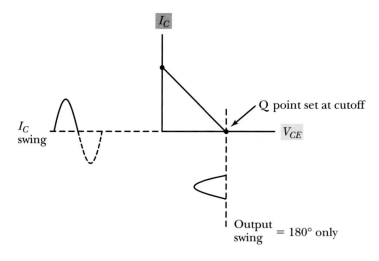

FIGURE 27–20 Q point set at cutoff for Class B amplifier operation

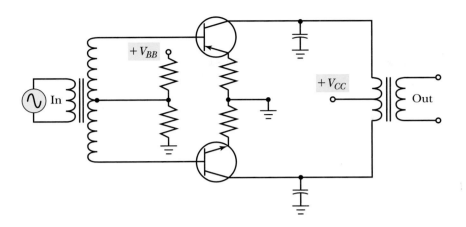

FIGURE 27–21 Two Class B amplifiers used in a push-pull amplifier circuit

Class AB amplifier operation

The Class AB amplifier is a compromise between Class A (low distortion, low efficiency) and Class B (high distortion, high efficiency). As long as the input signal isn't too large, a Class AB amplifier shows no signal distortion and it is more efficient than a Class A amplifier, but not quite as efficient as the Class B. Class AB amplifiers are commonly found in the lower-power sections of audio amplifier circuits.

Figure 27–22 shows that the Q point for a Class AB amplifier is about halfway between the Q points for the Class A and Class B amplifiers.

Class C amplifier operation

When a transistor amplifier stage is biased below cutoff value, Figure 27–23, collector current flows for less than 180° of the input cycle. Typical conduction time for Class C operation is about 120°.

Because Class C operation can be used in radio-frequency (RF) circuits where the other circuit components (such as a tuned LC circuit) complete the missing part of the output waveform, you frequently find this operation used in

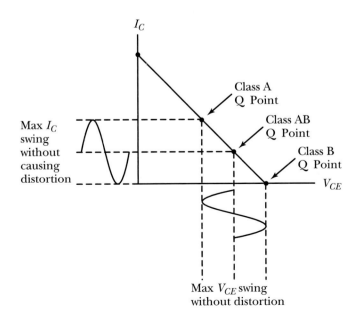

FIGURE 27–22 Class AB amplifier operation

FIGURE 27–23 Q point set below cutoff for Class C amplifier operation

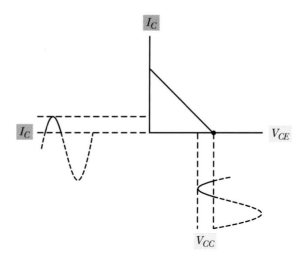

RF power amplifiers. Class C operation has high efficiency and large-signal handling capability.

IN-PROCESS LEARNING CHECK II

1. Class A amplifiers conduct during _____° of the input cycle; Class B amplifiers conduct for _____° of the input cycle; and Class C amplifiers conduct for approximately _____° of the input cycle.

2. A load line shows all operating points of a given amplifier from _____ to _____.

3. What is the amount of saturation current for a CE amplifier operated Class A when the collector resistance is 1 kΩ, the emitter resistance is zero (this resistor not used), and the supply voltage is 12 V? What is the value of the cutoff voltage for this circuit?

4. The Q point for a Class A amplifier is located at the _____ of the load line.

5. The Class _____ amplifier has the Q point located below the cutoff.

6. Collector current flows in a Class _____ amplifier, even when there is no signal applied to the input.

27–5 ANALYSIS OF BJT AMPLIFIERS

Analysis of BJT amplifier circuits can be crucial to understanding and troubleshooting electronic equipment. This section introduces a method for analyzing BJT amplifiers in a simple way and by approximations that are suitable for most technician's work. The examples are all for an NPN, common-emitter amplifier, but you can use them easily in other kinds of circuits you have studied.

Voltage and current calculations

As you have previously learned, to forward bias the emitter-base junction of a silicon transistor takes approximately 0.7 V difference of potential. Let's use that fact as a starting point for analyzing the CE circuit in Figure 27–24.

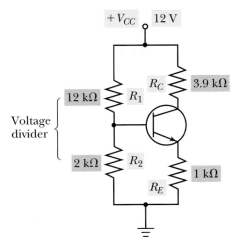

FIGURE 27–24 Analysis of a typical CE amplifier

Some further assumptions will be made for an easier analysis. We are able to use these assumptions because they introduce only a small percentage of error:

1. Assume the collector current is equal to the emitter current, $I_C = I_E$.

2. Assume the base current is much less than the voltage-divider current. Therefore, base current has negligible effect on the divider voltages, and so base current is not considered.

3. Assume in the divider branch that the voltages across the divider resistors equal the source voltage, $V_{R_1} + V_{R_2} = V_{CC}$. Therefore, the current through the divider is equal to $V_{CC}/(R_1 + R_2)$.

4. Assume in the transistor branch, the sum of the emitter-resistor voltage drop (V_E), transistor voltage drop (V_{CE}), and collector-resistor voltage drop (V_C) is equal to the source voltage: $V_{CC} = V_E + V_{CE} + V_C$. This also means that the voltage from the collector to ground (V_C) is equal to $V_{CC} - (I_C \times R_C)$. And $V_{CE} = V_{CC} - (V_C + V_E)$.

Example

Using the above information and Ohm's law, let's analyze the circuit parameters in Figure 27–24 so you can see how easy such analysis is.

1. Divider current:

$$I = V_{CC}/(R_1 + R_2) = 12 \text{ V}/14 \text{ k}\Omega = 0.857 \text{ mA}$$

2. Base voltage:

$$V_B = I \times R_2 = 0.857 \text{ mA} \times 2 \text{ k}\Omega = 1.71 \text{ V}$$

3. Emitter voltage:

$$V_E = V_B - 0.7 \text{ V} = 1.01 \text{ V}$$

4. Emitter current:

$$I_E = V_E/R_E = 1.01 \text{ V}/1 \text{ k}\Omega \approx 1 \text{ mA}$$

5. Collector voltage:

$$V_C = V_{CC} - (I_C \times R_C) = 12\text{ V} - 3.9\text{ V} = 8.1\text{ V}$$

6. Transistor voltage:

$$V_{CE} = V_{CC} - (V_{RC} + V_{RE}) = 12\text{ V} - 4.9\text{ V} = 7.1\text{ V}$$

PRACTICE PROBLEMS II

Assume that the circuit in Figure 27–24 has $R_1 = 10$ kΩ, $R_2 = 2$ kΩ, $R_E = 1$ kΩ, $R_C = 4.7$ kΩ, and $V_{CC} = 10$ V. Calculate the divider current, base voltage, emitter voltage, emitter current, collector voltage, and voltage across the transistor.

PRACTICAL NOTES

The critical parameters for the transistor stage are calculated with three of the above formulas.

1. $V_B = I \text{ (divider)} \times R_2$
2. $V_E = V_B - 0.7\text{ V}$
3. $I_E = V_E / R_E$

Using your approximations and knowing these parameters allows you to determine easily all the other parameters of interest.

Gain calculations

Gain, or voltage amplification (A_V), is calculated by comparing the voltage swing (peak-to-peak change) at the output to the voltage swing at the input.

FORMULA 27–2 $A_V = V_{out} / V_{in}$

Current gain (A_I) is computed in a similar fashion. That is, output change in current over input change in current.

FORMULA 27–3 $A_I = I_{out} / I_{in}$

Power gain (A_P) is calculated as the voltage gain times the current gain.

FORMULA 27–4 $A_P = A_V \times A_I$

Example

The input voltage to a CE amplifier is 50 mV and the output voltage is 1.2 V. The input current is 0.5 mA and the output current is 10 mA. What is the voltage gain? The current gain? The power gain?

Answers:

From Formula 27–2, the voltage gain is:

$$A_V = V_{out}/V_{in}$$

$$A_V = 1.2\ V/50\ mV = 24$$

From Formula 27–3, the current gain is:

$$A_I = I_{out}/I_{in}$$

$$A_I = 10\ mA/0.5\ mA = 20$$

From Formula 27–4, the power gain is:

$$A_P = A_V \times A_I$$

$$A_P = 24 \times 20 = 480$$

IN-PROCESS LEARNING CHECK III

1. The input signal to a CE amplifier is 4 mV and the output signal swings between +4 V and +5 V. What is the voltage gain of the amplifier? _____

2. A signal causes a common-emitter base current to swing by 0.03 mA and the collector current varies 1.5 mA peak-to-peak. What is the current gain of the amplifier? _____

3. A common-emitter amplifier has a voltage gain of 600 and a current gain of 30. What is the power gain? _____

SUMMARY

▶ The three basic transistor amplifier circuit configurations are the common-emitter (CE), common-collector (CC), and common-base (CB) circuits. Each has special features that are used in specific applications.

▶ For the common-emitter amplifier, the input is to the base and the output is from the collector. For the common-collector amplifier, the input is to the base and the output is from the emitter. For the common-base amplifier, the input is to the emitter, and the output is from the collector.

▶ The common-emitter circuit provides both current and voltage gain, and the phase of the amplified signal is 180° different from the input signal. That is, the signal is inverted using the amplification process. The common-base amplifier output is in phase with the input signal. The current gain of the common-base amplifier is approximately 1. The common-collector circuit features include a voltage gain of less than 1 and high current gain. It has a high input impedance and low output impedance, causing it to be used frequently for impedance matching or circuit isolation purposes.

▶ Regardless of the type of BJT amplifier configuration that is used (CE, CC, or CB), the basic requirements for biasing a BJT are strictly followed: the base-emitter P-N junction is forward biased and the collector-base P-N junction is reverse biased.

▶ Amplifier circuits can be composed of more than one stage connected one after the other. This enables the use of advantages of the three basic circuit configurations. A CC amplifier followed by a CE amplifier, for example provides high input impedance (CC advantage) and high voltage gain (CE advantage).

▶ A Darlington amplifier is composed of two stages of direct-coupled common-collector amplifiers. The circuit has an extremely high input impedance (the two beta values multiplied together), but a voltage gain that is less than unity.

▶ Small-signal amplifiers are designed to operate at small ac input signals (peak-to-peak ac current value is less than 0.1 times the amplifier's input bias current) and at power levels less than 1 W. Small-signal amplifiers are used where it is required to boost low voltage, low power signals.

▶ Large-signal amplifiers are designed for higher input current levels and output power levels (such as the output stages of an audio amplifier system).

▶ Four classes of operation for BJT amplifiers are Class A, Class B, Class AB, and Class C. In Class A amplifiers, collector current flows at all times, or during 360° of the input signal cycle. In Class AB amplifiers, collector current flows for more than 180° of the input cycle, but less than 360°. In Class B operation, current flows during half the input signal cycle (180°). And in Class C operation, collector current flows for approximately 120° of the input signal cycle.

▶ A load line is constructed on collector characteristic curves for a given transistor and circuit by drawing a line between the extremes of cutoff and saturation. In other words, the load line illustrates all possible operating points between those two extremes.

▶ The Q point, or quiescent operating point, of a transistor amplifier circuit is identified on the load line at the juncture of the base current and the load line for the given circuit. The "no-signal" values (quiescent dc values) for base current, collector current, and collector-to-emitter voltage are read from the collector curves/load line representation.

▶ If the Q point for a BJT amplifier stage is halfway between cutoff and saturation, the stage is operating as a Class A amplifier. If the Q point is set at cutoff, the stage is operating Class B. When the Q point is located somewhere between Class A and B, the circuit is said to be operating Class AB. And if the Q point is set below cutoff, the stage is operating Class C.

FORMULAS AND SAMPLE CALCULATOR SEQUENCES

FORMULA 27–1 $I_{C_{(sat)}} = V_{CC}/(R_C + R_E)$

V_{CC} value, ⊟, ⊡, R_C value, ⊞, R_E value, ⊡, ▱

FORMULA 27–2 $A_V = V_{out}/V_{in}$

V_{out} value, ⊟, V_{in} value, ▱

FORMULA 27–3 $A_I = I_{out}/I_{in}$

I_{out} value, ⊟, I_{in} value, ▱

FORMULA 27–4 $A_P = A_V \times A_I$

A_V value, ⊗, A_I value, ⊜

REVIEW QUESTIONS

1. In the common-emitter amplifier configuration, what transistor terminal is common to both input and output circuits?

2. Indicate between which junctions or electrodes the input signal is introduced, and between which electrodes the output signal is taken for the following circuit configurations:
   a. Common-emitter amplifier:
      input is between _____ and _____
      output is between _____ and _____
   b. Common-base amplifier:
      input is between _____ and _____
      output is between _____ and _____
   c. Common-collector amplifier:
      input is between _____ and _____
      output is between _____ and _____

3. For the following transistor amplifier configurations, what are the typical input and output impedance characteristics? (NOTE: Use "high," "low," or "medium" to describe them.)
   a. Common-emitter amplifier:
      input Z = _____
      output Z = _____
   b. Common-base amplifier:
      input Z = _____
      output Z = _____
   c. Common-collector amplifier:
      input Z = _____
      output Z = _____

4. Which transistor amplifier circuit configuration provides both voltage and current gain?

5. Which transistor amplifier circuit configuration causes 180° phase shift from input to output?

6. What is the name for a two-transistor, common-collector configuration that is noted for extremely high input impedance, but a voltage gain of less than 1?

7. Explain why a circuit would have a common-collector amplifier stage followed by a common-emitter stage.

8. What is the least popular of the three basic amplifier configurations?

9. Name one application for small-signal and one application for large-signal transistor amplifier stages.

10. Which amplifier class has the lowest efficiency?

11. Which amplifier class has the least distortion?

12. Which amplifier class has the highest efficiency?

13. Where is the Q point set to achieve Class A operation?

14. Where is the Q point set to achieve Class B operation?

15. A common-emitter circuit has a V_{CC} of 20 V. What is the approximate quiescent value of V_{CE} if the circuit operates as a Class A amplifier?

16. Briefly list four assumptions frequently used when analyzing a common-emitter circuit that uses voltage-divider bias.

17. For a voltage-divider bias circuit, V_{CC} equals 15 V, R_C equals 10 kΩ, and R_E equals 1 kΩ. (Assume $R_1 = 9.1$ kΩ and $R_2 = 1$ kΩ in the bias divider.) Find the following values:
    a. Divider current
    b. Base voltage
    c. Emitter voltage (assuming a silicon transistor)
    d. Collector voltage
    e. Collector-to-emitter voltage

18. If you need a transistor amplifier to provide 180° phase difference between the input and output signals, which circuit configuration would you choose?

19. A transistor amplifier circuit is needed that matches a high impedance at its input side to a low impedance connected to its output side. Which circuit configuration would you choose?

20. Which class of amplifier is used in a push-pull circuit?

PROBLEMS

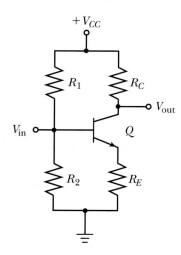

FIGURE 27–25

1. Referring to the circuit in Figure 27–25, let $+V_{CC} = 12$ V, $R_1 = 100$ kΩ, and $R_2 = 82$ kΩ. Determine the following:
   a. The current through the voltage bias divider
   b. The voltage across R_2
   c. The voltage across R_E

2. For the circuit in Figure 27–25, let $+V_{CC} = 12$ V, $R_1 = 82$ kΩ, $R_2 = 100$ kΩ, and $R_E = 100$ Ω. Determine the following:
   a. The current through R_1
   b. The voltage across R_2
   c. The voltage across R_E
   d. The current through R_E

3. In Figure 27–25, let $+V_{CC} = 9$ V, $R_1 = 100$ kΩ, $R_2 = 100$ kΩ, $R_E = 120$ Ω, and $R_C = 120$ Ω.
   a. $I_{R_1} = $ _____
   b. $V_{R_1} = $ _____
   c. $I_{R_2} = $ _____
   d. $V_{R_2} = $ _____
   e. V_{BE} (voltage measured between base and emitter) = _____
   f. $V_E = $ _____
   g. $I_E = $ _____
   h. $I_C = $ _____
   i. $V_{RC} = $ _____
   j. $V_{CE} = $ _____

4. If $+V_{CC} = 6$ V, $R_C = 1$ kΩ, and $R_E = 100$ Ω in Figure 27–25, determine the following:
   a. The saturation current for the circuit
   b. The cutoff voltage for the circuit

5. Referring to the circuit and values assigned in problem 3, determine the following:
   a. The saturation current for the circuit
   b. The cutoff voltage for the circuit

6. In a common-emitter BJT amplifier, such as the one shown in Figure 27–25, suppose $+V_{CC} = 10$ V and resistor R_1 is removed from the circuit. Determine the following:
   a. The current through R_2
   b. The voltage across R_E
   c. The voltage between the emitter and collector of the transistor
   d. The voltage across R_C
   e. The transistor's emitter current
   f. The class of operation
   g. The cutoff voltage

7. In the C_E amplifier of Figure 27–25, suppose $V_{CC} = +18$ V and resistor R_2 is shorted to common. Determine the following:
   a. The current through R_1
   b. The voltage across R_1

8. The input signal to a Class A amplifier is 120 mV, and the output signal level is 2.2 V. What is the voltage gain of the circuit?

9. There is a need for an audio amplifier that boosts the 50 mV signal from a crystal microphone to 7.1 V. What is the required voltage gain for the amplifier?

10. The voltage gain of a certain common-collector amplifier is 0.9, and its current gain is 2.2. What is the power gain?

ANALYSIS QUESTIONS

1. Figure 27–26 shows a family of collector characteristic curves and a load line for three different values of R_C: low resistance, medium resistance, and high resistance. Does this drawing indicate that to change the resistance value of R_C is to change the voltage gain of the amplifier? Explain your answer.

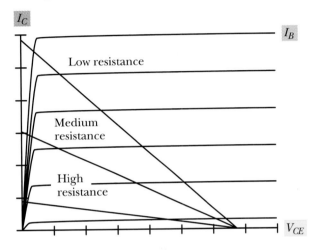

FIGURE 27–26

2. From the curves in Figure 27–27, determine the collector current value for a base current of 60 μA and a V_{CE} of 15 V.

3. What are the V_{CE} and I_C values at cutoff and saturation if a collector resistor with a value of 1 kΩ and an emitter resistor with a value of 220 Ω were present? (Use Figure 27–25 and assume $V_{CC} = 15$ V.)

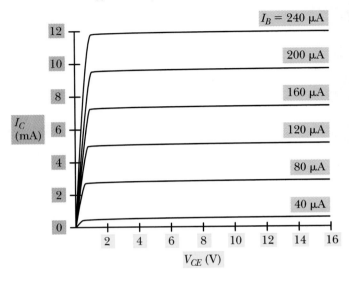

FIGURE 27–27

4. Using the information from question 3, what is the I_E value when the Q point is set for Class A operation? What is the V_E value?

5. For the conditions described in question 4, what are the divider resistor values that will set the Q point for Class A operation?

PERFORMANCE PROJECTS CORRELATION CHART

CHAPTER TOPIC	PERFORMANCE PROJECT	PROJECT NUMBER
The Common-Emitter (CE) Amplifier	Common-Emitter Amplifier	76
The Common-Collector Amplifier	Common-Collector Amplifier	77
Classification by Class of Operation	BJT Class A Amplifier	78
Class A Amplifier Operation		
Class B Amplifier Operation	BJT Class B Amplifier	79
Class C Amplifier Operation	BJT Class C Amplifier	80

NOTE: It is suggested that after completing the above projects, the student should be required to answer the questions in the "Summary" at the end of this section of projects in the Laboratory Manual.

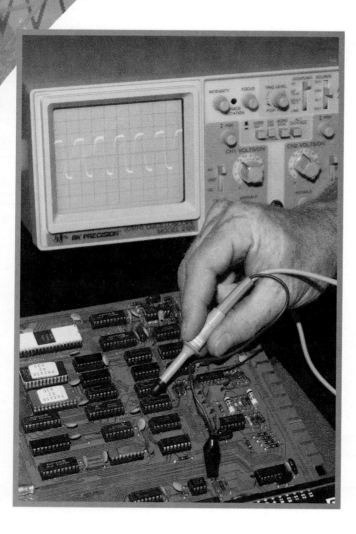

FIELD-EFFECT
TRANSISTORS
AND CIRCUITS

28

chapter

CHAPTER PREVIEW

The bipolar junction transistors (BJTs) you learned about earlier depend on current flow by means of two different charge carriers, namely, holes and electrons. The need for two different charge carriers is why such devices are called *bipolar*. By contrast, the transistor devices introduced in this lesson are *unipolar*. That is, current through the device is carried by just one type of charge carrier—either holes or electrons, depending on the type of semiconductor being used.

You will see how these unipolar devices control electron flow through them by means of applied electrical fields. Recall that BJTs control current flow by means of another path for current flow. Unipolar semiconductor devices that control their current flow by means of electrical fields are called **field-effect transistors** (abbreviated FETs).

You will learn about two important families of FETs: junction field-effect transistors and MOS field-effect transistors. You will see how each type is biased for use in basic types of amplifier circuits, and you will learn how the circuits operate.

OBJECTIVES

After studying this chapter, you should be able to:

1. Describe the **semiconductor structure** and identify the schematic symbols for N- and P-channel **JFETs, D-MOSFETs,** and **E-MOSFETs**

2. Determine the proper voltage polarities for operating **N-** and **P-channel FETs**

3. Explain the difference between depletion and enhancement modes of operation for FETs

4. Identify and explain the operation of **common-source, common-drain,** and **common-gate FET amplifier** circuits

5. Name some common practices for storing and handling MOSFET devices that ensure they are not destroyed by static electricity

28–1 JUNCTION FIELD-EFFECT TRANSISTORS

Figure 28–1 shows a simplified drawing of the semiconductor structure and symbols for a type of FET known as a **junction field-effect transistor**, or JFET. It is called a junction device because it contains a single P-N junction (which is always reverse biased under normal operating conditions).

The JFET has three external connections that are called the **gate** (G), **source** (S), and **drain** (D).

The source connection on a JFET is the connection that provides the source of charge carriers for the channel current. The drain connection acts as the place where the charge carriers are removed (or "drained") from the device. It doesn't matter whether you are working with an N-channel or a P-channel JFET, the charge carriers always flow from the source connection to the drain connection. Now, let's consider this important point in greater detail.

Remember from your earlier lessons on semiconductor materials that current flows through a solid block of N or P materials. The sample is neither a good

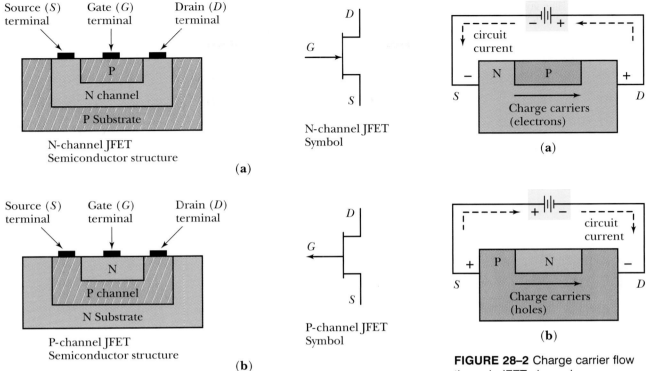

Source (S) terminal   Gate (G) terminal   Drain (D) terminal

P

N channel

P Substrate

N-channel JFET
Semiconductor structure

(a)

Source (S) terminal   Gate (G) terminal   Drain (D) terminal

N

P channel

N Substrate

P-channel JFET
Semiconductor structure

(b)

FIGURE 28–1 JFET semiconductor structures and symbols:
(a) N-channel FET; **(b)** P-channel FET

N-channel JFET
Symbol

P-channel JFET
Symbol

circuit current

− N   P +

S   D

Charge carriers
(electrons)

(a)

circuit current

+ P   N −

S   D

Charge carriers
(holes)

(b)

FIGURE 28–2 Charge carrier flow
through JFET channels:
(a) electron carriers in an N
channel; **(b)** hole carriers in a P
channel

conductor nor an insulator—it is a *semi*conductor. This is true for JFETs. And you can see from the structures in Figure 28–1 that the channel running between the source and drain connections are simply solid pieces of N or P semiconductor material. So if you just apply a negative voltage to the source terminal of an N-type JFET and a positive voltage to its drain terminal, the device will conduct current by means of electron carriers that flow from source to drain, Figure 28–2a. Or if you are working with a P-channel FET and you apply a positive voltage to the source connection and a negative voltage to the drain terminal, you will get current flow through the channel. In this case, holes carry the current charges, and the holes flow from source to drain, Figure 28–2b. The whole purpose behind the operation of JFETs is to control the flow of charge carriers through the channel material.

Figure 28–3 shows how an N-channel JFET is to be biased. Notice in the structure diagram that a dc voltage source, V_{DS}, is connected between the source and drain terminals. V_{DS} is connected so that its positive polarity is to the JFET's drain connection and its negative polarity is to the source connection. This lines up with what you just learned about the polarity and direction of current through an N-channel device. The second voltage source (V_{GS} connected between the source and gate terminals) is the one that controls the amount of current that flows through the channel.

One of the most important things to notice about V_{GS} is that it is connected with a polarity that reverse biases the P-N junction between the gate and source connections. This means there is nothing but reverse leakage current flowing be-

FIGURE 28–3 Bias voltages for an N-channel JFET

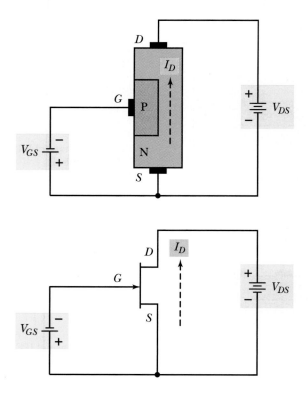

tween the source and gate (we usually consider there is no reverse current at all). More important, however, is the effect that the P-N depletion region has on the flow of charge carriers through the channel, Figure 28–4. The larger the amount of reverse bias applied to the gate-source junction, the farther the depletion region extends into the channel; and the farther the depletion region extends into the channel, the higher the resistance to current flow through the channel. The current flowing through the channel is called the *drain current*, and its symbol is I_D.

When there is no voltage applied to the gate-source junction, drain current is maximum. As V_{GS} increases, the effective width of the channel decreases until V_{GS} reaches a point where it stops the drain current altogether.

PRACTICAL NOTES

There are some important differences to notice between BJTs and JFETs:

1. BJTs use a current (I_B) to control a current (I_C), but JFETs use a voltage (V_{GS}) to control a current (I_D).

2. BJTs are turned off when there is zero base current ($I_B = 0$ A), but JFETs are turned on and maximum drain current flows when there is zero gate voltage ($V_{GS} = 0$).

BJTs and JFETs, however, are identical in appearance. They have three leads, they are provided in plastic and metal cases, and they have part designations that begin with 2N. To distinguish BJETs from JFETs, you must have access to a catalog, data book, or other listing of devices that specifies which 2N designations belong to BJTs and which belong to JFETs.

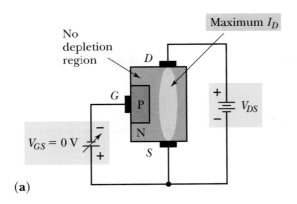

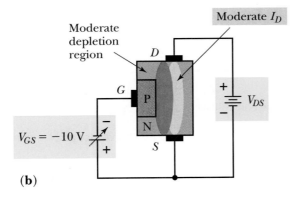

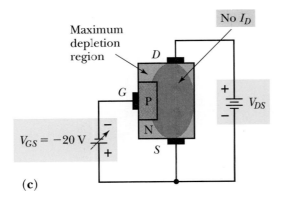

FIGURE 28–4 Depletion region in an N-channel JFET

JFET characteristic curves and ratings

Figure 28–5 is a family of drain characteristic curves for an N-channel JFET. Pick any one of the drain characteristic curves, and you can see that I_D increases with V_{DS} up to a certain point, then there is no more increase in I_D. What happens is that the channel becomes saturated with charge carriers, and the rate of flow changes very little as you increase or decrease the drain-source voltage. The value of V_{DS} where drain current goes into saturation is called the **pinch-off voltage**, V_P.

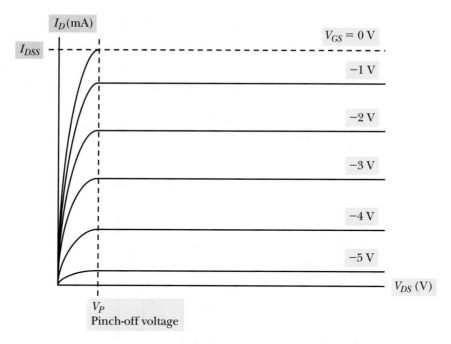

FIGURE 28–5 Drain characteristic curves for an N-channel JFET

Notice that the amount of drain current (I_D) becomes less as the gate-source voltage (V_{GS}) increases from 0 V to −5 V. The maximum amount of drain current flows through the channel when V_{GS} is at 0 V. This particular drain current level has a symbol of its own, I_{DSS}. The amount of gate-source voltage that is required for completely turning off the JFET in this example is about −5.3 V. This turn-off gate-source voltage level is denoted as $V_{GS_{(off)}}$.

Figure 28–6 shows a simple JFET amplifier. If the gate bias voltage (V_{GG}) is set to 0 V, the transistor will be normally conducting to its fullest extent. This

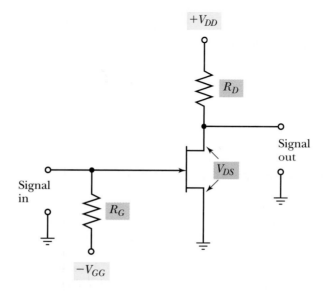

FIGURE 28–6 A simple N-channel JFET amplifier

means V_{DS} will be at its lowest voltage level, and I_D at its highest. A negative-going input signal, however, will tend to decrease the conduction of the transistor, thereby increasing the V_{DS} voltage. This is not a very efficient way to operate an amplifier because drain current is maximum when there is no input signal. A better approach is to set the bias level, V_{GG}, to a negative value that will tend to hold down the normal conduction of the FET. An input signal can then aid or oppose the V_{GG} bias level and thereby control the amount of drain current accordingly. In this circuit, the more negative the input, the more positive the output. (You will learn more about FET amplifier circuits later.)

You may notice that we have used N-channel JFETs in most of the examples. Generally speaking, N-channel versions are more commonly in use. You should have no trouble learning about and working with P-channel versions as long as you remember that the only difference is the polarity of the supply and bias voltages.

Although drain characteristic curves are useful for explaining the operation of a JFET, a transconductance curve shows some of the important features even more clearly. The sample of a transconductance curve in Figure 28–7 shows how the drain current (I_D) decreases as the gate voltage (V_{GS}) is made more negative. The curve also shows maximum drain current (I_{DSS}) when V_{GS} is zero, and how drain current is stopped when V_{GS} reaches the $V_{GS_{(off)}}$ level.

Another important specification for an FET is its **transconductance** (g_m). This value is based on the FET's transconductance curve and provides an indication of how much it amplifies signals. Transconductance for an FET serves much the same function as beta for a BJT. Transconductance is not a simple ratio like beta is, however. Formula 28–1 shows that transconductance for an FET is the ratio of a change in drain current to the corresponding change in gate-source voltage. (In mathematics, Δ symbolizes "a change in.") The unit of mea-

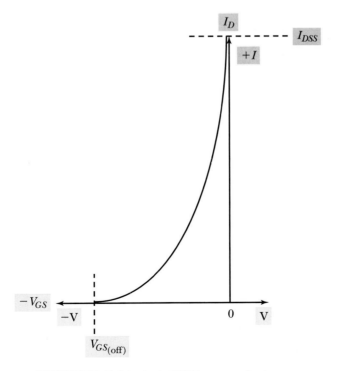

FIGURE 28–7 A typical JFET transconductance curve

surement for transconductance is the siemens. This is the same unit of measure for conductance you studied in an earlier lesson. And like conductance, the unit letter symbol for transconductance is *S*.

FORMULA 28–1 $g_m = \Delta I_D / \Delta V_{GS}$

The maximum ratings for JFETs enable the engineer to design circuits that will not allow the device to be destroyed by reverse voltage, excessive voltage, or excessive current flow. The most critical maximum JFET ratings follow:

- Maximum gate-source voltage (V_{GS}). This is the maximum reverse-bias voltage that can be applied continuously to the P-N junction. Typical values are between 30 V and 50 V. (Recall that the gate-source junction is normally operated with reverse bias.)

- Maximum gate-drain voltage (V_{GD}). This is the maximum reverse-bias voltage between the gate and drain sections that can be applied continuously. The value is usually identical with the maximum gate-source voltage rating, typically 30 V to 50 V.

- Maximum drain-source voltage (V_{DS}). This is the maximum forward voltage that can be controlled. Exceeding this rating causes the channel to conduct fully, regardless of the amount of applied gate voltage. This basically states the largest amount of supply voltage to be used with the device, and it is on the same order as the other maximum voltage levels, between 30 V and 50 V.

IN-PROCESS LEARNING CHECK I

Fill in the blanks as appropriate.

1. *JFET* is the abbreviation for _____.
2. The three terminals on a JFET are called _____, _____, and _____.
3. The arrow in the symbol for an N-channel JFET points _____ the source-drain connections, while the arrow for a P-channel JFET points _____ source-drain connections.
4. In an N-channel JFET, the source-drain voltage must be connected so that the positive polarity is applied to the _____ terminal of the JFET and the negative polarity is applied to the _____ terminal.
5. In a P-channel JFET, the source-drain voltage must be connected so that the positive polarity is applied to the _____ terminal of the JFET and the negative polarity is applied to the _____ terminal.
6. The charge carriers in the channel material of a JFET always flow from the _____ terminal to the _____ terminal.
7. In an N-channel JFET, the gate-source voltage must be connected so that the positive polarity is applied to the _____ terminal of the JFET and the negative polarity is applied to the _____ terminal.
8. In a P-channel JFET, the gate-source voltage must be connected so that the positive polarity is applied to the _____ terminal of the JFET and the negative polarity is applied to the _____ terminal.

9. V_{DS} stands for _____.
   V_{GS} stands for _____.
   I_D stands for _____.
   g_m stands for _____.

10. The greatest amount of current flows through the channel of a JFET when V_{GS} is at its _____ level, while the least amount of drain current flows when V_{GS} is at its _____ level.

Biasing JFET amplifiers

JFETS are somewhat more difficult to bias than BJTs are. A BJT is normally non-conducting, and a base bias of the same polarity as the collector supply voltage is applied in order to increase conduction. A JFET, however, is normally conducting, and a gate bias having the opposite polarity from the drain supply voltage is required to achieve biasing effects. The need for a biasing potential that has a polarity opposite the polarity required for the drain-current circuit complicates JFET biasing circuits.

Bear in mind the following facts and assumptions while analyzing the JFET biasing circuits:

1. As long as the gate terminal has the proper polarity, there is no current flowing in the gate circuit.

2. Source current and drain current are the same.

Gate bias

The most direct procedure for biasing a JFET amplifier is to apply a voltage to the gate that is opposite in polarity from the voltage applied to the drain. The drain supply voltage for an N-channel JFET is positive, for instance; so the amplifier can be biased by applying a negative voltage to the gate. This method is called the *gate bias* method. See the example in Figure 28–8.

The incoming ac signal passes through the input capacitor, where its voltage level is shifted by the value of V_{GG}, which increases and decreases the total volt-

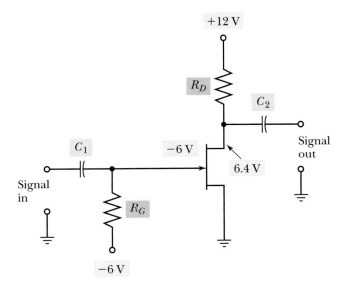

FIGURE 28–8 Gate-biased JFET amplifier

FIGURE 28–9 Self-biased
JFET amplifier

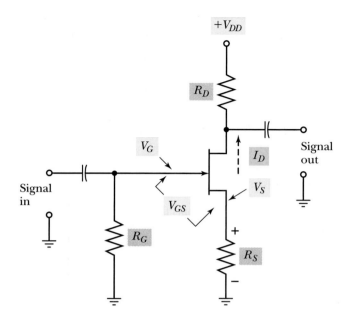

Self-bias

Figure 28–9 shows a JFET with the gate grounded through resistor R_G and the source grounded through resistor R_S. With no signal applied, the voltage at V_G will be zero (because of its connection to ground through R_G), and the drain current will be maximum. The voltage at the source terminal will be determined by the amount of drain current and the value of R_S. In any event, there will be a voltage at the source terminal of the JFET that is more positive than the ground potential. There will be a difference in potential between the source and gate: the gate voltage will be zero and the source voltage some positive level. The gate, in other words, will be less positive (more negative) than the source; and that is exactly what is required for biasing an N-channel JFET amplifier. Because this kind of bias is generated by the FET's own current, it is called **self-bias**.

One of the important advantages of using self-bias is that it does not require a separate negative-voltage power supply as does the gate-biasing method. A disadvantage of self-bias is that the amount of bias varies with the amount of drain current.

Voltage-divider bias

The voltage-divider method of bias is illustrated in Figure 28–10. Resistors R_1 and R_2 make up a voltage divider that applies a bias voltage to the gate that has the same polarity as the drain supply. Recall this is not the proper polarity for biasing a JFET. However, the voltage-divider potential is combined with the self-bias potential found across the source resistor (R_S), which has the proper polarity for biasing. This type of circuit is designed so that self-bias feature is balanced by the

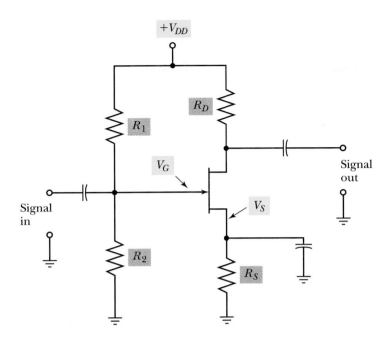

FIGURE 28–10 Voltage-divider biasing for JFET amplifiers

voltage-divider feature. The result is a stable biasing method that does not require a second power source that has a polarity opposite from V_{DD}.

JFET amplifier circuits

The basic configurations for JFET amplifiers are the same as for BJTs. The names are slightly different, but the features are generally the same: **common-source, common-drain**, and **common-gate**.

All of the JFET circuits shown in the lesson to this point have been common-source (CS) JFET amplifier circuits. The basic features of CS JFET amplifiers correspond to the basic features you have studied for the common-emitter BJT amplifier:

1. The input signal is introduced into the gate circuit, and the output is taken from the drain circuit (the source is common to the input and output).

2. The input circuit can have a high impedance (because of the reverse-biased source-drain P-N junction). The actual input impedance of the circuit is determined by the values of the resistors used in the gate biasing.

3. The output circuit is high impedance, determined by the value of the drain resistor.

4. The circuit provides voltage, current, and power gain. Current and power gains are extremely high.

5. There is a 180° phase reversal between the input and output signals.

Figure 28–11 shows the JFET version of a common-collector amplifier, which is called a **common-drain** (CD) amplifier. The basic features of practical CD amplifiers include:

1. The input signal is introduced into the gate, and the output is taken from the source (the drain is common to both the input and output).

FIGURE 28–11 A common-drain JFET amplifier

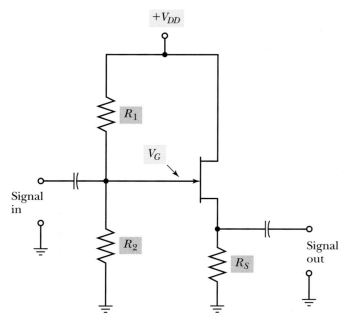

2. The input circuit can have a high impedance, but the actual input impedance of the circuit is determined by the values of the resistors used in the gate biasing.

3. The output circuit is low impedance, determined by the value of the source resistor and g_m of the FET.

4. The circuit provides voltage gain less than 1.

5. The circuit provides high current and power gains.

6. There is no phase reversal between the input and output signals.

The practical **common-gate** (CG) amplifier in Figure 28–12 has the following features:

1. The input signal is introduced into the source, and the output is taken from the drain (the gate is common to the input and output).

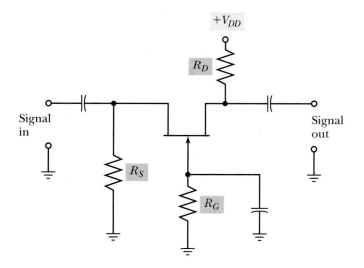

FIGURE 28–12 A common-gate JFET amplifier

2. The input impedance can be fairly low as determined by the values of R_S, R_D, and g_m.

3. The output circuit has high impedance.

4. The circuit provides good voltage and high power gains. Current gain is always less than 1.

5. There is no phase reversal between the input and output signals.

IN-PROCESS LEARNING CHECK II

Fill in the blanks as appropriate.

1. When a polarity opposite the drain polarity is applied to the gate of a JFET through a large-value resistor, the bias method is called _____ bias.

2. When the gate and source of JFET are both connected to ground through resistors, the bias method being used is called _____.

3. When the gate of a JFET is connected to a voltage divider and the source is connected through a resistor to ground, the bias method being used is called _____ bias.

4. The common-_____ FET amplifier has the input signal applied to the source terminal and the signal taken from the drain.

5. The common-_____ FET amplifier has the signal applied to the gate terminal and the output signal taken from the drain.

6. The common-_____ FET amplifier has the signal applied to the gate terminal and taken from the source terminal.

28–2 MOS FIELD-EFFECT TRANSISTORS

You have just learned how JFETs use the voltage of a reverse-biased P-N junction to control a current flow through a solid piece of semiconductor material. There is a second kind of FET technology that also uses a voltage to control a current. This is MOS technology. MOS stands for metal-oxide semiconductor. Field-effect transistors that operate according to this technology are called **metal-oxide semiconductor FETs,** or MOSFETs.

There are two types of MOSFETs. These two types of MOSFETs differ according to their structure and according to the way their controlling voltage affects the flow of current. The two types of MOSFETs are known as *enchancement type* (E-MOSFET) and *depletion type* (D-MOSFET). You will be studying both types of MOSFETs in this chapter. But first, let's look at a diagram that will aid your understanding of these families of MOSFET devices as you further your study, Figure 28–13.

The devices described in this chapter are all field-effect transistors, or FETs. The diagram shows that there are two different classifications of FETs: JFETs (which you have already studied) and MOSFETs (which you are about to study). Then the JFET family is further classified into N-channel and P-channel devices. The MOSFET family has an additional "generation." MOSFETs are basically divided into E-MOSFETs and D-MOSFETs—of which there are both N- and P-channel versions.

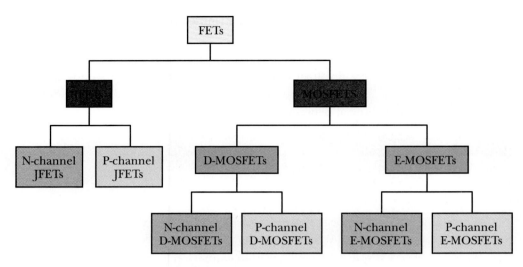

FIGURE 28–13 The "family tree" of FETs

There are other types of MOSFET devices available today, but they work by the same principles shown in this chapter. Once you know how the "family tree" of FETs works, you will have no trouble learning about the more specialized types that are not shown here.

D-MOSFETs

Figure 28–14 shows the elemental structure and symbol for depletion-type MOS-FETs. Notice that the source and drain terminals are connected to a "straight-through" piece of semiconductor material. In the case of an N-type D-MOSFET, the charges are carried by electrons in the N-type material. In the case of a P-type D-MOSFET, the charges are carried by holes through a P-type material.

The main difference between these MOSFETs and a typical JFET is the manner in which the gate terminal is separated from the channel. In a MOSFET, gate-channel separation is enabled by a very thin layer of a metal oxide known as silicon dioxide. Silicon dioxide is an extremely good insulator material. Thus, you can see that there will be no current flow between the gate terminal and the channel. However, as in a JFET, a voltage applied to the gate of a MOSFET greatly influences the amount of current flowing through the channel.

The electronic symbols for D-MOSFETs show the gate symbol entirely separated from the channel part of the symbol. This emphasizes that there is no possibility of current flow between the gate and the channel.

The arrow part of the symbols for D-MOSFETs indicates the nature of the base material, or substrate. You can see from the structure diagrams in Figure 28–14 that the substrate is not electrically connected to any of the terminals. It mainly serves as a foundation material for the channel, and it is always of the opposite type. For the N-type D-MOSFET, for instance, the substrate is a P-type material. The inward-pointing arrow in the symbol for an N-channel D-MOSFET indicates that the substrate is a P-type material (remember "*Pointing in = P*-type").

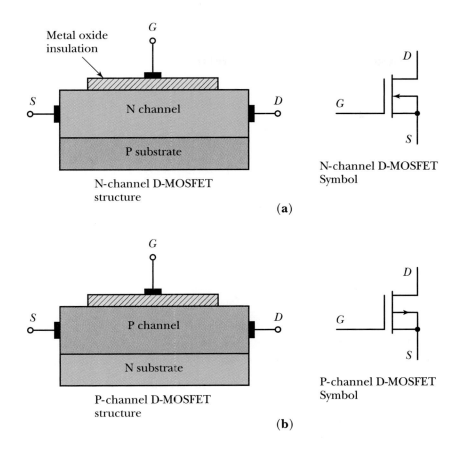

FIGURE 28–14 D-MOSFET structures and symbols

An outward-pointing arrow signifies a P-channel MOSFET that is built upon an N-type substrate (remember "*N*ot pointing in = *N*-type").

The channel portion of the device is indicated on the symbol as an unbroken line between the source and drain connections. This reminds us that the D-MOSFET uses an uninterrupted piece of one type of semiconductor material between the gate and drain.

Now let's consider how a D-MOSFET operates and where the "D" designation comes from.

D-mode of operation

D-MOSFETs are specifically designed to operate in the **depletion mode**. Figure 28–15 shows what is meant by this. Notice the polarity of the biasing voltages:

1. The V_{DS} polarity is determined by the fact that the device in this example has an N-type material for a channel. In order to satisfy the rule that charge carriers must flow from source to drain (and since the carriers in N-type semiconductors are electrons), it follows that the positive supply voltage is connected to the drain terminal and the negative to the source terminal.

2. The V_{GS} polarity is shown here as positive voltage to the source terminal and negative to the gate terminal. This is what determines the depletion mode of operation.

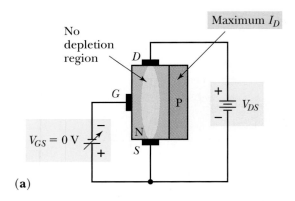

(a)

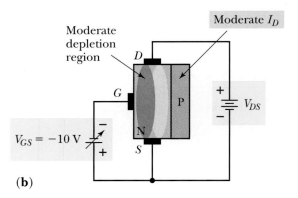

(b)

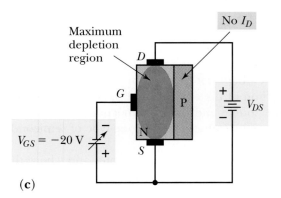

(c)

FIGURE 28–15 Action of an N-channel D-MOSFET in the depletion mode (D-mode) of operation: **(a)** V_{GS} is zero; **(b)** V_{GS} increased; **(c)** V_{GS} large enough to cut off channel current

While V_{GS} is at 0 V, Figure 28–15a, the charge carriers have a clear path through the channel, and drain current (I_D), flows freely. As V_{GS} is increased, however, electron carriers in the channel are forced away from the gate region, creating a depletion region of increasing size, Figure 28–15b. And as the size of this depletion region increases, the size of the charge-carrying channel decreases. Current through the channel can be completely cut off by making V_{GS}

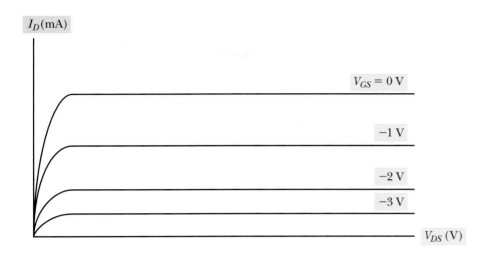

sufficiently large, Figure 28–15c. This is called the depletion mode of operation because V_{GS} controls channel current by depleting the number of available charge carriers.

Figure 28–16 shows the family of drain characteristic curves for a D-MOSFET operating in the depletion mode. The curves are virtually the same except for their drain current levels. Pick any one of them, and you can see that drain current increases rapidly with drain-source voltage at first. But the drain current soon reaches a saturation point where the only way to get more current flowing is to make the gate voltage less negative.

It is this depletion mode of operation that accounts for the "D" in the name D-MOSFET. And can you see that the operation of a D-MOSFET in its depletion mode works much like a JFET? Notice that drain current flows readily through the channel when there is zero gate voltage, but the current drops off dramatically as the gate voltage is taken more negative.

E-mode of operation

D-MOSFETs are very special in the sense that they can also be operated in an **enhancement mode**. Let's see what this means.

Figure 28–17 shows an N-channel D-MOSFET that looks like the basic circuit in Figure 28–15, but has the polarity of V_{GS} reversed. Instead of restricting current flow through the channel, V_{GS} in the enhancement mode actually increases the effective size of the channel and *enhances* the flow of current through the channel. The positive gate voltage actually allows the channel to widen into the substrate material by forcing substrate carriers (holes in the case of a P-type substrate) away from the channel region. Figure 28–18 shows the enhancement-mode portion of the drain characteristic curves.

Figure 28–19 is a complete set of drain curves for an N-channel D-MOSFET. Note that the range of gate voltages is from +2 V (maximum drain current) to −3 V (near cutoff). None of the other FETs discussed can be used in both a depletion and an enhancement mode. JFETs cannot be used in an enhancement mode because attempting to do so would simply forward bias the P-N junction

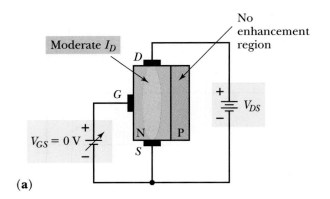

(a)

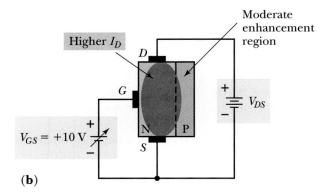

(b)

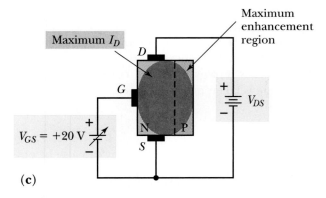

(c)

FIGURE 28–17 Action of an N-channel D-MOSFET in the enhancement mode (E-mode) of operation: **(a)** V_{GS} is zero; **(b)** V_{GS} increased; **(c)** V_{GS} further increased

and would not enable a voltage field that could control drain current. You will soon see that the structure of E-MOSFETs makes depletion-mode operation impossible. In this sense, the D-MOSFET is an extremely useful device, especially when you consider it can amplify true ac waveforms without the need for clamping the signal all positive or all negative.

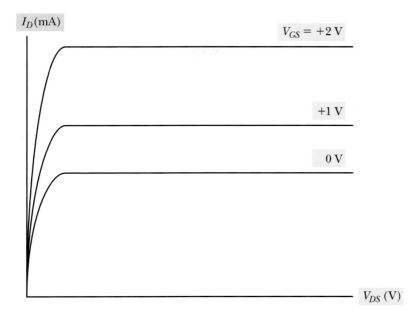

FIGURE 28–18 Family of drain characteristic curves for an N-channel D-MOSFET operating in its enhancement mode

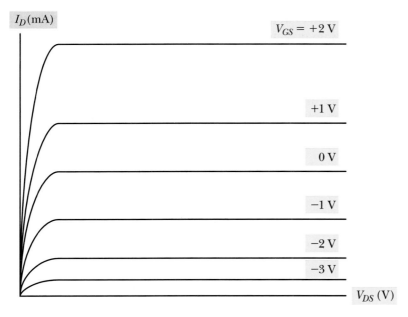

FIGURE 28–19 Complete family of drain characteristic curves for an N-channel D-MOSFET

The transconductance curve, Figure 28–20, for a D-MOSFET is very distinctive because the curve covers the $-V_{GS}$ part of the axis as well as the $+V_{GS}$ part of the axis. This emphasizes that a D-MOSFET can be operated in both the depletion ($-V_{GS}$ part of the axis) and enhancement ($+V_{GS}$ part of the axis) modes.

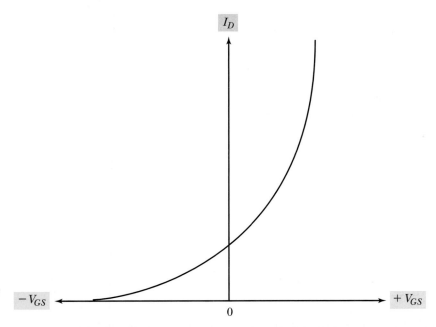

FIGURE 28–20 A typical D-MOSFET transconductance curve

A simple amplifier circuit using an N-channel D-MOSFET is shown in Figure 28–21. By having no dc gate bias, the transistor is set up for Class A operation. Recall this means the entire input waveform (both positive and negative phases) is amplified without distortion. In this amplifier, the positive portion of the input signal causes an increase in enhancement-mode drain current and a corresponding decrease in V_{DS} voltage. So as the input signal goes more positive, the output signal goes less positive. And when a negative portion of the input signal

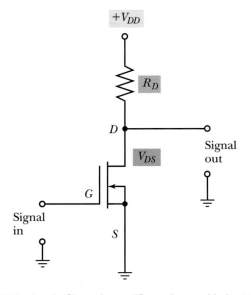

FIGURE 28–21 A simple Class-A amplifier using an N-channel D-MOSFET

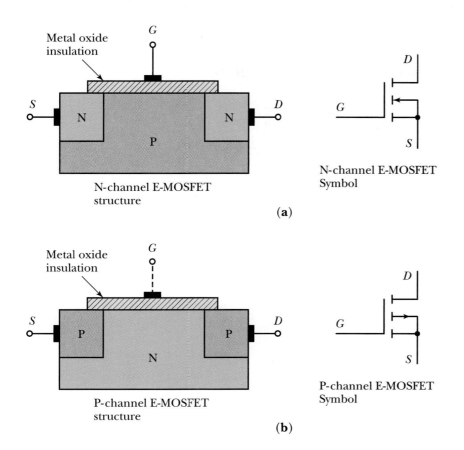

FIGURE 28–22 E-MOSFET structures and symbols

occurs, the depletion mode goes into effect and decreases the channel conduction. The result is an increase in V_{DS} voltage. As the input signal goes more negative, the output signal goes more positive.

Common-drain and common-gate D-MOSFET amplifiers are possible, and they have the same general characteristics as the JFET amplifiers you studied earlier. Also, the examples used in this lesson are all of N-channel D-MOSFETS. P-channel D-MOSFETs operate the same way as the N-channel versions, but with the polarities of V_{DS} and V_{GS} reversed.

E-MOSFETs

Figure 28–22 shows the basic structure and symbol for enhancement-type ("E") MOSFETs. The channel between the source and drain terminals is interrupted by a section of the substrate material. This interruption is noted on the schematic symbols by showing the channel broken into three segments: one for the source, one for the substrate (arrow), and one for the drain.

Current cannot possibly flow between source and drain on this device when there is no gate voltage applied, Figure 28–23a. There are two P-N junctions, and the middle one is far too wide to punch carriers through it (as is done with BJTs). But as shown in Figure 28–23b, a V_{GS} that is positive to the gate terminal causes holes to move away from the gate area and/or electrons to move into the gate area. This creates an *inversion layer* that completes a conductive channel between the source and drain. The more positive the gate becomes, the wider the inver-

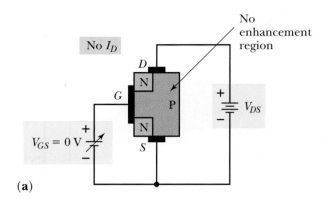

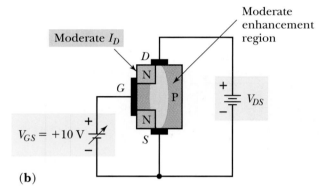

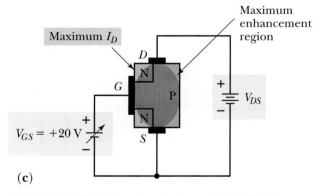

FIGURE 28–23 Action of an N-channel E-MOSFET:
(a) V_{GS} is zero; **(b)** V_{GS} increased; **(c)** V_{GS} further
increased

sion layer and the more conductive the channel, Figure 28–23c. The E-MOSFET
is the only FET that is non-conducting (switched off) when there is zero voltage
applied to the gate.

Figure 28–24 is a complete set of drain curves for an N-channel E-MOSFET.
Each of the curves has the same general shape as other FET curves. In the case
of E-MOSFETs, however, drain current is increased by increasing the amount of
V_{GS}. The minimum gate-source voltage that creates the inversion layer for an
E-MOSFET is called the *threshold gate-source voltage*, abbreviated $V_{GS_{(th)}}$. This is the
amount of gate voltage that is required for allowing any channel current to flow.

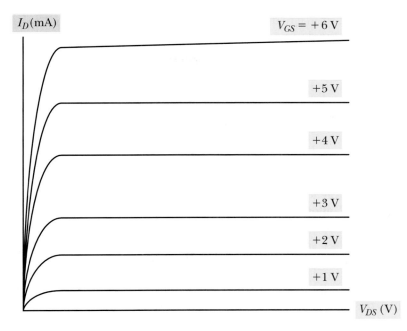

FIGURE 28–24 Family of drain characteristic curves for an N-channel E-MOSFET

The transconductance curve for an N-channel E-MOSFET is shown in Figure 28–25.

Consider the following features of E-MOSFETs that are common with D-MOSFETs:

1. Charge-carrier flow is from source to drain. For an N-channel E-MOSFET, this means electron flow is from source to drain—negative voltage is connected to the source and positive to the drain. For a P-channel version, hole

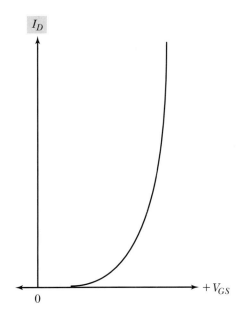

FIGURE 28–25 A typical E-MOSFET transconductance curve

flow is from source to drain—positive voltage is connected to the source and negative to the drain.

2. The type of semiconductor material used for the channel is opposite the type of material used for the substrate. An N-channel E-MOSFET uses a P-type semiconductor for its substrate. A P-channel version uses an N-type semiconductor for its substrate.

3. The arrow part of the schematic symbol indicates the type of material that is used for the substrate. An inward-pointing arrow signifies a P-type substrate material (therefore an N-channel device). An outward-pointing arrow represents an N-type substrate (and a P-channel device).

It is impossible to operate an E-MOSFET in a depletion mode. The structure of a D-MOSFET allows it to be operated in both modes, but the structure of an E-MOSFET prevents anything but enhancement-mode operation. The main reason an E-MOSFET cannot be used in a depletion mode is there is never a continuous semiconductor material where carriers can be depleted from.

The circuit in Figure 28–26 is an example of how an E-MOSFET can be used as a switch. Since there is no dc bias to the gate, this amplifier is operating Class B. When the input signal is zero, the transistor is turned off and the V_{DS} level is at $+V_{DD}$. Applying a positive voltage level as an input signal turns the transistor on. As a consequence, the V_{DS} level at the output drops to some minimum level.

E-MOSFETS are similar to BJTs in the sense that they are switched off when zero bias is applied. Therefore, the biasing methods are much the same. With no bias, for example, the circuit behaves like a Class B amplifier where there is no current through the device until the signal forward biases it. To operate as a Class A amplifier, the gate is biased with the same polarity that is applied to the drain. Figure 28–27 shows an example of forward biasing a common-source E-MOSFET by a procedure called *drain-feedback biasing*.

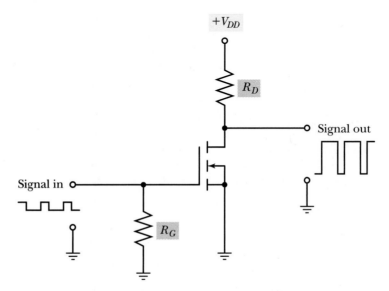

FIGURE 28–26 A simple switch circuit using an N-channel E-MOSFET

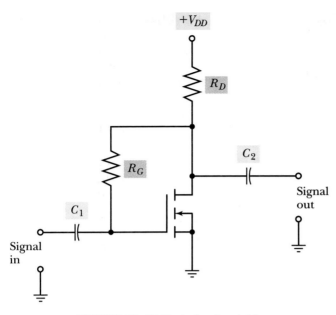

FIGURE 28–27 Drain-feedback bias

IN-PROCESS LEARNING CHECK III

Fill in the blanks as appropriate.

1. The term *MOS* stands for _____.

2. The gate terminal in a D-MOSFET is separated from the channel material by a thin layer of _____.

3. In a MOSFET, the direction of flow of charge carriers is always from the _____ terminal to the _____ terminal.

4. An outward-pointing arrow on a D-MOSFET symbol indicates a(n) _____ -type substrate and a(n) _____ channel. An inward-pointing arrow on a D-MOSFET symbol indicates a(n) _____-type substrate and a(n) _____ channel.

5. The proper polarity for V_{DS} of an N-channel D-MOSFET is _____ to the source and _____ to the drain. For a P-channel D-MOSFET, the proper polarity for V_{DS} is _____ to the source and _____ to the drain.

6. When operating a D-MOSFET in the depletion mode, the polarity of V_{GS} is _____ to the gate terminal. And when V_{GS} is at its minimum level, I_D is at its _____ level.

7. The _____ is the only FET that is nonconducting when the gate voltage is zero.

8. The proper polarity for V_{DS} of an N-channel E-MOSFET is _____ to the source and _____ to the drain. The proper polarity of V_{GS} is _____ to the gate terminal.

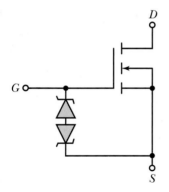

FIGURE 28–28 Diode clippers are sometimes used to prevent excessive static discharges in MOSFET devices.

28-3 FET HANDLING PRECAUTIONS

It was stated earlier that the silicon dioxide insulation between the gate and channel in MOSFETs is very thin. Although silicon dioxide is a good insulator, its thinness makes it prone to breakdown with voltages that are accidentally applied. In fact, the oxide layer can be destroyed by the simple electrical discharges we often feel as a result of brushing against certain materials in a dry atmosphere. P-N junctions (as in BJTs and JFETs) are not affected by electrical static discharges because they have relatively low impedances that harmlessly short out the static potential. The impedance of a MOSFET gate-channel connection is virtually infinite, so static potentials are not shorted out—at least not until the voltage punctures the insulation and creates a permanent short-circuit path!

Many MOSFET devices today (especially those found in integrated circuits) are protected to some extent by means of a bipolar zener clipper across the input terminals, Figure 28–28. Whenever the voltage at the input terminal attempts to exceed the breakdown voltage rating, the corresponding zener conducts and prevents the input voltage from rising above the clipping level.

Voltage-sensitive MOSFETs are packaged in some sort of conductive wrapping, usually a graphite-impregnated plastic foam. The graphite content of this foam makes it conductive, thereby keeping all the terminals of the MOSFET electrically shorted together. This prevents any difference in potential from building up between the terminals.

PRACTICAL NOTES

MOSFETs that are packed and shipped in a conductive foam or conductive wrapping should be kept in that protective environment until you are ready to use them.

Care should be exercised when handling MOSFETs. This applies to the time when you remove them from their conductive foam or wrapping to the time they are securely fastened into the circuit. It is important that you avoid picking up a static potential from your surroundings when you are about to handle a MOSFET. One way to make sure you are "discharged" is by briefly touching the metal case of a piece of plugged-in electronic test equipment (such as an oscilloscope) before you handle a MOSFET. The most reliable procedure is to use a conductive wrist band. This wrist band has a wire and connector that is fastened to ground, thus preventing any static buildup on your body. Soldering tools used for installing MOSFETs should have similar ground-wire connections on their tips.

///

SAFETY HINTS

Be sure your body and your soldering tools are not carrying a static electrical charge when you handle MOSFETs.

SUMMARY

▶ Field-effect transistors are unipolar devices. This means current through the devices is carried by just one type of charge carrier (either holes or electrons, depending on the type of semiconductor being used). This is in contrast to bipolar devices (BJTs), where holes carry the charge through one part and electrons carry the charge through a second part.

▶ The terminals on an FET are called gate, source, and drain. The flow of charge carriers is always from the source to the drain. The bias voltage that is used to control drain current is applied to the gate.

▶ FETs use a voltage (V_{GS}) to control the flow of current (I_D) through the device. This is in contrast to BJT, which uses a current (I_B) to control a current (I_C).

▶ The current flowing through the channel of an FET (I_D) rises sharply with increasing drain-source voltage (V_{DS}), but only to a certain point called the *pinch-off voltage*. Once the pinch-off voltage is reached, further increase in V_{DS} causes almost no increase in I_D.

▶ The family of drain characteristic curves and the transconductance curve for a JFET shows that it conducts fully (I_{DSS}) when no potential is applied to the gate-source junction ($V_{GS} = 0$ V). Increasing the amount of V_{GS} causes the amount of drain current (I_D) to decrease.

▶ MOSFETs differ from JFETs by the way the gate voltage is separated from the channel current. In JFETs, this separation is accomplished by means of a reverse-biased P-N junction. In MOSFETs, the gate voltage is separated from the channel current by means of an extremely good insulating material called silicon dioxide.

▶ The channel material in MOSFETs is opposite the type of material used as the substrate. So an N-channel MOSFET uses a P-type material as its substrate, and a P-channel MOSFET uses an N-type material as its substrate.

▶ The arrow portion of the symbols for MOSFETs indicates the type of substrate material used. An inward-pointing arrow indicates P-type substrate, and an outward-pointing arrow indicates an N-type substrate.

▶ The family of drain characteristic curves and the transconductance curve for a D-MOSFET show that it conducts fully when the maximum amount of enhancement-mode voltage is applied to the gate-source junction. For an N-channel D-MOSFET, the enhancement-mode polarity is positive; for a P-channel version, the polarity is negative. Reversing the polarity causes the D-MOSFET to enter its depletion-mode of operation, where increasing the amount of gate-source voltage causes drain current to decrease until cutoff occurs. A D-MOSFET can be operated in both the enhancement and depletion modes.

▶ A D-MOSFET is a natural choice for a Class A FET amplifier because the device operates about halfway between cutoff and saturation when there is no voltage applied to the gate.

▶ In an E-MOSFET, the source and drain are separated by two P-N junctions. Conduction is enabled only by creating a clear path for charge carriers (the inversion layer) between the source and drain segments of the structure.

▶ The family of drain characteristic curves and the transconductance curve for an E-MOSFET show that it is nonconducting when no voltage is applied to the gate-source junction. This means the device operates in the enhancement mode. For an N-channel E-MOSFET, increasing the positive polarity at the gate-source increases drain current. For a P-channel E-MOSFET, increasing the negative polarity at the gate-source terminals increases drain current.

▶ Drain-feedback bias can be used with E-MOSFETs because the drain polarity is the same as the polarity required for controlling gate bias.

▶ The gate-source insulation on MOSFETs is especially sensitive to voltage breakdown because of an accidental buildup of static electricity. Therefore, MOSFETs are usually shipped and stored in conductive plastic foam. Also, it is recommended that you ground your body and soldering tools when installing MOSFET devices in a circuit.

FORMULAS AND SAMPLE CALCULATOR SEQUENCES

FORMULA 28–1 $g_m = \Delta I_D / \Delta V_{GS}$

ΔI_D value, ➗ , ΔV_{GS} value, 🟰

REVIEW QUESTIONS

1. Explain the meaning of the following terms for JFETs: V_{DS}, V_{GS}, I_D, I_{DSS}, and $V_{GS_{(off)}}$.

2. Draw and properly label the schematic symbols for P-channel and N-channel JFETs, including the names of the terminals.

3. State the polarity of V_{DD} and V_{GG} for Class A, gate-biased N-channel and P-channel JFET amplifiers.

4. Draw a common-source, gate-biased amplifier that uses a P-channel JFET. Be sure to indicate the polarity of the V_{DD} and V_{GG} connections.

5. Describe the main differences between the operations of a JFET and a BJT.

6. Draw an N-channel JFET amplifier that is using self-bias. What is the value of V_{GS} for this amplifier when $V_S = +4$ V?

7. State, in words, the formula for transconductance.

8. Describe the difference between drain characteristic curves and transconductance curves for FETs.

9. Explain why a D-MOSFET makes an ideal Class A amplifier.

10. Draw the schematic symbols for N-channel and P-channel D-MOSFETs, and label the terminals.

11. Provide the polarity of V_{DD} and V_{GS} for the following D-MOSFET configurations:
    a. N-channel depletion mode
    b. P-channel depletion mode
    c. N-channel enhancement mode
    d. P-channel enhancement mode

12. Draw the schematic symbols for N-channel and P-channel E-MOSFETs, and label the terminals.

13. Explain the difference between depletion modes and enhancement modes of operation for FETs.

14. Provide the polarity of V_{DD} and V_G for N-channel and P-channel, Class A E-MOSFET amplifiers.

15. Explain why an E-MOSFET makes an ideal electronic switching circuit.

16. Define the term *threshold gate-source voltage* ($V_{GS_{(th)}}$) as it applies to E-MOSFET devices.

17. State the polarity of $V_{GS_{(off)}}$ for:
    a. N- and P-channel JFETs
    b. N- and P-channel D-MOSFETs
    c. N- and P-channel E-MOSFETs

18. List the types of FETs presented in this lesson that can be (or must be) operated in a depletion mode, and the types of FETs that can be (or must be) operated in an enhancement mode.

19. Explain the purpose of zener diodes that are internally connected between the gate and source terminals of some MOSFET devices.

20. Explain why MOSFETs are usually packaged and shipped in conductive foam.

PROBLEMS

1. For the circuit in Figure 28–29, let $-V_{GG} = 4$ V, $+V_{DD} = 12$ V, $R_G = 10$ mΩ, $R_D = 1$ kΩ, and $I_D = 8$ mA. Determine the following dc levels:
   a. Voltage measured between gate and common
   b. Voltage across resistor R_D
   c. Voltage measured between the source and drain
   d. Current through R_G

2. For the circuit in Figure 28–29, let $-V_{GG} = 5V$, $+V_{DD} = 12$ V, $R_G = 4.7$ MΩ, and $R_D = 2.2$ kΩ. If the dc voltage measured between the source and drain is 6 V, determine the following:
   a. The amount of dc drain current
   b. The dc voltage across R_D
   c. The dc voltage measured from the gate to common
   d. The dc current through R_G

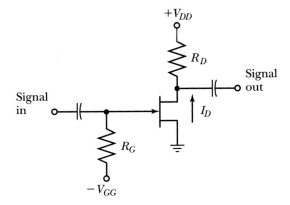

FIGURE 28–29

3. If a change in 2 V across R_G in Figure 28–29 causes the drain current to change 1 mA, determine the following:
   a. The value of transconductance
   b. The change in voltage between the source and drain

4. For the circuit in Figure 28–30, let $+V_{DD} = 12$ V, $R_G = 10$ MΩ, $R_S = 1$ kΩ, $R_D = 1.2$ kΩ, and $I_D = 5$ mA. Determine the following dc levels:
   a. V_{RS}
   b. V_{RD}
   c. V_{DS}
   d. V_{RG}
   e. V_{GS}
   f. I_{RG}

5. Referring to the circuit in Figure 28–30, let $+V_{DD} = 12$ V, $R_G = 4.7$ MΩ, $R_S = 1.2$ kΩ, $R_D = 2.2$ kΩ, and $V_{RS} = 2$ V. Determine the following dc levels:
   a. I_D
   b. V_{DS}
   c. Voltage measured between the gate and common
   d. V_{GS}
   e. V_{RG}
   f. I_{RG}

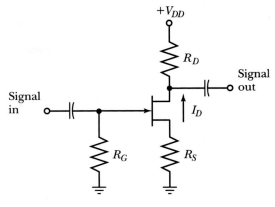

FIGURE 28–30

6. For the circuit in Figure 28–31, let $+V_{DD} = 18$V, $R_1 = 100$ kΩ, $R_2 = 10$ kΩ, $R_S = 330$ Ω, $R_D = 1$ kΩ, and $I_D = 10$ mA. Determine the following dc values:
   a. I_{R_2}
   b. V_{R_2}
   c. V_{R_1}
   d. V_{RS}
   e. V_{GS}
   f. I_{GS}
   g. V_{DS}

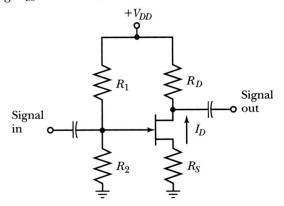

FIGURE 28–31

7. For the circuit in Figure 28–31, let $+V_{CC} = 12$V, $R_1 = 100$ kΩ, $R_2 = 33$ kΩ, $R_S = 470$ Ω, $R_D = 680$ Ω, and $V_{RS} = 4$ V_{DC}. Determine the following dc values:
   a. I_{R_2}
   b. V_{R_2}
   c. V_{R_1}
   d. V_{RS}
   e. V_{GS}
   f. I_{GS}
   g. V_{DS}

ANALYSIS QUESTIONS

1. Explain why you should never attempt to operate a JFET in an enhancement mode.

2. Research and describe the construction of a high frequency version of a D-MOSFET that is called a *dual-gate* MOSFET.

3. Explain why a D-MOSFET can be operated in both the depletion and enhancement modes while an E-MOSFET can be operated only in the enhancement mode.

4. Research and describe the operation of a high power version of an E-MOSFET device that is called a *VMOSFET.*

PERFORMANCE PROJECTS CORRELATION CHART

CHAPTER TOPIC	PERFORMANCE PROJECT	PROJECT NUMBER
JFET Characteristic Curves and Ratings Biasing JFET Amplifiers JFET Amplifier Circuits	JFET Biasing and Drain Characteristics Curves	81
	Common-Source JFET Amplifier	82

NOTE: It is suggested that after completing the above projects, the student should be required to answer the questions in the "Summary" at the end of this section of projects in the Laboratory Manual.

OPERATIONAL AMPLIFIERS

29

chapter

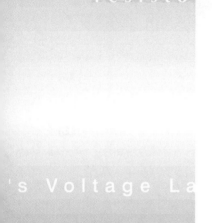

CHAPTER PREVIEW

The **operational amplifier (op-amp)** is very versatile and frequently used in modern electronic equipment of all types. For that reason, we feel it is important to provide an overview of this subject.

In this chapter you will learn about the basic features of op-amps. A variety of linear and nonlinear applications for these devices will be presented so you can understand their versatility. Additionally, you will learn that in many applications, a simple ideal op-amp approach can be used for amplifier analysis and design. Also, a brief discussion of some nonideal op-amp parameters will be presented.

This chapter is an introduction to op-amps. For this reason, you should pursue further studies on this topic in order to learn more about the operation, design, and applications of these unique and useful devices.

OBJECTIVES

After studying this chapter, you should be able to:

1. Explain the derivation of the term operational amplifier **(op-amp)**

2. Draw op-amp symbol(s)

3. Define the term **differential amplifier**

4. Draw a block diagram of typical circuits used in op-amps

5. List the key characteristics of an ideal op-amp

6. Identify linear and nonlinear applications circuits for op-amps

7. Distinguish between **inverting** and **noninverting op-amp circuits**

8. Perform voltage gain and resistance calculations for standard inverting and noninverting op-amp circuits

9. Describe the operation of op-amps in voltage amplifiers, voltage followers, comparators, and Schmitt trigger amplifiers

10. Describe the function of op-amps in circuits originally designed for analog computers: summing amplifiers, subtractor circuits, differentiators, and integrators

29-1 BACKGROUND INFORMATION

The name *operational amplifier* comes from the fact that this kind of circuit was once used for performing mathematical *operations* in analog computers. These early computer amplifier circuits were hand-wired circuits, using many discrete (individual) components, such as vacuum tubes, resistors, capacitors, and so on.

Today, the op-amp is a linear integrated-circuit (IC) device with all the components, including transistors, resistors, diodes, and their interconnections, fabricated on a single semiconductor substrate, Figure 29–1. The op-amp symbol and sample IC basing diagrams for a popular op-amp are shown in Figure 29–2.

It is easy to fabricate the IC transistors, resistors, and diodes, but more difficult to manufacture meaningful capacitor values on ICs. Therefore, we find the op-amp stages are direct coupled (without coupling capacitors between stages).

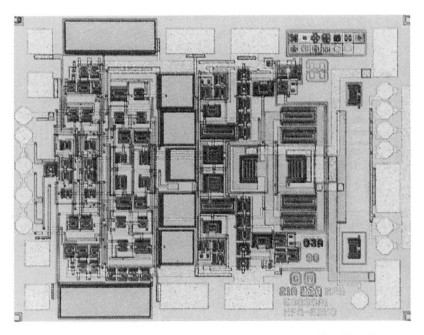

FIGURE 29–1 Microscopic view of a high-frequency amplifier IC chip *(Photo courtesy of Harris Semiconductor.)*

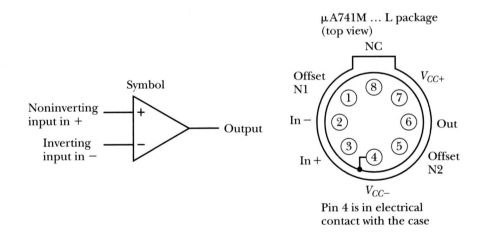

Symbol

Noninverting input in +

Inverting input in −

Output

μA741M … L package (top view)

NC

Offset N1

V_{CC+}

In −

Out

In +

Offset N2

V_{CC-}

Pin 4 is in electrical contact with the case

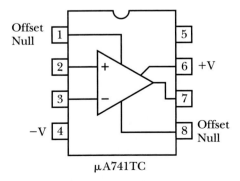

Offset Null

1

5

2

6 +V

3

7

−V 4

8 Offset Null

μA741TC

FIGURE 29–2 General-purpose operational amplifiers

One stage's output is coupled directly into the next stage's input. Because dc is not blocked between stages, sometimes these amplifiers are called *dc amplifiers,* as well as direct-coupled amplifiers.

By definition, an op-amp is a very high gain, dc amplifier. By controlling and using appropriate feedback from output to input, important amplifier characteristics are controlled by design. Characteristics such as the amplifier gain and bandwidth features are controlled by the external components connected to the op-amp IC.

Linear and nonlinear op-amp applications

Various **linear amplifier** applications (where the output waveform is similar to the input waveform) include **inverting amplifiers** (where the output is inverted from the input), **noninverting amplifiers**, voltage followers, summing amplifiers, differential amplifiers, amplifiers for instrumentation, stereo preamplifiers, logarithmic amplifiers, differentiators, integrators, sine-wave oscillators, and active filters.

Several **nonlinear amplifier** applications (where the output waveform is different from the input waveform) include comparators, multivibrators, hysteresis oscillators, and other special-purpose circuits.

Elemental op-amp information

Op-amps can have open-loop voltage gains (the gains that occur when there is no feedback from the output to the input of the op-amp) from several thousand up to about a million. The packaging requires few external connections. Look at the schematic symbol and typical connection points in Figure 29–3.

A simplified block diagram of the typical stages in an operational amplifier is shown in Figure 29–4.

As you can see from the preceding two figures, the schematic symbol, device connections, and block diagram for an op-amp are simple. However, the circuitry inside the device contains many circuits and connections shown in the typical IC op-amp diagram in Figure 29–5.

FIGURE 29–3 Op-amp symbols: **(a)** simplified symbol; **(b)** symbol showing *V* source connections

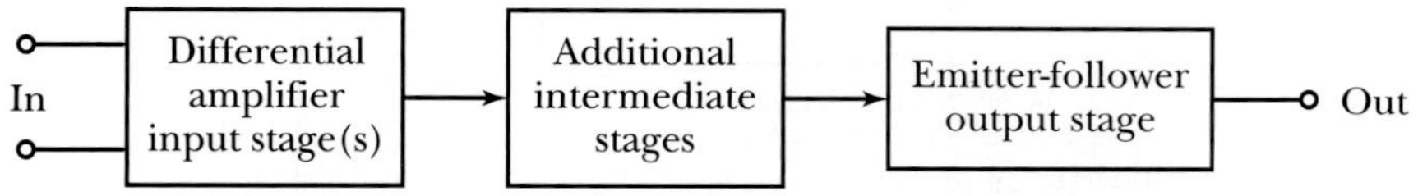

(a)

(b)

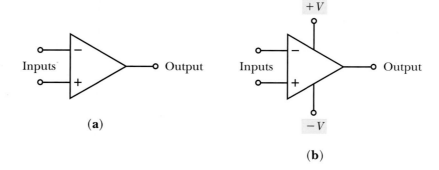

FIGURE 29–4 Simplified block diagram of an op-amp

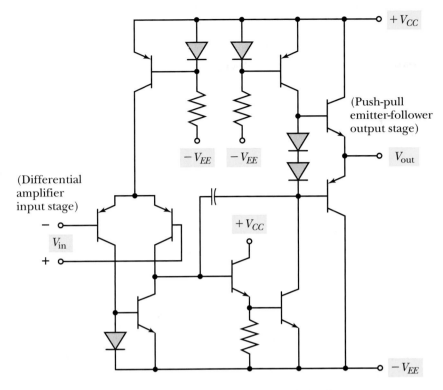

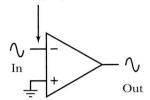

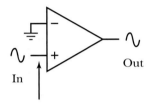

(Push-pull emitter-follower output stage)

V_{out}

$-V_{EE}$ $-V_{EE}$

(Differential amplifier input stage)

V_{in}

$+V_{CC}$

$-V_{EE}$

FIGURE 29–5 Typical op-amp internal circuitry

Inverting input

In

Out

Signal fed to inverting input; output is inverted

Out

In

Noninverting input

Signal fed to noninverting input; output is not inverted

FIGURE 29–6 Inverting and noninverting inputs

In Figure 29–6, the differential amplifier input stage of the op-amp has both an inverting input (at the negative input terminal) and a noninverting input (at the positive input terminal). If an input signal is applied to the inverting input with the noninverting input grounded, the output signal polarity is opposite the input signal polarity, Figure 29–7a. Conversely, if the input signal is fed to the positive input terminal while the negative input terminal is grounded, the output signal is the same polarity as the input, Figure 29–7b.

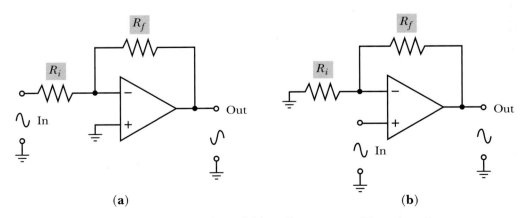

R_f

R_i

In

Out

(a)

R_f

R_i

In

Out

(b)

FIGURE 29–7 Op-amp connections: **(a)** inverting op-amp; **(b)** noninverting op-amp

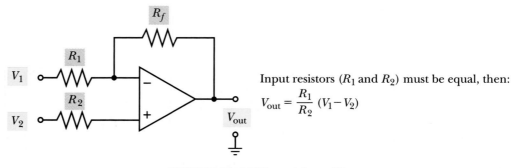

Input resistors (R_1 and R_2) must be equal, then:

$$V_{out} = \frac{R_1}{R_2}(V_1 - V_2)$$

FIGURE 29–8 Differential amplifier

The enormous gain figures of the basic op-amp device (i.e., open-loop gain) make it necessary to control this gain by use of external components. Notice in the op-amp circuit configuration in Figure 29–7, there is an input component (resistor R_i), and a feedback component (resistor R_f). This is the feedback that controls the amplifier gain.

You have seen that a signal fed to the inverting input of an op-amp enables an inverted version at the output. Conversely, a signal fed to the noninverting input enables a noninverted version at the output. What will happen if you apply signals to both inputs at the same time? The result will be a difference waveform; that is, a waveform that represents the noninverted input minus the inverted input. This amplifier configuration is important in many applications today. It is called a **differential amplifier** because it amplifies the *difference* between the two input waveforms. See Figure 29–8. If you apply the exact same signal to both inputs of an op-amp, the differential amplifier effect produces no signal output.

This differential effect introduces another term called common-mode rejection ratio. This is the degree the amplifier rejects (does not respond or produce output) for signals that are common to both inputs. This circuit feature provides useful rejection of undesired signals that can be present at both inputs. For example, undesired signals, such as 60-Hz pickup or other stray noise signals, may be present at the inputs. See Figure 29–9.

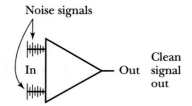

FIGURE 29–9 Noise signals common to both inputs cancel in the output

Ideal op-amp characteristics

With this background, let's now look at the characteristics of an ideal op-amp. An ideal op-amp would have:

1. infinite voltage gain;
2. infinite bandwidth;
3. infinite input impedance; and
4. zero output impedance.

Obviously these parameters are not achievable in a practical device. However, they are used to approximate operation when designing certain circuits using op-amps.

In practice, these devices have limitations similar to other semiconductor devices. That is, there are maximum current-carrying capabilities, maximum voltage ratings, and power limitations that must not be exceeded. Also, there is something less than a flat frequency response when attempting amplification from dc to radio frequencies.

As may be surmised from these discussions, real world op-amps cannot meet the absolute ideal parameters of infinite gain and so on. It is, however, realistic to assume these devices exhibit high voltage gain, high input impedance, low output impedance, and wide bandwidth amplifying characteristics. Using this information, let's look at some basic op-amp circuits.

PRACTICAL NOTES

Manufacturers produce explicit data sheets on every device they manufacture. These data sheets accurately describe each device's capabilities and limitations. Thus, it is critical that a technician learn how to find, use, and properly interpret the data found on these data sheets.

IN-PROCESS LEARNING CHECK I

Fill in the blanks as appropriate.

1. An op-amp is a high gain _____-coupled amplifier.
2. The input stage of an op-amp uses a _____ amplifier configuration.
3. The name operational amplifier derives from their early use to perform mathematical _____.
4. The open-loop gain of an op-amp is much _____ than the closed-loop gain.
5. To achieve an output that is the inversion of the input, the input signal is fed to the _____ input of the op-amp.
6. Input signals to the op-amp can be fed to the _____ input, the _____ input, or to both inputs.

29-2 AN INVERTING AMPLIFIER

In Figure 29–10 you see an amplifier with the input signal applied to inverting input through R_i. Part of the amplified and inverted output feeds back to the input through the feedback resistor, R_f. With the positive input grounded, the differential input signal equals the value of the input signal fed to the input.

If ideal (infinite input impedance) conditions are assumed, it can be shown that the closed-loop voltage gain (V_{out}/V_{in}) of this amplifier equals the ratio of the feedback resistor value to the input resistor value. The output V signal is, however, inverted from the input signal. This makes designing for a specific voltage gain easy.

FORMULA 29–1 $A_V = (R_f/R_i)$

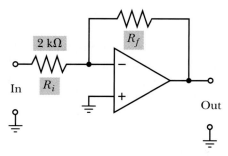

FIGURE 29–10 Inverting amplifier closed-loop gain

Example

For the circuit in Figure 29–10, if $R_f = 27$ kΩ and $R_i = 2$ kΩ, what is the voltage gain of the amplifier?

Answer:

From Formula 29–1:

$$A_V = (R_f/R_i)$$

$$A_V = 27 \text{ k}\Omega / 2 \text{ k}\Omega = 13.5$$

PRACTICAL NOTES

Be aware that there is phase inversion when the input signal is put into the negative input of the op-amp. This means, for example, that if +0.2 V were put into our example amplifier above, the output would be −2.7 V (i.e., 0.2 × 13.5).

PRACTICE PROBLEMS I

1. What is the voltage gain of an inverting op-amp if $R_i = 100$ kΩ and $R_f = 1$ MΩ?

2. For the circuit in Figure 29–10, what R_f value is needed to obtain a gain of 100?

3. What is the necessary value for R_i of an inverting op-amp if $R_f = 270$ kΩ and the gain is to be 20?

29–3 A NONINVERTING AMPLIFIER

The circuit for a noninverting op-amp is shown in Figure 29–11. As you can see, the junction of the feedback resistor (R_f) and input resistor (R_i) is still at the negative input terminal. However, the difference is that the input signal is fed to the positive input and R_i is grounded. R_f and R_i form a voltage divider for the feedback voltage. The differential voltage felt by the op-amp is the difference between the input voltage (V_{in}) fed to the positive input and the feedback voltage

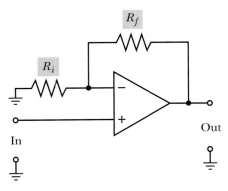

FIGURE 29–11 Noninverting amplifier closed-loop gain

felt at the negative input terminal. This differential input is then amplified by the gain of the op-amp and produces the voltage output. Without discussing the derivation of the noninverting amplifier gain, let us simply state that the gain of the noninverting amplifier equals the feedback resistance divided by the input resistance, plus 1.

FORMULA 29–2 $A_V \text{ (noninverting)} = (R_f/R_i) + 1$

Example

If the R_i and R_f values for a noninverting op-amp circuit are 1 kΩ and 12 kΩ, respectively, what is the voltage gain of the amplifier?

Answer:

From Formula 29–2:

$$A_V \text{ (noninverting)} = (R_f/R_i) + 1$$

$$A_V \text{ (noninverting)} = (12 \text{ k}\Omega/1 \text{ }\Omega) + 1 = 13$$

PRACTICE PROBLEMS II

1. What is the voltage gain of a noninverting op-amp if $R_f = 20{,}000$ Ω and $R_i = 2{,}000$ Ω?

2. For the circuit in Figure 29–11, what R_f value is needed to obtain a gain of 100 if $R_i = 10$ kΩ?

3. What is the value for R_i of a noninverting op-amp circuit if $R_f = 270$ kΩ and the gain is to be 20?

A special application of a noninverting op-amp is one where the gain is 1, or unity. Any amplifier that has a gain of 1 and has no phase shift between the input signal and output signal is called a *voltage follower.* An op-amp voltage follower consists of a noninverting amplifier configuration where the feedback resistance

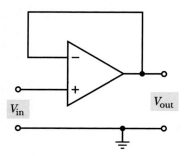

FIGURE 29–12 Voltage follower

is zero (the output is directly connected to the inverting input). See Figure 29–12.

One purpose of a voltage follower is to match a high impedance source to a low impedance load. The input impedance of an op-amp voltage follower is very close to the value of the input resistor, R_i. So if the circuit uses an input resistance of 1 MΩ, the input impedance is close to that value.

29–4 OP-AMP INPUT MODES

Op-amp circuits use one of three basic input modes of operation:

1. Single-ended
2. Double-ended (or difference)
3. Common-mode

Figure 29–13 shows each one of these input modes. Note that the **single-ended input** mode has one input connected to the input signal and the other connected to ground. The inputs of a **double-ended input** circuit, however, are connected to different signal sources. In the case of the **common-mode input,** both inputs are connected to the same signal source.

Single-ended input mode

The single-ended mode is the operation with one input grounded, Figure 29–13a. Sometimes the grounding is through a resistor, but that doesn't affect the mode of operation. This single-ended input mode is used in the examples of inverting and noninverting op-amp circuits you have studied so far in this chapter. Most op-amp circuits in modern electronic equipment use the single-ended input mode. Formulas 29–1 and 29–2 relate to the single-ended input mode.

Op-amps are not as ideal as our approximations usually presume. When there is zero input to an inverting op-amp circuit, for instance, there should be zero output. But that is not precisely the case. The specification is called *input-offset voltage*. Likewise, there is always a little bit of current flowing into the junction point between R_f and R_i, even when there is no input voltage. This current is called the *input-bias current*.

The double-ended (differential) input mode

The double-ended mode takes advantage of the fact that the op-amp is a differential amplifier. In this mode, signals are fed to both inputs at the same time.

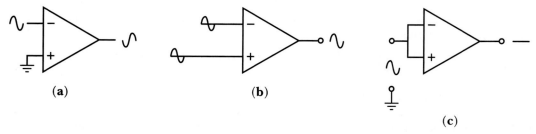

(a) (b) (c)

FIGURE 29–13 Modes of operation for op-amps: **(a)** single-ended; **(b)** double-ended; **(c)** common-mode

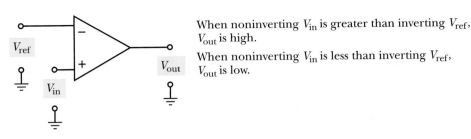

When noninverting V_{in} is greater than inverting V_{ref}, V_{out} is high.

When noninverting V_{in} is less than inverting V_{ref}, V_{out} is low.

FIGURE 29–14 Example of op-amp comparator

The result at the output is an amplified version of the difference between the two signals, Figure 29–13b.

Differential mode gain is the ratio of output voltage to the difference between the two signals fed to the two inputs of an op-amp that is functioning in the double-ended mode. Usually this gain figure is very high.

The comparator circuit

It often is useful to compare two voltages to see which is larger. An op-amp circuit that performs this function is shown in Figure 29–14. You can see that this is a basic double-ended input circuit. Because there is no feedback path, the gain of the circuit is extremely high. In fact, the output will normally be at one of the two extremes: close to $+V$ supply voltage or close to $-V$ supply voltage. Just a few millivolts of difference between V_{ref} and V_{in} can drive the output to one of those extremes. It is the polarity of the output, rather than the actual voltage level, that is important to the function of this circuit. With this circuit, if V_{in} (noninverting input) is greater than V_{ref} (inverting input), the output goes to $+V$ supply level. On the other hand, if V_{in} is less than V_{ref}, then the output goes to the $-V$ supply level.

Consider a specific application where V_{ref} is some dc voltage reference level and a sine-wave signal is fed to V_{in}. In this instance, the output is at its positive extreme only during the period when the sine-wave amplitude is greater than the dc reference value. During the remaining time, the output is at its low extreme. By changing the reference voltage level, the output can be varied from a square wave to a rectangular wave. One practical use of such a circuit is to convert a sine wave to a square wave.

A comparator circuit is an example of a nonlinear op-amp circuit. This means the waveform at the output does not necessarily resemble the waveforms at the input.

The subtractor circuit

A double-ended, or difference-mode, op-amp circuit with feedback is shown in Figure 29–15. This circuit performs the actual algebraic subtraction between two input voltage levels. It is an example of the mathematical computer applications of early operational amplifiers.

The output voltage (V_{out}) is related to the difference between the two input voltages (V_{in+} connected to the noninverting input and V_{in-} connected to the inverting input). The actual output voltage then depends on the amount of gain.

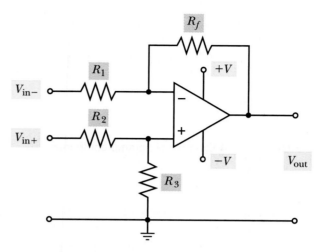

FIGURE 29–15 Op-amp as a difference amplifier

The values of the resistors in Figure 29–15 are all the same for reasons of simplicity. When all resistance values are the same for a difference amplifier, the gain is 1 and the output voltage is calculated as $V_{out} = V_{in+} - V_{in-}$.

Examples

1. Suppose the resistor values in Figure 29–15 all have a value of 10 kΩ. What is the output voltage when $V_{in-} = +2$ V and $V_{in+} = +5$ V?

Answer:

$$V_{out} = V_{in+} - V_{in-}$$

$$V_{out} = 5 - 2 = 3 \text{ V}$$

2. What is V_{out} for question 1 when $V_{in-} = +2$ V and $V_{in+} = -5$ V?

Answer:

$$V_{out} = -5 - (+2) = -7 \text{ V}$$

PRACTICAL NOTE

You have seen that the output of a difference amplifier is the algebraic difference between the two input voltage levels. It is a mathematical subtraction circuit. The output of a difference op-amp can go negative only if the device is supplied with both +V and −V power supplies as shown in the example offered here. Recall from an earlier lesson that many types of FET circuits also require dual-voltage power supplies.

The common-mode input

Although there is no practical application for an op-amp using the common-mode input, Figure 29–13c, the configuration is useful for defining the *common-*

mode gain of the op-amp. If you apply the same signal to both the inverting and noninverting inputs, you do not see any output. Why not? Because the differential amplifier at the inputs should cause the inverted and noninverted versions to cancel one another completely. That would be the ideal case. But no op-amp can be ideal, so there will always be a little bit of output signal.

The common-mode gain specification for an op-amp is usually less than 1. The lower the value, the better the quality of the device.

IN-PROCESS LEARNING CHECK II

Fill in the blanks as appropriate.

1. For an op-amp inverting amplifier the input signal is applied to the _____ input, and the feedback resistor is connected from the output to the _____ input.

2. The voltage gain of an inverting amplifier is a negative value in order to indicate _____ of the signal.

3. In the schematic for an op-amp inverting amplifier, resistor R_f is located between the _____ of the circuit and the _____ input. Resistor R_i is connected to the _____ terminal.

4. For an op-amp noninverting amplifier the input signal is applied to the _____ input, and the feedback resistor is connected from the output to the _____ input.

5. The special case of a noninverting op-amp circuit that has a voltage gain of 1 is called a(n) _____ circuit.

6. An op-amp circuit that has one signal input and one grounded input is called a(n) _____-ended circuit.

7. An op-amp circuit that has different signal sources connected to its inverting and noninverting inputs is called a(n) _____-ended circuit.

8. A comparator circuit is an example of an op-amp operating in the _____-ended input mode.

29–5 MORE OP-AMP APPLICATIONS

Summing amplifier

You have already been introduced to the op-amp circuit that provides a subtraction operation. The **summing amplifier** is one that provides the algebraic sum of the input voltage levels.

In Figure 29–16a, you see all the resistors have the same value. When that is the case, $V_{out} = V_1 + V_2$. In Figure 29–16b, you see the summing effect resulting with resistors of different values in the input and output circuits. Use Ohm's law to compute each input current. The sum of those currents passes through the feedback resistor (R_f), and the product of that current times the feedback resistor value provides the output voltage.

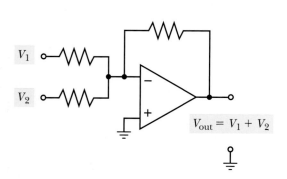

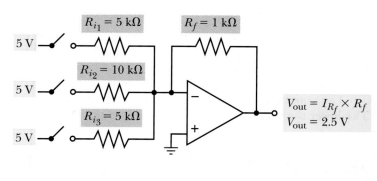

(a) Assuming all Rs equal:
$V_{out} = V_1 + V_2$

(b) With different R values and with all three switches closed:

$$I_{i_1} = \frac{5\ V}{5\ k\Omega} = 1\ mA$$

$$I_{i_2} = \frac{5\ V}{10\ k\Omega} = 0.5\ mA$$

$$I_{i_3} = \frac{5\ V}{5\ k\Omega} = 1\ mA$$

I through $R_f = 1 + 0.5 + 1 = 2.5\ mA$
$V_{out} = I_{R_f} \times R_f = 2.5\ mA \times 1\ k\Omega = 2.5\ V$

FIGURE 29–16 Examples of summing amplifiers

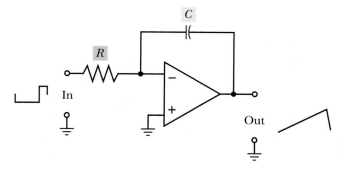

FIGURE 29–17 Example of op-amp integrator

Integrator circuit

The **integrator circuit** uses a capacitor, instead of a resistor, as the feedback component, Figure 29–17. It performs the mathematical (calculus) function of integration. This means the output is related to the "area under the curve" for the input waveform.

One popular application of this circuit is to produce a ramp-shaped voltage output when its input is a rectangular wave.

Differentiator circuit

The **differentiator circuit**, Figure 29–18, is another example of a calculus function. In this case, the output is proportional to the derivative of the input volt-

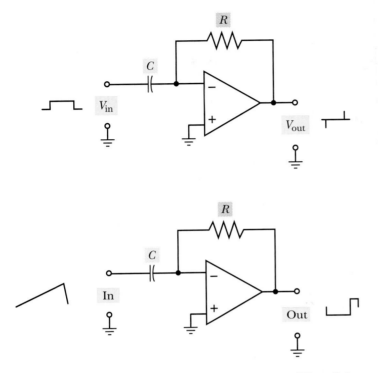

FIGURE 29–18 Examples of op-amp differentiators

age—the rate of change of voltage at the input. This operation is the opposite of integration, so you can see that a capacitor replaces the input resistor.

One application of this circuit is to detect pulsed edges (such as the leading or trailing edges of a rectangular wave). Another useful function of this circuit is to produce a rectangular output from a ramp-shaped input voltage.

Series-type linear voltage regulator

Figure 29–19 shows a simple series-type voltage regulator circuit that uses an op-amp as one critical element. This circuit regulates its voltage output for any

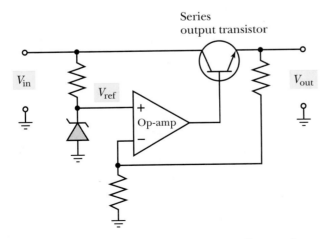

FIGURE 29–19 Example of op-amp series-type linear voltage regulator

changes in input voltage (V_{in}) or for changes in output load current demand (I_L). It essentially keeps the output voltage constant for these changes.

The voltage-divider resistors from the V_{out} line to ground sense changes in output voltage. Since one op-amp differential input is held constant by the zener diode reference voltage circuit, any change fed from the output through the R-divider network to the other op-amp input causes a differential input between the op-amp inputs. This, in turn, is amplified and controls the series output transistor conduction and, consequently, its voltage drop from collector to emitter. If the output voltage attempts to decrease (from a decrease in V_{in} or an increased load at V_{out}), there is less drop across the transistor, bringing the output back up to the desired level. Conversely, if the output voltage attempts to increase (from V_{in} increasing or the output load decreasing), the transistor voltage drop increases, bringing the output voltage back down to the desired regulated level.

An op-amp Schmitt trigger circuit

The circuit in Figure 29–20 is an example of one of the few op-amp circuits that uses positive feedback. Note that the feedback resistor (R_1) is connected between the output of the op-amp and the noninverting input. Operated in this way, the circuit is known as a **Schmitt trigger** circuit, or *hysteresis amplifier.*

This circuit uses a single-ended input mode, and it is further classified as a nonlinear application because the output waveform can be much different from the input waveform. There is some similarity in operation between the Schmitt trigger amplifier and the basic comparator circuit; see again Figure 29–14. Both types of circuits "square up" their input waveforms, but the use of external resistors in a Schmitt trigger amplifier enables more control over the on/off triggering levels.

When the input voltage is moving in a positive direction and reaches the *upper threshold point* (UTP), the output quickly switches (due to the positive feedback) to its negative output level. Then as the input waveform reaches its peak positive level and moves in the negative direction, it soon reaches the *lower threshold point* (LPT) where the output switches to its positive output level. With any Schmitt trigger circuit, there is always a difference between the UTP and LPT values. This difference is known as the **hysteresis** of the amplifier. Formulas for the threshold points refer to the circuit in Figure 29–20. V_{max} is the more positive of the dc supply voltages, and V_{min} is the more negative dc supply voltage.

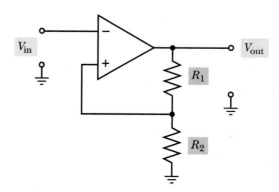

FIGURE 29–20 Op-amp as Schmitt trigger amplifier

FORMULA 29–3 $UTP = V_{max}[R_1/(R_1 + R_2)]$

FORMULA 29–4 $LTP = V_{min}[R_1/(R_1 + R_2)]$

Example

Suppose the circuit in Figure 29–20 has $V_{max} = +12$ V, $V_{min} = -12$ V, $R_1 = 10$ kΩ, and $R_2 = 100$ kΩ. Calculate the values of UTP, LTP, and the amount of hysteresis.

Answers:

From Formula 29–3:

$$UTP = V_{max}[R_1/(R_1 + R_2)]$$
$$UTP = 12[10 \text{ k}\Omega/(10 \text{ k}\Omega + 100 \text{ k}\Omega)] = 1.09 \text{ V}$$

From Formula 29–4:

$$LTP = V_{min}[R_1/(R_1 + R_2)]$$
$$LTP = -12[10 \text{ k}\Omega/(10 \text{ k}\Omega + 100 \text{ k}\Omega)]$$
$$= -1.09 \text{ V}$$

The hysteresis is the difference between UTP and LTP:

$$1.09 \text{ V} - (-1.09 \text{ V}) = 2.18 \text{ V}$$

SUMMARY

▶ Operational amplifiers were originally used in analog computers to perform mathematical operations.

▶ Integrated circuit op-amps are high gain, direct-coupled, dc amplifiers. Op-amps perform various useful functions; therefore they are applied in many electronic systems and subsystems. They are used as inverting and noninverting amplifiers for numerous purposes. They are also used in nonlinear applications such as comparators and Schmitt triggers.

▶ The open-loop gain of op-amp devices can be in the hundreds of thousands. The closed-loop gain of an op-amp circuit is much less than the open-loop gain. The closed-loop gain is controlled by the values and types of external components connected to the inputs and output terminals.

Using some ideal op-amp assumptions, the closed-loop gain is computed as the ratio of the feedback resistor value to the input resistor value.

▶ To produce an output that is inverted from the input, the input signal is connected to the inverting (negative) input of the op-amp. Conversely, to produce an output with the same phase or polarity as the input, the input signal is fed to the noninverting (positive) input.

▶ A circuit that produces a noninverted output (zero phase shift) and a gain of 1 (unity) is called a voltage follower. An op-amp voltage follower is created by using a feedback resistance of 0 Ω.

▶ The single-ended mode of operation for the input is achieved by grounding one input (sometimes through a resistor) and feeding the signal to the other input.

▶ The differential mode is used when different input signals are applied to the inverting and noninverting inputs.

▶ The common-mode operation of an op-amp is used when the same signal feeds both inputs. Although there is no practical application for this mode of operation, it provides the specification for common-mode rejection ratio.

▶ Operational amplifiers can perform mathematical operations on the values of input voltages. These operations are addition (summing amplifier), subtraction (subtractor circuit), differentiation (differentiator circuit), and integration (integrator circuit). Circuits such as signal mixers (summing amplifiers) and

waveform converters, for example, can perform their mathematical functions in ways that do not seem to be mathematical.

▶ A Schmitt trigger circuit provides a rectangular output waveform that switches state when the input voltage passes upward through a UTP (upper threshold point) and when the input voltage passes downward through a LTP (lower threshold point). These threshold voltage levels are determined by the values of the supply voltages and the values of external resistors. The difference between the UTP and LTP is the hysteresis of the amplifier. The Schmitt trigger is one of a very few op-amp circuits that uses positive feedback.

FORMULAS AND SAMPLE CALCULATOR SEQUENCES

FORMULA 29–1 $A_V = (R_f/R_i)$

R_f value, ⊟, R_i value, ⊟

FORMULA 29–2 $A_V \text{ (noninverting)} = (R_f/R_i) + 1$

R_f value, ⊟, R_i value, ⊟, ⊞, 1, ⊟

FORMULA 29–3 $\text{UTP} = V_{max}[R_1/(R_1 + R_2)]$

R_1 value, ⊟, ⟮, R_1, ⊞, R_2, ⟯, ⊟, ⊗, V_{max} value, ⊟

FORMULA 29–4 $\text{LTP} = V_{min}[R_1/(R_1 + R_2)]$

R_1 value, ⊟, ⟮, R_1, ⊞, R_2, ⟯, ⊟, ⊗, V_{min} value, ⊟

REVIEW QUESTIONS

1. Which amplifier circuit is typically used as the input stage for an operational amplifier?

2. Describe the difference between linear and nonlinear op-amp circuits and give an example of each.

3. List the key parameters to describe the ideal op-amp.

4. Define *open-loop* gain.

5. Give an example of an op-amp circuit that is operated in the open-loop gain mode.

6. What is the closed-loop gain of an inverting op-amp circuit when the ratio of feedback to input resistance (R_f/R_i) is 15? 10?

7. If R_i for an inverting op-amp is a variable resistor, increasing the value of R_i will (increase, decrease) _____ the gain of the amplifier.

8. What is the relationship between the amplitude and phase of the input and output signals from an inverting op-amp that has equal resistance values for R_i and R_f?

9. What is the closed-loop gain of a noninverting op-amp circuit when the ratio of feedback to input resistance (R_f/R_i) is 15? 10?

10. If R_i for a noninverting op-amp is a variable resistor, increasing the value of R_i will (increase, decrease) _____ the gain of the amplifier.

11. What is the relationship between the amplitude and phase of the input and output signals from an noninverting op-amp that has equal resistance values for R_i and R_f?

12. To produce a voltage follower with an op-amp what value resistance should be used as the feedback component if $R_i = 1$ MΩ?

13. For an op-amp operating in the single-ended mode, describe the output signal polarity when the positive input is grounded and the input signal feeds the negative input.

14. For an op-amp in common-mode operation, should the output level be large, small, or medium?

15. To produce integration in an op-amp, what type of component should be used as the feedback component? as the input component?

16. To produce differentiation in an op-amp, what type of component should be used as the feedback component? as the input component?

17. What type of op-amp circuit should be selected where it is required to change a rectangular waveform into a sawtooth (ramp) waveform?

18. What type of op-amp circuit should be selected where it is required to change a rectangular waveform into brief pulses?

19. Assuming equal resistors throughout the circuit, write the formula for the output of a summing amplifier having three input voltages.

20. For the circuit in Figure 29–19, briefly explain how a condition where V_{out} trying to increase could be compensated for by the regulator circuit. Explain in terms of the coordination between the feedback circuit, the op-amp, and the series output transistor.

PROBLEMS

1. For the circuit in Figure 29–21, the feedback resistor is 250 kΩ and an input resistor is 1,200 Ω. What is the closed-loop gain?

3. What value input resistor is required for an inverting op-amp if the feedback resistor is 1 MΩ and the closed-loop gain is to be 120?

4. For the circuit in Figure 29–22, the feedback resistor is 250 kΩ and an input resistor is 1,200 Ω. What is the closed-loop gain?

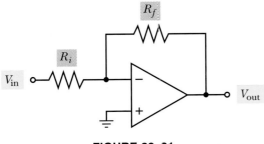

FIGURE 29–21

2. What is the closed-loop gain of the circuit in Figure 29–21 when $R_i = 1$ kΩ and $R_f = 1$ kΩ?

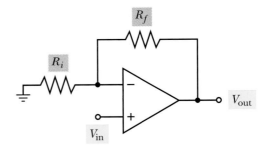

FIGURE 29–22

5. What is the closed-loop gain of the circuit in Figure 29–22 when $R_i = 1$ kΩ and $R_f = 1$ kΩ?

6. What value input resistor is required for an inverting op-amp if the feedback resistor is 1 MΩ and the closed-loop gain is to be 120?

7. All resistors in a certain difference-mode op-amp (see Figure 29–15) have the same value. Determine the output of the amplifier when the following is true:
   a. $V_{in+} = +2$ V and $V_{in-} = -3$ V
   b. $V_{in+} = -2$ V and $V_{in-} = -3$ V
   c. $V_{in+} = +2$ V and $V_{in-} = +3$ V
   d. $V_{in+} = -2$ V and $V_{in-} = +3$ V

8. All resistors in the summing op-amp of Figure 29–23 have the same value. Determine the output voltage, V_{out}, when the following is true:
   a. $V_1 = +2$ V and $V_2 = +3$ V
   b. $V_1 = -2$ V and $V_2 = +3$ V
   c. $V_1 = +2$ V and $V_2 = -3$ V
   d. $V_1 = -2$ V and $V_2 = -3$ V

9. For the summing op-amp of Figure 29–23, $R_{i_1} =$

10 kΩ, $R_{i_2} = 15$ kΩ, and $R_f = 20$ kΩ. Determine V_{out} when the following is true:
   a. $V_1 = +2$ V and $V_2 = +3$ V
   b. $V_1 = -2$ V and $V_2 = +3$ V
   c. $V_1 = +2$ V and $V_2 = -3$ V
   d. $V_1 = -2$ V and $V_2 = -3$ V

10. Calculate the UTP, LTP, and hysteresis for a Schmitt trigger circuit (Figure 29–20) where $R_1 = 100$ kΩ, $R_2 = 27$ kΩ, +V source = +12 V, and −V source = 0 V (grounded).

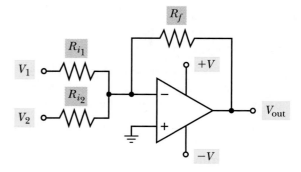

FIGURE 29–23

ANALYSIS QUESTIONS

1. Show how you can use the formula for the gain of a noninverting op-amp (Formula 29–2) to prove that the gain of a voltage follower, Figure 29–12, is unity.

2. Find and list the data handbook parameters typically shown for a 741-type operational amplifier IC device.

3. A sensitive electronic thermometer produces a voltage that changes 10 mV for every degree of

change in temperature. If the signal is connected to the input of an inverting op-amp circuit, what voltage gain is necessary so that the output of the amplifier produces a change of 2 V for every degree of change in temperature? If the input resistance to the op-amp is 2.2 kΩ, what is the value of the feedback resistor?

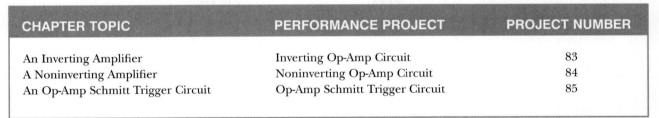

PERFORMANCE PROJECTS CORRELATION CHART

CHAPTER TOPIC	PERFORMANCE PROJECT	PROJECT NUMBER
An Inverting Amplifier	Inverting Op-Amp Circuit	83
A Noninverting Amplifier	Noninverting Op-Amp Circuit	84
An Op-Amp Schmitt Trigger Circuit	Op-Amp Schmitt Trigger Circuit	85

NOTE: It is suggested that after completing the above projects, the student should be required to answer the questions in the "Summary" at the end of this section of projects in the Laboratory Manual.

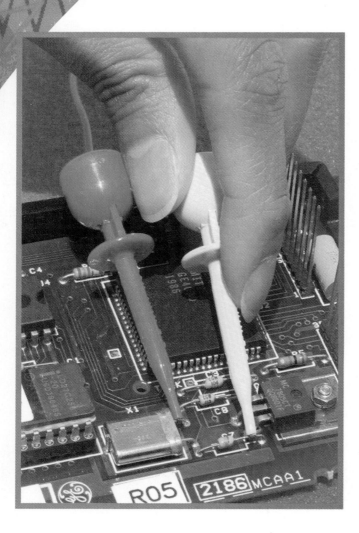

KEY TERMS

Astable multivibrator
Clapp oscillator
Colpitts oscillator
Crystal oscillator
Free-running multivibrator
Hartley oscillator
Monostable multivibrator
Multivibrators
One-shot multivibrator
Phase-shift oscillator
Piezoelectric effect
Positive feedback
Rectangular waveforms
Wien-bridge oscillator

OSCILLATORS AND MULTIVIBRATORS

30

chapter

CHAPTER PREVIEW

Someone once said that an oscillator is a "hot rod" or "turbo" amplifier. Let's examine that comment. An oscillator is a circuit that generates and sustains an output signal without an input signal supplied by another circuit or source. This capability enables oscillators to be used as signal sources for many circuit applications.

There are basically three kinds of electronic circuits: those that amplify signals, those that control the flow of signals, and those that generate signals. You have already studied amplifiers in previous chapters, and you will study control circuits in a later course of study. This chapter is about the oscillator circuits that generate signals.

You will find that the function of oscillators depends upon the qualities of *LC* tuned circuits or *RC* phase-shift circuits. A special type of oscillator, an astable (or free-running) multivibrator, generates its signal according to the charge and discharge time of *RC* circuits. You will also be introduced to a monostable (or one-shot) multivibrator.

OBJECTIVES

After studying this chapter, you should be able to:

1. Identify from schematic diagrams the BJT, FET, and op-amp versions of the **Hartley, Colpitts,** and **Clapp oscillators**

2. Identify the tuning components and describe the procedure for determining the oscillating frequency of the Hartley, Colpitts, and Clapp oscillators

3. Explain the operation of a **crystal oscillator**

4. Identify from schematic diagrams the **phase-shift** and **Wien-bridge oscillators**

5. Identify the tuning components and describe the procedure for determining the oscillating frequency of the phase-shift and Wien-bridge oscillators

6. Define the operation of a **monostable multivibrator** and calculate the duration of the output pulse

7. Define the operation of an **astable multivibrator** and determine the operating frequency for both symmetrical and nonsymmetrical output waveforms

30–1 INTRODUCTION TO OSCILLATORS

As stated earlier, an **oscillator** is a circuit that generates and sustains an output signal without an input signal supplied by another circuit or source. For example, an oscillator generates the waveform that produces the tones we hear from an electronic organ or keyboard. An oscillator generates the high frequency waveforms that carry radio and television signals around the earth and into space. Also, oscillators synchronize the operation of industrial controls and personal computers. Figure 30–1 shows the two most common kinds of waveforms—

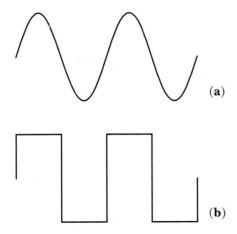

FIGURE 30–1 Standard waveforms
from oscillators and multivibrators:
(a) sinusoidal waveform;
(b) rectangular waveform

sinusoidal and rectangular—generated by oscillator circuits. These are the waveforms generated by the oscillator circuits discussed in this lesson.

Four conditions are absolutely required to begin and sustain the operation
of an oscillator:

1. A power source
2. A device or components that determine the frequency of oscillation
3. Amplification
4. **Positive feedback**

The power source for oscillator circuits is taken from batteries or a dc power
supply. Oscillator circuits sometimes create signals that find their way into the
wiring for the power supply. This is highly undesirable because the oscillations
can thus show up in circuits where they do not belong. At least one *decoupling capacitor* is used with oscillator circuits to bypass (short-circuit) any of these stray
oscillations from the power supply lines to ground. See the example in Figure
30–2. Decoupling capacitors generally have values on the order of 10 pF to 100
pF, and they are physically located very close to the oscillator circuit.

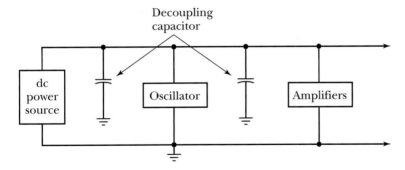

FIGURE 30–2 Power supply decoupling capacitors

The devices or components that control the frequency of oscillation include combinations of resistors, capacitors, inductors, and quartz crystals. You will soon find that the frequency determining components for some oscillators are *RC* circuits—combinations of resistors and capacitors. A number of different oscillators use *LC* tank circuits (a parallel circuit composed of inductors and capacitors) as the frequency determining components. One type of oscillator uses a quartz crystal to fix the rate of oscillation.

The amplification for oscillators can be provided by BJTs, FETs, and op-amps. Every oscillator must contain a device that provides signal amplification greater than 1.

Positive feedback contributes to the operation of an oscillator by feeding a portion of the amplifier's output back to its own input. The feedback signal must be phased in such a way that it enhances the amplifier's input/output signal (positive feedback). The amplifier part of the oscillator usually inverts the signal by 180°, then a phase-shifting component (capacitor or inductor) in the feedback path provides the necessary additional phase shift that is required for positive feedback. Positive feedback is also called *regenerative feedback.*

30–2 LC SINE-WAVE OSCILLATORS

An important and commonly used class of oscillators are those that use a parallel *LC* (tank) circuit to determine the operating frequency. As long as these circuits are operating properly, they produce sine-wave output waveforms. The frequency of these waveforms is the same as the resonant frequency of the tank circuit; and as you learned in an earlier chapter, you can compute the resonant frequency of a tank circuit by means of the formula shown here as Formula 30–1. In this formula, f_r is the resonant frequency (frequency of the oscillator), L is the value of the inductance, and C is the value of capacitance.

FORMULA 30–1 $f_r = 1/2\pi\sqrt{LC}$

LC sine-wave oscillators contain various combinations of two or more capacitors and inductors. You will find that the values of the inductors can be combined into a single *L* value and, likewise, the values of two or more capacitors can be combined into a single *C* value.

In all of these resonant circuits, increasing the value of capacitance decreases the operating frequency, while decreasing the capacitance increases operating frequency. This is also true for the inductances: increasing the inductance decreases frequency, while decreasing inductance increases frequency. We say that the operating frequency of an *LC* oscillator is inversely proportional to the square root of the product of the capacitance and inductance values in the oscillator's tank circuit.

Hartley Oscillators

The circuit in Figure 30–3 is known as a **Hartley oscillator**.

1. The power source is shown as $+V_{CC}$.

2. The frequency determining part of the oscillator is the *LC* tank circuit composed of L_{1_A}, L_{1_B}, and C_1.

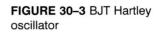

FIGURE 30–3 BJT Hartley oscillator

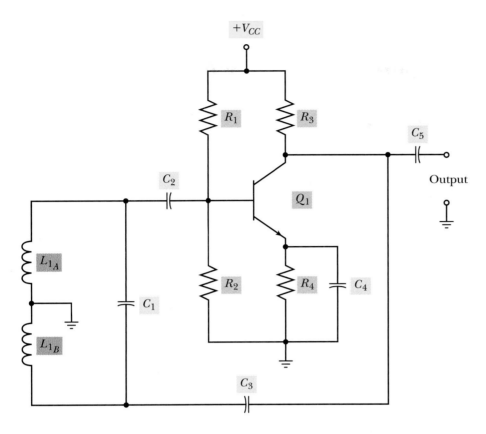

3. Amplification is provided by the NPN common-emitter transistor (Q_1) and its associated components.

4. Positive feedback is provided from the collector of the amplifier through capacitor C_3 to the tapped coil in the tank circuit.

The Hartley oscillator is characterized by having a tapped inductor in the LC resonant circuit. The tapped point for the inductor is grounded. You will sometimes see two separate inductors instead of a single, tapped inductor.

You can calculate the operating frequency of a Hartley oscillator by applying the basic parallel LC circuit formula for resonance:

FORMULA 30–2 $\quad f_r = 1/2\pi\sqrt{L_t C}$

In Formula 30–2, L_t is the total inductance and is found by:

FORMULA 30–3 $\quad L_t = L_{1_A} + L_{1_B}$

Example

What is the operating frequency of the Hartley oscillator shown in Figure 30–3 if $C = 100$ pF, $L_{1_A} = 20$ μH, and $L_{1_B} = 20$ μH?

FIGURE 30–4 Op-amp Hartley oscillator

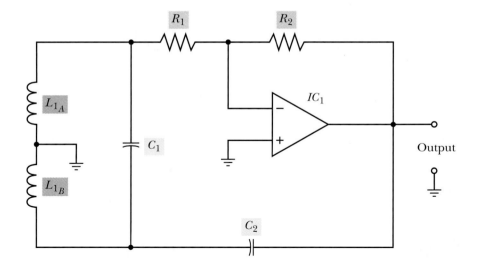

Answer:

From Formula 30–3:

$$L_t = L_{1_A} + L_{1_B} = 20 \ \mu H + 20 \ \mu H = 40 \ \mu H$$

From Formula 30–2:

$$f_r = 1/2\pi\sqrt{L_t C} = 1/2\pi\sqrt{(40 \ \mu H)(100 \ pF)}$$

$$f_r = 2.5 \ MHz$$

Figure 30–4 shows a version of the Hartley oscillator that uses an op-amp as the amplifier element. Resistor R_2 in this circuit provides the dc negative feedback that is necessary for the operation of an inverting op-amp. The regenerative feedback that is required for oscillation is from the output of IC_1, through coupling capacitor C_2, to the tank circuit, and through resistor R_1 to the inverting input of the op-amp.

A JFET Hartley oscillator is shown in Figure 30–5.

The formulas for finding the oscillating frequency of op-amp and FET versions of the Hartley oscillator are exactly the same as for the BJT versions. It is important you understand that, in *LC* oscillators, the type of amplifying device being used has little effect on the operating frequency. The operating frequency is determined by the values of the inductances and capacitors in the tank circuit. The amplifying device merely ensures continuous oscillation of the tank circuit.

PRACTICE PROBLEMS I

Refer to the JFET Hartley oscillator circuit in Figure 30–5 to answer the following questions.

1. Which component is the active amplifying device in this oscillator?

2. Which components make up the frequency determining part of the circuit?

3. Capacitor C_3 is part of the _____ portion of this oscillator circuit.

4. What is the total inductance of the tank circuit if $L_{1_A} = 0.1$ mH and $L_{1_B} = 0.1$ mH?

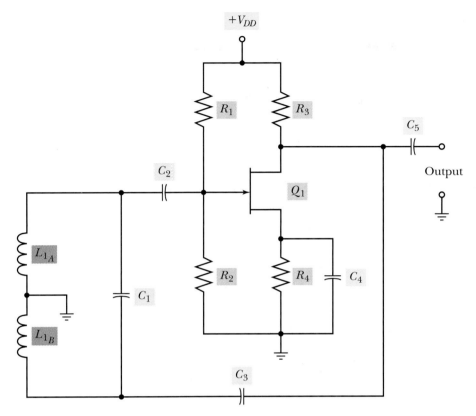

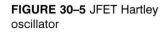

FIGURE 30–5 JFET Hartley oscillator

5. Suppose the inductances have the values indicated in question 4. Further suppose the value of the capacitor is 1,000 pF. What is the frequency of oscillation of this circuit?

6. What happens to the frequency of oscillation of this circuit if you decrease the value of C_1? Increase the value of L_{1_A}?

Colpitts oscillators

The circuit in Figure 30–6 is the BJT version of a **Colpitts oscillator**. It closely resembles a Hartley oscillator, Figure 30–3. The main difference is the location of the grounded tap in the resonant circuit. Recall that a Hartley oscillator has a grounded tap between two series inductors. The Colpitts oscillator has a grounded tap between a pair of series capacitors.

1. The dc power source is the $+V_{CC}$ connection.

2. The frequency determining element of the oscillator is the tank circuit comprised of L_1, C_1, and C_2.

3. Amplification is provided by the NPN common-emitter amplifier transistor circuit.

4. Positive feedback is accomplished by the connection from the collector of the amplifier transistor, through capacitor C_4, to the tank circuit.

FIGURE 30–6 BJT Colpitts
oscillator

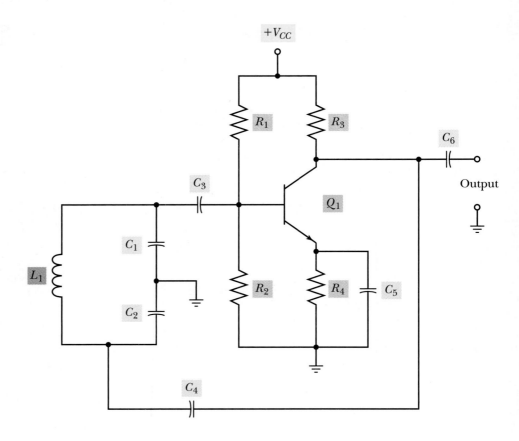

You can calculate the operating frequency of a Colpitts oscillator by applying a
slightly modified version of the resonance formula:

FORMULA 30–4 $f_r = 1/2\pi\sqrt{LC_T}$

Total capacitance (C_T) is determined by:

FORMULA 30–5 $C_T = C_1 C_2/(C_1 + C_2)$

Example

Referring to the Colpitts oscillator circuit in Figure 30–6, determine the operating frequency if $L = 100$ µH, $C_1 = 1,000$ pF, and $C_2 = 500$ pF.

Answer:

From Formula 30–5:

$$C_T = C_1 C_2/(C_1 + C_2) = 333 \text{ pF}$$

From Formula 30–4:

$$f_r = 1/2\pi\sqrt{LC_T}$$
$$f_r = 1/2\pi\sqrt{(100 \text{ µH})(333 \text{ pF})} = 872 \text{ kHz}$$

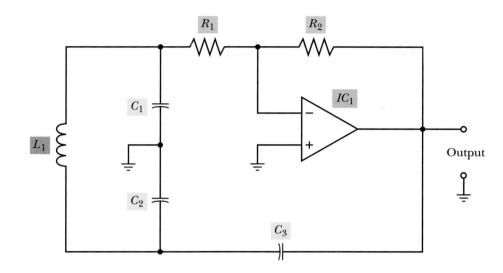

FIGURE 30–7 Op-amp
Colpitts oscillator

Figure 30–7 shows a version of the Colpitts oscillator that uses an op-amp as the amplifier element. This is basically an inverting amplifier where R_1 is the input resistor and R_2 is the feedback resistor. Proper phase inversion is automatic by means of this type of amplifier.

The Colpitts oscillator in Figure 30–8 is built around a JFET. Note transformer T_1 at the output. The stability of operation of Colpitts oscillators is often affected by changes in the amount of loading (impedance) across its output connections. Direct coupling or capacitive coupling to the next stage of the circuit is sometimes inadequate, so you will often find transformer coupling at the out-

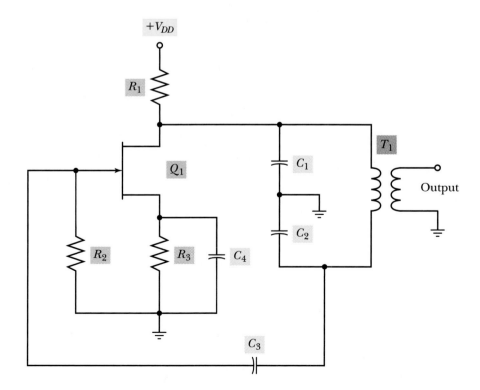

FIGURE 30–8 JFET Colpitts
oscillator

put of a Colpitts oscillator stage. Notice that the inductance value of the primary winding of coupling transformer T_1 is the inductance value of the oscillator's tank circuit.

PRACTICE PROBLEMS II

Refer to the JFET Colpitts oscillator circuit in Figure 30–8 to answer the following questions.

1. Which component is the active amplifying device in this oscillator?
2. Which components make up the frequency determining part of the circuit?
3. What is the total capacitance of the tank circuit if $C_1 = 100$ pF and $C_2 = 100$ pF?
4. What is the frequency of oscillation of this circuit if the total capacitance is the value found in question 3 and the inductor is rated at 750 μH?
5. What happens to the frequency of oscillation of a Colpitts oscillator when you decrease the total capacitance? increase the value of inductance?

PRACTICAL NOTES

You can remember that a Hartley oscillator has a divided inductance by recalling that H is the unit of measure for inductance (and *Hartley* begins with the letter *H*). Colpitts and Clapp oscillators, on the other hand, have split capacitances and begin with the letter *C* (also the first letter in *capacitance*).

Clapp oscillators

You have already learned that a Colpitts oscillator stage often uses transformer coupling at the output to stabilize operation when the input impedance of the next stage changes. Another way to help stabilize operation is to insert a third capacitor into the tank circuit, Figure 30–9. When this is done, the circuit is no longer called a Colpitts oscillator. It is called a **Clapp oscillator.**

1. The dc power source is the $+V_{CC}$ connection.
2. The frequency determining element of the oscillator is the tank circuit comprised of L_1, C_1, C_2, and C_3.
3. Amplification is provided by the NPN common-emitter amplifier transistor circuit.
4. Positive feedback is accomplished by the connection from the collector of the amplifier transistor to the tank circuit through C_5.

The main formula for finding the oscillating frequency for a Clapp oscillator is the same one used for a Colpitts oscillator (Formula 30–4). Because there

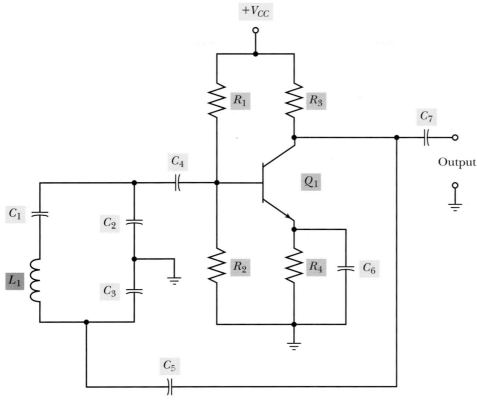

FIGURE 30–9 BJT Clapp oscillator

are three capacitors in the tank circuit for a Clapp oscillator, the formula for total capacitance (C_T) is different:

FORMULA 30–6 $\qquad C_T = 1/(1/C_1 + 1/C_2 + 1/C_3)$

Example

Suppose you find a Clapp oscillator, such as the one in Figure 30–9, that has $L = 10\ \mu H$, $C_1 = 0.01\ \mu F$, $C_2 = 0.01\ \mu F$, and $C_3 = 0.1\ \mu F$. Calculate the operating frequency of the circuit.

Answer:

From Formula 30–6:

$$C_T = 1/(1/C_1 + 1/C_2 + 1/C_3)$$
$$C_T = 4{,}762\ pF$$

From Formula 30–4:

$$f_r = 1/2\pi\sqrt{LC_T}$$
$$f_r = 1/2\pi\sqrt{(10\ \mu H)(4{,}762\ pF)} = 729.7\ kHz$$

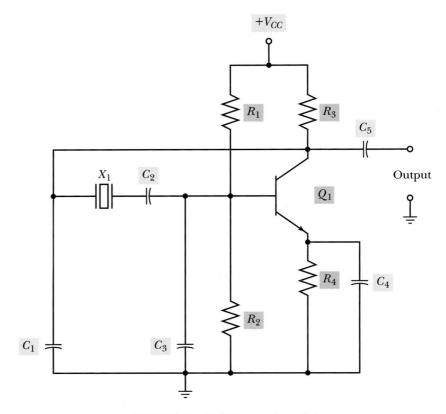

FIGURE 30–10 BJT crystal oscillator

30–3 CRYSTAL OSCILLATORS

Quartz crystals possess a piezoelectric quality that permits their use in very precise sine-wave oscillators. The **piezoelectric effect** is one where the application of a bending force to crystal material causes it to generate a small amount of voltage; then applying a voltage to the same material causes it to bend or warp as long as the voltage is applied. When a quartz crystal is connected to an amplifying device as shown in Figure 30–10, the amplifier can provide a burst of voltage that will cause the crystal to bend. Then when the voltage is removed and the crystal is allowed to return to its normal shape, the crystal generates a small voltage that is detected and amplified by the transistors. This impulse is then amplified and returned to the crystal to cause the crystal to bend again. This positive feedback cycle continues, thus producing oscillations determined by the rate that the crystal vibrates. The rate that the crystal vibrates is determined by its physical shape and size.

The crystal in a crystal oscillator replaces the tank circuit found in LC-type oscillators. The schematic symbol and equivalent RCL circuit are shown in Figure 30–11. In this equivalent circuit, R is the natural resistance of the crystal, L is the equivalent inductance, C_N is the "natural" capacitance, and C_M is the "mounting" capacitance (which includes the stray capacitance of the metal case, mounting terminals, etc.).

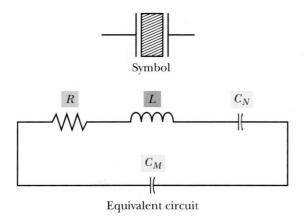

FIGURE 30–11 Symbol and equivalent circuit for a quartz crystal

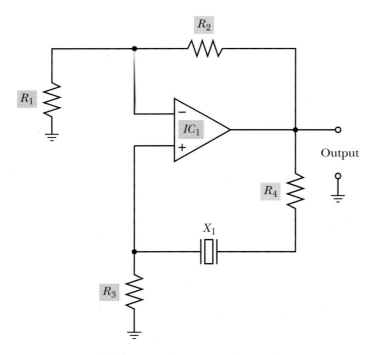

FIGURE 30–12 Op-amp crystal oscillator

Figure 30–12 shows an op-amp crystal oscillator with the following features:

1. The power source is the $+V$ and $-V$ connections.
2. The frequency determining element of the oscillator is crystal X_1.
3. Amplification is provided by the op-amp.
4. Positive feedback is accomplished by the connection from the output of the amplifier through the crystal to the noninverting input.

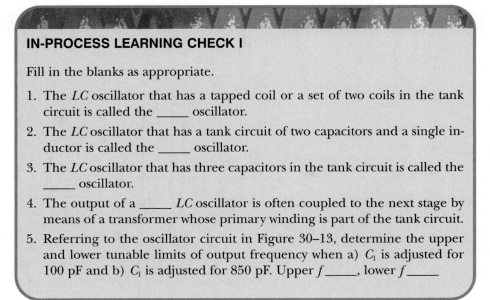

IN-PROCESS LEARNING CHECK I

Fill in the blanks as appropriate.

1. The LC oscillator that has a tapped coil or a set of two coils in the tank circuit is called the _____ oscillator.

2. The LC oscillator that has a tank circuit of two capacitors and a single inductor is called the _____ oscillator.

3. The LC oscillator that has three capacitors in the tank circuit is called the _____ oscillator.

4. The output of a _____ LC oscillator is often coupled to the next stage by means of a transformer whose primary winding is part of the tank circuit.

5. Referring to the oscillator circuit in Figure 30–13, determine the upper and lower tunable limits of output frequency when a) C_1 is adjusted for 100 pF and b) C_1 is adjusted for 850 pF. Upper f _____, lower f _____

TROUBLESHOOTING HINTS

In actual practice, LC oscillators tend to work properly or not at all. The best laboratory tool for checking the operation of an oscillator is an oscilloscope. An electronic frequency measuring instrument (sometimes called a "freq meter") is valuable for checking the operating frequency of an oscillator very accurately, but this meter can give you misleading information if you haven't determined that correct voltage levels exist by oscilloscope measurement.

The two most common reasons an LC oscillator stops operating are a defective amplifier element and excessive loading at the output. The best way to check for a defective amplifier element is to replace it with one that is known to be in good working order. As far as output loading is concerned, a trouble in the next stage in the circuit might place a low impedance across the output of the oscillator and stop it from running. You can check for this trouble by temporarily disconnecting the output of the oscillator from the next circuit along the line. If the oscillator begins functioning normally under these circumstances, the trouble is more likely in the next stage rather than in the oscillator circuit itself.

Avoid taking frequency and signal amplitude measurements while touching the components in a high frequency oscillator. The stray capacitance of your fingers and hand can shift the frequency and amplitude levels.

30–4 RC SINE-WAVE OSCILLATORS

A second major group of sine-wave oscillators is built around RC timing components rather than LC resonant circuits and quartz crystals. Whereas the operation of LC oscillators is based on resonant frequency conditions, the operation of RC oscillators is based on phase-shift conditions. The four basic elements and

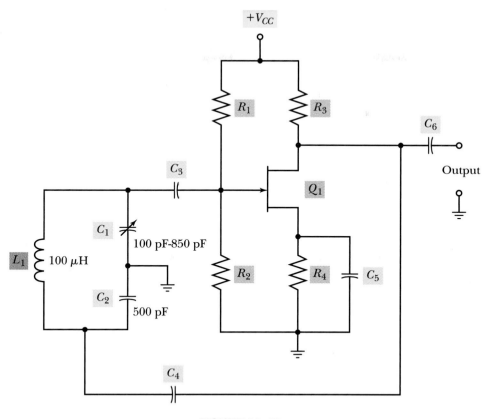

FIGURE 30–13

conditions for oscillation, however, are the same; that is, a power source, frequency determining elements, amplifier, and positive feedback.

RC phase-shift oscillator

The circuit in Figure 30–14 shows two versions of the **phase-shift oscillator**. You can always identify a phase-shift oscillator by means of its three-stage RC network. The amplifying device shown in these examples is an op-amp, but the three-stage RC network is readily identified on a schematic where the amplifying element is a BJT or FET.

1. The dc power source is the $+V_{CC}$ connection.
2. The frequency determining element of the oscillator is the RC phase-shift network comprised of R_1, C_1, R_2, C_2, R_3, and C_3.
3. Amplification is provided by the op-amp circuit.
4. Positive feedback is accomplished by the connection from the output of the op-amp, through the phase-shift network, to the inverting $(-)$ input of the op-amp.

As the name of this oscillator suggests, its operation is based on phase-shift principles. Recall that an RC circuit will shift the phase of an ac waveform by a

FIGURE 30–14 Phase-shift oscillators: **(a)** lead network; **(b)** lag network

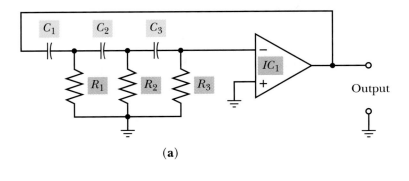

(a)

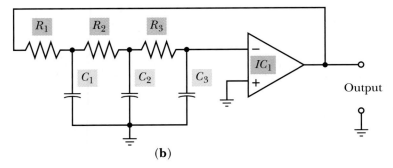

(b)

certain number of degrees, depending on the values of R and C and on the frequency. The RC network in a phase-shift oscillator is capable of producing a 180° phase shift at one particular frequency. When this 180° phase shift (which occurs at only one particular frequency) is combined with the normal 180° phase shift of the inverting amplifier, we have the positive feedback condition required for starting and maintaining oscillation. Oscillation occurs only at the frequency that causes a 180° phase shift through the RC network. At any other frequency, the feedback is some value other than the required 180° + 180° = 360° needed for positive feedback.

The principle of operation and formula for calculating the operating frequency are the same for the lead and lag versions. In the lead version, Figure 30–14a, the capacitors are in series with the feedback path, thus causing the feedback voltage developed across the Rs to lead the signal from the amplifier. In the lag version, Figure 30–14b, the resistors are in series with the feedback path and the Cs in parallel, thus enabling a lagging feedback voltage.

The formula for determining the operating frequency of a phase-shift oscillator is complicated unless you assume all of the capacitors in the RC network have the same value and all the resistors in the network have the same value. Assuming all three capacitors and all three resistors have the same value:

FORMULA 30–7 $f_r = 1/2\pi RC\sqrt{6}$

Example

What is the operating frequency of the phase-shift oscillator in Figure 30–14a if $R_1 = R_2 = R_3 = 3.3$ kΩ and $C_1 = C_2 = C_3 = 0.01$ μF?

Answer:

From Formula 30–7:

$$f_r = 1/2\pi RC\sqrt{6}$$

$$f_r = 1/2\pi(3.3 \text{ k}\Omega)(0.01 \text{ }\mu\text{F})\sqrt{6} = 1.97 \text{ kHz}$$

Phase-shift oscillators are known for their relative simplicity (no inductors), low cost, and reliability. Their main disadvantage is that it is difficult to fine-tune them. This means they are used only in circuits where a very exact frequency is not required. Another disadvantage is the loss in feedback signal level as it passes through the series of three *RC* combinations. To meet the voltage gain requirements for oscillation, the amplifier in a phase-shift oscillator must have a gain greater than 30.

Wien-bridge oscillator

The **Wien-bridge oscillator** in Figure 30–15 is a simple, stable, and popular *RC* oscillator configuration. Its identifying feature is a set of two *RC* combinations called a lead-lag network. A series *RC* combination is connected in series with the feedback path, and a parallel *RC* combination is connected in parallel with the feedback path. Another unusual feature is that it uses a noninverting amplifier instead of the usual inverting configuration.

1. The dc power source is the $+V_{CC}$ connection.
2. The frequency determining element of the oscillator is the lead-lag *RC* network comprised of C_1, R_1, C_2, and R_2.
3. Amplification is provided by the op-amp circuit.
4. Positive feedback is accomplished by the connection from the output of the op-amp, through the *RC* network, to the noninverting input (+) of the op-amp.

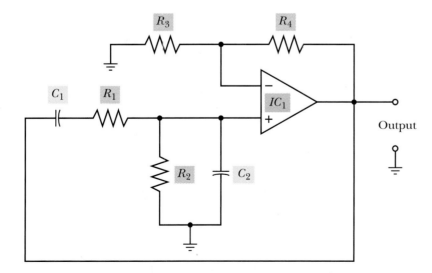

FIGURE 30–15 Wien-bridge oscillator

The frequency of oscillation is determined by the values of C_1, R_1, C_2, and R_2 in the lead-lag network. The important property of this kind of RC network is that the signal fed back from the output of the amplifier undergoes a phase shift of $0°$ at the oscillator frequency. Because the lead-lag network causes zero phase shift at the operating frequency, the amplifier must be a noninverting type to enable a $0°$ phase shift as well.

It is instructive to compare the operation of a Wien-bridge oscillator with the phase-shift oscillator you studied in the previous section of this chapter.

1. In the phase-shift oscillator, the RC section produces a $180°$ phase shift at the oscillation frequency. In the Wien-bridge oscillator, the RC section produces a $0°$ phase shift at the oscillation frequency.

2. The amplifier in a phase-shift oscillator is of the inverting type, thereby producing a $180°$ phase shift. The amplifier in a Wien-bridge oscillator is of the noninverting type, thereby producing zero phase shift between input signals and output signals.

3. Considering the combined RC network and amplifier phase shifts for a phase-shift oscillator:

$$\text{Total phase shift} = 180° + 180° = 360° = 0°$$

Considering the combined RC lead-lag and amplifier phase shifts for a Wien-bridge oscillator:

$$\text{Total phase shift} = 0° + 0° = 0°$$

You can see from this comparison that the phase-shift and Wien-bridge oscillators both provide a positive-feedback phase shift of $0°$ at the oscillating frequency.

As with the phase-shift oscillator, determining the oscillating frequency of the Wien-bridge oscillator is made simpler by making both capacitor values and both resistor values in the lead-lag network the same. In other words, let $C = C_1 = C_2$ and $R = R_1 = R_2$. Formula 30–8 provides the oscillating frequency of the Wien-bridge oscillator.

FORMULA 30–8 $f_r = 1/2\pi RC$

Example

Referring to the Wien-bridge circuit in Figure 30–15, determine the output frequency if $C_1 = C_2 = 0.1\ \mu\text{F}$, and $R_1 = R_2 = 100\ \text{k}\Omega$.

Answer:

From Formula 30–8:

$$f_r = 1/2\pi RC$$

$$f_r = 1/2\pi (100\ \text{k}\Omega)(0.1\ \mu\text{F}) = 15.9\ \text{Hz}$$

PRACTICE PROBLEMS III

Refer to the Wien-bridge oscillator circuit in Figure 30–15 for the following questions.

1. Determine the output frequency of the Wien-bridge oscillator when $C_1 = C_2 = 0.002$ μF, and $R_1 = R_2 = 4.7$ kΩ.

2. What happens to the output frequency if you double the value of the capacitors? Cut the resistor values in half?

3. Suppose $C_1 = C_2 = 0.1$ μF and you want to obtain a frequency of 1,000 Hz. What value should be used for the resistors?

IN-PROCESS LEARNING CHECK II

Fill in the blanks as appropriate.

1. Whereas _____ oscillators operate according to resonance, RC oscillators operate according to _____.

2. The four basic elements and conditions for starting and sustaining oscillation in RC sine-wave oscillators are _____, _____, _____, and _____.

3. The sine-wave oscillator that uses a three-stage RC network for achieving a 180° phase shift at the frequency of oscillation is the _____ oscillator.

4. The sine-wave oscillator that uses a lead-lag RC network to produce 0° phase shift at the frequency of oscillation is the _____ oscillator.

5. The amplifier element of a phase-shift oscillator must produce a phase shift of _____°, while the amplifier element of a Wien-bridge oscillator must produce a phase shift of _____°.

30–5 MULTIVIBRATORS

Multivibrators are special types of circuits that produce **rectangular waveforms** that switch approximately between ground and the supply voltage level. If the supply voltage level happens to be +5.2 V, for example, a multivibrator circuit will produce a rectangular waveform that switches between +0.3 V (close to ground potential) and about +5.0 V (close to the supply voltage). These voltage levels are those of a BJT or FET that is operating as a switch—either fully on (saturation) or switched off (cutoff).

Another characteristic of multivibrator circuits is that their timing and frequency determining components are always RC circuits. Do not expect to find inductors doing anything significant in a multivibrator circuit.

There are three types of multivibrator circuits: monostable multivibrators (or one-shot multivibrators), bistable multivibrators (or flip-flops), and astable multivibrators. We will discuss monostable and astable multivibrators in this chapter. You will learn about bistable multivibrators (flip-flops) in your later studies of digital electronics.

Monostable multivibrators

A **monostable multivibrator** is a circuit that produces a rectangular pulse of a fixed duration in response to a brief input trigger pulse. A monostable multivibrator does not run continuously as do oscillators (described earlier) and the astable multivibrator (described next). This circuit is also called a *one-shot* because one trigger pulse causes one output pulse to occur.

Figure 30–16 shows the schematic diagram for a BJT monostable multivibrator. The trigger pulse that starts the output timing interval is applied to the base of transistor Q_1 through resistor R_3. The output pulse is taken from the collector of transistor Q_2. The waveforms indicate the action of the circuit in response to a trigger pulse being applied at the *Trig in* point.

Monostable multivibrators have four states of operation: the resting state, the trigger-on state, the active state, and the return state. The following is a sum-

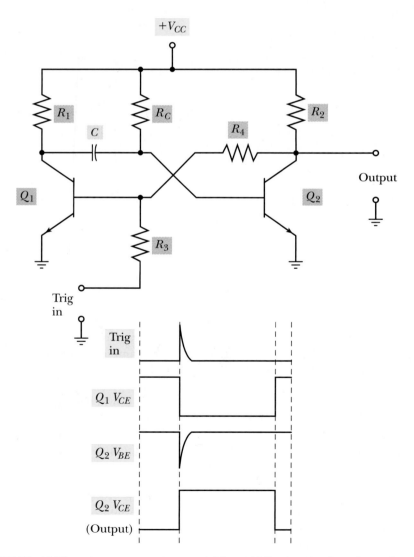

FIGURE 30–16 Discrete-component monostable multivibrator circuit and waveforms

mary of the four operating states for the monostable multivibrator shown in Figure 30–16.

1. The resting (quiescent) state is the standby state. It is where transistor Q_1 is turned off and Q_2 is turned on. You can see the effect in the waveforms where the V_{CE} for Q_1 is at the source voltage level ($+V_{CC}$) and the V_{CE} for Q_2 is close to ground potential. The circuit remains in this state until some external condition causes the positive-going spike to occur at the *Trig in* connection.

2. The trigger-on state begins when the *Trig in* terminal sees the positive-going edge of a trigger pulse waveform. This positive transition immediately biases transistor Q_1 to its ON state. This effectively places the Q_1 side of capacitor C at ground potential, which effectively grounds the base of transistor Q_2. This action turns off transistor Q_2. Note how the waveforms reflect these changes when the *Trig in* signal goes positive:

   a. The V_{CE} for Q_1 drops to 0 V. (Q_1 is switched on.)
   b. The base voltage (V_{BE}) for Q_2 immediately drops to 0 V. (This is what turns off Q_2.)
   c. The V_{CE} for Q_2 rises to $+V_{CC}$. (Q_2 is switched off.)

3. The active state begins as capacitor C starts charging with the collector current of switched-on transistor Q_1. As long as capacitor C is charging toward $+V_{CC}$ voltage level, transistor Q_2 is held off. And as long as Q_2 is held off, base current for transistor Q_1 flows through resistors R_4 and R_2. The waveforms show that the *Trig in* signal has returned to 0 V, Q_1 is still turned on, capacitor C is charging (through Q_1 and R_C), and transistor Q_2 is being held off.

4. The return state begins when capacitor C has charged very close to the $+V_{CC}$ level. When this happens, transistor Q_2 is biased on. The V_{CE} for Q_2 thus drops close to ground potential. The collector of Q_2 is direct-coupled to the base of Q_1; so when Q_2 is switched on, it switches transistor Q_1 off. Now everything has returned to the resting state where it remains until the next *Trig in* pulse is delivered to the base of Q_1 to turn it on again.

It is possible to estimate very closely the active timing interval for this monostable multivibrator. By studying this formula, you will find that the active timing interval is proportional to the values of the timing resistor (R_C) and timing capacitor (C).

FORMULA 30–9 $t_{on} = 0.69 R_C C$

Example

What is the time of the active state of a monostable multivibrator (Figure 30–16) when $R_C = 100$ kΩ and $C = 0.1$ μF?

Answer:

From Formula 30–9:

$$t_{on} = 0.69 R_C C$$

$$t_{on} = 0.69 \times 100 \text{ k}\Omega \times 0.1 \text{ }\mu\text{F} = 6.9 \text{ msec}$$

Monostable multivibrator circuits are used for performing electronic timing operations. The timing in such applications takes place from the moment the trigger waveform occurs to the end of the active output interval. You have already seen how you can closely estimate that time with the aid of Formula 30–9. A monostable multivibrator is used in the circuits for controlling the overhead light in modern automobiles. In this system, the overhead light goes on whenever you open the car door, but then it remains on for a fixed period from the moment the door is closed. In this example, the monostable multivibrator is triggered to its active state when the car door is closed. Barring any interruption (such as turning on the ignition key), the monostable multivibrator causes the overhead light to remain on for a preset period that is on the order of 1 minute.

Today's monostable multivibrators are rarely constructed from discrete components. Most are constructed from integrated circuits that replace both transistors and most of the resistors. Only the timing resistors and capacitors are external to the circuit. This makes sense when you consider that the timing will be different for different kinds of applications, and that we need to have direct access to the timing components. Figure 30–17 shows a monostable multivibrator circuit that uses a 555 timer integrated circuit.

Let's point out the sequence of events that transpires during a complete cycle of monostable operation.

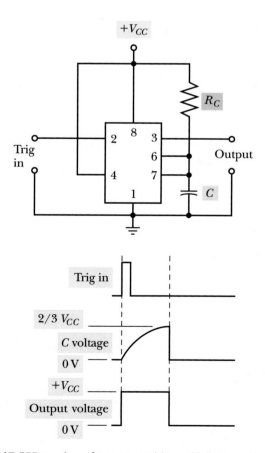

FIGURE 30–17 555 version of a monostable multivibrator circuit and waveforms

1. The quiescent state is the standby state. In this state, IC pin 7 is internally connected to ground so that capacitor C is held fully discharged in spite of the path to $+V_{CC}$ through R_C. Also, the output is pulled down very close to 0 V. The circuit remains in this state until the positive-going edge of a pulse occurs at the *Trig in* connection.

2. The active phase begins when the *Trig in* terminal at IC pin 2 sees the positive-going edge of a trigger pulse waveform. This positive transition immediately opens the internal ground connection at IC pin 7 and allows capacitor C to begin charging through R_C. Also, the triggering action switches the output to a voltage that is very close to $+V_{CC}$, where it remains until capacitor C is discharged again.

3. The timing interval ends when IC pin 6 senses that the voltage charge across capacitor C has reached 2/3 the amount of $+V_{CC}$. At that instant, IC pin 7 is internally shorted to ground. This shorting action immediately returns the output to 0 V and discharges capacitor C to ground. The circuit is now in its quiescent state until another trigger pulse occurs.

The timing interval of a 555-type monostable multivibrator is easily determined by Formula 30–10. In this formula, the output timing interval is in seconds, R_C is the value of the timing resistor, and C is the value of the timing capacitor.

FORMULA 30–10 $t_w = 1.1R_C C$

Example

What is the time of the active state of a 555-type monostable multivibrator, Figure 30–17, when $R_C = 1$ MΩ and $C = 10$ μF?

Answer:

From Formula 30–10:

$$t_w = 1.1R_C C$$

$$t_w = 1.1 \times 1 \text{ M}\Omega \times 10 \text{ }\mu\text{F} = 11 \text{ sec}$$

Astable multivibrators

An **astable multivibrator** is a type of RC oscillator that generates square or rectangular waveforms. The term **astable** means not stable. This type circuit has a no stable state, and it produces its output as a result of continually seeking a stable state. The frequency of oscillation is determined by the charge time of the RC networks. Astable multivibrators are also known as **free-running multivibrators** and/or relaxation oscillators.

A BJT version of an astable multivibrator is shown in Figure 30–18. Let's see how the circuit works, especially with regard to the fact that one of the two transistors is always in saturation while the other is in cutoff. The operation of the circuit is based on the saturation and cutoff states alternating between the two transistors.

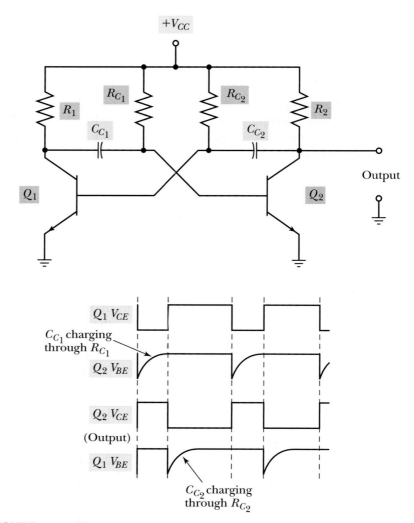

FIGURE 30–18 Discrete-component astable multivibrator circuit and waveforms

1. The active state for Q_1 begins when it is switched on. (How this happens will be shown in a moment.) As Q_1 is switched from cutoff (off) to saturation (on), its collector voltage drops from $+V_{CC}$ to about 0.7 V. This negative-going transition at the collector of Q_1 is coupled through capacitor C_{C_1} as a negative polarity to the base of Q_2. This reverse-bias voltage at the base of Q_2 turns it off.

2. During the active state of Q_1, capacitor C_{C_1} is charging through a path that begins with the ground connection to the emitter of Q_1, through the emitter and collector of Q_1, through C_{C_1} and R_{C_1} to $+V_{CC}$. The rate of charge of capacitor C_{C_1} depends on its own value and the value of resistor R_{C_1}. The RC time constant of R_{C_1} and C_{C_1} determines how long the active state of Q_1 persists.

3. The active state for Q_2 begins (and the active state for Q_1 ends) when C_{C_1} becomes charged through the path described in Step 2. When this happens, there is no longer a negative bias applied to the base of Q_2, so Q_2 begins to

conduct. Q_2 thus switches from cutoff to a state of saturation. Its collector voltage drops from $+V_{CC}$ to about 0.7 V, and this negative-going transition is coupled through capacitor C_{C_2} as a negative polarity to the base of Q_1. This action biases Q_1 off.

4. Q_2 remains in its active state for a period determined by the RC time constant of the circuit composed of C_{C_2} and R_{C_2}. When C_{C_2} becomes fully charged, the negative bias (due to the charging of C_{C_2}) disappears from the base of Q_1, and Q_1 is allowed to switch on. As Q_1 is switched on, this cycle returns to Step 1, where the switching-on action of Q_1 switches off Q_2.

The action described in these four states represents the operation of an astable multivibrator. The frequency of operation (or oscillating frequency) of this circuit depends on the RC time constant of R_{C_1} and C_{C_1}, and the RC time constant of R_{C_2} and C_{C_2}. The general formula for calculating either one of these time intervals is shown in Formula 30–11, where R is the value of the resistor and C is the value of the capacitor.

FORMULA 30–11 $t = 0.7RC$

As part of the procedure for determining the operating frequency of an astable multivibrator, you must solve Formula 30–11 for *both RC* paths in the circuit. If desired, you can plot the rectangular waveforms found at the output of the multivibrator.

Example

Referring to the circuit in Figure 30–18, determine the active time intervals for Q_1 and Q_2, and plot the waveform at the collector of Q_2 when:

$$R_{C_1} = 10 \text{ k}\Omega, \ C_{C_1} = 0.01 \ \mu\text{F}$$
$$R_{C_2} = 22 \text{ k}\Omega, \ C_{C_2} = 0.001 \ \mu\text{F}$$

Answer:

The active time for Q_1 is determined by the values of R_{C_1} and C_{C_1}. From Formula 30–11:

$$t_1 = 0.7R_{C_1}C_{C_1} = 0.7(10 \text{ k}\Omega)(0.01 \ \mu\text{F})$$
$$t_1 = 70 \ \mu\text{sec}$$

The active time for Q_2 is determined by the values of R_{C_2} and C_{C_2}. From Formula 30–11:

$$t_2 = 0.7R_{C_2}C_{C_2} = 0.7(22 \text{ k}\Omega)(0.001 \ \mu\text{F})$$
$$t_2 = 15 \ \mu\text{sec}$$

The waveform shown in the example represents the output taken from the collector of Q_2. The output of an astable multivibrator can also be taken from the collector of the opposite transistor, Q_1. The waveforms from either Q_1 or Q_2 are the same except they are inverted. That is, when one output is near 0 V, the other is near $+V_{CC}$.

The waveform in the example also shows that the output of an astable multivibrator is not necessarily symmetrical—the time interval of one of the states is not necessarily equal to the time of the opposite state. In this example, one interval is 70 μsec and the other is only 15 μsec.

Once you know the time interval of the two states, you can sum the intervals to arrive at the total period of the waveform: $T = t_1 + t_2$. And once you know the total period of the waveform, you can calculate frequency by the familiar period-to-frequency formula: $f = 1/T$.

Example

What is the frequency of oscillation of an astable multivibrator that has an output waveform with states of 100 μsec and 250 μsec?

Answer:

The total period is given by:

$$T = t_1 + t_2 = 100 \ \mu\text{sec} + 250 \ \mu\text{sec} = 350 \ \mu\text{sec}$$

The frequency is given by:

$$f = 1/T = 1/350 \ \mu\text{sec} = 2.86 \text{ kHz}$$

When the RC circuits in an astable multivibrator contain identical values of resistance and capacitance, the output waveform is symmetrical. This means that the two states have the same interval. This is often the case in actual practice, and the procedure for calculating the operating frequency is much simpler as indicated by Formula 30–12. In this formula, it is assumed that $R_{C_1} = R_{C_2} = R_C$, and $C_{C_1} = C_{C_2} = C$.

FORMULA 30–12 $f = 1/1.38R_C C$

Example

What is the frequency of oscillation of the astable multivibrator in Figure 30–18 if $C_{C_1} = C_{C_2} = 0.01 \ \mu\text{F}$ and $R_{C_1} = R_{C_2} = 47 \text{ k}\Omega$?

Answer:

From Formula 30–12:

$$f = 1/1.38R_C C$$

$$f = 1/(1.38)(47 \text{ k}\Omega)(0.01 \ \mu\text{F}) = 1.54 \text{ kHz}$$

Just as there is a simpler, integrated-circuit (IC) version of the monostable multivibrator, Figure 30–17 (page 996), there is also a simpler, IC version of the astable multivibrator. In fact both versions can be designed around the same IC device—the 555 timer IC. See the 555-type astable multivibrator in Figure 30–19.

1. The active state begins when the output is switched to its $+V_{CC}$ level. The instant this happens, capacitor C begins charging toward the $+V_{CC}$ level through resistors R_A and R_B.

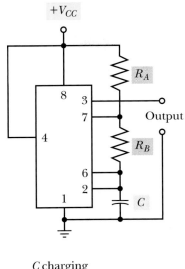

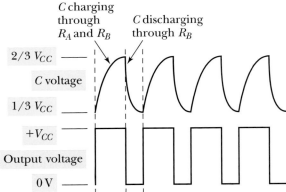

FIGURE 30–19 555 version of an astable multivibrator circuit and waveforms

2. During the active state of the output, capacitor C continues to charge until pin 6 senses a capacitor voltage that is $2/3$ times $+V_{CC}$. This marks the end of the active state.

3. At the end of the active state, the output voltage drops to zero. Also pin 7 is internally connected to ground, enabling the capacitor to begin discharging through resistor R_B.

4. The circuit remains in this zero-output state until pin 6 senses the capacitor voltage has discharged to $1/3$ times $+V_{CC}$. At that time, the output switches to its $+V_{CC}$ level, the internal ground connection to pin 7 is switched open, and capacitor C is again allowed to charge toward $+V_{CC}$ through resistors R_A and R_B.

The frequency of operation of this circuit depends on the RC time constant of $R_A + R_B$ and C during the charge (active) time, and the RC time constant of R_B and C during the discharge (zero-output) time. The formula for calculating the frequency of oscillation is given in Formula 30–13.

FORMULA 30–13 $f = 1/0.69C(R_A + 2R_B)$

Example

What is the frequency of operation of the 555-type astable multivibrator in Figure 30–19 if $R_A = 10$ kΩ, $R_B = 47$ kΩ, and $C = 0.001$ μF?

Answer:

From Formula 30–13:

$$f = 1/0.69C(R_A + 2R_B)$$

$$f = 1/0.69(0.001 \ \mu\text{F})[10 \ \text{k}\Omega + 2(47 \ \text{k}\Omega)] = 13.9 \ \text{kHz}$$

SUMMARY

▶ Four elements or conditions are absolutely required to begin and sustain the operation of an oscillator: a power source, frequency determining device or components, amplification greater than 1, and positive feedback.

▶ The Hartley oscillator is an *LC* sine-wave oscillator that has a tank circuit comprised of a capacitor and a tapped inductor (or two inductors in series) that is grounded at the tap (or common connection between the two inductors). The total inductance of the Hartley tank circuit is equal to the sum of the two inductance values.

▶ The Colpitts oscillator is an *LC* sine-wave oscillator that has a tank circuit comprised of a single inductor and two series-connected capacitors that are grounded at their common connection. The total capacitance of the Colpitts tank circuit is found by the product-over-sum rule as applied to the values of the two capacitors.

▶ Because of undesirable loading effects, the output of a Colpitts *LC* oscillator is often transformer coupled to the next stage. The primary winding of the coupling transformer serves as the inductance for the tank circuit.

▶ The Clapp oscillator is an *LC* sine-wave oscillator that is identical with the Colpitts oscillator except for the existence of three capacitors in the tank circuit. The extra capacitor provides greater stability against changes in the output loading. The total capacitance in the Clapp tank circuit is found by the formula used for the total capacitance of three capacitors in series.

▶ One way to remember the difference between Hartley and Colpitts oscillators is to remember: Hartley = H = Henrys (two inductance values), and Colpitts = C = Capacitance (two capacitance values).

▶ The frequency of a crystal-controlled oscillator is fixed by the mechanical shape and size of the crystal. The crystal is the equivalent of a parallel tuned (*LC* tank) circuit.

▶ The phase-shift oscillator is an *RC* sine-wave oscillator that uses a phase-shift network comprised of three *RC* elements to produce a 180° phase shift at one particular frequency. When capacitors are connected in series with the feedback path, the feedback voltage leads the output of the amplifier. When the resistors are connected in series with the feedback path, the feedback voltage lags the output of the amplifier.

▶ The Wien-bridge oscillator is an *RC* sine-wave oscillator that uses a lead-lag network comprised of two *RC* elements (one in series and one in parallel) to produce a 0° phase shift at the circuit's oscillating frequency. The amplifier is designed for noninverting operation. Thus the overall feedback at the oscillating frequency remains at 0°—as required for starting and sustaining oscillation.

▶ Multivibrators are characterized by generating rectangular waveforms and using only *RC* circuits for setting the timing or frequency of oscillation.

▶ A monostable multivibrator (also called a one-shot multivibrator) generates a single, *RC*-timed, output pulse in response to a single input trigger pulse.

▶ An astable multivibrator (also called a free-running multivibrator) generates a continuous rectangular waveform whose active and inactive times are determined by the charge times of capacitors in the circuit.

FORMULAS AND SAMPLE CALCULATOR SEQUENCES

FORMULA 30–1 $f_r = 1/2\pi\sqrt{LC}$

L value, [×], C value, [=], [√], [×], 6.28, [=], [1/x]

FORMULA 30–2 $f_r = 1/2\pi\sqrt{L_t C}$

(Use the keystrokes for Formula 30–1)

FORMULA 30–3 $L_t = L_{1_A} + L_{1_B}$

L_{1_A} value, [+], L_{1_B} value, [=]

FORMULA 30–4 $f_r = 1/2\pi\sqrt{LC_T}$

(Use the keystrokes for Formula 30–1)

FORMULA 30–5 $C_T = C_1 C_2/(C_1 + C_2)$

C_1 value, [×], C_2 value, [÷], [(], C_1 value, [+], C_2 value, [)], [=]

FORMULA 30–6 $C_T = 1/(1/C_1 + 1/C_2 + 1/C_3)$

C_1 value, [1/x], [+], C_2 value, [1/x], [+], C_3 value, [1/x], [=], [1/x]

FORMULA 30–7 $f_r = 1/2\pi RC\sqrt{6}$

6, [√], [×], 6.28, [×], R value, [×], C value, [=], [1/x]

FORMULA 30–8 $f_r = 1/2\pi RC$

R value, [×], C value, [×], 6.28, [=], [1/x]

FORMULA 30–9 $t_{\text{on}} = 0.69 R_C C$

0.69, ✕, R_C value, ✕, C value, =

FORMULA 30–10 $t_w = 1.1 R_C C$

1.1, ✕, R_C value, ✕, C value, =

FORMULA 30–11 $t = 0.7 R C$

0.7, ✕, R value, ✕, C value, =

FORMULA 30–12 $f = 1/1.38 R_C C$

1.38, ✕, R_C value, ✕, C value, =, 1/x

FORMULA 30–13 $f = 1/0.69 C (R_A + 2R_B)$

(, 2, ✕, R_B, +, R_A value,), ✕, 0.69, ✕, C value, =, 1/x

REVIEW QUESTIONS

1. List the four elements and conditions that are absolutely required to begin and sustain the operation of an oscillator.

2. Write the main formula that applies for finding the frequency of oscillation of all LC-type oscillators. Specify the meaning of each term.

3. Which type of LC oscillator uses a tapped coil? Two capacitors with a ground connection between them? Three capacitors and a single inductor in the tank circuit?

4. Draw the tank circuit portion of the following:
   a. A Hartley oscillator
   b. A Clapp oscillator
   c. A Colpitts oscillator

5. Suppose the components for the tank circuit of the Hartley oscillator you drew in question 4 have the following values: $L_1 = 10$ mH, $L_2 = 10$ mH, and $C = 0.1$ μF. What is the frequency of oscillation?

6. Assign values of your own choosing to the components for the tank circuit of the Clapp oscillator you drew in question 4. Calculate the frequency of oscillation. (Be sure to show all your work.)

7. Suppose the components for the tank circuit of the Colpitts oscillator you drew in question 4 has the following values: $L = 10$ mH, $C_1 = 0.1$ μF, and $C_2 = 0.1$ μF. What is the frequency of oscillation?

8. In any of the LC sine-wave oscillators studied in this chapter, you can increase the frequency of operation by _____ the value(s) of inductance or by _____ the value(s) of capacitance.

9. Which type of RC sine-wave oscillator studied in this chapter uses a network of three identical RC elements? Which uses a lead-lag network?

10. What element(s) of a crystal-controlled oscillator determine(s) its frequency of oscillation?

11. At the frequency of oscillation, the RC network in a phase-shift oscillator produces a phase shift of _____°, and the amplifier produces a phase shift of _____°.

12. Sketch the frequency determining RC network for (a) a phase-shift oscillator and (b) a Wien-bridge oscillator.

13. The capacitors in the RC network of a Wien-bridge oscillator have a value of 1 μF and the resistors have a value of 1 kΩ. What is the frequency of oscillation?

14. What is another common name for a monostable multivibrator?

15. In order to decrease the active time of a monostable multivibrator, you must _____ the value of the timing capacitor and/or _____ the value the timing resistors.

16. What is another common name for an astable multivibrator?

17. What is the frequency of oscillation of an astable multivibrator that produces a waveform near $+V_{CC}$ for 10 μsec and near 0 V for 8 μsec?

18. Explain why the $+V_{CC}$ time of the output waveform of a 555-type astable multivibrator is always longer than the 0-V time.

19. In a 555-type astable multivibrator, $R_A = 10$ kΩ, $R_B = 10$ kΩ, and $C = 0.1$ μF. What is the frequency of oscillation?

20. In order to increase the frequency of operation of the 555-type astable multivibrator, you must _____ the value of the timing capacitor, or you can _____ the value of either or both of the timing resistors.

PROBLEMS

1. Determine the resonant frequency of a parallel LC circuit where $L = 1$ mH and $C = 1$ nF.

2. A certain tank circuit is to resonate at 10.2 MHz. If you already know that $L = 20$ μH, determine the required value of C.

3. A BJT Harley oscillator (see Figure 30–3) uses inductances of 90 μH and 40 μH. If the capacitor in the tank circuit is 500 μF, what is the operating frequency of the circuit?

4. Determine the frequency of a Hartley oscillator that uses a tank circuit composed of a 1-mH center-tapped coil and a 100-nF capacitor.

5. An op-amp Colpitts oscillator (see Figure 30–7) uses capacitances of 90 pF and 40 pF. If the inductor in the tank circuit is 5 μH, what is the operating frequency of the circuit?

6. Referring to the FET Colpitts oscillator to Figure 30–8, replace C_2 with a variable capacitor that can be adjusted between 10 pF and 90 pF. If C_1 is fixed at 400 pF and the primary inductance of T_1 is 10 μH, calculate the lowest and highest operating frequencies for the circuit.

7. All three capacitors in the tank circuit for a Clapp oscillator (see Figure 30–9) are listed as 22 nF. What is the value for C_T?

8. If the operating frequency of a Clapp oscillator is to be 1.1 MHz when the inductor in the tank circuit is rated at 120 μH, what is the necessary value for C_T?

9. Calculate the frequency of a phase-shift oscillator (see Figure 30–14) where all the capacitors are rated as 1 μF and all the resistors at 1 MΩ.

10. A certain Wien-bridge oscillator (see Figure 30–15) is operating a 100 kHz when the timing resistor R_2 is equal to 1 MΩ. What is the value of the timing capacitor C_2?

ANALYSIS QUESTIONS

1. Draw a JFET and an op-amp version of the Clapp oscillator.

2. Research and briefly explain the operation of the Armstrong oscillator circuit.

3. Explain why the amplifier element of a phase-shift oscillator must be an inverting amplifier, while the amplifier element in a Wien-bridge oscillator must be a noninverting amplifier.

4. Research and draw the schematic diagram for a twin-T *RC* oscillator.

5. Explain why the amplifier element in a phase-shift oscillator must have a voltage gain that is greater than those for the other types of oscillators described in this lesson.

6. Research and explain the meanings of *interruptible* and *noninterruptible* monostable multivibrator.

7. Research and describe the function of a bistable multivibrator, or flip-flop.

8. Explain why the charge time for the timing capacitor of a 555 astable multivibrator (see Figure 30–19) is always longer than the discharge time. Further explain how this accounts for the fact that the time the output voltage is near $+V_{CC}$ is always longer than the time it is near 0 V.

9. Describe what happens to the operating frequency of a 555 astable multivibrator when the value of the timing capacitor is increased and when it is decreased.

10. Explain why the timing for 555 monostable and astable multivibrators is independent of the power supply voltage.

PERFORMANCE PROJECTS CORRELATION CHART

CHAPTER TOPIC	PERFORMANCE PROJECT	PROJECT NUMBER
Wien-Bridge Oscillator	Wien-Bridge Oscillator Circuit	86
Astable Multivibrators	555 Astable Multivibrator	87

NOTE: It is suggested that after completing the above projects, the student should be required to answer the questions in the "Summary" at the end of this section of projects in the Laboratory Manual.

THYRISTORS

31

CHAPTER PREVIEW

A **thyristor** is a four-layer semiconductor device that has only two states of operation: fully conducting or fully non-conducting. Recall that a junction diode is composed of two layers, P-N. Thyristors, with their four layers, are like two diodes connected in series: P-N-P-N. (It is important to realize, however, that two junction diodes connected in series cannot be made to perform like a thyristor.)

There are different types of thyristors that respond differently with regard to the ways they are turned on and off. The thyristors discussed in this chapter are the silicon-controlled rectifier (SCR), diac, and triac. Compared with the other semiconductor devices you have studied in previous chapters, thyristors are built for high current and voltage ratings. Whereas a typical BJT could have a maximum collector current of 100 mA at 30 V, a typical SCR could handle 5 or 10 A at 220 V.

Thyristors are found in consumer electronic devices, most notably in wall switch light dimmers and automobile electronic ignition systems. By far the greatest quantity and variety of thyristors, however, are used in industrial power control equipment.

OBJECTIVES

After studying this chapter, you should be able to:

1. Describe what a **thyristor** is

2. Describe in detail the way an **SCR** can be switched on and off

3. Explain the operation of simple SCR circuits including a power "on/off" push-button control circuit and an electronic "crowbar"

4. Identify the symbols for and describe the operation of the **gate-controlled switch, silicon-controlled switch,** and **light-activated SCR**

5. Identify and explain the purpose of thyristors connected in inverse parallel, or back-to-back

6. Identify the schematic symbol and explain the operation of a **diac**

7. Explain the details for starting and ending the conduction of a **triac**

8. Identify phase-control power circuits that use thyristors

9. Describe basic troubleshooting procedures for thyristors

31–1 SILICON-CONTROLLED RECTIFIERS (SCRS)

The **silicon-controlled rectifier (SCR)** is a simple four-layer **thyristor** device. The semiconductor structure shown in Figure 31–1 indicates the four layers and shows the standard schematic symbol. The terminals are called cathode *(K)*, anode *(A)*, and gate *(G)*.

If the gate terminal is left open or shorted directly to the cathode, the device cannot conduct current between the cathode and anode. The reason is that at least one of the three P-N junctions will be reverse biased, no matter which polarity of voltage you apply between the cathode and anode. The SCR can be made to conduct, however, by meeting two conditions:

1. A voltage is applied between the cathode and anode such that the anode is positive compared to the cathode.

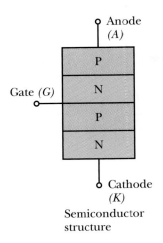

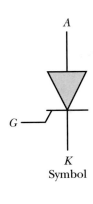

FIGURE 31–1 SCR
semiconductor structure and
schematic symbol

2. A voltage is applied to the gate such that the P-N junction between the gate and cathode is forward biased (at least momentarily).

Once an SCR begins conducting under the conditions just described, the only way to turn it off is by removing the voltage applied between the anode and cathode, or by reverse biasing the anode-cathode connection. You cannot turn off a conducting SCR by removing the gate voltage, and you cannot turn off an SCR by reversing the voltage applied to the gate. The gate of an SCR is intended to initiate conduction between the cathode and anode. Once this condition begins, the gate plays no further role in the operation.

Theory of SCR operation

The diagrams in Figure 31–2 should help you understand and remember the conditions for turning an SCR on and off. The four layers have been drawn to clarify the fact that they can be regarded as a set of two BJTs. The equivalent BJT that includes the anode and gate connections is a PNP transistor shown as Q_1 on

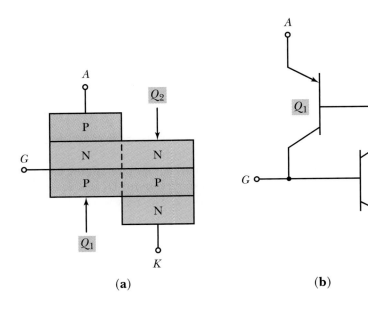

FIGURE 31–2 SCR as a set
of two BJTs:
(a) semiconductor structure;
(b) equivalent BJT circuit

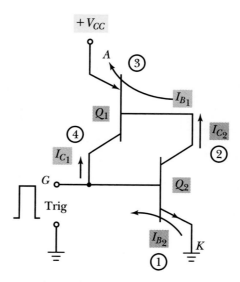

FIGURE 31–3 Turning on an SCR

the schematic. The equivalent BJT that includes the cathode is an NPN transistor shown as Q_2 on the schematic.

With the equivalent BJT circuit in Figure 31–2, it is impossible for just one BJT to conduct. Either they both conduct or neither one conducts. This is because forward-biasing base current for Q_1 cannot flow (to turn it on) unless Q_2 is turned on. The collector current for Q_2 provides the base current for Q_1. If Q_2 is *not* conducting, there is no base current for Q_1, so it cannot be conducting either.

Now suppose a positive pulse from an outside voltage source is applied to the gate terminal. The equivalent circuit in Figure 31–3 indicates the sequence of events that takes place. Observe the numbers (indicating the sequence of events) and paths for current flow:

Step 1. The positive pulse at the gate terminal causes forward-biasing base current (I_{B_2}) to flow at BJT Q_2.

Step 2. The forward-biasing base current for Q_2 turns on that transistor and causes collector current (I_{C_2}) to flow.

Step 3. The collector current for Q_2 provides forward-biasing base current (I_{B_1}) for BJT Q_1.

Step 4. The forward-biasing base current for Q_1 turns on that transistor and causes collector current (I_{C_1}) to flow.

By the time the process reaches Step 4, both transistors are in saturation. Even after the trigger potential at the gate terminal returns to zero, the transistors remain latched on. This is ensured by collector current I_{C_1} maintaining base current for Q_2. For all practical purposes, the collector current for Q_1 acts as a steady gating pulse that keeps both transistors latched on. The only way to interrupt this latching effect is by removing the supply voltage or reversing its polarity.

Figure 31–4 shows a simple SCR demonstration circuit. It indicates how an SCR can be latched into conduction and turned off by interrupting the source of power to it. The power source is a 12-V battery, the load to be controlled is a 12-V automobile lamp. The normally closed "off" push-button switch is con-

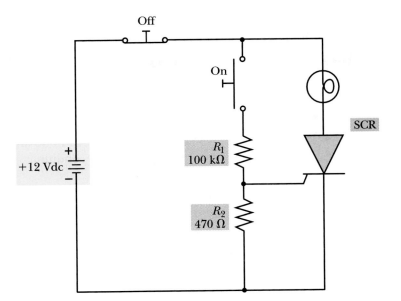

FIGURE 31–4 Simple SCR power "on/off" circuit

nected in series between the positive side of the dc power source and anode of the SCR. When this button is depressed, power is totally disconnected from the circuit. The normally open "on" push-button switch is connected in series with the positive side of the power source and the gate of the SCR. Whenever the "on" push button is depressed, a positive bias is applied to the gate of the SCR. This circuit is thus turned on by momentarily pressing the "on" button; and it is turned off by depressing the "off" button.

The most useful ratings for an SCR are listed here:

- *Peak forward blocking voltage:* the largest amount of forward-bias voltage the SCR can hold off without breaking down and conducting. The range of available values is between 30 V and 1,200 V.

- *Maximum forward current:* the largest amount of current, dc or rms, that the SCR can handle continuously without special heat sinking. The range of available values is between 0.8 A and 100 A, Figure 31–5.

FIGURE 31–5 80-ampere stud-mounted SCRs

- *Peak reverse voltage:* the largest amount of reverse voltage that can be applied between the cathode and anode without breaking into conduction. The value of the peak reverse voltage is often the same as the peak forward blocking voltage.

- *Holding current:* the smallest amount of cathode-anode forward current that can sustain conduction of the SCR. This value is on the order of 5 mA to 20 mA.

- *Forward "on" voltage:* the forward voltage drop between cathode and anode while the SCR is conducting. Under normal operating conditions, this voltage is on the order of 1 or 2 V.

There are also current and voltage ratings for the gate circuit, including forward cathode-gate voltage, maximum forward gate current, reverse breakdown voltage for the cathode-gate junction, and reverse gate leakage current.

Now that you have seen the wide range of SCR ratings, let's look at some further examples of SCR circuits.

SCR circuits

The circuit in Figure 31–6 provides an effective means for protecting sensitive electronic components (such as computer memory circuits) from sudden, unexpected surges of high voltage from a dc power supply. The power supply in this example is designed to provide +12 Vdc to the circuit being protected. Note that the zener diode is rated at 14 V, so you can realize that the zener does not normally conduct. In fact, the zener will not conduct at all unless the dc power supply voltage rises above 14 V (which occurs from a surge of high voltage). As long as the zener diode does not conduct, the SCR cannot conduct because the zener current is necessary for gating the SCR. But the moment the zener diode conducts due to a surge of high voltage, the SCR is gated on.

Once the SCR is gated on, it acts as a short circuit across the dc power supply. This causes a surge of relatively high current which blows the fuse to remove the dc power supply from the circuit being protected. This circuit is known as an electronic *crowbar.*

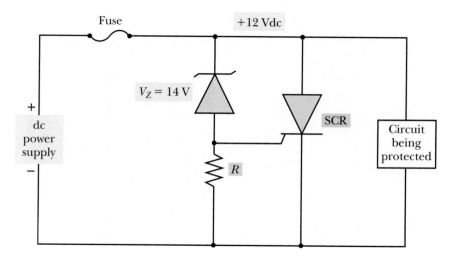

FIGURE 31–6 SCR used in a "crowbar" protection circuit

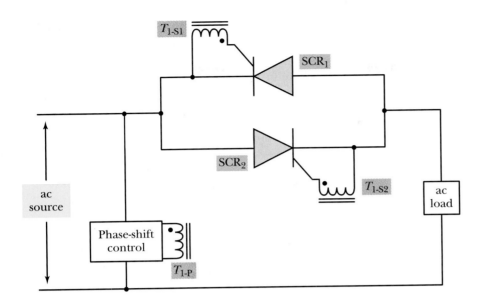

FIGURE 31–7 AC power control circuit using two SCRs connected in inverse parallel

The circuit in Figure 31–7 represents the most common application of SCRs. Such a circuit is used for smoothly adjusting the amount of ac power that is applied to a wide range of devices. For lamps, the circuit can be used to control their intensity from off to full brightness; for ac motors, the circuit can be used to control their speed; and for heating elements, the circuit can be used to control their amount of heating.

Before getting into the details of operation of this power control circuit, let's look at a few of its general features. First notice that the circuit is operated from an ac source. The ac power source is usually 120 Vac or 240 Vac as supplied by the local power utility. The ac load is the device that is to be operated from the ac source. Normally, the ac load runs at full power (light, speed, heat, etc.) at the rated ac voltage level; so an SCR control circuit of this type is expected to reduce the power level from the full-power level down to some other desired level.

Next, notice how the two SCRs (SCR$_1$ and SCR$_2$) are connected with respect to one another. Note that the cathode of one is connected to the anode of the other. This circuit arrangement is known as **inverse parallel** or back-to-back. Using two SCRs in inverse parallel is necessary in an ac power control circuit because it enables ac power to reach the load. If only one SCR is used, only half-wave pulsating dc could reach the load. By using the inverse parallel connection, SCR$_1$ can pass current to the load on one half-cycle, and SCR$_2$ can pass current to the load on the opposite half-cycle. The SCR that is switched on will be turned off at the end of its forward-biasing half-cycle because its anode voltage goes through zero and to the reverse-biasing polarity.

Finally, notice the arrangement of windings for transformer T_1. The primary of this transformer (T_{1-P}) is shown connected to the output of a phase-shift control circuit. The transformer has two separate secondaries, which are T_{1-S1} at the gate of SCR$_1$, and T_{1-S2} at the gate of SCR$_2$. The phasing dots on the transformer windings indicate the same phase of signal is applied to both SCRs.

Details of the phase-shift control circuit are not shown on the schematic diagram. It is sufficient for our purposes to describe the signal that is applied to T_{1-P}. The signal from the phase-shift control circuit consists of positive pulses that occur at twice the frequency of the ac source. If the ac source is 60 Hz, for example, the positive pulses from the phase-shift control circuit occur at 120 Hz. These pulses have a fixed amplitude (about 3 V at the secondary windings) and are applied to the gate of both SCRs. The occurrence of a gate pulse turns on the SCR that happens to be forward biased between its cathode and anode. The next gate pulse will occur on the next ac half-cycle and thus switch on the opposite SCR. The effect is that the SCRs conduct alternately.

Now you are ready to consider the phase-control waveforms shown in Figure 31–8. The trigger waveforms from the phase-control circuit are delayed by a vari-

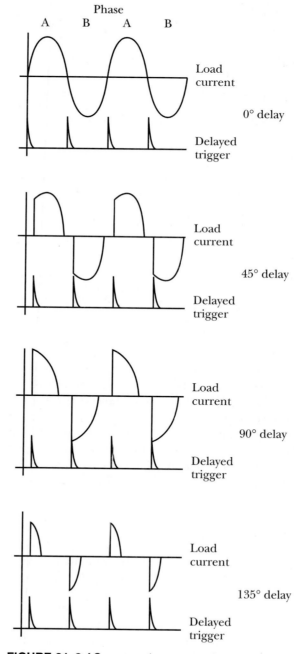

FIGURE 31–8 AC power phase control waveforms

able amount compared with the applied ac waveform. The longer the delay, the later in each half-cycle the SCRs are turned on and the lower the amount of power applied to the ac load. The amount of delay is measured in terms of degrees of the sinusoidal waveform. When the trigger pulses occur at 0°, for instance, there is no delay and the SCRs are triggered at the beginning of each ac half-cycle (maximum ac power to the load). When the trigger pulses are delayed 45°, however, the SCRs are not gated on until 45° into each half-cycle (about one quarter power applied to the load). And at a 90° delay, the SCRs conduct through the last half of each half-cycle. If you delayed to 180°, the SCRs would be gated at the end of each half-cycle, and no power would be applied to the ac load.

Other types of SCRs

The SCRs you have just studied are the most commonly used types. There are several other types of SCRs that are not quite as popular but that possess some features that are advantageous in certain applications.

Gate-controlled switch (GCS)

The **gate-controlled switch** is very similar to an SCR except that a GCS can be gated off as well as gated on. Recall the only way to stop the conduction of an SCR is by removing or reversing the voltage applied between the cathode and anode. A GCS can also be turned off by applying a reverse-bias voltage across the gate-cathode P-N junction. See Figure 31–9.

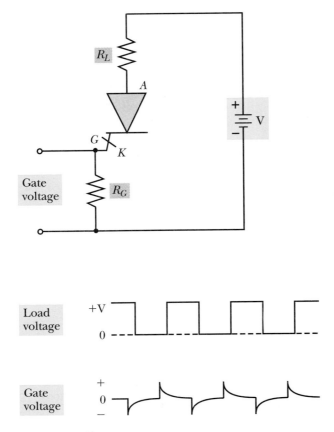

FIGURE 31–9 Gate-controlled switch (GCS) basic operation

The two conditions that must be met for switching on a GCS are the same as for an SCR:

1. A voltage is applied between the cathode and anode such that the anode is positive compared with the cathode.
2. A voltage is applied to the gate such that the P-N junction between the gate and cathode is forward biased (at least momentarily).

A GCS, however, can be turned off in two different ways:

1. The voltage applied between the cathode and anode is removed or reversed so that the cathode is positive with respect to the anode.
2. A voltage is applied to the gate such that the P-N junction between the gate and cathode is reverse biased (at least momentarily).

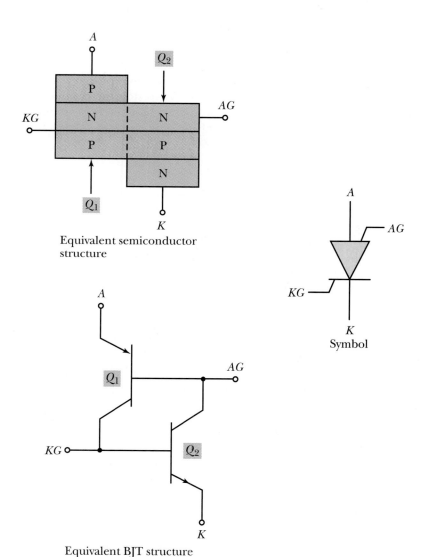

Equivalent semiconductor structure

Symbol

Equivalent BJT structure

FIGURE 31–10 Silicon-controlled switch (SCS) semiconductor structure, symbol, and equivalent BJT circuit

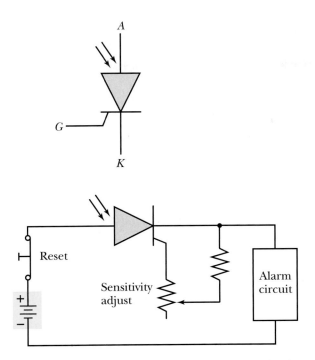

FIGURE 31–11 Light-activated SCR (LASCR) and alarm circuit application

The example circuit in Figure 31–9 is a chopper circuit that changes a dc current into a rectangular, or pulsating dc, waveform.

Silicon-controlled switch (SCS)

The **silicon-controlled switch** can be compared to an SCR that happens to have two gate terminals. Compare the equivalent BJT circuit for an SCS in Figure 31–10 to the SCR version in Figure 31–2. You can see the difference is that only one of the transistors for an SCR can be used for gating the device on, while either (or both) of the transistors for an SCS can be used for gating the device on.

Light-activated SCR (LASCR)

The **light-activated SCR** can be gated by applying a positive voltage to the gate terminal or by directing light through a lens. Once the LASCR is conducting, it is turned off only by removing the anode voltage or reversing it. See the symbol and demonstration circuit in Figure 31–11.

In the demonstration circuit, light falling onto the LASCR causes it to switch to its conducting state. Dc power is thus applied to the alarm circuit. The circuit is reset (switched off) by momentarily depressing the normally closed reset push button.

The greatest amount of sensitivity to light occurs when there is no connection to the gate terminal. The variable resistor in the diagram serves as a sensitivity adjustment. The greater the amount of resistance, the greater the sensitivity to light. Setting the variable resistor for minimum resistance provides minimum sensitivity.

IN-PROCESS LEARNING CHECK I

Fill in the blanks as appropriate.

1. A thyristor is a _____-layer device that has just _____ operating states.

2. An SCR is gated on by applying a _____-bias voltage to the _____ P-N junction. A _____ can also be gated on by shining a light onto its lens.

3. An SCR will conduct when it is gated and a voltage is applied between the cathode and anode such that the anode polarity is _____ and the cathode is _____.

4. The SCR rating that specifies the maximum forward voltage an SCR can handle is called the _____.

5. The SCR rating that specifies the maximum reverse voltage an SCR can handle is called the _____.

6. The smallest amount of cathode-anode forward current that can sustain conduction of an SCR is called the _____.

7. The _____ is a type of SCR that can be gated off as well as gated on.

8. The _____ is a type of SCR that has two gate terminals.

9. In order to control both cycles of ac power, two SCRs must be connected in reverse with respect to one another in an arrangement that is known as _____ or _____.

10. When a pair of SCRs are gated halfway through their respective half-cycles of the ac power waveform, the circuit is said to be firing at _____°.

31–2 DIACS

The **diac** (Figure 31–12) is another thyristor, or four-layer device. The semiconductor structure is identical with that of an SCR. The diac is different from an SCR in two important ways. First, the diac is a **bidirectional** device. This means that it conducts current in both directions. You can see this bidirectional nature by the symbol that resembles a pair of diodes connected in inverse parallel (back-to-back). The diac also differs from SCRs by having no gate terminal.

To understand exactly how a diac works, first consider two SCR ratings you have already studied: peak forward blocking voltage and peak reverse voltage. The peak forward blocking voltage is the largest amount of forward bias that can be applied to an SCR that is not gated on without the SCR breaking into conduction. The peak reverse voltage rating is the largest amount of reverse voltage that can be applied between the cathode and anode without breaking down the junctions. It is also important to recall that the forward blocking voltage and reverse breakdown voltage for an SCR are usually the same voltage levels. A diac has the same semiconductor structure as an SCR, so it follows that a diac also has forward blocking and reverse breakdown voltage ratings. In fact, the only way to get a diac to conduct is by exceeding its forward blocking and reverse breakdown ratings.

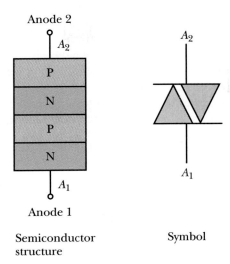

FIGURE 31–12 Diac semiconductor sturcture and symbol

Figure 31–13 shows a diac connected in series with an ac power source and resistive load. The voltage ratings for this version are 27 V. The waveforms show that no current flows through the circuit on either half-cycle until the applied voltage exceeds 27 V. The diac then remains conducting until the applied voltage waveform goes through zero. On one half-cycle, the diac's forward blocking voltage is exceeded; on the opposite half-cycle (and it conducts through the re-

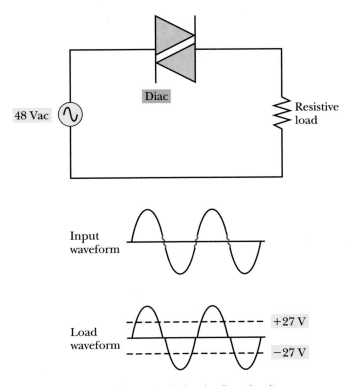

FIGURE 31–13 A simple diac circuit

mainder of the half-cycle), the diac's reverse breakdown voltage is exceeded (and it conducts through the remainder of the half-cycle).

The voltage ratings for diacs are typically between 25 V and 50 V. Current ratings are on the order of several amperes. Since diacs require only two terminals, they look much like rectifier diodes.

The most common commercial and industrial applications of diacs are as triggering devices for triacs as described in the next section.

31-3 TRIACS

The **triac** bidirectional thyristor works as a pair of SCRs connected in inverse parallel, Figure 31–14. A single gate terminal can switch on the triac in either direction. The triac was originally devised to replace sets of two inverse-parallel SCRs with a single device. It is said to be an "ac SCR," and the two conditions required for switching on a triac are as follows:

1. A voltage is applied between the cathode and anode (polarity is not important).
2. A forward-bias voltage is applied to the gate-cathode junction (at least momentarily).

A conducting triac is turned off by removing the anode voltage or reversing the present polarity applied to it.

Figure 31–15 shows one of the most popular thyristor circuits in the consumer marketplace. This circuit is typical of the circuits that are used for adjusting the brightness of the lights in a room. In this drawing, the lights would be the ac load. The ac source is the utility power supplied to the home. The diac (D_1) is connected to the gate of the triac (D_2)—the triac is gated each time the

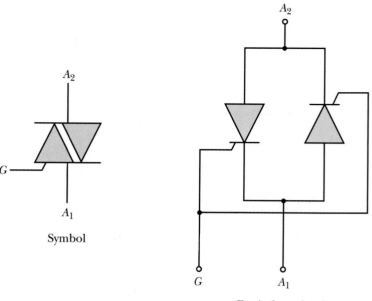

Symbol

Equivalent circuit

FIGURE 31–14 Triac symbol and equivalent SCR circuit

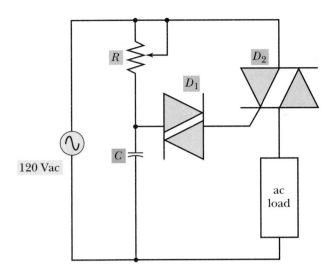

FIGURE 31–15 A light dimmer circuit using a diac and triac

diac breaks into conduction. The adjustable resistor (*R*) is the power control adjustment. Let's look at how this circuit works.

Resistor *R* and capacitor *C* form an ac phase-shift circuit. The ac waveform that is applied to D_1 is thus a phase-shifted, level-adjusted version of the ac source voltage. This means the diac will start conducting at a point on each half-cycle that can be varied from 0° to 180°. In turn, this means the triac is triggered into conduction at various points on each of its half-cycles, thereby enabling the control of power to the ac load, from minimum power to maximum power.

IN-PROCESS LEARNING CHECK II

Fill in the blanks as appropriate.

1. Thyristors that can conduct in two directions (such as a diac and triac) are said to be _____ devices.
2. A diac has _____ gate terminal(s), while a triac has _____ gate terminal(s).
3. The only way to get a diac to conduct is by exceeding its _____ blocking and reverse _____ ratings.
4. A triac operates like a pair of _____ connected back-to-back.
5. The most common commercial and industrial applications of _____ are as triggering devices for triacs.
6. In order to apply full ac power to a load, a triac should be fired at the _____° point on each half-cycle.

31–4 TROUBLESHOOTING THYRISTORS

When a thyristor fails, the trouble is usually due to one or more shorted P-N junctions. Most often, the short is between the anode and cathode, thereby allowing the device to conduct all the time. Also, gated thyristors (SCRs and triacs) can

FIGURE 31–16 Power thyristor mounted on a heat sink *(Photo courtesy of EG&G Wakefield Engineering.)*

fail due to a short circuit in one of the gate junctions. The way the device responds to a shorted gate depends on the nature of the short; sometimes the device is not gated on at all, sometimes it is gated on all the time, and sometimes it is gated on at the wrong time. Internal open circuits in a thyristor are fairly rare.

Thyristors are normally mounted on heat sinks to prevent over-heating, Figure 31–16.

One thing to bear in mind about thyristors is that they are by nature all-or-nothing devices. They are either switched completely on or switched completely off. So when you find a thyristor that is operating somewhere between cutoff and saturation, you may have found a shorted thyristor.

If possible, you should track down defective thyristors by noting their voltage waveforms while the circuit is turned on. (It is not always possible to check thyristors in a circuit that is turned on because a shorted thyristor might blow fuses and prevent you from making any meaningful oscilloscope checks.)

//

SAFETY HINTS

Thyristors (especially SCRs and triacs) are often used in electrical power control circuits. This means you can expect them to be handling lethal levels of current and voltage. Also, thyristors can be operated at temperatures high enough to cause painful burns when touched. So take special care when placing oscilloscope leads, test clips, etc., in an operating thyristor circuit.

When an SCR or triac is shorted, you will find low-level voltages between the anode and cathode, and nearly full power applied to the load. When observing SCRs connected in inverse parallel with an oscilloscope, consider that one shorted SCR will make the other appear shorted as well.

You should confirm a shorted SCR or triac by removing it from the circuit and checking the cathode-to-anode (SCR) or anode-to-anode (triac) resistance with an ohmmeter. A good SCR or triac will show high resistance in both directions (unlike a good diode that shows high resistance in one direction and low resistance in the other). A shorted SCR will indicate a low to medium resistance in both directions.

When an SCR or triac has a shorted gate, an oscilloscope check at the gate connection will show a very low gate signal level or no gate signal at all. Removed from the circuit and checked with an ohmmeter, the gate-cathode junction will test exactly like a rectifier diode. That is, a good junction will show high resistance in one direction and low resistance in the other. A shorted junction will show low to medium resistance in both directions.

SUMMARY

▶ A thyristor is a four-layer semiconductor device (P-N-P-N), which is either fully conducting or fully non-conducting. Three of the main types of thyristors are SCRs, diacs, and triacs. These thyristors generally have higher current and voltage ratings than other types of semiconductor devices.

▶ An SCR is a thyristor device that cannot conduct from cathode to anode until a positive voltage is applied to the gate. Once conduction begins, it cannot be stopped by removing or reversing the polarity of the gate voltage. The condition of an SCR can be stopped only by removing the voltage or reversing the cathode-anode voltage.

▶ The main ratings for SCRs are peak forward blocking voltage, maximum forward current, peak reverse voltage, holding current, and forward "on" voltage.

▶ A gate-controlled switch (GCS) is similar to an SCR except that a GCS can be gated off as well as gated on. The GCS is gated on by forward biasing the gate-cathode junction, and it is gated off by reverse biasing the gate-cathode junction. A GCS can also be turned off by reversing or eliminating

the voltage applied between the cathode and anode.

▶ A silicon-controlled switch (SCS) is an SCR with two separate gates. The SCS can be gated on by applying a positive gate signal to either or both gates. Like an ordinary SCR, however, the conduction of an SCS can be stopped only by removing or reversing the voltage applied between the cathode and anode.

▶ A light-activated SCR (LASCR) can be gated by a light source as well as a positive pulse to the gate.

▶ A diac is a bidirectional thyristor that has no gate terminals. It is turned on by exceeding the forward blocking or reverse breakdown voltage. Once conduction begins (in either direction), the diac is turned off by removing or reversing the polarity of the supply voltage.

▶ A triac functions like a pair of SCRs that are connected in inverse parallel (back-to-back). Triacs are designed mainly for phase-shift control of ac power.

▶ In an ac phase-shift power control circuit, the later the triac is fired in each half-cycle the lower the amount of power applied to the load.

REVIEW QUESTIONS

1. Specify the two possible states of operation of a thyristor.

2. Complete the following series of statements:
   a. A junction diode is a _____-layer device. An example is _____.
   b. A BJT is a _____-layer device. An example is _____.
   c. A thyristor is a _____-layer device. An example is _____.

3. Specify the conditions required for turning on an SCR, and describe how to turn it off.

4. Sketch the BJT equivalent circuit for an SCR and describe the process of gating the circuit on and show the paths for current that keep the circuit latched on.

5. Define *holding current* as it applies to thyristor devices, especially SCRs.

6. Define and compare the thyristor ratings of *peak forward blocking voltage* and *peak reverse voltage*.

7. Specify completely the meaning of the following abbreviations: GCS, LASCR, SCR, and SCS.

8. Demonstrate your understanding of inverse-parallel connections by drawing two junction diodes connected in inverse parallel. State another name for this type of connection.

9. Describe the most important difference between an SCR and a GCS.

10. Describe the most important differences between an SCR and an SCS.

11. Cite two different ways a LASCR can be switched on.

12. From the list of thyristors studied in this chapter (SCR, GCS, SCS, LASCR, diac, and triac), specify the following:
    a. Those having no gate terminal
    b. Those having one gate terminal
    c. Those having two gate terminals
    d. Those that are bidirectional

13. Define *bidirectional* in terms of current flow through electronic devices. Is a common resistor bidirectional? Is a junction diode bidirectional?

14. Describe how a diac is gated on and how it can be switched off.

15. Explain how the thyristor ratings (i.e., peak forward blocking voltage and peak reverse voltage), are extremely important to the theory of operation of a diac.

16. Describe how the schematic symbols for a diac and a triac are different.

17. Specify the conditions required for turning on a triac, and describe how to turn it off.

18. Explain the basic difference between the application of an SCR and a triac.

19. Describe the main purpose of an ac phase control circuit.

20. Sketch the output voltage waveform for an ac phase control circuit that is being fired at 90° on each half-cycle.

ANALYSIS QUESTIONS

1. Explain how an SCR can be considered a nonlinear power amplifier.

2. Research a parts catalog or thyristor data book, identify the part number, and list the ratings for the SCR having the lowest forward current rating and the SCR having the highest current rating.

3. Explain the similarities and differences between the action of a diac and a zener diode.

4. Research and explain the operation of a unijunction transistor (UJT) and show how it can be used to trigger SCRs or a triac in an ac phase control circuit.

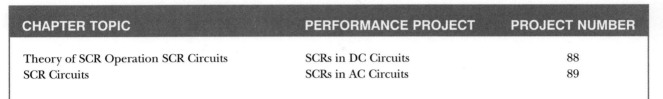

PERFORMANCE PROJECTS CORRELATION CHART

CHAPTER TOPIC	PERFORMANCE PROJECT	PROJECT NUMBER
Theory of SCR Operation SCR Circuits	SCRs in DC Circuits	88
SCR Circuits	SCRs in AC Circuits	89

NOTE: It is suggested that after completing the above projects, the student should be required to answer the questions in the "Summary" at the end of this section of projects in the Laboratory Manual.

Color Codes

A. RESISTOR COLOR CODE

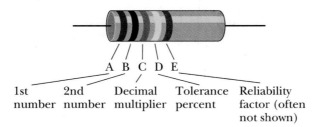

A	B	C	D	E
1st number	2nd number	Decimal multiplier	Tolerance percent	Reliability factor (often not shown)

COLOR SIGNIFICANCE CHART
(For resistors and capacitors)

COLOR	NUMBER COLOR REPRESENTS	DECIMAL MULTIPLIER	TOLERANCE PERCENT	VOLTAGE RATING	% CHANGE PER 1,000 HRS OPER.
Black	0	1	—	—	—
Brown	1	10	1*	100*	1%
Red	2	100	2*	200*	0.1%
Orange	3	1,000	3*	300*	0.01%
Yellow	4	10,000	4*	400*	0.001%
Green	5	100,000	5*	500*	—
Blue	6	1,000,000	6*	600*	—
Violet	7	10,000,000	7*	700*	—
Gray	8	100,000,000	8*	800*	—
White	9	1,000,000,000	9*	900*	—
Gold	—	0.1	5	1,000*	—
Silver	—	0.01	10	2,000*	—
No Color	—	—	20	500*	—

(*Applicable to capacitors only)

B. SAMPLE SURFACE-MOUNT TECHNOLOGY (SMT) "CHIP" RESISTOR CODING

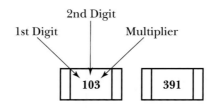

1st Digit	2nd Digit	Multiplier
	103	391

Code: 103 = 1, then 0, then 000 or 10,000 Ω   Code: 391 = 3, then 9, then 0, or 390 Ω

C. CAPACITOR COLOR CODES

5th and 6th band = V rating (multiply numbers by 100)

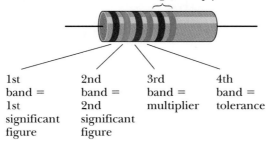

1st
band =
1st
significant
figure

2nd
band =
2nd
significant
figure

3rd
band =
multiplier

4th
band =
tolerance

See Color Significance Chart for meaning of colors

Band color coding system for tubular ceramic capacitors

1st
significant
figure

2nd
significant
figure

Multiplier

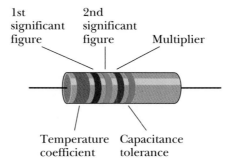

Temperature
coefficient

Capacitance
tolerance

D. SAMPLE CERAMIC DISK CAPACITORS CODED MARKINGS

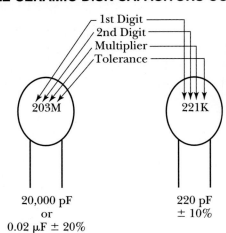

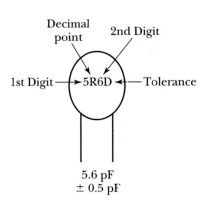

20,000 pF
or
0.02 μF ± 20%

220 pF
± 10%

5.6 pF
± 0.5 pF

Third-Digit Multiplier

Number	Multiply by
0	Nothing
1	10
2	100
3	1,000
4	10,000

Letter Tolerance Code

Under 10 pF Values:

Letter	Tolerance
B	± 0.1 pF
C	± 0.25 pF
D	± 0.5 pF
F	± 1.0 pF

Over 10 pF Values:

Letter	Tolerance
E	± 25%
F	± 1%
G	± 2%
H	± 2.5%
J	± 5%
K	± 10%
M	± 20%
P	− 0%, + 100%
S	− 20%, + 50%
W	− 0%, + 200%
X	− 20%, + 40%
Z	− 20%, + 80%

E. SURFACE-MOUNT TECHNOLOGY (SMT) "CHIP" CAPACITOR CODING

SMT Capacitor Significant Figures Letter Code

Character	Significant Figures	Character	Significant Figures
A	1.0	R	4.3
B	1.1	S	4.7
C	1.2	T	5.1
D	1.3	U	5.6
E	1.5	V	6.2
F	1.6	W	6.8
G	1.8	X	7.5
H	2.0	Y	8.2
J	2.2	Z	9.1
K	2.4	a	2.5
L	2.7	b	3.5
M	3.0	d	4.0
N	3.3	e	4.5
P	3.6	f	5.0
Q	3.9	m	6.0
		n	7.0
		t	8.0
		y	9.0

SMT Capacitor Multiplier Code

Number	Decimal Multiplier
0	1
1	10
2	100
3	1,000
4	10,000
5	100,000
6	1,000,000
7	10,000,000
8	100,000,000
9	0.1

(a)

Decoding Examples:

(C = 1.2)
(3 = × 1,000)

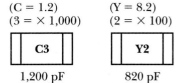

C3

1,200 pF

(Y = 8.2)
(2 = × 100)

Y2

820 pF

(Letter plus number = above 100 pF)

(Read directly)

39

39 pF

(Numbers only = below 100 pF)

(b)

F. SEMICONDUCTOR DIODE COLOR CODE(S)

(NOTE: Prefix "1N . . . " is understood.)

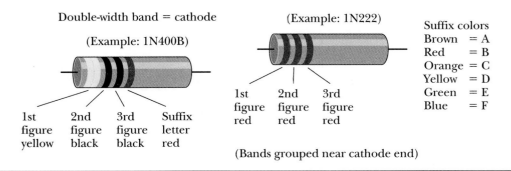

Double-width band = cathode

(Example: 1N400B)

1st figure yellow   2nd figure black   3rd figure black   Suffix letter red

(Example: 1N222)

1st figure red   2nd figure red   3rd figure red

Suffix colors
Brown   = A
Red   = B
Orange = C
Yellow   = D
Green   = E
Blue   = F

(Bands grouped near cathode end)

APPENDIX B

Answers to In-Process Learning Checks, Practice Problems, Odd-Numbered Review Questions, Problems, and Analysis Questions

CHAPTER 1

IN-PROCESS LEARNING CHECKS

I

1. Matter is anything that has *weight* and occupies *space.*

2. Three physical states of matter are *solid, liquid,* and *gas.*

3. Three chemical states of matter are *elements, compounds,* and *mixtures.*

4. The smallest particle that a compound can be divided into but retain its physical properties is the *molecule.*

5. The smallest particle that an element can be divided into but retain its physical properties is the *atom.*

6. The three parts of an atom that interest electronics students are the *electron, proton,* and *neutron.*

7. The atomic particle having a negative charge is the *electron.*

8. The atomic particle having a positive charge is the *proton.*

9. The atomic particle having a neutral charge is the *neutron.*

10. The *protons* and *neutrons* are found in the atom's nucleus.

11. The particle that orbits the nucleus of the atom is the *electron.*

II

1. An electrical system (circuit) consists of a source, a way to transport the electrical energy, and a *load.*

2. Static electricity is usually associated with *nonconductor*-type materials.

3. The basic electrical law is that *unlike* charges attract each other and *like* charges repel each other.

4. A positive sign or a negative sign often shows electrical *polarity.*

5. If two quantities are *directly* related, as one increases the other will *increase.*

6. If two quantities are *inversely* related, as one increases the other will *decrease.*

7. The unit of charge is the *coulomb.*

8. The unit of current is the *ampere.*

9. An ampere is an electron flow of one *coulomb* per second.

10. If two points have different electrical charge levels, there is a difference of *potential* between them.

11. The volt is the unit of *electromotive* force, or *potential* difference.

REVIEW QUESTIONS

1. Matter is anything that has weight and occupies space whether it is a solid, liquid, or gas.

3. Three chemical states of matter are elements, compounds, and mixtures. Copper is an element,

sugar is a compound, and sand and gold dust are a mixture.

5. A compound is a form of matter that can be chemically divided into simpler substances and that has two or more types of atoms. Examples are water (hydrogen and oxygen) and sugar (carbon, hydrogen, and oxygen).

7. I.D. of charges on particles (electron = negative; proton = positive).

9. Free electrons are electrons in the atomic structure that can be easily moved or removed from their original atom. These are outer-ring or valence electrons in conductor materials.

11. Chemical, mechanical, light, and heat.

13. Unlike charges attract and like charges repel each other.

15. Force of attraction or repulsion will be one-ninth the original force.

17. Unit of measure of electromotive force (emf) is the volt.

19. a. Unit of measure for current is the ampere.

    b. 4 amperes of current flow

21. A closed circuit has an unbroken path for current flow. An open circuit has discontinuity or a broken path; therefore, the circuit does not provide a continuous path for current.

23. − sign denotes a negative polarity point.

25. Point B is positive side of source.

ANALYSIS QUESTIONS

1. Drawing:

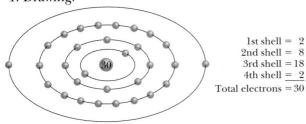

1st shell = 2
2nd shell = 8
3rd shell = 18
4th shell = 2
Total electrons = 30

3. Zinc

5. 30

7. Circuit diagram:

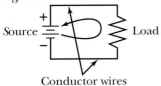

Conductor wires

9. Usually an undesired very low resistance path around part or all of a circuit.

CHAPTER 2

PRACTICE PROBLEMS

I

The circular-mil cross-sectional area $= 2.5^2 = 6.25$ CM.

II

The resistance of 300 feet of copper wire having a cross-sectional area of 2,048 CM $= 1.523 \, \Omega$.

III

1. 500 feet of #12 wire has a resistance of approximately $0.81 \, \Omega$.

2. The wire gage of wire having a cross-sectional area of 10,380 CM is #10 wire.

3. #10 wire has an approximate resistance of 1.018 Ω per thousand feet of length.

IV

1. 39 kΩ, 10% tolerance.

2. 18 kΩ, 5%, and 0.1%.

V

1. It is a TV receiver system.

2. The signal flows from left-to-right, as illustrated.

3. The receiver circuitry receives signals from the TV antenna.

4. Two types of outputs are picture information and audio sound information.

VI

1. There are three kinds of components (resistor, switch, battery).

2. The highest value resistor is 100 kΩ.

3. A SPST switch is used.

4. The voltage source is a battery.

VII

1. There are five resistors.

2. Resistor R_3. The lowest R value resistor is 1 kΩ.

3. There are two swtiches in the circuit.

4. The circuit applied voltage is 100 V.

5. S_2 is opened, and S_1 is closed.

IN-PROCESS LEARNING CHECKS

I

1. Charge is represented by the letter **Q**. The unit of measure is the **coulomb**, and the abbreviation is **C**.

2. The unit of potential difference is the **volt**. The symbol is **V**.

3. The abbreviation for current is **I**. The unit of measure for current is the **ampere**, and the symbol for this unit is **A**.

4. The abbreviation for resistance is **R**. The unit of measure for resistance is the **ohm**, and the symbol for this unit is Ω *(the lowercase Greek letter omega)*.

5. Conductance is the **ease** with which current flows through a component or circuit. The abbreviation is **G**. The unit of measure for conductance is the **siemens**, and the symbol for this unit is **S**.

6. 0.0000022 amperes = **2.2 microamperes (μA)**. Could be expressed as **2.2 × 10⁻⁶A**.

7. One-thousandth is represented by the metric prefix **milli**. When expressed as a power of 10, it is **10⁻³**.

8. 10,000 ohms might be expressed as **10 kΩ** (kilohms). Expressed as a whole number times a power of 10, **10 × 10³ ohms**.

II

1. The greater the resistivity of a given conductor, the **higher** its resistance will be per given length.

2. The circular-mil area of a conductor is equal to the square of its **diameter** in **mils (thousandths of an inch)**.

3. The smaller the diameter of a conductor, the **higher** its resistance will be per given length.

4. The higher the temperature in which a conductor must operate, the **higher** will be its resistance.

5. Wire tables show the AWG wire **size**, the cross-sectional area of wires in **circular mils**, the resistance per given length of copper wire, and the operating **temperature** at which these parameters hold true.

III

1. Negative lead connected to **Point B;** positive lead connected to **Point A**.

2. Negative lead connected to **Point B,** positive lead connected to **Point C**.

3. The first thing to do is to remove the source voltage from the circuit.

REVIEW QUESTIONS

1. 0.003 V

3. A micro unit is one-thousandth of a milli unit.

5. Schematic diagram of battery, two resistors, and a SPST switch:

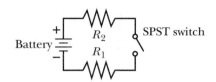

7. Resistor

9. Battery

11. Carbon or composition resistors and wirewound resistors. Carbon/composition resistors are composed of carbon or graphite mixed with powdered insulating material. Wirewound resistors are constructed with special resistance wire wrapped around a ceramic or insulator core.

13. Value, power rating, tolerance, and physical size

15. 222,000 Ω ±2%

17. Would likely use a wire-wound resistor

19. Would use the 5th band; expected failure rate = 0.01%

21. Two precautions: Use only one hand; insulate yourself from ground; use only tools with insulated handles.

23. Two dangers: electrical shock; physical injury

25. Causes of shock: Part of body provides current path across points having a potential difference.

ANALYSIS QUESTIONS

1. The resistor color code is necessary because many resistors are to small to enable printing their values on them. Also, color code provides a "universal" method, readable by anyone in the world, regardless of their language.

3. Generally, reliability refers to the ability of the component to perform as it is expected to, for the length of time intended, under normal operating conditions. Resistor reliability ratings typically indicate the percentage of failure for 1,000 hours of operation.

5. Block diagram of a simple power distribution system:

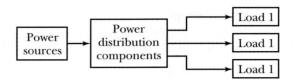

7. a. Brown, black, red, silver

b. Red, violet, orange

c. Brown, black, green, silver

d. Brown, black, black, gold

9. Two connection points

CHAPTER 3

PRACTICE PROBLEMS

I

1. $I = V/R = 150$ V/10 $\Omega = 15$ A

2. 7.5 A

3. a. Remain the same

b. 300 V/20 $\Omega = 15$ A

II

1. 25×10^3

2. 1.2335×10^1

3. 1.0×10^{-1}

4. 1.5×10^{-3}

5. 23,500

6. 1,000

7. 10,000

8. 0.015

III

1. 1,456,000

2. 0.0333

3. 1.5

4. 1,000,000

IV

1. $P = $ energy/time $= 25$ J/2 sec $= 12.5$ W

2. $P = $ energy/time $= 1$ J/100 sec $= 0.01$ W or 10 mW

V

1. Energy $= $ Power $\times$ Time $= 250$ W $\times 8$ hr $= 2,000$ Wh, or 2 kWh

VI

1. $P = V \times I = 150$ V $\times 10$ mA $= 1,500$ mW, or 1.5 W

2. a. I would decrease to half the original value.

b. Power diss. $= V \times I = 150 \times 5$ mA $= 750$ mW, or 0.75 W

VII

1. $V_A = I \times R = 1.5$ mA $\times 4.7$ k$\Omega = 7.05$ V

2. $I = V/R = 100$ V/27 k$\Omega = 3.704$ mA

3. $R = V/I = 100$ V/19.6 mA $= 5.102$ kΩ

VIII

1. $P = 40$ W; $I = 2$ A

2. $I = 1$ mA; $R = 100$ kΩ

3. $I = 12$ mA; $P = 1.44$ W (or 1,440 mW)

IN-PROCESS LEARNING CHECKS

I

1. $I = \textbf{1.5 A,}$ or $\textbf{1,500 mA.}$ ($I = V/R$; $I = 15$ V/10 $\Omega = 1.5$ A)

2. $V = \textbf{30 V.}$ ($V = I \times R$; $V = 2$ mA $\times 15$ k$\Omega = 30$ V)

3. $R = \textbf{27 k}\Omega.$ ($R = V/I$; $R = 135$ V/5 mA $= 27$ kΩ)

4. a. Current would decrease.

b. Voltage applied would be higher.

c. Circuit resistance must have halved.

II

1. negative, positive

2. positive, negative

3. electron

4. one, changes

5. expenditure, work, work, energy

6. joules, seconds

7. kilowatt hour

REVIEW QUESTIONS

1. a. $I = V/R$; $V = I \times R$; $R = V/I$

b. $P = V \times I$; $P = I^2 R$; $P = V^2/R$

3. a. 5.1×10^3

b. 47×10^3

c. 0.000001×10^3

d. 0.039×10^3

5. Energy is the ability to do work.

7. Work is measured in foot poundals, or foot pounds, or horsepower. Energy is measured in joules. Power (electrical) is measured in Watts.

9. Power unit = the watt

PROBLEMS

1. The new current will be four times the original current.

3. a. Schematic:

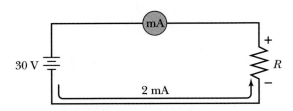

b. $R = V/I = 30$ V/2 mA = 15 kΩ; $P = V \times I = 30$ V $\times$ 2 mA = 60 mW

c. See diagram.

5. $V = I \times R = 3$ mA $\times$ 12 kΩ = 36 V

7. $R = V/I = 41$ V/50 mA = 820 Ω

9. $I = 3.162$ A

11. $R = V/I = 100$ V/8.5 mA = 11,765 kΩ

13. Power dissipated will increase ($P = V \times I$ and I remained the same while V doubled).

15. 2.8125 W

17. 20.83 mA

19. 33.85 Wh

21. 100 V

23. 100 V

25. a. 0.01

b. 10^{12}

c. 8.57×10^4 or 85,700

d. 6.1664×10^3 or 6.17×10^3

ANALYSIS QUESTIONS

1. The EE (exponent entry) key allows entering numbers with exponents, such as numbers raised to some power of 10.

3. Current flows only in one direction with the dc source. It alternates direction of flow with ac applied.

5. The direction a + (positive) charge would move.

7. V has doubled.

CHAPTER 4

PRACTICE PROBLEMS

I

1. Total resistance = 74 kΩ

2. Total resistance = 16 kΩ

II

1. $R_1 = 47$ Ω; $R_T = 100$ Ω

2. New $R_T = 10.5$ kΩ

3. $R_2 = 3.9$ kΩ

III

1. $V_A = 411$ V; $I_T = 3$ mA; $V_{R_2} = 81$ V; $V_{R_1} = 300$ V

2. $V_A = 342.5$ V; $I_T = 2.5$ mA; $V_{R_2} = 67.5$ V; $V_{R_1} = 250$ V

IV

1. $V_X = (R_X/R_T) \times V_T = (2.7$ kΩ/17.4 k$\Omega) \times 50$ V = $0.155 \times 50 = 7.75$ V (V_{R_1})

2. $(10$ kΩ/17.4 k$\Omega) \times 50$ V = $0.575 \times 50 = 28.75$ V (V_{R_3})

3. $7.75 + 13.5 + 28.75 = 50$ V (Yes, they add up correctly.)

V

$V_2 = 30$ V; $V_A = 75$ V

VI

$P_1 = 90$ W	$V_1 = 30$ V
$P_2 = 90$ W	$V_2 = 30$ V
$P_3 = 90$ W	$V_3 = 30$ V
$P_T = 270$ W	$R_3 = 33.3\%$ of R_T
$I_T = 3$ A	$P_3 = 33.3\%$ of P_T

VII

1. a. increase; b. decrease; c. decrease; d. increase; e. decrease

2. Yes

VIII

1. With the 47-kΩ resistor shorted the voltage drops are:
   10-kΩ resistor drops 14.6 V
   27-kΩ resistor drops 39.4 V
   47-kΩ resistor drops 0 V
   100-kΩ resistor drops 146 V

2. They increase.

IX

Givens: $R_1 = 20$ kΩ; $V_1 = \dfrac{2}{5} V_A$; $V_2 = \dfrac{1.5}{5} V_A$;

$V_A = 50$ V; $R_2 = 15$ kΩ; $R_3 = 15$ kΩ;
$V_2 = 15$ V; $V_3 = 15$ V; $P_T = 50$ mW; $P_1 = 20$
mW; $P_2 = 15$ mW; $P_3 = 15$ mW; $I = 1$ mA

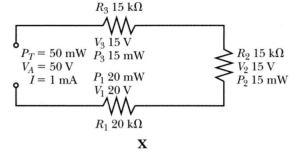

X

1. $V_{R_1} = 4.5$ V
2. $V_{R_2} = 1.5$ V

XI

D to C = 94 V; C to B = 54 V; B to A = 40 V; D to
B = 148 V

XII

1. Voltage across R_2 = 27.49 V (or approximately 27.5 V).
2. Voltage at top of R_2 is +27.5 V with respect to point C.
3. Voltage across R_1 = 9.995 V (or approximately 10 V).
4. Voltage at point C with respect to point A = −37.5 V.

XIII

1. Minimum power rating should be 2 times (60 V × 30 mA), or 2 × 1,800 mW = 3,600 mW (3.6 W).

2. R dropping equals $\dfrac{110 \text{ V}}{30 \text{ mA}}$ = 3.66 kΩ. Minimum power rating should be 2 times (110 V × 30 mA), or 2 × 3,300 mW = 6,600 mW (6.6 W).

IN-PROCESS LEARNING CHECKS

I

1. The primary identifying characteristic of a series circuit is that the **current** is the same throughout the circuit.

2. The total resistance in a series circuit must be greater than any one **resistance** in the circuit.

3. In a series circuit, the highest value voltage is dropped by the **largest or highest** value resistance,

and the lowest value voltage is dropped by the **smallest or lowest** value resistance.

4. V drop by the other resistor must be **100 V.**

5. Total resistance **increases.** Total current **decreases.** Adjacent resistor's voltage drop **decreases.**

6. Applied voltage is **130 V.**

REVIEW QUESTIONS

1. Current is the same throughout the circuit because there is only one path for current.

3. R_T equals the sum of all the resistor values in series.

5. b (e.g., In a series circuit the applied circuit voltage is equal to the sum of all the individual voltage drops around the circuit.)

7. a (e.g., The power distribution throughout a series circuit is directly related to the resistance distribution around the circuit.)

9. c (e.g., If a short is placed across part of a series circuit, R_T decreases.)

PROBLEMS

1. $R_T = 10/5 = 2$ Ω; each resistor = 1 Ω

3. Diagram, labeling, and calculations for a series circuit composed of a 50-, 40-, 30- and 20-Ω resistor.

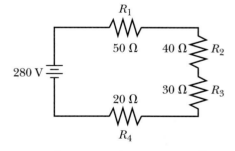

a. $R_T = 50 + 40 + 30 + 20 = 140$ Ω
b. If $I = 2$ A, then V applied = $IR = 2$ A × 140 Ω = 280 V
c. $V_1 = 100$ V; $V_2 = 80$ V; $V_3 = 60$ V; $V_4 = 40$ V
d. $P_T = I_T \times V_T = 2$ A × 280 V = 560 W
e. $P_2 = 2$ A × 80 V = 160 W; $P_4 = 2$ A × 40 V = 80 W
f. 1/7 V_A dropped by R_4
g. If R_3 increased in value:
   (1) total resistance would increase.
   (2) total current would decrease.

(3) V_1 would decrease; V_2 would decrease; and V_4 would decrease.

(4) P_T would decrease because I decreased and V stayed the same.

h. If R_2 shorted:

(1) total resistance would decrease.

(2) total current would increase.

(3) V_1 would increase; V_3 would increase; V_4 would increase (because of increased I with Rs staying the same).

5. Diagram of three sources to acquire 60 V if sources are 100 V, 40 V, and 120 V.

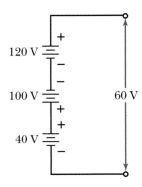

7. a. $V_A = 34$ V (4 V + 20 V + 10 V)

b. $R_3 = 2.5$ kΩ (to drop 5 V of the 10 V dropped by R_3 and R_4)

c. $R_2 = 10$ kΩ (to drop 20 V with 2 mA of current)

9. 17.22 V

11. 2.15 V

13. 3.87 V

15. Most power = R_1; least power = R_2

17. $V_3 = 70$ V

19. a. R_1 is set at 3.81 kΩ

b. $I_T = 4.2$ mA

c. R_1 dissipates 16% of the total power

d. I_T would equal 4 mA

e. Voltmeter would indicate 50 V

21. a. $P_{R_2} = 423$ mW

b. $P_T = 819$ mW

c. $I_T = 3$ mA

23. Circuit with series "dropping resistor" used to drop from a 100-V source to a 50-V source with 50-mA load current:

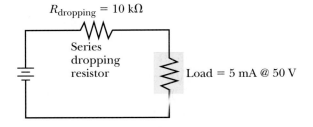

25. Decrease

27. $V_{applied} = 182$ V

ANALYSIS QUESTIONS

1. In the scientific mode, the results always indicate a number between 1 and 10 times some power of 10.

   In the engineering mode, the results show a number times a power of 10 with an exponent that is some multiple or submultiple of 3 (taking advantage of the metric system).

3. The "divide-and-conquer" or "split" rule indicates that where there is a linear, or series, flow it is often good to make the first check at the middle of the questionable area. This enables eliminating half the questionable area with just one test.

5. To identify "initial suspect area": Analyze all symptom information; determine all sections or components in the system that could cause the symptoms; then "bracket" the area of uncertainty, or suspect area in which to make tests and diagnoses.

7. Bracketing: enclosing the area of uncertainty, or "suspect area," either mentally or on paper by placing parenthesis, brackets, etc., to highlight the area in which you will troubleshoot.

9. Useful documentation includes block diagrams, schematic diagrams, and operator manuals.

11. If current drops to zero, an open condition has occurred somewhere in the series path.

13. The method suggested here is to use the "divide-and-conquer" (half-split) technique.

CHAPTER 5

PRACTICE PROBLEMS

I

$V_{R_1} = V_{R_2} = 2 \text{ mA} \times 27 \text{ k}\Omega = 54 \text{ V}$
$V_A = V_{R_1} = V_{R_2} = 54 \text{ V}$

II

$I_T = I_1 + I_2 + I_3 + I_4 + I_5$
The lowest R value branch passes the most current.
The highest R value branch passes the least current.

III

$V_2 = 50 \text{ V}; V_1 = 50 \text{ V}; V_A = 50 \text{ V}; I_1 = 5 \text{ A}; I_T = 6 \text{ A}$

IV

1. $R_1 = 15 \text{ k}\Omega; I_2 = 2.5 \text{ mA}; I_3 = 2.5 \text{ mA}; R_2 = 30 \text{ k}\Omega$
2. $R_1 = 27 \text{ k}\Omega; I_2 = 3.78 \text{ mA}; I_3 = 2.23 \text{ mA}; R_2 = 33$
   $\text{k}\Omega; R_T = 11.75 \text{ k}\Omega.$

V

1. $V_{R_1} = 30 \text{ V}; V_T = 30 \text{ V}$
2. $R_T = 1.67 \text{ k}\Omega$

VI

1. $R_T = 5.35 \text{ k}\Omega$
2. $R_T = 0.76 \text{ k}\Omega$ (or 760 Ω)

VII

Total resistance = 25 Ω

VIII

1. $R_T = 12 \Omega$
2. $R_T = 7.67 \text{ k}\Omega$

IX

$R_T = 27.15 \Omega$

X

$R_2 = 100 \Omega$

XI

$P_T = 2,250 \text{ mW}$ or 2.25 W

XII

$R_T = 10 \text{ k}\Omega$	$R_1 = 60 \text{ k}\Omega$	$R_2 = 30 \text{ k}\Omega$	$R_3 = 20 \text{ k}\Omega$
$V_A = 120 \text{ V}$	$V_1 = 120 \text{ V}$	$V_2 = 120 \text{ V}$	$V_3 = 120 \text{ V}$
$I_T = 12 \text{ mA}$	$I_1 = 2 \text{ mA}$	$I_2 = 4 \text{ mA}$	$I_3 = 6 \text{ mA}$
$P_T = 1.44 \text{ W}$	$P_1 = 240 \text{ mW}$	$P_2 = 480 \text{ mW}$	$P_3 = 720 \text{ mW}$

XIII

1. $I_1 = 1.85 \text{ mA}$ (through 27-kΩ resistor)

$I_2 = 1.07 \text{ mA}$ (through 47-kΩ resistor)
$I_3 = 0.50 \text{ mA}$ (through 100-kΩ resistor)

2. $R_T = 1.54 \Omega$
   $I_1 = 0.616 \text{ A}$ (through 10-Ω resistor)
   $I_2 = 1.1 \text{ A}$ (through 5.6-Ω resistor)
   $I_3 = 2.28 \text{ A}$ (through 2.7-Ω resistor)

XIV

1. $I_1 = 1.65 \text{ mA}$
   $I_2 = 0.35 \text{ mA}$
2. $I_1 = 16.4 \text{ mA}$
   $I_2 = 3.6 \text{ mA}$

REVIEW QUESTIONS

1. b
3. c
5. a
7. $1/R_T = 1/R_1 + 1/R_2 + \ldots 1/R_n$ and/or "Reciprocal of the reciprocals": $\dfrac{1}{1/R_1 + 1/R_2 + \ldots 1/R_n}$
9. $R_T = R_1 \times R_2/R_1 + R_2$
11. $R_u = R_k \times R_e/R_k - R_e$

PROBLEMS

1. $R_T = 17.15 \text{ k}\Omega$; product-over-the-sum method
3. $I_2 = 30 \text{ mA}$; Kirchhoff's current law
5. a. I_1 will RTS.   b. I_1 will D.
   I_2 will D. I_2 will D.
   I_3 will RTS. I_3 will I.
   I_T will D. I_T will I.
   P_T will D.
   V_1 will RTS.
   V_T will RTS.
7. $I_1 = 12 \text{ mA}$
   $I_2 = 6 \text{ mA}$ (since I branch is inverse to R branch)
9. $P_{R_1} = 0.5 \text{ A} \times 50 \text{ V} = 25 \text{ W}$
11. a. $M_1 = 2.13 \text{ mA}$
    b. $M_2 = 37 \text{ mA}$
    c. $M_3 = 39.13 \text{ mA}$
    d. $M_4 = 10 \text{ mA}$
    e. $M_5 = 49.13 \text{ mA}$
    f. $P_{R_1} = 213 \text{ mW}$
    g. $P_{R_2} = 3.7 \text{ W}$
    h. $P_{R_3} = 1 \text{ W}$
    i. $P_T = 4.91 \text{ W}$

j. New P_T = 1.21 W

13. R_X = 31.59 kΩ

15. a. R_4
    b. P_{R_4} = 71.7 mW

17. V_T = 18 V

19. I_{R_2} = 5.45 mA

21. R_1 dissipates the most power.

23. I_T would be 34 mA if R_2 were changed to a 1.5-kΩ resistor value.

25. Power dissipated by R_3 would not change as long as V_T and R_3 were the same values.

27. 120 V

29. R_e = 5.45 kΩ

ANALYSIS QUESTIONS

1. The purpose of performing "mental approximations" is to provide a rapid check as to whether you might have made a mistake in solving a problem. If your approximation comes out completely different from your other calculations, chances are you have a simple mathematical error that needs correcting.

3. a. 4.864864865 03
   b. 4.864864865 03

5. Circuit diagram of a three-branch parallel circuit.

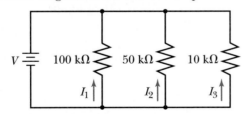

7. A method of isolating parallel components, or an individual branch, when troubleshooting a paral-

lel circuit is to "lift" one end of the branch to be tested from the remainder of the circuit so as to electrically isolate it. (NOTE: This may require unsoldering one lead of the component from the circuit for testing and then resoldering after testing.)

CHAPTER 6

PRACTICE PROBLEMS

I

1. The only component carrying total current is R_7.

2. The only single components in parallel with each other are R_2 and R_3.

3. The only single components that pass the same current are R_1, R_4, and R_5.

II

1. R_T for Figure 6–11 = 12.1 kΩ

2. R_T for Figure 6–12 = 7.63 kΩ

III

1. I_T = 2 A; I_1 = 1.33 A; I_2 = 0.67 A; I_3 = 0.33 A.

2. I_T = 6.17 mA; I_1 = 2.36 mA; I_2 = 1.38 mA; I_3 = 2.43 mA

IV

1. *Voltage across:* R_1 = 30 V; R_2 = 100 V; R_3 = 25 V; R_4 = 50 V; R_5 = 50 V; R_6 = 25 V.
   Current through: R_1 = 3 A; R_2 = 0.5 A; R_3 = 2.5 A; R_4 = 0.5 A; R_5 = 2 A; and R_6 = 2.5 A.

2. *Voltage across:* R_1 = 50 V; R_2 = 10.7 V; R_3 = 8.9 V; R_4 = 15.2 V; R_5 = 4.45 V; R_6 = 0.72 V; R_7 = 1.94 V; R_8 = 2.0 V; R_9 = 2.0 V.
   Current through: R_1 = 5 mA; R_2 = 2.28 mA; R_3 = 2.28 mA; R_4 = 2.72 mA; R_5 = 2.0 mA; R_6 = 0.72 mA; R_7 = 0.72 mA; R_8 = 0.36 mA; R_9 = 0.36 mA.

V

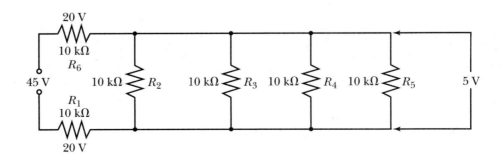

VI

1. Circuit diagram:

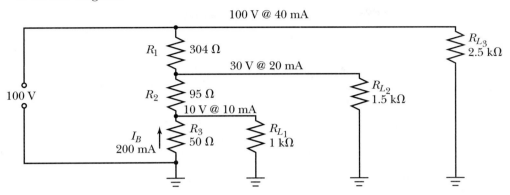

2. Load #1 = 1 kΩ
   Load #2 = 1.5 kΩ
   Load #3 = 2.5 kΩ
   R_1 divider resistor = 304 Ω
   R_2 divider resistor = 95 Ω
   R_3 divider resistor = 50 Ω

3. See labeling of diagram in answer 1.

VII

1. $V_{R_L} = 150$ V; $P_{R_L} = 300$ mW

2. $R_L = 75$ kΩ

3. $R_2 = 50$ kΩ

4. The change, as specified, means R_2 has decreased.

5. The change, as specified, means R_1 has increased.

6. V_{R_2} equals 166.66 V.

7. If R_1 opens, V_{R_L} drops to 0 V.
   If R_2 shorts, V_{R_L} drops to 0 V.

IN-PROCESS LEARNING CHECKS

I

For Figure 6–13a: $R_T = 10$ Ω

For Figure 6–13b: $R_T = 120$ Ω

II

$I_1 = 0.4$ mA; $I_T = 1$ mA; $V_T = 85$ V; $R_T = 85$ kΩ; $P_3 = 36$ mW; and the 100-kΩ resistor (R_3) is dissipating the most power.

III

1. Figure 6–25: The defective component is R_5. It has drastically increased in value from 3.3 kΩ to 17.3 kΩ.

2. Figure 6–26: (NOTE: Total circuit current *should be* 2 mA, meaning R_3 should drop 2 mA × 47 kΩ V, or 94 V. Instead it is dropping 121 − 17.6 = 103.4 V, which is higher than it should be.) Possible bad components could include R_1 or R_2 decreased in value, or R_3 increased in value.

3. Figure 6–27: If R_2 increases in value, V_{R_1} increases and V_{R_3} decreases. (This is due to total circuit resistance increasing, which decreases the total current through R_3, resulting in R_3 dropping less voltage. Kirchhoff's voltage law would indicate, therefore, that the voltage across parallel resistors R_1 and R_2 has to increase so the loop voltages equal V applied.)

ANSWERS TO SPECIAL "THINKING EXERCISE" (PAGE 207)

If R_2 opens:

V_T remains the same.

V_1 decreases due to lower total current through R_1.

V_2 increases because V_T stays the same, and V_1 and V_4 decrease due to lower total current through them.

V_3 increases for the same reason V_2 increases.

V_4 decreases due to lower I_T caused by higher R_T.

I_T decreases due to higher R_T with same applied voltage.

I through R_1 decreases due to higher circuit total R.

I through R_2 decreases to zero due to its opening.

I through R_3 increases due to higher V across it.

I through R_4 decreases due to lower I_T through it.

P_T decreases due to lower I_T with same V_T.

P_3 increases due to higher I through same R value.

P_4 decreases due to lower current through same R value.

REVIEW QUESTIONS

1. A series-parallel circuit has portions that are connected in series and portions that are connected in parallel. The series portions must be analyzed using series circuit principles, and the parallel portions must be analyzed using parallel circuit principles. The results of the analysis of each type circuit portion must be combined with results from other connected portions to solve for the total circuit quantities.

3. Series portions of series-parallel circuits are identified as those through which the same (common) current passes through the components and by the fact that the components are in "tandem."

5. A general approach for solving for a series-parallel circuit's current distribution is to start at the source end of the circuit and work out.

7. a

9. d

PROBLEMS

1. $R_T = 22\ \Omega$

3. R_T will D.
   R_4 will RTS.
   V_1 will I.
   V_3 will D.
   P_T will I.

5. Minimum voltage is 13 V.
   Maximum voltage is 18 V.

7. $R_4 = 6\ k\Omega$ (so that branches voltage equals 80 V). $I_2 = 10\ mA$ (since $I_T = 20\ mA$ and I_1 can be calculated as 10 mA, due to V across the 8-$k\Omega$ resistor having to be 80 V. That is, the 100 V applied minus the 20-V drop across the 1-$k\Omega$ resistor that has I_T passing through it).

9. Light gets dimmer, due to the additional current that passes through the resistor in series with the source. This drops the voltage across the remaining sections of the circuit.

11.

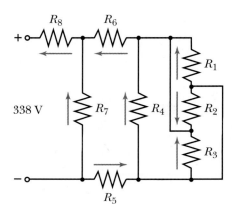

a. $R_T = 169\ k\Omega$
b. $I_T = 2\ mA$
c. $V_{R_1} = 14\ V$
d. $V_{R_2} = 14\ V$
e. $V_{R_3} = 14\ V$
f. $V_{R_4} = 14\ V$
g. $V_{R_5} = 62\ V$
h. $V_{R_6} = 62\ V$
i. $V_{R_7} = 138\ V$
j. $V_{R_8} = 200\ V$

13. Before connecting voltmeter, $V_{R_1} = 5\ V$.

15. Approximately 4 V

17. Yes, connecting the meter changes R_T and I_T. A higher I_T will pass through R_1, causing its voltage to increase from 5 V to 6 V. Less voltage will be dropped across R_2 than was present before the meter was connected.

19. Total current with the meter across one resistor = 60 mA.

21. If the meter were connected from the bottom of R_2 to the top of R_1, the voltages across the two resistors would not be affected.

23. $R_3 = 1\ k\Omega$; $R_2 = 495\ \Omega$; $R_1 = 490\ \Omega$.

25. R_1 would have to be decreased in value.

27. Bleeder current would become 102.8 mA.

29. Minimum rating should exceed 10.1 W for R_2.

31. P_T being supplied is 112 mA $\times$ 200 V = 22.4 watts.

ANALYSIS QUESTIONS

1. If R_2 decreased greatly in value:
   a. voltage across R_3 would I.
   b. voltage across R_1 would I.
   c. current through the load connected across R_4 would I.

3. If the sensor is connected across a balanced bridge circuit, when the bridge is unbalanced by a change in conditions, the sensor's output can then be fed to control elements, which respond to the changed parameter across the sensor, and make appropriate system corrections, based on this feedback.

5. $R_X = 200$ kΩ

7. 0.0555:1 $(R_1:R_2)$ or 18:1 $(R_2:R_1)$

9. First, set the calculator to the engineering mode. Next, put in one resistor's value, then press the **1/x** key, then the plus (**+**) key. Then, put in the second resistor's value, press the **1/x** key, then the **+** key. Then, input the third resistor's value, press the **1/x** key, then press the equal (**=**) key, then press the **1/x** key, one more time, for the answer.

CHAPTER 7

PRACTICE PROBLEMS

I

$I = 4$ A
$V_L = 16$ V
$P_T = 80$ W
$P_L = 64$ W
Efficiency = 80%

II

1. I through $R_1 = 0$ mA
   V across R_1 with respect to point A = 0 V
   I through $R_2 = 7.5$ mA
   V across R_2 with respect to point A = +75 V
   I through $R_3 = 7.5$ mA
   V across R_3 with respect to point A = −75 V

2. I through $R_1 = 10$ mA
   V across R_1 with respect to point A = +100 V
   I through $R_2 = 12.5$ mA
   V across R_2 with respect to point A = −125 V
   I through $R_3 = 2.5$ mA
   V across R_3 with respect to point A = +25 V

III

1. If $R_L = 175$ Ω, $I_L = 0.25$ A and $V_L = 43.75$ V (Figure 7–10)

2. If $I_L = 125$ mA, then $R_L = 375$ Ω (Figure 7–10)

3. $I_L = 1.16$ mA; $V_{R_L} = 0.55$ V

4. In Figure 7–13, $R_{TH} = 4$ kΩ and $V_{TH} = 50$ V. If $R_L = 16$ kΩ, then $V_L = 40$ V and $I_L = 2.5$ mA.

IV

If R_L is 60 Ω, $I_L = 0.588$ A and $V_L = 35.28$ V.

V

1. If $R_L = 100$ Ω in Figure 7–18a, $I_L = 0.4$ A, and $V_L = 40$ V.

2. If R_L changes from 25 Ω to 50 Ω, I_L decreases (from 1 A to 0.66 A), and V_L increases (from 25 V to 33.5 V).

IN-PROCESS LEARNING CHECKS

I

1. For maximum power transfer to occur between source and load, the load resistance should **equal** the source resistance.

2. The higher the efficiency of power transfer from source to load, the **greater** the percentage of total power is dissipated by the load.

3. Maximum power transfer occurs at **50%** efficiency.

4. If the load resistance is less than the source resistance, efficiency is **smaller** than the efficiency at maximum power transfer.

5. To analyze a circuit having two sources, the superposition theorem indicates that Ohm's law **can** be used.

6. What are two key observations needed in using the superposition theorem to analyze a circuit with more than one source? **Noting the direction of current flow and polarity of voltage drops.**

7. Using the superposition theorem, if the sources are considered "voltage" sources, are these sources considered shorted or opened during the analysis process? **Shorted.**

8. When using the superposition theorem when determining the final result of your analysis, the calculated parameters are combined, or superimposed, **algebraically.**

REVIEW QUESTIONS

1. Matching the impedance of an antenna system to the impedance of a transmitter output stage, and matching the output impedance of an audio amplifier to the impedance of a speaker.

3. $P_{out}/P_{in} \times 100$

5. To "Thevenize" a circuit:
   (a) Open or remove R_L.
   (b) Determine the open circuit V at points where R_L is to be connected. (This value is V_{TH}.)
   (c) Determine the resistance looking toward the source from the R_L connection points, assuming the source is shorted. (This value is R_{TH}.)
   (d) Draw the Thevenin equivalent circuit with V_{TH} as source, and R_{TH} in series with the source.
   (e) Calculate I_L and V_L for given values of R_L, using a two-resistor series circuit analysis technique.

7. The most important application of the superposition theorem is for analyzing circuits having more than one source.

9. c

11. d

PROBLEMS

1. 0 V

2. V_{TH} = 40.5 V (voltage at Points A and B with R_L removed)
   R_{TH} = 14.85 kΩ (R at Points A and B with source shorted)
   I_L = 1.63 mA (V_L/R_L in Thevenin equivalent circuit)
   V_L = 16.3 V (from simple series Thevenin equivalent circuit analysis)

5. Norton equivalent circuit is a constant current source with a parallel resistance, called R_N, and the load (R_L) in parallel with R_N and the source.
   To convert parameters:
   $R_N = R_{TH}$ = 14.85 kΩ
   I_N (constant current source) = V_{TH}/R_{TH} = 40.5 V/14.85 kΩ = 2.73 mA
   The circuit is shown below:

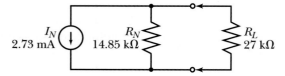

7. The maximum power that can be delivered to the load in Figure 7–22 is 1.25 W.

9. R_{TH} = 29.49 Ω; V_{TH} = 34.45 V

11. R_{TH} = 10 Ω; $V_{TH} = I_N \times R_N$ = 3 A × 10 Ω = 30 V

13. Efficiency = 83.2% (P_{out} = 277 W; P_{in} = 333 W; $P_{out}/P_{in} \times 100$ = 83.2%)

15. For maximum power transfer, R_L should match the r_{int} of source or should equal 5Ω.

ANALYSIS QUESTIONS

1. It is important for a voltage source to have as low an internal resistance (r_{int}) as possible so that source output voltage changes will be minimized as current demands for the connected circuit change. Ideally, a voltage source would have zero internal resistance; thus, it would not have any internal voltage drop for any current demand level.

3. Load resistance should be decreased.

5. Norton's theorem

7. R_{TH} is the fixed-value, Thevenin equivalent-circuit series resistor considered to be in series with whatever the R_L value is, for calculation purposes. Finding this value enables using a simple two-resistor series circuit analysis in calculating circuit parameters for various values of R_L. R_{TH} is determined by "looking back" into the network at the points R_L will be connected, and considering the voltage source to be shorted (or 0 Ω).

9. R_N is the fixed-value, Norton equivalent-circuit parallel resistor considered to be in parallel with the "constant current" source, and R_L. R_N (like R_{TH}) is also found by looking back into the network at the points R_L is to be connected and considering the voltage source as being replaced by its internal R (r_{int}), or 0 Ω in this case. Knowing the R_N value (along with the equivalent circuit I_N value) enables reducing circuit analysis to a simple two-resistor, parallel circuit, current-division approach.

11. a. Various answers possible (e.g., some IC circuits)
    b. Superposition theorem

13. Yes, it is possible to measure the circuit R_N. Since this is the same circuit (Figure 7–23), and R_N are determined from the same points, under the same conditions, you would use the same tech-

nique as described for measuring R_{TH} in the previous question.

15. Battery resistance would equal 12 V − 11 V/50 A; 1 V/50 A = 0.02 Ω.

CHAPTER 8

PRACTICE PROBLEMS

I

1. $I_A = 0.55$ A
2. $I_B = 0.25$ A
3. $V_1 = 9.6$ V
4. $V_2 = 8.25$ V
5. $V_3 = 12.1$ V
6. $V_4 = 8.25$ V
7. $V_5 = 11.75$ V
8. $V_{A-B} = $ close to 0 V

Considered from Points A and B with respect to the opposite ends of resistors R_2 and R_4, respectively: Point A is +8.25 V with respect to the common connection points of R_2 and R_4. Point B is +8.25 V with respect to the common connection points of R_2 and R_4. Thus, Point A and Point B are at the same potential. Considered from Points A and B with respect to the opposite ends of R_3 and R_5, respectively: Point A is −12.1 V with respect to the common connection points of R_3 and R_5. Point B is −11.75 V with respect to the common connection points of R_3 and R_5. Therefore, Point A is −0.6 V with respect to Point B, or Point B is +0.6 V with respect to point A. (NOTE: Variance is due to the fact that the math was not carried out to a large number of decimal places.)

II

1. $I_A = 0.714$ A
2. $I_B = 0.143$ A
3. $V_1 = 7.14$ V; current through $R_1 = 0.714$ A
4. $V_2 = 12.85$ V; current through $R_2 = 0.857$ A
5. $V_3 = 2.145$ V; current through $R_3 = 0.143$ A

III

1. $V_{R_1} = 18.5$ V
2. $V_{R_2} = 41.5$ V
3. $V_{R_3} = 6.5$ V

4. Current through $R_1 = 0.925$ A
5. Current through $R_2 = 4.15$ A
6. Current through $R_3 = 3.25$ A

Solution

$$I_1 + I_2 = I_3$$

Loop with 60-V source:

$$\left(\frac{V_{R_1}}{R_1}\right) + \left(\frac{V_{R_3}}{R_3}\right) = \frac{V_{R_2}}{R_2}$$
$$V_{R_1} + V_{R_2} = 60$$
$$V_{R_1} = 60 - V_{R_2}$$

Loop with 48-V source:

$$V_{R_3} + V_{R_2} = 48$$
$$V_{R_3} = 48 - V_{R_2}$$

Substituting these terms into $I_1 + I_2 = I_3$:

$$\left(60 - \frac{V_{R_2}}{20}\right) + \left(48 - \frac{V_{R_2}}{2}\right) = \frac{V_{R_2}}{10}$$

Multiplying each term by 20 to move and/or remove the denominators:

$$(60 - V_{R_2}) + 10(48 - V_{R_2}) = 2\,V_{R_2}$$
$$60 - V_{R_2} + 480 - 10\,V_{R_2} = 2\,V_{R_2}$$
$$540 = 13\,V_{R_2}$$
$$V_{R_2} = 41.5 \text{ V}$$
$$I_{R_2} = \frac{V_{R_2}}{10} = \frac{41.5 \text{ V}}{10 \text{ Ω}} = 4.15 \text{ A}$$
$$V_{R_3} = 48 - V_{R_2} = 48 - 41.5 = 6.5 \text{ V}$$
$$I_{R_3} = \frac{V_{R_3}}{2} = \frac{6.5 \text{ V}}{2 \text{ Ω}} = 3.25 \text{ A}$$
$$V_{R_1} = 60 - 41.5 = 18.5 \text{ V}$$
$$I_{R_1} = \frac{18.5}{20} = 0.925 \text{ A}$$

Verifying:

$$V_{R_1} + V_{R_2} = V_{S_1};\ 18.5 \text{ V} + 41.5 \text{ V} = 60 \text{ V}$$
$$V_{R_2} + V_{R_3} = V_{S_2};\ 41.5 \text{ V} + 6.5 \text{ V} = 48 \text{ V}$$

IV

1. $R_A = 31.82$ Ω; $R_B = 70$ Ω; $R_C = 46.67$ Ω
2. $R_1 = 22$ Ω; $R_2 = 10$ Ω; $R_3 = 15$ Ω
3. Yes

V

$(R_1 = 7\ \Omega;\ R_2 = 3.5\ \Omega;\ R_3 = 2.33\ \Omega)$

$R_T = (R_D + R_2)\|(R_E + R_3) + R_1$

$R_T = 13.44\ \Omega$

$I_T = V_T/R_T = 84\ V/13.44\ \Omega = 6.25\ A$

IN-PROCESS LEARNING CHECKS

I

1. Path
2. Mesh
3. + (positive)
4. +, −

II

1. Current
2. Major node
3. Reference node
4. Current

III

1. (a) **T** (tee) (b) π (pi)
2. AC (or electrical)
3. Current
4. is not

REVIEW QUESTIONS

1. A loop is a complete path in a circuit, from one side of the source, through a series of circuit components, back to the other side of the source.
3. Mesh currents are "assumed" currents (and current directions) used in conjunction with Kirchhoff's voltage law to help in analyzing networks. Mesh currents are different than actual currents in that they are assumed not to branch.
5. c
7. a
9. a
11. b

PROBLEMS

1. $I_A = 0.8\ A$
3. $V_{R_1} = 8\ V$
5. $V_{R_3} = 4\ V$
7. $I_{R_2} = 1.2\ A$

9. I_1 (current through R_1) = 2.13 A
11. I_3 (current through R_2) = 2.34 A
13. $V_{R_2} = 11\ V$
15. $R_A = 45\ \Omega$
17. $R_C = 30\ \Omega$

ANALYSIS QUESTIONS

1. Mesh or nodal analysis techniques are used in situations in which basic series and parallel analysis techniques might be cumbersome or very difficult.
3. In industrial settings, you might find delta and wye circuit configurations related to three-phase power, three-phase motors, etc.

CHAPTER 9

PRACTICE PROBLEMS

I

1. $A = \dfrac{1.5\ V - 1.25\ V}{0.1\ A} = \dfrac{0.25\ V}{0.1\ A} = 2.5\ \Omega$

2. Since we defined useful life voltage between 1.4 V and 1.6 V, this cell is *bad*.

II

1. It will deliver 7.5 amperes for 20 hours $\left(\dfrac{150\ \text{ampere-hours}}{20\ \text{time in hours}}\right)$.

2. C rating would be 60/12.5, meaning it could deliver 12.5 amperes for 4.8 hours.

IN-PROCESS LEARNING CHECKS

I

1. The carbon-zinc type dry cell is composed of a negative electrode made of *zinc* and a positive electrode made of *carbon.* The electrolyte solution in this type cell is *ammonium chloride,* called sal *ammoniac.*

2. The process which causes hydrogen bubbles to form on the positive electrode, and which is detrimental to the cell's operation, is called *polarization.*

3. Manganese dioxide is often added to the cell to prevent the process mentioned in question 2. This agent is sometimes known as a *depolarizer* agent.

4. Another problem that can occur in dry cells is that of little batteries being formed because of impurities suspended in the cell's electrolyte near the zinc electrode. This type activity is known as *local action.* To reduce this undesired activity, mercury is put on the zinc by a process called *amalgamation.*

5. The shelf life of dry cells is *hurt* by storing the cells in hot conditions.

6. Nominal cell terminal voltage of dry cells is about *1.5* V.

REVIEW QUESTIONS

1. A positive ion is an atom that has lost electron(s).

3. Amalgamation is the process of coating the inside of the zinc electrode with mercury in a carbon-zinc cell to minimize local action.

5. Manganese dioxide

7. Mercury cell

9. Used in applications in which there is cyclic rather than continuous use (e.g., cameras and portable transmitters)

PROBLEMS

1. 0.05 Ω

3. 100 ampere-hours

5. Two parallel branches of two cells in series for each parallel branch

ANALYSIS QUESTIONS

1. The characteristics of the charging action are higher charging currents at first (when the battery's series-opposing voltage is lowest). Charging current decreases as the battery charges and its terminal voltage approaches the source value.

3. The source's positive terminal is connected to the battery's positive terminal and the source's negative terminal is connected to the battery's negative terminal. This reverses the chemical action in the battery and recharges it.

5. a. AAA (carbon-zinc): $V = 1.5$ V (current rating not shown in the catalog)
   b. AA (carbon-zinc): $V = 1.5$ V (current rating not shown in the catalog)
   c. A (carbon-zinc): not a standard type

d. C (carbon-zinc): $V = 1.5$ V (current rating not shown in the catalog)

e. D (carbon-zinc): $V = 1.5$ V (current rating not shown in the catalog)

7. Current capability is primarily determined by the size of the active area of the electrodes.

9. If load R is increased, then load current decreases and the current demand from the cell decreases. This means that the battery or cell should be able to deliver normal output voltage for a longer period.

11. One key advantage of the lithium cell for applications mentioned in the preceding question is that the lithium cell has a long life; therefore, it does not need to be changed often.

CHAPTER 10

PRACTICE PROBLEMS

I

II

Figure 10–13a = Move away from each other.

Figure 10–13b = Move toward one another.

Figure 10–13c = Move away from each other.

III

Figure 10–17a = N at right end, S at left end; Figure 10–17b = N at left end, S at right end; and Figure 10–17c = N at left end, S at right end.

IN-PROCESS LEARNING CHECKS

I

1. Small magnets that are suspended and free to move align in a *North* and *South* direction.

2. Materials that lose their magnetism after the magnetizing force is removed are called *temporary magnets.* Materials that retain their magnetism after the magnetizing force is removed are called *permanent magnets.*

3. A wire carrying dc current *does* establish a magnetic field. Once a magnetic field is established, if the current level doesn't change, the field is *stationary.*

4. A law of magnetism is like poles **repel** each other and unlike poles **attract** each other.

5. A maxwell represents **one** line of force.

6. A weber represents 10^8 lines of force.

7. Lines of force are **continuous.**

8. Lines of force related to magnets exit the **North** pole of a magnet and enter the **South** pole.

9. Nonmagnetic materials do **not** stop the flow of magnetic flux lines through themselves.

10. Yes, magnetism **can** be induced from one magnetic object to another object, if it is a magnetic material.

II

1. For a given coil dimension and core material, what two factors primarily affect the strength of an electromagnet? **Number of turns** and **amount of current.**

2. The left-hand rule for determining the polarity of electromagnetics states that when the fingers of your left hand point **in the same direction** as the current passing through the coil, the thumb points toward the **North** pole of the electromagnet.

3. Adjacent current-carrying conductors that are carrying current in the same direction tend to **attract each other.**

4. If you grasp a current-carrying conductor so that the thumb of your left hand is in the direction of current through the conductor, the fingers "curled around the conductor" **will indicate** the direction of the magnetic field around the conductor.

5. When representing an end view of a current conductor pictorially, it is common to show current coming out of the paper via a **dot.**

III

1. A *B-H* curve is also known as a **magnetization** curve.

2. *B* stands for flux **density.**

3. *H* stands for magnetizing **intensity.**

4. The point where increasing current through a coil causes no further significant increase in flux density is called **saturation.**

5. The larger the area inside a "hysteresis loop" the **larger** the magnetic losses represented.

REVIEW QUESTIONS

1. b

3. b

5. d

7. Permeability

9. φ, Weber

11. MMF, Ampere-turns (NI)

13. 10^8

15. The shield simply diverts lines of flux through itself, preventing flux from reaching the items inside the shield enclosure.

17. Drawing of a hysteresis loop, with callouts for saturation, residual magnetism, and coercive force:

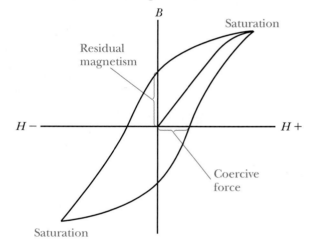

19. b

21. The direction of an induced voltage or current is such that it tends to oppose the change causing it.

23. Motor effect in a generator relates to the induced current causing an opposition to the motion that induced it.

25. Toroidal coil forms are efficient conductors of magnetic flux because of the high permeability materials used in them and because of their geometry not allowing much leakage flux.

PROBLEMS

1. Drawing of a typical field pattern for unlike poles facing each other and like poles facing each other:

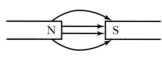 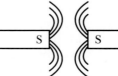

3. Labeling of North and South poles of the electromagnet circuit shown:

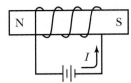

5. a. $10^8/10^6 = 100$ lines of flux
   b. Teslas = webers per square meter (Since there are 100 lines in 0.005 m², there are $1/0.005 \times 100$ lines in a square meter, or lines = $200 \times 100 = 20,000$.)
   Number of webers per square meter: teslas = $20,000/100,000,000 = 0.0002$ T

7. 1,750 gauss = $1,750 \times 10^{-4}$ teslas or 0.175 teslas

9. H = $3 \times 1,500/1 = 4,500$ ampere-turns per meter

ANALYSIS QUESTIONS

1. Magnetomotive force is the force by which a magnetic field is produced and relates to the ampere-turns involved in an electromagnet. Magnetic field intensity add the dimension of ampere-turns per a given length of the electromagnetic coil. In the SI system, field intensity is AT/meter.

3. Reluctance will triple.

5. Flux density will increase.

7. Earth's geographic N pole = magnetic S pole.

CHAPTER 11

PRACTICE PROBLEMS

I

$$R_S = \frac{V_m}{I_S}$$

$$R_S = \frac{100 \text{ mV}}{99 \text{ mA}}$$

$$R_S = 1.01 \ \Omega$$

II

$$R_{Mult} = \frac{V \text{ (range)}}{I \text{ (fs)}} - R_M$$

$$R_{Mult} = \frac{250}{50 \times 10^{-6}} - 2,000$$

$$R_{Mult} = 5,000,000 - 2,000$$
$$R_{Mult} = 4,998,000 \text{ or } 4,998 \text{ M}\Omega$$

III

1. Ohms-per volt rating of a 20-microampere movement is $\dfrac{1}{fs} = \dfrac{1}{20 \times 10^{-6}} = 50,000$ ohms per volt.

2. Full-scale current rating of a meter rated at 2,000-ohms per volt is $1/2,000 = 0.5$ mA movement.

IN-PROCESS LEARNING CHECK

I

1. Three essential elements in an analog meter movement include a **moving** element, an **indicating** means, and some form of **damping.**

2. The purpose of the permanent magnet in a d'Arsonval meter movement is to establish a fixed **magnetic field.**

3. The movable element in a moving coil meter moves due to the interaction of the fixed **magnetic field** and the magnetic field of the movable coil when current passes through the coil.

4. Electrical conduction of current in a movable meter movement is accomplished through the **springs,** which also provides a desired **mechanical** force against which the movable coil must act.

5. The element in a moving-coil movement producing electrical damping is the **coil frame.**

REVIEW QUESTIONS

1. Permanent magnet, pole pieces, soft iron core, movable coil, support frame for coil, spindle and bearings, springs, pointer, counterbalance, calibrated scale, and stops

3. c

5. b

7. Turn power off, break circuit, insert meter (observing polarity), power on, read meter.

9. a

11. Less loading effect; easier to read

13. a

15. Do not have to break circuit.

17. Electrical damping used to minimize overshoot and pointer oscillations.

19. c

21. c

PROBLEMS

1. Full-scale current rating = 2 mA

   R meter = 200 m V/2 mA = 100 Ω

3. $R_{mult} = VI(range)/I(fs) - R_m = 30 \quad V/3 \quad mA - 50 \Omega = 9,950 \Omega$

5. Voltmeter reading = 3.33 V

   Without voltmeter present, R_2 would drop half the applied voltage, or 5 V.

7. When 0–50 V range is reading 22 V, percentage fs = 44%.

   When 75% of fs is flowing, meter reads 37.5 V.

9. Midscale = 1,500 Ω

   Quarter scale = 4,500 Ω

11. 49.75 V

ANALYSIS QUESTIONS

1. A shorting type switch is where the movable contact makes contact with the next contact position before "leaving" the previous contact position. This type switch is often used in multiple-range ammeter circuits to ensure that the meter is never "unshunted."

3. Drawing of a three-resistor parallel circuit with dc source, showing how ammeters should be connected to measure I_T and the current through the middle branch:

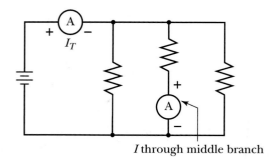

   I through middle branch

5. If an analog meter is connected into a dc circuit with the wrong polarity, the pointer would tend to bang against the stop pin on the left side of the scale. This can damage the pointer and possibly the meter movement as well.

CHAPTER 12

PRACTICE PROBLEMS

I

1. The sine of 35° = 0.5735.

2. Cycle 1 ends and cycle 2 begins at Point G.

3. Maximum rate of change for the waveform shown occurs at Points J, N, R, and T.

II

1.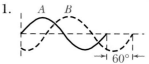

2.

III

1. 10 V

2. 9 V

3. 90%

4. No

5. Yes

6. a. 120.8 V
   b. 171 V
   c. 120.8 V
   d. 171 V
   e. 108.9 V

IV

1. a. $V_1 = 50$ V
   b. $V_2 = 50$ V
   c. $I_p = 7.07$ mA
   d. $P_p = 999.7$ mW
   e. $P_1 = 250$ mW

2. a. $I_1 = 10$ mA
   b. $I_2 = 10$ mA
   c. $I_p = 28.28$ mA
   d. $V_{p-p} = 282.8$ V

3. a. 66.7 V
   b. 33.3 V
   c. 33.3 V
   d. 444.89 mW

V

The duty cycle = Pulse width/Time for one period times 100. Duty cycle = (1 µs/15 µs) × 100 = 6.67% duty cycle.

VI

1. V_{avg} = Baseline value + (duty cycle × p-p amplitude)

   V_{avg} = 0 + (pulse width/period × p-p amplitude)
   V_{avg} = 0 + [(2 ms/30 ms) × 6 V] = 0 + 0.066 × 6 = 0.396 V

2. −2.604 V

IN-PROCESS LEARNING CHECKS

I

1. The basic difference is that dc (direct current) is in one direction and of one polarity. Alternating current (ac) periodically changes direction and polarity.

2. The reference, or 0° position is horizontally to the right when describing angular motion.

3. The y-axis is the vertical axis in the coordinate system.

4. The second quadrant in the coordinate system is that quadrant between 90° and 180°.

5. A vector represents a given quantity's magnitude and direction with respect to location in space.

6. A phasor represents a rotating vector's relative position with respect to time.

II

1. 10-kHz

2. 0.0025 seconds, or 2.5 msec

3. 50 μsec

4. Amplitude

5. 45° and 135°; 225° and 315°

6. 1 and 2

7. 0.067 μsec

8. 40-Hz

9. As frequency increases, T *decreases.*

10. The longer a given signal's period, the *longer* the time for each alternation.

REVIEW QUESTIONS

1. d

3. b

5. b

7. c

9. b

11. a

13. c

15. c

17. A "periodic wave" is one that repeats itself in time and form.

19. b

PROBLEMS

1. t = 1/f = 1/2,000 = 0.0005 seconds, or 0.5 msec

3. 1/120 of a second.

5. 400 Hz

7. The peak-to-peak value would also double.

9. The frequency has decreased to one-third its original frequency.

11. One alternation is half the time of a cycle; therefore, the time for one alternation = 1 ms.

13. Diagram:

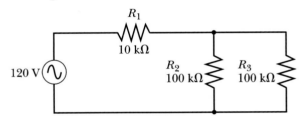

a. V_{R_1} = 20 V
b. I_{R_1} = 2 mA
c. P_{R_1} = 40 mW
d. I_{R_2} = 1 mA
e. V_{R_3} = 100 V
f. θ between V applied and I total = 0°

15. λ will increase to three times the original.

ANALYSIS QUESTIONS

1. Two differences between ac and dc are 1) ac changes polarity, periodically, dc does not; 2) ac continuously varies in amplitude, dc does not; 3) ac can be stepped up or down via transformer action, dc cannot.

3. If rms value = 100 V, then peak value = 141 V and peak-to-peak value = 282 V.

5. 70.7 V

7. Since A = 1/0.002 × 10^{-3} = 500 kHz, B = 250 kHz.

9. a. Graphic drawing of rectangular wave:

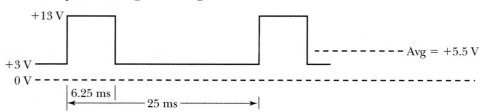

b. The V_{avg} = Baseline value + (Duty cycle × p-to-p amplitude)

V_{avg} = 3 + (Pulse width/period × p-to-p amplitude)

V_{avg} = 3 + [(6.25 ms/25 ms) × 10 V] = 3 + 0.25 × 10 = 5.5 V

CHAPTER 13

PRACTICE PROBLEMS

I

When the vertical frequency is four times the horizontal frequency, the waveform shows four cycles of the given waveform.

II

1. 3.5 V (rms: 5 V peak voltage)

2. a. 12 V (p-to-p) b. 4.24 V (rms)

III

1. +1 V dc

2. +0.8 V dc

IV

1. 30 V (p-to-p); 10.6 V (rms)

2. 1.76 V/div (rms)

3. Use the 10 position setting of the vertical volts/div control.

V

1. 1.5 V (p-to-p); 0.53 V (rms)

2. 3 V (p-to-p); 1.06 V (rms)

VI

72 degrees per major division

VII

1. a. Period (T) = 0.25 s

   b. Frequency = 4 Hz

2. a. Period (T) = 50 μs

   b. Frequency = 20 kHz

IN-PROCESS LEARNING CHECKS

I

1. The part of the oscilloscope producing the visual display is the ***cathode-ray tube.***

2. The scope control that influences the brightness of the display is the ***intensity*** control.

3. A waveform is moved up or down on the screen by using the ***vertical position*** control.

4. A waveform is moved left or right on the screen by using the ***horizontal position*** control.

5. For a signal fed to the scope's vertical input, the controls that adjust the number of cycles seen on the screen are the horizontal ***time/div.*** and the horizontal ***time variable*** controls.

6. The control(s) that help keep the waveform from moving or jiggling on the display are associated with the ***synchronization*** circuitry.

REVIEW QUESTIONS

1. Name is cathode-ray tube and abbreviation is CRT.

3. To protect and add life to coating on CRT screen

5. Vertical input terminal(s)

7. "Ext Trig" terminal (NOTE: Other names are also used.)

9. To retain proper calibration

PROBLEMS

1. 4 × 0.1 = 0.4 ms/cycle; $f = 1/t = 1/0.4$ ms = 2,500 Hz

3. 2 divisions

5. T = 0.004 s

7. 0.0004

9. p-to-p = 5 V; rms V = 1.77 V; $I = 1.77$ V/4.7 kΩ = 0.3765 mA

ANALYSIS QUESTIONS

1. List of control names should be based on the specific oscilloscope names students can observe on your equipment in your training setting:
   a. Control the trace focus _____.
   b. Control the trace intensity _____.
   c. Control(s) the vertical position _____.
   d. Control(s) the horizontal position _____.
   e. Vertical attenuator/amplifier controls _____.
   f. Horizontal sweep frequency controls _____.

3. Diagram B: vertical position and horizontal gain controls
   Diagram C: vertical position and horizontal gain controls
   Diagram D: vertical gain and horizontal position controls
   Diagram E: horizontal gain control
   Diagram F: vertical gain control
   Diagram G: horizontal sweep frequency controls (horizontal time/div. and time variable)
   Diagram H: vertical gain control
   Diagram I: vertical and horizontal gain controls
   Diagram J: horizontal sweep frequency controls (horizontal time/div. and time variable)
   Diagram K: horizontal gain control
   Diagram L: vertical position, vertical and horizontal gain controls
   Diagram M: vertical gain control

CHAPTER 14

PRACTICE PROBLEMS

I
Since 1 weber = 10^8 flux lines and cutting 1 Wb/sec induces 1 V, then cutting 2×10^8 flux lines (2 Wb) in 0.5 sec = 4 V.

II
$$V_L = L\frac{di}{dt}$$
$$V_L = 5\left(\frac{3}{1}\right) = 15 \text{ V}$$

III
10 mH + 6 mH + 100 μH = 16.1 mH total inductance

IV
1. Three equal value inductors (15 H) in parallel have a total equivalent inductance = 1/3 of one of them. L_T = 5 henrys.

2. L_T of 10 mH, 12 mH, 15 mH, and 20 mH inductors in parallel is 3.33 mH.

V
a. dc current through coil = 5 A
b. energy = $\frac{LI^2}{2} = \frac{1(25)}{2} = 12.5$ J

VI
Two seconds = approximately 2.4 TC.
After 2.4 TC, V_R is about 91% of V applied, or 9.1 V.
After 2.4 TC, V_L is about 9% of V applied, or 0.9 V.

VII
0.92 mA

IN-PROCESS LEARNING CHECKS

I
1. All other factors remaining the same, when an inductor's number of turns increases four times, the inductance value *increases* by a factor of *16 times.*

2. All other factors remaining the same, when an inductor's diameter triples, the inductance value *increases* by a factor of *9 times.* (NOTE: A is directly proportional to d^2.)

3. All other factors remaining the same, when an inductor's length increases, the inductance value *decreases.*

4. When the core of an air-core inductor is replaced by material having a permeability of ten, the inductance value *increases* by a factor of *10 times.*

REVIEW QUESTIONS

1. Inductance (or self-inductance) is the property of a circuit that opposes a *change* in current flow.

3. Number of turns on the coil; length of the coil; cross-sectional area of the coil; relative permeability of the core material

5. The amount of emf induced in a circuit is related to the number of magnetic flux lines cut or linked per unit time.

7. c

9. c

11. b

13. a

15. c

17. Be careful regarding the "high-voltage kick" produced when the test leads break connection with the inductor just tested. Use appropriate personal safety measures.

19. The 63.2% indicates the percentage of change that will occur from the present level toward the final (stable) level or value during the period of one time constant.

PROBLEMS

1. Inductance will increase to 400 mH (L varies as the square of the turns).

3. New inductance is 500×100 μH = 0.05 H.

5. $L_T = L_1 + L_2 + L_3 = 0.1$ H + 0.2 H + 0.001 H = 0.301 H, or 301 mH

7. 1 TC = $L/R = 250 \times 10^{-3}/100 = 0.0025$ sec, or 2.5 ms
   It will take 5 times 2.5 ms, or 12.5 ms for current to complete the change from one level to another.

9. Tapping a coil at midpoint yields an inductance of one-fourth the value of the total inductance of the coil. In the case of the 0.4 H coil, L equals about 0.1 H at the 250-turn tap.

11. V_L after 0.6 μs equals approximately 6.76 V.

ANALYSIS QUESTIONS

1. Inductance will decrease.

3. Total inductance will increase.

5. Since the inductance change is proportional to the square of the turns, $L_1/L_2 = (N_1)^2/(N_2)^2$; $4/6 = (1,000)^2/(N_2)^2$; $N_2 = 1224.74$ turns.

7. Because the voltage-per-turn is less than the varnish insulation breakdown voltage (The voltage-per-turn in this case is approximately 2 V/turn.)

9. The joules of energy stored by an inductor is directly related to the amount of inductance the inductor has because the strength of the magnetic field built-up by a given value of current through the coil is directly related to the inductance value. The energy is stored in the form of the magnetic field.

CHAPTER 15

PRACTICE PROBLEMS

I

1. 3 kΩ

2. 333.3 Ω

II

1. 6,280 Ω

2. 79.6 mH, or approximately 80 mH

3. 238.85 Hz, or approximately 239 Hz

III

1. 5 kΩ

2. $X_L = 5,024$ Ω, so $I \times X_L = 3$ mA $\times$ 5,025 Ω = approximately 15 V

3. 40 mA

IN-PROCESS LEARNING CHECKS

I

1. When current increases through an inductor, the cemf *hinders* the current increase.

2. When current decreases through an inductor, the cemf *hinders* the current decrease.

3. In a pure inductor, the *voltage* leads the *current* by 90°

4. What memory aid helps in recalling the relationship described in question 3? *"Eli"* the ice man.

5. The opposition that an inductor shows to ac is termed *inductive reactance.*

6. The opposition that an inductor shows to ac *increases* as inductance increases.

7. The opposition that an inductor shows to ac *decreases* as frequency decreases.

8. X_L is *directly* related to inductance value.

9. X_L is *directly* related to frequency.

REVIEW QUESTIONS

1. c

3. c

5. b

7. d

9. b

PROBLEMS

1. $L = X_L/2\pi f = 376.8/6.28 \times 100 = 376.8/628 = 0.6$ H; at 450 Hz, $X_L = 1695.6 \ \Omega$

3. Total inductance 2 mH × 6 mH/2 mH + 6 mH = 12/8 = 1.5 mH $f = X_L/2\pi L = 39.8$ kHz

5. $Q = X_L/R = 6.28 \times 20$ MHz × 10 μH/10 Ω = 125.6/10 = 12.56 Q not a dc parameter. Q will increase, because $Q = \dfrac{X_L}{R}$ and condition is $\dfrac{3 \ X_L}{2 \ R}$.

7. $V_T = 45$ V
   $V_{L_1} = 15$ V
   $I_T = 0.75$ mA

9. Approximately 6 V

11. L_T will RTS.
    X_{L_T} will I.
    I_T will D.
    V_{L_4} will RTS.
    I_1 will D.

13. $L = 0.318$ H

15. $X_{LT} = 28.89$ kΩ

17. I_T would decrease.

19. Voltage applied would have to decrease to one-third its original value.

ANALYSIS QUESTIONS

1. Six times greater

3. Circuit current will increase, because X_L will decrease.

5. $Q = X_L/R$ (as f decreases, X_L decreases while R remains the same); Q will decrease.

7. Less losses (less wasted energy)

9. b

CHAPTER 16

PRACTICE PROBLEMS

I

5 pounds at an angle of 36.9° from horizontal.

II

$V_S = \sqrt{V_R^2 + V_L^2} = \sqrt{25 + 25} = \sqrt{50} = 7.07$ V

III

1. $\cos \theta = \dfrac{\text{adj}}{\text{hyp}} = \dfrac{5}{10}$ and $\cos^{-1} = 60°$

2. Tangent of 45° = 1.0

3. Sine of 30° = 0.5

IV

V applied = 150 V

V

1.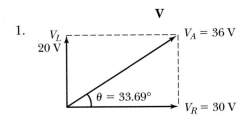

2. $\theta = 33.69°$ (Tan θ = opp/adj = 20/30 = 0.6666. Angle whose tangent = 0.6666 is 33.69°.)

3. Tangent function

4. Angle would have been greater (56.3°).

VI

1. $V_R = 35.35$ V (Cos θ = adj/hyp and cos 45° = 0.7071. Therefore, 0.7071 = adj/50 = 35.35.)

2. a. $V_R = 32.5$ V
   b. $V_L = 56.29$ V

3. a. Angle = 56.3°
   b. $V_T = 43.2$ V

VII

1.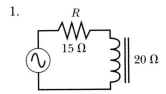

2.

3. $Z = \sqrt{15^2 + 20^2}$
   $Z = \sqrt{625}$
   $Z = 25 \ \Omega$

4. Tan $\theta = \dfrac{X_L}{R}$
   Tan $\theta = \dfrac{20}{15}$
   Tan $\theta = 1.33$
   Angle = 53.1°
   Yes, answer agrees.

VIII

1. $I_T = \sqrt{I_R^2 + I_L^2}$
   $I_T = \sqrt{5^2 + 3^2}$
   $I_T = \sqrt{34}$
   $I_T = 5.83$ A

2. $Z = \dfrac{V_T}{I_T}$

   $Z = \dfrac{300}{5.83}$

   $Z = 51.45\ \Omega$

3. $\theta -30.9°$ (Tan θ = opp/adj = 3 A/5 A = 0.6. Angle whose tangent equals 0.6 is 30.9°.)

IX

$Z = 47 \times 100/(\sqrt{47^2 + 100^2}) = 4{,}700/110.5 = 42.5\ \Omega$

IN-PROCESS LEARNING CHECKS

I

1. In a purely resistive ac circuit, the circuit voltage and current are **in** phase.

2. In a purely inductive ac circuit, the current through the inductance and the voltage across the inductance are **90** degrees out of phase. In this case, the **voltage** leads the **current.**

3. A quantity expressing both magnitude and direction is a **vector** quantity.

4. The length of the vector expresses the **magnitude** of a quantity.

REVIEW QUESTIONS

1. b
3. a
5. a
7. c
9. c
11. c
13. $Z = \sqrt{R^2 + X_L^2}$
15. $Z = R \times X_L/\sqrt{R^2 + X_L^2}$
17. b
19. a

PROBLEMS

1. a. 36 Ω
   b. Pythagorean result = 36 Ω
   c. Angle between R and Z = 56.3°

3. Diagrams for the specified series RL circuit are shown:

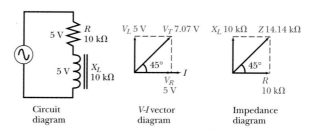

Circuit diagram V-I vector diagram   Impedance diagram

Since $\theta = 45°$, then R must $= X_L$, and V_R must $= V_L$.
$Z = R^2 + X_L^2 = 200 \times 10^6 = 14.14 \times 10^3 = 14.14$ kΩ

5. a. Z = vector sum of 10 kΩ R and approximate 20 kΩ X_L; $Z = 22.36$ kΩ
   b. $V_R = IR = (67\ \text{V}/22.36\ \text{k}\Omega) \times 10\ \text{k}\Omega$ = approximately 3 mA $\times$ 10 kΩ = 30 V
   c. θ = approximately 63.43°

7. $I = 5$ mA
   $Z = 40$ kΩ
   $V_L = 105.4$ V
   $X_L = 21.08$ kΩ
   $L = X_L/2\pi f = 56$ mH

9. Diagrams and calculations for the specified parallel RL circuit are shown:

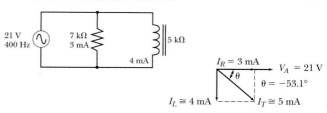

$X_L = 2\pi fL = 5$ kΩ (approximately)
$I_R = 21\ \text{V}/7\ \text{k}\Omega = 3$ mA
$I_L = 21\ \text{V}/5\ \text{k}\Omega = 4.2$ mA (or about 4 mA)
$I_T = I_R^2 + I_L^2 = 5.16$ mA (or about 5 mA)
$Z = V_T/I_T = 21\ \text{V}/5\ \text{mA} = 4.2$ kΩ
Cos θ = adj/hyp = 3 mA/5 mA = 0.6; $\theta = -53.1°$ (current lags V_T by 53.1°)

ANALYSIS QUESTIONS

1. a. R will RTS.
   b. L will RTS.
   c. X_L will I.
   d. Z will I.
   e. θ will D.
   f. I_T will D.
   g. V_T will RTS.

3. a. *R* will I.
   b. *L* will RTS.
   c. X_L will RTS.
   d. *Z* will I.
   e. θ will D.
   f. I_T will D.
   g. V_T will RTS.

5. a. *R* will I.
   b. *L* will RTS.
   c. X_L will RTS.
   d. *Z* will I.
   e. θ will I.
   f. I_T will D.
   g. V_T will RTS.

7. 57, sin

9. 34, cos

11. Not having to know the hypotenuse to calculate

CHAPTER 17

PRACTICE PROBLEMS

I

$M = k\sqrt{L_1 \times L_2}$
$M = 0.95\sqrt{10 \times 15}$
$M = 0.95\sqrt{150}$
$M = 0.95 \times 12.25 = 11.63$ H

II

$L_T = L_1 + L_2 - 2M$
$L_T = 10\text{ H} + 10\text{ H} - (2 \times 5\text{ H})$
$L_T = 20\text{ H} - 10\text{ H} = 10\text{ H}$

III

1. $L_T = \dfrac{1}{\dfrac{1}{L_1 + M} + \dfrac{1}{L_2 + M}}$

   $L_T = \dfrac{1}{\dfrac{1}{8 + 4} + \dfrac{1}{8 + 4}}$

   $L_T = \dfrac{1}{\dfrac{1}{12} + \dfrac{1}{12}}$

   $L_T = \dfrac{1}{\dfrac{1}{6}} = 1 \times \dfrac{6}{1} = 6$ H

2. $L_T = \dfrac{1}{\dfrac{1}{L_1 - M} + \dfrac{1}{L_2 - M}}$

$L_T = \dfrac{1}{\dfrac{1}{8 - 4} + \dfrac{1}{8 - 4}}$

$L_T = \dfrac{1}{\dfrac{1}{4} + \dfrac{1}{4}}$

$L_T = \dfrac{1}{\dfrac{1}{2}} = 1 \times \dfrac{2}{1} = 2$ H

IV

1. $\dfrac{N_p}{N_s} = \dfrac{V_p}{V_s}$; s-p turns ratio $= \dfrac{100}{300}$ or 1:3

2. Turns ratio equals voltage ratio. Since the transformer steps-up the voltage six times, the secondary must have six times the turns of the primary. Therefore, the p-s turns ratio is 1:6.

3. Since the p-s turns ratio is 1:5, the secondary has five times the voltage of the primary, or $5 \times 50\text{ V} = 250$ V.

4. Voltage ratio with p-s turns ratio of 4:1 (primary-to-secondary). This is a step-down transformer, since the primary voltage is four times that of the secondary.

V

Since the impedance is transformed in relation to the square of the turns ratio, when the secondary has twice the number of turns of the primary, the impedance at the primary looks like one-fourth the impedance across the secondary, or $2{,}000/4 = 500$ Ω.

VI

1. The *primary-to-secondary* impedance ratio is related to the square of the turns relationship of primary-to-secondary. Since the primary has five times as many turns as the secondary, the primary impedance is 5^2 times greater than secondary. Therefore, the p-s *Z* ratio is 25:1.

2. The impedance looking into the primary is $(16)^2 \times 4\text{ Ω} = 256 \times 4 = 1{,}024$ Ω.

IN-PROCESS LEARNING CHECKS

I

1. Producing voltage via a changing magnetic field is called electromagnetic *induction.*

2. Inducing voltage in one circuit by varying current in another circuit is called *mutual inductance.*

3. The fractional amount of the total flux that links two circuits is called the *coefficient* of coupling, which is represented by the letter **k**. When 100% of the flux links the two circuits, the *coefficient* of coupling has a value of *1*.

4. The closer coils are, the **higher** the coupling factor produced. Compared with parallel coils, perpendicular coils have a **lower** degree of coupling.

REVIEW QUESTIONS

1. b

3. b

5. b

7. a

9. c

11. c

13. b

15. c

17. b

PROBLEMS

1. $I_p = 200$ mA

3. $N_P/N_S = 20:1$

5. $M = 1.2$ H

7. $L_T = 2.5$ H

9. Volts-per-turn = 0.1 volts-per-turn

11. Turns = 2,000 turns

13. Primary-to-secondary turns ratio = 1:4.75

15. Voltage applied to primary = 62.5 V

ANALYSIS QUESTIONS

1. A "flyback" transformer (sometimes called a "horizontal transformer") is typically used in the horizontal deflection circuits of televisions. It primarily provides high voltage to the picture tube or CRT. It may also provide low filament voltage for the high voltage rectifier. The scanning circuit in which it operates uses a frequency of 15,750 Hz.

3. Primary winding should have 3535.5 turns.

5. Number of secondary turns = 200 turns

CHAPTER 18

PRACTICE PROBLEMS

I

1. $C = \dfrac{Q}{V} = \dfrac{100 \times 10^{-6}}{25} = 4 \times 10^{-6} = 4 \ \mu F$

2. $Q = CV = 10 \times 10^{-6} \times 250 = 2,500 \ \mu C$

3. $V = \dfrac{Q}{C} = \dfrac{50 \times 10^{-6}}{2 \times 10^{-6}} = 25$ V

II

$$E = \frac{1}{2}CV^2 = \frac{10 \times 10^{-6} \times (100)^2}{2} = \frac{100,000 \times 10^{-6}}{2}$$

$$= 50,000 \ \mu J \text{ or } 0.05 \text{ J}$$

III

1. $C = \dfrac{2.25 \text{ kA}}{10^7 \text{s}} \times (n - 1) = \dfrac{2.25 \times 10 \times 1}{10^7 \times 0.1}$

   $C = \dfrac{22.5}{10^6} \times (3) = 22.5 \times 10^{-6} \times 3 = 67.5 \ \mu F$

2. $C = \dfrac{8.85 \text{ kA}}{10^{12} \text{s}} = \dfrac{8.85 \times 1 \times 0.2}{10^{12} \times 0.005} = \dfrac{1.77 \times 10^{-12}}{0.005}$

   $= 354$ pF

IV

1. $C_T = \dfrac{C_1 \times C_2}{C_1 + C_2} = \dfrac{5 \times 10^{-6} \times 20 \times 10^{-6}}{25 \times 10^{-6}} =$

   $\dfrac{100 \times 10^{-12}}{25 \times 10^{-6}} = 4 \times 10^{-6}$, or $4 \ \mu$ F

2. Total capacitance of four 20-μF capacitors in series is equal to one of the equal value capacitors divided by 4. Thus, $20/4 = 5 \ \mu F$, $C_T = C_1/C_N$.

4. V_{C_1} = four-fifths V_T, or 80 V. V_{C_2} = one-fifth V_T, or 20 V.

   ($Q_T = C_T \times V_T$; 4 μF $\times$ 100 V = 400 μC)

   $\left(V_{C_1} = \dfrac{Q}{C_1}; \ = \dfrac{400 \ \mu C}{5 \ \mu F} = 80 \text{ V} \right)$

   $\left(V_{C_2} = \dfrac{Q}{C_2}; \ = \dfrac{400 \ \mu C}{20 \ \mu F} = 20 \text{ V} \right)$

V

1. C_T = sum of individual capacitances; therefore = 60 μF.

2. $Q_{C_1} = C_1 \times V = 12 \ \mu\text{F} \times 60 \ \text{V} = 720 \ \mu\text{C}$
$Q_{C_1} = Q_{C_2} = Q_{C_3} = Q_{C_4} = Q_{C_5}$
$Q_T = \text{sum of all the charges} = 5 \times 720 \ \mu\text{C} = 3{,}600 \ \mu\text{C}.$

VI

1. One time constant for circuit $= 0.001 \times 10^{-6} \times 10 \times 10^3$
   1 TC = 0.01 milliseconds (or 10 microseconds)
   Time allowed for charge = 20 μs, or 2 TC
   After 2 TC, capacitor is charged to 86.5% of V applied.
   In this case, $V_C = 86.5\%$ of 100 V, or 86.5 V, after 20 μs.

2. The voltage across the resistor 15 μs after the switch is closed will be equal to the percentage of V applied that V_R drops to after 1.5 TC.
   (15 μs = 1.5 $\times$ the 10-μs TC). According to the TC chart, the resistor voltage is approximately 22% of V applied, or 22 V. Charging current = 2.2 mA.

3. V_C 30 μs after the switch is closed $\cong$ 95 V.

4. V_R 25 μs after the switch is closed $\cong$ 8.2 V.

IN-PROCESS LEARNING CHECKS

I

1. A capacitor is an electrical component consisting of two conducting surfaces called **plates** that are separated by a nonconductor called the **dielectric.**

2. A capacitor is a device that stores electrical **charge** when voltage is applied.

3. Electrons **do not** travel through the capacitor dielectric.

4. During charging action, one capacitor plate collects **electrons,** making that plate **negative.** At the same time, the other plate is losing **electrons** to become **positive.**

5. Once the capacitor has charged to the voltage applied, it acts like an **open** circuit to dc. When the level of dc applied increases, the capacitor **charges** to reach the new level. When the level of dc applied decreases, the capacitor **discharges** to reach the new level.

6. When voltage is first applied to a capacitor, **maximum** charge current occurs.

7. As the capacitor becomes charged, current **decreases** through the circuit in series with the capacitor.

REVIEW QUESTIONS

1. b

3. a

5. a

7. d

9. b

11. c

13. Observe polarity.

15. Mica

17. Small size; very little lead length provides little stray undesired electrical parameters, such as inductance.

19. b

PROBLEMS

1. New capacitance is four times the original capacitance, since C is directly related to the area of the plates facing each other and inversely related to the dielectric thickness.

3. $C = \dfrac{Q}{V} = \dfrac{200 \ \text{pC}}{50} = 4 \ \text{pF}$

5. $C_T \ (\text{pF}) = \dfrac{1}{\dfrac{1}{C_1} + \dfrac{1}{C_2} + \dfrac{1}{C_3}} = \dfrac{1}{\dfrac{1}{100} + \dfrac{1}{400} + \dfrac{1}{1{,}000}}$
   $C_T = \dfrac{1}{0.01 + 0.0025 + 0.001} = \dfrac{1}{0.0135} = 74.07 \ \text{pF}$
   $C_T = 100 \ \text{pF} + 400 \ \text{pF} + 1{,}000 \ \text{pF} = 1{,}500 \ \text{pF}$

7. $Q_{C_1} = C_1 \times V_1 = 10 \ \mu\text{F} \times 100 \ \text{V} = 1{,}000 \ \mu\text{C}$
   $Q_{C_2} = C_2 \times V_2 = 20 \ \mu\text{F} \times 100 \ \text{V} = 2{,}000 \ \mu\text{C}$
   $Q_{C_3} = C_3 \times V_3 = 50 \ \mu\text{F} \times 100 \ \text{V} = 5{,}000 \ \mu\text{C}$
   $C_T = C_1 + C_2 + C_3 = 80 \ \mu\text{F}$
   $Q_T = Q_1 + Q_2 + Q_3 = 8{,}000 \ \mu\text{C}$

9. $v_R = V \text{ applied} \times \epsilon^{-t/\tau} = 200 \times 2.718^{-1.2} = 60.23 \ \text{V}$
   $v_C = V_A - V_R,$ or approximately $200 - 60 = 140 \ \text{V}$

11. Voltage across $C_5 = 8$ V

13. 600 V

15. $J = 12{,}800 \ \mu\text{J}$

17. $J = C(V^2)/2 = 30 \ \mu\text{F} \times (225)^2/2 = 1.518/2 = 0.759375 \ \text{J}$, or 759,375 μJ.

ANALYSIS QUESTIONS

1. Students should reflect the ranges and operational procedures for measuring capacitance with

a DMM in terms of the equipment or vendor data available to them.

3. Students should reflect the ranges and operational procedures for measuring capacitance with an LCR meter in terms of the equipment or vendor data available to them.

CHAPTER 19

PRACTICE PROBLEMS

I

1. Yes, the X_C formula verifies the values of X_C shown in Figure 19–3.

2. $X_C = 1/(2\pi fC) = 1/(6.28 \times 1,000 \times 10^{-6}) = 159\ \Omega$

3. X_C remains the same if f is doubled and C is halved simultaneously.

II

1. $f = 1/(2\pi C X_C) = 1/(6.28 \times 500 \times 10^{-12} \times 1,000) = 318\ \text{kHz}$

2. $C = 1/(2\pi fX_C) = 1/(6.28 \times 2,000 \times 3,180) = 0.025\ \mu\text{F}$

III

$X_{C_1} = 1/(2\pi fC) = 1/(6.28 \times 1,000 \times 2 \times 10^{-6}) = 79.5\ \Omega$

$X_{C_2} = 1/(2\pi fC) = 1/(6.28 \times 1,000 \times 4 \times 10^{-6}) = 39.75\ \Omega$

$X_{C_T} = X_{C_1} + X_{C_2} + X_{C_3} + X_{C_4} = 79.5 + 39.75 + 200 + 300 = 619.25\ \Omega$

IV

1. $X_{C_T} = 1/(2\pi fC_T) = 1/(6.28 \times 1,000 \times 6 \times 10^{-6}) = 26.5\ \Omega$

2. $X_{C_T} = 1/(2\pi fC_T) = 1/(6.28 \times 500 \times 25 \times 10^{-6}) = 12.7\ \Omega$

3. $X_{C_T} = 1,600/4 = 400\ \Omega$

V

1. $X_{C_1} = 1,000\ \Omega$; $X_{C_2} = 1,000\ \Omega$; $X_{C_3} = 2,000\ \Omega$; $X_{C_T} = 4,000\ \Omega$; $C_2 = 0.795\ \mu\text{F}$; $C_3 = 0.3975\ \mu\text{F}$; $V_2 = 25\ \text{V}$; $I_T = 25\ \text{mA}$

2. $C_1 = 0.5\ \mu\text{F}$; $X_{C_1} = 1,000\ \Omega$; $X_{C_2} = 1,000\ \Omega$; $X_{C_3} = 500\ \Omega$; $I_2 = 2\ \text{mA}$; $I_T = 8\ \text{mA}$

IN-PROCESS LEARNING CHECKS

I

1. For a capacitor to develop a potential difference between its plates, there must be a **charging** current.

2. For the potential difference between capacitor plates to decrease (once it is charged), there must be a **discharging** current.

3. When ac is applied to capacitor plates, the capacitor will alternately **charge** and **discharge**.

4. The value of instantaneous capacitor current directly relates to the value of **capacitance** and the rate of change of **voltage**.

5. The amount of charging or discharging current is maximum when the rate of change of voltage is **maximum**. For sine-wave voltage, this occurs when the sine wave is near **zero** points.

6. Capacitor current **leads** the voltage across the capacitor by **90** degrees. The expression that helps you remember this relationship is "Eli the **ice** man."

REVIEW QUESTIONS

1. c

3. b

5. b

7. b

9. c

11. a

13. c

15. b

17. b

19. b

PROBLEMS

1. $X_C = \dfrac{1}{6.28 fC} = \dfrac{1}{7.85 \times 10^{-6}} = 127.38\ \text{k}\Omega$

3. X_C decreases to one-third its original value, since X_C is inverse to frequency $\left(X_C = \dfrac{1}{2\pi fC} \right)$

5. For the circuit shown:
$C_T = \dfrac{C_1 \times C_2}{C_1 + C_2} = \dfrac{0.636 \times 1.59}{0.636 + 1.59} = 0.454\ \mu\text{F}$

$X_{C_T} = X_{C_1} + X_{C_2} = 25\ \text{k}\Omega + 10\ \text{k}\Omega = 35\ \text{k}\Omega$

$V_A = 70\ \text{V}$ (sum of V_{C_1} and V_{C_2})

$C_1 = \dfrac{1}{6.28 \times 10\ \text{Hz} \times X_C}$; where $X_C = \dfrac{V}{I} =$

$\dfrac{50\ \text{V}}{2\ \text{mA}} = 25\ \text{k}\Omega$. Thus, $C_1 =$

$\dfrac{1}{6.28 \times 10 \times 25\ \text{k}\Omega} = \dfrac{1}{1570 \times 10^3} = 0.636\ \mu\text{F}.$

$C_2 = \dfrac{1}{6.28 \times 10\ \text{Hz} \times X_C}$, where $X_C = \dfrac{V}{I} = \dfrac{20\ \text{V}}{2\ \text{mA}} =$

$10\ \text{k}\Omega$. Thus, $C_2 = \dfrac{1}{6.28 \times 10 \times 10\ \text{k}\Omega} =$

$\dfrac{1}{628 \times 10^3} = 1.59\ \mu\text{F}$

$X_{C_2} = 10\ \text{k}\Omega$

7. a. C_1 will RTS.
   b. C_2 will RTS.
   c. C_T will RTS.
   d. X_{C_1} will D.
   e. X_{C_2} will D.
   f. X_{C_3} will D.
   g. I will I.
   h. V_{C_1} will RTS.
   i. V_{C_2} will RTS.
   j. V_A will RTS.

9. Total current increases to 12.56 mA, if $C_2 = C_1$ and all other factors remain the same.

11. X_C of $C_1 = 1,000\ \Omega$

13. $C_T = 0.063\ \mu\text{F}$

15. Increase

ANALYSIS QUESTIONS

1. A capacitor can be said to be an "open circuit" with respect to dc because the insulating dielectric material between the capacitor's plates breaks the dc current path. Only very small "leakage" current can pass through the insulating material.

3. Capacitive voltage divider with the parameters specified as shown.

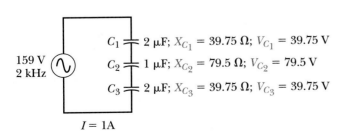

C_1 ⊣⊢ $2\ \mu\text{F}$; $X_{C_1} = 39.75\ \Omega$; $V_{C_1} = 39.75\ \text{V}$

$159\ \text{V}$
$2\ \text{kHz}$
C_2 ⊣⊢ $1\ \mu\text{F}$; $X_{C_2} = 79.5\ \Omega$; $V_{C_2} = 79.5\ \text{V}$

C_3 ⊣⊢ $2\ \mu\text{F}$; $X_{C_3} = 39.75\ \Omega$; $V_{C_3} = 39.75\ \text{V}$

$I = 1\text{A}$

CHAPTER 20

PRACTICE PROBLEMS

I

1. $V_T = \sqrt{V_R^2 + V_C^2} = \sqrt{80^2 + 60^2} = \sqrt{10,000}$
   $V_T = 100\ \text{V}$

   θ ; $V_R = 80\ \text{V}$; $V_C = 60\ \text{V}$; $V_T = 100\ \text{V}$

2. $Z = \sqrt{R^2 + X_C^2} = \sqrt{160^2 + 120^2} = \sqrt{40,000}$
   $Z = 200\ \Omega$

   θ ; $R = 160\ \Omega$; $X_C = 120\ \Omega$; $Z = 200\ \Omega$

II

1. $V_T = 80\ \text{V}$; $\theta = -48.5°$

   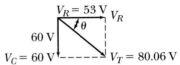

   $V_R = 53\ \text{V}$; V_R ; θ ; $60\ \text{V}$; $V_C = 60\ \text{V}$; $V_T = 80.06\ \text{V}$

2. $\theta = -48.54°$

   $R = 26.5\ \Omega$; $\theta\ -48.5°$; $X_C = 30\ \Omega$; $Z = 40\ \Omega$

III

Angle of output voltage "lead" equals arctan X_C/R.
With R set at $500\ \Omega$:
$X_C = 1/(2\pi fC) = 1/(6.28 \times 400 \times 0.5 \times 10^{-6}) = 796\ \Omega$
$R = 500\ \Omega$
$X_C/R = 796/500 = 1.59$
arctan $1.59 =$ angle of $57.8°$
Angle of lead $= 57.8°$

IV

Lag angle $= 90°$ minus circuit phase angle
Phase angle $=$ arctan X_C/R
$X_C = 1/(2\pi fC) = 1/(6.28 \times 400 \times 0.5 \times 10^{-6}) = 796\ \Omega$
$R = 750\ \Omega$
$X_C/R = 796/750 = 1.06$
arctan $1.06 =$ angle of $46.7°$
Angle of lag $= 90 - 46.7°$; Angle of lag $= 43.3°$

V

1. $I_T = 22.36\ \text{mA}$
2. $\theta = 63.43°$

3. $Z = 7.11 \text{ k}\Omega$

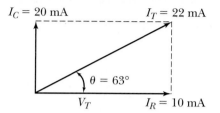

$I_C = 20 \text{ mA}$ $I_T = 22 \text{ mA}$

$\theta = 63°$

V_T $I_R = 10 \text{ mA}$

IN-PROCESS LEARNING CHECKS

I

1. In series RC circuits, the **current** is the reference vector, shown at $0°$.

2. In the $V\text{-}I$ vector diagram of a series RC circuit, the circuit voltage is shown **lagging** the circuit current.

3. In series RC circuits, an impedance diagram **can** be drawn.

4. Using the Pythagorean approach, the formula to find V_T in a series of RC circuit is $V_T = \sqrt{V_R^2 + V_C^2}$.

5. Using the Pythagorean approach, the formula to find Z in a series RC circuit is $Z = \sqrt{R^2 + X_C^2}$.

6. The trig function used to solve for phase angle when you do not know the hypotenuse value is the **tangent** function.

II

1. In parallel RC circuits, **voltage** is the reference vector in $V\text{-}I$ vector diagram.

2. In parallel RC circuits, current through the resistor(s) is plotted **out of phase** with the circuit current vector.

3. In parallel RC circuits, current through capacitor branch(es) is plotted at **+90 degrees** with respect to circuit applied voltage.

4. The Pythagorean formula to find circuit current in parallel RC circuits is $I_T = \sqrt{I_R^2 + I_C^2}$.

5. **No**

6. **Yes**

7. **No**

REVIEW QUESTIONS

1. d

3. b

5. b

7. b

9. b

11. b

13. b

15. b

17. b

19. Ohm's law and Pythagorean theorem ($I_T = V_T/Z$; and $I_T = \sqrt{I_R^2 + I_C^2}$

PROBLEMS

1. $Z = \sqrt{R^2 + X_C^2} = \sqrt{64 \times 10^6 + 225 \times 10^6} = \sqrt{289 \times 10^6} = 17 \text{ k}\Omega$

3. $C = \dfrac{1}{2\pi f X_C} = \dfrac{1}{6.28 \times 1{,}060 \times 15 \text{ k}\Omega} = \dfrac{1}{99.8 \times 10^6} \times 106 = 0.01 \ \mu\text{F}$

5. a. R will RTS.
   b. X_C will D.
   c. Z will D.
   d. I_T will I.
   e. θ will D.
   f. V_C will D.
   g. V_R will I.
   h. V_T will RTS.

7. $R = \dfrac{V_R}{I} = \dfrac{150 \text{ V}}{10 \text{ mA}} = 15 \text{ k}\Omega$

9. $f = \dfrac{1}{2\pi C X_C} = \dfrac{1}{6.28 \times 0.02 \times 10^{-6} \times 30 \times 10^3} = \dfrac{1}{3.77 \times 10^{-3}} = 265 \text{ Hz}$

11. $R_T = 20 \text{ k}\Omega + 18 \text{ k}\Omega + 30 \text{ k}\Omega = 68 \text{ k}\Omega$

13. $Z = \sqrt{R^2 + X_C^2} = \sqrt{(68 \times 10^3)^2 + (52 \times 10^3)^2} = \sqrt{7{,}328 \times 10^6} = 85.6 \text{ k}\Omega$

15. $f = \dfrac{1}{2\pi C X_C} = \dfrac{1}{6.28 \times 0.1 \times 10^{-6} \times 30 \times 10^3} = \dfrac{1}{0.0188} = 53 \text{ Hz}$

17. $I_C = \dfrac{320 \text{ V}}{16 \text{ k}\Omega} = 20 \text{ mA}$

19. $\text{Tan } \theta = \dfrac{I_C}{I_R} = \dfrac{20}{32} = 0.625$. Angle whose tangent equals 0.625 is $32°$. $\theta = 32°$

21. $R = 750 \ \Omega$; $X_C = 1{,}000 \ \Omega$; $I = 20 \text{ mA}$; $V_R = 15 \text{ V}$; $V_C = 20 \text{ V}$; $\theta = -53.1°$

ANALYSIS QUESTIONS

1. Two basic circuit operational differences between series RC circuits and series RL circuits are as follows:
   a. Circuit current leads circuit voltage in series RC circuits and lags circuit voltage in series RL circuits.
   b. As frequency of operation is increased, or C is increased in series RC circuits, the phase angle between circuit current and circuit voltage decreases. For a series RL circuit, as frequency increases, or L is increased, the circuit phase angle increases.

3. Yes, there are similar circuit effects when comparing a series RC circuit and a parallel RL circuit. For example, as frequency of operation is increased, a *series RC circuit* will cause phase angle to decrease. The same is true for a *parallel RL circuit;* that is, an increase in f will cause the phase angle to decrease.

CHAPTER 21

PRACTICE PROBLEMS

I

$R = 200 \ \Omega$
$X_{C_T} = 500 \ \Omega$
$X_L = 250 \ \Omega$
$Z = \sqrt{R^2 + X^2} = \sqrt{200^2 + 250^2} = \sqrt{102{,}500} = 320.15 \ \Omega$
$V_T = I \times Z = 0.25 \times 320.15 = 80.04 \ \text{V}$
$V_{C_2} = I \times X_{C_2} = 0.25 \times 250 = 62.5 \ \text{V}$
$V_L = I \times X_L = 0.25 \times 250 = 62.5 \ \text{V}$
$V_R = I \times R = 0.25 \times 200 = 50 \ \text{V}$
$\theta = \dfrac{\text{angle whose tangent equals } X(\text{net})}{R}$

Tan $\theta = 250/200 = 1.25$; therefore, $\theta = -51.34°$ (V with respect to I)

II

$I_R = \dfrac{V}{R} = \dfrac{300}{200} = 1.5 \ \text{A}$

$I_L = \dfrac{V}{X_L} = \dfrac{300}{200} = 1.5 \ \text{A}$

$I_C = \dfrac{V}{X_C} = \dfrac{300}{300} = 1 \ \text{A}$

$I_X = I_L - I_C = 1.5 - 1 = 0.5 \ \text{A}$

$I_T \sqrt{I_R^2 + I_X^2} = \sqrt{1.5^2 + 0.5^2} = \sqrt{2.5} = 1.58 \ \text{A}$

$Z = \dfrac{V}{I} = \dfrac{300}{1.58} = 189.8 \ \Omega$

$\theta = $ angle whose tan equals $I_X/I_R = 0.5/1.5 = 0.333$

Therefore, $\theta = -18.43°$ (I with respect to V)

III

1. $20 + j30$ in rectangular form converts to $36.05 \ \angle 56.31°$ in polar form.

2. $30 - j20$ in rectangular form converts to $36.05 \ \angle -33.7°$ in polar form.

IV

1. $66 \ \angle 60°$ in polar form converts to $33 + j57.15$ in rectangular form.

2. $75 \ \angle 48°$ converts to $50.2 + j55.7$ in rectangular form.

V

1. $\quad 20 + j5$
   $\quad \underline{35 - j6}$
   $\quad 55 - j1$

2. $\quad 34.9 - j10$
   $\quad \underline{15.4 + j12}$
   $\quad 50.3 + j2$

3. $\quad 35 - j6$ $\qquad\qquad\qquad 35 - j6$
   $\quad \underline{20 + j5 \ (\text{change sign})} \qquad \underline{-20 - j5}$
   $\qquad\qquad\qquad\qquad\qquad\qquad\quad 15 - j11$

4. $\quad 37.8 - j13$ $\qquad\qquad\qquad 37.8 - j13$
   $\quad \underline{22.9 - j7 \ (\text{change sign})} \quad \underline{-22.9 + j7}$
   $\qquad\qquad\qquad\qquad\qquad\qquad\quad 14.9 - j6$

5. $5 \ \angle 20° \times 25 \ \angle -15° = 125 \ \angle 5°$

6. $24.5 \ \angle 31° \times 41 \ \angle 23° = 1{,}004.5 \ \angle 54°$

7. $\dfrac{5 \ \angle 20°}{25 \ \angle -15°} = 0.2 \ \angle 35°$

8. $\dfrac{33 \ \angle 34°}{13 \ \angle 45°} = 2.54 \ \angle -11°$

9. $I = 12 \ \angle 25°$, then $I^2 = 144 \ \angle 50°$

10. If $I^2 = 49 \ \angle 66°$, then $I = 7 \ \angle 33°$

11. $\dfrac{5 \ \angle 50° + (30 + j40)}{2.6 \ \angle -15°} =$

    $\dfrac{(3.2 + j3.8) + (30 + j40)}{2.6 \ \angle -15°} = \dfrac{33.2 + j43.8}{2.6 \ \angle -15°} =$

    $\dfrac{54.9 \ \angle 52.8°}{2.6 \ \angle -15°} = 21.1 \ \angle 67.8°$

12. $\dfrac{10 \angle 53.1° + (25 - j25)}{4.8 \angle 36°} =$

$\dfrac{31 - j17}{4.8 \angle 36°} = \dfrac{35 \angle -28.7°}{4.8 \angle 36°} = 7.3 \angle -64.7°$

IN-PROCESS LEARNING CHECKS

I

1. When both capacitive and inductive reactance exist in the same circuit, the net reactance is **the difference between their reactances.**

2. When plotting reactances for a series circuit containing L and C, the X_L is plotted **upward** from the reference, and the X_C is plotted **downward** from the reference.

3. In a series circuit with both capacitive and inductive reactance, the capacitive voltage is plotted **180** degrees out of phase with the inductive voltage.

4. If inductive reactance is greater than capacitive reactance in a series circuit, the formula to find the circuit impedance is $Z = \sqrt{R^2 + (X_L - X_C)^2}$.

5. In a series RLC circuit, current is found by $\dfrac{V}{Z}$.

6. **False**

7. If the frequency applied to a series RLC circuit decreases while the voltage applied remains the same, the component(s) whose voltage will increase is/are the **capacitor(s).**

II

1. In parallel RLC circuits, capacitive and inductive branch currents are **180 degrees** out of phase with each other.

2. The formula to find total current in a parallel RLC circuit when I_C is greater than I_L is $I_T = \sqrt{I_R^2 + (I_C - I_L)^2}$.

3. **False**

4. If the frequency applied to a parallel RLC circuit increases while the applied voltage remains the same, the branch current(s) that will increase is/are the **capacitive** branch(es).

III

1. Power dissipated by a perfect capacitor over one whole cycle of input is **zero.**

2. The formula for apparent power is $S = V \times I$, and the unit expressing apparent power is **volt-amperes.**

3. True power in RLC circuits is dissipated by the **resistances.**

4. The formula for power factor is **p.f. = cos θ.**

5. If the power factor is one, the circuit is purely **resistive.**

6. The lower the power factor, the more **reactive** the circuit is acting.

7. In the power triangle, true power is on the **horizontal** axis, reactive power (VAR) is on the **vertical** axis, and apparent power is the **hypotenuse** of the right triangle.

REVIEW QUESTIONS

1. b
3. d
5. b
7. b
9. c
11. c
13. c
15. b
17. a
19. b

PROBLEMS

1. $Z = \sqrt{R^2 + X_T^2} = \sqrt{300^2 + 150^2} = \sqrt{112,500} = 335.4 \ \Omega$

$I = \dfrac{V_T}{Z} = \dfrac{10 \text{ V}}{0.3354 \text{ k}\Omega} = 29.8 \text{ mA}$

$V_C = I \times X_C = 29.8 \text{ mA} \times 250 \ \Omega = 7.45 \text{ V}$
$V_L = I \times X_L = 29.8 \text{ mA} \times 100 \ \Omega = 2.98 \text{ V}$
$V_R = I \times R = 29.8 \text{ mA} \times 300 \ \Omega = 8.94 \text{ V}$

$\text{Cos } \theta = \dfrac{\text{adj}}{\text{hyp}} = \dfrac{R}{Z} = \dfrac{300}{335.4} = 0.89$

$\theta = $ angle whose $\cos = 0.89 = 26.56°$ (I_T will lead V_T.)

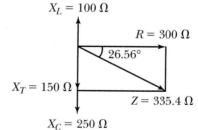

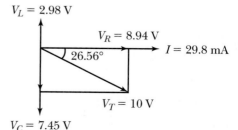

3. True power $= I^2R = 0.0298^2 \times 300 = 0.266$ W

Apparent power $= V \times I = 10$ V $\times 0.0298$ mA $= 0.298$ VA

Reactive power $= V \times I \times \sin \theta = 10$ V $\times 0.0298 \times 0.447 = 0.133$ VAR

Power factor $= \dfrac{\text{true power}}{\text{apparent power}} = \dfrac{0.266}{0.298} = 0.89$

(or p.f. $= \cos \theta = \cos 26.56° = 0.89$)

5. $Z = 20 + j40$

If the circuit were an RC circuit, it would be stated as

$Z = 20 - j40$

7. $I_R = \dfrac{20}{20} = 1$ A

$I_C = \dfrac{20}{20} = 1$ A

$I_T = 1.414$ A at an angle of $45°$

$I_T = 1.414 \angle 45°$

9. Converting $14.14 \angle 45°$ to rectangular form: $10 + j10$. That is, $R = Z \times \cos \theta = 14.14 \times \cos 45° = 10$, since the angle is $45°$; R and X must be equal; thus $Z = R + jX_L$

$14.14 = 10 + j10$

11. $Z = 61.43 \angle -75.96°$ kΩ (polar form)

$Z = 14,900 - j59,600$ (rectangular form)

13. V_C will decrease.

15. $Z = 1.79 \angle -63.4°$ Ω

17. $Z = 5.82 \angle 14°$ Ω

19. $I_R = 800$ V$/400$ Ω $= 2$ A

$I_C = 800$ V$/800$ Ω $= 1$ A

$I_L = 800$ V$/500$ Ω $= 1.6$ A

$I_{XT} = 1.6$ A $- 1$ A $= 0.6$ A

$I_T = 2^2 + 0.6^2 = 2.09$ A

$Z = V_T/I_T = 800$ V$/2.09$ A $= 383.8$ Ω

$\theta = 16.69°$

21. I_R will RTS.

I_C will I.

I_L will D.

I_T will I.

Z_T will D.

θ will I.

23. $1.414 \angle 45°$ A $= 1 + j1$ A in rectangular form.

25. $1.79 \angle -63.4°$Ω $= 0.8 - j1.6$ Ω in rectangular form.

27. $Z_T = 12.5 - j2.5$ Ω (rectangular)

29. $I_T = 3.84 + j0.8$ A (rectangular)

ANALYSIS QUESTIONS

1. A "real number" in math can be any rational or irrational number. In complex numbers related to the coordinate system, the real term in the complex number is plotted on the horizontal (or x) axis of the coordinate system. Imaginary numbers are complex numbers; they involve the square roots of negative numbers. The imaginary unit, or j operator, represents the square root of -1. For our purposes, the j operator is the vertical element used in ac analysis.

3. The reason simply measuring voltage with a voltmeter and the current with a current meter does not provide adequate information to find the actual power dissipated by the circuit is that in ac circuits containing reactances, currents and voltages are not generally in-phase with one another. Thus, taking the product of values that are not in-phase will not yield the actual power being dissipated.

CHAPTER 22

PRACTICE PROBLEMS

I

$f_r = \dfrac{1}{(2\pi\sqrt{LC})} =$

$\dfrac{1}{(6.28 \times \sqrt{250 \times 10^{-6} \times 500 \times 10^{-12}})} =$

450.4 kHz

II

$L = \dfrac{25,330}{f^2C} = 10.13$ μH

III

1. $Q = \dfrac{X_L}{R} = \dfrac{1,000}{50} = 20$

2. $V_C = I \times X_C$

$I = \dfrac{V_T}{Z} = \dfrac{12}{50} = 0.24$ A

$V_C = 0.24$ A $\times 1,000$ Ω $= 240$ V

3. If r_s doubles, $Q = \dfrac{1,000}{100} = 10$.

4. If r_s is made smaller, the skirts are steeper.

IV

1. $f_r = \dfrac{1}{(2\pi\sqrt{LC})} = 2.517$ MHz

2. $Q = \dfrac{X_L}{R} = 316.13$

3. $Z = Q \times X_L$ = approximately 999 kΩ

4. If r_s is 1 kΩ, the f_r is lower.

V

1. $f_1 = 2{,}494$ kHz and $f_2 = 2{,}506$ kHz

2. $f_r = 500$ kHz and bandpass = from 495 kHz to 505 kHz

3. $f_1 = 3{,}200$ kHz $- \dfrac{2 \text{ kHz}}{2} = 3{,}200$ kHz $- 1$ kHz = 3,199 kHz

IN-PROCESS LEARNING CHECKS

I

1. In circuits containing both inductance and capacitance, when the source frequency increases, inductive reactance *increases* and capacitive reactance *decreases.*

2. In a series *RLC* (or *LC*) circuit, when X_L equals X_C, the circuit is said to be at *resonance.*

3. In a series *RLC* (or *LC*) circuit, when X_L equals X_C, the circuit impedance is *minimum;* the circuit current is *maximum;* and the phase angle between applied voltage and current is *zero* degrees.

4. In a series *RLC* (or *LC*) circuit, when X_L equals X_C, the voltage across the inductor or the capacitor is *maximum.*

II

1. In a high-*Q* parallel resonant circuit, *Z* is *maximum.*

2. In a high-*Q* parallel circuit using ideal components, X_L *does* equal X_C at resonance.

3. The power factor of an ideal parallel resonant circuit is *1.*

4. Phase angle in an ideal parallel resonant circuit is *0 degrees.*

5. In a parallel resonant circuit, the current circulating in the tank circuit is *higher* than the line current.

REVIEW QUESTIONS

1. b
3. a
5. b
7. c
9. b
11. c
13. c
15. a
17. b, c
19. b

PROBLEMS

1. a. At resonance $Z = R$; therefore, $Z = 10$ Ω
   b. $f_r = 0.159/\sqrt{LC} = 15.9$ MHz
   c. $Q = X_L/R$;
   $X_L = 2\pi fL = 99.8$ Ω
   $Q = 99.8/10 = 9.98$
   d. $B.W. = 1.59$ MHz
   e. Bandpass = $15.105 - 16.695$ MHz
   f. Below f_r acts capacitively

3. a. $f_r = 5.486$ MHz
   b. $Q = 32.155$
   c. $BW = 0.17$ MHz
   d. Bandpass = 5.65 MHz
   e. At resonance, $Z = 31{,}018.6$ Ω
   f. Below resonance the circuit acts inductively.

5. Resonant frequence is 1,060 Hz.

7. $L = 45$ mH

9. *R* value must be 3.33 Ω.

11. Resonant frequency is 795 kHz.

13. Exhibited impedance equals 110 kΩ.

ANALYSIS QUESTIONS

1. The tuned circuit in a receiver input stage is typically a parallel *LC* circuit.

3. The circuits described would be considered "resonant filters."

5. *Selectivity* is the ability to select desired signal frequencies, while rejecting undesired signal frequencies. Referring to resonant circuits, selectivity indicates the sharpness of the response curve. A narrow and more pointed response curve re-

sults in a higher degree of frequency selectiveness, or selectivity.

Bandwidth is the frequency limits (total difference in frequency) between upper and lower frequencies over which a specified response level is shown by a frequency-sensitive circuit, such as a resonant circuit. The cutoff level is defined at 70.7% of peak response, sometimes called the half-power points on the response curve.

Bandpass is the band of frequencies that passes through a filter or circuit with minimal attenuation or degradation, often designated by specific frequencies. When specified, they are designated indicating that the lower frequency is at the lower frequency "half-power point" of the response curve and the upper frequency, which is at the upper "half-power point" on the response curve. Bandpass is often expressed as *the number of Hertz between these two frequency response limits.*

7. a. X_L will D.
   b. X_C will I.
   c. R will RTS.
   d. Z will I.
   e. I will D.
   f. θ will I.
   g. V_C will I.
   h. V_L will D.
   i. V_R will D.
   j. V_T will RTS.
   k. BW will RTS.
   l. Q will RTS (close to the same).
   m. I_C will D.
   n. I_L will D.
   o. I_T will D.

9. a. X_L will D.
   b. X_C will I.
   c. R will RTS.
   d. Z will D.
   e. I will I.
   f. θ will I.
   g. V_C will RTS.
   h. V_L will RTS.
   i. V_R will RTS.
   j. V_T will RTS.
   k. BW will RTS.
   l. Q will RTS.
   m. I_C will D.
   n. I_L will I.
   o. I_T will I.

11. The condition is such that $X_L = X_C$.

13. Circuit diagram of a bandpass filter using a series resonant and a parallel resonant circuit:

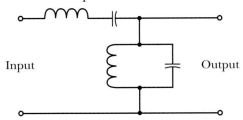

CHAPTER 23

IN-PROCESS LEARNING CHECKS

I

1. Before doping (in their pure form) semiconductor materials are sometimes called *intrinsic* semiconductors. Once they are doped with tiny amounts of an impurity atom, they are called *extrinsic* semiconductors.

2. N-type materials are formed by doping a *tetra*valent semiconductor material with a *penta*valent material. P-type materials are formed by doping a *tetra*valent semiconductor material with a *tri*valent material.

3. The doping atoms for N-type materials are called *donor* atoms because they donate extra electrons to the covalent bonds. The doping atoms for P-type materials are called *acceptor* atoms because they accept electrons that will fill the holes in the covalent bonds.

4. The majority carriers in an N-type material are *electrons,* and the minority carriers are *holes.* The majority carriers in a P-type material are *holes,* and the minority carriers are *electrons.*

REVIEW QUESTIONS

1. a. A *trivalent* atom has three valence electrons.
   b. A *tetravalent* atom has four valence electrons.
   c. A *pentavalent* atom has five valence electrons.

3. a. Electrons
   b. Holes

5. a. 0.7 V
   b. 0.3 V

7. b

9. P-N junction semiconductor showing an external dc power supply connected for forward biasing.

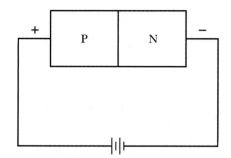

11. Forward-bias portion of the *I-V* curve for a P-N junction.

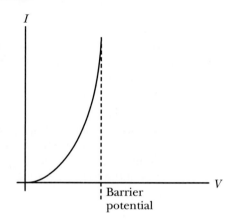

PROBLEMS

1. 32

2. 3

ANALYSIS QUESTIONS

1. Valence-band electrons are bound to their atom and are not free to move from one atom to another in the material. Electrons in the conduction band are free to move. Also, conduction-band electrons possess more energy than their corresponding valence-band electrons.

3. Increasing the amount of forward bias narrows the depletion region. Decreasing the amount of forward bias or applying a reverse bias increases the width of the depletion region.

5. In a semiconductor, holes flow from positive to negative, and electrons flow from negative to positive. Conventional current flow shows current flowing from positive to negative, and the electron-flow convention shows current flowing from negative to positive. So in a manner of speaking, hole flow is similar to conventional current flow, and electron flow is similar to the electron version of current flow.

CHAPTER 24

PRACTICE PROBLEMS

I

1. 12.6 V

2. 0 V

3. 0.7 V

4. 11.9 V

II

1. 0.7 V

2. 75.3 V

3. 753 µA

III

1. 0.9 V, 10.9 V

2. 9.1 V, 9.1 V

3. 0.9 mA, 10.9 mA

4. 8.1 mW, 99.19 mW

IV

1. 325 Ω

IN-PROCESS LEARNING CHECKS

I

1. To cause conduction in a semiconductor diode, the diode must be *forward* biased.

2. To reverse bias a P-N junction, connect the negative source voltage to the *P*-type material and the positive source voltage to the *N*-type material.

3. The anode of a junction diode corresponds to the *P*-type material, and the cathode corresponds to the *N*-type material.

4. The forward conduction voltage drop across a silicon diode is approximately 0.7 V.

5. When a diode is reverse biased in a series circuit, it acts like an *open* switch.

6. There will be no current flow through a series diode circuit when the diode is *reverse* biased.

7. When a diode is connected in series with a resistor, the voltage across the resistor is very nearly equal to the dc source voltage when the diode is *forward* biased.

8. The three most general diode ratings are *forward voltage drop, average forward current,* and *peak reverse voltage* (or *reverse-breakdown voltage*).

II

1. Rectifier diodes are used where it is necessary to change *ac* current power into *dc* current power.

2. Where additional cooling is necessary, a rectifier diode can be connected to a *heat sink* to dissipate heat more efficiently.

3. When an ac waveform is applied to a rectifier diode, the diode's *reverse breakdown voltage* rating must be greater than the peak voltage level.

4. The main current specification for rectifier diodes is *average forward current.*

5. To test the forward conduction of a diode with an ohmmeter, connect the *negative* lead of the meter to the cathode and the *positive* lead to the anode.

6. The forward resistance of a good diode should be much *less* than its reverse resistance.

7. Switching diodes have a *reverse recovery time* rating that is hundreds of times less than most rectifier diodes.

8. The type of diode circuit that removes the peaks from an input waveform is called a *clipper* circuit.

9. The type of diode circuit that changes the baseline level of an input waveform is called a *clamp* circuit.

III

1. An LED emits light energy when an electron in the *conduction* band falls to the *valence* band.

2. The choice of *semiconductor material* determines the color of the light from an LED.

3. The arrow on an LED symbol points *away from* the diode.

4. An LED emits light when it is *forward* biased.

5. The formula for calculating the value of the resistor to be used in series with an LED and its dc voltage source is $R = (V_S - V_D)/I_F$, where:
   V_S is *the voltage of the source*
   V_D is *the voltage across the diode*
   I_F is *the forward diode current*

6. The reverse-breakdown voltage of an LED is generally *lower* than that of a typical rectifier diode, and the forward voltage drop of an LED is *higher* than that of a rectifier diode.

7. Light falling onto the depletion region of a reverse-biased diode *increases* the amount of reverse leakage current.

8. The arrow on a photodiode symbol points *toward* the diode.

9. A photodiode is connected into the circuit so that it is *reverse* biased.

10. In a diode opto-isolator, an *LED* is the source and a *photodiode* is the detector of light energy.

REVIEW QUESTIONS

1.

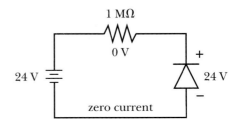

Reverse-biased diode   Forward-biased diode

3.

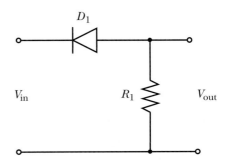

5.

7. A diode clamp circuit shifts the baseline of an input waveform from 0 V to a different level. Whether the baseline is shifted positive or nega-

tive depends on the direction the diode is connected into the circuit.

9. a. The reverse breakdown voltage is the maximum amount of reverse bias a diode can withstand indefinitely.
   b. The forward conduction voltage is the voltage drop across a diode when it is carrying forward-bias current. This value is fairly constant at 0.3 V for germanium diodes and 0.7 V for silicon diodes.
   c. The average forward current specification is the maximum amount of forward current a diode can carry for an indefinite time.
   d. The reverse recovery time is the time that is required for changing the condition of a diode from reverse bias to forward bias.

11. The voltage across the zener diode will be unchanged. The voltage across the resistor will increase. Increasing the applied voltage will cause the current through the circuit to increase.

13.

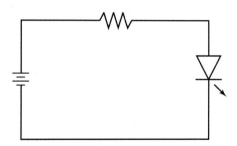

15.

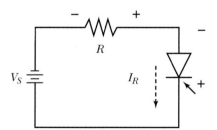

17. Light-emitting diode and photodiode.

19. Price-scanners in supermarkets, compact disk players (CDs), optical disk drives for computers, fiber-optic communications.

21. The negative-resistance region is the peculiar feature of the *I-V* curve for tunnel diodes.

23. Oscillator element and electronic switch.

25. Schottky diodes use a hot-carrier junction, which allows them to operate much faster than conventional diodes.

PROBLEMS

1. a. 5.3 mA
   b. 5.3 V
   c. 5.3 mA

3. 100.7 V

5. a. 3 V
   b. 25 mA
   c. 75 mW
   d. 225 mW

7. 25 mW

9. a. 25 mA
   b. 25 mA
   c. 113 mW

ANALYSIS QUESTIONS

1. Connecting a resistor in series with both diodes ensures that the voltage across each of the parallel branches exceeds the voltage required for forward conduction of both diodes.

3. Answer varies with the results of the research, but should be in the general range of 1.5 to 120 V.

5. Laser diodes provide the source of coherent light required for scanning the bar code.

7. Answers will vary.

CHAPTER 25

PRACTICE PROBLEMS

I

$V_{dc} = 0.45 \times V_{rms} = 0.45 \times 200 \text{ V} = 90 \text{ V}$

II

$V_{dc} = 0.45 \times V_{rms} = 0.45 \times 440 \text{ V} = 198 \text{ V}$

$PIV = 1.414 \times V_{rms} = 1.414 \times 440 \text{ V} = 622 \text{ V}$

Ripple frequency = same as ac input = 60 pps (pulse per second)

Polarity of output = positive with respect to ground reference

III

$V_{dc} = 0.9 \times V_{rms}$ of half the secondary $= 0.9 \times 150 \text{ V} = 135 \text{ V}$

Ripple frequency $= 2 \times$ ac input frequency $= 2 \times 400 = 800$ pps

$PIV = 1.414 \times V_{rms}$ full secondary $= 1.414 \times 300 \text{ V} = 424 \text{ V}$

IV

1. $V_{dc} = 0.9 \times V_{rms}$ input $= 0.9 \times 120 \text{ V} = 108 \text{ V}$

2. $I_{dc} = V_{dc}/R_L = 108 \text{ V}/50 \text{ } \Omega = 2.16 \text{ A}$

3. PIV (Bridge) = $1.414 \times V_{rms}$ input = $1.414 \times 120 = 169.7 \text{ V}$

4. Ripple frequency (Bridge) = $2 \times$ ac input frequency = $2 \times 60 \text{ Hz} = 120 \text{ pps}$

IN-PROCESS LEARNING CHECKS

I

1. The simplest rectifier circuit is the *half-wave* circuit.

2. The ripple frequency of a bridge rectifier circuit is *two times* the ripple frequency of a half-wave rectifier.

3. The circuit having the highest output voltage for a given transformer secondary voltage is the *bridge* rectifier.

4. For a given full transformer secondary voltage, the full-wave rectifier circuit unfiltered dc output voltage is *equal to* the dc output of a half-wave rectifier that is using the same transformer.

5. To find the average dc output (unfiltered) of a center-tapped, full-wave rectifier, multiply the full secondary rms voltage by *0.45.*

II

1. In a C-type filter, the output of the rectifier circuit is connected in *parallel* with a filter *capacitor* and the load resistance.

2. In an L-type filter, the output of the rectifier circuit is connected in *series* with a filter *choke (inductor)* and the load resistance.

3. In an L-type filter, energy is stored in the form of a *magnetic* field.

4. The five main performance characteristics used in the selection and quality of a power supply filter network are *output ripple, regulation, rectifier peak current limits, load current,* and *output voltage.*

5. The two main causes of swings in the output voltage of a power supply are changes in the *input* voltage level and changes in the *output* current demand.

6. The three terminals on an integrated-circuit voltage regulator are the unregulated *dc input,* regulated *dc output,* and *common ground.*

7. The *lower* the percent of ripple, the higher the quality of the power supply.

8. Full-load output voltage is taken when the current from the power supply is *maximum.*

9. The *higher* the percent of regulation, the higher the quality of the power supply.

REVIEW QUESTIONS

1.

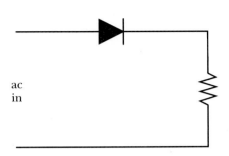

3. When you turn around a rectifier diode, the polarity of the output from a half-wave rectifier is reversed. Switching the connections to the ac source has no effect on the operation of the rectifier, however.

5.

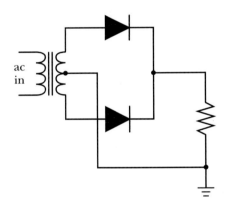

7. 120 pps

9.

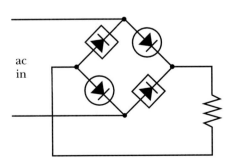

11. 800 Hz

13. Excessive ripple is most likely caused by an open capacitor or shorted inductor in the power supply filter network.

15.

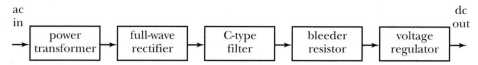

ac in → power transformer → full-wave rectifier → C-type filter → bleeder resistor → voltage regulator → dc out

17. A no-load condition occurs when no current is drawn from the power supply. A full-load condition occurs when the maximum allowable amount of current is drawn from the power supply.

19. The RMS input voltage is 212 V.

PROBLEMS

1. a. 54 Vdc
   b. 170 V (unfiltered)
   c. 60 Hz

3. 1.7 A

5. ≅ 15.45:1 step-down

7. 10%

9. 10.6% regulation

ANALYSIS QUESTIONS

1.

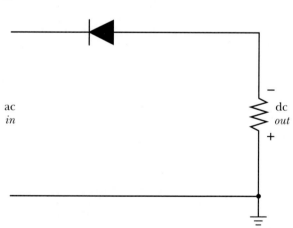

ac *in*   dc *out*

3. Capacitor discharges less between charging pulses.

5. Optimum inductance is that value of the first inductor where peak rectifier current is kept from exceeding the dc load current by more than 10% at maximum load levels.

7. A shorted diode allows ac voltage to reach the filter capacitor. Electrolytic capacitors can handle only dc voltages of a certain polarity. When ac voltage is applied to an electrolytic filter capacitor, when a rectifier diode shorts, for example, the reverse half-cycle of ac voltage shorts the capacitor and quickly destroys it.

CHAPTER 26

PRACTICE PROBLEMS

I

1. $I_E = I_C + I_B = 10$ mA $+ 100$ μA $= 10.1$ mA

2. $\beta_{dc} = I_C/I_B = 2$ mA/15 μA $= 133$ for both NPN and PNP transistors

3. $I_C = \beta_{dc} \times I_B = 100 \times 10$ μA $= 1$ mA
   $I_E = I_C + I_B = 1$ mA $+ 10$ μA $= 1.01$ mA
   $\alpha = I_C/I_E = 1$ mA/1.01 mA $= 0.99$

IN-PROCESS LEARNING CHECKS

I

1. Under normal operating conditions, the base-emitter junction of a BJT must be *forward* biased, while the collector-base junction is *reverse* biased.

2. Current must be flowing through the *emitter-base* junction of a BJT before current can flow between the emitter and collector.

3. The symbol for an NPN transistor shows the emitter element pointing *away from* the base element.

4. The symbol for a PNP transistor shows the emitter element pointing *in toward* the base element.

5. For BJTs:
   V_{BE} is the *base-emitter* voltage
   V_{CB} is the *collector-base* voltage
   V_{CE} is the *collector-emitter* voltage

6. For a typical BJT circuit:
   I_C stands for the *collector* current
   I_B is the *base* current
   I_E is the *emitter* current

7. Stated in words, the dc beta of a BJT is the ratio of *collector current* divided by *base current*.

8. Stated in words, the alpha of a BJT is the ratio of *collector current* divided by *emitter current*.

9. The α of a BJT is *less* than 1, while the β_{dc} is *greater* than 1.

II

1. BJTs use a small amount of base *current* to control a larger amount of collector *current*. It can be said that BJTs are *current* controllers.

2. When base current in a BJT is zero, the collector current is *zero*.

3. BJTs are basically used as *switches* and *amplifiers*.

4. When a BJT is being used as a switch (as in Figure 26-5), V_{CE} is maximum when I_B is *minimum,* and V_{CE} is minimum when I_B is *maximum.* In the same circuit I_C is maximum when I_B is *maximum,* and I_C is minimum when I_B is *minimum.*

5. When a BJT is being used as voltage amplifier (as in Figure 26-7), an increase in base current causes a *decrease* in V_{CE}.

6. On a family of collector characteristic curves, the horizontal axis represents the *collector voltage* and the vertical axis represents the *collector current.* Each curve in the family represents a different level of *base current.*

REVIEW QUESTIONS

1. Two, two

3. N, P

5. Negative to base; positive (or less negative) to emitter; negative to collector.

7. Approximately 96–99% of a BJT's emitter current appears in the collector circuit.

9. A reverse-biased P-N junction is one where the P-type material is connected to the negative side of the source and the N-type material is connected to the positive side of the source.

11. Reverse

13. No

15. Collector current (I_C) increases as base current (I_B) increases.

17. Increasing the forward-biasing current (I_B) in a typical BJT voltage amplifier causes the collector-emitter voltage drop (V_{CE}) to decrease.

19. Define the following ratings for a BJT:
   a. $I_{C_{(MAX)}}$ = maximum collector current; the greatest amount of current the transistor can handle for an indefinitely long period.
   b. $P_{D_{(MAX)}}$ = maximum power dissipation; greatest amount of power the transistor can dissipate without the help of a heat sink.

PROBLEMS

1. a. $I_C = 40.5$ mA
   b. $\beta_{dc} = 80$

3. $\beta_{dc} = 500$

5. $I_C = 1.6$ mA

7. α = 0.99

9. a. $I_C = 38.4$ mA
   b. $I_B = 1.6$ mA
   c. $\beta_{dc} = 25$

11. $\beta_{ac} = 100$

13. $P_d = 480$ mW

15. $\beta_{ac} \cong 100$

ANALYSIS QUESTIONS

1. Outline drawings for TO-3, TO-5, TO-92, TO-220, and SOT-89 transistor packages are shown.

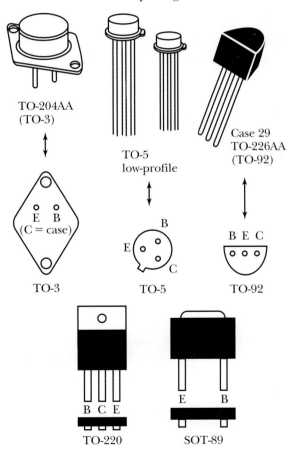

TO-204AA (TO-3)

TO-5 low-profile

Case 29 TO-226AA (TO-92)

TO-3

TO-5

TO-92

TO-220

SOT-89

3. The purpose of a heat sink is to draw heat away from a semiconductor device and to dissipate it into the air by means of heat convection. Heat sinks are made from extruded aluminum stock

that usually has rows of fins that greatly increase the effective area exposed to surrounding air. The semiconductor device to be called is securely bolted to the heat sink, sometimes with a thin layer of mica serving to isolate electrically the case of the semiconductor from the heat sink. Heat sinks are sometimes anodized with a flat black color to enhance heat dissipation according to the principle of black-body radiation.

5. Formula 26-1 shows that I_E is greater than I_C (unless all terms are zero): $I_E = I_C + I_B$. And since I_E is always greater than I_C, the quotient I_C/I_E has to be less than 1. Therefore, $\alpha = I_C/I_E$ (Formula 26-5) is always less than 1.

CHAPTER 27

PRACTICE PROBLEMS

I

Cutoff point is 5 V.
Saturation current is (from Formula 27-1):
$I_{C_{(SAT)}} = V_{CC}/(R_C + R_E) = 5 \text{ V}/517 \ \Omega = 9.67 \text{ mA}$

II

Divider current: $I = V_{CC}/(R_1 + R_2) = 10 \text{ V}/12 \text{ k}\Omega = 0.83 \text{ mA}$
Base voltage: $V_B = I \times R_2 = 0.83 \text{ mA} \times 2 \text{ k}\Omega = 1.67 \text{ V}$
Emitter voltage: $V_E = V_B - 0.7 \text{ V} = 0.97 \text{ V}$
Emitter current: $I_E = V_E/R_E = 0.97 \text{ V}/1 \text{ k}\Omega = 0.97 \text{ mA}$
Collector voltage: $V_C = V_{CC} - (I_C \times R_C) = 10 \text{ V} - 4.6 \text{ V} = 5.4 \text{ V}$
Transistor voltage: $V_{CE} = V_{CC} - (V_{RC} + V_{RE}) = 10 \text{ V} - 5.57 \text{ V} = 4.43 \text{ V}$

IN-PROCESS LEARNING CHECKS

I

1. The BJT amplifier circuit having the base terminal common to both the input and output circuits is the *common-base (CB)* amplifier.

2. The BJT amplifier circuit providing both voltage and current gain is the *common-emitter (CE)* amplifier.

3. The BJT amplifier circuit having 180° phase difference between input and output signals is the *common-emitter (CE)* amplifier.

4. The BJT amplifier circuit having a voltage gain of less than 1 is the *common-collector (CC)* amplifier.

5. The BJT amplifier circuit having an input to the base and output from the emitter is called a common-*collector* amplifier.

6. In an amplifier that uses more than one stage, the overall gain is found by *multiplying* the gains of the individual stages.

7. The voltage-follower amplifier is another name for a common-*collector* amplifier.

8. The Darlington amplifier is made up of two common-*collector* amplifiers.

II

1. Class A amplifiers conduct during *360°* of the input cycle; Class B amplifiers conduct for *180°* of the input cycle; and Class C amplifiers conduct for approximately *120°* of the input cycle.

2. A load line shows all operating points of a given amplifier from *cutoff* to *saturation.*

3. The amount of saturation current = *12 mA.* The value of the cutoff voltage for this circuit = *12 V.*

4. The Q point for a Class A amplifier is located at the *halfway point* on the load line (halfway between cutoff and saturation).

5. The Class *C* amplifier has the Q point located below the cutoff.

6. Collector current flows in a Class *A* amplifier, even when there is no signal applied to the input.

III

1. The voltage gain of the amplifier = *250.*

2. The current gain of the amplifier = *50.*

3. The power gain = *18,000.*

REVIEW QUESTIONS

1. Emitter

3. Typical impedance characteristics of the various amplifier configurations follow.
   a. Common-emitter: input Z = low, output Z = low to medium
   b. Common-base: input Z = low, output Z = fairly high
   c. Common-collector: input Z = high, output Z = low

5. Common-emitter

7. A common-collector amplifier stage followed by a common-emitter stage provides a high input impedance followed by high voltage gain.

9. Small-signal transistor amplifier stages may be used in the first stages of radio and TV receivers; large-signal transistor amplifier stages in audio power amplifier stages for driving loudspeakers.

11. Class A amplifiers have the least distortion.

13. The Q point is usually set at the midpoint of the load line for Class A operation.

15. For Class A operation of a common-emitter circuit, the quiescent value of V_{CE} is about one-half the value of V_{CC}, or 10 V.

17. a. Divider current: $I = V_{CC}/(R_1 + R_2) = 15$ V/ 10.1 kΩ = 1.49 mA

   b. Base voltage: $V_B = I \times R_2 = 1.49$ mA $\times$ 1 kΩ = 1.49 V

   c. Emitter voltage: $V_E = V_B - 0.7$ V = 1.49 V $-$ 0.7 V = 0.79 V

   d. Collector voltage: $V_C = V_{CC} - (I_C \times R_C) = 15$ V $-$ (0.79 mA $\times$ 10 kΩ) = 15 V $-$ 7.9 V = 7.1 V

   e. Collector-to-emitter voltage: $V_{CE} = V_{CC} - (V_{RC} + V_{RE}) = 7.1$ V $-$ 0.79 V = 6.31 V

19. A common-collector BJT amplifier circuit is needed to match a high impedance to a low impedance.

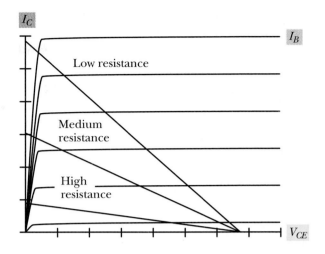

3. Cutoff voltage = 15 V; saturation current = 11.7 mA.

5. The ratio of $R_1:R_2$ should be approximately 6:1.

PROBLEMS

1. a. $I_{bias} = 65.9$ μA
   b. $V_{R_2} = 5.41$
   c. $V_{RE} = 4.71$ V

3. a. $I_{R_1} = 45$ μA
   b. $V_{R_1} = 4.5$ V
   c. $I_{R_2} = 45$ μA
   d. $V_{R_2} = 4.5$ V
   e. $V_{BE} = 3.8$ V
   f. $V_E = 3.8$ V
   g. $I_E = 31.7$ mA
   h. $I_C = 31.7$ mA
   i. $V_{RC} = 3.8$ V
   j. $V_{CE} = 1.4$ V

5. a. 37.5 mA
   b. 9 V

7. a. $I_{R_1} = 180$ μA
   b. $V_{R_1} = 18$ V

9. $A_V = 142$

ANALYSIS QUESTIONS

1. The steeper the slope of the load line, the less change in collector voltage for a given change in base current. The diagram shows that a steeper curve is created by lowering the collector resistance. You can conclude that the voltage gain of an amplifier changes in proportion to changes in the collector resistance.

CHAPTER 28

IN-PROCESS LEARNING CHECKS

I

1. *JFET* is the abbreviation for *junction field-effect transistor.*

2. The three terminals on a JFET are called *source, gate,* and *drain.*

3. The arrow in the symbol for an N-channel JFET points *in toward* the source-drain connections, while the arrow for a P-channel JFET points *away from* source-drain connections.

4. In an N-channel JFET, the source-drain voltage must be connected so that the positive polarity is applied to the *drain* terminal of the JFET and the negative polarity is applied to the *source* terminal.

5. In a P-channel JFET, the source-drain voltage must be connected so that the positive polarity is applied to the *source* terminal of the JFET and the negative polarity is applied to the *drain* terminal.

6. The charge carriers in the channel material of a JFET always flow from the *source* terminal to the *drain* terminal.

7. In an N-channel JFET, the gate-source voltage must be connected so that the positive polarity is applied to the *source* terminal of the JFET and the negative polarity is applied to the *gate* terminal.

8. In a P-channel JFET, the source-drain voltage must be connected so that the positive polarity is

applied to the **gate** terminal of the JFET and the negative polarity is applied to the **source** terminal.

9. V_{DS} stands for **voltage between the drain and source terminals.**
   V_{GS} stands for **voltage between the gate and source terminals.**
   I_D stands for **drain current.**
   g_m stands for **transconductance.**

10. The greatest amount of current flows through the channel of a JFET when V_{GS} is at its **zero** level, while the least amount of drain current flows when V_{GS} is at its **maximum** (or $V_{GS_{(OFF)}}$) level.

II

1. When a polarity opposite the drain polarity is applied to the gate of a JFET through a large-value resistor, the bias method is called **gate** bias.

2. When the gate and source of JFET are both connected to ground through resistors, the bias method being used is called **self bias.**

3. When the gate of a JFET is connected to a voltage divider and the source is connected through a resistor to ground, the bias method being used is called **voltage-divider** bias.

4. The common-**gate** FET amplifier has the input signal applied to the source terminal and the signal taken from the drain.

5. The common-**source** FET amplifier has the signal applied to the gate terminal and the output signal taken from the drain.

6. The common-**drain** FET amplifier has the signal applied to the gate terminal and output taken from the source terminal.

III

1. The term *MOS* stands for **metal-oxide semiconductor.**

2. The gate terminal in a D-MOSFET is separated from the channel material by a thin layer of **silicon dioxide** (or **metal-oxide insulation**).

3. In a MOSFET, the direction of flow of charge carriers is always from the **source** terminal to the **drain** terminal.

4. An outward-pointing arrow on a D-MOSFET symbol indicates an **N**-type substrate and a **P**-type channel. An inward-pointing arrow on a D-MOSFET symbol indicates a **P**-type substrate and an **N**-type channel.

5. The proper polarity for V_{DS} of an N-channel D-MOSFET is **negative** to the source and **positive** to

the drain. For a P-channel D-MOSFET, the proper polarity for V_{DS} is **positive** to the source and **negative** to the drain.

6. When operating a D-MOSFET in the depletion mode, the polarity of V_{GS} is **negative** to the gate terminal. And when V_{GS} is at its minimum level, I_D is at its **maximum** level.

7. The **E-MOSFET** is the only FET that is nonconducting when the gate voltage is zero.

8. The proper polarity for V_{DS} of an N-channel E-MOSFET is **negative** to the source and **positive** to the drain. The proper polarity of V_{GS} is **positive** to the gate terminal.

REVIEW QUESTIONS

1. V_{DS} = voltage measured between the drain and source terminals.
   V_{GS} = voltage measured between the gate and source terminals.
   I_D = current at the drain terminal (usually the same is found at the source terminal).
   I_{DSS} = drain current that flows when no bias voltage is applied to the gate.
   $V_{GS_{(OFF)}}$ = the amount of gate-source voltage required to turn off drain current.

3. For N-channel: V_{DD} is positive, and V_{GG} is negative. For P-channel: V_{DD} is negative, and V_{GG} is positive.

5. A JFET uses a voltage (V_{GS}) to control a current (I_D), whereas a BJT uses a current (I_B) to control a current (I_C). Also, a JFET is fully conducting when there is no bias applied to the gate, whereas a BJT is switched off when there is no bias applied to the base.

7. Transconductance is equal to a change in drain current divided by the corresponding change in gate-source voltage.

9. A D-MOSFET makes an ideal Class-A amplifier because the required gate-source bias voltage is zero. This greatly simplifies the required bias circuitry.

11. a. For the N-channel depletion mode: V_{DD} is positive and V_{GS} is negative.
    b. For the P-channel depletion mode: V_{DD} is negative and V_{GS} is positive.
    c. For the N-channel enhancement mode: V_{DD} is positive and V_{GS} is positive.
    d. For the P-channel enhancement mode: V_{DD} is negative and V_{GS} is negative.

13. In the depletion mode, increasing the amount of gate-source voltage depletes the channel of charge

carriers, thus tending to decrease the amount of drain current. In an enhancement mode, increasing the amount of gate-source voltage enhances the density of charge carriers in the channel, thus tending to increase the amount of drain current.

15. An E-MOSFET makes an ideal electronic switching circuit (Class-B amplifier) because it is off when there is no gate-source voltage applied, and it is turned on when the gate-source voltage is applied.

17. a. For an N-channel JFET, $V_{GS_{(OFF)}}$ is negative. For a P-channel JFET, $V_{GS_{(OFF)}}$ is positive.
    b. For an N-channel D-MOSFET, $V_{GS_{(OFF)}}$ is negative. For a P-channel D-MOSFET, $V_{GS_{(OFF)}}$ is positive.
    c. For an N-channel E-MOSFET, $V_{GS_{(OFF)}}$ is 0 V. For a P-channel E-MOSFET $V_{GS_{(OFF)}}$ is 0 V.

19. The zener diodes that are internally connected between the gate and source terminals of some MOSFET devices protect the device from damage that could otherwise be caused by an external build-up of static voltage.

PROBLEMS

1. a. $V_G = -4$ V
   b. $V_{RD} = 8$ V
   c. $V_{DS} = 4$ V
   d. $I_{RG} = 0$

3. a. $g_m = 0.5$ mS (or 500 uS)
   b. $\Delta V_{DS} = 2$ V

5. a. $I_D = 1.67$ mA
   b. $V_{DS} = 6.33$ V
   c. $V_G = 0$
   d. $V_{GS} = 2$ V
   e. $V_{RG} = 0$
   f. $I_{RG} = 0$

7. a. $I_{R_2} = 90.2$ μA
   b. $V_{R_2} = 2.98$ V
   c. $V_{R_1} = 9.02$ V
   d. $V_{RS} = 4$ V
   e. $V_{GS} = 1.02$ V
   f. $I_{GS} = 0$
   g. $V_{DS} = 2.21$ V

ANALYSIS QUESTIONS

1. Attempting to operate a JFET in an enhancement mode would forward bias the gate-source junction. This means the device could not operate as an FET, and also there would be a risk of over-

heating the gate junction due to excessive forward current flow.

3. A D-MOSFET has an uninterrupted channel of N-type or P-type material, which makes it possible to operate it in a depletion mode. It can also be operated in an enhancement mode because of a region of the opposite type of material that can become part of the channel when the gate-source voltage contributes charge carriers to the opposite-type material. An E-MOSFET, on the other hand, does not have an uninterrupted channel of a type of semiconductor material, so it is not possible to operate it in a depletion mode. An E-MOSFET can be operated in the enhancement mode because it has an opposite-type material that can take on charge carriers that form a conductive channel, or inversion layer.

CHAPTER 29

PRACTICE PROBLEMS

I

1. $A_V = (R_f/R_i) = (1 \text{ M}\Omega/100 \text{ k}\Omega) = 10$
2. $R_f = A_V \times R_i = 100 \times 2 \text{ k}\Omega = 200 \text{ k}\Omega$
3. $R_i = R_f/A_V = 270 \text{ k}\Omega/20 = 13.5 \text{ k}\Omega$

II

1. A_V (noninverting) $= (R_f/R_i) + 1 = (20{,}000 \ \Omega/2{,}000 \ \Omega) + 1 = 11$
2. $R_f = R_i(A_V - 1) = (10 \text{ k}\Omega)(100 - 1) = 990 \text{ k}\Omega$
3. $R_i = R_f/(A_V - 1) = 270 \text{ k}\Omega/19 = 14.2 \text{ k}\Omega$

IN-PROCESS LEARNING CHECKS

I

1. An op-amp is a high gain *direct*-coupled amplifier.
2. The input stage of an op-amp uses a *differential* amplifier configuration.
3. The name operational amplifier derives from their early use to perform mathematical *operations.*
4. The open-loop gain of an op-amp is much *higher* than the closed-loop gain.
5. To achieve an output that is the inversion of the input, the input signal is fed to the *negative* input of the op-amp.
6. Input signals to the op-amp can be fed to the *inverting* input, the *noninverting* input, or to both inputs.

II

1. For an op-amp inverting amplifier the input signal is applied to the *inverting* input, and the feed-

back resistor is connected from the output to the *inverting* input.

2. The voltage gain of an inverting amplifier is a negative value in order to indicate *inversion* of the signal.

3. In the schematic for an op-amp inverting amplifier, resistor R_f is located between the *output* of the circuit and the *inverting* input. Resistor R_i is connected to the *inverting* terminal.

4. For an op-amp noninverting amplifier the input signal is applied to the *noninverting* input, and the feedback resistor is connected from the output to the *inverting* input.

5. The special case of a noninverting op-amp circuit that has a voltage gain of 1 is called a *voltage follower* circuit.

6. An op-amp circuit that has one signal input and one grounded input is called a *single*-ended circuit.

7. An op-amp circuit that has different signal sources connected to its inverting and noninverting inputs is called a *double*-ended circuit.

8. A comparator circuit is an example of an op-amp operating in the *double*-ended input mode.

REVIEW QUESTIONS

1. The amplifier circuit typically used as the input stage for an operational amplifier is a *differential* amplifier.

3. The key parameters for the ideal op-amp are infinite voltage gain, infinite bandwidth, infinite input impedance, and zero output impedance.

5. A comparator circuit is an example of an op-amp circuit that is operated in the open-loop gain mode.

7. Decrease

9. 16, 11

11. The output voltage waveform is twice the input voltage and has the same phase.

13. The output signal has a phase that is opposite the input.

15. The feedback component is a capacitor, and the input component is a resistor.

17. Integrator

19. $V_{out} = V_1 + V_2 + V_3$

PROBLEMS

1. $A_V = 208$

3. $R_i = 8.33 \text{ k}\Omega$

5. $A_V = 2$

7. a. $V_{out} = +5 \text{ V}$
   b. $V_{out} = +1 \text{ V}$
   c. $V_{out} = -1 \text{ V}$
   d. $V_{out} = -5 \text{ V}$

9. a. $V_{out} = +8 \text{ V}$
   b. $V_{out} = 0 \text{ V}$
   c. $V_{out} = 0 \text{ V}$
   d. $V_{out} = -8 \text{ V}$

ANALYSIS QUESTIONS

1. From Formula 29-2: A_V (noninverting) $= (R_f/R_i) + 1$. For a voltage follower, $R_f = 0$. Substituting this into the formula, we get: A_V (noninverting) $= (0/R_i) + 1 = 0 + 1 = 1$.

   Thus, the voltage gain of the circuit is 1, or unity.

3. Voltage gain $= V_{out}/V_{in} = 2 \text{ V}/10 \text{ mV} = 200$ $R_f = A_V \times R_i = 200 \times 2.2 \text{ k}\Omega = 440 \text{ k}\Omega$

CHAPTER 30

PRACTICE PROBLEMS

I

1. JFET Q_1 is the active amplifying device.

2. L_{1_A}, L_{1_B}, and C_1 make up the frequency determining part of the circuit.

3. Capacitor C_3 is part of the *feedback* portion of this oscillator circuit.

4. From Formula 30-3:
   $L_T = L_{1_A} + L_{1_B} = 0.1 \text{ mH} + 0.1 \text{ mH} = 0.2 \text{ mH}$

5. From Formula 30-2:
   $f_r = 1/2\pi\sqrt{L_T C} = 1/2\pi\sqrt{(0.2 \text{ mH})(1,000 \text{ pF})}$

6. Decreasing the value of C_1 increases the oscillating frequency. Increasing the value of L_{1_A} decreases the oscillating frequency.

II

1. The active amplifying device is JFET Q_1.

2. C_1, C_2, and the inductance of the primary of transformer T_1 make up the frequency determining part of the circuit.

3. From Formula 30-5:
   $C_T = C_1 C_2/(C_1 + C_2) = 50 \text{ pF}$

4. From Formula 30-4:
$f_r = 1/2\pi\sqrt{LC_T}$
$f_r = 1/2\pi\sqrt{(750 \text{ }\mu\text{H})(50 \text{ pF})} = 822$ kHz

5. When the total capacitance is decreased, the frequency of oscillation increases. When the value of inductance increases, the frequency of oscillation decreases.

III

1. From Formula 30-8:
$f_r = 1/2\pi RC$
$f_r = 1/2\pi(4.7 \text{ k}\Omega)(0.002 \text{ }\mu\text{F}) = 16.9$ kHz

2. Doubling the values of the capacitors reduces the operating frequency by one-half. Cutting the resistor values in half causes the operating frequency to double.

3. Rearranging Formula 30-8 to solve for the value of R:
$R = 1/2\pi f_r C$
$R = 1/2\pi(1 \text{ kHz})(0.1 \text{ }\mu\text{F}) = 1.6 \text{ k}\Omega$

IN-PROCESS LEARNING CHECKS

I

1. The LC oscillator that has a tapped coil or a set of two coils in the tank circuit is called the *Hartley* oscillator.

2. The LC oscillator that has a tank circuit of two capacitors and a single inductor is called the *Colpitts* oscillator.

3. The LC oscillator that has three capacitors in the tank circuit is called the *Clapp* oscillator.

4. The output of a *Colpitts* LC oscillator is often coupled to the next stage by means of a transformer whose primary winding is part of the tank circuit.

5. When $C_1 = 100$ pF:
$C_T = (100 \text{ pF})(500 \text{ pF})/(100 \text{ pF} + 500 \text{ pf}) = 83.3$ pF
$f_r = 1/2\pi\sqrt{(100 \text{ }\mu\text{H})(83.8 \text{ pF})} = 1.74 \text{ MHz}$
When $C_1 = 850$ pF:
$C_T = (850 \text{ pF})(500 \text{ pF})/(850 \text{ pF} + 500 \text{ pF}) = 315$ pF
$f_r = 1/2\pi\sqrt{(100 \text{ }\mu\text{H})(315 \text{ pF})} = 897$ kHz
or 0.897 MHz

II

1. Whereas *LC* oscillators' frequency of operation depend on resonance, *RC* oscillators' frequency of operation depend on *phase-shift.*

2. The four basic elements and conditions for starting and sustaining oscillation in *RC* sine-wave oscillators are *power source, frequency determining elements, amplifier,* and *positive feedback.*

3. The sine-wave oscillator that uses a three-stage *RC* network for achieving a 180° phase shift at the frequency of oscillation is the *phase-shift* oscillator.

4. The sine-wave oscillator that uses a lead-lag *RC* network to produce 0° phase shift at the frequency of oscillation is the *Wien-bridge* oscillator.

5. The amplifier element of a phase-shift oscillator must produce a phase shift of *180°,* while the amplifier element of a Wien-bridge oscillator must produce a phase shift of *0°.*

REVIEW QUESTIONS

1. A power source, a device or components that determine the frequency of oscillation, amplification, and positive feedback.

3. The Hartley oscillator uses a tapped coil. The Colpitts oscillator has two capacitors with a ground connection between them. The Clapp oscillator uses three capacitors and a single inductor in the tank circuit.

5. From Formula 30-3:
$L_T = L_{1_A} + L_{1_B} = 20$ mH
From Formula 30-2:
$f_r = 1/2\pi\sqrt{L_T C} = 3.56$ kHz

7. From Formula 30-5:
$C_T = C_1 C_2/(C_1 + C_2) = 0.05 \text{ }\mu\text{F}$
From Formula 30-4:
$f_r = 1/2\pi\sqrt{LC_T} = 7.12$ kHz

9. The phase-shift oscillator uses a network of three identical *RC* elements. The Wien-bridge oscillator uses a lead-lag network.

11. At the frequency of oscillation, the *RC* network in a phase-shift oscillator produces a phase shift of *180°,* and the amplifier produces a phase shift of *180°.*

13. From Formula 30-8:
$f_r = 1/2\pi RC = 159$ Hz

15. In order to decrease the active time of a monostable multivibrator, you must *decrease* the value of the timing capacitor and/or *decrease* the value of the timing resistors.

17. $T = t_1 + t_2 = 10 \text{ }\mu\text{s} + 8 \text{ }\mu\text{s} = 18 \text{ }\mu\text{s}$
$f = 1/T = 55.6$ kHz

19. From Formula 30-12:
$$f = 1/0.69C(R_A + 2R_B) = 483 \text{ Hz}$$

PROBLEMS

1. 160 kHz
3. 625 Hz
5. 13.5 MHz
7. 7.33 nF
9. 65 Hz

ANALYSIS QUESTIONS

1.

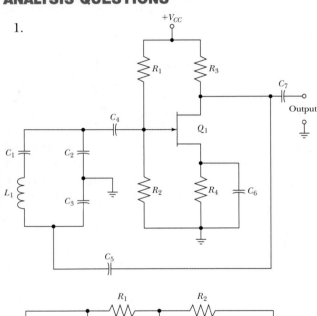

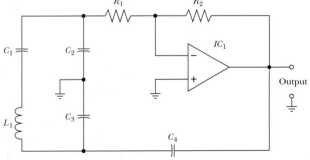

3. The total phase shift from the output to the input of the amplifier element in an oscillator must be 0° (or 360°) at the operating frequency to provide positive feedback. In a phase-shift oscillator, the *RC* network shifts the waveform by 180°, so the amplifier element must also shift the signal by 180° to satisfy the requirement for positive feedback. The *RC* network in a Wien-bridge oscillator, on the other hand, does not shift the phase of the waveform at the oscillator frequency; so the amplifier must not shift the waveform, either, in order to provide a total of 0° for positive feedback.

5. The feedback path through the three-section *RC* network of a phase-shift oscillator reduces the signal level a great deal by the time it reaches the input of the amplifier element. None of the other oscillators described in this chapter reduces the signal so much. The amplifier must have a high gain to overcome the unusual amount of signal loss through the *RC* network.

7. A *bistable multivibrator,* or flip-flop, is a type of multivibrator that is stable in both of its two states—on and off. It changes state each time it is triggered.

9. The operating frequency of a 555 stable multivibrator decreases as the value of the timing capacitor increases. The operating frequency increases as the value of the timing.

CHAPTER 31

IN-PROCESS LEARNING CHECKS

I

1. A thyristor is a ***four***-layered device that has just ***two*** operating states.

2. An SCR is gated on by applying a ***forward***-biasing voltage to the ***gate-cathode*** P-N junction. A ***LASCR*** (or ***light-activated SCR***) can also be gated on by shining a light onto its lens.

3. An SCR will conduct when it is gated and a voltage is applied between the cathode and anode such that the anode polarity is ***positive*** and the cathode is ***negative.***

4. The SCR rating that specifies the maximum forward voltage an SCR can handle is called the ***peak forward blocking voltage.***

5. The SCR rating that specifies the maximum reverse voltage an SCR can handle is called the ***peak reverse voltage.***

6. The smallest amount of cathode-anode forward current that can sustain conduction of an SCR is called the ***holding current.***

7. The ***GCS*** (or ***gate-controlled switch***) is a type of SCR that can be gated off as well as gated on.

8. The ***silicon-controlled switch*** is a type of SCR that has two gate terminals.

9. In order to control both cycles of ac power, two SCRs must be connected in reverse with respect to one another in an arrangement that is known as *inverse parallel* or *back-to-back.*

10. When a pair of SCRs are gated halfway through their respective half-cycles of the ac power waveform, the circuit is said to be firing at *90°.*

II

1. Thyristors that can conduct in two directions (such as a diac and triac) are said to be *bidirectional* devices.

2. A diac has *no* gate terminals, while a triac has *one* gate terminal.

3. The only way to get a diac to conduct is by exceeding its *forward* blocking and reverse *breakdown* ratings.

4. A triac operates like a pair of *SCRs* connected back-to-back.

5. The most common commercial and industrial applications of *diacs* are as triggering devices for triacs.

6. In order to apply full ac power to a load, a triac should be fired at the *0°* point on each half-cycle.

REVIEW QUESTIONS

1. The two possible states of operation of a thyristor are *fully conducting* and *fully nonconducting.*

3. The conditions required for turning on an SCR are 1) forward-biasing voltage between the cathode and anode, and 2) forward-biasing voltage between the cathode and gate. The SCR is turned off only by removing or reversing the cathode-anode voltage.

5. *Holding current* is the least amount of forward-conducting current a thyristor can conduct and remain conducting.

7. *GCS* = gate-controlled switch; *LASCR* = light-activated SCR; *SCR* = silicon-controlled rectifier; and *SCS* = silicon-controlled switch.

9. The most important difference between an SCR and a GCS is that a GCS can be turned on and off with gate pulses. An SCR can only be turned on with gate pulses.

11. A LASCR can be switched on by applying a positive gate pulse or by shining a light onto its lens assembly.

13. A bidirectional device is one that is capable of conducting current in both directions. A common resistor is bidirectional because it can conduct current in both directions. A junction diode is not bidirectional because it passes current in only one direction.

15. Since a diac is switched on only by exceeding the forward blocking voltage and peak reverse voltage, these ratings are extremely important to the theory of operation of a diac (a bidirectional device).

17. A triac is gated on by applying a voltage between the two anode connections (polarity is not relevant) and applying a forward-biasing voltage to the gate. A triac is turned off when the applied voltage is reduced to zero or reversed in polarity.

19. An ac phase control circuit is used for switching on a circuit at an adjustable point along each half-cycle of the ac waveform. Such circuits are used for controlling the amount of ac power applied to motors, heating elements, and lamps.

ANALYSIS QUESTIONS

1. An SCR can be considered a power amplifier because it uses a small gate voltage and current to switch on relatively large amounts of anode current and voltage. It can be considered a nonlinear amplifier because the output (load) waveform can be significantly different from the input (gate) waveform.

3. The only similarity between the action of a diac and a zener diode is that they begin conducting only after the applied voltage reaches a certain level. They are quite different in the way they are turned off. The zener diode stops conducting when the applied voltage falls below the zener level; however, the diac does not stop conducting until the applied voltage drops near zero.

APPENDIX C

Chapter Challenge Test Results

Find the number listed next to the test you chose and record the result.

1. 1.4 V
2. signal normal
3. 2 kΩ
4. 2 V
5. 1.5 V
6. high
7. 1.5 V
8. 0 V
9. 0 V
10. 3 V
11. 0 V
12. 7 V
13. ∞ Ω
14. 0 Ω
15. ≈0 V
16. 1 kΩ
17. 400 Ω
18. 1.5 V
19. ∞ Ω
20. 1.5 V
21. ⌐‾_
22. 10 V
23. no signal
24. circuit operates normally
25. 10 kΩ
26. slightly high
27. 0 V
28. 0 V
29. 0 Ω
30. greatly below max value

31. 5 V
32. 400 Ω
33. ∞ Ω
34. 6.4 V
35. noticeably high
36. 4 kΩ
37. 10 kΩ
38. 3.3 V
39. 14 V
40. 0 V
41. ≈1.5 mH
42. 3.0 V
43. 7 V
44. 1.5 V
45. high
46. 0 Ω
47. 0 V
48. 2.3 V
49. 14 V
50. 0.5 V
51. 1.5 V
52. 10 V
53. _⌣‾ (14 VAC)
54. 10 kΩ
55. max value
56. 10 V
57. signal normal
58. ∞ Ω
59. 1.5 kΩ
60. 275 pF

61. 14 V
62. 1 kΩ
63. ∞ Ω
64. 1.5 V
65. ≈6 VDC
66. 0 V
67. 7.5 V
68. 10 kΩ
69. low
70. signal normal
71. no change in operation
72. 0 Ω
73. 1.5 V
74. ‾⌣‾ (120 VAC)
75. 10 V
76. slightly low
77. 5.9 V
78. 3.1 V
79. 7.5 V
80. 7 V
81. 2 V
82. 0 V
83. 10 V
84. 0 Ω
85. slightly below max value
86. 1.5 V
87. 5 Ω
88. 14 V

89. 2.3 V
90. ∞ Ω
91. 0 V
92. normal
93. 10 kΩ
94. ≈5 V
95. 10 V
96. ≈11 V
97. 0 V
98. low
99. 1.5 V
100. ∞ Ω
101. 0 V
102. 0 V
103. 0 V
104. ≈5 kΩ
105. 0 V
106. no signal
107. 50 pF
108. ∞ Ω
109. 12 kΩ
110. 0 V
111. no change in operation
112. 1.5 V
113. signal normal
114. 297 Ω
115. 0 V
116. ≈5 V
117. within normal range

Using a Scientific Calculator

Sample problem: Find R_T, I_T, and V_T for a series circuit containing three 20-kΩ resistors, where $V_1 = 10$ V. See the diagram below.

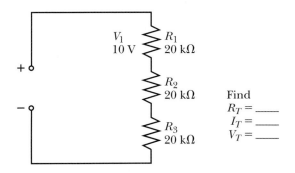

V_1
10 V

R_1
20 kΩ

R_2
20 kΩ

R_3
20 kΩ

Find
$R_T =$ ____
$I_T =$ ____
$V_T =$ ____

ELEMENTAL METHOD

Step 1: Turn calculator on and clear.

Step 2: Find R_T where $R_T = R_1 + R_2 + R_3$, by entering 20,000 + 20,000 + 20,000, then press the ⊜ button. The answer should be 60,000. (This represents 60,000 Ω, or 60 kΩ.)

Step 3: Find I by finding the I through R_1, where $I_1 = V_1/R_1$. Clear the calculator by pushing the **CE/C** button. Input as follows: Input 10 (for V_1), then the ÷ symbol button, then 20,000 (for R_1), then the ⊜ button. The answer should be 0.0005. (This represents 0.0005 A, or 0.5 mA.)

Step 4: Find V_T by using the formula $V_T = I_T \times R_T$. You may clear the calculator (or, leave the 0.0005 you had just calculated, since you need it in this calculation). At any rate, be sure that 0.0005 is showing, then push the ✕ symbol button, then input 60,000 (for the R_T value). Push the ⊜ button. The answer should be 30, representing 30 V.

NOTE: This approach is appropriate for many calculators. Some brands, depending on their notation approach, will not use the exact same sequence of key strokes.

Using the "elemental approach" just described, if you solve for the parameters asked for in the circuit below, your answers should match the ones shown in parenthesis after each of the blanks. Try using your calculator and the "elemental approach" to see if you get the answers, as shown.

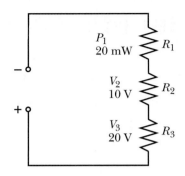

Find:
$V_T =$ _____ (40 V)
$P_3 =$ _____ (40 mW)
$R_1 =$ _____ (5 kΩ)
$R_2 =$ _____ (5 kΩ)
$R_3 =$ _____ (10 kΩ)

Now that you've used the elemental method of entering numeric data, let's look at three other ways you can enter data and use a scientific calculator to advantage. The three ways include:

1. using the **EE** or **EXP** key on a scientific calculator (along with powers of 10);

2. using the **EE** or **EXP** key on a scientific calculator that is in the "engineering mode" (along with powers of 10); and

3. using the **EE** or **EXP** key on a scientific calculator that is in the "scientific notation mode" (along with powers of 10).

All three of these methods will save you from having to enter numerous zeros for very large or very small numbers. For our particular example problem, these approaches save you from having to put in all the zeros for large resistance values or all the zeros after the decimal for the small current value. Any one of these approaches can be helpful for most of your electronic calculations. Once you have learned these techniques, use the one that suits your needs best.

The significant key you will use on the keyboard in conjunction with these modes may be called **EE** on some calculators, or, **EXP** on others. Let's see how this system works.

SPECIAL NOTATIONS REGARDING SCIENTIFIC CALCULATORS

Numerous brands and types of scientific calculators are on the market. Functions such as squaring a number, finding square roots, finding the arc tangent, and finding antilogarithms may differ slightly from calculator to calculator. For the types of calculations performed throughout the text, the chart below shows some common variations you may see or use. If you are using a different type of

TASK DESCRIPTION	CALCULATOR #1 KEYSTROKE(S) TI 36X SOLAR	CALCULATOR #2 KEYSTROKE(S) HP 20S	CALCULATOR #3 KEYSTROKE(S) (Others)
Finding the reciprocal	1/x	1/x	2nd, 1/x
Squaring a number	x^2	↰ , x^2	2nd, x^2
Taking a square root	$\sqrt{x}$	$\sqrt{x}$	$\sqrt{x}$
Finding the arctan	2nd, tan	↰ , ATAN	INV, tan
Finding the antilog of a natural logarithm (ϵ)	2nd, e^x	e^x	e^x
Converting from polar to rectangular form: Finding the x-axis and y-axis (j) values:	phasor value, X♦Y, angle value, 2nd (P ▸ R) (Displays x-axis value) Follow above sequence with X♦Y key again (Displays y-axis *[j]* value)	phasor value, INPUT, angle value, ⌐, → R (Displays y-axis *[j]* value) Follow above sequence with ↰, SWAP (Displays x-axis value)	phasor value, P-R key, angle value, = key (Displays x-axis value) Follow above sequence with X-Y key (Displays y-axis *[j]* value)
Converting from rectangular to polar form: Finding the phasor value and the angle	x-axis value, X♦Y, y-axis (j) value, 3rd, (R ▸ P) Displays the phasor value Press the X♦Y key again (Displays the angle value)	x-axis value, INPUT, y-axis (j) value, ↰, → P, (Displays angle value) Follow above sequence with ↰, SWAP (Displays the phasor value)	x-axis value, R-P key, y-axis (j) value, = (Displays phasor value) Follow the above sequence with X-Y key, (Displays the angle value)

scientific calculator than the one used in creating the sample calculator sequences in our text (Calculator #1), the equivalent keystrokes chart, shown above, may be helpful to you.

USING THE EE (OR EXP) KEY WITH POWERS OF 10 FOR METRIC UNITS

Using the same sample problem as before; that is, finding R_T, I_T, and V_T, where there are three series 20-kΩ resistors and $V_1 = 10$ V:

Step 1: Turn calculator on and clear by pressing the **AC/ON** key.

Step 2: Find R_T (where $R_T = R_1 + R_2 + R_3$) by entering the following:

input 20, **EE** (or **EXP**), 3,

(the display reads 20^{03} or 20. 03, meaning 20×10^3)

then input **+**,

(the display now reads 20000)

then input 20, **EE** (or **EXP**), 3, a second time

(the display once again reads 20^{03} or 20. 03, showing the number you just entered)

then input ➕,

(the display now reads 40000)

then input 20, **EE** (or **EXP**), 3, a third time

(the display will read 20^{03} or 20. 03, again representing the number you just entered)

then enter ▬,

(the display will read 60000. This indicates that the value of R_T is 60,000 Ω, or 60 kΩ)

Step 3: Find I by finding the I through R_1, (where $I_1 = V_1/R_1$):

Clear the calculator by pressing the **CE/C** key.

Enter 10 (for the V_1 value),

then enter ➗,

then enter 20, **EE** (or **EXP**), 3 (for the 20-kΩ R_1 value)

then press the ▬ key.

The display reads 0.0005. (This indicates that current is 0.5 mA.)

Step 4: Find V_T by using the formula $V_T = I_T \times R_T$.

Clear the calculator with the **CE/C** button.

Enter the value of I_T: 0.5, **EE** (or **EXP**), **+/−**, 3 (the display should read 0.5^{-03} or $0.5 - 03$)

then press the ✖ key,

then input 60, **EE** (or **EXP**), 3

then, press the ▬ key.

The display reads 30. (This indicates that $V_T = 30$ V.)

Now, you try this technique to see if you can match the answers for the circuit parameters called for in the circuit shown below.

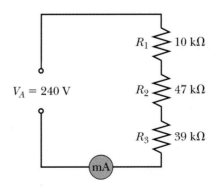

FIND:
R_T = _____ (96 kΩ)
I_T = _____ (2.5 mA)
V_{R_2} = _____ (117.5 V)

USING THE ENGINEERING MODE WITH POWERS OF 10 FOR METRIC UNITS

The engineering mode provides a notation that again uses powers of 10. Because most science and engineering calculations use metric measurements that typically have *exponents* that are in multiples or submultiples of three, the data input and answers appear as some number, times a power of 10 with a multiple of three exponent. For example, in electronics, kilohms and kilowatts are expressed in terms of 10^3. Millivolts and milliamperes are expressed in terms of 10^{-3}. Microamperes are stated as 10^{-6}. Nanoamperes are indicated as 10^{-9} and so on.

Using the earlier problem that asked for R_T, I_T, and V_T when the circuit was comprised of three series 20-kΩ resistors and $V_1 = 10$ V:

Step 1: Turn calculator on and clear by pressing the **AC/ON** key.

Step 2: Press any buttons required to get your calculator into the *engineering mode*. See your calculator's instruction manual for this. NOTE: For the TI 36X solar calculator, the key sequence to get into the engineering mode is **3rd**, ENG.

(When you are in this mode, the calculator readout will read 0.^{00} or 0.00 before new data are entered.)

Step 3: Find R_T in the *engineering mode* as follows:

Be sure your calculator is in the *engineering mode* and the display reads 0.^{00} or 0.00.

Enter 20, **EE** (or **EXP**), 3 (the readout reads 20.^{03} or 20.03),

then press **+** (the readout will still read 20.^{03} or 20.03),

then enter 20, **EE** (or **EXP**), 3 a second time (the readout will still read 20.^{03} or 20.03),

then enter **+** (the readout will read 40.^{03} or 40. 03),

then enter 20, **EE** (or **EXP**), 3 a third time (the readout will read 20.^{03} or 20.03),

then press the **=** sign (the readout now reads 60.^{03} or 60.03).

This indicates 60 $\times$ 10^3, or 60,000 Ω, or 60 kΩ.

Step 4: Find I_T in the *engineering mode* by entering the following (where $I_T = I_1$ and $I_1 = V_1/R_1$):

Be sure your calculator is in the *engineering mode* and the display reads 0.^{00} or 0.00.

Enter 10 (for the V_1 value),

then press the **÷** key,

then enter 20, **EE** (or **EXP**), 3,

then press the **=** key

The display should show 500.$^{-06}$ or 500.$-$06.

(This indicates 500 $\times$ 10^{-6}, or 0.0005 A, which is the same as 0.5 mA.

$I_T = 0.5$ mA.)

Step 5: Find V_T in the *engineering mode* by using the formula $V_T = I_T \times R_T$.

Be sure the calculator is in the *engineering mode* and the display reads 0.^{00} or 0.00.

Enter the I_T value by entering 0.5, **EE** (or **EXP**), **+/−**, 3, then press the **×** key, then enter 60, **EE** (or **EXP**), 3,

then, press the **=** key.

The display reads 30.^{00} or 30. 00 (This means 30×10^0, which means that $V_T = 30$ V.)

Get some practice in using the *engineering mode* by seeing if you can match the answers shown for the parameters specified.

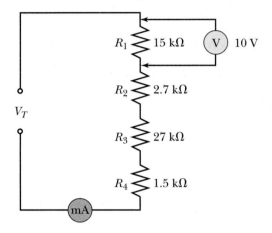

FIND:
$R_T =$ _____ (46.2 kΩ)
$I_T =$ _____ (0.67 mA)
$V_T =$ _____ (30.95 V)

USING THE SCIENTIFIC NOTATION MODE WITH POWERS OF 10 FOR METRIC UNITS

As you know, scientific notation expresses numerical values as a number between 1 and 10, times the appropriate power of 10. Performing the same calculations as we have been using, the sequence of keys are as follows.

(NOTE: Make sure the calculator is *cleared* and you are in the *scientific notation mode* for each of the following calculations.)

To find R_T for our example circuit:

Enter 20, **EE** (or **EXP**), 3, **+**, 20, **EE** (or **EXP**), 3, **+**,

20, **EE** (or **EXP**), 3, then press **=**.

The display reads 6.^{04} or 6. 04, meaning 6×10^4, or 60,000. Again, this is interpreted as 60,000 Ω, or 60 kΩ, for our circuit.

To find I_T (which is equal to I_1):

Enter 10, **÷**, 20, **EE** (or **EXP**), 3, then press **=**.

The display reads 5.$^{-04}$ or 5.−04, interpreted as 5×10^{-4}, or 0.5 mA.

To find V_T:

Enter 5, **EE** (or **EXP**), **+/−**, 4, **×**, 60, **EE** (or **EXP**), 3, then press **=**.

The display reads $3.^{01}$ or 3. 01, meaning 3×10^1, or 30 V.

Get some practice in using the *scientific notation mode* by seeing if you can match the answers shown for the parameters specified.

$R_T =$ _____ (25.9 kΩ)
$I_T =$ _____ (0.5 mA)
$V_T =$ _____ (12.95 V)

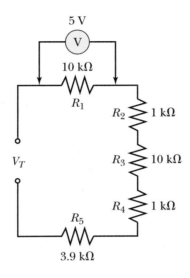

Powers of 10 Review

Rules for Moving Decimal Places Using Powers of 10

To express a small number as a larger one times a power of 10, move the decimal point to the right and count the number of places moved from the original point. The number of places counted (or moved) is the appropriate negative power of 10.

Example

The number 0.00253 can be expressed using powers of 10 in any of the following ways:
(1) 253×10^{-5} (2) 2530×10^{-6} (3) 25.3×10^{-4} and so on.

To express a large number as a smaller one times a power of 10, move the decimal point to the left and count the number of places moved from the original point. The number of places counted is the appropriate positive power of 10.

Example

The number 42,853 can be expressed using powers of 10 in any of the following ways:
(1) 4.2853×10^{4} (2) 42.853×10^{3} (5) 4285.3×10^{1} and so on.

Rule for Multiplication Using Powers of 10

When multiplying powers of 10, add the exponents (algebraically). (NOTE: Laws of exponents are applicable when using powers of 10.)

Example

$10^{2} \times 10^{2} = 10^{4}$
$2 \times 10^{4} \times 2 \times 10^{-3} = 4 \times 10^{1}$

Rule for Division Using Powers of 10

When dividing powers of 10, subtract the exponents (algebraically).

Example

$$\frac{4 \times 10^{-2} \text{ (numerator)}}{2 \times 10^{-4} \text{ (denominator)}} = 2 \times 10^{-2} \times 10^{4} = 2 \times 10^{2}$$

$$\frac{5 \times 10^{3}}{2.5 \times 10^{6}} = 2 \times 10^{3} \times 10^{-6} = 2 \times 10^{-3}$$

NOTE: Change the sign of the denominator power of 10 and add algebraically to the numerator power of 10 exponent. If the power of 10 exponent in the denominator is the same sign and same value as the numerator, they cancel.

Rules for Multiplication and Division Combined

The rules are:

1. multiply all factors in the numerator;
2. multiply all factors in the denominator; and
3. divide.

Example

$$\frac{15 \times 10^2 \times 10 \times 10^4}{2 \times 10^{-2} \times 12.5 \times 10^{-2}} = \frac{150 \times 10^6}{25 \times 10^{-4}} = 6 \times 10^{10}$$

Rule for Addition of Powers of 10

Change all factors to the same power of 10 and add the coefficients. The answer retains the same power of 10 as the added factors.

Example

$$(2 \times 10^2) + (5 \times 10^3) = (0.2 \times 10^3) + (5 \times 10^3) = 5.2 \times 10^3$$

Examples of coefficients

Rules for Subtraction of Powers of 10

Change all powers of 10 to the same power of 10 and subtract the coefficients.

Example

$$(5 \times 10^3) - (2 \times 10^2) = (5 \times 10^3) - (0.2 \times 10^3) = 4.8 \times 10^3$$

APPENDIX F

Copper-Wire Table

Wire Size A.W.G. (B&S)	Diam. in Mils[1]	Circular Mil Area	Turns per Linear Inch (25.4 mm)[2]			Cont.-duty current[3] single wire in open air	Cont.-duty current[3] wires or cables in conduits or bundles	Feet per Pound (0.45 kg) Bare	Ohms per 1000 ft. 25°C	Current Carrying Capacity[4] at 700 C.M. per Amp.	Diam. in mm.	Nearest British S.W.G. No.
			Enamel	S.C.E.	D.C.C.							
1	289.3	83,690	—	—	—	—	—	3.947	0.1264	119.6	7.348	1
2	257.6	66,370	—	—	—	—	—	4.977	0.1593	94.8	6.544	3
3	229.4	52,640	—	—	—	—	—	6.276	0.2009	75.2	5.827	4
4	204.3	41,740	—	—	—	—	—	7.914	0.2533	59.6	5.189	5
5	181.9	33,100	—	—	—	—	—	9.980	0.3195	47.3	4.621	7
6	162.0	26,250	—	—	—	—	—	12.58	0.4028	37.5	4.115	8
7	144.3	20,820	—	—	—	—	—	15.87	0.5080	29.7	3.665	9
8	128.5	16,510	7.6	—	7.1	73	46	20.01	0.6405	23.6	3.264	10
9	114.4	13,090	8.6	—	7.8	—	—	25.23	0.8077	18.7	2.906	11
10	101.9	10,380	9.6	9.1	8.9	55	33	31.82	1.018	14.8	2.588	12
11	90.7	8,234	10.7	—	9.8	—	—	40.12	1.284	11.8	2.305	13
12	80.8	6,530	12.0	11.3	10.9	41	23	50.59	1.619	9.33	2.053	14
13	72.0	5,178	13.5	—	12.8	—	—	63.80	2.042	7.40	1.828	15
14	64.1	4,107	15.0	14.0	13.8	32	17	80.44	2.575	5.87	1.628	16
15	57.1	3,257	16.8	—	14.7	—	—	101.4	3.247	4.65	1.450	17
16	50.8	2,583	18.9	17.3	16.4	22	13	127.9	4.094	3.69	1.291	18
17	45.3	2,048	21.2	—	18.1	—	—	161.3	5.163	2.93	1.150	18
18	40.3	1,624	23.6	21.2	19.8	16	10	203.4	6.510	2.32	1.024	19
19	35.9	1,288	26.4	—	21.8	—	—	256.5	8.210	1.84	0.912	20
20	32.0	1,022	29.4	25.8	23.8	11	7.5	323.4	10.35	1.46	0.812	21
21	28.5	810	33.1	—	26.0	—	—	407.8	13.05	1.16	0.723	22
22	25.3	642	37.0	31.3	30.0	—	5	514.2	16.46	0.918	0.644	23
23	22.6	510	41.3	—	37.6	—	—	648.4	20.76	0.728	0.573	24
24	20.1	404	46.3	37.6	35.6	—	—	817.7	26.17	0.577	0.511	25
25	17.9	320	51.7	—	38.6	—	—	1,031	33.00	0.458	0.455	26
26	15.9	254	58.0	46.1	41.8	—	—	1,300	41.62	0.363	0.405	27
27	14.2	202	64.9	—	45.0	—	—	1,639	52.48	0.288	0.361	29
28	12.6	160	72.7	54.6	48.5	—	—	2,067	66.17	0.228	0.321	30
29	11.3	127	81.6	—	51.8	—	—	2,607	83.44	0.181	0.286	31
30	10.0	101	90.5	64.1	55.5	—	—	3,287	105.2	0.144	0.255	33
31	8.9	80	101	—	59.2	—	—	4,145	132.7	0.114	0.227	34
32	8.0	63	113	74.1	61.6	—	—	5,227	167.3	0.090	0.202	36
33	7.1	50	127	—	66.3	—	—	6,591	211.0	0.072	0.180	37
34	6.3	40	143	86.2	70.0	—	—	8,310	266.0	0.057	0.160	38
35	5.6	32	158	—	73.5	—	—	10,480	335	0.045	0.143	38–39
36	5.0	25	175	103.1	77.0	—	—	13,210	423	0.036	0.127	39–40
37	4.5	20	198	—	80.3	—	—	16,660	533	0.028	0.113	41
38	4.0	16	224	116.3	83.6	—	—	21,010	673	0.022	0.101	42
39	3.5	12	248	—	86.6	—	—	26,500	848	0.018	0.090	43
40	3.1	10	282	131.6	89.7	—	—	33,410	1,070	0.014	0.080	44

[1]A mil is 0.001 inch. A circular mil is a square mil $\times \frac{\pi}{4}$. The circular mil (c.m.) area of a wire is the square of the mil diameter.

[2]Figures given are approximate only; insulation thickness varies with manufacturer.

[3]Max. wire temp. of 212°F (100°C) and max. ambient temp. of 135°F (57°C).

[4]700 circular mils per ampere is a satisfactory design figure for small transformers, but values from 500 to 1,000 c.m. are commonly used.

Useful Conversion Factors

To Convert	Into	Multiply By
ampere-turns/cm	amp-turns/inch	2.54
ampere-turns/inch	amp-turns/cm	0.3937
ampere-turns/inch	amp-turns/meter	39.37
ampere-turns/meter	amp-turns/inch	0.0254
Centigrade	Fahrenheit	$(°C \times 9/5) + 32$
centimeters	feet	3.281×10^{-2}
centimeters	inches	0.3937
centimeters/sec	feet/sec	0.03281
circular mils	sq mils	0.7854
circumference	radians	6.283
circular mils	sq inches	7.854×10^{-7}
coulombs	faradays	1.036×10^{5}
degrees	radians	0.01745
dynes	joules/cm	10^{-7}
dynes	joules/meter (newtons)	10^{-5}
ergs	foot-pounds	7.367×10^{-6}
ergs	joules	10^{-7}
farads	microfarads	10^{6}
faradays	ampere-hours	26.8
faradays	coulombs	9.649×10^{4}
feet	centimeters	30.48
foot-pounds	ergs	1.356×10^{7}
gausses	lines/sq inch	6.452
gausses	webers/sq meter	10^{-4}
gilberts	ampere-turns	0.7958
grams	dynes	980.7
grams	pounds	2.205×10^{-3}
henries	millihenries	1,000
horsepower	foot-lbs/sec	550
horsepower	kilowatts	0.7457
horsepower	watts	745.7
inches	centimeters	2.54
inches	millimeters	25.4
joules	ergs	10^{7}
joules	watt-hrs	2.778×10^{-4}
kilolines	maxwells	1,000
kilowatts	foot-lbs/sec	737.6
kilowatts	horsepower	1.341
kilowatts	watts	1,000
kilowatt-hrs	joules	3.6×10^{6}
lines/sq inch	webers/sq meter	$1,550 \times 10^{-5}$
maxwells	kilolines	0.001
maxwells	webers	10^{-8}
microfarad	farads	10^{-6}
microns	meters	1×10^{-6}
millihenries	henries	0.001
mils	inches	0.001
nepers	decibels	8.686
ohms	megohms	10^{-6}
quadrants	degrees	90
quadrants	radians	1.571
radians	degrees	57.3
square inches	sq cms	6.452
temperature (°F) -32	temperature (°C)	5/9
watts	horsepower	1.341×10^{-3}
watts (Abs.)	joules/sec	1
webers	maxwells	10^{8}
webers/sq meter	webers/sq inch	6.452×10^{-4}
yards	meters	0.9144

APPENDIX H

Schematic Symbols

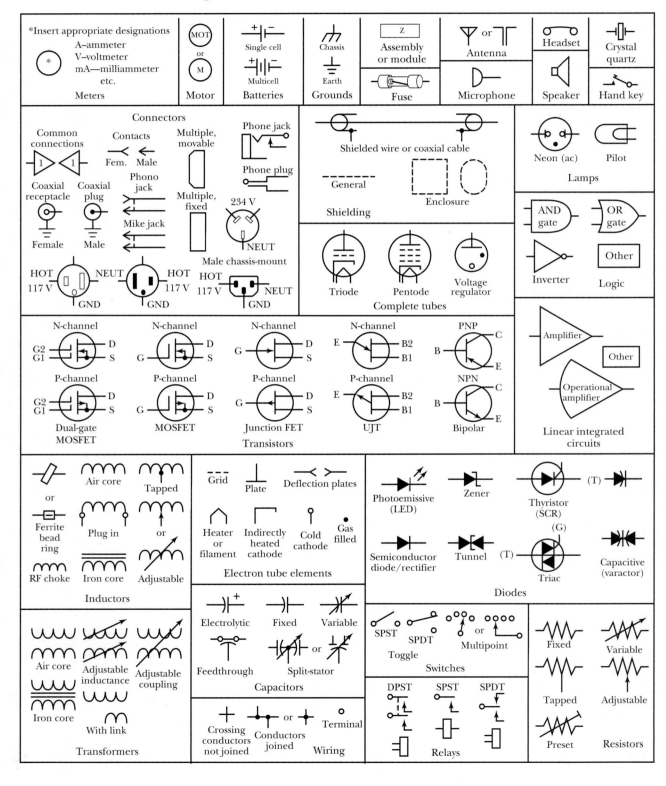

Glossary

ac: abbreviation for alternating current. The letters "ac" are also used as a prefix, or modifier to designate voltages, waveforms, and so forth that periodically alternate in polarity.

ac beta (β_{ac}): for a BJT, the amount of change in collector current divided by the corresponding amount of change in base current.

admittance (Y): the ability of a circuit with both resistance and reactance to pass ac current; the reciprocal of impedance.

alternation: one-half of a cycle. Alternations are often identified as the positive alternation or the negative alternation when dealing with ac quantities.

amalgamation: in carbon-zinc dry cell manufacturing, coating the inside of the zinc electrode with a thin layer of mercury to help prevent local action.

ampere (A): the basic unit of current flow. An ampere of current flow represents electron movement at a rate of one coulomb per second; that amount of current flowing through one ohm of resistance with one volt applied.

ampere-hour rating: a method of rating cells or batteries based on the amount of current they supply over a specified period with specified conditions.

analog operation: circuit operation where the outputs vary continuously over a range as a direct result of the changing input levels; also called linear operation.

AND gate: a digital logic gate where both inputs must be high for the output to be high.

anode: the terminal on a diode that is composed of a P-type semiconductor and must be made positive with respect to the cathode in order to cause conduction through a diode. Its symbol is a triangle or arrow.

apparent power (S): the product of V and I in an ac circuit, without regard to phase difference between V and I. The unit of measure is volt-amperes. In the power triangle it is the vector resulting from true and reactive power.

assumed voltage method: a method of finding total resistance in parallel circuits by assuming a circuit applied voltage value to find branch currents, which enables finding total current and total circuit resistance with less complex mathematics than other methods for certain situations.

astable multivibrator: a type of multivibrator that has no stable state. It is a form of RC-controlled oscillator that produces a rectangular waveform. It is also known as a free-running multivibrator.

AT: the abbreviation for ampere-turns or magneto-motive force.

atom: the basic building block of matter composed of different types of particles. Major atom particles are the electron, proton, and neutron.

attenuator: a controllable voltage divider network in oscilloscopes that helps adjust the amplitude of input signals to a useable size.

autotransformer: a special single-winding transformer where part of the winding is in both the primary and secondary circuit. These transformers can be made to step-up or step-down, depending on where on the winding the primary source and the secondary load are connected.

average value (for one-half cycle): in ac sinusoidal values, the average height of the curve above the zero axis, expressed as 0.637 times maximum value. (NOTE: The average height over an *entire* cycle is zero; however, the average height of one alternation is as defined above.)

B: the symbol representing flux density.

backoff ohmmeter scale: the reverse (right-to-left) scale characteristic of series-type ohmmeters. Zero ohms is indicated on the scale's right end and infinite ohms on the scale's left end. Also, this scale is a nonlinear scale. That is, calibration marks and values are not evenly spaced across the scale.

bandpass: the band of frequencies that passes through a filter with minimal attenuation or degradation; often designated by specific frequencies, where the lower frequency is the frequency at the low-frequency half-power point, and the upper frequency is the frequency at the upper half-power point on the resonance curve; also called pass-band.

bandpass filter: a filter that passes a selected band of frequencies with little loss but greatly attenuates frequencies either above or below the selected band.

bandstop filter: a filter that stops (greatly attenuates) a selected band of frequencies while allowing all other frequencies to pass with little attenuation; also called band-reject filter.

bandwidth: the frequency limits where a specified response level is shown by a frequency-sensitive circuit, such as a resonant circuit. The cutoff level is defined at 70.7% of peak response, sometimes called the half-power points on the response curve.

barrier potential: a small voltage that is developed across the depletion layer of a P-N junction. For silicon junctions, the barrier potential is about 0.7 V, and for germanium junctions it is about 0.3 V.

base: the center of the NPN or PNP sandwich in a BJT that separates the emitter and the collector transistor regions. The emitter-base junction is forward-biased and shows low impedance to current flow. The collector-base junction is reverse-biased and displays high resistance to current flow.

base current (I_B): the level of current that flows into or out of the base terminal on a BJT. This current is normally due to the forward biasing of the emitter-base P-N junction.

base-emitter voltage (V_{BE}): the level of voltage applied between the base and emitter terminals on a BJT. Forward base current flow is enabled when the polarity of V_{BE} is such that it forward biases the base-emitter junction.

battery: a dc voltage source containing a combination of cells, connected to produce higher voltage or current than a single cell produces alone.

bidirectional: the ability of a device to conduct current in two directions. Diacs and triacs are bidirectional, whereas SCRs and rectifier diodes are unidirectional (conduction current in only one direction).

bilateral resistance: any resistance having equal resistance in either direction; that is, R is the same for current passing either way through the component.

bipolar junction transistor (BJT): a three-terminal, bipolar semiconductor containing both majority and minority current carriers. The device has alternate layers of P- and N-type semiconductor materials (which form two PN junction diodes). This transistor requires the flow of both types of charge carriers (electrons and holes) to complete the path for current flow through it. Examples are the NPN and the PNP transistor.

bleeder: the resistive component assuring that the minimum current drawn from a power supply system is sufficient to achieve reasonable voltage regulation over the range of output currents demanded from the supply; serves to discharge the filter capacitors when the power supply is turned off, which adds safety for people working on electronic equipment fed by the power supply. This resistive component is also called bleeder resistor.

bleeder current: the fixed current through a resistive system that helps keep voltage constant under varying load conditions.

bleeder resistor: the resistor connected in parallel with power supply to draw a bleeder current.

block diagrams: graphic illustrations of systems or subsystems by means of blocks (or boxes) that contain information about the function of each block. These blocks are connected by lines that show direction of flow (signal, fluid, electrical energy, etc.) and how the blocks interact. Block diagrams can illustrate any kind of system: electrical, electronic, hydraulic, mechanical, and so forth. "Flow" generally moves from left to right through the diagram but is not a requirement.

bridge circuit: a special series-parallel circuit usually composed of two, two-resistor branches. Depending on the relative values of these resistors, the bridge output (between the center points of each branch) may be balanced or unbalanced.

bridge rectifier: a full-wave rectifier circuit that requires only four diodes arranged in series-parallel such that two of them conduct on one ac input half-cycle and the other two conduct on the opposite ac input half-cycle.

bridging or shunting: connecting a component, device, or circuit in parallel with an existing component, device, or circuit.

capacitance: the ability to store electrical energy in the form of an electrostatic field.

capacitive reactance (X_C): the opposition a capacitor gives to ac or pulsating dc current. The symbol used to represent capacitive reactance is X_C.

capacitor: a device consisting of two or more conductors, separated by nonconductor(s). When charged, a capacitor stores electrical energy in the form of an electrostatic field and blocks dc current.

cathode: the part of a diode that is composed of an N-type semiconductor and must be made negative with respect to the anode in order to cause conduction through the diode. Its symbol is a bar with a perpendicular connecting lead. (This bar symbol is also perpendicular to the anode arrow symbol.)

cell: (relates to voltage sources) a single (stand-alone) element or unit within a combination of similar units that convert chemical energy into electrical energy. The chemical action within a cell produces dc voltage at its output terminals or contacts.

cemf (counter-emf or back-emf): the voltage induced in an inductance that opposes a current change and is present due to current changes causing changing flux linkages with the conductor(s) making up the inductor.

charge (electrical): for our purposes, charge can be thought of as the electrical energy present where there is an accumulation of excess electrons, or a deficiency of electrons. For example, a capacitor stores electrical energy in the form of "charge" when one of its plates has more electrons than the other.

charging a capacitor: moving electrons to one plate of a capacitor and moving electrons off the other plate; resultant charge movement is called charging current.

chip: a semiconductor substrate where active devices (transistors) and passive devices (resistors and capacitors) form a complete circuit.

"chip" capacitor: a surface-mount device (SMD), sometimes called a "surface-mount" capacitor. Characterized by its small size, almost zero lead-length connection method, and its use in printed circuit board surface-mount applications.

"chip" resistor: a surface-mount device (SMD), sometimes called "surface-mount" resistor designed to be used in surface-mount technology applications. Typically, very small and rectangular in shape, with metallic end electrodes as the points of connection to the component.

circuit: a combination of elements or components that are connected to provide paths for current flow to perform some useful function.

clamp circuit: a circuit that offsets the zero baseline of an input waveform to a different positive or negative value.

Clapp oscillator: an LC resonant-circuit oscillator that requires a single inductor and three capacitors. One capacitor is in the inductor leg of the tank circuit, and two capacitors are connected in series and center-tapped to ground in the other leg of the tank circuit.

class A, B, and C: in reference to amplifier operation, the classifications that denote the operating range of the circuit. Class A operates halfway between cutoff and saturation. Class B operates at cutoff. Class C operates at below cutoff bias.

clipper circuit: a type of circuit that clips off the portion of a waveform that extends beyond a prescribed voltage level. Clipper circuits can be designed to clip positive or negative levels, or both.

coefficient of coupling (k): the amount or degree of coupling between two circuits. In inductively coupled circuits, it is expressed as the fractional amount of the total flux in one circuit that links the other circuit. If all the flux links, the coefficient of coupling is 1. The symbol for coefficient of coupling is k. When k is multiplied by 100, the percentage of total flux linking the circuits is expressed.

collector: the transistor electrode that receives current carriers that have transversed from the emitter through the base region; the electrode that removes carriers from the base-collector junction.

collector-base voltage (V_{CB}): the amount of voltage present between the collector and base of a BJT. For proper operation of the BJT, this voltage must have a polarity that reverse biases the P-N junction between the base and collector regions of the transistor.

collector characteristic curves: a family of curves that shows how BJT collector current increases with an increasing amount of emitter-collector voltage at several different base-current levels. These curves can be used for constructing load lines for BJT amplifier circuits.

collector current (I_C): the value of current measured at the collector terminal of a BJT.

collector-emitter voltage (V_{CE}): the amount of voltage present between the emitter and collector terminals of a BJT. For proper operation of the BJT, this voltage must have a polarity such that the flow of majority charge carriers is from emitter to collector.

Colpitts oscillator: an *LC* resonant-circuit oscillator that uses a single inductor in parallel with two series-connected capacitors that are grounded at their common point.

common-base: relating to transistor amplifiers, the circuit configuration where the base is common to both input and output circuits.

common-base amplifier (CB): a BJT amplifier configuration where the input signal is applied to the emitter and the output is taken from the collector.

common-collector: the transistor amplifier circuit configuration where the collector is common for input and output circuits.

common-collector amplifier (CC): a BJT amplifier configuration where the input signal is applied to the base and the output is taken from the emitter. Also known as an emitter-follower circuit.

common-drain amplifier (CD): an FET amplifier configuration where the input signal is applied to the gate and the output is taken from the source. Also known as a source-follower circuit.

common-emitter: in reference to transistor amplifiers, the circuit configuration where the emitter is common to both input and output circuits.

common-emitter amplifier (CE): a BJT amplifier configuration where the input signal is applied to the base and the output is taken from the collector.

common-gate amplifier (CG): an FET amplifier configuration where the input signal is applied to the source terminal and the output is taken from the drain.

common-mode gain: comparison (ratio) of the output voltage of a differential amplifier to the common-mode input voltage. Ideally, the output should be zero when the same input is fed to both differential amplifier inputs.

common-mode input: an input voltage to a differential amplifier that is common to both its inputs.

common-mode rejection ratio (CMRR): a measure expressing the amount of rejection a differential amplifier shows to common-mode inputs; the ratio between the differential-mode gain and the common-mode gain equals the CMRR.

common-source amplifier (CS): an FET amplifier configuration where the input signal is applied to the gate and the output is taken from the drain.

complex numbers: term that expresses a combination of a *real* and an *imaginary* term which must be added vectorially or with phasors. The real term is either a positive or negative number displayed on the horizontal axis of the rectangular coordinate system. The imaginary term, *j operator,* expresses a value on the vertical or *j*-axis. The value of the real term is expressed first. For example, $5 + j10$ means plus 5 units (to the right) on the horizontal axis and up 10 units on the vertical axis.

compound: a form of matter that can be chemically divided into simpler substances and that has two or more types of atoms.

conductance: the ease with which current flows through a component or circuit; the opposite of resistance.

conductor: a material that has many free electrons due to its atoms' outer rings having less than four electrons, which is less than the eight needed for chemical and/or electrical stability.

constant current source: a current source whose output current is constant even with varying load resistances. Its output voltage varies to enable constant output current, as appropriate.

constant voltage source: a voltage source where output voltage (voltage applied to circuitry connected to source output terminals) remains virtually constant even under varying current demand (load) situations.

coordinate system: in ac circuit analysis, a system to indicate phase angles within four, 90° quadrants. The vertical component is the "Y" axis. The horizontal component is the "X" axis.

coordinates: the vertical and horizontal axes that describe the position and magnitude of a vector.

copper losses: the I^2R losses in the copper wire of the transformer windings. Since current passes through the wire and the wire has resistance to current flow, an I^2R loss is created in the form of heat, rather than this energy being a useful part of energy transfer from primary source to secondary load.

core: soft iron piece to help strengthen magnetic field; helps produce a uniform air gap between the permanent magnet's pole pieces and itself.

core losses: energy dissipated in the form of heat by the transformer core that subtracts from the energy transferred from primary to secondary; basically comprised of eddy-current losses (due to spurious induced currents in the core material creating an I^2R loss) and hysteresis losses (due to energy used in constantly changing the polarity of the magnetic domains in the core material).

cosine: a trigonometric function of an angle of a right triangle; the ratio of the adjacent side to the hypotenuse.

cosine function: a trigonometric function used in ac circuit analysis; relationship of a specific angle of a right triangle to the side adjacent to that angle and the hypotenuse where cos θ = adjacent/hypotenuse.

coulomb: the basic unit of charge; the amount of electrical charge represented by 6.25×10^{18} electrons.

covalent bonding: the pairing of adjacent atom electrons to create a more stable atomic structure.

CRT (cathode-ray tube): the oscilloscope component where electrons strike a luminescent screen material causing light to be emitted. Waveforms are displayed on an oscilloscope's CRT.

crystal oscillator: an oscillator circuit that uses a special element (usually quartz-crystal material and holder) for the primary frequency determining element in the circuit.

current (I): the progressive movement of electrons through a conductor. Current is measured in amperes.

current divider circuit: any parallel circuit divides total current through its branches in inverse relationship to its branch R values.

current leading voltage: sometimes used to describe circuits where the circuit current leads the circuit voltage, or the circuit voltage lags the circuit current.

current ratio: the relationship of the current values of secondary and primary, or vice versa. In an ideal transformer, the current ratio is the *inverse* of the voltage and turns ratios; $I_P/I_S = N_S/N_P = V_S/V_P$.

cycle: a complete series of values of a periodic quantity that recur over time.

damping (electrical): the process of preventing an instrument's pointer from oscillating above and below the measured value. Electrical damping is due to the magnetic field produced by induced current opposing the motion that induced it.

dc beta (β_{dc}): for a BJT, the ratio of collector current (I_C) to the corresponding amount of base current (I_B).

delta (Δ): When used in mathematical equations, a term meaning "a small change in." When used in conjunction with an electrical network configuration, it describes the general triangular appearance of the schematic diagram layout, which depicts the electrical connections between components. Also, sometimes, "delta networks" are laid out in a fashion that they are called pi (π) networks.

delta network: an interconnection of components in a closed ring configuration that schematically can be represented as a triangular shape, similar to the Greek letter delta. A set of three branches connected so as to provide one closed mesh. Delta networks often appear in three-phase power systems, but also appear in other electrical and electronic circuits.

delta-wye conversion: the mathematical process of converting network component values, used in a delta network, to appropriate values to be used in a "Wye" network, and yet present the same impedance and load conditions to a source feeding such a network.

depletion mode: a mode of operation for FET devices where the gate bias tends to reduce the availability of charge carriers in the channel.

depletion region: the region near the junction of a P-N junction where the current carriers are depleted due to the combining effect of holes and electrons.

depolarizer: in a dry cell, the material to help prevent or lessen polarization. A common depolarizing material is manganese dioxide.

diac: a type of bidirectional thyristor that is switched on by exceeding the peak forward blocking voltage or peak reverse breakdown voltage. Diacs do not require gate terminals.

dielectric: nonconductive material separating capacitor plates.

dielectric constant: the relative permittivity of a given material compared to the permittivity of vacuum or air; the relative ability of a given material to store electrostatic energy (per given volume) to that of vacuum or air.

dielectric strength: the highest voltage a given dielectric material tolerates before electrically breaking down or rupturing; rated in volts per thousandths of an inch, V per mil.

differential amplifier: an amplifier circuit configuration having two inputs and one output; produces an output that is related to the difference (or differential) between the two inputs.

differential-mode gain: the gain of a differential amplifier when operated in the differential mode; the ratio of its output voltage to the differential mode (difference between the two input voltages) input voltage.

differentiator circuit: an op-amp configuration that produces an output voltage proportional to the rate of change of the input signal.

digital operation: circuit operation where the output varies in discrete steps or in an on-off, high-low, 1-0 category of operation.

discharging a capacitor: balancing (or neutralizing) the charges on capacitor plates by allowing excess electrons on the negative charged plate to move through a circuit to the positive charged plate. When opposite capacitor plates have the same charge, the voltage between them is zero and the capacitor is fully discharged.

doping: introducing specific amounts of P- or N-type elements (called impurities) into a semiconductor crystal (lattice) structure to aid current conduction through the material.

doping material: a material that is added to a pure semiconductor material that enables the existence of charge carriers.

double-ended input: a type of op-amp configuration that uses two inputs.

drain: the portion of an FET that receives charge carriers from the channel.

dropping resistor: a resistor connected in series with a given load that drops the circuit applied voltage to the level required by the load when passing the rated load current through it. (Recall a *voltage drop* is the difference in voltage between two points caused by a loss of pressure (or loss of emf) as current flows through a component offering opposition to current flow.)

effective value (rms): the value of ac voltage or current that produces the same heating effect in a resistor as that produced by a dc voltage or current of equal value. In ac, this value is 0.707 times maximum value.

efficiency: percentage, or ratio, of useful output (such as power) delivered compared to amount of input required to deliver that output. For example, in electrical circuits, the amount of power input to a device, component stage, or system compared to the amount of power delivered to the load. Generic formula is Efficiency $(\%) = (P_{out}/P_{in}) \times 100$.

electrical charge: an excess or deficiency of electrons compared to the normal neutral condition of having an equal number of electrons and protons in each atom of the material.

electrode: generally, one of the electrical terminals that is a conducting path for electrons in or out of a cell, battery, or other electrical or electronic device.

electrolyte: a liquid, paste, or substance through which electrical conduction occurs due to chemical processes of the electrolyte and other materials immersed or exposed to it.

electromagnetism: the property of a conductor to produce a magnetic field when current is passing through it.

electron: the negatively charged particle in an atom orbiting the atom's nucleus.

electrostatic (electric) field: the region surrounding electrically charged bodies; area where an electrically charged particle experiences force.

element: a form of matter that cannot be chemically divided into simpler substances and that has only one type of atom.

emitter: the portion of a BJT that emits charge carriers into the device.

emitter current (I_E): the amount of current found at the emitter terminal of a BJT.

energy: the ability to do work, or that which is expended in doing work. The basic unit of energy is the erg, where 980 ergs = 1 gram cm of work. A common unit in electricity/electronics is the joule, which equals 10^7 ergs.

enhancement mode: a mode of operation for FET devices where the gate bias tends to enhance the availability of charge carriers in the channel.

ε: the Greek letter epsilon, used to signify the mathematical value of 2.71828. This number, raised to various exponents that relate to time constants, can be used to determine electrical circuit parameter values in circuits where exponential changes occur.

equation: a mathematical statement of equality between two quantities or sets of quantities. Example, $A = B + C$.

equivalent resistance: the total resistance of two or more resistances in parallel.

Faraday's law: the law states the amount of induced voltage depends on the rate of cutting magnetic flux lines. In other words, greater number of flux lines cut per unit time produces greater induced voltage.

feedback: with signal amplifying or signal generating circuits, that small portion of the output signal that is coupled back to the input circuit of the stage.

field effect transistor (FET): a unipolar transistor that depends on a voltage field to control the flow of current through a channel.

filter: in power supplies, the component or combination of components that reduce the ac component of the pulsating dc output of the rectifier prior to the output fed to a load.

flux: relating to magnetism; the magnetic lines of force.

flux density (B): the magnetic flux per unit cross-sectional area; unit of measure is the tesla (T) representing 1 Wb per square meter.

focus control: a scope control that adjusts the narrowness of the electron stream striking the screen to create a clear and sharp presentation.

force: in electrical charges, the amount of attraction or repulsion between charges. In electromotive force, the potential difference in volts.

forward bias: the external voltage applied to a semiconductor junction that causes forward conduction, for example, positive to P-type material and negative to N-type material.

frame/movable coil: the parts in a moving-coil meter movement that move in response to the interaction between the fixed magnetic field and the field of the current-carrying, movable coil.

free electrons: the electrons in the outer ring of conductor materials that are easily moved from their original atom. Sometimes termed valence electrons.

free-running multivibrator: a type of multivibrator that runs continuously. It is a form of an *RC*-controlled oscillator circuit that produces a rectangular waveform. It is technically known as an astable multivibrator.

frequency (f): the number of cycles occurring in a unit of time. Generally, frequency is the number of cycles per second, called hertz (Hz), which means cycles per second. In formulas, frequency is expressed as f, where $f = 1/T$.

full-scale current: the value of current that causes full-scale, or full-range deflection of the pointer or indicator device in a current meter movement;

the maximum current that should pass through the moving-coil in a current meter movement.

full-wave rectifier: a type of rectifier that produces a dc output for both ac half-cycles of input.

gain controls: oscilloscope controls that adjust the horizontal and vertical size of the image displayed.

gate: (1) the terminal on a thyristor, such as an SCR or triac, that is used for the voltage that turns on the device. (2) The terminal on an FET where the voltage is applied that will control the amount of current flowing through the channel.

gate-controlled switch (GCS): a type of SCR that can be gated off as well as gated on.

ground reference: a common reference point for electrical/electronic circuits; a common line or conducting surface in electrical circuits used as the point where electrical measurements are made; a common line or conductor where many components make connection to one side of a power source. Typically, ground reference is the chassis, or a common printed circuit "bus" (or line) called chassis ground. In power line circuits, may also be "earth ground."

H: the symbol representing magnetic field intensity.

half-wave rectifier: a type of rectifier that produces a dc output for only one of the half-cycles of input.

Hartley oscillator: an *LC* resonant-circuit oscillator that uses a single capacitor in parallel with two series-connected inductors that are grounded at their common point.

heat sink: a mounting surface for semiconductor devices that is intended to carry away excessive heat from the semiconductor, by dissipating it into the surrounding air.

henry: the basic unit of inductance, abbreviated "H". One henry of inductance is that amount of inductance having one volt induced cemf when the rate of current change is one ampere per second.

Hertz (Hz): the unit of frequency expressing a frequency of one cycle per second.

high-pass filter: a filter that passes frequencies *above* a specified cutoff frequency with little attenuation. Frequencies below the cutoff are greatly attenuated.

holding current: the minimum amount of forward current that a thyristor can carry and still remain switched on.

horizontal (X) amplifier: the scope amplifier circuitry that amplifies signals fed to the horizontal deflection circuitry.

hybrid circuits: generally, circuits comprised of components created by more than one technique, such as thin-film, discrete transistors, and so on.

hysteresis: the difference in voltage between the amount of turn-on voltage and turn-off voltage for an amplifier that is used as a switch. A Schmitt-trigger amplifier, for example, might switch on when the input exceeds $+2$ V, but not switch off until the input drops below $+1.25$ V. The amount of hysteresis in this case is 0.75 V.

I_m or I_{fs}: full-scale current value of the meter movement.

imaginary numbers: complex numbers; the imaginary part of a complex number that is a real number multiplied by the square root of minus one.

impedance: the total opposition an ac circuit displays to current at a given frequency; results from both resistance and reactance. Symbol representing impedance is Z and unit of measure is the ohm.

impedance ratio: the relationship of impedances for the transformer windings; directly related to the *square* of the turns ratio; $Z_P/Z_S = N_P^2/N_S^2$. Reflected impedance is the impedance reflected back to the primary from the secondary. This value depends on the square of the turns ratio and value of the load impedance connected across the secondary.

inductance (self-inductance): the property of a circuit to oppose a change in current flow.

induction: the ability to induce voltage in a conductor; ability to induce magnetic properties from one object to another by magnetically linking the objects.

inductive reactance: the opposition an inductance gives to ac or pulsating current.

in-line: typically means in series with the source.

instantaneous value: the value of a sinusoidal quantity at a specific moment, expressed with lower case letters, such as *e* or *v* for instantaneous voltage and *i* for instantaneous current.

insulator: a material in which the atoms' outermost ring electrons are tightly bound and not free to move from atom to atom, as in a conductor. Typically, these materials have close to the eight outer shell electrons required for chemical/electrical stability.

integrated circuit (IC): a combination of electronic circuit elements. In relation to semiconductor ICs, this term generally indicates that all of these components are fabricated on a single substrate or piece of semiconductor material.

integrator circuit: an op-amp configuration that produces an output voltage proportional to the "area under the curve" for the input waveform.

intensity control: a scope control that adjusts the number of electrons striking the screen per unit time; thus controlling the brightness of the screen display.

inverse parallel: a way of interconnecting two diodes or triacs such that they are in parallel with the anode of one connected to the cathode of the other. Also known as a back-to-back connection.

inverter: a digital logic gate where the output is the opposite of the input. If the input is low, the output is high. If the input is high, the output is low.

inverting amplifiers: amplifier configurations (especially in reference to op-amps) that produce an output waveform having the opposite polarity of the input waveform.

ion: atoms that have gained or lost electrons and are no longer electrically balanced, or neutral atoms. Atoms that lose electrons become positive ions. Atoms that gain electrons become negative ions.

I_S: current through the shunt resistance.

isolation transformer: a 1:1 turns ratio transformer providing electrical isolation between circuits connected to its primary and secondary windings.

joule: the unit of energy. One joule of energy is the amount of energy moving one coulomb of charge between two points with a potential difference of one volt of emf; the work performed by a force of one newton acting through a distance of one meter equals one joule of energy. Also, 3.6×10^6 joules = 1 kilowatt hour.

junction field effect transistor (JFET): FETs that use the voltage field around a reverse-biased P-N junction to control the flow of charge carriers through the channel region.

Kirchhoff's current law: the value of current entering a point must equal the value of current leaving that same point in the circuit.

Kirchhoff's voltage law: the arithmetic sum of voltage drops around a closed loop equals V applied; the algebraic sum of voltages around the entire closed loop, including the source, equals zero.

L, L_T: L is the abbreviation representing inductance or inductor. L_T is the abbreviation for total inductance. L_1, L_2, and so on represent specific individual inductances.

L/C ratio: the ratio of inductance to capacitance in an LC resonant circuit. Generally, the higher this ratio is, the higher the Q of the circuit and the more selective its response.

large-signal amplifiers: amplifiers designed to operate effectively with relatively large ac input signals; e.g., power levels greater than 1 watt.

laser diode: a light-emitting diode that produces light of a single frequency (coherent light) when it is forward conducting.

leading in phase: in ac circuits the electrical quantity that first reaches its maximum positive point is specified as leading the electrical quantity that reaches that point later.

left-hand rule: related to magnetism; when the thumb and fingers of the left hand determine the direction of magnetic fields of current-carrying conductors to find the polarity of an electromagnet's poles.

Lenz's law: the law states the direction of an induced voltage or current opposes the change causing it.

light-activated SCR (LASCR): an SCR that can be gated on by directing light energy onto its gate-cathode junction.

light-emitting diode (LED): a junction diode that emits light energy when it is forward biased and conducting.

linear amplifier: an amplifier that produces an output waveform that has the same general shape as the input waveform.

linear network: a circuit whose electrical behavior does not change with different voltage or current values.

lines: in magnetic circuits, lines represent flux.

lines of force (flux): imaginary lines representing direction and location of the magnetic influence of a magnetic field relative to a magnet.

Lissajous pattern(s): scope patterns on the CRT screen when sine-wave signals are simultaneously applied to the horizontal and vertical deflection systems. These patterns are useful to determine phase and frequency relationships between the two signals.

load: the amount of current or power drain required from the source by a component, device, or circuit.

loaded voltage divider: a network of resistors designed to create various voltage levels of output from one source voltage; where parallel loads, which demand current, are connected.

loading effect: changing the electrical parameters of an existing circuit when a load component, device, or circuit is connected. With voltmeters, the effect of the meter circuit's resistance when in parallel with the portion of the tested circuit; thus, causing a change in the circuit's operation and some change in the electrical quantity values throughout the circuit.

load line: a straight line drawn between the points of cutoff voltage and saturation current on a family of collector characteristic curves.

local action: in carbon-zinc cells, the formation of multiple little local cells created by zinc impurities that are suspended in electrolyte near the zinc electrode. Local action causes the cell to be eaten away and degenerates the cell's chemical capability, even when the cell is not in use.

long time-constant circuit: when the circuit conditions are such that the t/RC ratio = 0.1 (or less). This means the RC time required to charge or discharge the capacitor is long compared with the time allowed during each alternation or cycle.

loop: in electrical networks, a complete electrical circuit or current path.

low-pass filter: a filter that permits all frequencies *below* a specified cutoff frequency to pass with little attenuation. Frequencies above the cutoff frequency are attenuated greatly.

***L/R* time constant:** the time required for an exponentially changing quantity to change by 63.2% of the total value of change to be achieved. For a circuit containing inductance, it is the time (in seconds) required for the current to acquire a level that is 63.2% of its final value. One time constant (in seconds) in a series RL or inductor circuit equals L/R, where L is in henrys and R is in ohms.

magnetic field: the field of influence or area where magnetic effects are observed surrounding magnets; a region of space where magnetic effects are observed.

magnetic field intensity (H): the amount of magnetomotive force per unit length. In the MKSA system, it is the ampere-turns per meter (AT/meter).

magnetic polarity: the relative direction of a magnet's flux with respect to its ends or poles; defined as the North-seeking pole and the South-seeking pole of a magnet.

magnetism: the property causing forces of attraction or repulsion in certain materials; property causing voltage to be induced in nearby conductors when there is relative motion between the magnetic object's field and the conductor.

magnetomotive force (mmf or F): the force causing magnetic lines of force through a medium, thus establishing a magnetic field; developed by current through a coil, where the amount of mmf relates to the amount of current and the number of turns on the coil ($F = AT$).

magnitude: the comparative size or amount of one quantity with respect to another quantity of the same type.

majority carriers: the electrons or holes in extrinsic semiconductor materials that enable current flow. In N-type materials, the majority carriers are electrons. In P-type materials, the majority carriers are holes.

matter: anything that has weight and occupies space, whether it is a solid, liquid, or gas.

maximum power transfer theorem: a theorem that states when the resistance (or impedance) of the load is properly matched to the internal resistance of the power source, maximum power transfer can take place between the source and the load.

Maxwell (Mx): one line of force, or one flux line.

mesh: sometimes called loops. In electrical networks, a set of branches that form a complete electrical path. If any branch is omitted, the remainder of the circuit does not form a complete path.

metal-oxide semiconductor FET (MOSFET): a unipolar silicon transistor that uses a very thin layer of metal-oxide insulation between the gate and channel.

meter shunt (ammeter): the parallel resistance used with ammeter movements to allow them to measure currents higher than their full-scale current rating; a percentage of the total current being measured is shunted, or bypassed around the meter by the meter shunt.

microprocessor: typically, a VLSI chip that acts as the central processing unit (CPU) for a computer system.

minority carriers: the electrons in P-type material and the holes in N-type material; called minority since there are fewer electrons in P-type material than holes and fewer holes in N-type material than electrons.

mode: related to semiconductor ICs; the type of operation for which the IC is designed. The linear mode is commonly used for amplifiers. The digital mode is often used for gates.

molecule: the smallest particle of a compound that resembles the compound substance itself.

monolithic: one piece as in a monolithic IC formed on one semiconductor substrate (or one stone).

monostable multivibrator: a type of multivibrator that is stable in only one state. An input trigger pulse sets the circuit to its unstable state where it remains for a period of time determined by an RC circuit. This circuit is also known as a one-shot multivibrator.

MOS: abbreviation for metal-oxide semiconductor; relates to the fabrication method.

moving-coil (d'Arsonval) movement: a type of meter movement. A coil is suspended within a permanent magnetic field and caused to move when current passes through it. Movement is due to the interaction of its magnetic field with the permanent magnet's field.

μ_r: the symbol for relative permeability.

multimeter: an instrument designed to measure several types of electrical quantities, for example, current, voltage, and resistance; commonly found in the forms of the VOM and the digital multimeter (DMM).

multiple-source circuit: a circuit containing more than one voltage or current source.

multiplier resistor(s) (voltmeter): the current-limiting series resistance(s) used with current meter movements, enabling them to be used as voltmeters that measure voltages much greater than the basic movement's voltage drop at full-scale current. The value of the multiplier resistance is such that its resistance (plus the meter's resistance) limits current to the full-scale value at the desired voltage-range level.

multivibrator: a circuit that produces a waveform that represents a condition of being fully switched on or fully switched off. The three basic types of multivibrators are the monostable (one-shot), the bistable, and the astable (free-running) multivibrators.

mutual inductance (M or L_M): the property that enables a current change in one conductor or coil to induce a voltage in another conductor or coil, and vice versa. This induced voltage results from magnetic flux of one conductor or coil linking the other conductor or coil. The symbol for mutual inductance is M or L_M and the unit of measure is the henry.

NAND gate: an AND gate with inverted output. That is, if both inputs are high, the output is low. For all other input conditions, the output is high.

network: a combination of electrically connected components.

neutron: a particle in the nucleus of the atom that displays no charge. Its mass is approximately equal to the proton.

node: a current junction or branching point in a circuit.

noninverting amplifiers: amplifiers that produce an output that has the same polarity or phase as their input.

nonlinear amplifier: an amplifier that produces an output waveform that is different from its input waveform.

NOR gate: an OR gate with inverted output. That is, if either or both inputs are high, the output is low. Only when both inputs are low is output high.

Norton's theorem: a theorem regarding circuit networks stating that the network can be replaced (and/or analyzed) by using an equivalent circuit consisting of a single "constant-current" source (called I_N) and a single shunt resistance (called R_N).

NPN transistor: a transistor made of P material sandwiched between two outside N material areas. Then, going from end-to-end, the transistor is an NPN type. The symbol for the NPN transistor shows the emitter arrow going away from the base.

N-type material: a semiconductor material that has an excess of electrons as compared to holes.

ohm (Ω): the basic unit of resistance; the amount of electrical resistance limiting the current to one ampere with 1 V applied.

Ohm's law: a mathematical statement describing the relationships among current, voltage, and resistance in electrical circuits. Common equations are $V = I \times R$, $I = V/R$, and $R = V/I$.

ohms-per-volt rating: a voltmeter rating indicating how many ohms of resistance are needed to limit the current through the meter to its full-scale value when one volt is applied to the meter circuit.

one-shot multivibrator: a popular name for a monostable multivibrator. (*See* monostable multivibrator.)

op-amp: abbreviation for operational amplifier; a versatile amplifier circuit with very high gain, high input impedance, and low output impedance.

open circuit: any break in the current path that is undesired, such as a broken wire or component, or designed, such as open switch contacts in a lighting circuit.

operational amplifier (op amp): a versatile differential amplifier circuit with very high gain, high input impedance, and low output impedance.

optoelectronic device: a group of electronic devices that generate or respond to light.

OR gate: a digital logic gate where if either input is high, or if both inputs are high, the output is high.

oscilloscope: a device that visually displays signal waveforms so time and amplitude parameters are easily determined.

parallel branch: a single current path within a circuit having two or more current paths where each of the paths are connected to the same voltage points; a single current path within a parallel circuit.

parallel circuit: a circuit with two or more paths for current flow where all components are connected between the same voltage points.

peak-inverse-voltage (PIV): the maximum voltage across a diode in the reverse (nonconducting) direction. The maximum voltage that a diode tolerates in this direction is the peak-inverse-voltage rating.

peak-to-peak value (p-p): the difference between the positive peak value and the negative peak value of a periodic waveform; computed for the sine wave as 2.828 times effective value.

peak value (V_{pk} or maximum value) (V_{max}): the maximum positive or negative value that a sinusoidal quantity reaches, sometimes referred to as peak value. It is calculated as 1.414 times effective value.

pentavalent atoms: an atom normally having five electrons in the valence shell. Such materials are often used as a donor impurity in the manufacture of N-type semiconductor materials. Examples are phosphorus, arsenic, and antimony.

period (T): the time required for one complete cycle. In formulas, it is expressed as T, where $T = 1/f$.

permeability (mu or μ): the ability of a material to pass, conduct, or concentrate magnetic flux.

phase angle (θ): the difference in time or angular degrees between ac electrical quantities (or periodic functions) when the two quantities reach a certain point in their periodic function. This difference is expressed by time difference (the fraction of a period involved), or by angular difference (expressed as degrees or radians). (NOTE: A radian = 57.3°. $2\pi \times$ a radian = 360°.) An example is the difference in phase between circuit voltage and circuit current in ac circuits that contain reactances, such as inductive reactance.

phase-shift oscillator: an RC sine-wave oscillator that uses a sequence of three RC circuits to achieve a total phase shift of 180° at the frequency of oscillation.

phasor: a quantity that expresses position relative to time.

photodiode cell: a device where resistance between terminals is changed by light striking its light-sensitive element; a light-controlled variable resistance. Typically, the more light striking it, the lower its terminal resistance is.

piezoelectric effect: an effect enabled by certain kinds of materials whereby applying a voltage to the material causes it to change size and shape, and bending the material causes it to generate a voltage.

pinch-off voltage (V_P): the source-drain voltage level (of an FET) where a further increase in source-drain voltage causes virtually no increase in drain current.

P-N junction: the location in a semiconductor where the P- and N-type materials join; where the transition from one type material to the other type occurs.

P-N junction diode: a semiconductor diode that consists of a single P-N junction.

PNP transistor: a transistor where the middle section is N-type material and the two outside sections are composed of P-type material, thus called a PNP transistor. The symbol for the PNP transistor shows the emitter arrow going toward the base.

pointer: the element in a meter that provides the indicating function of how much movement has taken place by the movable coil.

polar form notation: expressing circuit parameters with respect to polar coordinates; expressing a point with respect to distance and angle (direction) from a fixed reference point on the polar axis. For example, $25 \angle 45°$.

polarity: in an electrical circuit, a means of designating differences in the electrical charge condition of two points, or a means of designating direction of current flow. In a magnetic system, a means of designating the magnetic differences between two points or locations.

polarization: relating to cells or batteries, the undesired formation of hydrogen bubbles around the positive electrode that degenerates the cell's operation.

pole pieces: special meter pieces used to help linearize the magnetic field in a moving-coil movement.

position control(s) (vertical and horizontal): scope controls that move the display vertically or horizontally; thus enabling centering the display on the screen.

positive feedback: a portion of the output of an amplifier system that is fed back to the input with the same phase or polarity as the original input signal.

potential: in electrical circuits, the potential, or ability, to move electrons. In electromotive force, the difference in charge levels at two points. This potential difference is measured in volts.

power: the rate of doing work. In mechanical terms, a horsepower equals work done at a rate of 550 foot pounds per second. In terms of electrical power, it is electrical work at a rate of one joule per second.

power factor (p.f.): the ratio of true power to apparent power. It is equal to the cosine of the phase angle (θ).

powers of 10: working with small or large numbers can be simplified by converting the unit to a more manageable number times a power of ten. An example is 0.000006 ampere expressed as 6×10^{-6} amperes. Some of the frequently used powers of ten are 10^{-12} for pico (p) units, 10^{-6} for micro (μ) units, 10^{-3} for milli units, 10^3 for kilo units, and 10^6 for mega units.

power supply: a circuit that converts ac power into dc power required for the operation of most kinds of electronic systems.

product-over-the-sum method: a method of finding total resistance in parallel circuits by using two resistances at a time in the formula $R_T = R_1 \times R_2/ R_1 + R_2$.

proton: the positively charged particle in an atom located in the nucleus of the atom. Its weight is approximately 1,836 times that of the electron, and approximately the same weight as the neutron.

P-type material: a semiconductor material that has an excess number of holes compared to electrons.

pulsating dc: a varying voltage that is one polarity, as opposed to reversing polarity as ac does.

push-pull amplifier: an amplifier circuit that uses two Class B amplifiers, one to amplify one alternation of the ac input waveform and a second to amplify the opposite alternation. The circuit has the advantage of Class-B efficiency and the low distortion of a Class-A amplifier.

Pythagorean theorem: a theorem expressing the mathematical relationship of the sides of a right triangle, where the square of the hypotenuse is equal to the sum of the squares of the other two sides. That is, $c^2 = a^2 + b^2$. Equation used in electronics is $c = \sqrt{a^2 + b^2}$.

Q (figure of merit): a number indicating the ratio of the amount of stored energy in the inductor magnetic field to the amount of energy dissipated. Found by using the ratio of the inductance's inductive reactance (X_L) to the resistance of the inductor (R). That is, $Q = X_L/R$.

Q factor: the relative quality or figure of merit of a given component or circuit; often related to the ratio of X_L/R of the inductor in the resonant circuit. Other circuit losses can also enter into the effective Q of the circuit, such as the loading that a load coupled to the resonant circuit causes and the effect of R introduced in the circuit. The less the losses are, the higher the Q is. The sharpness of the resonant circuit response curve is directly related to the Q of the circuit.

Q point: the operating point of an amplifier when there is no input signal.

quadrants: segments of the coordinate system representing 90° angles, starting at a 0° reference plane.

range select (switch): the switch used with meter movements that connects or disconnects various shunt resistances values (for current or ohmmeter measurements), or multiplier resistances for the voltage measurement mode.

RC time constant: the time, in seconds, represented by the product of R in ohms and C in farads; time required for a capacitor to charge or discharge 63.2% of the change in voltage level applied. Five (5) RC time constants are needed for a capacitor to completely charge or discharge to a changed voltage level.

reactance: the opposition a capacitor or inductor gives to alternating current or pulsating direct current.

reactive power (Q): the vertical component of the ac power triangle that equals $VI \sin \theta$; expressed as VAR (voltampere-reactive).

real numbers: in math, any rational or irrational number.

rectangular form notation: expressing a set of parameters with the complex number system. For example, $R \pm jX$ indicates a combination of resistance and reactance. Thus, $5 + j10$ means 5 ohms of resistance in series with 10 ohms of inductive reactance.

rectangular waveforms: a "square" waveform that has only two voltage levels, often one level that is close to the main supply voltage and one that is close to zero volts.

rectifier circuit: a circuit that converts an ac waveform to pulsating dc.

rectifier diode: a junction diode that is designed for the higher current and higher reverse-voltage conditions encountered in power supply rectifier circuits.

rectify (rectification): the process of changing ac to pulsating dc.

regulator: a section of some power supplies that keeps the dc output voltage at a desired level in spite of changes in the input voltage level and/or changes in the output loading.

relative permeability (μ_r): comparing a material's permeability with air or a vacuum; relative permeability is not a constant, since it varies for a given material depending on the degree of magnetic field intensity present; relative permeability

$$(\mu_r) = \frac{\text{flux density with given core}}{\text{flux density with vacuum core}}.$$

reluctance ($\Re$): the opposition of a given path to the flow of magnetic lines of force through it.

residual magnetism: the magnetism remaining in the core material of an electromagnet after the magnetizing force is removed.

resistance (R): in an electrical circuit, the opposition to electron movement or current flow. Its value is affected by many factors, including the dimensions and material of the conductors, the physical makeup and types of the components in the circuit, and temperature.

resistor: an electrical component that provides resistance in ohms and that controls or limits current or distributes or divides voltage.

resistor color code: a system to display the ohmic value and tolerance of a resistor by means of colored stripes or bands around the body of the resistor.

resonance: a special circuit condition existing when opposite types of reactances in a circuit are equal and have cancelling effects (i.e., $X_L = X_C$).

reverse bias: the external voltage applied to a semiconductor junction that causes the semiconductor not to conduct; for example, negative to P-type material and positive to N-type material.

reverse breakdown voltage: a rating for semiconductors that specifies the maximum amount of reverse voltage that a P-N junction can withstand without breaking down and conducting in the reverse direction.

reverse Polish notation (RPN): an operational system, based on a technique derived from the work of a Polish scientist, used in some calculators to simplify and minimize the number of input steps required to solve certain complex mathematical problems. Since the notation method used is somewhat reversed from the scientist's original technique, the system, or mode, is sometimes called reverse Polish notation (RPN).

ripple: the ac component in a pulsating dc signal, such as the output of a power supply.

R_M: meter movement resistance value.

R_{Mult}: resistance value of the resistor used that enables a basic current meter to measure voltage by its current-limiting effect.

saturation: in magnetism, the condition when increasing the magnetizing force does not increase the flux density in a magnetic material.

scalar: a quantity expressing only quantity, such as a "real number."

scale: the face of a meter containing calibration marks that interpret the pointer's position in appropriate units of measure.

schematic diagrams: graphic illustrations of circuits or subcircuits and/or systems or subsystems. Schematics are more detailed than block diagrams and give specific information regarding how components are connected. Schematics show the types, values, and ratings of component parts, and in many cases show the electrical values of voltage, current, and so forth at various points throughout the circuit.

Schmitt trigger: a nonlinear amplifier circuit that converts an ac input waveform into a rectangular waveform. The output waveform goes to its high level when the input exceeds a certain voltage level, and the output goes to its low level when the input falls below a certain voltage level.

SCR: abbreviation for silicon-controlled rectifier; often used to control the time during an ac cycle the load receives power from the power source. A four-layer, P-N-P-N device where conduction is initiated by a voltage to the gate. Conduction doesn't cease until anode voltage is reversed, removed, or greatly reduced although the initiating signal has been removed.

selectivity: referring to resonant circuits, selectivity indicates the sharpness of the response curve. A narrow and more pointed response curve results in a higher degree of frequency selectiveness or selectivity.

self-bias: a method of biasing an amplifier such that the bias voltage is generated by the main flow of current through the device, itself. The method is usually used where the biasing voltage has a polarity opposite the main supply voltage (as with depletion-mode FETs).

semiconductor: a material with four valence (outermost ring) electrons. It is neither a good conductor, nor a good insulator.

semiconductor diode: a semiconductor P-N junction that passes current easily in one direction while not passing current easily in the other direction.

sensitivity rating: a means of rating meters, or meter movements to indicate how much current is required to cause full-scale deflection; how much voltage is dropped by the meter movement at full-scale current value. Also, voltmeters are rated by the ohms-per-volt required to limit current to full-scale value. (This is inversely related to the current sensitivity rating. That is, the reciprocal of the full-scale current value yields the ohms-per-volt rating.)

series circuit: a circuit in which components are connected in tandem or in a string providing only one path for current flow.

series-parallel circuit: a circuit comprised of a combination of both series-connected and parallel-connected components and/or subcircuits.

series-type ohmmeter circuit: the ohmmeter circuit where the meter, its zero-adjust resistance, and its internal voltage source are in series. When the test leads are connected across an unknown resistance to be measured, that resistance is placed in series with the meter circuit.

short circuit: generally, an undesired very low resistance path across two points in a circuit. It may be across one component, several components, or across the entire circuit. There can be designed shorts in circuits, such as purposeful "jumpers" or closed switch contacts.

short time-constant circuit: describes a circuit where the frequency is such that the capacitor has at least ten times one RC time constant to charge or discharge. A short time-constant circuit is sometimes described as one where: $t/RC = 10$ (or greater).

siemens: the unit used to measure or quantify conductance; the reciprocal of the ohm ($G = 1/R$).

silicon-controlled rectifier (SCR): four-layer (P-N-P-N) thyristor device that is gated on by forward-biasing the gate-cathode junction and is turned off only by eliminating or reversing the polarity of the cathode-anode voltage.

silicon-controlled switch (SCS): an SCR that has two gate terminals that can be used for switching on the device.

sine: a trigonometric function of an angle of a right triangle; the ratio of the opposite side to the hypotenuse.

sine function: a trigonometric function used in ac circuit analysis; expresses the relationship of a spec-

ified angle of a right triangle to the side opposite that angle and the hypotenuse where $\sin \theta =$ opposite/hypotenuse.

sine wave: a wave or waveform expressed in terms of the sine function relative to time.

single-ended input: a type of op-amp configuration that uses a single input.

sinusoidal quantity: a quantity that varies as a function of the sine or cosine of an angle.

skirts: term typically used to describe the portions of a resonance response curve that descend down from the peak response point. The steepness of the skirts reflects the selectivity of the circuit.

small-signal amplifiers: amplifiers designed to operate effectively with small ac input signals; that is, power levels less than 1 watt, and peak-to-peak ac current values less than 0.1 times the amplifier's input bias current.

source: in electrical circuits, the source is the device supplying the circuit applied voltage, or the electromotive force causing current to flow through the circuit. Sources range from simple flashlight cells to complex electronic circuits.

source: the portion of an FET that emits charge carriers into the channel.

source/gate/drain: the electrodes in a field effect transistor device; gate controls current flow between the source and drain.

springs: the elements in a moving-coil meter movement that conduct current to the movable coil and provide a mechanical force against which the moving coil must operate.

SSI, MSI, LSI, and VLSI: SSI is small-scale integration (less than 10 gates on a single chip). MSI is medium-scale integration (less than 100 gates on a single chip). LSI is large-scale integration (more than 100 gates on a single chip). VLSI is very-large-scale integration (over 1000 gates on a single chip).

steady-state condition: the condition where circuit values and conditions are stable or constant; opposite of transient conditions, such as switch-on or switch-off conditions, when values are changing from one state to another.

stops: pegs that limit the left and right movement of the meter's pointer and related moving elements.

subcircuit: a part or portion of a larger circuit.

summing amplifier: an op-amp configuration that mathematically adds (or sums) the voltage levels found at two or more inputs.

superconductivity: the flow of electrical current through an electrical conductor whose resistance has decreased to almost 0 Ω due to the conductor material and the super-cold temperature at which it is being operated. The discovery was made in 1911 that this "superconductivity" phenomenon appears in certain materials that are operated below a critical specific cold temperature level. (Usually close to absolute zero.)

superposition theorem: superimposing one set of values on another to create a set of values different from either set of the superimposed values alone; the algebraic sum of the superimposed sets of values.

surface mount technology: the technology in which discrete electrical components or devices (such as resistors and capacitors) are manufactured in a way that allows miniaturization of the component elements and minimizing of lead lengths. These components are designed to be mounted and soldered directly to the "surface" of printed circuit boards.

susceptance (B): the relative ease that ac current passes through a pure reactance; the reciprocal of reactance; the imaginary component of admittance.

sweep circuits: circuits in an oscilloscope that cause the horizontal trace and the rapid retrace of the electron beam across the CRT.

sweep frequency control(s): scope controls that set the number of times the horizontal trace occurs per unit time. Frequently, this control is calibrated in time/div. to measure the time per division traced on the display. (NOTE: The formula $f = 1/T$ is then used to determine frequency.)

switches: devices for making or breaking continuity in a circuit.

switching diode: a junction diode that is designed to allow a very short recovery time when the applied voltage changes polarity.

synchronization: the process of bringing two entities into time coherence. In a scope, the circuitry and process that cause the beginning of the horizontal trace to be synchronized with the vertical input signal.

synchronization control(s): scope controls that control the timing of the horizontal sweep in synchronization with the signal(s) being tested.

tangent function: a trigonometric function used in ac circuit analysis; expresses the relationship of a specified angle of a right triangle to the side opposite and side adjacent to that angle where tan θ = opposite/adjacent.

TC: the abbreviation often used to represent time constant.

tesla: a unit of flux density representing one weber per square meter in the MKSA system of units.

tetravalent atoms: an atom normally having four electrons in the outer, or valence, shell. Such materials are often used as the base material for semiconductors. Examples include silicon and germanium.

Thevenin's theorem: a theorem used to simplify analysis of circuit networks because the network can be replaced by a simplified equivalent circuit consisting of a single voltage source (called V_{TH}) and a single series resistance (called R_{TH}).

thin-film process: a process where thin layers of conductive or nonconductive materials are used to form integrated circuits.

thyristor: a semiconductor device that is composed of four layers, P-N-P-N. The conduction is either all or nothing; it is either switched fully on or turned fully off.

transconductance (g_m): for FETs, the ratio of a change in drain current to the corresponding change in gate voltage. Unit of measurement is siemens.

transformers: devices that transfer energy from one circuit to another through electromagnetic (mutual) induction.

triac: a special semiconductor device, sometimes called a 5-layer N-P-N-P-N device, that controls power over an entire cycle, rather than half the cycle, like the SCR device; equivalent of two SCRs in parallel connected in opposite directions but having a common gate.

trivalent materials: materials that have three valence electrons; used in doping semiconductor material to produce P-type materials where holes are the majority current carriers. Examples include boron, aluminum, gallium, and indium.

true (or real) power (P): the power expended in the ac circuit resistive parts that equals $VI \cos \theta$, or apparent power $\times \cos \theta$. The unit of measure is watts.

tunnel diode: a special PN junction diode that has been doped to exhibit high-speed switching capability and a special negative-resistance over part of its operating range; sometimes used in oscillator and amplifier circuits.

turns ratio: the number of turns on the separate windings of a transformer; expressed as the number of turns on the secondary versus the number of turns on the primary (N_S/N_P); or frequently stated as the ratio of primary-to-secondary turns.

universal (Ayrton or ring) shunt: a method of connecting and selectively tapping resistances so various series and shunt resistance values are connected to the meter circuit, thus affording multiple current ranges. Circuitry assures that the meter is never without some shunt resistance value; thus, providing some measure of meter protection.

valence electrons: the electrons in the outer shell of an atom

varactor diode: a silicon, voltage-controlled semiconductor capacitor where capacitance is varied by changing the junction reverse bias; used in various applications where voltage-controlled, variable capacitance can be advantageous.

vector: a quantity that illustrates both magnitude and direction. A vector's direction represents position relative to space.

vertical (Y) amplifier: the scope amplifier circuitry that amplifies signals fed to the vertical input jack(s).

vertical volts/div. control(s): scope controls that adjust the amplification or attenuation of the input signal fed to the vertical deflection system. Generally, this control has positions that provide calibrated vertical deflections in ranges of millivolts per division or volts per division; thus enabling measurement of waveform voltages.

volt (V): the basic unit of potential difference or electromotive force (emf); the amount of emf causing a current flow of one ampere through a resistance of one ohm.

voltage: the amount of emf available to move a current.

voltage divider action: the dividing of a specific source voltage into various levels of voltages by means of series resistors. The distribution of voltages at various points in the circuit is dependent on what proportion each resistor is of the total resistance in the circuit.

voltage follower: an amplifier configuration where the output voltage is slightly less than the input voltage level, and the output has the same phase or polarity as the input.

voltage rate of change (dV/dt): the amount of voltage level change per unit time. For example, a rate of change of one volt per second, 10 V per μ second, and so forth.

voltage ratio: in an ideal transformer with 100% coupling ($k = 1$), the voltage ratio is the same as the turns ratio, since the voltage induced in each turn of the secondary and the primary windings is the same for a given rate of flux change; $V_S/V_P = N_S/N_P$.

voltampere-reactive (VAR): the unit of reactive power in contrast to real power in watts; 1 VAR is equal to one reactive volt-ampere.

watt: the basic unit of electrical power. The basic formula for electrical power is P (in watts) equals V (in volts) times I (in amperes).

weber: the unit of magnetic flux indicating 10^8 maxwells (flux lines).

Wheatstone bridge circuit: a special application of the bridge circuit commonly used to measure unknown resistances. It utilizes a sensitive current or voltage meter to determine when the bridge is balanced.

Wien-bridge oscillator: a type of RC sine-wave oscillator that uses a lead-lag network to maintain a zero-degree phase shift of the feedback signal at the frequency of oscillation.

work: the expenditure of energy, where work = force $\times$ distance. The basic unit of work is the foot poundal. NOTE: At sea level, 32.16 foot poundals = one foot pound.

wye network: also sometimes called a "T" network. A wye network consists of three branches whose connections all join at one end of each branch, and are separate electrical points at the other end of each branch. This electrical connection system can be schematically represented in the shape of a Y or a T, hence the names Wye and Tee are used in describing this electrical network.

wye-delta conversion: the mathematical process of converting network component values, used in a wye network, to appropriate values to be used in a delta network, and yet present the same impedance and load conditions to a source feeding such a network.

zener diode: a junction diode that is designed to allow reverse breakdown at a certain voltage level, thereby acting as a voltage regulator device.

zero-adjust control: the adjustable series resistor in an ohmmeter used to regulate current through the meter movement so that with the test probes across zero ohms (or touching each other), there is full-scale current through the meter, and the pointer is located above the zero-ohms mark on the scale.

INDEX

Page numbers in *italic* indicate figures.